QMW Library
23 1164881 5

KU-406-637

OPTICS

Third Edition

Eugene Hecht

Adelphi University

An imprint of Addison Wesley Longman, Inc.

Reading, Massachusetts • Menlo Park, California • New York • Harlow, England
Don Mills, Ontario • Sydney • Mexico City • Madrid • Amsterdam

Sponsoring Editor: *Julia Berrisford*
Production Supervisor: *Helen Wythe*
Manufacturing Supervisor: *Hugh Crawford*
Cover Designer: *Eileen R. Hoff*
Production Service: *HRS Electronic Text Management*
Text Design: *HRS Electronic Text Management*
Composition: *HRS Electronic Text Management*
Illustration: *Oxford Illustrators and HRS Electronic Text Management*

Library of Congress Cataloging-in-Publication Data

Hecht, Eugene
Optics / Eugene Hecht; — 3rd ed.
p. cm.
Includes index.
ISBN 0-201-83887-7
1. Optics. II. Title.
QC355.2.H42 1998
535—dc20

Printed in the United States.

1 2 3 4 5 6 7 8 9 0 - MA - 00999897

Preface

To *Ca, b. w. l.*

This third edition has been guided by three distinct imperatives: to fine-tune the pedagogy, modernize the discourse, and update the content.

During the past decade, using the second edition in the classroom, a number of small but significant pedagogical refinements have evolved and these have all been incorporated throughout the revised text. The intent, as ever, is to be responsive to the students' changing needs.

This edition continues the program of gradually modernizing the treatment. In this, there are several goals, vis-à-vis the student: to impart an appreciation of the central role of atomic scattering in almost every aspect of Optics; to provide an understanding, as early as possible, of the insightful perspective offered by Fourier Theory; and to make clear, from the outset, the underlying quantum-mechanical nature of light.

Technological advances have been made in a wide range of areas from lenses and lasers to telescopes and fibers, and the computer continues to extend our perception of it all. This third edition treats (even if sometimes only in introductory fashion) all of the significant advances that today's students should be aware of.

Chapter 2 (*The Mathematics of Wave Motion*), which lays the theoretical groundwork for wave theory, has been revised in order to both make the existing material clearer and to provide a broader foundation for what's to come. It now contains a more complete, albeit low-level, discussion of the *Superposition Principle* and a new introductory section on *Phasors and the Addition of Waves*. These will better prepare students for early qualitative discussion of both scattering and QED.

Chapter 3 (*Electromagnetic Theory, Photons, and Light*) has a new section called *Averaging Harmonic Functions* (p. 47). It provides a detailed look at the averaging process and introduces the sinc function in the course of the exposition. In keeping with the commitment to integrate quantum-mechanical ideas into the basic discussion, *Photons* (p. 50) and *Photon Counting* (p. 53) are qualitatively considered early on. Some experimental results give the section on dispersion (p. 70) a more practical quality than it had before. The chapter ends with a few remarks about Quantum Field Theory (p. 80).

The Propagation of Light (Chapter 4) now begins with a section on *Rayleigh Scattering* that leads to a more realistic discussion of the transmission of light (p. 88), reflection, and refraction all from a scattering perspective. The chapter ends with a very simple discussion of reflection via Quantum Electrodynamics à la Feynman (p. 138).

Geometrical Optics (Chapter 5) contains a number of new drawings (e.g., Figs. 5.1, 5.2, 5.16*a*, 5.26, 5.28, 5.32, 5.35, 5.41, 5.42, 5.86, 5.88, 5.98, 5.125, 5.127, 5.128, 5.129, 5.132, and 5.133) and some great photographs (e.g., pp. 159 and 237). There are several new sections: one called *QED and the Lens* (p. 173) and another, *Capillary Optics* (p. 202). The discussion of telescopes has been extended (see *Aplanatic Reflectors*, p. 228) and now treats the Hubble Space Telescope. Section 5.8 (*Wavefront Shaping*) contains new discussions of *Adaptive Optics* (5.8.1) and *Phase Conjugation* (5.8.2).

Geometrical Optics continues in Chapter 6 and here attention has been given to improving the discussion of *Matrix Methods* (6.2.1). Two new sections, *Matrix Analysis of Mirrors* and *Flat Mirrors and the Planar Optical Cavity*, enhance the treatment. A discussion of the HST and the COSTAR module gives spherical aberration a little prominence (p. 260). Several nice computer-generated figures brighten the sections on aberrations (Figs. 6.23, 6.27, etc.). Section 6.4 (*GRIN Systems*) is a fairly extensive introduction to gradient index optics.

In Chapter 7 (*The Superposition of Waves*) the analysis of standing waves has been made more accessible with the inclusion of several diagrams and photographs (Figs. 7.7, 7.8, 7.9, 7.10, and 7.12). The treatments of the concepts of *Group Velocity* (p. 296) and *Coherence Length* (p. 308) have both been elaborated and made more effective. The new sections, *The Discrete Fourier Transform* and *Fourier Analysis and*

Diffraction, serve to introduce a number of important ideas relating to the spatial frequency perspective much earlier than was done in the last edition.

No substantial changes were made in Chapter 8 (*Polarization*), although, as elsewhere, the prose was tightened up and the analysis clarified here and there; fresh diagrams were developed, and a new selection of problems added.

Chapter 9 (*Interference*) now contains an early introduction to coherence in Section 9.2.1 (*Temporal and Spatial Coherence*) and the ideas are immediately carried into the discussion of *Young's Experiment* (p. 385). To underscore the quantum-mechanical texture of interference, many optical interference photographs are now accompanied by equivalent material particle fringe patterns (e.g., Fig. 9.13). Here too the *Fourier Perspective* (p. 389) is woven into the discussion much earlier than previously.

Chapters 10 (*Diffraction*), 11 (*Fourier Optics*), and 12 (*Basics of Coherence Theory*) have undergone a line-by-line fine tuning, but no major overhaul.

Chapter 13, which is now *Modern Optics: Lasers and Other Topics*, is the result of an extensive reorganization and revision. The greatly extended coverage of lasers begins with two new sections on *Radiant Energy and Matter in Equilibrium* (13.1.1) and *Stimulated Emission* (13.1.2).

This third edition continues the agenda of unifying the discourse, as much as possible, within the framework of a few grand ideas. Thus the concept of interference, which is one of the premier notions in Optics (and not surprisingly in Quantum Mechanics, as well), is used qualitatively to understand propagation phenomena long before it's studied formally in Chapter 9. Among other benefits, this approach of presenting advanced concepts in simplified form early in the exposition allows the student to develop a cohesive perspective.

At the request of users, I have added about 125 new problems. Most of these were designed to develop needed analytic skills and are of the "easy-to-intermediate" variety. As in previous editions, the complete solutions to many of the problems (those without asterisks) can be found at the back of the book.

Over the years, I have received comments, articles, and photographs from hundreds of colleagues, and I most sincerely thank them all. I am especially grateful to Professors M. W Coffey of the University of Colorado and H. Fearn of California State University for their many suggestions. Anyone interested in the discipline who wishes to exchange ideas can contact me at Adelphi University, Physics Department, Garden City, N.Y. 11530.

Fortunately, my editor for this edition was Julia Berrisford—I thank her for her professionalism, tenacity, patience, and inexhaustible good humor. The book was produced by HRS Electronic Text Management, which did a brilliant job of getting it all together. Lorraine Burke watched over every aspect of the process with incredible skill; Ed Burke designed a beautiful book; Karen Mahakian was the perfect compositor; and Pat Hannagan produced incomparable art. All have my deepest respect and appreciation. Finally, I thank my dear friend Carolyn Eisen Hecht for coping with one more edition of one more book.

Freeport, New York E.H.

Contents

1 A Brief History

1.1 Prolegomenon

In chapters to come we will evolve a formal treatment of much of the science of Optics, with particular emphasis on aspects of contemporary interest. The subject embraces a vast body of knowledge accumulated over roughly three thousand years of the human scene. Before embarking on a study of the modern view of things optical, let's briefly trace the road that led us there, if for no other reason than to put it all in perspective.

1.2 In the Beginning

The origins of optical technology date back to remote antiquity. Exodus 38:8 (ca. 1200 B.C.E.) recounts how Bezaleel, while preparing the ark and tabernacle, recast "the looking-glasses of the women" into a brass laver (a ceremonial basin). Early mirrors were made of polished copper, bronze, and later on of speculum, a copper alloy rich in tin. Specimens have survived from ancient Egypt—a mirror in perfect condition was unearthed along with some tools from the workers' quarters near the pyramid of Sesostris II (ca. 1900 B.C.E.) in the Nile valley. The Greek philosophers Pythagoras, Democritus, Empedocles, Plato, Aristotle, and others developed several theories of the nature of light (that of the last named being quite similar to the aether theory of the nineteenth century). The rectilinear propagation of light (p. 88) was known, as was the Law of Reflection (p. 96) enunciated by Euclid (300 B.C.E.) in his book *Catoptrics*. Hero of Alexandria attempted to explain both these phenomena by asserting that light traverses the shortest allowed path between two points. The burning glass (a positive lens used to start fires) was alluded to by Aristophanes in his comic play *The Clouds* (424 B.C.E.). The apparent bending of objects partly immersed in water (p. 102) is mentioned in Plato's *Republic*. Refraction was studied by Cleomedes (50 A.D.) and later by Claudius Ptolemy (130 A.D.) of Alexandria, who tabulated fairly precise measurements of the angles of incidence and refraction for several media (p. 100). It is clear from the accounts of the historian Pliny (23–79 A.D.) that the Romans also possessed burning glasses. Several glass and crystal spheres have been found among Roman ruins, and a planar convex lens was recovered in Pompeii. The Roman philosopher Seneca (3 B.C.E.–65 A.D.) pointed out that a glass globe filled with water could be used for magnifying purposes. And it is certainly possible that some Roman artisans may have used magnifying glasses to facilitate very fine detailed work.

After the fall of the Western Roman Empire (475 A.D.), which roughly marks the start of the Dark Ages, little or no scientific progress was made in Europe for a great while. The dominance of the Greco-Roman-Christian culture in the lands embracing the Mediterranean soon gave way by conquest to the rule of Allah. The center of scholarship shifted to the Arab world, and Optics was studied and extended, especially by Alhazen (ca. 1000 A.D.). He elaborated on the Law of Reflection, putting the angles of incidence and reflection in the same plane normal to the interface (p. 101); he studied spherical and parabolic mirrors and gave a detailed description of the human eye (p. 204).

By the latter part of the thirteenth century, Europe was only beginning to rouse from its intellectual stupor. Alhazen's work was translated into Latin, and it had a great effect on the writings of Robert Grosseteste (1175–1253), Bishop of Lincoln, and on the Polish mathematician Vitello (or Witelo), both of whom were influential in rekindling the study of Optics. Their works were known to the Franciscan Roger Bacon (1215–

1294), who is considered by many to be the first scientist in the modern sense. He seems to have initiated the idea of using lenses for correcting vision and even hinted at the possibility of combining lenses to form a telescope. Bacon also had some understanding of the way in which rays traverse a lens. After his death, Optics again languished. Even so, by the mid-1300s, European paintings were depicting monks wearing eyeglasses. And alchemists had come up with a liquid amalgam of tin and mercury that was rubbed onto the back of glass plates to make mirrors. Leonardo da Vinci (1452–1519) described the *camera obscura* (p. 219), later popularized by the work of Giovanni Battista Della Porta (1535–1615), who discussed multiple mirrors and combinations of positive and negative lenses in his *Magia naturalis* (1589).

This, for the most part, modest array of events constitutes what might be called the first period of Optics. It was undoubtedly a beginning—but on the whole a humble one. The whirlwind of accomplishment and excitement was to come later, in the seventeenth century.

FIGURE 1.1 Johannes Kepler (1571–1630). (Courtesy Burndy Library.)

1.3 From the Seventeenth Century

It is not clear who actually invented the refracting telescope, but records in the archives at The Hague show that on October 2, 1608, Hans Lippershey (1587–1619), a Dutch spectacle maker, applied for a patent on the device. Galileo Galilei (1564–1642), in Padua, heard about the invention and within several months had built his own instrument (p. 171), grinding the lenses by hand. The compound microscope was invented at just about the same time, possibly by the Dutchman Zacharias Janssen (1588–1632). The microscope's concave eyepiece was replaced with a convex lens by Francisco Fontana (1580–1656) of Naples, and a similar change in the telescope was introduced by Johannes Kepler (1571–1630). In 1611, Kepler published his *Dioptrice*. He had discovered total internal reflection (p. 121) and arrived at the small angle approximation to the Law of Refraction, in which case the incident and transmission angles are proportional. He evolved a treatment of first-order Optics for thin-lens systems and in his book describes the detailed operation of both the Keplerian (positive eyepiece) and Galilean (negative eyepiece) telescopes. Willebrord Snell (1591–1626), professor at Leyden, empirically discovered the long-hidden *Law of Refraction* (p. 100) in 1621—this was one of the great moments in Optics. By learning precisely how rays of light are redirected on traversing a boundary between two media, Snell in one swoop swung open the door to modern applied Optics. René Descartes (1596–1650) was the first to publish the now familiar formulation of the Law of Refraction in terms of sines. Descartes deduced the law using a model in which light was viewed as a pressure transmitted by an elastic medium; as he put it in his *La Dioptrique* (1637)

> …recall the nature that I have attributed to light, when I said that it is nothing other than a certain motion or an action conceived in a very subtle matter, which fills the pores of all other bodies….

FIGURE 1.2 René Descartes by Frans Hals (1596–1650). (© Musées Nationaux.)

The universe was a plenum. Pierre de Fermat (1601–1665), taking exception to Descartes's assumptions, rederived the Law of Reflection (p. 105) from his own *Principle of Least Time* (1657).

The phenomenon of diffraction, that is, the deviation from rectilinear propagation that occurs when light advances beyond an obstruction (p. 433), was first noted by Professor Francesco Maria Grimaldi (1618–1663) at the Jesuit College in Bologna. He had observed bands of light within the shadow of a rod illuminated by a small source. Robert Hooke (1635–1703), curator of experiments for the Royal Society, London, later also observed diffraction effects. He was the first to study the colored interference patterns (p. 393) generated by thin films (*Micrographia*, 1665). He proposed the idea that light was a rapid vibratory motion of the medium propagating at a very great speed. Moreover, "every pulse or vibration of the luminous body will generate a sphere"—this was the beginning of the wave theory. Within a year of Galileo's death, Isaac Newton (1642–1727) was born. The thrust of Newton's scientific effort was to build on direct observation and avoid speculative hypotheses. Thus he remained ambivalent for a long while about the actual nature of light. Was it corpuscular—a stream of particles, as some maintained? Or was light a wave in an all-pervading medium, the aether? At the age of 23, he began his now famous experiments on dispersion.

> I procured me a triangular glass prism to try therewith the celebrated phenomena of colours.

Newton concluded that white light was composed of a mixture of a whole range of independent colors (p. 77). He maintained that the corpuscles of light associated with the various colors excited the aether into characteristic vibrations. Even though his work simultaneously embraced both the wave and emission (corpuscular) theories, he did become more committed to the latter as he grew older. His main reason for rejecting the wave theory as it stood then was the daunting problem of explaining rectilinear propagation in terms of waves that spread out in all directions.

After some all-too-limited experiments, Newton gave up trying to remove chromatic aberration from refracting telescope lenses. Erroneously concluding that it could not be done, he turned to the design of reflectors. Sir Isaac's first reflecting telescope, completed in 1668, was only 6 inches long and 1 inch in diameter, but it magnified some 30 times.

At about the same time that Newton was emphasizing the emission theory in England, Christiaan Huygens (1629–1695), on the continent, was greatly extending the wave theory. Unlike Descartes, Hooke, and Newton, Huygens correctly concluded that light effectively slowed down on entering more dense media. He was able to derive the laws of reflection and refraction and even explained the double refraction of calcite (p. 332), using his wave theory. And it was while working with calcite that he discovered the phenomenon of *polarization* (p. 319).

> As there are two different refractions, I conceived also that there are two different emanations of the waves of light....

Thus light was either a stream of particles or a rapid undulation of aethereal matter. In any case, it was generally agreed that its speed was exceedingly large. Indeed, many believed that light propagated instantaneously, a notion that went back at least as far as Aristotle. The fact that it was finite was determined by the Dane Ole Christensen Römer (1644–1710). Jupiter's nearest moon, Io, has an orbit about that planet that is nearly in the plane of Jupiter's own orbit around the Sun. Römer made a careful study of the eclipses of Io as it moved through the shadow behind Jupiter. In 1676 he predicted that on November 9th Io would emerge from the dark some 10 minutes later than would have been expected on the basis of its yearly averaged motion. Precisely on schedule, Io per-

FIGURE 1.3 Sir Isaac Newton (1642–1727) by Sir Godfrey Kneller (1702). (Courtesy of the National Portrait Gallery.)

FIGURE 1.4 Christianne Huygens (1629–1695). (Rijksmuseum voor de geschiedenis der natuurwetenschappen courtesy AIP Emilio Segré Visual Archives.)

formed as predicted, a phenomenon Römer correctly explained as arising from the finite speed of light. He was able to determine that light took about 22 minutes to traverse the diameter of the Earth's orbit around the Sun—a distance of about 186 million miles. Huygens and Newton, among others, were quite convinced of the validity of Römer's work. Independently estimating the Earth's orbital diameter, they assigned values to c equivalent to 2.3×10^8 m/s and 2.4×10^8 m/s, respectively.*

The great weight of Newton's opinion hung like a shroud over the wave theory during the eighteenth century, all but stifling its advocates. Despite this, the prominent mathematician Leonhard Euler (1707–1783) was a devotee of the wave theory, even if an unheeded one. Euler proposed that the undesirable color effects seen in a lens were absent in the eye (which is an erroneous assumption) because the different media present negated dispersion. He suggested that achromatic lenses (p. 272) might be constructed in a similar way. Enthused by this work, Samuel Klingenstjerna (1698–1765), a professor at Upsala, reperformed Newton's experiments on achromatism and determined them to be in error. Klingenstjerna was in communication with a London optician, John Dollond (1706–1761), who was observing similar results. Dollond finally, in 1758, combined two elements, one of crown and the other of flint glass, to form a single achromatic lens. Incidentally, Dollond's invention was actually preceded by the unpublished work of the amateur scientist Chester Moor Hall (1703–1771) in Essex.

*A. Wróblewski, *Am. J. Phys.* **53** (7), July 1985, p. 620.

1.4 The Nineteenth Century

The wave theory of light was reborn at the hands of Dr. Thomas Young (1773–1829), one of the truly great minds of the century. In 1801, 1802, and 1803, he read papers before the Royal Society extolling the wave theory and adding to it a new fundamental concept, the so-called *Principle of Interference* (p. 377):

> When two undulations, from different origins, coincide either perfectly or very nearly in direction, their joint effect is a combination of the motions belonging to each.

He was able to explain the colored fringes of thin films and determined wavelengths of various colors using Newton's data. Even though Young, time and again, maintained that his conceptions had their very origins in the research of Newton, he was severely attacked. In a series of articles, probably written by Lord Brougham, in the *Edinburgh Review*, Young's papers were said to be "destitute of every species of merit."

Augustin Jean Fresnel (1788–1827), born in Broglie, Normandy, began his brilliant revival of the wave theory in France, unaware of the efforts of Young some 13 years earlier. Fresnel synthesized the concepts of Huygens's wave description and the interference principle (p. 434). The mode of propagation of a primary wave was viewed as a succession of spherical secondary wavelets, which overlapped and interfered to reform the advancing primary wave as it would appear an instant later. In Fresnel's words:

> The vibrations of a luminous wave in any one of its points may be considered as the sum of the elementary movements conveyed to it at the same moment, from the separate action of all the portions of the unobstructed wave considered in any one of its anterior positions.

These waves were presumed to be longitudinal, in analogy with sound waves in air. Fresnel was able to calculate the dif-

FIGURE 1.5 Augustin Jean Fresnel (1788–1827). (Cultural Service of the French Embassy.)

fraction patterns arising from various obstacles and apertures (p. 433) and satisfactorily accounted for rectilinear propagation in homogeneous isotropic media, thus dispelling Newton's main objection to the undulatory theory. When finally apprised of Young's priority to the interference principle, a somewhat disappointed Fresnel nonetheless wrote to Young telling him that he was consoled by finding himself in such good company—the two great men became allies.

Huygens was aware of the phenomenon of polarization arising in calcite crystals, as was Newton. Indeed, the latter in his *Opticks* stated,

> Every Ray of Light has therefore two opposite Sides....

It was not until 1808 that Étienne Louis Malus (1775–1812) discovered that this two-sidedness of light also arose upon reflection (p. 342); the phenomenon was not inherent to crystalline media. Fresnel and Dominique François Arago (1786–1853) then conducted a series of experiments to determine the effect of polarization on interference, but the results were utterly inexplicable within the framework of their longitudinal wave picture—this was a dark hour indeed. For several years Young, Arago, and Fresnel wrestled with the problem until finally Young suggested that the aethereal vibration might be *transverse* as is a wave on a string. The two-sidedness of light was then simply a manifestation of the two orthogonal vibrations of the aether, transverse to the ray direction. Fresnel went on to evolve a mechanistic description of aether oscillations, which led to his now famous formulas for the amplitudes of reflected and transmitted light (p. 111). By 1825 the emission (or corpuscular) theory had only a few tenacious advocates.

The first terrestrial determination of the speed of light was performed by Armand Hippolyte Louis Fizeau (1819–1896) in 1849. His apparatus, consisting of a rotating toothed wheel and a distant mirror (8633 m), was set up in the suburbs of Paris from Suresnes to Montmartre. A pulse of light leaving an opening in the wheel struck the mirror and returned. By adjusting the known rotational speed of the wheel, the returning pulse could be made either to pass through an opening and be seen or to be obstructed by a tooth. Fizeau arrived at a value of the speed of light equal to 315 300 km/s. His colleague Jean Bernard Léon Foucault (1819–1868) was also involved in research on the speed of light. In 1834 Charles Wheatstone (1802–1875) had designed a rotating-mirror arrangement in order to measure the duration of an electric spark. Using this scheme, Arago had proposed to measure the speed of light in dense media but was never able to carry out the experiment. Foucault took up the work, which was later to provide material for his doctoral thesis. On May 6, 1850, he reported to the Academy of Sciences that the speed of light in water was *less* than that in air. This result was in direct conflict with Newton's formulation of the emission theory and a hard blow to its few remaining devotees.

While all of this was happening in Optics, quite independently, the study of electricity and magnetism was also bearing fruit. In 1845 the master experimentalist Michael Faraday (1791–1867) established an interrelationship between electromagnetism and light when he found that the polarization direction of a beam could be altered by a strong magnetic field applied to the medium. James Clerk Maxwell (1831–1879) brilliantly summarized and extended all the empirical knowledge on the subject in a single set of mathematical equations. Beginning with this remarkably succinct and beautifully symmetrical synthesis, he was able to show, purely theoretically, that the electromagnetic field could propagate as a transverse wave in the luminiferous aether (p. 43).

Solving for the speed of the wave, Maxwell arrived at an expression in terms of electric and magnetic properties of the medium ($c = 1/\sqrt{\epsilon_0\mu_0}$). Upon substituting known empirical-

FIGURE 1.6 James Clerk Maxwell (1831–1879). (AIP Emilio Segré Visual Archives.)

ly determined values for these quantities, he obtained a numerical result equal to the measured speed of light! The conclusion was inescapable—*light was "an electromagnetic disturbance in the form of waves" propagated through the aether*. Maxwell died at the age of 48, eight years too soon to see the experimental confirmation of his insights and far too soon for physics. Heinrich Rudolf Hertz (1857–1894) verified the existence of long electromagnetic waves by generating and detecting them in an extensive series of experiments published in 1888.

The acceptance of the wave theory of light seemed to necessitate an equal acceptance of the existence of an all-pervading substratum, the luminiferous aether. If there were waves, it seemed obvious that there must be a supporting medium. Quite naturally, a great deal of scientific effort went into determining the physical nature of the aether, yet it would have to possess some rather strange properties. It had to be so tenuous as to allow an apparently unimpeded motion of celestial bodies. At the same time it could support the exceedingly high-frequency ($\sim 10^{15}$ Hz) oscillations of light traveling at 186 000 miles per second. That implied remarkably strong restoring forces within the aethereal substance. The speed at which a wave advances through a medium is dependent on the characteristics of the disturbed substratum and not on any motion of the source. This is in contrast to the behavior of a stream of particles whose speed with respect to the source is the essential parameter.

Certain aspects of the nature of aether intrude when studying the optics of moving objects, and it was this area of research, evolving quietly on its own, that ultimately led to the next great turning point. In 1725 James Bradley (1693–1762), then Savilian Professor of Astronomy at Oxford, attempted to measure the distance to a star by observing its orientation at two different times of the year. The position of the Earth changed as it orbited around the Sun and thereby provided a large base line for triangulation on the star. To his surprise, Bradley found that the "fixed" stars displayed an apparent systematic movement related to the direction of motion of the Earth in orbit and not dependent, as had been anticipated, on the Earth's position in space. This so-called *stellar aberration* is analogous to the well-known falling-raindrop situation. A raindrop, although traveling vertically with respect to an observer at rest on the Earth, will appear to change its incident angle when the observer is in motion. Thus a corpuscular model of light could explain stellar aberration rather handily. Alternatively, the wave theory also offers a satisfactory explanation provided that *the aether remains totally undisturbed as the Earth plows through it.*

In response to speculation as to whether the Earth's motion through the aether might result in an observable difference between light from terrestrial and extraterrestrial sources, Arago set out to examine the problem experimentally. He found that there were no such observable differences. Light behaved just as if the Earth were at rest with respect to the aether. To explain these results, Fresnel suggested in effect that light was partially dragged along as it traversed a transparent medium in motion. Experiments by Fizeau, in which light beams passed down moving columns of water, and by Sir George Biddell Airy (1801–1892), who used a water-filled telescope in 1871 to examine stellar aberration, both seemed to confirm Fresnel's drag hypothesis. Assuming an aether at *absolute rest*, Hendrik Antoon Lorentz (1853–1928) derived a theory that encompassed Fresnel's ideas.

In 1879 in a letter to D. P. Todd of the U.S. Nautical Almanac Office, Maxwell suggested a scheme for measuring the speed at which the solar system moved with respect to the luminiferous aether. The American physicist Albert Abraham Michelson (1852–1931), then a naval instructor, took up the idea. Michelson, at the tender age of 26, had already established a favorable reputation by performing an extremely precise determination of the speed of light. A few years later, he began an experiment to measure the effect of the Earth's motion through the aether. Since the speed of light in aether is constant and the Earth, in turn, presumably moves in relation to the aether (orbital speed of 67 000 mi/h), the speed of light measured with respect to the Earth should be affected by the planet's motion. In 1881 he published his findings. There was no detectable motion of the Earth with respect to the aether—the aether was stationary. But the decisiveness of this surprising result was blunted somewhat when Lorentz pointed out an oversight in the calculation. Several years later Michelson, then professor of physics at Case School of Applied Science in Cleveland, Ohio, joined with Edward Williams Morley (1838–1923), a well-known professor of chemistry at Western Reserve, to redo the experiment with considerably greater precision. Amazingly enough, their results, published in 1887, once again were negative:

> It appears from all that precedes reasonably certain that if there be any relative motion between the earth and the luminiferous aether, it must be small; quite small enough entirely to refute Fresnel's explanation of aberration.

Thus, whereas an explanation of stellar aberration within the context of the wave theory required the existence of a relative motion between Earth and aether, the Michelson–Morley Experiment refuted that possibility. Moreover, the findings of Fizeau and Airy necessitated the inclusion of a partial drag of light due to motion of the medium.

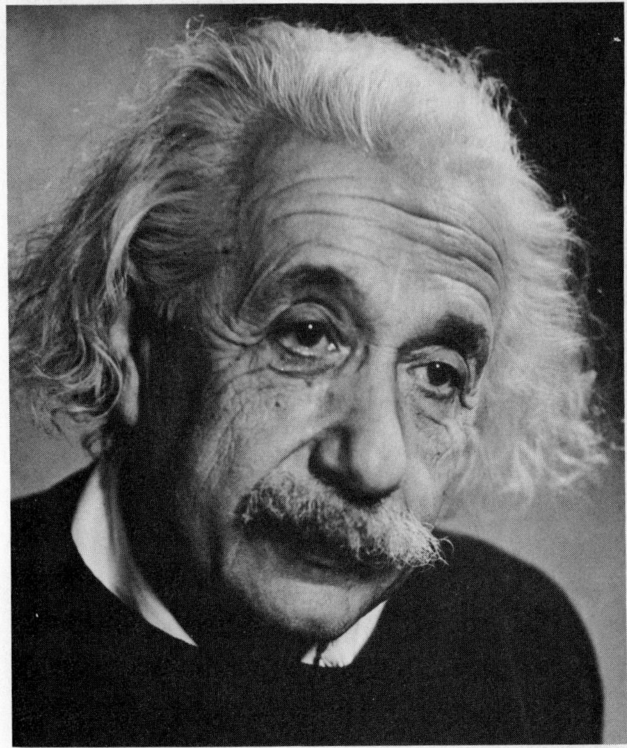

FIGURE 1.7 Albert Einstein (1879–1955).

1.5 Twentieth-Century Optics

Jules Henri Poincaré (1854–1912) was perhaps the first to grasp the significance of the experimental inability to observe any effects of motion relative to the aether. In 1899 he began to make his views known, and in 1900 he said:

> Our aether, does it really exist? I do not believe that more precise observations could ever reveal anything more than *relative* displacements.

In 1905 Albert Einstein (1879–1955) introduced his *Special Theory of Relativity*, in which he too, quite independently, rejected the aether hypothesis.

> The introduction of a "luminiferous aether" will prove to be superfluous inasmuch as the view here to be developed will not require an "absolutely stationary space."

He further postulated:

> light is always propagated in empty space with a definite velocity c which is independent of the state of motion of the emitting body.

The experiments of Fizeau, Airy, and Michelson–Morley were then explained quite naturally within the framework of Einstein's relativistic kinematics.* Deprived of the aether, physicists simply had to get used to the idea that electromagnetic waves could propagate through free space—there was no

*See, for example, *Special Relativity* by French, Chapter 5.

alternative. Light was now envisaged as a self-sustaining wave with the conceptual emphasis passing from aether to field. The electromagnetic wave became an entity in itself.

On October 19, 1900, Max Karl Ernst Ludwig Planck (1858–1947) read a paper before the German Physical Society in which he introduced the beginnings of what was to become yet another great revolution in scientific thought—*Quantum Mechanics*, a theory embracing submicroscopic phenomena (p. 50). In 1905, building on these ideas, Einstein proposed a new form of corpuscular theory in which he asserted that light consisted of globs or "particles" of energy. Each such quantum of radiant energy or *photon*,* as it came to be called, had an energy proportional to its frequency ν, that is, $\mathscr{E} = h\nu$, where h is known as Planck's constant (Fig. 1.8). By the end of the 1920s, through the efforts of Bohr, Born, Heisenberg, Schrödinger, De Broglie, Pauli, Dirac, and others, Quantum Mechanics had become a well-verified theory. It gradually became evident that the concepts of particle and wave, which in the macroscopic world seem so obviously mutually exclusive, must be merged in the submicroscopic domain. The mental image of an atomic particle (e.g., electrons and neutrons) as a minute localized lump of matter would no longer suffice. Indeed, it was found that these "particles" could generate interference and diffraction patterns in precisely the same way as would light (p. 433). Thus photons, protons, electrons, neutrons, and so forth—the whole lot—have both particle and wave manifestations. Still, the matter was by no means settled. "Every physicist thinks that he knows what a photon is," wrote

*The word *photon* was coined by G. N. Lewis, *Nature*, December 18, 1926.

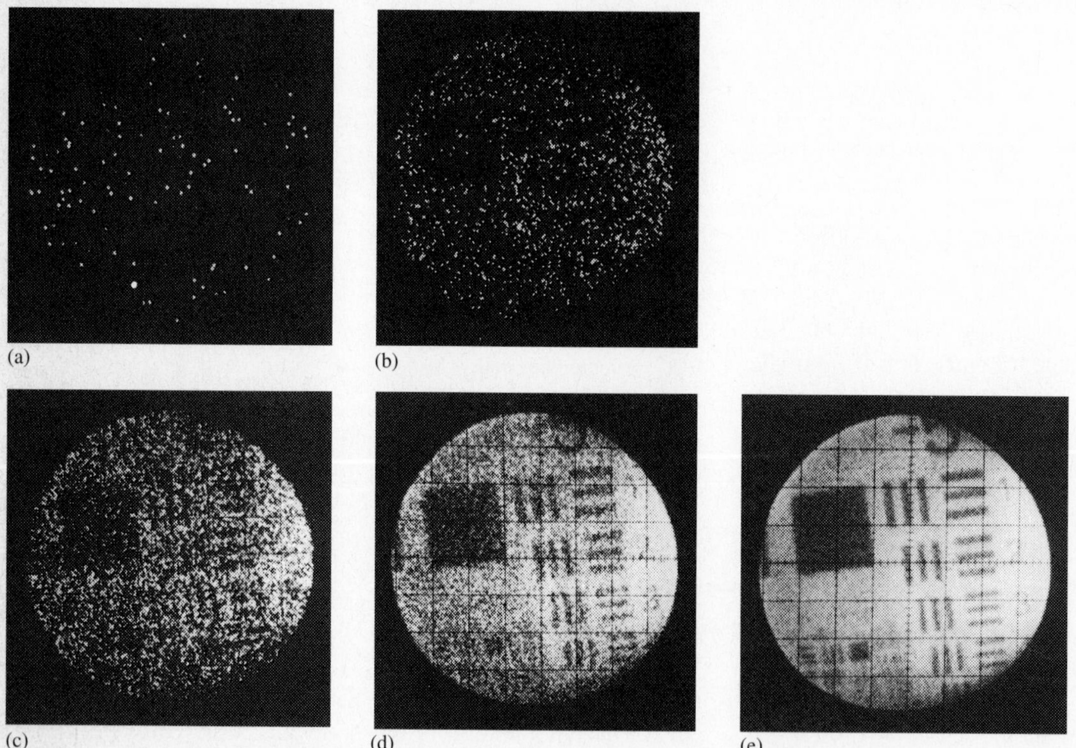

FIGURE 1.8 A rather convincing illustration of the particle nature of light. This sequence of photos was made using a position-sensing photomultiplier tube illuminated by an (8.5×10^3 count-per-second) image of a bar chart. The exposure times were (*a*) 8 ms, (*b*) 125 ms, (*c*) 1 s, (*d*) 10 s, and (*e*) 100 s. Each dot can be interpreted as the arrival of a single photon. (Photos courtesy of ITT Corporation, Electro-Optical Products Division, Tube and Sensor Laboratories, Fort Wayne, Indiana.)

Einstein. "I spent my life to find out what a photon is and I still don't know it."

Relativity liberated light from the aether and showed the kinship between mass and energy (via $\mathscr{E}_0 = mc^2$). What seemed to be two almost antithetical quantities now became interchangeable. Quantum Mechanics went on to establish that a particle* of momentum p had an associated wavelength λ, such that $p = h/\lambda$. The neutrino, a neutral particle presumably having zero rest mass, was postulated for theoretical reasons in 1930 by Wolfgang Pauli (1900–1958) and verified experimentally in the 1950s. The easy images of submicroscopic specks of matter became untenable, and the wave-particle dichotomy dissolved into a duality.

Quantum Mechanics also treats the manner in which light is absorbed and emitted by atoms (p. 63). Suppose we cause a gas to glow by heating it or passing an electrical discharge through it. The light emitted is characteristic of the very structure of the atoms constituting the gas. Spectroscopy, which is the branch of Optics dealing with spectrum analysis (p. 73), developed from the research of Newton. William Hyde Wollaston (1766–1828) made the earliest observations of the dark lines in the solar spectrum (1802). Because of the slit-shaped aperture generally used in spectroscopes, the output consisted of narrow colored bands of light, the so-called *spectral lines*. Working independently, Joseph Fraunhofer (1787–1826) greatly extended the subject. After accidentally discovering the double line of sodium (p. 273), he went on to study sunlight and made the first wavelength determinations using diffraction gratings (p. 465). Gustav Robert Kirchhoff (1824–1887) and Robert Wilhelm Bunsen (1811–1899), working together at Heidelberg, established that each kind of atom had its own signature in a characteristic array of spectral lines. And in 1913 Niels Henrik David Bohr (1885–1962) set forth a precursory quantum theory of the hydrogen atom, which was able to predict the wavelengths of its emission spectrum. The light emitted by an atom is now understood to arise from its outermost electrons (p. 63). The process is the domain of modern quantum theory, which describes the most minute details with incredible precision and beauty.

The flourishing of applied Optics in the second half of the twentieth century represents a renaissance in itself. In the 1950s several workers began to inculcate Optics with the mathematical techniques and insights of communications theory. Just as the idea of momentum provides another dimension in which to visualize aspects of mechanics, the concept of spatial frequency offers a rich new way of appreciating a broad range of optical phenomena. Bound together by the mathematical formalism of Fourier analysis (p. 300), the outgrowths of this contemporary emphasis have been far-reaching. Of particular interest are the theory of image formation and evaluation (p. 523), the *transfer functions* (p. 543), and the idea of *spatial filtering* (p. 313).

The advent of the high-speed digital computer brought with it a vast improvement in the design of complex optical systems. Aspherical lens elements (p. 149) took on renewed practical significance, and the *diffraction-limited* system with an appreciable field of view became a reality. The technique of ion bombardment polishing, in which one atom at a time is chipped away, was introduced to meet the need for extreme precision in the preparation of optical elements. The use of single and multilayer thin-film coatings (reflecting, antireflecting, etc.) became commonplace (p. 418). Fiberoptics evolved into a practical communications tool (p. 195), and thin-film light guides continued to be studied. A great deal of attention was paid to the infrared end of the spectrum (surveillance systems, missile guidance, etc.), and this in turn stimulated the development of infrared materials. Plastics began to be used extensively in Optics (lens elements, replica gratings, fibers, aspherics, etc.). A new class of partially vitrified glass ceramics with exceedingly low thermal expansion was developed. A resurgence in the construction of astronomical observatories (both terrestrial and extraterrestrial) operating across the whole spectrum was well under way by the end of the 1960s and vigorously sustained in the 1980s and 1990s (p. 221).

The first laser was built in 1960, and within a decade laserbeams spanned the range from infrared to ultraviolet. The availability of high-power coherent sources led to the discovery of a number of new optical effects (harmonic generation, frequency mixing, etc.) and thence to a panorama of marvelous new devices. The technology needed to produce a practicable optical communications system developed rapidly. The sophisticated use of crystals in devices such as second-harmonic generators (p. 636), electro-optic and acousto-optic modulators, and the like spurred a great deal of contemporary research in crystal optics. The wavefront reconstruction technique known as *holography* (p. 617), which produces magnificent three-dimensional images, was found to have numerous additional applications (nondestructive testing, data storage, etc.).

*Perhaps it might help if we just called them all *wavicles*.

The military orientation of much of the developmental work in the 1960s continued in the 1970s, 1980s, and the 1990s with added vigor. That technological interest in Optics ranges across the spectrum from "smart bombs" and spy satellites to "death rays" and infrared gadgets that see in the dark. But economic considerations coupled with the need to improve the quality of life have brought products of the discipline into the consumer marketplace as never before. Today lasers are in use everywhere: reading videodiscs in living rooms, cutting steel in factories, scanning labels in supermarkets, and performing surgery in hospitals. Millions of optical display systems on clocks and calculators and computers are blinking all around the world. The almost exclusive use, for the last one hundred years, of electrical signals to handle and transmit data is now rapidly giving way to more efficient optical techniques. A far-reaching revolution in the methods of processing and communicating information is quietly taking place, a revolution that will continue to change our lives in the years ahead.

Profound insights are slow in coming. What few we have took over three thousand years to glean, even though the pace is ever quickening. It is marvelous indeed to watch the answer subtly change while the question immutably remains—*what is light*?*

*For more reading on the history of Optics, see F. Cajori, *A History of Physics*, and V. Ronchi, *The Nature of Light*. Excerpts from a number of original papers can conveniently be found in W. F. Magie, *A Source Book in Physics*, and in M. H. Shamos, *Great Experiments in Physics*.

2 Wave Motion

The issue of the actual nature of light is central to a complete treatment of Optics, and we will struggle with it throughout this work. The straightforward question "Is light a wave phenomenon or a particle phenomenon?" is far more complicated than it might at first seem. For example, the essential feature of a particle is its localization; it exists in a well-defined, "small" region of space. Practically, we tend to take something familiar like a ball or a pebble and shrink it down in imagination until it becomes vanishingly small, and that's a "particle," or at least the basis for the concept of "particle." But a ball interacts with its environment; it has a gravitational field that interacts with the Earth (and the Moon, and Sun, etc.). This field, which spreads out into space—whatever *it* is—cannot be separated from the ball; it is an inextricable part of the ball just as it is an inextricable part of the definition of "particle." Real particles interact via fields, and, in a sense, the field is the particle and the particle is the field. That little conundrum is the domain of Quantum Field Theory, a discipline we'll talk more about later (p. 138). Suffice it to say now that if light is a stream of submicroscopic particles (photons), they are by no means "ordinary" miniball classical particles.

On the other hand, the essential feature of a wave is its non-localization. *A classical traveling wave is a self-sustaining disturbance of a medium, which moves through space transporting energy and momentum.* We tend to think of the ideal wave as a continuous entity that exists over an extended region. But when we look closely at real waves (such as waves on strings), we see composite phenomena comprised of vast numbers of particles moving in concert. The media supporting these waves are atomic (i.e., particulate), and so the waves are not continuous entities in and of themselves. The only possible exception might be the electromagnetic wave. Conceptually, the classical electromagnetic wave (p. 43) is supposed to be a continuous entity, and *it* serves as the model for the very notion of wave as distinct from particle. But in this century we have found that the energy of an electromagnetic wave is *not* distributed continuously. The classical formulation of the electromagnetic theory of light, however wonderful it is on a macroscopic level, is profoundly wanting on a microscopic level. Einstein was the first to suggest that the electromagnetic wave, which we perceive macroscopically, is the statistical manifestation of a fundamentally granular underlying microscopic phenomenon (p. 50). In the subatomic domain, the classical concept of a physical wave is an illusion. Still, in the large-scale regime that we ordinarily work in, electromagnetic waves seem real enough and classical theory applies superbly well.

Because both the classical and quantum-mechanical treatments of light make use of the mathematical description of waves, this chapter lays out the basics of what both formalisms will need. The ideas we develop here will apply to all physical waves from a surface tension ripple in a cup of tea to a pulse of light reaching us from some distant galaxy.

2.1 One-Dimensional Waves

An essential aspect of a traveling wave is that it is a self-sustaining disturbance of the medium through which it propagates. The most familiar waves, and the easiest to visualize (Fig. 2.1), are the mechanical waves, among which are waves on strings, surface waves on liquids, sound waves in the air, and compression waves in both solids and fluids. Sound waves are **longitudinal**—*the medium is displaced in the direction of motion of the wave*. Waves on a string (and electromagnetic

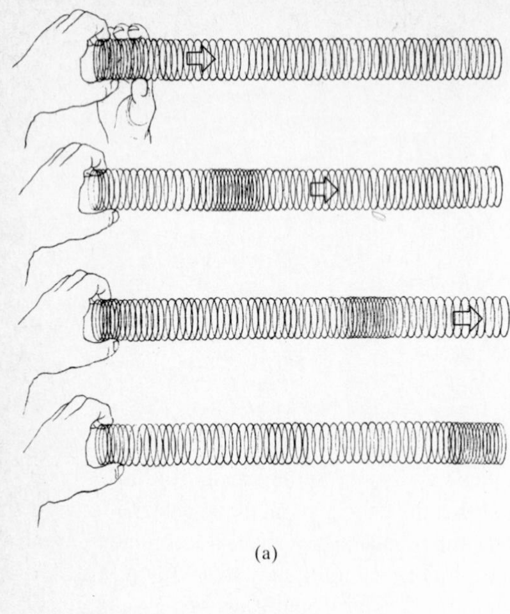

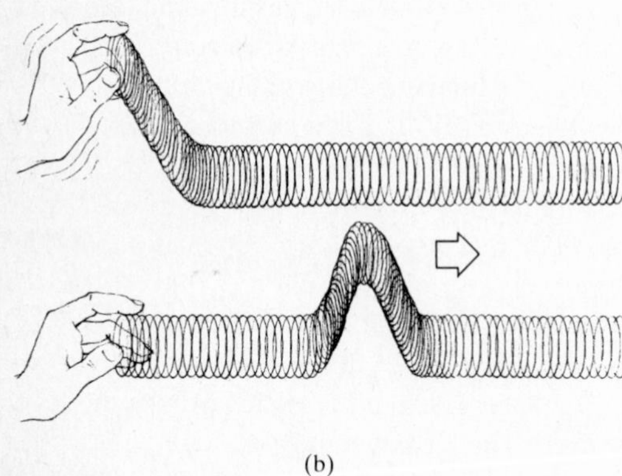

FIGURE 2.1 *(a)* A longitudinal wave in a spring. *(b)* A transverse wave in a spring.

waves) are **transverse**—*the medium is displaced in a direction perpendicular to that of the motion of the wave*. In all cases, although the energy-carrying disturbance advances through the medium, the individual participating atoms remain in the vicinity of their equilibrium positions: *the disturbance advances, not the material medium*. That's one of several crucial features of a wave that distinguishes it from a stream of particles. The wind blowing across a field sets up "waves of grain" that sweep by, even though each stalk only sways in place. Leonardo da Vinci seems to have been the first person to recognize that a wave does not transport the medium through which it travels, and it is precisely this property that allows waves to propagate at very great speeds.

Envision some such disturbance ψ moving in the positive x-direction with a constant speed v. The specific nature of the disturbance is at the moment unimportant. It might be the vertical displacement of the string in Fig. 2.2 or the magnitude of an electric or magnetic field associated with an electromagnetic wave (or even the quantum-mechanical probability amplitude of a matter wave).

Since the disturbance is moving, it must be a function of both position and time and can therefore be written as

$$\psi = f(x, t) \tag{2.1}$$

The shape of the disturbance at any instant, say $t = 0$, can be found by holding time constant at that value. In this case,

$$\psi(x, t)\big|_{t=0} = f(x, 0) = f(x) \tag{2.2}$$

represents the shape or **profile** of the wave at that time. For example, if $f(x) = e^{-ax^2}$, where a is a constant, the profile has the shape of a bell; that is, it is a **Gaussian function**. The process is analogous to taking a "photograph" of the pulse as

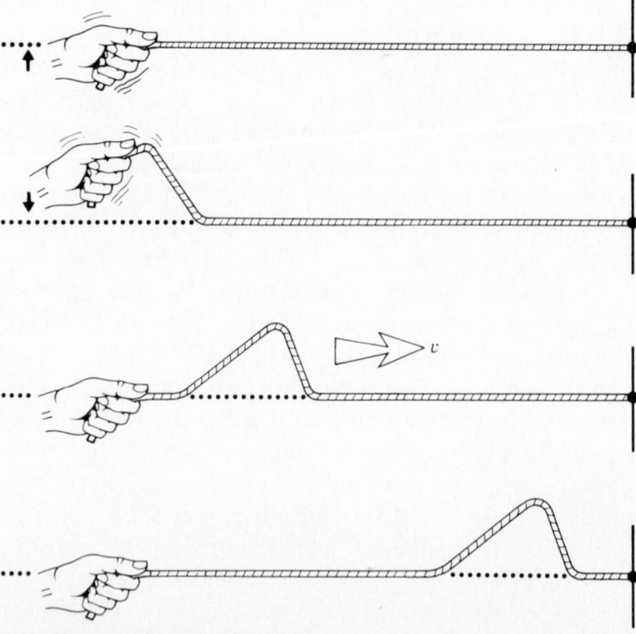

FIGURE 2.2 A wave on a string.

it travels by. For the moment we limit ourselves to a wave that *does not change its shape* as it progresses through space. Figure 2.3 is a "double exposure" of such a disturbance taken at the beginning and end of a time interval t. The pulse has moved along the x-axis a distance vt, but in all other respects it remains unaltered. We now introduce a coordinate system S', which travels along with the pulse at the speed v. In this system ψ is no longer a function of time, and as we move along with S', we see a stationary constant profile with the same functional form as Eq. (2.2). Here, the coordinate is x' rather than x, so that

$$\psi = f(x') \tag{2.3}$$

The disturbance looks the same at any value of t in S' as it did at $t = 0$ in S when S and S' had a common origin. It follows from Fig. 2.3 that

$$x' = x - vt \tag{2.4}$$

so that ψ can be written in terms of the variables associated with the stationary S system as

$$\psi(x, t) = f(x - vt) \tag{2.5}$$

This then represents the most general form of the one-dimensional **wavefunction**. To be more specific, we have only to choose a shape, Eq. (2.2), and then substitute $(x - vt)$ for x in $f(x)$. ***The resulting expression describes a wave having the desired profile, moving in the positive x-direction with a speed*** v. Thus, $\psi(x, t) = e^{-a(x-vt)^2}$ is a bell-shaped wave.

To see how this all works in a bit more detail, let's unfold the analysis for a specific pulse, for example, $\psi(x) = 3/[10x^2 + 1] = f(x)$. That profile is plotted in Fig. 2.4*a*, and if it was a wave on a rope, ψ would be the vertical displacement

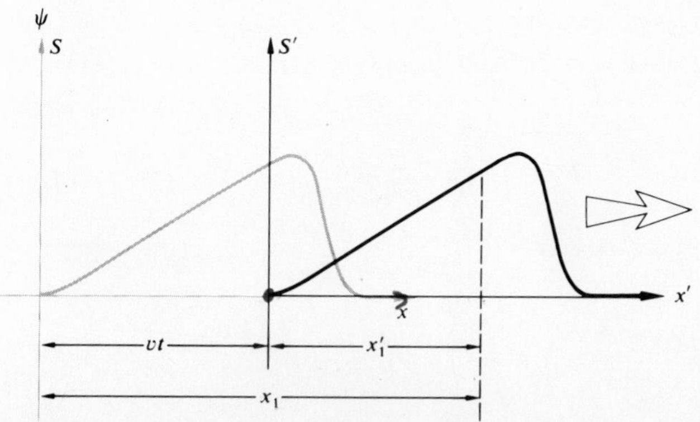

FIGURE 2.3 Moving reference frame.

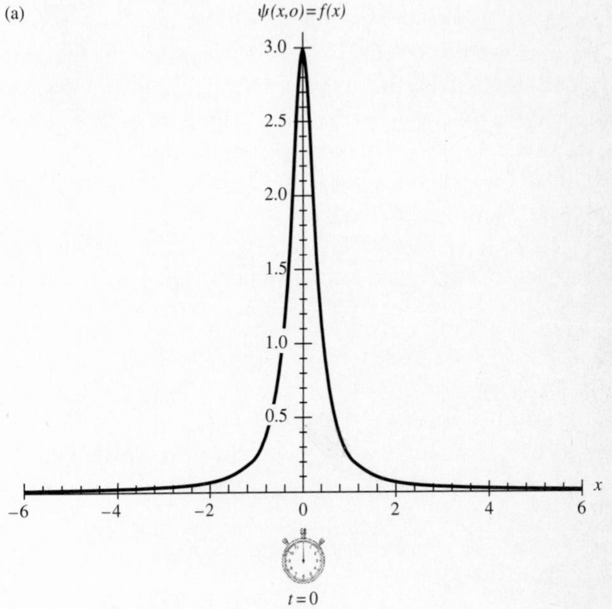

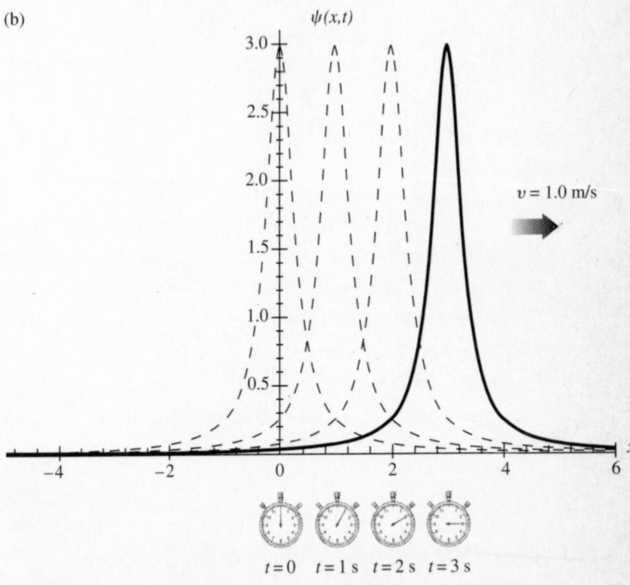

FIGURE 2.4 (*a*) The profile of a pulse given by the function $f(x) = 3/(10x^2 + 1)$. (*b*) The profile shown in (*a*) is now moving as a wave, $\psi(x, t) = 3/[10(x - vt)^2 + 1]$, to the right. It has a speed of 1 m/s and advances in the positive x-direction.

and we might even replace it by the symbol y. Whether ψ represents displacement or pressure or electric field, we now have the profile of the disturbance. To turn $f(x)$ into $\psi(x, t)$, that is, to turn it into the description of a wave moving in the positive x-direction at a speed v, we replace x wherever it appears in $f(x)$ by $(x - vt)$, thereby yielding $\psi(x, t) = 3/[10(x - vt)^2 + 1]$. If v is arbitrarily set equal to, say, 1.0 m/s and the function is plotted successively at $t = 0$, $t = 1$ s, $t = 2$ s, and $t = 3$ s, we get Fig. 2.4*b*, which shows the pulse sailing off to the right at 1.0 m/s, just the way it's supposed to. Incidentally, had we substituted $(x + vt)$ for x in the profile function, the resulting wave would move off to the left.

If we check the form of Eq. (2.5) by examining ψ after an increase in time of Δt and a corresponding increase of $v\,\Delta t$ in x, we find

$$f[(x + v\,\Delta t) - v(t + \Delta t)] = f(x - vt)$$

and the profile is unaltered.

Similarly, if the wave was traveling in the negative x-direction, that is, to the left, Eq. (2.5) would become

$$\psi = f(x + vt), \quad \text{with} \quad v > 0 \tag{2.6}$$

We may conclude therefore that, regardless of the shape of the disturbance, the variables x and t must appear in the function as a unit, that is, as a single variable in the form $(x \mp vt)$. Equation (2.5) is often expressed equivalently as some function of $(t - x/v)$, since

$$f(x - vt) = F\left(-\frac{x - vt}{v}\right) = F(t - x/v) \tag{2.7}$$

The pulse shown in Fig. 2.2 and the disturbance described by Eq. (2.5) are spoken of as *one-dimensional*, because the waves sweep over points lying on a line—it takes only one space variable to specify them. Don't be confused by the fact that in this particular case the rope happens to rise up into a second dimension. In contrast, a two-dimensional wave propagates out across a surface, like the ripples on a pond, and can be described by two space variables.

2.1.1 The Differential Wave Equation

In 1747 Jean Le Rond d'Alembert introduced partial differential equations into the mathematical treatment of physics. That same year, he wrote an article on the motion of vibrating strings in which the so-called *differential wave equation* appears for the first time. This linear, second-order, partial differential equation is usually taken as the defining expression for all types of physical waves. The reason it's a *partial* differential equation is that the wave must be a function of several independent variables, namely, those of space and time. A *linear* differential equation is essentially one consisting of two or more terms, each composed of a constant multiplying a function $\psi(x)$ or its derivatives. The relevant point is that each such term must appear only to the first power; nor can there be any cross products of ψ with its derivatives, or of its derivatives. Recall that the *order* of a differential equation equals the order of the highest derivative in that equation. Furthermore, if a differential equation is of order N, the solution will contain N arbitrary constants. Accordingly, we now derive the one-dimensional form of the wave equation guided by the foreknowledge (p. 15) that the simplest of waves traveling at a fixed speed requires two constants (amplitude and frequency or wavelength) to specify it, and this suggests second derivatives. In any event, take the partial derivative of $\psi(x, t)$ with respect to x, holding t constant. Using $x' = x \mp vt$, we have

$$\frac{\partial \psi}{\partial x} = \frac{\partial f}{\partial x'}\frac{\partial x'}{\partial x} = \frac{\partial f}{\partial x'} \quad \text{since} \quad \frac{\partial x'}{\partial x} = 1 \tag{2.8}$$

Holding x constant, the partial derivative with respect to time is

$$\frac{\partial \psi}{\partial t} = \frac{\partial f}{\partial x'}\frac{\partial x'}{\partial t} = \mp v \frac{\partial f}{\partial x'} \tag{2.9}$$

Combining Eqs. (2.8) and (2.9) yields

$$\frac{\partial \psi}{\partial t} = \mp v \frac{\partial \psi}{\partial x} \tag{2.10}$$

This says that the rate of change of ψ with t and with x are equal, to within a multiplicative constant, as shown in Fig. 2.5. The second partial derivatives of Eqs. (2.8) and (2.9) are

$$\frac{\partial^2 \psi}{\partial x^2} = \frac{\partial^2 f}{\partial x'^2}$$

and

$$\frac{\partial^2 \psi}{\partial t^2} = \frac{\partial}{\partial t}\left(\mp v \frac{\partial f}{\partial x'}\right) = \mp v \frac{\partial}{\partial x'}\left(\frac{\partial f}{\partial t}\right)$$

Since

$$\frac{\partial \psi}{\partial t} = \frac{\partial f}{\partial t}$$

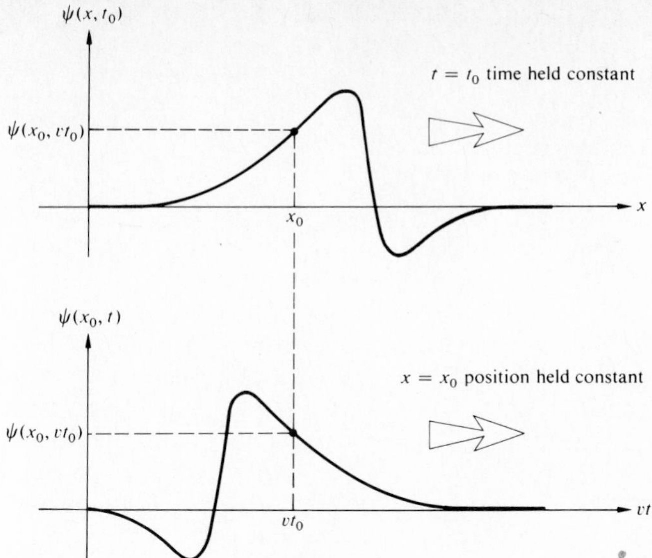

FIGURE 2.5 Variation of ψ with x and t.

it follows, using Eq. (2.9), that

$$\frac{\partial^2 \psi}{\partial t^2} = v^2 \frac{\partial^2 f}{\partial x'^2}$$

Combining these equations, we obtain

$$\frac{\partial^2 \psi}{\partial x^2} = \frac{1}{v^2} \frac{\partial^2 \psi}{\partial t^2} \tag{2.11}$$

which is the desired one-dimensional **differential wave equation**. As a rule, partial differential equations arise when the system being described is continuous. The fact that time is one of the independent variables reflects the continuity of temporal change in the process under analysis. Field theories, in general, treat continuous distributions of quantities in space and time and so take the form of partial differential equations. Maxwell's formulation of electromagnetism, which is a field theory, yields a variation of Eq. (2.11), and from that the concept of the electromagnetic wave arises in a completely natural way (p. 44).

We began this discussion with the special case of waves that have a constant shape as they propagate, even though, as a rule, waves don't maintain a fixed profile. Still, that simple assumption has led us to the general formulation, the differential wave equation. If a function is a solution of that equation, it represents a wave. As we've seen, it will at the same time be a function of $(x \mp vt)$—specifically, one that is twice differentiable (in a nontrivial way) with respect to both x and t.

2.2 Harmonic Waves

Let's now examine the simplest wave form, one for which the profile is a sine or cosine curve. These are variously known as sinusoidal waves, simple harmonic waves, or more succinctly as **harmonic waves**. We shall see in Chapter 7 that any wave shape can be synthesized by a superposition of harmonic waves, and they therefore take on a special significance.

Choose as the profile the simple function

$$\psi(x, t)\big|_{t=0} = \psi(x) = A \sin kx = f(x) \tag{2.12}$$

where k is a positive constant known as the **propagation number**. It's necessary to introduce the constant k simply because we cannot take the sine of a quantity that has physical units. The sine is the ratio of two lengths and is therefore unitless. Accordingly, kx is properly in radians, which is not a real physical unit. The sine varies from $+1$ to -1 so that the maximum value of $\psi(x)$ is A. This maximum disturbance is known as the **amplitude** of the wave (Fig. 2.6). To transform Eq. (2.12) into a *progressive wave* traveling at speed v in the positive x-direction, we need merely replace x by $(x - vt)$, in which case

$$\psi(x, t) = A \sin k(x - vt) = f(x - vt) \tag{2.13}$$

This is clearly (see Problem 2.14) a solution of the differential wave equation. Holding either x or t fixed results in a sinusoidal disturbance; the wave is periodic in both space and time. The **spatial period** is known as the **wavelength** and is denoted by λ. Wavelength is *the number of units of length per wave*. The customary measure of λ is the nanometer, where 1 nm $= 10^{-9}$ m; although the micron (1 μm $= 10^{-6}$ m) is often used, and the older angstrom (1 Å $= 10^{-10}$ m) can still be found in the literature. An increase or decrease in x by the amount λ should leave ψ unaltered, that is,

$$\psi(x, t) = \psi(x \pm \lambda, t) \tag{2.14}$$

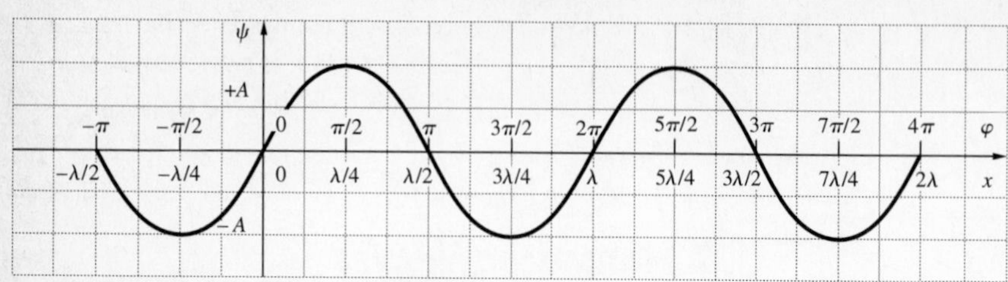

FIGURE 2.6 A harmonic function, which serves as the profile of a harmonic wave. One wavelength corresponds to a change in phase φ of 2π rad.

In the case of a harmonic wave, this is equivalent to altering the argument of the sine function by $\pm 2\pi$. Therefore,

$$\sin k(x - vt) = \sin k[(x \pm \lambda) - vt] = \sin [k(x - vt) \pm 2\pi]$$

and so

$$|k\lambda| = 2\pi$$

or, since both k and λ are positive numbers,

$$k = 2\pi/\lambda \tag{2.15}$$

Figure 2.6 shows how to plot the profile given by Eq. (2.12) in terms of λ. Here φ is the argument of the sine function, also called the **phase**. Notice that $\psi(x) = 0$ whenever $\sin \varphi = 0$, which happens when $\varphi = 0, \pi, 2\pi, 3\pi$, and so on. That occurs at $x = 0, \lambda/2, \lambda$, and $3\lambda/2$, respectively.

In an analogous fashion to the above discussion of λ, we now examine the **temporal period**, τ. This is the amount of time it takes for one complete wave to pass a stationary observer. In this case, it is the repetitive behavior of the wave in time that is of interest, so that

$$\psi(x, t) = \psi(x, t \pm \tau) \tag{2.16}$$

and

$$\sin k(x - vt) = \sin k[x - v(t \pm \tau)]$$

$$= \sin [k(x - vt) \pm 2\pi]$$

Therefore,

$$|kv\tau| = 2\pi$$

But these are all positive quantities; hence

$$kv\tau = 2\pi \tag{2.17}$$

or

$$\frac{2\pi}{\lambda} v\tau = 2\pi$$

from which it follows that

$$\tau = \lambda/v \tag{2.18}$$

The period is *the number of units of time per wave* (Fig. 2.7), the inverse of which is the **temporal frequency** ν, or *the number of waves per unit of time* (i.e., per second). Thus,

$$\nu \equiv 1/\tau$$

in units of cycles per second or Hertz. Equation (2.18) then becomes

$$v = \nu\lambda \tag{2.19}$$

Imagine that you are at rest and a harmonic wave on a string is progressing past you. The number of waves that sweep by per second is ν, and the length of each is λ. In 1.0 s, the overall length of the disturbance that passes you is the product $\nu\lambda$. If, for example, each wave is 2.0 m long and they come at a rate of 5.0 per second, then in 1.0 s, 10 m of wave fly by. This is just what we mean by the speed of the wave (v)—the rate, in m/s, at which it advances. Said slightly differently, because a length of wave λ passes by in a time τ, its speed must equal $\lambda/\tau = \nu\lambda$. Incidentally, Newton derived this relationship in the *Principia* (1687) in a section called "To find the velocity of waves."

Two other quantities are often used in the literature of wave motion. One is the **angular temporal frequency**

$$\omega \equiv 2\pi/\tau = 2\pi\nu \tag{2.20}$$

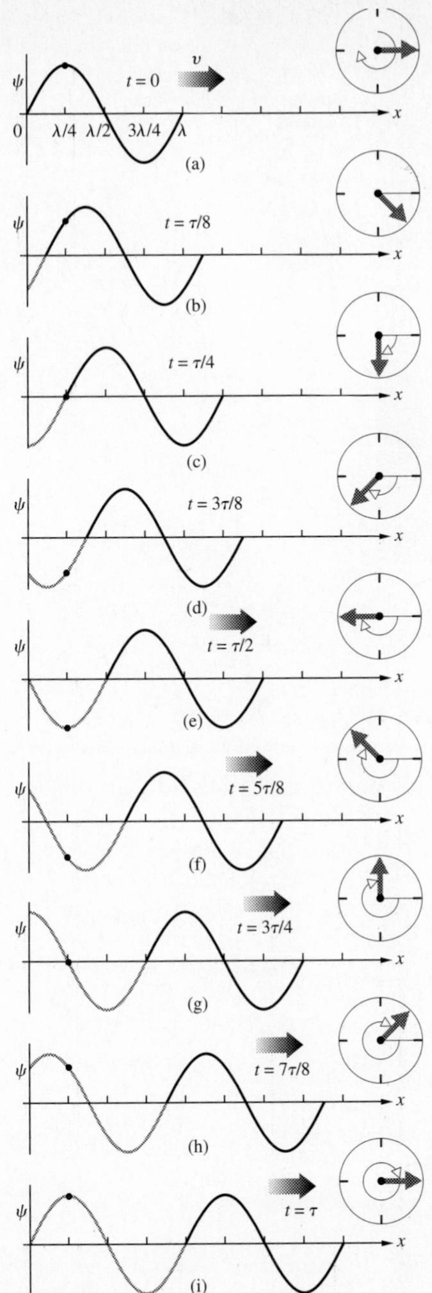

FIGURE 2.7 A harmonic wave moving along the *x*-axis during a time of one period. Note that if this is a picture of a rope any one point on it only moves vertically. We'll discuss the significance of the rotating arrow in Section 2.6.

given in units of radians per second. The other, which is important in spectroscopy, is the **wave number** or **spatial frequency**

$$\kappa \equiv 1/\lambda \tag{2.21}$$

measured in inverse meters. In other words, κ is *the number of waves per unit of length* (i.e., per meter). All of these quantities apply equally well to waves that are not harmonic, as long as each such wave is made up of a single regularly repeated **profile-element** (Fig. 2.8).

Using the above definitions a number of equivalent expressions can be written for the traveling harmonic wave:

$$\psi = A \sin k(x \mp vt) \qquad [2.13]$$

$$\psi = A \sin 2\pi\left(\frac{x}{\lambda} \mp \frac{t}{\tau}\right) \tag{2.22}$$

$$\psi = A \sin 2\pi(\kappa x \mp \nu t) \tag{2.23}$$

$$\psi = A \sin (kx \mp \omega t) \tag{2.24}$$

$$\psi = A \sin 2\pi\nu\left(\frac{x}{v} \mp t\right) \tag{2.25}$$

Of these, Eqs. (2.13) and (2.24) will be encountered most frequently. Note that all these idealized waves are of infinite extent; that is, for any fixed value of t, there is no mathematical limitation on x, which varies from $-\infty$ to $+\infty$. Each such wave has a single constant frequency and is therefore **monochromatic** or, even better, **monoenergetic**. Real waves are never monochromatic. Even a perfect sinusoidal generator cannot have been operating forever. Its output will unavoidably contain a range of frequencies, albeit a small one, just because the wave does not extend back to $t = -\infty$. Thus all waves comprise a band of frequencies, and when that band is narrow the wave is said to be **quasimonochromatic**.

Before we move on, let's put some numbers into Eq. (2.13) and see how to deal with each term. To that end, arbitrarily let $v = 1.0$ m/s and $\lambda = 2.0$ m. Then the wavefunction

$$\psi = A \sin \frac{2\pi}{\lambda}(x - vt)$$

in SI units becomes

$$\psi = A \sin \pi (x - t)$$

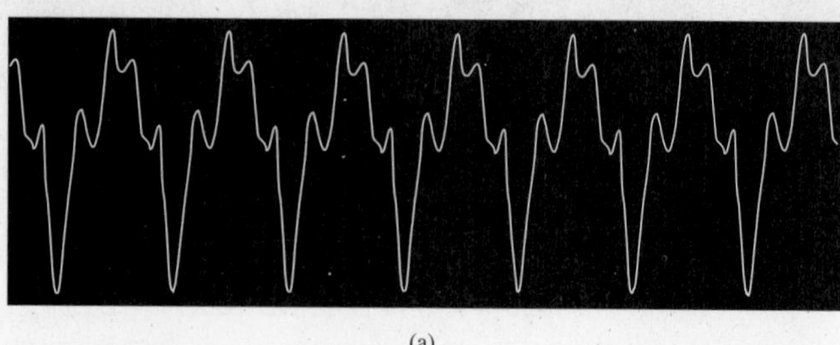

(a)

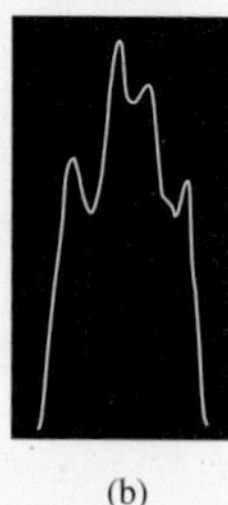

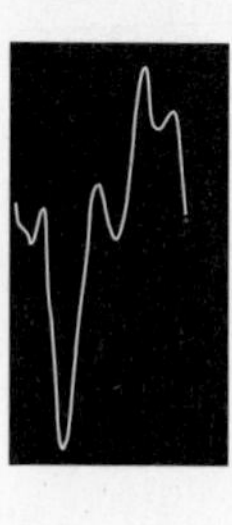

(b)

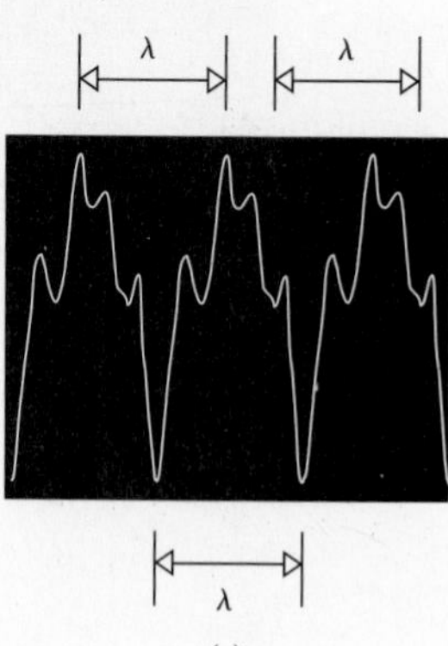

(c)

FIGURE 2.8 (*a*) The waveform produced by a saxophone. Imagine any number of profile-elements (*b*) that, when repeated, create the waveform (*c*). The distance over which the wave repeats itself is called the wavelength, λ.

Figure 2.9 shows how the wave progresses to the right at 1.0 m/s as the time goes from $t = 0$ [whereupon $\psi = A \sin \pi x$] to $t = 1.0$ s [whereupon $\psi = A \sin \pi(x - 1.0)$] to $t = 2.0$ s [whereupon $\psi = A \sin \pi(x - 2.0)$].

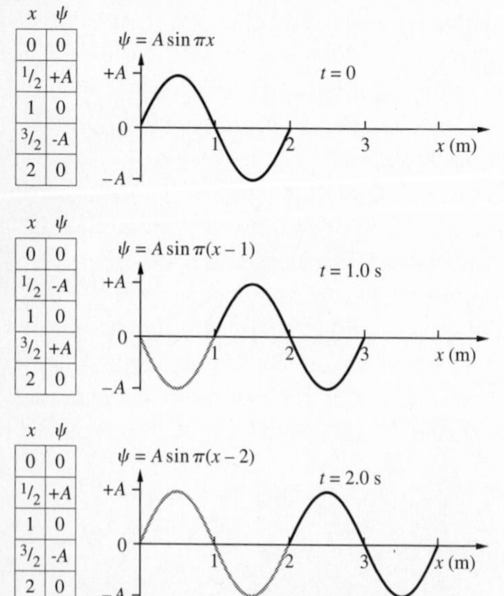

FIGURE 2.9 A progressive wave of the form $\psi(x, t) = A \sin k(x - vt)$, moving to the right at a speed of 1.0 m/s.

2.3 Phase and Phase Velocity

Examine any one of the harmonic wavefunctions, such as

$$\psi(x, t) = A \sin (kx - \omega t) \tag{2.26}$$

The entire argument of the sine is the phase φ of the wave, where

$$\varphi = (kx - \omega t) \tag{2.27}$$

At $t = x = 0$,

$$\psi(x, t)\big|_{\substack{x=0 \\ t=0}} = \psi(0, 0) = 0$$

which is certainly a special case. More generally, we can write

$$\psi(x, t) = A \sin (kx - \omega t + \varepsilon) \tag{2.28}$$

where ε is the **initial phase**. To get a sense of the physical meaning of ε, imagine that we wish to produce a progressive harmonic wave on a stretched string, as in Fig. 2.10. In order to generate harmonic waves, the hand holding the string would have to move such that its vertical displacement y was proportional to the negative of its acceleration, that is, in simple harmonic motion (see Problem 2.15). But at $t = 0$ and $x = 0$, the hand certainly need not be on the x-axis about to move downward, as in Fig. 2.10. It could, of course, begin its motion on an upward swing, in which case $\varepsilon = \pi$, as in Fig. 2.11. In this latter case,

$$\psi(x, t) = y(x, t) = A \sin (kx - \omega t + \pi)$$

which is equivalent to

$$\psi(x, t) = A \sin (\omega t - kx) \tag{2.29}$$

or
$$\psi(x, t) = A \cos \left(\omega t - kx - \frac{\pi}{2}\right)$$

The initial phase angle is just the constant contribution to the phase arising at the generator and is independent of how far in space, or how long in time, the wave has traveled.

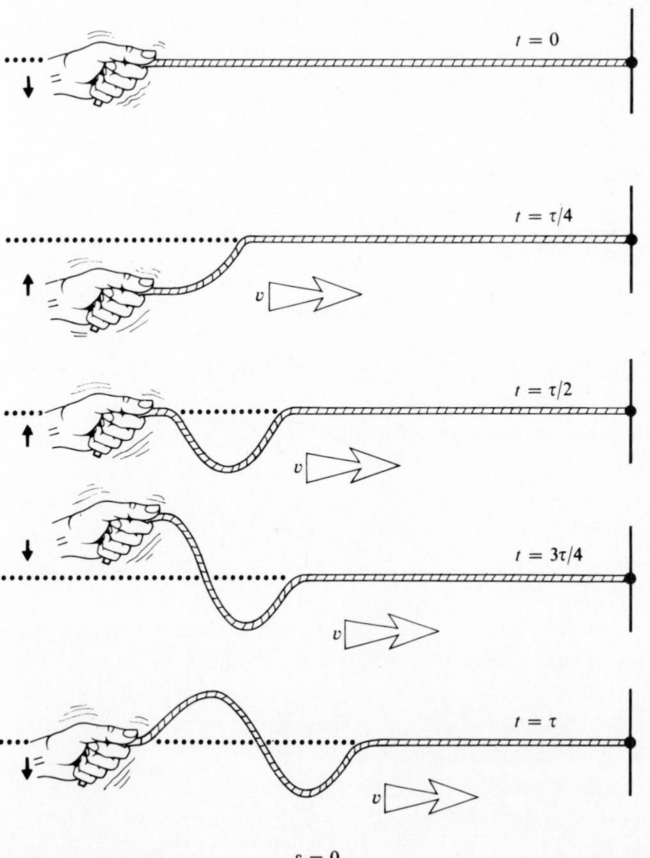

FIGURE 2.10 With $\varepsilon = 0$ note that at $x = 0$ and $t = \tau/4 = \pi/2\omega$, $y = A \sin(-\pi/2) = -A$.

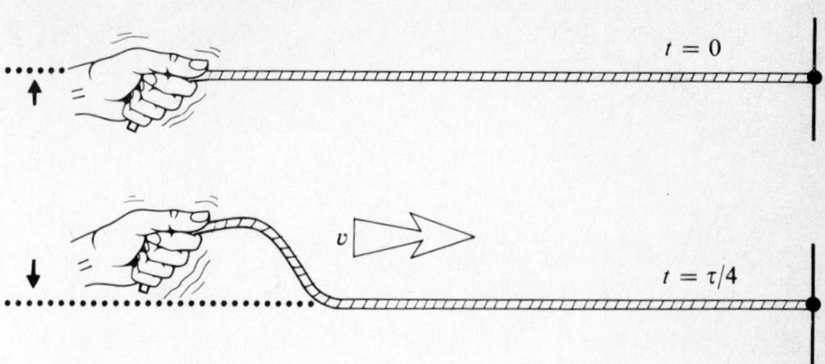

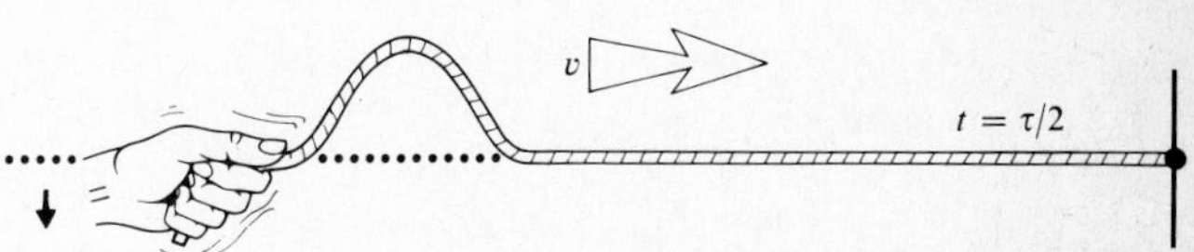

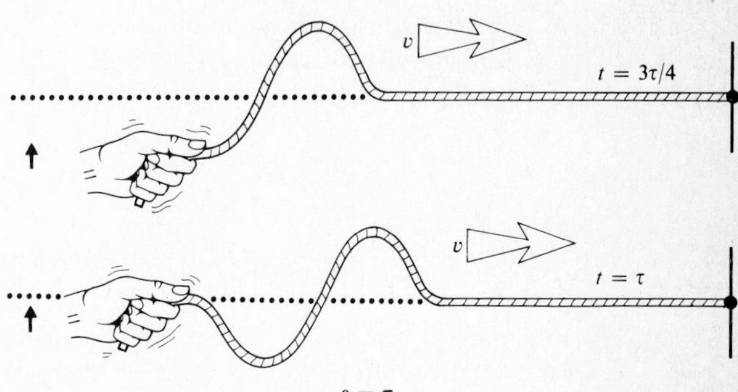

FIGURE 2.11 With $\varepsilon = \pi$ note that at $x = 0$ and $t = \tau/4$, $y = A \sin(\pi/2) = A$.

The phase in Eq. (2.26) is $(kx - \omega t)$, whereas in Eq. (2.29) it's $(\omega t - kx)$. Nonetheless, both of these equations describe waves moving in the positive x-direction that are otherwise identical except for a relative phase difference of π. As is often the case, when the initial phase is of no particular significance in a given situation, either Eq. (2.26) or (2.29) or, if you like, a cosine function can be used to represent the wave. Even so, in some situations one expression for the phase may be mathematically more appealing than another; the literature abounds with both, and so we will use both.

The phase of a disturbance such as $\psi(x, t)$ given by Eq. (2.28) is

$$\varphi(x, t) = (kx - \omega t + \varepsilon)$$

and is obviously a function of x and t. In fact, the partial derivative of φ with respect to t, holding x constant, is the *rate-of-change of phase with time*, or

$$\left|\left(\frac{\partial\varphi}{\partial t}\right)_x\right| = \omega \tag{2.30}$$

The rate-of-change of phase at any fixed location is the angular frequency of the wave, the rate at which a point on the rope in Fig. 2.10 oscillates up and down. That point must go through the same number of cycles per second as the wave. For each cycle, φ changes by 2π.

Similarly, the *rate-of-change of phase with distance*, holding t constant, is

$$\left|\left(\frac{\partial\varphi}{\partial x}\right)_t\right| = k \tag{2.31}$$

These two expressions should bring to mind an equation from the theory of partial derivatives, one used frequently in Thermodynamics, namely,

$$\left(\frac{\partial x}{\partial t}\right)_\varphi = \frac{-\,(\partial\varphi/\partial t)_x}{(\partial\varphi/\partial x)_t} \tag{2.32}$$

The term on the left represents the *speed of propagation of the condition of constant phase*. Imagine a harmonic wave and choose any point on the profile, for example, a crest of the wave. As the wave moves through space, the displacement y of the crest remains fixed. Since the only variable in the harmonic wavefunction is the phase, it too must be constant for that moving point. That is, the phase is fixed at such a value as to yield the constant y corresponding to the chosen point. The point moves along with the profile at the speed v, and so too does the condition of constant phase.

Taking the appropriate partial derivatives of φ as given, for example, by Eq. (2.29) and substituting them into Eq. (2.32), we get

$$\left(\frac{\partial x}{\partial t}\right)_\varphi = \pm\frac{\omega}{k} = \pm v \tag{2.33}$$

This is the *speed* at which the profile moves and is known commonly as the **phase velocity** of the wave. The phase velocity is accompanied by a positive sign when the wave moves in the direction of increasing x and a negative one in the direction of decreasing x. This is consistent with our development of v as the magnitude of the wave velocity: $v > 0$.

Consider the idea of the propagation of constant phase and how it relates to any one of the harmonic wave equations, say

$$\psi = A \sin k(x \mp vt)$$

with
$$\varphi = k(x - vt) = \text{constant}$$

As t increases, x must increase. Even if $x < 0$ so that $\varphi < 0$, x must increase (i.e., become less negative). Here, then, the condition of constant phase moves in the direction of increasing x. As long as the two terms in the phase subtract from each other, the wave travels in the positive x-direction. On the other hand, for

$$\varphi = k(x + vt) = \text{constant}$$

as t increases x can be positive and decreasing or negative and becoming more negative. In either case, the constant-phase condition moves in the decreasing x-direction.

Any point on a harmonic wave having a fixed magnitude moves such that $\varphi(x, t)$ is constant in time, in other words, $d\varphi(x, t)/dt = 0$, or alternatively, $d\psi(x, t)/dt = 0$. This is true for all waves, periodic or not, and it leads (Problem 2.21) to the expression

$$\pm v = -\left(\frac{\partial\psi}{\partial t}\right)_x \bigg/ \left(\frac{\partial\psi}{\partial x}\right)_t \tag{2.34}$$

which can be used to conveniently provide v when we have $\psi(x, t)$. Note that because v is always a positive number, when the ratio on the right turns out negative the motion is in the negative x-direction.

Figure 2.12 depicts a source producing hypothetical two-dimensional waves on the surface of a liquid. The essentially sinusoidal nature of the disturbance, as the medium rises and falls, is evident in the diagram. But there is another useful way to envision what's happening. *The curves connecting all the points with a given phase form a set of concentric circles*. Furthermore, given that A is everywhere constant at any one distance from the source, if φ is constant over a circle, ψ too must be constant over that circle. In other words, all the corresponding peaks and troughs fall on circles, and we speak of these as *circular waves*, each of which expands outward at the speed v.

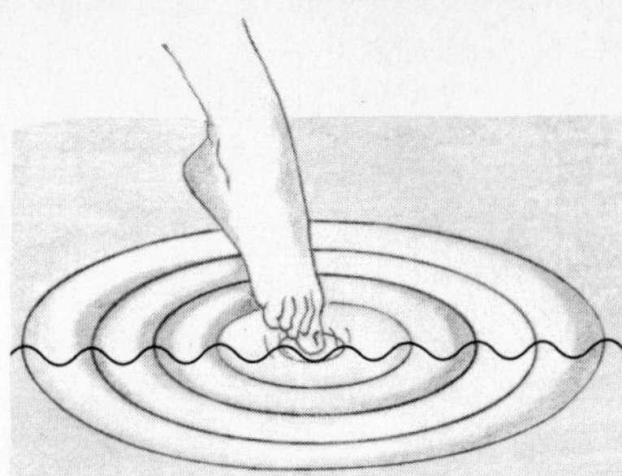

FIGURE 2.12 Idealized circular waves. (Photo by E.H.)

2.4 The Superposition Principle

The form of the differential wave equation [Eq. (2.11)] reveals an intriguing property of waves, one that is quite unlike the behavior of a stream of classical particles. Suppose that the wavefunctions ψ_1 and ψ_2 are each separate solutions of the wave equation; it follows that $(\psi_1 + \psi_2)$ is also a solution. This is known as the **Superposition Principle**, and it can easily be proven since it must be true that

$$\frac{\partial^2 \psi_1}{\partial x^2} = \frac{1}{v^2}\frac{\partial^2 \psi_1}{\partial t^2} \quad \text{and} \quad \frac{\partial^2 \psi_2}{\partial x^2} = \frac{1}{v^2}\frac{\partial^2 \psi_2}{\partial t^2}$$

Adding these yields

$$\frac{\partial^2 \psi_1}{\partial x^2} + \frac{\partial^2 \psi_2}{\partial x^2} = \frac{1}{v^2}\frac{\partial^2 \psi_1}{\partial t^2} + \frac{1}{v^2}\frac{\partial^2 \psi_2}{\partial t^2}$$

and so

$$\frac{\partial^2}{\partial x^2}(\psi_1 + \psi_2) = \frac{1}{v^2}\frac{\partial^2}{\partial t^2}(\psi_1 + \psi_2)$$

which establishes that $(\psi_1 + \psi_2)$ is indeed a solution. What this means is that when two separate waves arrive at the same place in space wherein they overlap, they will simply add to (or subtract from) one another without permanently destroying or disrupting either wave. ***The resulting disturbance at each point in the region of overlap is the algebraic sum of the individual constituent waves at that location*** (Fig. 2.13). Once having past through the region where the two waves coexist, each will move out and away unaffected by the encounter.

Keep in mind that we are talking about a *linear* superposition of waves, a process that's widely valid and the most commonly encountered. Nonetheless, it is also possible for the wave amplitudes to be large enough to drive the medium in a nonlinear fashion (p. 634). For the time being we'll concentrate on the linear differential wave equation, which results in a linear Superposition Principle.

Much of Optics involves the superposition of waves in one way or another. Even the basic processes of reflection and refraction are manifestations of the scattering of light from countless atoms (p. 85), a phenomenon that can only be treated satisfactorily in terms of the overlapping of waves. It therefore becomes crucial that we understand the process, at least qualitatively, as soon as possible. Consequently, carefully examine

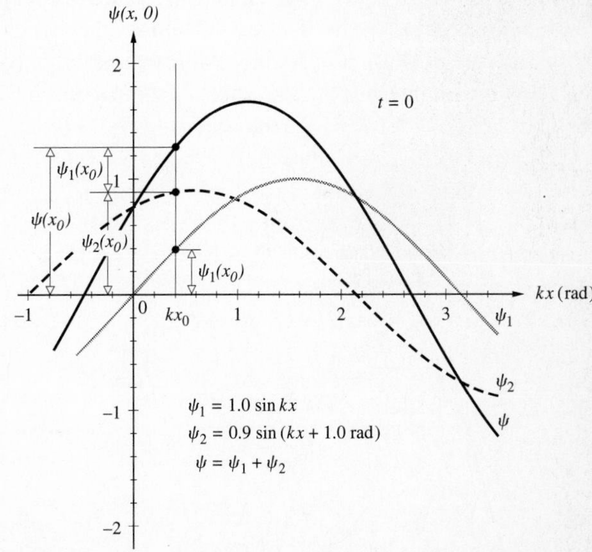

FIGURE 2.13 The superposition of two equal-wavelength sinusoids ψ_1 and ψ_2. The resultant is a sinusoid with the same wavelength, which at every point equals the algebraic sum of the constituent sinusoids. Thus at $x = x_0$, $\psi(x_0) = \psi_1(x_0) + \psi_2(x_0)$; the magnitudes add.

the two coexisting waves in Fig. 2.13. At every point (i.e., every value of kx) we simply add ψ_1 and ψ_2, either of which could be positive or negative. As a quick check, keep in mind that wherever either constituent wave is zero (e.g., $\psi_1 = 0$), the resultant disturbance equals the value of the other nonzero constituent wave ($\psi = \psi_2$), and those two curves cross at that location (e.g., at $kx = 0$ and $+3.14$ rad). On the other hand, $\psi = 0$ wherever the two constituent waves have equal magnitudes and opposite signs (e.g., at $kx = +2.67$ rad). Incidentally, notice how a relative *positive* phase difference of 1.0 rad between the two curves shifts ψ_2 to the *left* with respect to ψ_1 by 1.0 rad.

Developing the illustration a bit further, Fig. 2.14 shows how the resultant arising from the superposition of two nearly equal-amplitude waves depends on the *phase-angle difference* between them. In Fig. 2.14*a* the two constituent waves have the same phase; that is, their phase-angle difference is zero, and they are said to be **in-phase**; they rise and fall in-step, reinforcing each other. The composite wave, which then has a substantial amplitude, is sinusoidal with the same frequency and wavelength as the component waves (p. 289). Following the sequence of the drawings, we see that the resultant amplitude diminishes as the phase-angle difference increases until, in Fig. 2.14*d*, it almost vanishes when that difference equals π. The waves are then said to be 180° **out-of-phase**. The fact that waves which are out-of-phase tend to diminish each other has given the name **interference** to the whole phenomenon.

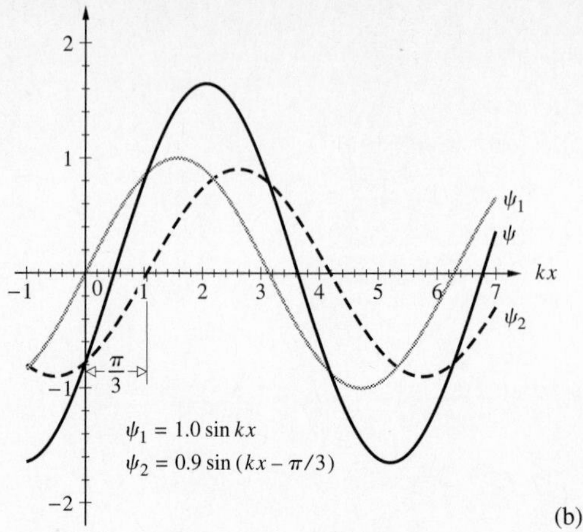

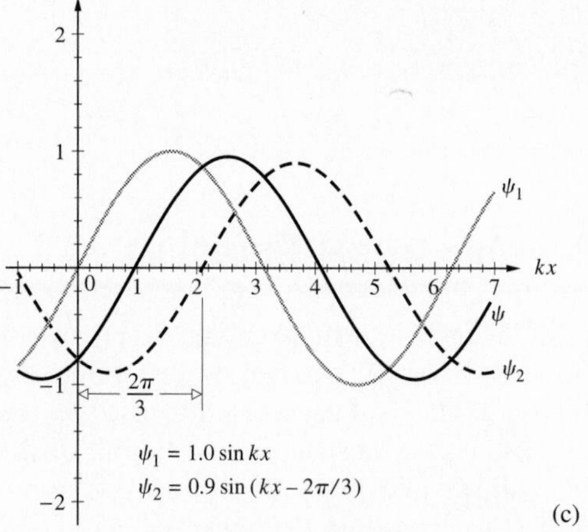

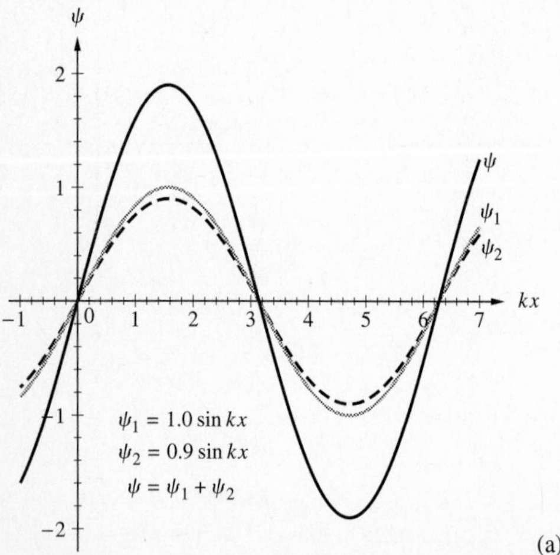

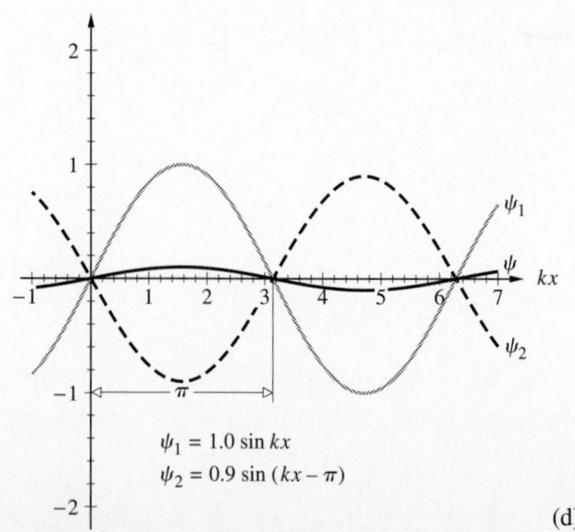

FIGURE 2.14a. The superposition of two sinusoids with amplitudes of 1.0 and 0.9. In (*a*) they are in-phase. In (*b*) ψ_1 leads ψ_2 by $\pi/3$. In (*c*) ψ_1 leads ψ_2 by $2\pi/3$. And (*d*) ψ_1 and ψ_2 are out-of-phase by π and almost cancel each other.

2.5 The Complex Representation

As we develop the analysis of wave phenomena, it will become evident that the sine and cosine functions that describe harmonic waves can be somewhat awkward for our purposes. The expressions formulated will sometimes be rather involved and the trigonometric manipulations required to cope with them will be even more unattractive. The complex-number representation offers an alternative description that is mathematically simpler to process. In fact, complex exponentials are used extensively in both Classical and Quantum Mechanics, as well as in Optics.

The complex number $\tilde{z}$ has the form

$$\tilde{z} = x + iy \tag{2.35}$$

where $i = \sqrt{-1}$. The real and imaginary parts of $\tilde{z}$ are respectively x and y, where both x and y are themselves real numbers. This is illustrated graphically in the Argand diagram in Fig. 2.15*a*. In terms of polar coordinates (r, θ),

$$x = r\cos\theta, \qquad y = r\sin\theta$$

and

$$\tilde{z} = x + iy = r(\cos\theta + i\sin\theta)$$

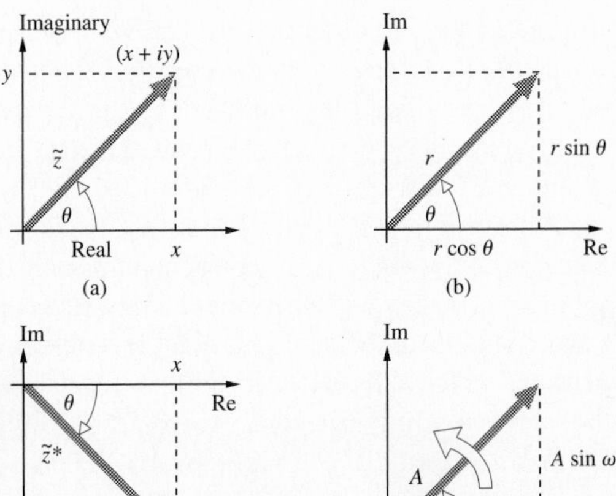

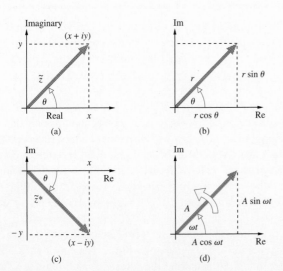

FIGURE 2.15 An Argand diagram is a representation of a complex number in terms of its real and imaginary components. This can be done using either (*a*) x and y or (*b*) r and θ. Moreover, when θ is a constantly changing function of time (*d*), the arrow rotates at a rate ω.

The *Euler formula**

$$e^{i\theta} = \cos\theta + i\sin\theta$$

leads to the expression $e^{-i\theta} = \cos\theta - i\sin\theta$ and adding and subtracting these two equations yields

$$\cos\theta = \frac{e^{i\theta} + e^{-i\theta}}{2}$$

and

$$\sin\theta = \frac{e^{i\theta} - e^{-i\theta}}{2i}$$

Moreover, the Euler formula allows us (Fig. 2.15*b*) to write

$$\tilde{z} = re^{i\theta} = r\cos\theta + ir\sin\theta$$

where r is the *magnitude* of $\tilde{z}$ and θ is the *phase angle* of $\tilde{z}$, in radians. The magnitude is often denoted by $|\tilde{z}|$ and referred to as the *modulus* or *absolute value* of the complex number. The *complex conjugate*, indicated by an asterisk (Fig. 2.15*c*), is found by replacing i wherever it appears, with $-i$, so that

$$\tilde{z}^* = (x + iy)^* = (x - iy)$$

$$\tilde{z}^* = r(\cos\theta - i\sin\theta)$$

and

$$\tilde{z}^* = re^{-i\theta}$$

The operations of addition and subtraction are quite straightforward:

$$\tilde{z}_1 \pm \tilde{z}_2 = (x_1 + iy_1) \pm (x_2 + iy_2)$$

and therefore

$$\tilde{z}_1 \pm \tilde{z}_2 = (x_1 \pm x_2) + i(y_1 \pm y_2)$$

Notice that this process is very much like the component addition of vectors.

Multiplication and division are most simply expressed in polar form

$$\tilde{z}_1\tilde{z}_2 = r_1r_2e^{i(\theta_1+\theta_2)}$$

*If you have any doubts about this identity, take the differential of $\tilde{z} = \cos\theta + i\sin\theta$, where $r = 1$. This yields $d\tilde{z} = i\tilde{z}\,d\theta$, and integration gives $\tilde{z} = \exp(i\theta)$.

and
$$\frac{\tilde{z}_1}{\tilde{z}_2} = \frac{r_1}{r_2} e^{i(\theta_1 - \theta_2)}$$

A number of facts that will be useful in future calculations are well worth mentioning at this point. It follows from the ordinary trigonometric addition formulas (Problem 2.31) that

$$e^{\tilde{z}_1 + \tilde{z}_2} = e^{\tilde{z}_1} e^{\tilde{z}_2}$$

and so, if $\tilde{z}_1 = x$ and $\tilde{z}_2 = iy$,

$$e^{\tilde{z}} = e^{x+iy} = e^x e^{iy}$$

The modulus of a complex quantity is given by

$$r = |\tilde{z}| \equiv (\tilde{z}\tilde{z}^*)^{1/2}$$

and
$$|e^{\tilde{z}}| = e^x$$

Inasmuch as $\cos 2\pi = 1$ and $\sin 2\pi = 0$,

$$e^{i2\pi} = 1$$

Similarly,

$$e^{i\pi} = e^{-i\pi} = -1 \quad \text{and} \quad e^{\pm i\pi/2} = \pm i$$

The function $e^{\tilde{z}}$ is periodic; that is, it repeats itself every $i2\pi$:

$$e^{\tilde{z}+i2\pi} = e^{\tilde{z}} e^{i2\pi} = e^{\tilde{z}}$$

Any complex number can be represented as the sum of a real part Re $(\tilde{z})$ and an imaginary part Im (z)

$$\tilde{z} = \text{Re}\,(\tilde{z}) + i\,\text{Im}\,(\tilde{z})$$

such that

$$\text{Re}\,(\tilde{z}) = \tfrac{1}{2}(\tilde{z} + \tilde{z}^*) \quad \text{and} \quad \text{Im}\,(\tilde{z}) = \frac{1}{2i}(\tilde{z} - \tilde{z}^*)$$

Both of these expressions follow immediately from the Argand diagram, Fig. 2.15*a* and *c*. For example, $\tilde{z} + \tilde{z}^* = 2x$ because the imaginary parts cancel, and so Re $(\tilde{z}) = x$.

From the polar form where

$$\text{Re}\,(\tilde{z}) = r\cos\theta \quad \text{and} \quad \text{Im}\,(\tilde{z}) = r\sin\theta$$

it is clear that either part could be chosen to describe a harmonic wave. It is customary, however, to choose the real part, in which case a harmonic wave is written as

$$\psi(x, t) = \text{Re}\,[Ae^{i(\omega t - kx + \varepsilon)}] \tag{2.36}$$

which is, of course, equivalent to

$$\psi(x, t) = A\cos(\omega t - kx + \varepsilon)$$

Henceforth, wherever it's convenient, we shall write the wavefunction as

$$\psi(x, t) = Ae^{i(\omega t - kx + \varepsilon)} = Ae^{i\varphi} \tag{2.37}$$

and utilize this complex form in the required computations. This is done to take advantage of the ease with which complex exponentials can be manipulated. Only after arriving at a final result, and then only if we want to represent the actual wave, must we take the real part. It has, accordingly, become quite common to write $\psi(x, t)$, as in Eq. (2.37), where it is understood that the actual wave is the real part.

2.6 Phasors and the Addition of Waves

The arrow in the Argand diagram (Fig. 2.15*d*) is set rotating at a frequency ω by letting the angle equal ωt. This suggests a scheme for representing (and ultimately adding) waves that we will introduce here qualitatively and develop later (p. 290) quantitatively. Figure 2.16 depicts a harmonic wave of amplitude A traveling to the left. The arrow in the diagram has a length A and revolves at a constant rate such that the changing angle it makes with the reference x-axis is ωt. This rotating arrow and its associated phase angle together constitute a **phasor**, which tells us everything we need to know about the corresponding harmonic wave. It's common to express a phasor in terms of its amplitude, A, and phase, φ, as $A\angle\varphi$.

To see how this works, let's first examine each part of Fig. 2.16 separately. The phasor in Fig. 2.16*a* has a zero phase angle; that is, it lies along the reference axis; the associated sine function can also serve as a reference. In Fig. 2.16*b* the phasor has a phase angle of $+\pi/3$ rad, and the sine curve is shifted to the left by $\pi/3$ rad. That sine curve reaches its first

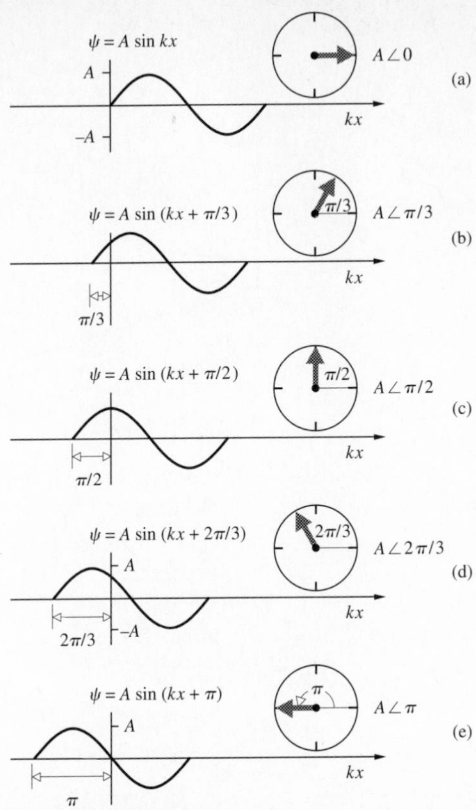

FIGURE 2.16 A plot of the function $\psi = A \sin (kx + \omega t)$ and the corresponding phasor diagrams. In (*a*), (*b*), (*c*), (*d*), and (*e*), the values of ωt are 0, $\pi/3$, $\pi/2$, $2\pi/3$, and π, respectively.

peak at a smaller value of kx than does the reference curve in part (*a*), and therefore it *leads* the reference by $\pi/3$ rad. In parts (*c*), (*d*), and (*e*) of Fig. 2.16, the phase angles are $+\pi/2$ rad, $+2\pi/3$ rad, and $+\pi$ rad, respectively. The entire sequence of curves can be seen as a wave, $\psi = A \sin (kx + \omega t)$, traveling to the left. It is equivalently represented by a phasor rotating counterclockwise such that its phase angle at any moment is ωt. Much the same thing happens in Fig. 2.7, but there the wave advances to the right and the phasor rotates clockwise.

When wavefunctions are combined, we are usually interested in the resulting amplitude and phase. With that in mind, reexamine the way waves add together in Fig. 2.14. Apparently, for disturbances that are in-phase (Fig. 2.14*a*) the amplitude of the resultant wave, A, is the sum of the constituent amplitudes: $A = A_1 + A_2 = 1.0 + 0.9 = 1.9$. This is the same thing we would get if we added two colinear vectors pointing in the same direction. Similarly (Fig. 2.14*d*), when the component waves are 180° out-of-phase $A = A_1 - A_2 = 1.0 - 0.9 = 0.1$ as if two colinear oppositely directed vectors were added. Although phasors are not vectors, they do add in a similar way. Later, we'll prove that two arbitrary phasors, $A_1\angle\varphi_1$ and $A_2\angle\varphi_2$, combine tip-to-tail, as vectors would (Fig. 2.17), to produce a resultant $A\angle\varphi$. Because both phasors rotate together at a rate ω, we can simply freeze them at $t = 0$ and not worry about their time dependence, which makes them a lot easier to draw.

The four phasor diagrams in Fig. 2.18 correspond to the four wave combinations taking place sequentially in Fig. 2.14. When the waves are in-phase (as in Fig. 2.14*a*), we take the phases of both wave-1 and wave-2 to be zero (Fig. 2.18*a*) and position the corresponding phasors tip-to-tail along the zero-φ reference axis. When the waves differ in phase by $\pi/3$ (as in Fig. 2.14*b*), the phasors have a relative phase (Fig. 2.18*b*) of $\pi/3$. The resultant, which has an appropriately reduced amplitude, has a phase φ that is between 0 and $\pi/3$, as can be seen in both Figs. 2.14*b* and 2.18*b*. When the two waves differ in phase by $2\pi/3$ (as in Fig. 2.14*c*), the corresponding phasors almost form an equilateral triangle in Fig. 2.18*c* (but for the fact that $A_1 > A_2$) and so A now lies between A_1 and A_2. Finally, when the phase-angle difference for the two waves (and the two phasors) is π rad (i.e., 180°), they almost cancel and the resulting amplitude is a minimum. Notice (in Fig. 2.18*d*) that the resultant phasor points along the reference axis and so has the same phase (i.e., zero) as $A_1\angle\varphi_1$. Thus it is 180° out-of-phase with $A_2\angle\varphi_2$; the same thing is true of the corresponding waves in Fig. 2.14*d*.

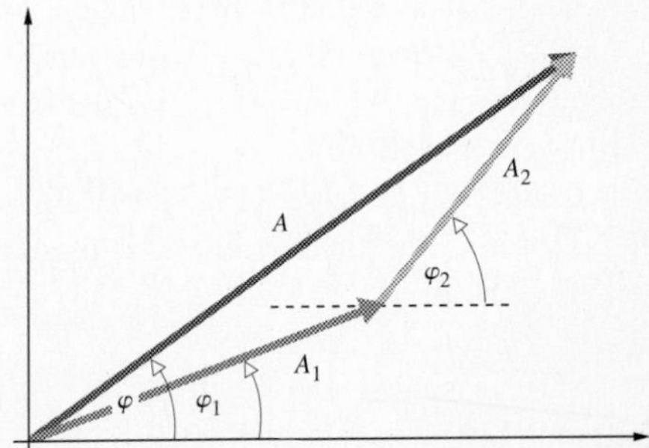

FIGURE 2.17 The sum of two phasors $A_1\angle\varphi_1$ and $A_2\angle\varphi_2$ equals $A\angle\varphi$.

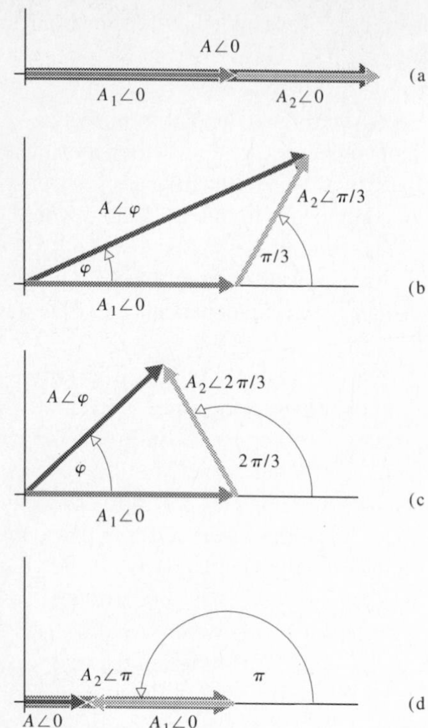

FIGURE 2.18 The addition of phasors.

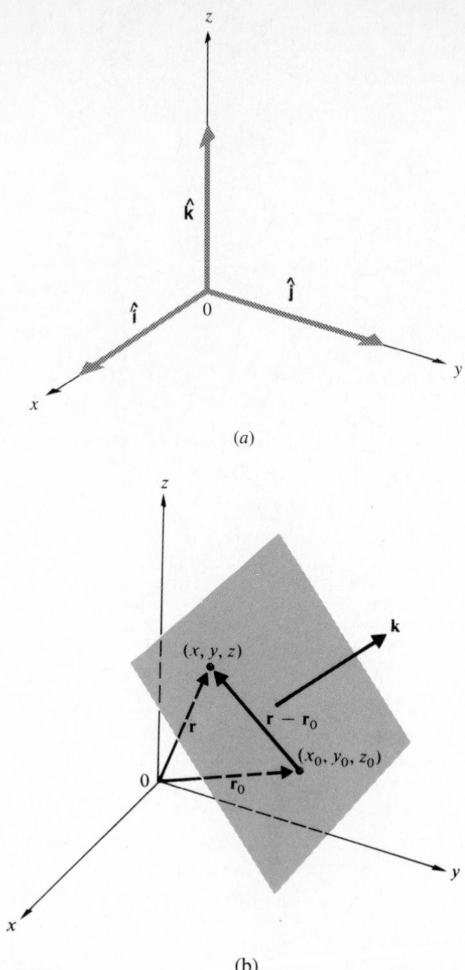

FIGURE 2.19 (*a*) The Cartesian unit basis vectors. (*b*) A plane wave moving in the **k**-direction.

2.7 Plane Waves

The plane wave is perhaps the simplest example of a three-dimensional wave. It exists at a given time, when all the surfaces on which a disturbance has constant phase form a set of planes, each generally perpendicular to the propagation direction. There are quite practical reasons for studying this sort of disturbance, one of which is that by using optical devices, we can readily produce light resembling plane waves.

The mathematical expression for a plane that is perpendicular to a given vector **k** and that passes through some point (x_0, y_0, z_0) is rather easy to derive (Fig. 2.19). First we write the position vector in Cartesian coordinates in terms of the unit basis vectors (Fig. 2.19*a*),

$$\mathbf{r} = x\hat{\mathbf{i}} + y\hat{\mathbf{j}} + z\hat{\mathbf{k}}$$

It begins at some arbitrary origin O and ends at the point (x, y, z), which can, for the moment, be anywhere in space. Similarly,

$$(\mathbf{r} - \mathbf{r}_0) = (x - x_0)\hat{\mathbf{i}} + (y - y_0)\hat{\mathbf{j}} + (z - z_0)\hat{\mathbf{k}}$$

By setting

$$(\mathbf{r} - \mathbf{r}_0)\cdot\mathbf{k} = 0 \tag{2.38}$$

we force the vector $(\mathbf{r} - \mathbf{r}_0)$ to sweep out a plane perpendicular to **k**, as its endpoint (x, y, z) takes on all allowed values. With

$$\mathbf{k} = k_x\hat{\mathbf{i}} + k_y\hat{\mathbf{j}} + k_z\hat{\mathbf{k}} \tag{2.39}$$

Equation (2.38) can be expressed in the form

$$k_x(x - x_0) + k_y(y - y_0) + k_z(z - z_0) = 0 \qquad (2.40)$$

or as

$$k_x x + k_y y + k_z z = a \qquad (2.41)$$

where

$$a = k_x x_0 + k_y y_0 + k_z z_0 = \text{constant} \qquad (2.42)$$

The most concise form of the equation of a plane perpendicular to **k** is then just

$$\mathbf{k} \cdot \mathbf{r} = \text{constant} = a \qquad (2.43)$$

The plane is the locus of all points whose position vectors each have the same projection onto the **k**-direction.

We can now construct a set of planes over which $\psi(\mathbf{r})$ varies in space sinusoidally, namely,

$$\psi(\mathbf{r}) = A \sin (\mathbf{k} \cdot \mathbf{r}) \qquad (2.44)$$

$$\psi(\mathbf{r}) = A \cos (\mathbf{k} \cdot \mathbf{r}) \qquad (2.45)$$

or

$$\psi(\mathbf{r}) = Ae^{i\mathbf{k} \cdot \mathbf{r}} \qquad (2.46)$$

For each of these expressions $\psi(\mathbf{r})$ is constant over every plane defined by $\mathbf{k} \cdot \mathbf{r}$ = constant. Since we are dealing with harmonic functions, they should repeat themselves in space after a displacement of λ in the direction of **k**. Figure 2.20 is a rather humble representation of this kind of expression. We have drawn only a few of the infinite number of planes, each having a different $\psi(\mathbf{r})$. The planes should also have been drawn with an infinite spatial extent, since no limits were put on **r**. The disturbance clearly occupies all of space.

The spatially repetitive nature of these harmonic functions can be expressed by

$$\psi(\mathbf{r}) = \psi\left(\mathbf{r} + \frac{\lambda \mathbf{k}}{k}\right) \qquad (2.47)$$

where k is the magnitude of **k** and $\mathbf{k}/k$ is a unit vector parallel to it (Fig. 2.21). In the exponential form, this is equivalent to

$$Ae^{i\mathbf{k} \cdot \mathbf{r}} = Ae^{i\mathbf{k} \cdot (\mathbf{r} + \lambda \mathbf{k}/k)} = Ae^{i\mathbf{k} \cdot \mathbf{r}} e^{i\lambda k}$$

For this to be true, we must have

$$e^{i\lambda k} = 1 = e^{i2\pi}$$

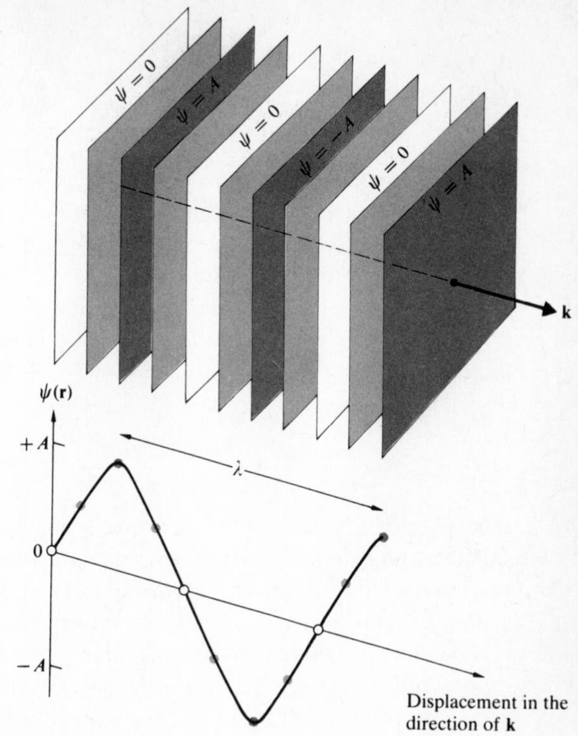

FIGURE 2.20 Wavefronts for a harmonic plane wave.

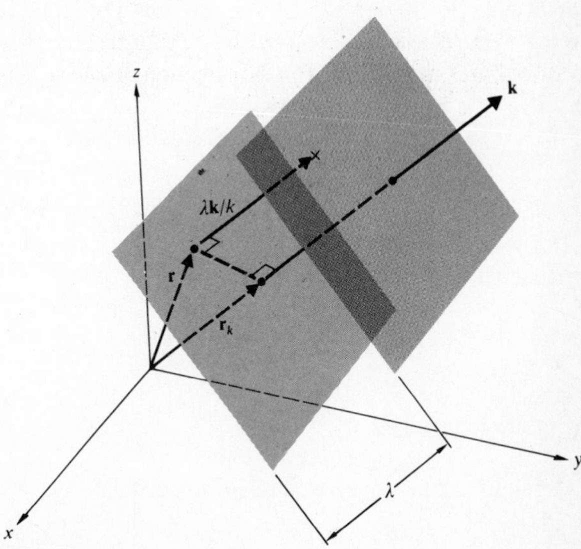

FIGURE 2.21 Plane waves.

Therefore,

$$\lambda k = 2\pi$$

and

$$k = 2\pi/\lambda$$

The vector **k**, whose magnitude is the *propagation number k* (already introduced), is called the **propagation vector**.

At any fixed point in space where **r** is constant, the phase is constant as is $\psi(\mathbf{r})$; in short, the planes are motionless. To get things moving, $\psi(\mathbf{r})$ must be made to vary in time, something we can accomplish by introducing the time dependence in an analogous fashion to that of the one-dimensional wave. Here then

$$\psi(\mathbf{r}, t) = Ae^{i(\mathbf{k}\cdot\mathbf{r} \mp \omega t)} \tag{2.48}$$

with A, ω, and k constant. As this disturbance travels along in the **k**-direction, we can assign a phase corresponding to it at each point in space and time. At any given time, ***the surfaces joining all points of equal phase are known as* wavefronts**. Note that the wavefunction will have a constant value over the wavefront only if the amplitude A has a fixed value at every point on the wavefront. In general, A is a function of **r** and may not be constant over all space or even over a wavefront. In the latter case, the wave is said to be *inhomogeneous*. We will not be concerned with this sort of disturbance until later, when we consider laserbeams and total internal reflection.

The phase velocity of a plane wave given by Eq. (2.48) is equivalent to the propagation velocity of the wavefront. In Fig. 2.21, the scalar component of **r** in the direction of **k** is r_k. The disturbance on a wavefront is constant, so that after a time dt, if the front moves along **k** a distance dr_k, we must have

$$\psi(\mathbf{r}, t) = \psi(r_k + dr_k, t + dt) = \psi(r_k, t) \tag{2.49}$$

In exponential form, this is

$$Ae^{i(\mathbf{k}\cdot\mathbf{r} \mp \omega t)} = Ae^{i(kr_k + k\,dr_k \mp \omega t \mp \omega\, dt)} = Ae^{i(kr_k \mp \omega t)}$$

and so it must be that $k\,dr_k = \pm\omega\,dt$

The magnitude of the wave velocity, dr_k/dt, is then

$$\frac{dr_k}{dt} = \pm\frac{\omega}{k} = \pm v \tag{2.50}$$

We could have anticipated this result by rotating the coordinate system in Fig. 2.21 so that **k** was parallel to the x-axis. For that orientation

$$\psi(\mathbf{r}, t) = Ae^{i(kx \mp \omega t)}$$

since $\mathbf{k}\cdot\mathbf{r} = kr_k = kx$. The wave has thereby been effectively reduced to the one-dimensional disturbance already discussed.

Now consider the two waves in Fig. 2.22; both have the same wavelength λ such that $k_1 = k_2 = k = 2\pi/\lambda$. Wave-1 propagating along the z-axis can be written as

$$\psi_1 = A_1 \cos\left(\frac{2\pi}{\lambda} z - \omega t\right)$$

where, because $\mathbf{k}_1$ and **r** are parallel, $\mathbf{k}_1\cdot\mathbf{r} = kz = (2\pi/\lambda)z$. Similarly for wave-2, $\mathbf{k}_2\cdot\mathbf{r} = k_z z + k_y y = (k\cos\theta)z + (k\sin\theta)y$ and

$$\psi_2 = A_2 \cos\left[\frac{2\pi}{\lambda}(z\cos\theta + y\sin\theta) - \omega t\right]$$

We'll return to these expressions and what happens in the region of overlap when we consider interference in more detail.

The plane harmonic wave is often written in Cartesian coordinates as

$$\psi(x, y, z, t) = Ae^{i(k_x x + k_y y + k_z z \mp \omega t)} \tag{2.51}$$

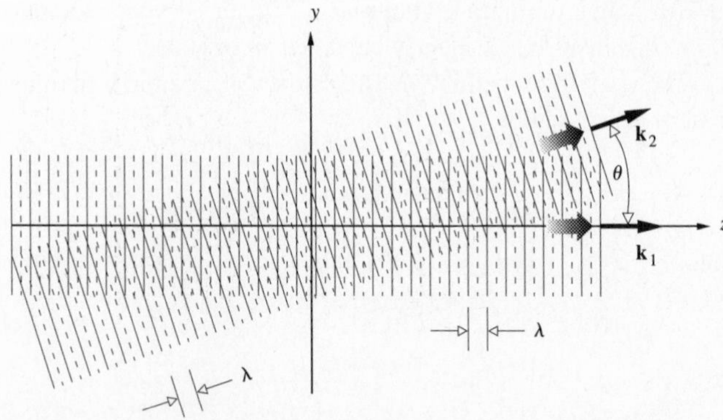

FIGURE 2.22 Two overlapping waves of the same wavelength traveling in different directions.

or

$$\psi(x, y, z, t) = Ae^{i[k(\alpha x+\beta y+\gamma z)\mp\omega t]} \tag{2.52}$$

where α, β, and γ are the direction cosines of $\mathbf{k}$ (see Problem 2.33). In terms of its components, the magnitude of the propagation vector is

$$|\mathbf{k}| = k = (k_x^2 + k_y^2 + k_z^2)^{1/2} \tag{2.53}$$

and of course

$$\alpha^2 + \beta^2 + \gamma^2 = 1 \tag{2.54}$$

We have examined plane waves with a particular emphasis on harmonic functions. The special significance of these waves is twofold: first, physically, sinusoidal waves can be generated relatively simply by using some form of harmonic oscillator; second, ***any three-dimensional wave can be expressed as a combination of plane waves***, each having a distinct amplitude and propagation direction.

We can certainly imagine a series of plane waves like those in Fig. 2.20 where the disturbance varies in some fashion other than harmonically (Fig. 2.23). It will be seen in the next section that harmonic plane waves are, indeed, a special case of a more general plane-wave solution.

FIGURE 2.23 The image of a single collimated laser pulse caught as it swept along the surface of a ruler. This ultrashort burst of light corresponded to a portion of a plane wave. It extended in time for 300×10^{-15} s and was only a fraction of a millimeter long. (Photo courtesy J. Valdmanis and N.H. Abramson.)

2.8 The Three-Dimensional Differential Wave Equation

Of all the three-dimensional waves, only the plane wave (harmonic or not) can move through space with an unchanging profile. Clearly, the idea of a wave as a disturbance whose profile is *unaltered* is somewhat lacking. Alternatively, we can define a wave as any solution of the differential wave equation. What we need now is a three-dimensional wave equation. This should be rather easy to obtain, since we can guess at its form by generalizing from the one-dimensional expression, Eq. (2.11). In Cartesian coordinates, the position variables x, y, and z must certainly appear symmetrically* in the three-dimensional equation, a fact to be kept in mind. The wavefunction $\psi(x, y, z, t)$ given by Eq. (2.52) is a particular solution of the differential equation we are looking for. In analogy with the derivation of Eq. (2.11), we compute the following partial derivatives from Eq. (2.52):

$$\frac{\partial^2\psi}{\partial x^2} = -\alpha^2k^2\psi \tag{2.55}$$

$$\frac{\partial^2\psi}{\partial y^2} = -\beta^2k^2\psi \tag{2.56}$$

$$\frac{\partial^2\psi}{\partial z^2} = -\gamma^2k^2\psi \tag{2.57}$$

and

$$\frac{\partial^2\psi}{\partial t^2} = -\omega^2\psi \tag{2.58}$$

Adding the three spatial derivatives and utilizing the fact that $\alpha^2 + \beta^2 + \gamma^2 = 1$, we obtain

$$\frac{\partial^2\psi}{\partial x^2} + \frac{\partial^2\psi}{\partial y^2} + \frac{\partial^2\psi}{\partial z^2} = -k^2\psi \tag{2.59}$$

Combining this with the time derivative, Eq. (2.58), and remembering that $v = \omega/k$, we arrive at

$$\frac{\partial^2\psi}{\partial x^2} + \frac{\partial^2\psi}{\partial y^2} + \frac{\partial^2\psi}{\partial z^2} = \frac{1}{v^2}\frac{\partial^2\psi}{\partial t^2} \tag{2.60}$$

*There is no distinguishing characteristic for any one of the axes in Cartesian coordinates. We should therefore be able to change the names of, say, x to z, y to x, and z to y (keeping the system right-handed) without altering the differential wave equation.

the *three-dimensional differential wave equation*. Note that x, y, and z do appear symmetrically, and the form is precisely what one might expect from the generalization of Eq. (2.11).

Equation (2.60) is usually written in a more concise form by introducing the *Laplacian* operator

$$\nabla^2 \equiv \frac{\partial^2}{\partial x^2} + \frac{\partial^2}{\partial y^2} + \frac{\partial^2}{\partial z^2} \tag{2.61}$$

whereupon it becomes simply

$$\nabla^2 \psi = \frac{1}{v^2} \frac{\partial^2 \psi}{\partial t^2} \tag{2.62}$$

Now that we have this most important equation, let's briefly return to the plane wave and see how it fits into the scheme of things. A function of the form

$$\psi(x, y, z, t) = Ae^{ik(\alpha x + \beta y + \gamma z \mp vt)} \tag{2.63}$$

is equivalent to Eq. (2.52) and, as such, is a solution of Eq. (2.62). It can also be shown (Problem 2.34) that

$$\psi(x, y, z, t) = f(\alpha x + \beta y + \gamma z - vt) \tag{2.64}$$

and

$$\psi(x, y, z, t) = g(\alpha x + \beta y + \gamma z + vt) \tag{2.65}$$

are both plane-wave solutions of the differential wave equation. The functions f and g, which are twice differentiable, are otherwise arbitrary and certainly need not be harmonic. A linear combination of these solutions is also a solution, and we can write this in a slightly different manner as

$$\psi(\mathbf{r}, t) = C_1 f(\mathbf{r} \cdot \mathbf{k}/k - vt) + C_2 g(\mathbf{r} \cdot \mathbf{k}/k + vt) \tag{2.66}$$

where C_1 and C_2 are constants.

Cartesian coordinates are particularly suitable for describing plane waves. However, as various physical situations arise, we can often take better advantage of existing symmetries by making use of some other coordinate representations.

2.9 Spherical Waves

Toss a stone into a tank of water. The surface ripples that emanate from the point of impact spread out in two-dimensional circular waves. Extending this imagery to three dimensions, envision a small pulsating sphere surrounded by a fluid. As the source expands and contracts, it generates pressure variations that propagate outward as spherical waves.

Consider now an idealized point source of light. The radiation emanating from it streams out radially, uniformly in all directions. The source is said to be *isotropic*, and the resulting wavefronts are again concentric spheres that increase in diameter as they expand out into the surrounding space. The obvious symmetry of the wavefronts suggests that it might be more convenient to describe them in terms of spherical polar coordinates (Fig. 2.24). In this representation the Laplacian operator is

$$\nabla^2 \equiv \frac{1}{r^2}\frac{\partial}{\partial r}\left(r^2 \frac{\partial}{\partial r}\right) + \frac{1}{r^2 \sin\theta}\frac{\partial}{\partial \theta}\left(\sin\theta \frac{\partial}{\partial \theta}\right) + \frac{1}{r^2 \sin^2\theta}\frac{\partial^2}{\partial \phi^2} \tag{2.67}$$

where r, θ, ϕ are defined by

$$x = r\sin\theta\cos\phi, \qquad y = r\sin\theta\sin\phi, \qquad z = r\cos\theta$$

Remember that we are looking for a description of spherical waves, waves that are spherically symmetrical (i.e., ones that do not depend on θ and ϕ) so that

$$\psi(\mathbf{r}) = \psi(r, \theta, \phi) = \psi(r)$$

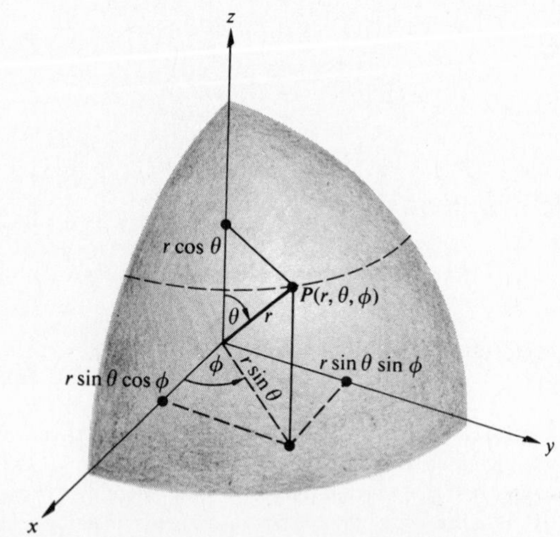

FIGURE 2.24 The geometry of spherical coordinates.

The Laplacian of $\psi(r)$ is then simply

$$\nabla^2\psi(r) = \frac{1}{r^2}\frac{\partial}{\partial r}\left(r^2\frac{\partial\psi}{\partial r}\right) \tag{2.68}$$

We can obtain this result without being familiar with Eq. (2.67). Start with the Cartesian form of the Laplacian, Eq. (2.61), operate on the spherically symmetrical wavefunction $\psi(r)$, and convert each term to polar coordinates. Examining only the x-dependence, we have

$$\frac{\partial\psi}{\partial x} = \frac{\partial\psi}{\partial r}\frac{\partial r}{\partial x}$$

and

$$\frac{\partial^2\psi}{\partial x^2} = \frac{\partial^2\psi}{\partial r^2}\left(\frac{\partial r}{\partial x}\right)^2 + \frac{\partial\psi}{\partial r}\frac{\partial^2 r}{\partial x^2}$$

since

$$\psi(\mathbf{r}) = \psi(r)$$

Using

$$x^2 + y^2 + z^2 = r^2$$

we have

$$\frac{\partial r}{\partial x} = \frac{x}{r}, \quad \frac{\partial^2 r}{\partial x^2} = \frac{1}{r}\frac{\partial}{\partial x}(x) + x\frac{\partial}{\partial x}\left(\frac{1}{r}\right) = \frac{1}{r}\left(1 - \frac{x^2}{r^2}\right)$$

and

$$\frac{\partial^2\psi}{\partial x^2} = \frac{x^2}{r^2}\frac{\partial^2\psi}{\partial r^2} + \frac{1}{r}\left(1 - \frac{x^2}{r^2}\right)\frac{\partial\psi}{\partial r}$$

Now having $\partial^2\psi/\partial x^2$, we form $\partial^2\psi/\partial y^2$ and $\partial^2\psi/\partial z^2$, and on adding get

$$\nabla^2\psi(r) = \frac{\partial^2\psi}{\partial r^2} + \frac{2}{r}\frac{\partial\psi}{\partial r}$$

which is equivalent to Eq. (2.68). This result can be expressed in a slightly different form:

$$\nabla^2\psi = \frac{1}{r}\frac{\partial^2}{\partial r^2}(r\psi) \tag{2.69}$$

The differential wave equation can then be written as

$$\frac{1}{r}\frac{\partial^2}{\partial r^2}(r\psi) = \frac{1}{v^2}\frac{\partial^2\psi}{\partial t^2} \tag{2.70}$$

Multiplying both sides by r, yields

$$\frac{\partial^2}{\partial r^2}(r\psi) = \frac{1}{v^2}\frac{\partial^2}{\partial t^2}(r\psi) \tag{2.71}$$

Notice that this expression is now just the one-dimensional differential wave equation, Eq. (2.11), where the space variable is r and the wavefunction is the product $(r\psi)$. The solution of Eq. (2.71) is then simply

$$r\psi(r, t) = f(r - vt)$$

or

$$\psi(r, t) = \frac{f(r - vt)}{r} \tag{2.72}$$

This represents a spherical wave progressing radially outward from the origin, at a constant speed v, and having an arbitrary functional form f. Another solution is given by

$$\psi(r, t) = \frac{g(r + vt)}{r}$$

and in this case the wave is converging toward the origin.* The fact that this expression blows up at $r = 0$ is of little practical concern.

A special case of the general solution

$$\psi(r, t) = C_1\frac{f(r - vt)}{r} + C_2\frac{g(r + vt)}{r} \tag{2.73}$$

is the *harmonic spherical wave*

$$\psi(r, t) = \left(\frac{\mathcal{A}}{r}\right)\cos k(r \mp vt) \tag{2.74}$$

or

$$\psi(r, t) = \left(\frac{\mathcal{A}}{r}\right)e^{ik(r \mp vt)} \tag{2.75}$$

wherein the constant $\mathcal{A}$ is called the *source strength*. At any fixed value of time, this represents a cluster of concentric spheres filling all space. Each wavefront, or surface of constant phase, is given by

$$kr = \text{constant}$$

Notice that the amplitude of any spherical wave is a function of r, where the term r^{-1} serves as an attenuation factor. Unlike the plane wave, a spherical wave decreases in amplitude, thereby changing its profile, as it expands and moves out from

*Other more complicated solutions exist when the wave is not spherically symmetrical. See C. A. Coulson, *Waves*, Chapter 1.

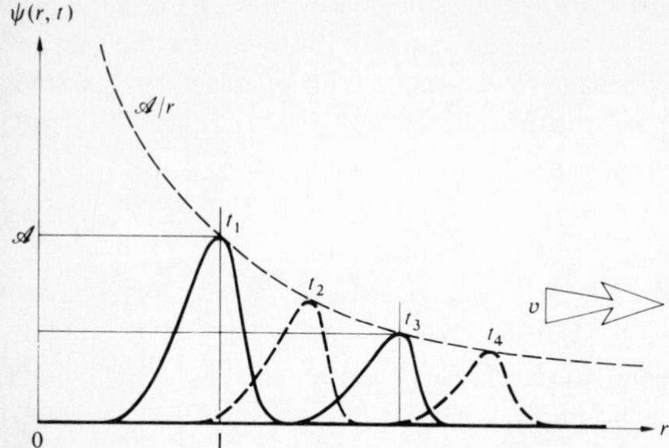

FIGURE 2.25 A "quadruple exposure" of a spherical pulse.

the origin.* Figure 2.25 illustrates this graphically by showing a "multiple exposure" of a spherical pulse at four different times. The pulse has the same extent in space at any point along any radius r; that is, the width of the pulse along the r-axis is a constant. Figure 2.26 is an attempt to relate the diagrammatic representation of $\psi(r, t)$ in the previous figure to its actual form as a spherical wave. It depicts half the spherical pulse at two different times, as the wave expands outward. Remember that these results would obtain regardless of the direction of r, because of the spherical symmetry. We could also have drawn a harmonic wave, rather than a pulse, in Figs. 2.25 and 2.26. In this case, the sinusoidal disturbance would have been bounded by the curves

$$\psi = \mathscr{A}/r \quad \text{and} \quad \psi = -\mathscr{A}/r$$

*The attenuation factor is a direct consequence of energy conservation. Chapter 3 contains a discussion of how these ideas apply specifically to electromagnetic radiation.

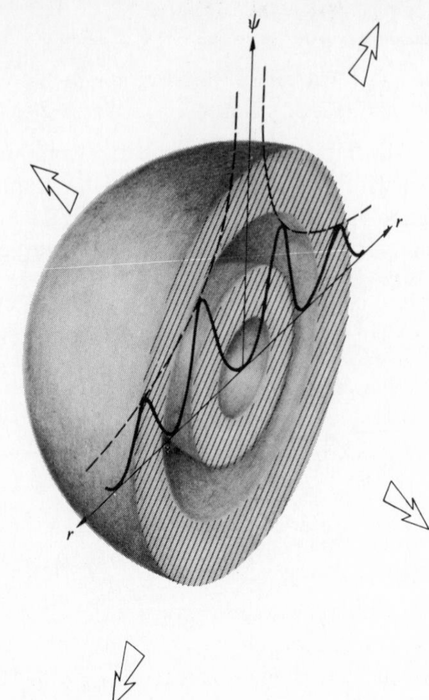

FIGURE 2.26 Spherical wavefronts.

The outgoing spherical wave emanating from a point source and the incoming wave converging to a point are idealizations. In actuality, light only approximates spherical waves, as it also only approximates plane waves.

As a spherical wavefront propagates out, its radius increases. Far enough away from the source, a small area of the wavefront will closely resemble a portion of a plane wave (Fig. 2.27).

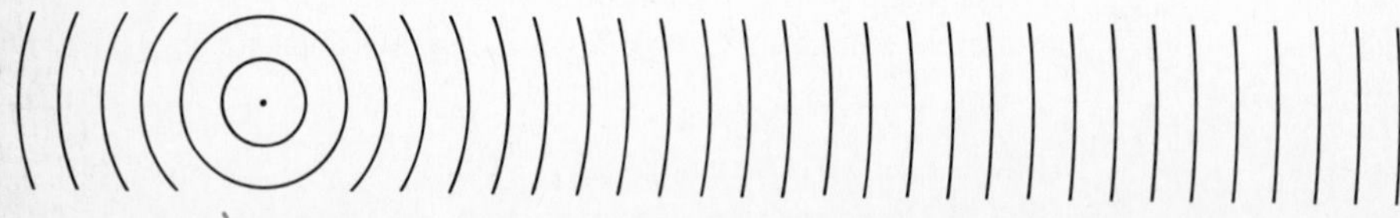

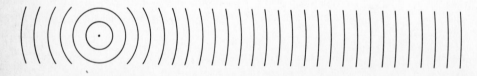

FIGURE 2.27 The flattening of spherical waves with distance.

2.10 Cylindrical Waves

We will now briefly examine another idealized waveform, the infinite circular cylinder. Unfortunately, a precise mathematical treatment is far too involved to do here. We shall, instead, just outline the procedure. The Laplacian of ψ in cylindrical coordinates (Fig. 2.28) is

$$\nabla^2\psi = \frac{1}{r}\frac{\partial}{\partial r}\left(r\frac{\partial\psi}{\partial r}\right) + \frac{1}{r^2}\frac{\partial^2\psi}{\partial\theta^2} + \frac{\partial^2\psi}{\partial z^2} \tag{2.76}$$

where

$$x = r\cos\theta,\quad y = r\sin\theta,\quad \text{and}\quad z = z$$

The simple case of cylindrical symmetry requires that

$$\psi(\mathbf{r}) = \psi(r, \theta, z) = \psi(r)$$

The θ-independence means that a plane perpendicular to the z-axis will intersect the wavefront in a circle, which may vary in r, at different values of z. In addition, the z-independence further restricts the wavefront to a right circular cylinder centered on the z-axis and having infinite length. The differential wave equation becomes

$$\frac{1}{r}\frac{\partial}{\partial r}\left(r\frac{\partial\psi}{\partial r}\right) = \frac{1}{v^2}\frac{\partial^2\psi}{\partial t^2} \tag{2.77}$$

After a bit of manipulation, in which the time dependence is separated out, Eq. (2.77) becomes something called Bessel's equation. The solutions of Bessel's equation for large values of r gradually approach simple trigonometric forms. When r is sufficiently large,

$$\psi(r, t) \approx \frac{\mathscr{A}}{\sqrt{r}}\, e^{ik(r \mp vt)}$$

$$\psi(r, t) \approx \frac{\mathscr{A}}{\sqrt{r}}\cos k(r \mp vt) \tag{2.78}$$

This represents a set of coaxial circular cylinders filling all space and traveling toward or away from an infinite line source. No solutions in terms of arbitrary functions can now be found as there were for both spherical [Eq. (2.73)] and plane [Eq. (2.66)] waves.

A plane wave impinging on the back of a flat opaque screen containing a long, thin slit will result in the emission, from that slit, of a disturbance resembling a cylindrical wave (see Fig. 2.29). Extensive use has been made of this technique to generate cylindrical lightwaves (p. 386).

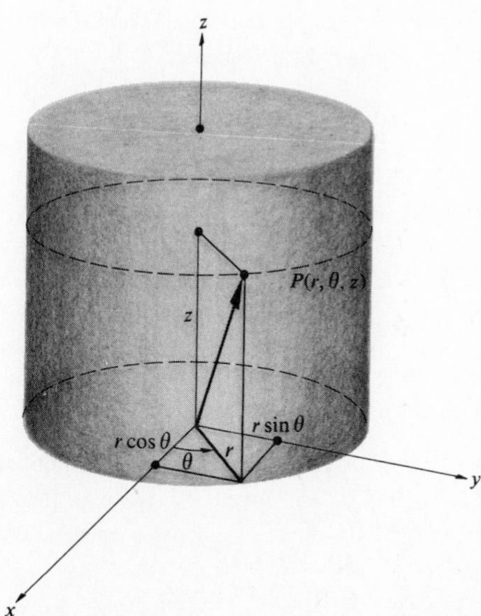

FIGURE 2.28 The geometry of cylindrical coordinates.

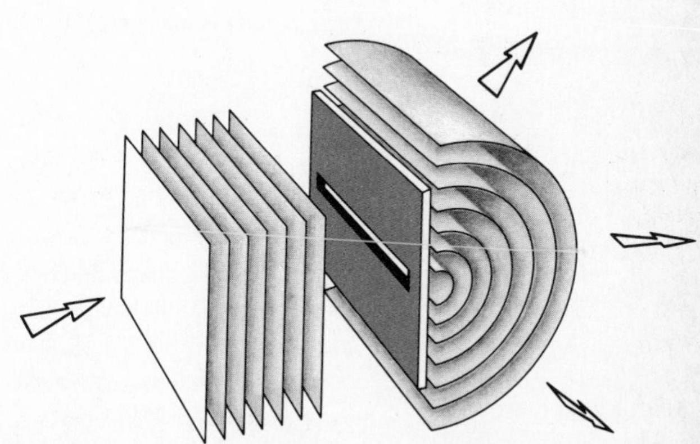

FIGURE 2.29 Cylindrical waves emerging from a long, narrow slit.

PROBLEMS

2.1 How many "yellow" lightwaves ($\lambda = 580$ nm) will fit into a distance in space equal to the thickness of a piece of paper (0.003 in)? How far will the same number of microwaves ($\nu = 10^{10}$ Hz, i.e., 10 GHz, and $v = 3 \times 10^8$ m/s) extend?

2.2* The speed of light in vacuum is approximately 3×10^8 m/s. Find the wavelength of red light having a frequency of 5×10^{14} Hz. Compare this with the wavelength of a 60-Hz electromagnetic wave.

2.3* It is possible to generate ultrasonic waves in crystals with wavelengths similar to light (5×10^{-5} cm) but with lower frequencies (6×10^8 Hz). Compute the corresponding speed of such a wave.

2.4* A youngster in a boat on a lake watches waves that seem to be an endless succession of identical crests passing with a half-second interval between each. If every disturbance takes 1.5 s to sweep straight along the length of her 4.5-m-long boat, what are the frequency, period, and wavelength of the waves?

2.5* A vibrating hammer strikes the end of a long metal rod in such a way that a periodic compression wave with a wavelength of 4.3 m travels down the rod's length at a speed of 3.5 km/s. What was the frequency of the vibration?

2.6 A violin is submerged in a swimming pool at the wedding of two scuba divers. Given that the speed of compression waves in pure water is 1498 m/s, what is the wavelength of an A-note of 440 Hz played on the instrument?

2.7* A wavepulse travels 10 m along the length of a string in 2.0 s. A harmonic disturbance of wavelength 0.50 m is then generated on the string. What is its frequency?

2.8* Show that for a periodic wave $\omega = (2\pi/\lambda)v$.

2.9* Make up a table with columns headed by values of θ running from $-\pi/2$ to 2π in intervals of $\pi/4$. In each column place the corresponding value of $\sin\theta$, beneath those the values of $\cos\theta$, beneath those the values of $\sin(\theta - \pi/4)$, and similarly with the functions $\sin(\theta - \pi/2)$, $\sin(\theta - 3\pi/4)$, and $\sin(\theta + \pi/2)$. Plot each of these functions, noting the effect of the phase shift. Does $\sin\theta$ lead or lag $\sin(\theta - \pi/2)$; in other words, does one of the functions reach a particular magnitude at a smaller value of θ than the other and therefore lead the other (as $\cos\theta$ leads $\sin\theta$)?

*There are complete solutions in the back of the book to all problems—except those with an asterisk.

2.10* Make up a table with columns headed by values of kx running from $x = -\lambda/2$ to $x = +\lambda$ in intervals of x of $\lambda/4$—of course, $k = 2\pi/\lambda$. In each column place the corresponding values of $\cos(kx - \pi/4)$ and beneath that the values of $\cos(kx + 3\pi/4)$. Next plot the functions $15\cos(kx - \pi/4)$ and $25\cos(kx + 3\pi/4)$.

2.11* Make up a table with columns headed by values of ωt running from $t = -\tau/2$ to $t = +\tau$ in intervals of t of $\tau/4$—of course, $\omega = 2\pi/\tau$. In each column place the corresponding values of $\sin(\omega t + \pi/4)$ and $\sin(\pi/4 - \omega t)$ and then plot these two functions.

2.12* The profile of a harmonic wave, traveling at 1.2 m/s on a string, is given by

$$y = (0.02\text{ m})\sin(157\text{ m}^{-1})\,x$$

Determine its amplitude, wavelength, frequency, and period.

2.13 Given the wavefunctions

$$\psi_1 = 4\sin 2\pi(0.2x - 3t)$$

and

$$\psi_2 = \frac{\sin(7x + 3.5t)}{2.5}$$

determine in each case the values of (a) frequency, (b) wavelength, (c) period, (d) amplitude, (e) phase velocity, and (f) direction of motion. Time is in seconds and x is in meters.

2.14* Show that

$$\psi(x, t) = A\sin k(x - vt) \qquad [2.13]$$

is a solution of the differential wave equation.

2.15 Show that if the displacement of the string in Fig. 2.7 is given by

$$y(x, t) = A\sin[kx - \omega t + \varepsilon]$$

then the hand generating the wave must be moving vertically in simple harmonic motion.

2.16 Write the expression for a harmonic wave of amplitude 10^3 V/m, period 2.2×10^{-15} s, and speed 3×10^8 m/s. The wave is propagating in the negative x-direction and has a value of 10^3 V/m at $t = 0$ and $x = 0$.

2.17 Consider the pulse described in terms of its displacement at $t = 0$ by

$$y(x, t)|_{t=0} = \frac{C}{2 + x^2}$$

where C is a constant. Draw the wave profile. *Write* an expression for the wave, having a speed v in the negative x-direction, as a function of time t. If $v = 1\ \text{m/s}$, sketch the profile at $t = 2\ \text{s}$.

2.18* Please determine the magnitude of the wavefunction $\psi(z, t) = A \cos [k(z + vt) + \pi]$ at the point $z = 0$, when $t = \tau/2$ and when $t = 3\tau/4$.

2.19 Does the following function, in which A is a constant,

$$\psi(y, t) = (y - vt)A$$

represent a wave? Explain your reasoning.

2.20* Use Eq. (2.33) to calculate the speed of the wave whose representation in SI units is

$$\psi(y, t) = A \cos \pi(3 \times 10^6 y + 9 \times 10^{14} t)$$

2.21 Beginning with the following theorem: If $z = f(x, y)$ and $x = g(t)$, $y = h(t)$, then

$$\frac{dz}{dt} = \frac{\partial z}{\partial x}\frac{dx}{dt} + \frac{\partial z}{\partial y}\frac{dy}{dt}$$

Derive Eq. (2.34).

2.22 Using the results of the previous problem, show that for a harmonic wave with a phase $\varphi(x, t) = k(x - vt)$ we can determine the speed by setting $d\varphi/dt = 0$. Apply the technique to Problem 2.20 to find the speed of that wave.

2.23* A Gaussian wave has the form $\psi(x, t) = Ae^{-a(bx + ct)^2}$. Use the fact that $\psi(x, t) = f(x \mp vt)$ to determine its speed and then verify your answer using Eq. (2.34).

2.24 Create an expression for the *profile* of a harmonic wave traveling in the z-direction whose magnitude at $z = -\lambda/12$ is 0.866, at $z = +\lambda/6$ is $1/2$, and at $z = \lambda/4$ is 0.

2.25 Which of the following expressions correspond to traveling waves? For each of those, what is the speed of the wave? The quantities a, b, and c are positive constants.

(a) $\psi(z, t) = (az - bt)^2$

(b) $\psi(x, t) = (ax + bt + c)^2$

(c) $\psi(x, t) = 1/(ax^2 + b)$

2.26* Determine which of the following describe traveling waves:

(a) $\psi(y, t) = e^{-(a^2y^2 + b^2t^2 - 2abty)}$

(b) $\psi(z, t) = A \sin (az^2 - bt^2)$

(c) $\psi(x, t) = A \sin 2\pi\left(\dfrac{x}{a} + \dfrac{t}{b}\right)^2$

(d) $\psi(x, t) = A \cos^2 2\pi(t - x)$

Where appropriate, draw the profile and find the speed and direction of motion.

2.27 Given the traveling wave $\psi(x, t) = 5.0 \exp (-ax^2 - bt^2 - 2\sqrt{ab}\, xt)$, determine its direction of propagation. Calculate a few values of ψ and make a sketch of the wave at $t = 0$, taking $a = 25\ \text{m}^{-2}$ and $b = 9.0\ \text{s}^{-2}$. What is the speed of the wave?

2.28* Imagine a sound wave with a frequency of 1.10 kHz propagating with a speed of 330 m/s. Determine the phase difference in radians between any two points on the wave separated by 10.0 cm.

2.29 Consider a lightwave having a phase velocity of 3×10^8 m/s and a frequency of 6×10^{14} Hz. What is the shortest distance along the wave between any two points that have a phase difference of 30°? What phase shift occurs at a given point in 10^{-6} s, and how many waves have passed by in that time?

2.30 Write an expression for the wave shown in Fig. P.2.30. Find its wavelength, velocity, frequency, and period.

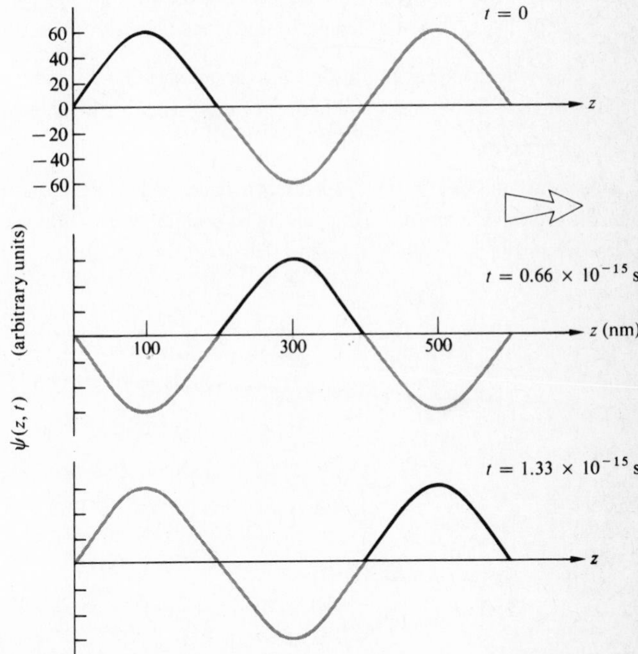

FIGURE P.2.30 A harmonic wave.

2.31* Working with exponentials directly, show that the magnitude of $\psi = Ae^{i\omega t}$ is A. Then rederive the same result using Euler's formula. Prove that $e^{i\alpha}e^{i\beta} = e^{i(\alpha+\beta)}$.

2.32* Show that the imaginary part of a complex number $\tilde{z}$ is given by $(\tilde{z} - \tilde{z}^*)/2i$.

2.33 Beginning with Eq. (2.51), verify that

$$\psi(x, y, z, t) = Ae^{i[k(\alpha x + \beta y + \gamma z) \mp \omega t]}$$

and that
$$\alpha^2 + \beta^2 + \gamma^2 = 1$$

Draw a sketch showing all the pertinent quantities.

2.34* Show that Eqs. (2.64) and (2.65), which are plane waves of arbitrary form, satisfy the three-dimensional differential wave equation.

2.35 De Broglie's hypothesis states that every particle has associated with it a wavelength given by Planck's constant ($h = 6.6 \times 10^{-34}$ J . s) divided by the particle's momentum. Compare the wavelength of a 6.0-kg stone moving at a speed of 1.0 m/s with that of light.

2.36 Write an expression in Cartesian coordinates for a harmonic plane wave of amplitude A and frequency ω propagating in the direction of the vector $\mathbf{k}$, which in turn lies on a line drawn from the origin to the point (4, 2, 1). *Hint:* First determine $\mathbf{k}$ and then dot it with $\mathbf{r}$.

2.37* Write an expression in Cartesian coordinates for a harmonic plane wave of amplitude A and frequency ω propagating in the positive x-direction.

2.38 Show that $\psi(\mathbf{k} \cdot \mathbf{r}, t)$ may represent a plane wave where $\mathbf{k}$ is normal to the wavefront. *Hint:* Let $\mathbf{r}_1$ and $\mathbf{r}_2$ be position vectors drawn to any two points on the plane and show that $\psi(\mathbf{r}_1, t) = \psi(\mathbf{r}_2, t)$.

2.39* Make up a table with columns headed by values of θ running from $-\pi/2$ to 2π in intervals of $\pi/4$. In each column place the corresponding value of sin θ, and beneath those the values of 2 sin θ. Next add these, column by column, to yield the corresponding values of the function sin θ + 2 sin θ. Plot each of these three functions, noting their relative amplitudes and phases.

2.40* Make up a table with columns headed by values of θ running from $-\pi/2$ to 2π in intervals of $\pi/4$. In each column place the corresponding value of sin θ, and beneath those the values of $\sin(\theta - \pi/2)$. Next add these, column by column, to yield the corresponding values of the function $\sin\theta + \sin(\theta - \pi/2)$. Plot each of these three functions, noting their relative amplitudes and phases.

2.41* With the last two problems in mind, draw a plot of the three functions (a) $\sin\theta$, (b) $\sin(\theta - 3\pi/4)$, and (c) $\sin\theta + \sin(\theta - 3\pi/4)$. Compare the amplitude of the combined function (c) in this case with that of the previous problem.

2.42* Make up a table with columns headed by values of kx running from $x = -\lambda/2$ to $x = +\lambda$ in intervals of x of $\lambda/4$. In each column place the corresponding values of cos kx and beneath that the values of $\cos(kx + \pi)$. Next plot the three functions cos kx, $\cos(kx + \pi)$, and $\cos kx + \cos(kx + \pi)$.

3 Electromagnetic Theory, Photons, and Light

The work of J. Clerk Maxwell and subsequent developments since the late 1800s have made it evident that light is most certainly electromagnetic in nature. Classical electrodynamics, as we shall see, unalterably leads to the picture of a continuous transfer of energy by way of electromagnetic waves. In contrast, the more modern view of Quantum Electrodynamics (p. 80) describes electromagnetic interactions and the transport of energy in terms of massless elementary "particles" known as *photons*. The quantum nature of radiant energy is not always apparent, nor is it always of practical concern in Optics. There is a range of situations in which the detecting equipment is such that it is impossible, and desirably so, to distinguish individual quanta.

If the wavelength of light is small in comparison to the size of the apparatus, one may use, as a first approximation, the techniques of *Geometrical Optics*. A somewhat more precise treatment, which is applicable as well when the dimensions of the apparatus are small, is that of *Physical Optics*. In Physical Optics the dominant property of light is its wave nature. It is even possible to develop most of the treatment without ever specifying the kind of wave one is dealing with. Certainly, as far as the classical study of Physical Optics is concerned, it will suffice admirably to treat light as an electromagnetic wave.

We can think of light as the most tenuous form of matter. Indeed, one of the basic tenets of Quantum Mechanics is that both light and material particles display similar wave-particle properties. As Erwin C. Schrödinger (1887–1961), one of the founders of quantum theory, put it:

> In the new setting of ideas the distinction [between particles and waves] has vanished, because it was discovered that all particles have also wave properties, and *vice versa*. Neither of the two concepts must be discarded, they must be amalgamated. Which aspect obtrudes itself depends not on the physical object, but on the experimental device set up to examine it.*

The quantum-mechanical treatment associates a wave equation with a particle, be it a photon, electron, proton, or whatever. In the case of material particles, the wave aspects are introduced by way of the field equation known as Schrödinger's Equation. For light we have a representation of the wave nature in the form of the classical electromagnetic field equations of Maxwell. With these as a starting point one can construct a quantum-mechanical theory of photons and their interaction with charges. The dual nature of light is evidenced by the fact that it propagates through space in a wavelike fashion and yet displays particlelike behavior during the processes of emission and absorption. Electromagnetic radiant energy is created and destroyed in quanta or photons and not continuously as a classical wave. Nonetheless, its motion through a lens, a hole, or a set of slits is governed by wave characteristics. If we're unfamiliar with this kind of behavior in the macroscopic world, it's because the wavelength of a material object varies inversely with its momentum (p. 56), and even a grain of sand (which is barely moving) has a wavelength so small as to be indiscernible in any conceivable experiment.

The photon has zero mass, and therefore exceedingly large numbers of low-energy photons can be envisioned as present in a beam of light. Within that model, dense streams of photons act on the average to produce well-defined classical fields (p. 52). We can draw a rough analogy with the flow of commuters through a train station during rush hour. Each one pre-

*Erwin C. Schrödinger, *Science Theory and Man.*

sumably behaves individually as a quantum of humanity, but all have the same intent and follow fairly similar trajectories. To a distant, myopic observer there is a seemingly smooth and continuous flow. The behavior of the stream *en masse* is predictable from day to day, and so the precise motion of each commuter is unimportant, at least to the observer. The energy transported by a large number of photons is, *on the average*, equivalent to the energy transferred by a classical electromagnetic wave. It is for these reasons that the field representation of electromagnetic phenomena has been, and will continue to be, so useful.

Quite pragmatically, then, we can consider light to be a classical electromagnetic wave, keeping in mind that there are situations for which this description is woefully inadequate.

3.1 Basic Laws of Electromagnetic Theory

Our intent in this section is to review and develop some of the ideas needed to appreciate the concept of electromagnetic waves.

We know from experiments that charges, even though separated in vacuum, experience a mutual interaction. Recall the familiar electrostatics demonstration in which a pith ball somehow senses the presence of a charged rod without actually touching it. As a possible explanation we might speculate that each charge emits (and absorbs) a stream of undetected particles (*virtual photons*). The exchange of these particles among the charges may be regarded as the mode of interaction. Alternatively, we can take the classical approach and imagine instead that every charge is surrounded by something called an electric field. We then need only suppose that each charge interacts directly with the electric field in which it is immersed. Thus if a point charge $q_{\cdot}$ experiences a force $\mathbf{F}_E$, the *electric field* $\mathbf{E}$ at the position of the charge is defined by $\mathbf{F}_E = q_{\cdot}\mathbf{E}$. In addition, we observe that a moving charge may experience another force $\mathbf{F}_M$, which is proportional to its velocity $\mathbf{v}$. We are thus led to define yet another field, namely, the *magnetic induction* or just the *magnetic field* $\mathbf{B}$, such that $\mathbf{F}_M = q_{\cdot}\mathbf{v} \times \mathbf{B}$. If forces $\mathbf{F}_E$ and $\mathbf{F}_M$ occur concurrently, the charge is moving through a region pervaded by both electric and magnetic fields, whereupon $\mathbf{F} = q_{\cdot}\mathbf{E} + q_{\cdot}\mathbf{v} \times \mathbf{B}$.

As we shall see, electric fields are generated by both electric charges and by *time-varying magnetic fields*. Similarly, magnetic fields are generated by electric currents and by *time-varying electric fields*. This interdependence of $\mathbf{E}$ and $\mathbf{B}$ is a key point in the description of light.

3.1.1 Faraday's Induction Law

Michael Faraday made a number of major contributions to electromagnetic theory. One of the most significant was his discovery that a time-varying magnetic flux passing through a closed conducting loop results in the generation of a current around that loop. The flux of the magnetic field through any open area A bounded by the conducting loop (Fig. 3.1) is given by

$$\Phi_M = \iint_A \mathbf{B} \cdot d\mathbf{S} \tag{3.1}$$

The induced *electromotive force*, or *emf*, developed around the loop is then

$$\text{emf} = -\frac{d\Phi_M}{dt} \tag{3.2}$$

We should not, however, get too involved with the image of wires and current and emf. Our present concern is with the electric and magnetic fields themselves. The emf exists only

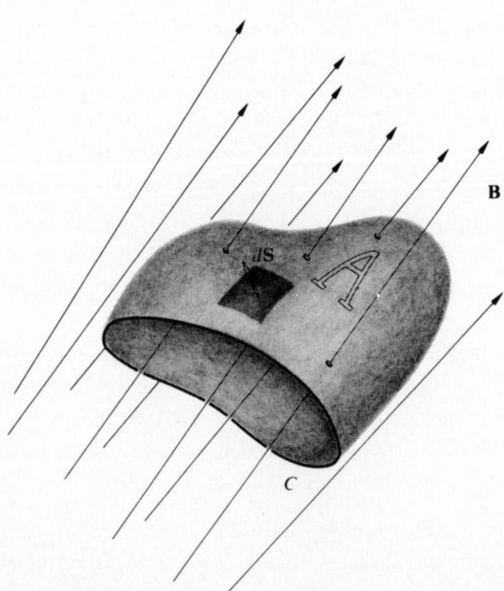

FIGURE 3.1 B-field through an open area A.

as a result of the presence of an electric field given by

$$\text{emf} = \oint_C \mathbf{E} \cdot d\mathbf{l} \tag{3.3}$$

taken around the closed curve C, corresponding to the loop. Equating Eqs. (3.2) and (3.3), and making use of Eq. (3.1), we get

$$\oint_C \mathbf{E} \cdot d\mathbf{l} = -\frac{d}{dt} \iint_A \mathbf{B} \cdot d\mathbf{S} \tag{3.4}$$

We began this discussion by examining a conducting loop and have arrived at Eq. (3.4); this expression, except for the path C, contains no reference to the physical loop. In fact, the path can be chosen arbitrarily and need not be within, or anywhere near, a conductor. The electric field in Eq. (3.4) arises not from the presence of electric charges but rather from the time-varying magnetic field. With no charges to act as sources or sinks, the field lines close on themselves, forming loops (Fig. 3.2). For the case in which the path is fixed in space and unchanging in shape, the *Induction Law* (Eq. 3.4) can be

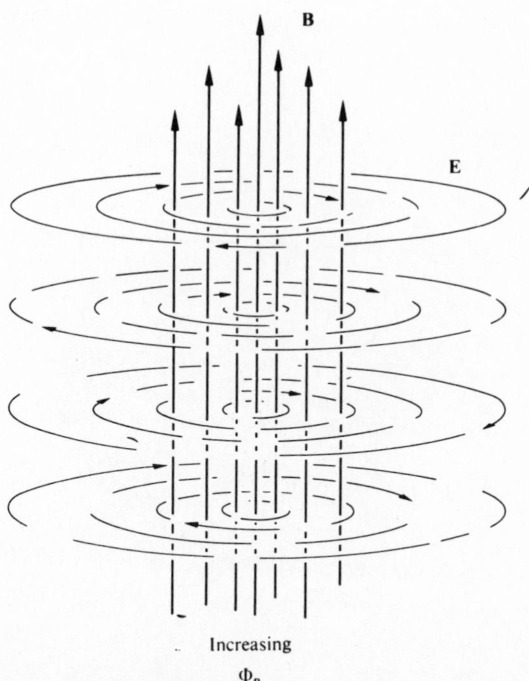

FIGURE 3.2 A time-varying **B**-field. Surrounding each point where Φ_M is changing, the **E**-field forms closed loops.

rewritten as

$$\oint_C \mathbf{E} \cdot d\mathbf{l} = -\iint_A \frac{\partial \mathbf{B}}{\partial t} \cdot d\mathbf{S} \tag{3.5}$$

This, in itself, is a rather fascinating expression, since it indicates that ***a time-varying magnetic field will have an electric field associated with it***.

Another less mathematical way to appreciate this is as follows: A charge changing its location in a fixed **B**-field will generally experience a force. But since motion is relative, a stationary charge in a moving (changing) **B**-field will also experience a force. Yet whenever a charge experiences a force there must be an **E**-field present. Hence, $\partial \mathbf{B}/\partial t$ must be accompanied by an **E**.

3.1.2 Gauss's Law—Electric

Another fundamental law of electromagnetism is named after the German mathematician Karl Friedrich Gauss (1777–1855). It relates the flux of electric field intensity through a closed surface A

$$\Phi_E = \oiint_A \mathbf{E} \cdot d\mathbf{S} \tag{3.6}$$

to the total enclosed charge. The circled double integral is meant to serve as a reminder that the surface is closed. The vector $d\mathbf{S}$ is in the direction of an *outward normal*, as shown in Fig. 3.3. If the volume enclosed by A is V, and if within it there is a continuous charge distribution of density ρ, then Gauss's Law is

$$\oiint_A \mathbf{E} \cdot d\mathbf{S} = \frac{1}{\epsilon} \iiint_V \rho \, dV \tag{3.7}$$

The integral on the left is the difference between the amount of flux flowing into and out of any closed surface A. If there is a difference, it will be due to the presence of sources or sinks of the electric field within A. Clearly then, the integral must be proportional to the total enclosed charge, inasmuch as charges are the sources (+) and sinks (−) of the electric field.

The constant ϵ is known as the **electric permittivity** of the medium. For the special case of a vacuum, the *permittivity of free space* is given by $\epsilon_0 = 8.854\,2 \times 10^{-12}\ \text{C}^2/\text{N}\cdot\text{m}^2$. One function of the ϵ in Eq. (3.7) is, of course, to balance out the units, but the concept is basic to the description of the parallel plate capacitor (see Section 3.1.4). There it's the medium-dependent proportionality constant between the device's

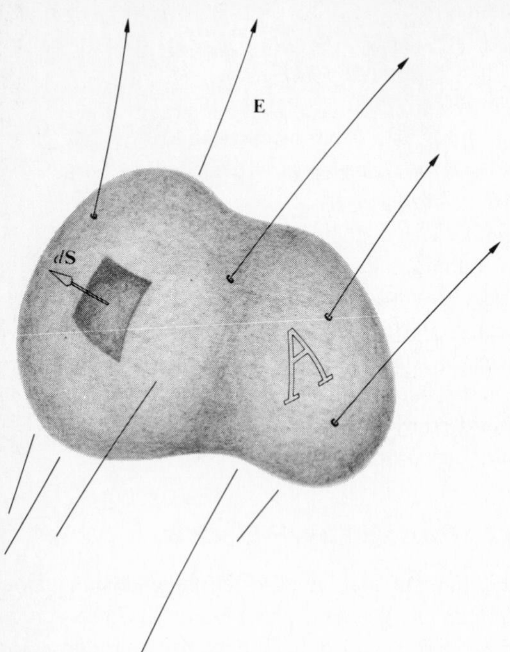

FIGURE 3.3 E-field through a closed area A.

capacitance and its geometric characteristics. Indeed ϵ is often measured by a procedure in which the material under study is placed within a capacitor. Conceptually, the permittivity embodies the electrical behavior of the medium: in a sense, it is a measure of the degree to which the material is permeated by the electric field in which it is immersed.

In the early days of the development of the subject, people in various areas worked in different systems of units, a state of affairs leading to some obvious difficulties. This necessitated the tabulation of numerical values for ϵ in each of the different systems, which was, at best, a waste of time. Recall that the same problem regarding densities was neatly avoided by using specific gravity (i.e., density ratios). Thus it was advantageous to tabulate values not of ϵ but of a new related quantity that is independent of the system of units being used. Accordingly, we define K_E as ϵ/ϵ_0. This is the **dielectric constant** (or *relative permittivity*), and it is appropriately unitless. The permittivity of a material can then be expressed in terms of ϵ_0 as

$$\epsilon = K_E \epsilon_0 \tag{3.8}$$

Our interest in K_E anticipates the fact that the permittivity is related to the speed of light in dielectric materials, such as glass, air, and quartz.

3.1.3 Gauss's Law—Magnetic

There is no known magnetic counterpart to the electric charge; that is, no isolated magnetic poles have ever been found, despite extensive searching, even in lunar soil samples. Unlike the electric field, the magnetic field **B** does not diverge from or converge toward some kind of magnetic charge (a monopole source or sink). Magnetic fields can be described in terms of current distributions. Indeed we might envision an elementary magnet as a small current loop in which the lines of **B** are continuous and closed. Any closed surface in a region of magnetic field would accordingly have an equal number of lines of **B** entering and emerging from it (Fig. 3.4). This situation arises from the absence of any monopoles within the enclosed volume. The flux of magnetic field Φ_M through such a surface is zero, and we have the magnetic equivalent of Gauss's Law:

$$\Phi_M = \oint\!\!\!\oint_A \mathbf{B} \cdot d\mathbf{S} = 0 \tag{3.9}$$

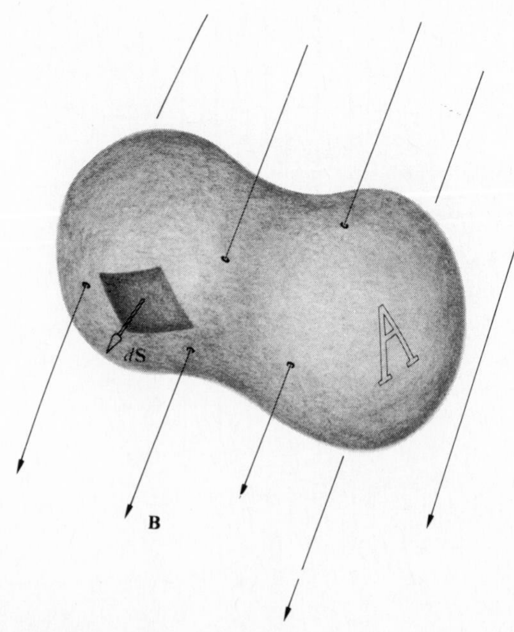

FIGURE 3.4 B-field through a closed area A.

3.1.4 Ampère's Circuital Law

Another equation that will be of great interest is due to André Marie Ampère (1775–1836). Known as the *Circuital Law*, it relates a line integral of **B** tangent to a closed curve C, with the total current i passing within the confines of C:

$$\oint_C \mathbf{B}\cdot d\mathbf{l} = \mu \iint_A \mathbf{J}\cdot d\mathbf{S} = \mu i \tag{3.10}$$

The open surface A is bounded by C, and J is the current per unit area (Fig. 3.5). The quantity μ is called the **permeability** of the particular medium. For a vacuum $\mu = \mu_0$ (the *permeability of free space*), which is defined as $4\pi \times 10^{-7}$ $\text{N}\cdot\text{s}^2/\text{C}^2$. As in Eq. (3.8),

$$\mu = K_M \mu_0 \tag{3.11}$$

with K_M being the dimensionless *relative permeability*.

Equation (3.10), though often adequate, is not the whole truth. Ampère's Law is not particular about the area used, provided it's bounded by the curve C, which makes for an obvious problem when charging a capacitor, as shown in Fig. 3.6*a*. If flat area A_1 is used, a net current of i flows through it and there is a **B**-field along curve C—the right side of Eq. (3.10) is nonzero, so the left side is nonzero. But if area A_2 is used instead to encompass C, no net current passes through it and the field must now be zero, even though nothing physical has actually changed. Something is obviously wrong!

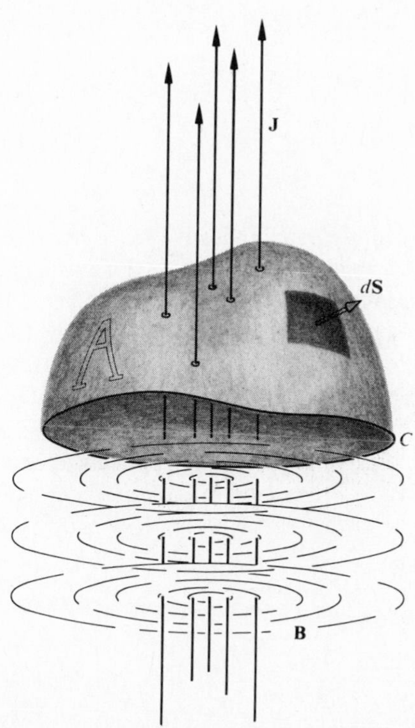

FIGURE 3.5 Current density through an open area A.

Moving charges are not the only source of a magnetic field. While charging or discharging a capacitor, one can measure a **B**-field in the region between its plates (Fig. 3.6*b*), which is indistinguishable from the field surrounding the leads, even though no electric current actually traverses the capacitor. Notice, however, that if A is the area of each plate and Q the charge on it,

$$E = \frac{Q}{\epsilon A}$$

As the charge varies, the electric field changes, and taking the derivative of both sides yields

$$\epsilon \frac{\partial E}{\partial t} = \frac{i}{A}$$

which is effectively a current density. James Clerk Maxwell hypothesized the existence of just such a mechanism, which he called the *displacement current density*,* defined by

$$\mathbf{J}_D \equiv \epsilon \frac{\partial \mathbf{E}}{\partial t} \tag{3.12}$$

The restatement of Ampère's Law as

$$\oint_C \mathbf{B}\cdot d\mathbf{l} = \mu \iint_A \left(\mathbf{J} + \epsilon \frac{\partial \mathbf{E}}{\partial t}\right)\cdot d\mathbf{S} \tag{3.13}$$

was one of Maxwell's greatest contributions. It points out that even when $\mathbf{J} = 0$, ***a time-varying E-field will be accompanied by a B-field*** (Fig. 3.7).

*Maxwell's own words and ideas concerning this mechanism are examined in an article by A. M. Bork, *Am. J. Phys.* **31**, 854 (1963). Incidentally, Clerk is pronounced *clark*.

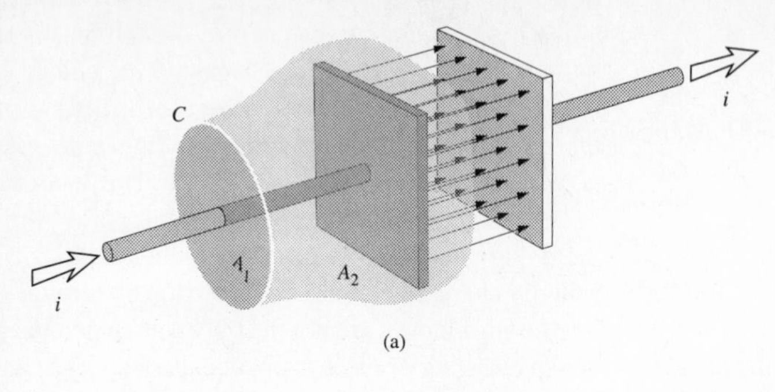

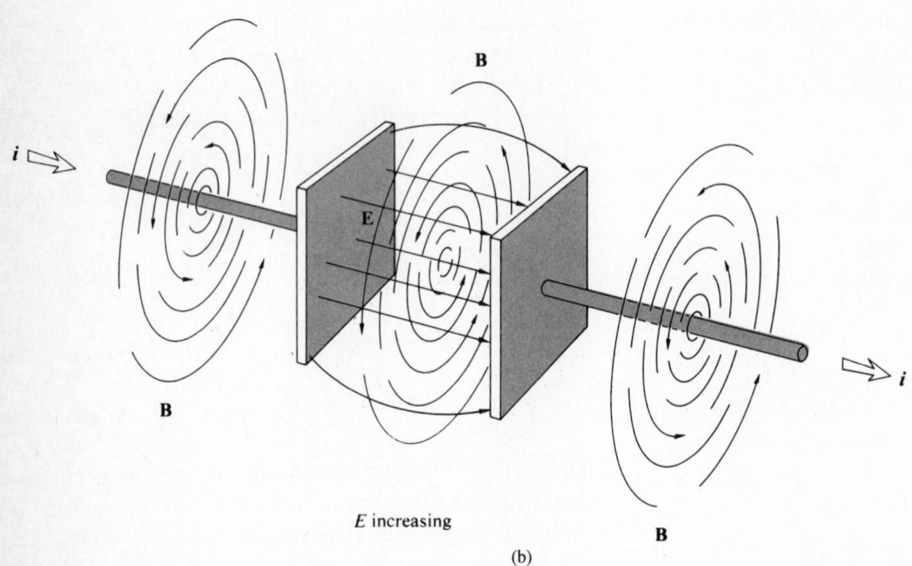

FIGURE 3.6 (a) Ampère's Law is indifferent to which area A_1 or A_2 is bounded by the path C. Yet a current passes through A_1 and not through A_2, and that means something is very wrong. (b) **B**-field concomitant with a time-varying **E**-field in the gap of a capacitor.

3.1.5 Maxwell's Equations

The set of integral expressions given by Eqs. (3.5), (3.7), (3.9), and (3.13) have come to be known as Maxwell's Equations. Remember that these are generalizations of experimental results. The simplest statement of Maxwell's Equations applies to the behavior of the electric and magnetic fields in free space, where $\epsilon = \epsilon_0$, $\mu = \mu_0$, and both ρ and $\mathbf{J}$ are zero. In that instance,

$$\oint_C \mathbf{E} \cdot d\mathbf{l} = -\iint_A \frac{\partial \mathbf{B}}{\partial t} \cdot d\mathbf{S} \tag{3.14}$$

$$\oint_C \mathbf{B} \cdot d\mathbf{l} = \mu_0\epsilon_0 \iint_A \frac{\partial \mathbf{E}}{\partial t} \cdot d\mathbf{S} \tag{3.15}$$

$$\oiint_A \mathbf{B} \cdot d\mathbf{S} = 0 \tag{3.16}$$

$$\oiint_A \mathbf{E} \cdot d\mathbf{S} = 0 \tag{3.17}$$

Observe that except for a multiplicative scalar, the electric and magnetic fields appear in the equations with a remarkable symmetry. However **E** affects **B**, **B** will in turn affect **E**. The mathematical symmetry implies a good deal of physical symmetry.

Maxwell's Equations can be written in a differential form, which will be somewhat more useful for our purposes. The appropriate calculation is carried out in Appendix 1, and the consequent equations for *free space*, in Cartesian coordinates, are as follows:

$$\begin{aligned} \frac{\partial \mathbf{E}_z}{\partial y} - \frac{\partial \mathbf{E}_y}{\partial z} &= -\frac{\partial \mathbf{B}_x}{\partial t} && \text{(i)} \\ \frac{\partial \mathbf{E}_x}{\partial z} - \frac{\partial \mathbf{E}_z}{\partial x} &= -\frac{\partial \mathbf{B}_y}{\partial t} && \text{(ii)} \\ \frac{\partial \mathbf{E}_y}{\partial x} - \frac{\partial \mathbf{E}_x}{\partial y} &= -\frac{\partial \mathbf{B}_z}{\partial t} && \text{(iii)} \end{aligned} \tag{3.18}$$

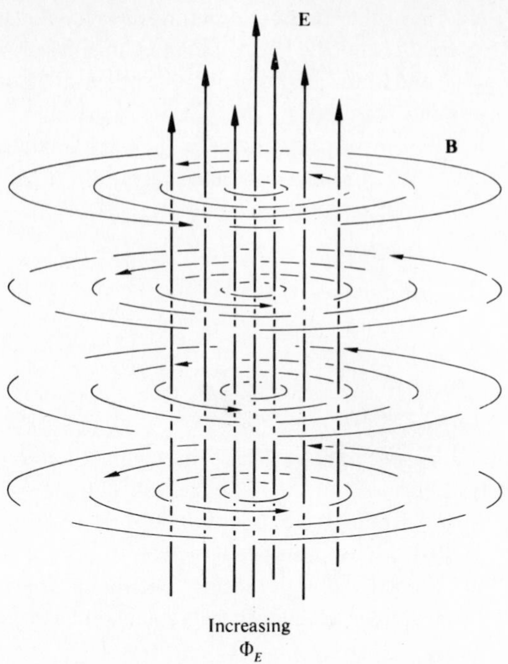

FIGURE 3.7 A time-varying **E**-field. Surrounding each point where Φ_E is changing, the **B**-field forms closed loops.

$$\frac{\partial \mathbf{B}_z}{\partial y} - \frac{\partial \mathbf{E}_y}{\partial z} = \mu_0 \epsilon_0 \frac{\partial \mathbf{E}_x}{\partial t} \quad \text{(i)}$$

$$\frac{\partial \mathbf{B}_x}{\partial z} - \frac{\partial \mathbf{B}_z}{\partial x} = \mu_0 \epsilon_0 \frac{\partial \mathbf{E}_y}{\partial t} \quad \text{(ii)} \qquad (3.19)$$

$$\frac{\partial \mathbf{B}_y}{\partial x} - \frac{\partial \mathbf{B}_x}{\partial y} = \mu_0 \epsilon_0 \frac{\partial \mathbf{E}_z}{\partial t} \quad \text{(iii)}$$

$$\frac{\partial \mathbf{B}_x}{\partial x} + \frac{\partial \mathbf{B}_y}{\partial y} + \frac{\partial \mathbf{B}_z}{\partial z} = 0 \qquad (3.20)$$

$$\frac{\partial \mathbf{E}_x}{\partial x} + \frac{\partial \mathbf{E}_y}{\partial y} + \frac{\partial \mathbf{E}_z}{\partial z} = 0 \qquad (3.21)$$

The transition has thus been made from the formulation of Maxwell's Equations in terms of integrals over *finite regions* to a restatement in terms of derivatives at *points* in space.

We now have all that is needed to comprehend the magnificent process whereby electric and magnetic fields, inseparably coupled and mutually sustaining, propagate out into space as a single entity, free of charges and currents, sans matter, sans aether.

3.2 Electromagnetic Waves

We have relegated to Appendix 1 a complete and mathematically elegant derivation of the electromagnetic wave equation. Here the focus is on the equally important task of developing a more intuitive appreciation of the physical processes involved. Three observations, from which we might build a qualitative picture, are readily available to us: the general perpendicularity of the fields, the symmetry of Maxwell's Equations, and the interdependence of **E** and **B** in those equations.

In studying electricity and magnetism, one soon becomes aware that a number of relationships are described by vector cross-products or, if you like, right-hand rules. In other words, an occurrence of one sort produces a related, perpendicularly directed response. Of immediate interest is the fact that a time-varying **E**-field generates a **B**-field, which is everywhere perpendicular to the direction in which **E** changes (Fig. 3.7). In the same way, a time-varying **B**-field generates an **E**-field, which is everywhere perpendicular to the direction in which **B** changes (Fig. 3.2). We might, consequently, anticipate the general transverse nature of the **E**- and **B**-fields in an electromagnetic disturbance.

Consider a charge that is somehow caused to *accelerate* from rest. When the charge is motionless, it has associated with it a constant radial **E**-field extending in all directions presumably to infinity. At the instant the charge begins to move, the **E**-field is altered in the vicinity of the charge, and this alteration propagates out into space at some finite speed. The time-varying electric field induces a magnetic field by means of Eq. (3.15) or (3.19). If the charge's velocity is constant, the rate-of-change of the **E**-field is steady, and the resulting **B**-field is constant. But here the charge is accelerating, $\partial \mathbf{E}/\partial t$ is itself not constant, so the induced **B**-field is time-dependent. The time-varying **B**-field generates an **E**-field, (3.14) or (3.18), and the process continues, with **E** and **B** coupled in the form of a pulse. As one field changes, it generates a new field that extends a bit farther, and the pulse moves out from one point to the next through space.

We can draw an overly mechanistic but rather picturesque analogy, if we imagine the electric field lines as a dense radial distribution of strings (p. 58). When somehow plucked, each string is distorted, forming a kink that travels outward from the source. All these kinks combine at any instant to yield a three-dimensional expanding pulse in the continuum of the electric field.

The **E**- and **B**-fields can more appropriately be considered as two aspects of a single physical phenomenon, the *electromagnetic field*, whose source is a moving charge. The disturbance, once it has been generated in the electromagnetic field, is an untethered wave that moves beyond its source and independently of it. Bound together as a single entity, the time-varying electric and magnetic fields regenerate each other in an endless cycle. The electromagnetic waves reaching us from the relatively nearby Andromeda galaxy (which can be seen with the naked eye) have been on the wing for 2 200 000 years.

We have not yet considered the direction of wave propagation with respect to the constituent fields. Notice, however, that the high degree of symmetry in Maxwell's Equations for free space suggests that the disturbance will propagate in a direction that is symmetrical to both **E** and **B**. That implies that an electromagnetic wave cannot be purely longitudinal (i.e., as long as **E** and **B** are not parallel). Let's now replace conjecture with a bit of calculation.

Appendix 1 shows that Maxwell's Equations for free space can be manipulated into the form of two extremely concise vector expressions:

$$\nabla^2 \mathbf{E} = \epsilon_0 \mu_0 \frac{\partial^2 \mathbf{E}}{\partial t^2} \qquad \text{[A1.26]}$$

and

$$\nabla^2 \mathbf{B} = \epsilon_0 \mu_0 \frac{\partial^2 \mathbf{B}}{\partial t^2} \qquad \text{[A1.27]}$$

The Laplacian,* ∇^2, operates on each component of **E** and **B**, so that the two vector equations actually represent a total of six scalar equations. In Cartesian coordinates,

$$\frac{\partial^2 E_x}{\partial x^2} + \frac{\partial^2 E_x}{\partial y^2} + \frac{\partial^2 E_x}{\partial z^2} = \epsilon_0 \mu_0 \frac{\partial^2 E_x}{\partial t^2} \qquad (3.22)$$

$$\frac{\partial^2 E_y}{\partial x^2} + \frac{\partial^2 E_y}{\partial y^2} + \frac{\partial^2 E_y}{\partial z^2} = \epsilon_0 \mu_0 \frac{\partial^2 E_y}{\partial t^2} \qquad (3.23)$$

and there are four other equations with precisely the same form for E_z, B_x, B_y, and B_z. Expressions of this sort, which relate the space and time variations of some physical quantity, had been studied long before Maxwell's work and were known to describe wave phenomena (p. 14). Each and every component of the electromagnetic field (E_x, E_y, E_z, B_x, B_y, B_z) obeys the scalar differential wave equation

$$\frac{\partial^2 \psi}{\partial x^2} + \frac{\partial^2 \psi}{\partial y^2} + \frac{\partial^2 \psi}{\partial z^2} = \frac{1}{v^2} \frac{\partial^2 \psi}{\partial t^2} \qquad [2.60]$$

provided that

$$v = 1/\sqrt{\epsilon_0 \mu_0} \qquad (3.24)$$

To evaluate v Maxwell made use of the results of electrical experiments performed in 1856 in Leipzig by Wilhelm Weber (1804–1891) and Rudolph Kohlrausch (1809–1858). Equivalently, nowadays μ_0 is assigned a value of $4\pi \times 10^{-7}$ m · kg/C^2 in SI units, and until recently one might determine ϵ_0 directly from simple capacitor measurements. In any event, in modern units

$$\epsilon_0 \mu_0 \approx (8.85 \times 10^{-12}\ \text{s}^2 \cdot \text{C}^2/\text{m}^3 \cdot \text{kg})(4\pi \times 10^{-7}\ \text{m} \cdot \text{kg}/\text{C}^2)$$

or

$$\epsilon_0 \mu_0 \approx 11.12 \times 10^{-18}\ \text{s}^2/\text{m}^2$$

And now the moment of truth—in free space, the predicted speed of all electromagnetic waves would then be

$$v = \frac{1}{\sqrt{\epsilon_0 \mu_0}} \approx 3 \times 10^8\ \text{m/s}$$

This theoretical value was in remarkable agreement with the previously measured speed of light (315 300 km/s) determined by Fizeau. The results of Fizeau's experiments, performed in 1849 with a rotating toothed wheel, were available to Maxwell and led him to comment:

> This velocity [i.e., his theoretical prediction] is so nearly that of light, that it seems we have strong reason to conclude that light itself (including radiant heat, and other radiations if any) is an electromagnetic disturbance in the form of waves propagated through the electromagnetic field according to electromagnetic laws.

This brilliant analysis was one of the great intellectual triumphs of all time. It has become customary to designate the speed of light in vacuum by the symbol c, which comes from

*In Cartesian coordinates,

$$\nabla^2 \mathbf{E} = \hat{\mathbf{i}} \nabla^2 E_x + \hat{\mathbf{j}} \nabla^2 E_y + \hat{\mathbf{k}} \nabla^2 E_z$$

the Latin word *celer*, meaning fast. In 1983 the 17th Conférence Générale des Poids et Mesures in Paris adopted a new definition of the meter and thereby fixed the speed of light in vacuum as exactly

$$c = 2.997\,924\,58 \times 10^8 \text{ m/s}$$

3.2.1 Transverse Waves

The experimentally verified transverse character of light must now be explained within the context of electromagnetic theory. To that end, consider the fairly simple case of a plane wave propagating in vacuum in the positive x-direction. The electric field intensity is a solution of Eq. (A1.26), where $\mathbf{E}$ is constant over each of an infinite set of planes perpendicular to the x-axis. It is therefore a function only of x and t; that is, $\mathbf{E} = \mathbf{E}(x, t)$. We now refer back to Maxwell's Equations, and in particular to Eq. (3.21), which is generally read as *the divergence of* **E** *equals zero*. Since $\mathbf{E}$ is not a function of either y or z, the equation can be reduced to

$$\frac{\partial E_x}{\partial x} = 0 \tag{3.25}$$

If E_x is not zero—that is, if there is some component of the field in the direction of propagation—this expression tells us that it does not vary with x. At any given time E_x is constant for all values of x, but of course, this possibility cannot therefore correspond to a traveling wave advancing in the positive x-direction. Alternatively, it follows from Eq. (3.25) that for a wave, $E_x = 0$; the electromagnetic wave has no electric field component in the direction of propagation. The **E**-field associated with the plane wave is then exclusively *transverse*.

The fact that the **E**-field is transverse means that in order to completely specify the wave we will have to specify the moment-by-moment direction of **E**. Such a description corresponds to the **polarization** of the light and it will be treated in Chapter 8. Without any loss of generality, we deal here with *plane* or *linearly polarized waves*, for which the direction of the vibrating **E**-vector is fixed. Thus we orient our coordinate axes so that the electric field is parallel to the y-axis, whereupon

$$\mathbf{E} = \hat{\mathbf{j}} E_y(x, t) \tag{3.26}$$

Returning to Eq. (3.18), it follows that

$$\frac{\partial E_y}{\partial x} = -\frac{\partial B_z}{\partial t} \tag{3.27}$$

Therefore B_x and B_y are constant and of no interest at present. The time-dependent **B**-field can only have a component in the z-direction. Clearly then, ***in free space, the plane electromagnetic wave is transverse*** (Fig. 3.8). Except in the case of normal incidence, such waves propagating in real material media are generally not transverse—a complication arising from the fact that the medium may be dissipative or contain free charge.

We have not specified the form of the disturbance other than to say that it is a plane wave. Our conclusions are therefore quite general, applying equally well to both pulses and continuous waves. We have already pointed out that harmonic functions are of particular interest, because any waveform can be expressed in terms of sinusoidal waves using Fourier techniques (p. 300). We therefore limit the discussion to harmonic waves and write $E_y(x, t)$ as

$$E_y(x, t) = E_{0y} \cos\left[\omega(t - x/c) + \varepsilon\right] \tag{3.28}$$

the speed of propagation being c. The associated magnetic flux density can be found by directly integrating Eq. (3.27), that is,

$$B_z = -\int \frac{\partial E_y}{\partial x}\, dt$$

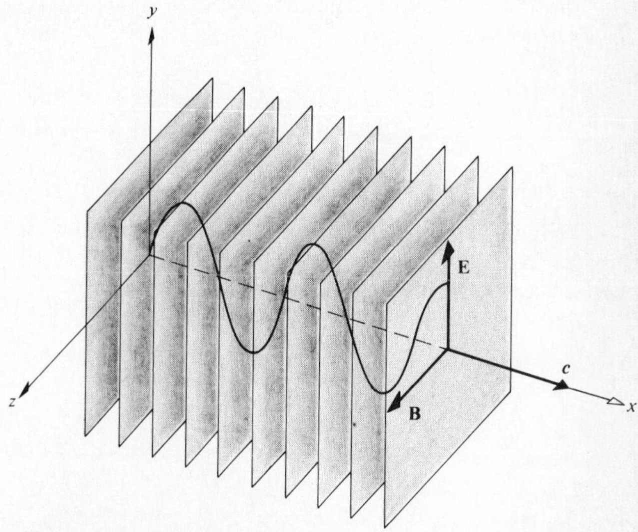

FIGURE 3.8 The field configuration in a plane harmonic electromagnetic wave.

Using Eq. (3.28), we obtain

$$B_z = -\frac{E_{0y}\omega}{c}\int \sin\left[\omega(t - x/c) + \varepsilon\right]dt$$

or

$$B_z(x, t) = \frac{1}{c}\,E_{0y}\cos\left[\omega(t - x/c) + \varepsilon\right] \tag{3.29}$$

The constant of integration, which represents a time-independent field, has been disregarded. Comparison of this result with Eq. (3.28) makes it evident that

$$E_y = cB_z \tag{3.30}$$

Since E_y and B_z differ only by a scalar, and so have the same time-dependence, **E** and **B** are *in-phase* at all points in space. Moreover, $\mathbf{E} = \hat{\mathbf{j}}E_y(x, t)$ and $\mathbf{B} = \hat{\mathbf{k}}B_z(x, t)$ are *mutually perpendicular*, and their cross-product, $\mathbf{E} \times \mathbf{B}$, points in the propagation direction, $\hat{\mathbf{i}}$ (Fig. 3.9).

Plane waves, though important, are not the only solutions to Maxwell's Equations. As we saw in the previous chapter, the differential wave equation allows many solutions, among which are cylindrical and spherical waves (Fig. 3.10).

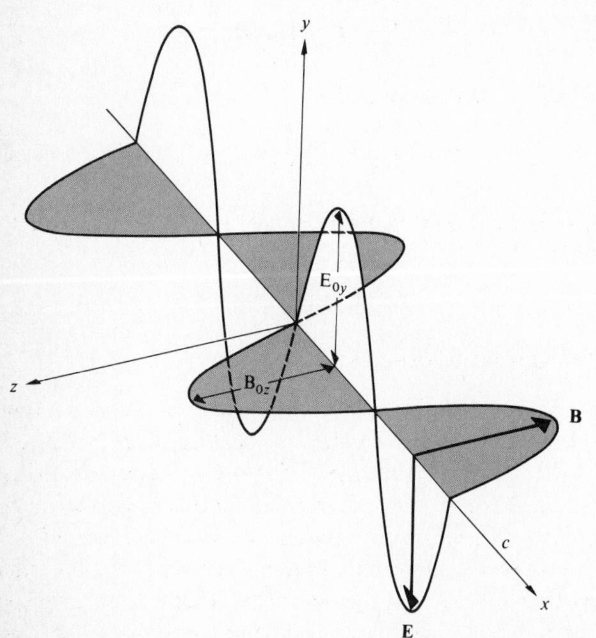

FIGURE 3.9 Orthogonal harmonic **E**- and **B**-fields for a plane polarized wave.

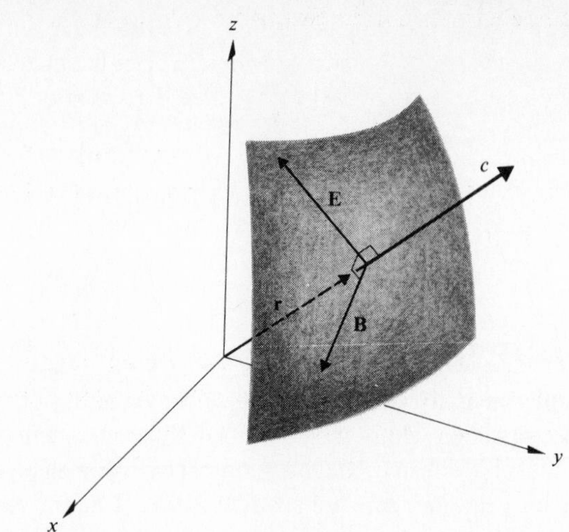

FIGURE 3.10 Portion of a spherical wavefront far from the source.

3.3 Energy and Momentum

One of the most significant properties of the electromagnetic wave is that it transports energy and momentum. The light from even the nearest star beyond the Sun travels 25 million million miles to reach the Earth, yet it still carries enough energy to do work on the electrons within your eye.

3.3.1 The Poynting Vector

Any electromagnetic wave exists within some region of space, and it is therefore natural to consider the *radiant energy per unit volume*, or **energy density**, u. We suppose that the electric field itself can somehow store energy. This is a major logical step since it imparts to the field the attribute of physical reality—if the field has energy it is a thing-in-itself. Moreover, inasmuch as the classical field is continuous, its energy is continuous. Let's assume as much and see where it leads. We can compute (Problem 3.6) the energy density of the **E**-field (e.g., between the plates of a charged capacitor) to be

$$u_E = \frac{\epsilon_0}{2}E^2 \tag{3.31}$$

Similarly, the energy density of the B-field alone (as it might be computed within a current-carrying toroid) is

$$u_B = \frac{1}{2\mu_0} B^2 \tag{3.32}$$

The relationship $E = cB$ was derived specifically for plane waves; nonetheless it's quite general in its applicability. Using $c = 1/\sqrt{\epsilon_0\mu_0}$, it follows that

$$u_E = u_B \tag{3.33}$$

The energy streaming through space in the form of an electromagnetic wave is shared equally between the constituent electric and magnetic fields. Inasmuch as

$$u = u_E + u_B \tag{3.34}$$

$$u = \epsilon_0 E^2 \tag{3.35}$$

or equivalently,

$$u = \frac{1}{\mu_0} B^2 \tag{3.36}$$

To represent the flow of electromagnetic energy associated with a traveling wave, let S symbolize the transport of energy per unit time (the power) across a unit area. In the SI system it has units of W/m^2. Figure 3.11 depicts an electromagnetic wave traveling with a speed c through an area A. During a very small interval of time Δt, only the energy contained in the cylindrical volume, $u(c\,\Delta t\,A)$, will cross A. Thus

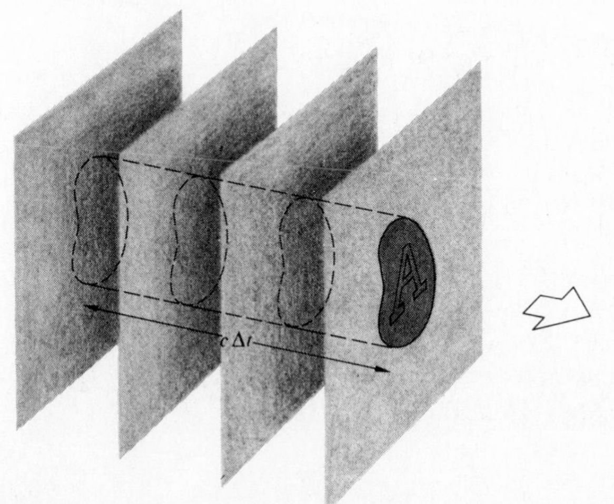

FIGURE 3.11 The flow of electromagnetic energy.

$$S = \frac{uc\,\Delta t\,A}{\Delta t\,A} = uc \tag{3.37}$$

or, using Eq. (3.35),

$$S = \frac{1}{\mu_0} EB \tag{3.38}$$

We now make the reasonable assumption (for isotropic media) that the energy flows in the direction of the propagation of the wave. The corresponding *vector* **S** is then

$$\mathbf{S} = \frac{1}{\mu_0} \mathbf{E} \times \mathbf{B} \tag{3.39}$$

or

$$\mathbf{S} = c^2\epsilon_0 \mathbf{E} \times \mathbf{B} \tag{3.40}$$

The magnitude of **S** is the power per unit area crossing a surface whose normal is parallel to **S**. Named after John Henry Poynting (1852–1914), it has come to be known as the **Poynting vector**. Let's now apply these considerations to the case of a harmonic, linearly polarized (the directions of the **E**- and **B**-fields are fixed) plane wave traveling through free space in the direction of **k**:

$$\mathbf{E} = \mathbf{E}_0 \cos(\mathbf{k}\cdot\mathbf{r} - \omega t) \tag{3.41}$$

$$\mathbf{B} = \mathbf{B}_0 \cos(\mathbf{k}\cdot\mathbf{r} - \omega t) \tag{3.42}$$

Using Eq. (3.40), we find

$$\mathbf{S} = c^2\epsilon_0 \mathbf{E}_0 \times \mathbf{B}_0 \cos^2(\mathbf{k}\cdot\mathbf{r} - \omega t) \tag{3.43}$$

This is the instantaneous flow of energy per unit area per unit time.

Averaging Harmonic Functions

It should be evident that **E** × **B** cycles from maxima to minima. At optical frequencies ($\approx 10^{15}$ Hz), **S** is an extremely rapidly varying function of time (indeed, twice as rapid as the fields, since cosine-squared has double the frequency of cosine); therefore, its instantaneous value would be an impractical quantity to measure directly. This suggests that we employ an averaging procedure. That is to say, we absorb the radiant energy during some finite interval of time using, for example, a photocell, a film plate, or the retina of a human eye.

The specific form of Eq. (3.43), and the central role played by harmonic functions, suggest that we take a moment to study the average values of such functions. The time-averaged value of some function $f(t)$ over an interval T is written as $\langle f(t)\rangle_T$ and given by

$$\langle f(t)\rangle_T = \frac{1}{T}\int_{t-T/2}^{t+T/2} f(t)\,dt$$

The resulting value of $\langle f(t)\rangle_T$ very much depends on T. To find the average of a harmonic function, evaluate

$$\langle e^{i\omega t}\rangle_T = \frac{1}{T}\int_{t-T/2}^{t+T/2} e^{i\omega t}dt = \frac{1}{i\omega T}e^{i\omega t}\Big|_{t-T/2}^{t+T/2}$$

$$\langle e^{i\omega t}\rangle_T = \frac{1}{i\omega T}(e^{i\omega(t+T/2)} - e^{i\omega(t-T/2)})$$

and $$\langle e^{i\omega t}\rangle_T = \frac{1}{i\omega T}e^{i\omega t}(e^{i\omega T/2} - e^{-i\omega T/2})$$

The parenthetical term should remind us (p. 23) of $\sin \omega T/2$. Hence

$$\langle e^{i\omega t}\rangle_T = \left(\frac{\sin \omega T/2}{\omega T/2}\right)e^{i\omega t}$$

The ratio in brackets is so common and important in Optics that it's given its own name; $(\sin u)/u$ is called (sinc u). Taking the real and imaginary parts of the above expression yields

$$\langle \cos \omega t\rangle_T = (\text{sinc } u)\cos \omega t$$

and $$\langle \sin \omega t\rangle_T = (\text{sinc } u)\sin \omega t$$

The average of the cosine is itself a cosine, oscillating with the same frequency but having a sinc-function amplitude that drops off from its initial value of 1.0 very rapidly (Fig. 3.12 and Table 1 in the Appendix). Since sinc $u = 0$ at $u = \omega T/2 = \pi$, which happens when $T = \tau$, it follows that $\cos \omega t$ averaged over an interval T equal to one period equals zero. Similarly, $\cos \omega t$ averages to zero over any whole number of periods, as does $\sin \omega t$. That's reasonable in that each of these functions encompasses as much positive area above the axis as negative area below the axis, and that's what the defining integral corresponds to. After an interval of several periods, the sinc term will be so small that the fluctuations around zero will be negligible: $\langle \cos \omega t\rangle_T$ and $\langle \sin \omega t\rangle_T$ are then essentially zero.

It's left for Problem 3.8 to show that $\langle \cos^2 \omega t\rangle_T = \frac{1}{2}[1 + \text{sinc } \omega T \cos 2\omega t]$, which oscillates about a value of $1/2$ at a frequency of 2ω and rapidly approaches $1/2$ as T increases beyond a few dozen periods. In the case of light $\tau \approx 10^{-15}$ s and so averaging over even a microsecond corresponds to $T \approx 10^9\tau$, far more than enough to drive the sinc function to some totally negligible value, whereupon $\langle \cos^2 \omega t\rangle_T = 1/2$. Figure 3.13 suggests the same result; we chop off the humps above

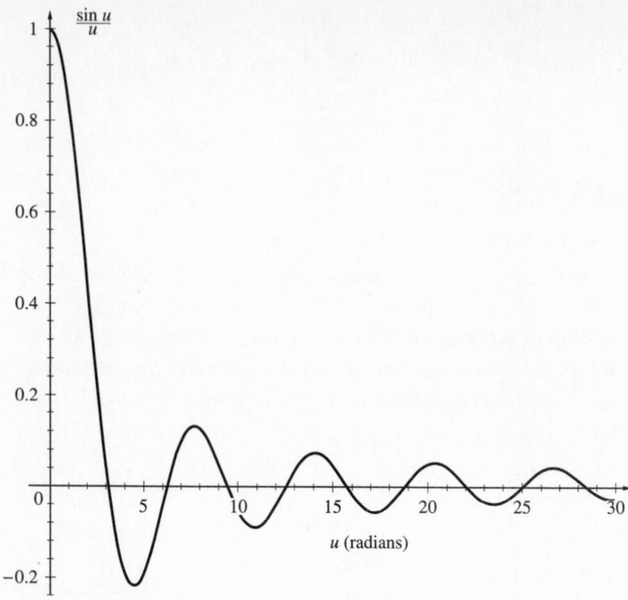

FIGURE 3.12: sinc u. Notice how the sinc function has a value of zero at $u = \pi$, 2π, 3π, and so forth.

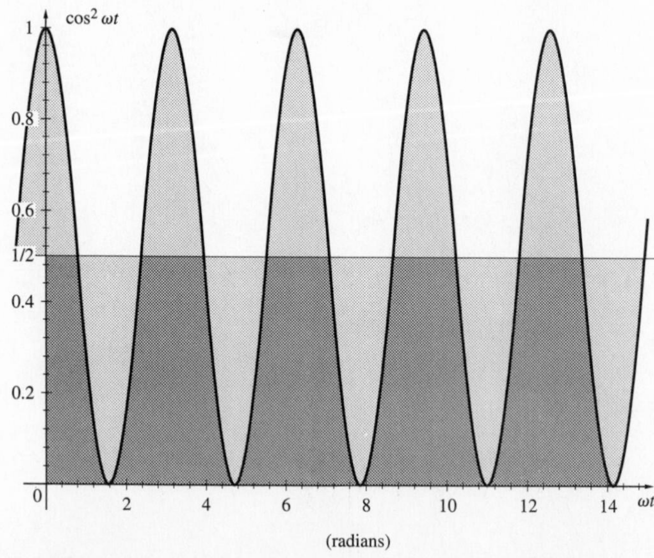

FIGURE 3.13 Using the peaks above the $\frac{1}{2}$ line to fill the troughs beneath it suggests that the average is $\frac{1}{2}$.

the 1/2 line and use them to fill in the missing areas beneath the line. After enough cycles, the area under the $f(t)$ curve divided by T, which is $\langle f(t)\rangle_T$, approaches 1/2.

3.3.2 Irradiance

When we talk about the "amount" of light illuminating a surface, we are referring to something called the **irradiance**,* denoted by I—the ***average energy per unit area per unit time***. Any kind of light-level detector has an entrance window that admits radiant energy through some fixed area A. The dependence on the size of that particular window is removed by dividing the total energy received by A. Furthermore, since the power arriving cannot be measured instantaneously, the detector must integrate the energy flux over some finite time, T. If the quantity to be measured is the *net* energy per unit area received, it depends on T and is therefore of limited utility. Someone else making a similar measurement under the same conditions can get a different result using a different T. If, however, the T is now divided out, a highly practical quantity results, one that corresponds to the average energy per unit area per unit time, namely, I.

The time-averaged value ($T >> \tau$) of the magnitude of the Poynting vector, symbolized by $\langle S\rangle_T$, is a measure of I. In the specific case of harmonic fields and Eq. (3.43),

$$\langle S\rangle_T = c^2\epsilon_0|\mathbf{E}_0 \times \mathbf{B}_0|\langle\cos^2(\mathbf{k}\cdot\mathbf{r} - \omega t)\rangle$$

Because $\langle\cos^2(\mathbf{k}\cdot\mathbf{r} - \omega t)\rangle_T = \frac{1}{2}$ for $T >> \tau$ (see Problem 3.7)

$$\langle S\rangle_T = \frac{c^2\epsilon_0}{2}|\mathbf{E}_0 \times \mathbf{B}_0|$$

or

$$I \equiv \langle S\rangle_T = \frac{c\epsilon_0}{2}E_0^2 \tag{3.44}$$

The irradiance is proportional to the square of the amplitude of the electric field. Two alternative ways of saying the same thing are simply

$$I = \frac{c}{\mu_0}\langle B^2\rangle_T \tag{3.45}$$

and

$$I = \epsilon_0 c\langle E^2\rangle_T \tag{3.46}$$

*In the past physicists generally used the word *intensity* to mean the flow of *energy per unit area per unit time*. By international, if not universal, agreement, that term is slowly being replaced in Optics by the word *irradiance*.

Within a linear, homogeneous, isotropic dielectric, the expression for the irradiance becomes

$$I = \epsilon v\langle E^2\rangle_T \tag{3.47}$$

Since, as we have learned, **E** is considerably more effective at exerting forces and doing work on charges than is **B**, we shall refer to **E** as the **optical field** and use Eq. (3.46) almost exclusively.

The time rate of flow of radiant energy is the **optical power** P or **radiant flux**, generally expressed in watts. If we divide the radiant flux incident on or exiting from a surface by the area of the surface, we have the **radiant flux density** (W/m^2). In the former case, we speak of the *irradiance*, in the latter the *exitance*, and in either instance the **flux density**. The irradiance is a measure of the *concentration* of power.

The Inverse Square Law

We saw earlier that the spherical-wave solution of the differential wave equation has an amplitude that varies inversely with r. Let's now examine this same feature within the context of energy conservation. Consider an isotropic point source in free space, emitting energy equally in all directions (i.e., emitting spherical waves). Surround the source with two concentric imaginary spherical surfaces of radii r_1 and r_2, as shown in Fig. 3.14. Let $E_0(r_1)$ and $E_0(r_2)$ represent the amplitudes of the

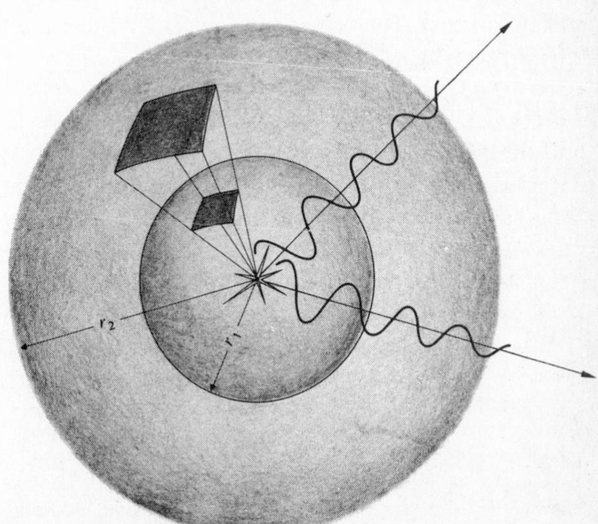

FIGURE 3.14 The geometry of the inverse square law.

waves over the first and second surfaces, respectively. If energy is to be conserved, the total amount of energy flowing through each surface per second must be equal, since there are no other sources or sinks present. Multiplying I by the surface area and taking the square root, we get

$$r_1 E_0(r_1) = r_2 E_0(r_2)$$

Inasmuch as r_1 and r_2 are arbitrary, it follows that

$$rE_0(r) = \text{constant},$$

and the amplitude must drop off inversely with r. The irradiance from a point source is proportional to $1/r^2$. This is the well-known **Inverse Square Law**, which is easily verified with a point source and a photographic exposure meter.

3.3.3 Photons

Light is absorbed and emitted in tiny discrete bursts, in particles of electromagnetic "stuff," known as photons. That much has been confirmed and is well established, but the question of whether or not light is "really" a stream of photons is far from settled, and that issue will be revisited several times to come.*

Ordinarily, a light beam delivers so many minute energy quanta that its inherent granular nature is totally hidden and a continuous phenomenon is observed macroscopically. That sort of thing is commonplace in Nature; the forces exerted by the individual gas molecules in a wind blend into what seems a continuous pressure, but it obviously isn't. Indeed, that analogy between a gas and a flow of photons is one we will come back to presently.

As the great French physicist Louis de Broglie put it, "Light is, in short, the most refined form of matter," and all matter, including light, is quantized; at base it comes in minute elementary units—quarks, leptons, and photons. That overarching unity is among the most appealing reasons to embrace the photon as particle.

*Thus whenever, for brevity's sake, an expression such as "a beam of photons" is used, the reader is advised to keep in mind that although the light-*is*-corpuscular model has wide acceptance (especially in high-energy physics) and is a crucial part of the contemporary discourse, like almost everything else in physics, it is not yet established beyond all doubt. See, for example, the summary article by R. Kidd, J. Ardini, and A. Anton, "Evolution of the modern photon," *Am. J. Phys.* **57** (1), 27 (1989).

The Failure of Classical Theory

In 1900 Max Planck produced a rather tentative, and somewhat erroneous, analysis of a process known as *blackbody radiation* (p. 575). Nonetheless, the expression he came up with beautifully fit all the existing experimental data, a feat no other formulation had even come close to. Basically, he considered electromagnetic (EM) waves in equilibrium within an isothermal chamber (or cavity). All the EM-radiation within the cavity is emitted and absorbed by the walls of the enclosure—none enters from outside. This ensures that its spectral composition will match that emanating from an ideal black surface. The goal was to predict the spectrum of the radiation that would emerge from a small opening in the cavity. Totally stymied by the problem, as a last resort, Planck turned to the classical statistical analysis of Maxwell and Boltzmann, which was developed as the basis of the kinetic theory of gases. Philosophically, this is a completely deterministic treatment that assumes one can follow, at least in principle, every atom as it moves around in the system. Consequently, each atom is taken to be recognizable, independent, and enumerable. For purely computational reasons, Planck hypothesized that each one of the oscillators lining the walls of the chamber could absorb and emit only discrete amounts of energy proportional to its oscillatory frequency, ν. These energy jolts were equal to whole number multiples of $h\nu$, where h, now called **Planck's Constant**, was found to be 6.626×10^{-34} J·s. Being a rather traditional man, Planck otherwise held fast to the classical wave picture of light, insisting that only the oscillators were quantized.

Prophetically, J. J. Thomson (1903)—the discoverer of the electron—extended the idea, suggesting that electromagnetic waves might actually be radically different from other waves; perhaps local concentrations of radiant energy truly existed. Thomson had observed that when a beam of high-frequency EM-radiation (X-rays) was shone onto a gas, only certain of the atoms, here and there, were ionized. It was as if the beam had "hot spots" rather than having its energy distributed continuously over the wavefront (Fig. 3.15).

The concept of the photon in its modern incarnation came into being in 1905 by way of Einstein's brilliant theoretical work on the Photoelectric Effect. When a metal is bathed in EM-radiation, it emits electrons. The details of that process had been studied experimentally for decades, but it defied analysis via classical Electromagnetic Theory. Einstein's startling treatment established that ***the electromagnetic field itself is quantized***. Each constituent photon has an energy given by

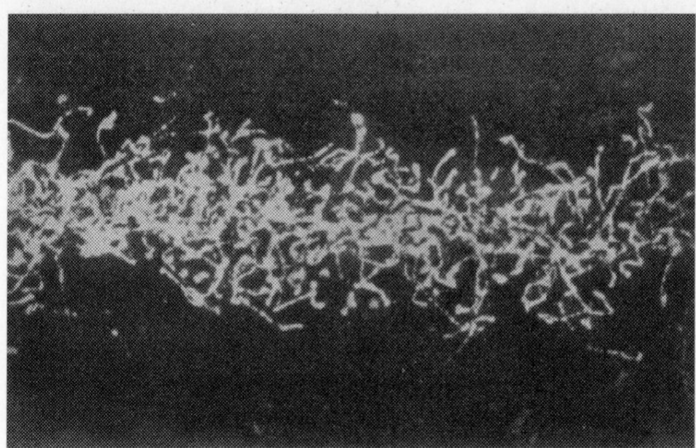

FIGURE 3.15 A beam of X-rays enters a cloud chamber on the left. The tracks are made by electrons emitted via either the Photoelectric Effect (these tend to leave long tracks at large angles to the beam) or the Compton Effect (short tracks more in the forward direction). Although classically, the X-ray beam has its energy uniformly distributed along transverse wavefronts, the scattering seems discrete and random. (From the Smithsonian Report, 1915.)

the product of Planck's Constant and the frequency of the radiation field:

$$\mathscr{E} = h\nu \tag{3.48}$$

Photons are stable, chargeless, massless elementary particles that exist only at the speed c.

In 1924 Satyendra N. Bose formulated a new and rigorous proof of Planck's blackbody equation using statistical methods applied to light quanta. The cavity was envisioned to be filled with a "gas" of photons, ***which were taken to be totally indistinguishable, one from the other***. That was a crucial feature of this quantum-mechanical treatment. It meant that the microparticles were completely interchangeable, and this had a profound effect on the statistical formulation. In a mathematical sense, each particle of this quantum "gas" is related to every other particle, and no one of them can be taken as statistically independent of the system as a whole. That's very different from the independent way classical microparticles behave in an ordinary gas. The quantum-mechanical probability function that describes the statistical behavior of thermal light is now known as the Bose-Einstein distribution. The photon, whatever it is, became an indispensable tool of theoretical physics.

Unlike ordinary objects, photons cannot be seen directly; what is known of them comes from observing the results of their being either created or annihilated. Light is never *seen* just sailing along through space. A photon is observed by detecting the effect it has on its surroundings, and it only has an observable effect when it either comes into, or goes out of, existence. Photons begin and end on charged particles; most often they are emitted from and absorbed by electrons. And these are usually the electrons circulating in the clouds around atoms. A number of experiments have directly confirmed the quantal nature of the emission process. For example, imagine a very dim source surrounded, at equal distances, by identical photodetectors each capable of measuring a minute amount of light. If the emission, *no matter how faint*, is a continuous wave, as is maintained classically, all the detectors should register each emitted pulse in coincidence. That doesn't happen; instead, counts are registered by detectors independently, one at a time, in clear agreement with the idea that atoms emit localized light quanta in random directions.

Furthermore, it has been confirmed that when an atom emits light (i.e., a photon), it recoils in the opposite direction, just as a pistol recoils when it fires a bullet. In Fig. 3.16 atoms pumped up with excess energy (i.e., excited, p. 64) are formed into a narrow beam. They soon spontaneously radiate photons in random directions and are themselves kicked backward,

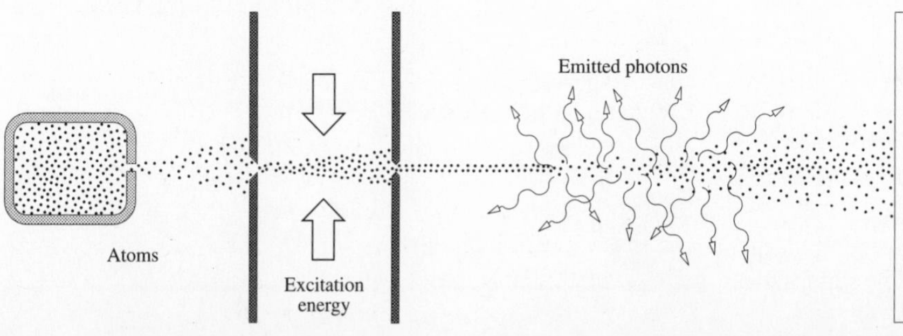

FIGURE 3.16 When so-called excited atoms forming a narrow beam radiate photons, they recoil laterally and the beam spreads out. Alternatively if the beam is formed of atoms that are not excited (i.e., they are in their ground states), it remains narrow all the way to the screen.

often laterally away from the beam. The resulting spread of the beam is a quantum-mechanical effect inconsistent with the classical picture of the emission of a symmetrical wave.

A Barrage of Photons

When we analyze phenomena involving the activity of immense numbers of participants, the use of statistical techniques is often the only practical way to proceed. In addition to the classical Maxwell-Boltzmann statistics (for distinguishable particles), there are two kinds of quantum-mechanical statistics (for indistinguishable particles): Bose-Einstein and Fermi-Dirac. The first treats particles that *are not* subject to the Pauli Exclusion Principle (i.e., particles that have zero or integer spins). Fermi-Dirac statistics treats particles that *are* subject to the Pauli Exclusion Principle (i.e., those that have spins that are odd integer multiples of $\frac{1}{2}$). Photons are called **bosons**, they are spin-1 particles, and the manner in which they group together obeys Bose-Einstein statistics. Similarly, electrons are **fermions**; they are spin-$\frac{1}{2}$ particles that obey Fermi-Dirac statistics.

Microparticles have defining physical characteristics such as mass, charge, and spin—characteristics that do not change. When these are given, we have completely specified the kind of particle being considered. Alternatively, there are alterable properties of any given microparticle that describe its momentary condition such as energy, momentum, and spin orientation. When all of these alterable quantities are given, we have specified the particular **state** the particle happens to be in at the moment.

Fermions are committed loners; *only one fermion can occupy any given state*. By comparison, bosons are gregarious; *any number of them can occupy the same state, and moreover, they actually tend to cluster close together*. ***When a very large number of photons occupy the same state, the inherent granularity of the light beam essentially vanishes and the electromagnetic field appears as the continuous medium of an electromagnetic wave***. Thus we can associate a monochromatic (monoenergetic) plane wave with a stream of photons having a high population density, all progressing in the same state (with the same energy, same frequency, same momentum, same direction). ***Different monochromatic plane waves represent different photon states***.

Unlike the photon, because electrons are fermions large numbers of them cannot tightly cluster in the same state, and a monoenergetic beam of electrons does not manifest itself on a macroscopic scale as a continuous classical wave. In that regard, EM-radiation is quite distinctive.

For a uniform monochromatic light beam of frequency ν, the quantity $I/h\nu$ is the average number of photons impinging on a unit area (normal to the beam) per unit time, namely, the photon flux density. More realistically, if the beam is quasi-monochromatic (p. 17) with an average frequency ν_0, its **mean photon flux density** is $I/h\nu_0$. Given that an incident quasi-monochromatic beam has a cross-sectional area A, its **mean photon flux** is

$$\Phi = AI/h\nu_0 = P/h\nu_0 \tag{3.49}$$

where P is the **optical power** of the beam in watts. The mean photon flux is the average number of photons arriving per unit of time (Table 3.1). For example, a small 1.0 mW He-Ne laserbeam with a mean wavelength of 632.8 nm delivers a mean photon flux of $P/h\nu_0 = (1.0 \times 10^{-3}\ \text{W})/[(6.626 \times 10^{-34}\ \text{J}\cdot\text{s})(2.998 \times 10^8\ \text{m/s})/(632.8 \times 10^{-9}\ \text{m})] = 3.2 \times 10^{15}$ photons per second.

Imagine a uniform beam of light having a constant irradiance (and therefore a constant mean photon flux) incident on a screen. The energy of the beam is deposited on the screen in a random flurry of minute bursts. Individually, the incoming photons register at locations on the plane that are totally unpredictable, and arrive at moments in time that are equally unpredictable. It looks as if the beam is composed of a random stream of photons, but that conclusion, however tempting, goes beyond the observation. What *can* be said is that the light delivers its energy in a staccato of impacts that are random in space and time across the beam.

TABLE 3.1 The Mean Photon Flux Density for a Sampling of Common Sources

Light Source	Mean Photon Flux Density Φ/A in units of (photons/s·m^2)
Laserbeam (10 mW, He-Ne, focused to 20 μm)	10^{26}
Laserbeam (1 mW, He-Ne)	10^{21}
Bright sunlight	10^{18}
Indoor light level	10^{16}
Twilight	10^{14}
Moonlight	10^{12}
Starlight	10^{10}

Suppose that we project a light pattern onto the screen; it might be a set of interference fringes or the image of a woman's face. The barrage of photons forming the image is a statistical tumult; we cannot predict when a photon will arrive at any given location. But we can determine the likelihood of one or more photons hitting any particular point during a substantial time interval. ***At any location on the screen, the measured (or classically computed) value of the irradiance is proportional to the probability of detecting a photon at that location*** (p. 137).

Figure 1.8, which is a pictorial record of the arrival of individual photons, was produced using a special kind of photomultiplier tube. To underscore the inherent photonic nature of radiant energy, let's now use an entirely different and more straightforward photographic approach to record the incidence of light. A photographic emulsion contains a distribution of microscopic ($\approx 10^{-6}$ m) silver halide crystals, each comprising approximately 10^{10} Ag atoms. A single photon can interact with such a crystal, disrupting a silver-halogen bond and freeing up an Ag atom. One or more silver atoms then serve as a development center on the exposed crystal. The film is developed using a chemical reducing agent. It dissolves each exposed crystal, depositing at that site, all of its Ag atoms as a single clump of the metal.

Figure 3.17 shows a series of photographs taken with increasing amounts of illumination. Using extremely dim light, a few thousand photons, the first picture is composed of roughly as many silver clumps, making a pattern that only begins to suggest an overall image. As the number of participating photons goes up (roughly by a factor of 10 for each successive picture), the image becomes increasingly smooth and recognizable. When there are tens of millions of photons forming the image, the statistical nature of the process is lost and the picture assumes a familiar continuous appearance.

PHOTON COUNTING

What, if anything, can be said about the statistical nature of the barrage of photons delivered as a beam of light? To answer that question, researchers within the last 30 years or so have conducted experiments in which they literally counted indi-

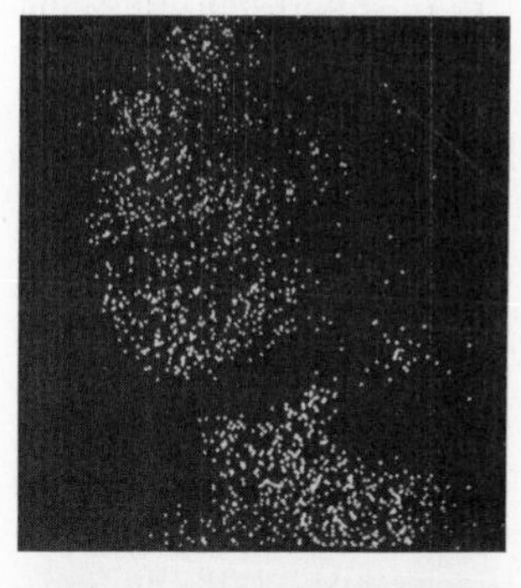
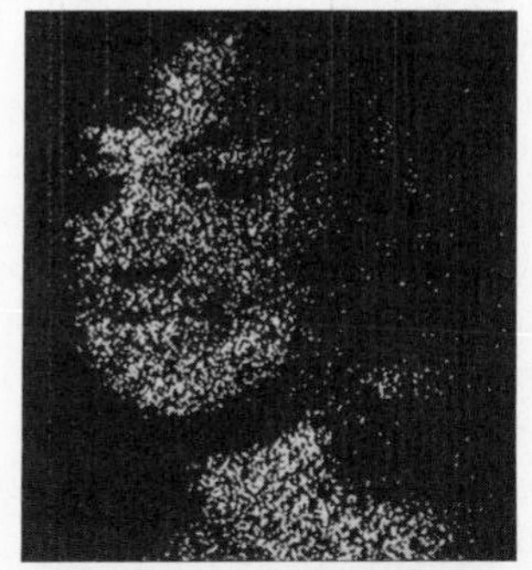

FIGURE 3.17 These photographs (which were electronically enhanced), are a compelling illustration of the granularity displayed by light in its interaction with matter. Under exceedingly faint illumination, the pattern (each spot corresponding to one photon) seems almost random, but as the light level increases the quantal character of the process gradually becomes obscured. (See *Advances in Biological and Medical Physics* V, 1957, 211–242.) (Photos courtesy Radio Corporation of America.)

vidual photons. What they found was that the pattern of arrival of photons was characteristic of the type of source.* We cannot go into the theoretical details here, but it is informative at least to look at the results for the two extreme cases of what is often called *coherent* and *chaotic* light.

Consider an ideal continuous laserbeam of *constant irradiance*; remember that irradiance is a time-averaged quantity via Eq. (3.46). The beam has a constant optical power P—which is also a time-averaged quantity—and, from Eq. (3.49), a corresponding mean photon flux Φ. Figure 3.18 depicts the random arrival of photons on a time scale that is short compared to the interval over which the irradiance is averaged. Thus the macroscopic quantity P can be measured to be constant, even though there is an underlying discontinuous transfer of energy.

Now pass the beam through a shutter that stays open for a short sampling time T (which might be in the range from about 10 μs to perhaps 10 ms), and count the number of photons arriving at a photodetector during that interval. After a brief pause repeat the procedure, and do it again and again, tens of thousands of times. The results are presented in a histogram (Fig. 3.19), where the number of trials in which N photons were counted is plotted against N. Few trials register either very few photons or very many photons. On average, the number of photons per trial is $N_{av} = \Phi T = PT/h\nu_0$. The shape of the data plot, which can be derived using probability theory, closely approximates the well-known *Poisson distribution.* It represents a graph of the probability that the detector (during a trial interval lasting a time T) will record zero photons, one photon, two photons, and so forth.

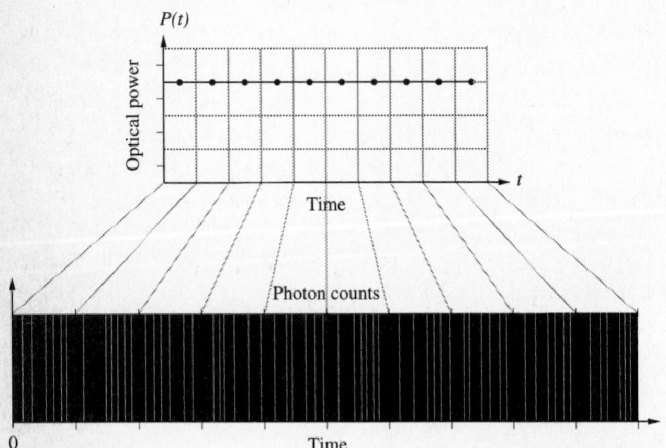

FIGURE 3.18 A constant optical power and the corresponding random set of photon counts. The arrival of each photon is an independent event.

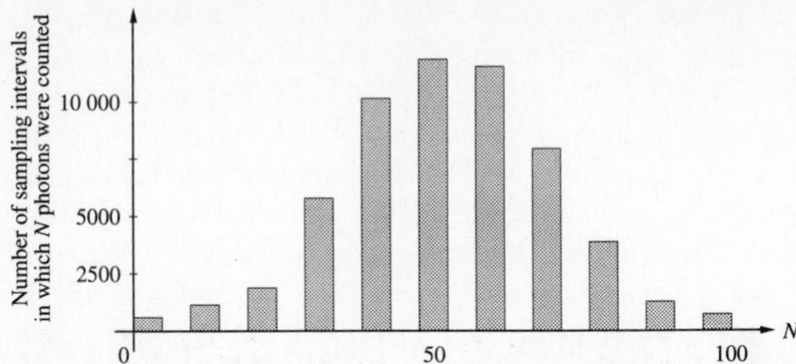

FIGURE 3.19 A typical histogram showing the probability or photon-count distribution for a beam of constant irradiance.

The Poisson distribution is the same symmetrical curve one gets when counting either the number of particles randomly emitted by a long-lived radioactive sample, or the number of raindrops randomly descending on an area in a steady shower. It's also the curve of the probability of getting a head, plotted against the number of heads occurring (N), for a coin tossed more than about 20 times. Thus with $N_{max} = 20$ the highest probability occurs near the average value N_{av}, namely, at $\frac{1}{2}N_{max}$ or 10 and the lowest at $N = 0$ and $N = 20$. The most probable value will be 10 heads out of 20 tosses, and the likelihood of getting either no heads or all heads is vanishingly small. It would seem that however an ideal laser produces light, it generates a stream of photons whose individual arrival is random and statistically independent. For reasons that will be explored later, an ideal monoenergetic beam—a monochromatic plane wave—is the epitome of what is known as **coherent light**.

Not surprisingly, ***the statistical distribution of the number of photons arriving at a detector depends on the nature of the source***; it is fundamentally different for an ideal *coherent source* at one extreme, as compared to an equally idealized completely incoherent or *chaotic source* at the other extreme. A stabilized laser resembles a coherent source, and an ordinary thermal source such as a lightbulb or a star or a gas discharge lamp more closely resembles a chaotic source. In the case of ordinary light, there are inherent fluctuations in the irradiance and therefore in the optical power (p. 49). These

*See P. Koczyk, P. Wiewior, and C. Radzewicz, "Photon counting statistics—Undergraduate experiment," *Am. J. Phys.* **64** (3), 240 (1996).

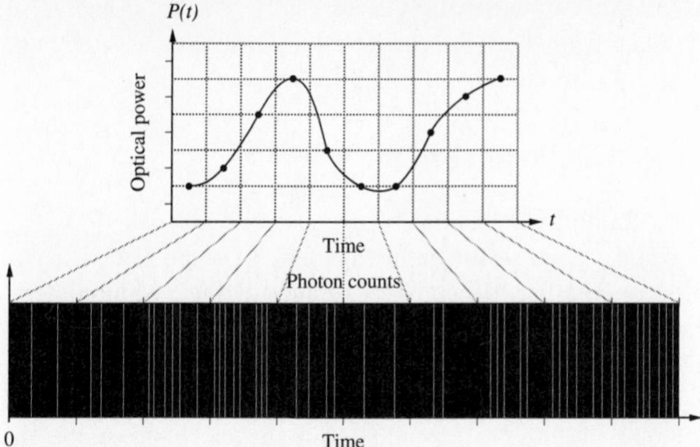

FIGURE 3.20 A time-varying optical power and the corresponding random set of photon counts. Now there are fluctuations that are correlated, and the photon arrivals are no longer independent.

fluctuations are correlated, and the associated number of emitted photons, though random in time, is correspondingly also correlated (Fig. 3.20). The greater the optical power, the greater the number density of photons. Because the arrival of photons at the detector is not a succession of independent events, Bose-Einstein statistics apply (Fig. 3.21). Here the most likely number of counts per interval is zero, whereas, ideally, for laser light the most likely number of photons to be measured during a sampling interval equals the average number recorded. Thus even if a beam of laser light and a beam of ordinary light have the same average irradiance and the same frequency spectrum, they are still inherently distinguishable—a result that extends beyond classical theory.

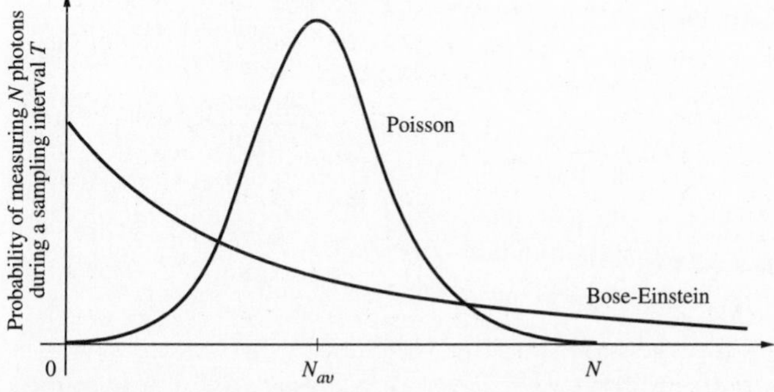

FIGURE 3.21 Poisson and Bose-Einstein photon counting distributions.

3.3.4 Radiation Pressure and Momentum

As long ago as 1619, Johannes Kepler proposed that it was the pressure of sunlight that blew back a comet's tail so that it always pointed away from the Sun. That argument particularly appealed to the later proponents of the corpuscular theory of light. After all, they envisioned a beam of light as a stream of particles, and such a stream would obviously exert a force as it bombarded matter. For a while it seemed as though this effect might at last establish the superiority of the corpuscular over the wave theory, but all the experimental efforts to that end failed to detect the force of radiation, and interest slowly waned.

Ironically, it was Maxwell in 1873 who revived the subject by establishing theoretically that waves do indeed exert pressure. "In a medium in which waves are propagated," wrote Maxwell, "there is a pressure in the direction normal to the waves, and numerically equal to the energy in a unit of volume."

When an electromagnetic wave impinges on some material surface, it interacts with the charges that constitute bulk matter. Regardless of whether the wave is partially absorbed or reflected, it exerts a force on those charges and hence on the surface itself. For example, in the case of a good conductor, the wave's electric field generates currents, and its magnetic field generates forces on those currents.

It's possible to compute the resulting force via Electromagnetic Theory, whereupon Newton's Second Law (which maintains that force equals the time rate-of-change of momentum) suggests that the *wave itself carries momentum*. Indeed, whenever we have a flow of energy, it's reasonable to expect that there will be an associated momentum—the two are the related time and space aspects of motion.

As Maxwell showed, the **radiation pressure**, $\mathcal{P}$, equals the energy density of the electromagnetic wave. From Eqs. (3.31) and (3.32), for a vacuum, we know that

$$u_E = \frac{\epsilon_0}{2} E^2 \text{ and } u_M = \frac{1}{2\mu_0} B^2$$

Since $\mathcal{P} = u = u_E + u_M$,

$$\mathcal{P} = \frac{\epsilon_0}{2} E^2 + \frac{1}{2\mu_0} B^2$$

Alternatively, using Eq. (3.37), we can express the pressure in terms of the magnitude of the Poynting vector, namely,

$$\mathcal{P}(t) = \frac{S(t)}{c} \tag{3.50}$$

Notice that this equation has the units of power divided by area, divided by speed—or equivalently, force times speed divided by area and speed, or just force over area. This is the instantaneous pressure that would be exerted on a perfectly absorbing surface by a normally incident beam.

Inasmuch as the **E**- and **B**-fields are rapidly varying, $S(t)$ is rapidly varying, so it is eminently practical to deal with the average radiation pressure, namely,

$$\langle \mathcal{P}(t) \rangle_{\mathrm{T}} = \frac{\langle S(t) \rangle_{\mathrm{T}}}{c} = \frac{I}{c} \tag{3.51}$$

expressed in newtons per square meter. This same pressure is exerted on a source that itself is radiating energy.

Referring back to Fig. 3.11, if p is momentum, the force exerted by the beam on an absorbing surface is

$$A\mathcal{P} = \frac{\Delta p}{\Delta t} \tag{3.52}$$

If p_{V} is the *momentum per unit volume of the radiation*, then an amount of momentum $\Delta p = p_{\mathrm{V}}(c\,\Delta t\,A)$ is transported to A during each time interval Δt, and

$$A\mathcal{P} = \frac{p_{\mathrm{V}}(c\,\Delta t\,A)}{\Delta t} = A\,\frac{S}{c}$$

Hence the volume density of electromagnetic momentum is

$$p_{\mathrm{V}} = \frac{S}{c^2} \tag{3.53}$$

When the surface under illumination is perfectly reflecting, the beam that entered with a velocity $+c$ will emerge with a velocity $-c$. This corresponds to twice the change in momentum that occurs on absorption, and hence

$$\langle \mathcal{P}(t) \rangle_{\mathrm{T}} = 2\,\frac{\langle S(t) \rangle_{\mathrm{T}}}{c}$$

Notice, from Eqs. (3.50) and (3.52), that if some amount of energy $\mathscr{E}$ is transported per square meter per second, then there will be a corresponding momentum $\mathscr{E}/c$ transported per square meter per second.

In the photon picture, each quantum, has an energy $\mathscr{E} = h\nu$. We can then expect a photon to carry a momentum

$$p = \frac{\mathscr{E}}{c} = \frac{h}{\lambda} \tag{3.54}$$

Its vector momentum would be

$$\mathbf{p} = \hbar\mathbf{k}$$

where **k** is the propagation vector and $\hbar \equiv h/2\pi$. This all fits in rather nicely with Special Relativity, which relates the mass m, energy, and momentum of a particle by

$$\mathscr{E} = [(cp)^2 + (mc^2)^2]^{1/2}$$

For a photon $m = 0$ and $\mathscr{E} = cp$.

These quantum-mechanical ideas have been confirmed experimentally utilizing the Compton Effect, which detects the energy and momentum transferred to an electron upon interaction with an individual X-ray photon.

The average flux density of electromagnetic energy from the Sun impinging normally on a surface just outside the Earth's atmosphere is about 1400 $\mathrm{W/m^2}$. Assuming complete absorption, the resulting pressure would be $4.7 \times 10^{-6}\,\mathrm{N/m^2}$, or 1.8×10^{-9} ounce/$\mathrm{cm^2}$, as compared with, say, atmospheric pressure of about $10^5\,\mathrm{N/m^2}$. The pressure of solar radiation at the Earth is tiny, but it is still responsible for a substantial planetwide force of roughly 10 tons. Even at the very surface of the Sun, radiation pressure is relatively small (see Problem 3.25). As one might expect, it becomes appreciable within the blazing body of a large bright star, where it plays a significant part in supporting the star against gravity. Despite the Sun's modest flux density, it nonetheless can produce appreciable effects over long-acting times. For example, had the pressure of sunlight exerted on the Viking spacecraft during its journey been neglected, it would have missed Mars by about 15 000 km. Calculations show that it is even feasible to use the pressure of sunlight to propel a space vehicle among the inner planets.* Ships with immense reflecting sails driven by solar radiation pressure may some day ply the dark sea of local space.

The pressure exerted by light was actually measured as long ago as 1901 by the Russian experimenter Pyotr Nikolaievich Lebedev (1866–1912) and independently by the Americans Ernest Fox Nichols (1869–1924) and Gordon Ferrie Hull (1870–1956). Their accomplishments were formidable, considering the light sources available at the time. Nowadays, with the advent of the laser, light can be focused down to a spot size approaching the theoretical limit of about one wavelength in radius. The resulting irradiance, and therefore the pressure, is appreciable, even with a laser rated at just

*The charged-particle flux called the "solar wind" is 1000 to 100 000 times less effective in providing a propulsive force than is sunlight.

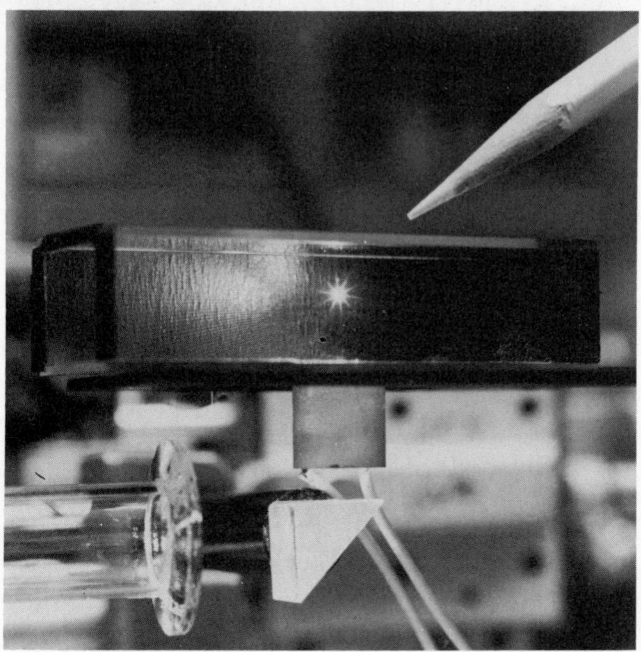

FIGURE 3.22 The tiny starlike speck is a minute (one-thousandth of an inch diameter) transparent glass sphere suspended in midair on an upward 250 mW laserbeam. (Photo courtesy Bell Laboratories.)

a few watts. It has thus become practical to consider radiation pressure for all sorts of applications, such as separating isotopes, accelerating particles, and even optically levitating small objects (Fig. 3.22).

Light can also transport angular momentum, but that raises a number of issues that will be treated later (p. 324).

3.4 Radiation

Electromagnetic radiation comes in a broad range of wavelengths and frequencies, although in vacuum it all travels at the same speed. Despite the fact that we distinguish different regions of the spectrum with names like radiowaves, microwaves, infrared, and so forth, there is only one entity, one essence of electromagnetic wave. Maxwell's Equations are independent of wavelength and so suggest no fundamental differences in kind. Accordingly, it is reasonable to look for a common source-mechanism for all EM-radiation. What we find is that the various types of radiant energy seem to have a common origin in that they are all associated with *nonuniformly moving charges*. We are, of course, dealing with waves in the electromagnetic field, and charge *is* that which gives rise to field, so this is not altogether surprising.

A stationary charge has a constant **E**-field, no **B**-field, and hence produces no radiation—where would the energy come from if it did? A uniformly moving charge has both an **E**- and a **B**-field, but it does not radiate. If you traveled along with the charge, the current would thereupon vanish, hence **B** would vanish, and we would be back at the previous case, uniform motion being relative. That's reasonable, since it would make no sense at all if the charge stopped radiating just because you started walking along next to it. That leaves *nonuniformly moving charges*, which assuredly do radiate. In the photon picture this is underscored by the conviction that the fundamental interactions between substantial matter and radiant energy are between photons and charges.

We know in general that free charges (those not bound within an atom) emit electromagnetic radiation when accelerated. That much is true for charges changing speed along a straight line within a linear accelerator, sailing around in circles inside a cyclotron, or simply oscillating back and forth in a radio antenna—***if a charge moves nonuniformly, it radiates***. A free charged particle can spontaneously absorb or emit a photon, and an increasing number of important devices, ranging from the free-electron laser (1977) to the synchrotron radiation generator, utilize this mechanism on a practical level.

3.4.1 Linearly Accelerating Charges

Consider a charge moving at a constant speed. It essentially has attached to it an unchanging radial electric field and a surrounding circular magnetic field. Although at any point in space the **E**-field changes from moment to moment, at any instant its value can be determined by supposing that the field lines move along, fixed to the charge. Thus the field does not disengage from the charge, and there is no radiation.

The electric field of a charge at rest can be represented, as in Fig. 3.23, by a uniform, radial distribution of straight field lines. For a charge moving at a constant velocity **v**, the field lines are still radial and straight, but they are no longer uniformly distributed. The nonuniformity becomes evident at high speeds and is usually negligible when $v << c$.

In contrast, Fig. 3.24 shows the field lines associated with an electron accelerating uniformly to the right. The points O_1, O_2, O_3, and O_4 are the positions of the electron after equal

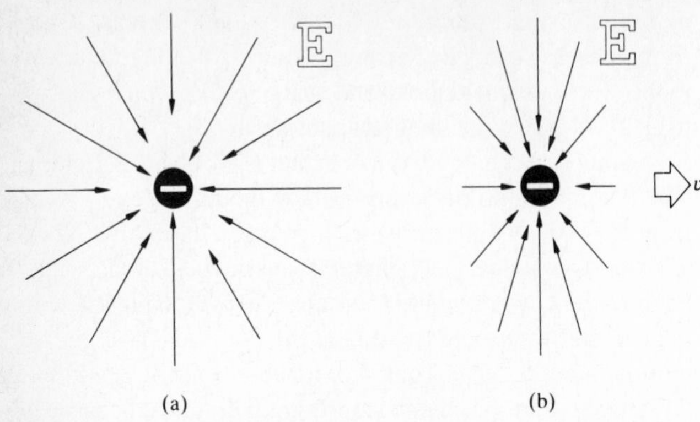

FIGURE 3.23 (*a*) Electric field of a stationary electron. (*b*) Electric field of a moving electron.

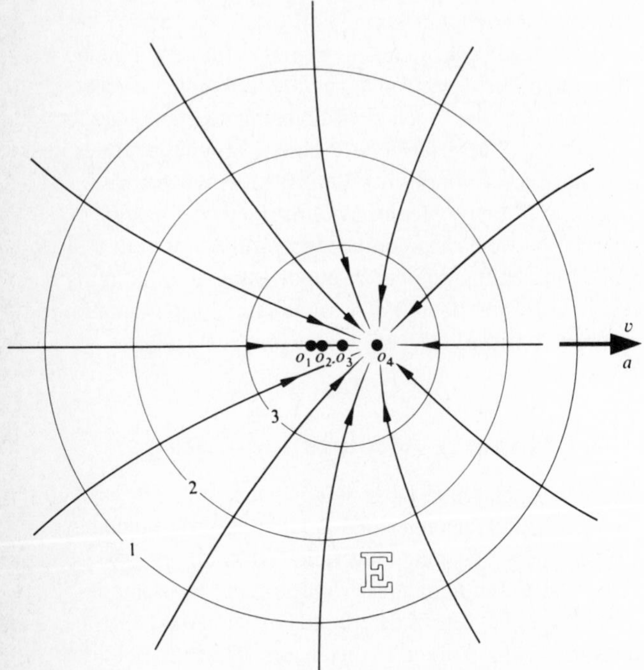

FIGURE 3.24 Electric field of a uniformly accelerating electron.

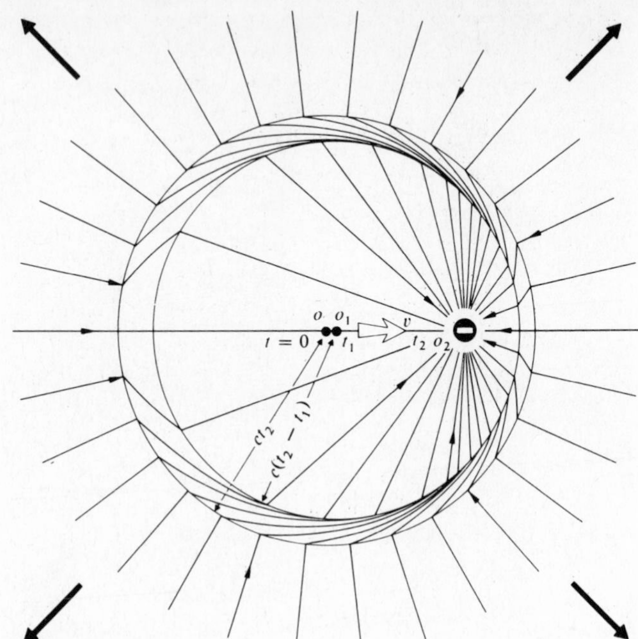

FIGURE 3.25 A kink in the E-field lines.

time intervals. The field lines are now curved, and this is a significant difference. As a further contrast, Fig. 3.25 depicts the field of an electron at some arbitrary time t_2. Before $t = 0$ the particle was always at rest at the point O. The charge was then uniformly accelerated until time t_1, reaching a speed v, which was maintained constant thereafter. We can anticipate that the surrounding field lines will somehow carry the information that the electron has accelerated. We have ample reason to assume that this "information" will propagate at the speed c. If, for example, $t_2 = 10^{-8}$ s, no point beyond 3 m from O would be aware of the fact that the charge had even moved. All the lines in that region would be uniform, straight, and centered on O, as if the charge were still there. At time t_2 the electron is at point O_2 moving with a constant speed v. In the vicinity of O_2 the field lines must then resemble those in Fig. 3.22*b*. Gauss's Law requires that the lines outside the sphere of radius ct_2 connect to those within the sphere of radius $c(t_2 - t_1)$, since there are no charges between them. It is now apparent that during the interval when the particle accelerated, the field lines became distorted and a kink appeared. The exact shape of the lines within the region of the kink is of little interest here. What is significant is that there now exists a *transverse component* of the electric field $\mathbf{E}_T$, which propagates outward as a pulse. At some point in space the transverse electric field will be a function of time, and it will therefore be accompanied by a magnetic field.

The radial component of the electric field drops off as $1/r^2$, while the transverse component goes as $1/r$. At large distances from the charge, the only significant field will be the $\mathbf{E}_T$-component of the pulse, which is known as the *radiation field*.* For a positive charge moving slowly ($v << c$), the electric and magnetic radiation fields can be shown to be proportional to $\mathbf{r} \times (\mathbf{r} \times \mathbf{a})$ and $(\mathbf{a} \times \mathbf{r})$, respectively, where $\mathbf{a}$ is the acceleration. For a negative charge the reverse occurs, as shown in Fig. 3.26. Observe that the irradiance is a function of θ and that $I(0) = I(180°) = 0$ while $I(90°) = I(270°)$ is a maximum.

The energy that is radiated out into the surrounding space is supplied to the charge by some external agent. That agent is responsible for the accelerating force, which in turn does work on the charge.

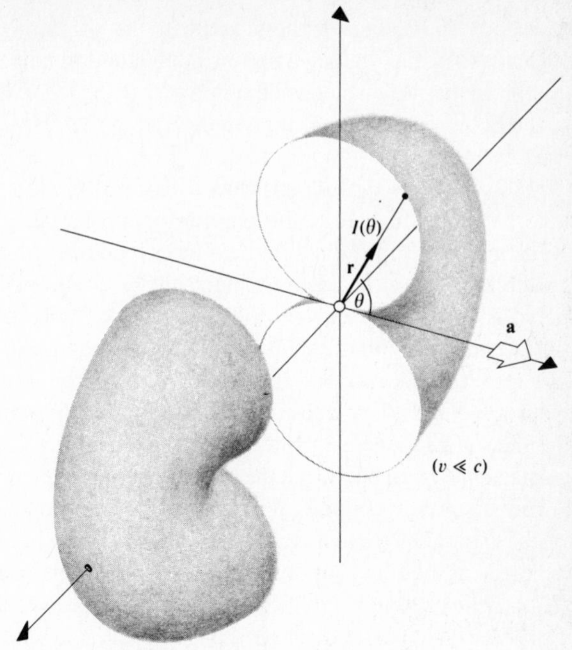

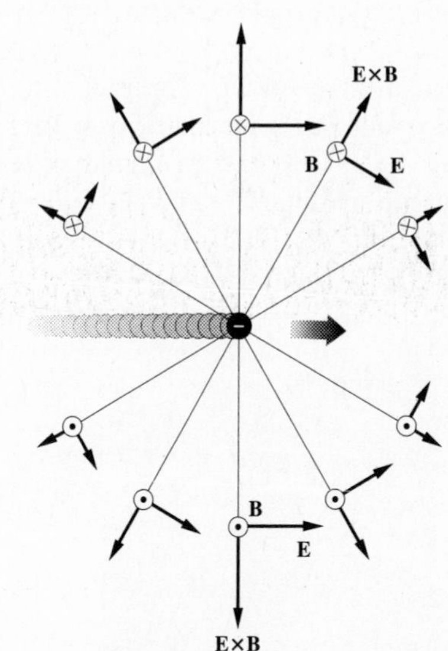

FIGURE 3.26 The toroidal radiation pattern of a linearly accelerating charge (split to show cross section).

3.4.2 Synchrotron Radiation

A free charged particle traveling on any sort of curved path is accelerating and will radiate. This provides a powerful mechanism for producing radiant energy, both naturally and in the laboratory. The synchrotron radiation generator, a research tool developed in the 1970s, does just that. Clumps of charged particles, usually electrons or positrons, interacting with an applied magnetic field are made to revolve around a large, essentially circular track at a precisely controlled speed. The frequency of the orbit determines the fundamental frequency of the emission (which also contains higher harmonics), and that's continuously variable, more or less, as desired. Incidentally, it's necessary to use clumps of charge; *a uniform loop of current does not radiate.*

A charged particle slowly revolving in a circular orbit radiates a doughnut-shaped pattern similar to the one depicted in Fig. 3.26. Again the distribution of radiation is symmetrical around **a**, which is now the centripetal acceleration acting inward along the radius drawn from the center of the circular orbit to the charge. The higher the speed, the more an observer at rest in the laboratory will "see" the backward lobe of the radiation pattern shrink while the forward lobe elongates in the direction of motion. At speeds approaching c, the particle

*The details of this calculation using J. J. Thomson's method of analyzing the kink can be found in J. R. Tessman and J. T. Finnell, Jr., "Electric Field of an Accelerating Charge." *Am. J. Phys.* **35**, 523 (1967). As a general reference for radiation, see, for example, Marion and Heald, *Classical Electromagnetic Radiation*, Chapter 7.

beam (usually with a diameter comparable to that of a straight pin) radiates essentially along a narrow cone pointing tangent to the orbit in the instantaneous direction of **v** (Fig. 3.27). For $v \approx c$ the radiation will be strongly polarized in the plane of the motion.

This "searchlight," often less than a few millimeters in diameter, sweeps around as the particle clumps circle the machine, much like the headlight on a train rounding a turn. With each revolution the beam momentarily ($<\frac{1}{2}$ ns) flashes through one of many windows in the device. As we will learn (p. 308) when a signal has a short duration it must comprise a broad range of frequencies. The result is a tremendously intense source of rapidly pulsating radiation, tunable over a wide range of frequencies, from infrared to light to X-rays. When magnets are used to make the circulating electrons wiggle in and out of their circular orbits, bursts of high-frequency X-rays of unparalleled intensity can be created. These beams are hundreds of thousands of times more powerful than a dental X-ray (which is roughly a fraction of a watt) and can easily burn a finger-sized hole through a 3-mm-thick lead plate.

Although this technique was first used to produce light in an electron synchrotron as long ago as 1947, it took several decades to recognize that what was an energy-robbing nuisance to the accelerator people might be a major research tool in itself (Fig. 3.28).

In the astronomical realm, we can expect that some regions exist that are pervaded by magnetic fields. Charged particles trapped in these fields will move in circular or helical orbits, and if their speeds are high enough, they will emit synchrotron radiation. Figure 3.29 shows five photographs of the extragalactic Crab Nebula.* Radiation emanating from the nebula extends over the range from radio frequencies to the extreme ultraviolet. Assuming the source to be trapped circulating charges, we can anticipate strong polarization effects. These are evident in the first four photographs, which were taken through a polarizing filter. The direction of the electric field vector is indicated in each picture. Since in synchrotron radiation, the emitted **E**-field is polarized in the orbital plane, we can conclude that each photograph corresponds to a par-

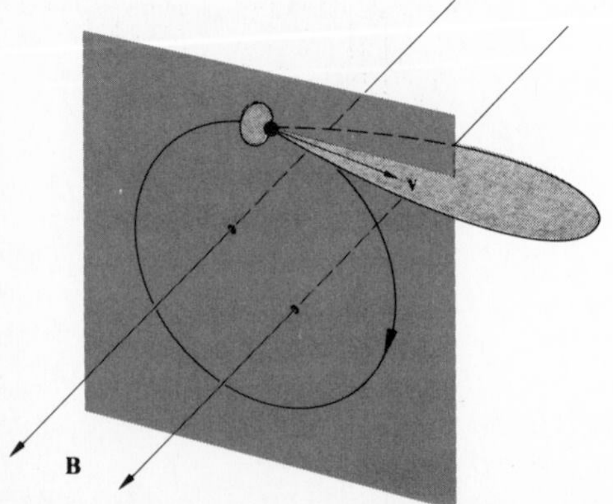

FIGURE 3.27 Radiation pattern for an orbiting charge.

FIGURE 3.28 The first beam of "light" from the National Synchrotron Light Source (1982) emanating from its ultraviolet electron storage ring.

*The Crab Nebula is believed to be expanding debris left over after the cataclysmic death of a star. From its rate of expansion, astronomers calculated that the explosion took place in A.D. 1050. This was subsequently corroborated when a study of old Chinese records the (chronicles of the Beijing Observatory) revealed the appearance of an extremely bright star, in the same region of the sky, in A.D. 1054.

> In the first year of the period Chihha, the fifth moon, the day Chi-chou [i.e., July 4, 1054], a great star appeared.... After more than a year, it gradually became invisible.

There is little doubt that the Crab Nebula is the remnant of that supernova.

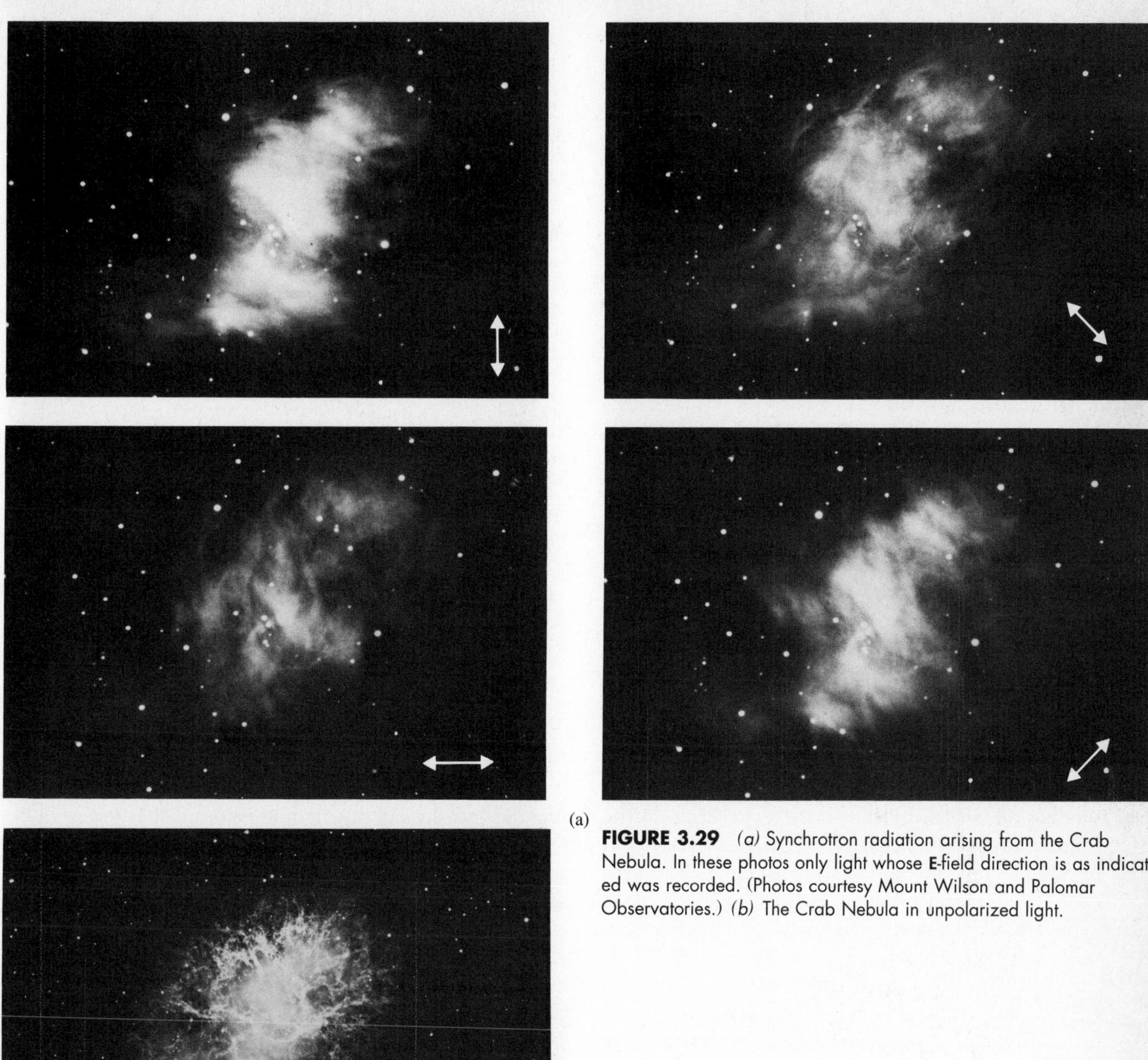

(a)

FIGURE 3.29 *(a)* Synchrotron radiation arising from the Crab Nebula. In these photos only light whose **E**-field direction is as indicated was recorded. (Photos courtesy Mount Wilson and Palomar Observatories.) *(b)* The Crab Nebula in unpolarized light.

(b)

ticular uniform magnetic field orientation normal to the orbits and to **E**.

It is believed that a majority of the low-frequency radiowaves reaching the Earth from outer space have their origin in synchrotron radiation. In 1960 radio astronomers used these long-wavelength emissions to identify a class of objects known as quasars. In 1955 bursts of polarized radiowaves were discovered emanating from Jupiter. Their origin is now attributed to spiraling electrons trapped in radiation belts surrounding the planet.

3.4.3 Electric Dipole Radiation

Perhaps the simplest electromagnetic wave-producing mechanism to visualize is the oscillating dipole—two charges, one plus and one minus, vibrating to and fro along a straight line. And yet this arrangement is surely the most important of all.

Both light and ultraviolet radiation arise primarily from the rearrangement of the outermost, or weakly bound, electrons in atoms and molecules. It follows from the quantum-mechanical analysis that the electric dipole moment of the atom is the major source of this radiation. The rate of energy emission from a material system, although a quantum-mechanical process, can be envisioned in terms of the classical oscillating electric dipole. This mechanism is therefore central to understanding the way atoms, molecules, and even nuclei emit and absorb electromagnetic waves. Figure 3.30 schematically depicts the electric field distribution in the region of an electric dipole. In this configuration, a negative charge oscillates linearly in simple harmonic motion about an equal stationary positive charge. If the angular frequency of the oscillation is ω, the time-dependent dipole moment $\wp(t)$ has the scalar form

$$\wp = \wp_0 \cos \omega t \tag{3.55}$$

Note that $\wp(t)$ could represent the collective moment of the oscillating charge distribution on the atomic scale or even an oscillating current in a linear television antenna.

At $t = 0$, $\wp = \wp_0 = qd$, where d is the initial maximum separation between the centers of the two charges (Fig. 3.30a). The dipole moment is actually a vector in the direction from $-q$ to $+q$. The figure shows a sequence of field line patterns as the displacement, and therefore the dipole moment decreases, then goes to zero, and finally reverses direction. When the charges effectively overlap, $\wp = 0$ and the field lines must close on themselves.

Very near the atom, the **E**-field has the form of a static electric dipole. A bit farther out, in the region where the closed loops form, there is no specific wavelength. The detailed treatment shows that the electric field is composed of five different terms, and things are fairly complicated. Far from the dipole, in what is called the *wave* or *radiation zone*, the field configuration is much simpler. In this zone a fixed wavelength has been established; **E** and **B** are transverse, mutually perpendicular, and in phase. Specifically,

$$E = \frac{\wp_0 k^2 \sin \theta}{4\pi\epsilon_0} \frac{\cos (kr - \omega t)}{r} \tag{3.56}$$

and $B = E/c$, where the fields are oriented as in Fig. 3.31. The Poynting vector $\mathbf{S} = \mathbf{E} \times \mathbf{B}/\mu_0$ always points radially outward in the wave zone. There, the **B**-field lines are circles concentric with, and in a plane perpendicular to, the dipole axis. This is understandable, since **B** can be considered to arise from the time-varying oscillator current.

The irradiance (radiated radially outward from the source) follows from Eq. (3.44) and is given by

$$I(\theta) = \frac{\wp_0^2 \omega^4}{32\pi^2 c^3 \epsilon_0} \frac{\sin^2 \theta}{r^2} \tag{3.57}$$

again an inverse-square-law dependence on distance. The angular flux density distribution is toroidal, as in Fig. 3.26. The axis along which the acceleration takes place is the symmetry axis of the radiation pattern. Notice the dependence of the irradiance on ω^4—*the higher the frequency, the stronger the radiation*; that feature will be important when we consider scattering.

It's not difficult to attach an AC generator between two conducting rods and send currents of free electrons oscillating up and down that "transmitting antenna." Figure 3.32 shows the arrangement carried to its logical conclusion—a fairly standard AM radio tower. An antenna of this sort will function most efficiently if its length corresponds to the wavelength being transmitted or, more conveniently, to $\frac{1}{2}\lambda$. The radiated wave is then formed at the dipole in synchronization with the oscillating current producing it. AM radiowaves are unfortunately several hundred meters long. Consequently, the antenna shown in the figure has half the $\frac{1}{2}\lambda$-dipole essentially buried

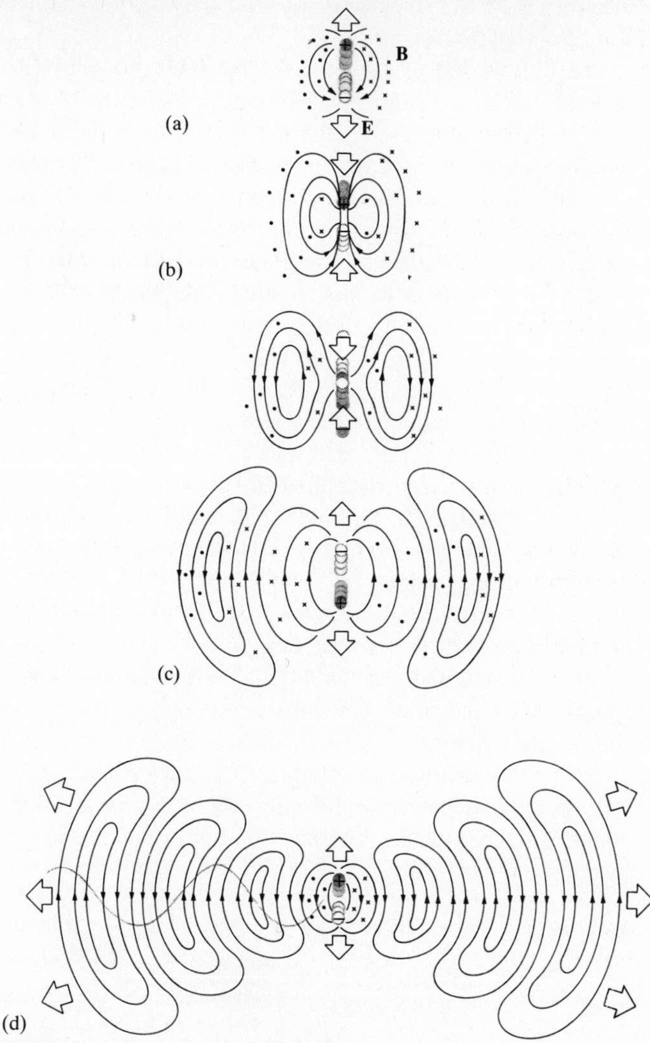

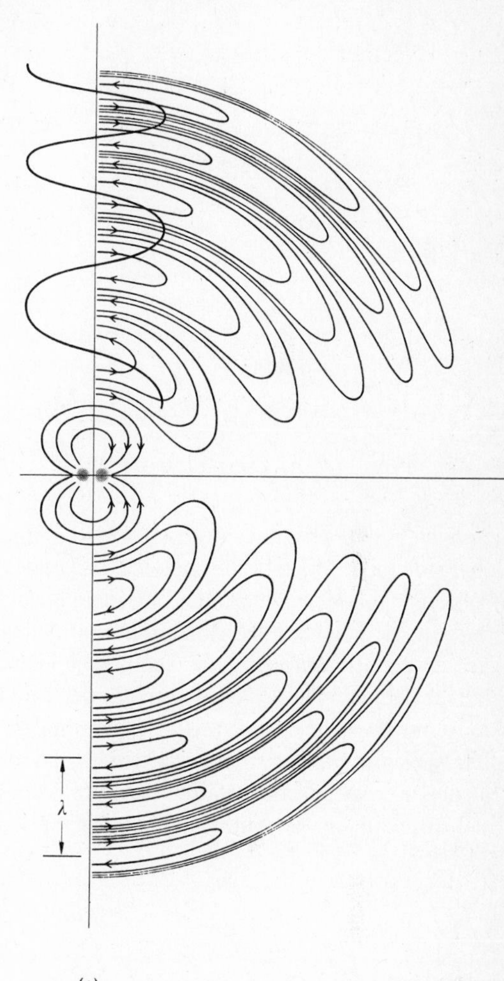

FIGURE 3.30 The E-field of an oscillating electric dipole.

in the Earth. That at least saves some height, allowing the device to be built only $\frac{1}{4}\lambda$ tall. Moreover, this use of the Earth also generates a so-called *ground wave* that hugs the planet's surface, where most people with radios are likely to be located. A commercial station usually has a range somewhere between 25 and 100 miles.

3.4.4 The Emission of Light from Atoms

Surely the most significant mechanism responsible for the natural emission and absorption of radiant energy—especially of light—is the *bound charge*, electrons confined within atoms. These minute negative particles, which surround the massive

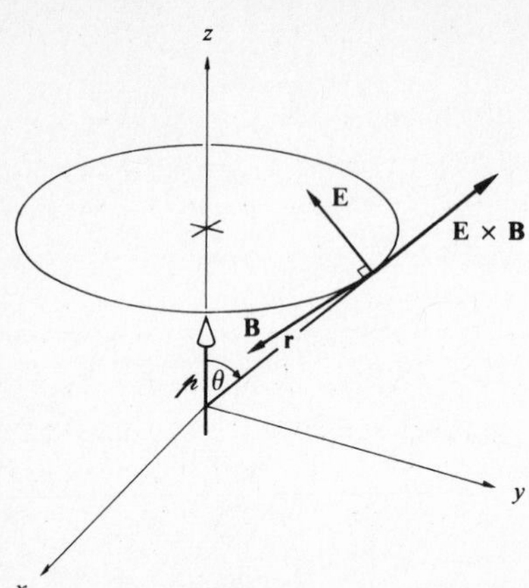

FIGURE 3.31 Field orientations for an oscillating electric dipole.

positive nucleus of each atom, constitute a kind of distant, tenuous charged cloud. Much of the chemical and optical behavior of ordinary matter is determined by its outer or valence electrons. The remainder of the cloud is ordinarily formed into "closed," essentially unresponsive, shells around and tightly bound to the nucleus. These closed or filled shells are made up of specific numbers of electron pairs. Even though it is not completely clear what occurs internally when an atom radiates, we do know with some certainty that light is emitted during readjustments in the outer charge distribution of the electron cloud. This mechanism is ultimately the predominant source of light in the world.

Usually, an atom exists with its clutch of electrons arranged in some stable configuration that corresponds to their lowest energy distribution or *level*. Every electron is in the lowest possible energy state available to it, and the atom as a whole is in its so-called **ground state** configuration. There it will likely remain indefinitely, if left undisturbed. Any mechanism that pumps energy into the atom will alter the ground state. For instance, a collision with another atom, an electron, or a photon can affect the atom's energy state profoundly. An atom can exist with its electron cloud in only certain specific configurations corresponding to only certain values of energy. In addition to the ground state, there are higher energy levels, the **excited states**, each associated with a specific cloud configuration and a specific well-defined energy. When one or more electrons occupies a level higher than its ground-state level, the atom is said to be **excited**—a condition that is inherently unstable and temporary.

At low temperatures, atoms tend to be in their ground state; at progressively higher temperatures, more and more of them will become excited through atomic collisions. This sort of mechanism is indicative of a class of relatively gentle excitations—glow discharge, flame, spark, and so forth—which energize only the outermost unpaired valence electrons. We will initially concentrate on these outer electron transitions, which give rise to the emission of light, and the nearby infrared and ultraviolet.

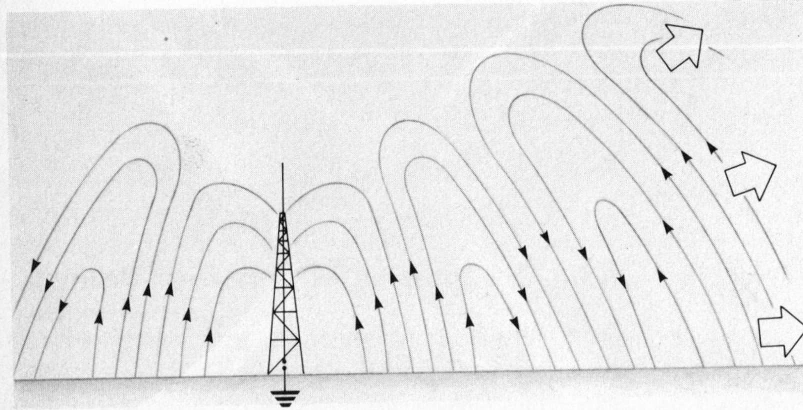

FIGURE 3.32 Electromagnetic waves from a transmitting tower.

When enough energy is imparted to an atom (typically to the valence electron), whatever the cause, the atom can react by suddenly ascending from a lower to a higher energy level (Fig. 3.33). The electron will make a very rapid transition, a **quantum jump**, from its ground-state orbital configuration to one of the well-delineated excited states, one of the quantized rungs on its energy ladder. As a rule, ***the amount of energy taken up in the process equals the energy difference between the initial and final states, and since that is specific and well defined, the amount of energy that can be absorbed by an atom is quantized*** (i.e., limited to specific amounts). This state of atomic excitation is a short-lived resonance phenomenon. Usually, after about 10^{-8} s or 10^{-9} s, the excited atom spontaneously relaxes back to a lower state, most often the ground state, losing the excitation energy along the way. This energy readjustment can occur by way of the emission of light or (especially in dense materials) by conversion to thermal energy through interatomic collisions within the medium.

If the atomic transition is accompanied by the emission of light (as it is in a rarefied gas), ***the energy of the photon exactly matches the quantized energy decrease of the atom***. That corresponds to a specific frequency, by way of $\Delta\mathscr{E} = h\nu$, a frequency associated with both the photon and the atomic transition between the two particular states. This is said to be a **resonance frequency**, one of several (each with its own likelihood of occurring) at which the atom very efficiently absorbs and emits energy. The atom radiates a quantum of energy that presumably is created spontaneously, on the spot, by the shifting electron.

Even though what occurs during the atom-transition interval of 10^{-8} s is far from clear, it can be helpful to imagine the orbital electron somehow making its downward energy transition via a gradually damped oscillatory motion at the specific resonance frequency. The radiated light can then be envisioned in a semiclassical way as emitted in a short oscillatory directional pulse, or **wavetrain**, lasting less than roughly 10^{-8} s—a picture that is in agreement with certain experimental observations (see Section 7.4.2, Fig. 7.25). It is useful to think of this electromagnetic pulse as associated in some inextricable fashion with the photon. In a way, the pulse is a semiclassical representation of the manifest wave nature of the photon. But the two are *not* equivalent in all respects: the electromagnetic wavetrain is a classical creation that describes the propagation and spatial distribution of light extremely well, yet its energy is not quantized, and that is an essential characteristic of the photon. So when we consider photon wavetrains, keep in mind that there is more to the idea than just a classical oscillatory pulse of electromagnetic wave. Of course, the reason even to introduce the notion of the emission of wavetrains is to have a basis for talking about the frequency of the light. This is perhaps the central problem in any naïve photon model: What agency manifests the frequency?

The emission spectra of single atoms or low-pressure gases, whose atoms do not interact appreciably, consist of sharp

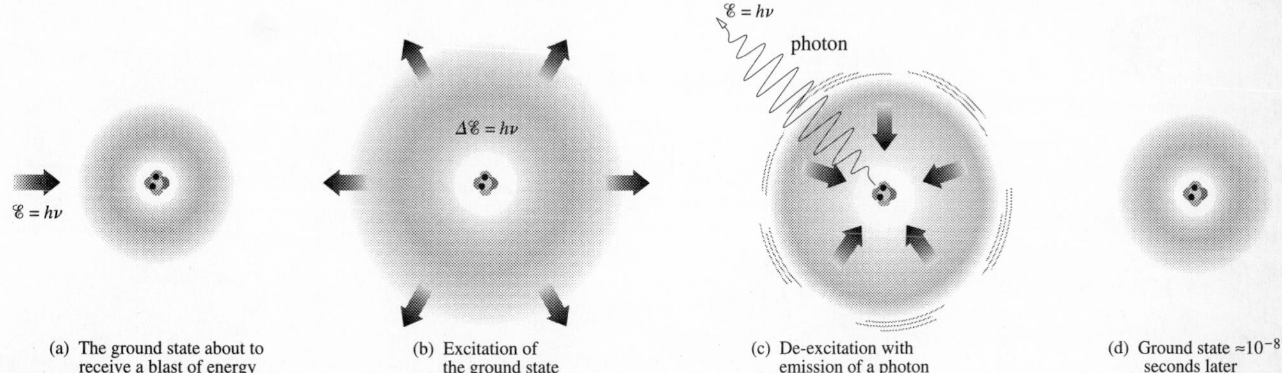

$\mathscr{E} = h\nu$
$\Delta\mathscr{E} = h\nu$
$\mathscr{E} = h\nu$
photon
(a) The ground state about to receive a blast of energy
(b) Excitation of the ground state
(c) De-excitation with emission of a photon
(d) Ground state ≈10^{-8} seconds later

FIGURE 3.33 The excitation of an atom. (*a*) Energy in the amount $h\nu$ is delivered to the atom. (*b*) Since this matches the energy needed to reach an excited state, the atom absorbs the energy and attains a higher energy level. (*c*) With the emission of a photon it drops back (*d*) and returns to the ground state in about 10^{-8} s.

"lines," that is, fairly well-defined frequencies characteristic of the atoms. There is always some frequency broadening of that radiation due to atomic motion, collisions, and so forth; hence it's never precisely monochromatic. Generally, however, the atomic transition from one level to another is characterized by the emission of a well-defined narrow range of frequencies. On the other hand, the spectra of solids and liquids, in which the atoms are interacting with one another, are broadened into wide frequency bands. When two atoms are brought close together, the result is a slight shift in their respective energy levels, because they act on each other. The many interacting atoms in a solid create a tremendous number of such shifted levels, in effect spreading out each of their original levels, blurring them into essentially continuous bands. Materials of this nature emit and absorb over broad ranges of frequencies.

3.5 Light in Matter

The response of dielectric or nonconducting materials to electromagnetic fields is of special concern in Optics. We will, of course, be dealing with transparent dielectrics in the form of lenses, prisms, plates, films, and so forth, not to mention the surrounding sea of air.

The net effect of introducing a homogeneous, isotropic dielectric into a region of free space is to change ϵ_0 to ϵ and μ_0 to μ in Maxwell's Equations. The phase speed in the medium now becomes

$$v = 1/\sqrt{\epsilon\mu} \tag{3.58}$$

The ratio of the speed of an electromagnetic wave in vacuum to that in matter is known as the **absolute index of refraction** n:

$$n \equiv \frac{c}{v} = \sqrt{\frac{\epsilon\mu}{\epsilon_0\mu_0}} \tag{3.59}$$

In terms of the relative permittivity and relative permeability of the medium, n becomes

$$n = \sqrt{K_E K_M} \tag{3.60}$$

There are magnetic substances that are transparent in the infrared and microwave regions of the spectrum. But we are primarily interested in materials that are transparent in the visible, and these are all essentially "nonmagnetic." Indeed, K_M generally doesn't deviate from 1 by any more than a few parts in 10^4 (e.g., for diamond $K_M = 1 - 2.2 \times 10^{-5}$). Setting $K_M = 1$ in the formula for n results in an expression known as *Maxwell's Relation*, namely,

$$n = \sqrt{K_E} \tag{3.61}$$

wherein K_E is presumed to be the *static dielectric constant*. As indicated in Table 3.2, this relationship seems to work well only for some simple gases. The difficulty arises because K_E and therefore n are actually *frequency-dependent*. The dependence of n on the wavelength (or color) of light is a well-known effect called **dispersion**. It arises on a microscopic level, and so Maxwell's Equations are quite oblivious to it. Sir Isaac Newton used prisms to disperse white light into its constituent colors over three hundred years ago, and the phenomenon was well known, if not well understood, even then.

TABLE 3.2 Maxwell's Relation

Substance	$\sqrt{K_E}$	n
Gases at 0°C and 1 atm		
Air	1.000294	1.000293
Helium	1.000034	1.000036
Hydrogen	1.000131	1.000132
Carbon dioxide	1.00049	1.00045
Liquids at 20°C		
Benzene	1.51	1.501
Water	8.96	1.333
Ethyl alcohol (ethanol)	5.08	1.361
Carbon tetrachloride	4.63	1.461
Carbon disulfide	5.04	1.628
Solids at room temperature		
Diamond	4.06	2.419
Amber	1.6	1.55
Fused silica	1.94	1.458
Sodium chloride	2.37	1.50

Values of K_E correspond to the lowest possible frequencies, in some cases as low as 60 Hz, whereas n is measured at about 0.5×10^{15} Hz. Sodium D light was used ($\lambda = 589.29$ nm).

Scattering and Absorption

What is the physical basis for the frequency dependence of n? The answers to that question can be found by examining the interaction of an incident electromagnetic wave with the array of atoms constituting a dielectric material. An atom can react to incoming light in two different ways, depending on the incident frequency or equivalently on the incoming photon energy ($\mathscr{E} = h\nu$). Generally, the atom will "scatter" the light, redirecting it without otherwise altering it. On the other hand, if the photon's energy matches that of one of the excited states, the atom will "absorb" the light, making a quantum jump to that higher energy level. In the dense atomic landscape of ordinary gases (at pressures of about 10^2 Pa and up), solids, and liquids, it's very likely that this excitation energy will rapidly be transferred, via collisions, to random atomic motion, thermal energy, before a photon can be emitted. This commonplace process (the taking up of a photon and its conversion into thermal energy) was at one time widely known as "absorption," but nowadays that word is more often used to refer just to the "taking up" aspect, regardless of what then happens to the energy. Consequently, it's now better referred to as **dissipative absorption**.

In contrast to this excitation process, *ground-state* or **nonresonant scattering** occurs with incoming radiant energy of other frequencies—that is, lower than the resonance frequencies. Imagine an atom in its lowest state and suppose that it interacts with a photon whose energy is too small to cause a transition to any of the higher, excited states. Despite that, the electromagnetic field of the light can be supposed to drive the electron cloud into oscillation. There is no resulting atomic transition; the atom remains in its ground state while the cloud vibrates ever so slightly at the frequency of the incident light. Once the electron cloud starts to vibrate with respect to the positive nucleus, the system constitutes an oscillating dipole and so presumably will *immediately* begin to radiate at that same frequency. The resulting scattered light consists of a photon that sails off in some direction carrying the same amount of energy as did the incident photon—***the scattering is elastic***. In effect, the atom resembles a little dipole oscillator, a model employed by Hendrik Antoon Lorentz (1878) in order to extend Maxwell's Theory, in a classical way, to the atomic domain. If the incident light is unpolarized, the atomic oscillators scatter in random directions.

When an atom is irradiated with light, the process of excitation and spontaneous emission is rapidly repeated. In fact, with an emission lifetime of $\approx 10^{-8}$ s an atom could emit upward of 10^8 photons per second in a situation in which there was enough energy to keep reexciting it. Atoms have a very strong tendency to interact with resonant light (they have a large *absorption cross section*). This means that the saturation condition, in which the atoms of a low-pressure gas are constantly emitting and being reexcited, occurs at a modest value of irradiance ($\approx 10^2$ W/m^2). So it's not very difficult to get atoms firing out photons at a rate of 100 million per second.

Generally, we can imagine that in a medium illuminated by an ordinary beam of light, each atom behaves as though it was a "source" of a tremendous number of photons (scattered either elastically or resonantly) that fly off in all directions. A stream of energy like this resembles a classical spherical wave. ***Thus we imagine an atom (even though it is simplistic to do so) as a point source of spherical electromagnetic waves***—provided we keep in mind Einstein's admonition that "outgoing radiation in the form of spherical waves does not exist."

When a material with no resonances in the visible is bathed in light, nonresonant scattering occurs and it gives each participating atom the appearance of being a tiny source of spherical wavelets. As a rule, the closer the frequency of the incident beam is to an atomic resonance, the more strongly will the interaction occur and, in dense materials, the more energy will be dissipatively absorbed. It is precisely this mechanism of *selective absorption* (see Section 4.9) that creates much of the visual appearance of things. It is primarily responsible for the color of your hair, skin, and clothing, the color of leaves and apples and paint.

3.5.1 Dispersion

Maxwell's Theory treats matter as continuous, representing its electric and magnetic responses to applied **E**- and **B**-fields in terms of constants, ϵ and μ. Consequently, K_E and K_M are also constant, and n is therefore unrealistically independent of frequency. To deal theoretically with dispersion (the frequency dependence of the refractive index), it is necessary to incorporate the atomic nature of matter and to exploit some frequency-dependent aspect of that nature. Following H. A. Lorentz, the contributions of large numbers of atoms can be averaged to represent the behavior of an isotropic dielectric medium.

When a dielectric is subjected to an applied electric field, the internal charge distribution is distorted. This corresponds to the generation of electric dipole moments, which in turn contribute to the total internal field. More simply stated, the external field separates positive and negative charges in the medium (each pair of which is a dipole), and these then contribute an additional field component. The resultant dipole moment per unit volume is called the **electric polarization P**. For most materials **P** and **E** are proportional and can satisfactorily be related by

$$(\epsilon - \epsilon_0)\mathbf{E} = \mathbf{P} \tag{3.62}$$

The redistribution of charge and the consequent polarization can occur via the following mechanisms. There are molecules that have a permanent dipole moment as a result of unequal sharing of valence electrons. These are known as *polar molecules*; the nonlinear water molecule is a fairly typical example (Fig. 3.34). Each hydrogen–oxygen bond is polar covalent, with the H-end positive with respect to the O-end. Thermal agitation keeps the molecular dipoles randomly oriented. With the introduction of an electric field, the dipoles align themselves, and the dielectric takes on an **orientational polarization**. In the case of *nonpolar molecules* and *atoms*, the applied field distorts the electron cloud, shifting it relative to the nucleus, thereby producing a dipole moment. In addition to this **electronic polarization**, there is another process that's applicable specifically to molecules, for example, the ionic crystal NaCl. In the presence of an electric field, the positive and negative ions undergo a shift with respect to each other. Dipole moments are therefore induced, resulting in what is called **ionic** or **atomic polarization**.

If the dielectric is subjected to an incident harmonic electromagnetic wave, its internal charge structure will experience time-varying forces and/or torques. These will be proportional to the electric field component of the wave.* For fluids that are polar dielectrics the molecules actually undergo rapid rotations, aligning themselves with the $\mathbf{E}(t)$-field. But these molecules are relatively large and have appreciable moments of inertia. At high driving frequencies ω, polar molecules will be unable to follow the field alternations. Their contributions to **P** will decrease, and K_E will drop markedly. The relative permittivity of water is fairly constant at approximately 80, up to about 10^{10} Hz, after which it falls off quite rapidly.

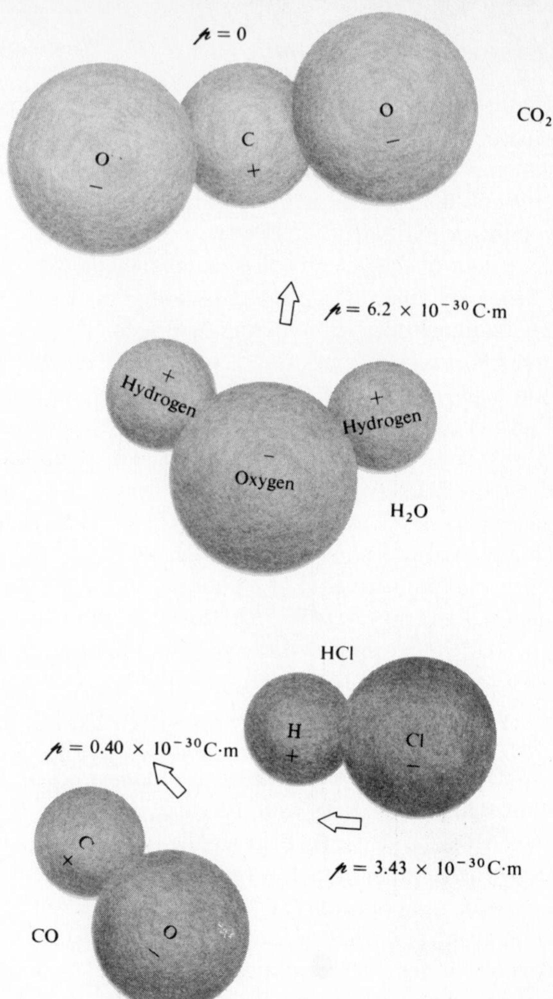

FIGURE 3.34 Assorted molecules and their dipole moments.

In contrast, electrons have little inertia and can continue to follow the field contributing to $K_E(\omega)$ even at optical frequencies (of about 5×10^{14} Hz). Thus the dependence of n on ω is governed by the interplay of the various electric polarization mechanisms contributing at the particular frequency. With this in mind, it is possible to derive an analytical expression for $n(\omega)$ in terms of what's happening within the medium on an atomic level.

The electron cloud of the atom is bound to the positive nucleus by an attractive electric force that sustains it in some sort of equilibrium configuration. Without knowing much

*Forces arising from the magnetic component of the field have the form $\mathbf{F}_M = q_e\mathbf{v} \times \mathbf{B}$ in comparison to $\mathbf{F}_E = q_e\mathbf{E}$ for the electric component; but $v << c$, so it follows from Eq. (3.30) that $\mathbf{F}_M$ is generally negligible.

more about the details of all the internal atomic interactions, we can anticipate that, like other stable mechanical systems which are not totally disrupted by small perturbations, a net force, F, must exist that returns the system to equilibrium. Moreover, we can reasonably expect that for very small displacements, x, from equilibrium (where $F = 0$), the force will be linear in x. In other words, a plot of $F(x)$ versus x will cross the x-axis at the equilibrium point ($x = 0$) and will be a straight line very close on either side. Thus for small displacements it can be supposed that the restoring force has the form $F = -kx$. Once somehow momentarily disturbed, an electron bound in this way will oscillate about its equilibrium position with a **natural** or **resonant frequency** given by $\omega_0 = \sqrt{k/m_e}$, where m_e is its mass. This is the oscillatory frequency of the *undriven* system.

A material medium is envisioned as an assemblage, in vacuum, of a very great many polarizable atoms, each of which is small (by comparison to the wavelength of light) and close to its neighbors. When a lightwave impinges on such a medium, each atom can be thought of as a classical **forced oscillator** being driven by the time-varying electric field $E(t)$ of the wave, which is assumed here to be applied in the x-direction. Figure 3.35*b* is a mechanical representation of just such an oscillator in an *isotropic medium* where the negatively charged shell is fastened to a stationary positive nucleus by identical springs. Even under the illumination of bright sunlight, the amplitude of the oscillations will be no greater than about 10^{-17} m. The force (F_E) exerted on an electron of charge q_e by the $E(t)$ field of a harmonic wave of frequency ω is of the form

$$F_E = q_e E(t) = q_e E_0 \cos \omega t \tag{3.63}$$

Consequently, Newton's Second Law provides the equation of motion; that is, the sum of the forces equals the mass times the acceleration:

$$q_e E_0 \cos \omega t - m_e \omega_0^2 x = m_e \frac{d^2x}{dt^2} \tag{3.64}$$

The first term on the left is the driving force, and the second is the opposing restoring force. To satisfy this expression, x will have to be a function whose second derivative isn't very much different from x itself. Furthermore, we can anticipate that the electron will oscillate at the same frequency as $E(t)$, so we "guess" at the solution

$$x(t) = x_0 \cos \omega t$$

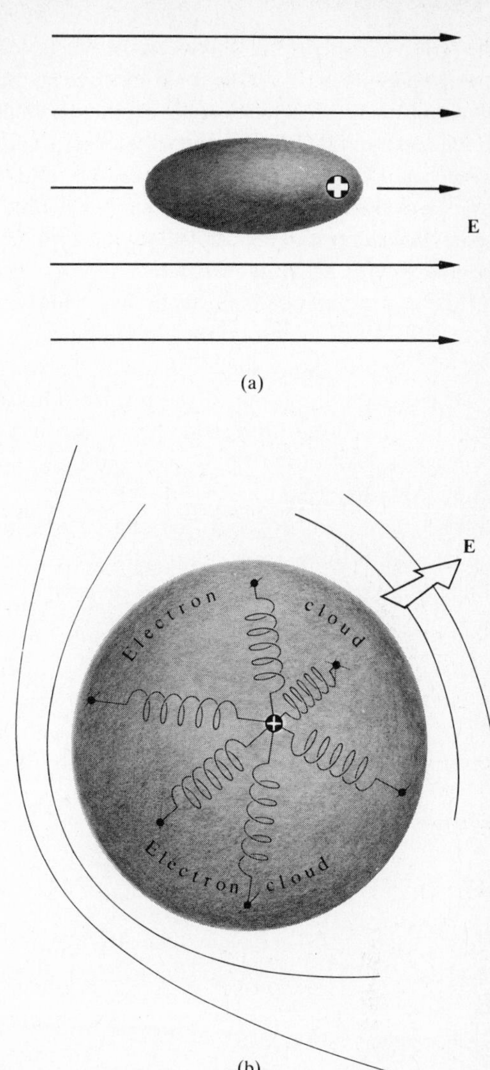

FIGURE 3.35 (*a*) Distortion of the electron cloud in response to an applied **E**-field. (*b*) The mechanical oscillator model for an isotropic medium—all the springs are the same, and the oscillator can vibrate equally in all directions.

and substitute it in the equation to evaluate the amplitude x_0. In this way we find that

$$x(t) = \frac{q_e/m_e}{(\omega_0^2 - \omega^2)} E_0 \cos \omega t \tag{3.65}$$

or

$$x(t) = \frac{q_e/m_e}{(\omega_0^2 - \omega^2)} E(t) \tag{3.66}$$

This is the relative displacement between the negative cloud and the positive nucleus. It's traditional to leave q_e positive and speak about the displacement of the oscillator. Without a driving force (no incident wave) the oscillator will vibrate at its *resonance frequency* ω_0. In the presence of a field whose frequency is less than ω_0, $E(t)$ and $x(t)$ have the same sign, which means that the oscillator can follow the applied force (i.e., is in-phase with it). However, when $\omega > \omega_0$, the displacement $x(t)$ is in a direction opposite to that of the instantaneous force $q_eE(t)$ and therefore 180° out-of-phase with it. Remember that we are talking about oscillating dipoles where for $\omega_0 > \omega$, the relative motion of the *positive* charge is a vibration in the direction of the field. Above resonance the positive charge is 180° out of phase with the field, and the dipole is said to lag by π rad.

The dipole moment is equal to the charge q_e times its displacement, and if there are N contributing electrons per unit volume, the electric polarization, or density of dipole moments, is

$$P = q_e x N \tag{3.67}$$

Hence

$$P = \frac{q_e^2 NE/m_e}{(\omega_0^2 - \omega^2)} \tag{3.68}$$

and from Eq. (3.62)

$$\epsilon = \epsilon_0 + \frac{P(t)}{E(t)} = \epsilon_0 + \frac{q_e^2 N/m_e}{(\omega_0^2 - \omega^2)} \tag{3.69}$$

Using the fact that $n^2 = K_E = \epsilon/\epsilon_0$, we can arrive at an expression for n as a function of ω, which is known as a **dispersion equation**:

$$n^2(\omega) = 1 + \frac{Nq_e^2}{\epsilon_0 m_e}\left(\frac{1}{\omega_0^2 - \omega^2}\right) \tag{3.70}$$

At frequencies increasingly above resonance, $(\omega_0^2 - \omega^2) < 0$, and the oscillator undergoes displacements that are approximately 180° out-of-phase with the driving force. The resulting electric polarization will therefore be similarly out-of-phase with the applied electric field. Hence the dielectric constant and therefore the index of refraction will both be less than 1. At frequencies increasingly below resonance, $(\omega_0^2 - \omega^2) > 0$, the electric polarization will be nearly in-phase with the applied electric field. The dielectric constant and the corresponding index of refraction will then both be greater than 1. This kind of behavior, which actually represents only part of what happens, is nonetheless generally observed in all sorts of materials.

We can test the utility of the analysis using a dispersive prism (p. 189) made of the sample material under study, but first we rewrite Eq. (3.70), as is done in Problem 3.46:

$$(n^2 - 1)^{-1} = -C\lambda^{-2} + C\lambda_0^{-2}$$

where, since $\omega = 2\pi c/\lambda$, the multiplicative constant is given by $C = 4\pi^2 c^2 \epsilon_0 m_e/Nq_e^2$. Figure 3.36 is a plot of $(n^2 - 1)^{-1}$ versus λ^{-2} using data from a student experiment. A crown-glass prism was illuminated with the various wavelengths from a He discharge tube, and the index of refraction was measured for each one (Table 3.3). The resulting curve is indeed a straight line; its slope (using $y = mx + b$) equals $-C$, and its y-intercept corresponds to $C\lambda_0^{-2}$. From this it follows that the resonant frequency is 2.95×10^{15} Hz, which is properly in the ultraviolet.

As a rule, any given substance will actually undergo several transitions from $n > 1$ to $n < 1$ as the illuminating frequency is made to increase. The implication is that instead of a single frequency ω_0 at which the system resonates, there apparently are several such frequencies. It would seem reasonable to generalize matters by supposing that there are N molecules per unit volume, each with f_j oscillators having natural frequencies ω_{0j}, where $j = 1, 2, 3, \ldots$. In that case,

$$n^2(\omega) = 1 + \frac{Nq_e^2}{\epsilon_0 m_e}\sum_j\left(\frac{f_j}{\omega_{0j}^2 - \omega^2}\right) \tag{3.71}$$

This is essentially the same result as that arising from the quantum-mechanical treatment, with the exception that some

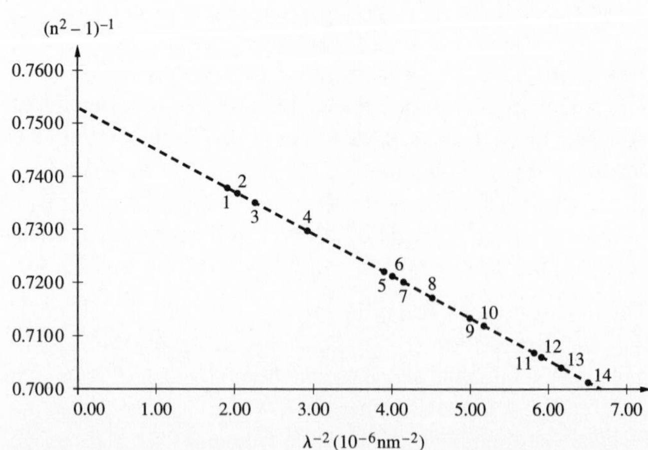

FIGURE 3.36 Graph of $(n^2 - 1)^{-1}$ vs λ^{-2} for the data shown in Table 3.3. See N. Gauthier, *Phys. Teach.*, **25**, 502 (1987).

TABLE 3.3 Dispersion of Crown Glass. The wavelengths are those of a He discharge tube. The corresponding indices were measured.

	Wavelength λ (nm)	Index of Refraction n
1.	728.135	1.5346
2.	706.519, 706.570	1.5352
3.	667.815	1.53629
4.	587.562, 587.587	1.53954
5.	504.774	1.54417
6.	501.567	1.54473
7.	492.193	1.54528
8.	471.314	1.54624
9.	447.148	1.54943
10.	438.793	1.55026
11.	414.376	1.55374
12.	412.086	1.55402
13.	402.619	1.55530
14.	388.865	1.55767

of the terms must be reinterpreted. Accordingly, the quantities ω_{0j} would then be the characteristic frequencies at which an atom may absorb or emit radiant energy. The f_j terms, which satisfy the requirement that $\Sigma_j f_j = 1$, are weighting factors known as *oscillator strengths*. They reflect the emphasis that should be placed on each one of the modes. Since they measure the likelihood that a given atomic transition will occur, the f_j terms are also known as *transition probabilities*.

A similar reinterpretation of the f_j terms is even required classically, since agreement with the experimental data demands that they be less than unity. This is obviously contrary to the definition of the f_j that led to Eq. (3.71). One then supposes that a molecule has many oscillatory modes but that each of these has a distinct natural frequency and strength.

Notice that when ω equals any of the characteristic frequencies, n is discontinuous, contrary to actual observation. This is simply the result of having neglected the damping term, which should have appeared in the denominator of the sum. Incidentally, the damping, in part, is attributable to energy lost when the forced oscillators reradiate. In solids, liquids, and gases at high pressure ($\approx 10^3$ atm), the interatomic distances are roughly 10 times less than those of a gas at standard temperature and pressure. Atoms and molecules in this relatively close proximity experience strong interactions and a resulting "frictional" force. The effect is a damping of the oscillators and a dissipation of their energy within the substance in the form of "heat" (random molecular motion).

Had we included a damping force proportional to the speed (of the form $m_e\gamma\, dx/dt$) in the equation of motion, the dispersion equation (3.71) would have been

$$n^2(\omega) = 1 + \frac{Nq_e^2}{\epsilon_0 m_e}\sum_j \frac{f_j}{\omega_{0j}^2 - \omega^2 + i\gamma_j\omega} \tag{3.72}$$

Although this expression is fine for rarefied media such as gases, there is another complication that must be overcome if the equation is to be applied to dense substances. Each atom interacts with the local electric field in which it is immersed. Yet unlike the isolated atoms considered above, those in a dense material will also experience the induced field set up by their brethren. Consequently, an atom "sees" in addition to the applied field $E(t)$ another field,* namely, $P(t)/3\epsilon_0$. Without going into the details here, it can be shown that

$$\frac{n^2 - 1}{n^2 + 2} = \frac{Nq_e^2}{3\epsilon_0 m_e}\sum_j \frac{f_j}{\omega_{0j}^2 - \omega^2 + i\gamma_j\omega} \tag{3.73}$$

Thus far we have been considering electron-oscillators almost exclusively, but the same results would have been applicable to ions bound to fixed atomic sites as well. In that instance m_e would be replaced by the considerably larger ion mass. Thus, although electronic polarization is important over the entire optical spectrum, the contributions from ionic polarization significantly affect n only in regions of resonance ($\omega_{0j} = \omega$).

The implications of a complex index of refraction will be considered later, in Section 4.8. At the moment we limit the discussion, for the most part, to situations in which absorption is negligible (i.e., $\omega_{0j}^2 - \omega^2 >> \gamma_j\omega$) and n is real, so that

$$\frac{n^2 - 1}{n^2 + 2} = \frac{Nq_e^2}{3\epsilon_0 m_e}\sum_j \frac{f_j}{\omega_{0j}^2 - \omega^2} \tag{3.74}$$

Colorless, transparent materials have their characteristic frequencies outside the visible region of the spectrum (which is why they are, in fact, colorless and transparent). In particular, glasses have effective natural frequencies above the visible in the ultraviolet, where they become opaque. In cases for

*This result, which applies to isotropic media, is derived in almost any text on Electromagnetic Theory.

which $\omega_{0j}^2 >> \omega^2$, by comparison, ω^2 may be neglected in Eq. (3.74), yielding an essentially constant index of refraction over that frequency region. For example, the important characteristic frequencies for glasses occur at wavelengths of about 100 nm. The middle of the visible range is roughly five times that value, and there, $\omega_{0j}^2 >> \omega^2$. Notice that as ω increases toward ω_{0j}, $(\omega_{0j}^2 - \omega^2)$ decreases and *n gradually increases with frequency*, as is clearly evident in Fig. 3.37. This is called **normal dispersion**. In the ultraviolet region, as ω approaches a natural frequency, the oscillators will begin to resonate. Their amplitudes will increase markedly, and this will be accompanied by damping and a strong absorption of energy from the incident wave. When $\omega_{0j} = \omega$ in Eq. (3.73), the damping term obviously becomes dominant. The regions immediately surrounding the various ω_{0j} in Fig. 3.38 are called **absorption bands**. There $dn/d\omega$ is negative, and the process is spoken of as **anomalous** (i.e., abnormal) **dispersion**. When white light passes through a glass prism, the blue constituent has a higher index than the red and is therefore deviated through a larger angle (see Section 5.5.1). In contrast, when we use a liquid-cell prism containing a dye solution with an absorption band in the visible, the spectrum is altered markedly (see Problem 3.43). All substances possess absorption bands somewhere within the electromagnetic frequency spectrum, so that the term *anomalous dispersion*, being a carryover from the late 1800s, is certainly a misnomer.

As we have seen, atoms within a molecule can also vibrate

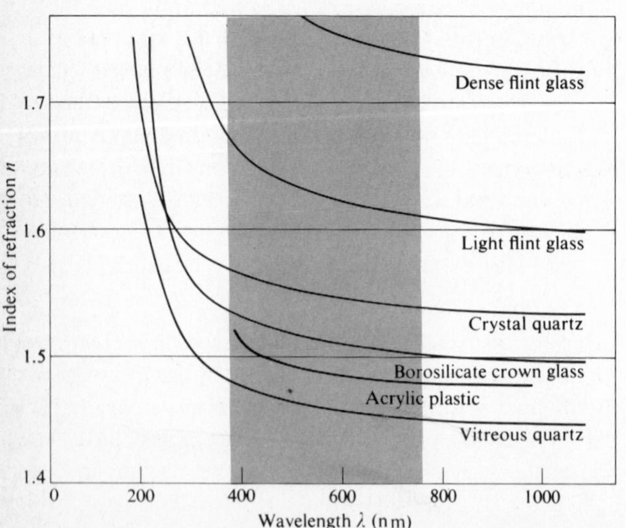

FIGURE 3.37 The wavelength dependence of the index of refraction for various materials.

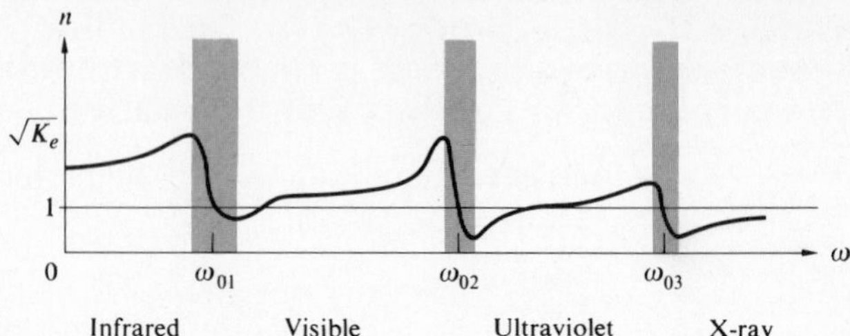

FIGURE 3.38 Refractive index versus frequency.

about their equilibrium positions. But the nuclei are massive, and so the natural oscillatory frequencies are low, in the infrared. Molecules such as H_2O and CO_2 have resonances in both the infrared and ultraviolet. When water is trapped within a piece of glass during its manufacture, these molecular oscillators are available, and an infrared absorption band exists. The presence of oxides also results in infrared absorption. Figure 3.39 shows the $n(\omega)$ curves (ranging from the ultraviolet to the infrared) for a number of important optical crystals. Note how they rise in the ultraviolet and fall in the infrared. At the even lower frequencies of radiowaves, glass is again transparent. In comparison, a piece of stained glass evidently has a resonance in the visible where it absorbs out a particular range of frequencies, transmitting the complementary color.

As a final point, notice that if the driving frequency is greater than any of the ω_{0j} terms, then $n^2 < 1$ and $n < 1$. Such a situation can occur, for example, if we beam X-rays onto a glass plate. This is an intriguing result, since it leads to $v > c$, in seeming contradiction to Special Relativity. We will consider this behavior again later on, when we discuss the group velocity (Section 7.6).

In partial summary then, over the visible region of the spectrum, electronic polarization is the operative mechanism determining $n(\omega)$. Classically, one imagines electron-oscillators vibrating at the frequency of the incident wave. When the wave's frequency is appreciably different from a characteristic or natural frequency, the oscillations are small, and there is little dissipative absorption. At resonance, however, the oscillator amplitudes are increased, and the field does an increased amount of work on the charges. Electromagnetic energy removed from the wave and converted into mechanical energy is dissipated thermally within the substance, and one speaks of an absorption peak or band. The material, although essentially transparent at other frequencies, is fairly opaque to incident radiation at its characteristic frequencies (Fig. 3.40).

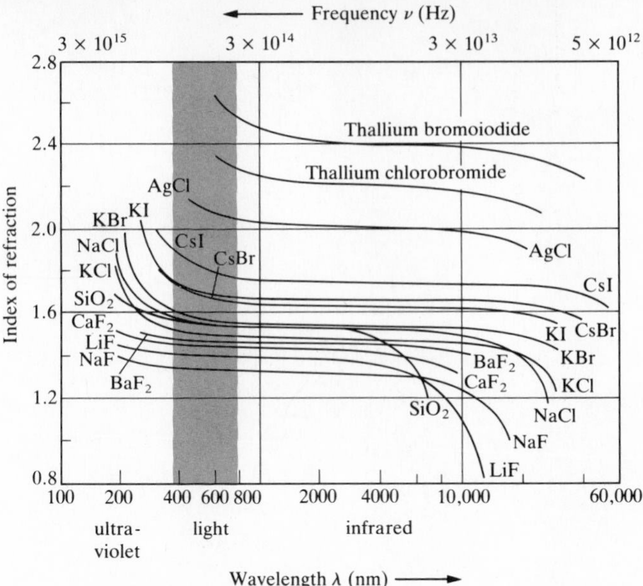

FIGURE 3.39 Index of refraction versus wavelength and frequency for several important optical crystals. (Adapted from data published by The Harshaw Chemical Co.)

FIGURE 3.40 A group of semiconductor lenses made from ZnSe, CdTe, GaAs, and Ge. These materials are particularly useful in the infrared (2 μm to 30 μm), where they are highly transparent despite the fact that they are quite opaque in the visible region of the spectrum. (Photo courtesy Two-Six Incorporated.)

3.6 The Electromagnetic-Photon Spectrum

In 1867, when Maxwell published the first extensive account of his electromagnetic theory, the frequency band was only known to extend from the infrared, across the visible, to the ultraviolet. Although this region is of major concern in Optics, it is a small segment of the vast electromagnetic spectrum (see Fig. 3.41). This section enumerates the main categories (there is actually some overlapping) into which the spectrum is usually divided.

3.6.1 Radiofrequency Waves

In 1887, eight years after Maxwell's death, Heinrich Hertz, then professor of physics at the Technische Hochschule in Karlsruhe, Germany, succeeded in generating and detecting electromagnetic waves.* His transmitter was essentially an oscillatory discharge across a spark gap (a form of oscillating electric dipole). For a receiving antenna, he used an open loop of wire with a brass knob on one end and a fine copper point on the other. A small spark visible between the two ends marked the detection of an incident electromagnetic wave. Hertz focused the radiation, determined its polarization, reflected and refracted it, caused it to interfere setting up standing waves, and then even measured its wavelength (on the order of a meter). As he put it:

> I have succeeded in producing distinct rays of electric force, and in carrying out with them the elementary experiments which are commonly performed with light and radiant heat. ... We may perhaps further designate them as rays of light of very great wavelength. The experiments described appear to me, at any rate, eminently adapted to remove any doubt as to the identity of light, radiant heat, and electromagnetic wave motion.

The waves used by Hertz are now classified in the *radiofrequency* range, which extends from a few hertz to about 10^9 Hz (λ, from many kilometers to 0.3 m or so). These are generally emitted by an assortment of electric circuits. For example, the 60-Hz alternating current circulating in power lines radiates

*David Hughes may well have been the first person who actually performed this feat, but his experiments in 1879 went unpublished and unnoticed for many years.

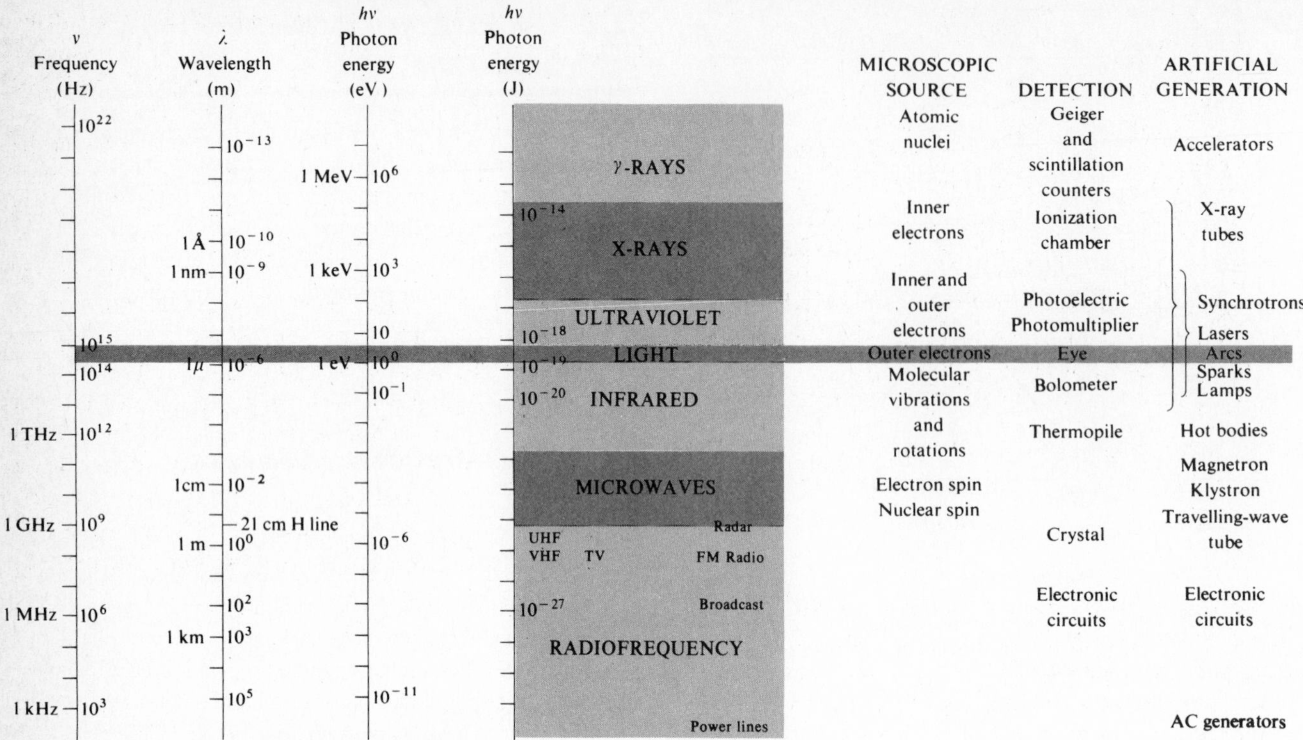

FIGURE 3.41 The electromagnetic-photon spectrum.

with a wavelength of 5×10^6 m, or about 3×10^3 miles. There is no upper limit to the theoretical wavelength; one could leisurely swing the proverbial charged pith ball and, in so doing, produce a rather long, if not very strong, wave. Indeed, waves more than 18 million miles long have been detected streaming down toward Earth from outer space. The higher frequency end of the band is used for television and radio broadcasting.

At 1 MHz (10^6 Hz) a radiofrequency photon has an energy of 6.62×10^{-28} J or 4×10^{-9} eV, a very small quantity by any measure. The granular nature of the radiation is generally obscured, and only a smooth transfer of radiofrequency energy is apparent.

3.6.2 Microwaves

The microwave region extends from about 10^9 Hz up to about 3×10^{11} Hz. The corresponding wavelengths go from roughly 30 cm to 1.0 mm. Radiation capable of penetrating the Earth's atmosphere ranges from less than 1 cm to about 30 m. Microwaves are therefore of interest in space-vehicle communications, as well as radio astronomy. In particular, neutral hydrogen atoms, distributed over vast regions of space, emit 21-cm (1420-MHz) microwaves. A good deal of information about the structure of our own and other galaxies has been gleaned from this particular emission.

Molecules can absorb and emit energy by altering the state of motion of their constituent atoms—they can be made to vibrate and rotate. Again, the energy associated with either motion is quantized, and molecules possess rotational and vibrational energy levels in addition to those due to their electrons. Only polar molecules will experience forces via the **E**-field of an incident electromagnetic wave that will cause them to rotate into alignment, and only they can absorb a photon and make a rotational transition to an excited state. Since massive molecules are not able to swing around easily, we can anticipate that they will have low-frequency rotational resonances (far IR, 0.1 mm, to microwave, 1 cm). For instance, water molecules are polar (see Fig. 3.34), and if exposed to an electro-

FIGURE 3.42 Microwave antennae on the top of the Eiffel Tower in Paris.

magnetic wave, they will swing around, trying to stay lined up with the alternating **E**-field. This will occur with particular vigor at any one of its rotational resonances. Consequently, water molecules efficiently and dissipatively absorb microwave radiation at or near such a frequency. The microwave oven (12.2 cm, 2.45 GHz) is an obvious application. On the other hand, nonpolar molecules, such as carbon dioxide, hydrogen, nitrogen, oxygen, and methane, cannot make rotational transitions by way of the absorption of photons.

Nowadays microwaves are used for everything from transmitting telephone conversations (Fig. 3.42) and interstation television to cooking hamburgers, from guiding planes and catching speeders (by radar) to studying the origins of the Universe, opening garage doors, and viewing the surface of the planet (Fig. 3.43). They are also quite useful for studying Physical Optics with experimental arrangements that are scaled up to convenient dimensions.

Photons in the low-frequency end of the microwave spectrum have little energy, and one might expect their sources to be electric circuits exclusively. Emissions of this sort can, however, arise from atomic transitions, if the energy levels involved are quite near each other. The apparent ground state of the cesium atom is a good example. It is actually a pair of closely spaced energy levels, and transitions between them involve an energy of only 4.14×10^{-5} eV. The resulting microwave emission has a frequency of $9.192\,631\,77 \times 10^{9}$ Hz. This is the basis for the well-known cesium clock, the standard of frequency and time.

3.6.3 Infrared

The infrared region, which extends roughly from 3×10^{11} Hz to about 4×10^{14} Hz, was first detected by the renowned astronomer Sir William Herschel (1738–1822) in 1800. As the name implies, this band of EM-radiation lies just beneath red

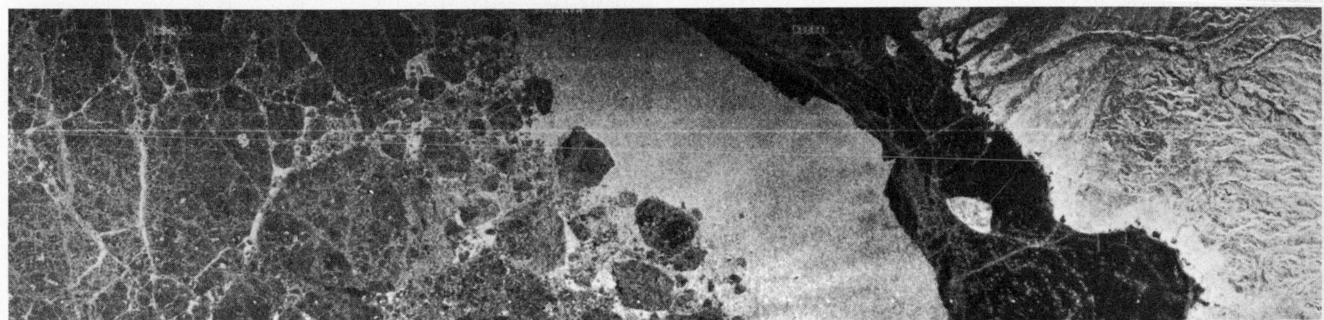

FIGURE 3.43 A photograph of an 18 by 75 mile area northeast of Alaska. It was taken by the *Seasat* satellite 800 kilometers (500 miles) above the Earth. The overall appearance is somewhat strange because this is actually a radar or microwave picture. The wrinkled gray region on the right is Canada. The small, bright shell shape is Banks Island, embedded in a black band of shore-fast, first-year sea ice. Adjacent to that is open water, which appears smooth and gray. The dark gray blotchy area at the far left is the main polar ice pack. There are no clouds because the radar "sees" right through them.

light. The infrared, or IR, is often subdivided into four regions: the *near IR*, that is, near the visible, (780–3000 nm), the *intermediate IR* (3000–6000 nm), the *far IR* (6000–15000 nm), and the *extreme IR* (15000 nm–1.0 mm). This is again a rather loose division, and there is no universality in the nomenclature. Radiant energy at the long-wavelength extreme can be generated by either microwave oscillators or incandescent sources (i.e., molecular oscillators). Indeed, any material will radiate and absorb IR via thermal agitation of its constituent molecules.

The molecules of any object at a temperature above absolute zero ($-273°C$) will radiate IR, even if only weakly (see Section 13.1.1). On the other hand, infrared is copiously emitted in a continuous spectrum from hot bodies, such as electric heaters, glowing coals, and ordinary house radiators. Roughly half the electromagnetic energy from the Sun is IR, and the common lightbulb actually radiates far more IR than light. Like all warm-blooded creatures, we too are infrared emitters. The human body radiates IR quite weakly, starting at around 3000 nm, peaking in the vicinity of 10 000 nm, and trailing off from there into the extreme IR and, negligibly, beyond. This emission is exploited by see-in-the-dark sniperscopes, as well as by some rather nasty "heat"-sensitive snakes (Crotalidae, pit vipers, and Boidae, constrictors) that tend to be active at night.

Besides rotating, a molecule can vibrate in several different ways, with its atoms moving in various directions with respect to one another. The molecule need not be polar, and even a linear system such as CO_2 has three basic vibrational modes and a number of energy levels, each of which can be excited by photons. The associated vibrational emission and absorption spectra are, as a rule, in the IR (1000 nm to 0.1 mm). Many molecules have both vibrational and rotational resonances in the IR and are good absorbers, which is one reason IR is often misleadingly called "heat waves"—just put your face in the sunshine and feel the resulting buildup of thermal energy.

Infrared radiant energy is generally measured with a device that responds to the heat generated on absorption of IR by a blackened surface. There are, for example, thermocouple, pneumatic (e.g., Golay cells), pyroelectric, and bolometer detectors. These in turn depend on temperature-dependent variations in induced voltage, gas volume, permanent electric polarization and resistance, respectively. The detector can be coupled by way of a scanning system to a cathode ray tube to produce an instantaneous televisionlike IR picture (Fig. 3.44) known as a thermograph (which is quite useful for diagnosing all sorts of problems, from faulty transformers to faulty people). Photographic films sensitive to near IR (< 1300 nm) are also available. There are IR spy satellites that look out for rocket launchings, IR resource satellites that look out for crop diseases, and IR astronomical satellites that look out into space; there are "heat-seeking" missiles guided by IR, and IR lasers and telescopes peering into the heavens.

Small differences in the temperatures of objects and their surroundings result in characteristic IR emission, that can be used in many ways, from detecting brain tumors and breast cancers to spotting a lurking burglar. The CO_2 laser, because it is a convenient source of continuous power at appreciable levels of 100 W and more, is widely used in industry, especially in precision cutting and heat treating. Its extreme-IR emissions (18.3 μm–23.0 μm) are readily absorbed by human tissue, making the laserbeam an effective bloodless scalpel that cauterizes as it cuts.

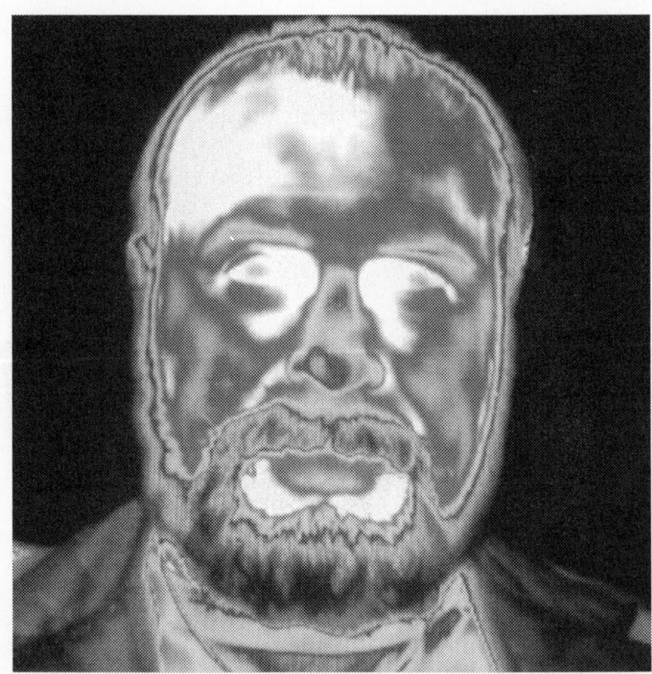

FIGURE 3.44 Thermograph of the author. Note the cool beard and how far the hairline has receded since the first edition of this book. (Photo courtesy E.H.)

3.6.4 Light

Light corresponds to the electromagnetic radiation in the narrow band of frequencies from about 3.84×10^{14} Hz to roughly 7.69×10^{14} Hz (see Table 3.4). It is generally produced by a rearrangement of the outer electrons in atoms and molecules. (Don't forget synchrotron radiation, which is a different mechanism.)*

In an incandescent material, a hot, glowing metal filament, or the solar fireball, electrons are randomly accelerated and undergo frequent collisions. The resulting broad emission spectrum is called **thermal radiation**, and it is a major source of light. In contrast, if we fill a tube with some gas and pass an electric discharge through it, the atoms therein will become excited and radiate. The emitted light is characteristic of the particular energy levels of those atoms, and it is made up of a series of well-defined frequency bands or lines. Such a device is known as a gas discharge tube. When the gas is the krypton 86 isotope, the lines are particularly narrow (zero nuclear spin, therefore no hyperfine structure). The orange-red line of Kr 86, whose vacuum wavelength is 605.780 210 5 nm, has a width (at half height) of only 0.000 47 nm, or about 400 MHz. Accordingly, until 1983 it was the international standard of length (1 650 763.73 wavelengths equaled a meter).

TABLE 3.4 Approximate Frequency and Vacuum Wavelength Ranges for the Various Colors

Color	λ_0 (nm)	ν (THz)*
Red	780–622	384–482
Orange	622–597	482–503
Yellow	597–577	503–520
Green	577–492	520–610
Blue	492–455	610–659
Violet	455–390	659–769

*1 terahertz (THz) = 10^{12} Hz, 1 nanometer (nm) = 10^{-9} m.

Newton was the first to recognize that **white light** is actually a mixture of all the colors of the visible spectrum, that the prism does not create color by altering white light to different degrees, as had been thought for centuries, but simply fans out the light, separating it into its constituent colors. Not surprisingly, the very concept of *whiteness* seems dependent on our perception of the Earth's daylight spectrum—a broad frequency distribution that falls off more rapidly in the violet than in the red (Fig. 3.45). The human eye-brain detector perceives as white a wide mix of frequencies, usually with about the same amount of energy in each portion. That is what we mean when we speak about "white light"—much of the color of the spectrum, with no region predominating. Nonetheless, many different distributions will appear more or less white. We recognize a piece of paper to be white whether it's seen indoors under incandescent light or outside under skylight, even though those whites are quite different. In fact, there are many pairs of colored light beams (e.g., 656-nm red and 492-nm cyan) that will produce the sensation of whiteness, and the eye cannot always distinguish one white from another; it cannot frequency analyze light into its harmonic components the way the ear can analyze sound (see Section 7.3).

Colors are the subjective human physiological and psychological responses, primarily, to the various frequency regions extending from about 384 THz for red, through orange, yellow, green, and blue, to violet at about 769 THz (Table 3.4). Color is not a property of the light itself but a manifestation of the electrochemical sensing system—eye, nerves, brain. To be more precise, we should not say "yellow light" but rather

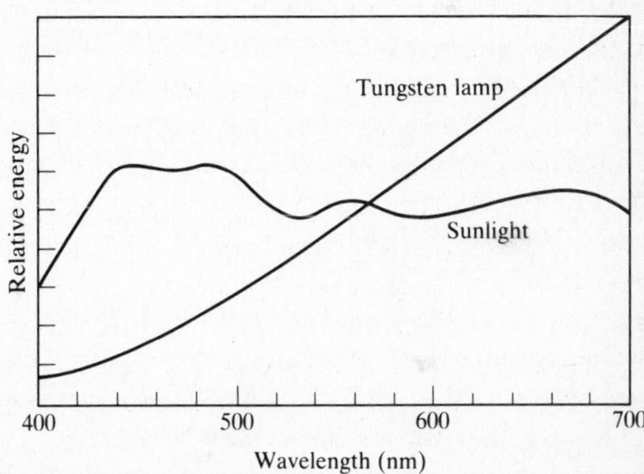

FIGURE 3.45 A graph of sunlight compared with the light from a tungsten lamp.

*There is no need here to define light in terms of human physiology. On the contrary, there is plenty of evidence to indicate that this would not be a very good idea. For example, see T. J. Wang, "Visual Response of the Human Eye to X Radiation." *Am. J. Phys.* **35**, 779 (1967).

"light that is seen as yellow." Remarkably, a variety of different frequency mixtures can evoke the same color response from the eye-brain sensor. A beam of red light (peaking at, say, 430 THz) overlapping a beam of green light (peaking at, say, 540 THz) will result, believe it or not, in the perception of *yellow* light, even though no frequencies are actually present in the so-called yellow band. Apparently, the eye-brain averages the input and "sees" yellow (Section 4.9). That's why a color television screen can manage with only three phosphors: red, green, and blue.

In a flood of bright sunlight where the photon flux density might be 10^{21} photons/$m^2 \cdot s$, we can generally expect the quantum nature of the energy transport to be thoroughly masked. However, in very weak beams, since photons in the visible range ($h\nu \approx$ 1.6 eV to 3.2 eV) are energetic enough to produce effects on a distinctly individual basis, the granularity will become evident. Research on human vision indicates that as few as 10 light photons, and possibly even 1, may be detectable by the eye.

3.6.5 Ultraviolet

Adjacent to and just beyond light in the spectrum is the ultraviolet region (approximately 8×10^{14} Hz to about 3.4×10^{16} Hz), discovered by Johann Wilhelm Ritter (1776–1810). Photon energies therein range from roughly 3.2 eV to 100 eV. Ultraviolet, or UV, rays from the Sun will thus have more than enough energy to ionize atoms in the upper atmosphere and in so doing create the ionosphere. These photon energies are also of the order of the magnitude of many chemical reactions, and ultraviolet rays become important in triggering those reactions. Fortunately, ozone (O_3) in the atmosphere absorbs what would otherwise be a lethal stream of solar UV. At wavelengths less than around 290 nm, UV is germicidal (i.e., it kills microorganisms). The particlelike aspects of radiant energy become increasingly evident as the frequency rises.

Humans cannot see UV very well, because the cornea absorbs it, particularly at the shorter wavelengths, while the eye lens absorbs most strongly beyond 300 nm. A person who has had a lens removed because of cataracts can see UV ($\lambda >$ 300 nm). In addition to insects, such as honeybees, a fair number of other creatures can visually respond to UV. Pigeons, for example, are capable of recognizing patterns illuminated by UV and probably employ that ability to navigate by the Sun even on overcast days.

An atom emits a UV photon when an electron makes a long jump down from a highly excited state. For example, the outermost electron of a sodium atom can be raised to higher and higher energy levels until it is ultimately torn loose altogether at 5.1 eV, and the atom is ionized. If the ion subsequently recombines with a free electron, the latter will rapidly descend to the ground state, most likely in a series of jumps, each resulting in the emission of a photon. It is possible, however, for the electron to make one long plunge to the ground state, radiating a single 5.1-eV UV photon. Even more energetic UV can be generated when the inner, tightly bound electrons of an atom are excited.

The unpaired valence electrons of isolated atoms can be an important source of colored light. But when these same atoms combine to form molecules or solids, the valence electrons are ordinarily paired in the process of creating the chemical bonds that hold the thing together. Consequently, the electrons are often more tightly bound, and their molecular-excited states are higher up in the UV. Molecules in the atmosphere, such as N_2, O_2, CO_2, and H_2O, have just this sort of electronic resonance in the UV.

Nowadays there are ultraviolet photographic films and microscopes, UV orbiting celestial telescopes, synchrotron sources, and ultraviolet lasers (Fig. 3.46).

3.6.6 X-rays

X-rays were rather fortuitously discovered in 1895 by Wilhelm Conrad Röntgen (1845–1923). Extending in frequency from roughly 2.4×10^{16} Hz to 5×10^{19} Hz, they have extremely short wavelengths; most are smaller than an atom. Their photon energies (100 eV to 0.2 MeV) are large enough so that X-ray quanta can interact with matter one at a time in a clearly granular fashion, almost like bullets of energy. One of the most practical mechanisms for producing X-rays is the rapid deceleration of high-speed charged particles. The resulting broad-frequency *bremsstrahlung* (German for "braking radiation") arises when a beam of energetic electrons is fired at a material target, such as a copper plate. Collisions with the Cu nuclei produce deflections of the beam electrons, which in turn radiate X-ray photons.

In addition, the atoms of the target may become ionized during the bombardment. Should that occur through removal of an inner electron strongly bound to the nucleus, the atom will emit X-rays as the electron cloud returns to the ground state. The resulting quantized emissions are specific to the target atom, revealing its energy level structure, and accordingly are called *characteristic* radiation.

Traditional medical film-radiography generally produces little more than simple shadow castings, rather than photographic images in the usual sense; it has not been possible to fabricate useful X-ray lenses. But modern focusing methods using mirrors (see Section 5.4) have begun an era of X-ray imagery, creating detailed pictures of all sorts of things, from imploding fusion pellets to celestial sources, such as the Sun (Fig. 3.47), distant quasars, and black holes—objects at temperatures of millions of degrees that emit predominantly in the X-ray region. Orbiting X-ray telescopes have given us an exciting new eye on the Universe. There are X-ray microscopes, picosecond X-ray streak cameras, X-ray diffraction gratings, and interferometers, and work continues on X-ray holography. In 1984 a group at the Lawrence Livermore National Laboratory succeeded in producing laser radiation at a wavelength of 20.6 nm. Although this is more accurately in the extreme ultraviolet (XUV), it's close enough to the X-ray region to qualify as the first soft X-ray laser.

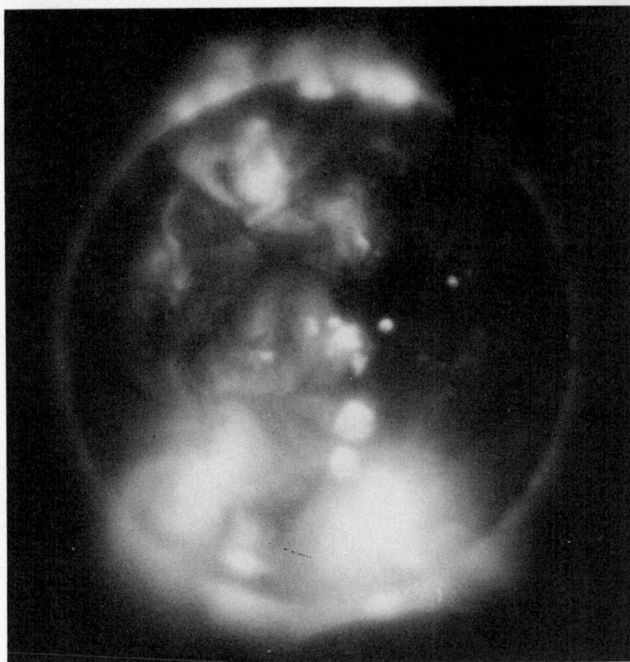

FIGURE 3.47 X-ray photograph of the Sun taken March 1970. The limb of the Moon is visible in the southeast corner. (Courtesy Dr. G. Vaiana and NASA.)

FIGURE 3.46 An ultraviolet photograph of Venus taken by *Mariner 10*.

3.6.7 Gamma Rays

These are the highest energy (10^4 eV to about 10^{19} eV), shortest wavelength electromagnetic radiations. They are emitted by particles undergoing transitions within the atomic nucleus. A single gamma-ray photon carries so much energy that it can be detected with little difficulty. At the same time its wavelength is so small that it is now extremely difficult to observe any wavelike properties.

We have gone full cycle from the radiofrequency wavelike response to gamma-ray particlelike behavior. Somewhere, not far from the (logarithmic) center of the spectrum, there is light. As with all electromagnetic radiation, its energy is quantized, but here in particular what we "see" will depend on how we "look."

3.7 Quantum Field Theory

A charged particle exerts forces on other charged particles. It creates a web of electromagnetic interaction around itself that extends out into space. That imagery leads to the concept of the electric field, which is a representation of the way the electromagnetic interaction reveals itself on a macroscopic level. The static electric field is, in effect, a spatial conception summarizing the interaction among charges. Through Faraday's vision, the idea of the field was extended, and it became appropriate to imagine that one charge sets up an **E**-field in space and another charge, immersed in that field, interacts directly with it, and vice versa. What began as a mapping of the force distribution (whatever its cause) became a thing, a field, capable itself of exerting force. Still the picture seems straightforward, even if many questions come to mind. Does the static **E**-field have a physical reality in-and-of itself? If it does, does it fill space with energy and how exactly does that happen? Is anything actually flowing? How does the field produce a force on a charge? Does it take time to exert its influence?

FIGURE 3.48 An ultrashort pulse of green light from a neodymium-doped glass laser. The pulse passed through a water cell whose wall is marked in millimeters. During the 10-picosecond exposure the pulse moved about 2.2 mm. (Photo courtesy Bell Laboratories.)

Once the electromagnetic field became a reality, physicists could imagine disturbances of that tenuous medium which so conveniently spans the void of space; ***light was an electromagnetic wave in the electromagnetic field***. Although it's easy enough to envision a wave sweeping through an existing field (p. 57), it's not so obvious how a localized pulse launched into space, like the one shown in Fig. 3.48, might be conceptualized. There is no static field filling space, extending out in front of the pulse; if the pulse advances through the medium of the electromagnetic field, it must first create that medium itself as it progresses. That's not impossible to imagine on some level, but it's hardly what one would call a classical wave. For any traditional wave, a medium in equilibrium is the fundamental starting point; it exists at any location before and after the wave passes. So this idea of an electromagnetic wave, which is so beautiful mathematically, is not quite so transparent conceptually.

As early as 1905, Einstein already considered the classical equations of Electromagnetic Theory to be descriptions of the average values of the quantities being considered. "To me it seems absurd," he wrote to Planck, "to have energy continuously distributed in space without assuming an aether. . . . While Faraday's representation was useful in the development of electrodynamics, it does not follow in my opinion that this view must be maintained in all its detail." Classical theory wonderfully accounted for everything being measured, but it was oblivious to the exceedingly fine granular structure of the phenomenon. Using thermodynamic arguments, Einstein proposed that electric and magnetic fields were quantized, that they are particulate rather than continuous. After all, classical theory evolved decades before the electron was even discovered. If charge (the fundamental source of electromagnetism) is quantized, shouldn't the theory reflect that in some basic way?

Today, we are guided by Quantum Mechanics, a highly mathematical theory that provides tremendous computational and predictive power but is nonetheless disconcertingly abstract. In particular, the subdiscipline that treats microparticles and their interactions, Quantum Field Theory (QFT), in its various forms, is the most fundamental and arguably the most successful of all physical theories. Light quanta come out of the theory in a completely natural way by quantizing the electromagnetic field. The apparent implication of this is that all microparticles originate in the same way from their own individual fields: **the field's the thing**, as it were. Thus the electron is the quantum of the electron field, the proton is the quantum of the proton field, and so forth. Filling in the details has been the business of field theorists for much of this century.

There are two distinct philosophical currents in contemporary QFT: the field-centered and the particle-centered. In the field-centered view ***fields are the fundamental entities, particles are just the quanta of the fields***. In the particle-centered view, ***particles are the fundamental entities, fields are just the macroscopic coherent states of particles***. The field tradition goes back to L. de Broglie (1923), E. Schrödinger, P. Jordan, and W. Pauli, whose research laid the foundations of the quantum-mechanical variant sometimes called Wave Mechanics. The particle tradition began with the early work of W. Heisenberg (1925), although its spiritual mentor was P.A.M. Dirac who set the particle agenda with his theory of the electron-positron pair. The particular offshoot of QFT that strives to provide a relativistic quantum-mechanical treatment of the electromagnetic interaction is called Quantum Electrodynamics (QED), and it too has its particle-centered and field-centered proponents. Some of the basic ideas of QED have been made accessible on this level by R. P. Feynman, and insofar as they illuminate Optics we'll explore them later in this text (p. 137).

Contemporary physics by way of QFT holds that all fields are quantized; that each of the fundamental Four Forces (Gravitational, Electromagnetic, Strong, and Weak) is mediated by a special kind of field particle. These *messenger bosons* are continuously absorbed and emitted by the interacting material particles (electrons, protons, etc.). This ongoing exchange *is* the interaction. The mediating particle of the electric field is the **virtual photon**. This massless messenger travels at the speed of light and transports momentum and energy. When two electrons repel one another, or an electron and proton attract, it is by emitting and absorbing virtual photons and thereby transferring momentum from one to the other, that transfer being a measure of the action of force. The messenger particles of the electromagnetic force are called *virtual* photons because they are bound to the interaction. Virtual photons can never escape to be detected directly by some instrument, however unsettling that is philosophically and however hard that makes it to establish their existence. Indeed, virtual photons (as distinct from real photons) exist only as the means of interaction. They are creatures of theory whose metaphysical status is yet to be determined.*

On a macroscopic level, messenger particles can manifest themselves as a continuous field provided they can group in very large numbers. Fundamental particles have an intrinsic angular momentum, or *spin*, that determines their grouping characteristics. Quantum Theory tells us that the desired field behavior can occur only if forces are mediated by messenger particles having angular momenta equal to integer multiples of $h/2\pi$ (i.e., 0, $1h/2\pi$, $2h/2\pi$, $3h/2\pi$, ...). The angular momentum of the virtual photon is $1(h/2\pi)$; it's a spin-1 particle. The exceedingly important class of interactions that have spin-1 messengers are known as **gauge forces**, and the electromagnetic force is the model for all the gauge forces. Today, the magic of action-at-a-distance is understood via the no less mysterious exchange of virtual particles, but at least now a highly predictive mathematical theory is in place that describes the phenomenon.

*For a discussion of the issues being struggled with, see H. R. Brown and R. Harré, *Philosophical Foundations of Quantum Field Theory*.

PROBLEMS

3.1 Consider the plane electromagnetic wave (in SI units) given by the expressions $E_x = 0$, $E_y = 2\cos[2\pi \times 10^{14}(t - x/c) + \pi/2]$, and $E_z = 0$.

(a) What are the frequency, wavelength, direction of motion, amplitude, initial phase angle, and polarization of the wave?

(b) Write an expression for the magnetic flux density.

3.2 Write an expression for the **E**- and **B**-fields that constitute a plane harmonic wave traveling in the $+z$-direction. The wave is linearly polarized with its plane of vibration at 45° to the yz-plane.

3.3* Considering Eq. (3.30), show that the expression

$$\mathbf{k} \times \mathbf{E} = \omega \mathbf{B}$$

is correct as it applies to a plane wave for which the direction of the electric field is constant.

3.4* An electromagnetic wave is specified (in SI units) by the following function:

$$\mathbf{E} = (-3\hat{\mathbf{i}} + 3\sqrt{3}\hat{\mathbf{j}})(10^4\ \text{V/m})e^{i[\frac{1}{3}(\sqrt{5}x + 2y)\pi \times 10^7 - 9.42 \times 10^{15}t]}$$

Find (a) the direction along which the electric field oscillates, (b) the scalar value of amplitude of the electric field, (c) the direction of propagation of the wave, (d) the propagation number and wavelength, (e) the frequency and angular frequency, and (f) the speed.

3.5 The electric field of an electromagnetic wave traveling in the positive x-direction is given by

$$\mathbf{E} = E_0\hat{\mathbf{j}} \sin \frac{\pi z}{z_0} \cos (kx - \omega t)$$

(a) Describe the field verbally. (b) Determine an expression for k. (c) Find the phase speed of the wave.

3.6* Calculate the energy input necessary to charge a parallel plate capacitor by carrying charge from one plate to the other. Assume the energy is stored in the field between the plates and compute the energy per unit volume, u_E, of that region, that is, Eq. (3.31). *Hint:* since the electric field increases throughout the process, either integrate or use its average value $E/2$.

3.7 The time average of some function $f(t)$ taken over an interval T is given by

$$\langle f(t) \rangle_T = \frac{1}{T}\int_t^{t+T} f(t')dt'$$

where t' is just a dummy variable. If $\tau = 2\pi/\omega$ is the period of a harmonic function, show that

$$\langle \sin^2(\mathbf{k} \cdot \mathbf{r} - \omega t) \rangle = \tfrac{1}{2}$$

$$\langle \cos^2(\mathbf{k} \cdot \mathbf{r} - \omega t) \rangle = \tfrac{1}{2}$$

and

$$\langle \sin(\mathbf{k} \cdot \mathbf{r} - \omega t) \cos(\mathbf{k} \cdot \mathbf{r} - \omega t) \rangle = 0$$

when $T = \tau$ and when $T >> \tau$.

3.8* Show that a more general formulation of the previous problem yields

$$\langle \cos \omega t \rangle_T = \tfrac{1}{2}[1 + \text{sinc}\ \omega T \cos 2\omega t]$$

for any interval T.

3.9* With the previous problem in mind prove that

$$\langle \sin^2 \omega t \rangle_T = \tfrac{1}{2}[1 - \text{sinc}\ \omega T \cos 2\omega t]$$

for any interval T.

3.10* Consider a linearly polarized plane electromagnetic wave traveling in the $+x$-direction in free space having as its plane of vibration the xy-plane. Given that its frequency is 10 MHz and its amplitude is $E_0 = 0.08$ V/m,

(a) Find the period and wavelength of the wave.

(b) Write an expression for $E(t)$ and $B(t)$.

(c) Find the flux density, $\langle S \rangle$, of the wave.

3.11 A linearly polarized harmonic plane wave with a scalar amplitude of 10 V/m is propagating along a line in the xy-plane at 45° to the x-axis with the xy-plane as its plane of vibration. Please write a vector expression describing the wave assuming both k_x and k_y are positive. Calculate the flux density taking the wave to be in vacuum.

3.12 Pulses of UV lasting 2.00 ns each are emitted from a laser which has a beam of diameter 2.5 mm. Given that each burst carries an energy of 6.0 J, (a) determine the length in space of each wavetrain, and (b) find the average energy per unit volume for such a pulse.

3.13* A laser provides pulses of EM-radiation in vacuum lasting 10^{-12} s. If the radiant flux density is 10^{20} W/m^2, determine the amplitude of the electric field of the beam.

3.14 A 1.0-mW laser has a beam diameter of 2 mm. Assuming the divergence of the beam to be negligible, compute its energy density in the vicinity of the laser.

3.15* A cloud of locusts having a density of 100 insects per cubic meter is flying north at a rate of 6 m/min. What is the flux density of locusts, that is, how many cross an area of 1 m^2 perpendicular to their flight path per second?

3.16 Imagine that you are standing in the path of an antenna which is radiating plane waves of frequency 100 MHz and flux density 19.88×10^{-2} W/m^2. Compute the photon flux density, that is, the number of photons per unit time per unit area. How many photons, on the average, will be found in a cubic meter of this region?

3.17* How many photons per second are emitted from a 100 W yellow lightbulb if we assume negligible thermal losses and a quasi-monochromatic wavelength of 550 nm? In actuality only about 2.5% of the total dissipated power emerges as visible radiation in an ordinary 100 W lamp.

3.18 A 3.0-V flashlight bulb draws 0.25 A, converting about 1.0% of the dissipated power into light ($\lambda \approx 550$ nm). If the beam has a cross-sectional area of 10 cm^2, and is approximately cylindrical,

(a) How many photons are emitted per second?

(b) How many photons occupy each meter of the beam?

(c) What is the flux density of the beam as it leaves the flashlight?

3.19* An isotropic quasimonochromatic point source radiates at a rate of 100 W. What is the flux density at a distance of 1 m? What are the amplitudes of the **E**- and **B**-fields at that point?

3.20 Using energy arguments, show that the amplitude of a cylindrical wave must vary inversely with $\sqrt{r}$. Draw a diagram indicating what's happening.

3.21* What is the momentum of a 10^{19}-Hz X-ray photon?

3.22 Consider an electromagnetic wave impinging on an electron. It is easy to show kinematically that the average value of the time rate-of-change of the electron's momentum **p** is proportional to the average value of the time rate-of-change of the work, W, done on it by the wave. In particular,

$$\left\langle \frac{d\mathbf{p}}{dt} \right\rangle = \frac{1}{c}\left\langle \frac{dW}{dt} \right\rangle \hat{\mathbf{i}}$$

Accordingly, if this momentum change is imparted to some completely absorbing material, show that the pressure is given by Eq. (3.51).

3.23* Derive an expression for the radiation pressure when the normally incident beam of light is totally reflected. Generalize this result to the case of oblique incidence at an angle θ with the normal.

3.24 A completely absorbing screen receives 300 W of light for 100 s. Compute the total linear momentum transferred to the screen.

3.25 The average magnitude of the Poynting vector for sunlight arriving at the top of Earth's atmosphere (1.5×10^{11} m from the Sun) is about 1.4 kW/m^2.

(a) Compute the average radiation pressure exerted on a metal reflector facing the Sun.

(b) Approximate the average radiation pressure at the surface of the Sun whose diameter is 1.4×10^9 m.

3.26* A surface is placed perpendicular to a beam of light of constant irradiance (I). Suppose that the fraction of the irradiance absorbed by the surface is α. Show that the pressure on the surface is given by

$$\mathcal{P} = (2 - \alpha)I/c$$

3.27 What force on the average will be exerted on the (40 m $\times$ 50 m) flat, highly reflecting side of a space station wall if it's facing the Sun while orbiting Earth?

3.28 A parabolic radar antenna with a 2-m diameter transmits 200-kW pulses of energy. If its repetition rate is 500 pulses per second, each lasting 2 μs, determine the average reaction force on the antenna.

3.29 Consider the plight of an astronaut floating in free space with only a 10-W lantern (inexhaustibly supplied with power). How long will it take to reach a speed of 10 m/s using the radiation as propulsion? The astronaut's total mass is 100 kg.

3.30 Consider the uniformly moving charge depicted in Fig. 3.23*b*. Draw a sphere surrounding it and show via the Poynting vector that the charge does not radiate.

3.31* A plane, harmonic, linearly polarized light wave has an electric field intensity given by

$$E_z = E_0 \cos \pi\, 10^{15}\left(t - \frac{x}{0.65c}\right)$$

while traveling in a piece of glass. Find

(a) The frequency of the light.

(b) Its wavelength.

(c) The index of refraction of the glass.

3.32* What is the speed of light in diamond if the index of refraction is 2.42?

3.33* Given that the wavelength of a lightwave in vacuum is 540 nm, what will it be in water, where $n = 1.33$?

3.34* Determine the index of refraction of a medium if it is to reduce the speed of light by 10% as compared to its speed in vacuum?

3.35 If the speed of light (the phase speed) in Fabulite ($SrTiO_3$) is 1.245×10^8 m/s, what is its index of refraction?

3.36* What is the distance that yellow light travels in water (where $n = 1.33$) in 1.00 s?

3.37* A 500-nm lightwave in vacuum enters a glass plate of index 1.60 and propagates perpendicularly across it. How many waves span the glass if it's 1.00 cm thick?

3.38* Yellow light from a sodium lamp ($\lambda_0 = 589$ nm) traverses a tank of glycerin (of index 1.47), which is 20.0 m long, in a time t_1. If it takes a time t_2 for the light to pass through the same tank when filled with carbon disulfide (of index 1.63), determine the value of $t_2 - t_1$.

3.39* A lightwave travels from point A to point B in vacuum. Suppose we introduce into its path a flat glass plate ($n_g = 1.50$) of thickness $L = 1.00$ mm. If the vacuum wavelength is 500 nm, how many waves span the space from A to B with and without the glass in place? What phase shift is introduced with the insertion of the plate?

3.40 The low-frequency relative permittivity of water varies from 88.00 at 0°C to 55.33 at 100°C. Explain this behavior. Over the same range in temperature, the index of refraction (λ = 589.3 nm) goes from roughly 1.33 to 1.32. Why is the change in n so much smaller than the corresponding change in K_E?

3.41 Show that for substances of low density, such as gases, which have a single resonant frequency ω_0, the index of refraction is given by

$$n \approx 1 + \frac{Nq_e^2}{2\epsilon_0 m_e(\omega_0^2 - \omega^2)}$$

3.42* In the next chapter, Eq. (4.47), we'll see that a substance reflects radiant energy appreciably when its index differs most from the medium in which it is embedded.

(a) The dielectric constant of ice measured at microwave frequencies is roughly 1, whereas that for water is about 80 times greater—why?

(b) How is it that a radar beam easily passes through ice but is considerably reflected when encountering a dense rain?

3.43 Fuchsin is a strong (aniline) dye, which in solution with alcohol has a deep red color. It appears red because it absorbs the green component of the spectrum. (As you might expect, the surfaces of crystals of fuchsin reflect green light rather strongly.) Imagine that you have a thin-walled hollow prism filled with this solution. What will the spectrum look like for incident white light? By the way, anomalous dispersion was first observed in about 1840 by Fox Talbot, and the effect was christened in 1862 by Le Roux. His work was promptly forgotten, only to be rediscovered eight years later by C. Christiansen.

3.44* Take Eq. (3.71) and check out the units to make sure that they agree on both sides.

3.45 The resonant frequency of lead glass is in the UV fairly near the visible, whereas that for fused silica is far into the UV. Use the dispersion equation to make a rough sketch of n versus ω for the visible region of the spectrum.

3.46* Show that Eq. (3.70) can be rewritten as

$$(n^2 - 1)^{-1} = -C\lambda^{-2} + C\lambda_0^{-2}$$

where $C = 4\pi^2c^2\epsilon_0 m_e/Nq_e^2$.

3.47 Augustin Louis Cauchy (1789–1857) determined an empirical equation for $n(\lambda)$ for substances that are transparent in the visible. His expression corresponded to the power series relation

$$n = C_1 + C_2/\lambda^2 + C_3/\lambda^4 + \cdots$$

where the Cs are all constants. In light of Fig. 3.38, what is the physical significance of C_1?

3.48 Referring to the previous problem, realize that there is a region between each pair of absorption bands for which the Cauchy Equation (with a new set of constants) works fairly well. Examine Fig. 3.37: what can you say about the various values of C_1 as ω decreases across the whole spectrum? Dropping all but the first two terms, use Fig. 3.38 to determine approximate values for C_1 and C_2 for borosilicate crown glass in the visible.

3.49* Crystal quartz has refractive indexes of 1.557 and 1.547 at wavelengths of 410.0 nm and 550.0 nm, respectively. Using only the first two terms in Cauchy's Equation, calculate C_1 and C_2 and determine the index of refraction of quartz at 610.0 nm.

3.50* In 1871 Sellmeier derived the equation

$$n^2 = 1 + \sum_j \frac{A_j\lambda^2}{\lambda^2 - \lambda_{0j}^2}$$

where the A_j terms are constants and each λ_{0j} is the vacuum wavelength associated with a natural frequence ν_{0j}, such that $\lambda_{0j}\nu_{0j} = c$. This formulation is a considerable practical improvement over the Cauchy Equation. Show that where $\lambda >> \lambda_{0j}$, Cauchy's Equation is an approximation of Sellmeier's. *Hint:* Write the above expression with only the first term in the sum; expand it by the binomial theorem; take the square root of n^2 and expand again.

3.51* If an ultraviolet photon is to dissociate the oxygen and carbon atoms in the carbon monoxide molecule, it must provide 11 eV of energy. What is the minimum frequency of the appropriate radiation?

4 The Propagation of Light

4.1 Introduction

Our present concern is with the basic phenomena of *transmission* (p. 88), *reflection* (p. 95) and *refraction* (p. 100). These will be described classically in two ways: first, via the general notions of waves and rays and then from the more specific perspective of Electromagnetic Theory (p. 109). After that, we'll turn to a highly simplified treatment of Quantum Electrodynamics (QED) for a modern interpretation of what's happening (p. 137).

Most students have already studied these fundamental propagation phenomena in some introductory way and found ideas like the Laws of Reflection and Refraction to be straightforward and simple. But that's only because such treatments are from a macroscopic perspective that tends to be misleadingly superficial. For instance, reflection, which looks as obvious as light "bouncing off a surface," is a wonderfully subtle affair usually involving the coordinated behavior of countless atoms. The more deeply we explore these processes, the more challenging they become. Beyond that, many fascinating questions need to be addressed: How does light move through a material medium? What happens to it as it does? Why does light appear to travel at a speed other than c when photons can only exist at c?

Each encounter of light with bulk matter can be viewed as a cooperative event arising when a stream of photons sails through, and interacts with, an array of atoms suspended (via electromagnetic fields) in the void. The details of that journey determine why the sky is blue and blood is red, why your cornea is transparent and your hand opaque, why snow is white and rain is not. At its core, this chapter is about **scattering**, in particular, the absorption and prompt re-emission of EM-radiation by electrons associated with atoms and molecules. ***The processes of transmission, reflection, and refraction are macroscopic manifestations of scattering occurring on a submicroscopic level.***

To begin the analysis, let's first consider the propagation of radiant energy through various homogeneous media.

4.2 Rayleigh Scattering

Imagine a narrow beam of sunlight having a broad range of frequencies advancing through empty space. As it progresses, the beam spreads out very slightly, but apart from that, all the energy continues forward at c. There is no scattering, and the beam cannot be seen from the side. Nor does the light tire or diminish in any way. When a star in a nearby galaxy 1.7×10^5 light-years away was seen to explode in 1987, the flash of light that reached Earth had been sailing through space for 170 000 years before it got here. ***Photons are timeless***.

Now, suppose we mix a wisp of air into the void—some molecules of nitrogen, oxygen, and so forth. These molecules have no resonances in the visible, no one of them can be raised into an excited state by absorbing a quantum of light, and the gas is therefore transparent. Instead, each molecule behaves as a little oscillator whose electron cloud can be driven into a ground-state vibration by an incoming photon. Immediately upon being set vibrating, the molecule initiates the re-emission of light. A photon is absorbed, and without delay another photon of the same frequency (and wavelength) is emitted; the light is *elastically scattered*. The molecules are randomly oriented, and photons scatter out every which way (Fig. 4.1). Even when the light is fairly dim, the number of photons is immense, and it looks as if the molecules are scattering little classical spherical wavelets (Fig. 4.2)—energy streams out in every direction. Still, the scattering process is quite weak and

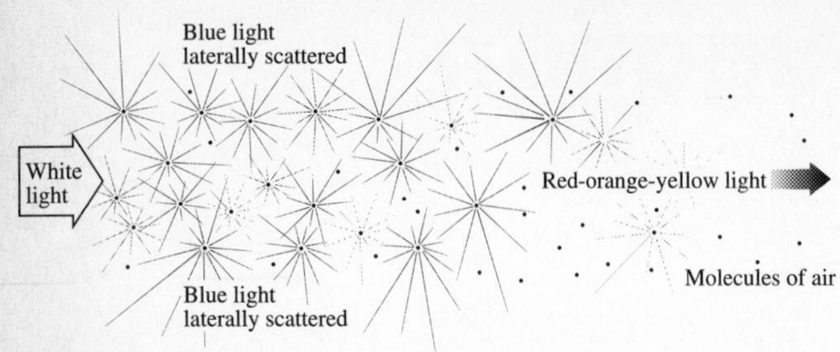

FIGURE 4.1 Sunlight traversing a region of widely spaced air molecules. The light laterally scattered is mostly blue, and that's why the sky is blue. The unscattered light, which is rich in red, is viewed only when the Sun is low in the sky at sunrise and sunset.

the gas tenuous, so the beam is very little attenuated unless it passes through a tremendous volume of air.

The amplitudes of these ground-state vibrations, and therefore the amplitudes of the scattered light, increase with frequency because all the molecules have electronic resonances in the UV. The closer the driving frequency is to a resonance, the more vigorously the oscillator responds. So, violet light is strongly scattered laterally out of the beam, as is blue to a slightly lesser degree, as is green to a considerably lesser

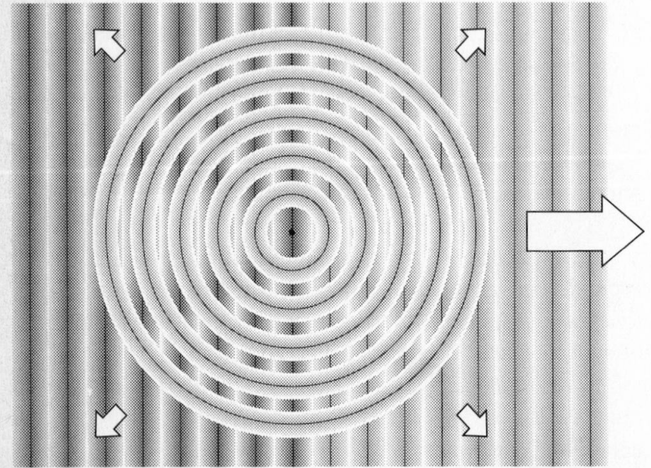

FIGURE 4.2 A plane wave, incident from the left, sweeps across an atom and spherical wavelets are scattered. The process is continuous, and hundreds of millions of photons per second stream out of the scattering atom in all directions.

degree, as is yellow to a still lesser degree, and so on. The beam that traverses the gas will thus be richer in the red end of the spectrum, while the light scattered out (sunlight not having very much violet in it in the first place) will be richer in blue. That, in part, is why the sky is blue.

Long before Quantum Mechanics, Lord Rayleigh (1871) analyzed scattered sunlight in terms of molecular oscillators. Using a simple argument based on dimensional analysis (see Problem 4.1), he correctly concluded that the intensity of the scattered light was proportional to $1/\lambda^4$, and therefore increases with ν^4. Before this work, it was widely believed that the sky was blue because of scattering from minute dust particles. Since that time, scattering involving particles smaller than a wavelength (i.e., less than about $\lambda/15$) has been referred to as Rayleigh Scattering. Atoms and ordinary molecules fit the bill since they are a few tenths of a nanometer in diameter, whereas light has a wavelength of around 500 nm. A human's blue eyes, a bluejay's feathers, the blue-tailed skinks's blue tail, and the baboon's blue buttocks are all colored via Rayleigh Scattering. Indeed, in the animal kingdom scattering is the cause of almost all the blue, much of the green, and even some of the purple coloration. Scattering from the tiny alveolar cells in the barbs of the jay's feathers make it blue, whereas a parrot's green is a blend of yellow arising from preferential absorption (p. 133) and blue via scattering.

As we will see in a moment, a dense uniform substance will not appreciably scatter laterally, and that applies to much of the lower atmosphere. After all, if blue light were strongly scattered out at sea level, a far-off mountain would appear reddish and that's not the case even over distances of tens of kilometers. Even in the middle regions of the atmosphere, the density is great enough to suppress Rayleigh Scattering; something else must be contributing to the blue of the sky. What happens in the mid-atmosphere is that thermal motion of the air results in rapidly changing *density fluctuations* on a local scale. These momentary, fairly random fluctuations cause more molecules to be in one place than another and to radiate more in one direction than another. M. Smoluchowski (1908) and A. Einstein (1910) independently provided the basic ideas for the theory of scattering from these fluctuations, which gives similar results to those of Rayleigh. Scattering from inhomogeneities in density is of interest whenever light travels great distances in a medium, such as the glass fiber of a communications link (p. 199).

Sunlight streaming into the atmosphere from one direction is scattered in all directions. Without an atmosphere, the daytime sky would be as black as the void of space, as black as the Moon sky. When the Sun is low over the horizon, its light passes through a great thickness of air (far more so than it does at noon). With the blue-end appreciably attenuated, the reds and yellows propagate along the line-of-sight from the Sun to produce Earth's familiar fiery sunsets.

4.2.1 Scattering and Interference

In dense media, a tremendous number of close-together atoms or molecules contribute an equally tremendous number of scattered electromagnetic wavelets. These wavelets overlap and interfere in a way that does not occur in a tenuous medium. As a rule, ***the denser the substance through which light advances, the less the lateral scattering***, and to understand why that's so, we must examine the interference taking place.

Interference has already been discussed (p. 22) and will be treated in further detail in Chapters 7 and 9; here, the basics suffice. Recall that interference is *the superposition of two or more waves producing a resultant disturbance that is the sum of the overlapping wave contributions*. Figure 2.14 shows two harmonic waves of the same frequency traveling in the same direction. When they are precisely in-phase (Fig. 2.14*a*), the resultant at every point is the sum of the two wave-height values. This extreme case is called ***total constructive interference***. When the phase difference reaches 180°, the waves tend to cancel, and we have the other extreme, called ***total destructive interference*** (Fig. 2.14*d*).

The theory of Rayleigh Scattering has independent molecules randomly arrayed in space so that the phases of the secondary wavelets scattered off to the side have no particular relationship to one another and there is no sustained pattern of interference. That situation occurs when the separation between the molecular scatterers is roughly a wavelength or more, as it is in a tenuous gas. In Fig. 4.3*a* a parallel beam of light is incident from the left. This so-called *primary light field* (in this instance composed of plane waves) illuminates a group of widely spaced molecules. A continuing progression of primary wavefronts sweep over and successively energize and re-energize each molecule, which, in turn, scatters light in all directions, and in particular out to some lateral point *P*. Because the lengths of their individual paths to *P* differ greatly in comparison to λ, some of the wavelets arriving at *P* are ahead of others while some are behind, and that by substantial fractions of a wavelength (Fig. 4.3*b*). In other words, the phases of the wavelets at *P* differ greatly. (Remember that the molecules are also moving around, and that changes the phases as well.) At any moment some wavelets interfere constructively, some destructively, and the shifting random hodgepodge of overlapping wavelets effectively averages away the interfer-

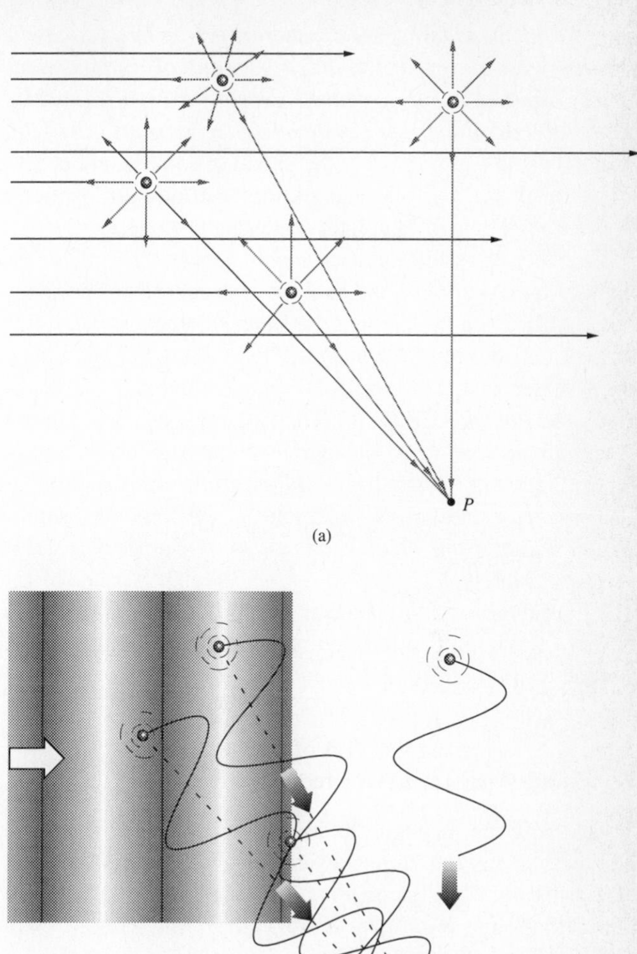

FIGURE 4.3 (*a*) The scattering of light from a widely spaced distribution of molecules. (*b*) The waves arriving at a point *P* off to the side, have a jumble of different phases and tend not to interfere in a sustained fashion; a good deal of light is therefore scattered laterally.

ence. ***Random, widely spaced scatterers driven by an incident primary wave emit wavelets that are essentially independent of one another in all directions except forward. Laterally scattered light, unimpeded by interference, streams out of the beam.*** This is approximately the situation existing about 100 miles up in the Earth's tenuous high-altitude atmosphere, where a good deal of blue-light scattering takes place.

That the scattered irradiance should depend on $1/\lambda^4$ is easily seen by returning to the concept of dipole radiation (Section 3.4.3). Each molecule is taken as an electron oscillator driven into vibration by the incident field. Being far apart, they are assumed to be independent of one another and each radiates in accord with Eq. (3.56). The scattered electric fields are essentially independent, and there is no interference laterally. Accordingly, the net irradiance at P is the algebraic sum of the scattered irradiances from each molecule (p. 289). For an individual scatterer the irradiance is given by Eq. (3.57), and it varies with ω^4.

The advent of the laser has made it relatively easy to observe Rayleigh Scattering directly in low-pressure gases, and the results confirm the theory.

Forward Propagation

To see why the forward direction is special, why the wave advances in any medium, consider Fig. 4.4. Notice that for a forward point P all the different paths taken by the light are about the same length; scattering alters the various path lengths by very little. (The scattered wavelets arrive at P more or less in-phase and essentially interfere constructively.) A more detailed description is provided by Fig. 4.5. It depicts a sequence in time showing two molecules A and B, interacting with an incoming primary plane wave—a solid arc represents a secondary wavelet peak (a positive maximum); a dashed arc corresponds to a trough (a negative maximum). In (*a*), the primary wavefront impinges on molecule A, which begins to scatter a spherical wavelet. For the moment, suppose the wavelet is 180° out-of-phase with the incident wave. (A driven oscillator is usually out-of phase with the driver: p. 92.) Thus A begins to radiate a trough (a negative E-field) in response to being driven by a peak (a positive E-field). Part (*b*) shows the spherical wavelet and the plane wave overlapping, marching out-of-step but marching together. The incident wavefront impinges on B, and it, in turn, begins to reradiate a wavelet, which must also be out-of-phase by 180°. In (*c*) and (*d*), we see the point of all of this, namely, that both wavelets are moving forward—they are in-phase with each other. That condition would be true for all such wavelets regardless of both how many molecules there were and how they were distributed. Because of the asymmetry introduced by the beam itself, ***all the scattered wavelets add constructively with each other in the forward direction.***

4.2.2 The Transmission of Light Through Dense Media

Now, suppose the amount of air in the region under consideration is increased. In fact, imagine that each little cube of air, one wavelength on a side, contains a great many molecules, whereupon it is said to have an appreciable *optical density*. (This usage probably derives from the fact that early experiments on gases indicated that an increase in density is accompanied by a proportionate increase in the index of refraction.) At the wavelengths of light, the Earth's atmosphere at STP has about 3 million molecules in such a λ^3-cube. The scattered wavelets ($\lambda \approx 500$ nm) radiated by sources so close together

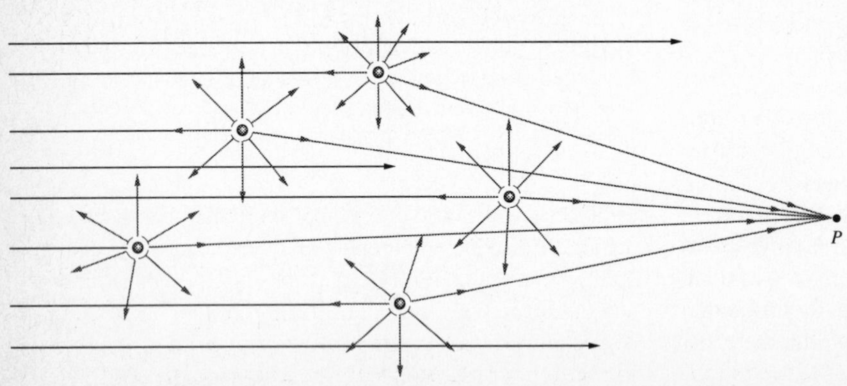

FIGURE 4.4 Scattering in the forward direction doesn't change the light paths very much, and the waves all arrive at P pretty much in-phase.

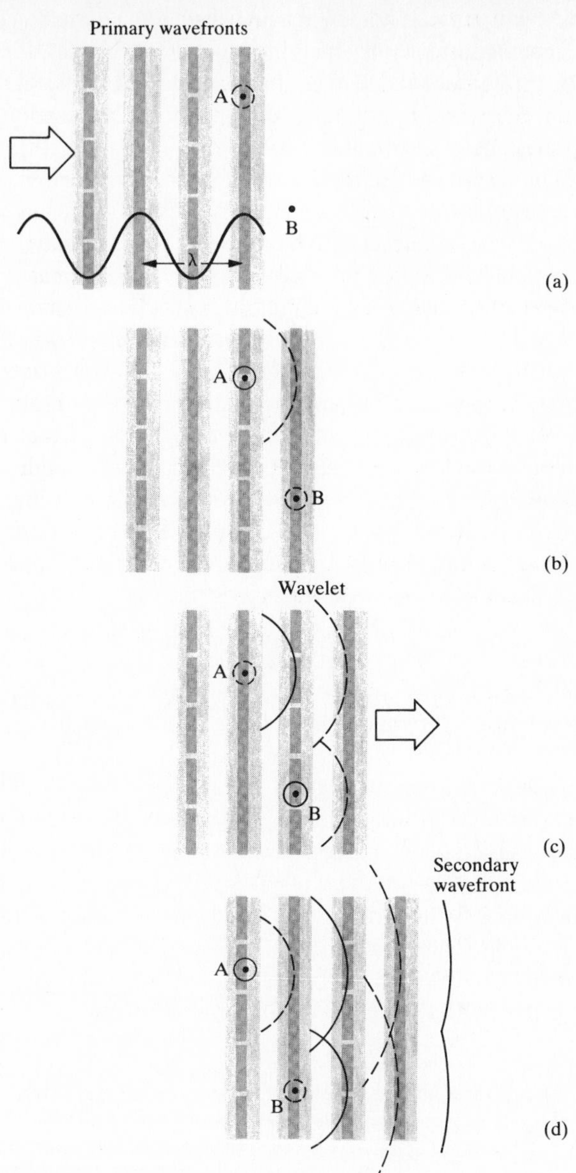

FIGURE 4.5 In the forward direction the scattered wavelets arrive in-phase on planar wavefronts—trough with trough, peak with peak.

($\approx$3 nm) cannot properly be assumed to arrive at some point P with random phases—interference will be important. This is equally true in liquids and solids where the atoms are 10 times closer and arrayed in a far more orderly fashion. In such cases, the light beam effectively encounters a uniform medium with no discontinuities to destroy the symmetry. Again, the scattered wavelets interfere constructively in the forward direction (that much is independent of the arrangement of the molecules), but now destructive interference predominates in all other directions. ***Little or no light ends up scattered laterally or backwards in a dense homogeneous medium.***

To illustrate the phenomenon, Fig. 4.6 shows a beam moving through an ordered array of close-together scatterers. All along wavefronts throughout the beam, sheets of molecules are energized in-phase, radiate, and are re-energized, over and over again as the light sweeps past. Thus some molecule A radiates spherically out of the beam, but because of the ordered close arrangement, there will be a molecule B, a distance $\approx\lambda/2$ away, such that both wavelets cancel in the transverse direction. Here, where λ is thousands of times larger than the scatterers and their spacing, there will likely always

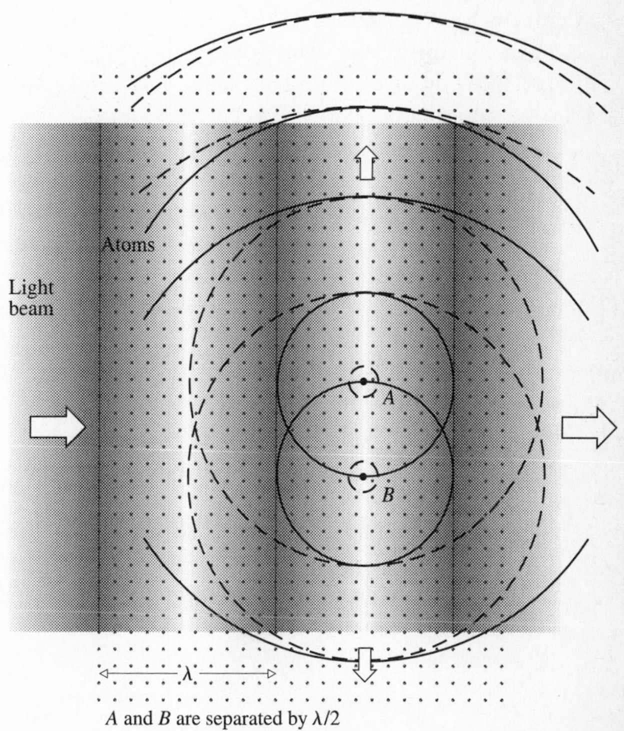

FIGURE 4.6 A plane wave impinging from the left. The medium is composed of many closely spaced atoms. Among countless others, a wavefront stimulates two atoms, A and B, that are very nearly one-half wavelength apart. The wavelets they emit interfere destructively. Trough overlaps crest, and they completely cancel each other in the direction perpendicular to the beam. That process happens over and over again, and little or no light is scattered laterally.

be pairs of molecules that tend to negate each other's wavelets in any given lateral direction. Even if the medium is not perfectly ordered, the net electric field at a point in any transverse direction will be the sum of a great many *tiny* scattered fields, each somewhat out-of-phase with the next, so that the sum (which will be different from point to point) will always be small (Fig. 4.7). This makes sense from the perspective of conservation of energy—we can't have constructive interference in every direction. ***Interference produces a redistribution of energy, out of the regions where it's destructive into the regions where it's constructive.***

The more dense, uniform, and ordered the medium is (the more nearly homogeneous), the more complete will be the lateral destructive interference and the smaller the amount of nonforward scattering. Thus most of the energy will go into the forward direction, and the beam will advance essentially undiminished (Fig. 4.8).

Scattering on a per-molecule basis is extremely weak. In order to have half its energy scattered, a beam of green light will have to traverse $\approx$150 km of atmosphere. Since about 1000 times more molecules are in a given volume of liquid than in the same volume of vapor (at atmospheric pressure), we can expect to see an increase in scattering. Still, the liquid is a far more ordered state with much less pronounced density fluctuations, and that should suppress the nonforward scattering appreciably. Accordingly, an increased scattering per unit volume is observed in liquids, but it's more like 5 to 50 times as much rather than 1000 times. Molecule for molecule, liquids scatter substantially less than gases. Put a few drops of milk in a tank of water and illuminate it with a bright flashlight beam. A faint but unmistakable blue haze will scatter out laterally, and the direct beam will emerge decidedly reddened.

Transparent amorphous solids, such as glass and plastic, will also scatter light laterally, but very weakly. Good crystals, like quartz and mica, with their almost perfectly ordered structures, scatter even more faintly. Of course, imperfections of all sorts (dust and bubbles in liquids, flaws and impurities in solids) will serve as scatterers, and when these are small, as in the gem moonstone, the emerging light will be bluish.

In 1869 John Tyndall experimentally studied the scattering produced by small particles. He found that as the size of the particles increased (from a fraction of a wavelength), the amount of scattering of the longer wavelengths increased proportionately. Ordinary clouds in the sky testify to the fact that relatively large droplets of water scatter white light with no appreciable coloration. The same is true of the microscopic globules of fat and protein in milk.

When the number of molecules in a particle is small, they are all close to one another and act in unison; their wavelets interfere constructively, and the scattering is strong. As the size of the particle approaches a wavelength, the atoms at its extremities no longer radiate wavelets that are necessarily in-phase and the scattering begins to diminish. This happens first at the short wavelengths (blue), and so as the particle size increases, it scatters proportionately more of the red end of the spectrum (and it does so increasingly in the forward direction). The theoretical analysis of scattering from spherical particles of any size was first published by Gustav Mie in 1908. Mie Scattering depends only weakly on wavelength and becomes independent of it (white light in, white light out) when the particle size exceeds λ. Reasonably enough, Rayleigh Scattering is the small-size limiting case of Mie Scattering.

On an overcast day, the sky looks hazy gray because of large water droplets. In the same way, some inexpensive plastic food containers and white garbage-bag plastic look pale blue-white in scattered light and are distinctly orange in transmitted light. The garbage bags, in order to be made opaque, contain (2 to 2.5%) clear TiO_2 spheres (n = 2.76) about 200 nm in diameter, and these Mie scatter bluish-white.*

*It has only recently been observed (and that was by chance) that inhomogeneous opaque materials, such as milk and white paint, can reduce the effective speed of light to as little as one-tenth the value anticipated for the medium. See S. John, "Localization of light," *Phys. Today* **44**, 32 (1991).

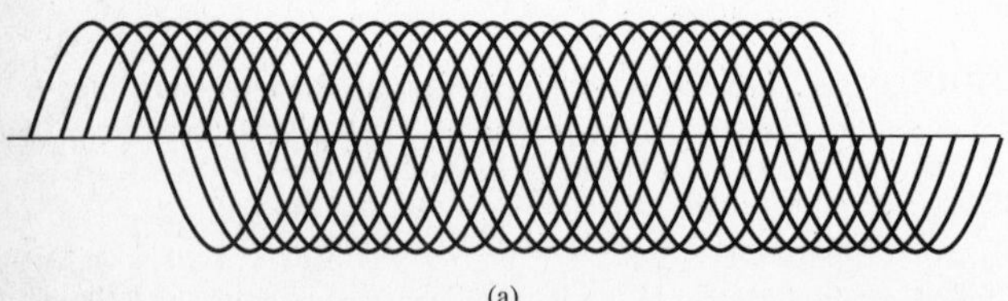

FIGURE 4.7 (*a*) When a great many tiny slightly shifted waves arrive at a point in space, there is generally as much positive *E*-field as negative, and the resultant disturbance is nearly zero. (*b*) The tiny phasors representing those waves form a very small circular figure, and the resultant (which oscillates in a way that depends on the number of waves) is never large.

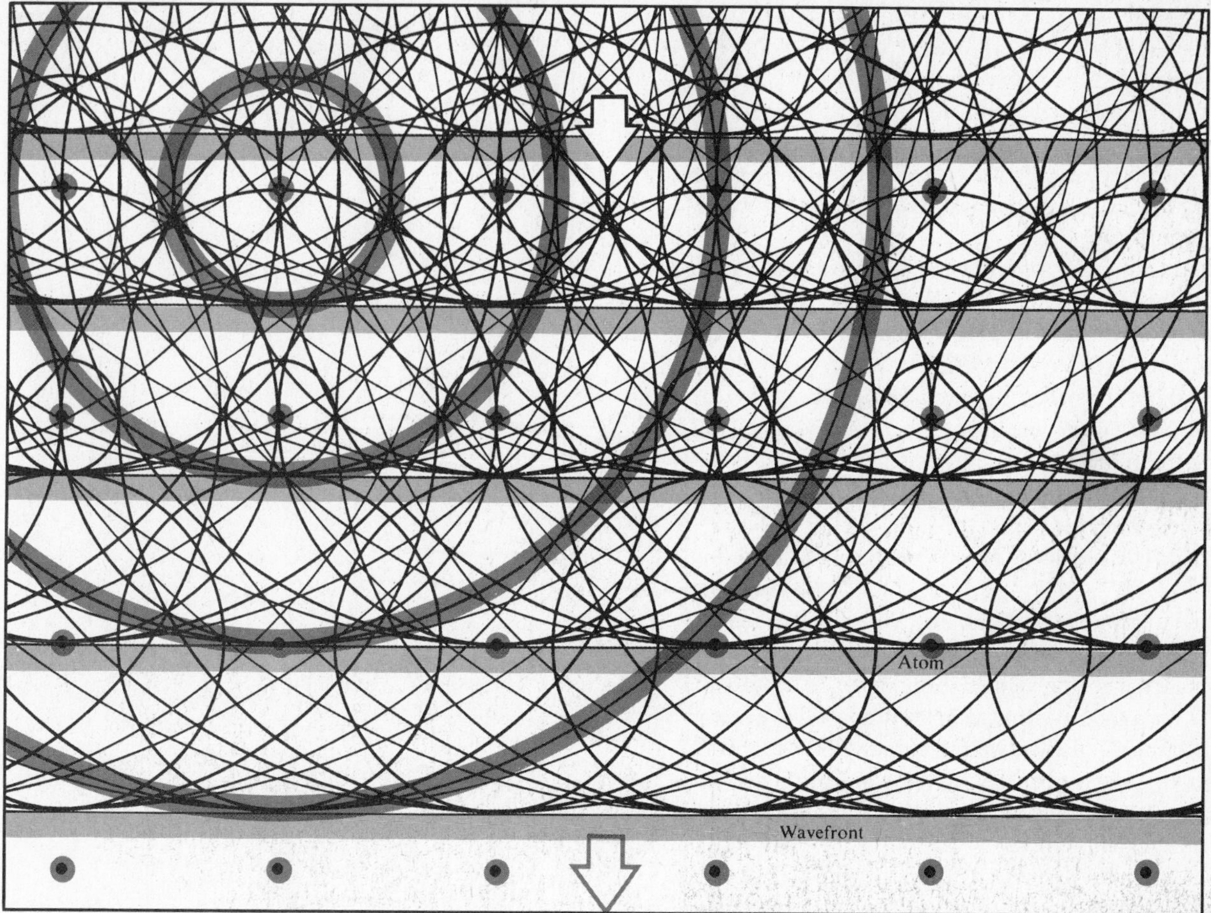

FIGURE 4.8 A downward plane wave incident on an ordered array of atoms. Wavelets scatter in all directions and overlap to form an ongoing secondary plane wave traveling downward.

4.2.3 Transmission and the Index of Refraction

The transmission of light through a homogeneous medium is an ongoing repetitive process of scattering and rescattering. Each such event introduces a phase shift into the light field, which ultimately shows up as a shift in the apparent phase velocity of the transmitted beam from its nominal value of c. That corresponds to an index of refraction for the medium ($n = c/v$) which is other than one, even though **photons exist only at a speed** c.

To see how this comes about, return to Fig. 4.5. Recall that the scattered wavelets all combine in-phase in the forward direction to form what might best be called the *secondary wave*. For empirical reasons alone we can anticipate that the secondary wave will combine with what is left of the primary wave to yield the only observed disturbance within the medium, namely, the **transmitted wave**. *Both the primary and secondary electromagnetic waves propagate through the interatomic void with the speed c.* Yet the medium can certainly possess an index of refraction other than 1. The refracted wave may appear to have a phase velocity less than, equal to, or even greater than c. The key to this apparent contradiction resides in the phase relationship between the secondary and primary waves.

The classical model predicts that the electron-oscillators will be able to vibrate almost completely in-phase with the driving force (i.e., the primary disturbance) only at relatively low frequencies. As the frequency of the electromagnetic field increases, the oscillators will fall behind, lagging in phase by a proportionately larger amount. A detailed analysis reveals that at resonance the phase lag will reach 90°, increasing thereafter to almost 180°, or half a wavelength, at frequencies well above the particular characteristic value. Problem 4.4 explores this phase lag for a damped driven oscillator, and Fig. 4.9 summarizes the results.

In addition to these lags there is another effect that must be considered. When the scattered wavelets recombine, the resultant secondary wave* itself lags the oscillators by 90°.

The combined effect of both these mechanisms is that at frequencies below resonance, the secondary wave lags the primary (Fig. 4.10) by some amount between approximately 90° and 180°, and at frequencies above resonance, the lag ranges from about 180° to 270°. But a phase lag of $\delta \geq 180°$ is equivalent to a phase lead of $360° - \delta$, [e.g., $\cos(\theta - 270°) = \cos(\theta + 90°)$]. This much can be seen on the right side of Fig. 4.9*b*.

Within the transparent medium, the primary and secondary waves overlap and, depending on their amplitudes and relative phase, generate the net transmitted disturbance. Except for the fact that it is weakened by scattering, the primary wave travels into the material just as if it were traversing free space. By comparison to this free-space wave, which initiated the process, the resultant transmitted wave is phase shifted, and this phase difference is crucial.

When the secondary wave lags (or leads) the primary, the resultant transmitted wave must also lag (or lead) it by some amount (Fig. 4.11). This qualitative relationship will serve our purposes for the moment, although it should be noted that the phase of the resultant also depends on the amplitudes of the interacting waves [see Eq. (7.10)]. At frequencies below ω_0 the transmitted wave lags the free-space wave, whereas at frequencies above ω_0 it leads the free-space wave. For the special case in which $\omega = \omega_0$ the secondary and primary waves are out-of-phase by 180°; the former works against the latter, so that the refracted wave is appreciably reduced in amplitude although unaffected in phase.

*This point will be made more plausible when we consider the predictions of the Huygens–Fresnel theory in the diffraction chapter. Most texts on E & M treat the problem of radiation from a sheet of oscillating charges, in which case the 90° phase lag is a natural result (see Problem 4.5).

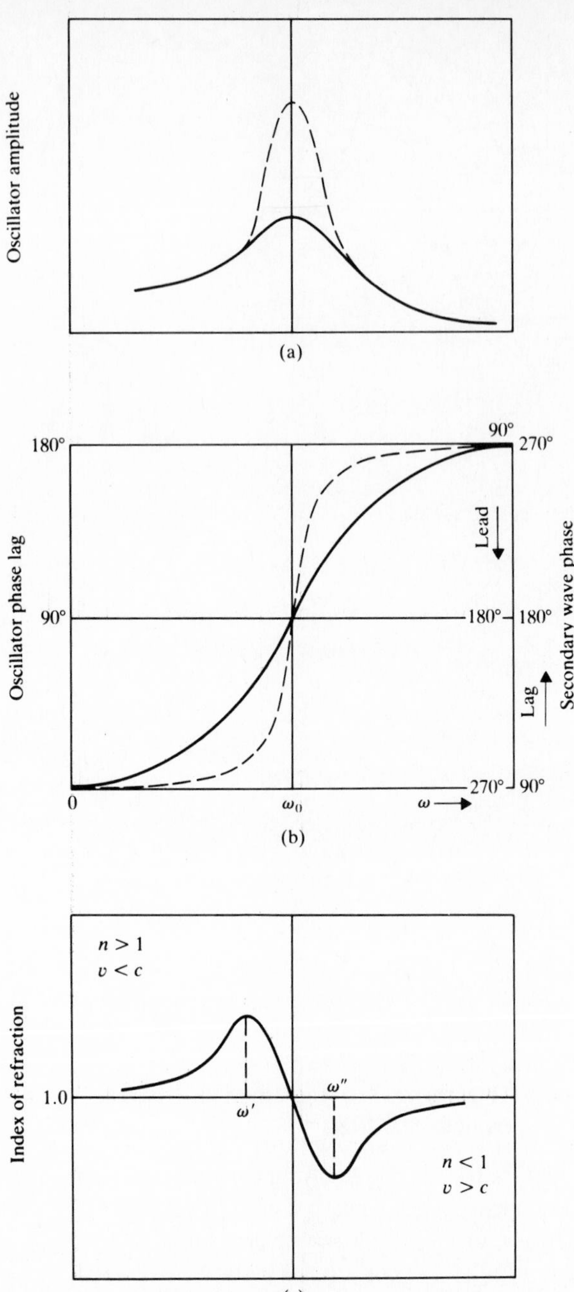

FIGURE 4.9 A schematic representation of (*a*) amplitude and (*b*) phase lag versus driving frequency for a damped oscillator. The dashed curves correspond to decreased damping. The corresponding index of refraction is shown in (*c*).

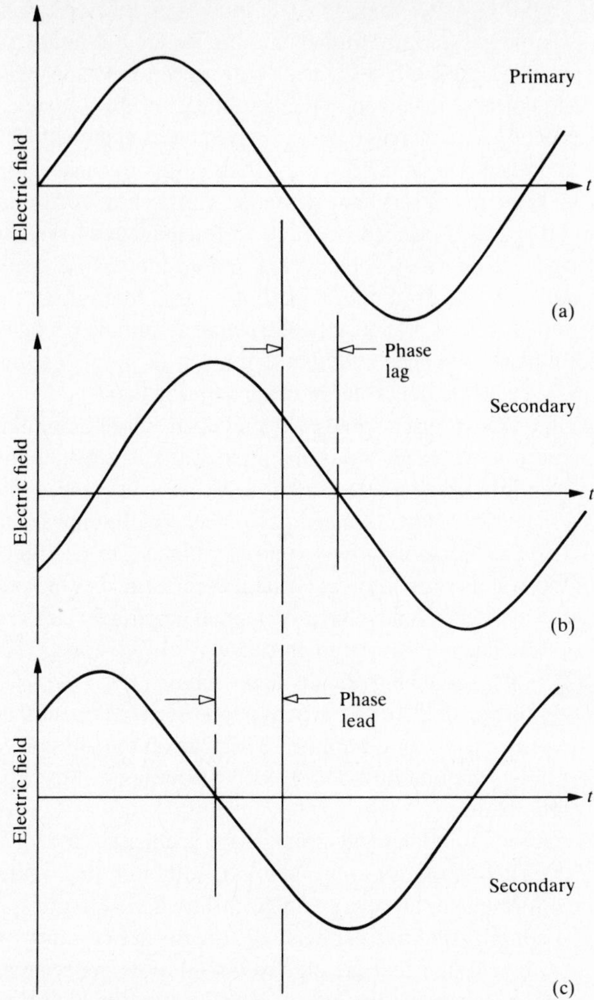

FIGURE 4.10 A primary wave (*a*) and two possible secondary waves. In (*b*) the secondary lags the primary—it takes longer to reach any given value. In (*c*) the secondary wave reaches any given value before (at an earlier time than) the primary; that is, it leads.

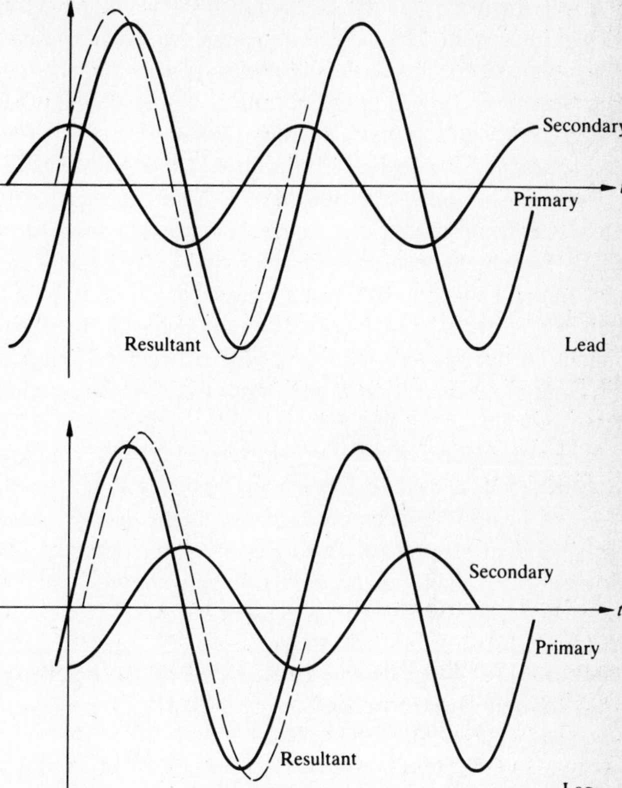

FIGURE 4.11 If the secondary leads the primary, the resultant will also lead it.

As the transmitted wave advances through the medium, scattering occurs over and over again. Light traversing the substance is progressively retarded (or advanced) in phase. Evidently, since the speed of the wave is the rate of advance of the condition of constant phase, a change in the phase should correspond to a change in the speed.

We now wish to show that a phase shift is indeed tantamount to a difference in phase velocity. In free space, the disturbance at some point *P* may be written as

$$E_P(t) = E_0 \cos \omega t \tag{4.1}$$

If *P* is surrounded by a dielectric, there will be a cumulative phase shift ε_P, which was built up as the wave moved through the medium to *P*. At ordinary levels of irradiance the medium will behave linearly, and the frequency in the dielectric will be the same as that in vacuum, even though the wavelength and speed may differ. Once again, but this time in the medium, the disturbance at *P* is

$$E_P(t) = E_0 \cos (\omega t - \varepsilon_P) \tag{4.2}$$

where the subtraction of ε_P corresponds to a phase lag. An observer at *P* will have to wait a longer time for a given crest to arrive when she is in the medium than she would have had to wait in vacuum. That is, if you imagine two parallel waves of the same frequency, one in vacuum and one in the material, the vacuum wave will pass P a time ε_P/ω before the other wave. Clearly then, *a phase lag of ε_P corresponds to a reduction in speed*, $v < c$ and $n > 1$. Similarly, *a phase lead yields an increase in speed*, $v > c$ and $n < 1$. Again, the scattering process is a continuous one, and the cumulative phase shift builds as the light penetrates the medium. That is to say, ε is a function of the length of dielectric traversed, as it must be if v is to be constant (see Problem 4.5). In the vast majority of situations encountered in Optics $v < c$ and $n > 1$; see Table 4.1. The important exception is the case of X-ray propagation where $\omega > \omega_0$, $v > c$, and $n < 1$.

The overall form of $n(\omega)$, as depicted in Fig. 4.9c, can now be understood, as well. At frequencies far below ω_0 the amplitudes of the oscillators and therefore of the secondary waves are very small, and the phase angles are approximately 90°. Consequently, the refracted wave lags only slightly, and n is only slightly greater than 1. As ω increases, the secondary waves have greater amplitudes and lag by greater amounts. The result is a gradually decreasing wave speed and an increasing value of $n > 1$. Although the amplitudes of the secondary waves continue to increase, their relative phases approach 180° as ω approaches ω_0. Consequently, their ability to cause a further increase in the resultant phase lag diminishes. A turning point ($\omega = \omega'$) is reached where the refracted wave begins to experience a decreasing phase lag and an increasing speed ($dn/d\omega < 0$). That continues until $\omega = \omega_0$, whereupon the transmitted wave is appreciably reduced in amplitude but unaltered in phase and speed. At that point, $n = 1$, $v = c$, and we are more or less at the center of the absorption band.

At frequencies just beyond ω_0 the relatively large-amplitude secondary waves lead; the transmitted wave is advanced in phase, and its speed exceeds c ($n < 1$). As ω increases, the whole scenario is played out again in reverse (with some asymmetry due to frequency-dependent asymmetry in oscillator amplitudes and scattering). At even higher frequencies the secondary waves, which now have very small amplitudes, lead by nearly 90°. The resulting transmitted wave is advanced very slightly in phase, and n gradually approaches 1.

The precise shape of a particular $n(\omega)$ curve depends on the specific oscillator damping, as well as on the amount of absorption, which in turn depends on the number of oscillators participating.

A rigorous solution to the propagation problem is known as the *Ewald–Oseen Extinction Theorem*. Although the mathematical formalism, involving integrodifferential equations, is far too complicated to treat here, the results are certainly of interest. It is found that the electron-oscillators generate an electromagnetic wave having essentially two terms. One of these precisely cancels the primary wave within the medium. The other, and only remaining disturbance, moves through the dielectric at a speed $v = c/n$ as the transmitted wave.* ***Henceforth we shall simply assume that a lightwave propagating through any substantive medium travels at a speed $v \neq c$.***

Apparently, any quantum-mechanical model we construct will somehow have to associate a wavelength with the photon. That's easily done mathematically via the expression $p = h/\lambda$, even if it's not clear at this point what is doing the waving. Still the wave nature of light seems inescapable; it will have to

TABLE 4.1 Approximate Indices of Refraction of Various Substances*

Substance	Index
Air	1.000 29
Ice	1.31
Water	1.333
Ethyl alcohol (C_2H_5OH)	1.36
Fused quartz (SiO_2)	1.4584
Carbon tetrachloride (CCl_4)	1.46
Turpentine	1.472
Benzene (C_6H_6)	1.501
Plexiglass	1.51
Crown glass	1.52
Sodium chloride (NaCl)	1.544
Light flint glass	1.58
Polystyrene	1.59
Carbon disulfide (CS_2)	1.628
Dense flint glass	1.66
Lanthanum flint glass	1.80
Zircon ($ZrO_2 \cdot SiO_2$)	1.923
Fabulite ($SrTiO_3$)	2.409
Diamond (C)	2.417
Rutile (TiO_2)	2.907
Gallium phosphide	3.50

*Values vary with physical conditions—purity, pressure, etc. These correspond to a wavelength of 589 nm

*For a discussion of the Ewald–Oseen theorem, see *Principles of Optics* by Born and Wolf, Section 2.4.2; this is heavy reading. Also look at Reali, "Reflection from dielectric materials." *Am. J. Phys.* **50**, 1133 (1982).

be infused into the theory one way or another. And once we have the idea of a photon-wavelength, it's natural to bring in the concept of relative phase. Thus ***the index of refraction arises when the absorption and emission process advances or retards the phases of the scattered photons, even as they travel at speed*** *c*.

4.3 Reflection

When a beam of light impinges on the surface of a transparent material, such as a sheet of glass, the wave "sees" a vast array of closely spaced atoms that will somehow scatter it. Remember that the wave may be ≈500 nm long, whereas the atoms and their separations (≈0.2 nm) are thousands of times smaller. In the case of transmission through a dense medium, the scattered wavelets cancel each other in all but the forward direction and just the ongoing beam is sustained. But that can only happen if there are no discontinuities. This is not the case at an interface between two different transparent media (such as air and glass), which is a jolting discontinuity. When a beam of light strikes such an interface, some light is always scattered backward, and we call this phenomenon **reflection**.

If the transition between two media is gradual—that is, if the dielectric constant (or the index of refraction) changes from that of one medium to that of the other over a distance of a wavelength or more—there will be very little reflection; the interface effectively vanishes. On the other hand, a transition from one medium to the other over a distance of 1/4 wavelength or less behaves very much like a totally discontinuous change.

Internal and External Reflection

Imagine that light is traveling across a large homogeneous block of glass (Fig. 4.12). Now, suppose that the block is sheared in half perpendicular to the beam. The two segments are then separated, exposing the smooth flat surfaces depicted in Fig. 4.12*b*. Just before the cut was made, there was no lightwave traveling to the left inside the glass—we know the beam only advances. Now there must be a wave (beam-I) moving to the left, reflected from the surface of the right-hand block. The implication is that a region of scatterers on and beneath the exposed surface of the right-hand block is now "unpaired," and the backward radiation they emit can no longer be canceled. The region of oscillators that was adjacent to these, prior to the cut, is now on the section of the glass that is to the left. When the two sections were together, these scatterers presumably also emitted wavelets in the backward direction that were 180° out-of-phase with, and canceled, beam-I. Now they produce reflected beam-II. Each molecule scatters light in the backward direction, and, in principle, ***each and every molecule contributes to the reflected wave***. Nonetheless, in practice, it is a thin layer ($\approx\lambda/2$ deep) of unpaired atomic oscillators near the surface that is effectively responsible for the reflection. For an air–glass interface, about 4% of the energy of an incident beam falling perpendicularly *in* air *on* glass will be reflected straight back out by this layer of unpaired scatterers (p. 89). And that's true whether the glass is 1.0 mm thick or 1.00 m thick.

Beam-I reflects off the right-hand block, and because light was initially traveling from a less to a more optically dense medium, this is called **external reflection**. In other words, the index of the incident medium (n_i) is less than the index of the transmitting medium (n_t). Since the same thing happens to the unpaired layer on the section that was moved to the left, it, too, reflects backwards. With the beam incident perpendicularly *in* glass *on* air, 4% must again be reflected, this time as beam-II. This process is referred to as **internal reflection** because $n_i > n_t$. If the two glass regions are made to approach one another increasingly closely (so that we can imagine the gap to be a thin film of, say, air), the reflected light will diminish until it ultimately vanishes as the two faces merge and disappear and the block becomes continuous again. Remember this ***180° rel-***

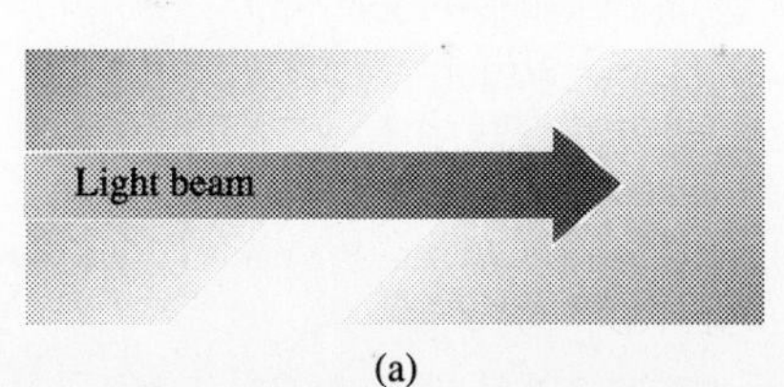

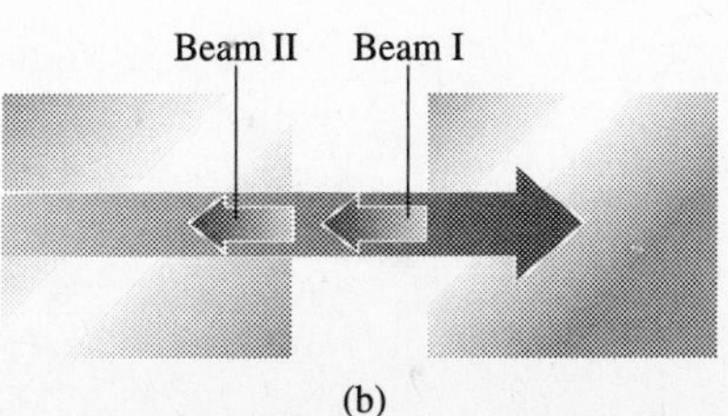

FIGURE 4.12 (*a*) A light beam propagating through a dense homogeneous medium such as glass. (*b*) When the block of glass is cut and parted, the light is reflected backward at the two new interfaces. Beam-I is externally reflected, and beam-II is internally reflected. Ideally, when the two pieces are pressed back together, the two reflected beams cancel one another.

ative phase shift between internally and externally reflected light (see Section 4.10 for a more rigorous treatment)—we will come back to it later on.

Experience with the common mirror makes it obvious that white light is reflected as white—it certainly isn't blue. To see why, first remember that the layer of scatterers responsible for the reflection is effectively about $\lambda/2$ thick (as per Fig. 4.6). Thus the larger the wavelength, the deeper the region contributing (typically upward of a thousand atom layers), and the more scatterers there are acting together. This tends to balance out the fact that each scatterer is less efficient as λ increases (remember $1/\lambda^4$). The combined result is that ***the surface of a transparent medium reflects all wavelengths about equally and doesn't appear colored in any way***. That, as we will see, is why this page looks white under white-light illumination.

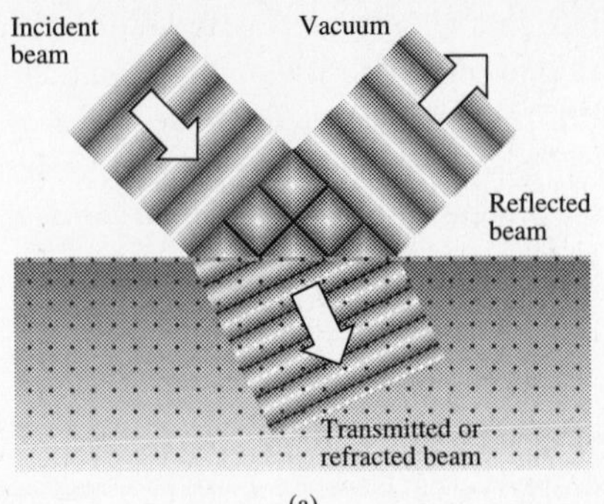

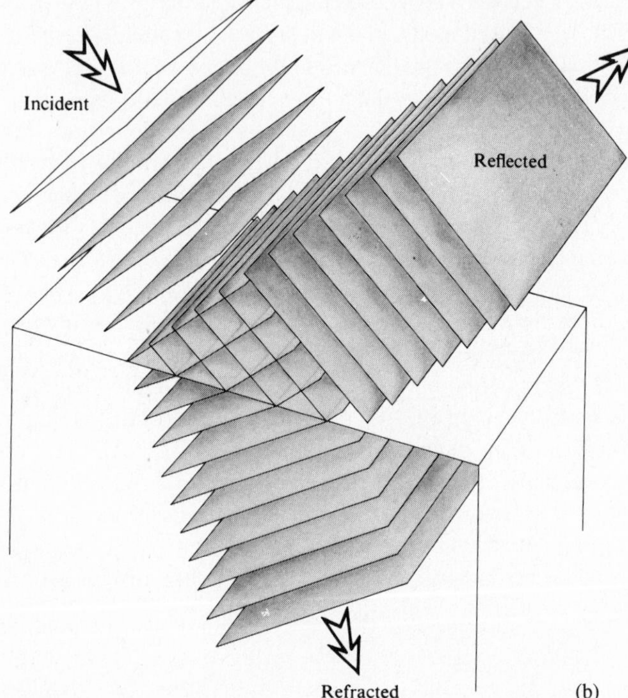

FIGURE 4.13 A beam of plane waves incident on a distribution of molecules constituting a piece of clear glass or plastic. Part of the incident light is reflected and part refracted.

4.3.1 The Law of Reflection

Figure 4.13 shows a beam composed of plane wavefronts impinging at some angle on the smooth, flat surface of an optically dense medium (let it be glass). Assume that the surrounding environment is vacuum. Follow one wavefront as it sweeps in and across the molecules on the surface. For the sake of simplicity, in Figs. 4.14 and 4.15 we have omitted everything but one molecular layer at the interface. As the wavefront descends, it energizes and re-energizes one scatterer after another, each of which radiates a stream of photons that can be regarded as a hemispherical wavelet in the incident medium. Because the wavelength is so much greater than the separation between the molecules, the wavelets emitted back into the incident medium advance together and add constructively in only one direction, and there is one well-defined *reflected* beam. That would not be true if the incident radiation was short-wavelength X-rays, in which circumstance there would be several reflected beams. And it would not be true if the scatterers were far apart compared to λ, as they are for a diffraction grating (p. 465), in which case there would also be several reflected beams. The direction of the reflected beam is determined by the constant phase difference between the atomic scatterers. That, in turn, is determined by the angle made by the incident wave and the surface, the so-called **angle-of-incidence**.

In Fig. 4.16, the line $\overline{AB}$ lies along an incoming wavefront, while $\overline{CD}$ lies on an outgoing wavefront—in effect, $\overline{AB}$ transforms on reflection into $\overline{CD}$. With Fig. 4.15 in mind, we see that the wavelet emitted from A will arrive at C in-phase with the wavelet just being emitted from D (as it is stimulated by

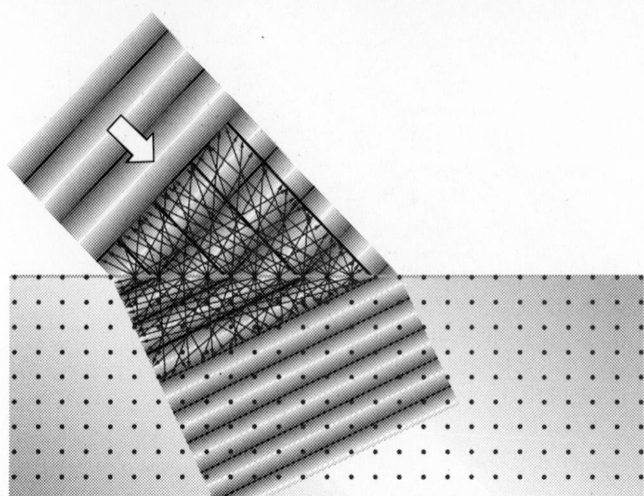

FIGURE 4.14 A plane wave sweeps in stimulating atoms across the interface. These radiate and reradiate almost continuously, thereby giving rise to both the reflected and transmitted waves.

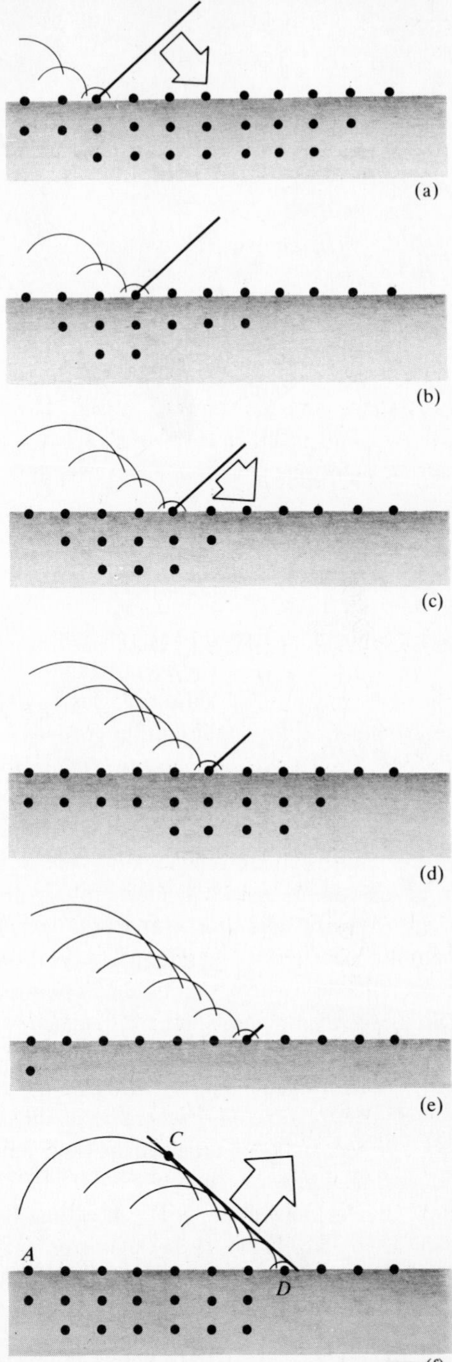

FIGURE 4.15 The reflection of a wave as the result of scattering.

B), as long as the distances $\overline{AC}$ and $\overline{BD}$ are equal. In other words, if all the wavelets emitted from all the surface scatterers are to overlap in-phase and form a single reflected plane wave, it must be that $\overline{AC} = \overline{BD}$. Then, since the two triangles have a common hypotenuse

$$\frac{\sin \theta_i}{\overline{BD}} = \frac{\sin \theta_r}{\overline{AC}}$$

All the waves travel in the incident medium with the same speed v_i. It follows that in the time (Δt) it takes for point B on the wavefront to reach point D on the surface, the wavelet emitted from A reaches point C. In other words, $\overline{BD} = v_i \Delta t = \overline{AC}$, and so from the above equation, $\sin \theta_i = \sin \theta_r$, which means that

$$\theta_i = \theta_r \tag{4.3}$$

The angle-of-incidence equals the angle-of-reflection. This equation is the first part of the **Law of Reflection**. It initially appeared in the book *Catoptrics*, which was purported to have been written by Euclid. We say that a beam is *normally incident* when $\theta_i = 0°$, in which case $\theta_r = 0°$ and for a mirror the beam reflects back on itself. Similarly, *glancing incidence* corresponds to $\theta_i \approx 90°$ and perforce $\theta_r \approx 90°$.

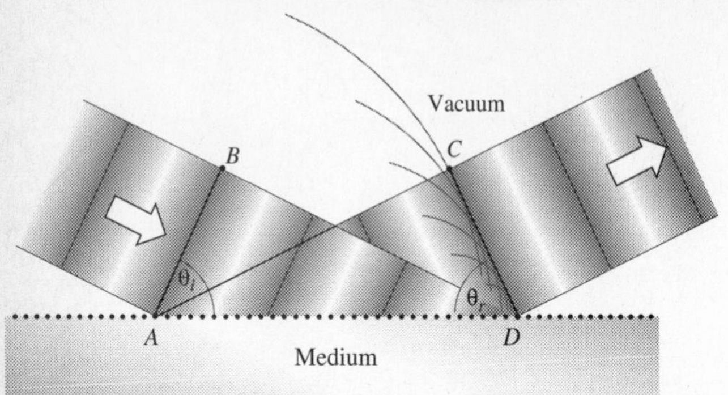

FIGURE 4.16 Plane waves enter from the left and are reflected off to the right. The reflected wavefront $\overline{CD}$ is formed of waves scattered by the atoms on the surface from A to D. Just as the first wavelet arrives at C from A, the atom at D emits, and the wavefront along $\overline{CD}$ is completed.

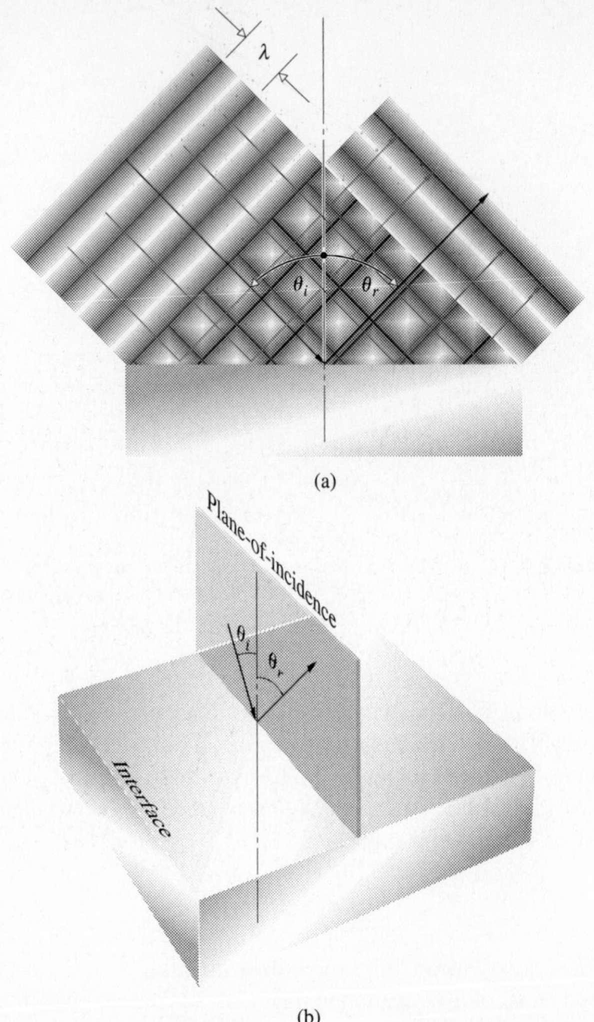

FIGURE 4.17 (*a*) Select one ray to represent the beam of plane waves. Both the angle-of-incidence θ_i and the angle-of-reflection θ_r are measured from a perpendicular drawn to the reflecting surface. (*b*) The incident ray and the reflected ray define the *plane-of-incidence*, perpendicular to the reflecting surface.

RAYS

Drawing wavefronts can get things a bit cluttered, so we introduce another convenient scheme for visualizing the progression of light. The imagery of antiquity was in terms of straight-line streams of light, a notion that got into Latin as "radii" and reached English as "rays." ***A ray is a line drawn in space corresponding to the direction of flow of radiant energy***. It is a mathematical construct and not a physical entity. In a medium that is uniform (homogeneous), rays are straight. If the medium behaves in the same manner in every direction (isotropic), ***the rays are perpendicular to the wavefronts***. Thus for a point source emitting spherical waves, the rays, which are perpendicular to them, point radially outward from the source. Similarly, the rays associated with plane waves are all parallel. Rather than sketching bundles of rays, we can simply draw one incident ray and one reflected ray (Fig. 4.17*a*). *All the angles are now measured from the perpendicular (or normal) to the surface*, and θ_i and θ_r have the same numerical values as before (Fig. 4.16).

The ancient Greeks knew the Law of Reflection. It can be deduced by observing the behavior of a flat mirror, and nowadays that observation can be done most simply with a flashlight or, even better, a low-power laser. The second part of the Law of Reflection maintains that ***the incident ray, the perpendicular to the surface, and the reflected ray all lie in a plane*** called the **plane-of-incidence** (Fig. 4.17*b*)—this is a three-dimensional business. Try to hit some target in a room with a flashlight beam by reflecting it off a stationary mirror, and the importance of this second part of the law becomes obvious!

Figure 4.18*a* shows a beam of light incident upon a reflecting surface that is smooth (one for which any irregularities are small compared to a wavelength). In that case, the light reemitted by millions upon millions of atoms will combine to form a single well-defined beam in a process called **specular**

Specular

Diffuse

(a)

(b)

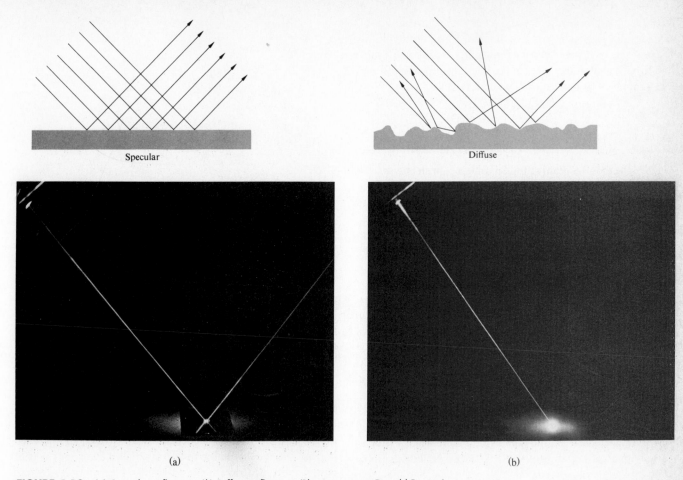

FIGURE 4.18 (*a*) Specular reflection. (*b*) Diffuse reflection. (Photos courtesy Donald Dunitz.)

reflection (from the word for a common mirror alloy in ancient times, speculum). Provided the ridges and valleys are small compared to λ, the scattered wavelets will still arrive more or less in-phase when $\theta_i = \theta_r$. This is the situation assumed in Figs. 4.13, 4.15, 4.16, and 4.17. On the other hand, when the surface is rough in comparison to λ, although the angle-of-incidence will equal the angle-of-reflection for each ray, the whole lot of rays will emerge every which way, constituting what is called **diffuse reflection** (Fig. 4.19). Both of

FIGURE 4.19 The cruiser *Aurora*, which played a key role in the Communist Revolution (1917), docked in St. Petersburg. Where the water is still, the reflection is specular. The image blurs where the water is rough and the reflection diffuse.

these conditions are extremes; the reflecting behavior of most surfaces lies somewhere between them. Thus, although the paper of this page was deliberately manufactured to be a fairly diffuse scatterer, the cover of the book reflects in a manner that is somewhere between diffuse and specular.

4.4 Refraction

Figure 4.13 shows a beam of light impinging on an interface at some angle ($\theta_i \neq 0$). The interface corresponds to a major inhomogeneity, and the atoms that compose it scatter light both backward, as the reflected beam, and forward, as the transmitted beam. The fact that the incident rays are bent or "turned out of their way," as Newton put it, is called **refraction**.

Examine the transmitted or refracted beam. Speaking classically, each energized molecule on the interface radiates wavelets into the glass that expand out at speed c. These can be imagined as combining into a secondary wave that then recombines with the unscattered remainder of the primary wave, to form the net transmitted wave. The process continues over and over again as the wave advances in the transmitting medium.

However we visualize it, immediately on entering the transmitting medium, there is a single net field, a single net wave. As we have seen, this transmitted wave usually propagates with an effective speed $v_t < c$. It's essentially as if the atoms at the interface scattered "slow wavelets" into the glass that combine to form the "slow transmitted wave." We'll come back to this imagery when we talk about Huygens's Principle. In any event, because the cooperative phenomenon known as the transmitted electromagnetic wave *is* slower than the incident electromagnetic wave, the transmitted wavefronts are refracted, displaced (turned with respect to the incident wavefronts), and the beam bends.

4.4.1 The Law of Refraction

Figure 4.20 picks up where we left off with Figs. 4.13 and 4.16. The diagram depicts several wavefronts, all shown at a single instant in time. Remember that each wavefront is a surface of constant phase, and, to the degree that the phase of the net field is retarded by the transmitting medium, each wavefront is held back, as it were. The wavefronts "bend" as they

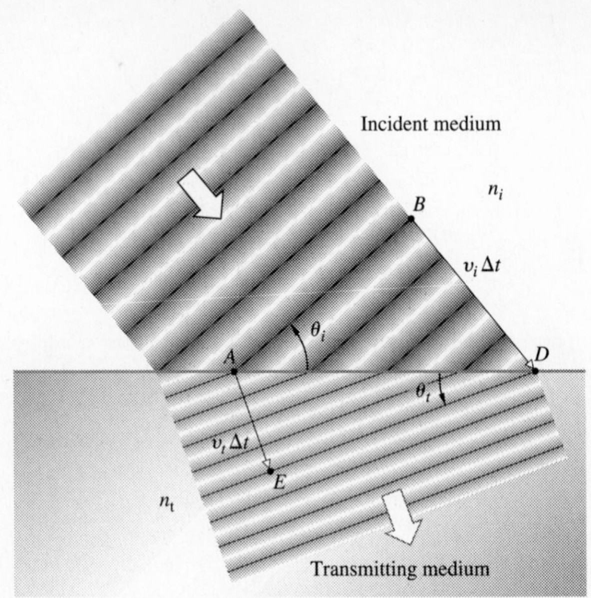

FIGURE 4.20 The refraction of waves. The atoms in the region of the surface of the transmitting medium reradiate wavelets that combine constructively to form a refracted beam.

cross the boundary because of the speed change. Alternatively, we can envision Fig. 4.20 as a multiple-exposure picture of a single wavefront showing it after successive equal intervals of time. Notice that in the time Δt, which it takes for point B on a wavefront (traveling at speed v_i) to reach point D, the transmitted portion of that same wavefront (traveling at speed v_t) has reached point E. If the glass ($n_t = 1.5$) is immersed in an incident medium that is vacuum ($n_i = 1$) or air ($n_i =$ 1.000 3) or anything else where $n_t > n_i$, $v_t < v_i$ and $\overline{AE} < \overline{BD}$, the wavefront bends. The refracted wavefront extends from E to D, making an angle with the interface of θ_t. As before, the two triangles ABD and AED in Fig. 4.19 share a common hypotenuse ($\overline{AD}$), and so

$$\frac{\sin \theta_i}{\overline{BD}} = \frac{\sin \theta_t}{\overline{AE}}$$

where $\overline{BD} = v_i\,\Delta t$ and $\overline{AE} = v_t\,\Delta t$. Hence

$$\frac{\sin \theta_i}{v_i} = \frac{\sin \theta_t}{v_t}$$

Multiply both sides by c, and since $n_i = c/v_i$ and $n_t = c/v_t$

$$n_i \sin \theta_i = n_t \sin \theta_t \tag{4.4}$$

This equation is the first portion of the **Law of Refraction**, also known as **Snell's Law** after the man who proposed it, Willebrord Snel van Royen (1591–1626). At first the indices of refraction were simply experimentally determined constants of the physical media. Later, Newton was actually able to derive Snell's Law using his own corpuscular theory. By then, the significance of n as a measure of the speed of light was evident. Still later, Snell's Law was shown to be a natural consequence of Maxwell's Electromagnetic Theory (p. 111).

It is again convenient to transform the diagram into a ray representation (Fig. 4.21) wherein all the angles are measured from the perpendicular. Along with Eq. (4.4), there goes the understanding that ***the incident, reflected, and refracted rays all lie in the plane-of-incidence***. In other words, the respective unit propagation vectors $\hat{\mathbf{k}}_i$, $\hat{\mathbf{k}}_r$, and $\hat{\mathbf{k}}_t$ are coplanar (Fig. 4.22).

When $n_i < n_t$ (that is, when the light is initially traveling within the lower-index medium), it follows from Snell's Law that $\sin \theta_i > \sin \theta_t$, and since the same function is everywhere positive between 0° and 90°, then $\theta_i > \theta_t$. Rather than going straight through, ***the ray entering a higher-index medium bends toward the normal*** (Fig. 4.23*a*). The reverse is also true (Fig. 4.23*b*); that is, ***on entering a medium having a lower index, the ray, rather than going straight through, will bend away from the normal*** (Fig. 4.24). Notice that this implies that the rays will traverse the same path going either way, into or out of either medium. The arrows can be reversed and the resulting picture is still true.

Snell's Law can be rewritten in the form

$$\frac{\sin\theta_i}{\sin\theta_t} = n_{ti} \tag{4.5}$$

where $n_{ti} \equiv n_t/n_i$ is the **relative index of refraction** of the two media.

Let $\hat{\mathbf{u}}_n$ be a unit vector normal to the interface pointing in the direction from the incident to the transmitting medium

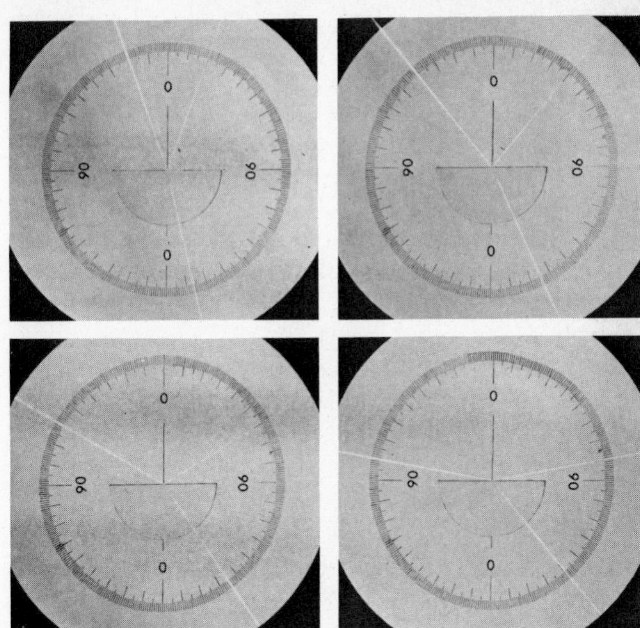

FIGURE 4.22 Refraction at various angles of incidence. Notice that the bottom surface is cut circular so that the transmitted beam within the glass always lies along a radius and is normal to the lower surface in every case. (Photos courtesy *PSSC College Physics*, © 1965 Education Development Center Inc.)

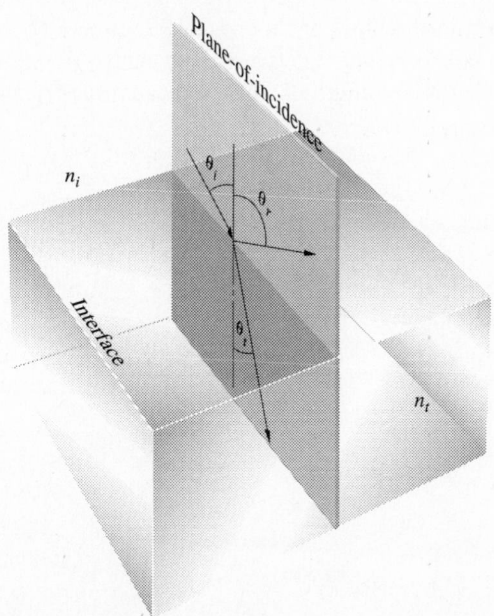

FIGURE 4.21 The incident, reflected, and transmitted beams each lie in the plane-of-incidence.

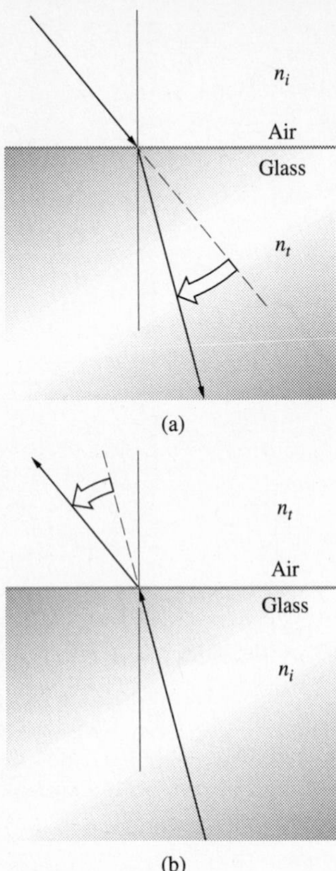

FIGURE 4.23 The bending of rays at an interface. (a) When a beam of light enters a more optically dense medium, one with a greater index of refraction ($n_i < n_t$), it bends toward the perpendicular. (b) When a beam goes from a more dense to a less dense medium ($n_i > n_t$), it bends away from the perpendicular.

FIGURE 4.24 Rays from the submerged portion of the pencil bend on leaving the water as they rise toward the viewer. See Problem 4.16.

(Fig. 4.25). As you will have the opportunity to prove in Problem 4.25, the complete statement of the Law of Refraction can be written vectorially as

$$n_i(\hat{\mathbf{k}}_i \times \hat{\mathbf{u}}_n) = n_t(\hat{\mathbf{k}}_t \times \hat{\mathbf{u}}_n) \tag{4.6}$$

or alternatively,

$$n_t\hat{\mathbf{k}}_t - n_i\hat{\mathbf{k}}_i = (n_t \cos \theta_t - n_i \cos \theta_i)\, \hat{\mathbf{u}}_n \tag{4.7}$$

Fig. 4.20 illustrates the three important changes that occur in the beam traversing the interface. (1) It changes direction. Because the leading portion of the wavefront in the glass slows down, the part still in the air advances more rapidly, sweeping past and bending the wave toward the normal. (2) The beam in the glass has a broader cross section than the beam in the air; hence, the transmitted energy is spread thinner. (3) The wavelength decreases because the frequency is

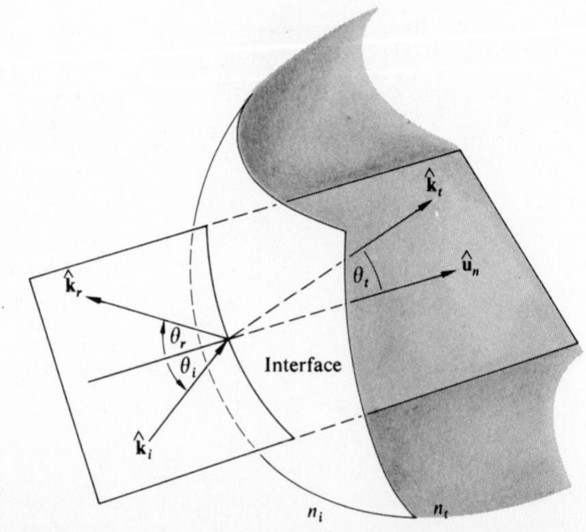

FIGURE 4.25 The ray geometry.

unchanged while the speed decreases; $\lambda = v/\nu = c/n\nu$ and

$$\lambda = \frac{\lambda_0}{n} \tag{4.8}$$

This latter notion suggests that the color aspect of light is better thought of as associated with its frequency (or energy, $\mathscr{E} = h\nu$) than its wavelength, since the frequency changes with the medium through which the light moves. Color is so much a physio-psychological phenomenon (p. 131) that it must be treated rather gingerly. Still, even though it's a bit simplistic, it's useful to remember that blue photons are more energetic than red photons. When we talk about wavelengths and colors, we should always be referring to **vacuum wavelengths** (henceforth to be given as λ_0).

In all the situations treated thus far, it was assumed that the reflected and refracted beams always had the same frequency as the incident beam, and that's ordinarily a reasonable assumption. Light of frequency ν impinges on a medium and presumably drives the molecules into simple harmonic motion. That's certainly the case when the amplitude of the vibration is fairly small, as it is when the electric field driving the molecules is small. The E-field for bright sunlight is only about 1000 V/m (while the B-field is less than a tenth of the Earth's surface field). This isn't very large compared to the fields keeping a crystal together, which are of the order of 10^{11} V/m—just about the same magnitude as the cohesive field holding the electron in an atom. We can usually expect the oscillators to vibrate in simple harmonic motion, and so the frequency will remain constant—the medium will ordinarily respond linearly. That will not be true, however, if the incident beam has an exceedingly large-amplitude E-field, as can be the case with a high-power laser. So driven, at some frequency ν the medium can behave in a nonlinear fashion, resulting in reflection and refraction of harmonics (2ν, 3ν, etc.) in addition to ν. Nowadays, second-harmonic generators (p. 636) are available commercially. You shine red light (694.3 nm) into an appropriately oriented transparent nonlinear crystal (of, for example, potassium dihydrogen phosphate, KDP, or ammonium dihydrogen phosphate, ADP) and out will come a beam of UV (347.15 nm).

One feature of the above treatment merits some further discussion. It was reasonably assumed that each point on the interface in Fig. 4.13 coincides with a particular point on each of the incident, reflected, and transmitted waves. In other words, there is a fixed phase relationship between each of the waves at all points along the interface. As the incident front sweeps across the interface, every point on it in contact with the interface is also a point on both a corresponding reflected front and a corresponding transmitted front. This situation is known as **wavefront continuity**, and it will be justified in a more mathematically rigorous treatment in Section 4.6.1. Interestingly, Sommerfeld* has shown that the laws of reflection and refraction (independent of the kind of wave involved) can be derived directly from the requirement of wavefront continuity and the solution to Problem 4.22 demonstrates as much.

4.4.2 Huygens's Principle

Suppose that light passes through a nonuniform sheet of glass, as in Fig. 4.26, so that the wavefront Σ is distorted. How can we determine its new form Σ'? Or for that matter, what will Σ' look like at some later time, if it is allowed to continue unobstructed?

A preliminary step toward the solution of this problem appeared in print in 1690 in the work entitled *Traité de la Lumière*, which had been written 12 years earlier by the Dutch physicist Christiaan Huygens. It was there that he enunciated what has since become known as **Huygens's Principle:** ***every point on a propagating wavefront serves as the source of spherical secondary wavelets, such that the wavefront at some later time is the envelope of these wavelets.***

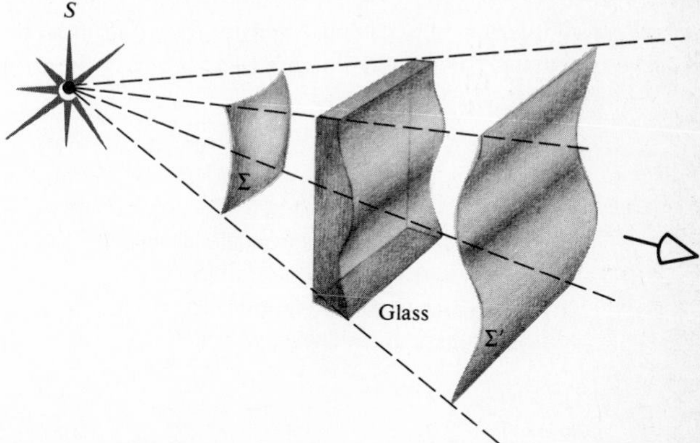

FIGURE 4.26 Distortion of a portion of a wavefront on passing through a material of nonuniform thickness.

*A. Sommerfeld, *Optics*, p. 151. See also J. J. Sein, *Am. J. Phys.* **50**, 180 (1982).

A further crucial point is that ***if the propagating wave has a frequency ν, and is transmitted through the medium at a speed v_t, then the secondary wavelets have that same frequency and speed.*** Huygens was a brilliant scientist, and this is the basis of a remarkably insightful, though quite naive, scattering theory. It's a very early treatment and naturally has several shortcomings, one of which is that it doesn't overtly incorporate the concept of interference, and perforce cannot deal with lateral scattering. Moreover, the idea that the secondary wavelets propagate at a speed determined by the medium (a speed that may even be anisotropic, e.g., p. 335) is a happy guess. Nonetheless, Huygens's Principle can be used to arrive at Snell's Law in a way that's similar to the treatment that led to Eq. (4.4). It's probably best not to fuss over the physical details (such as how to rationalize propagation in vacuum) and just use the Principle as a tool—a highly useful fiction that works. After all, if Einstein is right, there are only scattered photons; the wavelets themselves are a theoretical construct.

If the medium is homogeneous, the wavelets may be constructed with finite radii, whereas if it is inhomogeneous, the wavelets must have infinitesimal radii. Figure 4.27 should make this fairly clear; it shows a view of a wavefront Σ, as well as a number of spherical secondary wavelets, which, after a time t, have propagated out to a radius of vt. The envelope of all these wavelets is then asserted to correspond to the advanced wave Σ'. It is easy to visualize the process in terms of mechanical vibrations of an elastic medium. Indeed this is the way that Huygens envisioned it within the context of an all-pervading aether, as is evident from this comment by him:

> We have still to consider, in studying the spreading out of these waves, that each particle of matter in which a wave proceeds not only communicates its motion to the next particle to it, which is on the straight line drawn from the luminous point, but that it also necessarily gives a motion to all the others which touch it and which oppose its motion. The result is that around each particle there arises a wave of which this particle is a center.

Fresnel, in the 1800s, successfully modified Huygens's Principle mathematically adding in the concept of interference. A little later on, Kirchhoff showed that the *Huygens–Fresnel Principle* was a direct consequence of the differential wave equation [Eq. (2.60)], thereby putting it on a firm mathematical base. That there was a need for a reformulation of the principle is evident from Fig. 4.27, where we deceptively only drew hemispherical wavelets.* Had we drawn them as spheres, there would have been a *backwave* moving toward the source—something that is not observed. Since this difficulty was taken care of theoretically by Fresnel and Kirchhoff, we need not be disturbed by it.

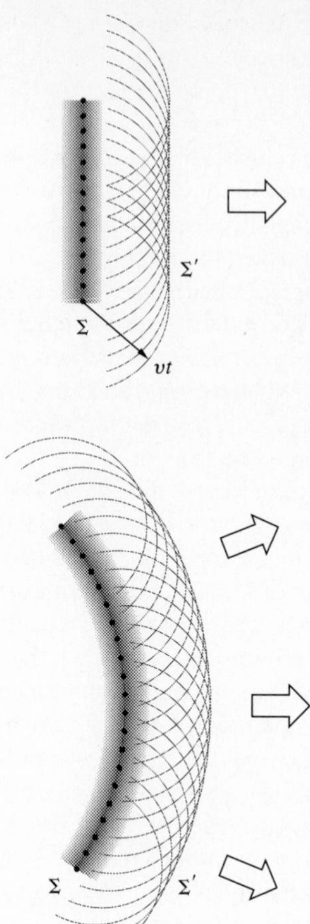

FIGURE 4.27 According to Huygens's Principle, a wave propagates as if the wavefront were composed of an array of point sources, each emitting a spherical wave.

4.4.3 Light Rays and Normal Congruence

In practice, one can produce very narrow *beams* or *pencils* of light (e.g., a laserbeam), and we might imagine a ray to be the

*See E. Hecht, *Phys. Teach.* 18, 149 (1980).

unattainable limit on the narrowness of such a beam. Bear in mind that in an *isotropic medium* (i.e., one whose properties are the same in all directions) ***rays are orthogonal trajectories of the wavefronts***. That is to say, *they are lines normal to the wavefronts at every point of intersection.* Evidently, *in such a medium a ray is parallel to the propagation vector* **k**. As you might suspect, this is not true in *anisotropic* substances, which we will consider later (see Section 8.4.1). *Within homogeneous isotropic materials, rays will be straight lines*, since by symmetry they cannot bend in any preferred direction, there being none. Moreover, because the speed of propagation is identical in all directions within a given medium, the spatial separation between two wavefronts, measured along rays, must be the same everywhere.* Points where a single ray intersects a set of wavefronts are called *corresponding points*, for example, A, A', and A'' in Fig. 4.28. *Evidently the separation in time between any two corresponding points on any two sequential wavefronts is identical.* If wavefront Σ is transformed into Σ'' after a time t'', the distance between corresponding points on any and all rays will be traversed in that same time t''. This will be true even if the wavefronts pass from one homogeneous isotropic medium into another. This just means that each point on Σ can be imagined as following the path of a ray to arrive at Σ'' in the time t''.

If a group of rays is such that we can find a surface that is orthogonal to each and every one of them, they are said to form a **normal congruence**. For example, the rays emanating from a point source are perpendicular to a sphere centered at the source and consequently form a normal congruence.

We can now briefly consider a scheme that will also allow us to follow the progress of light through various isotropic media. The basis for this approach is the ***Theorem of Malus and Dupin*** (introduced in 1808 by E. Malus and modified in 1816 by C. Dupin), according to which ***a group of rays will preserve its normal congruence after any number of reflections and refractions*** (as in Fig. 4.28). From our present vantage point of the wave theory, this is equivalent to the statement that rays remain orthogonal to wavefronts throughout all propagation processes in isotropic media. As shown in Problem 4.24, the theorem can be used to derive the Law of Reflection as well as Snell's Law. It is often most convenient to carry out a ray trace through an optical system and then reconstruct the wavefronts using the idea of equal transit times between corresponding points and the orthogonality of the rays and wavefronts.

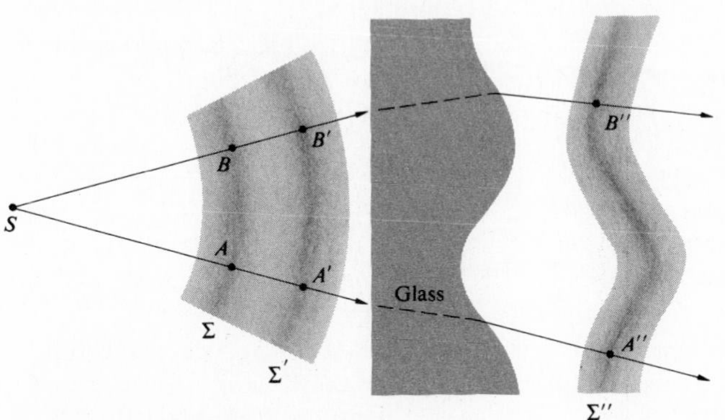

FIGURE 4.28 Wavefronts and rays.

*When the material is inhomogeneous or when there is more than one medium involved, it will be the *optical path length* (see Section 4.5) between the two wavefronts that is the same.

4.5 Fermat's Principle

The laws of reflection and refraction, and indeed the manner in which light propagates in general, can be viewed from an entirely different and intriguing perspective afforded us by **Fermat's Principle**. The ideas that will unfold presently have had a tremendous influence on the development of physical thought in and beyond the study of Classical Optics.

Hero of Alexandria, who lived some time between 150 B.C.E. and A.D. 250, was the first to propose what has since become known as a *variational principle*. In his treatment of reflection, he asserted that *the path taken by light in going from some point S to a point P via a reflecting surface was the shortest possible one*. This can be seen rather easily in Fig. 4.29, which depicts a point source S emitting a number of rays that are then "reflected" toward P. Presumably, only one of these paths will have any physical reality. If we draw the rays as if they emanated from S' (the image of S), none of the distances to P will have been altered (i.e., $SAP = S'AP$, $SBP = S'BP$, etc.). But obviously the straight-line path $S'BP$, which corresponds to $\theta_i = \theta_r$, is the shortest possible one. The same kind of reasoning (Problem 4.27) makes it evident that points S, B, and P must lie in what has previously been defined as the plane-of-incidence. For over fifteen hundred years Hero's curious observation stood alone, until in 1657 Fermat propounded his celebrated *Principle of Least Time*, which encom-

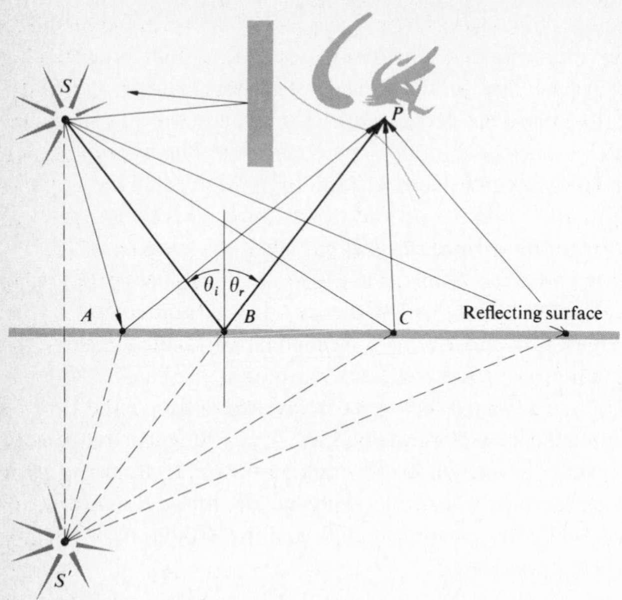

FIGURE 4.29 Minimum path from the source *S* to the observer's eye at *P*.

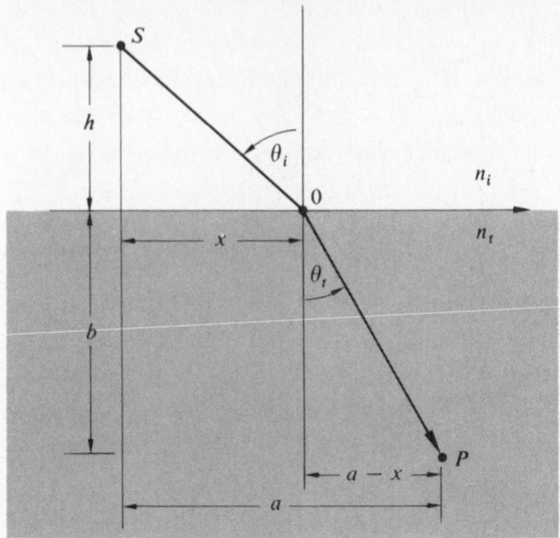

FIGURE 4.30 Fermat's Principle applied to refraction.

passed both reflection and refraction. A beam of light traversing an interface does not take a straight line or *minimum spatial path* between a point in the incident medium and one in the transmitting medium. Fermat consequently reformulated Hero's statement to read: *the actual path between two points taken by a beam of light is the one that is traversed in the least time*. As we shall see, even this form of the statement is incomplete and a bit erroneous at that. For the moment then, let us embrace it but not passionately.

As an example of the application of the principle to the case of refraction, refer to Fig. 4.30, where we minimize t, the transit time from S to P, with respect to the variable x. In other words, changing x shifts point O, changing the ray from S to P. The smallest transit time will then presumably coincide with the actual path. Hence

$$t = \frac{\overline{SO}}{v_i} + \frac{\overline{OP}}{v_t}$$

or

$$t = \frac{(h^2 + x^2)^{1/2}}{v_i} + \frac{[b^2 + (a - x)^2]^{1/2}}{v_t}$$

To minimize $t(x)$ with respect to variations in x, we set $dt/dx = 0$, that is,

$$\frac{dt}{dx} = \frac{x}{v_i(h^2 + x^2)^{1/2}} + \frac{-(a - x)}{v_t[b^2 + (a - x)^2]^{1/2}} = 0$$

Using the diagram, we can rewrite the expression as

$$\frac{\sin\theta_i}{v_i} = \frac{\sin\theta_t}{v_t}$$

which is no less than Snell's Law (Eq. 4.4). If a beam of light is to advance from S to P in the least possible time, it must comply with the Law of Refraction.

Suppose that we have a stratified material composed of m layers, each having a different index of refraction, as in Fig. 4.31. The transit time from S to P will then be

$$t = \frac{s_1}{v_1} + \frac{s_2}{v_2} + \cdots + \frac{s_m}{v_m}$$

or

$$t = \sum_{i=1}^{m} s_i/v_i$$

where s_i and v_i are the path length and speed, respectively, associated with the ith contribution. Thus

$$t = \frac{1}{c}\sum_{i=1}^{m} n_i s_i \tag{4.9}$$

in which the summation is known as the **optical path length** (OPL) traversed by the ray, in contrast to the spatial path length $\sum_{i=1}^{m} s_i$. Clearly, for an inhomogeneous medium where

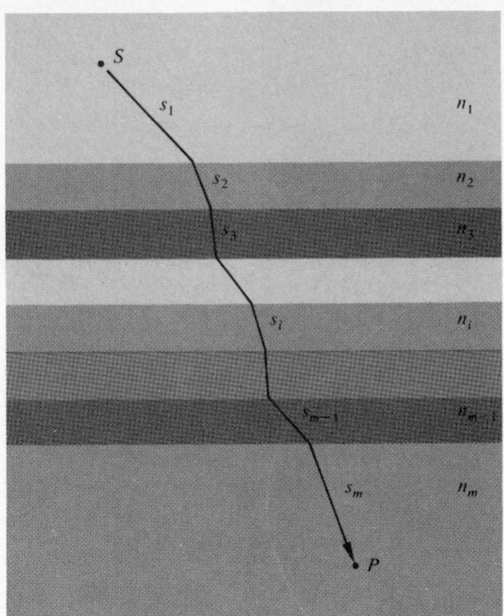

FIGURE 4.31 A ray propagating through a layered material.

n is a function of position, the summation must be changed to an integral:

$$OPL = \int_S^P n(s)\, ds \tag{4.10}$$

The optical path length corresponds to the distance in vacuum equivalent to the distance traversed (s) in the medium of index n. That is, the two will correspond to the same number of wavelengths, $(OPL)/\lambda_0 = s/\lambda$, and the same phase change as the light advances.

Inasmuch as $t = (OPL)/c$, we can restate Fermat's Principle: *light, in going from point S to P, traverses the route having the smallest optical path length*. Accordingly, when light rays from the Sun pass through the inhomogeneous atmosphere of the Earth, as shown in Fig. 4.32*a*, they bend so as to traverse the lower, denser regions as abruptly as possible, minimizing the *OPL*. Ergo, one can still see the Sun after it has actually passed below the horizon. In the same way, a road viewed at a glancing angle, as in Fig. 4.32*b*, appears to reflect the environs as if it were covered with a sheet of water. The air near the roadway is warmer and less dense than that farther above it. Rays bend upward, taking the shortest optical path, and in so doing they appear to be reflected from a mirrored surface. The effect is particularly easy to see on long modern highways. The only requirement is that you look at the road at near glancing incidence, because the rays bend very gradually.

The original statement of Fermat's *Principle of Least Time* has some serious failings and is in need of alteration. To that end, recall that if we have a function, say $f(x)$, we can determine the specific value of the variable x that causes $f(x)$ to have a *stationary* value by setting $df/dx = 0$ and solving for x. By a stationary value, we mean one for which the slope of $f(x)$ versus x is zero or equivalently where the function has a maximum, minimum, or a point of inflection with a horizontal tangent.

Fermat's Principle in its modern form reads: ***a light ray in going from point S to point P must traverse an optical path length that is stationary with respect to variations of that path.*** In essence what that means is that the curve of the *OPL* versus x will have a somewhat flattened region in the vicinity of where the slope goes to zero. The zero-slope point corresponds to the actual path taken. In other words, the *OPL* for the true trajectory will equal, to a first approximation, the *OPL*

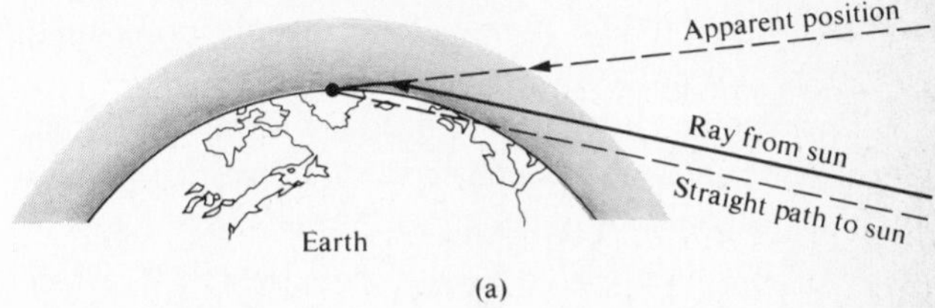

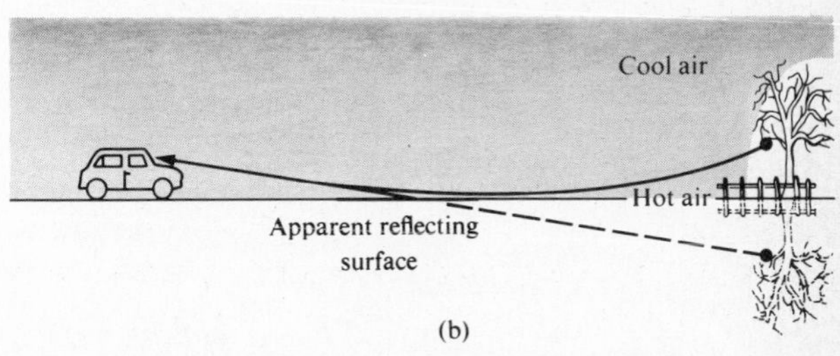

FIGURE 4.32 The bending of rays through inhomogeneous media. (*a*) Because the rays bend as they pass through the atmosphere the Sun appears higher in the sky. (*b*) At very low angles the rays appear to be coming from beneath the road as if reflected in a puddle. (*c*) The effect can be understood via sound waves where the speed, and therefore the wavelength increases in the less dense medium. That bends the wavefronts and the rays.

of paths immediately adjacent to it.* For example, in a situation where the *OPL* is a minimum, as with the refraction illustrated in Fig. 4.30, the *OPL* curve will look something like Fig. 4.33. A small change in x in the vicinity of O has little effect on the *OPL*, but a similar change in x anywhere well away from O results in a substantial change in *OPL*. Thus there will be many paths neighboring the actual one that would take nearly the same time for the light to traverse. This latter insight makes it possible to begin to understand how light manages to be so clever in its meanderings.

Suppose that a beam of light advances through a homogeneous isotropic medium (Fig. 4.34) so that a ray passes from points S to P. Atoms within the material are driven by the incident disturbance, and they reradiate in all directions. Wavelets progressing along paths in the immediate vicinity of a stationary straight-line path will reach P by routes that differ only slightly in *OPL*. They will therefore arrive nearly in-phase and reinforce each other. Wavelets taking other paths away from the stationary one will arrive at P appreciably out-of-phase with each other and will therefore tend to cancel. That being the case, energy will effectively propagate along that ray from S to P that satisfies Fermat's Principle. And this is true whether we're talking about interfering electromagnetic waves or photon probability amplitudes (p. 137).

To see that the *OPL* for a ray need not always be a minimum, examine Fig. 4.35, which depicts a segment of a hollow three-dimensional ellipsoidal mirror. If the source S and the observer P are at the foci of the ellipsoid, then by definition the length SQP will be constant, regardless of where on the perimeter Q happens to be. It is also a geometrical property of the ellipse that $\theta_i = \theta_r$ for any location of Q. All optical paths from S to P via a reflection are therefore precisely equal—none is a minimum, and the *OPL* is clearly stationary with respect to variations. Rays leaving S and striking the mirror will arrive at the focus P. From another viewpoint we can say that radiant energy emitted by S will be scattered by electrons in the mirrored surface such that the wavelets will substantially reinforce each other only at P, where they have traveled the same distance and have the same phase. In any case, if a plane mirror was tangent to the ellipse at Q, the exact same path SQP traversed by a ray would then be a relative minimum. At the other extreme, if the mirrored surface conformed to a curve lying within the ellipse, like the dashed one shown, that same ray along SQP would now negotiate a relative maximum *OPL*. This is true even though other unused paths (where $\theta_i \neq \theta_r$)

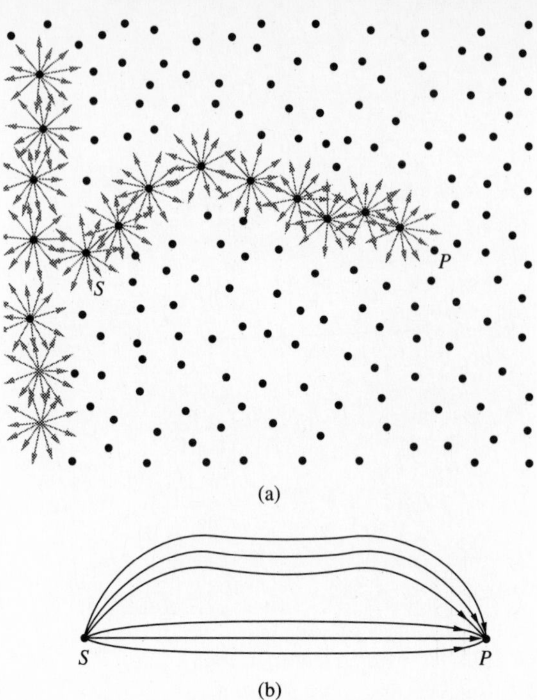

FIGURE 4.34 (*a*) Light can presumably take any number of paths from S to P, but it apparently takes only the one that corresponds to a stationary *OPL*. All other routes effectively cancel out. (*b*) For example, if some light takes each of the three upper paths in the diagram, it arrives at P with three very different phases and interferes more or less destructively.

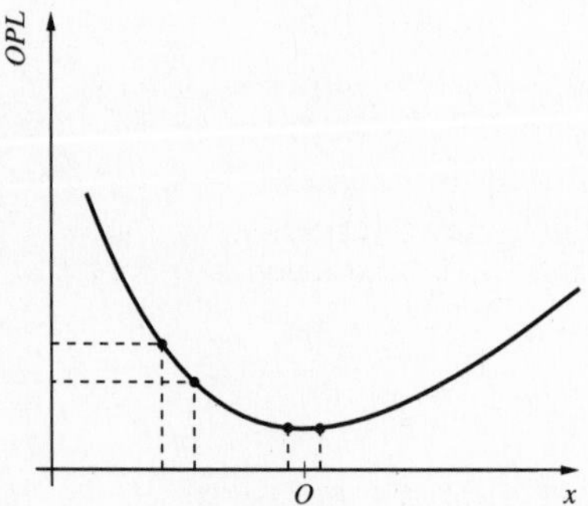

FIGURE 4.33 In the situation shown in Fig. 4.30 the actual location of point O corresponds to a path of minimum *OPL*.

*The first derivative of the *OPL* vanishes in its Taylor series expansion, since the path is stationary.

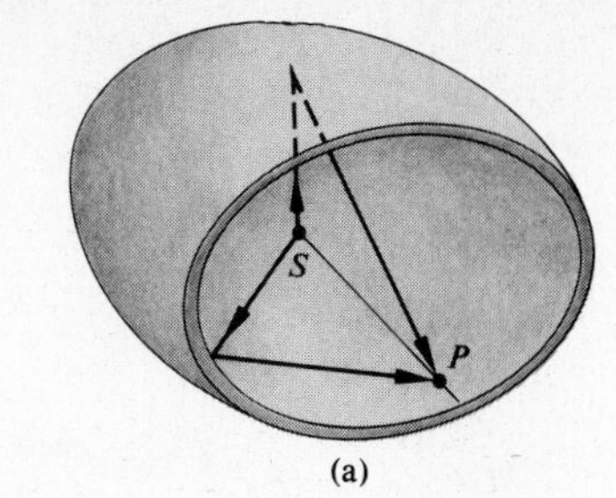

(a)

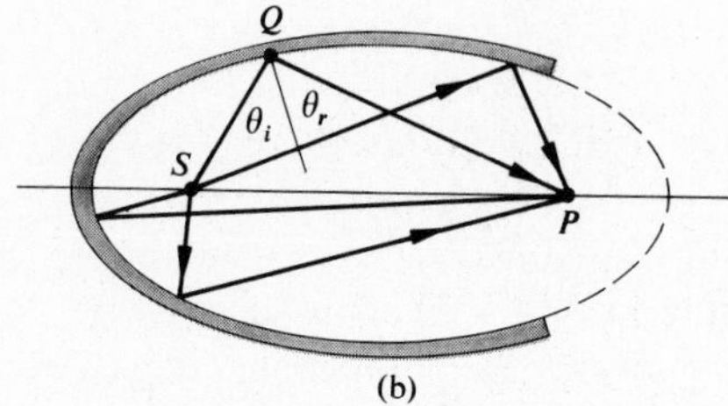

(b)

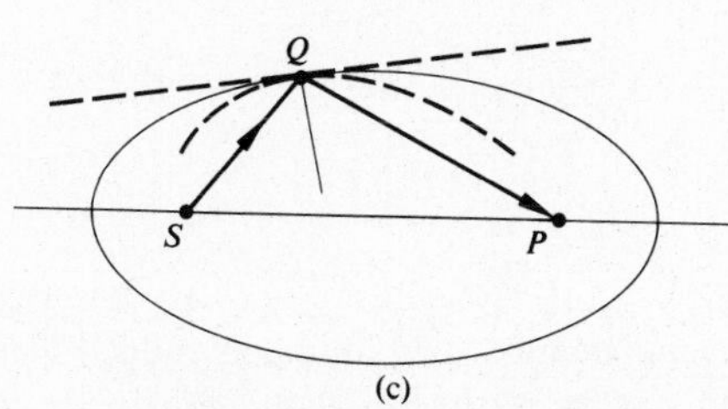

(c)

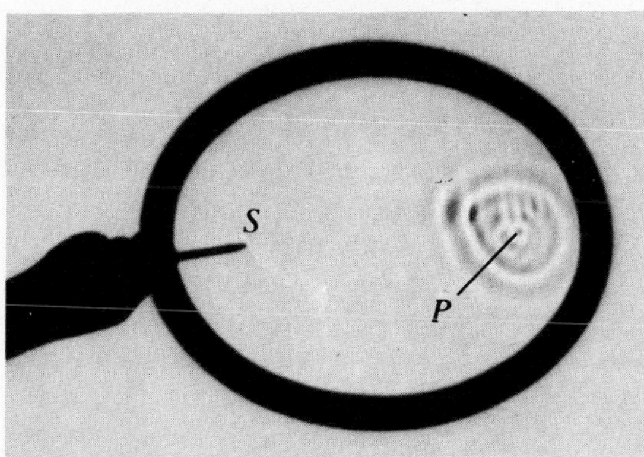

FIGURE 4.35 Reflection off an ellipsoidal surface. Observe the reflection of waves using a frying pan filled with water. Even though these are usually circular, it is well worth playing with. (Photo courtesy *PSSC College Physics*, © 1965 Education Development Center Inc.)

would actually be shorter (i.e., apart from inadmissible curved paths). Thus in all cases the rays travel a stationary *OPL* in accord with the reformulated Fermat's Principle. Note that since the Principle speaks only about the path and not the direction along it, a ray going from *P* to *S* will trace the same route as one from *S* to *P*. This is the very useful *Principle of Reversibility*.

Fermat's achievement stimulated a great deal of effort to supersede Newton's laws of mechanics with a similar variational formulation. The work of many men, notably Pierre de Maupertuis (1698–1759) and Leonhard Euler, finally led to the mechanics of Joseph Louis Lagrange (1736–1813) and hence to the *Principle of Least Action*, formulated by William Rowan Hamilton (1805–1865). The striking similarity between the principles of Fermat and Hamilton played an important part in Schrödinger's development of Quantum Mechanics. In 1942 Richard Phillips Feynman (1918–1988) showed that Quantum Mechanics can be fashioned in an alternative way using a variational approach. The continuing evolution of variational principles brings us back to Optics via the modern formalism of Quantum Optics.

Fermat's Principle is not so much a computational device as it is a concise way of thinking about the propagation of light. It is a statement about the grand scheme of things without any concern for the contributing mechanisms, and as such it will yield insights under a myriad of circumstances.

4.6 The Electromagnetic Approach

Thus far, we have studied reflection and refraction from the perspectives of Scattering Theory, the Theorem of Malus and Dupin, and Fermat's Principle. Yet another and even more powerful approach is provided by Electromagnetic Theory. Unlike the previous techniques, which say nothing about the incident, reflected, and transmitted radiant flux densities (i.e., I_i, I_r, I_t, respectively), Electromagnetic Theory treats these within the framework of a far more complete description.

4.6.1 Waves at an Interface

Suppose that the incident monochromatic lightwave is planar, so that it has the form

$$\mathbf{E}_i = \mathbf{E}_{0i} \exp\,[i(\mathbf{k}_i \cdot \mathbf{r} - \omega_i t)] \tag{4.11}$$

or, more simply,

$$\mathbf{E}_i = \mathbf{E}_{0i} \cos\,(\mathbf{k}_i \cdot \mathbf{r} - \omega_i t) \tag{4.12}$$

Assume that $\mathbf{E}_{0i}$ is constant in time; that is, the wave is linearly or plane polarized. We'll find in Chapter 8 that any form of light can be represented by two orthogonal linearly polarized waves, so that this doesn't actually represent a restriction. Note that just as the origin in time, $t = 0$, is arbitrary, so too is the origin O in space, where $\mathbf{r} = 0$. Thus, making no assumptions about their directions, frequencies, wavelengths, phases, or amplitudes, we can write the reflected and transmitted waves as

$$\mathbf{E}_r = \mathbf{E}_{0r} \cos(\mathbf{k}_r \cdot \mathbf{r} - \omega_r t + \varepsilon_r) \tag{4.13}$$

and

$$\mathbf{E}_t = \mathbf{E}_{0t} \cos(\mathbf{k}_t \cdot \mathbf{r} - \omega_t t + \varepsilon_t) \tag{4.14}$$

Here ε_r and ε_t are *phase constants* relative to $\mathbf{E}_i$ and are introduced because the position of the origin is not unique. Figure 4.36 depicts the waves in the vicinity of the planar interface between two homogeneous lossless dielectric media of indices n_i and n_t.

The laws of Electromagnetic Theory (Section 3.1) lead to certain requirements that must be met by the fields, and they are referred to as the *boundary conditions*. Specifically, one of these is that the component of the electric field **E** that is tangent to the interface must be continuous across it (the same is true for **H**). In other words, the total tangential component of **E** on one side of the surface must equal that on the other

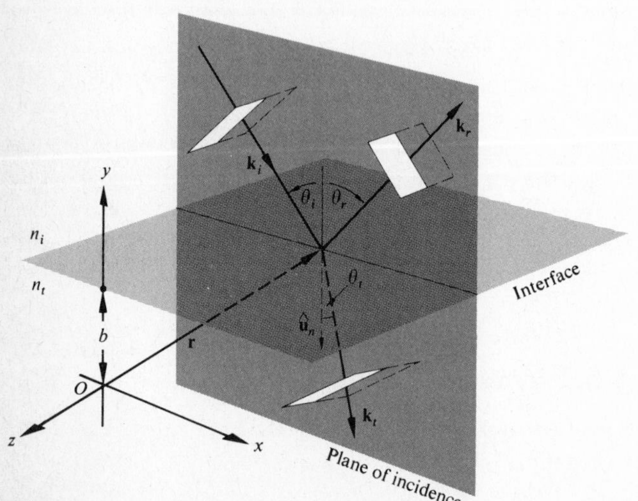

FIGURE 4.36 Plane waves incident on the boundary between two homogeneous, isotropic, lossless dielectric media.

(Problem 4.33). Thus since $\hat{\mathbf{u}}_n$ is the unit vector normal to the interface, regardless of the direction of the electric field within the wavefront, the cross-product of it with $\hat{\mathbf{u}}_n$ will be perpendicular to $\hat{\mathbf{u}}_n$ and therefore tangent to the interface. Hence

$$\hat{\mathbf{u}}_n \times \mathbf{E}_i + \hat{\mathbf{u}}_n \times \mathbf{E}_r = \hat{\mathbf{u}}_n \times \mathbf{E}_t \tag{4.15}$$

or

$$\hat{\mathbf{u}}_n \times \mathbf{E}_{0i} \cos(\mathbf{k}_i \cdot \mathbf{r} - \omega_i t) + \hat{\mathbf{u}}_n \times \mathbf{E}_{0r} \cos(\mathbf{k}_r \cdot \mathbf{r} - \omega_r t + \varepsilon_r) = \hat{\mathbf{u}}_n \times \mathbf{E}_{0t} \cos(\mathbf{k}_t \cdot \mathbf{r} - \omega_t t + \varepsilon_t) \tag{4.16}$$

This relationship must obtain at any instant in time and at any point on the interface ($y = b$). Consequently, $\mathbf{E}_i$, $\mathbf{E}_r$, and $\mathbf{E}_t$ must have precisely the same functional dependence on the variables t and r, which means that

$$(\mathbf{k}_i \cdot \mathbf{r} - \omega_i t)|_{y=b} = (\mathbf{k}_r \cdot \mathbf{r} - \omega_r t + \varepsilon_r)|_{y=b} = (\mathbf{k}_t \cdot \mathbf{r} - \omega_t t + \varepsilon_t)|_{y=b} \tag{4.17}$$

With this as the case, the cosines in Eq. (4.16) cancel, leaving an expression independent of t and r, as indeed it must be. Inasmuch as this has to be true for all values of time, the coefficients of t must be equal, to wit

$$\omega_i = \omega_r = \omega_t \tag{4.18}$$

Recall that the electrons within the media are undergoing (linear) forced vibrations at the frequency of the incident wave. Whatever light is scattered has that same frequency. Furthermore,

$$(\mathbf{k}_i \cdot \mathbf{r})|_{y=b} = (\mathbf{k}_r \cdot \mathbf{r} + \varepsilon_r)|_{y=b} = (\mathbf{k}_t \cdot \mathbf{r} + \varepsilon_t)|_{y=b} \tag{4.19}$$

wherein **r** terminates on the interface. The values of ε_r and ε_t correspond to a given position of O, and thus they allow the relation to be valid regardless of that location. (For example, the origin might be chosen such that **r** was perpendicular to $\mathbf{k}_i$ but not to $\mathbf{k}_r$ or $\mathbf{k}_t$.) From the first two terms we obtain

$$[(\mathbf{k}_i - \mathbf{k}_r) \cdot \mathbf{r}]_{y=b} = \varepsilon_r \tag{4.20}$$

Recalling Eq. (2.43), this expression simply says that the endpoint of **r** sweeps out a plane (which is of course the interface) perpendicular to the vector $(\mathbf{k}_i - \mathbf{k}_r)$. To phrase it slightly differently, $(\mathbf{k}_i - \mathbf{k}_r)$ is parallel to $\hat{\mathbf{u}}_n$. Notice, however, that since the incident and reflected waves are in the same medium, $k_i = k_r$. From the fact that $(\mathbf{k}_i - \mathbf{k}_r)$ has no component in the plane of the interface, that is, $\hat{\mathbf{u}}_n \times (\mathbf{k}_i - \mathbf{k}_r) = 0$, we conclude that

$$k_i \sin \theta_i = k_r \sin \theta_r$$

Hence we have the Law of Reflection, that is,

$$\theta_i = \theta_r$$

Furthermore, since $(\mathbf{k}_i - \mathbf{k}_r)$ is parallel to $\hat{\mathbf{u}}_n$ all three vectors, $\mathbf{k}_i$, $\mathbf{k}_r$, and $\hat{\mathbf{u}}_n$, are in the same plane, the plane of incidence. Again, from Eq. (4.19)

$$[(\mathbf{k}_i - \mathbf{k}_t) \cdot \mathbf{r}]_{y=b} = \varepsilon_t \tag{4.21}$$

and therefore $(\mathbf{k}_i - \mathbf{k}_t)$ is also normal to the interface. Thus $\mathbf{k}_i$, $\mathbf{k}_r$, $\mathbf{k}_t$, and $\hat{\mathbf{u}}_n$ are all coplanar. As before, the tangential components of $\mathbf{k}_i$ and $\mathbf{k}_t$ must be equal, and consequently

$$k_i \sin \theta_i = k_t \sin \theta_t \tag{4.22}$$

But because $\omega_i = \omega_t$, we can multiply both sides by c/ω_i to get

$$n_i \sin \theta_i = n_t \sin \theta_t$$

which is Snell's Law. Finally, if we had chosen the origin O to be in the interface, it is evident from Eqs. (4.20) and (4.21) that ε_r and ε_t would both have been zero. That arrangement, though not as instructive, is certainly simpler, and we'll use it from here on.

4.6.2 The Fresnel Equations

We have just found the relationship that exists among the phases of $\mathbf{E}_i(\mathbf{r}, t)$, $\mathbf{E}_r(\mathbf{r}, t)$, and $\mathbf{E}_t(\mathbf{r}, t)$ at the boundary. There is still an interdependence shared by the amplitudes $\mathbf{E}_{0i}$, $\mathbf{E}_{0r}$, and $\mathbf{E}_{0t}$, which can now be evaluated. To that end, suppose that a plane monochromatic wave is incident on the planar surface separating two isotropic media. Whatever the polarization of the wave, we shall resolve its **E**- and **B**-fields into components parallel and perpendicular to the plane-of-incidence and treat these constituents separately.

Case 1: E perpendicular to the plane-of-incidence. Assume that **E** is perpendicular to the plane-of-incidence and that **B** is parallel to it (Fig. 4.37). Recall that $E = vB$, so that

$$\hat{\mathbf{k}} \times \mathbf{E} = v\mathbf{B} \tag{4.23}$$

and,

$$\hat{\mathbf{k}} \cdot \mathbf{E} = 0 \tag{4.24}$$

(i.e., **E**, **B**, and the unit propagation vector $\hat{\mathbf{k}}$ form a right-handed system). Again, making use of the continuity of the tangential components of the **E**-field, we have at the boundary at any time and any point

$$\mathbf{E}_{0i} + \mathbf{E}_{0r} = \mathbf{E}_{0t} \tag{4.25}$$

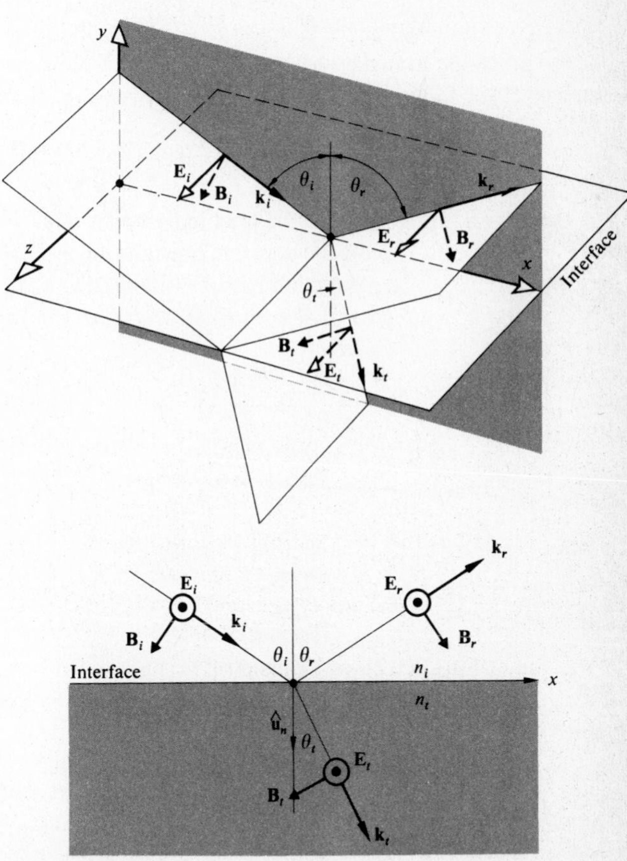

FIGURE 4.37 An incoming wave whose E-field is normal to the plane-of-incidence.

where the cosines cancel. Realize that the field vectors as shown really ought to be envisioned at $y = 0$ (i.e., at the surface), from which they have been displaced for the sake of clarity. Note too that although $\mathbf{E}_r$ and $\mathbf{E}_t$ must be normal to the plane of incidence by symmetry, *we are guessing that they point outward* at the interface when $\mathbf{E}_i$ does. The directions of the **B**-fields then follow from Eq. (4.23).

We will need to invoke another of the boundary conditions in order to get one more equation. The presence of material substances that become electrically polarized by the wave has a definite effect on the field configuration. Thus, although the tangential component of **E** is continuous across the boundary, its normal component is not. Instead, the normal component of the product $\epsilon\mathbf{E}$ is the same on either side of the interface. Similarly, the normal component of **B** is continuous, as is the tangential component of $\mu^{-1}\mathbf{B}$. Here the magnetic effect of the two media appears via their permeabilities μ_i and μ_t. This boundary condition will be the simplest to use, particularly as applied to reflection from the surface of a conductor.* Thus the continuity of the tangential component of $\mathbf{B}/\mu$ requires that

$$-\frac{B_i}{\mu_i}\cos\theta_i + \frac{B_r}{\mu_i}\cos\theta_r = -\frac{B_t}{\mu_t}\cos\theta_t \tag{4.26}$$

where the left and right sides are the total magnitudes of $\mathbf{B}/\mu$ parallel to the interface in the incident and transmitting media, respectively. The positive direction is that of increasing x, so that the scalar components of $\mathbf{B}_i$ and $\mathbf{B}_t$ appear with minus signs. From Eq. (4.23) we have

$$B_i = E_i/v_i \tag{4.27}$$

$$B_r = E_r/v_r \tag{4.28}$$

and

$$B_t = E_t/v_t \tag{4.29}$$

Since $v_i = v_r$ and $\theta_i = \theta_r$, Eq. (4.26) can be written as

$$\frac{1}{\mu_i v_i}(E_i - E_r)\cos\theta_i = \frac{1}{\mu_t v_t}E_t\cos\theta_t \tag{4.30}$$

Making use of Eqs. (4.12), (4.13), and (4.14) and remembering that the cosines therein equal one another at $y = 0$, we obtain

$$\frac{n_i}{\mu_i}(E_{0i} - E_{0r})\cos\theta_i = \frac{n_t}{\mu_t}E_{0t}\cos\theta_t \tag{4.31}$$

*In keeping with our intent to use only the E- and B-fields, at least in the early part of this exposition, we have avoided the usual statements in terms of H, where

$$\mathbf{H} = \mu^{-1}\mathbf{B} \tag{A1.14}$$

Combined with Eq. (4.25), this yields

$$\left(\frac{E_{0r}}{E_{0i}}\right)_\perp = \frac{\dfrac{n_i}{\mu_i}\cos\theta_i - \dfrac{n_t}{\mu_t}\cos\theta_t}{\dfrac{n_i}{\mu_i}\cos\theta_i + \dfrac{n_t}{\mu_t}\cos\theta_t} \tag{4.32}$$

and

$$\left(\frac{E_{0t}}{E_{0i}}\right)_\perp = \frac{2\dfrac{n_i}{\mu_i}\cos\theta_i}{\dfrac{n_i}{\mu_i}\cos\theta_i + \dfrac{n_t}{\mu_t}\cos\theta_t} \tag{4.33}$$

The $\perp$ subscript serves as a reminder that we are dealing with the case in which **E** is perpendicular to the plane of incidence. These two expressions, *which are completely general statements applying to any linear, isotropic, homogeneous media*, are two of the **Fresnel Equations**. Most often one deals with dielectrics for which $\mu_i \approx \mu_t \approx \mu_0$; consequently, the common form of these equations is simply

$$r_\perp \equiv \left(\frac{E_{0r}}{E_{0i}}\right)_\perp = \frac{n_i\cos\theta_i - n_t\cos\theta_t}{n_i\cos\theta_i + n_t\cos\theta_t} \tag{4.34}$$

and

$$t_\perp \equiv \left(\frac{E_{0t}}{E_{0i}}\right)_\perp = \frac{2n_i\cos\theta_i}{n_i\cos\theta_i + n_t\cos\theta_t} \tag{4.35}$$

Here $r_\perp$ denotes the **amplitude reflection coefficient**, and $t_\perp$ is the **amplitude transmission coefficient**.

Case 2: E parallel to the plane of incidence. A similar pair of equations can be derived when the incoming **E**-field lies in the plane-of-incidence, as shown in Fig. 4.38. Continuity of the tangential components of **E** on either side of the boundary leads to

$$E_{0i}\cos\theta_i - E_{0r}\cos\theta_r = E_{0t}\cos\theta_t \tag{4.36}$$

In much the same way as before, continuity of the tangential components of $\mathbf{B}/\mu$ yields

$$\frac{1}{\mu_i v_i}E_{0i} + \frac{1}{\mu_r v_r}E_{0r} = \frac{1}{\mu_t v_t}E_{0t} \tag{4.37}$$

Using the fact that $\mu_i = \mu_r$ and $\theta_i = \theta_r$, we can combine these formulas to obtain two more of the *Fresnel Equations*:

$$r_\parallel \equiv \left(\frac{E_{0r}}{E_{0i}}\right)_\parallel = \frac{\dfrac{n_t}{\mu_t}\cos\theta_i - \dfrac{n_i}{\mu_i}\cos\theta_t}{\dfrac{n_i}{\mu_i}\cos\theta_t + \dfrac{n_t}{\mu_t}\cos\theta_i} \tag{4.38}$$

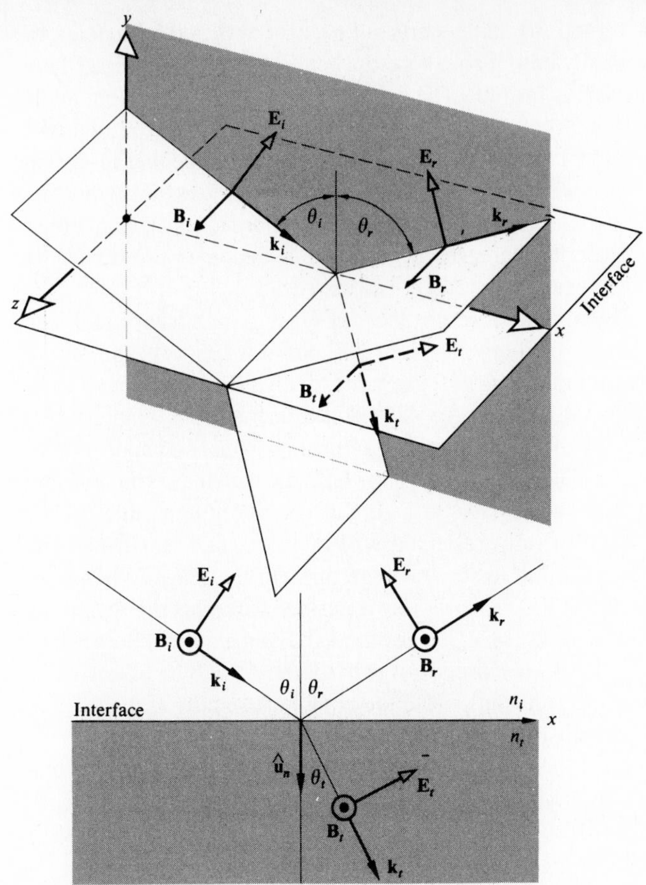

FIGURE 4.38 An incoming wave whose **E**-field is in the plane-of-incidence.

and

$$t_{\parallel} \equiv \left(\frac{E_{0t}}{E_{0i}}\right)_{\parallel} = \frac{2\dfrac{n_i}{\mu_i}\cos\theta_i}{\dfrac{n_i}{\mu_i}\cos\theta_t + \dfrac{n_t}{\mu_t}\cos\theta_i} \tag{4.39}$$

When both media forming the interface are dielectrics that are essentially "nonmagnetic" (p. 66), the amplitude coefficients become

$$r_{\parallel} = \frac{n_t \cos\theta_i - n_i \cos\theta_t}{n_i \cos\theta_t + n_t \cos\theta_i} \tag{4.40}$$

and

$$t_{\parallel} = \frac{2n_i \cos\theta_i}{n_i \cos\theta_t + n_t \cos\theta_i} \tag{4.41}$$

One further notational simplification can be made using Snell's Law, whereupon the Fresnel Equations for dielectric media become (Problem 4.35)

$$r_{\perp} = -\frac{\sin(\theta_i - \theta_t)}{\sin(\theta_i + \theta_t)} \tag{4.42}$$

$$r_{\parallel} = +\frac{\tan(\theta_i - \theta_t)}{\tan(\theta_i + \theta_t)} \tag{4.43}$$

$$t_{\perp} = +\frac{2\sin\theta_t \cos\theta_i}{\sin(\theta_i + \theta_t)} \tag{4.44}$$

$$t_{\parallel} = +\frac{2\sin\theta_t \cos\theta_i}{\sin(\theta_i + \theta_t)\cos(\theta_i - \theta_t)} \tag{4.45}$$

A note of caution must be introduced here. Bear in mind that the directions (or more precisely, the phases) of the fields in Figs. 4.37 and 4.38 were selected rather arbitrarily. For example, in Fig. 4.37 we could have assumed that $\mathbf{E}_r$ pointed inward, whereupon $\mathbf{B}_r$ would have had to be reversed as well. Had we done that, the sign of $r_{\perp}$ would have turned out to be positive, leaving the other amplitude coefficients unchanged. The signs appearing in Eqs. (4.42) through (4.45), which are positive except for the first, correspond to the particular set of field directions selected. The minus sign in Eq. (4.42), as we will see, just means that we didn't guess correctly concerning $\mathbf{E}_r$ in Fig. 4.37. Nonetheless, be aware that the literature is not standardized, and all possible sign variations have been labeled the *Fresnel Equations*—to avoid confusion *they must be related to the specific field directions from which they were derived.*

4.6.3 Interpretation of the Fresnel Equations

This section examines the physical implications of the Fresnel Equations. In particular, we are interested in determining the fractional amplitudes and flux densities that are reflected and refracted. In addition we shall be concerned with any possible phase shifts that might be incurred in the process.

Amplitude Coefficients

Let's briefly examine the form of the amplitude coefficients over the entire range of θ_i values. At nearly normal incidence ($\theta_i \approx 0$) the tangents in Eq. (4.43) are essentially equal to sines, in which case

$$[r_\parallel]_{\theta_i=0} = [-r_\perp]_{\theta_i=0} = \left[\frac{\sin(\theta_i - \theta_t)}{\sin(\theta_i + \theta_t)}\right]_{\theta_i=0}$$

We will come back to the physical significance of the minus sign presently. After expanding the sines and using Snell's Law, this expression becomes

$$[r_\parallel]_{\theta_i=0} = [-r_\perp]_{\theta_i=0} = \left[\frac{n_t \cos\theta_i - n_i \cos\theta_t}{n_t \cos\theta_i + n_i \cos\theta_t}\right]_{\theta_i=0} \tag{4.46}$$

which follows as well from Eqs. (4.34) and (4.40). In the limit, as θ_i goes to 0, $\cos\theta_i$ and $\cos\theta_t$ both approach one, and consequently

$$[r_\parallel]_{\theta_i=0} = [-r_\perp]_{\theta_i=0} = \frac{n_t - n_i}{n_t + n_i} \tag{4.47}$$

This equality of the reflection coefficients arises because the plane-of-incidence is no longer specified when $\theta_t = 0$. Thus, for example, at an air ($n_i = 1$) glass ($n_t = 1.5$) interface at nearly normal incidence, the amplitude reflection coefficients equal ± 0.2. (See Problem 4.38.)

When $n_t > n_i$ it follows from Snell's Law that $\theta_i > \theta_t$, and $r_\perp$ is negative for all values of θ_i (Fig. 4.39). In contrast, Eq. (4.43) tells us that $r_\parallel$ starts out positive at $\theta_i = 0$ and decreases gradually until it equals zero when $(\theta_i + \theta_t) = 90°$, since there $\tan \pi/2$ is infinite. The particular value of the incident angle for which this occurs is denoted by θ_p and referred to as the **polarization angle** (see Section 8.6.1). Notice that $r_\parallel \to 0$ at θ_p, just when the phase shifts 180°. That means we won't see the **E**-field do any flipping when θ_i approaches θ_p from either side. As θ_i increases beyond θ_p, $r_\parallel$ becomes progressively more negative, reaching -1.0 at 90°.

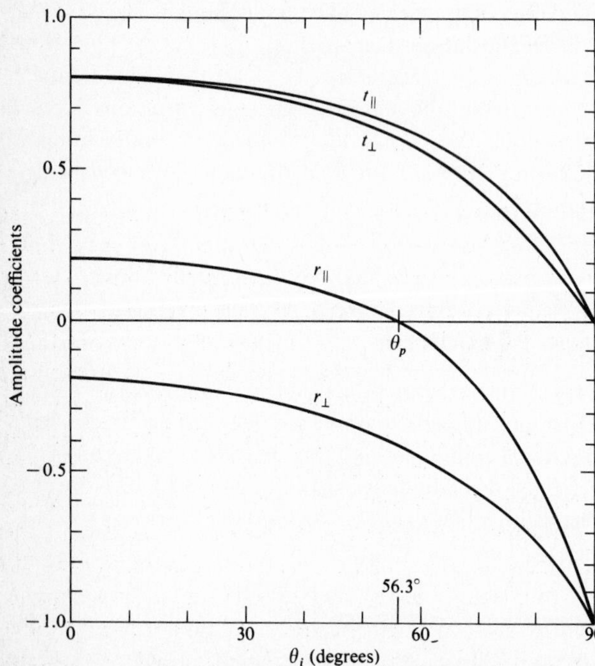

FIGURE 4.39 The amplitude coefficients of reflection and transmission as a function of incident angle. These correspond to external reflection $n_t > n_i$ at an air–glass interface ($n_{ti} = 1.5$).

If you place a single sheet of glass, a microscope slide, on this page and look straight down into it ($\theta_i = 0$), the region beneath the glass will seem decidedly grayer than the rest of the paper, because the slide will reflect at both its interfaces, and the light reaching and returning from the paper will be diminished appreciably. Now hold the slide near your eye and again view the page through it as you tilt it, increasing θ_i. The amount of light reflected will increase, and it will become more difficult to see the page through the glass. When $\theta_i \approx 90°$ the slide will look like a perfect mirror as the reflection coefficients (Fig. 4.39) go to -1.0. Even a rather poor surface (Fig. 4.40), such as the cover of this book, will be mirrorlike at glancing incidence. Hold the book horizontally at the level of the middle of your eye and face a bright light; you will see the source reflected rather nicely in the cover. This suggests that even X-rays could be mirror-reflected at glancing incidence (p. 246), and modern X-ray telescopes are based on that very fact.

At normal incidence Eqs. (4.35) and (4.41) lead rather straightforwardly to

$$[t_\parallel]_{\theta_i=0} = [t_\perp]_{\theta_i=0} = \frac{2n_i}{n_i + n_t} \tag{4.48}$$

It will be shown in Problem 4.43 that the expression

$$t_\perp + (-r_\perp) = 1 \tag{4.49}$$

holds for all θ_i, whereas

$$t_\parallel + r_\parallel = 1 \tag{4.50}$$

is true only at normal incidence.

The foregoing discussion, for the most part, was restricted to the case of **external reflection** (i.e., $n_t > n_i$). The opposite situation of **internal reflection**, in which the incident medium is the more dense ($n_i > n_t$), is of interest as well. In that instance $\theta_t > \theta_i$, and $r_\perp$, as described by Eq. (4.42), will always be positive. Figure 4.41 shows that $r_\perp$ increases from its initial value [Eq. (4.47)] at $\theta_i = 0$, reaching $+1$ at what is called the **critical angle**, θ_c. Specifically, θ_c is the special value of the incident angle (p. 121) for which $\theta_t = \pi/2$. Like-

FIGURE 4.40 At near-glancing incidence the walls and floor are mirrorlike—this despite the fact that the surfaces are rather poor reflectors at $\theta_i = 0°$.

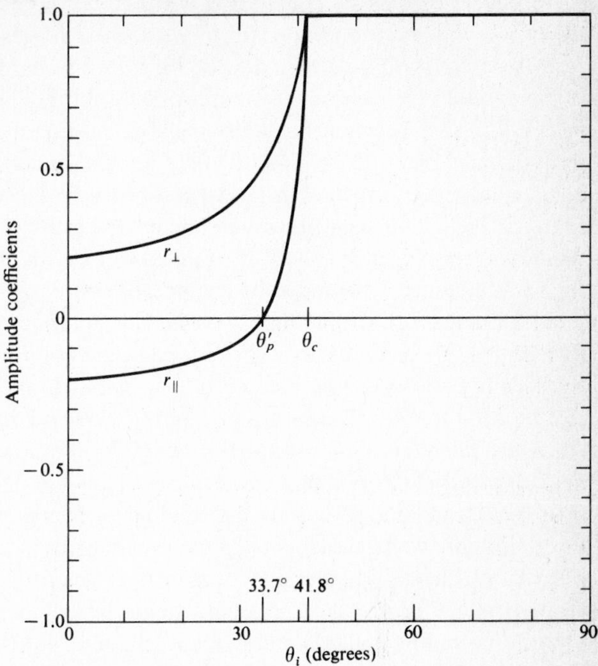

FIGURE 4.41 The amplitude coefficients of reflection as a function of incident angle. These correspond to internal reflection $n_t < n_i$ at an air–glass interface ($n_{ti} = 1/1.5$).

wise, $r_\parallel$ starts off negatively [Eq. (4.47)] at $\theta_i = 0$ and thereafter increases, reaching $+1$ at $\theta_i = \theta_c$, as is evident from the Fresnel Equation (4.40). Again, $r_\parallel$ passes through zero at the *polarization angle* θ'_p. It is left for Problem 4.57 to show that the polarization angles θ'_p and θ_p for internal and external reflection at the interface between the same media are simply the complements of each other. We will return to internal reflection in Section 4.7, where it will be shown that $r_\perp$ and $r_\parallel$ are complex quantities for $\theta_i > \theta_c$.

Phase Shifts

It should be evident from Eq. (4.42) that $r_\perp$ is negative regardless of θ_i when $n_t > n_i$. Yet we saw earlier that had we chosen $[\mathbf{E}_r]_\perp$ in Fig. 4.37 to be in the opposite direction, the first Fresnel Equation (4.42) would have changed signs, causing $r_\perp$ to become a positive quantity. The sign of $r_\perp$ is associated with the relative directions of $[\mathbf{E}_{0i}]_\perp$ and $[\mathbf{E}_{0r}]_\perp$. Bear in mind that a reversal of $[\mathbf{E}_{0r}]_\perp$ is tantamount to introducing a phase shift, $\Delta\varphi_\perp$, of π radians into $[\mathbf{E}_r]_\perp$. Hence at the boundary $[\mathbf{E}_i]_\perp$ and $[\mathbf{E}_r]_\perp$ will be antiparallel and therefore π out-of-phase with each other, as indicated by the negative value of $r_\perp$. When we consider components normal to the plane of incidence, there is no confusion as to whether two fields are in-phase or π radians out-of-phase: if parallel, they're in-phase; if antiparallel, they're π out-of-phase. In summary, then, ***the component of the electric field normal to the plane-of-incidence undergoes a phase shift of π radians upon reflection when the incident medium has a lower index than the transmitting medium***. Similarly, $t_\perp$ and $t_\parallel$ are always positive and $\Delta\varphi = 0$. *Further-*

more, when $n_i > n_t$ no phase shift in the normal component results on reflection, that is, $\Delta\varphi_\perp = 0$ so long as $\theta_i < \theta_c$.

Things are a bit less obvious when we deal with $[\mathbf{E}_i]_\parallel$, $[\mathbf{E}_r]_\parallel$, and $[\mathbf{E}_t]_\parallel$. It now becomes necessary to define more explicitly what is meant by *in-phase*, since the field vectors are coplanar but generally not colinear. The field directions were chosen in Figs. 4.37 and 4.38 such that if you looked down any one of the propagation vectors toward the direction from which the light was coming, **E**, **B**, and **k** would appear to have the same relative orientation whether the ray was incident, reflected, or transmitted. We can use this as the required condition for two **E**-fields to be in-phase. Equivalently, but more simply, ***two fields in the incident plane are in-phase if their y-components are parallel and are out-of-phase if the components are antiparallel.*** Notice that when two **E**-fields are out-of-phase so too are their associated **B**-fields and vice versa. With this definition we need only look at the vectors normal to the plane of incidence, whether they be **E** or **B**, to determine the relative phase of the accompanying fields in the incident plane. Thus in Fig. 4.42*a* $\mathbf{E}_i$ and $\mathbf{E}_t$ are in-phase, as are $\mathbf{B}_i$ and $\mathbf{B}_t$, whereas $\mathbf{E}_i$ and $\mathbf{E}_r$ are out-of-phase, along with $\mathbf{B}_i$ and $\mathbf{B}_r$. Similarly, in Fig. 4.42*b* $\mathbf{E}_i$, $\mathbf{E}_r$, and $\mathbf{E}_t$ are in-phase, as are $\mathbf{B}_i$, $\mathbf{B}_r$, and $\mathbf{B}_t$.

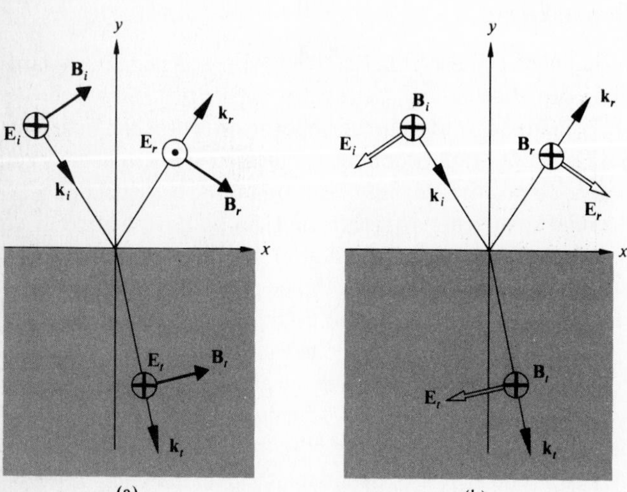

FIGURE 4.42 Field orientations and phase shifts.

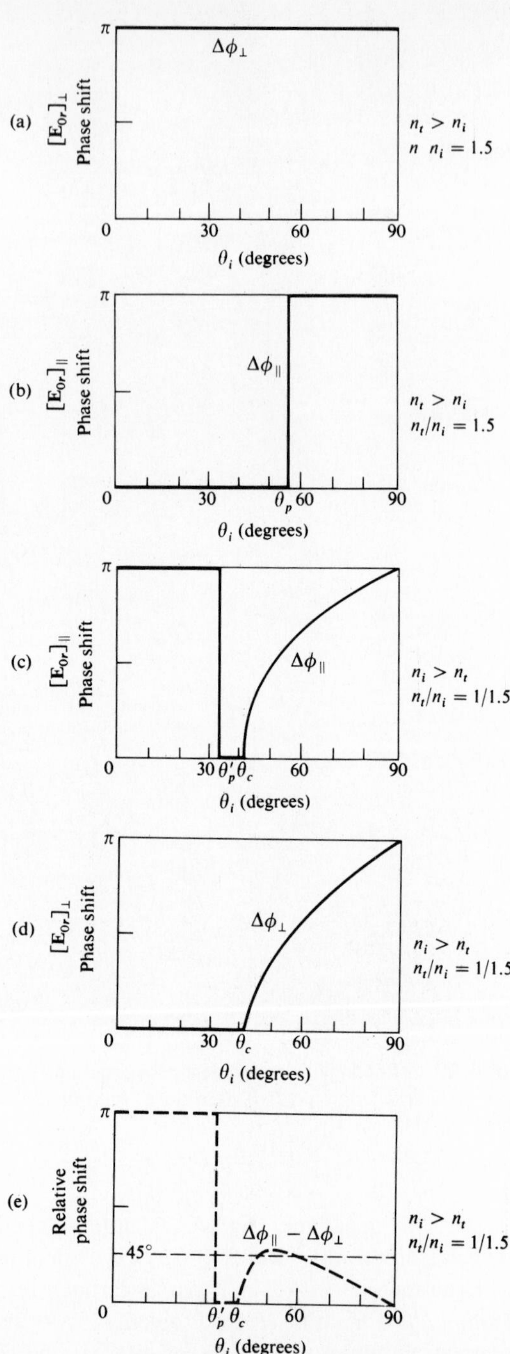

FIGURE 4.43 Phase shifts for the parallel and perpendicular components of the E-field corresponding to internal and external reflection.

Now, the amplitude reflection coefficient for the parallel component is given by

$$r_{\parallel} = \frac{n_t \cos \theta_i - n_i \cos \theta_t}{n_t \cos \theta_i + n_i \cos \theta_t}$$

which is positive ($\Delta\varphi_{\parallel} = 0$) as long as

$$n_t \cos \theta_i - n_i \cos \theta_t > 0$$

that is, if

$$\sin \theta_i \cos \theta_i - \cos \theta_t \sin \theta_t > 0$$

or equivalently

$$\sin (\theta_i - \theta_t) \cos (\theta_i + \theta_t) > 0 \tag{4.51}$$

This will be the case for $n_i < n_t$ if

$$(\theta_i + \theta_t) < \pi/2 \tag{4.52}$$

and for $n_i > n_t$ when

$$(\theta_i + \theta_t) > \pi/2 \tag{4.53}$$

Thus when $n_i < n_t$, $[\mathbf{E}_{0r}]_{\parallel}$ and $[\mathbf{E}_{0i}]_{\parallel}$ will be in-phase ($\Delta\varphi_{\parallel} = 0$) until $\theta_i = \theta_p$ and out-of-phase by π radians thereafter. The transition is not actually discontinuous, since $[\mathbf{E}_{0r}]_{\parallel}$ goes to zero at θ_p. In contrast, for internal reflection $r_{\parallel}$ is negative until θ_p', which means that $\Delta\varphi_{\parallel} = \pi$. From θ_p' to θ_c, $r_{\parallel}$ is positive and $\Delta\varphi_{\parallel} = 0$. Beyond θ_c, $r_{\parallel}$ becomes complex, and $\Delta\varphi_{\parallel}$ gradually increases to π at $\theta_i = 90°$.

Figure 4.43, which summarizes these conclusions, will be of continued use to us. The actual functional form of $\Delta\varphi_{\parallel}$ and $\Delta\varphi_{\perp}$ for internal reflection in the region where $\theta_i > \theta_c$ can be found in the literature,* but the curves depicted here will suffice for our purposes. Figure 4.43*e* is a plot of the relative phase shift between the parallel and perpendicular components, that is, $\Delta\varphi_{\parallel} - \Delta\varphi_{\perp}$. It is included here because it will be useful later on (e.g., when we consider polarization effects). Finally, many of the essential features of this discussion are illustrated in Figs. 4.44 and 4.45. The amplitudes of the reflected vectors are in accord with those of Figs. 4.39 and 4.41 (for an air–glass interface), and the phase shifts agree with those of Fig. 4.43.

Many of these conclusions can be verified with the simplest experimental equipment, namely, two linear polarizers, a

*Born and Wolf, *Principles of Optics*, p. 49.

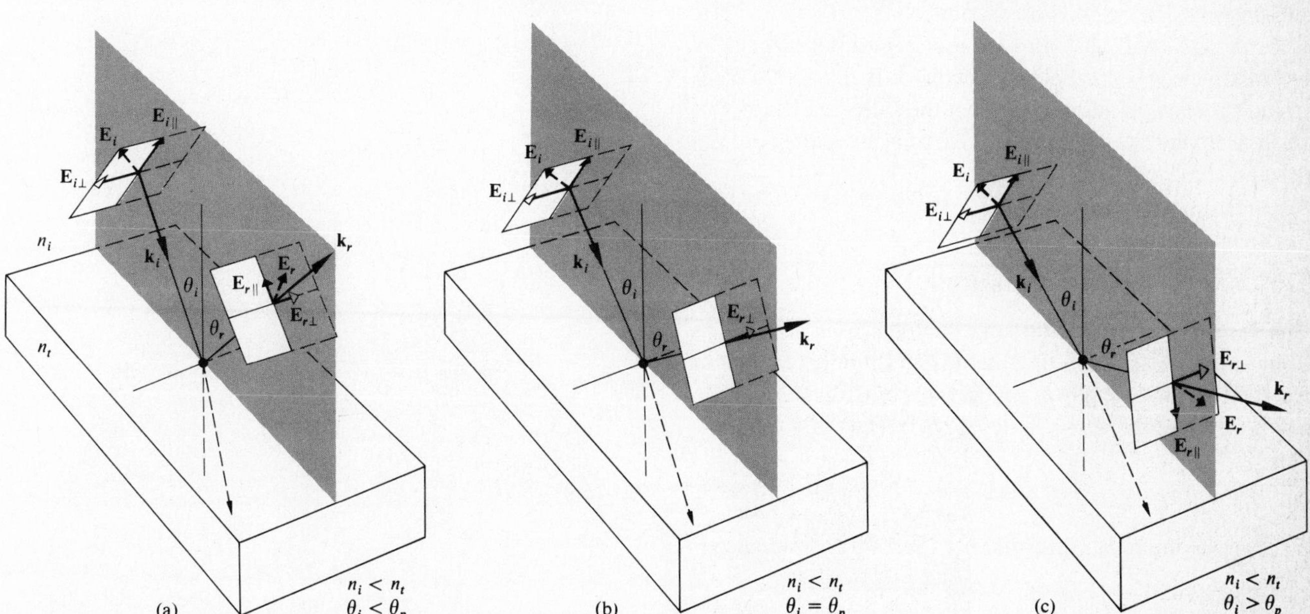

FIGURE 4.44 The reflected E-field at various angles concomitant with external reflection.

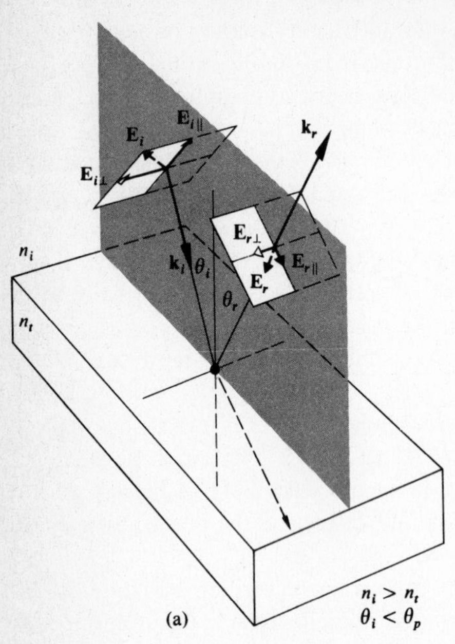

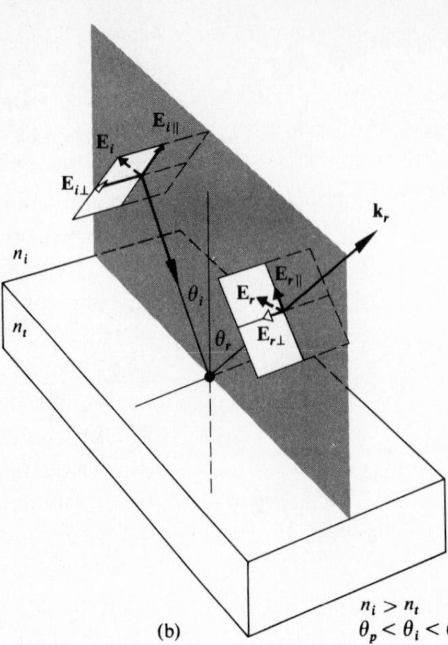

FIGURE 4.45 The reflected E-field at various angles concomitant with internal reflection.

piece of glass, and a small source, such as a flashlight or high-intensity lamp. By placing one polarizer in front of the source (at 45° to the plane of incidence), you can easily duplicate the conditions of Fig. 4.44. For example, when $\theta_i = \theta_p$ (Fig. 4.44*b*) no light will pass through the second polarizer if its transmission axis is parallel to the plane-of-incidence. In comparison, at near-glancing incidence the reflected beam will vanish when the axes of the two polarizers are almost normal to each other.

Reflectance and Transmittance

Consider a circular beam of light incident on a surface, as shown in Fig. 4.46, such that there is an illuminated spot of area A. Recall that the power per unit area crossing a surface in vacuum whose normal is parallel to **S**, the Poynting vector, is given by

$$\mathbf{S} = c^2\epsilon_0 \mathbf{E} \times \mathbf{B} \qquad [3.40]$$

Furthermore, the radiant flux density (W/m^2) or irradiance is

$$I = \langle S \rangle_T = \frac{c\epsilon_0}{2} E_0^2 \qquad [3.44]$$

This is the average energy per unit time crossing a unit area normal to **S** (in isotropic media **S** is parallel to **k**). In the case at hand (Fig. 4.46), let I_i, I_r, and I_t be the incident, reflected, and transmitted flux densities, respectively. The cross-sectional areas of the incident, reflected, and transmitted beams are, respectively, $A \cos \theta_i$, $A \cos \theta_r$, and $A \cos \theta_t$. Accordingly, the

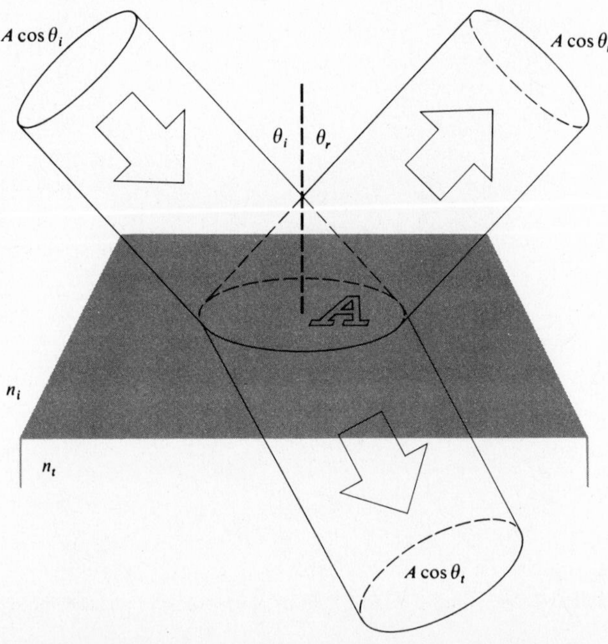

FIGURE 4.46 Reflection and transmission of an incident beam.

incident power is $I_i A \cos\theta_i$; this is the energy per unit time flowing in the incident beam, and it's therefore the power arriving on the surface over A. Similarly, $I_r A \cos\theta_r$ is the power in the reflected beam, and $I_t A \cos\theta_t$ is the power being transmitted through A. We define the **reflectance** R to be the ratio of the reflected power (or flux) to the incident power:

$$R \equiv \frac{I_r A \cos\theta_r}{I_i A \cos\theta_i} = \frac{I_r}{I_i} \tag{4.54}$$

In the same way, the **transmittance** T is defined as the ratio of the transmitted to the incident flux and is given by

$$T \equiv \frac{I_t \cos\theta_t}{I_i \cos\theta_i} \tag{4.55}$$

The quotient I_r/I_i equals $(v_r\epsilon_r E_{0r}^2/2)/(v_i\epsilon_i E_{0i}^2/2)$, and since the incident and reflected waves are in the same medium, $v_r = v_i$, $\epsilon_r = \epsilon_i$, and

$$R = \left(\frac{E_{0r}}{E_{0i}}\right)^2 = r^2 \tag{4.56}$$

In like fashion (assuming $\mu_i = \mu_t = \mu_0$),

$$T = \frac{n_t \cos\theta_t}{n_i \cos\theta_i}\left(\frac{E_{0t}}{E_{0i}}\right)^2 = \left(\frac{n_t \cos\theta_t}{n_i \cos\theta_i}\right) t^2 \tag{4.57}$$

where use was made of the fact that $\mu_0\epsilon_t = 1/v_t^2$ and $\mu_0 v_t \epsilon_t = n_t/c$. Notice that at normal incidence, which is a situation of great practical interest, $\theta_t = \theta_i = 0$, and the transmittance [Eq. (4.55)], like the reflectance [Eq. (4.54)], is then simply the ratio of the appropriate irradiances. Since $R = r^2$, we need not worry about the sign of r in any particular formulation, and that makes reflectance a convenient notion. Observe that in Eq. (4.57) T is not simply equal to t^2, for two reasons. First, the ratio of the indices of refraction must be there, since the speeds at which energy is transported into and out of the interface are different, in other words, $I \propto v$, from Eq. (3.47). Second, the cross-sectional areas of the incident and reflected beams are different. The energy flow per unit area is affected accordingly, and that manifests itself in the presence of the ratio of the cosine terms.

Let's now write an expression representing the conservation of energy for the configuration depicted in Fig. 4.46. In other words, the total energy flowing into area A per unit time must equal the energy flowing outward from it per unit time:

$$I_i A \cos\theta_i = I_r A \cos\theta_r + I_t A \cos\theta_t \tag{4.58}$$

When both sides are multiplied by c this expression becomes

$$n_i E_{0i}^2 \cos\theta_i = n_i E_{0r}^2 \cos\theta_i + n_t E_{0t}^2 \cos\theta_t$$

or

$$1 = \left(\frac{E_{0r}}{E_{0i}}\right)^2 + \left(\frac{n_t \cos\theta_t}{n_i \cos\theta_i}\right)\left(\frac{E_{0t}}{E_{0i}}\right)^2 \tag{4.59}$$

But this is simply

$$R + T = 1 \tag{4.60}$$

where there was no absorption. It is convenient to use the component forms, that is,

$$R_\perp = r_\perp^2 \tag{4.61}$$

$$R_\parallel = r_\parallel^2 \tag{4.62}$$

$$T_\perp = \left(\frac{n_t \cos\theta_t}{n_i \cos\theta_i}\right) t_\perp^2 \tag{4.63}$$

and

$$T_\parallel = \left(\frac{n_t \cos\theta_t}{n_i \cos\theta_i}\right) t_\parallel^2 \tag{4.64}$$

which are illustrated in Fig. 4.47. Furthermore, it can be shown (Problem 4.62) that

$$R_\parallel + T_\parallel = 1 \tag{4.65}$$

and

$$R_\perp + T_\perp = 1 \tag{4.66}$$

When $\theta_i = 0$, the incident plane becomes undefined, and any distinction between the parallel and perpendicular components of R and T vanishes. In this case Eqs. (4.61) through (4.64), along with (4.47) and (4.48), lead to

$$R = R_\parallel = R_\perp = \left(\frac{n_t - n_i}{n_t + n_i}\right)^2 \tag{4.67}$$

and

$$T = T_\parallel = T_\perp = \frac{4n_t n_i}{(n_t + n_i)^2} \tag{4.68}$$

Thus 4% of the light incident normally on an air–glass ($n_g = 1.5$) interface will be reflected back, whether internally, $n_i > n_t$, or externally, $n_i < n_t$ (Problem 4.63). This will be of concern to anyone who is working with a complicated lens system, which might have 10 or 20 such air–glass boundaries. Indeed, if you look perpendicularly into a stack of about 50 microscope slides (cover-glass sliders are much thinner and easier to handle in large quantities), most of the light will be reflected. The stack will look very much like a mirror (Fig.

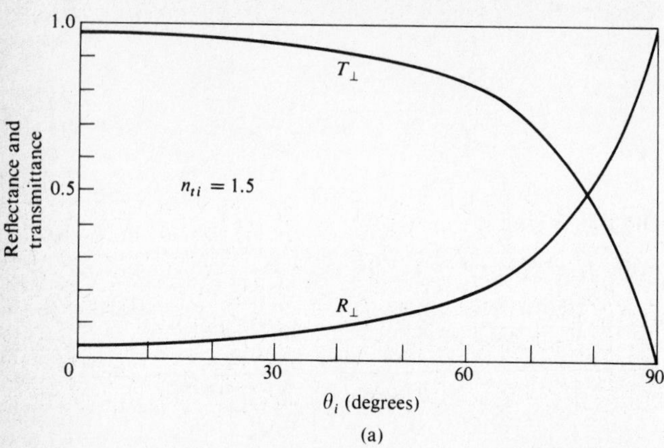

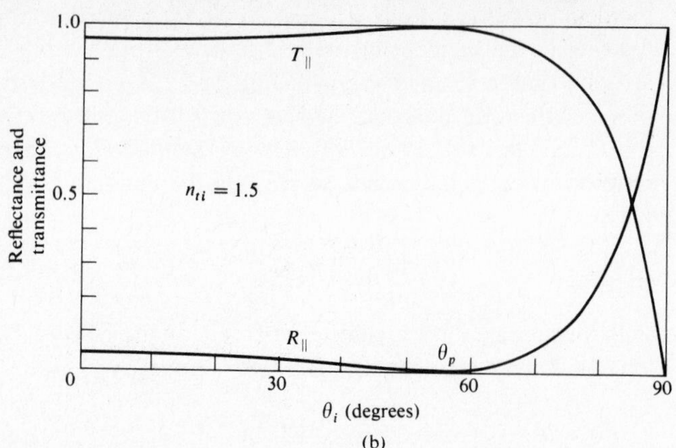

FIGURE 4.47 Reflectance and transmittance versus incident angle.

4.48). Roll up a thin sheet of clear plastic into a multiturned cylinder and it too will look like shiny metal. The many interfaces produce a large number of closely spaced *specular* reflections that send much of the light back into the incident medium, more or less, as if it had undergone a single frequency-independent reflection. A smooth gray-metal surface does pretty much the same thing—it has a large, frequency-independent specular reflectance—and looks shiny (that's what "shiny" is). If the reflection is diffuse, the surface will appear gray or even white if the reflectance is large enough.

Figure 4.49 is a plot of the reflectance at a single interface, assuming normal incidence for various transmitting media in air. Figure 4.50 depicts the corresponding dependence of the transmittance at normal incidence on the number of interfaces and the index of the medium. Of course, this is why you can't see through a roll of "clear" smooth-surfaced plastic tape, and it's also why the many elements in a periscope must be coated with antireflection films (Section 9.9.2).

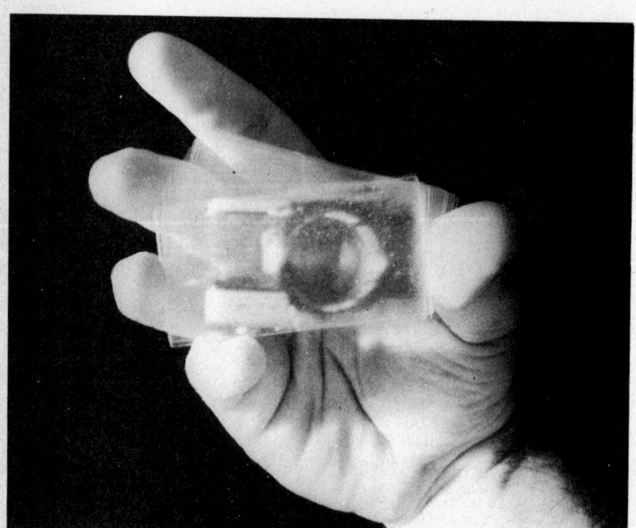

FIGURE 4.48 Near normal reflection off a stack of microscope slides. You can see the image of the camera that took the picture. (Photo by E.H.)

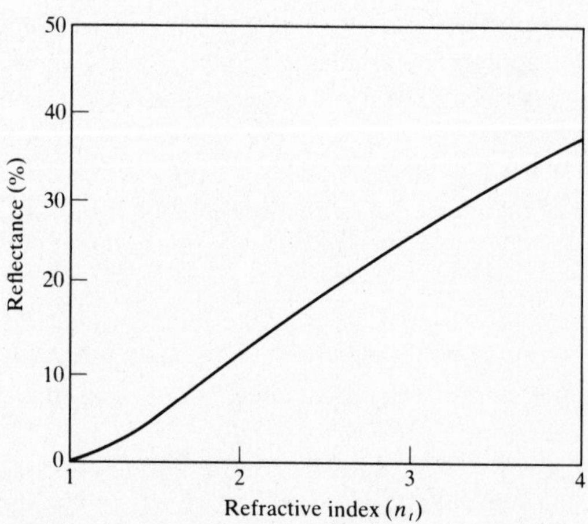

FIGURE 4.49 Reflectance at normal incidence in air (n_i = 1.0) at a single interface.

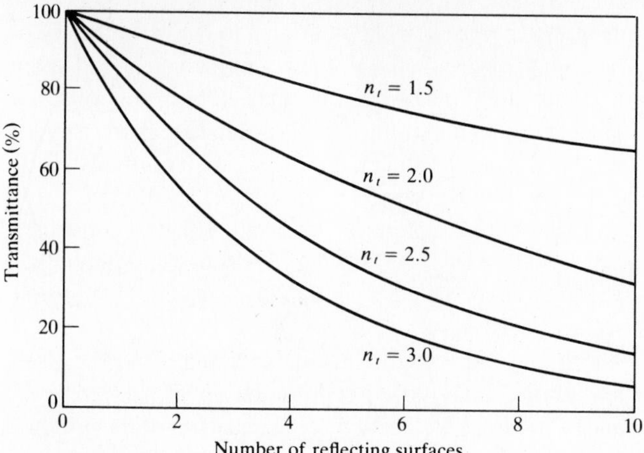

FIGURE 4.50 Transmittance through a number of surfaces in air ($n_i = 1.0$) at normal incidence.

4.7 Total Internal Reflection

In the previous section it was evident that something rather interesting was happening in the case of internal reflection ($n_i > n_t$) when θ_i was equal to or greater than θ_c, the so-called **critical angle**. Let's now return to that situation for a closer look. Suppose that we have a source embedded in an optically dense medium, and we allow θ_i to increase gradually, as indicated in Fig. 4.51. We know from the preceding section (Fig. 4.41) that $r_{\parallel}$ and $r_{\perp}$ increase with increasing θ_i, and therefore $t_{\parallel}$ and $t_{\perp}$ both decrease. Moreover $\theta_t > \theta_i$, since

$$\sin \theta_i = \frac{n_t}{n_i} \sin \theta_t$$

and $n_i > n_t$, in which case $n_{ti} < 1$. Thus as θ_i becomes larger, the transmitted ray gradually approaches tangency with the boundary, and as it does more and more of the available energy appears in the reflected beam. Finally, when $\theta_t = 90°$, $\sin \theta_t = 1$ and

$$\sin \theta_c = n_{ti} \tag{4.69}$$

As noted earlier, *the critical angle is that special value of θ_i for which $\theta_t = 90°$*. The larger n_i is, the smaller n_{ti} is, and the smaller θ_c is. For incident angles greater than or equal to θ_c, all the incoming energy is reflected back into the incident medium in the process known as **total internal reflection** (Fig. 4.52).

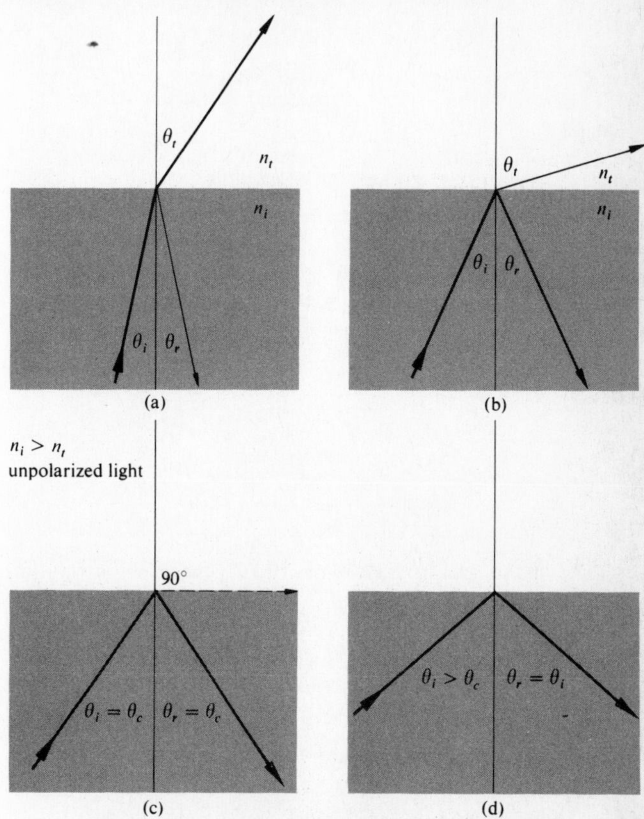

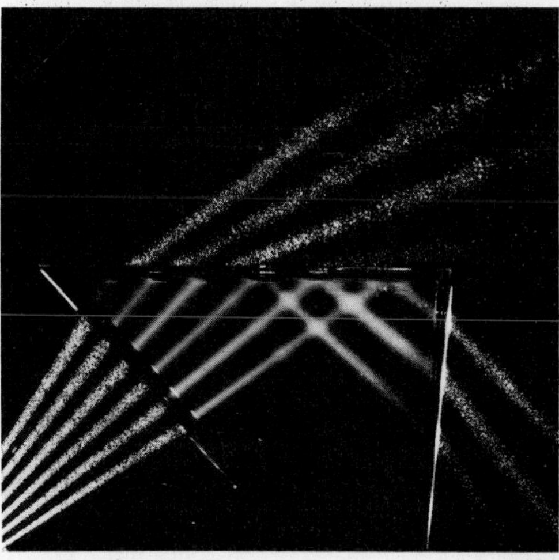

FIGURE 4.51 Internal reflection and the critical angle. (Photo courtesy of Educational Service, Inc.)

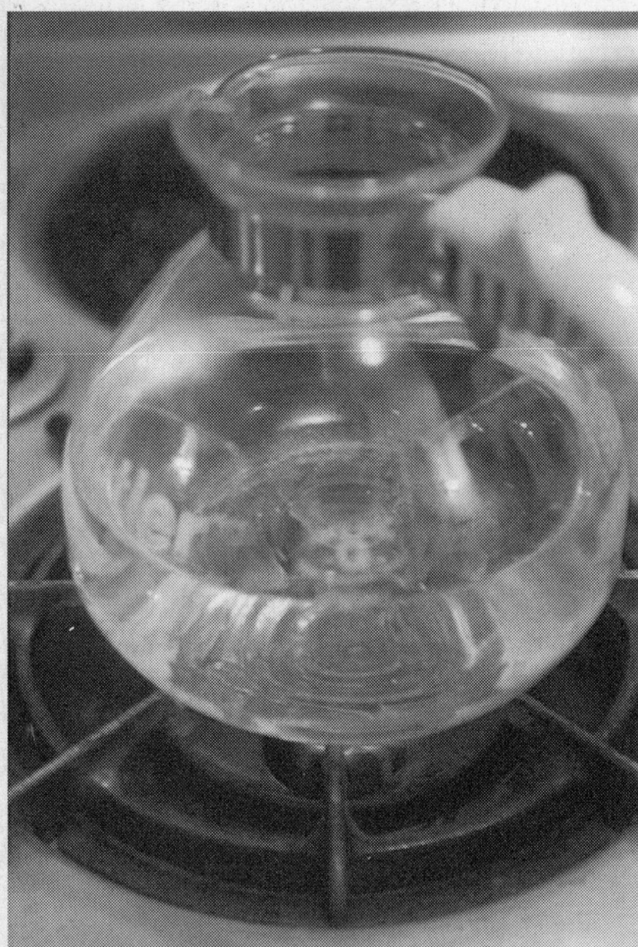

FIGURE 4.52 An everyday instance of total internal reflection. Here the bottom of the teapot is seen through the water via total internal reflection. Notice that no light from the front flame is transmitted across the bottom to the camera.

It should be stressed that the transition from the conditions depicted in Fig. 4.51*a* to those of 4.51*d* takes place without any discontinuities. As θ_i becomes larger, the reflected beam grows stronger and stronger while the transmitted beam grows weaker, until the latter vanishes and the former carries off all the energy at $\theta_r = \theta_c$. It's an easy matter to observe the diminution of the transmitted beam as θ_i is made larger. Just place a glass microscope slide on a printed page, this time blocking out any specularly reflected light. At $\theta_i \approx 0$, θ_t is roughly zero, and the page as seen through the glass is fairly bright and clear. But if you move your head, allowing θ_t (the angle at which you view the interface) to increase, the region of the printed page covered by the glass will appear darker and darker, indicating that T has indeed been markedly reduced.

The critical angle for our air–glass interface is roughly 42° (see Table 4.2). Consequently, a ray incident normally on the left face of either of the prisms in Fig. 4.53 will have a $\theta_i >$ 42° and therefore be internally reflected. This is a convenient way to reflect nearly 100% of the incident light without having to worry about the deterioration that can occur with metallic surfaces (Fig. 4.54).

Another useful way to view the situation is via Fig. 4.55, which shows a simplified representation of scattering off atomic oscillators. We know that the net effect of the presence of the homogeneous isotropic media is to alter the speed of the light from c to v_i and v_t, respectively (p. 91). The resultant wave is the superposition of these wavelets propagating at the appropriate speeds. In Fig. 4.55*a* an incident wave results in the emission of wavelets successively from scattering centers

TABLE 4.2 Critical Angles

n_{it}	θ_c (degrees)	θ_c (radians)	n_{it}	θ_c (degrees)	θ_c (radians)
1.30	50.2849	0.8776	1.50	41.8103	0.7297
1.31	49.7612	0.8685	1.51	41.4718	0.7238
1.32	49.2509	0.8596	1.52	41.1395	0.7180
1.33	48.7535	0.8509	1.53	40.8132	0.7123
1.34	48.2682	0.8424	1.54	40.4927	0.7067
1.35	47.7946	0.8342	1.55	40.1778	0.7012
1.36	47.3321	0.8261	1.56	39.8683	0.6958
1.37	46.8803	0.8182	1.57	39.5642	0.6905
1.38	46.4387	0.8105	1.58	39.2652	0.6853
1.39	46.0070	0.8030	1.59	38.9713	0.6802
1.40	45.5847	0.7956	1.60	38.6822	0.6751
1.41	45.1715	0.7884	1.61	38.3978	0.6702
1.42	44.7670	0.7813	1.62	38.1181	0.6653
1.43	44.3709	0.7744	1.63	37.8428	0.6605
1.44	43.9830	0.7676	1.64	37.5719	0.6558
1.45	43.6028	0.7610	1.65	37.3052	0.6511
1.46	43.2302	0.7545	1.66	37.0427	0.6465
1.47	42.8649	0.7481	1.67	36.7842	0.6420
1.48	42.5066	0.7419	1.68	36.5296	0.6376
1.49	42.1552	0.7357	1.69	36.2789	0.6332

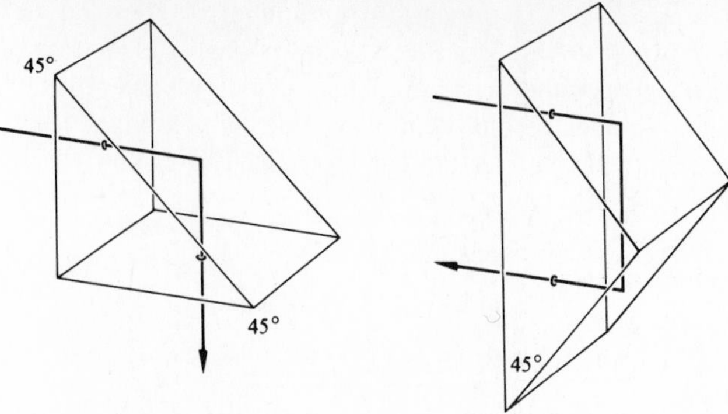

FIGURE 4.53 Total internal reflection.

A and B. These overlap to form the transmitted wave. The reflected wave, which comes back down into the incident medium as usual ($\theta_i = \theta_r$), is not shown. In a time t the incident front travels a distance $v_i t = \overline{CB}$, while the transmitted front moves a distance $v_t t = \overline{AD} > \overline{CB}$. Since one wave moves from A to E in the same time that the other moves from C to B, and since they have the same frequency and period, they must change phase by the same amount in the process. Thus the disturbance at point E must be in-phase with that at

FIGURE 4.54 The prism behaves like a mirror and reflects a portion of the pencil (reversing the lettering on it). The operating process is total internal reflection.

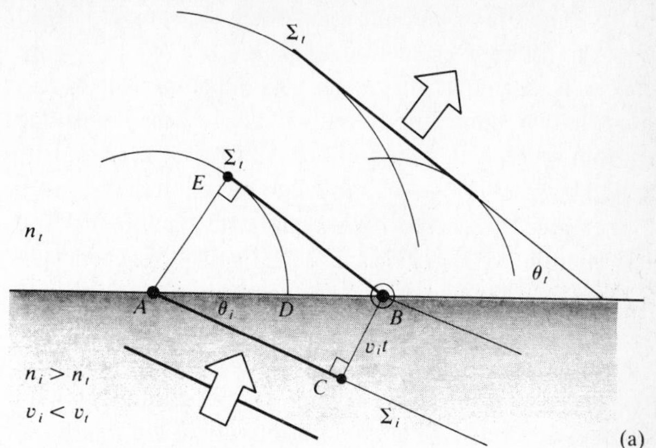

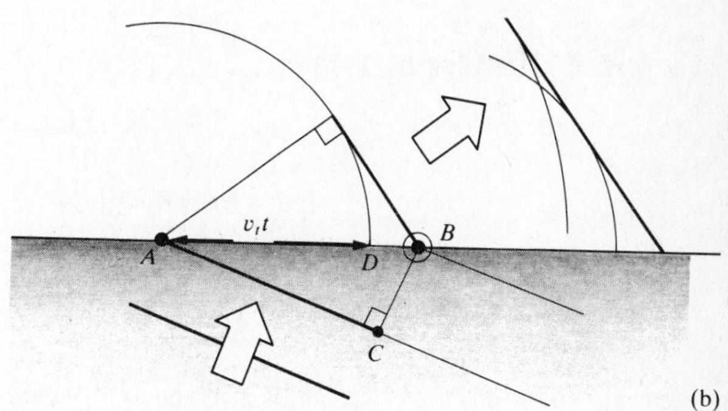

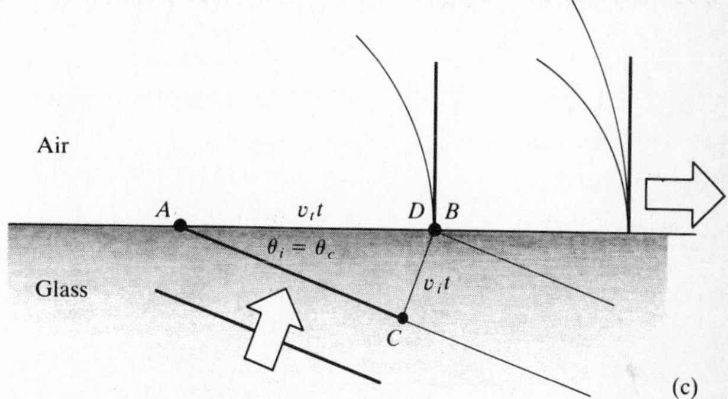

FIGURE 4.55 An examination of the transmitted wave in the process of total internal reflection from a scattering perspective. Here we keep θ_i and n_i constant and in successive parts of the diagram decrease n_t, thereby increasing v_t. The reflected wave ($\theta_r = \theta_i$) is not drawn.

point B; both of these points must be on the same transmitted wavefront (remember Section 4.4.2).

It can be seen that the greater v_t is in comparison to v_i, the more tilted the transmitted front will be (i.e., the larger θ_t will be). That much is depicted in Fig. 4.55*b*, where n_{ti} has been taken to be smaller by assuming n_t to be smaller. The result is a higher speed v_t, increasing $\overline{AD}$ and causing a greater transmission angle. In Fig. 4.55*c* a special case is reached: $\overline{AD} = \overline{AB} = v_t t$, and the wavelets will overlap in-phase *only along the line of the interface*, $\theta_t = 90°$. From triangle ABC, $\sin\theta_i = v_i t/v_t t = n_t/n_i$, which is Eq. (4.69). For the two given media (i.e., for the particular value of n_{ti}), the direction in which the scattered wavelets will add constructively in the transmitting medium is along the interface. The resulting disturbance ($\theta_t = 90°$) is known as a *surface wave*.

4.7.1 The Evanescent Wave

Because the frequency of X-rays is higher than the resonance frequencies of the atoms of the medium, Eq. (3.70) suggests, and experiments confirm, that the index of refraction of X-rays is less than 1.0. Thus the wave velocity of X-rays (i.e., the phase speed) in matter exceeds its value (c) in vacuum, although it usually does so by less than 1 part in 10 000, even in the densest solids. When X-rays traveling in air enter a dense material like glass the beam bends ever so slightly *away* from the normal rather than toward it. With the above discussion of total internal reflection in mind, we should expect that X-rays will be **totally "externally" reflected** when, for example, $n_i = n_{air}$ and $n_t = n_{glass}$. This is the way it's often spoken of in the literature, but that's a misnomer; since for X-rays $n_{air} > n_{glass}$ and therefore $n_i > n_t$ (even though glass is physically more dense than air), the process is actually still *internal* reflection. In any event, because n_t is less than, but very nearly equal to, 1 the index ratio $n_{ti} \approx 1$ and $\theta_c \approx 90°$.

In 1923 A. H. Compton reasoned that even though X-rays incident on a sample at ordinary angles are not specularly reflected, they should be totally "externally" reflected at glancing incidence. He shined 0.128 nm X-rays on a glass plate and got a critical angle of about 10 minutes of arc (0.167°) measured with respect to the surface. That yielded an index of refraction for glass that differed from 1 by -4.2×10^{-6}.

We'll come back to some important practical applications of both total internal and total "external" reflection later on (p. 202).

If we assume in the case of total internal reflection that there is no transmitted wave, it becomes impossible to satisfy the boundary conditions using only the incident and reflected waves—things are not at all as simple as they might seem. Furthermore, we can reformulate Eqs. (4.34) and (4.40) (Problem 4.66) such that

$$r_\perp = \frac{\cos\theta_i - (n_{ti}^2 - \sin^2\theta_i)^{1/2}}{\cos\theta_i + (n_{ti}^2 - \sin^2\theta_i)^{1/2}} \tag{4.70}$$

and

$$r_\parallel = \frac{n_{ti}^2\cos\theta_i - (n_{ti}^2 - \sin^2\theta_i)^{1/2}}{n_{ti}^2\cos\theta_i + (n_{ti}^2 - \sin^2\theta_i)^{1/2}} \tag{4.71}$$

Since $\sin\theta_c = n_{ti}$ when $\theta_i > \theta_c$, $\sin\theta_i > n_{ti}$, and both $r_\perp$ and $r_\parallel$ become complex quantities. Despite this (Problem 4.67), $r_\perp r_\perp^* = r_\parallel r_\parallel^* = 1$ and $R = 1$, which means that $I_r = I_i$ and $I_t = 0$. Thus, although there must be a transmitted wave, it cannot, on the average, carry energy across the boundary. We shall not perform the complete and rather lengthy computation needed to derive expressions for all the reflected and transmitted fields, but we can get an appreciation of what's happening in the following way. The wavefunction for the transmitted electric field is

$$\mathbf{E}_t = \mathbf{E}_{0t}\exp i(\mathbf{k}_t\cdot\mathbf{r} - \omega t)$$

where

$$\mathbf{k}_t\cdot\mathbf{r} = k_{tx}x + k_{ty}y$$

there being no z-component of $\mathbf{k}$. But

$$k_{tx} = k_t\sin\theta_t$$

and

$$k_{tx} = k_t\cos\theta_t$$

as seen in Fig. 4.56. Once again using Snell's Law,

$$k_t\cos\theta_t = \pm k_t\left(1 - \frac{\sin^2\theta_i}{n_{ti}^2}\right)^{1/2} \tag{4.72}$$

or, since we are concerned with the case where $\sin\theta_i > n_{ti}$,

$$k_{ty} = \pm ik_t\left(\frac{\sin^2\theta_i}{n_{ti}^2} - 1\right)^{1/2} \equiv \pm i\beta$$

and

$$k_{tx} = \frac{k_t}{n_{ti}}\sin\theta_i$$

Hence

$$\mathbf{E}_t = \mathbf{E}_{0t}e^{\mp\beta y}e^{i(k_t x\sin\theta_i/n_{ti} - \omega t)} \tag{4.73}$$

Neglecting the positive exponential, which is physically untenable, we have a wave whose amplitude drops off expo-

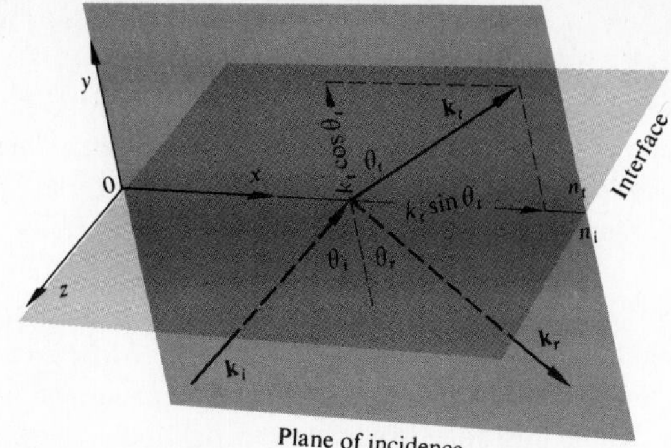

FIGURE 4.56 Propagation vectors for internal reflection.

nentially as it penetrates the less dense medium. The disturbance advances in the x-direction as a surface or **evanescent wave**. Notice that the wavefronts or surfaces of constant phase (parallel to the yz-plane) are perpendicular to the surfaces of constant amplitude (parallel to the xz-plane), and as such the wave is *inhomogeneous* (p. 28). Its amplitude decays rapidly in the y-direction, becoming negligible at a distance into the second medium of only a few wavelengths.

If you are still concerned about the conservation of energy, a more extensive treatment would have shown that energy actually circulates back and forth across the interface, resulting on the average in a zero net flow through the boundary into the second medium. Yet one puzzling point remains, inasmuch as there is still a bit of energy to be accounted for, namely, that associated with the evanescent wave that moves along the boundary in the plane-of-incidence. Since this energy could not have penetrated into the less dense medium under the present circumstances (so long as $\theta_i \geq \theta_c$), we must look elsewhere for its source. Under actual experimental conditions the incident beam would have a finite cross section and therefore would obviously differ from a true plane wave. This deviation gives rise (via diffraction) to a slight transmission of energy across the interface, which is manifested in the evanescent wave.

Incidentally, it is clear from (c) and (d) in Fig. 4.43 that the incident and reflected waves (except at $\theta_i = 90°$) do not differ in phase by π and cannot therefore cancel each other. It follows from the continuity of the tangential component of **E** that there must be an oscillatory field in the less dense medium, with a component parallel to the interface having a frequency ω (i.e., the evanescent wave).

The exponential decay of the surface wave, or *boundary wave*, as it is also sometimes called, has been confirmed experimentally at optical frequencies.*

Imagine that a beam of light traveling within a block of glass is internally reflected at a boundary. Presumably, if you pressed another piece of glass against the first, the air–glass interface could be made to vanish, and the beam would then propagate onward undisturbed. Furthermore, you might expect this transition from total to no reflection to occur gradually as the air film thinned out. In much the same way, if you hold a drinking glass or a prism, you can see the ridges of your fingerprints in a region that, because of total internal reflection, is otherwise mirrorlike. In more general terms, when the evanescent wave extends with appreciable amplitude across the rare medium into a nearby region occupied by a higher-index material, energy may flow through the gap in what is known as **frustrated total internal reflection** (FTIR). The evanescent wave, having traversed the gap, is still strong enough to drive electrons in the "frustrating" medium; they in turn will generate a wave that significantly alters the field configuration, thereby permitting energy to flow. Figure 4.57 is a

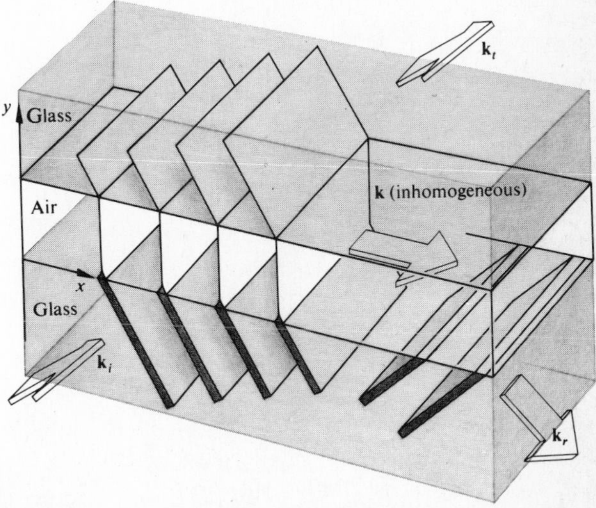

FIGURE 4.57 Frustrated total internal reflection.

*Take a look at the fascinating article by K. H. Drexhage, "Monomolecular layers and light." *Sci. Am.* **222**, 108 (1970).

schematic representation of FTIR: the width of the lines depicting the wavefronts decreases across the gap as a reminder that the amplitude of the field behaves in the same way. The process as a whole is remarkably similar to the quantum-mechanical phenomenon of *barrier penetration* or *tunneling*, which has numerous applications in contemporary physics.

One can demonstrate FTIR with the prism arrangement of Fig. 4.58 in a manner that is fairly self-evident. Moreover, if the hypotenuse faces of both prisms are made planar and parallel, they can be positioned so as to transmit and reflect any desired fraction of the incident flux density. Devices that perform this function are known as ***beamsplitters***. A *beamsplitter*

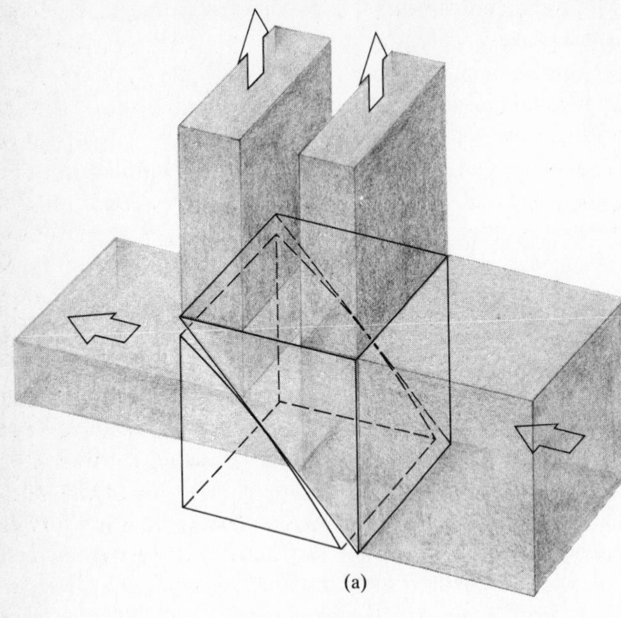
(a)

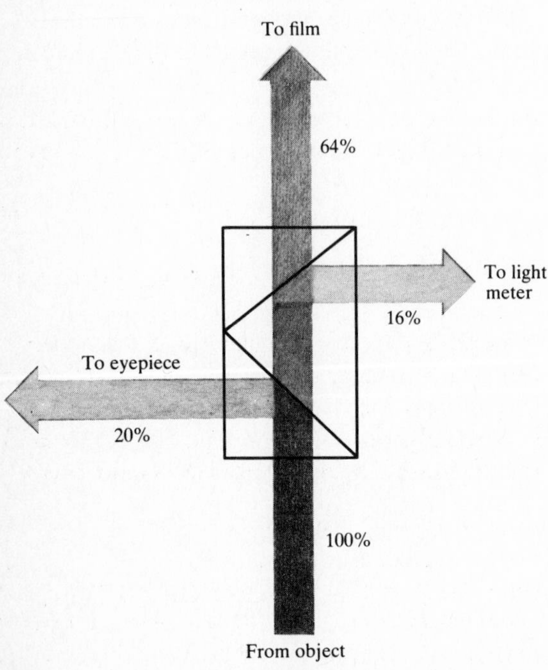

(b)

(c)

FIGURE 4.58 (*a*) A beamsplitter utilizing FTIR. (*b*) A typical modern application of FTIR: a conventional beamsplitter arrangement used to take photographs through a microscope. (*c*) Beamsplitter cubes. (Photo courtesy Melles Griot.)

cube can be made rather conveniently by using a thin, low-index transparent film as a precision spacer. Low-loss reflectors whose transmittance can be controlled by frustrating internal reflection are of considerable practical interest. FTIR can also be observed in other regions of the electromagnetic spectrum. Three-centimeter microwaves are particularly easy to work with, inasmuch as the evanescent wave will extend roughly 10^5 times farther than it would at optical frequencies. One can duplicate the above optical experiments with solid prisms made of paraffin or hollow ones of acrylic plastic filled with kerosene or motor oil. Any one of these would have an index of about 1.5 for 3-cm waves. It then becomes an easy matter to measure the dependence of the field amplitude on y.

4.8 Optical Properties of Metals

The characteristic feature of conducting media is the presence of a number of free electric charges (free in the sense of being unbound, i.e., able to circulate within the material). For metals these charges are of course electrons, and their motion constitutes a current. The current per unit area resulting from the application of a field **E** is related by means of Eq. (A1.15) to the conductivity of the medium σ. For a dielectric there are no free or conduction electrons and $\sigma = 0$, whereas for metals σ is nonzero and finite. In contrast, an idealized "perfect" conductor would have an infinite conductivity. This is equivalent to saying that the electrons, driven into oscillation by a harmonic wave, would simply follow the field's alternations. There would be no restoring force, no natural frequencies, and no absorption, only re-emission. In real metals the conduction electrons undergo collisions with the thermally agitated lattice or with imperfections and in so doing irreversibly convert electromagnetic energy into joule heat. The absorption of radiant energy by a material is a function of its conductivity.

Waves in a Metal

If we visualize the medium as continuous, Maxwell's Equations lead to

$$\frac{\partial^2 \mathbf{E}}{\partial x^2} + \frac{\partial^2 \mathbf{E}}{\partial y^2} + \frac{\partial^2 \mathbf{E}}{\partial z^2} = \mu\epsilon \frac{\partial^2 \mathbf{E}}{\partial t^2} + \mu\sigma \frac{\partial \mathbf{E}}{\partial t} \tag{4.74}$$

which is Eq. (A1.21) in Cartesian coordinates. The last term, $\mu\sigma\, \partial\mathbf{E}/\partial t$, is a first-order time derivative, like the damping force in the oscillator model (p. 71). The time rate-of-change of **E** generates a voltage, currents circulate, and since the material is resistive, light is converted to thermal energy—ergo absorption. This expression can be reduced to the unattenuated wave equation, if the permittivity is reformulated as a complex quantity. This in turn leads to a complex index of refraction, which, as we saw earlier (p. 72), is tantamount to absorption. We then need only substitute the complex index

$$\tilde{n} = n_R - in_I \tag{4.75}$$

(where the real and imaginary indices n_R and n_I are both real numbers) into the corresponding solution for a nonconducting medium. Alternatively, we can utilize the wave equation and appropriate boundary conditions to yield a specific solution. In either event, it is possible to find a simple sinusoidal plane-wave solution applicable within the conductor. Such a wave propagating in the y-direction is ordinarily written as

$$\mathbf{E} = \mathbf{E}_0 \cos(\omega t - ky)$$

or as a function of n,

$$\mathbf{E} = \mathbf{E}_0 \cos \omega(t - \tilde{n}y/c)$$

but here the refractive index must be taken as complex. Writing the wave as an exponential and using Eq. (4.75) yields

$$\mathbf{E} = \mathbf{E}_0 e^{(-\omega n_I y/c)} e^{i\omega(t - n_R y/c)} \tag{4.76}$$

or

$$\mathbf{E} = \mathbf{E}_0 e^{-\omega n_I y/c} \cos \omega(t - n_R y/c) \tag{4.77}$$

The disturbance advances in the y-direction with a speed c/n_R, precisely as if n_R were the more usual index of refraction. As the wave progresses into the conductor, its amplitude, $\mathbf{E}_0$ exp $(-\omega n_I y/c)$, is exponentially attenuated. Inasmuch as irradiance is proportional to the square of the amplitude, we have

$$I(y) = I_0 e^{-\alpha y} \tag{4.78}$$

where $I_0 = I(0)$; that is, I_0 is the irradiance at $y = 0$ (the interface), and $\alpha \equiv 2\omega n_I/c$ is called the *absorption coefficient* or (even better) the **attenuation coefficient**. The flux density will drop by a factor of $e^{-1} = 1/2.7 \approx \frac{1}{3}$ after the wave has propagated a distance $y = 1/\alpha$, known as the **skin** or **penetration depth**. For a material to be transparent the penetration depth must be large in comparison to its thickness. The penetration depth for metals, however, is exceedingly small. For example, copper at ultraviolet wavelengths ($\lambda_0 \approx 100$ nm) has a miniscule penetration depth, about 0.6 nm, while it is still only about 6 nm in the infrared ($\lambda_0 \approx 10\,000$ nm). This accounts for the generally observed opacity of metals, which nonetheless can become partly transparent when formed into extremely thin films (e.g., in the case of partially silvered two-way mirrors). The familiar metallic sheen of conductors corresponds to a high reflectance, which exists because the incident wave cannot effectively penetrate the material. Relatively few electrons in the metal "see" the transmitted wave, and therefore, although each absorbs strongly, little total energy is dissipated by them. Instead, most of the incoming energy reappears as the reflected wave. The majority of metals, including the less common ones (e.g., sodium, potassium, cesium, vanadium, niobium, gadolinium, holmium, yttrium, scandium, and osmium) have a silvery gray appearance like that of aluminum, tin, or steel. They reflect almost all the incident light (roughly 85 to 95%) regardless of wavelengths and are therefore essentially colorless.

Equation (4.77) is certainly reminiscent of Eq. (4.73) and FTIR. In both cases there is an exponential decay of the amplitude. Moreover, a complete analysis would show that the transmitted waves are not strictly transverse, there being a component of the field in the direction of propagation in both instances.

The representation of metal as a continuous medium works fairly well in the low-frequency, long-wavelength domain of the infrared. Yet we certainly might expect that as the wavelength of the incident beam decreased the actual granular nature of matter would have to be reckoned with. Indeed, the continuum model shows large discrepancies from experimental results at optical frequencies. And so we again turn to the classical atomistic picture initially formulated by Hendrik Lorentz, Paul Karl Ludwig Drude (1863–1906), and others. This simple approach will provide qualitative agreement with the experimental data, but the ultimate treatment requires quantum theory.

The Dispersion Equation

Envision the conductor as an assemblage of driven, damped oscillators. Some correspond to free electrons and will therefore have zero restoring force, whereas others are bound to the atom, much like those in the dielectric media of Section 3.5.1. The conduction electrons are, however, the predominant contributors to the optical properties of metals. Recall that the displacement of a vibrating electron was given by

$$x(t) = \frac{q_e/m_e}{(\omega_0^2 - \omega^2)} E(t) \tag{3.66}$$

With no restoring force, $\omega_0 = 0$, the displacement is opposite in sign to the driving force $q_e E(t)$ and therefore 180° out-of-phase with it. This is unlike the situation for transparent dielectrics, where the resonance frequencies are above the visible and the electrons oscillate in-phase with the driving force (Fig. 4.59). Free electrons oscillating out-of-phase with the incident light will reradiate wavelets that tend to cancel the incoming disturbance. The effect, as we have already seen, is a rapidly decaying refracted wave.

Assuming that the average field experienced by an electron moving about within a conductor is just the applied field $\mathbf{E}(t)$, we can extend the dispersion equation of a rare medium [Eq.

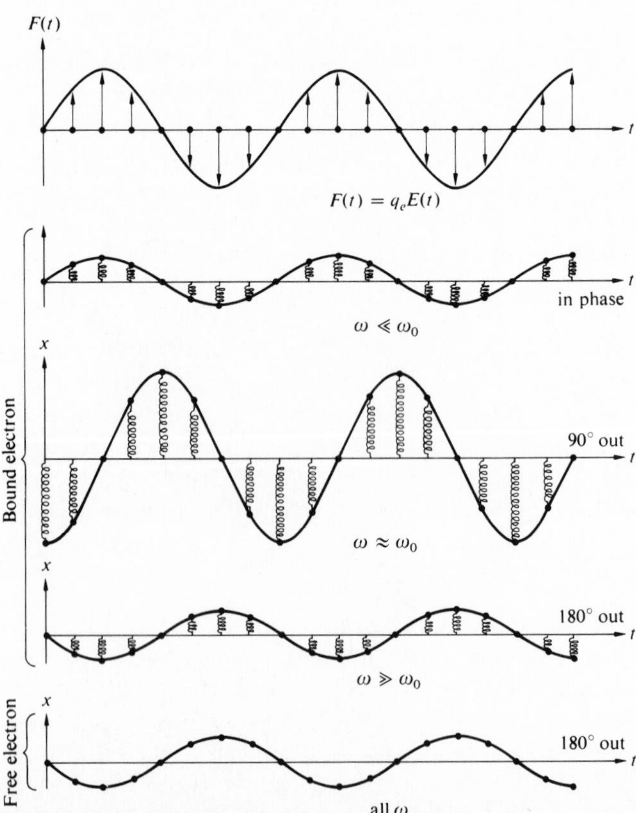

FIGURE 4.59 Oscillations of bound and free electrons.

(3.72)] to read

$$n^2(\omega) = 1 + \frac{Nq^2{}_e}{\epsilon_0 m_e}\left[\frac{f_e}{-\omega^2 + i\gamma_e\omega} + \sum_j \frac{f_j}{\omega_{0j}^2 - \omega^2 + i\gamma_j\omega}\right] \tag{4.79}$$

The first bracketed term is the contribution from the free electrons, wherein N is the number of atoms per unit volume. Each of these has f_e conduction electrons, which have no natural frequencies. The second term arises from the bound electrons and is identical to Eq. (3.72). It should be noted that if a metal has a particular color, it indicates that the atoms are partaking of selective absorption by way of the bound electrons, in addition to the general absorption characteristic of the free electrons. Recall that a medium that is very strongly absorbing at a given frequency doesn't actually absorb much of the incident light at that frequency but rather *selectively reflects* it. Gold and copper are reddish yellow because n_I increases with wavelength, and the larger values of λ are reflected more strongly. Thus, for example, gold should be fairly opaque to the longer visible wavelengths. Consequently, under white light, a gold foil less than roughly 10^{-6} m thick will indeed transmit predominantly greenish blue light.

We can get a rough idea of the response of metals to light by making a few simplifying assumptions. Accordingly, neglect the bound electron contribution and assume that γ_e is also negligible for very large ω, whereupon

$$n^2(\omega) = 1 - \frac{Nq_e^2}{\epsilon_0 m_e \omega^2} \tag{4.80}$$

The latter assumption is based on the fact that at high frequencies the electrons will undergo a great many oscillations between each collision. Free electrons and positive ions within a metal may be thought of as a plasma whose density oscillates at a natural frequency ω_p, the **plasma frequency**. This in turn can be shown to equal $(Nq_e^2/\epsilon_0 m_e)^{1/2}$, and so

$$n^2(\omega) = 1 - (\omega_p/\omega)^2 \tag{4.81}$$

The plasma frequency serves as a critical value below which the index is complex and the penetrating wave drops off exponentially [Eq. (4.77)] from the boundary; at frequencies above ω_p, n is real, absorption is small, and the conductor is transparent. In the latter circumstance n is less than 1, as it was for dielectrics at very high frequencies (v can be greater than c—see p.72). Hence we can expect metals in general to be fairly transparent to X-rays. Table 4.3 lists the plasma frequencies for some of the alkali metals that are transparent even to ultraviolet.

TABLE 4.3 Critical Wavelengths and Frequencies for Some Alkali Metals

Metal	λ_p (observed) nm	λ_p (calculated) nm	$\nu_p = c/\lambda_p$ (observed) Hz
Lithium (Li)	155	155	1.94×10^{15}
Sodium (Na)	210	209	1.43×10^{15}
Potassium (K)	315	287	0.95×10^{15}
Rubidium (Rb)	340	322	0.88×10^{15}

The index of refraction for a metal will usually be complex, and the impinging wave will suffer absorption in an amount that is frequency dependent. For example, the outer visors on the Apollo space suits were overlaid with a very thin film of gold (Fig. 4.60). The coating reflected about 70% of the incident light and was used under bright conditions, such as low and forward Sun angles. It was designed to decrease the ther-

FIGURE 4.60 Edwin Aldrin Jr. at Tranquility Base on the Moon. The photographer, Neil Armstrong, is reflected in the gold-coated visor. (Photo courtesy NASA.)

mal load on the cooling system by strongly reflecting radiant energy in the infrared while still transmitting adequately in the visible. Inexpensive metal-coated sunglasses which are quite similar in principle are also available commercially, and they're well worth having just to experiment with.

The ionized upper atmosphere of the Earth contains a distribution of free electrons that behave very much like those confined within a metal. The index of refraction of such a medium will be real and less than 1 for frequencies above ω_p. In July of 1965 the *Mariner IV* spacecraft made use of this effect to examine the ionosphere of the planet Mars, 216 million kilometers from Earth.*

If we wish to communicate between two distant terrestrial points, we might bounce low-frequency waves off the Earth's ionosphere. To speak to someone on the Moon, however, we should use high-frequency signals, to which the ionosphere would be transparent.

Reflection from a Metal

Imagine that a plane wave initially in air impinges on a conducting surface. The transmitted wave advancing at some angle to the normal will be inhomogeneous. But if the conductivity of the medium is increased, the wavefronts will become aligned with the surfaces of constant amplitude, whereupon $\mathbf{k}_t$ and $\hat{\mathbf{u}}_n$ will approach parallelism. In other words, in a good conductor the transmitted wave propagates in a direction normal to the interface regardless of θ_i.

Let's now compute the reflectance, $R = I_r/I_i$, for the simplest case of normal incidence on a metal. Taking $n_i = 1$ and $n_t = \tilde{n}$ (i.e., the complex index), we have from Eq. (4.47) that

$$R = \left(\frac{\tilde{n}-1}{\tilde{n}+1}\right)\left(\frac{\tilde{n}-1}{\tilde{n}+1}\right)^* \tag{4.82}$$

and therefore, since $\tilde{n} = n_R - in_I$,

$$R = \frac{(n_R-1)^2 + n_I^2}{(n_R+1)^2 + n_I^2} \tag{4.83}$$

If the conductivity of the material goes to zero, we have the case of a dielectric, whereupon in principle the index is real ($n_I = 0$), and the attenuation coefficient, α, is zero. Under those circumstances, the index of the transmitting medium n_t is n_R, and the reflectance [Eq. (4.83)] becomes identical with that of Eq. (4.67). If instead n_I is large while n_R is comparatively small, R in turn becomes large (Problem 4.72). In the unattainable limit where $\tilde{n}$ is purely imaginary, 100% of the incident flux density would be reflected ($R = 1$). Notice that it is possible for the reflectance of one metal to be greater than that of another even though its n_I is smaller. For example, at $\lambda_0 = 589.3$ nm the parameters associated with solid sodium are roughly $n_R = 0.04$, $n_I = 2.4$, and $R = 0.9$; and those for bulk tin are $n_R = 1.5$, $n_I = 5.3$, and $R = 0.8$; whereas for a gallium single crystal $n_R = 3.7$, $n_I = 5.4$, and $R = 0.7$.

The curves of $R_\parallel$ and $R_\perp$ for oblique incidence shown in Fig. 4.61 are somewhat typical of absorbing media. Thus, although R at $\theta_i = 0$ is about 0.5 for gold, as opposed to nearly 0.9 for silver in white light, the two metals have reflectances that are quite similar in shape, approaching 1.0 at $\theta_i = 90°$. Just as with dielectrics (Fig. 4.47), $R_\parallel$ drops to a minimum at what is now called the *principal angle-of-incidence*, but here that minimum is nonzero. Figure 4.62 illustrates the spectral reflectance at normal incidence for a number of evaporated metal films under ideal conditions. Observe that although gold transmits fairly well in and below the green region of the spectrum, silver, which is highly reflective across the visible, becomes transparent in the ultraviolet at about 316 nm.

Phase shifts arising from reflection off a metal occur in both components of the field (i.e., parallel and perpendicular to the plane of incidence). These are generally neither 0 nor π, with a notable exception at $\theta_i = 90°$, where, just as with a dielectric, both components shift phase by 180° on reflection.

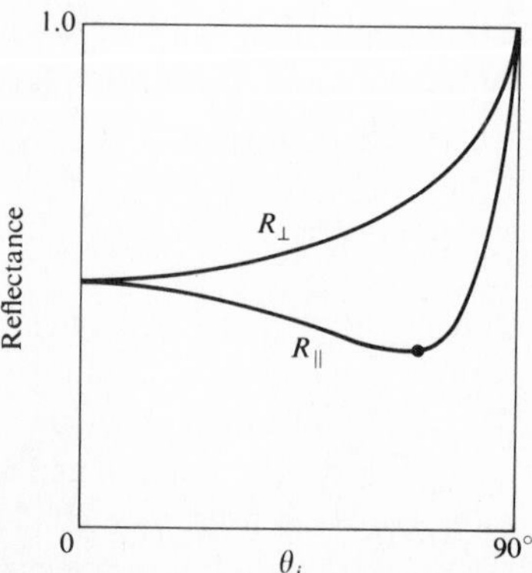

FIGURE 4.61 Typical reflectance for a linearly polarized beam of white light incident on an absorbing medium.

*R. Von Eshelman, *Sci. Am.* 220, 78 (1969).

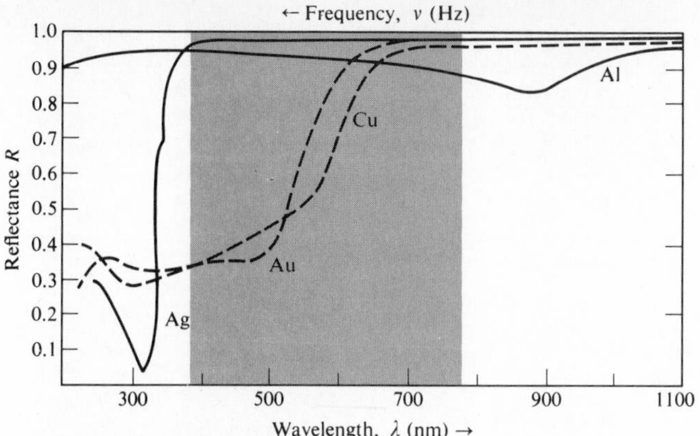

FIGURE 4.62 Reflectance versus wavelength for silver, gold, copper, and aluminum.

4.9 Familiar Aspects of the Interaction of Light and Matter

Let's now examine some of the phenomena that paint the everyday world in a marvel of myriad colors.

As we saw earlier (p. 77), light that contains a roughly equal amount of every frequency in the visible region of the spectrum is perceived as white. A broad source of white light (whether natural or artificial) is one for which every point on its surface can be imagined as sending out a stream of light of every visible frequency. Given that we evolved on this planet, it's not surprising that a source appears white when its emission spectrum resembles that of the Sun. Similarly, a reflecting surface that accomplishes essentially the same thing will also appear white: a highly reflecting, frequency-independent, *diffusely* scattering object will be perceived as white under white light illumination.

Although water is essentially transparent, water vapor appears white, as does ground glass. The reason is simple enough—if the grain size is small but larger than the wavelengths involved, light will enter each transparent particle, be reflected and refracted, and emerge. There will be no distinction among any of the frequency components, so the reflected light reaching the observer will be white. This is the mechanism accountable for the whiteness of things like sugar, salt, paper, cloth, clouds, talcum powder, snow, and paint, each grain or fiber of which is actually transparent.

Similarly, a wadded-up piece of crumpled clear plastic wrap will appear whitish, as will an ordinarily transparent material filled with small air bubbles (e.g., shaving cream or beaten egg white). Even though we usually think of paper, talcum powder, and sugar as each consisting of some sort of opaque white substance, it's an easy matter to dispel that misconception. Cover a printed page with a few of these materials (a sheet of white paper, some grains of sugar, or talcum) and illuminate it from behind. You'll have little difficulty seeing through them. In the case of white paint, one simply suspends colorless transparent particles, such as the oxides of zinc, titanium, or lead, in an equally transparent vehicle, for example, linseed oil or acrylics. Obviously, if the particles and vehicle have the same index of refraction, there will not be any reflections at the grain boundaries. The particles will simply disappear into the conglomeration, which itself remains clear. In contrast, if the indices are markedly different, there will be a good deal of reflection at all wavelengths (Problem 4.65), and the paint will appear white and opaque [take another look at Eq. (4.67)]. To color paint one need only dye the particles so that they absorb all frequencies except the desired range.

Carrying the logic in the reverse direction, if we reduce the relative index, n_{ti}, at the grain or fiber boundaries, the particles of material will reflect less, thereby decreasing the overall whiteness of the object. Consequently, a wet white tissue will have a grayish, more transparent look. Wet talcum powder loses its sparkling whiteness, becoming a dull gray, as does wet white cloth. In the same way, a piece of dyed fabric soaked in a clear liquid (e.g., water, gin, or benzene) will lose its whitish haze and become much darker, the colors then being deep and rich like those of a still-wet watercolor painting.

A diffusely reflecting surface that absorbs somewhat—uniformly across the spectrum—will reflect a bit less than a white surface and so appear mat gray. The less it reflects, the darker the gray, until it absorbs almost all the light and appears black. A surface that reflects perhaps 70% or 80% or more, but does so specularly, will appear the familiar shiny gray of a typical metal. Metals possess tremendous numbers of free electrons (p. 128) that scatter light very effectively, independent of frequency: they are not bound to the atoms and have no associated resonances. Moreover, the amplitudes of the vibrations are an order of magnitude larger than they were for the bound electrons. The incident light cannot penetrate into the metal any more than a fraction of a wavelength or so before it's canceled completely. There is little or no refracted light; most of the energy is reflected out, and only the small remainder is absorbed. Note that the primary difference between a gray surface and a mirrored surface is one of diffuse versus specular reflection. An artist paints a picture of a polished "white" metal, such as silver or aluminum, by "reflecting" images of things in the room on top of a gray surface.

When the distribution of energy in a beam of light is not effectively uniform across the spectrum, the light appears colored. Figure 4.63 depicts typical frequency distributions for what would be perceived as red, green, and blue light. These curves show the predominant frequency regions, but there can be a great deal of variation in the distributions, and they will still provoke the responses of red, green, and blue. In the early 1800s Thomas Young showed that a broad range of colors could be generated by mixing three beams of light, provided their frequencies were widely separated. When three such beams combine to produce white light, they are called **primary colors**. There is no single unique set of these primaries, nor do they have to be quasimonochromatic. Since a wide range of colors can be created by mixing red (R), green (G), and blue (B), these tend to be used most frequently. They are the three components (emitted by three phosphors) that generate the whole gamut of hues seen on a color television set.

Figure 4.64 summarizes the results when beams of these three primaries are overlapped in a number of different combinations: Red light plus blue light is seen as *magenta* (M), a reddish purple; blue light plus green light is seen as *cyan* (C), a bluish green or turquoise; and perhaps most surprising, red light plus green light is seen as *yellow* (Y). The sum of all three primaries is white:

$$\mathrm{R} + \mathrm{B} + \mathrm{G} = \mathrm{W}$$

$$\mathrm{M} + \mathrm{G} = \mathrm{W}, \text{ since } \mathrm{R} + \mathrm{B} = \mathrm{M}$$

$$\mathrm{C} + \mathrm{R} = \mathrm{W}, \text{ since } \mathrm{B} + \mathrm{G} = \mathrm{C}$$

$$\mathrm{Y} + \mathrm{B} = \mathrm{W}, \text{ since } \mathrm{R} + \mathrm{G} = \mathrm{Y}$$

Any two colored light beams that together produce white are said to be **complementary**, and the last three symbolic statements exemplify that situation. Thus

$$\mathrm{R} + \mathrm{B} + \mathrm{G} = \mathrm{W}$$

$$\mathrm{R} + \mathrm{B} \qquad = \mathrm{W} - \mathrm{G} = \mathrm{M}$$

$$\mathrm{B} + \mathrm{G} = \mathrm{W} - \mathrm{R} = \mathrm{C}$$

$$\mathrm{R} + \qquad \mathrm{G} = \mathrm{W} - \mathrm{B} = \mathrm{Y}$$

which means, for example, that a filter that absorbs blue out of white light passes yellow.

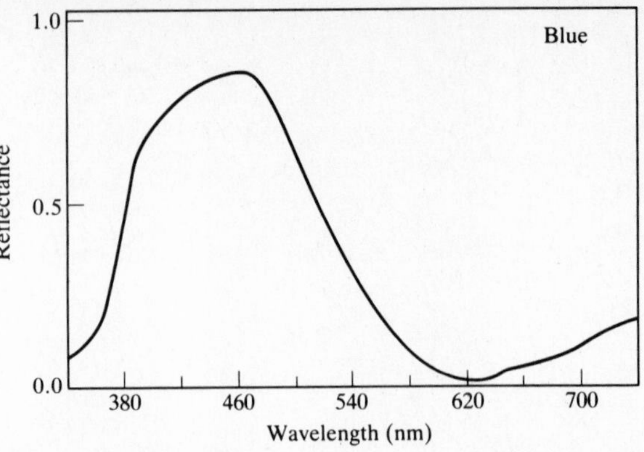

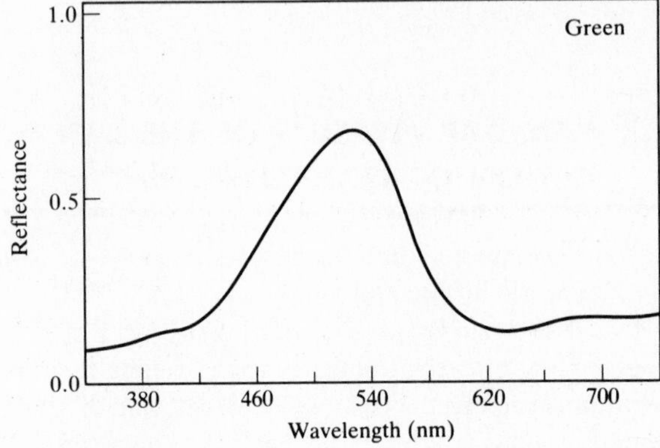

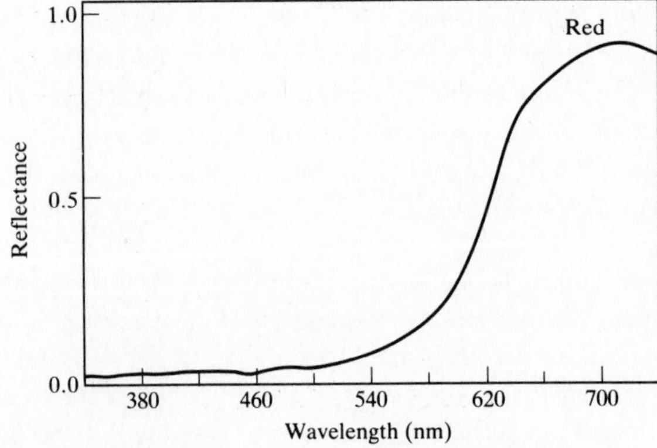

FIGURE 4.63 Reflection curves for blue, green, and red pigments. These are typical, but there is a great deal of possible variation among the colors.

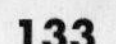

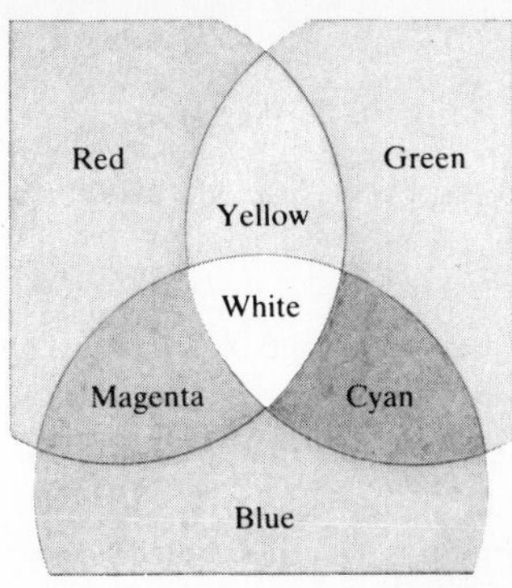

FIGURE 4.64 Three overlapping beams of colored light. A color television set uses these same three primary light sources—red, green, and blue.

Suppose we overlap beams of magenta and yellow light:

$$M + Y = (R + B) + (R + G) = W + R$$

The result is a combination of red and white, or pink. That raises another point: we say a color is **saturated**, that it is deep and intense, when it does not contain any white light. As Fig. 4.65 shows, pink is unsaturated red—red superimposed on a background of white.

The mechanism responsible for the yellowish red hue of gold and copper is, in some respects, similar to the process that causes the sky to appear blue. Putting it rather succinctly, the molecules of air have resonances in the ultraviolet and will be driven into larger-amplitude oscillations as the frequency of the incident light increases toward the ultraviolet. They effectively take energy from and re-emit the blue component of sunlight in all directions, transmitting the complementary red end of the spectrum with little alteration. This is analogous to the selective reflection or scattering of yellow-red light that takes place at the surface of a gold film and the concomitant transmission of blue-green light.

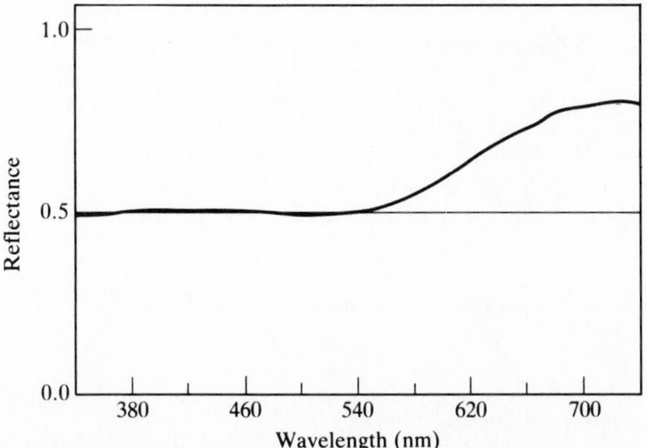

FIGURE 4.65 Spectral reflection of a pink pigment.

The characteristic colors of most substances have their origin in the phenomenon of **selective** or **preferential absorption**. For example, water has a very faint green-blue tint because of its absorption of red light. That is, the H_2O molecules have a broad resonance in the infrared, which extends somewhat into the visible. The absorption isn't very strong, so there is no accentuated reflection of red light at the surface. Instead it is transmitted and gradually absorbed out until at a depth of about 30 m of seawater, red is almost completely removed from the sunlight. This same process of selective absorption is responsible for the colors of brown eyes and butterflies, of birds and bees and cabbages and kings. Indeed, the great majority of objects in nature appear to have characteristic colors as the result of preferential absorption by pigment molecules. In contrast with most atoms and molecules, which have resonances in the ultraviolet and infrared, the pigment molecules must obviously have resonances in the visible. Yet visible photons have energies of roughly 1.6 eV to 3.2 eV, which, as you might expect, are on the low side for ordinary electron excitation and on the high side for excitation via molecular vibration. Despite this, *there are atoms where the bound electrons form incomplete shells* (gold, for example) and variations in the configuration of these shells provide a mode for low-energy excitation. In addition, there is the large group of organic dye molecules, which evidently also have resonances in the visible. All such substances, whether natural or synthetic, consist of long-chain molecules made up of regularly alternating single and double bonds in what is called a conjugated system. This structure is typified by the carotene molecule $C_{40}H_{56}$ (Fig. 4.66). The carotenoids range in color from yellow to red and are found in carrots, tomatoes, daffodils, dandelions, autumn leaves, and people. The chlorophylls are another group of familiar natural pigments, but here a portion of the long chain is turned around on itself to form a ring. In any event, conjugated systems of this sort contain a number of particularly mobile electrons known as *pi electrons.* They are not bound to specific atomic sites but instead can range over the relatively large dimensions of the molecular chain or ring. In the phraseology of Quantum Mechanics, we say that these are long-wavelength, low-frequency, and therefore low-energy, electron states. The energy required to raise a pi electron to an excited state is comparatively low, corresponding to that of visible photons. In effect, the molecule can

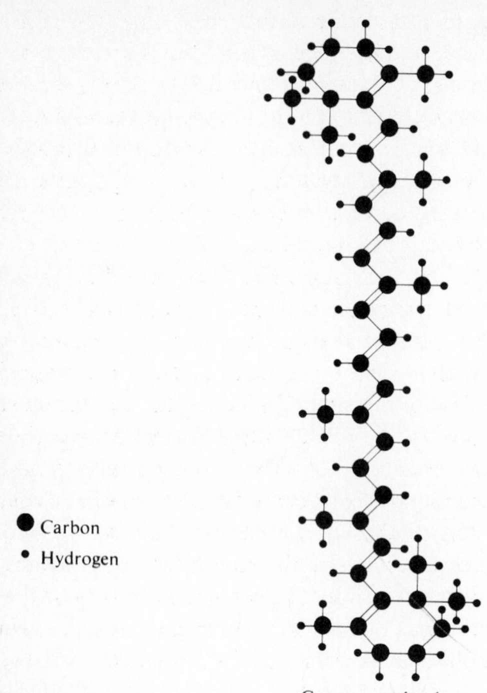

FIGURE 4.66 The carotene molecule.

be imagined as an oscillator having a resonance frequency in the visible.

The energy levels of an individual atom are precisely defined; that is, the resonances are very sharp. With solids and liquids, however, the proximity of the atoms results in a broadening of the energy levels into wide bands. The resonances spread over a broad range of frequencies. Consequently, we can expect that a dye will not absorb just a narrow portion of the spectrum; indeed if it did, it would reflect most frequencies and appear nearly white.

Imagine a piece of stained glass with a resonance in the blue where it strongly absorbs. If you look through it at a white-light source composed of red, green, and blue, the glass will absorb blue, passing red and green, which is yellow (Fig. 4.67). The glass looks yellow: yellow cloth, paper, dye, paint, and ink all selectively absorb blue. If you peer at something that is a pure blue through a yellow filter, one that passes yellow and absorbs blue, the object will appear black. Here the filter colors the light yellow by removing blue, and we speak of the process as **subtractive coloration**, as opposed to **additive coloration**, which results from overlapping beams of light.

In the same way, fibers of a sample of white cloth or paper are essentially transparent, but when dyed each fiber behaves as if it were a chip of colored glass. The incident light penetrates the paper, emerging for the most part as a reflected beam only after undergoing numerous reflections and refractions within the dyed fibers. The exiting light will be colored to the extent that it lacks the frequency component absorbed by the dye. This is precisely why a leaf appears green, or a banana yellow.

A bottle of ordinary blue ink looks blue in either reflected or transmitted light. But if the ink is painted on a glass slide and the solvent evaporates, something rather interesting happens. The concentrated pigment absorbs so effectively that it preferentially reflects at the resonant frequency, and we are back to the idea that a strong absorber (larger n_I) is a strong reflector. Thus, concentrated blue-green ink reflects red, whereas red-blue ink reflects green. Try it with a felt marker (overhead projector pens are best), but you must use reflected light, being careful not to inundate the sample with unwanted light from below. The most convenient way to accomplish that is to put colored ink onto a black surface that isn't very absorbant. For example, smear red ink over a black area on a glossy printed page and it will glow green in reflected light. Gentian violet, which you can buy in any drugstore, works

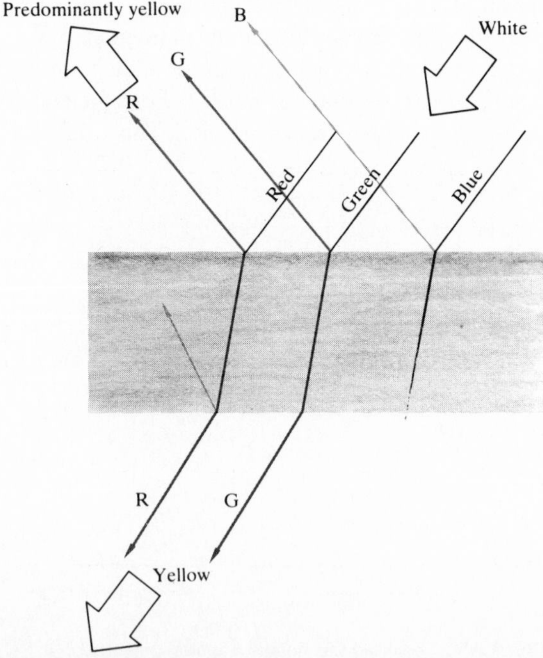

FIGURE 4.67 Yellow stained glass.

beautifully. Put some on a glass slide and let it dry in a thick coat. Examine both the reflected and transmitted light—they will be complementary.

The whole range of colors (including red, green, and blue) can be produced by passing white light through various combinations of magenta, cyan, and yellow filters (Fig. 4.68).

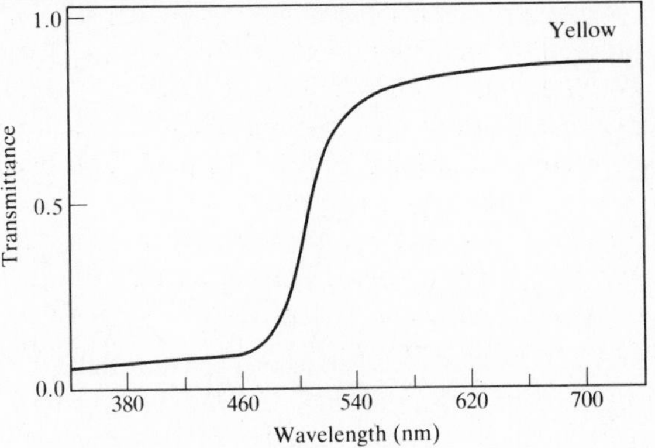

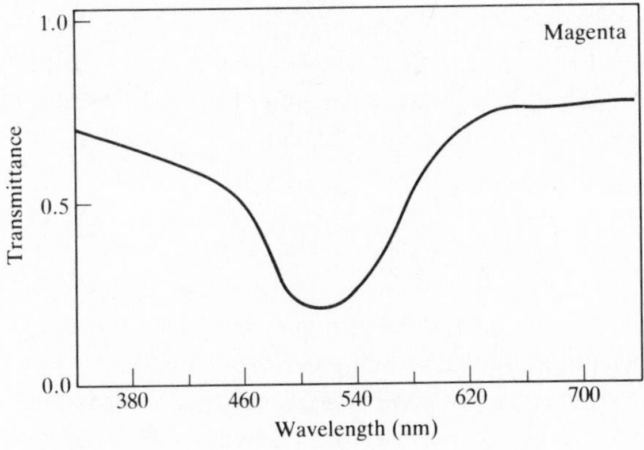

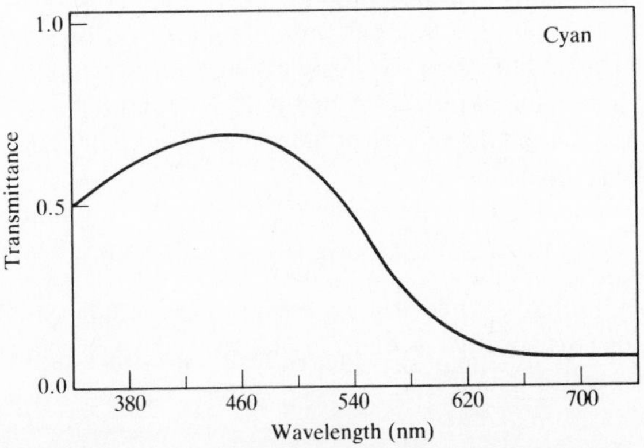

FIGURE 4.68 Transmission curves for colored filters.

These are the primary colors of subtractive mixing, the primaries of the paint box, although they are often mistakenly spoken of as red, blue, and yellow. They are the basic colors of the dyes used to make photographs and the inks used to print them. Ideally, if you mix all the subtractive primaries together (either by combining paints or by stacking filters), you get no color, no light—black. Each removes a region of the spectrum, and together they absorb it all.

If the range of frequencies being absorbed spreads across the visible, the object will appear black. That is not to say that there is no reflection at all—you obviously can see a reflected image in a piece of black patent leather, and a rough black surface reflects also, only diffusely. If you still have those red and blue inks, mix them, add some green, and you'll get black.

In addition to the above processes specifically related to reflection, refraction, and absorption, there are other color-generating mechanisms, which we shall explore later on. For example, scarabaeid beetles mantle themselves in the brilliant colors produced by diffraction gratings on their wing cases, and wavelength-dependent interference effects contribute to the color patterns seen on oil slicks, mother-of-pearl, soap bubbles, peacocks, and hummingbirds.

4.10 The Stokes Treatment of Reflection and Refraction

A rather elegant and novel way of looking at reflection and transmission at a boundary was developed by the British physicist Sir George Gabriel Stokes (1819–1903). Suppose that we have an incident wave of amplitude E_{0i} impinging on the planar interface separating two dielectric media, as in Fig. 4.69*a*. As we saw earlier in this chapter, because r and t are the fractional amplitudes reflected and transmitted, respectively (where $n_i = n_1$ and $n_t = n_2$), then $E_{0r} = rE_{0i}$ and $E_{0t} = tE_{0i}$. Again we are reminded that Fermat's Principle led to the Principle of Reversibility, which implies that the situation depicted in Fig. 4.69*b*, where all the ray directions are reversed, must also be physically possible. With the one proviso that there be no energy dissipation (no absorption), a wave's meanderings must be reversible. Equivalently, in the idiom of modern physics one speaks of *time-reversal invariance*, that is, if a process occurs, the reverse process can also occur. Thus if we take a hypothetical motion picture of the wave incident on, reflecting from, and transmitting through the interface, the behavior depicted when the film is run backward must also be

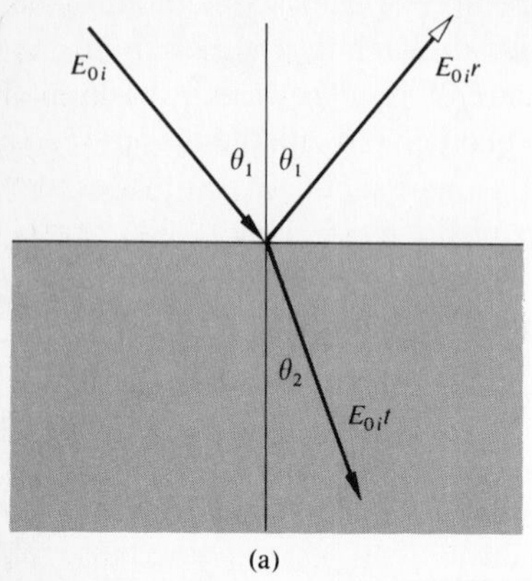

(a)

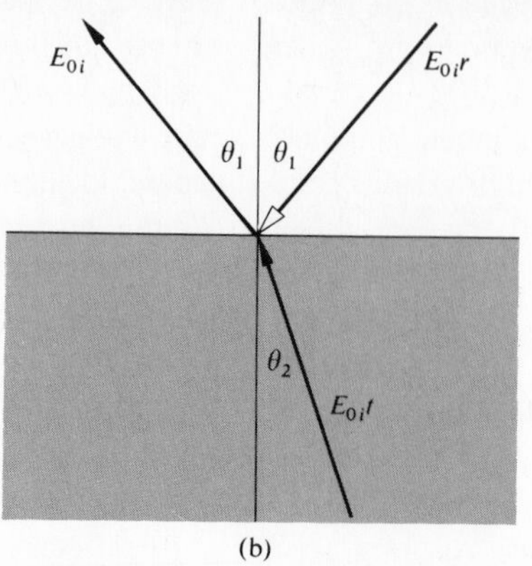

(b)

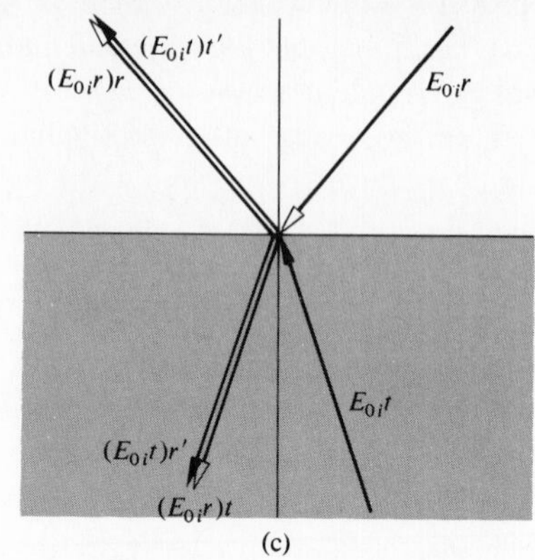

(c)

FIGURE 4.69 Reflection and refraction via the Stokes treatment.

physically realizable. Accordingly, examine Fig. 4.69*c*, where there are now two incident waves of amplitudes $E_{0i}r$ and $E_{0i}t$. A portion of the wave whose amplitude is $E_{0i}t$ is both reflected and transmitted at the interface. Without making any assumptions, let r' and t' be the amplitude reflection and transmission coefficients, respectively, for a wave incident from below (i.e., $n_i = n_2$, $n_t = n_1$). Consequently, the reflected portion is $E_{0i}tr'$, and the transmitted portion is $E_{0i}tt'$. Similarly, the incoming wave whose amplitude is $E_{0i}r$ splits into segments of amplitude $E_{0i}rr$ and $E_{0i}rt$. If the configuration in Fig. 4.69*c* is to be identical with that in Fig. 4.69*b*, then obviously

$$E_{0i}tt' + E_{0i}rr = E_{0i} \tag{4.84}$$

and

$$E_{0i}rt + E_{0i}tr' = 0 \tag{4.85}$$

Hence

$$tt' = 1 - r^2 \tag{4.86}$$

and

$$r' = -r \tag{4.87}$$

the latter two equations being known as the Stokes Relations. This discussion calls for a bit more caution than is usually granted it. It must be pointed out that *the amplitude coefficients are functions of the incident angles*, and therefore the Stokes Relations might better be written as

$$t(\theta_1)t'(\theta_2) = 1 - r^2(\theta_1) \tag{4.88}$$

and

$$r'(\theta_2) = -r(\theta_1) \tag{4.89}$$

where $n_1 \sin \theta_1 = n_2 \sin \theta_2$. The second equation indicates, by virtue of the minus sign, that *there is a 180° phase difference between the waves internally and externally reflected*. It is most important to keep in mind that here θ_1 and θ_2 are pairs of angles that are related by way of Snell's Law. Note as well that we never did say whether n_1 was greater or less than n_2, so Eqs. (4.88) and (4.89) apply in either case. Let's return for a moment to one of the Fresnel Equations:

$$r_\perp = -\frac{\sin(\theta_i - \theta_t)}{\sin(\theta_i + \theta_t)} \tag{4.42}$$

If a ray enters from above, as in Fig. 4.69*a*, and we assume $n_2 > n_1$, $r_\perp$ is computed by setting $\theta_i = \theta_1$ and $\theta_t = \theta_2$ (external reflection), the latter being derived from Snell's Law. If, on the other hand, the wave is incident at that same angle from below (in this instance internal reflection), $\theta_i = \theta_1$ and we again substitute in Eq. (4.42), but here θ_t is not θ_2, as before. The values of $r_\perp$ for internal and external reflection *at the same incident angle* are obviously different. Now suppose, in this case of internal reflection, that $\theta_i = \theta_2$. Then $\theta_t = \theta_1$, the ray directions are the reverse of those in the first situation, and Eq. (4.42) yields

$$r'_\perp(\theta_2) = -\frac{\sin(\theta_2 - \theta_1)}{\sin(\theta_2 + \theta_1)}$$

Although it may be unnecessary we once again point out that this is just the negative of what was determined for $\theta_i = \theta_1$ and external reflection, that is,

$$r'_\perp(\theta_2) = -r_\perp(\theta_1) \tag{4.90}$$

The use of primed and unprimed symbols to denote the amplitude coefficients should serve as a reminder that we are once more dealing with angles related by Snell's Law. In the same way, interchanging θ_i and θ_t in Eq. (4.43) leads to

$$r'_{\parallel}(\theta_2) = -r_{\parallel}(\theta_1) \tag{4.91}$$

The 180° phase difference between each pair of components is evident in Fig. 4.43, but keep in mind that when $\theta_i = \theta_p$, $\theta_t = \theta'_p$ and vice versa (Problem 4.69). Beyond $\theta_i = \theta_c$ there is no transmitted wave, Eq. (4.89) is not applicable, and as we have seen, the phase difference is no longer 180°.

It is common to conclude that both the parallel and perpendicular components of the externally reflected beam change phase by π radians while the internally reflected beam undergoes no phase shift at all. This is quite incorrect (compare Figs. 4.44*a* and 4.45*a*).

4.11 Photons, Waves, and Probability

Much of the theoretical grounding of Optics is predicated on wave theory. We take for granted both that we understand the phenomenon and that it's "real." As one example out of the many that will be encountered, the process of scattering seems to be understandable only in terms of interference; classical particles simply do not interfere. When a beam propagates through a dense medium, interference in the forward direction is constructive, whereas in all other directions it's almost completely destructive. Thus nearly all the light energy advances in the forward direction. But this raises interesting questions about the basic nature of interference and the usual interpretation of what's happening. ***Interference is a nonlocalized phenomenon; it cannot happen at only one single point in space***, even though we often talk about the interference at a point P. The principle of Conservation of Energy makes it clear that if there is constructive interference at one point, the "extra" energy at that location must have come from elsewhere. There must therefore be destructive interference somewhere else. ***Interference takes place over an extended region of space in a coordinated fashion that leaves the total amount of radiant energy unchanged.***

Now imagine a light beam traversing a dense medium, as in Fig. 4.6. Do real energy-carrying electromagnetic wavelets (which are never actually measured) propagate out laterally, only to interfere destructively everywhere beyond the beam? If so, these wavelets cancel and the energy they transport outward is inexplicably returned to the beam since, in the end, there is no net lateral scattering. That's true no matter how far away P is. Moreover, this applies to *all* interference effects (Chapter 9). If two or more electromagnetic waves arrive at point P out-of-phase and cancel, "What does that mean as far as their energy is concerned?" Energy can be redistributed, but *it* doesn't cancel out. We've learned from Quantum Mechanics that at base interference is one of the most fundamental mysteries in physics.

Remembering Einstein's admonition that there are no spherical wavelets emitted by atoms, perhaps we're being too literal in our interpretation of the classical wave field. After all, strictly speaking, the classical electromagnetic wave with its continuous distribution of energy does not actually exist. Perhaps we should think of the wavelets and the overall pattern they produce (rather than being a *real* wave field) as a theoretical device that, wonderfully enough, tells us where the light will end up. In any event, Maxwell's Equations provide a means of calculating the macroscopic distribution of electromagnetic energy in space.

Moving ahead in a semiclassical way, imagine a distribution of light given by some function of the off-axis angle θ. For example, consider the irradiance on a screen placed far beyond a slit-shaped aperture (p. 442) such that $I(\theta) = I(0)\,\text{sinc}^2\beta(\theta)$. Suppose that instead of observing the pattern by eye a detector composed of a diaphragm followed by a photomultiplier tube is used. Such a device could be moved around from one point to another, and over a constant time interval, it could measure the number of photons arriving at each location, $N(\theta)$. Taking a great many such measurements, a spatial distribution of the number of photon counts would emerge that would be of the very same form as that for the irradiance, namely, $N(\theta) = N(0)\text{sinc}^2\beta(\theta)$: the number of photons detected is proportional to the irradiance. A countable quantity like this lends itself to statistical analysis, and we can talk about the probability of detecting a photon at any point on the screen. That is, a probability distribution can be constructed, reminiscent of Fig. 3.19. Because the space variables (θ, x, y, or z) are continuous, it's necessary to introduce a **probability density**, let it be $\wp(\theta)$. Then $\wp(\theta)\,d\theta$ is the probability that a photon will be found in the infinitesimal range from θ to $\theta + d\theta$. In this case $\wp(\theta) = \wp(0)\,\text{sinc}^2\beta(\theta)$.

The square of the net electric field amplitude at every point in space corresponds to the irradiance (which can be measured directly), and that's equivalent to the likelihood of finding photons at any point. Accordingly, let's tentatively define the **probability amplitude** as that quantity whose absolute value squared equals the probability density. Thus the net E_0 at P can be interpreted as being proportional to a **semiclassical** probability amplitude inasmuch as ***the probability of detecting a photon at some point in space depends on the irradiance at that location and*** $I \propto E_0^2$. This is in accord with Einstein's conception of the light field, which Max Born (who initiated the statistical interpretation of Quantum Mechanics) described as a *Gespensterfeld*, or phantom field. In that view the waves of that field reveal how the photons distribute in space in the sense that the square of the absolute value of the wave amplitude somehow relates to the probability density of arriving photons. In the formal treatment of Quantum Mechanics, *the probability amplitude is generally a complex quantity* whose absolute value squared corresponds to the probability density (e.g., the Schrödinger wavefunction is a probability amplitude). Thus, however reasonable it was to consider E_0 as equivalent to a semiclassical probability amplitude, that usage cannot be carried over, as is, into quantum theory.

Still, all of this suggests that we might take the scattering process, considered in terms of probabilities, as the basis for a computational scheme. Each scattered wavelet is then a measure of the probability amplitude for light taking a particular route from one point to another, and the net electric field at P is the sum of all the scattered fields arriving via all possible routes. A quantum-mechanical methodology analogous to this was devised by Feynman, Schwinger, Tomonaga, and Dyson in the course of their development of Quantum Electrodynamics. In brief, the final observable outcome of an event is determined by the superposition of all the various probability amplitudes associated with each and every possible way that the event can occur. In other words, each "route" along which an event can take place, each way it can happen, is given an abstract mathematical representation, a complex probability amplitude. All of these then combine—and interfere, as complex quantities are wont to do—to produce a net probability amplitude for the event to take place.

What follows is a greatly simplified version of that analysis.

4.11.1 QED

Feynman was rather unequivocal in his stance regarding the nature of light:

> I want to emphasize that light comes in this form—particles. It is very important to know that light behaves like particles, especially for those of you who have gone to school, where you were probably told something about light behaving like waves. I'm telling you the way it *does* behave—like particles.

For him "light is made of particles (as Newton originally thought)"; it's a stream of photons whose behavior *en masse* can be determined statistically. For example, if 100 photons are incident perpendicularly on a piece of glass in air, on average 4 will be reflected backward from the first surface they encounter. Which 4 cannot be known, and in fact how those particular 4 photons are selected is a mystery. What can be deduced and confirmed experimentally is that 4% of the incident light will be reflected (p. 119).

Feynman's analysis proceeds from a few general computational rules, with the ultimate justification being that it works; the scheme makes accurate predictions. (1) ***The probability amplitude associated with the occurrence of an event is the "sum" of the constituent probability amplitudes correspond-***

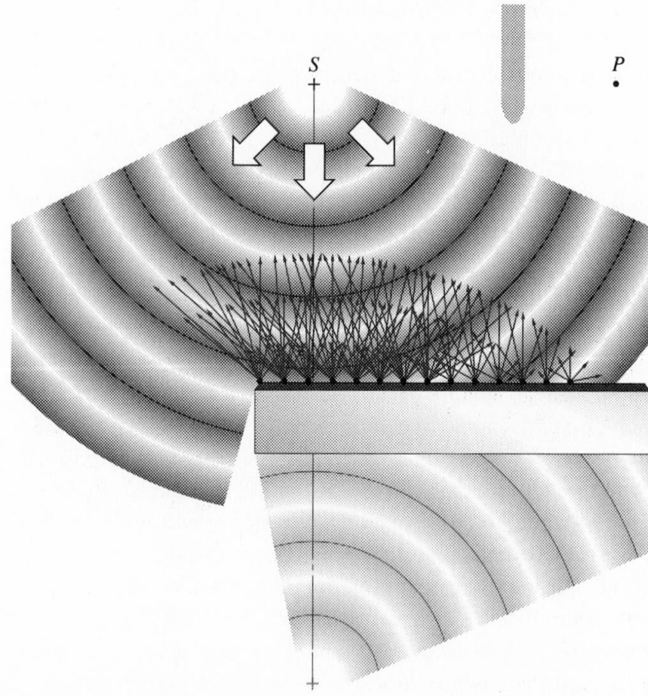

FIGURE 4.70 A schematic representation of reflection. A wave from *S* sweeps down and spreads across the surface of the mirror. Every atom on the interface subsequently scatters light back in all upward directions. And some of it ultimately arrives at *P* having come from every scatterer on the surface.

ing to each and every possible way the event can occur. (2) ***Each such constituent probability amplitude is generally expressible as a complex quantity***. Rather than analytically combining these constituent probability amplitudes, we can use the phasor representation (p. 24) to approximate the summation and thereby arrive at a resultant probability amplitude. (3) ***The probability of occurrence of the event as a whole is proportional to the absolute square of the resultant probability amplitude***.

We can appreciate how all of this comes together by treating the reflection pictured in Fig. 4.70; a point source S illuminates a mirror, and light is subsequently scattered upward in every direction from every point on the mirror. We wish to determine the probability of a detector at P, recording the arrival of a photon. Here the classical perspective, with its familiar wavelet model, can be used as an analogue to provide guidance (and perhaps a little intellectual comfort, if you still believe in classical EM waves).

For simplicity, take the mirror to be a narrow strip (which is essentially one-dimensional); that doesn't change things conceptually. Divide it into a number of equal-sized lengths (Fig. 4.71*a*), each of which establishes a possible path to P. (Of course, every atom on the surface is a scatterer, and so there are a multitude of paths, but the several we have drawn will do.) Classically, we know that every route from S to the mirror to P corresponds to the path of a scattered wavelet, and that the amplitude (E_{0j}) and phase of each such wavelet at P will determine the net resultant amplitude, E_0. As we saw with Fermat's Principle (p. 105), the optical path length from S to the mirror to P establishes the phase of each wavelet arriving at P. Moreover, the greater the *path length* is, the more the light spreads out (via the Inverse Square Law) and the smaller is the amplitude of the wavelet arriving at P.

Figure 4.71*b* is a plot of the *OPL* with its minimum at the observed path (*S-I-P*), for which $\theta_i = \theta_r$. A large change in *OPL*, as between (*S-A-P*) and (*S-B-P*) is accompanied by a large phase difference and a correspondingly large rotation of the phasors drawn in Fig. 4.71*c*. Going from A to B to C and so on to I, the optical path lengths decrease less and less rapidly, and each phasor leads the previous one by a smaller angle (set by the slope of the curve). In effect, the phasors to the left of I rotate counterclockwise from A to I. Since the *OPL* is a minimum at I, the phasors from that region are large and differ very little in phase angle. Going from I to J to K and so on to Q, the optical path lengths increase more and more rapidly, and each phasor lags the previous one by a larger angle. In effect, the phasors to the right of I rotate clockwise from I to Q.

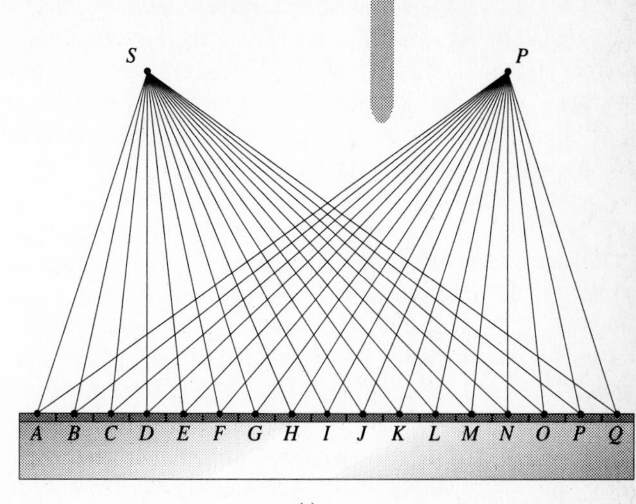

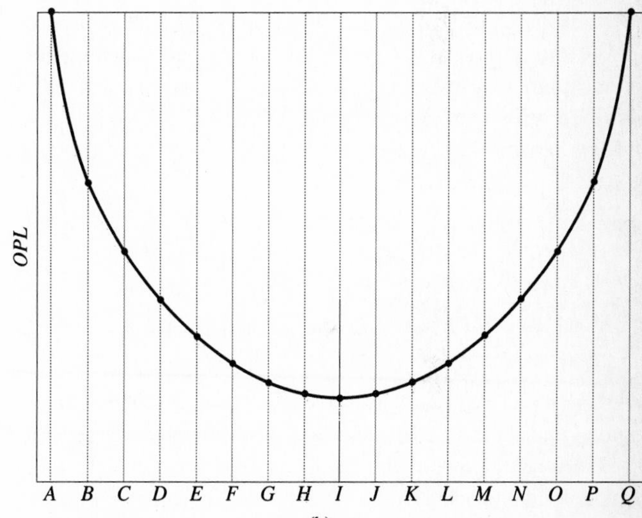

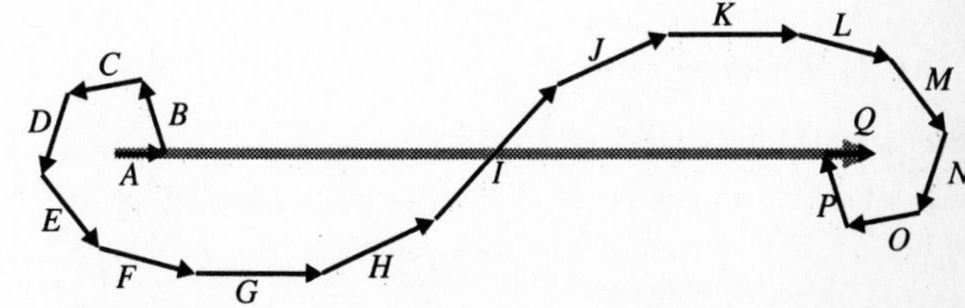

FIGURE 4.71 (*a*) Feynman's analysis of the problem of reflection via QED. (*a*) A number of paths from S to the mirror to P. (*b*) The *OPL* for light going from S to P along the paths depicted in (*a*). Each path has a probability amplitude associated with it. These add to produce a net amplitude.

In Fig. 4.71*c* the resultant amplitude is drawn from the starting tail to the ending tip, and classically it corresponds to the net electric field amplitude at *P*. The irradiance, *I*, is proportional to the square of the net field amplitude, and that, in turn, should be a measure of the likelihood of finding a photon when a detector is placed at *P*.

Now let's move beyond the classical ideas of scattered wavelets and electric fields (nonetheless being guided by them) and construct a quantum-mechanical treatment. Photons can go from *S* to the mirror to *P* along each of an innumerable number of distinct paths. It's reasonable to assume that each such path makes a specific contribution to the end result; an exceedingly long route out to the very edge of the mirror and back to *P* should contribute differently than a more direct route. Following Feynman, we associate some (as yet unspecified) complex quantity, a constituent **quantum-mechanical (QM) probability amplitude**, with each possible path. Each such constituent QM probability amplitude can be represented as a phasor whose angle is determined by the total time of flight from *S* to the mirror to *P*, and whose size is determined by the path length traversed. (Of course, this is just what obtained with each phasor in Fig. 4.71*c*. Still there are convincing reasons why the classical *E*-field cannot be the QM probability amplitude.) The total QM probability amplitude is the sum of all such phasors corresponding to all possible paths, and that is analogous to the resultant phasor in Fig. 4.71*c*.

Now relabel Fig. 4.71*c* so that it represents the quantum-mechanical formulation. Clearly, ***most of the length of the resultant QM probability amplitude arises from contributions in the immediate vicinity of path S-I-P, where the constituent phasors are large and nearly in-phase***. Most of the accumulated probability for light to go from *S* to *P* via reflection arises along, and immediately adjacent to, path *S-I-P*. The regions at the ends of the mirror contribute very little because the phasors from those areas form tight spirals at both extremes (Fig. 4.71*c*). Covering the ends of the mirror will have little effect on the length of the resultant amplitude and therefore little effect on the amount of light reaching *P*. Keep in mind that this diagram is rather crude; instead of 17 routes from *S* to *P*, there are billions of possible paths, and the phasors on both ends of the spiral wind around countless times.

QED predicts that light emitted by a point source *S* reflects out to *P* from all across the mirror, but that the most likely route is *S-I-P*, in which case $\theta_i = \theta_r$. With your eye at *P* looking into the mirror, you see one sharp image of *S*.

PROBLEMS

4.1 Work your way through an argument using dimensional analysis to establish the λ^{-4} dependence of the percentage of light scattered in Rayleigh Scattering. Let E_{0i} and E_{0s} be the incident and scattered amplitudes, the latter at a distance r from the scatterer. Assume $E_{0s} \propto E_{0i}$ and $E_{0s} \propto 1/r$. Furthermore, plausibly assume that the scattered amplitude is proportional to the volume, V, of the scatterer; within limits this is reasonable. Determine the units of the constant of proportionality.

4.2* A white floodlight beam crosses a large volume containing a tenuous molecular gas mixture of mostly oxygen and nitrogen. Compare the relative amount of scattering occurring for the yellow (580 nm) component with that of the violet (400 nm) component.

4.3* Figure P.4.3 depicts light emerging from a point source. It shows three different representations of radiant energy streaming outward. Identify each one and discuss its relationship to the others.

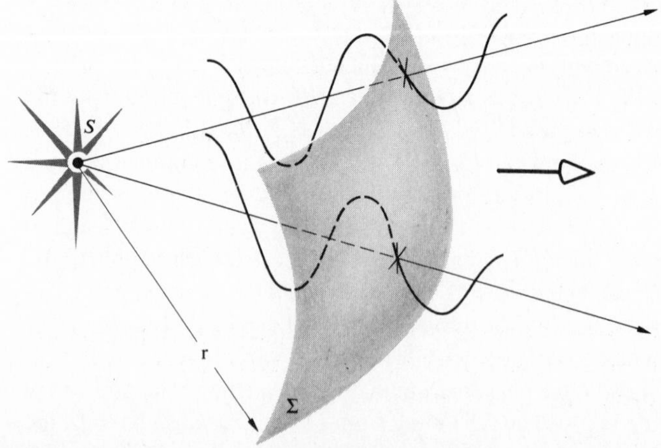

FIGURE P.4.3 A segment of a spherical wave.

4.4 The equation for a driven damped oscillator is

$$m_e\ddot{x} + m_e\gamma\dot{x} + m_e\omega_0^2 x = q_e E(t)$$

(a) Explain the significance of each term.

(b) Let $E = E_0 e^{i\omega t}$ and $x = x_0 e^{i(\omega t - \alpha)}$, where E_0 and x_0 are real quantities. Substitute into the above expression and show that

$$x_0 = \frac{q_e E_0}{m_e} \frac{1}{[(\omega_0^2 - \omega^2)^2 + \gamma^2\omega^2]^{1/2}}$$

(c) Derive an expression for the phase lag, α, and discuss how α varies as ω goes from $\omega \ll \omega_0$ to $\omega = \omega_0$ to $\omega \gg \omega_0$.

4.5 Imagine that we have a nonabsorbing glass plate of index n and thickness Δy, which stands between a source S and an observer P.

(a) If the unobstructed wave (without the plate present) is $E_u = E_0 \exp i\omega(t - y/c)$, show that with the plate in place the observer sees a wave

$$E_p = E_0 \exp i\omega[t - (n - 1)\,\Delta y/c - y/c]$$

(b) Show that if either $n \approx 1$ or Δy is very small, then

$$E_p = E_u + \frac{\omega(n - 1)\,\Delta y}{c} E_u e^{-i\pi/2}$$

The second term on the right may be envisioned as the field arising from the oscillators in the plate.

4.6* A very narrow laserbeam is incident at an angle of 58° on a horizontal mirror. The reflected beam strikes a wall at a spot 5.0 m away from the point of incidence where the beam hit the mirror. How far horizontally is the wall from that point of incidence?

4.7* On entering the tomb of FRED the Hero of Nod, you find yourself in a dark closed chamber with a small hole in a wall 3.0 m up from the floor. Once a year, on FRED's birthday, a beam of sunlight enters via the hole, strikes a small polished gold disk on the floor 4.0 m from the wall and reflects off it, lighting up a great diamond embedded in the forehead of a glorious statue of FRED, 20 m from the wall. Roughly how tall is the statue?

4.8 Calculate the transmission angle for a ray incident in air at 30° on a block of crown glass ($n_g = 1.52$).

4.9* A ray of yellow light from a sodium discharge lamp falls on the surface of a diamond in air at 45°. If at that frequency $n_d = 2.42$, compute the angular deviation suffered upon transmission.

4.10* Given an interface between water ($n_w = \frac{4}{3}$) and glass ($n_g = \frac{3}{2}$), compute the transmission angle for a beam incident in the water at 45°. If the transmitted beam is reversed so that it impinges on the interface, show that $\theta_t = 45°$.

4.11 A beam of 12-cm planar microwaves strikes the surface of a dielectric at 45°. If $n_{ti} = \frac{4}{3}$, compute (a) the wavelength in the transmitting medium, and (b) the angle θ_t.

4.12* Light of wavelength 600 nm in vacuum enters a block of glass where $n_g = 1.5$. Compute its wavelength in the glass. What color would it appear to someone embedded in the glass (see Table 3.4)?

4.13* A laserbeam impinges on an air–liquid interface at an angle of 55°. The refracted ray is observed to be transmitted at 40°. What is the refractive index of the liquid?

4.14* An underwater swimmer shines a beam of light up toward the surface. It strikes the air–water interface at 35°. At what angle will it emerge into the air?

4.15 Make a plot of θ_i versus θ_t for an air–glass boundary where $n_{ga} = 1.5$.

4.16* Prove that to someone looking straight down into a swimming pool, any object in the water will appear to be at 3/4 of its true depth.

4.17* A laserbeam impinges on the top surface of a 2.00-cm-thick parallel glass ($n = 1.50$) plate at an angle of 35°. How long is the actual path through the glass?

4.18* Light is incident in the air on an air–glass interface. If the index of refraction of the glass is 1.70, find the incident angle such that the transmission angle is to equal $\frac{1}{2}\theta_i$.

4.19* Suppose that you focus a camera with a close-up bellows attachment directly down on a letter printed on this page. The letter is then covered with a 1.00-mm-thick microscope slide ($n = 1.55$). How high must the camera be raised in order to keep the letter in focus?

4.20* A prism, ABC, is configured such that angle $BCA = 90°$ and angle $CBA = 45°$. What is the minimum value of its index of refraction if, while immersed in air, a beam traversing face AC is to be totally internally reflected from face BC?

4.21* A coin is resting on the bottom of a tank of water ($n_W = 1.33$) 1.00 m deep. On top of the water floats a layer of benzene ($n_b = 1.50$), which is 20.0 cm thick. Looking down nearly perpendicularly, how far beneath the topmost surface does the coin appear? Draw a ray diagram.

4.22 In Fig. P.4.22 the wavefronts in the incident medium match the fronts in the transmitting medium everywhere on the interface—a concept known as *wavefront continuity*. Write expressions for the number of waves per unit length along the interface in terms of θ_i and λ_i in one case and θ_t and λ_t in the other. Use these to derive Snell's Law. Do you think Snell's Law applies to sound waves? Explain.

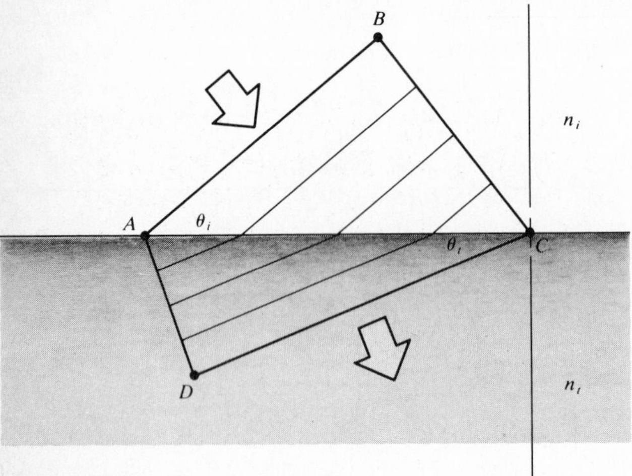

FIGURE P.4.22

4.23* With the previous problem in mind, return to Eq. (4.19) and take the origin of the coordinate system in the plane-of-incidence and on the interface (Fig. 4.37). Show that that equation is then equivalent to equating the x-components of the various propagation vectors. Show that it is also equivalent to the notion of wavefront continuity.

4.24 Making use of the ideas of equal transit times between corresponding points and the orthogonality of rays and wavefronts, derive the law of reflection and Snell's Law. The ray diagram of Fig. P.4.24 should be helpful.

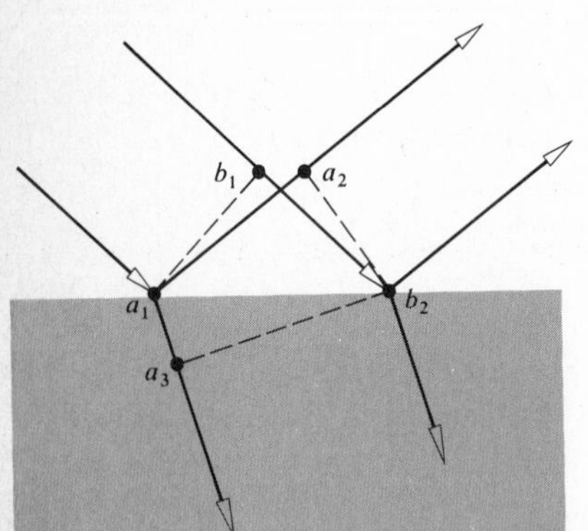

FIGURE P.4.24

4.25 Starting with Snell's Law, prove that the vector refraction equation has the form

$$n_t\hat{\mathbf{k}}_t - n_i\hat{\mathbf{k}}_i = (n_t \cos\theta_t - n_i \cos\theta_i)\hat{\mathbf{u}}_n \qquad [4.7]$$

4.26 Derive a vector expression equivalent to the Law of Reflection. As before, let the normal go from the incident to the transmitting medium, even though it obviously doesn't really matter.

4.27 In the case of reflection from a planar surface, use Fermat's Principle to prove that the incident and reflected rays share a common plane with the normal $\hat{\mathbf{u}}_n$, namely, the plane-of-incidence.

4.28* Derive the Law of Reflection, $\theta_i = \theta_r$, by using the calculus to minimize the transit time, as required by Fermat's Principle.

4.29* According to the mathematician Hermann Schwarz, there is one triangle that can be inscribed within an acute triangle such that it has a minimal perimeter. Using two planar mirrors, a laserbeam, and Fermat's Principle, explain how you can show that this inscribed triangle has its vertices at the points where the altitudes of the acute triangle intersect its corresponding sides.

4.30 Show analytically that a beam entering a planar transparent plate, as in Fig. P.4.30, emerges parallel to its initial direction. Derive an expression for the lateral displacement of the beam. Incidentally, the incoming and outgoing rays would be parallel even for a stack of plates of different material.

FIGURE P.4.30

4.31* Show that the two rays that enter the system in Fig. P.4.31 parallel to each other emerge from it being parallel.

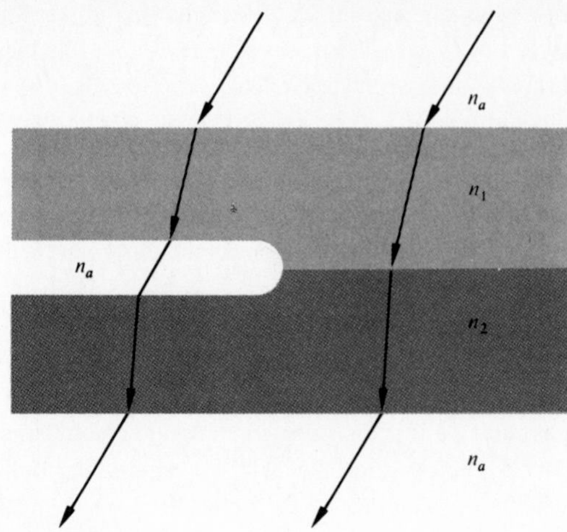

FIGURE P.4.31

4.32 Discuss the results of Problem 4.30 in the light of Fermat's Principle; that is, how does the relative index n_{21} affect things? To see the lateral displacement, look at a broad source through a thick piece of glass ($\approx\frac{1}{4}$ inch) or a stack (four will do) of microscope slides *held at an angle*. There will be an obvious shift between the region of the source seen directly and the region viewed through the glass.

4.33 Show that even in the nonstatic case the tangential component of the electric field intensity **E** is continuous across an interface. [*Hint:* Using Fig. P.4.33 and Eq. (3.5), shrink sides *FB* and *CD*, thereby letting the area bounded go to zero.]

4.34 Suppose a lightwave that is linearly polarized in the plane-of-incidence impinges at 30° on a crown-glass ($n_g = 1.52$) plate in air. Compute the appropriate amplitude reflection and transmission coefficients at the interface. Compare your results with Fig. 4.39.

4.35 Derive Eqs. (4.42) through (4.45) for $r_\perp$, $r_\parallel$, $t_\perp$, and $t_\parallel$.

4.36* A beam of light in air strikes the surface of a smooth piece of plastic having an index of refraction of 1.55 at an angle with the normal of 20.0°. The incident light has component *E*-field amplitudes parallel and perpendicular to the plane-of-incidence of 10.0 V/m and 20.0 V/m, respectively. Determine the corresponding reflected field amplitudes.

4.37* A laserbeam is incident on the interface between air and some dielectric of index *n*. For small values of θ_i show that $\theta_t = \theta_i/n$. Use this and Eq. (4.42) to establish that at near-normal incidence $[-r_\perp]_{\theta_i \approx 0} = (n-1)/(n+1)$.

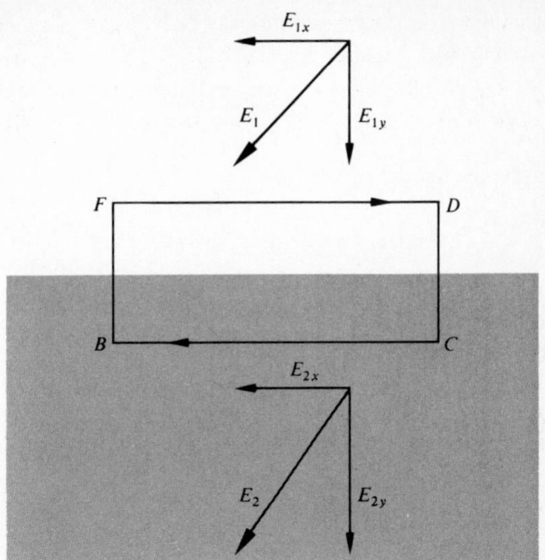

FIGURE P.4.33

4.38* Using the results of the previous problem, compare the amplitude reflection coefficients for an air–water ($n_w = 4/3$) interface with that of an air–crown glass ($n_g = 3/2$) interface, both at near-normal incidence. What are the corresponding ratios of the reflected to the incident irradiances?

4.39* Use Eq. (4.42) and the power series expansion of the sine function to establish that at near-normal incidence we can obtain a better approximation than the one in Problem 4.37, which is $[-r_\perp]_{\theta_i \approx 0} = (n-1)/(n+1)$, namely

$$[-r_\perp]_{\theta_i \approx 0} = \left(\frac{n-1}{n+1}\right)\left(1 + \frac{\theta_i^2}{n}\right)$$

4.40* Establish that at near-normal incidence the equation

$$[r_\parallel]_{\theta_i \approx 0} = \left(\frac{n-1}{n+1}\right)\left(1 - \frac{\theta_i^2}{n}\right)$$

is a good approximation. [*Hint:* Use the results of the previous problem, Eq. (4.43), and the power series expansions of the sine and cosine functions.]

4.41* Prove that for a vacuum-dielectric interface at glancing incidence $r_\perp \to -1$, as in Fig. 4.39.

4.42* In Fig. 4.39 the curve of $r_\perp$ approaches -1.0 as the angle-of-incidence approaches 90°. Prove that if $\alpha_\perp$ is the angle the curve makes with the vertical at $\theta_i = 90°$, then

$$\tan \alpha_\perp = \frac{\sqrt{n^2 - 1}}{2}$$

[*Hint:* First show that $d\theta_t/d\theta_i = 0$.]

4.43 Prove that

$$t_\perp + (-r_\perp) = 1 \qquad [4.49]$$

for all θ_i, first from the boundary conditions and then from the Fresnel Equations.

4.44* Verify that

$$t_\perp + (-r_\perp) = 1 \qquad [4.49]$$

for $\theta_i = 30°$ at a crown-glass and air interface ($n_{ti} = 1.52$).

4.45* Calculate the critical angle beyond which there is total internal reflection at an air–glass ($n_g = 1.5$) interface. Compare this result with that of Problem 4.15.

4.46* What is the critical angle for total internal reflection for diamond? What, if anything, does the critical angle have to do with the luster of a well-cut diamond?

4.47* Using a block of a transparent, unknown material, it is found that a beam of light inside the material is totally internally reflected at the air–block interface at an angle of 48.0°. What is its index of refraction?

4.48* A fish looking straight up toward the smooth surface of a pond receives a cone of rays and sees a circle of light filled with the images of sky and birds and whatever else is up there. This bright circular field is surrounded by darkness. Explain what is happening and compute the cone-angle.

4.49* A glass block having an index of 1.55 is covered with a layer of water of index 1.33. For light traveling in the glass, what is the critical angle at the interface?

4.50 Derive an expression for the speed of the evanescent wave in the case of internal reflection. Write it in terms of c, n_i, and θ_i.

4.51 Light having a vacuum wavelength of 600 nm, traveling in a glass ($n_g = 1.50$) block, is incident at 45° on a glass–air interface. It is then totally internally reflected. Determine the distance into the air at which the amplitude of the evanescent wave has dropped to a value of $1/e$ of its maximum value at the interface.

4.52 Figure P.4.52 depicts a laserbeam incident on a wet piece of filter paper atop a sheet of glass whose index of refraction is to be measured—the photograph shows the resulting light pattern. Explain what is happening and derive an expression for n_i in terms of R and d.

4.53 Consider the common mirage associated with an inhomogeneous distribution of air situated above a warm roadway. Envision the bending of the rays as if it were instead a problem in total internal reflection. If an observer, at whose head $n_a = 1.000\,29$, sees an apparent wet spot at $\theta_i \geq 88.7°$ down the road, find the index of the air immediately above the road.

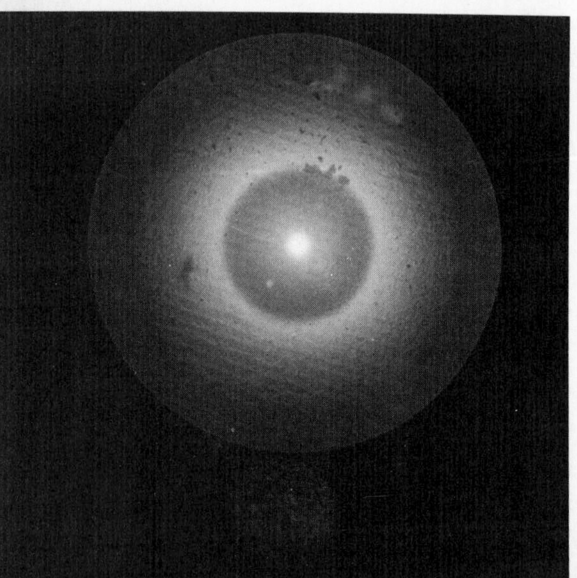

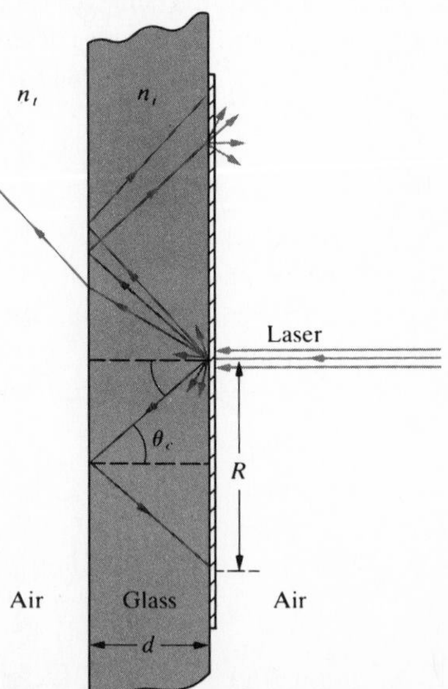

FIGURE P.4.52 (Photo and diagram courtesy S. Reich, The Weizmann Institute of Science, Israel.)

4.54* Use the Fresnel Equations to prove that light incident at $\theta_p = \frac{1}{2}\pi - \theta_t$ results in a reflected beam that is indeed polarized.

4.55 Show that $\tan\theta_p = n_t/n_i$ and calculate the polarization angle for external incidence on a plate of crown glass ($n_g = 1.52$) in air.

4.56* Beginning with Eq. (4.38), show that for two dielectric media, in general $\tan\theta_p = [\epsilon_t(\epsilon_t\mu_i - \epsilon_i\mu_t)/\epsilon_i(\epsilon_t\mu_t - \epsilon_i\mu_i)]^{1/2}$.

4.57 Show that the polarization angles for internal and external reflection at a given interface are complementary, that is, $\theta_p + \theta_p' = 90°$ (see Problem 4.55).

4.58 It is often useful to work with the *azimuthal angle* γ, which is defined as the angle between the plane of vibration and the plane-of-incidence. Thus for linearly polarized light,

$$\tan\gamma_i = [E_{0i}]_\perp/[E_{0i}]_\parallel \tag{4.92}$$

$$\tan\gamma_t = [E_{0t}]_\perp/[E_{0t}]_\parallel \tag{4.93}$$

and

$$\tan\gamma_r = [E_{0r}]_\perp/[E_{0r}]_\parallel \tag{4.94}$$

Figure P.4.58 is a plot of γ_r versus θ_i for internal and external reflection at an air–glass interface ($n_{ga} = 1.51$), where $\gamma_i = 45°$. Verify a few of the points on the curves and in addition show that

$$\tan\gamma_r = -\frac{\cos(\theta_i - \theta_t)}{\cos(\theta_i + \theta_t)}\tan\gamma_i \tag{4.95}$$

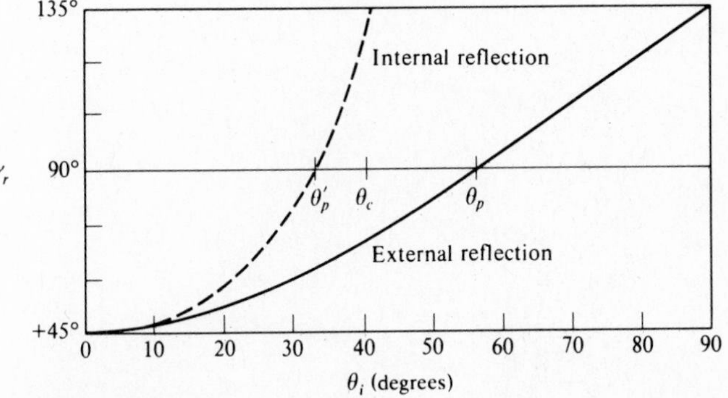

FIGURE P.4.58

4.59* Making use of the definitions of the azimuthal angles in Problem 4.58, show that

$$R = R_\parallel\cos^2\gamma_i + R_\perp\sin^2\gamma_i \tag{4.96}$$

and

$$T = T_\parallel\cos^2\gamma_i + T_\perp\sin^2\gamma_i \tag{4.97}$$

4.60 Make a sketch of $R_\perp$ and $R_\parallel$ for $n_i = 1.5$ and $n_t = 1$ (i.e., internal reflection).

4.61 Show that

$$T_\parallel = \frac{\sin 2\theta_i \sin 2\theta_t}{\sin^2(\theta_i + \theta_t)\cos^2(\theta_i - \theta_t)} \tag{4.98}$$

and

$$T_\perp = \frac{\sin 2\theta_i \sin 2\theta_t}{\sin^2(\theta_i + \theta_t)} \tag{4.99}$$

4.62* Using the results of Problem 4.61, that is, Eqs. (4.98) and (4.99), show that

$$R_\parallel + T_\parallel = 1 \qquad [4.65]$$

and

$$R_\perp + T_\perp = 1 \qquad [4.66]$$

4.63 Suppose that we look at a source perpendicularly through a stack of N microscope slides. The source seen through even a dozen slides will be noticeably darker. Assuming negligible absorption, show that the total transmittance of the stack is given by

$$T_t = (1 - R)^{2N}$$

and evaluate T_t for three slides in air.

4.64 Making use of the expression

$$I(y) = I_0 e^{-\alpha y} \qquad [4.78]$$

for an absorbing medium, we define a quantity called the *unit transmittance* T_1. At normal incidence, Eq. (4.55), $T = I_t/I_i$, and thus when $y = 1$, $T_1 \equiv I(1)/I_0$. If the total thickness of the slides in the previous problem is d and if they now have a transmittance per unit length T_1, show that

$$T_t = (1 - R)^{2N}(T_1)^d$$

4.65 Show that at normal incidence on the boundary between two dielectrics, as $n_{ti} \to 1$, $R \to 0$, and $T \to 1$. Moreover, prove that as $n_{ti} \to 1$, $R_\parallel \to 0$, $R_\perp \to 0$, $T_\parallel \to 1$, and $T_\perp \to 1$ for all θ_i. Thus as the two media take on more similar indices of refraction, less and less energy is carried off in the reflected wave. It should be obvious that when $n_{ti} = 1$ there will be no interface and no reflection.

4.66* Derive the expressions for $r_\perp$ and $r_\parallel$ given by Eqs. (4.70) and (4.71).

4.67 Show that when $\theta_i > \theta_c$ at a dielectric interface, $r_\parallel$ and $r_\perp$ are complex and $r_\perp r_\perp^* = r_\parallel r_\parallel^* = 1$.

4.68 Figure P.4.68 depicts a ray being multiply reflected by a transparent dielectric plate (the amplitudes of the resulting fragments are indicated). As in Section 4.5, we use the primed coefficient notation, because the angles are related by Snell's Law.

(a) Finish labeling the amplitudes of the last four rays.

(b) Show, using the Fresnel Equations, that

$$t_{\parallel}t'_{\parallel} = T_{\parallel} \tag{4.100}$$

$$t_{\perp}t'_{\perp} = T_{\perp} \tag{4.101}$$

$$r_{\parallel}^2 = r_{\parallel}'^2 = R_{\parallel} \tag{4.102}$$

and

$$r_{\perp}^2 = r_{\perp}'^2 = R_{\perp} \tag{4.103}$$

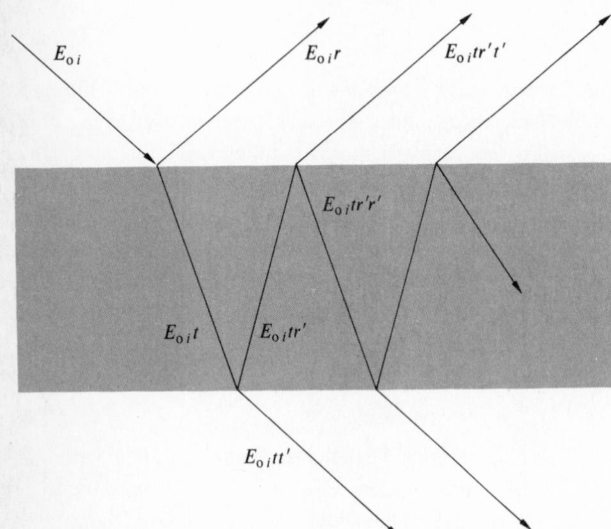

FIGURE P.4.68

4.69* A wave, linearly polarized in the plane-of-incidence, impinges on the interface between two dielectric media. If $n_i > n_t$ and $\theta_i = \theta'_p$, there is no reflected wave, that is, $r'_{\parallel}(\theta'_p) = 0$. Using Stokes's technique, start from scratch to show that $t_{\parallel}(\theta_p)t'_{\parallel}(\theta'_p) = 1$, $r_{\parallel}(\theta_p) = 0$, and $\theta_t = \theta_p$ (Problem 4.57). How does this compare with Eq. (4.100)?

4.70 Making use of the Fresnel Equations, show that $t_{\parallel}(\theta_p)t'_{\parallel}(\theta'_p) = 1$, as in the previous problem.

4.71 Figure P.4.71 depicts a glass cube surrounded by four glass prisms in very close proximity to its sides. Sketch in the paths that will be taken by the two rays shown and discuss a possible application for the device.

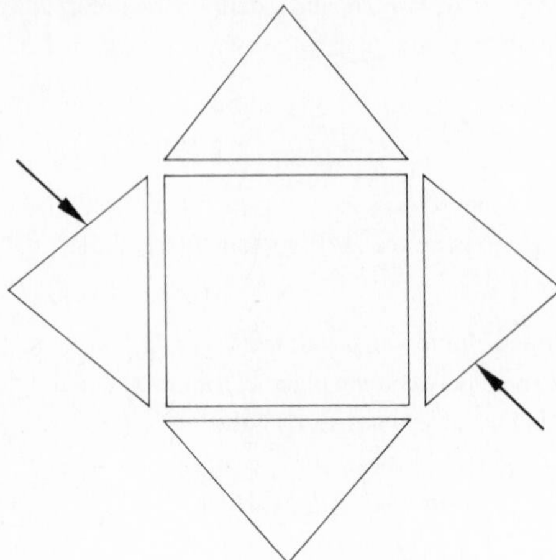

FIGURE P.4.71

4.72 Figure P.4.72 is a plot of n_I and n_R versus λ for a common metal. Identify the metal by comparing its characteristics with those considered in the chapter and discuss its optical properties.

FIGURE P.4.72

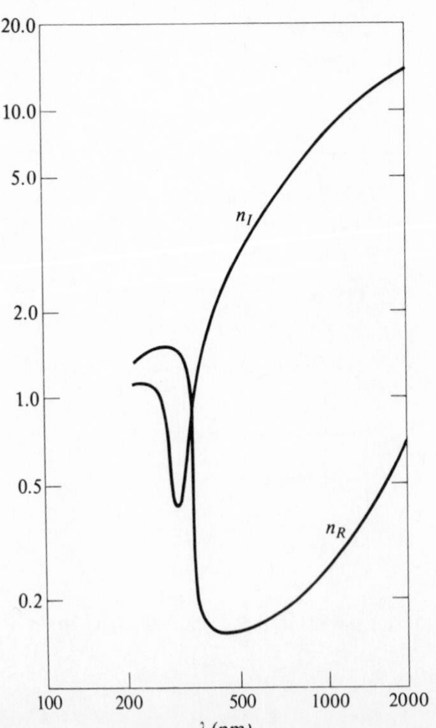

4.73 Figure P.4.73 shows a prism-coupler arrangement developed at the Bell Telephone Laboratories. Its function is to feed a laserbeam into a thin (0.000 01-inch) transparent film, which then serves as a sort of waveguide. One application is that of thin-film laserbeam circuitry—a kind of integrated optics. How do you think it works?

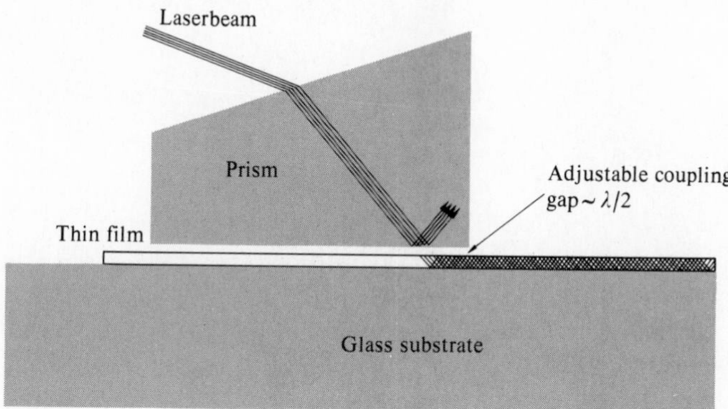

FIGURE P.4.73

5 Geometrical Optics

5.1 Introductory Remarks

The surface of an object that is either self-luminous or externally illuminated behaves as if it consisted of a very large number of radiating point sources. Each of these emits spherical waves; rays emanate radially in the direction of energy flow, that is, in the direction of the Poynting vector. In this case, the rays *diverge* from a given point source S, whereas if the spherical wave were collapsing to a point, the rays would of course be *converging*. Generally, one deals only with a small portion of a wavefront. ***A point from which a portion of a spherical wave diverges, or one toward which the wave segment converges, is known as a focus of the bundle of rays.***

Figure 5.1 depicts a point source in the vicinity of some arrangement of reflecting and refracting surfaces representing an *optical system*. Of the infinity of rays emanating from S, generally speaking, only one will pass through an arbitrary point in space. Even so, it is possible to arrange for an infinite number of rays to arrive at a certain point P, as in Fig. 5.1. If for a cone of rays coming from S there is a corresponding cone of rays passing through P, the system is said to be **stigmatic** for these two points. The energy in the cone (apart from some inadvertent losses due to reflection, scattering, and absorption) reaches P, which is then referred to as a **perfect image** of S. The wave could conceivably arrive to form a finite patch of light, or **blur spot**, about P; it would still be an image of S but no longer a perfect one.

It follows from the Principle of Reversibility (p. 109) that a point source placed at P would be equally well imaged at S, and accordingly the two are spoken of as **conjugate points**. In an *ideal optical system* every point of a three-dimensional region will be perfectly (or stigmatically) imaged in another region, the former being the **object space**, the latter the **image space**.

Most commonly, the function of an optical device is to collect and reshape a portion of the incident wavefront, often with the ultimate purpose of forming an image of an object. Notice that inherent in realizable systems is the limitation of being unable to collect all the emitted light; a system generally accepts only a segment of the wavefront. As a result, there will always be an apparent deviation from rectilinear propagation even in homogeneous media—the waves will be *diffracted*. The attainable degree of perfection of a real imaging optical system will be **diffraction-limited** (there will always be a blur spot, p. 459). As the wavelength of the radiant energy decreases in comparison to the physical dimensions of the optical system, the effects of diffraction become less significant. In the conceptual limit as $\lambda_0 \to 0$, rectilinear propagation obtains in homogeneous media, and we have the idealized domain of **Geometrical Optics**.* Behavior that is specifically attributable to the wave nature of light (e.g., interference and diffraction) would no longer be observable. In many situations, the great simplicity arising from the approximation of Geometrical Optics more than compensates for its inaccuracies. In short, *the subject treats the controlled manipulation of wavefronts (or rays) by means of the interpositioning of reflecting and/or refracting bodies, neglecting any diffraction effects.*

---

* *Physical Optics* deals with situations in which the nonzero wavelength of light must be reckoned with. Analogously, when the de Broglie wavelength of a material object is negligible, we have *Classical Mechanics*; when it is not, we have the domain of *Quantum Mechanics*.

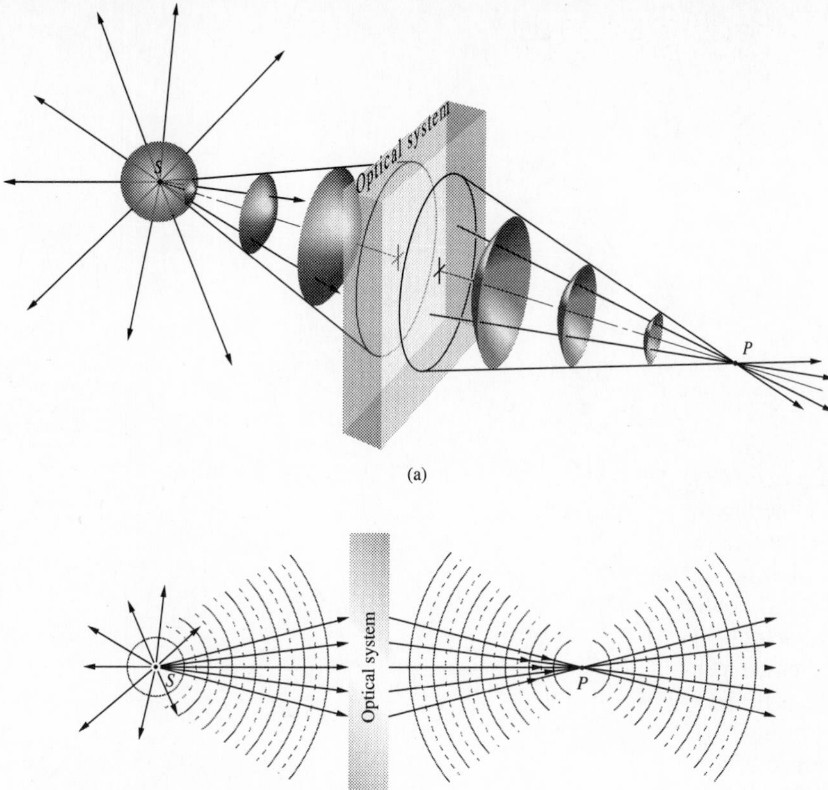

FIGURE 5.1 Conjugate foci. (*a*) A point source *S* sends out spherical waves. A cone of rays enters an optical system that inverts the wavefronts, causing them to converge on point *P*. (*b*) In cross section rays diverge from *S*, and a portion of them converge to *P*. If nothing stops the light at *P*, it continues on.

5.2 Lenses

The lens is no doubt the most widely used optical device, and that's not even considering the fact that we see the world through a pair of them. Human-made lenses date back at least to the burning-glasses of antiquity, which, as the name implies, were used to start fires long before the advent of matches. In the most general terms, **a lens is a refracting device (i.e., a discontinuity in the prevailing medium) that reconfigures a transmitted energy distribution**. That much is true whether we are dealing with UV, lightwaves, IR, microwaves, radiowaves, or even sound waves.

The configuration of a lens is determined by the required reshaping of the wavefront it is designed to perform. Point sources are basic, and so it is often desirable to convert diverging spherical waves into a beam of plane waves. Flashlights, projectors, and searchlights all do this in order to keep the beam from spreading out and weakening as it progresses. In just the reverse, it's frequently necessary to collect incoming parallel rays and bring them together at a point, thereby focusing the energy, as is done with a burning-glass or a telescope lens. Moreover, since the light reflected from someone's face scatters out from billions of point sources, a lens that causes each diverging wavelet to converge could form an image of that face (Fig. 5.2).

5.2.1 Aspherical Surfaces

To see how a lens works, imagine that we interpose in the path of a wave a transparent substance in which the wave's speed is different than it was initially. Figure 5.3*a* presents a cross-sec-

FIGURE 5.2 A person's face, like everything else we ordinarily see in reflected light, is covered with countless atomic scatterers.

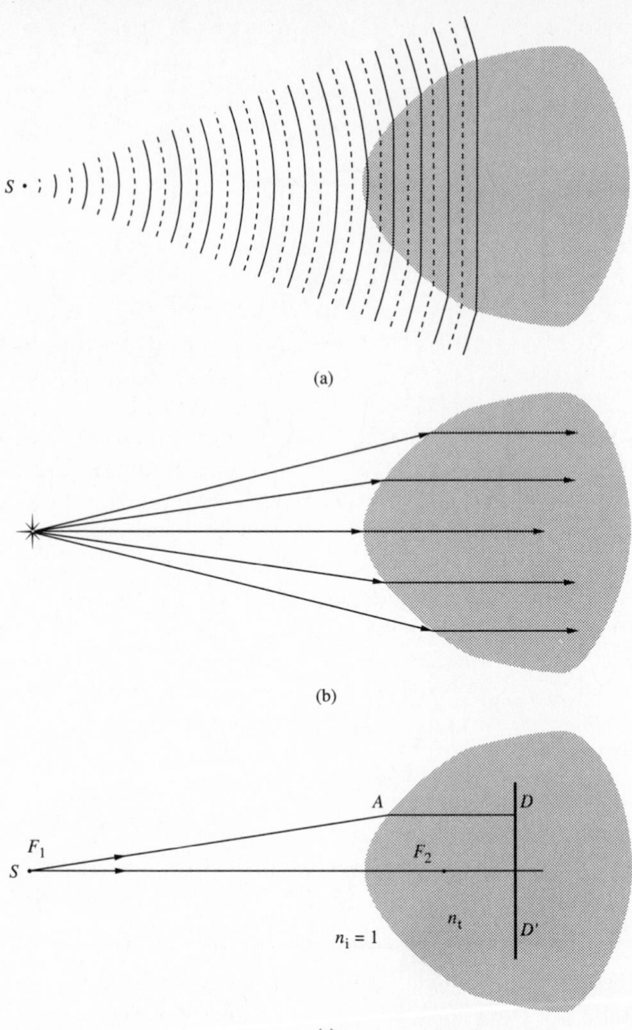

FIGURE 5.3 A hyperbolic interface between air and glass. (*a*) The wavefronts bend and straighten out. (*b*) The rays become parallel. (*c*) The hyperbola is such that the optical path from *S* to *A* to *D* is the same no matter where *A* is.

tional view of a diverging spherical wave traveling in an incident medium of index n_i impinging on the curved interface of a transmitting medium of index n_t. When n_t is greater than n_i, the wave slows upon entering the new substance. The central area of the wavefront travels more slowly than its outer extremities, which are still moving quickly through the incident medium. These extremities overtake the midregion, continuously flattening the wavefront. If the interface is properly configured, the spherical wavefront bends into a plane wave. The alternative ray representation is shown in Fig. 5.3*b*; the rays simply bend toward the local normal upon entering the more dense medium, and if the surface configuration is just right, the rays emerge parallel.

To find the required shape of the interface, refer to Fig. 5.3*c*, wherein point *A* can lie anywhere on the boundary. One wavefront is transformed into another, provided the paths along which the energy propagates are all "equal," thereby maintaining the phase of the wavefront (p. 28). A little spherical surface of constant phase emitted from *S* must evolve into a flat surface of constant phase at $\overline{DD'}$. Whatever path the light takes from *S* to $\overline{DD'}$, it must always be the same number of wavelengths long, so that the disturbance begins and ends in-phase. Radiant energy leaving *S* as a single wavefront must arrive at the plane $\overline{DD'}$, having traveled for the same amount of time to get there, no matter what the actual route taken by any particular ray. In other words, $\overline{F_1A}/\lambda_i$ (the number of wavelengths along the arbitrary ray from F_1 to *A*) plus $\overline{AD}/\lambda_t$ (the number of wavelengths along the ray from *A* to *D*) must be constant regardless of where on the interface *A* happens to be. Now, adding these and multiplying by λ_0, yields

$$n_i\,(\overline{F_1A}) + n_t\,(\overline{AD}) = \text{constant} \tag{5.1}$$

Each term on the left is the length traveled in a medium multiplied by the index of that medium, and, of course, each represents the optical path length *OPL* traversed. The optical path lengths from *S* to $\overline{DD'}$ are all equal. If Eq. (5.1) is divided by *c*, the first term becomes the time it takes to travel from *S* to *A* and the second term, the time from *A* to *D*; the right side remains constant (not the same constant, but constant). Equation (5.1) is equivalent to saying that all paths from S to $\overline{DD'}$ must take the same amount of time to traverse.

Let us return to finding the shape of the interface. Divide Eq. (5.1) by n_i, and it becomes

$$\overline{F_1A} + \left(\frac{n_t}{n_i}\right)\overline{AD} = \text{constant} \tag{5.2}$$

This is the equation of a hyperbola in which the eccentricity (e), which measures the bending of the curve, is given by $(n_t/n_i) > 1$; that is, $e = n_{ti} > 1$. The greater the eccentricity, the flatter the hyperbola (the larger the difference in the indices, the less the surface need be curved). When a point source is located at the focus F_1 and the interface between the two media is hyperbolic, plane waves are transmitted into the higher index material. It's left for Problem 5.3 to establish that when $(n_t/n_i) < 1$, the interface must be ellipsoidal. In each case pictured in Fig. 5.4, the rays either diverge from or converge toward a focal point, F. Furthermore, the rays can be reversed so that they travel either way; if a plane wave is incident (from the right) on the interface in Fig. 5.4*c*, it will converge (off to the left) at the farthest focus of the ellipsoid.

The first person to suggest using conic sections as surfaces for lenses and mirrors was Johann Kepler (1611), but he wasn't able to go very far with the idea without Snell's Law. Once that relationship was discovered, Descartes (1637) using his Analytic Geometry could develop the theoretical foundations of the optics of aspherical surfaces. The analysis presented here is in essence a gift from Descartes.

It's an easy matter now to construct lenses such that both the object and image points (or the incident and emerging light) will be outside of the medium of the lens. In Fig. 5.5*a* diverging incident spherical waves are made into plane waves at the first interface via the mechanism of Fig. 5.4*a*. These plane waves within the lens strike the back face perpendicularly and emerge unaltered: $\theta_i = 0$ and $\theta_t = 0$. Because the rays are reversible, plane waves incoming from the right will converge to point F_1, which is known as the focal point of the lens. Exposed on its flat face to the parallel rays from the Sun, our rather sophisticated lens would serve nicely as a burning-glass.

In Fig. 5.5*b*, the plane waves within the lens are made to converge toward the axis by bending at the second interface. Both of these lenses are thicker at their midpoints than at their edges and are therefore said to be **convex** (from the Latin *convexus*, meaning arched). Each lens causes the incoming beam to converge somewhat, to bend a bit more toward the central axis; therefore, they are referred to as **converging lenses**.

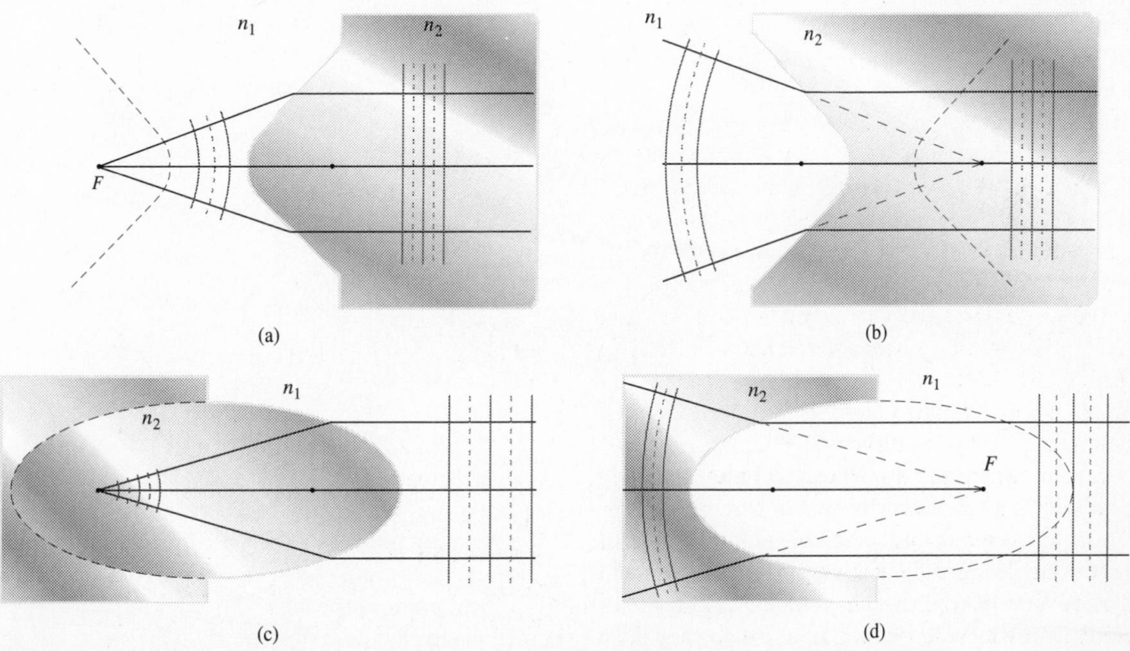

FIGURE 5.4 (*a*) and (*b*) Hyperboloidal and (*c*) and (*d*) ellipsoidal refracting surfaces ($n_2 > n_1$) in cross section.

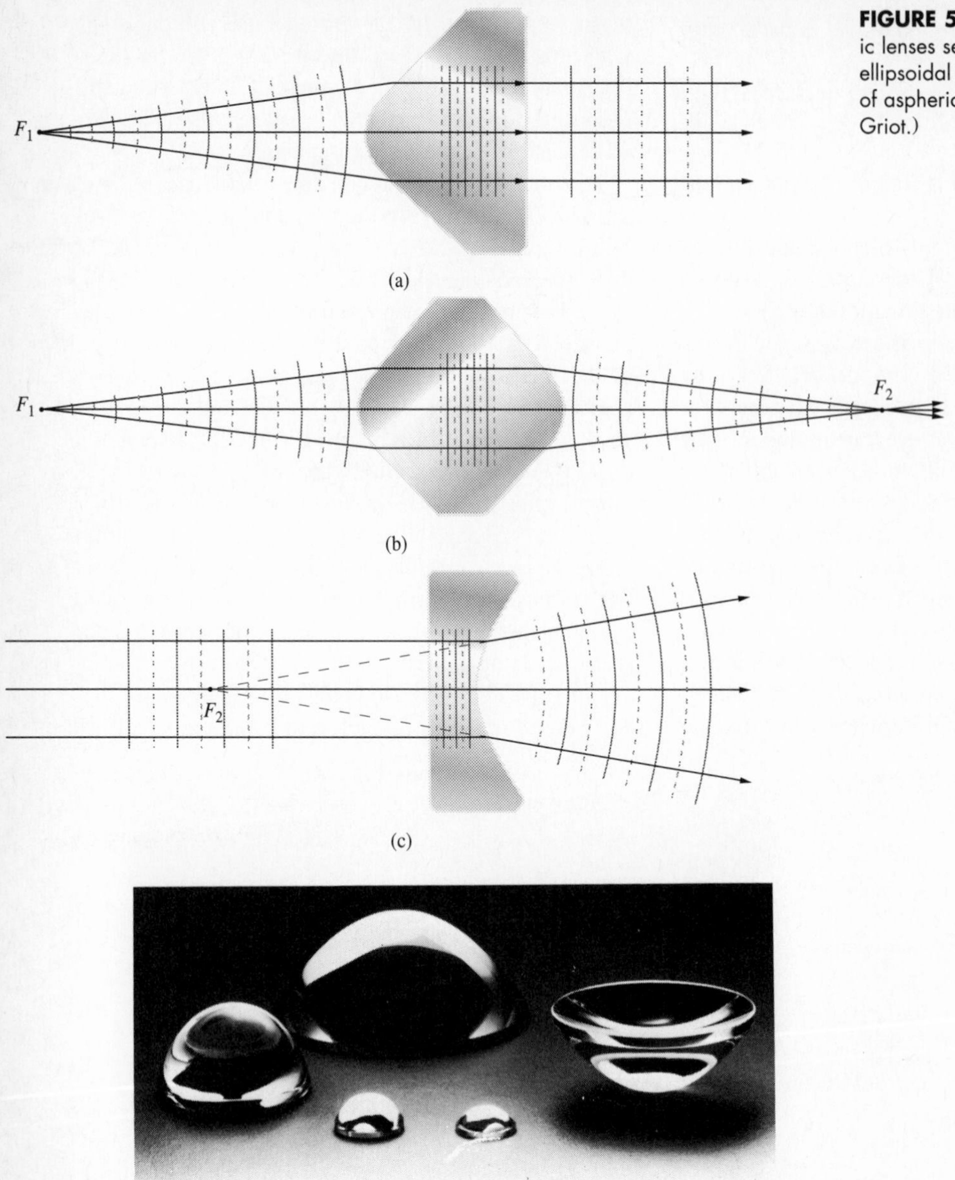

FIGURE 5.5 (*a*), (*b*), and (*c*) Several hyperbolic lenses seen in cross section. To explore an ellipsoidal lens see Problem 5.4. (*d*) A selection of aspherical lenses. (Photo courtesy Melles Griot.)

In contrast, a **concave** lens (from the Latin *concavus*, meaning hollow—and most easily remembered because it contains the word cave) is thinner in the middle than at the edges, as is evident in Fig. 5.5*c*. It causes the rays that enter as a parallel bundle to diverge. All such devices that turn rays outward away from the central axis (and in so doing add divergence to the beam) are called **diverging lenses**. In Fig. 5.5*c*, parallel rays enter from the left and, on emerging, seem to diverge from F_2; still, that point is taken as a focal point. **When a parallel bundle of rays passes through a converging lens, the point to which it converges (or when passing through a diverging lens, the point from which it diverges) is a focal point of the lens.**

If a point source is positioned on the central or optical axis at the point F_1 in front of the lens in Fig. 5.5*b*, rays will *converge* to the conjugate point F_2. A luminous image of the source would appear on a screen placed at F_2, an image that is therefore said to be **real**. On the other hand, in Fig. 5.5*c* the point source is at infinity, and the rays emerging from the system this time are *diverging*. They appear to come from a point F_2, but no actual luminous image would appear on a screen at that location. The image here is spoken of as **virtual**, as is the familiar image generated by a plane mirror.

Optical elements (lenses and mirrors) of the sort we have talked about, with one or both surfaces neither planar nor spherical, are referred to as *aspherics*. Although their operation is easy to understand and they perform certain tasks exceedingly well, they are still difficult to manufacture with great accuracy. Nonetheless, where the costs are justifiable or the required precision is not restrictive or the volume produced is large enough, aspherics are being used and will surely have an increasingly important role. The first quality glass aspheric to be manufactured in great quantities (tens of millions) was a lens for the Kodak disk camera (1982). Today aspherical lenses are frequently used as an elegant means of correcting imaging errors in complicated optical systems.

A new generation of computer-controlled machines, aspheric generators, is producing elements with tolerances (i.e., departures from the desired surface) of better than 0.5 μm (0.000 020 inch). This is still about a factor of 10 away from the generally required tolerance of $\lambda/4$ for quality optics, but that will surely come in time. Nowadays aspherics made in plastic and glass can be found in all kinds of instruments across the whole range of quality, including telescopes, projectors, cameras, and reconnaissance devices.

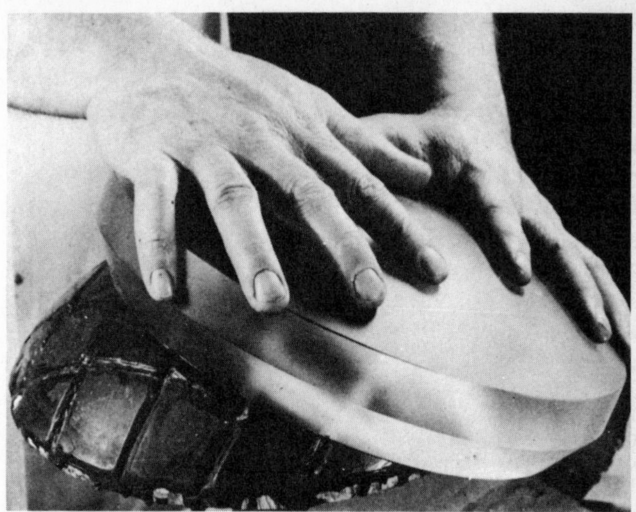

FIGURE 5.6 Polishing a spherical lens. (Photo courtesy Optical Society of America.)

5.2.2 Refraction at Spherical Surfaces

Imagine two pieces of material, one with a concave and the other a convex spherical surface, both having the same radius. It is a unique property of the sphere that such pieces will fit together in intimate contact regardless of their mutual orientation. Thus, if we take two roughly spherical objects of suitable curvature, one a grinding tool and the other a disk of glass, separate them with some abrasive, and then randomly move them with respect to each other, we can anticipate that any high spots on either object will wear away. As they wear, both pieces will gradually become more spherical (Fig. 5.6). Such surfaces are commonly generated in batches by automatic grinding and polishing machines.

Not surprisingly, the vast majority of quality lenses in use today have surfaces that are segments of spheres. Our intent here is to establish techniques for using such surfaces to simultaneously image a great many object points in light composed of a broad range of frequencies. Image errors, known as **aberrations**, will occur, but it is possible with the present technology to construct high-quality spherical lens systems whose aberrations are so well controlled that image fidelity is limited only by diffraction.

Figure 5.7 depicts a wave from the point source S impinging on a spherical interface of radius R centered at C. The

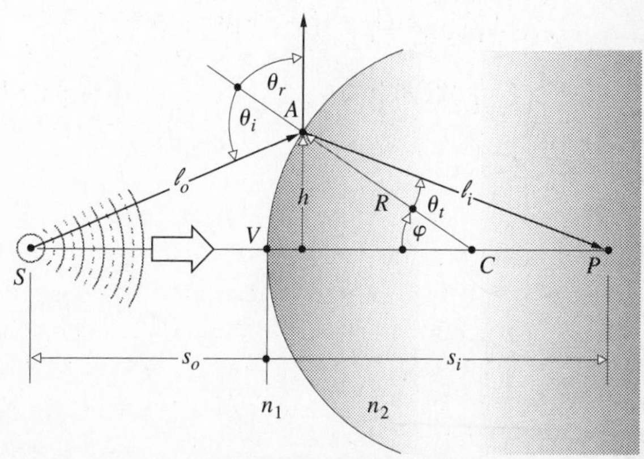

FIGURE 5.7 Refraction at a spherical interface. Conjugate foci.

point V is called the **vertex** of the surface. The length $s_o = \overline{SV}$ is known as the **object distance**. The ray $\overline{SA}$ will be refracted at the interface toward the local normal ($n_2 > n_1$) and therefore toward the central or **optical axis**. Assume that at some point P the ray will cross the axis, as will all other rays incident at the same angle θ_i (Fig. 5.8). The length $s_i = \overline{VP}$ is the **image distance**. Fermat's Principle maintains that the optical path length OPL will be stationary; that is, its derivative with respect to the position variable will be zero. For the ray in question,

$$OPL = n_1\ell_o + n_2\ell_i \tag{5.3}$$

Using the law of cosines in triangles SAC and ACP along with the fact that $\cos\varphi = -\cos(180° - \varphi)$, we get

$$\ell_o = [R^2 + (s_o + R)^2 - 2R(s_o + R)\cos\varphi]^{1/2}$$

and $$\ell_i = [R^2 + (s_i - R)^2 + 2R(s_i - R)\cos\varphi]^{1/2}$$

The OPL can be rewritten as

$$OPL = n_1[R^2 + (s_o + R)^2 - 2R(s_o + R)\cos\varphi]^{1/2}$$
$$+ n_2[R^2 + (s_i - R)^2 + 2R(s_i - R)\cos\varphi]^{1/2}$$

All the quantities in the diagram (s_i, s_o, R, etc.) are positive numbers, and these form the basis of a *sign convention* which is gradually unfolding and to which we shall return time and

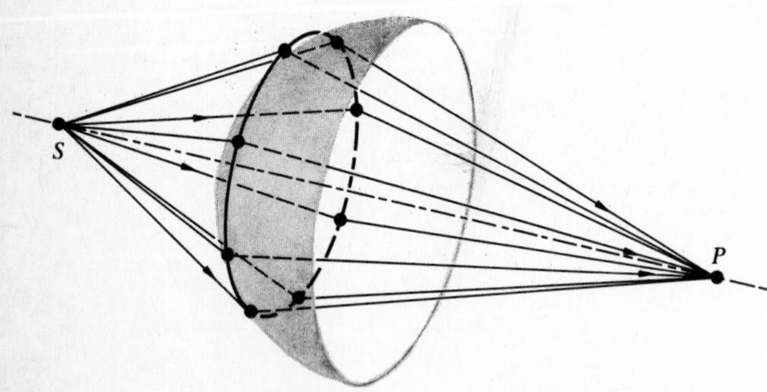

FIGURE 5.8 Rays incident at the same angle

Table 5.1 Sign Convention for Spherical Refracting Surfaces and Thin Lenses* (Light Entering from the Left)

s_o, f_o	+ left of V
x_o	+ left of F_o
s_i, f_i	+ right of V
x_i	+ right of F_i
R	+ if C is right of V
y_o, y_i	+ above optical axis

*This table anticipates the imminent introduction of a few quantities not yet spoken of.

again (see Table 5.1). Inasmuch as the point A moves at the end of a fixed radius (i.e., R = constant), φ is the position variable, and thus setting $d(OPL)/d\varphi = 0$, via Fermat's Principle we have

$$\frac{n_1R(s_o + R)\sin\varphi}{2\ell_o} - \frac{n_2R(s_i - R)\sin\varphi}{2\ell_i} = 0 \tag{5.4}$$

from which it follows that

$$\frac{n_1}{\ell_o} + \frac{n_2}{\ell_i} = \frac{1}{R}\left(\frac{n_2 s_i}{\ell_i} - \frac{n_1 s_o}{\ell_o}\right) \tag{5.5}$$

This is the relationship that must hold among the parameters for a ray going from S to P by way of refraction at the spherical interface. Although this expression is exact, it is rather complicated. If A is moved to a new location by changing φ, the new ray will not intercept the optical axis at P. (See Problem 5.1 concerning the Cartesian oval which is the interface configuration that would bring any ray, regardless of φ, to P.) The approximations that are used to represent ℓ_o and ℓ_i, and thereby simplify Eq. (5.5), are crucial in all that is to follow. Recall that

$$\cos\varphi = 1 - \frac{\varphi^2}{2!} + \frac{\varphi^4}{4!} - \frac{\varphi^6}{6!} + \cdots \tag{5.6}$$

and $$\sin\varphi = \varphi - \frac{\varphi^3}{3!} + \frac{\varphi^5}{5!} - \frac{\varphi^7}{7!} + \cdots \tag{5.7}$$

If we assume small values of φ (i.e., A close to V), $\cos\varphi \approx 1$.

Consequently, the expressions for ℓ_o and ℓ_i yield $\ell_o \approx s_o$, $\ell_i \approx s_i$, and to that approximation

$$\frac{n_1}{s_o} + \frac{n_2}{s_i} = \frac{n_2 - n_1}{R} \tag{5.8}$$

We could have begun this derivation with Snell's Law rather than Fermat's Principle (Problem 5.5), in which case small values of φ would have led to $\sin\varphi \approx \varphi$ and Eq. (5.8) once again. This approximation delineates the domain of what is called *first-order theory*; we'll examine *third-order theory* ($\sin\varphi \approx \varphi - \varphi^3/3!$) in the next chapter. Rays that arrive at shallow angles with respect to the optical axis (such that φ and h are appropriately small) are known as **paraxial rays**. The *emerging wavefront segment corresponding to these paraxial rays is essentially spherical and will form a "perfect" image at its center P located at* s_i. Notice that Eq. (5.8) is independent of the location of A over a small area about the symmetry axis, namely, the *paraxial region*. Gauss, in 1841, was the first to give a systematic exposition of the formation of images under the above approximation, and the result is variously known as *first-order*, *paraxial*, or **Gaussian Optics**. It soon became the basic theoretical tool by which lenses would be designed for several decades to come. If the optical system is well corrected, an incident spherical wave will emerge in a form very closely resembling a spherical wave. Consequently, as the perfection of the system increases, it more closely approaches first-order theory. Deviations from that of paraxial analysis will provide a convenient measure of the quality of an actual optical device.

If the point F_o in Fig. 5.9 is imaged at infinity ($s_i = \infty$), we have

$$\frac{n_1}{s_o} + \frac{n_2}{\infty} = \frac{n_2 - n_1}{R}$$

That special object distance is defined as the **first focal length** or the **object focal length**, $s_o \equiv f_o$, so that

$$f_o = \frac{n_1}{n_2 - n_1} R \tag{5.9}$$

The point F_o is known as the **first** or **object focus**. Similarly, the **second** or **image focus** is the axial point F_i, where the image is formed when $s_o = \infty$; that is,

$$\frac{n_1}{\infty} + \frac{n_2}{s_i} = \frac{n_2 - n_1}{R}$$

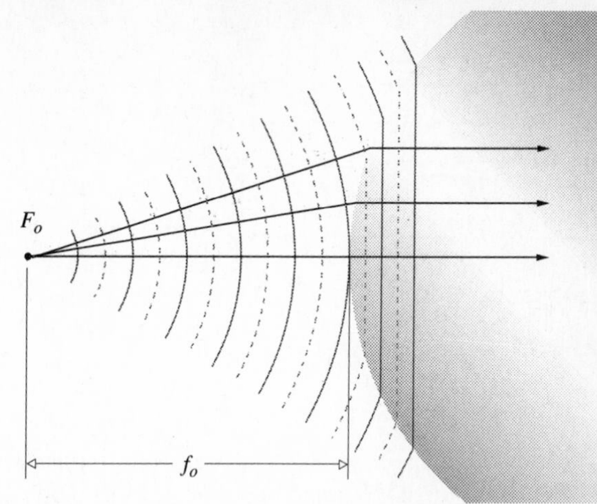

FIGURE 5.9 Plane waves propagating beyond a spherical interface—the object focus.

Defining the **second** or **image focal length** f_i as equal to s_i in this special case (Fig. 5.10), we have

$$f_i = \frac{n_2}{n_2 - n_1} R \tag{5.10}$$

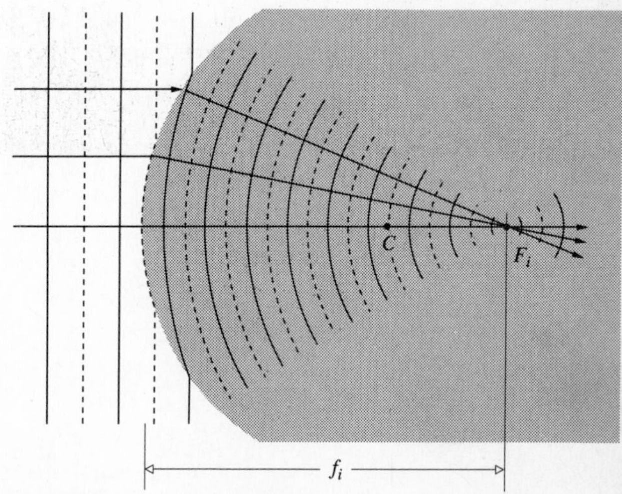

FIGURE 5.10 The reshaping of plane into spherical waves at a spherical interface—the image focus.

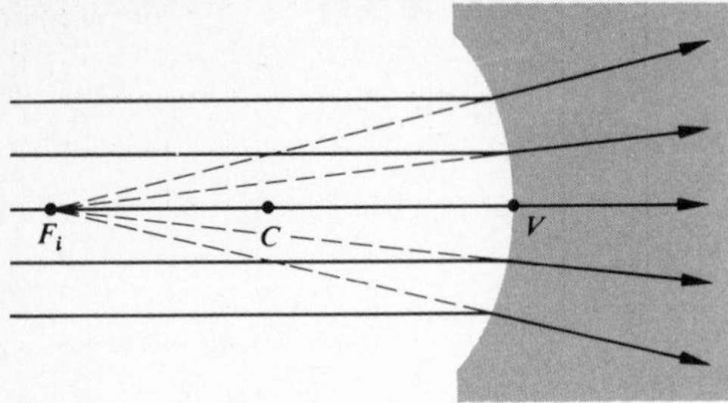

FIGURE 5.11 A virtual image point.

Recall that an image is virtual when the rays diverge from it (Fig. 5.11). Analogously, ***an object is virtual when the rays converge toward it*** (Fig. 5.12). Observe that the virtual object is now on the right-hand side of the vertex, and therefore s_o will be a negative quantity. Moreover, the surface is concave, and its radius will also be negative, as required by Eq. (5.9), since f_o would be negative. In the same way, the virtual image distance appearing to the left of V is negative.

5.2.3 Thin Lenses

Lenses are made in a wide range of forms; for example, there are acoustic and microwave lenses; some of the latter are made of glass or wax in easily recognizable shapes, whereas others are far more subtle in appearance (Fig. 5.13). Most often a lens has two or more refracting interfaces, and at least one of these is curved. Generally, the nonplanar surfaces are centered on a common axis. These surfaces are most frequently spherical segments and are often coated with thin dielectric films to control their transmission properties (see Section 9.9).

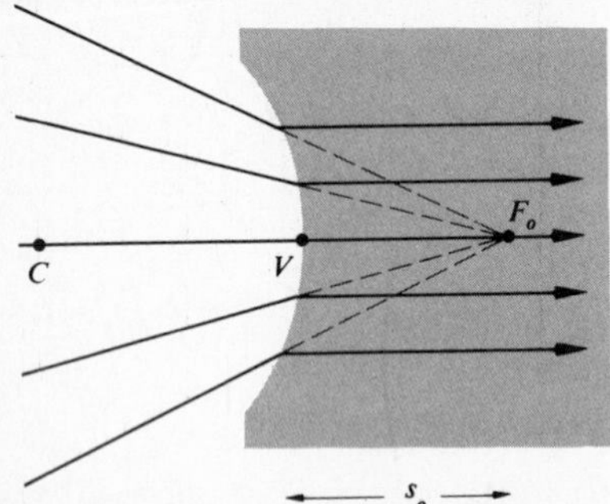

FIGURE 5.12 A virtual object point.

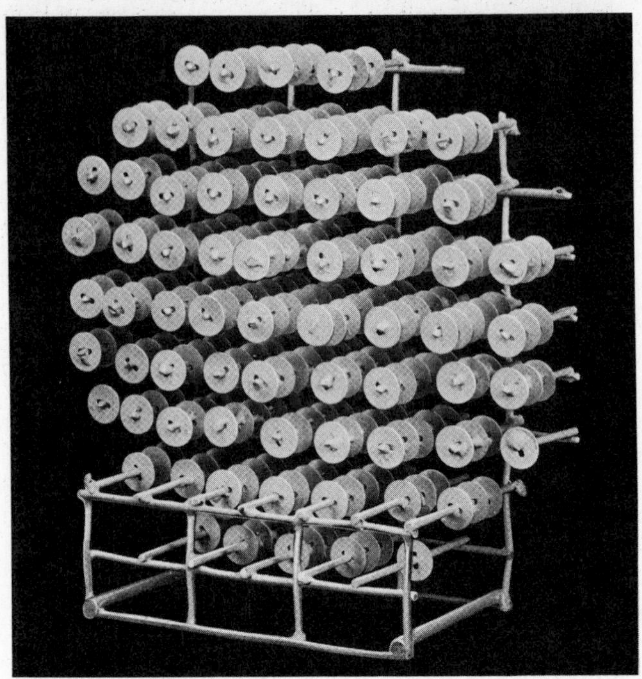

FIGURE 5.13 A lens for short-wavelength radiowaves. The disks serve to refract these waves much as rows of atoms refract light. (Photo courtesy Optical Society of America.)

A lens that consists of one element (i.e., it has only two refracting surfaces) is a *simple lens*. The presence of more than one element makes it a *compound lens*. A lens is also classified as to whether it is *thin* or *thick*—that is, whether or not its thickness is effectively negligible. We will limit ourselves, for the most part, to *centered systems* (for which all surfaces are rotationally symmetric about a common axis) of spherical surfaces. Under these restrictions, the simple lens can take the forms shown in Fig. 5.14.

Lenses that are variously known as ***convex***, ***converging***, or ***positive*** are thicker at the center and so tend to decrease the radius of curvature of the wavefronts. In other words, the wave converges more as it traverses the lens, assuming, of course, that the index of the lens is greater than that of the media in which it is immersed. ***Concave***, ***diverging***, or ***negative*** lenses, on the other hand, are thinner at the center and tend to advance that portion of the wavefront, causing it to diverge more than it did upon entry.

CONVEX	CONCAVE
$R_1 > 0$ $R_2 < 0$ Bi-convex	$R_1 < 0$ $R_2 > 0$ Bi-concave
$R_1 = \infty$ $R_2 < 0$ Planar convex	$R_1 = \infty$ $R_2 > 0$ Planar concave
$R_1 > 0$ $R_2 > 0$ Meniscus convex	$R_1 > 0$ $R_2 > 0$ Meniscus concave

FIGURE 5.14 Cross sections of various centered spherical simple lenses. The surface on the left is #1 since it is encountered first. Its radius is R_1. (Photo courtesy of Melles Griot.)

THIN-LENS EQUATIONS

Return to the discussion of refraction at a single spherical interface, where the location of the conjugate points S and P is given by

$$\frac{n_1}{s_o} + \frac{n_2}{s_i} = \frac{n_2 - n_1}{R} \qquad [5.8]$$

When s_o is large for a fixed $(n_2 - n_1)/R$, s_i is relatively small. As s_o decreases, s_i moves away from the vertex; that is, both θ_i and θ_t increase until finally $s_o = f_o$ and $s_i = \infty$. At that point, $n_1/s_o = (n_2 - n_1)/R$, so that if s_o gets any smaller, s_i will have to be negative, if Eq. (5.8) is to hold. In other words, the image becomes virtual (Fig. 5.15).

Let's now locate the conjugate points for a lens of index n_l surrounded by a medium of index n_m, as in Fig. 5.16, where another end has simply been ground onto the piece in Fig. 5.15c. This certainly isn't the most general set of circumstances, but it is the most common, and even more cogently, it is the simplest.* We know from Eq. (5.8) that the paraxial rays issuing from S at s_{o1} will meet at P', a distance, which we now call s_{i1}, from V_1, given by

$$\frac{n_m}{s_{o1}} + \frac{n_l}{s_{i1}} = \frac{n_l - n_m}{R_1} \qquad (5.11)$$

Thus, as far as the second surface is concerned, it "sees" rays coming toward it from P', which serves as its object point a distance s_{o2} away. Furthermore, the rays arriving at that second surface are in the medium of index n_l. Thus the object space for the second interface that contains P' has an index n_l. Note that the rays from P' to that surface are indeed straight lines. Considering the fact that

$$|s_{o2}| = |s_{i1}| + d$$

*See Jenkins and White, *Fundamentals of Optics*, p. 57, for a derivation containing three different indices.

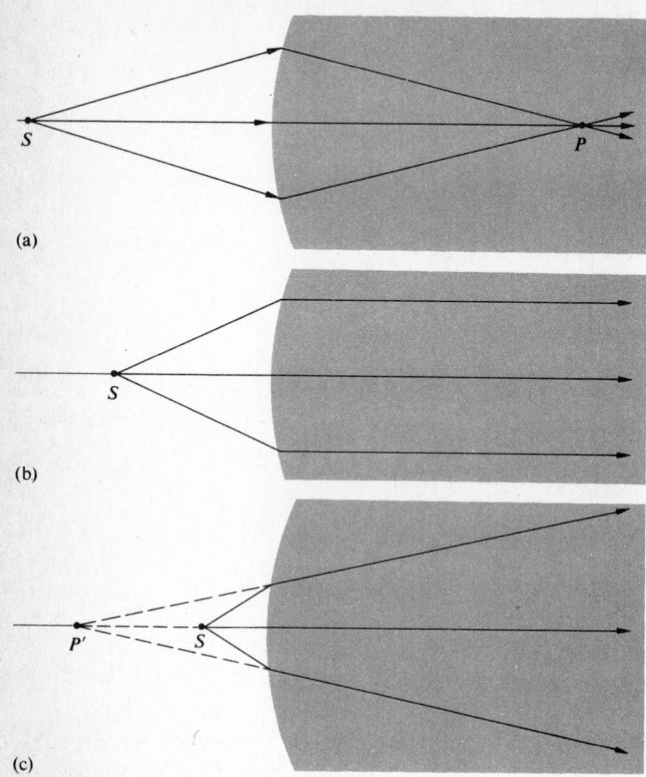

FIGURE 5.15 Refraction at a spherical interface between two transparent media shown in cross section.

since s_{o2} is on the left and therefore positive, $s_{o2} = |s_{o2}|$, and s_{i1} is also on the left and therefore negative, $-s_{i1} = |s_{i1}|$, we have

$$s_{o2} = -s_{i1} + d \tag{5.12}$$

Thus at the second surface Eq. (5.8) yields

$$\frac{n_l}{(-s_{i1} + d)} + \frac{n_m}{s_{i2}} = \frac{n_m - n_l}{R_2} \tag{5.13}$$

Here $n_l > n_m$ and $R_2 < 0$, so that the right-hand side is positive. Adding Eqs. (5.11) and (5.13), we have

$$\frac{n_m}{s_{o1}} + \frac{n_m}{s_{i2}} = (n_l - n_m)\left(\frac{1}{R_1} - \frac{1}{R_2}\right) + \frac{n_l d}{(s_{i1} - d)s_{i1}} \tag{5.14}$$

If the lens is thin enough ($d \to 0$), the last term on the right is effectively zero. As a further simplification, assume the surrounding medium to be air (i.e., $n_m \approx 1$). Accordingly, we

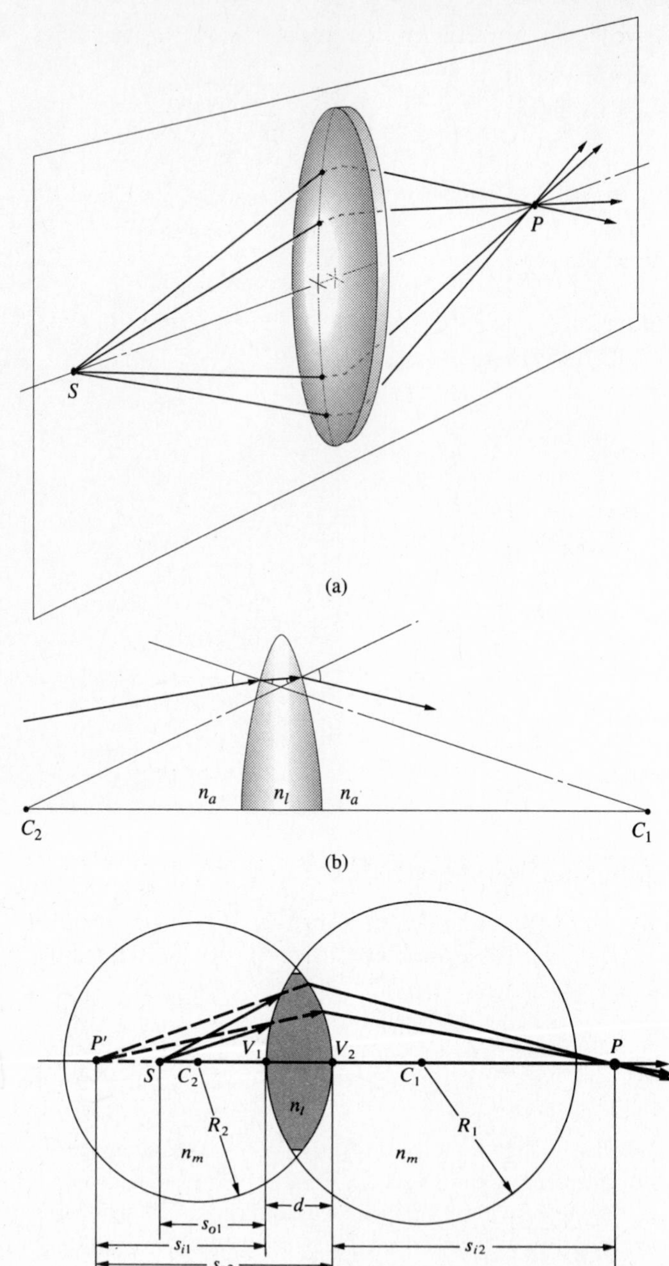

FIGURE 5.16 A spherical lens. (*a*) Rays in a vertical plane passing through a lens. Conjugate foci. (*b*) Refraction at the interfaces. The radius drawn from C_1 is normal to the first surface, and as the ray enters the lens it bends down *toward* that normal. The radius from C_2 is normal to the second surface; and as the ray emerges, since $n_l > n_a$, the ray bends down *away* from that normal. (*c*) The geometry.

have the very useful **Thin-Lens Equation**, often referred to as the **Lensmaker's Formula**:

$$\frac{1}{s_o} + \frac{1}{s_i} = (n_l - 1)\left(\frac{1}{R_1} - \frac{1}{R_2}\right) \tag{5.15}$$

where we let $s_{o1} = s_o$ and $s_{i2} = s_i$. The points V_1 and V_2 tend to coalesce as $d \to 0$, so that s_o and s_i can be measured from either the vertices or the lens center.

Just as in the case of the single spherical surface, if s_o is moved out to infinity, the image distance becomes the focal length f_i, or symbolically,

$$\lim_{s_o \to \infty} s_i = f_i$$

Similarly

$$\lim_{s_i \to \infty} s_o = f_o$$

It is evident from Eq. (5.15) that for a thin lens $f_i = f_o$, and consequently we drop the subscripts altogether. Thus

$$\frac{1}{f} = (n_l - 1)\left(\frac{1}{R_1} - \frac{1}{R_2}\right) \tag{5.16}$$

and

$$\frac{1}{s_o} + \frac{1}{s_i} = \frac{1}{f} \tag{5.17}$$

which is the famous **Gaussian Lens Formula** (Fig. 5.17).

As an example of how these expressions might be used, let's compute the focal length in air of a thin planar-convex lens having a radius of curvature of 50 mm and an index of 1.5. With light entering on the planar surface ($R_1 = \infty$, $R_2 = -50$),

$$\frac{1}{f} = (1.5 - 1)\left(\frac{1}{\infty} - \frac{1}{-50}\right)$$

whereas if instead it arrives at the curved surface ($R_1 = +50$, $R_2 = \infty$),

$$\frac{1}{f} = (1.5 - 1)\left(\frac{1}{+50} - \frac{1}{\infty}\right)$$

and in either case $f = 100$ mm. If an object is alternately placed at distances 600 mm, 200 mm, 150 mm, 100 mm, and 50 mm from the lens on either side, we can find the image points from Eq. (5.17):

$$s_i = \frac{s_o f}{s_o} - f = \frac{(600)(100)}{600 - 100}$$

and $s_i = 120$ mm. Similarly, the other image distances are 200 mm, 300 mm, ∞, and -100 mm, respectively.

Interestingly enough, when $s_o = \infty$, $s_i = f$; as s_o decreases, s_i increases positively until $s_o = f$ and s_i is negative thereafter. You can qualitatively check this out with a simple convex lens and a small electric light—the high-intensity variety is proba-

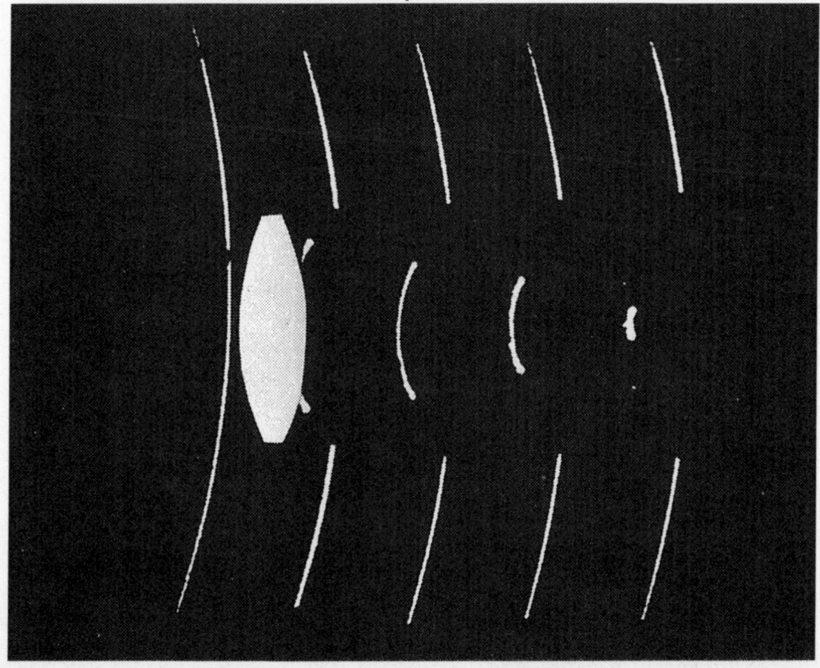

FIGURE 5.17 The actual wavefronts of a diverging lightwave partially focused by a lens. The photo shows five exposures, each separated by about 100 ps (i.e., 100×10^{-12} s), of a spherical pulse 10 ps long as it swept by and through a converging lens. The picture was made by Nils Abramson using a holographic technique.(Courtesy of N.H. Abramson)

bly the most convenient. Standing as far as you can from the source, project a clear image of it onto a white sheet of paper. You should be able to see the lamp quite clearly and not just as a blur. That image distance approximates f. Now move the lens in toward S, adjusting s_i to produce a clear image. It will surely increase. As $s_o \rightarrow f$, a clear image of the lamp can be projected, but only on an increasingly distant screen. For $s_o < f$, there will just be a blur where the farthest wall intersects the diverging cone of rays—the image is virtual.

Focal Points and Planes

Figure 5.18 summarizes pictorially some of the situations described analytically by Eq. 5.16. Observe that if a lens of index n_l is immersed in a medium of index n_m,

$$\frac{1}{f} = (n_{lm} - 1)\left(\frac{1}{R_1} - \frac{1}{R_2}\right) \tag{5.18}$$

The focal lengths in (a) and (b) of Fig. 5.18 are equal, because the same medium exists on either side of the lens. Since $n_l > n_m$, it follows that $n_{lm} > 1$. In both cases $R_1 > 0$ and $R_2 < 0$, so that each focal length is positive. We have a real object in (a) and a real image in (b). In (c), $n_l < n_m$, and consequently f is negative. In (d) and (e), $n_{lm} > 1$ but $R_1 < 0$, whereas $R_2 > 0$, so f is again negative, and the object in one case and the image in the other are virtual. In (f) $n_{lm} < 1$, yielding an $f > 0$.

Notice that in each instance it is particularly convenient to draw a ray through the center of the lens, which, because it is perpendicular to both surfaces, is undeviated. Suppose, instead, that an off-axis paraxial ray emerges from the lens parallel to its incident direction, as in Fig. 5.19. We maintain that all such rays will pass through the point defined as the **optical center** O of the lens. To see this, draw two parallel planes, one on each side tangent to the lens at any pair of points A and B. This can easily be done by selecting A and B

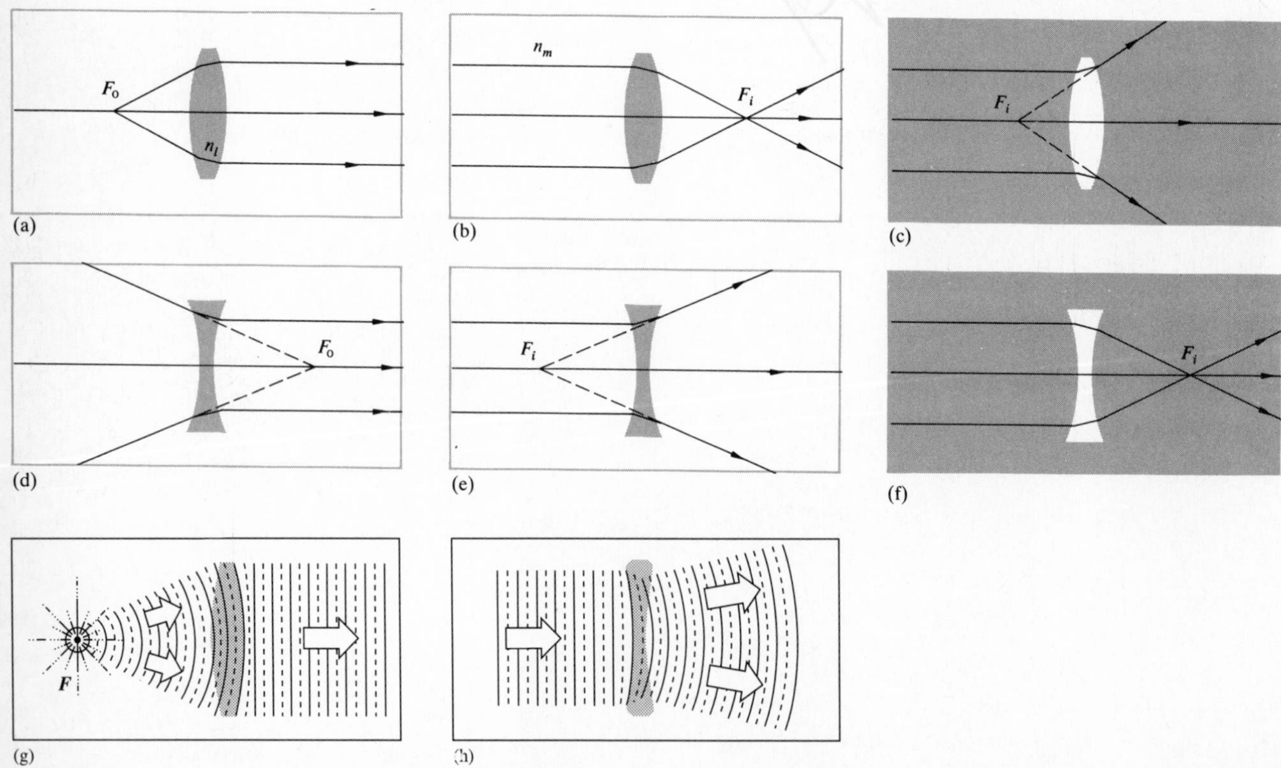

FIGURE 5.18 Focal lengths for converging and diverging lenses.

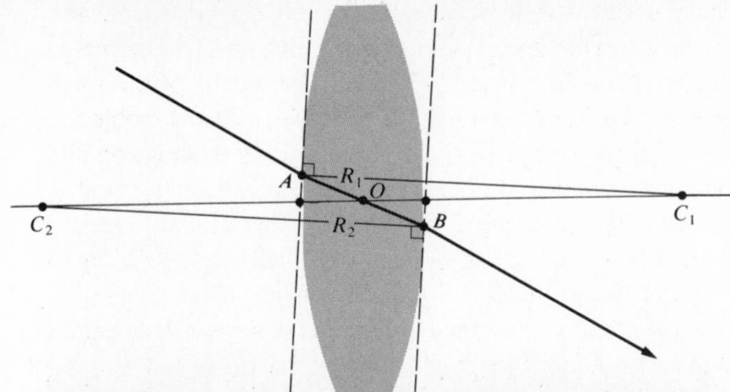

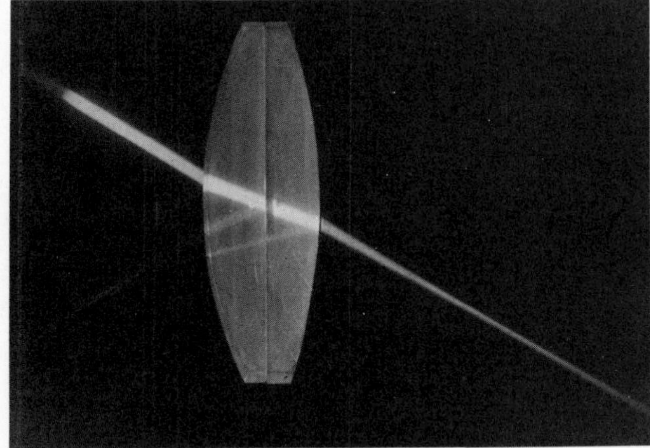

FIGURE 5.19 The optical center of a lens. (E.H.)

such that the radii $\overline{AC_1}$ and $\overline{BC_2}$ are themselves parallel. It is to be shown that the paraxial ray traversing $\overline{AB}$ enters and leaves the lens in the same direction. It's evident from the diagram that triangles AOC_1 and BOC_2 are similar, in the geometric sense, and therefore their sides are proportional. Hence, $|R_1|(\overline{OC_2}) = |R_2|(\overline{OC_1})$, and since the radii are constant, the location of O is constant, independent of A and B. As we saw earlier (Problem 4.30 and Fig. P.4.30), a ray traversing a medium bounded by parallel planes will be displaced laterally but will suffer no angular deviation. This displacement is proportional to the thickness, which for a thin lens is negligible. ***Rays passing through O may, accordingly, be drawn as straight lines***. It is customary when dealing with thin lenses simply to place O midway between the vertices.

Recall that a bundle of parallel paraxial rays incident on a spherical refracting surface comes to a focus at a point on the optical axis (Fig. 5.11). As shown in Fig. 5.20, this implies that several such bundles entering in a narrow cone will be focused on a spherical segment σ, also centered on C. The undeviated rays normal to the surface, and therefore passing through C, locate the foci on σ. Since the ray cone must indeed be narrow, σ can satisfactorily be represented as a plane normal to the symmetry axis and passing through the image focus. It is known as a **focal plane**. In the same way, limiting ourselves to paraxial theory, a lens will focus all incident parallel bundles of rays* onto a surface called the **second** or **back focal plane**, as in Fig. 5.21. Here each point on σ is located by the undeviated ray through O. Similarly, the **first** or **front focal plane** contains the object focus F_o.

Finite Imagery

Thus far we've treated the mathematical abstraction of a single-point source. Now let's deal with the fact that a great many such points combine to form a continuous finite object (Fig. 5.2). For the moment, imagine the object to be a segment of a

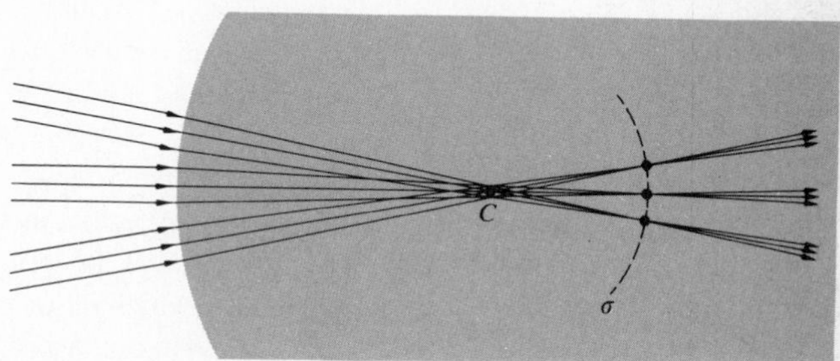

FIGURE 5.20 Focusing of several ray bundles.

*Perhaps the earliest literary reference to the focal properties of a lens appears in Aristophanes' play, *The Clouds*, which dates back to 423 B.C.E. In it Strepsiades plots to use a burning-glass to focus the Sun's rays onto a wax tablet and thereby melt out the record of a gambling debt.

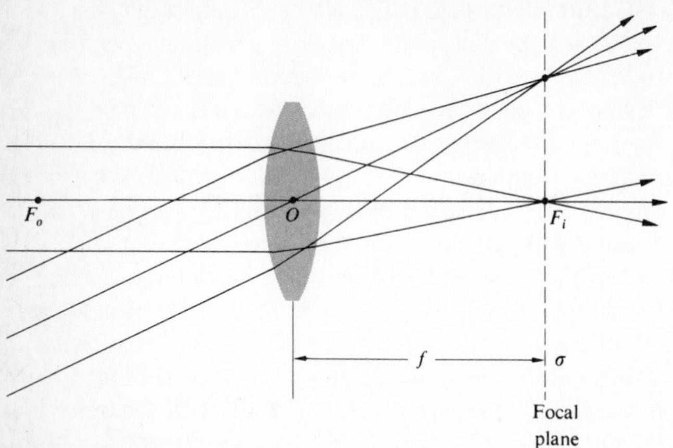

FIGURE 5.21 The focal plane of a lens.

sphere, σ_o, centered on C, as in Fig. 5.22. If σ_o is close to the spherical interface, point S will have a virtual image P ($s_i < 0$ and therefore on the left of V). With S farther away, its image will be real ($s_i > 0$ and therefore on the right-hand side). In either case, each point on σ_o has a conjugate point on σ_i lying on a straight line through C. Within the restrictions of paraxial theory, these surfaces can be considered planar. Thus a small planar object normal to the optical axis will be imaged into a small planar region also normal to that axis. Note that if σ_o is moved out to infinity, the cone of rays from each source point will become **collimated** (i.e., parallel), and the image points will lie on the focal plane (Fig. 5.21).

By cutting and polishing the right side of the piece depicted in Fig. 5.22, we can construct a thin lens. Once again, the image (σ_i in Fig. 5.22) formed by the first surface of the lens will serve as the object for the second surface, which in turn will generate a final image. Suppose then that σ_i in Fig. 5.22*a* is the object for the second surface, which is assumed to have a negative radius. We already know what will happen—the situation is identical to Fig. 5.22*b* with the ray directions reversed. The *final image formed by a lens of a small planar object normal to the optical axis will itself be a small plane normal to that axis.*

The location, size, and orientation of an image produced by a lens can be determined, particularly simply, with ray diagrams. To find the image of the object in Fig. 5.23, we must locate the image point corresponding to each object point. Since all rays issuing from a source point in a paraxial cone will arrive at the image point any two such rays will suffice to fix that point. Because we know the positions of the focal points, there are three rays that are especially easy to apply. Two of these make use of the fact that a ray passing through the focal point will emerge from the lens parallel to the central axis and vice versa; the third is the undeviated ray through the center of the lens O. Figure 5.24 shows how any *two* of these three rays locate the image of a point on the object. Incidentally, this technique dates back to the work of Robert Smith as long ago as 1738.

This graphical procedure can be made even simpler by replacing the thin lens with a plane passing through its center (Fig. 5.25). Presumably, if we were to extend every incoming ray forward a little and every outgoing ray backward a bit, each pair would meet on this plane. The total deviation of any ray can be envisaged as occurring all at once on that plane. This is equivalent to the actual process consisting of two separate angular shifts, one at each interface. (As we'll see later,

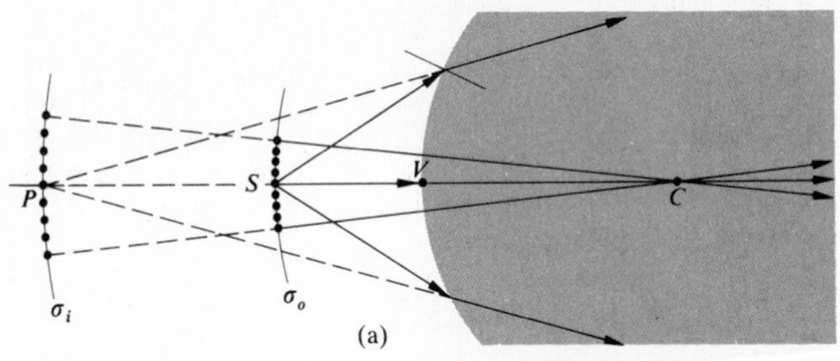

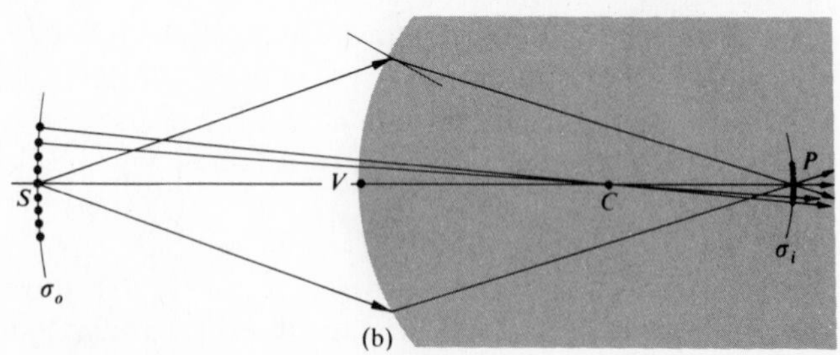

FIGURE 5.22 Finite imagery.

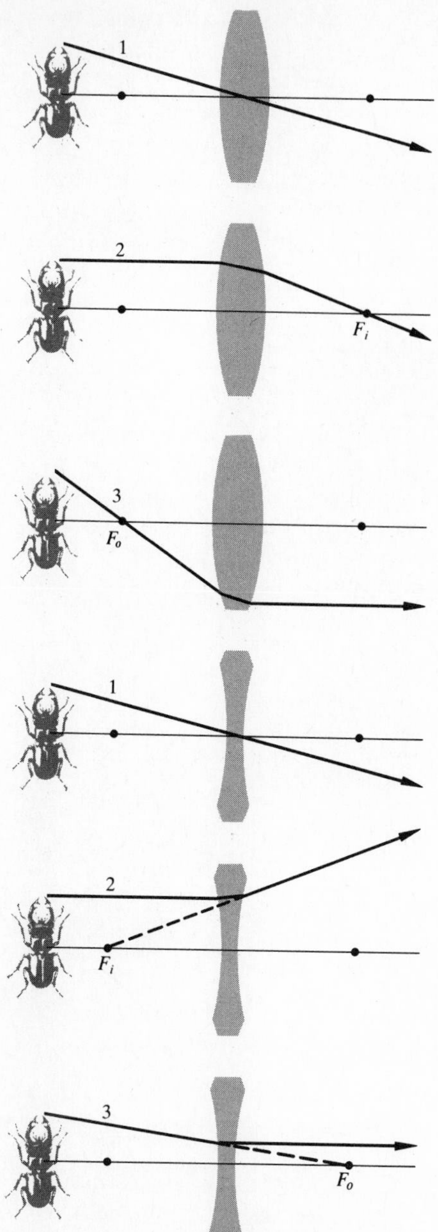

FIGURE 5.23 Tracing a few key rays through a positive and negative lens.

this is tantamount to saying that the two principal planes of a thin lens coincide.)

In accord with convention, transverse distances above the optical axis are taken as positive quantities, and those below the axis are given negative numerical values. Therefore in Fig. 5.25 $y_o > 0$ and $y_i < 0$. Here the image is said to be **inverted**, whereas if $y_i > 0$ when $y_o > 0$, it is **right-side-up** or **erect**. Observe that triangles AOF_i and $P_2P_1F_i$ are similar. Ergo

$$\frac{y_o}{|y_i|} = \frac{f}{(s_i - f)} \tag{5.19}$$

Similarly, triangles S_2S_1O and P_2P_1O are similar, and

$$\frac{y_o}{|y_i|} = \frac{s_o}{s_i} \tag{5.20}$$

where all quantities other than y_i are positive. Hence

$$\frac{s_o}{s_i} = \frac{f}{(s_i - f)} \tag{5.21}$$

and

$$\frac{1}{f} = \frac{1}{s_o} + \frac{1}{s_i}$$

which is, of course, the Gaussian Lens Equation [Eq. (5.17)]. Furthermore, triangles $S_2S_1F_o$ and BOF_o are similar and

$$\frac{f}{(s_o - f)} = \frac{|y_i|}{y_o} \tag{5.22}$$

Using the distances measured from the focal points and combining this information with Eq. (5.19), leads to

$$x_o x_i = f^2 \tag{5.23}$$

This is the **Newtonian form** of the lens equation, the first statement of which appeared in Newton's *Opticks* in 1704. The signs of x_o and x_i are reckoned with respect to their concomitant foci. By convention, x_o is taken to be positive left of F_o, whereas x_i is positive on the right of F_i. It is evident from Eq. (5.23) that x_o and x_i have like signs, which means that ***the object and image must be on opposite sides of their respective focal points***. This is a good thing for the neophyte to remember when making those hasty freehand ray diagrams for which he or she is already infamous.

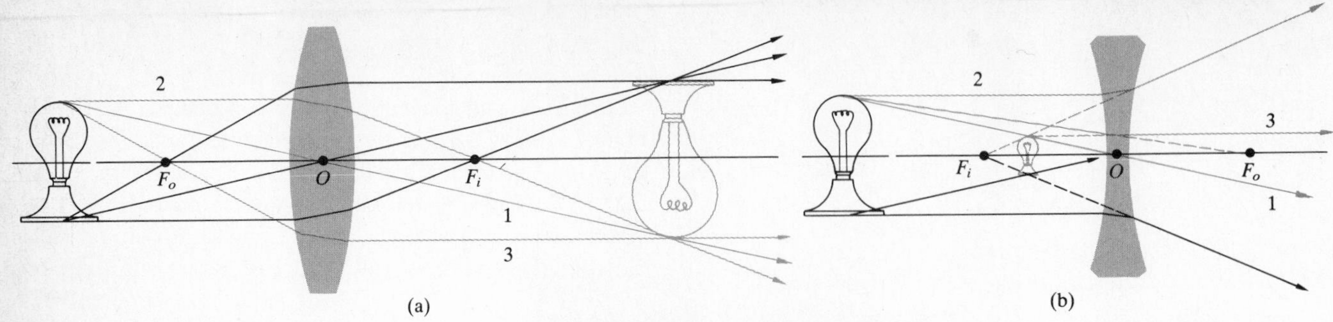

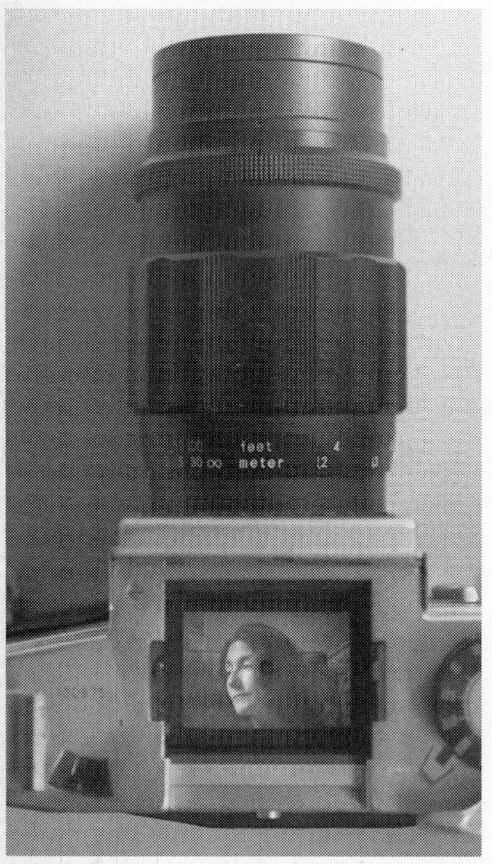

(c)

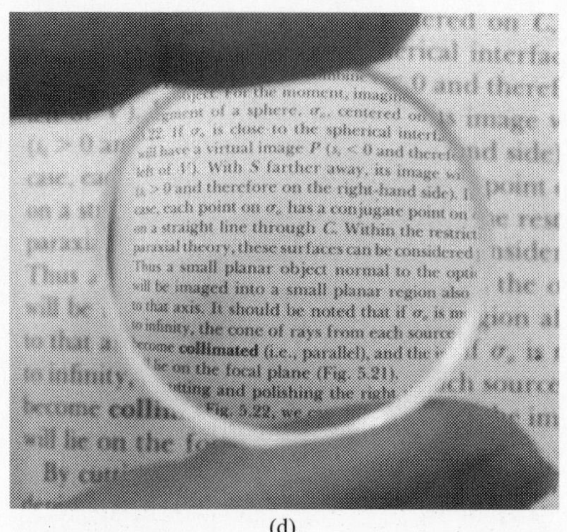

(d)

FIGURE 5.24 (*a*) A real object and a positive lens. (*b*) A real object and a negative lens. (*c*) A real image projected on the viewing screen of a 35-mm camera, much as the eye projects its image on the retina. Here a prism has been removed so you can see the image directly. (E.H.) (*d*) The minified, right-side-up, virtual image formed by a negative lens.

The ratio of the transverse dimensions of the final image formed by any optical system to the corresponding dimension of the object is defined as the *lateral* or **transverse magnification**, M_T, that is,

$$M_T \equiv \frac{y_i}{y_o} \tag{5.24}$$

Or from Eq. (5.20)

$$M_T = -\frac{s_i}{s_o} \tag{5.25}$$

Thus ***a positive M_T connotes an erect image, while a negative value means the image is inverted*** (see Table 5.2). Bear in

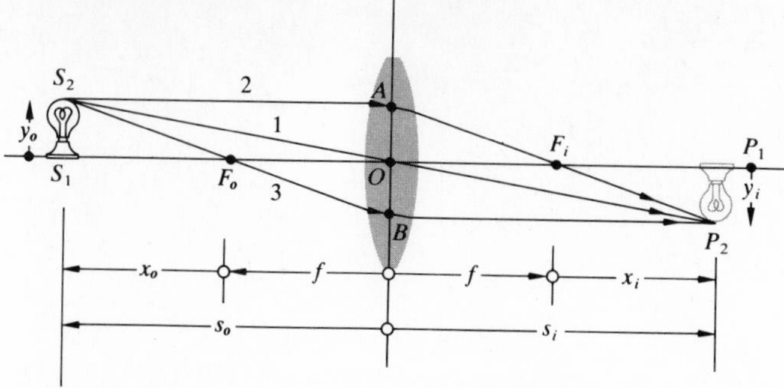

FIGURE 5.25 Object and image location for a thin lens.

mind that s_i and s_o are both positive for real objects and images. Clearly, then, ***all real images formed by a single thin lens will be inverted***. The Newtonian expression for the magnification follows from Eqs. (5.19) and (5.22) and Fig. 5.25:

$$M_T = -\frac{x_i}{f} = -\frac{f}{x_o} \tag{5.26}$$

The term *magnification* is a misnomer, since the magnitude of M_T can certainly be less than 1, in which case the image is smaller than the object. We have $M_T = -1$ when the object and image distances are positive and equal, and that happens [Eq. (5.17)] only when $s_o = s_i = 2f$. This turns out to be the configuration in which the object and image are as close together as they can possibly get (i.e., a distance $4f$ apart; see Problem 5.11). Table 5.3 summarizes a number of image configurations resulting from the juxtaposition of a thin lens and a real object.

Table 5.3 Images of Real Objects Formed by Thin Lenses

Convex				
Object	**Image**			
Location	Type	Location	Orientation	Relative Size
$\infty > s_o > 2f$	Real	$f < s_i < 2f$	Inverted	Minified
$s_o = 2f$	Real	$s_i = 2f$	Inverted	Same size
$f < s_o < 2f$	Real	$\infty > s_i > 2f$	Inverted	Magnified
$s_o = f$		$\pm\infty$		
$s_o < f$	Virtual	$\|s_i\| > s_o$	Erect	Magnified

Concave				
Object	**Image**			
Location	Type	Location	Orientation	Relative Size
Anywhere	Virtual	$\|s_i\| < \|f\|$,	Erect	Minified

Table 5.2 Meanings Associated with the Signs of Various Thin Lens and Spherical Interface Parameters

Quantity	Sign +	Sign −
s_o	Real object	Virtual object
s_i	Real image	Virtual image
f	Converging lens	Diverging lens
y_o	Erect object	Inverted object
y_i	Erect image	Inverted image
M_T	Erect image	Inverted image

We are now in a position to understand the entire range of behavior of a single convex or concave lens. To that end, suppose that a distant point source sends out a cone of light that is intercepted by a positive lens (Fig. 5.26). If the source is at infinity (i.e., so far away that it might just as well be infinity), rays coming from it entering the lens are essentially parallel (Fig. 5.26*a*) and will be brought together at the focal point F_i. If the source point S_1 is closer (Fig. 5.26*b*), but still fairly far away, the cone of rays entering the lens is narrow, and the rays come in at shallow angles to the surface of the lens. Because the rays do not diverge greatly, the lens bends each one into convergence, and they arrive at point P_1. As the source moves closer, the entering rays diverge more, and the resulting image point moves farther to the right. Finally, when the source point is at F_o, the rays are diverging so strongly that the lens can no longer bring them into convergence, and they emerge parallel to the central axis. Moving the source point closer results in rays that diverge so much on entering the lens that they still

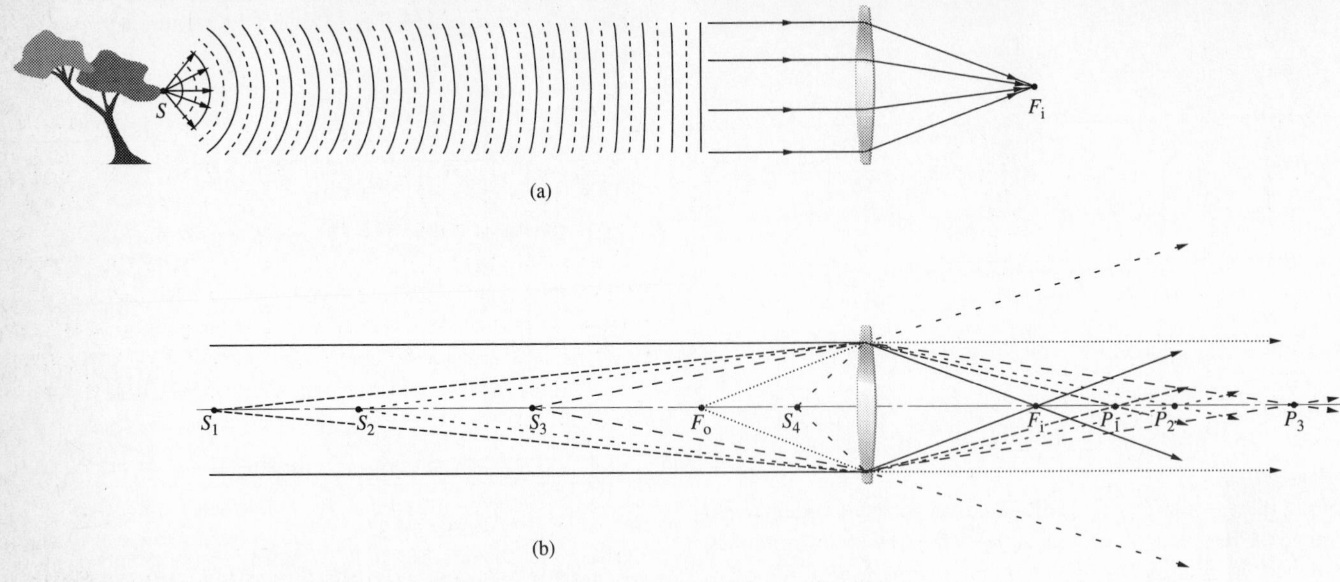

FIGURE 5.26 (*a*) The waves from a distant object flatten out as they expand, and the radii get larger and larger. Viewed from far away the rays from any point are essentially parallel, and the lens causes them to converge at F_i. (*b*) As a point source moves closer, the rays diverge more and the image point moves out away from the lens. The emerging rays no longer converge once the object reaches the focal point; nearer in still, they diverge.

diverge on leaving. The image point is now virtual—*there are no real images of objects at or closer in than f.* Figure 5.27 illustrates the behavior pictorially. **As the object approaches the lens, the real image moves away from it.**

It's useful to remember that *the ray entering the lens parallel to the central axis fixes the height of the real image* (Fig. 5.28). Because that ray diverges from the central axis, the size of the image increases rapidly as the object approaches F. Note, too, that the transformation from object to image space is not linear; all of the object space from $2f$ out to infinity, on the left of the lens, is compressed in the image space between f and $2f$, on the right of the lens. Figure 5.28 suggests that the image space is distorted, in the sense that advancing the object uniformly toward the lens has the effect of changing the image differently along and transverse to the central axis. The axial image intervals increase much more rapidly than the corresponding successive changes in the height of the image. This relative "flattening" of distant-object space is easily observable using a telescope (i.e., a long focal-length lens). You've probably seen the effect in a motion picture shot through a telephoto lens. Always staying far away, the hero vigorously runs a great distance toward the camera, but psychologically he seems to make no progress because his perceived size increases very little despite all his effort.

Presumably, the image of a three-dimensional object will itself occupy a three-dimensional region of space. The optical system can apparently affect both the transverse and longitudinal dimensions of the image. The **longitudinal magnification**, M_L, which relates to the axial direction, is defined as

$$M_L \equiv \frac{dx_i}{dx_o} \tag{5.27}$$

This is the ratio of an infinitesimal axial length in the region of the image to the corresponding length in the region of the

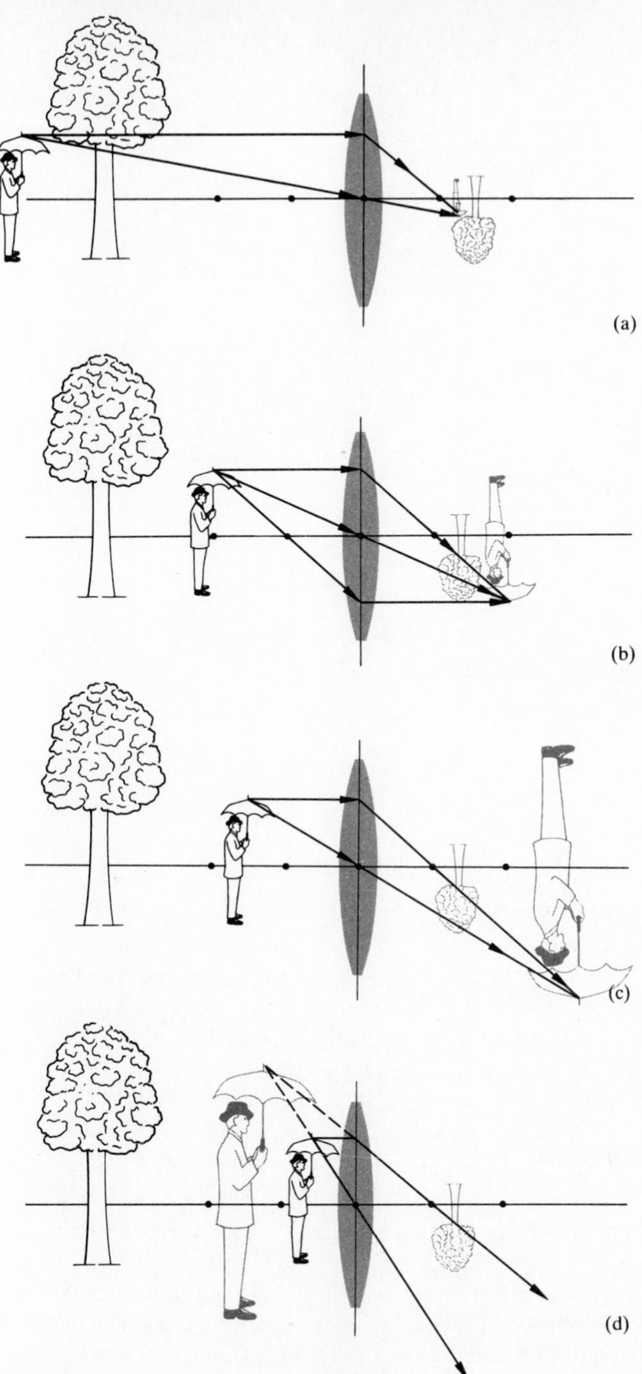

FIGURE 5.27 The image-forming behavior of a thin positive lens.

object. Differentiating Eq. (5.23) leads to

$$M_L = -\frac{f^2}{x_o^2} = -M_T^2 \tag{5.28}$$

for a thin lens in a single medium (Fig. 5.29). Evidently, $M_L < 0$, which implies that a positive dx_o corresponds to a negative dx_i and vice versa. In other words, a finger pointing toward the lens is imaged pointing away from it (Fig. 5.30).

Form the image of a window on a sheet of paper, using a simple convex lens. Assuming a lovely arboreal scene, image the distant trees on the screen. Now move the paper *away* from the lens, so that it intersects a different region of the image space. The trees will fade while the nearby window itself comes into view.

Thin-Lens Combinations

Our purpose here is not to become proficient in the intricacies of modern lens design, but rather to gain the familiarity necessary to utilize, and adapt those lens systems already available commercially.

In constructing a new optical system, one generally begins by sketching out a rough arrangement using the quickest approximate calculations. Refinements are then added as the designer goes on to the prodigious and more exact ray-tracing techniques. Nowadays these computations are most often carried out by computers. Even so, the simple thin-lens concept provides a highly useful basis for preliminary calculations in a broad range of situations.

No lens is actually a thin lens in the strict sense of having a thickness that approaches zero. Yet many simple lenses, for all practical purposes, function in a fashion equivalent to that of a thin lens (i.e., one that is thin in comparison to its diameter). Almost all spectacle lenses (which, by the way, have been used at least since the thirteenth century) are in this category. When the radii of curvature are large and the lens diameter is small, the thickness will usually be small as well. A lens of this sort would generally have a large focal length, compared with which the thickness would be quite small; many early telescope objectives fit that description perfectly.

We will now derive expressions for parameters associated with thin-lens combinations. The approach will be fairly simple, leaving the more elaborate traditional treatment for those tenacious enough to pursue the matter into the next chapter.

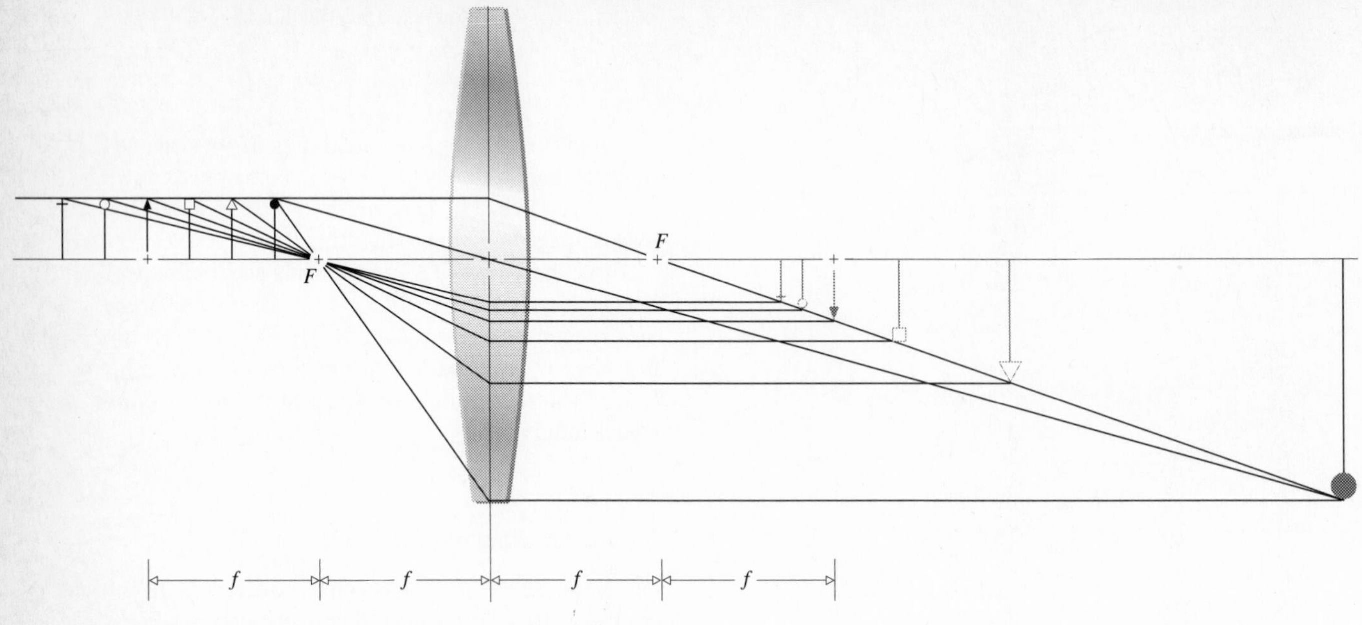

FIGURE 5.28 The number-2 ray entering the lens parallel to the central axis limits the image height.

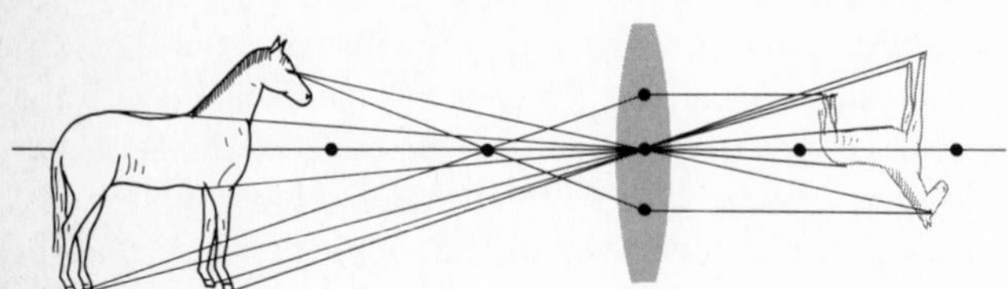

FIGURE 5.29 The transverse magnification is different from the longitudinal magnification.

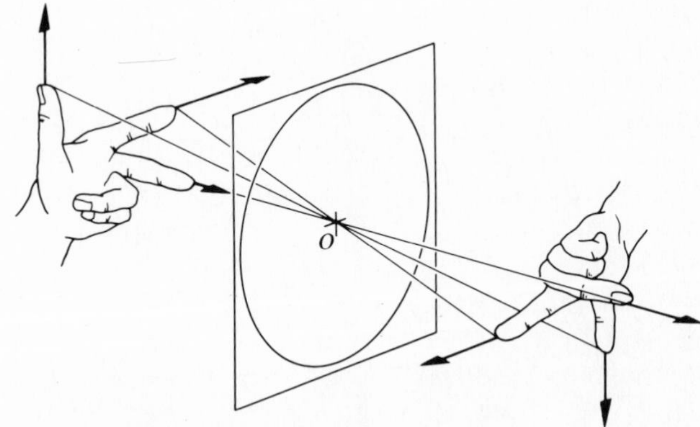

FIGURE 5.30 Image orientation for a thin lens.

Consider two thin positive lenses L_1 and L_2 separated by a distance d, which is smaller than either focal length, as in Fig. 5.31. The resulting image can be located graphically as follows. Overlooking L_2 for a moment, construct the image formed exclusively by L_1 using rays 1 and 3. As usual, these pass through the lens object and image foci, F_{o1} and F_{i1}, respectively. The object is in a normal plane, so that two rays determine its top, and a perpendicular to the optical axis finds its bottom. Ray 2 is then constructed running backward from P_1' through O_2. Insertion of L_2 has no effect on ray 2, whereas ray 3 is refracted through the image focus F_{i2} of L_2. The intersection of rays 2 and 3 fixes the image, which in this particular case is real, minified, and inverted. When the two lenses are close together, as they are here, the presence of L_2 essen-

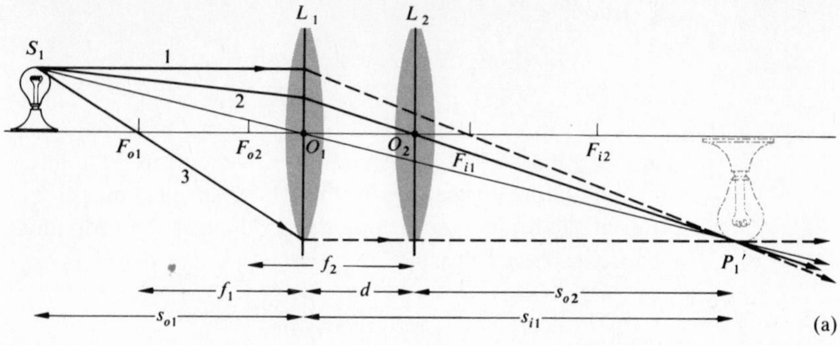

FIGURE 5.31 Two thin lenses separated by a distance smaller than either focal length.

tially adds convergence ($f_2 > 0$) or divergence ($f_2 < 0$) to the bundle of rays emerging from L_1; see Fig. 5.32.

A similar pair of lenses is illustrated in Fig. 5.33, in which the separation has been increased. Once again rays 1 and 3 through F_{i1} and F_{o1} fix the position of the intermediate image generated by L_1 alone. As before, ray 2 is drawn backward from O_2 to P_1' to S_1. The intersection of rays 2 and 3, as the latter is refracted through F_{i2}, locates the final image. This time it is real and erect. Notice that if the focal length of L_2 is increased with all else constant, the size of the image increases as well.

Analytically, for L_1

$$\frac{1}{s_{i1}} = \frac{1}{f_1} - \frac{1}{s_{o1}} \tag{5.29}$$

or

$$s_{i1} = \frac{s_{o1} f_1}{s_{o1} - f_1} \tag{5.30}$$

This is positive, and the intermediate image is to the right of L_1, when $s_{o1} > f_1$ and $f_1 > 0$. For L_2

$$s_{o2} = d - s_{i1} \tag{5.31}$$

and if $d > s_{i1}$, the object for L_2 is real (as in Fig. 5.33), whereas if $d < s_{i1}$, it is virtual ($s_{o2} < 0$, as in Fig. 5.31). In the former instance the rays approaching L_2 are diverging from P_1', whereas in the latter they are converging toward it. Furthermore,

$$\frac{1}{s_{i2}} = \frac{1}{f_2} - \frac{1}{s_{o2}}$$

or

$$s_{i2} = \frac{s_{o2} f_2}{s_{o2} - f_2}$$

Using Eq. (5.31), we obtain

$$s_{i2} = \frac{(d - s_{i1}) f_2}{(d - s_{i1} - f_2)} \tag{5.32}$$

In this same way we could compute the response of any number of thin lenses. It will often be convenient to have a single expression, at least when dealing with only two lenses, so substituting for s_{i1} from Eq. (5.29),

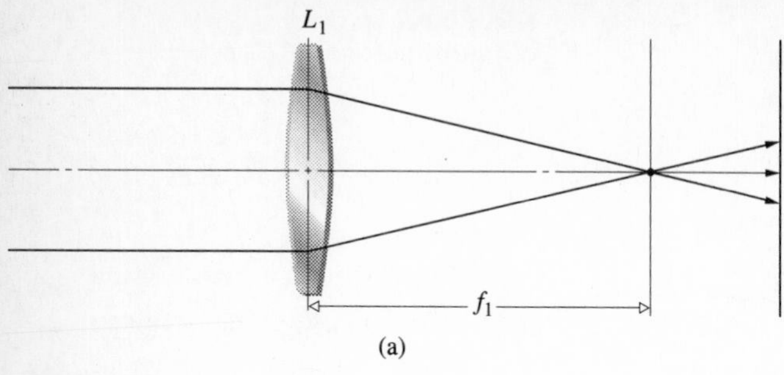

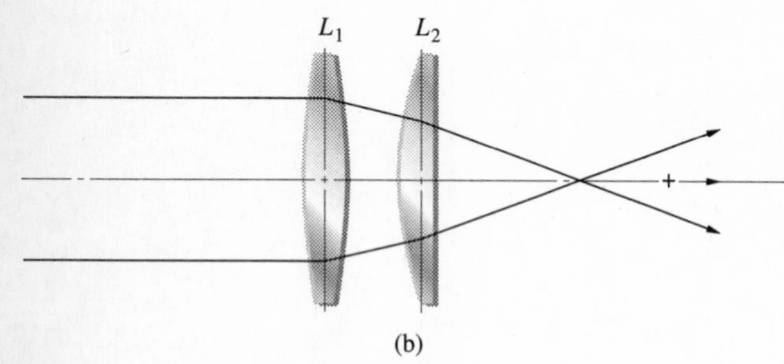

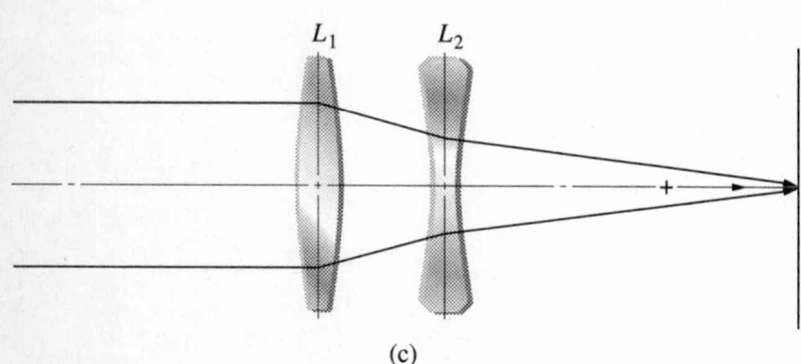

FIGURE 5.32 (a) The effect of placing a second lens, L_2, within the focal length of a positive lens, L_1. (b) When L_2 is positive, its presence adds convergence to the ray bundle. (c) When L_2 is negative, it adds divergence to the ray bundle.

$$s_{i2} = \frac{f_2 d - f_2 s_{o1} f_1/(s_{o1} - f_1)}{d - f_2 - s_{o1} f_1/(s_{o1} - f_1)} \tag{5.33}$$

Here s_{o1} and s_{i2} are the object and image distances, respectively, of the compound lens. As an example, let's compute the image distance associated with an object placed 50.0 cm from the first of two positive lenses. These in turn are separated by 20.0 cm and have focal lengths of 30.0 cm and 50.0 cm, respectively. By direct substitution

$$s_{i2} = \frac{50(20) - 50(50)(30)/(50-30)}{20 - 50 - 50(30)/(50-30)} = 26.2 \text{ cm}$$

and the image is real. Inasmuch as L_2 "magnifies" the intermediate image formed by L_1, the total transverse magnification of the compound lens is the product of the individual magnifications, that is,

$$M_T = M_{T1} M_{T2}$$

It is left as Problem 5.31 to show that

$$M_T = \frac{f_1 s_{i2}}{d(s_{o1} - f_1) - s_{o1} f_1} \tag{5.34}$$

In the above example

$$M_T = \frac{30(26.2)}{20(50-30) - 50(30)} = -0.72$$

and just as we should have guessed from Fig. 5.31, the image is minified and inverted.

The distance from the last surface of an optical system to the second focal point of that system as a whole is known as the **back focal length**, or b.f.l. Similarly, the distance from the vertex of the first surface to the first or object focus is the **front local length**, or f.f.l. Consequently, if we let $s_{i2} \to \infty$, s_{o2} approaches f_2, which combined with Eq. (5.31) tells us that $s_{i1} \to d - f_2$. Hence from Eq. (5.29)

$$\left.\frac{1}{s_{o1}}\right|_{s_{i2}=\infty} = \frac{1}{f_1} - \frac{1}{(d - f_2)} = \frac{d - (f_1 + f_2)}{f_1(d - f_2)}$$

But this special value of s_{o1} is the f.f.l.:

$$\text{f.f.l.} = \frac{f_1(d - f_2)}{d - (f_1 + f_2)} \tag{5.35}$$

In the same way, letting $s_{o1} \to \infty$ in Eq. (5.33), $(s_{o1} - f_1) \to s_{o1}$, and since s_{i2} is then the b.f.l., we have

$$\text{b.f.l.} = \frac{f_2(d - f_1)}{d - (f_1 + f_2)} \tag{5.36}$$

To see how this works numerically, let's find both the b.f.l. and f.f.l. for the thin-lens system in Fig. 5.34*a*, where $f_1 = -30$ cm and $f_2 = +20$ cm. Then

$$\text{b.f.l.} = \frac{20[10 - (-30)]}{10 - (-30 + 20)} = 40 \text{ cm}$$

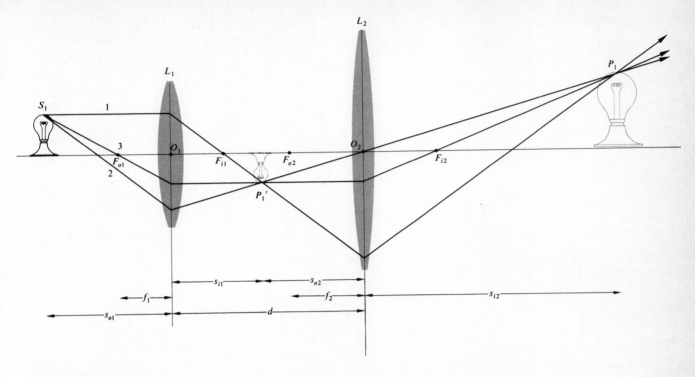

FIGURE 5.33 Two thin lenses separated by a distance greater than the sum of their focal lengths. Because the intermediate image is real, you could start with point P_i' and treat it as if it were a real object point for L_2. Thus a ray from P_i' through F_{o2} would arrive at P_1.

and similarly f.f.l. = 15 cm. Incidentally, notice that if $d = f_1 + f_2$, plane waves entering the compound lens from either side will emerge as plane waves (Problem 5.34), as in telescopic systems.

Observe that if $d \rightarrow 0$, that is, the lenses are brought into contact, as in the case of some achromatic doublets,

$$\text{b.f.l.} = \text{f.f.l.} = \frac{f_2 f_1}{f_2 + f_1} \tag{5.37}$$

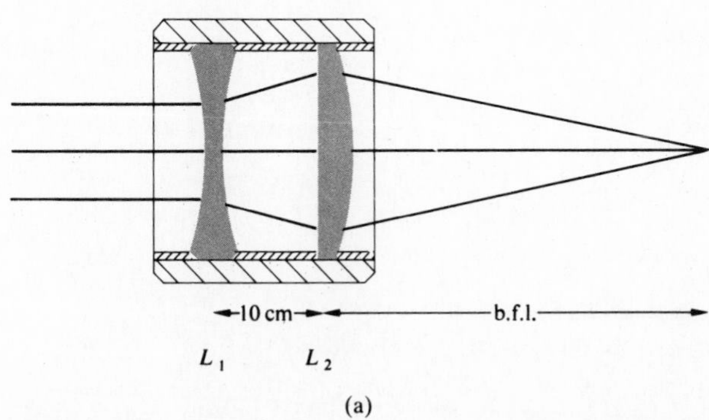

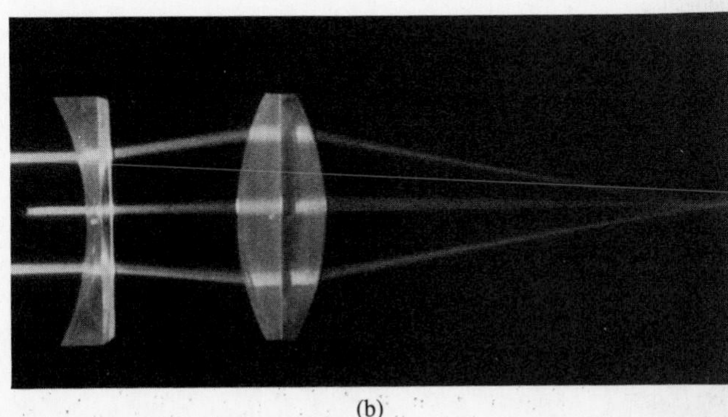

(b)

FIGURE 5.34 (a) A positive and negative thin-lens combination, (b) photo (E.H.)

The resultant thin lens has an *effective focal length*, f, such that

$$\frac{1}{f} = \frac{1}{f_1} + \frac{1}{f_2} \tag{5.38}$$

This implies that if there are N such lenses in contact,

$$\frac{1}{f} = \frac{1}{f_1} + \frac{1}{f_2} + \cdots + \frac{1}{f_N} \tag{5.39}$$

Many of these conclusions can be verified, at least qualitatively, with a few simple lenses. Figure 5.31 is easy to duplicate, and the procedure should be self-evident, whereas Fig. 5.33 requires a bit more care. First, determine the focal lengths of the two lenses by imaging a distant source. Then hold one of the lenses (L_2) at a fixed distance *slightly greater than its focal length* from the plane of observation (i.e., a piece of white paper). Now comes the maneuver that requires some effort if you don't have an optical bench. Move the second lens (L_1) toward the source, keeping it reasonably centered. Without any attempts to block out light entering L_2 directly,

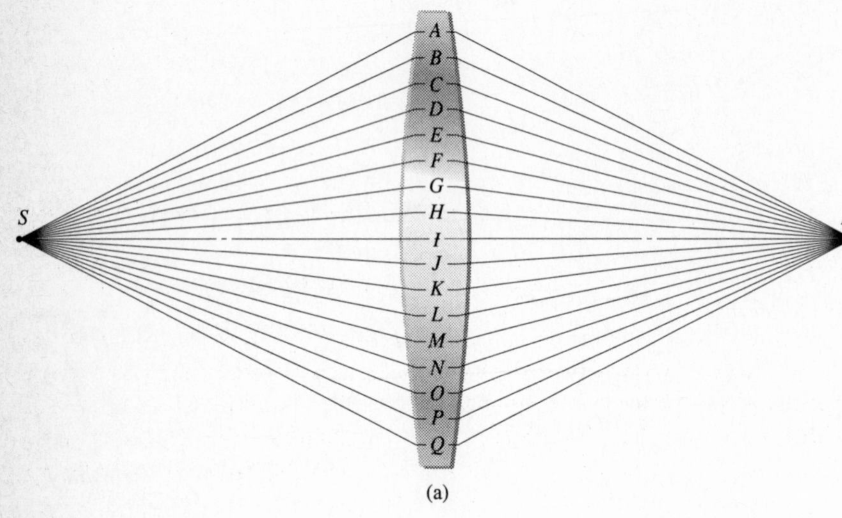

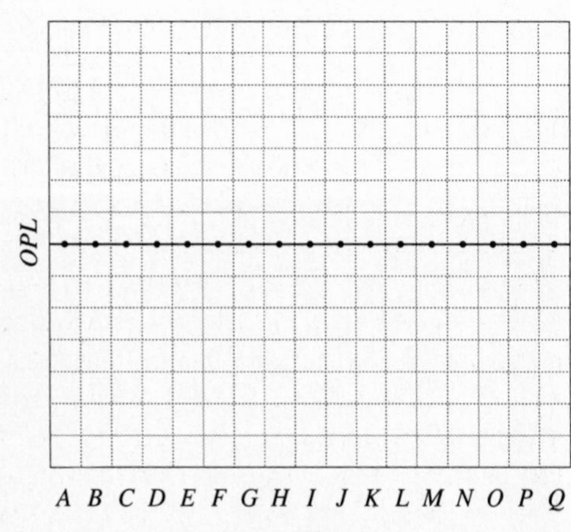

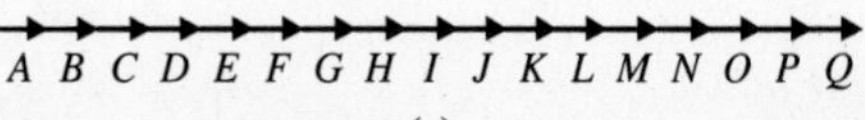

FIGURE 5.35 Feynman's analysis of the thin lens via QED. (*a*) A number of possible paths from *S* to *P*. (*b*) The *OPL* for light along each path. (*c*) The corresponding probability-amplitude phasors all adding in-phase.

you will probably see a blurred image of your hand holding L_1. Position the lenses so that the region on the screen corresponding to L_1 is as bright as possible. The scene spread across L_1 (i.e., its image within the image) will become clear and erect, as in Fig. 5.33.

QED and the Lens

One excellent reason for deriving the basic equations of this chapter from Fermat's Principle is that it keeps us thinking in terms of optical path lengths, and that naturally leads to the Feynman treatment of Quantum Electrodynamics. Keep in mind that many physicists consider their theories to be nothing more than the conceptual machinery for calculating the results of observations. And no matter how sophisticated a theory is, it must be in agreement with even the most "ordinary" observation. Thus to see how the operation of a lens fits into the QED worldview, return to Fig. 4.71 and the mirror for a brief review.

Light goes from point S to the mirror to point P along a tremendous number of possible routes. Classically, we note that the *OPL*s are different as, therefore, are the traversal times. In QED, each path has an associated probability amplitude (which has a phase angle proportional to the traversal time). When these are all summed, the most effective contribution to the overall probability of light arriving at P is seen to come from the paths immediately adjacent to the one that has the minimum *OPL*.

For a lens (Fig. 5.35) the situation is very different. We can again approximate things by dividing the device into a manageable number of segments with a possible light path, and therefore a tiny probability amplitude, corresponding to each one. Of course, there should be a lot more than 17 paths, so think of each of these as representing a cluster of billions of neighboring trajectories—the logic doesn't change. Each path has a little probability-amplitude phasor associated with it. Because the lens was designed specifically to make all the *OPL*s equal, a plot of *OPL* (or equivalently the transit times) against distance across the breadth of the lens is a straight line. Consequently, a photon takes the same time to traverse any one path; all the phasors (each assumed to be the same size) have the same phase angle. Thus, they all contribute equally to the likelihood of a photon arriving at P. Putting the phasors tip-to-tail results in a very large net amplitude, which when squared yields a very high probability of light reaching P via the lens. In the language of QED, ***a lens focuses light, by causing all the constituent probability amplitudes to have the same phase angle***.

For other points in the plane containing P that are close to the optical axis, the phase angles will differ proportionately. The phasors placed tip-to-tail will gradually spiral, and the net probability amplitude will initially diminish quickly, but not discontinuously so. Notice that the probability distribution is not a single infinitesimally narrow spike; the light cannot be focused to a point. The phasors for off-axis points cannot all at once add to zero; what happens, happens gradually and continuously. The resulting circularly symmetric probability distribution, $I(r)$, is known as the Airy pattern (p. 459).

5.3 Stops

5.3.1 Aperture and Field Stops

The intrinsically finite nature of all lenses demands that they collect only a fraction of the energy emitted by a point source. The physical limitation presented by the periphery of a simple lens therefore determines which rays shall enter the system to form an image. In that respect, the unobstructed or *clear diameter* of the lens functions as an aperture into which energy flows. Any element, be it the rim of a lens or a separate diaphragm, that determines the amount of light reaching the image is known as the **aperture stop** (abbreviated A.S.). The adjustable leaf diaphragm that is usually located behind the first few elements of a compound camera lens is just such an aperture stop. Evidently, it determines the light-gathering capability of the lens as a whole. As shown in Fig. 5.36, highly oblique rays can still enter a system of this sort. Usually, however, they are deliberately restricted in order to control the quality of the image. The element limiting the size or angular breadth of the object that can be imaged by the system is called the **field stop** or F.S.—it determines the field of view of the instrument. In a camera, the edge of the film itself bounds the image plane and serves as the field stop. Thus, while (Fig. 5.36) the aperture stop controls the number of rays from an object point reaching the conjugate image point, it is the field stop that will or will not obstruct those rays *in toto*. Neither the region above the top nor the region below the bottom of the object in Fig. 5.36 passes the field stop. Opening the circular

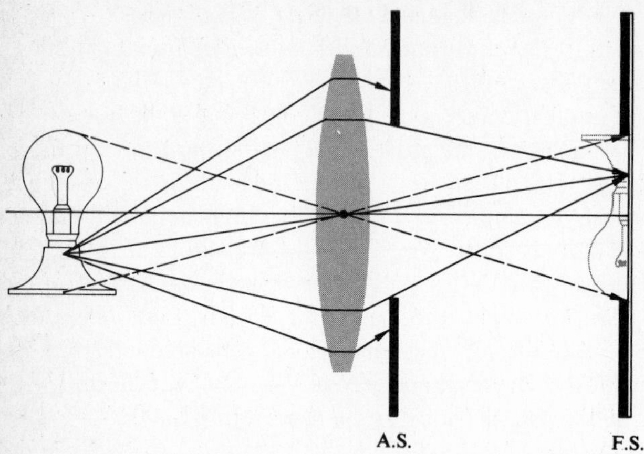

FIGURE 5.36 Aperture stop and field stop.

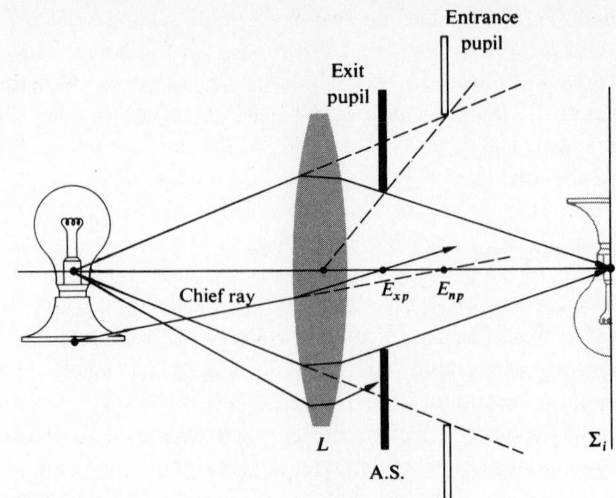

FIGURE 5.37 Entrance pupil and exit pupil.

aperture stop would cause the system to accept a larger energy cone and in so doing increase the irradiance at each image point. In contrast, opening the field stop would allow the regions beyond the extremities of the object, which were previously blocked, to be imaged.

5.3.2 Entrance and Exit Pupils

Another concept, useful in determining whether or not a given ray will traverse the entire optical system, is the *pupil*. This is simply an *image of the aperture stop*. The **entrance pupil** of a system is the ***image of the aperture stop as seen from an axial point on the object through those elements preceding the stop***. If there are no lenses between the object and the A.S., the latter itself serves as the entrance pupil. To illustrate the point, examine Fig. 5.37, which is a lens with a *rear aperture stop*. The image of the aperture stop in L is virtual (see Table 5.3) and magnified. It can be located by sending a few rays out from the edges of the A.S. in the usual way. In contrast, the **exit pupil** is the ***image of the A.S. as seen from an axial point on the image plane through the interposed lenses, if there are any***. In Fig. 5.37 there are no such lenses, so the aperture stop itself serves as the exit pupil. Notice that all of this just means that the cone of light actually entering the optical system is determined by the entrance pupil, whereas the cone leaving it is controlled by the exit pupil. No rays from the source point proceeding outside of either cone will make it to the image plane.

To use a telescope or a monocular as a camera lens, you might attach an external *front aperture stop* to control the amount of incoming light for exposure purposes. Figure 5.38 represents a similar arrangement in which the entrance and exit pupil locations should be self-evident. The last two diagrams include a ray labeled the **chief ray**. It is defined to be *any ray from an off-axis object point that passes through the center of the aperture stop. The chief ray enters the optical system along a line directed toward the midpoint of the*

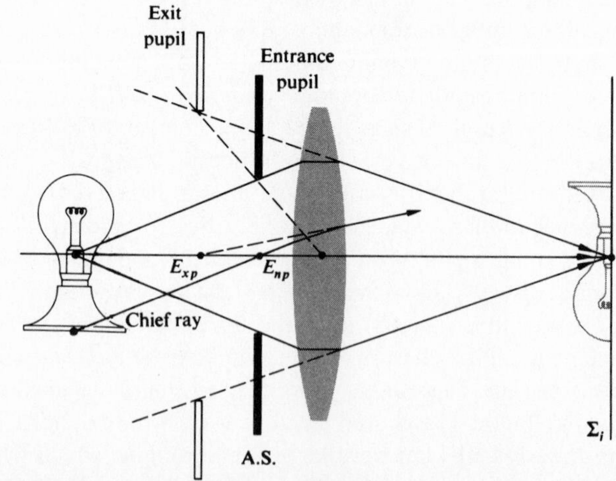

FIGURE 5.38 A front aperture stop.

entrance pupil, E_{np}, and leaves the system along a line passing through the center of the exit pupil, E_{xp}. The chief ray, associated with a conical bundle of rays from a point on the object, effectively behaves as the central ray of the bundle and is representative of it. Chief rays are of particular importance when the aberrations of a lens design are being corrected.

Figure 5.39 depicts a somewhat more involved arrangement. The two rays shown are those that are usually traced through an optical system. One is the chief ray from a point on the periphery of the object that is to be accommodated by the system. The other is called a **marginal ray**, since it goes from the axial object point to the rim or margin of the entrance pupil (or aperture stop).

In a situation where it is not clear which element is the actual aperture stop, each component of the system must be imaged by the remaining elements to its left. *The image that subtends the smallest angle at the axial object point is the entrance pupil.* The element whose image is the entrance pupil is then the aperture stop of the system for that object point. Problem 5.38 deals with just this kind of calculation.

Notice how the cone of rays, in Fig. 5.40, that can reach the image plane becomes narrower as the object point moves off-axis. The effective aperture stop, which for the axial bundle of rays was the rim of L_1, has been markedly reduced for the off-axis bundle. The result is a gradual fading out of the image at points near its periphery, a process known as **vignetting**.

The locations and sizes of the pupils of an optical system are of considerable practical importance. In visual instruments, the observer's eye is positioned at the center of the exit pupil. The pupil of the eye itself will vary from 2 mm to about 8 mm, depending on the general illumination level. Thus a telescope or binocular designed primarily for evening use might have an exit pupil of at least 8 mm. (You may have heard the term *night glasses*—they were quite popular on roofs during the Second World War.) In contrast, a daylight version will suffice with an exit pupil of 3 or 4 mm. The larger the exit pupil, the easier it is to align your eye properly with the instrument. Obviously, a telescopic sight for a high-powered rifle should have a large exit pupil located far enough behind the scope so as to avoid injury from recoil.

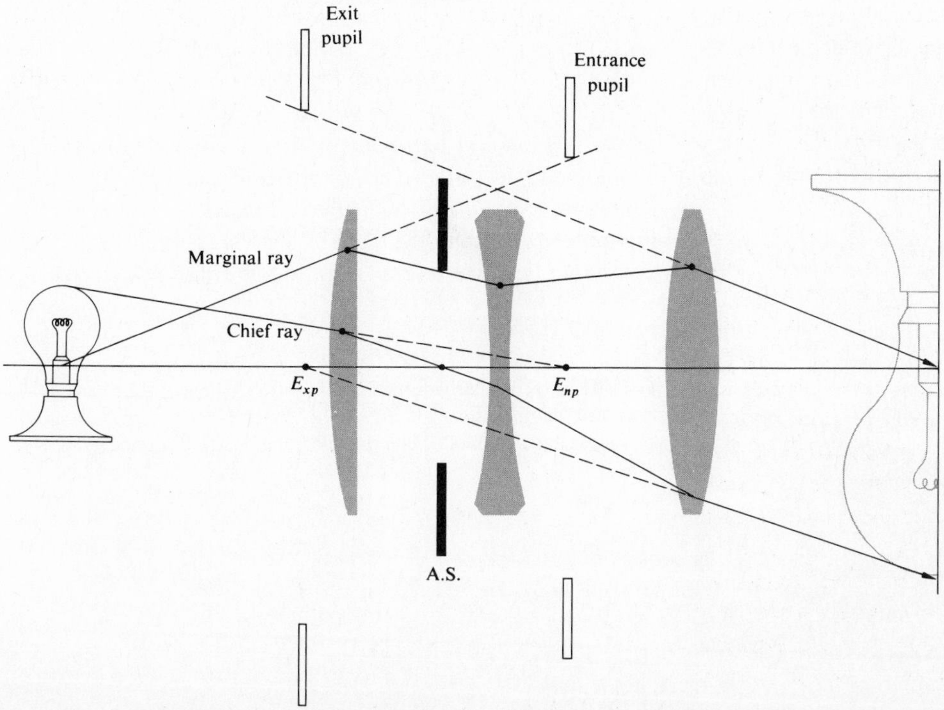

FIGURE 5.39 Pupils and stops for a three-lens system.

L_1 L_2
Effective aperture stop

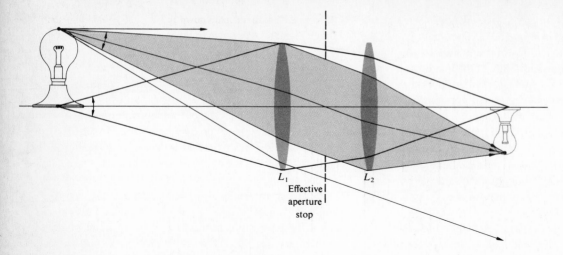

FIGURE 5.40 Vignetting.

5.3.3 Relative Aperture and *f*-Number

Suppose we collect the light from an extended source and form an image of it using a lens (or mirror). The amount of energy gathered by the lens (or mirror) from some small region of a distant source will be directly proportional to the area of the lens or, more generally, to the area of the entrance pupil. A large *clear aperture* will intersect a large cone of rays. Obviously, if the source were a laser with a very narrow beam, this would not necessarily be true. If we neglect losses due to reflection, absorption, and so forth, the incoming energy will be spread across a corresponding region of the image (Fig. 5.41). The energy per unit area per unit time (i.e., the flux density or irradiance) will be inversely proportional to the image area.

The entrance pupil area, if circular, varies as the square of its radius and is therefore proportional to the square of its diameter D. Furthermore, the image area will vary as the square of its lateral dimension, which in turn [Eqs. (5.24) and (5.26)] is proportional to f^2. (Keep in mind that we are talking about an extended object rather than a point source. In the latter case, the image would be confined to a very small area independent of f.) Thus the flux density at the image plane varies as $(D/f)^2$. The ratio D/f is known as the *relative aperture*, and its inverse is the **focal ratio** or ***f*-number**, often written $f/\#$, that is,

$$f/\# \equiv \frac{f}{D} \tag{5.40}$$

where $f/\#$ should be understood as a single symbol. For example, a lens with a 25-mm aperture and a 50-mm focal length has an *f-number* of 2, which is usually designated $f/2$. Figure 5.42 illustrates the point by showing a thin lens behind a variable iris diaphragm operating at either $f/2$ or $f/4$. A

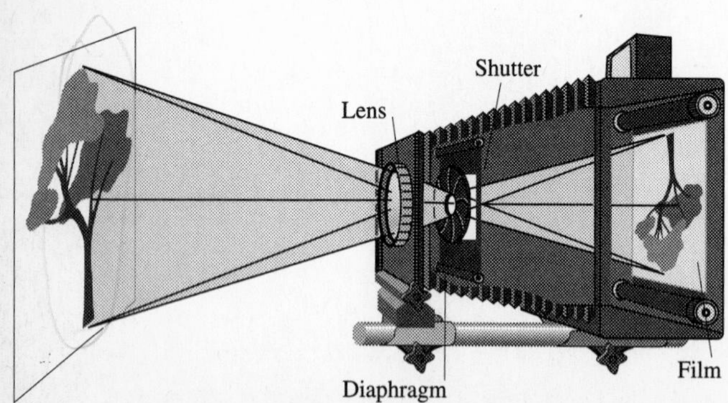

FIGURE 5.41 A large-format camera usually consists of a lens, followed by an adjustable diaphragm, a shutter that can rapidly open and close, and a sheet of film on which the image is formed.

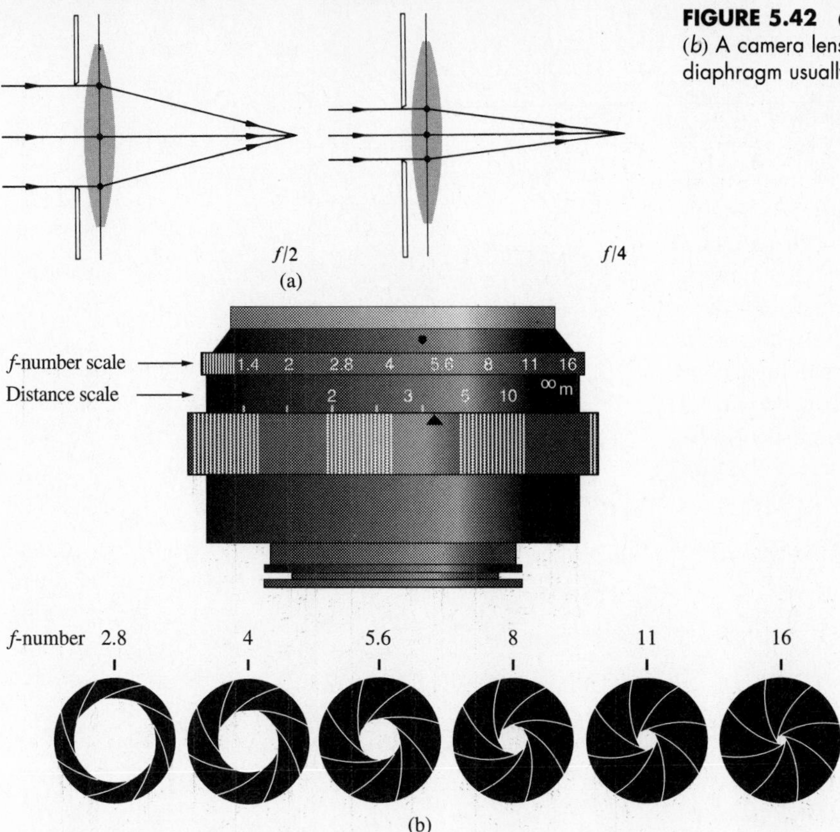

FIGURE 5.42 (*a*) Stopping down a lens to change the *f-number*. (*b*) A camera lens showing possible settings of the variable diaphragm usually located within the lens.

smaller *f-number* clearly permits more light to reach the image plane.

Camera lenses are usually specified by their focal lengths and largest possible apertures; for example, you might see "50 mm, $f/1.4$" on the barrel of a lens. Since the photographic exposure time is proportional to the square of the *f-number*, the latter is sometimes spoken of as the **speed** of the lens. An $f/1.4$ lens is said to be twice as fast as an $f/2$ lens. Usually, lens diaphragms have *f-number* markings of 1, 1.4, 2, 2.8, 4, 5.6, 8, 11, 16, 22, and so on. The largest relative aperture in this case corresponds to $f/1$, and that's a fast lens—$f/2$ is more typical. Each consecutive diaphragm setting increases the *f-number* by a multiplicative factor of $\sqrt{2}$ (numerically rounded off). This corresponds to a decrease in relative aperture by a multiplicative factor of $1/\sqrt{2}$ and therefore a decrease in flux density by one half. Thus, the same amount of light will reach the film whether the camera is set for $f/1.4$ at 1/500th of a second, $f/2$ at 1/250th of a second, or $f/2.8$ at 1/125th of a second.

The largest refracting telescope in the world, located at the Yerkes Observatory of the University of Chicago, has a 40-inch diameter lens with a focal length of 63 feet and therefore an *f-number* of 18.9. The entrance pupil and focal length of a mirror will, in exactly the same way, determine its *f-number*. Accordingly, the 200-inch diameter mirror of the Mount Palomar telescope, with a prime focal length of 666 inches, has an *f-number* of 3.33.

5.4 Mirrors

Mirror systems are increasingly being used, particularly in the X-ray, ultraviolet, and infrared regions of the spectrum. Although it is relatively simple to construct a reflecting device that will perform satisfactorily across a broad-frequency band, the same cannot be said of refracting systems. For example, a silicon or germanium lens designed for the infrared will be completely opaque in the visible (Fig. 3.40). As we will see later (p. 260), mirrors have other attributes that also contribute to their usefulness.

A mirror might simply be a piece of black glass or a finely polished metal surface. In the past mirrors were usually made by coating glass with silver, which was chosen because of its high efficiency in the UV and IR (see Fig. 4.62). Vacuum-evaporated coatings of aluminum on highly polished substrates have become the accepted standard for quality mirrors. Protective coatings of silicon monoxide or magnesium fluoride are often layered over the aluminum as well. In special applications (e.g., in lasers), where even the small losses due to metal surfaces cannot be tolerated, mirrors formed of multilayered dielectric films (see Section 9.9) are indispensable.

A new generation of lightweight precision mirrors continues to be developed for use in large-scale orbiting telescopes; the technology is by no means static.

5.4.1 Planar Mirrors

As with all mirror configurations, those that are planar can be either front- or back-surfaced. The latter type are most commonly found in everyday use because it allows the metallic reflecting layer to be completely protected behind glass. In contrast, the majority of mirrors designed for more critical technical usage are front-surfaced (Fig. 5.43).

From Section 4.3.1, it's an easy matter to determine the image characteristics of a planar mirror. Examining the point source and mirror arrangement of Fig. 5.43, we can quickly show that $|s_o| = |s_i|$; that is, the image P and object S are equidistant from the surface. To wit, $\theta_i = \theta_r$, from the Law of Reflection; $\theta_i + \theta_r$ is the exterior angle of triangle SPA and is therefore equal to the sum of the alternate interior angles, $\measuredangle VSA + \measuredangle VPA$. But $\measuredangle VSA = \theta_i$, and therefore $\measuredangle VSA = \measuredangle VPA$. This makes triangles VAS and VPA congruent, in which case $|s_o| = |s_i|$.

We are now faced with the problem of determining a sign convention for mirrors. Whatever we choose, and you should certainly realize that there is a choice, we need only be faithful unto it for all to be well. One obvious dilemma with respect to the convention for lenses is that now the virtual image is to the right of the interface. The observer sees P to be positioned behind the mirror, because the eye (or camera) cannot perceive the actual reflection; it merely interpolates the rays backward along straight lines. The rays from P in Fig. 5.43 are diverging, and no light can be cast on a screen located at P—the image is certainly virtual. Clearly, it is a matter of taste whether s_i should be defined as positive or negative in this instance. Since we rather like the idea of virtual object and

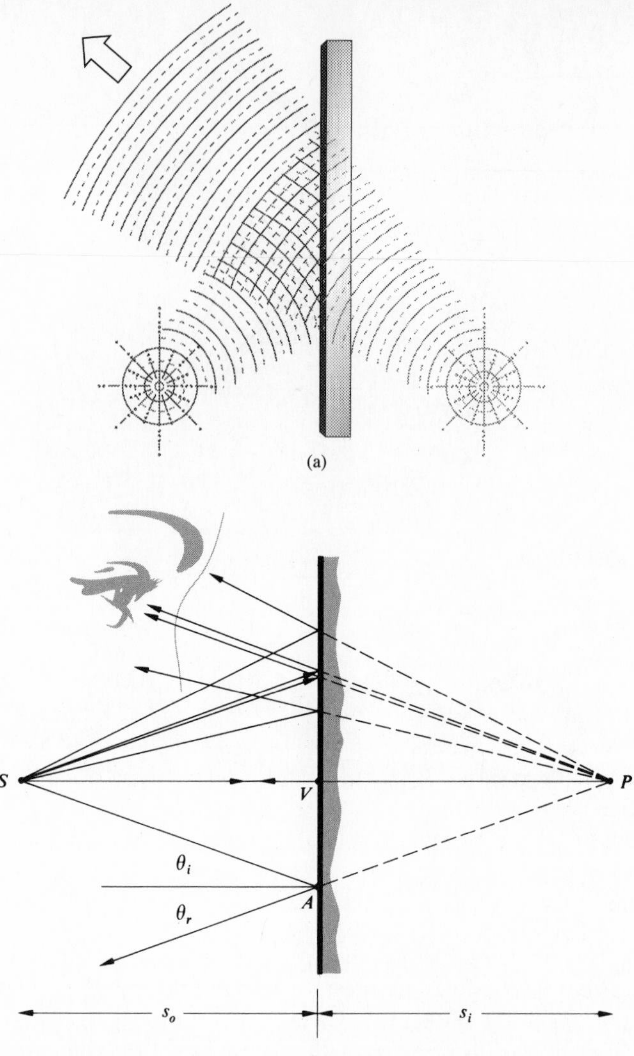

FIGURE 5.43 A planar mirror. (*a*) Reflection of waves. (*b*) Reflection of rays.

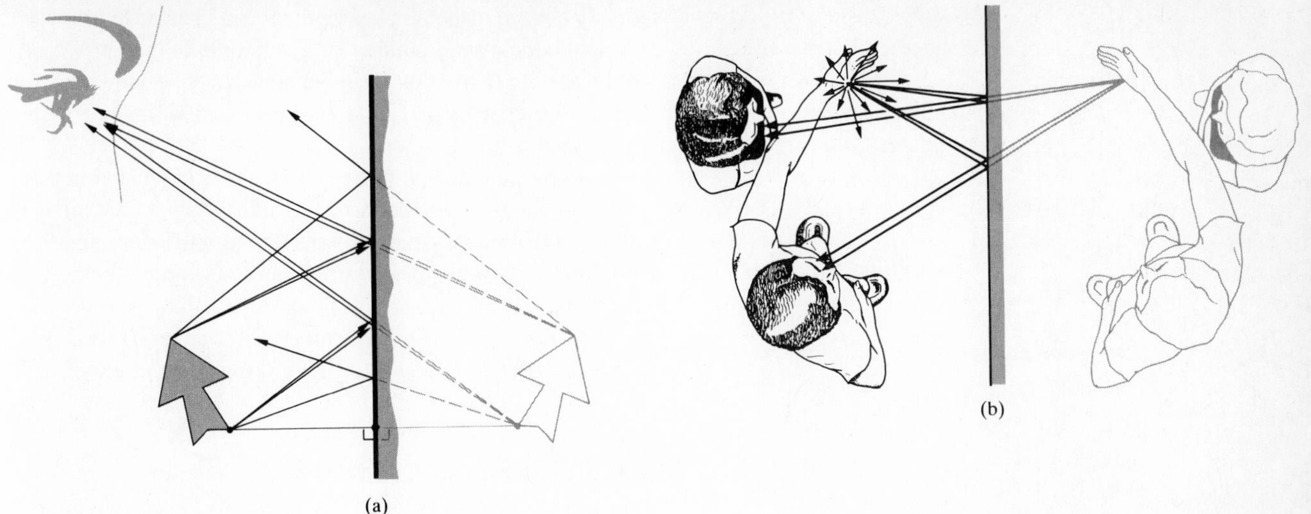

FIGURE 5.44 (*a*) The image of an extended object in a planar mirror. (*b*) Images in a planar mirror.

image distances being negative, we define s_o *and* s_i *as negative when they lie to the right of the vertex V.* This will have the added benefit of yielding a mirror formula identical to the Gaussian Lens Equation [Eq. (5.17)]. Evidently, the same definition of the transverse magnification [Eq. (5.24)] holds, where now, as before, $M_T = +1$ indicates a *life-size*, virtual, erect image.

Each point of the extended object in Fig. 5.44, a perpendicular distance s_i from the mirror, is imaged that same distance behind the mirror. In this way, the entire image is built up point by point. This is considerably different from the way a lens locates an image. The object in Fig. 5.30 was a left hand, and the image formed by the lens was also a left hand. To be sure, it might have been distorted ($M_L \neq M_T$), but it was still a left hand. The only evident change was a 180° rotation about the optical axis—an effect known as *reversion*. Contrarily, the mirror image of the left hand, determined by dropping perpendiculars from each point, is a right hand (Fig. 5.45). Such an image is sometimes said to be *perverted*. In deference to the more usual lay connotation of the word, its use in optics is happily waning. The process that converts a right-handed coordinate system in the object space into a left-handed one in the image space is known as *inversion*. Sys-

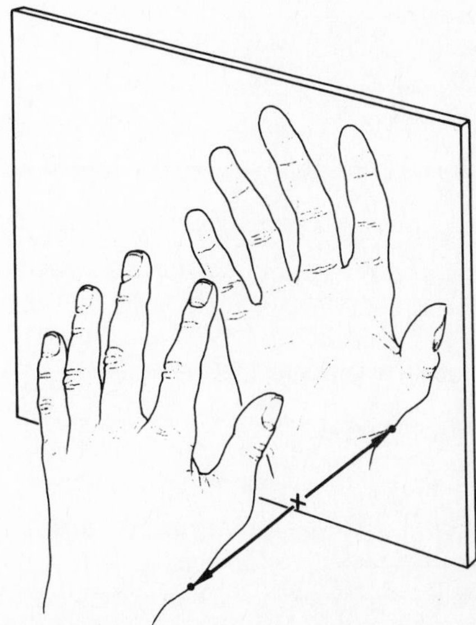

FIGURE 5.45 Mirror images—inversion.

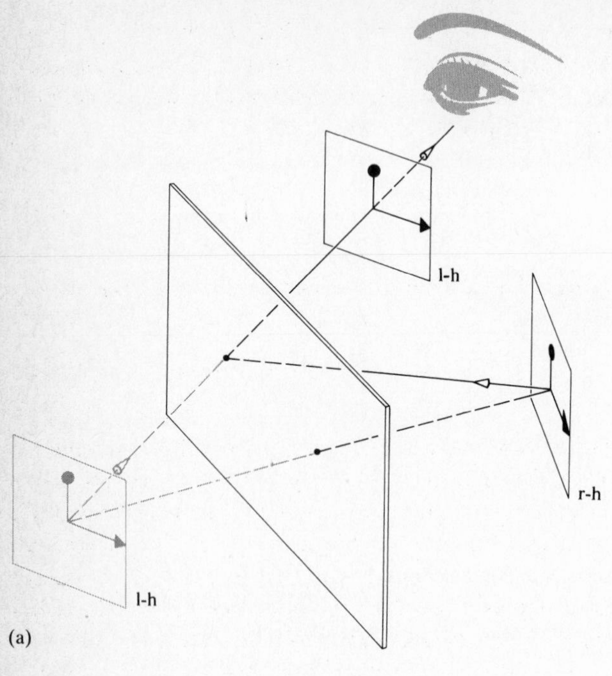

tems with more than one planar mirror can be used to produce either an odd or even number of inversions. In the latter case a right-handed (r-h) object will generate a right-handed image (Fig. 5.46), whereas in the former instance, the image will be left-handed (l-h).

A number of practical devices utilize rotating planar mirror systems—for example, choppers, beam deflectors, and image rotators. Mirrors are frequently used to amplify and measure the slight rotations of certain laboratory apparatus (galvanometers, torsion pendulums, current balances, etc.). As Fig. 5.47 shows, if the mirror rotates through an angle α, the reflected beam or image will move through an angle of 2α.

5.4.2 Aspherical Mirrors

Curved mirrors that form images very much like those of lenses or curved refracting surfaces have been known since the

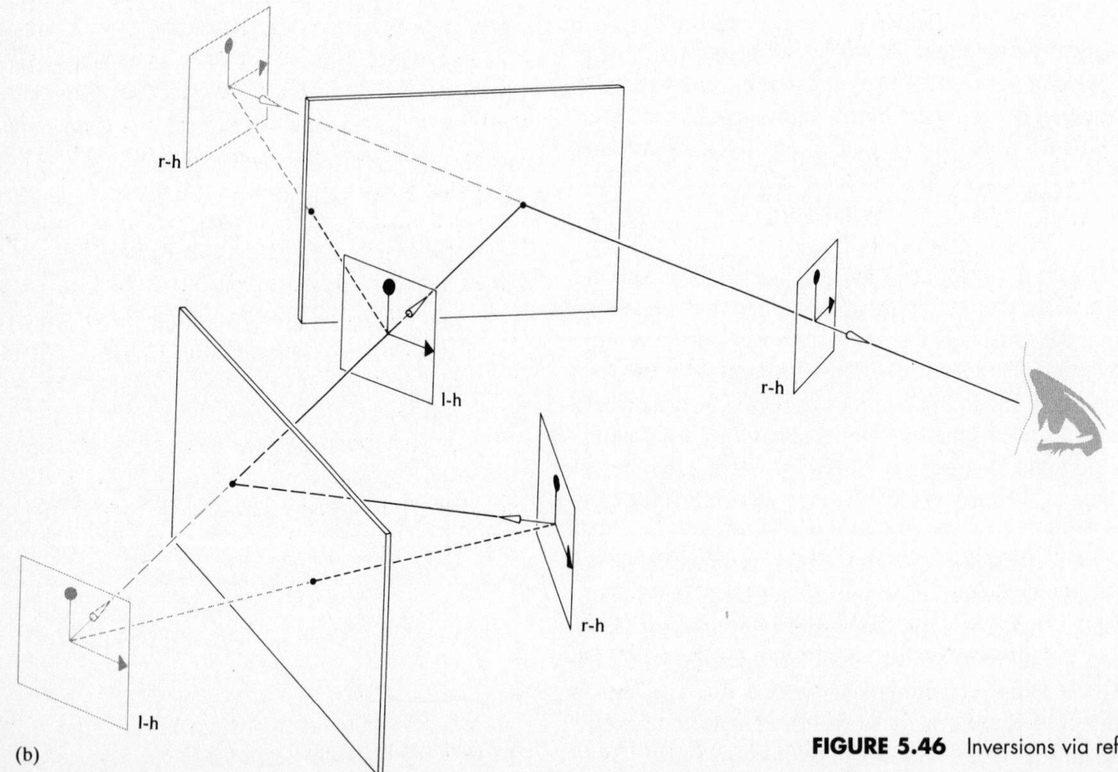

FIGURE 5.46 Inversions via reflection.

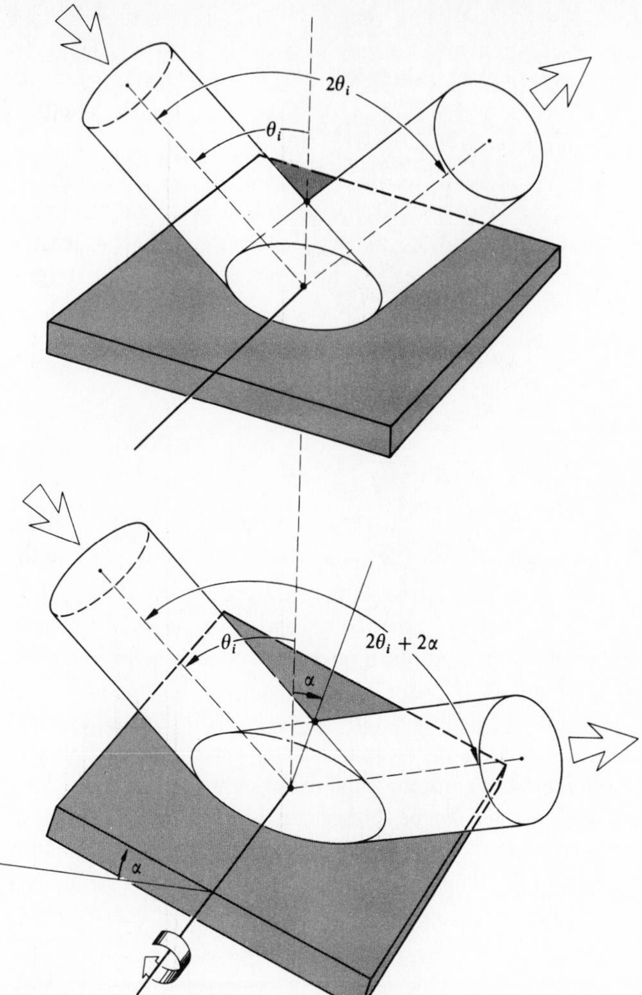

FIGURE 5.47 Rotation of a mirror and the concomitant angular displacement of a beam.

time of the ancient Greeks. Euclid, who is presumed to have authored the book entitled *Catoptrics*, discusses in it both concave and convex mirrors.* Fortunately, the conceptual basis for designing such mirrors was developed earlier when we studied Fermat's Principle as applied to imagery in refracting systems. Accordingly, let's determine the configuration a mirror must have if an incident plane wave is to be reformed upon reflection into a converging spherical wave (Fig. 5.48). Because the plane wave is ultimately to converge on point F, the optical path lengths for all rays must be equal; accordingly, for arbitrary points A_1 and A_2

$$OPL = \overline{W_1A_1} + \overline{A_1F} = \overline{W_2A_2} + \overline{A_2F} \qquad (5.41)$$

Since the plane Σ is parallel to the incident wavefronts,

$$\overline{W_1A_1} + \overline{A_1D_1} = \overline{W_2A_2} + \overline{A_2D_2} \qquad (5.42)$$

Equation (5.41) will therefore be satisfied for a surface for which $\overline{A_1F} = \overline{A_1D_1}$ and $\overline{A_2F} = \overline{A_2D_2}$ or, more generally, one for which $\overline{AF} = \overline{AD}$ for any point A on the mirror. In general, $\overline{AF} = e(\overline{AD})$, where e is the eccentricity of a **conic section.** Earlier (Section 5.2.1) the figure studied was a hyperbola for which $e = n_{ti} > 1$. In Problem 5.3 the figure is an ellipse and $e = n_{ti} < 1$. Here the second medium is identical to the first, $n_t = n_i$, and $e = n_{ti} = 1$; in other words, the surface is a paraboloid with F as its focus and Σ as its directrix. The rays could equally well be reversed (i.e., a point source at the focus of a paraboloid would result in the emission of plane waves from the system). Paraboloidals are used in a great variety of applications from flashlight and automobile headlight reflectors to giant radiotelescope antennas (Fig. 5.49), from microwave horns and acoustical dishes to optical telescope mirrors and Moon-based communications antennas. The convex paraboloidal mirror is also possible but is far less widely in use. Applying what we already know, it should be evident from Fig. 5.50 that an incident parallel bundle of rays will form a virtual image at F when the mirror is convex and a real image when it's concave.

There are other aspherical mirrors of interest, namely, the ellipsoid ($e < 1$) and hyperboloid ($e > 1$). Both produce perfect imagery between a pair of conjugate axial points corresponding to their two foci (Fig. 5.51). As we'll see presently,

**Dioptrics* denotes the optics of refracting elements, whereas *catoptrics* denotes the optics of reflecting surfaces.

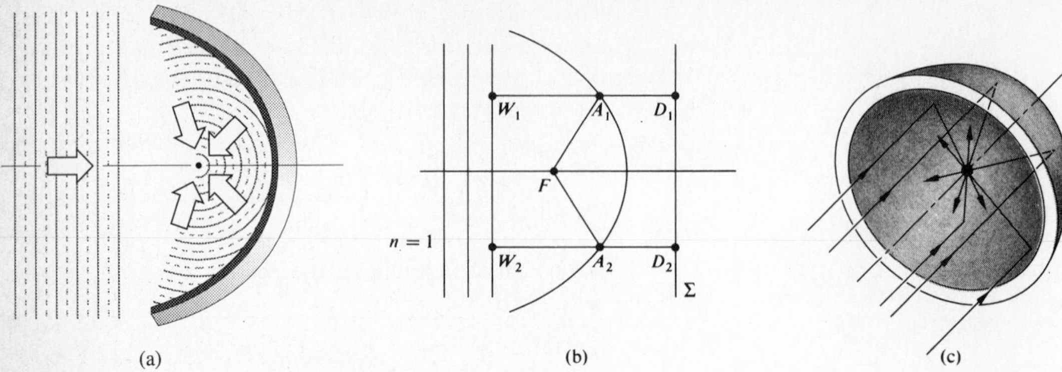

FIGURE 5.48 A paraboloidal mirror.

FIGURE 5.49 A classic paraboloidal radio antenna. (Photo courtesy of the Australian News and Information Bureau.)

the Cassegrain and Gregorian telescope configurations utilize convex secondary mirrors that are hyperboloidal and ellipsoidal, respectively. Like many new instruments, the primary mirror of the Hubble Space Telescope is hyperboloidal (Fig. 5.52).

A variety of aspherical mirrors are readily available commercially. In fact, one can purchase *off-axis elements*, in addition to the more common centered systems. Thus in Fig. 5.53 the focused beam can be further processed without obstructing the mirror. Incidentally, this geometry also obtains in large microwave horn antennas.

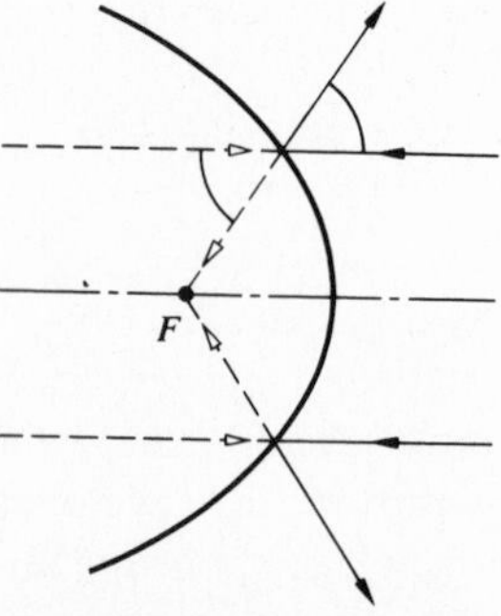

FIGURE 5.50 Real and virtual images for a paraboloidal mirror.

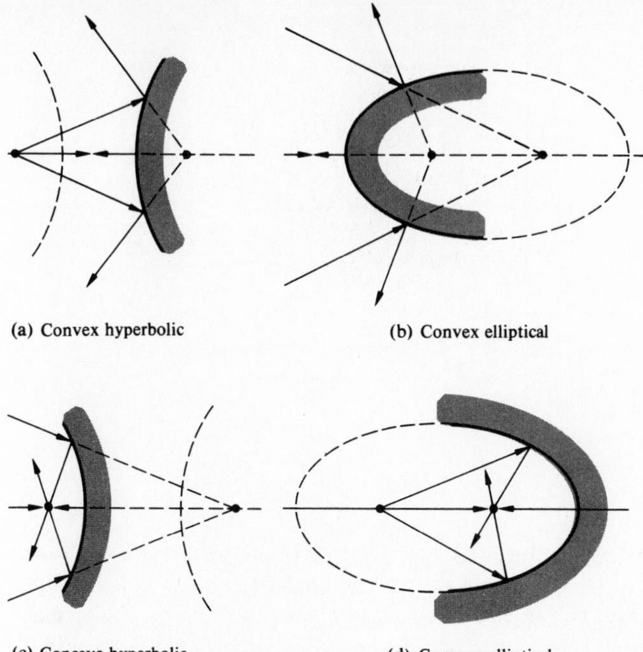

FIGURE 5.51 Hyperbolic and elliptical mirrors.

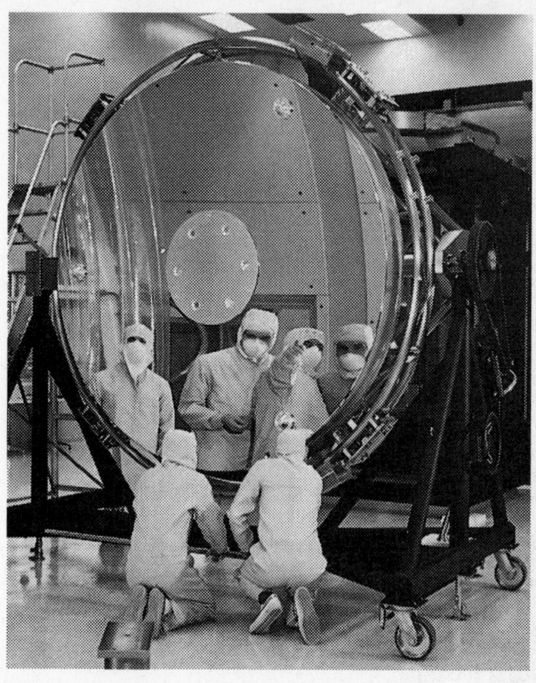

FIGURE 5.52 The 2.4-m-diameter hyperboloidal primary mirror of the Hubble Space Telescope. (Courtesy of NASA)

5.4.3 Spherical Mirrors

Precise aspheric surfaces are considerably more difficult to fabricate than are spherical ones, and, not surprisingly, they're considerably more expensive. Accordingly, we again turn to the spherical configuration to determine the circumstances under which it might perform adequately.

The Paraxial Region

The well-known equation for the circular cross section of a sphere (Fig. 5.54*a*) is

$$y^2 + (x - R)^2 = R^2 \tag{5.43}$$

where the center C is shifted from the origin O by one radius

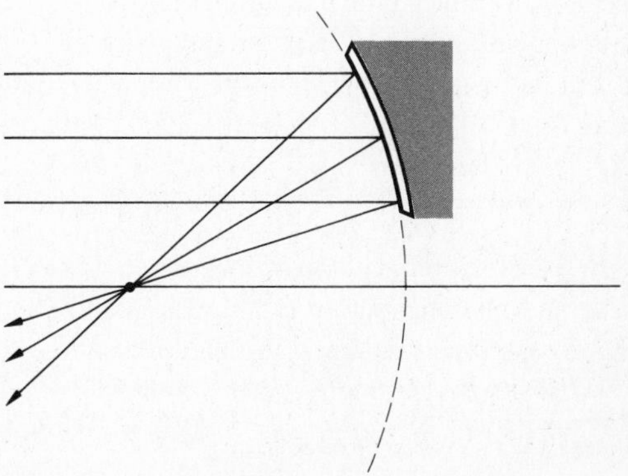

FIGURE 5.53 An off-axis parabolic mirror element.

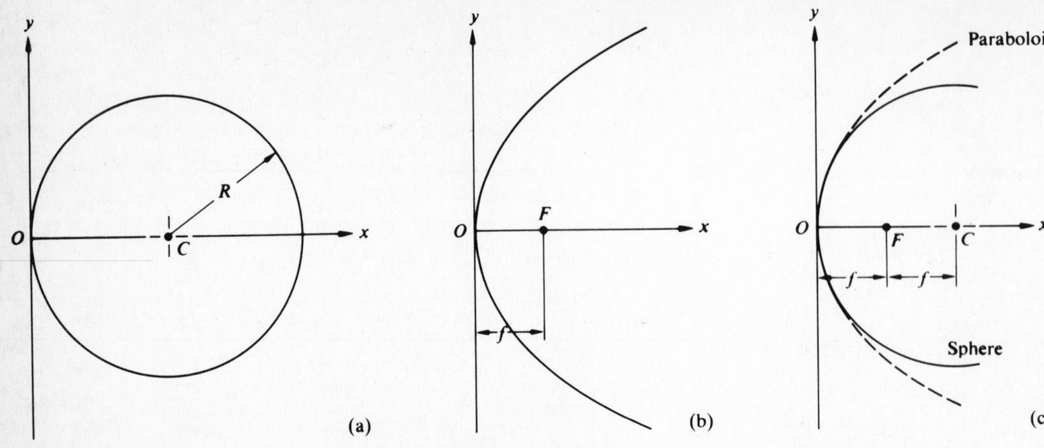

FIGURE 5.54 Comparison of spherical and paraboloidal mirrors.

R. After writing this as

$$y^2 - 2Rx + x^2 = 0$$

we can solve for x:

$$x = R \pm (R^2 - y^2)^{1/2} \tag{5.44}$$

Let's just concern ourselves with values of x less than R; that is, we'll study a hemisphere, open on the right, corresponding to the minus sign in Eq. (5.44). After expansion in a binomial series, x takes the form

$$x = \frac{y^2}{2R} + \frac{1y^4}{2^2 2! R^3} + \frac{1 \cdot 3y^6}{2^3 3! R^5} + \cdots \tag{5.45}$$

This expression becomes quite meaningful as soon as we realize that the standard equation for a parabola with its vertex at the origin and its focus a distance f to the right (Fig. 5.54*b*) is simply

$$y^2 = 4fx \tag{5.46}$$

By comparing these two formulas, we see that if $4f = 2R$ (i.e., if $f = R/2$), the first contribution in the series can be thought of as parabolic, and the remaining terms represent the deviation. If that deviation is Δx, then

$$\Delta x = \frac{y^4}{8R^3} + \frac{y^6}{16R^5} + \cdots$$

Evidently, this difference will be appreciable only when y is relatively large (Fig. 5.54*c*) in comparison to R. *In the paraxial region, that is, in the immediate vicinity of the central axis, these two configurations will be essentially indistinguishable.* If we stay within the paraxial theory of spherical mirrors as a first approximation, the conclusions drawn from our study of the stigmatic imagery of paraboloids are again applicable. In actual use, however, y will not be so limited, and aberrations will appear. Moreover, aspherical surfaces produce perfect images only for pairs of axial points—they too will suffer aberrations.

The Mirror Formula

The paraxial equation that relates conjugate object and image points to the physical parameters of a spherical mirror can be derived with the help of Fig. 5.55. To that end, observe that since $\theta_i = \theta_r$, the $\measuredangle SAP$ is bisected by $\overline{CA}$, which therefore divides the side $\overline{SP}$ of triangle SAP into segments proportional to the remaining two sides; that is,

$$\frac{\overline{SC}}{\overline{SA}} = \frac{\overline{CP}}{\overline{PA}} \tag{5.47}$$

Furthermore,

$$\overline{SC} = s_o - |R| \quad \text{and} \quad \overline{CP} = |R| - s_i$$

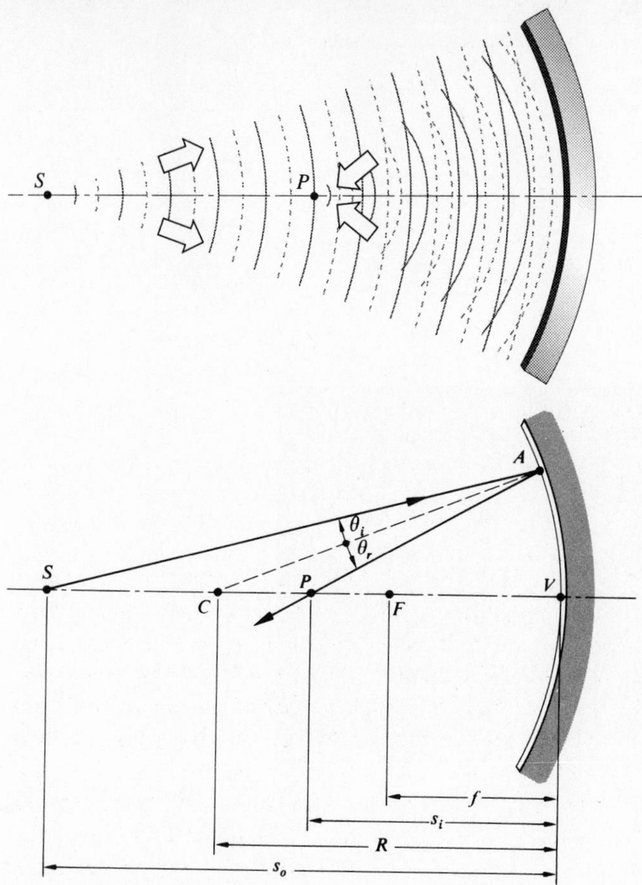

FIGURE 5.55 A concave spherical mirror. Conjugate foci.

where s_o and s_i are on the left and therefore positive. Using the same sign convention as we did with refraction, R will be negative because C is to the left of V (i.e., the surface is concave). Thus $|R| = -R$ and

$$\overline{SC} = s_o + R \quad \text{and} \quad \overline{CP} = -(s_i + R)$$

In the paraxial region $\overline{SA} \approx s_o$, $\overline{PA} \approx s_i$, and Eq. (5.47) becomes

$$\frac{s_o + R}{s_o} = -\frac{s_i + R}{s_i}$$

or

$$\frac{1}{s_o} + \frac{1}{s_i} = -\frac{2}{R} \tag{5.48}$$

which is the **Mirror Formula**. It's equally applicable to concave ($R < 0$) and convex ($R > 0$) mirrors. The *primary* or *object focus* is again defined by

$$\lim_{s_i \to \infty} s_o = f_o$$

and the *secondary* or *image focus* corresponds to

$$\lim_{s_o \to \infty} s_i = f_i$$

Consequently, from Eq. (5.48)

$$\frac{1}{f_o} + \frac{1}{\infty} = \frac{1}{\infty} + \frac{1}{f_i} = -\frac{2}{R}$$

to wit,

$$f_o = f_i = -\frac{R}{2} \tag{5.49}$$

as can be seen in Fig. 5.54*c*. Dropping the subscripts on the focal lengths yields

$$\frac{1}{s_o} + \frac{1}{s_i} = \frac{1}{f} \tag{5.50}$$

Observe that f will be positive for concave mirrors ($R < 0$) and negative for convex mirrors ($R > 0$). In the latter instance, the image is formed behind the mirror and is virtual (Figs. 5.56 and 5.57).

Finite Imagery

The remaining mirror properties are so similar to those of lenses and spherical refracting surfaces that we need only mention them briefly, without repeating the entire logical development of each item. Within the restrictions of paraxial theory, any parallel off-axis bundle of rays will be focused to a point on the *focal plane* passing through F normal to the optical axis. Likewise, a finite planar object perpendicular to the optical axis will be imaged (to a first approximation) in a plane similarly oriented; each object point will have a corresponding

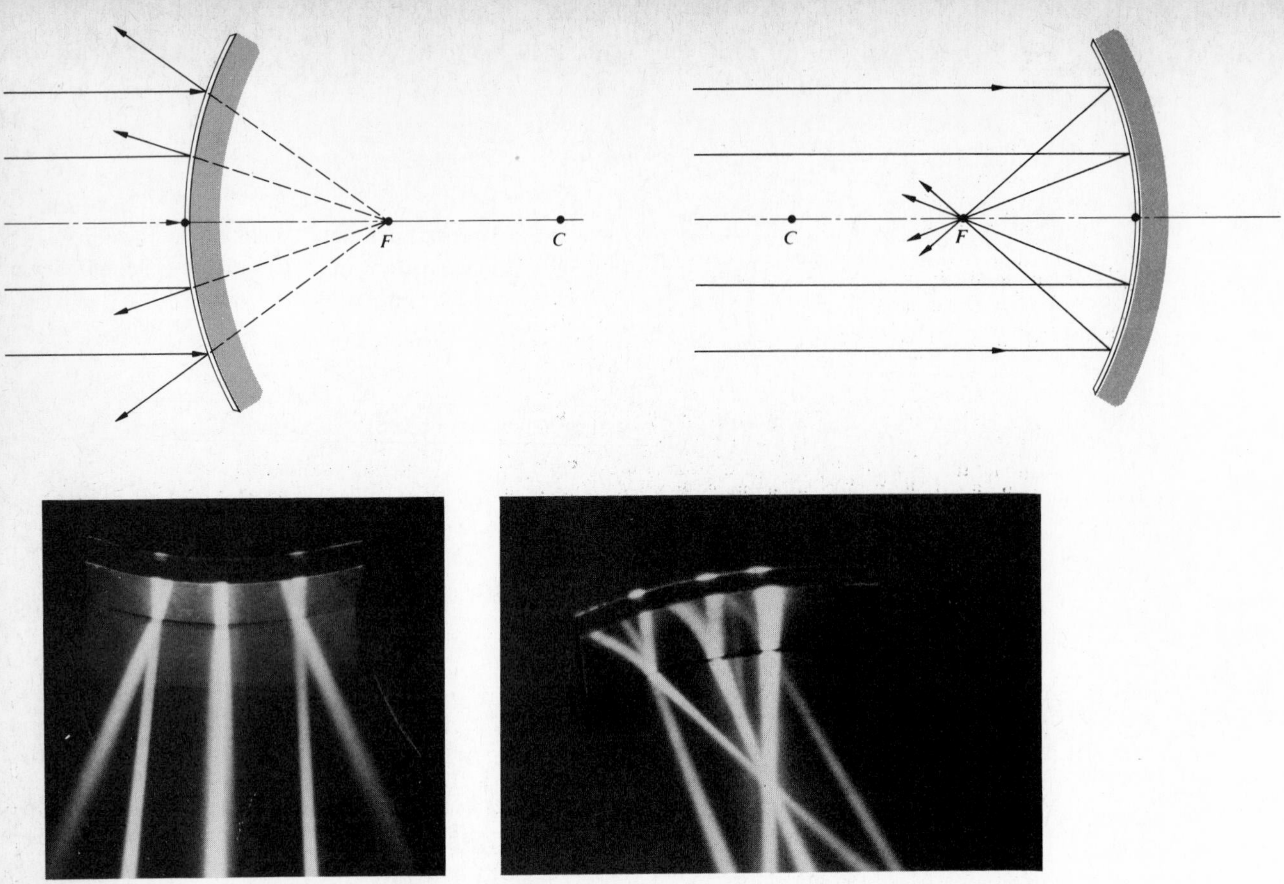

FIGURE 5.56 Focusing of rays via a spherical mirror. (E.H.)

FIGURE 5.57 A convex spherical mirror forming a virtual, right-side-up, minified image. See if you can locate the image of the author.

image point in that plane. This is certainly true for a plane mirror, but it only approximates the case for other configurations.

If a spherical mirror is used in a restricted fashion, the reflected waves arising from each object point will closely approximate spherical waves. Under such circumstances good finite images of extended objects can be formed.

Just as each image point produced by a thin lens lies along a straight line through the optical center O, each image point for a spherical mirror will lie on a ray passing through both the center of curvature C and the object point (Fig. 5.58). As with the thin lens (Fig. 5.24), the process for graphically locating the image is straightforward (Fig. 5.59). The top of the image is fixed at the intersection of two rays, one initially parallel to the axis and passing through F after reflection, and the other going straight through C (Fig. 5.60). The ray from any off-axis object point to the vertex forms equal angles with the central axis on reflection and is therefore particularly convenient to construct. So too is the ray that first passes through the focus and after reflection emerges parallel to the axis.

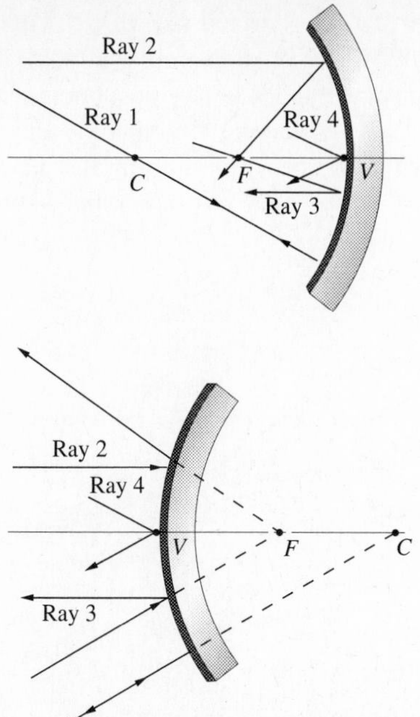

FIGURE 5.58 Four easy rays to draw. Ray 1 heads toward C and reflects back along itself. Ray 2 comes in parallel to the central axis and reflects toward (or away from) F. Ray 3 passes through (or heads toward) F and reflects off parallel to the axis. Ray 4 strikes point V and reflects such that $\theta_i = \theta_r$.

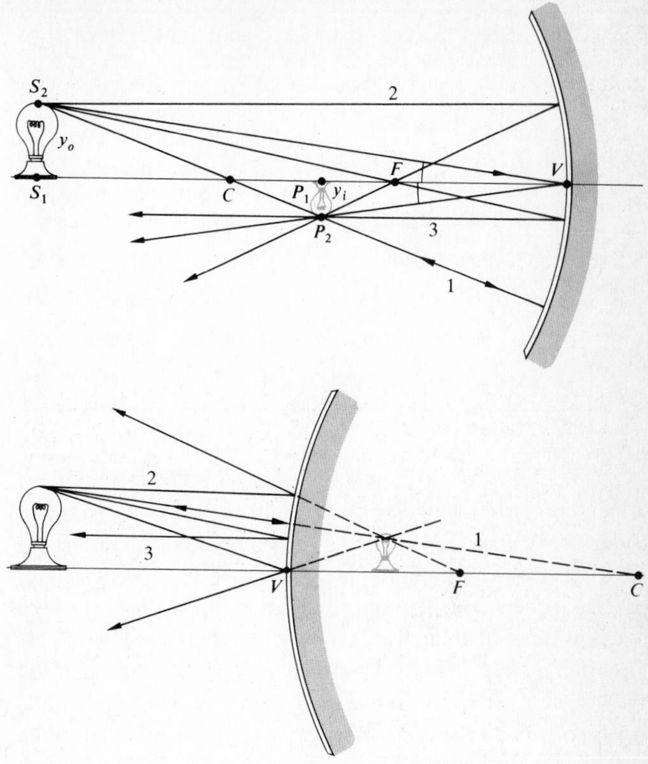

FIGURE 5.59 Finite imagery with spherical mirrors.

Notice that triangles S_1S_2V and P_1P_2V in Fig. 5.59*a* are similar, and hence their sides are proportional. Taking y_i to be negative, as we did before, since it's below the axis, $y_i/y_o = -s_i/s_o$, which is equal to M_T. This is the *transverse magnification*, just as it was for the lens [Eq. (5.25)].

The only equation that contains information about the

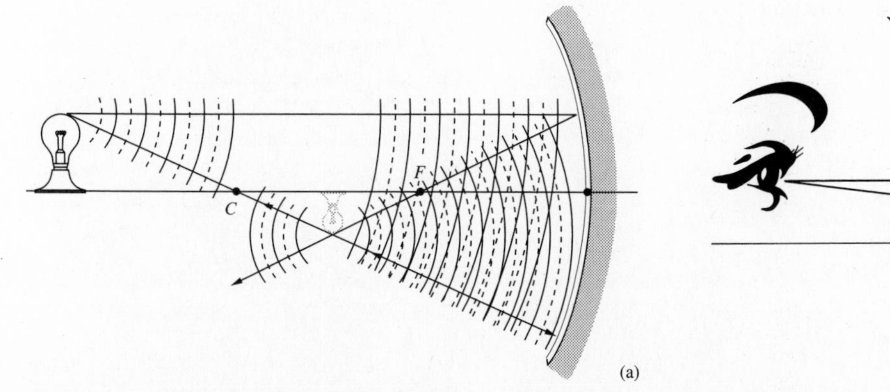

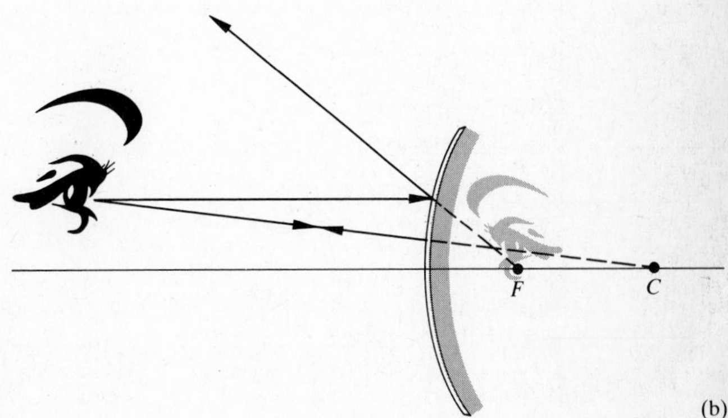

FIGURE 5.60 (*a*) Reflection from a concave mirror. (*b*) Reflection from a convex mirror.

Table 5.4 Sign Convention for Spherical Mirrors

Quantity	Sign +	Sign −
s_o	Left of V, real object	Right of V, virtual object
s_i	Left of V, real image	Right of V, virtual image
f	Concave mirror	Convex mirror
R	C right of V, convex	C left of V, concave
y_o	Above axis, erect object	Below axis, inverted object
y_i	Above axis, erect image	Below axis, inverted image

structure of the optical element (n, R, etc.) is that for f, and so, understandably, it differs for the thin lens [Eq. (5.16)] and spherical mirror [Eq. (5.49)]. The other functional expressions that relate s_o, s_i, and f or y_o, y_i, and M_T are, however, precisely the same. The only alteration in the previous sign convention appears in Table 5.4, where s_i on the left of V is now taken as positive. The striking similarity between the properties of a concave mirror and a convex lens on one hand and a convex mirror and a concave lens on the other are quite evident from a comparison of Tables 5.3 and 5.5, which are identical in all respects.

Table 5.5 Images of Real Objects Formed by Spherical Mirrors

Concave				
Object	**Image**			
Location	Type	Location	Orientation	Relative Size
$\infty > s_o > 2f$	Real	$f < s_i < 2f$	Inverted	Minified
$s_o = 2f$	Real	$s_i = 2f$	Inverted	Same size
$f < s_o < 2f$	Real	$\infty > s_i > 2f$	Inverted	Magnified
$s_o = f$		$\pm\infty$		
$s_o < f$	Virtual	$\|s_i\| > s_o$	Erect	Magnified
Convex				
Object	**Image**			
Location	Type	Location	Orientation	Relative Size
Anywhere	Virtual	$\|s_i\| < \|f\|$, $s_o > \|s_i\|$	Erect	Minified

The properties summarized in Table 5.5 and depicted in Fig. 5.61 can easily be verified empirically. If you don't have a spherical mirror at hand, a fairly crude but functional one can be made by carefully shaping aluminum foil over a spherical form, such as the end of a lightbulb (in that particular case R and therefore f will be small). A rather nice qualitative exper-

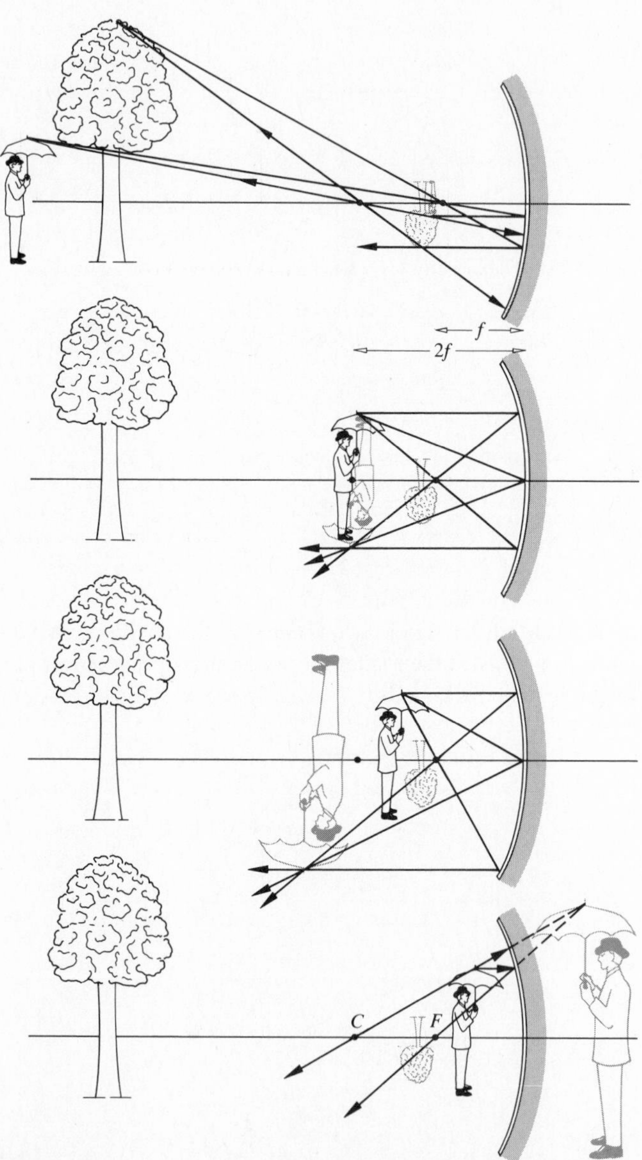

FIGURE 5.61 The image-forming behavior of a concave spherical mirror.

iment involves examining the image of some small object formed by a short focal-length concave mirror. As you move it toward the mirror from beyond a distance of $2f = R$, the image will gradually increase, until at $s_o = 2f$ it will appear inverted and life-size. Bringing it closer will cause the image to increase even more, until it fills the entire mirror with an unrecognizable blur. As s_o becomes smaller, the now erect, magnified image will continue to decrease until the object finally rests on the mirror, where the image is again life-size. If you are not moved by all of this to jump up and make a mirror, you might try examining the image formed by a shiny spoon—either side will be interesting.

5.5 Prisms

Prisms play many different roles in Optics; there are prism combinations that serve as beamsplitters (p. 000), polarizing devices (see Section 8.4.3), and interferometers. Despite this diversity, the vast majority of applications make use of only one of two main prism functions. First, a prism can serve as a dispersive device, as it does in a variety of spectrum analyzers (p. 191). As such it is capable of separating, to some extent, the constituent frequency components in a polychromatic light beam. Recall that the term *dispersion* was introduced earlier (p. 67) in connection with the frequency dependence of the index of refraction, $n(\omega)$, for dielectrics. In fact, the prism provides a highly useful means of measuring $n(\omega)$ over a wide range of frequencies and for a variety of materials (including gases and liquids).

Its second and more common function is to effect a change in the orientation of an image or in the direction of propagation of a beam. Prisms are incorporated in many optical instruments, often simply to fold the system into a confined space. There are inversion prisms, reversion prisms, and prisms that deviate a beam without inversion or reversion—and all of this without dispersion.

FIGURE 5.62 Prisms. (Photo courtesy Melles Griot.)

5.5.1 Dispersing Prisms

Prisms come in many sizes and shapes and perform a variety of functions (Fig. 5.62). Let's first consider the group known as **dispersing prisms**. Typically, a ray entering a dispersing prism, as in Fig. 5.63, will emerge having been deflected from its original direction by an angle δ known as the **angular devi-**

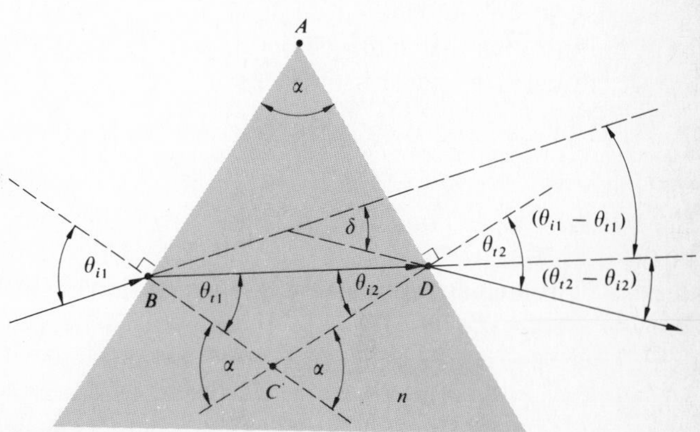

FIGURE 5.63 Geometry of a dispersing prism.

ation. At the first refraction the ray is deviated through an angle $(\theta_{i1} - \theta_{t1})$, and at the second refraction it is further deflected through $(\theta_{t2} - \theta_{i2})$. The total deviation is then

$$\delta = (\theta_{i1} - \theta_{t1}) + (\theta_{t2} - \theta_{i2})$$

Since the polygon *ABCD* contains two right angles, $\measuredangle BCD$ must be the supplement of the **apex angle** α. As the exterior angle to triangle *BCD*, α is also the sum of the alternate interior angles, that is,

$$\alpha = \theta_{t1} + \theta_{i2} \tag{5.51}$$

Thus

$$\delta = \theta_{i1} + \theta_{t2} - \alpha \tag{5.52}$$

We would like to write δ as a function of both the angle-of-incidence for the ray (i.e., θ_{i1}) and the prism angle α; these presumably would be known. If the prism index is n and it's immersed in air ($n_a \approx 1$), it follows from Snell's Law that

$$\theta_{t2} = \sin^{-1}(n \sin \theta_{i2}) = \sin^{-1}[n \sin(\alpha - \theta_{t1})]$$

Upon expanding this expression, replacing $\cos \theta_{t1}$ by $(1 - \sin^2 \theta_{t1})^{1/2}$, and using Snell's Law we have

$$\theta_{t2} = \sin^{-1}[(\sin \alpha)(n^2 - \sin^2 \theta_{i1})^{1/2} - \sin \theta_{i1} \cos \alpha]$$

The deviation is then

$$\delta = \theta_{i1} + \sin^{-1}[(\sin \alpha)(n^2 - \sin^2 \theta_{i1})^{1/2} - \sin \theta_{i1} \cos \alpha] - \alpha \tag{5.53}$$

Apparently, δ increases with n, which is itself a function of frequency, so we might designate the deviation as $\delta(\nu)$ or $\delta(\lambda)$. For most transparent dielectrics of practical concern, $n(\lambda)$ decreases as the wavelength increases across the visible [refer back to Fig. 3.38 for a plot of $n(\lambda)$ versus λ for various glasses]. Clearly, then, $\delta(\lambda)$ will be less for red light than it is for blue.

Missionary reports from Asia in the early 1600s indicated that prisms were well known and highly valued in China because of their ability to generate color. A number of scientists of the era, particularly Marci, Grimaldi, and Boyle, had made some observations using prisms, but it remained for the great Sir Isaac Newton to perform the first definitive studies of dispersion. On February 6, 1672, Newton presented a classic paper to the Royal Society entitled "A New Theory about Light and Colours.'' He had concluded that white light consisted of a mixture of various colors and that the process of refraction was color-dependent.

Returning to Eq. (5.53), it's evident that the deviation suffered by a monochromatic beam on traversing a given prism (i.e., n and α are fixed) is a function only of the incident angle at the first face, θ_{i1}. A plot of the results of Eq. (5.53) as applied to a typical glass prism is shown in Fig. 5.64. The smallest value of δ is known as the **minimum deviation**, δ_m, and it is of particular interest for practical reasons. The value of δ_m can be determined analytically by differentiating Eq. (5.53) and then setting $d\delta/d\theta_{i1} = 0$, but a more indirect route will certainly be simpler. Differentiating Eq. (5.52) and setting it equal to zero yields

$$\frac{d\delta}{d\theta_{i1}} = 1 + \frac{d\theta_{t2}}{d\theta_{i1}} = 0$$

or $d\theta_{t2}/d\theta_{i1} = -1$. Taking the derivative of Snell's Law at each interface, we get

$$\cos \theta_{i1}\, d\theta_{i1} = n \cos \theta_{t1}\, d\theta_{t1}$$

and

$$\cos \theta_{t2}\, d\theta_{t2} = n \cos \theta_{i2}\, d\theta_{i2}$$

Note as well, on differentiating Eq. (5.51), that $d\theta_{t1} = -d\theta_{i2}$, since $d\alpha = 0$. Dividing the last two equations and substituting for the derivatives leads to

$$\frac{\cos \theta_{i1}}{\cos \theta_{t2}} = \frac{\cos \theta_{t1}}{\cos \theta_{i2}}$$

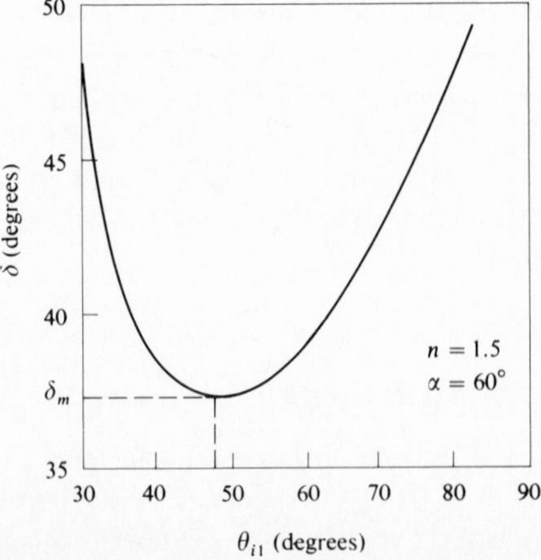

FIGURE 5.64 Deviation versus incident angle.

Making use of Snell's Law once again, we can rewrite this as

$$\frac{1 - \sin^2 \theta_{i1}}{1 - \sin^2 \theta_{t2}} = \frac{n^2 - \sin^2 \theta_{i1}}{n^2 - \sin^2 \theta_{t2}}$$

The value of θ_{i1} for which this is true is the one for which $d\delta/d\theta_{i1} = 0$. Inasmuch as $n \neq 1$, it follows that

$$\theta_{i1} = \theta_{t2}$$

and therefore

$$\theta_{t1} = \theta_{i2}$$

This means that *the ray for which the deviation is a minimum traverses the prism symmetrically, that is, parallel to its base.* Incidentally, there is a lovely argument for why θ_{i1} must equal θ_{t2}, which is neither as mathematical nor as tedious as the one we have evolved. In brief, suppose a ray undergoes a minimum deviation and $\theta_{i1} \neq \theta_{t2}$. Then if we reverse the ray, it will retrace the same path, so δ must be unchanged (i.e., $\delta = \delta_m$). But this implies that there are two different incident angles for which the deviation is a minimum, and this we know is not true—ergo $\theta_{i1} = \theta_{t2}$.

In the case when $\delta = \delta_m$, it follows from Eqs. (5.51) and (5.52) that $\theta_{i1} = (\delta_m + \alpha)/2$ and $\theta_{t1} = \alpha/2$, whereupon Snell's Law at the first interface leads to

$$n = \frac{\sin [(\delta_m + \alpha)/2]}{\sin \alpha/2} \tag{5.54}$$

This equation forms the basis of one of the most accurate techniques for determining the refractive index of a transparent substance. Effectively, one fashions a prism out of the material in question, and then, measuring α and $\delta_m(\lambda)$, $n(\lambda)$ is computed employing Eq. (5.54) at each wavelength of interest. Hollow prisms whose sides are fabricated of plane-parallel glass can be filled with liquids or gases under high pressure; the glass plates will not result in any deviation of their own.

Figures 5.65 and 5.66 show two examples of **constant-deviation dispersing prisms**, which are important primarily in spectroscopy. The **Pellin–Broca** prism is probably the most common of the group. Albeit a single block of glass, it can be envisaged as consisting of two 30°–60°–90° prisms and one 45°–45°–90° prism. Suppose that in the position shown a single monochromatic ray of wavelength λ traverses the component prism *DAE* symmetrically, thereafter to be reflected at 45° from face *AB*. The ray will then traverse prism *CDB* symmetrically, having experienced a total deviation of 90°. The ray can be thought of as having passed through an ordinary 60° prism (*DAE* combined with *CDB*) at minimum deviation. All other wavelengths present in the beam will emerge at other angles. If the prism is now rotated slightly about an axis normal to the paper, the incoming beam will have a new incident angle. A different wavelength component, say λ_2, will now undergo a minimum deviation, which is again 90°—hence the name *constant deviation*. With a prism of this sort, one can

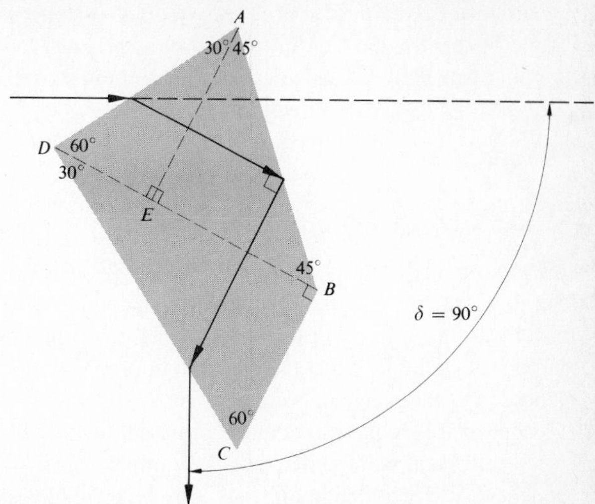

FIGURE 5.65 The Pellin–Broca prism.

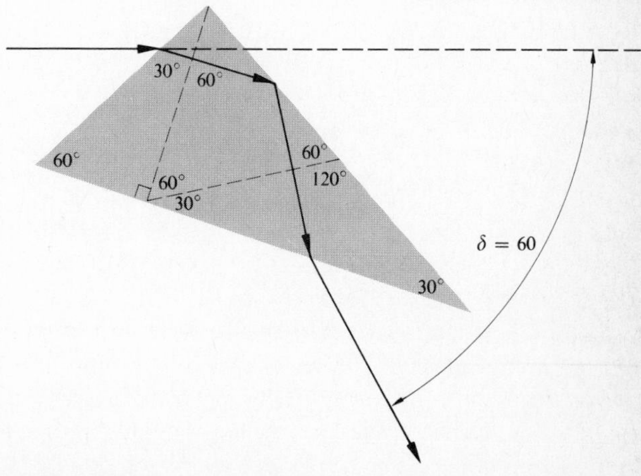

FIGURE 5.66 The Abbe prism.

conveniently set up the light source and viewing system at a fixed angle (here 90°), and then simply rotate the prism to look at a particular wavelength. The device can be calibrated so that the prism-rotating dial reads directly in wavelength.

5.5.2 Reflecting Prisms

We now examine **reflecting prisms**, in which dispersion is not desirable. In this case, the beam is introduced in such a way that at least one internal reflection takes place, for the specific purpose of changing either the direction of propagation or the orientation of the image or both.

Let's first establish that it is actually possible to have such an internal reflection without dispersion. In other words, is δ independent of λ? The prism in Fig. 5.67 is assumed to have as its profile an isosceles triangle—this happens to be a rather common configuration in any event. The ray refracted at the first interface is later reflected from face *FG*. As we saw earlier (Section 4.7), this will occur when the internal incident angle is greater than the critical angle θ_c, defined by

$$\sin \theta_c = n_{ti} \qquad [4.69]$$

For a glass–air interface, this requires that θ_i be greater than roughly 42°. To avoid any difficulties at smaller angles, let's further suppose that the base of our hypothetical prism is silvered as well—certain prisms do in fact require silvered faces. The angle of deviation between the incoming and outgoing rays is

$$\delta = 180° - \measuredangle BED \qquad (5.55)$$

From the polygon *ABED* it follows that

$$\alpha + \measuredangle ADE + \measuredangle BED + \measuredangle ABE = 360°$$

Moreover, at the two refracting surfaces

$$\measuredangle ABE = 90° + \theta_{i1}$$

and

$$\measuredangle ADE = 90° + \theta_{t2}$$

Substituting for $\measuredangle BED$ in Eq. (5.55) leads to

$$\delta = \theta_{i1} + \theta_{t2} + \alpha \qquad (5.56)$$

Since the ray at point *C* has equal angles of incidence and reflection, $\measuredangle BCF = \measuredangle DCG$. Thus, because the prism is isosceles, $\measuredangle BFC = \measuredangle DGC$, and triangles *FBC* and *DGC* are similar. It follows that $\measuredangle FBC = \measuredangle CDG$, and therefore $\theta_{t1} =$

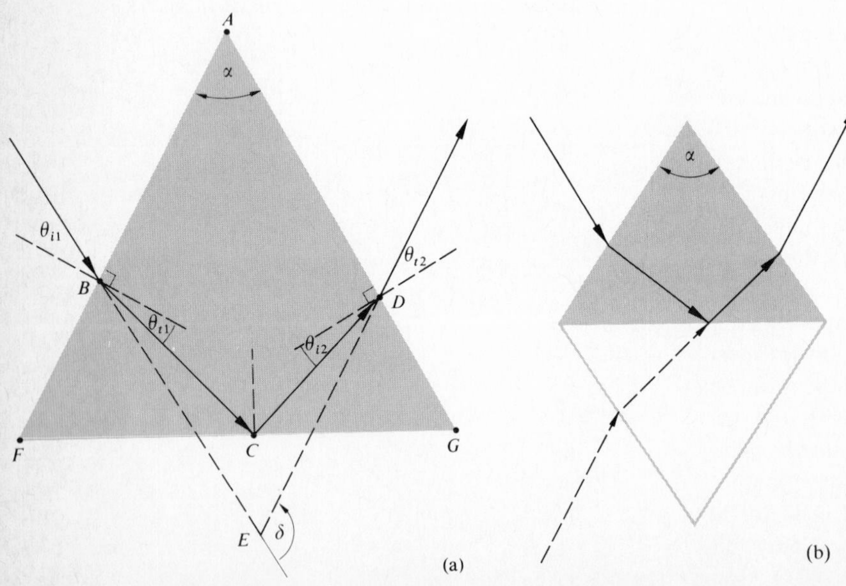

FIGURE 5.67 Geometry of a reflecting prism.

θ_{i2}. From Snell's Law we know that this is equivalent to $\theta_{i1} = \theta_{t2}$, whereupon the deviation becomes

$$\delta = 2\theta_{i1} + \alpha \tag{5.57}$$

which is certainly independent of both λ and n. The reflection will occur without any color preferences, and the prism is said to be **achromatic**. Unfolding the prism, that is, drawing its image in the reflecting surface FG, as in Fig. 5.67*b*, we see that it is equivalent in a sense to a parallelepiped or thick planar plate. The image of the incident ray emerges parallel to itself, regardless of wavelength.

A few of the many widely used reflecting prisms are shown in the next several figures. These are often made from BSC-2 or C-1 glass (see Table 6.2). For the most part, the illustrations are self-explanatory, so the descriptive commentary will be brief.

The **right-angle prism** (Fig. 5.68) deviates rays normal to the incident face by 90°. Notice that the top and bottom of the image have been interchanged; that is, the arrow has been flipped over, but the right and left sides have not. It is therefore an inversion system with the top face acting like a plane mirror. (To see this, imagine that the arrow and lollypop are vectors and take their cross-product. The resultant, arrow × lollypop, was initially in the propagation direction but is reversed by the prism.)

The **Porro prism** (Fig. 5.69) is physically the same as the right-angle prism but is used in a different orientation. After two reflections, the beam is deviated by 180°. Thus, if it enters right-handed, it leaves right-handed.

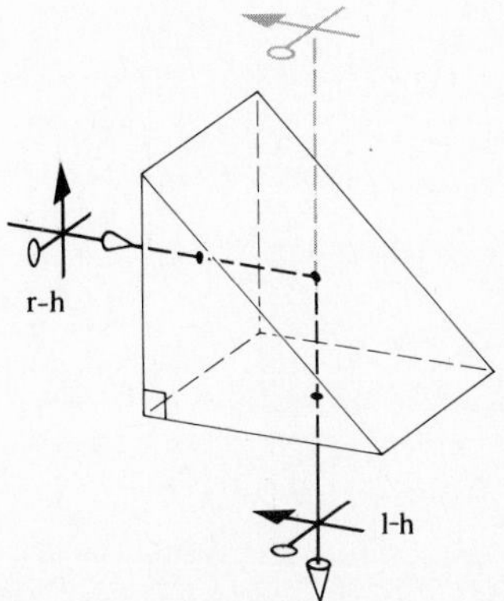

FIGURE 5.68 The right-angle prism.

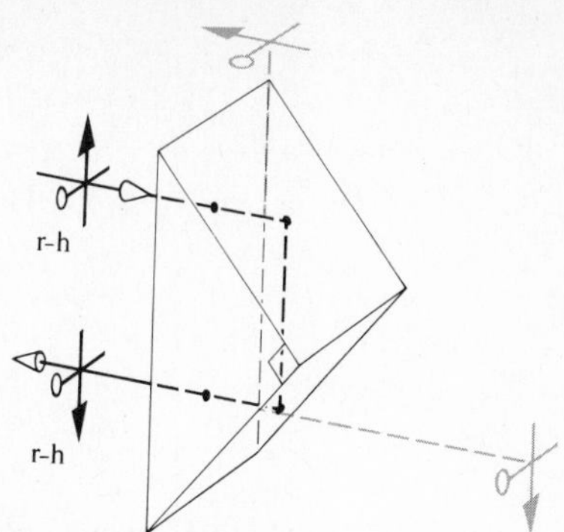

FIGURE 5.69 The Porro prism.

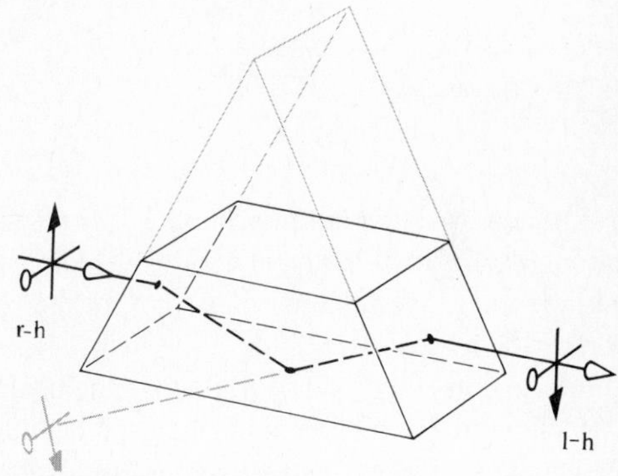

FIGURE 5.70 The Dove prism.

The **Dove prism** (Fig. 5.70) is a truncated version (to reduce size and weight) of the right-angle prism, used almost exclusively in collimated light. It has the interesting property (Problem 5.61) of rotating the image twice as fast as it is itself rotated about the longitudinal axis.

The **Amici prism** (Fig. 5.71) is essentially a truncated right-angle prism with a roof section added on to the hypotenuse face. In its most common use it has the effect of splitting the image down the middle and interchanging the right and left portions.* These prisms are expensive, because

*You can see how it actually works by placing two plane mirrors at right angles and looking directly into the combination. If you wink your *right* eye, the image will wink its *right* eye. Incidentally, if your eyes are equally strong, you will see two seams (images of the line where the mirrors meet), one running down the middle of each eye, with your nose presumably between them. If one eye is stronger, there will be only one seam, down the middle of that eye. If you close it, the seam will jump over to the other eye. This must be tried to be appreciated.

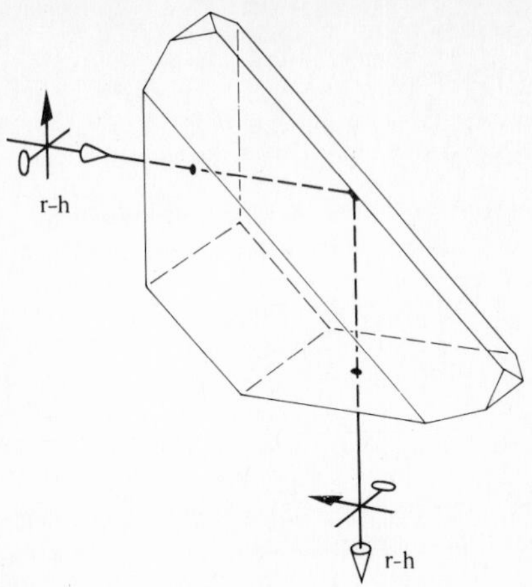

FIGURE 5.71 The Amici prism.

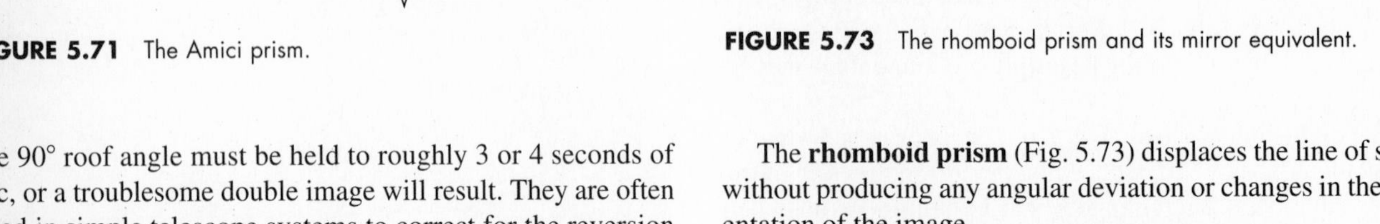

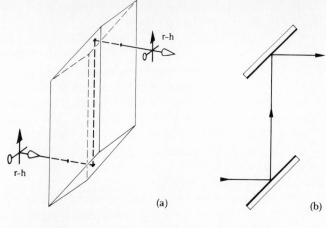

FIGURE 5.73 The rhomboid prism and its mirror equivalent.

the 90° roof angle must be held to roughly 3 or 4 seconds of arc, or a troublesome double image will result. They are often used in simple telescope systems to correct for the reversion introduced by the lenses.

The **penta prism** (Fig. 5.72) will deviate the beam by 90° without affecting the orientation of the image. Note that two of its surfaces must be silvered. These prisms are often used as end reflectors in small range finders.

The **rhomboid prism** (Fig. 5.73) displaces the line of sight without producing any angular deviation or changes in the orientation of the image.

The **Leman–Springer prism** (Fig. 5.74) also has a 90° roof. Here the line of sight is displaced without being deviated, but the emerging image is right-handed and rotated through 180°. The prism can therefore serve to erect images in telescope systems, such as gun sights and the like.

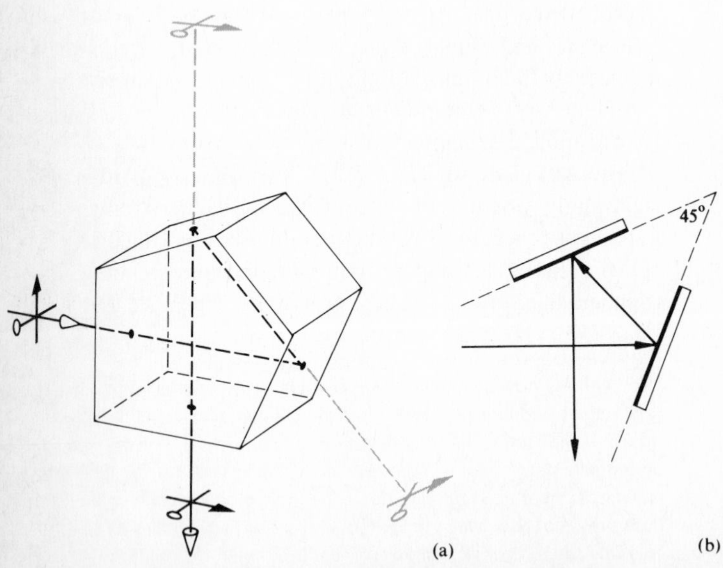

FIGURE 5.72 The penta prism and its mirror equivalent.

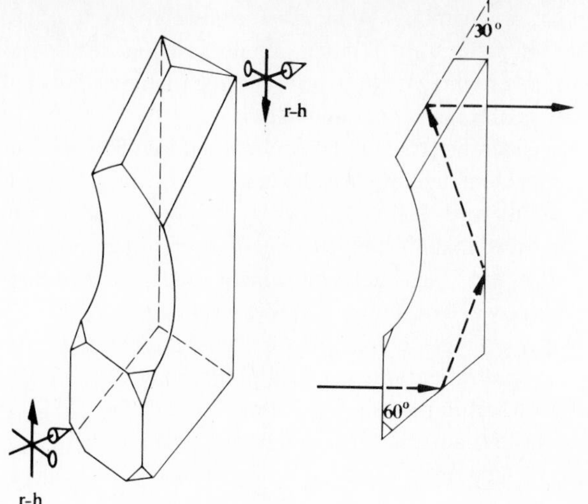

FIGURE 5.74 The Leman–Springer prism.

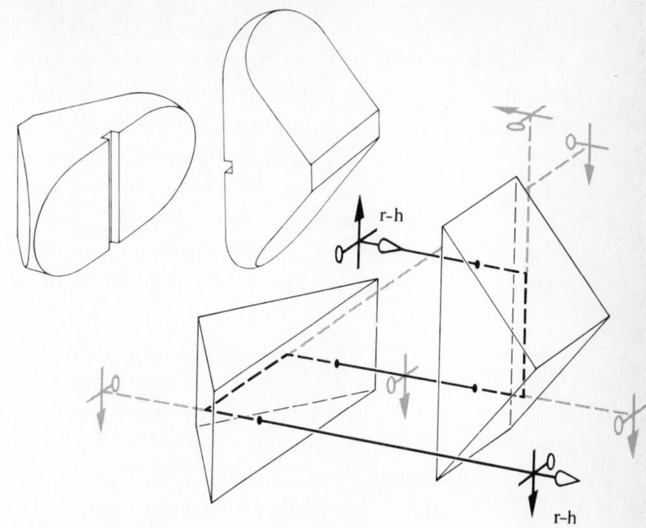

FIGURE 5.75 The double Porro prism.

Many more reflecting prisms perform specific functions. For example, if one cuts a cube so that the piece removed has three mutually perpendicular faces, it is called a **corner-cube prism**. It has the property of being retrodirective; that is, it will reflect all incoming rays back along their original directions. One hundred of these prisms are sitting in an 18-inch square array 240 000 miles from here, having been placed on the Moon during the Apollo 11 flight.*

The most common erecting system consists of two Porro prisms, as illustrated in Fig. 5.75. These are relatively easy to manufacture and are shown here with rounded corners to reduce weight and size. Since there are four reflections, the exiting image will be right-handed. A small slot is often cut in the hypotenuse face to obstruct rays that are internally reflected at glancing angles. Finding these slots after dismantling the family's binoculars is often an inexplicable surprise.

*J. E. Foller and E. J. Wampler, "The Lunar Laser Reflector." *Sci. Am.*, March 1970, p. 38.

5.6 Fiberoptics

The concept of channeling light within a long narrow dielectric (via total internal reflection) has been around for quite a while. John Tyndall (1870) showed that light could be contained within and guided along a thin stream of water. Soon after that, glass "light pipes" and, later, threads of fused quartz were used to further demonstrate the effect. But it wasn't until the early 1950s that serious work was done to transport images along bundles of short glass fibers.

After the advent of the laser (1960), there was an immediate appreciation of the potential benefits of sending information from one place to another using light, as opposed to electric currents or even microwaves. At those high optical frequencies (of the order of 10^{15} Hz), one hundred thousand times more information can be carried than with microwaves. Theoretically, that's the equivalent of sending tens of millions of television programs all at once on a beam of light. It wasn't long (1966) before the possibility of coupling lasers with fiberoptics for long-distance communications was pointed out. Thus began a tremendous technological transformation that's still roaring along today.

In 1970 researchers at the Corning Glass Works produced a silica fiber with a signal-power transmission of better than 1%

over a distance of 1 km (i.e., an attenuation of 20 dB/km), which was comparable to existing copper electrical systems. During the next two decades, the transmission rose to about 96% over 1 km (i.e., an attenuation of only 0.16 dB/km).

Because of its low-loss transmission, high-information-carrying capacity, small size and weight, immunity to electromagnetic interference, unparalleled signal security, and the abundant availability of the required raw materials (i.e., ordinary sand), ultrapure glass fibers have become the premier communications medium.

As long as the diameter of these fibers is large compared with the wavelength of the radiant energy, the inherent wave nature of the propagation is of little importance, and the process obeys the familiar laws of Geometrical Optics. On the other hand, if the diameter is of the order of λ, the transmission closely resembles the manner in which microwaves advance along waveguides. Some of the propagation modes are evident in the photomicrographic end views of fibers shown in Fig. 5.76. Here the wave nature of light must be reckoned with and this behavior resides in the domain of Physical Optics. Although optical waveguides, particularly of the thin-film variety, are of increasing interest, this discussion will be limited to the case of relatively large diameter fibers, those about the thickness of a human hair.

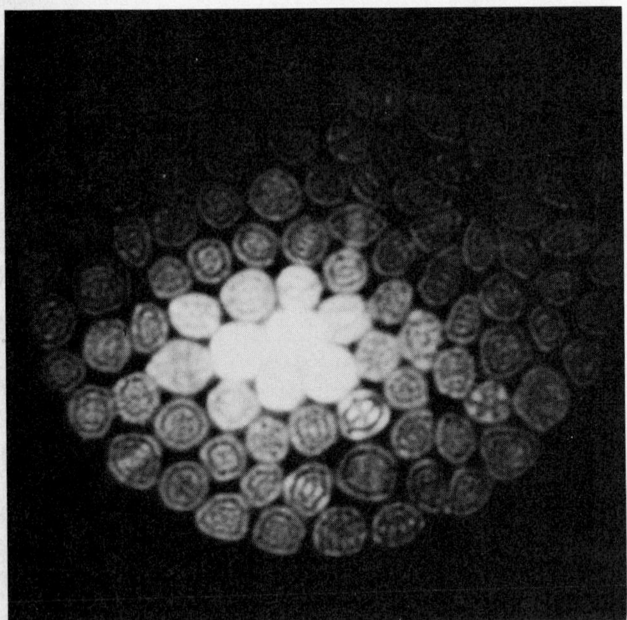

FIGURE 5.76 Optical waveguide mode patterns seen in the end faces of small-diameter fibers. (Photo courtesy of Narinder S. Kapany, AMP Fellow)

Consider the straight glass cylinder of Fig. 5.77 surrounded by an incident medium of index n_i—let it be air, $n_i = n_a$. Light striking its walls from within will be totally internally reflected, provided that the incident angle at each reflection is greater than $\theta_c = \sin^{-1} n_a/n_f$, where n_f is the index of the cylinder or fiber. As we will show, a *meridional ray* (i.e., one that is coplanar with the central or optical axis) might undergo several thousand reflections per foot as it bounces back and forth along a fiber, until it emerges at the far end (Fig. 5.78). If the fiber has a diameter D and a length L, the path length ℓ traversed by the ray will be

$$\ell = L/\cos\theta_t \tag{5.58}$$

or from Snell's Law

$$\ell = n_f L(n_f^2 - \sin^2\theta_i)^{-1/2} \tag{5.59}$$

The number of reflections N is then given by

$$N = \frac{\ell}{D/\sin\theta_t} \pm 1$$

or

$$N = \frac{L\sin\theta_i}{D(n_f^2 - \sin^2\theta_i)^{1/2}} \pm 1 \tag{5.60}$$

rounded off to the nearest whole number. The ± 1, which depends on where the ray strikes the end face, is of no signifi-

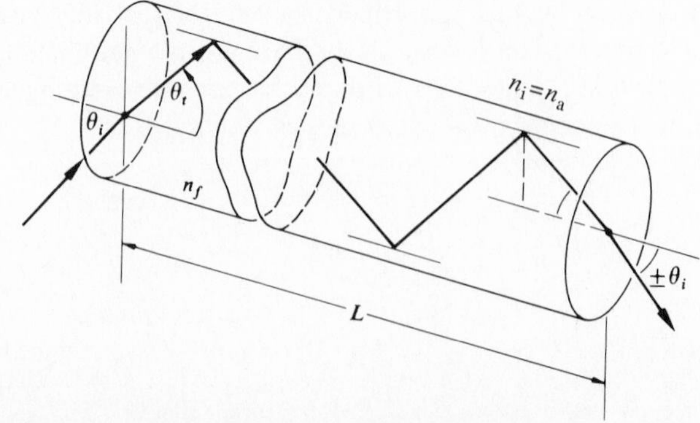

FIGURE 5.77 Rays reflected within a dielectric cylinder.

FIGURE 5.78 Light emerging from the ends of a loose bundle of glass fibers. (E.H.)

cance when N is large, as it is in practice. Thus, if D is 50 μm (i.e., 50 microns where 1 μm $= 10^{-6}$ m $= 39.37 \times 10^{-6}$ in), which is about 2×10^{-3} in (a hair from the head of a human is roughly 50 μm in diameter), and if $n_f = 1.6$ and $\theta_i = 30°$, N turns out to be approximately 2000 reflections per foot. Fibers are available in diameters as small as 2 μm or so but are seldom used in sizes much less than about 10 μm. Extremely thin glass (or plastic) filaments are quite flexible and can even be woven into fabric.

The smooth surface of a single fiber must be kept clean (of moisture, dust, oil, etc.), if there is to be no leakage of light (via frustrated total internal reflection). Similarly, if large numbers of fibers are packed in close proximity, light may leak from one fiber to another in what is known as *cross-talk*. For these reasons, it is customary to enshroud each fiber in a transparent sheath of lower index called a **cladding**. This layer need only be thick enough to provide the desired isolation, but for other reasons it generally occupies about one-tenth of the cross-sectional area. Although references in the literature to simple light pipes go back 100 years, the modern era of fiberoptics began with the introduction of clad fibers in 1953.

Typically, a fiber core might have an index (n_f) of 1.62, and the cladding an index (n_c) of 1.52, although a range of values is available. A clad fiber is shown in Fig. 5.79. Notice that there is a maximum value θ_{max} of θ_i, for which the internal ray will impinge at the critical angle, θ_c. Rays incident on the face at angles greater than θ_{max} will strike the interior wall at angles less than θ_c. They will be only partially reflected at each such encounter with the core–cladding interface and will quickly leak out of the fiber. Accordingly, θ_{max}, which is known as the acceptance angle, defines the half-angle of the acceptance cone of the fiber. To determine it start with

$$\sin\theta_c = n_c/n_f = \sin(90° - \theta_t)$$

Thus

$$n_c/n_f = \cos\theta_t \tag{5.61}$$

or

$$n_c/n_f = (1 - \sin^2\theta_t)^{1/2}$$

Making use of Snell's Law and rearranging terms, we have

$$\sin\theta_{max} = \frac{1}{n_i}(n_f^2 - n_c^2)^{1/2} \tag{5.62}$$

The quantity $n_i \sin\theta_{max}$ is defined as the **numerical aperture**, or NA. Its square is a measure of the light-gathering power of the system. The term originates in microscopy, where the equivalent expression describes the corresponding capabilities of the objective lens. It should clearly relate to the *speed* of the system, and, in fact,

$$f/\# = \frac{1}{2(\text{NA})} \tag{5.63}$$

Thus for a fiber

$$\text{NA} = (n_f^2 - n_c^2)^{1/2} \tag{5.64}$$

The left-hand side of Eq. (5.62) cannot exceed 1, and in air ($n_a = 1.000\,28 \approx 1$) that means that the largest value of NA is 1. In this case, the half-angle θ_{max} equals 90°, and the fiber totally internally reflects all light entering its face (Problem

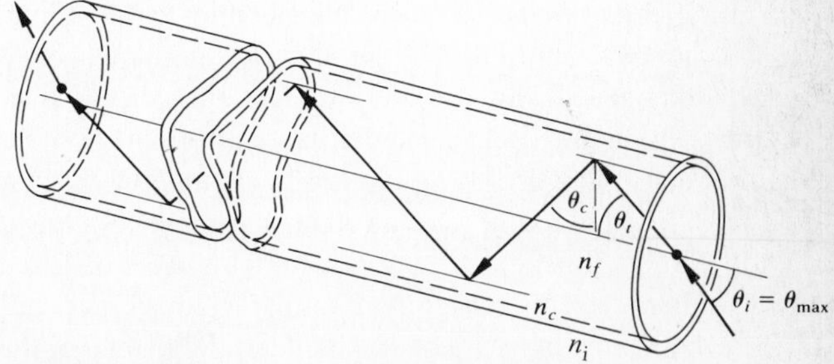

FIGURE 5.79 Rays in a clad optical fiber.

5.62). Fibers with a wide variety of numerical apertures, from about 0.2 up to and including 1.0, are commercially obtainable.

Bundles of free fibers whose ends are bound together (e.g., with epoxy), ground, and polished form flexible light guides. If no attempt is made to align the fibers in an ordered array, they form an *incoherent bundle*. This unfortunate use of the term *incoherent* (which should not be confused with coherence theory) just means, for example, that the first fiber in the top row at the entrance face may have its terminus anywhere in the bundle at the exit face. These *flexible light carriers* are, for that reason, relatively easy to make and inexpensive. Their primary function is simply to conduct light from one region to another. Conversely, when the fibers are carefully arranged so that their terminations occupy the same relative positions in both of the bound ends of the bundle, it is said to be *coherent*. Such an arrangement is capable of transmitting images and is consequently known as a *flexible image carrier*.

Coherent bundles are frequently fashioned by winding fibers on a drum to make ribbons, which are then carefully layered. When one end of such a device is placed face down flat on an illuminated surface, a point-by-point image of whatever is beneath it will appear at the other end (Fig. 5.80). These bundles can be tipped off with a small lens, so that they need not be in contact with the object under examination. Nowadays it is common to use fiberoptic instruments to poke into all sorts of unlikely places, from nuclear reactor cores and jet engines to stomachs and reproductive organs. When a device is used to examine internal body cavities, it's called an *endoscope*. This category includes bronchoscopes, colonoscopes, gastroscopes, and so forth, all of which are generally less than about 200 cm in length. Similar industrial instruments are usually two or three times as long and often contain 5000 to 50 000 fibers, depending on the required image resolution and the overall diameter that can be accommodated. An additional incoherent bundle incorporated into the device usually supplies the illumination.

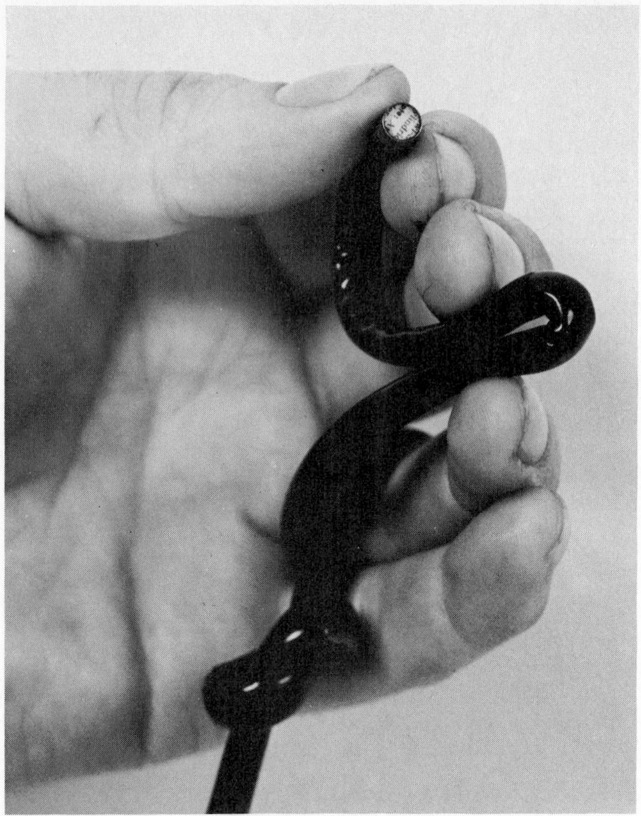

FIGURE 5.80 A coherent bundle of 10 μm glass fibers transmitting an image even though knotted and sharply bent. (Photo courtesy of American ACMI Div., American Hospital Supply Corp.)

Not all fiberoptic arrays are made flexible; for example, fused, rigid, coherent fiber faceplates, or *mosaics*, are used to replace homogeneous low-resolution sheet glass on cathode-ray tubes, vidicons, image intensifiers, and other devices. Mosaics consisting of literally millions of fibers with their claddings fused together have mechanical properties almost identical to homogeneous glass. Similarly, a sheet of fused *tapered* fibers can either magnify or minify an image, depending on whether the light enters the smaller or larger end of the fiber. The compound eye of an insect such as the housefly is effectively a bundle of tapered fiber optical filaments. The rods and cones that make up the human retina may also channel light through total internal reflection. Another common application of mosaics involving imaging is the *field flattener*. If the image formed by a lens system resides on a curved surface, it is often desirable to reshape it into a plane, for example, to match a film plate. A mosaic can be ground and polished on one of its end surfaces to correspond to the contour of the image and on the other to match the detector. Incidentally, a naturally occurring fibrous crystal known as *ulexite*, when polished, responds surprisingly like a fiberoptic mosaic. (Hobby shops often sell it for use in making jewelry.)

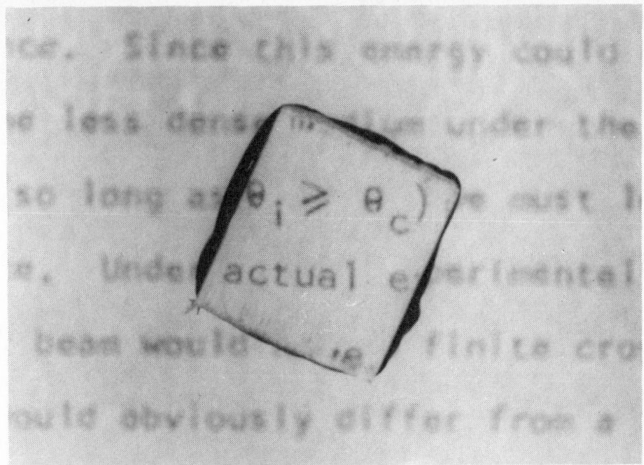

FIGURE 5.81 A stack of cover-glass slides held together by a rubber band serves as a coherent light guide. (E.H.)

If you have never seen the kind of light conduction we've been talking about, try looking down the edges of a stack of microscope slides. Even better are the much thinner (0.18-mm) cover-glass slides. Figure 5.81 shows the way light is conveyed to the upper surface of a stack of a few hundred of these slides held together by a rubber band.

Today fiberoptics has three very different applications: it is used for the direct (short-distance) transmission of images and illumination, it provides a variety of remarkable waveguides used in telecommunications, and it serves as the core of a new family of sensors. Transmitting images over distances of a few meters with coherent bundles, however beautiful and however useful, is a rather unsophisticated business that doesn't start to utilize the full potential inherent in fiberoptics. During the past few decades the application of lightguides to telecommunications has begun to replace copper wires and electricity as the primary information pathway. Worldwide, over 100 million kilometers of fibers have been installed since 1970. Even more recently, fiberoptic sensors—devices that measure pressure, sound, temperature, voltage, current, liquid levels, electric and magnetic fields, rotations, and so forth—have become the latest manifestation of the versatility of fibers.

5.6.1 Fiberoptic Communications Technology

The high frequencies of light allow for an incredible data-handling capacity. For example, with sophisticated transmitting techniques, a pair of copper telephone wires can be made to carry about two dozen simultaneous conversations. That should be compared with a single, ongoing, simple television transmission, which is equivalent to about 1300 simultaneous telephone conversations, and that, in turn, is roughly the equal of sending some 2500 typewritten pages each second. Clearly, at present it's quite impractical to attempt to send television over copper telephone lines. Yet by the mid-1980s it was already possible to transmit in excess of 12 000 simultaneous conversations over a *single pair* of fibers—that's more than nine television channels. Each such fiber has a line rate of about 400 million bits of information per second (400 Mb/s), or 6000 voice circuits. Fibers of this sort (with repeaters spaced every 40 km or so) formed the world's new intercity long-haul telecommunications grids. In the early 1990s researchers used **solitons**—carefully shaped pulses that travel without changing—to attain transmission rates of upwards of 4 Gb/s. This is the equivalent of 70 simultaneous color TV channels sent more than a million kilometers.

The accomplishments of the last few decades are impressive. For example, the first fiberoptic transatlantic cable TAT-8 was designed, using some clever data-handling techniques, to carry 40 000 conversations at once over just two pairs of glass fibers. TAT-1, a copper cable installed in 1956, could carry a mere 51 conversations, and the last of the bulky copper versions, TAT-7 (1983), can handle only about 8000. The TAT-8, which began operations in 1988, functions at 296 Mb/s (using single-mode 1300-nm fibers—see p. 200). It has regenerators or repeaters (to boost the signal strength) every 50 km (30 mi) or more. That should be compared with the copper TAT-7, which has amplifiers every 10 km or so. This feature is tremendously important in long-distance communications. Ordinary wire systems require repeaters roughly every kilometer; electrical coaxial networks extend that range to about 2 to 6 km; even radio transmissions through the atmosphere need regeneration every 30 to 50 km. It is anticipated that high-performance fiber systems will extend the repeater separation to upward of 150 km.

A major determining factor in the spacing of repeaters is the power loss due to attenuation of the signal as it propagates down the line. The decibel (dB) is the customary unit used to designate the ratio of two power levels, and as such it can provide a convenient indication of the power-out (P_o) with respect to the power-in (P_i). The number of dB = $-10 \log (P_o/P_i)$, and hence a ratio of 1:10 is 10 dB, 1:100 is 20 dB, 1:1000 is 30 dB, and so on. The attenuation (α) is usually specified in decibels per kilometer (dB/km) of fiber length (L). Thus $-\alpha L/10 = \log (P_o/P_i)$, and if we raise 10 to the power of both sides,

$$P_o/P_i = 10^{-\alpha L/10} \tag{5.65}$$

As a rule, reamplification of the signal is necessary when the power has dropped by a factor of about 10^{-5}. Commercial optical glass, the kind of material available for fibers in the mid-1960s, has an attenuation of about 1000 dB/km. Light, after being transmitted 1 km through the stuff, would drop in power by a factor of 10^{-100}, and regenerators would be needed every 50 m (which is little better than communicating with a string and two tin cans). By 1970 α was down to about 20 dB/km for fused silica (quartz, SiO_2), and it was reduced to as little as 0.16 dB/km in 1982. This tremendous decrease in attenuation was achieved mostly by removing impurities (especially the ions of iron, nickel, and copper) and reducing contamination by OH groups, largely accomplished by scrupulously eliminating any traces of water in the glass (p. 72).

Figure 5.82 depicts the three major fiber configurations used in communications today. In (a) the core is relatively wide, and the indices of core and cladding are both constant throughout. This is the so-called **stepped-index fiber**, with a homogeneous core of 50 to 150 μm or more and cladding with an outer diameter of roughly 100 to 250 μm. The oldest of the three types, the stepped-index fiber was widely used in first-generation systems (1975–1980). The comparatively large central core makes it rugged and easily infused with light, as well as easily terminated and coupled. It's the least expensive but also, the least effective of the lot, and for long-range applications, it has some serious drawbacks.

Depending on the launch angle into the fiber, there can be hundreds, even thousands, of different ray paths or modes by which energy can propagate down the core (Fig. 5.83). This then is a **multimode fiber**, wherein each mode corresponds to a slightly different transit time. Higher-angle rays travel longer paths; reflecting from side to side, they take longer to get to the end of the fiber than do rays moving along the axis. This is loosely spoken of as **intermodal dispersion** (or often just *modal dispersion*), even though it has nothing to do with a frequency-dependent index of refraction. Information to be trans-

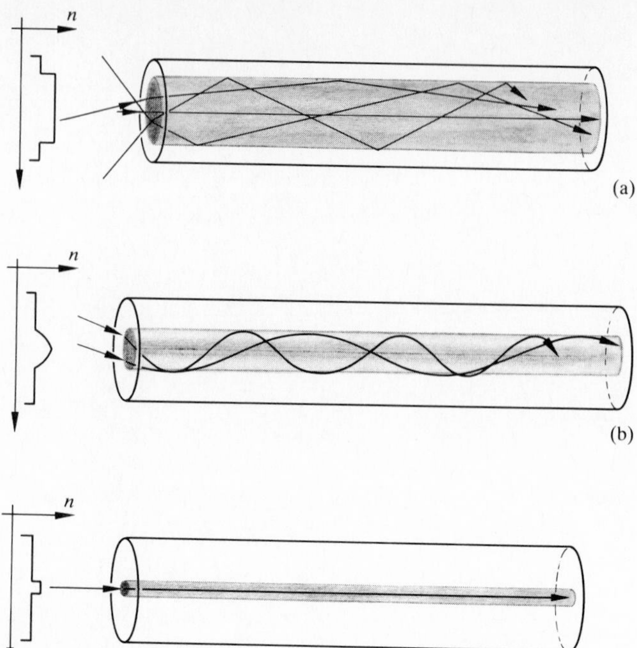

FIGURE 5.82 The three major fiberoptic configurations and their index profiles. (a) Multimode step-index fiber. (b) Multimode graded-index fiber. (c) Single-mode step-index fiber.

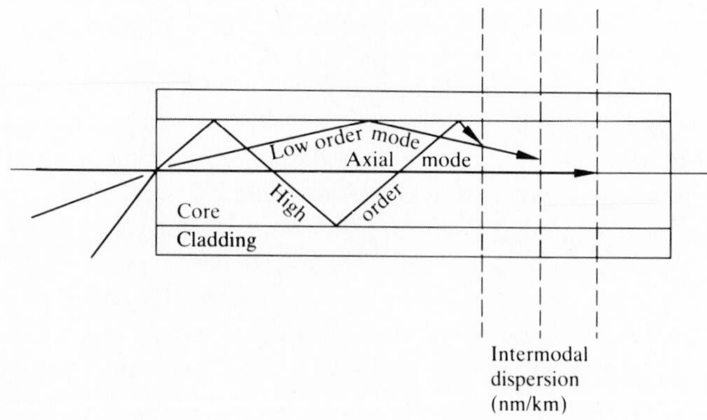

FIGURE 5.83 Intermodal dispersion in a stepped-index multimode fiber.

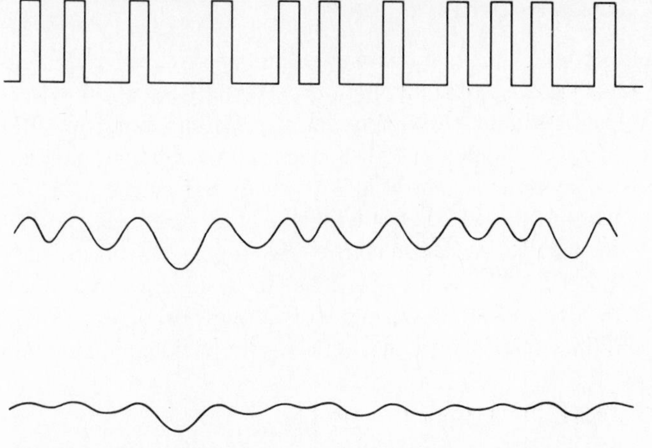

FIGURE 5.84 Rectangular pulses of light smeared out by increasing amounts of dispersion. Note how the closely spaced pulses degrade more quickly.

mitted is usually digitized in some coded fashion and then sent along the fibers as a flood of millions of pulses or bits per second. The different transit times have the undesirable effect of changing the shape of the pulses of light that represent the signal. What started as a sharp rectangular pulse can smear out, after traveling a few kilometers within the fiber, into an unrecognizable blur (Fig. 5.84).

The total time delay between the arrival of the axial ray and the slowest ray, the one traveling the longest distance, is $\Delta t = t_{\max} - t_{\min}$. Here, referring back to Fig. 5.79, the minimum time of travel is just the axial length L divided by the speed of light in the fiber:

$$t_{\min} = \frac{L}{v_f} = \frac{L}{c/n_f} = \frac{Ln_f}{c} \tag{5.66}$$

The nonaxial route (ℓ), given by Eq. (5.58), is longest when the ray is incident at the critical angle, whereupon Eq. (5.61) holds. Combining these two, we get $\ell = Ln_f/n_c$, and so

$$t_{\max} = \frac{\ell}{v_f} = \frac{Ln_f/n_c}{c/n_f} = \frac{Ln_f^2}{cn_c} \tag{5.67}$$

Thus it follows that, subtracting Eq. (5.66) from Eq. (5.67), we get

$$\Delta t = \frac{Ln_f}{c}\left(\frac{n_f}{n_c} - 1\right) \tag{5.68}$$

As an example, suppose $n_f = 1.500$ and $n_c = 1.489$. The delay, $\Delta t/L$, then turns out to be 37 ns/km. In other words, a sharp pulse of light entering the system will be spread out in time some 37 ns for each kilometer of fiber traversed. Moreover, traveling at a speed $v_f = c/n_f = 2.0 \times 10^8$ m/s, it will spread in space over a length of 7.4 m/km. To make sure that the transmitted signal will still be easily readable, we might require that the spatial (or temporal) separation be at least twice the spread-out width (Fig. 5.85). Now imagine the line to be 1.0 km long. In that case the output pulses are 7.4 m wide on emerging from the fiber and so must be separated by 14.8 m. This means that the input pulses must be at least 14.8 m apart; they must be separated in time by 74 ns and so cannot

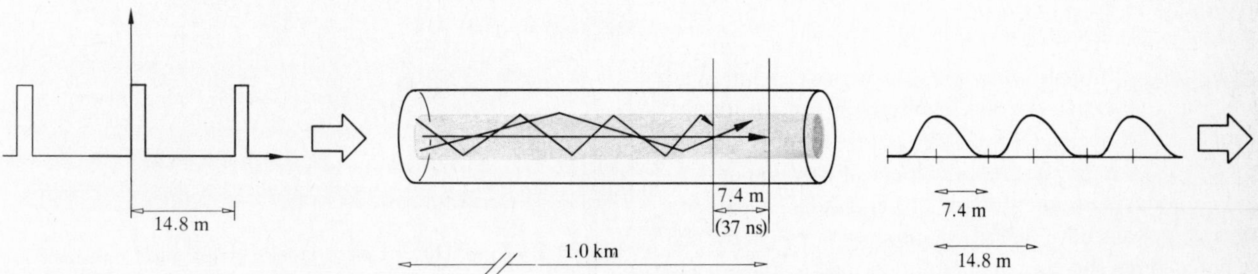

FIGURE 5.85 The spreading of an input signal due to intermodal dispersion.

come any faster than one every 74 ns, which is a rate of 13.5 million pulses per second. In this way the intermodal dispersion (which is typically 15 to 30 ns/km) limits the frequency of the input signal, thereby dictating the rate at which information can be fed through the system.

This problem of delay differences can be reduced as much as a hundredfold by gradually varying the refractive index of the core, decreasing it radially outward to the cladding (Fig. 5.82*b*). Instead of following sharp zigzag paths, the rays then smoothly spiral around the central axis. Because the index is higher along the center, rays taking shorter paths are slowed down by proportionately greater amounts, and rays spiralling around near the cladding move more swiftly over longer paths. The result is that all the rays tend to stay more or less together in these multimode **graded-index fibers**. Typically, a graded-index fiber has a core diameter of about 20 μm to 90 μm and an intermodal dispersion of only around 2 ns/km. They are intermediate in price and have been widely used in medium-distance intercity applications.

Multimode fibers with core diameters of 50 μm or more are often fed by *light-emitting diodes* (LEDs). These are comparatively inexpensive and are commonly used over relatively short spans at low transmission rates. The problem with them is that they emit a fairly broad range of frequencies. As a result, ordinary *material* or *spectral dispersion*, the fact that the fiber index is a function of frequency, becomes a limiting factor. That difficulty is essentially avoided by using spectrally pure laserbeams. Alternatively, the fibers can be operated at wavelengths near 1.3 μm, where silica glass (see Figs. 3.38 and 3.39) has little dispersion.

The last, and best, solution to the problem of intermodal dispersion is to make the core so narrow (less than 10 μm) that it will provide only one mode wherein the rays travel parallel to the central axis (Fig. 5.82*c*). Such **single-mode fibers** of ultrapure glass (both stepped-index and the newer graded-index) provide the best performance. Typically having core diameters of only 2 μm to 9 μm (around 10 wavelengths), they essentially eliminate intermodal dispersion. Although they are relatively expensive and require laser sources, these single-mode fibers operated at 1.55 μm (where the attenuation is about 0.2 dB/km, not far from the ideal silica value of 0.1 dB/km) are today's premier long-haul lightguides. A pair of such fibers may someday connect your home to a vast network of communications and computer facilities, making the era of the copper wire seem charmingly primitive.

Capillary Optics

Fiberoptics works by having radiant energy (of a relatively low frequency, namely, light or IR) totally internally reflect off a high-index/low-index interface within a narrow solid waveguide. Similarly, high-frequency EM-radiation (especially X-rays) can also be totally internally reflected (p. 121) off an air–glass interface (rather than a glass–air interface). The critical angle, *measured up from the surface*, is typically only about 0.2° for 10 keV ($\approx$0.12 nm) X-rays. Figure 5.86 shows how a beam follows the curve of a hollow capillary tube via multiple grazing-incidence reflection at the internal air–glass

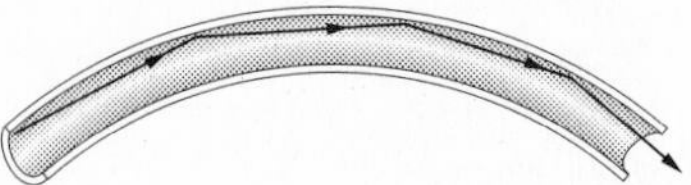

FIGURE 5.86 Multiple grazing-incidence reflections of X-rays within a hollow glass fiber.

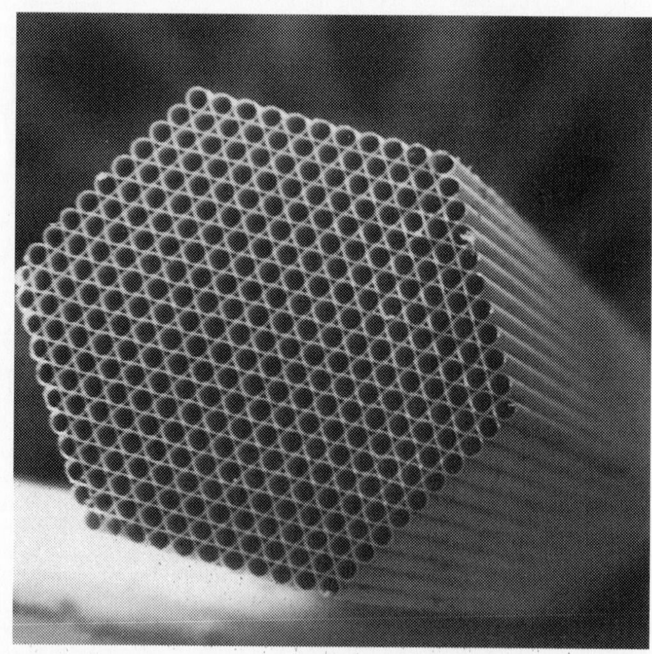

FIGURE 5.87 A scanning electron micrograph of a single multichannel thread containing hundreds of hollow capillaries. (Courtesy X-Ray Optical Systems, Inc. Albany, NY, USA)

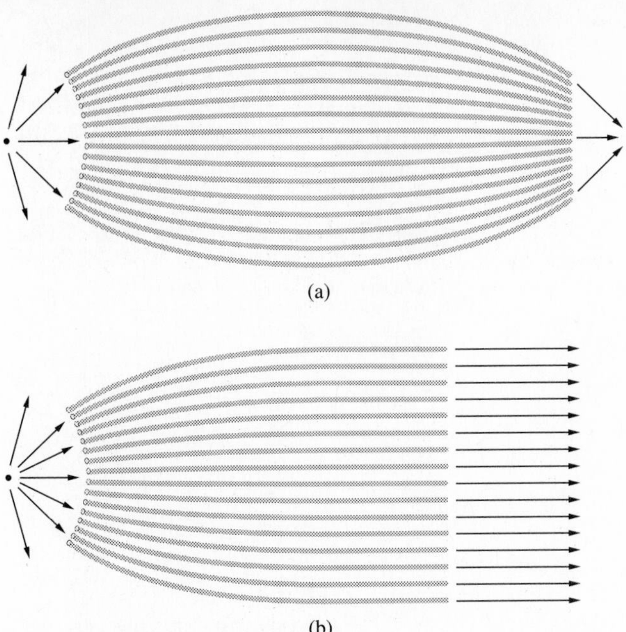

FIGURE 5.88 A bundle of multicapillary threads used to (*a*) focus or (*b*) collimate the X-rays from a point source.

interface. Bending the path of X-rays is otherwise a daunting business.

A single glass thread with a diameter of 300 to 600 μm can be fabricated so that it contains thousands of fine capillary channels each from 3 to 50 μm in diameter (Fig. 5.87). Thousands of such multichannel threads (Fig. 5.88) are then used together to conveniently focus or collimate X-ray beams in a way never before possible.

5.7 Optical Systems

We have developed paraxial theory to a point where it is possible to appreciate the principles underlying the majority of practical optical systems. To be sure, the subtleties involved in controlling aberrations are extremely important and still beyond this discussion. Even so, one could build, for example, a telescope (admittedly not a very good one, but a telescope nonetheless) using the conclusions already drawn from first-order theory.

What better starting point for a discussion of optical instruments than the most common of all—the eye?

5.7.1 Eyes

For our purposes, three main groupings of eyes can readily be distinguished: those that gather radiant energy and form images via a single-centered lens system, those that utilize a multifaceted arrangement of tiny lenses (feeding into channels resembling optical fibers), and the most rudimentary, those that simply function with a small lensless hole (p. 219). In addition to light-sensing eyes, the rattlesnake has infrared pinhole "eyes" called pits, which might be included in this last group.

Visual lens systems of the first type have evolved independently and remarkably similarly in at least three distinct kinds of organisms. Some of the more advanced mollusks (e.g., the octopus), certain spiders (e.g., the avicularia), and the vertebrates, ourselves included, possess eyes that each form a single continuous real image on a light-sensitive screen or retina. By comparison, the multifaceted compound eye (Fig. 5.89) developed independently among arthropods, the creatures with articulated bodies and limbs (e.g., insects and crayfish). It produces a mosaic sensory image composed of many small-field-of-view spot contributions, one from each tiny segment of the eye (as if one were looking at the world through a tightly packed bundle of exceedingly fine tubes). Like a television picture made up of different-intensity dots, the compound eye divides and digitizes the scene being viewed. There is no real image formed on a retinal screen; the synthesis takes place electrically in the nervous system. The horsefly has about 7000 such segments, and the predatory dragonfly, an especially fast flyer, gets a better view with 30 000, as compared with some ants that manage with only about 50. The more facets, the more image dots, and the better the resolution, the sharper the composite picture. This may well be the oldest of eye types: trilobites, the little sea creatures of 500 million years ago had well-developed compound eyes. Remarkably, however different the optics, the chemistry of the image-sensing mechanisms in all Earth animals is quite similar.

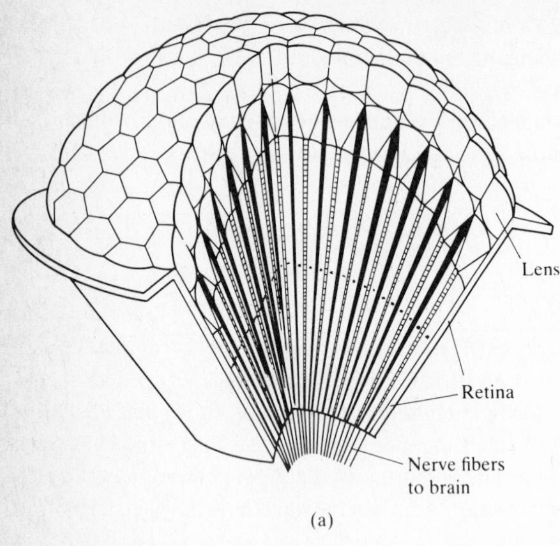

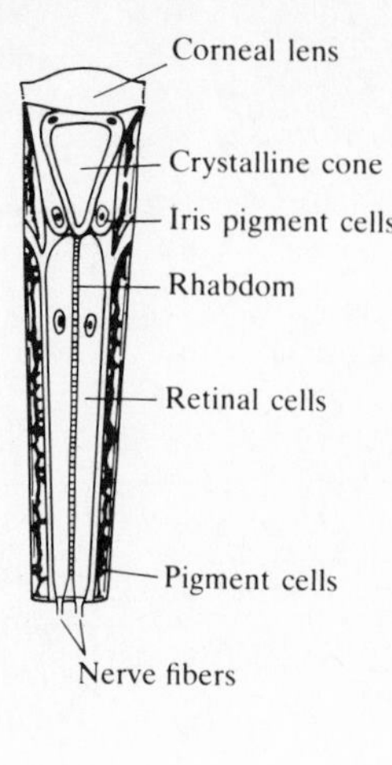

FIGURE 5.89 (*a*) The compound eye made up of many ommatidia. (*b*) An ommatidium, the little individual eye that each "sees" a small region in a particular direction. The corneal lens and crystalline cone channel the light into the sensing structure, the clear, rod-shaped rhabdom. Each of these is surrounded by retinal cells that lead via nerve fibers to the brain. (From Ackerman et al., *Biophysical Science*, © 1962, 1979. Englewood Cliffs, NJ: Prentice-Hall, Inc., p. 31. After R. Bushman, *Animals Without Backbones*.)

Structure of the Human Eye

The human eye can be thought of as a positive double-lens arrangement that casts a real image on a light-sensitive surface. That notion, in a rudimentary form, was apparently proposed by Kepler (1604), who wrote "Vision, I say, occurs when the image of the … external world … is projected onto the … concave retina." This insight gained wide acceptance only after a lovely experiment was performed in 1625 by the German Jesuit Christopher Scheiner (and independently, about five years later, by Descartes). Scheiner removed the coating on the back of an animal's eyeball and, peering through the nearly transparent retina from behind, was able to see a minified, inverted image of the scene beyond the eye. Although it resembles a simple camera (p. 176), the seeing system (eye, optic nerve, and visual cortex) functions much more like a closed-circuit computerized television unit.

The eye (Fig. 5.90) is an almost spherical (24 mm long by about 22 mm across) jellylike mass contained within a tough flexible shell, the **sclera**. Except for the front portion, or **cornea**, which is transparent, the sclera is white and opaque. Bulging out from the body of the sphere, the cornea's curved surface (which is slightly flattened, thereby cutting down on spherical aberration) serves as the first and strongest convex element of the lens system. Indeed, most of the bending imparted to a bundle of rays takes place at the air–cornea interface. Incidentally, one of the reasons you can't see very well under water ($n_W \approx 1.33$) is that its index is too close to that of the cornea ($n_C \approx 1.376$) to allow for adequate refraction.

Light emerging from the cornea passes through a chamber filled with a clear watery fluid called the **aqueous humor** ($n_{ah} \approx 1.336$). It nourishes the anterior portion of the eye. A ray that is strongly refracted toward the optical axis at the air–cornea interface will be only slightly redirected at the cornea–aqueous humor interface because of the similarity of their indices. Immersed in the aqueous is a diaphragm known as the **iris**, which serves as the aperture stop controlling the amount of light entering the eye through the hole, or **pupil**. It is the iris (from the Greek word for rainbow) that gives the eye its characteristic blue, brown, gray, green, or hazel color. Made up of circular and radial muscles, the iris can expand or contract the pupil over a range from about 2 mm in bright light to roughly 8 mm in darkness. In addition to this function, it is also linked to the focusing response and will contract to increase image sharpness when doing close work.

Immediately behind the iris is the **crystalline lens**. The name, which is somewhat misleading, dates back to about A.D. 1000 and the work of Abû 'Alî al Hasan ibn al Hasan ibn al Haitham, alias Alhazen of Cairo, who described the eye as partitioned into three regions that were watery, crystalline, and glassy, respectively. The lens, which has both the size and shape of a small bean (9 mm in diameter and 4 mm thick), is a

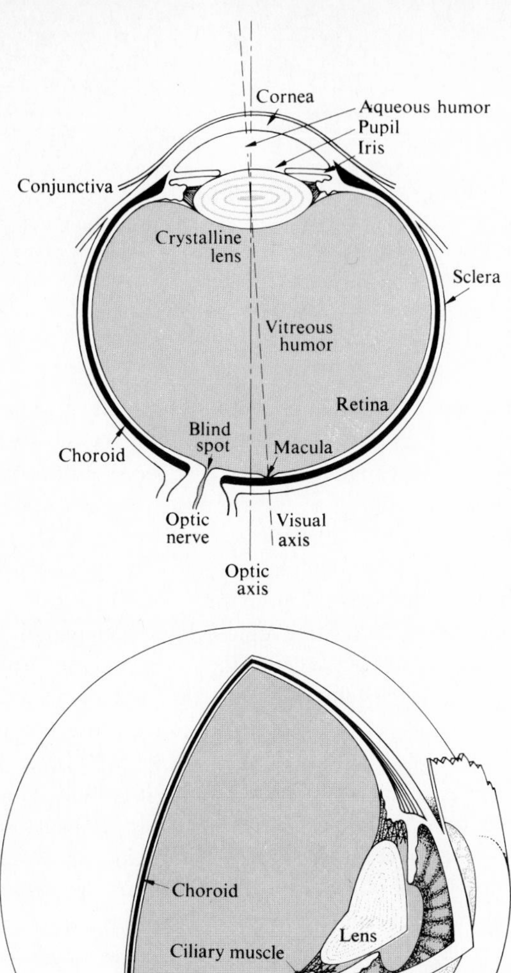

FIGURE 5.90 The human eye.

complex layered fibrous mass surrounded by an elastic membrane. In structure it is somewhat like a transparent onion, formed of roughly 22 000 very fine layers. It has some remarkable characteristics that distinguish it from man-made lenses, in addition to the fact that it continues to grow in size. Because of its laminar structure, rays traversing it will follow paths made up of minute, discontinuous segments. The lens as a whole is quite pliable, albeit less so with age. Moreover, its index of refraction ranges from about 1.406 at the inner core to approximately 1.386 at the less dense cortex and, as such it represents a gradient-index or GRIN system (p. 277). The crystalline lens provides the needed fine-focusing mechanism through changes in its shape; that is, it has a variable focal length—a feature we'll come back to presently.

The refracting components of the eye, the cornea and crystalline lens, can be treated as forming an effective double-element lens with an object focus of about 15.6 mm in front of the anterior surface of the cornea and an image focus of about 24.3 mm behind it on the retina. To simplify things a little, we can take the combined lens to have an optical center 17.1 mm in front of the retina, which falls just at the rear edge of the crystalline lens.

Behind the lens is another chamber filled with a transparent gelatinous substance made of collagen (a protein polymer) and hyaluronic acid (a protein concentrate). Known as the **vitreous humor** ($n_{vh} \approx 1.337$), this thick gel gives support to the eyeball. As an aside, it should be noted that the vitreous humor contains microscopic particles of cellular debris floating freely about. You can easily see their shadows, outlined with diffraction fringes, within your own eye by squinting at a light source or looking at the sky through a pinhole—strange little amoebalike objects (*muscae volitantes*) will float across the field of view. Incidentally, a marked increase in one's perception of these floaters may be indicative of retinal detachment. While you're at it, squint at the source again (a broad diffuse fluorescent light works well). Closing your lids almost completely, you'll actually be able to see the near circular periphery of your own pupil, beyond which the glare of light will disappear into blackness. If you don't believe it, block and then unblock some of the light; the glare circle will visibly expand and contract, respectively. You are seeing the shadow cast by the iris from the inside! Seeing internal objects like this is known as entoptic perception.

Within the tough sclerotic wall is an inner shell, the choroid. It is a dark layer, well supplied with blood vessels and richly pigmented with melanin. The choroid is the absorber of stray light, as is the coat of black paint on the inside of a camera. A thin layer (about 0.5 mm to 0.1 mm thick) of light receptor cells covers much of the inner surface of the choroid—this is the **retina** (from the Latin *rete*, meaning net).

The focused beam of light is absorbed via electrochemical reactions in this pinkish multilayered structure.

The human eye contains two kinds of photoreceptor cells: **rods** and **cones** (Fig. 5.91). Roughly 125 million of them are intermingled nonuniformly over most of the retina. The ensemble of rods (each about 0.002 mm in diameter) in some respects has the characteristics of a high-speed, black-and-white film (such as Tri-X). It is exceedingly sensitive, performing in light too dim for the cones to respond to, yet it is unable to distinguish color, and the images it relays are not well defined. In contrast, the ensemble of 6 or 7 million cones (each about 0.006 mm in diameter) can be imagined as a separate, but overlapping, low-speed color film. It performs in bright light, giving detailed colored views, but it is fairly insensitive at low light levels.

The normal wavelength range of human vision is roughly 390 nm to 780 nm (Table 3.4). However, studies have extended these limits down to about 310 nm in the ultraviolet and up to roughly 1050 nm in the infrared. Indeed, people have reported "seeing" X-radiation. The limitation on ultraviolet transmission in the eye is set by the crystalline lens, which absorbs in the UV. People who have had a lens removed surgically have greatly improved UV sensitivity.

FIGURE 5.91 An electron micrograph of the retina of a salamander (*Necturus Maculosus*). Two visual cones appear in the foreground and several rods behind them. Photo from E. R. Lewis, Y. Y. Zeevi, and F. S. Werblin, *Brain Research* **15**, 559 (1969).

FIGURE 5.92 To verify the existence of the blind spot, close one eye and, at a distance of about 10 inches, look directly at the X—the 2 will disappear. Moving closer will cause the 2 to reappear while the 1 vanishes.

The area of exit of the optic nerve from the eye contains no receptors and is insensitive to light; accordingly, it is known as the **blind spot** (see Fig. 5.92). The optic nerve spreads out over the back of the interior of the eye in the form of the retina.

Just about at the center of the retina is a small depression from 2.5 to 3 mm in diameter known as the yellow spot, or **macula**. It is composed of more than twice as many cones as rods. There is a tiny rod-free region about 0.3 mm in diameter at the center of the macula called the **fovea centralis**. (In comparison, the image of the full Moon on the retina is about 0.2 mm in diameter—Problem 5.66.) Here the cones are thinner (with diameters of 0.003 0 mm to 0.001 5 mm) and more densely packed than anywhere else in the retina. Since the fovea provides the sharpest and most detailed information, the eyeball is continuously moving, so that light coming from the area on the object of primary interest falls on this region. An image is constantly shifted across different receptor cells by these normal eye movements. If such movements did not occur and the image was kept stationary on a given set of photoreceptors, it would actually tend to fade out. Without the fovea the eye would lose 90 to 95% of its capability, retaining only peripheral vision.

Another fact that indicates the complexity of the sensing system is that the rods are multiply connected to nerve fibers, and a single such fiber can be activated by any one of about a hundred rods. By contrast, cones in the fovea are individually connected to nerve fibers. The actual perception of a scene is constructed by the eye–brain system in a continuous analysis of the time-varying retinal image. Just think how little trouble the blind spot causes, even with one eye closed.

Between the nerve-fiber layer of the retina and the humor is a network of large retinal blood vessels, which can be observed entoptically. One way is to close your eye and place a bright small source against the lid. You'll "see" a pattern of shadows (*Purkinje figures*) cast by the blood vessels on the sensitive retinal layer.

Accommodation

The fine focusing, or **accommodation**, of the human eye is a function performed by the crystalline lens. The lens is suspended in position behind the iris by ligaments that are connected to a circular yoke composed of the **ciliary muscles**. Ordinarily, these muscles are relaxed, and in that state they pull outward radially on the network of fine fibers holding the rim of the lens. This draws the pliable lens into a fairly flat configuration, increasing its radii, which in turn increases its focal length [Eq. (5.16)]. With the muscles completely relaxed, the light from an object at infinity will be focused on the retina (Fig. 5.93). As the object moves closer to the eye, the ciliary muscles contract, relieving the external tension on the periphery of the lens, which then bulges slightly under its own elastic forces. In so doing the focal length decreases such that s_i is kept constant. As the object comes still closer, the yoke of ciliary muscles becomes more tensely contracted, the circular region they encompass gets still smaller, and the lens surfaces take on even smaller radii. The closest point on which the eye can focus is known as the **near point**. In a normal eye it might be about 7 cm for a teenager, 12 cm or so for a young adult, roughly 28 to 40 cm in the middle-aged, and about 100 cm by 60 years of age. Visual instruments are designed with this in mind, so that the eye need not strain unnecessarily. Clearly, the eye cannot focus on two different objects at once. This will be made obvious if, while looking through a piece of glass, you try to focus on it and the scene beyond at the same time.

Mammals generally accommodate by varying the lens curvature, but there are other means. Fish move only the lens itself toward or away from the retina, just as the camera lens is moved to focus. Some mollusks accomplish the same thing by contracting or expanding the whole eye, thus altering the relative distance between lens and retina. For birds of prey, which must keep a rapidly moving object in constant focus over a wide range of distances as a matter of survival, the accommodation mechanism is quite different. They accommodate by greatly changing the curvature of the cornea.

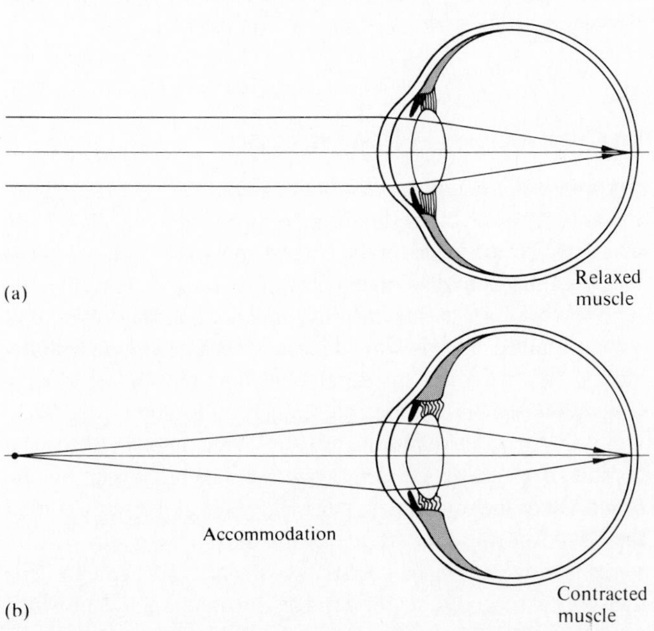

FIGURE 5.93 Accommodation—changes in the lens configuration.

5.7.2 Eyeglasses

Spectacles were probably invented some time in the late thirteenth century, possibly in Italy. A Florentine manuscript of the period (1299), which no longer exists, spoke of "spectacles recently invented for the convenience of old men whose sight has begun to fail." These were biconvex lenses, little more than variations on the handheld magnifying or reading glasses, and polished gemstones were no doubt employed as lorgnettes long before that. Roger Bacon (ca. 1267) wrote about negative lenses rather early on, but it was almost another two hundred years before Nicholas Cusa first discussed their use in eyeglasses and a hundred years more before such glasses ceased to be a novelty, in the late 1500s. Amusingly, it was considered improper to wear spectacles in public even as late as the eighteenth century, and we see few users in the paintings up until that time. In 1804 Wollaston, recognizing that traditional (fairly flat, biconvex, and concave) eyeglasses provided good vision only while one looked through their centers, patented a new, deeply curved lens. This was the forerunner of modern-day meniscus (from the Greek *meniskos*, the diminutive for moon, i.e., crescent) lenses, which allow the turning eyeball to see through them from center to margin without significant distortion.

It is customary and quite convenient in physiological optics to speak about the **dioptric power**, $\mathcal{D}$, of a lens, which is simply the reciprocal of the focal length. When f is in meters, the unit of power is the inverse meter, or *diopter*, symbolized by D: $1\ \text{m}^{-1} = 1$ D. For example, if a converging lens has a focal length of $+1$ m, its power is $+1$ D; with a focal length of -2 m (a diverging lens), $\mathcal{D} = -\frac{1}{2}$ D; for $f = +10$ cm, $\mathcal{D} = 10$ D. Since a thin lens of index n_l in air has a focal length given by

$$\frac{1}{f} = (n_l - 1)\left(\frac{1}{R_1} - \frac{1}{R_2}\right) \qquad [5.16]$$

its power is

$$\mathcal{D} = (n_l - 1)\left(\frac{1}{R_1} - \frac{1}{R_2}\right) \qquad (5.69)$$

You can get a sense of the direction in which we are moving by considering, in rather loose terms, that each surface of a lens bends the incoming rays—the more bending, the stronger the surface. A convex lens that strongly bends the rays at both surfaces has a short focal length and a large dioptric power. We already know that the focal length for two thin lenses in contact is given by

$$\frac{1}{f} = \frac{1}{f_1} + \frac{1}{f_2} \qquad [5.38]$$

This means that the combined power is the sum of the individual powers, that is,

$$\mathcal{D} = \mathcal{D}_1 + \mathcal{D}_2$$

Thus a convex lens with $\mathcal{D}_1 = +10$ D in contact with a negative lens of $\mathcal{D}_2 = -10$ D results in $\mathcal{D} = 0$; the combination behaves like a parallel sheet of glass. Furthermore, we can imagine a lens, for example, a double convex lens, as being composed of two planar-convex lenses in intimate contact, back to back. The power of each of these follows from Eq. (5.69); thus for the first planar-convex lens ($R_2 = \infty$),

$$\mathcal{D}_1 = \frac{(n_l - 1)}{R_1} \qquad (5.70)$$

and for the second,

$$\mathcal{D}_2 = \frac{(n_l - 1)}{-R_2} \qquad (5.71)$$

These expressions may be equally well defined as giving the *powers of the respective surfaces* of the initial double convex lens. In other words, *the power of any thin lens is equal to the sum of the powers of its surfaces*. Because R_2 for a convex lens is a negative number, both $\mathcal{D}_1$ and $\mathcal{D}_2$ will be positive in that case. The power of a surface, defined in this way, is not generally the reciprocal of its focal length, although it is when immersed in air. Relating this terminology to the generally used model for the human eye, we note that the power of the crystalline lens *surrounded by air* is about $+19$ D. The cornea provides roughly $+43$ of the total $+58.6$ D of the intact unaccommodated eye.

A normal eye, despite the connotation of the word, is not as common as one might expect. By the term *normal*, or its synonym *emmetropic*, we mean an eye that is capable of focusing parallel rays on the retina while in a relaxed condition—that is, one whose second focal point lies on the retina. For the unaccommodated eye, we define the object point whose image lies on the retina to be the **far point**. Thus for the normal eye the most distant point that can be brought to a focus on the retina, the far point, is located at infinity (which for all practical purposes is anywhere beyond about 5 m). In contrast, when the focal point does not lie on the retina, the eye is *ametropic* (e.g., it suffers hyperopia, myopia, or astigmatism). This can arise either because of abnormal changes in the refracting mechanism (cornea, lens, etc.) or because of alterations in the length of the eyeball that change the distance between the lens and the retina. The latter is by far the more common cause. Just to put things in proper perspective, note that about 25% of young adults require ± 0.5 D or less of eyeglass correction, and perhaps as many as 65% need only ± 1.0 D or less.

Nearsightedness—Negative Lenses

Myopia is the condition in which parallel rays are brought to focus in front of the retina; the power of the lens system as configured is too large for the anterior-posterior axial length of the eye. Images of distant objects fall in front of the retina, the far point is closer in than infinity, and all points beyond it will appear blurred. This is why myopia is often called **nearsightedness**; an eye with this defect sees nearby objects clearly (Fig. 5.94). To correct the condition, or at least its symptoms, we place an additional lens in front of the eye such that the combined spectacle–eye lens system has its focal point on the retina. Since the myopic eye can clearly see objects closer than the far point, the spectacle lens must cast relatively nearby images of distant objects. Hence we introduce a negative lens that will diverge the rays a bit. Resist the temptation to suppose that we are merely reducing the power of the system. In point of fact, the power of the lens–eye combination is most

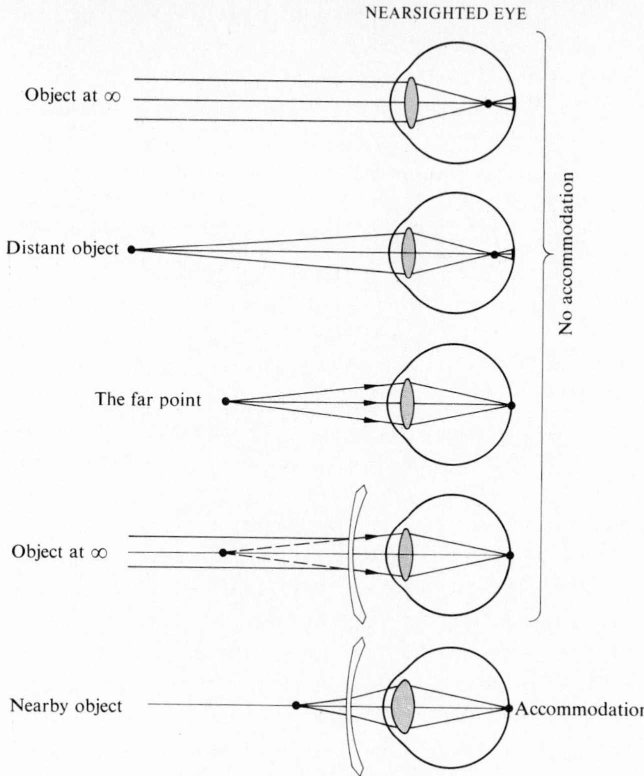

FIGURE 5.94 Correction of the nearsighted eye.

often made to equal that of the unaided eye. If you are wearing glasses to correct myopia, take them off; the world gets blurry, but it doesn't change size. Try casting a real image on a piece of paper using your glasses—it can't be done.

Suppose an eye has a far point of 2 m. All would be well if the spectacle lens appeared to bring more distant objects in closer than 2 m. If the virtual image of an object at infinity is formed by a concave lens at 2 m, the eye will see the object clearly with an unaccommodated lens. Thus using the thin-lens approximation (eyeglasses are generally thin to reduce weight and bulk), we have

$$\frac{1}{f} = \frac{1}{s_o} + \frac{1}{s_i} = \frac{1}{\infty} + \frac{1}{-2} \qquad [5.17]$$

and $f = -2$ m while $\mathcal{D} = -\frac{1}{2}$ D. Notice that the far-point distance, measured from the correction lens, equals its focal length (Fig. 5.95). The eye views the right-side-up virtual images of all objects formed by the correction lens, and those images are located between its far and near points. Incidentally, the near point also moves away a little, which is why myopes often prefer to remove their spectacles when threading needles or reading small print; they can then bring the material closer to the eye, thereby increasing the magnification.

The calculation we have just performed overlooks the separation between the correction lens and the eye—in effect, it applies to contact lenses more than to spectacles. The separation is usually made equal to the distance of the first focal point of the eye (≈16 mm) from the cornea, so that no magnification of the image over that of the unaided eye occurs. Many people have unequal eyes, yet both yield the same magnification. A change in M_T for one and not the other would be a disaster. Placing the correcting lens at the eye's first focal point avoids the problem completely, regardless of the power of that lens [take a look at Eq. (6.8)]. To see this, just draw a ray from the top of some object through that focal point. The ray will enter the eye and traverse it parallel to the optic axis, thus establishing the height of the image. Yet, since this ray is unaffected by the presence of the spectacle lens, whose center is at the focal point, the image's location may change on insertion of such a lens, but its height and therefore M_T will not [see Eq. (5.24)].

The question now becomes: What is the equivalent power of a spectacle lens at some distance d from the eye (i.e., equivalent to that of a contact lens with a focal length f_c that equals the far-point distance)? It will do for our purposes to approximate the eye by a single lens and take d from that lens to the spectacle as roughly equal to the cornea–eyeglass distance, around 16 mm. Given that the focal length of the correction lens is f_l and the focal length of the eye is f_e, the combination has a focal length provided by Eq. (5.36), that is,

$$\text{b.f.l.} = \frac{f_e(d - f_l)}{d - (f_l + f_e)} \qquad (5.72)$$

This is the distance from the eye-lens to the retina. Similarly,

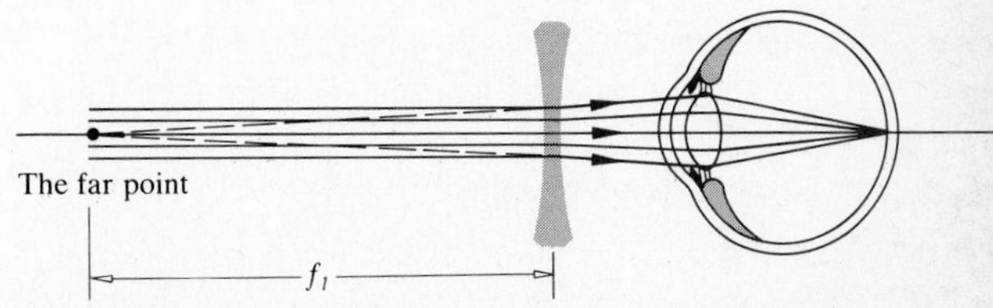

FIGURE 5.95 The far-point distance equals the focal length of the correction lens.

the equivalent contact lens combined with the eye-lens has a focal length given by Eq. (5.38):

$$\frac{1}{f} = \frac{1}{f_c} + \frac{1}{f_e} \tag{5.73}$$

where f = b.f.l. Inverting Eq. (5.72), setting it equal to Eq. (5.73), and simplifying, we obtain the result $1/f_c = 1/(f_l - d)$, independent of the eye itself. In terms of power,

$$\mathfrak{D}_c = \frac{\mathfrak{D}_l}{1 - \mathfrak{D}_l d} \tag{5.74}$$

A spectacle lens of power $\mathfrak{D}_l$ a distance d from the eye-lens has an effective power the same as that of a contact lens of power $\mathfrak{D}_c$. Notice that since d is measured in meters and thus is quite small, unless $\mathfrak{D}_l$ is large, as it often is, $\mathfrak{D}_c \approx \mathfrak{D}_l$. Usually, the point on your nose where you choose to rest your eyeglasses has little effect, but that's certainly not always the case; an improper value of d has resulted in many a headache.

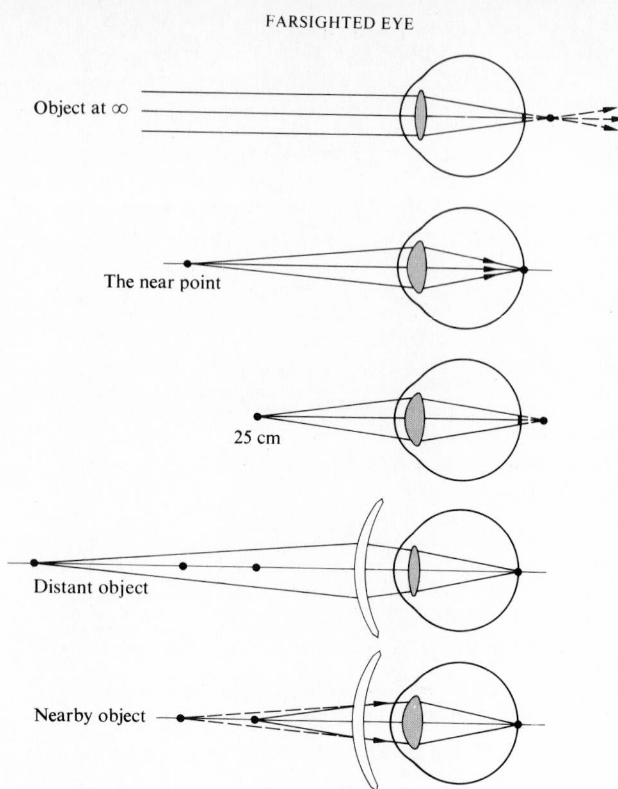

FIGURE 5.96 Correction of the farsighted eye.

Farsightedness—Positive Lenses

Hyperopia (or *hypermetropia*) is the defect that causes the second focal point of the unaccommodated eye to lie behind the retina (Fig. 5.96). **Farsightedness**, as you might have guessed it would be called, is often due to a shortening of the anteroposterior axis of the eye—the lens is too close to the retina. To increase the bending of the rays, a positive spectacle lens is placed in front of the eye. The hyperopic eye can and must accommodate to see distant objects distinctly, but it will be at its limit to do so for a near point, which is much farther away than it would be normally (this we take as 254 mm or just 25 cm). It will consequently be unable to see nearby objects clearly. A converging corrective lens with positive power will effectively move a close object out beyond the near point where the eye has adequate acuity; that is, it will form a distant virtual image, which the eye can then see clearly. Suppose that a hyperopic eye has a near point of 125 cm. For an object at +25 cm to have its image at $s_i = -125$ cm so that it can be seen as if through a normal eye, the focal length must be

$$\frac{1}{f} = \frac{1}{(-1.25)} + \frac{1}{0.25} = \frac{1}{0.31}$$

or $f = 0.31$ m and $\mathfrak{D} = +3.2$ D. This is in accord with Table 5.3, where $s_o < f$. These spectacles will cast real images—try it if you're hyperopic.

As shown in Fig. 5.97, the correcting lens allows the relaxed eye to view objects at infinity. In effect, it creates an image on its focal "plane" (passing through F), which then serves as a virtual object for the eye. The point (whose image lies on the retina) is once again the *far point*, and it's a distance f_l *behind* the lens. The hyperope can comfortably "see" the far point, and any lens located anywhere in front of the eye that has an appropriate focal length will serve that purpose.

Very gentle finger pressure on the lids above and below the

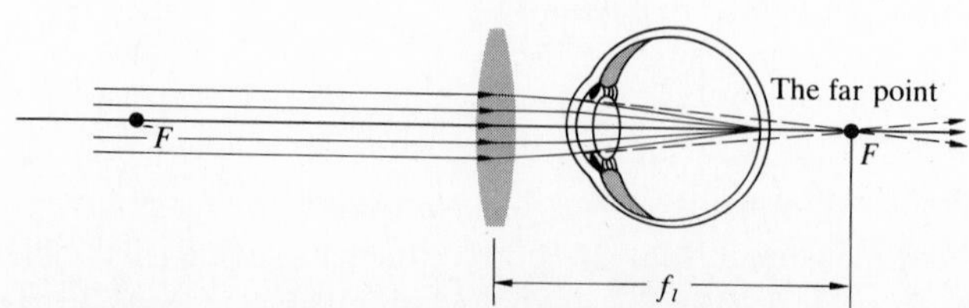

FIGURE 5.97 Again the far-point distance equals the focal length of the correction lens.

cornea will temporarily distort it, changing your vision from blurred to clear and vice versa.

Astigmatism—Anamorphic Lenses

Perhaps the most common eye defect is **astigmatism**. It arises from an uneven curvature of the cornea. In other words, the cornea is asymmetric. Suppose we pass two meridional planes (one containing the optical axis) through the eye such that the (curvature or) power is maximal on one and minimal on the other. If these planes are perpendicular, the *astigmatism* is *regular* and correctible; if not, it is *irregular* and not easily corrected. Regular astigmatism can take different forms; the eye can be emmetropic, myopic, or hyperopic in various combinations and degrees on the two perpendicular meridional planes. Thus, as a simple example, the columns of a checkerboard might be well focused while the rows are blurred due to myopia or hyperopia. Obviously, these meridional planes need not be horizontal and vertical (Fig. 5.98).

The great astronomer Sir George B. Airy used a concave sphero-cylindrical lens to ameliorate his own myopic astigmatism in 1825. This was probably the first time astigmatism had been corrected. But it was not until the publication in 1862 of a treatise on cylindrical lenses and astigmatism by the Dutchman Franciscus Cornelius Donders (1818–1889) that ophthalmologists were moved to adopt the method on a large scale.

Any optical system that has a different value of M_T or $\mathcal{D}$ in two principal meridians is said to be *anamorphic*. Thus, for example, if we rebuilt the system depicted in Fig. 5.34, this time using cylindrical lenses (Fig. 5.99), the image would be

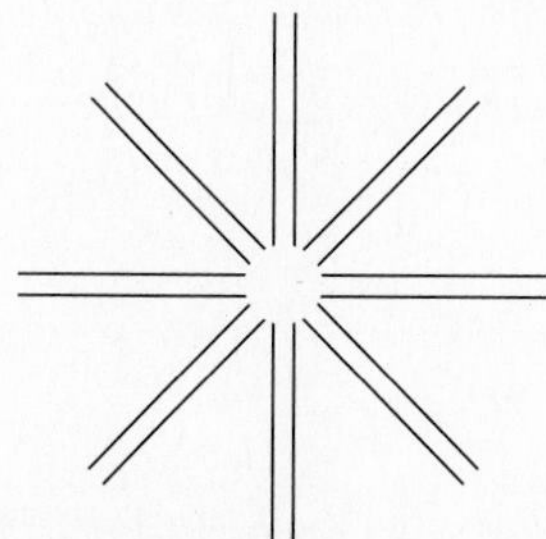

FIGURE 5.98 A test for astigmatism of the eye. View this figure through one unaided eye. If one set of lines appears bolder than the others, you have astigmatism. Hold the figure close to your eye; move it away slowly and note which set of lines comes into focus first. If two sets seem to be equally clear, rotate the figure until only one set is in focus. If all sets are clear you don't have astigmatism.

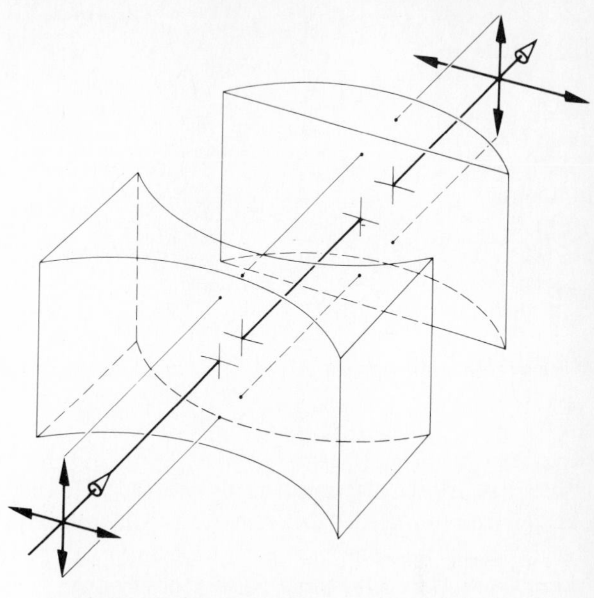

(a)

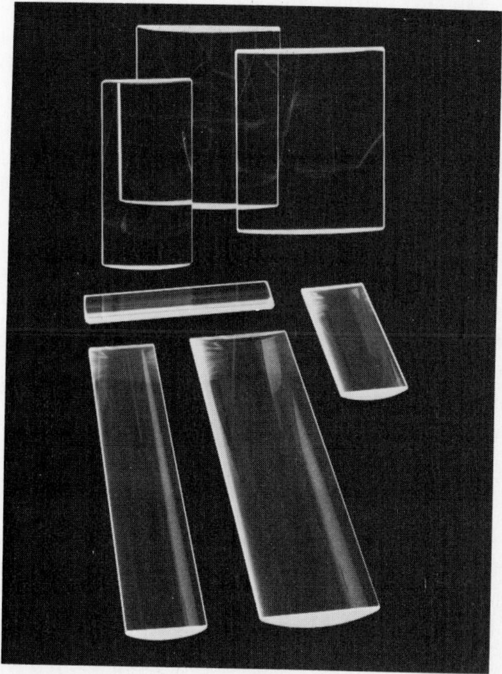

(b)

FIGURE 5.99 *(a)* An anamorphic system. *(b)* Cylindrical lenses (Photo courtesy Melles Griot.)

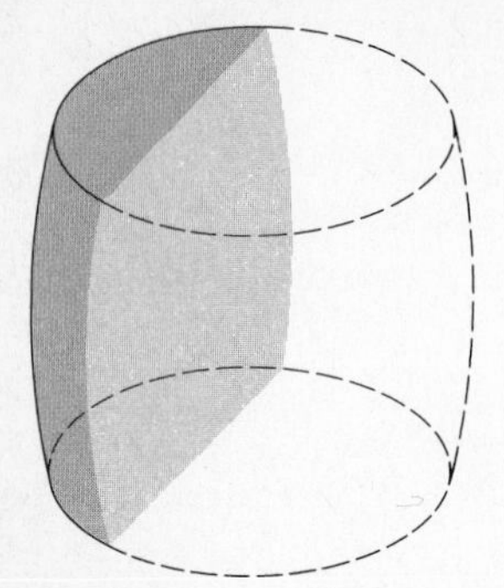
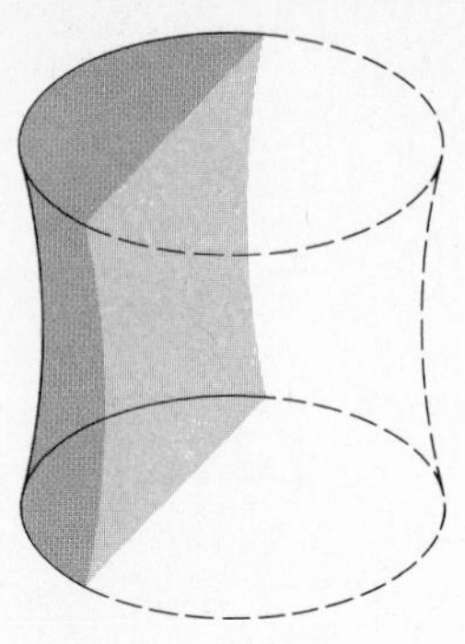

FIGURE 5.100 Toric surfaces.

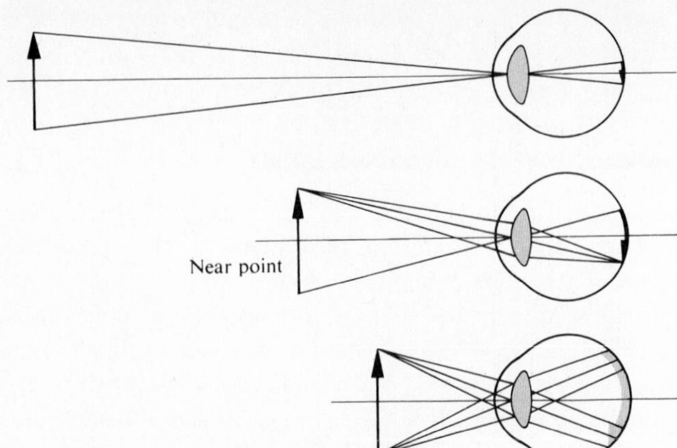

FIGURE 5.101 Images in relation to the near point.

distorted, having been magnified in only one plane. This is just the sort of distortion needed to correct for astigmatism when a defect exists in only one meridian. An appropriate planar cylindrical spectacle lens, either positive or negative, would restore essentially normal vision. When both perpendicular meridians require correction, the lens may be *sphero-cylindrical* or even *toric* as in Fig. (5.100).

Just as an aside, we note that anamorphic lenses are used in other areas, as for example, in the making of wide-screen motion pictures, where an extra-large horizontal field of view is compacted onto the regular film format. When shown through a special lens, the distorted picture spreads out again. On occasion a television station will show short excerpts without the special lens—you may have seen the weirdly elongated result.

5.7.3 The Magnifying Glass

An observer can cause an object to appear larger, for the purpose of examining it in detail, by simply bringing it closer to her eye. As the object is brought nearer and nearer, its retinal image increases, remaining in focus until the crystalline lens can no longer provide adequate accommodation. Should the object come closer than this *near point*, the image will blur (Fig. 5.101). A single positive lens can be used, in effect to add refractive power to the eye, so that the object can be brought still closer and yet be in focus. The lens so used is referred to variously as a **magnifying glass**, a *simple magnifier*, or a *simple microscope*. In any event, its function is *to provide an image of a nearby object that is larger than the image seen by the unaided eye* (Fig. 5.102). Devices of this sort have been around for a long time. In fact, a quartz convex lens ($f \approx$ 10 cm), which may have served as a magnifier, was unearthed in 1885 among the ruins of the palace of King Sennacherib (705–681 B.C.) of Assyria.

Evidently, it would be desirable for the lens to form a magnified, erect image. Furthermore, the rays entering the normal eye should not be converging. Table 5.3 (p. 165) immediately

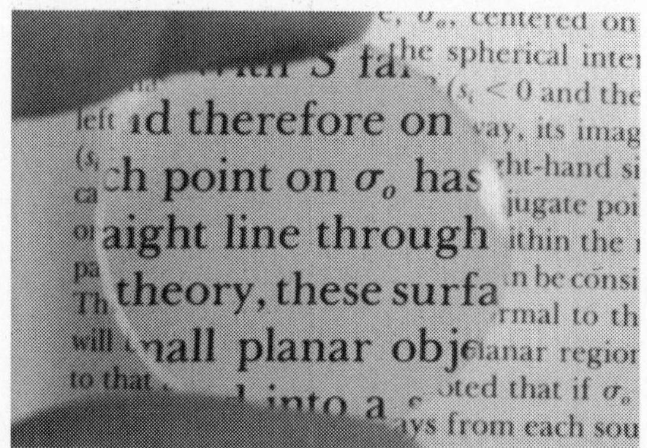

FIGURE 5.102 A positive lens used as a magnifying glass. (E.H.)

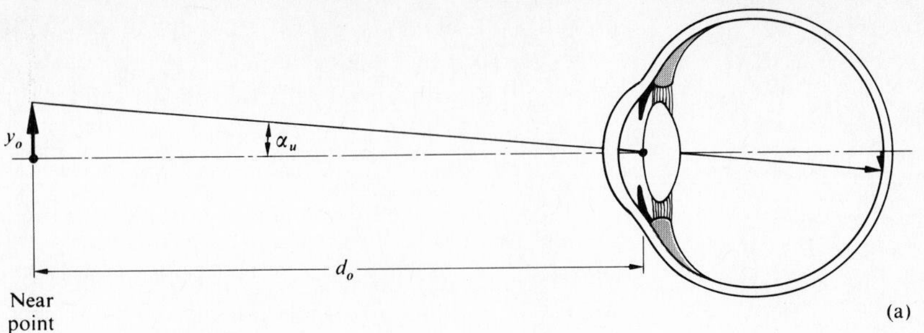

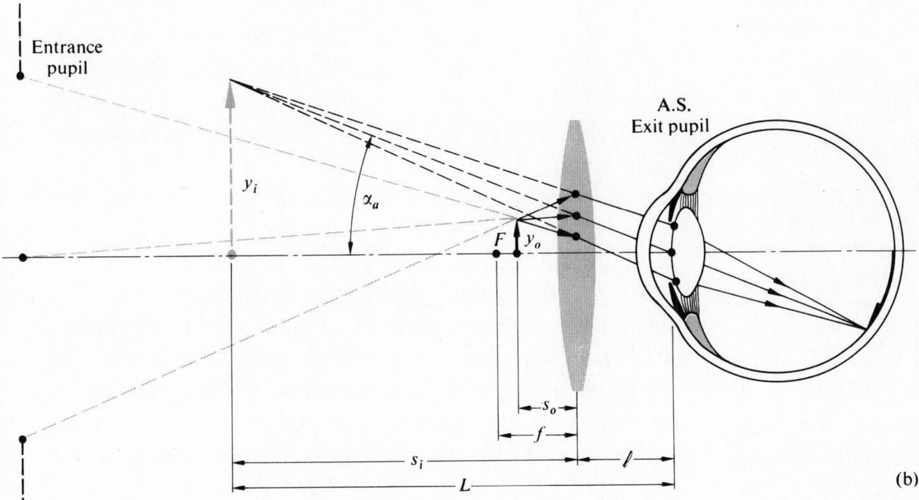

FIGURE 5.103 (*a*) An unaided view of an object. (*b*) The aided view through a magnifying glass. (*c*) A positive lens used as a magnifying glass. The object is less than one focal length from the lens.

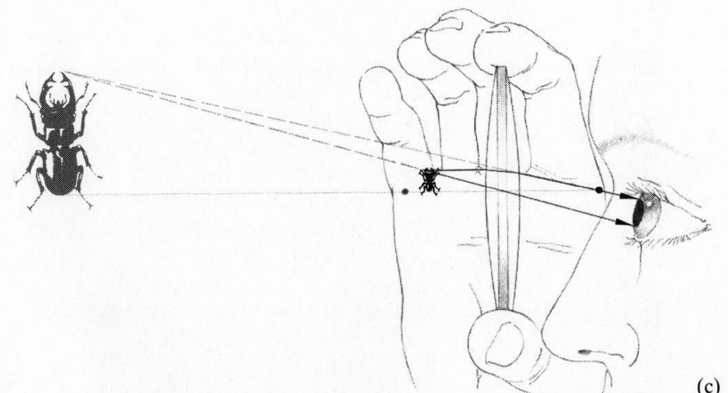

suggests placing the object within the focal length (i.e., $s_o < f$). The result is shown in Fig. 5.103. Because of the relatively tiny size of the eye's pupil, it will almost certainly always be the aperture stop, and as in Fig. 5.37 (p. 174), it will also be the exit pupil.

The **magnifying power**, *MP*, or equivalently, the **angular magnification**, M_A, of a visual instrument is defined as *the ratio of the size of the retinal image as seen through the instrument over the size of the retinal image as seen by the unaided eye at normal viewing distance*. The latter is generally taken as the distance to the near point, d_o. The ratio of angles α_a and α_u (which are made by chief rays from the top of the object in the instance of the aided and unaided eye, respectively) is equivalent to MP, that is,

$$\text{MP} = \frac{\alpha_a}{\alpha_u} \tag{5.75}$$

Keeping in mind that we are restricted to the paraxial region,

$\tan \alpha_a = y_i/L \approx \alpha_a$ and $\tan \alpha_u = y_o/d_o \approx \alpha_u$, so

$$\text{MP} = \frac{y_i d_o}{y_o L}$$

wherein y_i and y_o are above the axis and positive. If we make d_o and L positive quantities, MP will be positive, which is quite reasonable. When we use Eqs. (5.24) and (5.25) for M_T along with the Gaussian Lens Formula, the expression becomes

$$\text{MP} = -\frac{s_i d_o}{s_o L} = \left(1 - \frac{s_i}{f}\right)\frac{d_o}{L}$$

Inasmuch as the image distance is negative, $s_i = -(L - \ell)$, and

$$\text{MP} = \frac{d_o}{L}[1 + \mathfrak{D}(L - \ell)] \tag{5.76}$$

$\mathfrak{D}$ of course being the power of the magnifier $(1/f)$. There are three situations of particular interest: (1) When $\ell = f$ the magnifying power equals $d_o\mathfrak{D}$. (2) When ℓ is effectively zero,

$$[\text{MP}]_{\ell=0} = d_o\left(\frac{1}{L} + \mathfrak{D}\right)$$

In that case the largest value of MP corresponds to the smallest value of L, which, if vision is to be clear, must equal d_o. Thus

$$[\text{MP}]_{\substack{\ell=0\\L=d_o}} = d_o\mathfrak{D} + 1 \tag{5.77}$$

Taking $d_o = 0.25$ m for the standard observer, we have

$$[\text{MP}]_{\substack{\ell=0\\L=d_o}} = 0.25\mathfrak{D} + 1 \tag{5.78}$$

As L increases, MP decreases, and similarly as ℓ increases, MP decreases. If the eye is very far from the lens, the retinal image will indeed be small. (3) This last is perhaps the most common situation. Here we position the object at the focal point $(s_o = f)$, in which case the virtual image is at infinity $(L = \infty)$. Thus from Eq. (5.76)

$$[\text{MP}]_{L=\infty} = d_o\mathfrak{D} \tag{5.79}$$

for all practical values of ℓ. Because the rays are parallel, the eye views the scene in a relaxed, unaccommodated configuration, a highly desirable feature. Notice that $M_T = -s_i/s_o$ approaches infinity as $s_o \to f$, whereas in marked contrast, M_A merely decreases by 1 under the same circumstances.

A magnifier with a power of 10 D has a focal length $(1/\mathfrak{D})$ of 0.1 m and a MP equal to 2.5 when $L = \infty$. This is conventionally denoted as 2.5×, which means that the retinal image is 2.5 times larger with the object at the focal length of the lens than it would be were the object at the near point of the unaided eye (where the largest clear image is possible). The simplest single-lens magnifiers are limited by aberrations to roughly 2× or 3×. A large field of view generally implies a large lens; for practical reasons, this usually dictates a fairly small curvature of the surfaces. The radii are large, as is f, and therefore MP is small. The reading glass, the kind Sherlock Holmes made famous, is a typical example. The watchmaker's eye loupe is frequently a single-element lens, also of about 2× or 3×. Figure 5.104 shows a few more complicated magnifiers designed to operate in the range from roughly 10× to 20×. The double lens is quite common in a number of configurations. Although not particularly good, they perform satisfactorily, for example, in high-powered loupes. The Coddington is essentially a sphere with a slot cut in it to allow an aperture smaller than the pupil of the eye. A clear marble (any small sphere of glass qualifies) will also greatly magnify—but not without a good deal of distortion.

The relative refractive index of a lens and the medium in

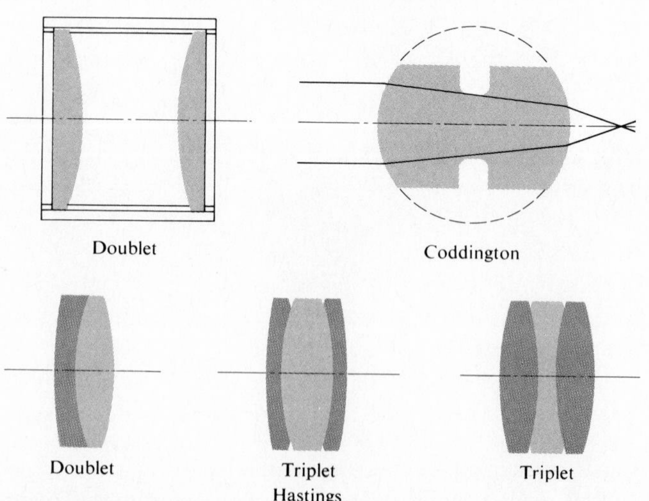

FIGURE 5.104 Magnifiers.

which it is immersed, n_{lm}, is wavelength dependent. But since the focal length of a simple lens varies with $n_{lm}(\lambda)$, this means that f is a function of wavelength, and the constituent colors of white light will focus at different points in space. The resultant defect is known as *chromatic aberration*. In order that the image be free of this coloration, positive and negative lenses made of different glasses are combined to form *achromates* (see Section 6.3.2). Achromatic, cemented, doublet, and triplet lenses are comparatively expensive and are usually found in small, highly corrected, high-power magnifiers.

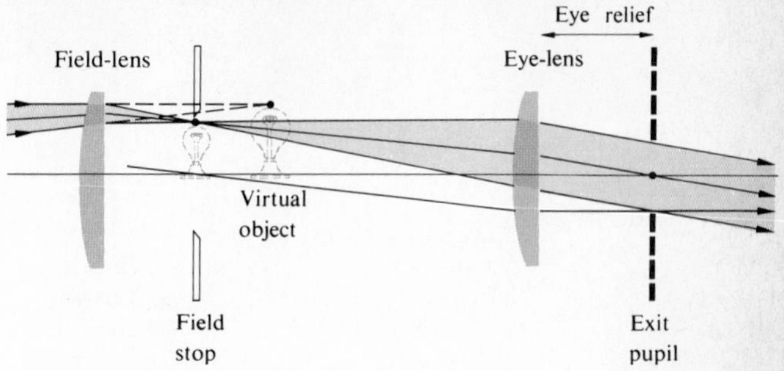

FIGURE 5.105 The Huygens eyepiece.

5.7.4 Eyepieces

The **eyepiece**, or **ocular**, is a visual optical instrument. Fundamentally a magnifier, it views not an actual object but the intermediate image of that object as formed by a preceding lens system. In effect, the eye looks into the ocular, and the ocular looks into the optical system—be it a spotting scope, compound microscope, telescope, or binocular. A single lens could serve the purpose, but poorly. If the retinal image is to be more satisfactory, the ocular cannot have extensive aberrations. The eyepiece of a special instrument, however, might be designed as part of the complete system, so that its lenses can be utilized in the overall scheme to balance out aberrations. Even so, standard eyepieces are used interchangeably on most telescopes and compound microscopes. Moreover, eyepieces are quite difficult to design, and the usual, and perhaps most fruitful, approach is to incorporate or slightly modify one of the existing designs.

The ocular must provide a virtual image (of the intermediate image), most often located at or near infinity, so that it can be comfortably viewed by a normal, relaxed eye. Furthermore, it must position the center of the exit pupil or *eye point* at which the observer's eye is placed at some convenient location, preferably at least 10 mm or so from the last surface. As before, ocular magnification is the product $d_o\mathfrak{D}$, or as it is often written, MP $= (250 \text{ mm})/f$.

The **Huygens** ocular, which dates back over 250 years, is still in wide use today (Fig. 5.105), particularly in microscopy. The lens adjacent to the eye is known as the **eye-lens**, and the first lens in the ocular is the **field-lens**. The distance from the eye-lens to the eye point is known as the **eye relief**, and for the Huygens ocular, it's only an uncomfortable 3 mm or so. Notice that this ocular requires the incoming rays to be converging so as to form a virtual object for the eye-lens. Clearly then, the Huygens eyepiece cannot be used as an ordinary magnifier. Its contemporary appeal rests in its low purchase price (see Section 6.3.2). Another old standby is the **Ramsden** eyepiece (Fig. 5.106). This time the principal focus is in front of the field-lens, so the intermediate image will appear there in easy access. This is where you would place a *reticle* (or *reticule*), which might contain a set of cross hairs, precision scales, or angularly divided circular grids. (When these are formed on a transparent plate, they are often called *graticules*.) Since the reticle and intermediate image are in the same plane, both will be in focus at the same time. The roughly 12-mm eye relief is an advantage over the previous ocular. The Ramsden is relatively popular and fairly inexpensive (see Problem 6.2). The **Kellner** eyepiece represents a definite increase in image qual-

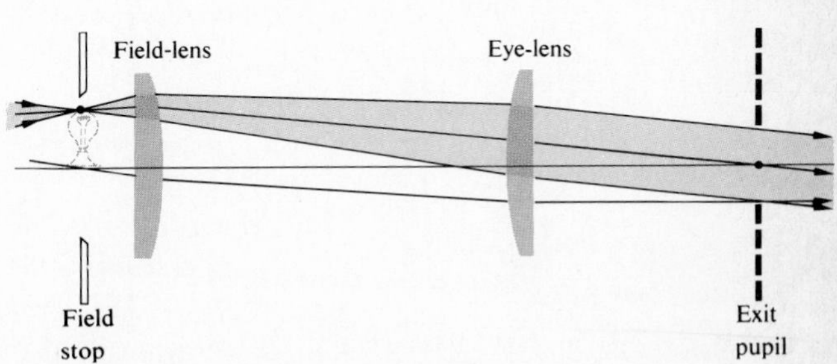

FIGURE 5.106 The Ramsden eyepiece.

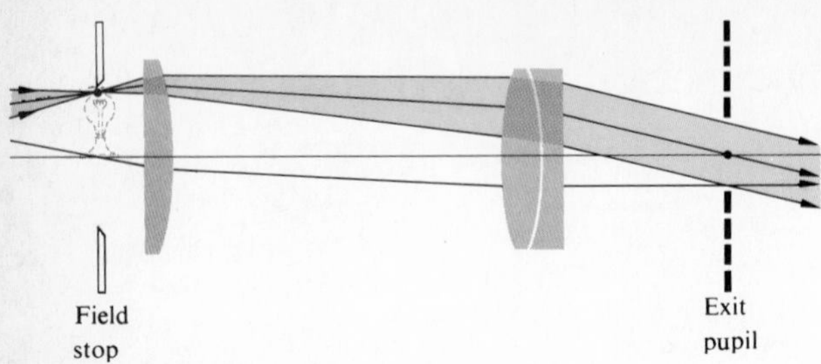

FIGURE 5.107 The Kellner eyepiece.

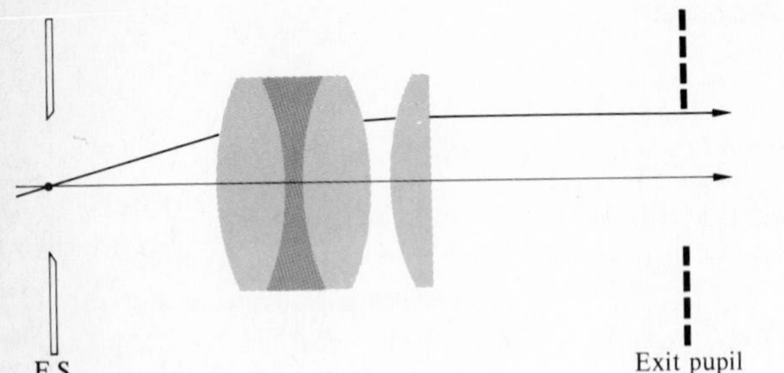

FIGURE 5.108 The orthoscopic eyepiece.

ity, although eye relief is between that of the previous two devices. The Kellner is essentially an achromatized Ramsden (Fig. 5.107). It is most commonly used in moderately wide-field telescopic instruments. The **orthoscopic** eyepiece (Fig. 5.108) has a wide field, high magnification, and long eye relief (≈20 mm). The **symmetrical (Plössl)** eyepiece (Fig. 5.109) has characteristics similar to those of the orthoscopic ocular but is generally somewhat superior to it. The **Erfle** (Fig. 5.110) is probably the most common wide-field (roughly ±30°) eyepiece. It is well corrected for all aberrations and comparatively expensive.*

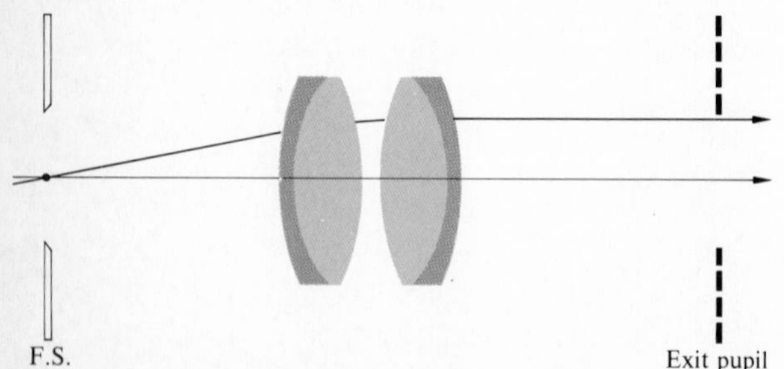

FIGURE 5.109 The symmetrical (Plössl) eyepiece.

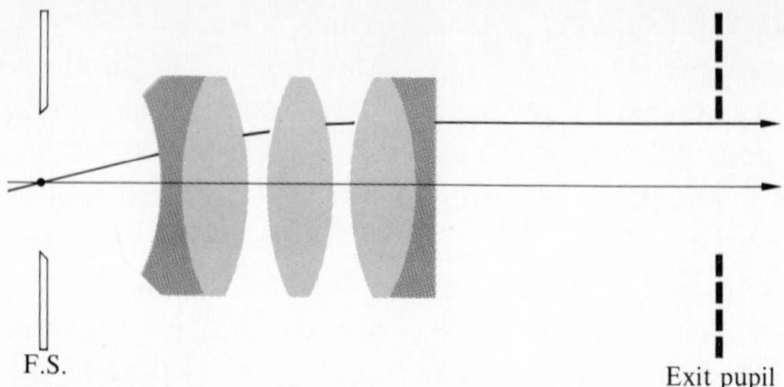

FIGURE 5.110 The Erfle eyepiece.

Although there are many other eyepieces, including variable-power *zoom* devices and ones with aspherical surfaces, those discussed here are representative. They are the devices you will ordinarily find on telescopes and microscopes and on long lists in the commercial catalogs.

5.7.5 The Compound Microscope

The compound microscope goes a step beyond the simple magnifier by providing higher angular magnification (greater than about 30×) of *nearby* objects. Its invention, which may have occurred as early as 1590, is generally attributed to a Dutch spectacle maker, Zacharias Janssen of Middleburg. Galileo runs a close second, having announced his invention of a compound microscope in 1610. A simple version, which is closer to these earliest devices than it is to a modern laboratory microscope, is depicted in Fig. 5.111. The lens system, here a singlet, closest to the object is referred to as the **objective**. It forms a real, inverted, and usually magnified image of the object. This image resides in space on the plane of the field stop of the eyepiece and has to be small enough to fit inside

*Detailed designs of these and other oculars can be found in the *Military Standardization Handbook—Optical Design*, MIL-HDBK-141.

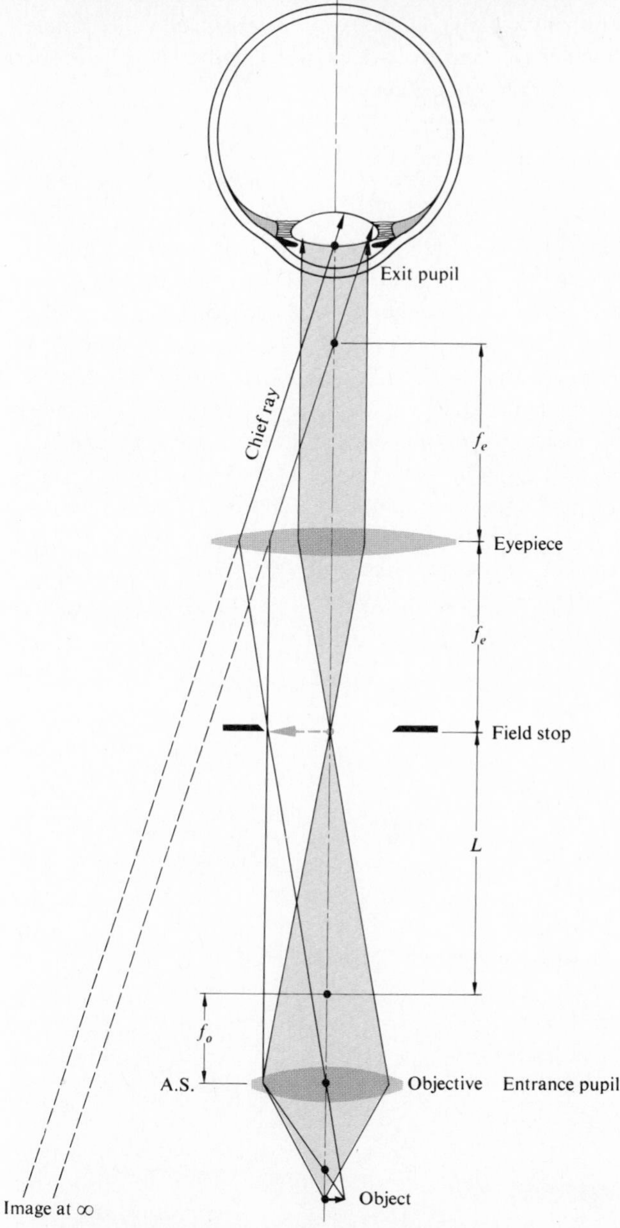

FIGURE 5.111 A rudimentary compound microscope. The objective forms a real image of a nearby object. The eyepiece, functioning like a magnifying glass, enlarges this intermediate image. The final virtual image can be bigger than the barrel of the device since it needn't fit inside.

the barrel of the device. Rays diverging from each point of this image will emerge from the eye-lens (which in this simple case is the eyepiece itself) parallel to each other, as noted in the previous section. The ocular magnifies the intermediate image still further. Thus the magnifying power of the entire system is the product of the transverse linear magnification of the objective, M_{To}, and the angular magnification of the eyepiece, M_{Ae}, that is,

$$\text{MP} = M_{To}M_{Ae} \tag{5.80}$$

The objective magnifies the object and brings it up in the form of a real image, where it can be examined as if through a magnifying glass.

Recall that $M_T = -x_i/f$, Eq. (5.26). With this in mind most, but not all, manufacturers design their microscopes such that the distance (corresponding to x_i) from the second focus of the objective to the first focus of the eyepiece is standardized at 160 mm. This distance, known as the **tube length**, is denoted by L in the figure. (Some authors define tube length as the image distance of the objective.) Hence, with the final image at infinity [Eq. (5.79)] and the standard near point taken as 254 mm (10 inches),

$$\text{MP} = \left(-\frac{160}{f_o}\right)\left(\frac{254}{f_e}\right) \tag{5.81}$$

Here the focal lengths are in millimeters, and the image is inverted (MP < 0). Accordingly, the barrel of an objective with a focal length f_o of, say, 32 mm will be engraved with the marking 5× (or ×5), indicating a power of 5. Combined with a 10× eyepiece ($f_e = 1$ inch), the microscope MP would then be 50×.

To maintain the distance relationships among the objective, field stop, and ocular, while a focused intermediate image of the object is positioned in the first focal plane of the eyepiece, all three elements are moved as a single unit.

The objective itself functions as the aperture stop and entrance pupil. Its image, formed by the eyepiece, is the exit pupil into which the eye is positioned. The field stop, which limits the extent of the largest object that can be viewed, is fabricated as part of the ocular. The image of the field stop formed by the optical elements following it is called the *exit window*, and the image formed by the optical elements preceding it is the *entrance window*. The cone angle subtended at the center of the exit pupil by the periphery of the exit window is said to be the *angular field of view in image space*.

A modern microscope objective can be roughly classified as one of three different kinds. It might be designed to work best with the object positioned below a cover glass, with no cover glass (metallurgical instruments), or with the object immersed in a liquid that is in contact with the objective. In some cases, the distinction is not critical, and the objective may be used with or without a cover glass. Four representative objectives are shown in Fig. 5.112 (see Section 6.3.1). In addition, the ordinary low-power (about 5×) cemented doublet achromate is quite common. Relatively inexpensive medium-power (10× or 20×) achromatic objectives, because of their short focal lengths, can conveniently be used when expanding and spatially filtering laserbeams.

There is one other characteristic quantity of importance, which must be mentioned here even if only briefly. The brightness of the image is, in part, dependent on the amount of light gathered in by the objective. The *f-number* is a useful parameter for describing this quantity, particularly when the object is a distant one (see Section 5.3.3). However, for an instrument working at *finite conjugates* (s_i and s_o both finite), the numerical aperture, NA, is more appropriate (see Section 5.6). In the present instance

$$\text{NA} = n_i \sin \theta_{\max} \tag{5.82}$$

where n_i is the refractive index of the immersing medium (air, oil, water, etc.) adjacent to the objective lens, and $\theta_{\max}$ is the half-angle of the maximum cone of light picked up by that lens (Fig. 5.112*b*). In other words, $\theta_{\max}$ is the angle made by a marginal ray with the axis. The numerical aperture is usually the second number etched in the barrel of the objective. It ranges from about 0.07 for low-power objectives to 1.4 or so for high-power (100×) ones. Of course, if the object is in the air, the numerical aperture cannot be greater than 1.0. Incidentally, Ernst Abbe (1840–1905), while working in the Carl Zeiss microscope workshop, introduced the concept of the numerical aperture. It was he who recognized that the minimum

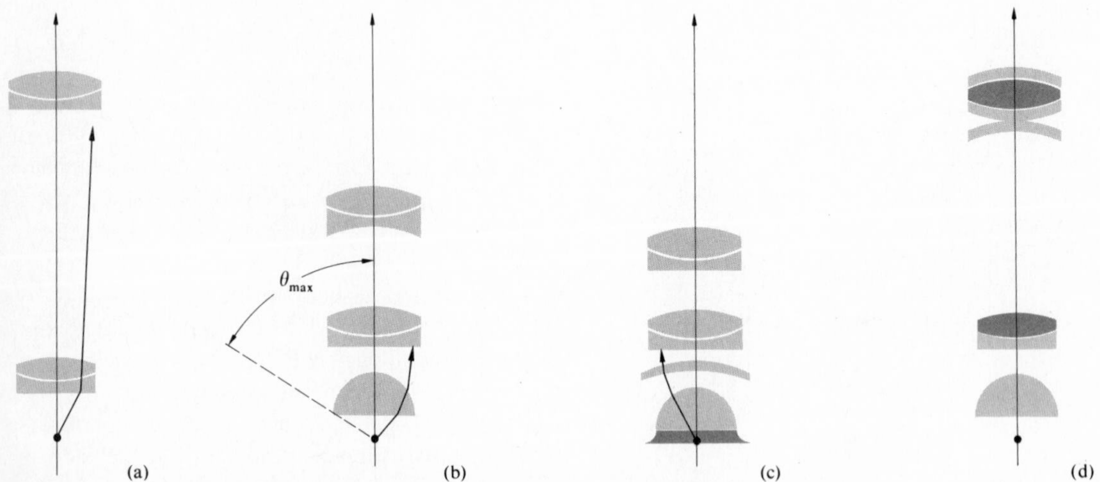

FIGURE 5.112 Microscope objectives: (*a*) Lister objective, 10×, NA = 0.25, f = 16 mm (two cemented achromates). (*b*) Amici objective, from 20×, NA = 0.5, f = 8 mm to 40×, NA = 0.8, f = 4 mm. (*c*) Oil-immersion objective, 100×, NA = 1.3, f = 1.6 mm (see Figure 6.18). (*d*) Apochromatic objective, 55×, NA = 0.95, f = 3.2 (contains two fluorite lenses).

transverse distance between two object points that can be resolved in the image, that is, the *resolving power*, varied directly as λ and inversely as the NA.

5.7.6 The Camera

The prototype of the modern photographic camera* was a device known as the *camera obscura*, the earliest form of which was simply a dark room with a small hole in one wall. Light entering the hole cast an inverted image of the sunlit outside scene on an inside screen. The principle was known to Aristotle, and his observations were preserved by Arab scholars throughout Europe's long Dark Ages. Alhazen utilized it to examine solar eclipses indirectly over eight hundred years ago. The notebooks of Leonardo da Vinci contain several descriptions of the obscura, but the first detailed treatment appears in *Magia naturalis* (*Natural Magic*) by Giovanni della Porta. He recommended it as a drawing aid, a function to which it was soon quite popularly put. Johannes Kepler, the renowned astronomer, had a portable tent version, which he used while surveying in Austria. By the latter part of the 1600s, the small hand-held camera obscura was commonplace. Note that the eye of the Nautilus, a little cuttlefish, is literally an open pinhole obscura, that simply fills with seawater on immersion.

By replacing the viewing screen with a photosensitive surface, such as a film plate, the obscura becomes a camera in the modern sense of the word. The first permanent photograph was made in 1826 by Joseph Nicéphore Niépce (1765–1833), who used a box camera with a small convex lens, a sensitized pewter plate, and roughly an eight-hour exposure.

The lensless pinhole camera (Fig. 5.113) is by far the least complicated device for the purpose, yet it has several endearing and, indeed, remarkable virtues. It can form a well-defined, practically undistorted image of objects across an extremely wide angular field (due to great depth of focus) and over a large range of distances (great depth of field). If initial-

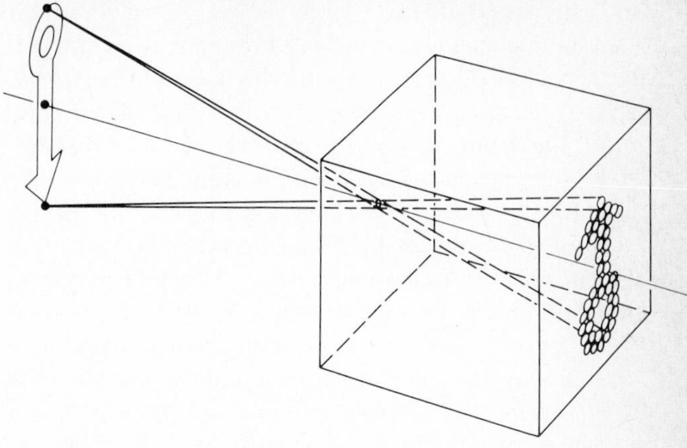

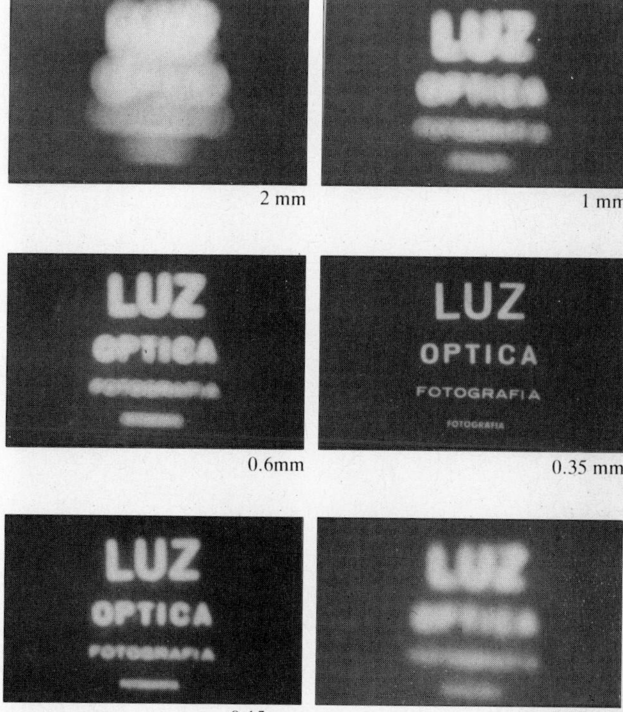

FIGURE 5.113 The pinhole camera. Note the variation in image clarity as the hole diameter decreases. (Photos courtesy Dr. N. Joel, UNESCO.)

*See W. H. Price, "The Photographic Lens," *Sci. Am.* (August 1976) p. 72.

ly the entrance pupil is very large, no image results. As it is decreased in diameter, the image forms and grows sharper. After a point, further reduction in the hole size causes the image to blur again, and one quickly finds that the aperture size for maximum sharpness is proportional to its distance from the image plane. (A hole with a 0.5-mm diameter at 0.25 m from the film plate is convenient and works well.) There is no focusing of the rays at all, so no defects in that mechanism are responsible for the drop-off in clarity. The problem is actually one of diffraction, as we shall see later on (Section 10.2.5). In most practical situations, the pinhole camera's one overriding drawback is that it is insufferably slow (roughly $f/500$). This means that exposure times will generally be far too long, even with the most sensitive films. The obvious exception is a stationary subject, such as a building (Fig. 5.114), for which the pinhole camera excels.

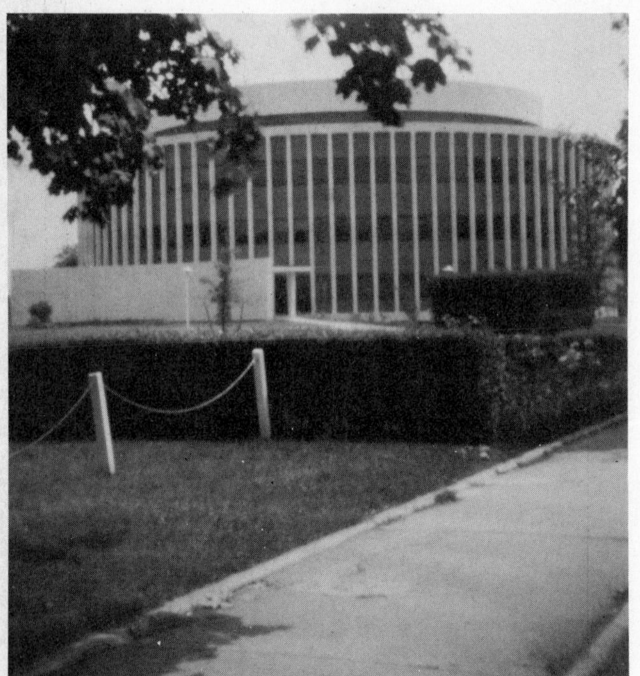

FIGURE 5.114 Photograph taken with a pinhole camera (Science Building, Adelphi University). Hole diameter 0.5 mm, film plane distance 24 cm, A.S.A. 3000, shutter speed 0.25 s. Note depth of field. (E.H.)

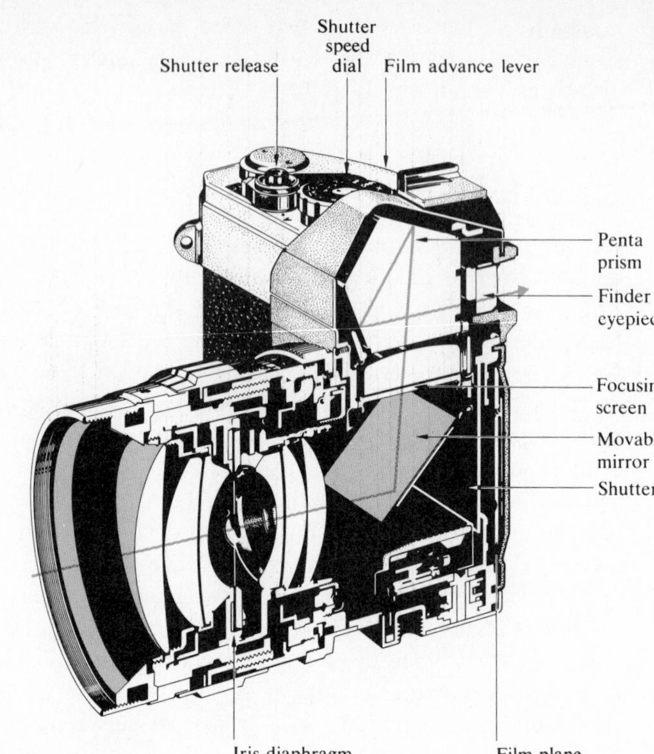

FIGURE 5.115 A single-lens reflex camera.

Figure 5.115 depicts the essential components of a popular and representative modern camera—the single-lens reflex, or SLR. Light traversing the first few elements of the lens then passes through an iris diaphragm, used in part to control the exposure time or, equivalently, the *f-number*; it is in effect a variable-aperture stop. On emerging from the lens, light strikes a movable mirror tilted at 45°, then goes up through the focusing screen to the penta prism and out the finder eyepiece. When the shutter release is pressed, the diaphragm closes down to a preset value, the mirror swings up out of the way, and the focal-plane shutter opens, exposing the film. The shutter then closes, the diaphragm opens fully, and the mirror drops back in place. Nowadays SLR systems have any one of a number of built-in light-meter arrangements, which are automatically coupled to the diaphragm and shutter, but those components are excluded from the diagram for the sake of simplicity.

To focus the camera, the entire lens is moved toward or away from the film plane. Since its focal length is fixed, as s_o varies, so too must s_i. The *angular field of view* can loosely be thought of as relating to the fraction of the scene included in the photograph. It is furthermore required that the entire photograph surface correspond to a region of satisfactory image quality. More precisely, the angle subtended at the lens, by a circle encompassing the film area, is the angular field of view φ (Fig. 5.116). As a rough but reasonable approximation of a common arrangement, take the diagonal distance across the film to equal the focal length. Thus $\varphi/2 \approx \tan^{-1}\frac{1}{2}$; that is, $\varphi \approx 53°$. If the object comes in from infinity, s_i must increase. The lens is then backed away from the film plate to keep the image in focus, and the field of view, as recorded on the film whose periphery is the field stop, decreases. A **standard** SLR lens has a focal length in the range of about 50 to 58 mm and a field of view of 40° to 50°. With the film size kept constant, a reduction of f results in a wider field angle. Accordingly, **wide-angle** SLR lenses range from $f \approx 40$ mm down to about 6 mm, and φ goes from about 50° to a remarkable 220° (the latter being a special-purpose lens wherein distortion is unavoidable). The *telephoto* has a long focal length, roughly 80 mm or more. Consequently, its field of view drops off rapidly, until it is only a few degrees at $f \approx 1000$ mm.

The standard photographic objective must have a large relative aperture, $1/(f/\#)$, to keep exposure times short. Moreover, the image is required to be flat and undistorted, and the lens should have a wide angular field of view as well. The evolution of a modern lens still begins with a creative insight that leads to a promising new form. In the past, these were laboriously perfected relying on intuition, experience, and, of course trial and error with a succession of developmental lenses. Today, for the most part, the computer serves this function without the need of numerous prototypes.

Many contemporary photographic objectives are variations of well-known successful forms. Figure 5.117 illustrates the general configuration of several important lenses, roughly progressing from wide angle to telephoto. Particular specifications are not given, because variations are numerous. The *Aviogon* and *Zeiss Orthometer* are wide-angle lenses, whereas the *Tessar* and *Biotar* are often standard lenses. The *Cooke triplet*, described in 1893 by H. Dennis Taylor of Cooke and Sons, is still being made (note the similarity with the Tessar). It contains the smallest number of elements by which all seven third-order aberrations can essentially be made to vanish. Even earlier (ca. 1840), Josef Max Petzval designed what was then a rapid (portrait) lens for Voightländer and Son. Its modern offshoots are myriad.

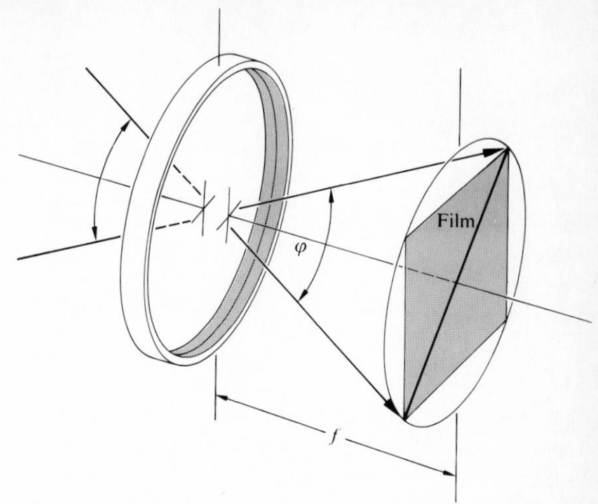

FIGURE 5.116 Angular field of view when focused at infinity.

5.7.7 The Telescope

It is not at all clear who actually invented the telescope. In point of fact, it was probably invented and reinvented many times. Recall that by the seventeenth century spectacle lenses had been in use in Europe for about three hundred years. During that long span of time, the fortuitous juxtapositioning of two appropriate lenses to form a telescope seems almost inevitable. In any event, it is most likely that a Dutch optician, possibly even the ubiquitous Zacharias Jenssen of microscope fame, first constructed a telescope and in addition had inklings of the value of what he was peering into. The earliest indisputable evidence of the discovery, however, dates to October

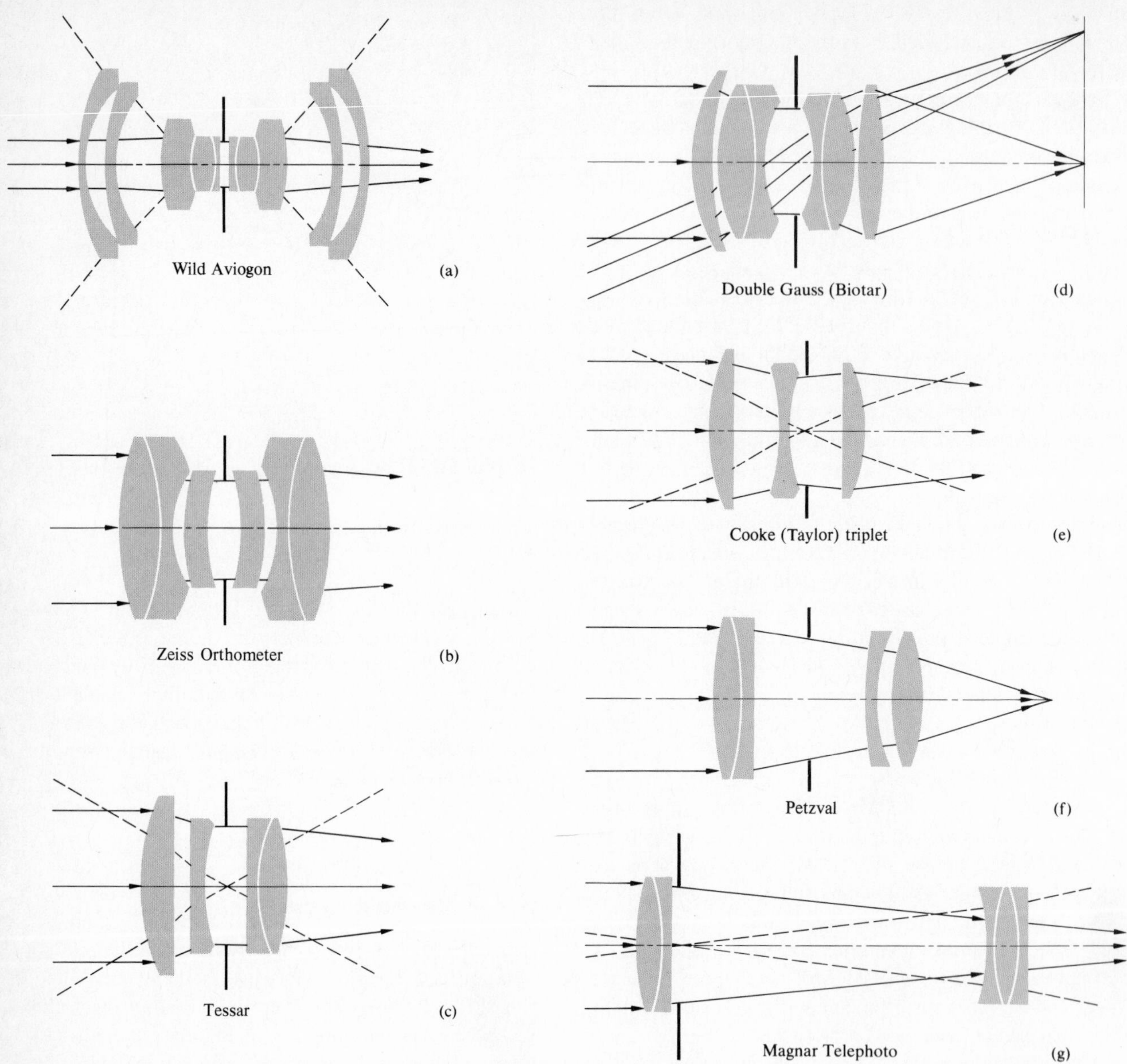

FIGURE 5.117 Camera lenses.

2, 1608, when Hans Lippershey petitioned the States-General of Holland for a patent on a device for seeing at a distance (which is what *teleskopos* means in Greek). As you might have guessed, its military possibilities were immediately recognized. His patent was therefore not granted; instead the government purchased the rights to the instrument, and he received a commission to continue research. Galileo heard of this work, and by 1609 he had fashioned a telescope of his own, using two lenses and an organ pipe as a tube. It was not long before he had constructed a number of greatly improved instruments and was astounding the world with the astronomical discoveries for which he is famous.

Refracting Telescopes

A simple **astronomical telescope** is shown in Fig. 5.118. Unlike the compound microscope, which it closely resembles, its primary function is to enlarge the retinal image of a *distant* object. In the illustration, the object is at a finite far distance from the objective, so that the real intermediate image is formed just beyond its second focal point. This image will be the object for the next lens system, that is, the ocular. It follows from Table 5.3 (p. 165) that if the eyepiece is to form a virtual magnified final image (within the range of normal accommodation), the object distance must be less than or equal to the focal length, f_e. In practice, *the position of the intermediate image is fixed, and only the eyepiece is moved to focus the instrument*. Notice that *the final image is inverted*, but as long as the scope is used for astronomical observations, this is of little consequence, especially since most work is photographic.

At great object distances the incident rays are effectively parallel—the intermediate image resides at the second focus of the objective. Usually, the eyepiece is located so that its first focus overlaps the second focus of the objective, in which case rays diverging from a point on the intermediate image will leave the ocular parallel to each other. A normal viewing eye can then focus the rays in a relaxed configuration. If the eye is nearsighted or farsighted, the ocular can be moved in or out so that the rays diverge or converge a bit to compensate. (If you are astigmatic, you'll have to keep your glasses on when using ordinary visual instruments.) We saw earlier (Section 5.2.3) that both the back and front focal lengths of a thin-lens combination go to infinity when the two lenses are separated by a

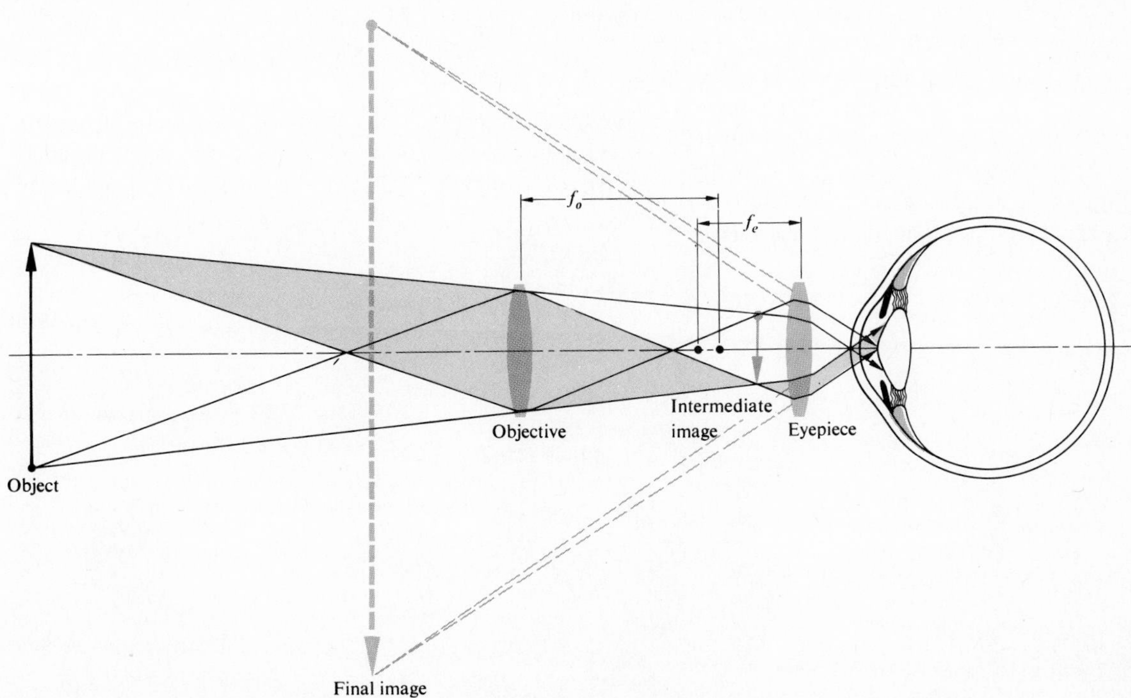

FIGURE 5.118 Keplerian astronomical telescope (accommodating eye).

distance d equal to the sum of their focal lengths (Fig. 5.119). The astronomical telescope in this configuration of *infinite conjugates* is said to be *afocal*, that is, without a focal length. As a side note, if you shine a collimated (parallel rays, i.e., plane waves) narrow laserbeam into the back end of a scope focused at infinity, it will emerge still collimated but with an increased cross section. It is often desirable to have a broad, quasimonochromatic, plane-wave beam, and specific devices of this sort are now available commercially.

The periphery of the objective is the aperture stop, and it encompasses the entrance pupil as well, there being no lenses to the left of it. If the telescope is trained directly on some distant galaxy, the visual axis of the eye will presumably be colinear with the central axis of the scope. The entrance pupil of the eye should then coincide in space with the exit pupil of the scope. However, the eye is not immobile. It will move about scanning the entire field of view, which quite often contains many points of interest. In effect, the eye examines different regions of the field by rotating so that rays from a particular area fall on the fovea centralis. The direction established by the chief ray through the center of the entrance pupil to the fovea centralis is the *primary line of sight*. The axial point, fixed in reference to the head, through which the primary line of sight always passes, regardless of the orientation of the eyeball, is called the *sighting intersect*. When it is desirable to have the eye surveying the field, the sighting intersect should be positioned at the center of the telescope's exit pupil. In that case, the primary line of sight will always correspond to a chief ray through the center of the exit pupil, however the eye moves.

Suppose that the margin of the visible object subtends a half-angle of α at the objective (Fig. 5.120). This is essentially the same as the angle α_u, which would be subtended at the unaided eye. As in previous sections, the angular magnification is

$$\text{MP} = \frac{\alpha_a}{\alpha_u} \qquad [5.75]$$

Here α_u and α_a are measures of the field of view in object and image space, respectively. The first is the half-angle of the actual cone of rays collected, and the second relates to the apparent cone of rays. If a ray arrives at the objective with a negative slope, it will enter the eye with a positive slope and vice versa. To make the sign of MP positive for erect images, and therefore consistent with previous usage (Fig. 5.103), either α_u or α_a must be taken to be negative—we choose the former because the ray has a negative slope. Observe that the ray passing through the first focus of the objective passes through the second focus of the eyepiece; that is, F_{o1} and F_{e2} are conjugate points. In the paraxial approximation $\alpha \approx \alpha_u \approx \tan \alpha_u$ and $\alpha_a \approx \tan \alpha_a$. The image fills the region of the field stop, and half its extent equals the distance $\overline{BC} = \overline{DE}$. Thus, from triangles $F_{o1}BC$ and $F_{e2}DE$, the ratio of the tangents yields

$$\text{MP} = -\frac{f_o}{f_e} \qquad (5.83)$$

Another convenient expression for the MP comes from considering the transverse magnification of the ocular. Inasmuch as the exit pupil is the image of the objective (Fig. 5.120), we have

$$M_{Te} = -\frac{f_e}{x_o} = -\frac{f_e}{f_o}$$

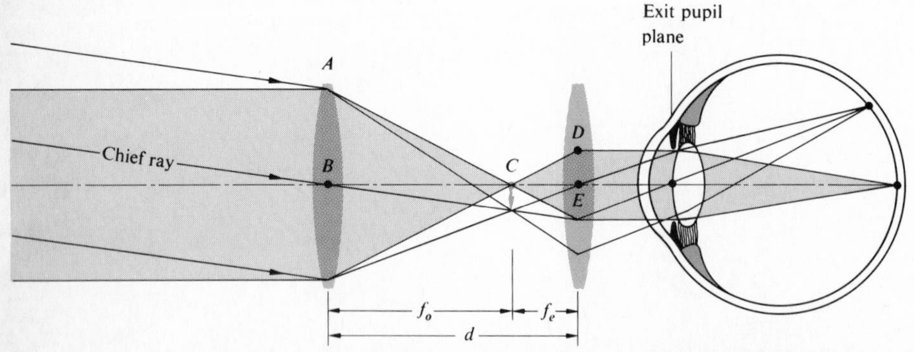

FIGURE 5.119 Astronomical telescope—infinite conjugates.

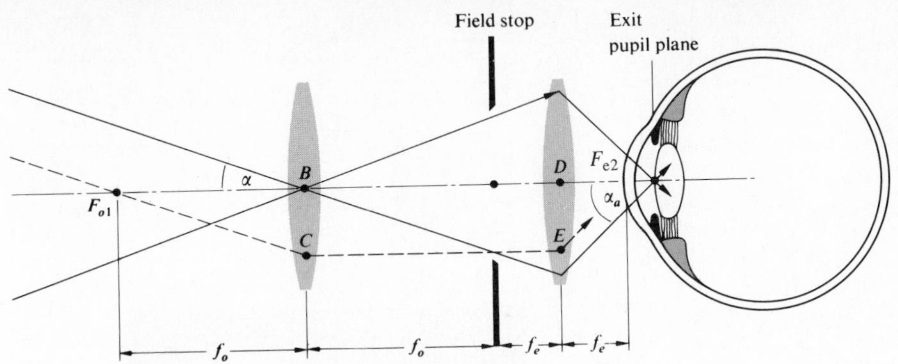

FIGURE 5.120 Ray angles for a telescope.

Furthermore, if D_o is the *diameter of the objective* and D_{ep} is the *diameter of its image, the exit pupil*, then $M_{Te} = D_{ep}/D_o$. These two expressions for M_{Te} compared with Eq. (5.83) yield

$$\text{MP} = \frac{D_o}{D_{ep}} \tag{5.84}$$

Here D_{ep} is actually a negative quantity, since the image is inverted. It is an easy matter to build a simple refracting scope by holding a lens with a long focal length in front of one with a short focal length and making sure that $d = f_o + f_e$. But again, well-corrected telescopic instruments generally have multi-element objectives, usually doublets or triplets.

To be useful when the orientation of the object is of importance, a scope must contain an additional **erecting system**; such an arrangement is known as a **terrestrial telescope**. A single erecting lens or lens system is usually located between the ocular and objective, with the result that the image is right side up. Figure 5.121 shows one with a cemented doublet objective and a Kellner eyepiece. It will obviously have to have a long draw tube, the picturesque kind that comes to mind when you think of wooden ships and cannonballs.

For that reason, **binoculars** (binocular telescopes) generally utilize erecting prisms, which accomplish the same thing in less space and also increase the separation of the objectives, thereby enhancing the stereoscopic effect. Most often these are double Porro prisms, as in Fig. 5.122 (notice the involved modified Erfle eyepiece, the wide field stop, and the achromatic doublet objective). Binoculars customarily bear several numerical markings, for example, 6×30, 7×50, or 20×50. The initial number is the magnification, here $6\times$, $7\times$, or $20\times$. The second number is the entrance-pupil diameter or, equivalently, the clear aperture of the objective, expressed in millimeters. It follows from Eq. (5.84) that the exit-pupil diameter will be the second number divided by the first, or in this case

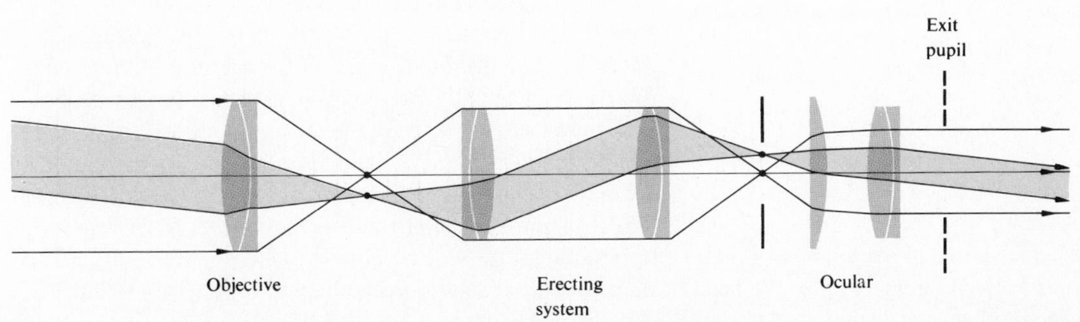

FIGURE 5.121 A terrestrial telescope.

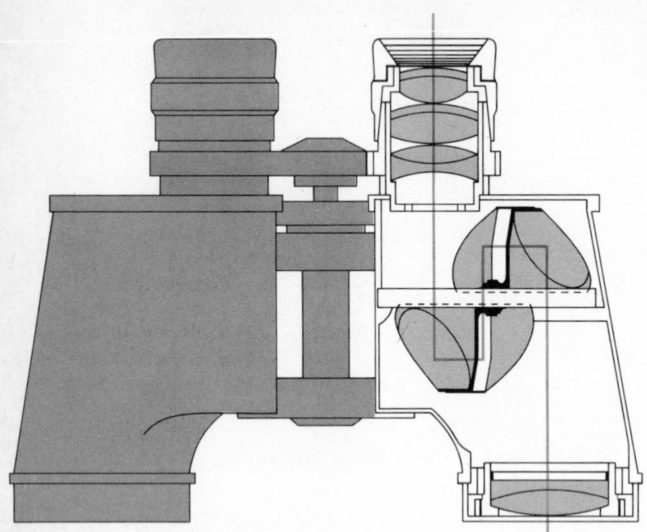

FIGURE 5.122 A binocular.

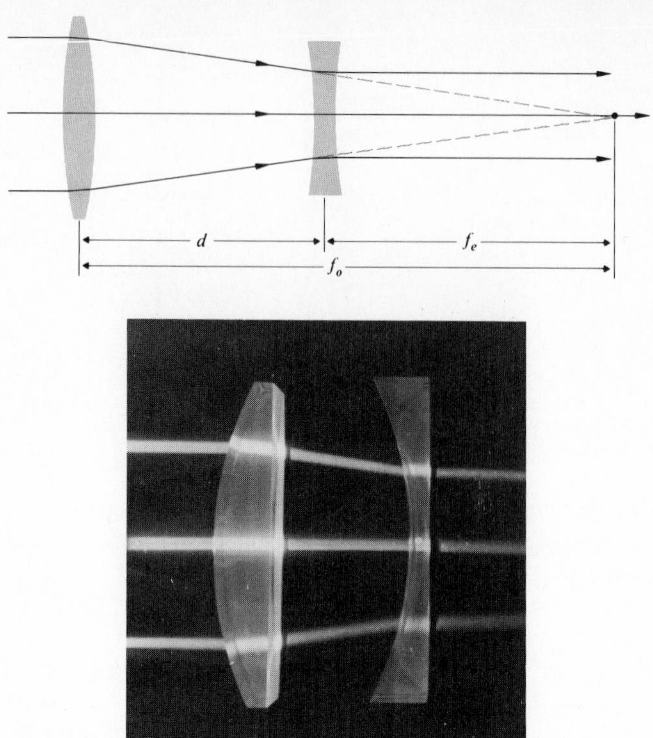

FIGURE 5.123 The Galilean telescope. Galileo's first scope had a planar-convex objective (5.6 cm in diameter, f = 1.7 m, R = 93.5 cm) and a planar-concave eyepiece, both of which he ground himself. It was 3× in contrast to his last scope, which was 32×. (E.H.)

5, 7.1, and 2.5, all in millimeters. You can hold the instrument away from your eye and see the bright circular exit pupil surrounded by blackness. To measure it, focus the device at infinity, point it at the sky, and observe the emerging sharp disk of light, using a piece of paper as a screen. Determine the eye relief while you're at it.

By the way, as long as $d = f_o + f_e$, the scope will be afocal, even if the eyepiece is negative (i.e., $f_e < 0$). The telescope built by Galileo (Fig. 5.123) had just such a negative lens as an eyepiece and therefore formed an erect image [$f_e < 0$ and MP > 0 in Eq. (5.83)]. As a telescope, the system is now mainly of historical and pedagogical interest, although one can still purchase two such scopes mounted side by side to form a Galilean field glass. It is quite useful, however, as a laserbeam expander, because it has no internal focal points where a high power beam would otherwise ionize the surrounding air.

Reflecting Telescopes

Put rather simply, a telescope should allow us to see things *clearly* that are far away, and often extremely *faint*. We need to be able to resolve fine details, that is, to distinguish separate features that are small and close together, such as the two stars in a binary system. A spy satellite that can spot people walking around is highly desirable, but one that will identify their military service from the markings on their uniforms is even better. The measure of that ability is the **resolution**, and it increases with the diameter (D) of the aperture admitting light into the system. All other things being equal (under ideal seeing conditions), a large-diameter telescope will have better resolution than a small-diameter telescope. There's another even more compelling reason to increase the size of the aperture: to improve the **light-gathering power**. A telescope with a large aperture will be able to collect more light and see fainter more distant objects than an otherwise identical but smaller one.

The difficulties inherent in making large lenses are underscored by the fact that the largest refracting instrument is the 40-inch Yerkes telescope in Williams Bay, Wisconsin, whereas the reflector on Mount Palomar in southwestern California is 200 inches in diameter. The problems are evident; a lens must be transparent and free of internal flaws such as bubbles. A front-surfaced mirror obviously need not be; indeed, it need not even be transparent. A lens can be supported only by its rim and may sag under its own weight; a mirror can be supported by its rim and back as well. Furthermore, since there is no refraction and therefore no effect on the focal length due to the wavelength dependence of the index, mirrors suffer no chromatic aberration. For these and other reasons (e.g., their frequency response), reflectors predominate in large telescopes.

Invented by the Scotsman James Gregory (1638–1675) in 1661, the reflecting telescope was first successfully constructed by Newton in 1668, and only became an important research tool in the hands of William Herschel a century later. Figure 5.124 depicts a number of reflector arrangements, each having a concave paraboloidal primary mirror. The venerable 200-inch Hale telescope is so large that a little enclosure, where an observer can sit, is positioned at the prime focus (Fig. 5.124*a*). In the Newtonian version (Fig. 5.124*b*) a plane mirror or prism brings the beam out at right angles to the axis of the scope, where it can be photographed, viewed, spectrally analyzed, or photoelectrically processed. In the classical Gregorian arrangement (Fig. 5.124*c*), which is not particularly popular, a concave ellipsoidal secondary mirror reinverts the image, returning the beam through a hole in the primary. The classical Cassegrain system (Fig. 5.124*d*) utilizes a convex hyperboloidal secondary mirror to increase the effective focal length (refer back to Fig. 5.51, p. 183). It functions as if the primary mirror had the same aperture but a larger focal length or radius of curvature.

The simple single-mirror paraboloidal telescope (Fig. 5.124*a*) was designed to function with rays entering along its optical axis. But there will always be objects of interest elsewhere in the field of view other than at its direct center. When a parallel bundle of off-axis rays are reflected by a paraboloid, they do not all meet at the same point. The image of a distant off-axis point (e.g., a star) is an off-axis asymmetric blur caused by the combined aberrations of *coma* (p. 261) and *astigmatism* (p. 264). This blurring becomes unacceptable rather quickly as the object moves off axis; that's especially

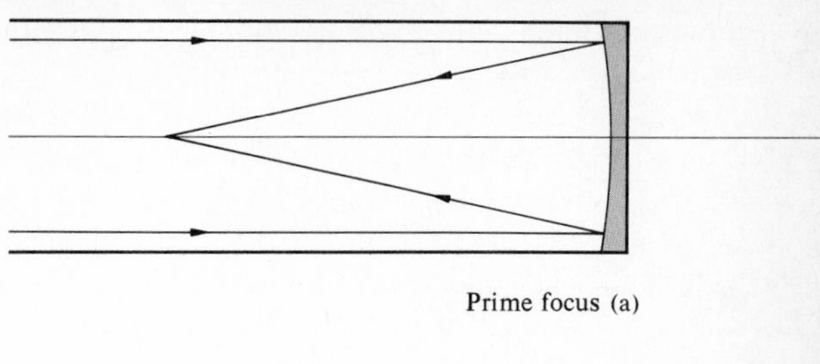

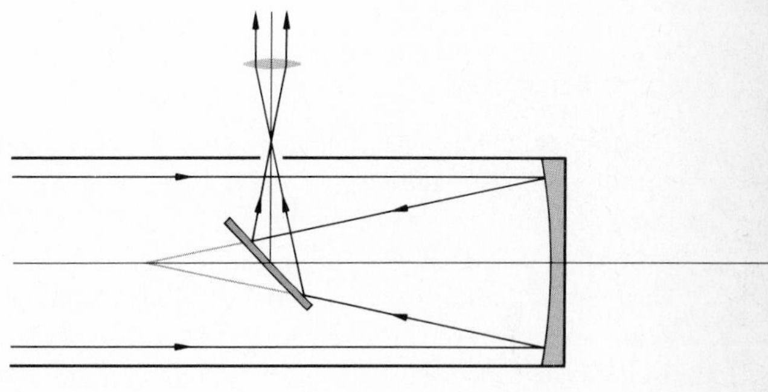

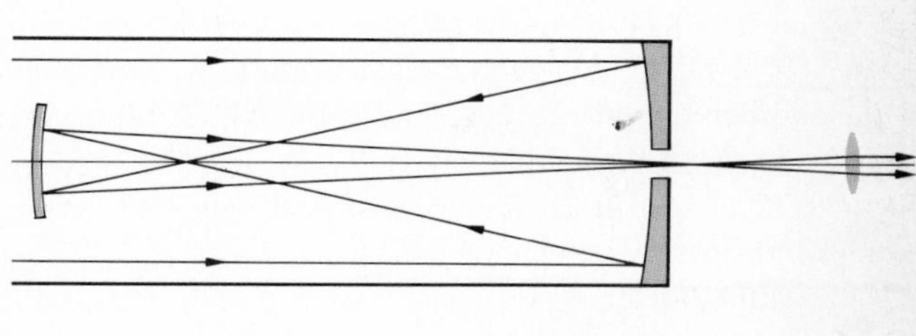

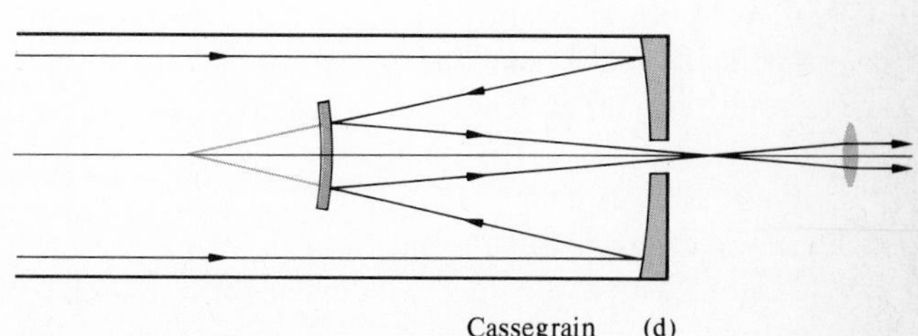

FIGURE 5.124 Reflecting telescopes.

true for the contribution due to coma, and it ends up limiting the acceptable field of view to something quite narrow. Even for a slow $f/10$ system, the angular radius of the acceptable field of view is only about 9 arc-minute off-axis, and it drops to a mere 1.4 arc-minutes at $f/4$. The classical two-mirror telescopes (Figs. 5.124*b*, *c*, and *d*) are similarly severely limited in their fields of view by coma.

Aplanatic Reflectors

An optical system that has negligible amounts of both spherical aberration (p. 258) and coma is called an **aplanat**, and there are aplanatic versions of both the Cassegrain and the Gregorian scopes. The Ritchey-Chrétien telescope is an aplanatic Cassegrain having a hyperboloidal primary and secondary. In recent times, this configuration has become the leading choice for devices with apertures of 2 m or more. Perhaps the best known example of its kind is the 2.4-m Hubble Space Telescope (HST), pictured in Fig. 5.125. Only telescopes in space (i.e., above the absorbing atmosphere) can work efficiently in the ultraviolet—which, for example, is where one would like to examine hot young stars. With its newly installed (1993) charge-coupled-device (CCD) arrays, the HST can "see" from about 1 μm in the IR to 121.6 nm in the UV. This complements ground-based telescopes that can provide diffraction-limited imaging in the wavelength range greater than 10 μm. (Incidentally, CCDs have a sensitivity about 50 times greater than otherwise comparable photographic film; the era of dropping film packs out of spy satellites is over.)

With little or no coma, the field of view of the Ritchey-Chrétien is limited by astigmatism. Thus an $f/10$ instrument will have an acceptable angular field radius of about 18 arc-minutes, twice that of an equivalent paraboloidal telescope. In comparison to the aplanatic Gregorian, the Ritchey-Chrétien has a smaller secondary and therefore blocks less light, and is substantially shorter in length; both features make it much more desirable.

Because an instrument can only collect a portion of the incident wavefront to be reformed into an image, there will always be diffraction: the light will deviate from straight-line propagation and spread out somewhat in the image plane. When an optical system with a circular aperture receives plane waves, rather than there being an image "point" (whatever that means), the light actually spreads out into a tiny circular spot (called an Airy disk, containing about 84% of the energy) surrounded by very faint rings. The radius of the Airy disk determines the overlapping of neighboring images and therefore the resolution. That's why an imaging system that is as perfect as possible is referred to as **diffraction limited**.

For a perfect instrument, the ideal theoretical angular resolution is given by Eq. (10.59), namely, the radius of the Airy disk, $1.22\lambda/D$ radians. Here D is the diameter of the instrument in the same units as λ. Another way to present the angular resolution is in arc-seconds, in which case it equals $2.52 \times 10^5\lambda/D$. Because of atmospheric distortions, ground-based telescopes, regardless of their size, seldom have angular resolutions better than about 1 arc-second. That is, the images of two stars separated by an angle of less than 1 arc-second blend into an undecipherable blur. By comparison, the HST, high above the atmosphere, for which $D = 2.4$ m, has a diffraction-limited angular resolution at $\lambda = 500 \times 10^{-9}$ m, of about 0.05 arc-second.

The world's two largest telescopes are the twin Keck aplanatic Cassegrains. Separated by 85 m, these two giant telescopes are perched atop the extinct volcano Mauna Kea in Hawaii, at an altitude of 13 600 feet. Each has a 10-m hyperboloidal primary composed of 36 hexagonal elements. These are deeply curved so that the $f/1.75$ system has a focal length of only 17.5 m. This is indicative of the new generation of large telescopes that tend to have fast mirrors (less than $f/2$) with relatively small focal lengths. Short telescopes are more economical to build and house, and are more stable and accurately steered.

The technology now exists for interferometrically combining the images from several separate optical telescopes, thereby tremendously increasing the overall effective aperture. A new generation of ground-based optical telescope arrays is destined to contribute significantly to the way we see the Universe.

Catadioptric Telescopes

A combination of reflecting (*catoptric*) and refracting (*dioptric*) elements is called a *catadioptric* system. The best known of these, although not the first, is the classic *Schmidt optical*

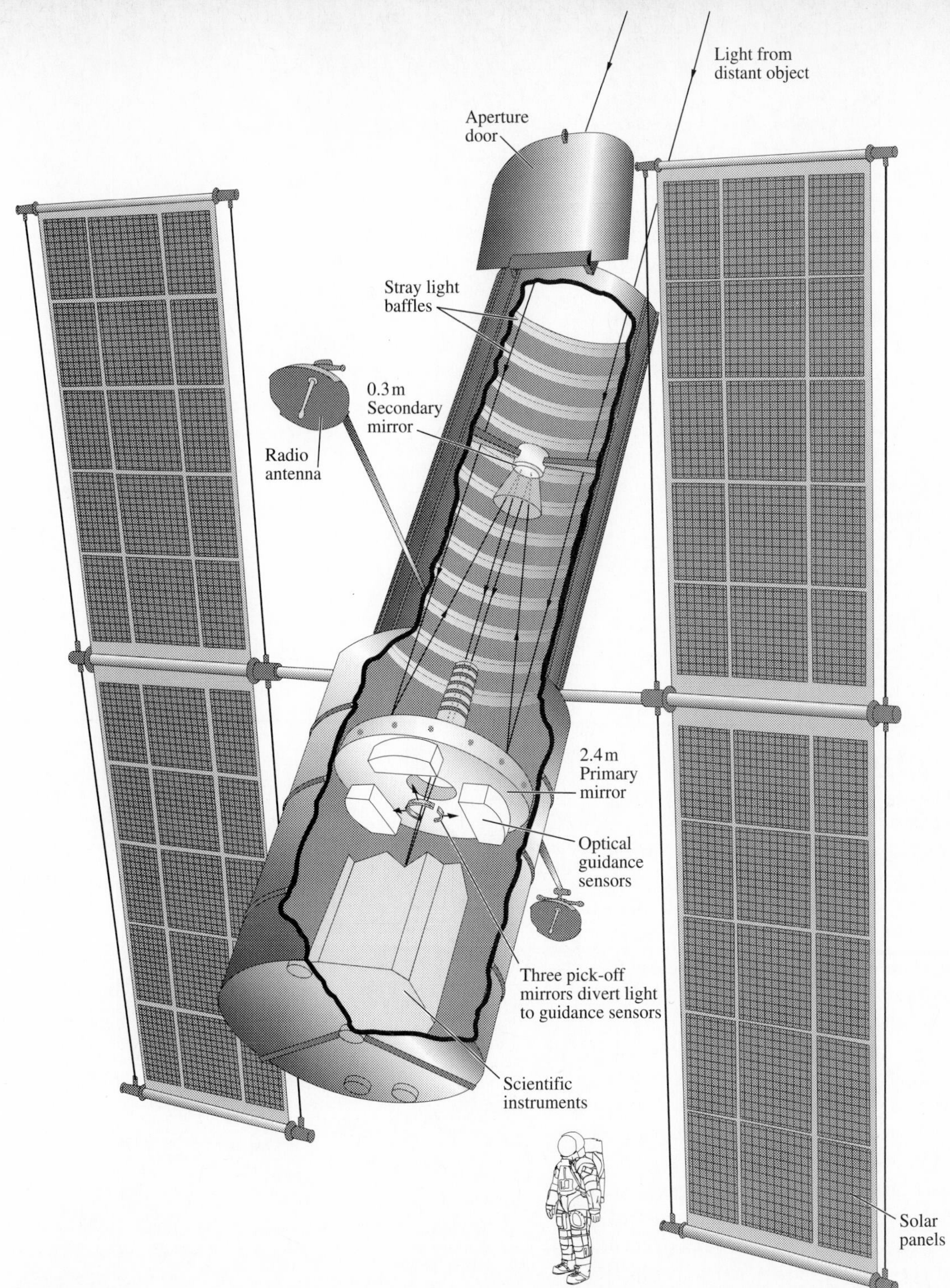

FIGURE 5.125 The Hubble Space Telescope. The spacecraft is 13 m (43 ft) long (about 16 ft from the primary to the secondary) and has a mass of 11 600 kg. It's in a 599-km by 591-km orbit with a period of 96 minutes. The primary mirror of the HST is pictured in Fig. 5.52.

system. We must treat it here, even if only briefly, because it represents an important approach to the design of large-aperture, extended-field reflecting systems. As seen in Fig. 5.126, bundles of parallel rays reflecting off a spherical mirror will form images, let's say of a field of stars, on a spherical image surface, the latter being a curved film plate in practice. The only problem with such a scheme is that although it is free of other aberrations (astigmatism and coma, see Section 6.3.1), we know that rays reflected from the outer regions of the mirror will not arrive at the same focus as those from the paraxial region. In other words, the mirror is a sphere, not a paraboloid, and it suffers *spherical aberration* (Fig. 5.126*b*). If this could be corrected, the system (in theory at least) would be capable of perfect imagery over a wide field of view. Since there is no one central axis, there are, in effect, no off-axis points. Recall that the paraboloid forms perfect images only at axial points, the image deteriorating rapidly off axis.

One evening in 1929, while sailing on the Indian Ocean (returning from an eclipse expedition to the Philippines), Bernhard Voldemar Schmidt (1879–1935) showed a colleague a sketch of a system he had designed to cope with the spherical aberration of a spherical mirror. He would use a thin glass corrector plate on whose surface would be ground a very shallow toroidal curve (Fig. 5.126*c*). Light rays traversing the outer regions would be deviated by just the amount needed to be sharply focused on the image sphere. The corrector must overcome one defect without introducing appreciable amounts of other aberrations. This first system was built in 1930, and in 1949 the famous 48-inch Schmidt telescope of the Palomar Observatory was completed. It is a fast ($f/2.5$), wide-field device, ideal for surveying the night sky. A single photograph could encompass a region the size of the bowl of the Big Dipper—this compared with roughly 400 photographs by the 200-inch reflector to cover the same area.

Major advances in the design of catadioptric instrumentation have occurred since the introduction of the original Schmidt system. There are now catadioptric satellite and missile tracking instruments, meteor cameras, compact commercial telescopes, telephoto objectives, and missile-homing guidance systems. Innumerable variations on the theme exist; some replace the correcting plate with concentric meniscus lens arrangements (Bouwers–Maksutov), and others use solid thick mirrors. One highly successful approach utilizes a triplet aspheric lens array (Baker).

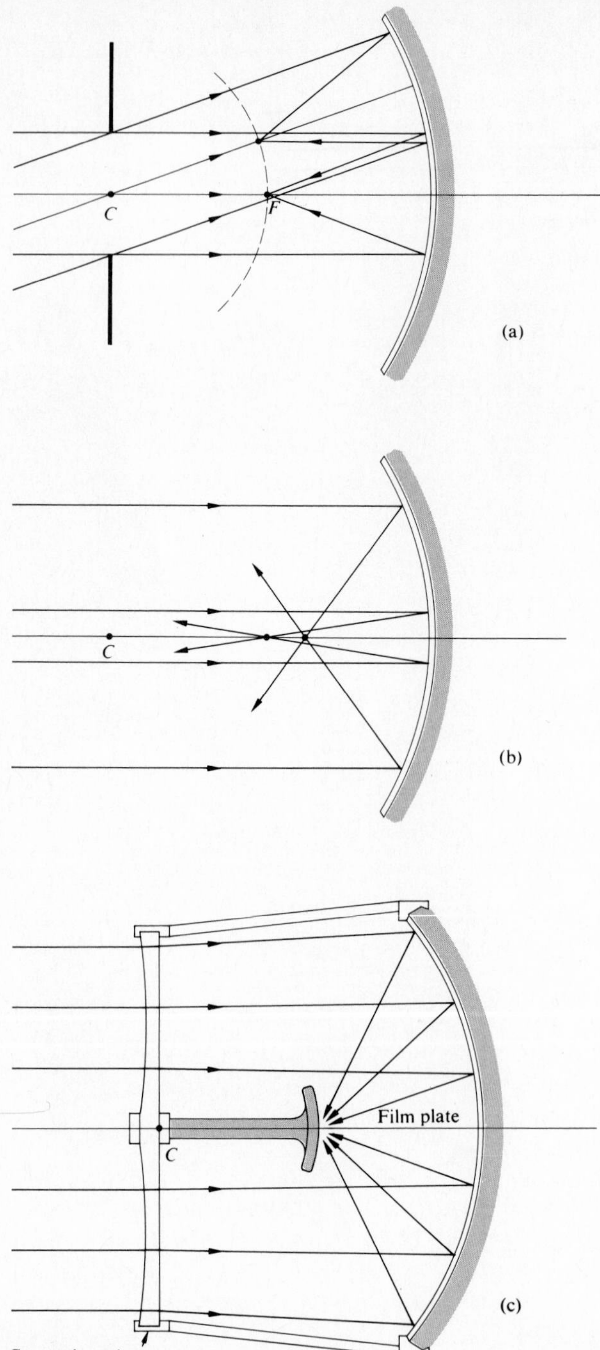

FIGURE 5.126 The Schmidt optical system.

5.8 Wavefront Shaping

This chapter has been about reshaping wavefronts in one way or another, but the changes introduced by traditional lenses and mirrors are global, affecting the whole processed portion of the wavefront in more-or-less the same way. By contrast, for the first time it is now possible to take an incoming wavefront and reconfigure every portion of it differently to fit specific needs.

Consider a plane wave passing either through some inhomogeneous medium of index $n(\mathbf{r})$ or through a medium of nonuniform thickness—a piece of shower-door glass will do (Fig. 5.127*a*). The wavefronts are essentially held back in proportion to the *OPL* and distort accordingly. When, for example, such a wrinkled wave reflects from an ordinary planar mirror, it goes off reversed in direction but otherwise unchanged (Fig. 5.127*b*). The leading and trailing wavefront regions remain leading and trailing, with only the direction of propagation reversed; the wavefront remains distorted. The scene beyond a crinkled-glass shower door is equally blurred whether you look at it directly or in a mirror.

If a more sophisticated mirror could be devised that could reshape the reflected wavefronts, we might be able to get rid of undesirable distortions that are unavoidably introduced in a variety of situations. This section explores two different state-of-the-art techniques for accomplishing just that.

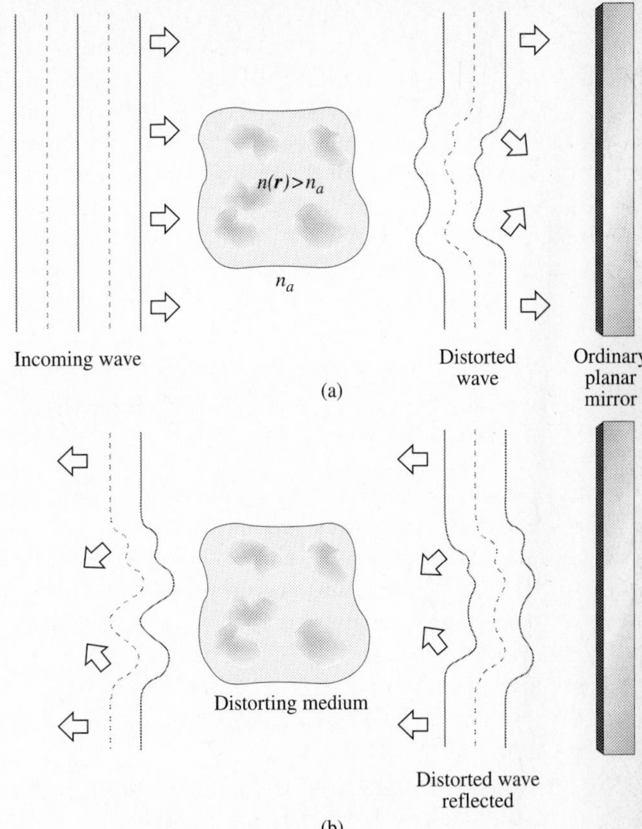

FIGURE 5.127 (*a*) A plane wave becomes distorted on passing through an inhomogeneous medium. (*b*) When such a wrinkled wave reflects off a traditional mirror, it changes direction. Regions that were originally leading or trailing remain that way as the wave, still wrinkled, moves off in a new direction. Passing through the inhomogeneous medium a second time increases the distortion.

5.8.1 Adaptive Optics

One of the most significant breakthroughs in telescope technology to occur in recent times is called **adaptive optics**, and it has already begun to provide a way to deal with the daunting problem of atmospheric distortion. As Newton put it, "If the Theory of making Telescopes could at length be fully brought into Practice, yet there would be certain Bounds beyond which telescopes could not perform. For the Air through which we look upon the Stars, is in perpetual Tremor; as may be seen by the tremulous Motion of Shadows cast from high Towers, and by the twinkling of the fix'd Stars." Adaptive optics is a methodology used to control that "perpetual Tremor"—first by measuring the turbulence-induced distortions of the incident light, and then by using that information to reconfigure the lightwave, bringing it back to a pristine condition as if it had never traversed the swirling tumult of the atmosphere (Fig. 5.128).

Driven by thermal energy from the Sun, the Earth's atmosphere is a shifting sea of turbulent air. Variations in density are accompanied by variations in the index of refraction and therefore in optical path length. Wavefronts streaming down from a point on a distant star arrive at the atmosphere almost precisely as plane waves (with a wavelength in the mid-visible of about 0.5 μm). As they sweep through the 100 miles or so of

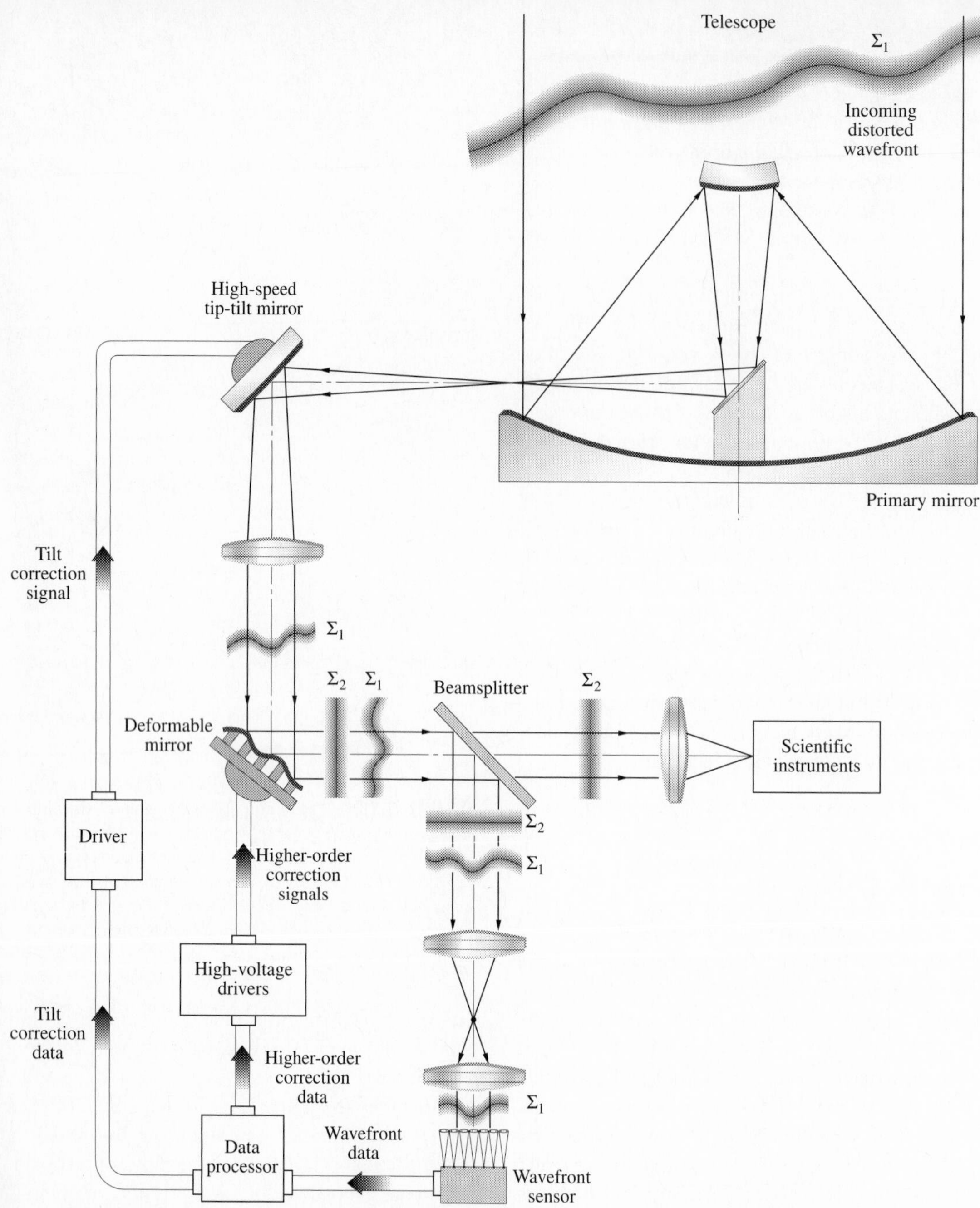

FIGURE 5.128 An adaptive-optics system. The distorted wavefront Σ_1 is analyzed and reconfigured. The corrected planar wavefront is sent on to the scientific instruments.

shifting air, path length differences of a few micrometers are introduced, and the wavefronts distort into a bumpy undulated surface. What reaches ground level is a succession of broadly wrinkled wavefronts shaped much like what you would get if you spread 10-cm tiles down on a floor that had first been randomly strewn with tiny tough beetles; that is, contiguous tiles, slightly tilted every which way. The turbulence changes unpredictably on a time scale of milliseconds, and the progression of wavefronts traversing it is continuously bent and buckled anew (as if the beetles were aimlessly walking around under the tiles, lifting and shifting them).

The tile imagery, however weird, is useful because in 1966 David L. Fried showed that the optical results of atmospheric turbulence could be modeled in a fairly simple way. In effect (because the speed of light is so great), one can assume that at any moment the atmosphere behaved as if it were compressed into a horizontal array of small, contiguous, wedge-shaped refracting regions or stable cells. At any given ground site, the local portion of a stellar wavefront is composed of many randomly tilted, small, fairly flat areas (each analogous to a single tile). In someone's backyard, these areas are typically about 10 cm across, although under the very best conditions (e.g., on a astronomical mountaintop) they might reach as much as 20 or occasionally 30 cm "when the seeing is good." Over each such **isoplanatic region**, the wavefront is fairly smooth and has little curvature: the difference between leading bumps and trailing depressions is about $\lambda/17$, and it's a rule of thumb that if wave distortions are less than $\lambda/10$ the image quality will be very good. The more turbulence there is, the smaller the stable cells are, and the smaller are the corresponding isoplanatic regions of the wavefront.

The effect of turbulence on the image formed by a telescope, one trained on a star, depends strongly on the size of its aperture. If the instrument has an aperture of only a few centimeters, the small admitted portion of a wavefront (having traversed only a part of a stable cell) will likely be quite flat. Turbulence will primarily alter the tilt of that otherwise planar incoming wavefront section. That means that a sharp Airy image can momentarily be formed via that section, but the Airy-image spot will wander around as the atmosphere changes and each successive planar wavefront section arrives at a different angle (our mythical beetles keep moving). By contrast, for a large-diameter telescope, several meters across, the large admitted wavefront section is a mosaic of many flat, tilted regions. The image is then a simultaneous superposition of numerous shifting Airy spots, and the result is a shimmering blur. Clearly, increasing the aperture will collect more light, but it will not proportionately improve the resolution.

The critical aperture size at which blurring becomes appreciable is a measure of the turbulence. It's called the **Fried parameter**, and it's almost universally represented by r_0; this is an unfortunate choice of symbol since this is not a radius. Pronounced "*r naught*," it corresponds to the size of the region over which the incoming wavefront can be taken to be essentially planar. On those rare occasions when r_0 exceeds 30 cm, a very distant star will be "perfectly" imaged as an Airy disk. As the turbulence increases, r_0 decreases; moreover, as the wavelength increases, r_0 increases: $r_0 \propto \lambda^{1.2}$. It follows that the angular resolution of a large ground-based telescope is actually $1.22\lambda/r_0$ and since r_0 is rarely better than 20 cm, the most powerful Earthbound instrument has little more resolution than a humble 6-inch telescope!

When there's a wind above a telescope, it, in effect, blows the isoplanatic regions past the aperture. A 5-m/s breeze will carry an $r_0 = 10$-cm isoplanatic region by in 20 ms. To monitor and ultimately respond to such atmospheric changes, an electro-optical-mechanical control system should operate 10 or 20 times faster, sampling the data at upwards of 1000 times per second.

Figure 5.128 is a schematic drawing of a typical astronomical adaptive optics system. In this simple arrangement, the telescope is pointed at a star that will serve both as the object of attention and as a beacon for correcting distortions. Before anything clever is done, the large beam from the primary mirror is reduced to several centimeters in diameter so that it can be handled more conveniently. In the process, each isoplanatic region at the primary becomes focused down to a correspondingly small region in the reduced beam.

The first step is to analyze the distorted wavefront, Σ_1, transmitted by the telescope and now present in miniature in the reduced beam. This is done with a **wavefront sensor**, of which there are several types. The one considered here is a Hartmann sensor (Fig. 5.129), which consists of a compact array of thousands of independent detectors tightly grouped side-by-side. Light incident on the sensor first encounters a sheet of closely packed tiny identical lenslets, at whose focal plane there is a CCD array (Fig. 5.129*a*). The device is located in the beam in such a way that a lenslet is about the size of

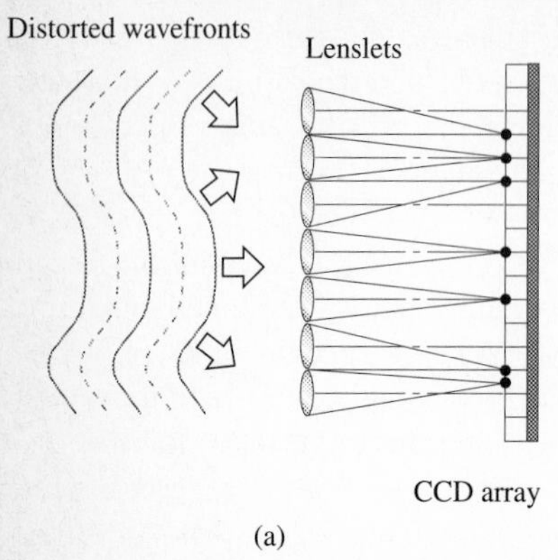

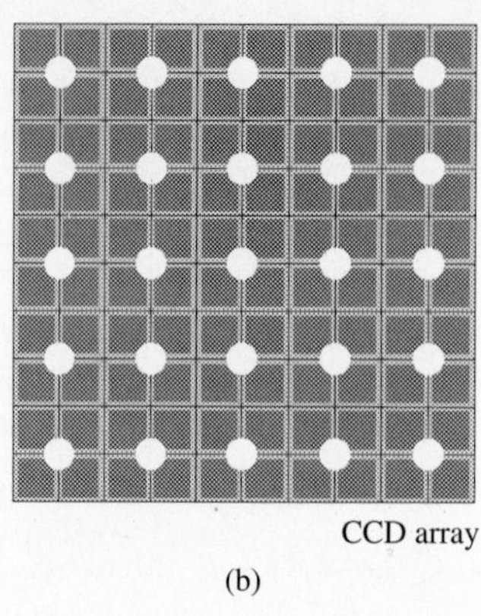

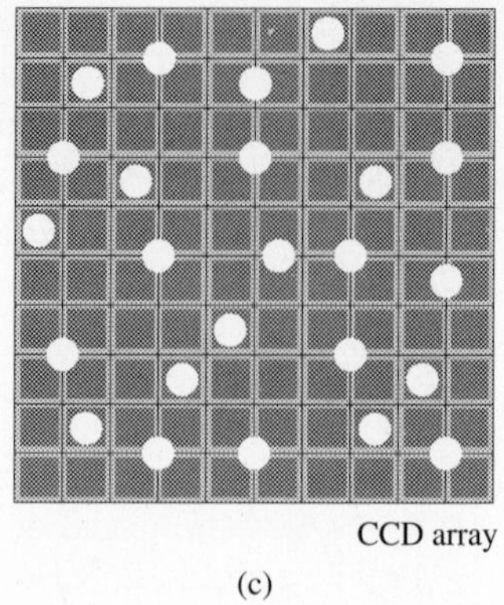

FIGURE 5.129 The Hartmann wavefront sensor. (*a*) Lenslets focus light down to a CCD array. Each square cluster of four CCD elements forms a detector. (*b*) When the incident wave is planar, Airy-image spots form at null points at the centers of each four-element detector. (*c*) When the wavefront is distorted, Airy-image spots are shifted from the null positions.

an isoplanatic region. Each lenslet then forms a minute image of the star on a cluster of four CCD pixel elements grouped around its optical axis. If the overall wavefront were perfectly flat, that is, if every isoplanatic region had zero tilt and all were parallel, each lenslet would produce an Airy-image spot at a null position between its own four pixel elements (Fig. 5.129*b*). But when any isoplanatic region is tilted, the corresponding image spot shifts and the four CCD elements record an unbalanced signal that indicates the exact displacement (Fig. 5.129*c*). The output from all of these minute detectors is computer analyzed, Σ_1 is theoretically reconstructed, and the corrections necessary to flatten the wavefront are calculated.

If an overall tilt of the wavefront is detected, a signal is sent to the fast-steering flat mirror, which initially receives the light from the primary, and that tilt is counteracted. The now untilted, but still wrinkled, wavefront heads toward a "rubber mirror," a flexible reflector that can rapidly and precisely be deformed. It might, for example, be composed of a thin faceplate reflector mounted on hundreds of actuators that rapidly push and pull it into the desired shape. Driven by signals from the computer, the mirror is bent into an inverse configuration to that of the wavefront. In effect, wavefront bumps impinge on matching mirror depressions, and vice versa. The result is to reflect a beam of distortion-free wavefronts, Σ_2, that correspond to the condition of the starlight before it entered the atmosphere. A small fraction of the radiant energy goes back into the sensor-computer-mirror control loop to continuously maintain the correction process, while the remainder travels on to the scientific instruments.

Because many objects of interest to astronomers—planets, galaxies, nebulae, and so on—are imaged as extended bodies, using these as an adaptive-optics beacon is precluded. Still, if you wish to examine a galaxy, you could use a nearby star as a beacon. Unfortunately, however, there will frequently not be any stars in the vicinity that are bright enough for the purpose. One way to get around this limitation is to use a laserbeam to create an artificial guide star (Fig. 5.130). To date, this has successfully been done in two different ways. In one, a laser pulse, focused at altitudes in the range from around 10 to 40 km, is projected up through the telescope. A portion of that light is backscattered downward from air molecules via Rayleigh Scattering. Alternatively, there is a layer of sodium atoms (probably deposited by meteors) at an altitude of 92 km, well above most of the atmospheric turbulence. A laser tuned to 589 nm can excite the sodium, thereby producing a small bright yellow beacon anywhere in the sky.

The results (Fig. 5.131) have been so encouraging* that most of the world's major telescopes are already, or soon will be, using adaptive optics.

*See L. A. Thompson, "Adaptive Optics in Astronomy," *Phys. Today* **47**, 24 (1994); J. W. Hardy, "Adaptive Optics," *Sci. Am.* **60** (June 1994); R. Q. Fugate and W. J. Wild, "Untwinkling the Stars—Part I," *Sky & Telescope* **24** (May 1994); W. J. Wild and R. Q. Fugate, "Untwinkling the Stars—Part II," *Sky & Telescope* **20** (June 1994).

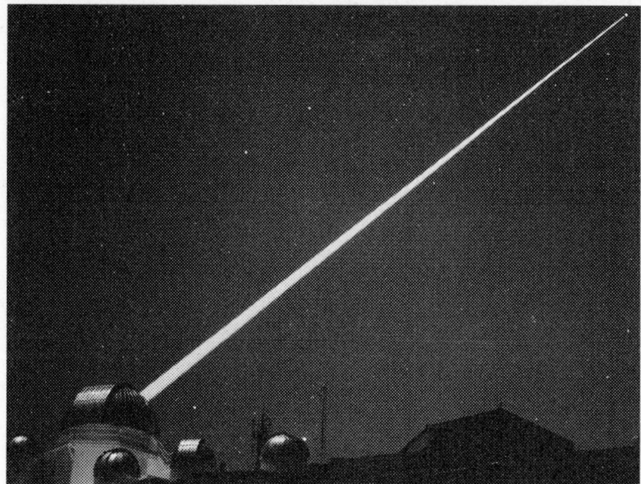

FIGURE 5.130 The creation of a laserguide star at the Phillips Laboratory, Kirtland Air Force Base, New Mexico. (Courtesy of Phillips Laboratory, Department of the US Air Force)

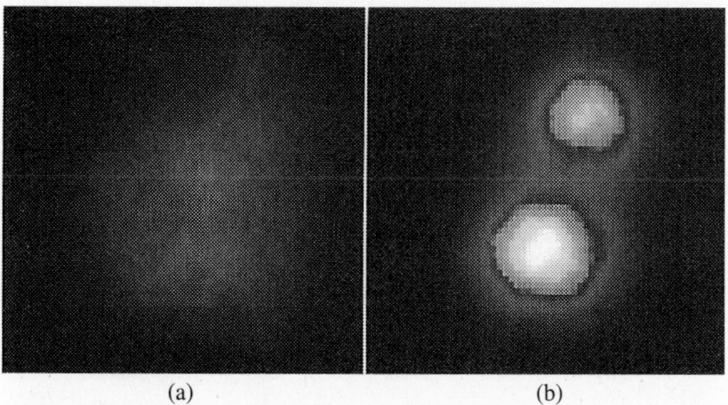

FIGURE 5.131 A 1-second exposure of 53ξ Ursa Major using a 1.5-m telescope at the Phillips Laboratory. (Courtesy of Phillips Laboratory, Department of the US Air Force) (*a*) The ordinary uncompensated image is undecipherable. (*b*) Using adaptive optics, the image is improved dramatically.

5.8.2 Phase Conjugation

Another important new technology for reshaping wavefronts is known as **phase conjugation**; here the wave is turned inside-out during a special kind of reflection.

Imagine a stream of plane waves traveling to the right in the positive z-direction impinging perpendicularly on an ordinary flat mirror. The incident wave (p. 28) is expressible as $E_i = E_0 \cos(kz - \omega t)$, or in complex form as $\tilde{E}_i = E_0 e^{i(kz-\omega t)} = E_0 e^{ikz} e^{-i\omega t} = \tilde{E}(z) e^{-i\omega t}$, where the space and time parts have been separated. For this simple geometry, the reflected waves ride right back over the incident waves; ***they are identical except for the direction of propagation***. The reflected wave is $E_r = E_0 \cos(-kz - \omega t)$, or $\tilde{E}_r = E_0 e^{-ikz} e^{-i\omega t} = E^*(z) e^{-i\omega t}$. Changing the sign of the space part of the phase changes the direction of the wave, and the same thing is accomplished by taking the complex conjugate in the exponential formulation. For this reason, the reflected wave is also called a **phase-conjugated wave** or just a **conjugate wave**. A situation of this sort is characterized by the fact that we could, in principle, take a motion picture of it, which when shown forwards or backwards, would be indistinguishable. Consequently, a phase-conjugated wave is said to be **time reversed**. For monochromatic waves, changing the sign of the time part (i.e., time reversal) is equivalent to reversing the direction of propagation: $\cos[kz - \omega(-t)] = \cos(kz + \omega t) = \cos(-kz - \omega t)$.

A very simple, phase-conjugated reflection occurs when there is a point source at the center of curvature of a concave spherical mirror. The waves flow, expanding out to the mirror, and on reflection, contract back on themselves to the source point. Presumably, a conventional reflecting surface could be made to exactly match any particular wavefront and thereby reflect a conjugate for that specific incoming wave (Fig. 5.132). That's a rather impractical approach, especially if you can't anticipate the shape of the wavefront or if it changes from moment to moment.

Fortunately, in 1972 a team of Russian scientists discovered a method for producing phase conjugation for *any* incident wavefront using Brillouin Scattering. They directed an intense beam of laser light into a tube containing high-pressure methane gas. At power levels of about a million watts, pressure-density variations occur, and the medium becomes a remarkable kind of mirror, reflecting back almost all the incoming light. What surprised the investigators was that the beam scattered back out of the gas was phase conjugated. The medium, in this case the methane, adjusts itself to the presence of the electromagnetic field in just such a way as to turn the backscattered wave inside-out, so that regions that were originally leading were now trailing. Today there are several means to the same end, all using media that produce nonlinear optical

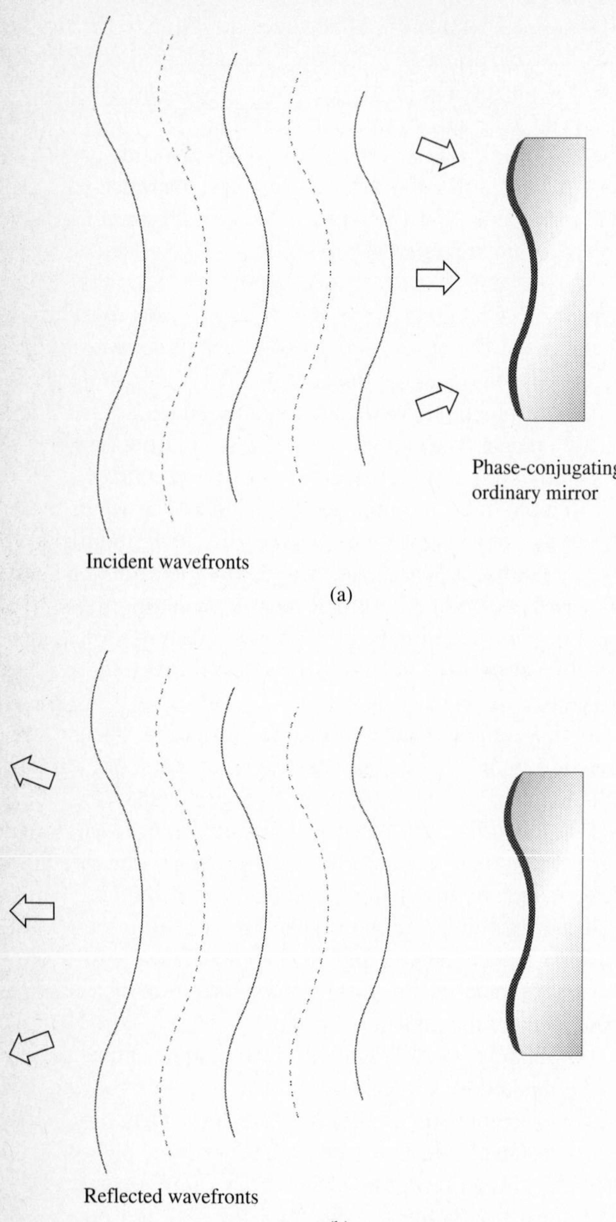

FIGURE 5.132 The operation of a rather limited phase-conjugating mirror. It only works for the incoming wavefronts shown in (*a*).

effects. There are myriad potential applications from tracking satellites to improving laserbeam quality.*

As an example of the kinds of things that can be done, consider the following: If a beam that has been distorted by passing through an inhomogeneous medium (Fig. 5.127) is reflected from an ordinary mirror and made to retraverse that medium, the beam will become even more distorted. By contrast, if the same thing is done using a phase-conjugating mirror, on passing back through the distorting medium for a second time the beam will be restored to its pristine condition. Figure 5.133 illustrates the technique, and Fig. 5.134 shows

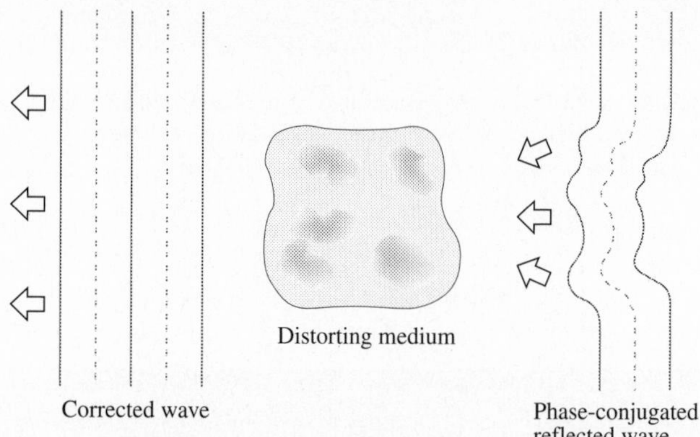

FIGURE 5.133 When the distorted wave in Fig. 5.127 is reflected by a phase-conjugating mirror, it's turned inside-out, or conjugated. Compare it to the conventionally reflected wave in Fig. 5.127*b*. On traversing the inhomogeneous medium a second time, regions of the wavefront that are now leading will be held back, and vice versa. The wave that emerges after a round-trip will be identical to the one that originally entered (Fig. 5.127*a*).

*See D. M. Pepper, "Applications of Optical Phase Conjugation," *Sci. Am.* **74** (January 1986) and V. V. Shkunov and B. Ya. Zel'dovich, "Optical Phase Conjugation," *Sci. Am.* **54** (December 1985).

(a)

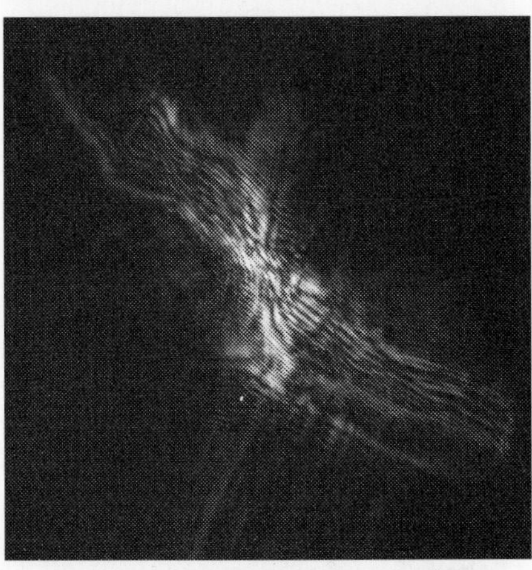

(b)

(c)

FIGURE 5.134 Using phase conjugation to remove distortion. (*a*) Image of a cat reflected from a mirror—no introduced distortion. (*b*) The same cat wave after twice traversing an inhomogeneous medium. (*c*) After passing through the inhomogeneous medium, the wave was phase conjugated and returned through the medium a second time. Most of the image distortion vanished. (Photos courtesy of Jack Feinberg, University of Southern California.)

the results of an actual experiment. The image of a cat was impressed on a collimated argon-ion laserbeam (λ = 514.5 nm) by simply passing the beam through a photographic transparency of the cat. As a reference standard, the image-carrying wave was sent, via a beamsplitter, to an ordinary mirror where it was reflected back through the beamsplitter and onto a ground-glass screen so it could be photographed (Fig. 5.134*a*). Next, a phase distorter (e.g., a piece of shower-door glass) was introduced between the beamsplitter and the mirror so that the wave traversed it twice. The image returned from the mirror was unrecognizable (Fig. 5.134*b*). Finally, the conventional mirror was removed and replaced by a phase conjugator. Even though the wave again passed twice through the distorter, the image was restored to its original clarity (Fig. 5.134*c*).

PROBLEMS

5.1 The shape of the interface pictured in Fig. P.5.1 is known as a Cartesian oval after René Descartes who studied it in the early 1800s. It's the perfect configuration to carry any ray from S to the interface to P. Prove that the defining equation is

$$\ell_o n_1 + \ell_i n_2 = \text{constant}$$

Show that this is equivalent to

$$n_1(x^2 + y^2)^{1/2} + n_2[y^2 + (s_o + s_i - x^2)]^{1/2} = \text{constant}$$

where x and y are the coordinates of point A.

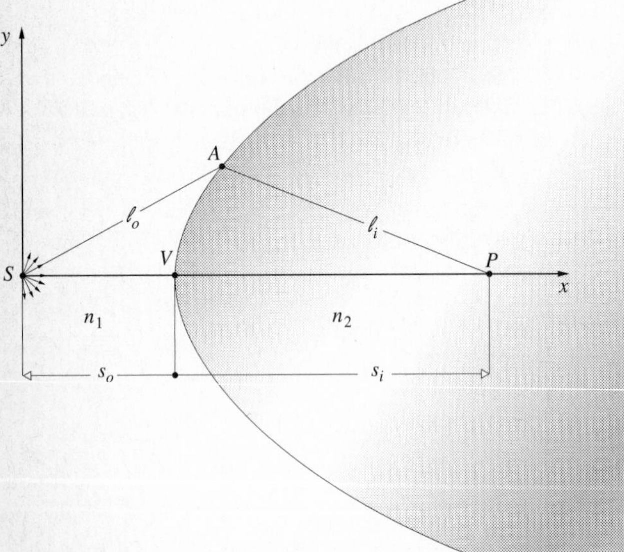

FIGURE P.5.1

5.2 Construct a Cartesian oval such that the conjugate points will be separated by 11 cm when the object is 5 cm from the vertex. If $n_1 = 1$ and $n_2 = \frac{3}{2}$, draw several points on the required surface.

5.3* Use Fig. P.5.3 to show that if a point source is placed at the focus F_1 of the ellipsoid, plane waves will emerge from the far side. Remember that the defining requirement for an ellipse is that the net distance from one focus to the curve and back to the other focus is constant.

5.4 Diagrammatically construct both a sphero-elliptic positive lens and an ellipto-spheric negative lens, showing rays and wavefronts as they pass through the lens. Do the same for an oval-spheric positive lens.

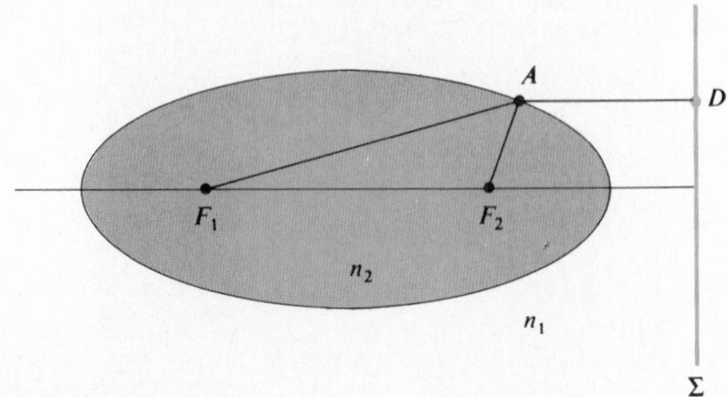

FIGURE P.5.3

5.5* Making use of Fig. P.5.5, Snell's Law, and the fact that in the paraxial region $\alpha = h/s_o$, $\varphi \approx h/R$, and $\beta \approx h/s_i$, derive Eq. (5.8).

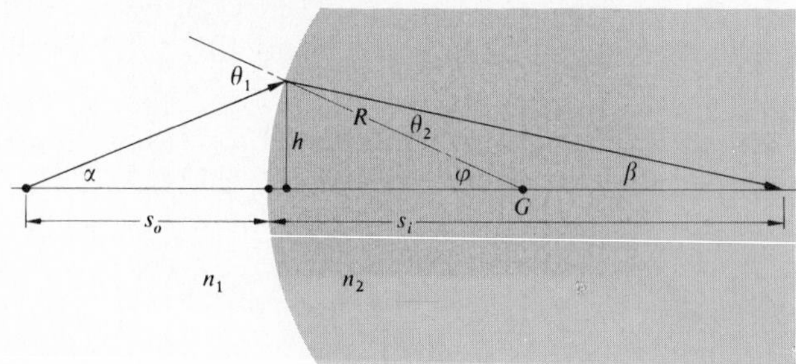

FIGURE P.5.5

5.6* Show that, in the paraxial domain, the magnification produced by a single spherical interface between two continuous media, as shown in Fig. P.5.6, is given by

$$M_T = -\frac{n_1 s_i}{n_2 s_o}$$

Use the small-angle approximation for Snell's Law and approximate the angles by their tangents.

5.7* Imagine a hemispherical interface, with a radius of curvature of radius 5.00 cm, separating two media: air on the left, water on the right. A 3.00-cm-tall toad is on the central axis, in air, facing the convex interface and 30.0 cm from its vertex. Where in the water will it

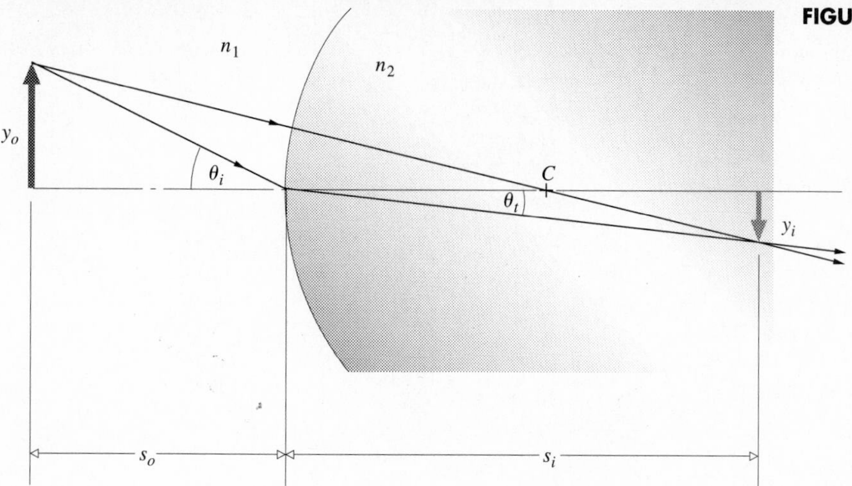

FIGURE P.5.6

be imaged? How big will it appear to a fish in the water? Use the results of the previous problem, even though our frog is pushing the paraxial approximation.

5.8 Locate the image of an object placed 1.2 m from the vertex of a gypsy's crystal ball, which has a 20-cm diameter (n = 1.5). Make a sketch (of the rays, not the gypsy).

5.9* Return to Problem 5.7 and suppose we cut off the medium on the right forming a thick water biconvex lens, with each surface having a radius of curvature of 5.00 cm. If the lens is 10.0 cm thick, determine the total magnification and everything you can about the toad's image.

5.10 A biconcave lens (n_l = 1.5) has radii of 20 cm and 10 cm and an axial thickness of 5 cm. Describe the image of an object 1-inch tall placed 8 cm from the first vertex. Use the thin-lens equation to see how far off it is in determining the final-image location.

5.11 Prove that the minimum separation between conjugate *real* object and image points for a thin positive lens is $4f$.

5.12 An object 2 cm high is positioned 5 cm to the right of a positive thin lens with a focal length of 10 cm. Describe the resulting image completely, using *both* the Gaussian and Newtonian equations.

5.13 Make a rough graph of the Gaussian Lens Equation; that is, plot s_i versus s_o, using unit intervals of f along each axis. (Get both segments of the curve.)

5.14* A parallel bundle of rays from a very distant point source is incident on a thin negative lens having a focal length of −50.0 cm. The rays make an angle of 6.0° with the optical axis of the lens. Locate the image of the source.

5.15 What must the focal length of a thin negative lens be for it to form a virtual image 50 cm away of an ant that is 100 cm away? Given that the ant is to the right of the lens, locate and describe its image.

5.16* Compute the focal length in air of a thin biconvex lens (n_l = 1.5) having radii of 20 and 40 cm. Locate and describe the image of an object 40 cm from the lens.

5.17 Determine the focal length of a planar-concave lens (n_l = 1.5) having a radius of curvature of 10 cm. What is its power in diopters?

5.18* Determine the focal length in air of a thin spherical planar-convex lens having a radius of curvature of 50.0 mm and an index of 1.50. What, if anything, would happen to the focal length if the lens were placed in a tank of water?

5.19* We would like to place an object 45 cm in front of a lens and have its image appear on a screen 90 cm behind the lens. What must be the focal length of the appropriate positive lens?

5.20 The horse in Fig. 5.29 is 2.25 m tall, and it stands with its face 15.0 m from the plane of the thin lens whose focal length is 3.00 m.

(a) Determine the location of the image of the equine nose.
(b) Describe the image in detail—type, orientation, and magnification.
(c) How tall is the image?
(d) If the horse's tail is 17.5 m from the lens, how long, nose-to-tail, is the image of the beast?

5.21* A candle that is 6.00 cm tall is standing 10 cm from a thin concave lens whose focal length is −30 cm. Determine the location of the image and describe it in detail. Draw an appropriate ray diagram.

5.22* The image projected by an equiconvex lens ($n = 1.50$) of a frog 5.0 cm tall and 0.60 m from a screen is to be 25 cm high. Please compute the necessary radii of the lens.

5.23* A thin piece of wire 4.00 mm long is located in a plane perpendicular to the optical axis and 60.0 cm in front of a thin lens. The sharp image of the wire formed on a screen is 2.00 mm long. What is the focal length of the lens? When the screen is moved farther from the lens by 10.0 mm, the image blurs to a width of 0.80 mm. What is the diameter of the lens? [*Hint:* Image a source point on the axis.]

5.24 A thin double convex glass lens (with an index of 1.56) while surrounded by air has a 10-cm focal length. If it is placed under water (having an index of 1.33) 100 cm beyond a small fish, where will the guppy's image be formed?

5.25 Consider a homemade television projection system that uses a large positive lens to cast the image of the TV screen onto a wall. The projected picture is enlarged three times, and although dim, it's nice and clear. If the lens has a focal length of 60 cm, what should be the distance between the screen and the wall? Why use a large lens? How should we mount the set with respect to the lens?

5.26 Write an expression for the focal length (f_w) of a thin lens immersed in water ($n_w = \frac{4}{3}$) in terms of its focal length when it's in air (f_a).

5.27* Observe the three vectors **A**, **B**, and **C** in Fig. P.5.27, each of which has a length of $0.10f$ where f is the focal length of the thin positive lens. The plane formed by **A** and **B** is at a distance of $1.10f$ from the lens. Describe the image of each vector.

5.28* A convenient way to measure the focal length of a positive lens makes use of the following fact. If a pair of conjugate object and (real) image points (S and P) are separated by a distance $L > 4f$, there will be two locations of the lens, a distance d apart, for which the same pair of conjugates obtain. Show that

$$f = \frac{L^2 - d^2}{4L}$$

Note that this avoids measurements made specifically from the vertex, which are generally not easy to do.

5.29* Two positive lenses with focal lengths of 0.30 m and 0.50 m are separated by a distance of 0.20 m. A small butterfly rests on the central axis 0.50 m in front of the first lens. Locate the resulting image with respect to the second lens.

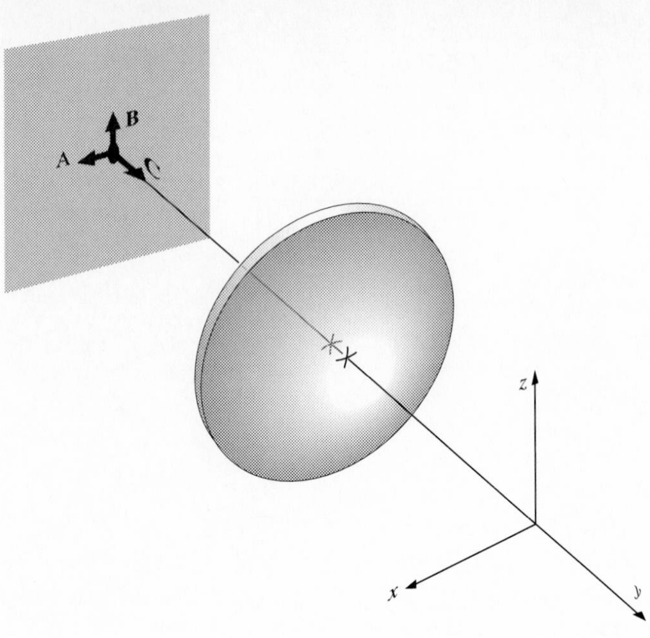

FIGURE P.5.27

5.30 In the process of constructing a doublet, an equiconvex thin lens L_1 is positioned in intimate contact with a thin negative lens, L_2, such that the combination has a focal length of 50 cm in air. If their indices are 1.50 and 1.55, respectively, and if the focal length of L_2 is −50 cm, determine all the radii of curvature.

5.31 Verify Eq. (5.34), which gives M_T for a combination of two thin lenses.

5.32 Compute the image location and magnification of an object 30 cm from the front doublet of the thin-lens combination in Fig. P.5.32. Do the calculation by finding the effect of each lens separately. Make a sketch of appropriate rays.

5.33* Two thin lenses having focal lengths of +15.0 cm and −15.0 cm are positioned 60.0 cm apart. A page of print is held 25.0 cm in front of the positive lens. Describe, in detail, the image of the print (i.e., insofar as it's paraxial).

5.34* Draw a ray diagram for the combination of two positive lenses wherein their separation equals the sum of their respective focal

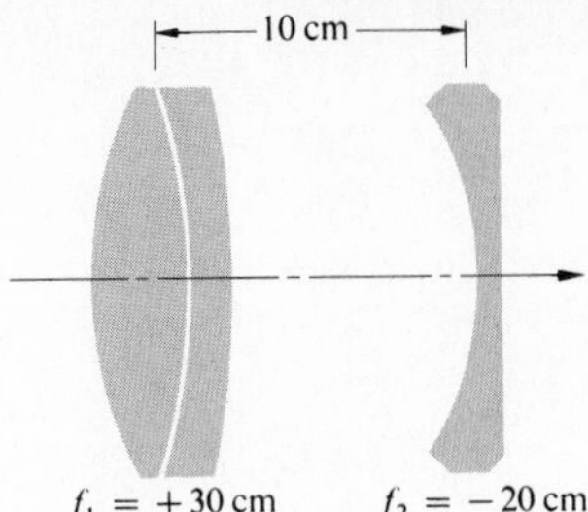

FIGURE P.5.32

lengths. Do the same thing for the case in which one of the lenses is negative.

5.35 Redraw the ray diagram for a compound microscope (Fig. 5.111), but this time treat the intermediate image as if it were a real object. This approach should be a bit simpler.

5.36* Consider a thin positive lens L_1, and using a ray diagram, show that if a second lens L_2 is placed at the focal point of L_1, the magnification does not change. That's a good reason to wear eyeglasses, whose lenses are different, at the correct distance from the eye.

5.37* Figures P.5.37*a* and P.5.37*b* are taken from an introductory physics book. What's wrong with them?

5.38 Consider the case of two positive thin lenses, L_1 and L_2, separated by 5 cm. Their diameters are 6 and 4 cm, respectively, and their focal lengths are $f_1 = 9$ cm and $f_2 = 3$ cm. If a diaphragm with a hole 1 cm in diameter is located between them, 2 cm from L_2, find (a) the aperture stop and (b) the locations and sizes of the pupils for an axial point, S, 12 cm in front of (to the left of) L_1.

5.39 Make a sketch roughly locating the aperture stop and entrance and exit pupils for the lens in Fig. P.5.39.

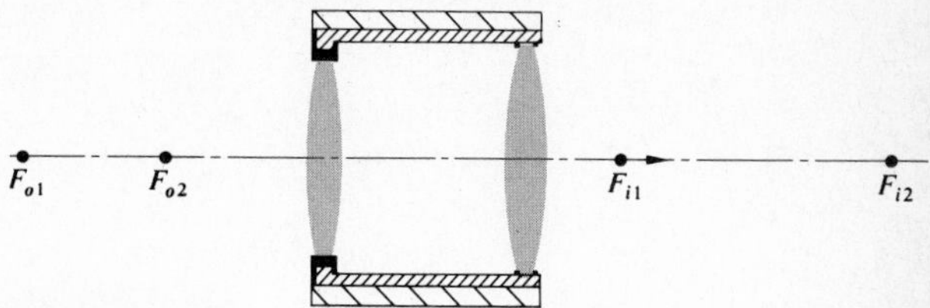

FIGURE P.5.39

5.40 Make a sketch roughly locating the aperture stop and entrance and exit pupils for the lens in Fig. P.5.40, assuming the object point to be beyond (to the left of) F_{o1}.

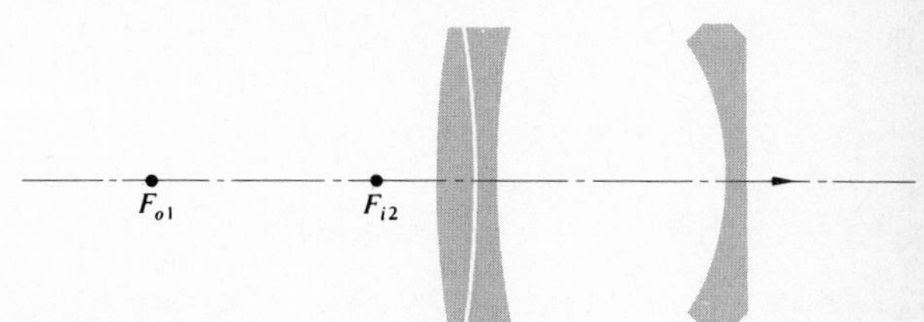

FIGURE P.5.40

5.41 Figure P.5.41 shows a lens system, an object, and the appropriate pupils. Diagrammatically locate the image.

5.42 Draw a ray diagram locating the images of a point source as formed by a pair of mirrors at 90° (Fig. P.5.42*a*). Now create a ray diagram locating the images of the arrow shown in Fig. P.5.42*b*.

5.43 Examine Velasquez's painting of *Venus and Cupid* (Fig. P.5.43). Is Venus looking at herself in the mirror? Explain.

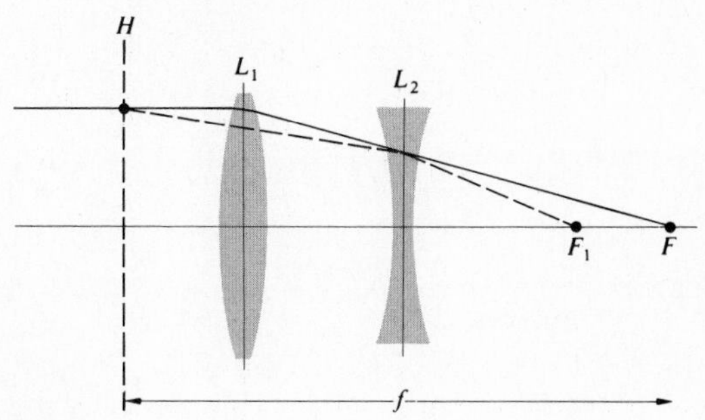

FIGURE P.5.37*a*

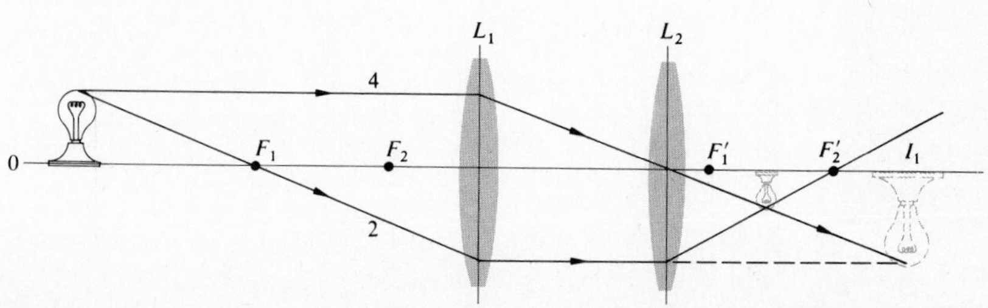

FIGURE P.5.37*b*

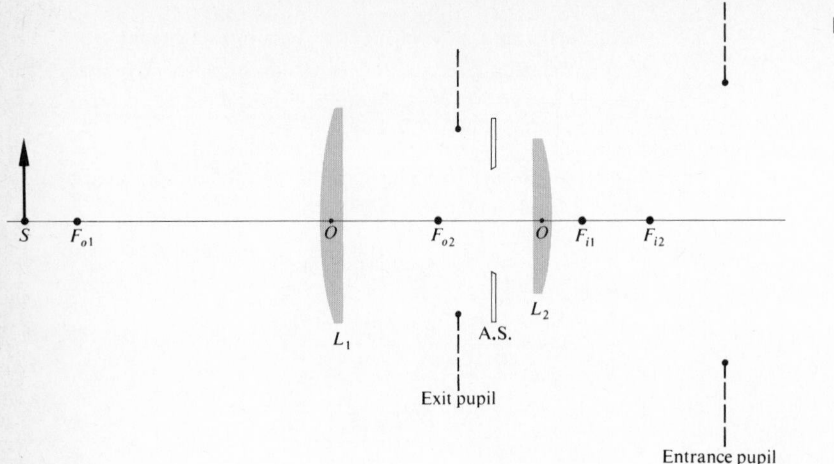

FIGURE P.5.41

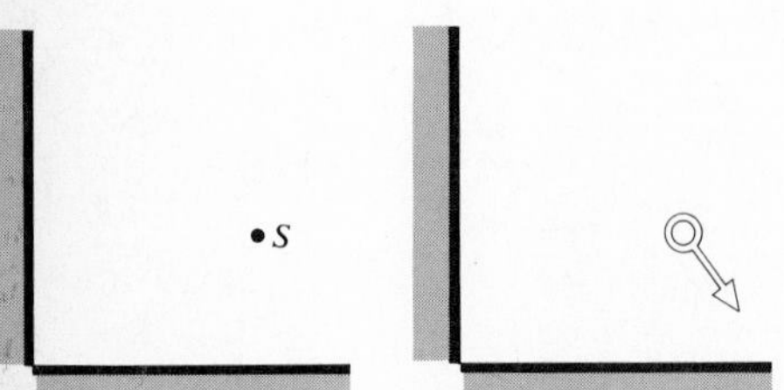

FIGURE P.5.42*a* **FIGURE P.5.42*b***

FIGURE P.5.43 *The Toilet of Venus* by Diego Rodriguez de Silva y Valazquez (Reproduced by courtesy of the Trustees, The National Gallery, London)

5.44 Manet's painting *The Bar at the Folies Bergères* (Fig. P.5.44) shows a girl standing in front of a large planar mirror. Reflected in it is her back and a man in evening dress with whom she appears to be talking. It would seem that Manet's intent was to give the uncanny feeling that the viewer is standing where that gentleman must be. From the laws of Geometrical Optics, what's wrong?

FIGURE P.5.44 *The Bar at the Folies Bergere* by Edouard Manet—(Courtauld Institute Galleries, London. Courtauld Collection.)

5.45 Show that Eq. (5.48) for a spherical surface is equally applicable to a plane mirror.

5.46 Locate the image of a paperclip 100 cm away from a convex spherical mirror having a radius of curvature of 80 cm.

5.47* Imagine that you are standing 5 feet from, and looking directly toward, a brass ball 1 foot in diameter hanging in front of a pawn shop. Describe the image you would see in the ball.

5.48 The image of a red rose is formed by a concave spherical mirror on a screen 100 cm away. If the rose is 25 cm from the mirror, determine its radius of curvature.

5.49 From the image configuration determine the shape of the mirror hanging on the back wall in van Eyck's painting of *John Arnolfini and His Wife* (Fig. P.5.49).

5.50* There are several varieties of retro-reflector that are commercially available; one type is comprised of transparent spheres, the backs of which are silvered. Light is refracted at the front surface, focused onto the rear surface, and there reflected back out in the direction it came. Determine the necessary index of refraction of the spheres. Assume the incident light is collimated.

5.51* Design an eye for a robot using a concave spherical mirror such that the image of an object 1.0 m tall and 10 m away fills its 1.0-cm-square photosensitive detector (which is movable for focusing purposes). Where should this detector be located with respect to the mirror? What should be the focal length of the mirror? Draw a ray diagram.

5.52 Design a little dentist's mirror to be fixed at the end of a shaft for use in the mouth of some happy soul. The requirements are (1) that the image be erect as seen by the dentist and (2) that when held 1.5 cm from a tooth the mirror produces an image twice life-size.

5.53 An object is located at a distance s_o from a spherical mirror of radius R. Show that the resulting image will be magnified by an amount

$$M_T = \frac{R}{2s_o + R}$$

5.54* A device used to measure the radius of curvature of the cornea of the eye is called a keratometer. This is useful information when fitting contact lenses. In effect, an illuminated object is placed a known distance from the eye, and the image reflected off the cornea is observed. The instrument allows the operator to measure the size of that virtual image. If the magnification is found to be 0.037× when the object distance is set at 100 mm, what is the radius of curvature?

FIGURE P.5.49 van Eyck's *The Marriage of Giovanni Arnolfini and Giovanni Cenami* (Reproduced by courtesy of the Trustees, The National Gallery, London)

5.55* Considering the operation of a spherical mirror, prove that the locations of the object and image are given by

$$s_o = f(M_T - 1)/M_T \quad \text{and} \quad s_i = -f(M_T - 1)$$

5.56 A man whose face is 25 cm away looks into the bowl of a soupspoon and sees his image reflected with a magnification of −0.064. Determine the radius of curvature of the spoon.

5.57* In an amusement park a large upright convex spherical mirror is facing a plane mirror 10.0 m away. A girl 1.0 m tall standing midway between the two sees herself twice as tall in the plane mirror as in the spherical one. In other words, the angle subtended at the observer by the image in the plane mirror is twice the angle subtended by the image in the spherical mirror. What is the focal length of the latter?

5.58* A homemade telephoto "lens" (Fig. P.5.58) consists of two spherical mirrors. The radius of curvature is 2.0 m for the primary and 60 cm for the secondary. How far from the smaller mirror should the film plane be located if the object is a star? What is the effective focal length of the system?

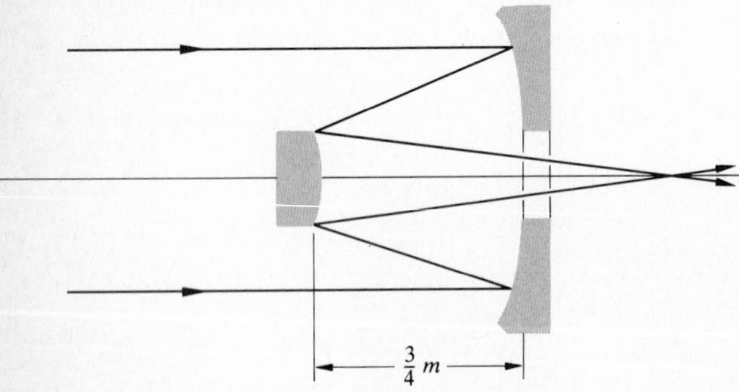

FIGURE P.5.58

5.59* Suppose you have a concave spherical mirror with a focal length of 10 cm. At what distance must an object be placed if its image is to be erect and one and a half times as large? What is the radius of curvature of the mirror? Check with Table 5.5.

5.60 Describe the image that would result for an object 3 inches tall placed 20 cm from a spherical concave shaving mirror having a radius of curvature of −60 cm.

5.61 Referring to the dove prism in Fig. 5.70, rotate it through 90° about an axis along the ray direction. Sketch the new configuration and determine the angle through which the image is rotated.

5.62 Determine the numerical aperture of a single clad optical fiber, given that the core has an index of 1.62 and the clad 1.52. When immersed in air, what is its maximum acceptance angle? What would happen to a ray incident at, say, 45°?

5.63 Given a fused silica fiber with an attenuation of 0.2 dB/km, how far can a signal travel along it before the power level drops by half?

5.64 The number of modes in a stepped-index fiber is provided by the expression

$$N_m \simeq \tfrac{1}{2}(\pi D\,\text{NA}/\lambda_0)^2$$

Given a fiber with a core diameter of 50 μm and $n_c = 1.482$ and $n_f = 1.500$, determine N_m when the fiber is illuminated by an LED emitting at a central wavelength of 0.85 μm.

5.65* Determine the intermodal delay (in ns/km) for a stepped-index fiber with a cladding of index 1.485 and a core of index 1.500.

5.66 Using the information on the eye in Section 5.7.1, compute the approximate size (in millimeters) of the image of the Moon as cast on the retina. The Moon has a diameter of 2160 miles and is roughly 230 000 miles from here, although this, of course, varies.

5.67* Figure P.5.67 shows an arrangement in which the beam is deviated through a constant angle σ, equal to twice the angle β between the plane mirrors, regardless of the angle-of-incidence. Prove that this is indeed the case.

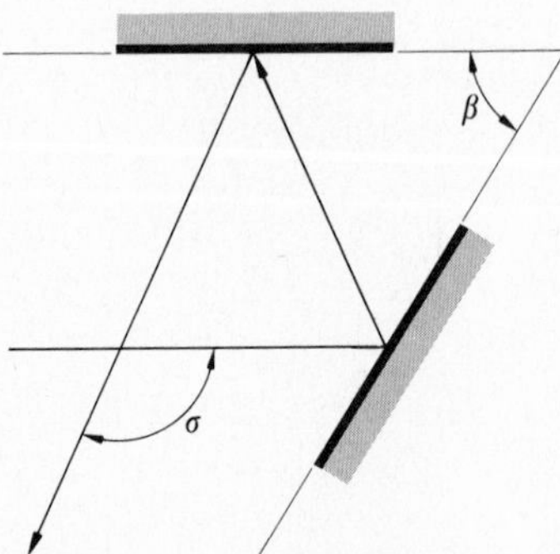

FIGURE P.5.67

5.68 An object 20 m from the objective (f_o = 4 m) of an astronomical telescope is imaged 30 cm from the eyepiece (f_o = 60 cm). Find the total linear magnification of the scope.

5.69* Figure P.5.69, which purports to show an erecting lens sys-

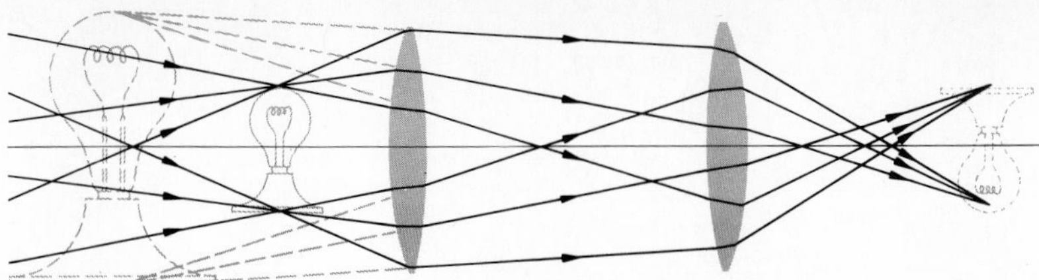

FIGURE P.5.69

tem, is taken from an old, out-of-print optics text. What's wrong with it?

5.70* Figure P.5.70 shows a pin hole in an opaque screen being used for something practical. Explain what's happening and why it works. Try it.

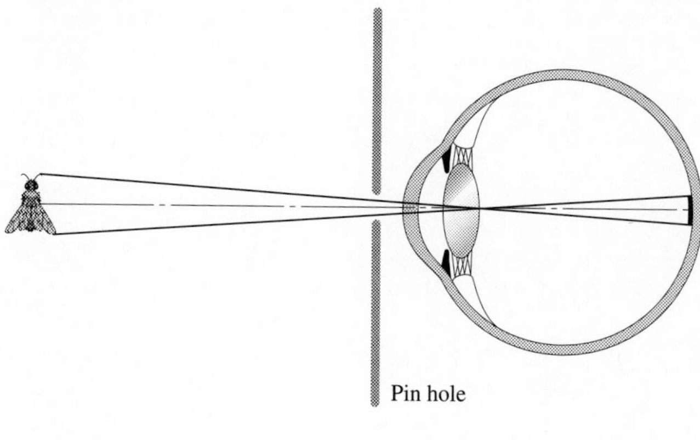

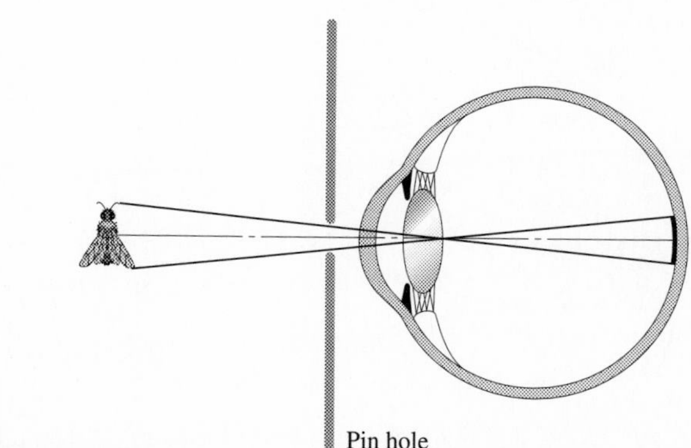

FIGURE P.5.70

5.71* If a photograph of a moving merry-go-round is perfectly exposed, but blurred, at $\frac{1}{30}$ s and $f/11$, what must the diaphragm setting be if the shutter speed is raised to $\frac{1}{120}$ s in order to "stop" the motion?

5.72 The field of view of a simple two-element astronomical telescope is restricted by the size of the eye-lens. Make a ray sketch showing the vignetting that arises.

5.73 A *field-lens*, as a rule, is a positive lens placed at (or near) the intermediate image plane in order to collect the rays that would otherwise miss the next lens in the system. In effect, it increases the field of view without changing the power of the system. Redraw the ray diagram of the previous problem to include a field-lens. Show that as a consequence the eye relief is reduced somewhat.

5.74* Describe completely the image that results when a bug sits at the vertex of a thin positive lens. How does this relate directly to the manner in which a field-lens works? (See Problem 5.73.)

5.75* It is determined that a patient has a near point at 50 cm. If the eye is approximately 2.0 cm long,

(a) How much power does the refracting system have when focused on an object at infinity? when focused at 50 cm?
(b) How much accommodation is required to see an object at a distance of 50 cm?
(c) What power must the eye have to see clearly an object at the standard near-point distance of 25 cm?
(d) How much power should be added to the patient's vision system by a correcting lens?

5.76* An optometrist finds that a farsighted person has a near point at 125 cm. What power will be required for contact lenses if they are effectively to move that point inward to a more workable distance of 25 cm so that a book can be read comfortably? Use the fact that if the object is imaged at the near point, it can be seen clearly.

5.77 A farsighted person can see very distant mountains with relaxed eyes while wearing +3.2-D contact lenses. Prescribe spectacle lenses that will serve just as well when worn 17 mm in front of the cornea. Locate and compare the far point in both cases.

5.78* A jeweler is examining a diamond 5.0 mm in diameter with a loupe having a focal length of 25.4 mm.

(a) Determine the maximum angular magnification of the loupe.
(b) How big does the stone appear through the magnifier?
(c) What is the angle subtended by the diamond at the unaided eye when held at the near point?
(d) What angle does it subtend at the aided eye?

5.79 Suppose we wish to make a microscope (that can be used with

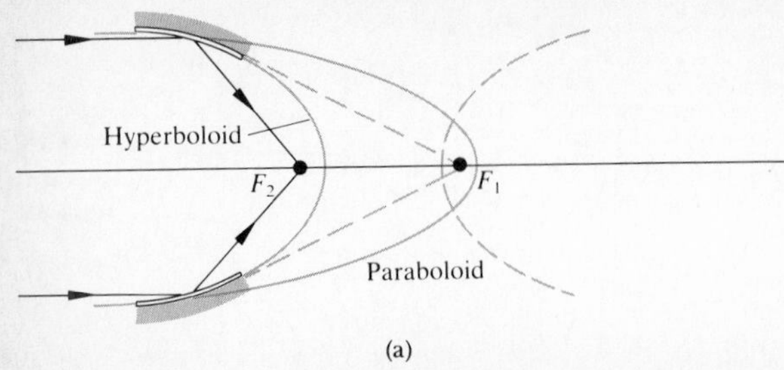

(a)

(b)

FIGURE P.5.80 a,b

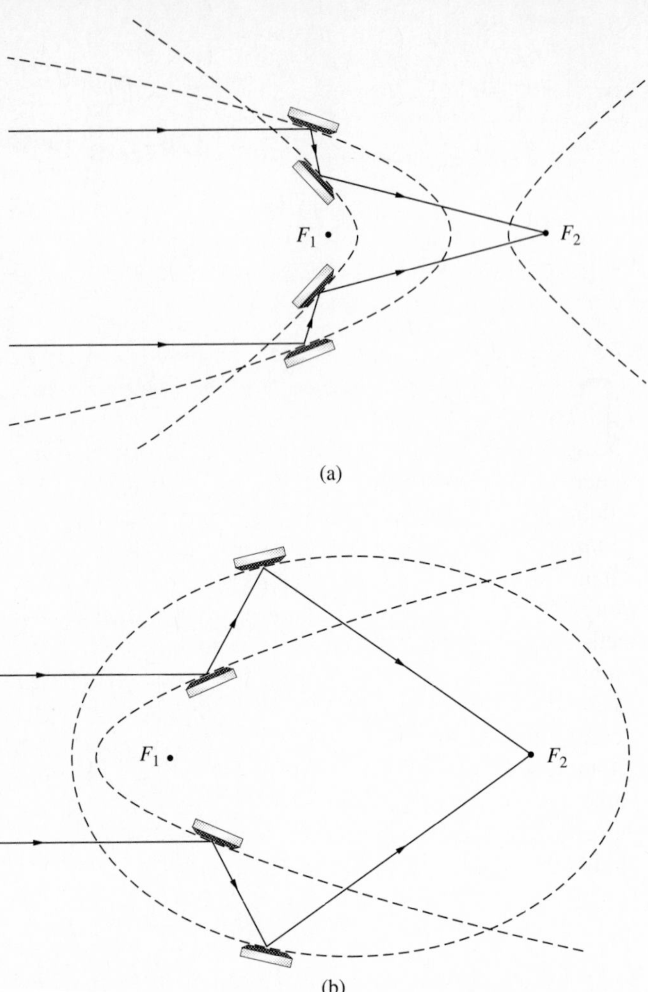

FIGURE P.5.81a,b

a relaxed eye) out of two positive lenses, both with a focal length of 25 mm. Assuming the object is positioned 27 mm from the objective, (a) how far apart should the lenses be, and (b) what magnification can we expect?

5.80* Figure P.5.80 shows a glancing-incidence X-ray focusing system designed in 1952 by Hans Wolter. How does it work? Microscopes with this type of system have been used to photograph, in X-rays, the implosion of fuel pellet targets in laser fusion research. Similar X-ray optical arrangements have been used in astronomical telescopes (Fig. 3.47).

5.81* The two glancing-incidence aspherical mirror systems depicted in Fig. P.5.81 are designed to focus X-rays. Explain how each works: identify the shapes of the mirrors, discuss the locations of their various foci, and so on.

5.82* The orbiting Hubble Space Telescope has a 2.4-m primary, which we will assume to be diffraction limited. Suppose we wanted to use it to read the print on the side of a distant Russian satellite. Assuming that a resolution of 1.0 cm at the satellite will do, how far away could it be from the HST?

6 More on Geometrical Optics

The preceding chapter, for the most part, dealt with paraxial theory as applied to thin spherical lens systems. The two predominant approximations were, rather obviously, that we had *thin* lenses and that first-order theory was sufficient for their analysis. Neither of these assumptions can be maintained throughout the design of a precision optical system, but, taken together, they provide the basis for a first rough solution. This chapter carries things a bit further by examining thick lenses and aberrations; even at that, it is only a beginning. The advent of computerized lens design requires a certain shift in emphasis—there is little need to do what a computer can do better. Moreover, the sheer wealth of existing material developed over centuries demands a bit of judicious pruning to avoid a plethora of pedantry.

6.1 Thick Lenses and Lens Systems

Figure 6.1 depicts a thick lens (i.e., one whose thickness is by no means negligible). As we shall see, it could equally well be envisioned more generally as an optical system, allowing for the possibility that it consists of a number of simple lenses, not merely one. The first and second focal points, or if you like, the object and image foci, F_o and F_i, can conveniently be measured from the two (outermost) vertices. In that case we have the familiar front and back focal lengths denoted by f.f.l. and b.f.l. When extended, the incident and emerged rays will meet at points, the locus of which forms a curved surface that may or may not reside within the lens. The surface, approximating a plane in the paraxial region, is termed the **principal plane** (see Section 6.3.1). Points where the primary and secondary principal planes (as shown in Fig. 6.1) intersect the optical axis are known as the **first** and **second principal points**, H_1 and H_2, respectively. They provide a set of very useful references from which to measure several of the system parameters. We saw earlier (Fig. 5.19, p. 161) that a ray traversing the

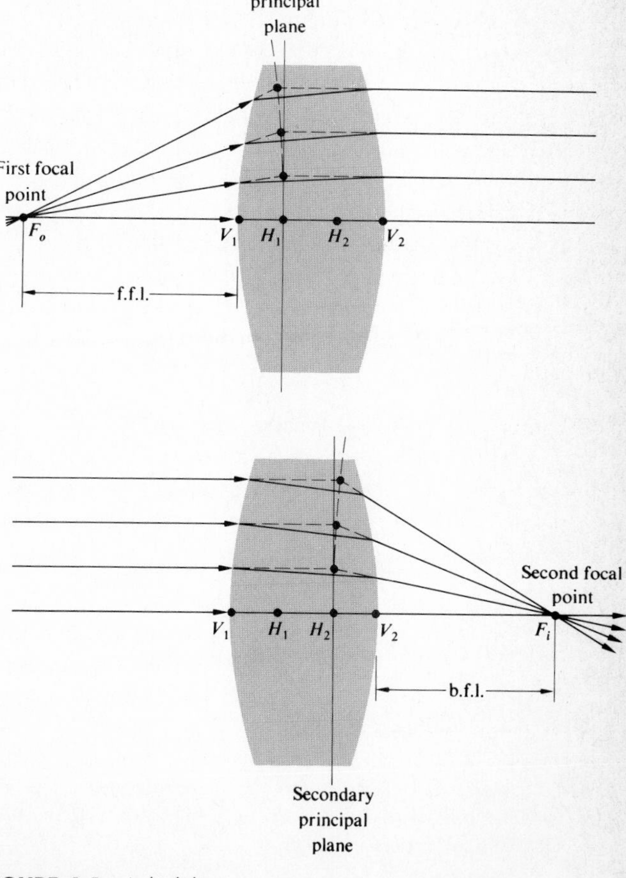

FIGURE 6.1 A thick lens.

lens through its optical center emerges parallel to the incident direction. Extending both the incoming and outgoing rays until they cross the optical axis locates what are called the **nodal points**, N_1 and N_2 in Fig. 6.2. ***When the lens is surrounded on both sides by the same medium, generally air, the nodal and principal points will be coincident.*** The six points, two focal, two principal, and two nodal, constitute the **cardinal points** of the system. As shown in Fig. 6.3, the principal planes can lie completely outside the lens system. Here, though differently configured, each lens in either group has the same power. Observe that in the symmetrical lens the principal planes are, quite reasonably, symmetrically located. In the case of either the planar-concave or planar-convex lens, one principal plane is tangent to the curved surface—as should be expected from the definition (applied to the paraxial region). In contrast, the principal points can be external for meniscus lenses. One often speaks of this succession of shapes with the same power as exemplifying *lens bending*. A rule of thumb for ordinary glass lenses in air is that the separation $\overline{H_1H_2}$ roughly equals one-third the lens thickness $\overline{V_1V_2}$.

The thick lens can be treated as consisting of two spherical refracting surfaces separated by a distance d_l between their vertices, as in Section 5.2.3, where the thin-lens equation was derived. After a great deal of algebraic manipulation,* wherein d_l is not negligible, one arrives at a very interesting result for the thick lens immersed in air. The expression for the conjugate points once again can be put in Gaussian form,

$$\frac{1}{s_o} + \frac{1}{s_i} = \frac{1}{f} \tag{6.1}$$

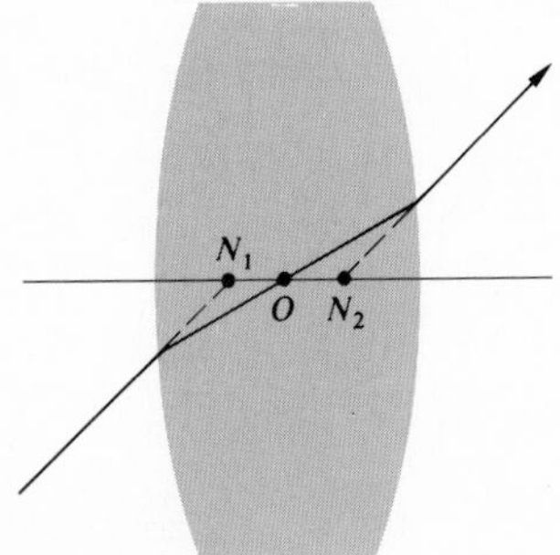

FIGURE 6.2 Nodal points.

*For the complete derivation, see Morgan, *Introduction to Geometrical and Physical Optics*, p. 57. We will be deriving much of this material using matrices in Section 6.2.1.

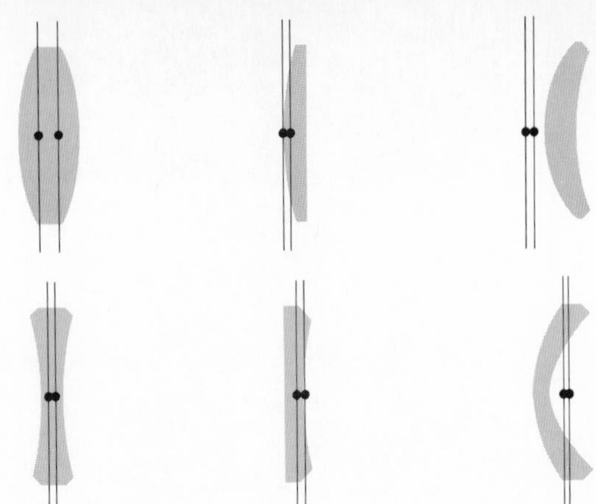

FIGURE 6.3 Lens bending.

provided that both these object and image distances are measured from the first and second principal planes, respectively. Moreover, the *effective focal length*, or simply the *focal length*, *f*, is also reckoned with respect to the principal planes and is given by

$$\frac{1}{f} = (n_l - 1)\left[\frac{1}{R_1} - \frac{1}{R_2} + \frac{(n_l - 1)d_l}{n_l R_1 R_2}\right] \tag{6.2}$$

The principal planes are located at distances of $\overline{V_1H_1} = h_1$ and $\overline{V_2H_2} = h_2$, *which are positive when the planes lie to the right of their respective vertices*. Figure 6.4 illustrates the arrangement of the various quantities. The values of h_1 and h_2 are (Problem 6.18) given by

$$h_1 = -\frac{f(n_l - 1)d_l}{R_2 n_l} \tag{6.3}$$

and

$$h_2 = -\frac{f(n_l - 1)d_l}{R_1 n_l} \tag{6.4}$$

In the same way the Newtonian form of the lens equation holds, as is evident from the similar triangles in Fig. 6.4. Thus

$$x_o x_i = f^2 \tag{6.5}$$

so long as *f* is given the present interpretation. And from the same triangles

$$M_T = \frac{y_i}{y_o} = -\frac{x_i}{f} = -\frac{f}{x_o} \tag{6.6}$$

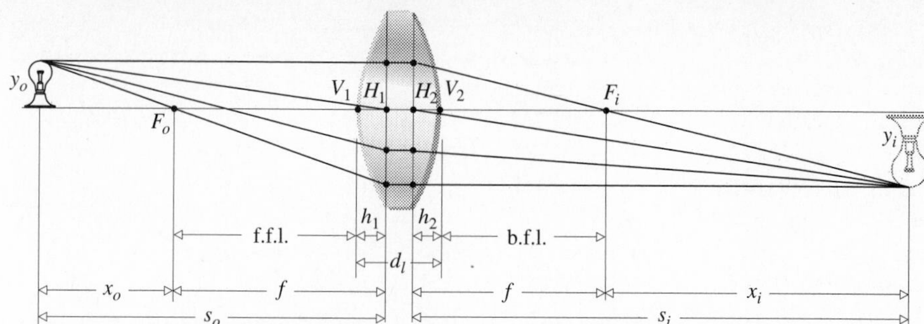

FIGURE 6.4 Thick lens geometry.

Obviously if $d_l \rightarrow 0$, Eqs. (6.1), (6.2), and (6.5) are transformed into the thin-lens expressions Eqs. (5.17), (5.16), and (5.23). As a numerical example, let's find the image distance for an object positioned 30 cm from the vertex of a double convex lens having radii of 20 cm and 40 cm, a thickness of 1 cm, and an index of 1.5. From Eq. (6.2) the focal length (in centimeters) is

$$\frac{1}{f} = (1.5 - 1)\left[\frac{1}{20} - \frac{1}{-40} + \frac{(1.5 - 1)1}{1.5(20)(-40)}\right]$$

and $f = 26.8$ cm. Furthermore,

$$h_1 = -\frac{26.8(0.50)1}{-40(1.5)} = +0.22 \text{ cm}$$

and

$$h_2 = -\frac{26.8(0.5)1}{20(1.5)} = -0.44 \text{ cm}$$

which means that H_1 is to the right of V_1, and H_2 is to the left of V_2. Finally, $s_o = 30 + 0.22$, whereupon

$$\frac{1}{30.2} + \frac{1}{s_i} = \frac{1}{26.8}$$

and $s_i = 238$ cm, measured from H_2.

The principal points are conjugate to each other. In other words, since $f = s_o s_i/(s_o + s_i)$, when $s_o = 0$, s_i must be zero, because f is finite and thus a point at H_1 is imaged at H_2. Furthermore, an object in the first principal plane ($x_o = -f$) is imaged in the second principal plane ($x_i = -f$) with unit magnification ($M_T = 1$). It is for this reason that they are sometimes spoken of as *unit planes*. Any ray directed toward a point on the first principal plane will emerge from the lens as if it originated at the corresponding point (the same distance above or below the axis) on the second principal plane.

Suppose we now have a compound lens consisting of two thick lenses, L_1 and L_2 (Fig. 6.5). Let s_{o1}, s_{i1}, and f_1 and s_{o2}, s_{i2}, and f_2 be the object and image distances and focal lengths for the two lenses, all measured with respect to their own principal planes. We know that the transverse magnification is the product of the magnifications of the individual lenses, that is,

$$M_T = \left(-\frac{s_{i1}}{s_{o1}}\right)\left(-\frac{s_{i2}}{s_{o2}}\right) = -\frac{s_i}{s_o} \tag{6.7}$$

where s_o and s_i are the object and image distances for the com-

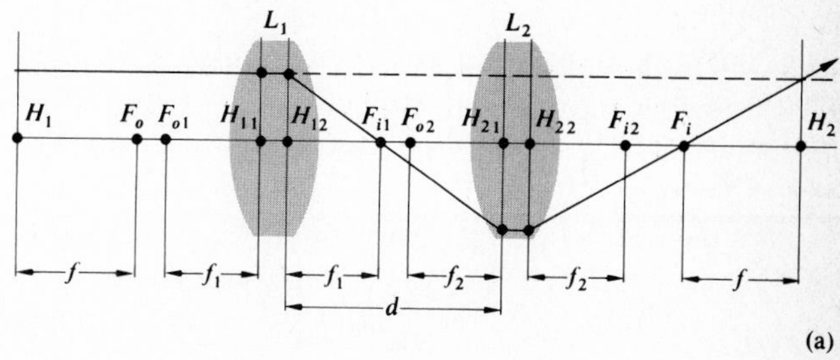

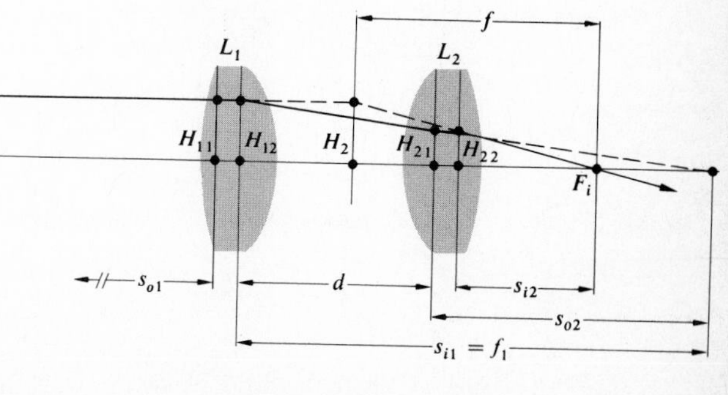

FIGURE 6.5 A compound thick lens.

bination as a whole. When s_o is equal to infinity $s_o = s_{o1}$, $s_{i1} = f_1$, $s_{o2} = -(s_{i1} - d)$, and $s_i = f$. Since

$$\frac{1}{s_{o2}} + \frac{1}{s_{i2}} = \frac{1}{f_2}$$

it follows (Problem 6.1), upon substituting into Eq. (6.7), that

$$-\frac{f_1 s_{i2}}{s_{o2}} = f$$

or

$$f = -\frac{f_1}{s_{o2}}\left(\frac{s_{o2} f_2}{s_{o2} - f_2}\right) = \frac{f_1 f_2}{s_{i1} - d + f_2}$$

Hence

$$\frac{1}{f} = \frac{1}{f_1} + \frac{1}{f_2} - \frac{d}{f_1 f_2} \tag{6.8}$$

This is the effective focal length of the combination of two thick lenses where all distances are measured from principal planes. The principal planes for the system as a whole are located using the expressions

$$\overline{H_{11}H_1} = \frac{fd}{f_2} \tag{6.9}$$

and

$$\overline{H_{22}H_2} = \frac{fd}{f_1} \tag{6.10}$$

which will not be derived here (see Section 6.2.1). We have in effect found an equivalent thick-lens representation of the compound lens. Note that if the component lenses are thin, the pairs of points H_{11}, H_{12}, and H_{21}, H_{22} coalesce, whereupon d becomes the center-to-center lens separation, as in Section 5.2.3. For example, returning to the thin lenses of Fig. 5.34 where $f_1 = -30$ cm, $f_2 = 20$ cm, and $d = 10$ cm, as in Fig. 6.6,

$$\frac{1}{f} = \frac{1}{-30} + \frac{1}{20} - \frac{10}{(-30)(20)}$$

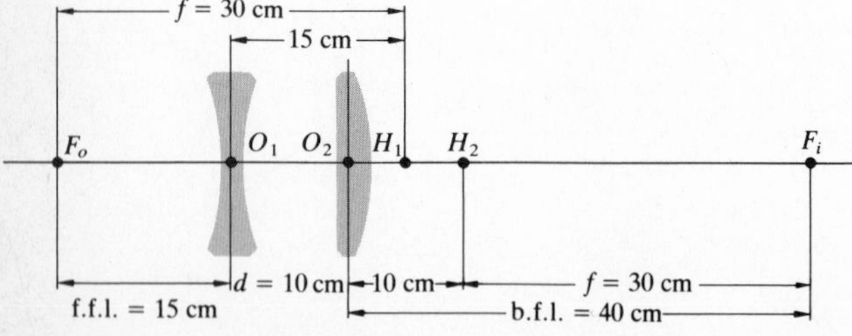

FIGURE 6.6 A compound lens.

so $f = 30$ cm. We found earlier (p. 170) that b.f.l. = 40 cm and f.f.l. = 15 cm. Moreover, since these are thin lenses, Eqs. (6.9) and (6.10) can be written as

$$\overline{O_1H_1} = \frac{30(10)}{20} = +15 \text{ cm}$$

and

$$\overline{O_2H_2} = -\frac{30(10)}{-30} = +10 \text{ cm}$$

Both are positive, and therefore the planes lie to the right of O_1 and O_2, respectively. Both computed values agree with the results depicted in the diagram. If light enters from the right, the system resembles a telephoto lens that must be placed 15 cm from the film plane, yet has an effective focal length of 30 cm.

The same procedures can be extended to three, four, or more lenses. Thus

$$f = f_1\left(-\frac{s_{i2}}{s_{o2}}\right)\left(-\frac{s_{i3}}{s_{o3}}\right)\cdots \tag{6.11}$$

Equivalently, the first two lenses can be envisioned as combined to form a single thick lens whose principal points and focal length are calculated. It, in turn, is combined with the third lens, and so on with each successive element.

6.2 Analytical Ray Tracing

Ray tracing is unquestionably one of the designer's chief tools. Having formulated an optical system on paper, one can mathematically shine rays through it to evaluate its performance. Any ray, paraxial or otherwise, can be traced through the system exactly. Conceptually, it's a simple matter of applying the refraction equation

$$n_i(\hat{\mathbf{k}}_i \times \hat{\mathbf{u}}_n) = n_t(\hat{\mathbf{k}}_t \times \hat{\mathbf{u}}_n) \tag{4.6}$$

at the first surface, locating where the transmitted ray then strikes the second surface, applying the equation once again, and so on all the way through. At one time ***meridional rays*** (those in the plane of the optical axis) were traced almost exclusively, because nonmeridional or ***skew rays*** (which do not intersect the axis) are considerably more complicated to deal with mathematically. The distinction is of less importance to a computer (Fig. 6.7), which simply takes a trifle longer to

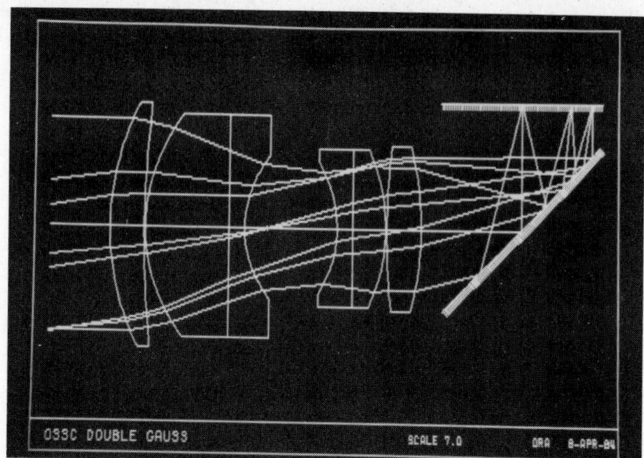

FIGURE 6.7 Computer ray tracing. (Photo courtesy of Optical Research Associates, Pasadena, California.)

make the trace. Thus, whereas it would probably take 10 or 15 minutes for a skilled person with a calculator to evaluate the trajectory of a single skew ray through a single surface, a computer would require less than a thousandth of a second for the same job, and equally important, it would be ready for the next calculation with undiminished enthusiasm.

The simplest case that will serve to illustrate the ray-tracing process is that of a paraxial, meridional ray traversing a thick spherical lens. Applying Snell's Law in Fig. 6.8 at point P_1 yields

$$n_{i1}\theta_{i1} = n_{t1}\theta_{t1}$$

or

$$n_{i1}(\alpha_{i1} + \alpha_1) = n_{t1}(\alpha_{t1} + \alpha_1)$$

Inasmuch as $\alpha_1 = y_1/R_1$, this becomes

$$n_{i1}(\alpha_{i1} + y_1/R_1) = n_{t1}(\alpha_{t1} + y_1/R_1)$$

Rearranging terms yields

$$n_{t1}\alpha_{t1} = n_{i1}\alpha_{i1} - \left(\frac{n_{t1} - n_{i1}}{R_1}\right) y_1$$

but as we saw in Section 5.7.2, the power of a single refracting surface is

$$\mathscr{D}_1 = \frac{(n_{t1} - n_{i1})}{R_1}$$

Hence

$$n_{t1}\alpha_{t1} = n_{i1}\alpha_{i1} - \mathscr{D}_1 y_1 \tag{6.12}$$

This is often called the **refraction equation** pertaining to the first interface. Having undergone refraction at point P_1, the ray advances through the homogeneous medium of the lens to point P_2 on the second interface. The height of P_2 can be expressed as

$$y_2 = y_1 + d_{21}\alpha_{t1} \tag{6.13}$$

on the basis that $\tan \alpha_{t1} \approx \alpha_{t1}$. This is known as the **transfer equation**, because it allows us to follow the ray from P_1 to P_2. Recall that the angles are positive if the ray has a positive slope. Since we are dealing with the paraxial region $d_{21} \approx \overline{V_2V_1}$ and y_2 is easily computed. Equations (6.12) and (6.13) are then used successively to trace a ray through the entire system. Of course, these are meridional rays and because of the lenses' symmetry about the optical axis, such a ray remains in the same meridional plane throughout its sojourn. The process

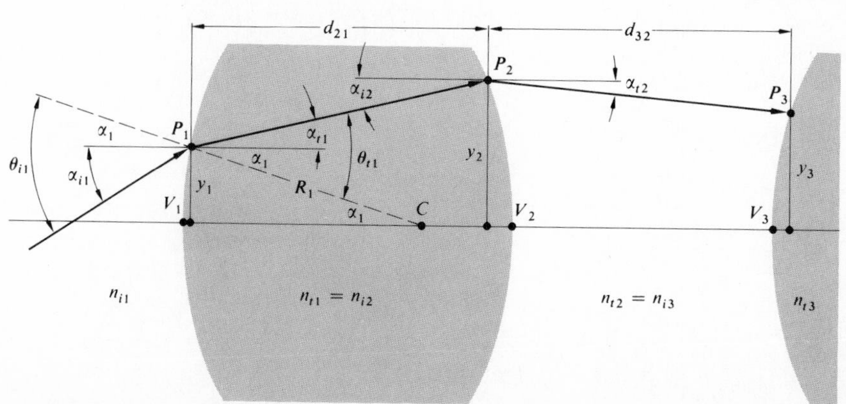

FIGURE 6.8 Ray geometry.

is two-dimensional; there are two equations and two unknowns, α_{t1} and y_2. In contrast, a skew ray would have to be treated in three dimensions.

6.2.1 Matrix Methods

In the beginning of the 1930s, T. Smith formulated an interesting way of handling the ray-tracing equations. The simple linear form of the expressions and the repetitive manner in which they are applied suggested the use of matrices. The processes of refraction and transfer might then be performed mathematically by matrix operators. These initial insights were not widely appreciated for almost 30 years. However, the early 1960s saw a rebirth of interest in this approach.* We shall only outline some of the salient features of the method, leaving a more detailed study to the references.

Matrix Analysis of Lenses

Let's begin by writing the formulas

$$n_{t1}\alpha_{t1} = n_{i1}\alpha_{i1} - \mathscr{D}_1 y_{i1} \tag{6.14}$$

and

$$y_{t1} = 0 + y_{i1} \tag{6.15}$$

which are not very insightful, since we merely replaced y_1 in Eq. (6.12) by the symbol y_{i1} and then let $y_{t1} = y_{i1}$. This last bit of business is for purely cosmetic purposes, as you will see in a moment. In effect, it simply says that the height of reference point P_1 above the axis in the incident medium (y_{i1}) equals its height in the transmitting medium (y_{t1})—which is obvious. But now the pair of equations can be recast in matrix form as

$$\begin{bmatrix} n_{t1}\alpha_{t1} \\ y_{t1} \end{bmatrix} = \begin{bmatrix} 1 & -\mathscr{D}_1 \\ 0 & 1 \end{bmatrix} \begin{bmatrix} n_{i1}\alpha_{i1} \\ y_{i1} \end{bmatrix} \tag{6.16}$$

This could equally well be written as

$$\begin{bmatrix} \alpha_{t1} \\ y_{t1} \end{bmatrix} = \begin{bmatrix} n_{i1}/n_{t1} & -\mathscr{D}_1/n_{t1} \\ 0 & 1 \end{bmatrix} \begin{bmatrix} \alpha_{i1} \\ y_{i1} \end{bmatrix} \tag{6.17}$$

so that the precise form of the 2×1 column matrices is actually a matter of preference. In any case, these can be envisioned as rays on either side of P_1, one before and the other after refraction. Accordingly, using $\mathscr{r}_{t1}$ and $\mathscr{r}_{i1}$ for the two rays, we can write

$$\mathscr{r}_{t1} \equiv \begin{bmatrix} n_{t1}\alpha_{t1} \\ y_{t1} \end{bmatrix} \quad \text{and} \quad \mathscr{r}_{i1} \equiv \begin{bmatrix} n_{i1}\alpha_{i1} \\ y_{i1} \end{bmatrix} \tag{6.18}$$

The 2×2 matrix is the **refraction matrix**, denoted as

$$\mathscr{R}_1 \equiv \begin{bmatrix} 1 & -\mathscr{D}_1 \\ 0 & 1 \end{bmatrix} \tag{6.19}$$

and Eq. (6.16) can be stated concisely as

$$\mathscr{r}_{t1} = \mathscr{R}_1 \mathscr{r}_{i1} \tag{6.20}$$

which just says that $\mathscr{R}_1$ transforms the rays $\mathscr{r}_{i1}$ into the ray $\mathscr{r}_{t1}$ during refraction at the first interface. Notice that the way we arranged the terms in Eqs. (6.14) and (6.15) determined the form of the refraction matrix. Accordingly, several equivalent variations of the matrix can be found in the literature.

From Fig. 6.8 we have $n_{i2}\alpha_{i2} = n_{t1}\alpha_{t1}$, that is,

$$n_{i2}\alpha_{i2} = n_{t1}\alpha_{t1} + 0 \tag{6.21}$$

and

$$y_{i2} = d_{21}a_{t1} + y_{t1} \tag{6.22}$$

where $n_{i2} = n_{t1}$, $\alpha_{i2} = \alpha_{t1}$, and use was made of Eq. (6.13), with y_2 rewritten as y_{i2} to make things pretty. Thus

$$\begin{bmatrix} n_{i2}\alpha_{i2} \\ y_{i2} \end{bmatrix} = \begin{bmatrix} 1 & 0 \\ d_{21}/n_{t1} & 1 \end{bmatrix} \begin{bmatrix} n_{t1}\alpha_{t1} \\ y_{t1} \end{bmatrix} \tag{6.23}$$

The **transfer matrix**

$$\mathscr{T}_{21} \equiv \begin{bmatrix} 1 & 0 \\ d_{21}/n_{t1} & 1 \end{bmatrix} \tag{6.24}$$

takes the transmitted ray at P_1 (i.e., $\mathscr{r}_{t1}$) and transforms it into the incident ray at P_2:

$$\mathscr{r}_{i2} \equiv \begin{bmatrix} n_{i2}\alpha_{i2} \\ y_{i2} \end{bmatrix}$$

Hence Eqs. (6.21) and (6.22) become simply

$$\mathscr{r}_{i2} = \mathscr{T}_{21}\mathscr{r}_{t1} \tag{6.25}$$

*For further reading, see K. Hallbach, "Matrix Representation of Gaussian Optics," *Am. J. Phys.* **32**, 90 (1964); W. Brouwer, *Matrix Methods in Optical Instrument Design*; E. L. O'Neill, *Introduction to Statistical Optics*; or A. Nussbaum, *Geometric Optics*.

If use is made of Eq. (6.20), this becomes

$$\mathbf{r}_{i2} = \mathscr{T}_{21}\mathscr{R}_1\mathbf{r}_{i1} \tag{6.26}$$

The 2 × 2 matrix formed by the product of the transfer and refraction matrices $\mathscr{T}_{21}\mathscr{R}_1$ will carry the ray incident at P_1 into the ray incident at P_2. Notice that the determinant of $\mathscr{T}_{21}$, denoted by $|\mathscr{T}_{21}|$, equals 1; that is, $(1)(1) - (0)(d_{21}/n_{t1}) = 1$. Similarly $|\mathscr{R}_1| = 1$, and since the determinant of a matrix product equals the product of the individual determinants, $|\mathscr{T}_{21}\mathscr{R}_1| = 1$. This provides a quick check on the computations. Carrying the procedure through the second interface (Fig. 6.8) of the lens, which has a refraction matrix $\mathscr{R}_2$, it follows that

$$\mathbf{r}_{t2} = \mathscr{R}_2\mathbf{r}_{i2} \tag{6.27}$$

where

$$\mathscr{R}_2 \equiv \begin{bmatrix} 1 & -\mathscr{D}_2 \\ 0 & 1 \end{bmatrix}$$

and the power of the second surface is

$$\mathscr{D}_2 = \frac{(n_{t2} - n_{i2})}{R_2}$$

From Eq. (6.26)

$$\mathbf{r}_{t2} = \mathscr{R}_2\mathscr{T}_{21}\mathscr{R}_1\mathbf{r}_{i1} \tag{6.28}$$

The **system matrix** $\mathscr{A}$ is then defined as

$$\mathscr{A} \equiv \mathscr{R}_2\mathscr{T}_{21}\mathscr{R}_1 \tag{6.29}$$

and has the form

$$\mathscr{A} = \begin{bmatrix} a_{11} & a_{12} \\ a_{21} & a_{22} \end{bmatrix} \tag{6.30}$$

Inasmuch as

$$\mathscr{A} = \begin{bmatrix} 1 & -\mathscr{D}_2 \\ 0 & 1 \end{bmatrix}\begin{bmatrix} 1 & 0 \\ d_{21}/n_{t1} & 1 \end{bmatrix}\begin{bmatrix} 1 & -\mathscr{D}_1 \\ 0 & 1 \end{bmatrix}$$

or

$$\mathscr{A} = \begin{bmatrix} 1 & -\mathscr{D}_2 \\ 0 & 1 \end{bmatrix}\begin{bmatrix} 1 & -\mathscr{D}_1 \\ \dfrac{d_{21}}{n_{t1}} & 1 - \dfrac{\mathscr{D}_1 d_{21}}{n_{t1}} \end{bmatrix}$$

it follows that

$$\mathscr{A} = \begin{bmatrix} 1 - \dfrac{\mathscr{D}_2 d_{21}}{n_{t1}} & -\mathscr{D}_1 - \mathscr{D}_2 + \dfrac{\mathscr{D}_2\mathscr{D}_1 d_{21}}{n_{t1}} \\ \dfrac{d_{21}}{n_{t1}} & \dfrac{\mathscr{D}_1 d_{21}}{n_{t1}} \end{bmatrix}$$

and again $|\mathscr{A}| = 1$ (Problem 6.17). Because we are working with only one lens, let's simplify the notation a little letting $d_{21} = d_l$ and $n_{t1} = n_l$. Consequently,

$$\begin{bmatrix} a_{11} & a_{12} \\ a_{21} & a_{22} \end{bmatrix} = \begin{bmatrix} 1 - \dfrac{\mathscr{D}_2 d_l}{n_l} & -\mathscr{D}_1 - \mathscr{D}_2 + \dfrac{\mathscr{D}_1\mathscr{D}_2 d_l}{n_l} \\ \dfrac{d_l}{n_l} & 1 - \dfrac{\mathscr{D}_1 d_l}{n_l} \end{bmatrix} \tag{6.31}$$

The value of each element in $\mathscr{A}$ is expressed in terms of the physical lens parameters, such as thickness, index, and radii (via $\mathscr{D}$). Thus the cardinal points, which are properties of the lens determined solely by its makeup, should be deducible from $\mathscr{A}$. The system matrix in this case, Eq. (6.31), transforms an incident ray at the first surface to an emerging ray at the second surface; as a reminder, we will write it as $\mathscr{A}_{21}$.

The concept of image formation enters rather directly (Fig. 6.9) after introduction of appropriate object and image planes. Consequently, the first operator $\mathscr{T}_{1O}$ transfers the reference point from the object (i.e., P_O to P_1). The next operator $\mathscr{A}_{21}$ then carries the ray through the lens, and a final transfer $\mathscr{T}_{I2}$ brings it to the image plane (i.e., P_I). Thus the ray at the image point ($\mathbf{r}_I$) is given by

$$\mathbf{r}_I = \mathscr{T}_{I2}\mathscr{A}_{21}\mathscr{T}_{1O}\mathbf{r}_O \tag{6.32}$$

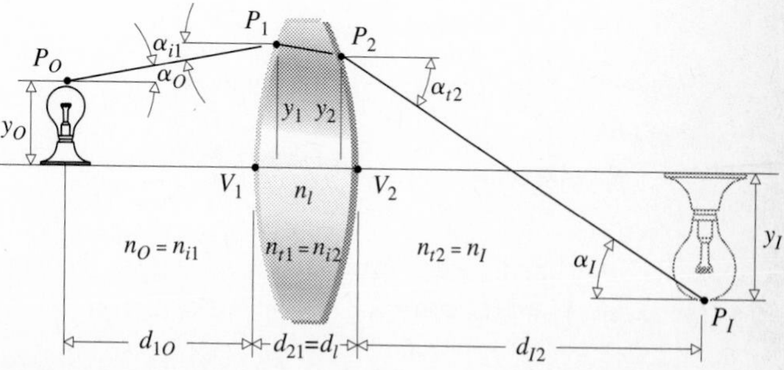

FIGURE 6.9 Image geometry.

where $\boldsymbol{r}_O$ is the ray at P_O. In component form this is

$$\begin{bmatrix} n_I\alpha_I \\ y_I \end{bmatrix} = \begin{bmatrix} 1 & 0 \\ d_{I2}/n_I & 1 \end{bmatrix}\begin{bmatrix} a_{11} & a_{12} \\ a_{21} & a_{22} \end{bmatrix} \times \begin{bmatrix} 1 & 0 \\ d_{1O}/n_O & 1 \end{bmatrix}\begin{bmatrix} n_O\alpha_O \\ y_O \end{bmatrix} \tag{6.33}$$

Notice that $\mathcal{T}_{1O}\boldsymbol{r}_O = \boldsymbol{r}_{i1}$ and that $\mathcal{A}_{21}\boldsymbol{r}_{i1} = \boldsymbol{r}_{t2}$, hence $\mathcal{T}_{I2}\boldsymbol{r}_{t2} = \boldsymbol{r}_I$. The subscripts O, 1, 2, …, I correspond to reference points P_O, P_1, P_2, and so on, and subscripts i and t denote the side of the reference point (i.e., whether incident or transmitted). Operation by a refraction matrix will change i to t but not the reference point designation. On the other hand, operation by a transfer matrix obviously does change the latter.

Ordinarily, the physical significances of the components of $\mathcal{A}$ are found by expanding out Eq. (6.33), but this is too involved to do here. Instead, let's return to Eq. (6.31) and examine several of the terms. For example,

$$-a_{12} = \mathfrak{D}_1 + \mathfrak{D}_2 - \mathfrak{D}_1\mathfrak{D}_2 d_l/n_l$$

If we suppose, for the sake of simplicity, that the lens is in air, then

$$\mathfrak{D}_1 = \frac{n_l - 1}{R_1} \quad \text{and} \quad \mathfrak{D}_2 = \frac{n_l - 1}{-R_2}$$

as in Eqs. (5.70) and (5.71). Hence

$$-a_{12} = (n_l - 1)\left[\frac{1}{R_1} - \frac{1}{R_2} + \frac{(n_l - 1)d_l}{R_1 R_2 n_l}\right]$$

But this is the expression for the focal length of a thick lens [Eq. (6.2)]; in other words,

$$a_{12} = -1/f \tag{6.34}$$

Thus the power of the lens as a whole is given by

$$-a_{12} = \mathfrak{D}_l = \mathfrak{D}_1 + \mathfrak{D}_2 - \frac{\mathfrak{D}_1\mathfrak{D}_2 d_l}{n_l}$$

If the embedding media were different on each side of the lens (Fig. 6.10), this would become

$$a_{12} = -\frac{n_{i1}}{f_o} = -\frac{n_{t2}}{f_i} \tag{6.35}$$

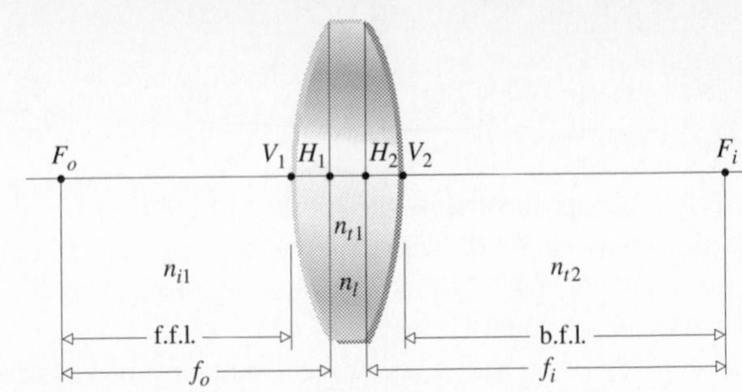

FIGURE 6.10 Principal planes and focal lengths.

Similarly, it is left as a problem to verify that

$$\overline{V_1H_1} = \frac{n_{i1}(1 - a_{11})}{-a_{12}} \tag{6.36}$$

and

$$\overline{V_2H_2} = \frac{n_{t2}(a_{22} - 1)}{-a_{12}} \tag{6.37}$$

which locate the principal points.

As an example of how the technique can be used, let's apply it, at least in principle, to the Tessar lens* shown in Fig. 6.11. The system matrix has the form

$$\mathcal{A}_{71} = \mathcal{R}_7\mathcal{T}_{76}\mathcal{R}_6\mathcal{T}_{65}\mathcal{R}_5\mathcal{T}_{54}\mathcal{R}_4\mathcal{T}_{43}\mathcal{R}_3\mathcal{T}_{32}\mathcal{R}_2\mathcal{T}_{21}\mathcal{R}_1$$

where

$$\mathcal{T}_{21} = \begin{bmatrix} 1 & 0 \\ \dfrac{0.357}{1.6116} & 1 \end{bmatrix} \qquad \mathcal{T}_{32} = \begin{bmatrix} 1 & 0 \\ \dfrac{0.189}{1} & 1 \end{bmatrix}$$

$$\mathcal{T}_{43} = \begin{bmatrix} 1 & 0 \\ \dfrac{0.081}{1.6053} & 1 \end{bmatrix}$$

*This particular example was chosen primarily because Nussbaum's book *Geometric Optics* contains a simple Fortran computer program written specifically for this lens. It would be almost silly to evaluate the system matrix by hand.

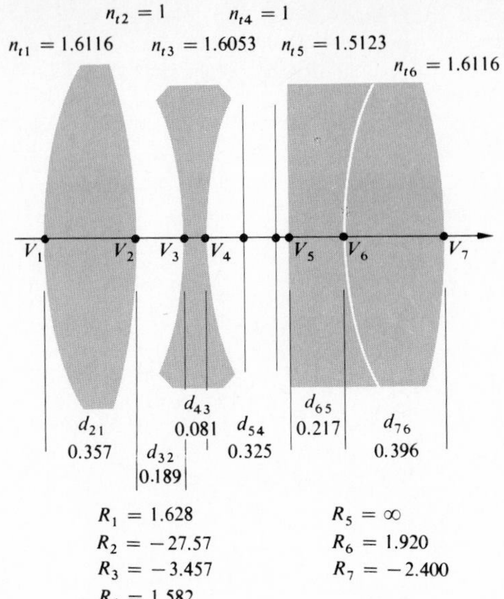

FIGURE 6.11 A Tessar.

and so forth. Furthermore,

$$\mathcal{R}_1 = \begin{bmatrix} 1 & -\dfrac{1.6116 - 1}{1.628} \\ 0 & 1 \end{bmatrix} \quad \mathcal{R}_2 = \begin{bmatrix} 1 & -\dfrac{1 - 1.6116}{-27.57} \\ 0 & 1 \end{bmatrix}$$

$$\mathcal{R}_3 = \begin{bmatrix} 1 & -\dfrac{1.6053 - 1}{-3.457} \\ 0 & 1 \end{bmatrix}$$

and so on. Multiplying out the matrices, in what is obviously a horrendous, though conceptually simple, calculation, one presumably will get

$$\mathcal{A}_{71} = \begin{bmatrix} 0.848 & -0.198 \\ 1.338 & 0.867 \end{bmatrix}$$

and from that, $f = 5.06$, $\overline{V_1H_1} = 0.77$, and $\overline{V_7H_2} = -0.67$.

Thin Lenses

As a last point, it is often convenient to consider a system of thin lenses using the matrix representation. To that end, return to Eq. (6.31). It describes the system matrix for a single lens, and if we let $d_l \to 0$, it corresponds to a thin lens. This is equivalent to making $\mathcal{T}_{21}$ a unit matrix; thus

$$\mathcal{A} = \mathcal{R}_2\mathcal{R}_1 = \begin{bmatrix} 1 & -(\mathcal{D}_1 + \mathcal{D}_2) \\ 0 & 1 \end{bmatrix}$$

But as we saw in Section 5.7.2, the power of a thin lens $\mathcal{D}$ is the sum of the powers of its surfaces. Hence

$$\mathcal{A} = \begin{bmatrix} 1 & -\mathcal{D} \\ 0 & 1 \end{bmatrix} = \begin{bmatrix} 1 & -1/f \\ 0 & 1 \end{bmatrix}$$

In addition, for two thin lenses separated by a distance d, in air, the system matrix is

$$\mathcal{A} = \begin{bmatrix} 1 & -1/f_2 \\ 0 & 1 \end{bmatrix}\begin{bmatrix} 1 & 0 \\ d & 1 \end{bmatrix}\begin{bmatrix} 1 & -1/f_1 \\ 0 & 1 \end{bmatrix}$$

or

$$\mathcal{A} = \begin{bmatrix} 1 - d/f_2 & -1/f_1 + d/f_1f_2 - 1/f_2 \\ d & -d/f_1 + 1 \end{bmatrix}$$

Clearly then,

$$-a_{12} = \frac{1}{f} = \frac{1}{f_1} + \frac{1}{f_2} - \frac{d}{f_1f_2}$$

and from Eqs. (6.36) and (6.37)

$$\overline{O_1H_1} = fd/f_2 \qquad \overline{O_2H_2} = -fd/f_1$$

all of which should be quite familiar by now. Note how easy it would be with this approach to find the focal length and principal points for a compound lens composed of three, four, or more thin lenses.

Matrix Analysis of Mirrors

To derive the appropriate matrix for reflection, consult Fig. 6.12, which depicts a concave spherical mirror, and write down two equations that describe the incident and reflected rays. Again, the final form of the matrix depends on how we arrange these two equations and the signs we assign to the various quantities. What's needed is an expression relating the ray angles and another relating their heights at the point of interaction with the mirror.

First let's consider the ray angles. The Law of Reflection is $\theta_i = \theta_r$; therefore from the geometry $\tan(\alpha_i - \theta_i) = y_i/R$, and

$$(\alpha_i - \theta_i) \approx y_i/R \tag{6.38}$$

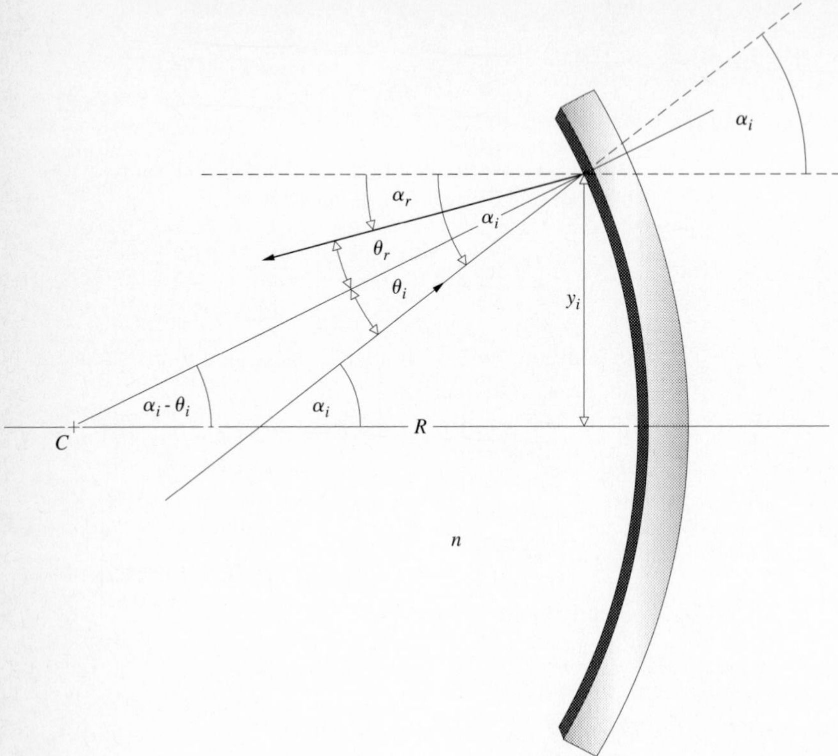

FIGURE 6.12 The geometry for reflection from a mirror. The ray angles α_i and α_r are measured from the direction of the optical axis.

Taking these angles to be positive, y is positive, but R isn't, and this equation will be in error as soon as we enter a negative value for the radius. Therefore rewrite it as $(\alpha_i - \theta_i) = -y_i/R$. Now to get α_r into the analysis, note that $\alpha_i = \alpha_r + 2\theta_i$ and $\theta_i = (\alpha_i - \alpha_r)/2$. Substituting this into Eq. (6.38) yields $\alpha_r = -\alpha_i - 2y_i/R$, and multiplying by n, the index of the surrounding medium (where usually $n = 1$), leads to

$$n\alpha_r = -n\alpha_i - 2ny_i/R$$

The second necessary equation is simply $y_r = y_i$ and so

$$\begin{bmatrix} n\alpha_r \\ y_r \end{bmatrix} = \begin{bmatrix} -1 & -2n/R \\ 0 & 1 \end{bmatrix} \begin{bmatrix} n\alpha_i \\ y_i \end{bmatrix}$$

Thus the mirror matrix $\mathcal{M}$ for a spherical configuration is given by

$$\mathcal{M}_{\circ} = \begin{bmatrix} -1 & -2n/R \\ 0 & 1 \end{bmatrix} \tag{6.39}$$

remembering from Eq. (5.49) that $f = -2/R$.

Flat Mirrors and the Planar Optical Cavity

For a flat mirror ($R \rightarrow \infty$) in air ($n = 1$), the matrix is

$$\mathcal{M}_{|} = \begin{bmatrix} -1 & 0 \\ 0 & 1 \end{bmatrix}$$

where the minus sign in the first position reverses the ray upon reflection. Figure 6.13 shows two planar mirrors facing each other forming an **optical cavity** (p. 585). Light leaves point O, traverses the gap in the positive direction, is reflected by mirror-1, retraces the gap in the negative direction, and is reflected by mirror-2. The system matrix is

$$\mathcal{A} = \mathcal{M}_{|2}\mathcal{T}_{21}\mathcal{M}_{|1}\mathcal{T}_{12}$$

$$\mathcal{A} = \begin{bmatrix} -1 & 0 \\ 0 & 1 \end{bmatrix} \begin{bmatrix} 1 & 0 \\ -d & 1 \end{bmatrix} \begin{bmatrix} -1 & 0 \\ 0 & 1 \end{bmatrix} \begin{bmatrix} 1 & 0 \\ d & 1 \end{bmatrix}$$

and

$$\mathcal{A} = \begin{bmatrix} 1 & 0 \\ 2d & 1 \end{bmatrix}$$

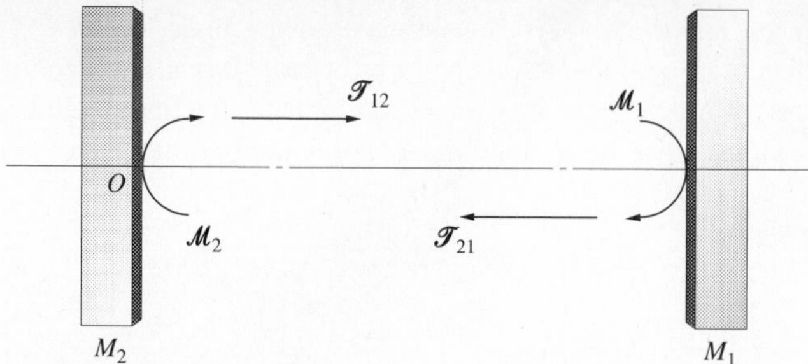

FIGURE 6.13 A schematic representation of a planar cavity formed by mirrors M_1 and M_2.

where again the determinant of the system matrix is one: $|\mathscr{A}| = 1$. Presumably, if the initial ray was axial ($\alpha = 0$), the system matrix should bring it back to its starting point so that the final ray $\mathscr{r}_f$ is identical to the initial ray $\mathscr{r}_i$. That is,

$$\mathscr{A}\mathscr{r}_i = \mathscr{r}_f = \mathscr{r}_i$$

This is a special kind of mathematical relationship known as an ***eigenvalue equation*** where, a bit more generally,

$$\mathscr{A}\mathscr{r}_i = a\mathscr{r}_i$$

and a is a constant. In other words,

$$\begin{bmatrix} 1 & 0 \\ 2d & 1 \end{bmatrix} \begin{bmatrix} \alpha_i \\ y_i \end{bmatrix} = a \begin{bmatrix} \alpha_i \\ y_i \end{bmatrix}$$

If $\alpha_i = 0$ and the initial ray is launched axially, then $y_i = ay_i$ and it follows that $a = 1$. The system matrix functions like a unit matrix that carries $\mathscr{r}_i$ into $\mathscr{r}_i$ after two reflections. Axial rays of light travel back and forth across the so-called ***resonant cavity*** without escaping.

Cavities can be constructed in a number of different ways using a variety of mirrors (Fig. 13.12, p. 588). If after traversing a cavity some number of times the light ray returns to its original location and orientation, the beam will be trapped and the cavity is said to be *stable*; that's why the eigenvalue discussion is important. To analyze the *confocal cavity* composed of two concave spherical mirrors facing each other, see Problem 6.24.

6.3 Aberrations

To be sure, we already know that first-order theory is no more than a good approximation—an exact ray trace or even measurements performed on a prototype system would certainly reveal inconsistencies with the corresponding paraxial description. Such departures from the idealized conditions of Gaussian Optics are known as **aberrations**. There are two main types: **chromatic aberrations** (which arise from the fact that n is actually a function of frequency or color) and **monochromatic aberrations**. The latter occur even with light that is quasimonochromatic, and they in turn fall into two subgroupings. There are monochromatic aberrations such as *spherical aberration, coma*, and *astigmatism* that deteriorate the image, making it unclear. In addition, there are aberrations that deform the image, for example, *Petzval field curvature* and *distortion*.

We have known all along that spherical surfaces in general would yield perfect imagery only in the paraxial region. Now we must determine the kind and extent of deviations that result simply from using those surfaces with finite apertures. By the judicious manipulation of a system's physical parameters (e.g., the powers, shapes, thicknesses, glass types, and separations of the lenses, as well as the locations of stops), these aberrations can indeed be minimized. In effect, one cancels out the most undesirable faults by a slight change in the shape of a lens here, or a shift in the position of a stop there (very much like trimming up a circuit with small variable capacitors, coils, and pots). When it's all finished, the unwanted deformations of the wavefront incurred as it passes through one surface will, it is hoped, be negated as it traverses some other surfaces farther down the line.

As early as 1950, ray-tracing programs were being developed for the new digital computers, and by 1954 efforts were already under way to create lens-designing software. In the early 1960s, computerized lens design was a tool of the trade used by manufacturers worldwide. Today there are elaborate computer programs for "automatically" designing and analyzing the performance of all sorts of complicated optical systems.

6.3.1 Monochromatic Aberrations

The paraxial treatment was based on the assumption that sin φ, as in Fig. 5.7, could be represented satisfactorily by φ alone; that is, the system was restricted to operating in an extremely narrow region about the optical axis. Obviously, if rays from the periphery of a lens are to be included in the formation of an image, the statement $\sin\varphi \approx \varphi$ is somewhat unsatisfactory. Recall that we also occasionally wrote Snell's Law simply as $n_i\theta_i = n_t\theta_t$, which again would be inappropriate. In any event, if the first two terms in the expansion

$$\sin\varphi = \varphi - \frac{\varphi^3}{3!} + \frac{\varphi^5}{5!} - \frac{\varphi^7}{7!} + \cdots \qquad [5.7]$$

are retained as an improved approximation, we have the so-called *third-order theory*. Departures from first-order theory that then result are embodied in the five *primary aberrations* (spherical aberration, coma, astigmatism, field curvature, and distortion). These were first studied in detail by Ludwig von Seidel (1821–1896) in the 1850s. Accordingly, they are frequently spoken of as the ***Seidel aberrations***. In addition to the first two contributions, the series obviously contains many other terms, smaller to be sure, but still to be reckoned with. Thus there are most certainly *higher-order aberrations*. The difference between the results of exact ray tracing and the computed primary aberrations can therefore be thought of as the sum of all contributing higher-order aberrations. We shall restrict this discussion exclusively to the primary aberrations.

Spherical Aberration

Let's return for a moment to Section 5.2.2 (p. 153), where we computed the conjugate points for a single refracting spherical interface. We found that for the paraxial region,

$$\frac{n_1}{s_o} + \frac{n_2}{s_i} = \frac{n_2 - n_1}{R} \qquad [5.8]$$

If the approximations for ℓ_o and ℓ_i are improved a bit (Problem 6.26), we get the third-order expression:

$$\frac{n_1}{s_o} + \frac{n_2}{s_i} = \frac{n_2 - n_1}{R} + h^2\left[\frac{n_1}{2s_o}\left(\frac{1}{s_o} + \frac{1}{R}\right)^2 + \frac{n_2}{2s_i}\left(\frac{1}{R} - \frac{1}{s_i}\right)^2\right] \qquad (6.40)$$

The additional term, which varies approximately as h^2, is clearly a measure of the deviation from first-order theory. As shown in Fig. 6.14, rays striking the surface at greater distances above the axis (h) are focused nearer the vertex. In brief, spherical aberration, or SA, corresponds to a dependence of focal length on aperture for nonparaxial rays. Similarly, for a converging lens, as in Fig. 6.15, the marginal rays will, in effect, be bent too much, being focused in front of the

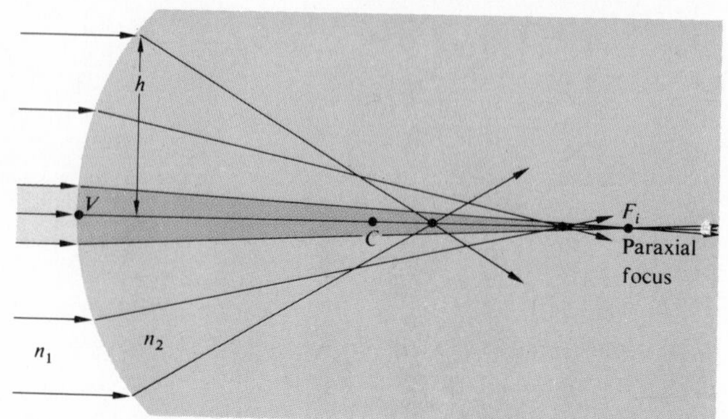

FIGURE 6.14 Spherical aberration resulting from refraction at a single interface.

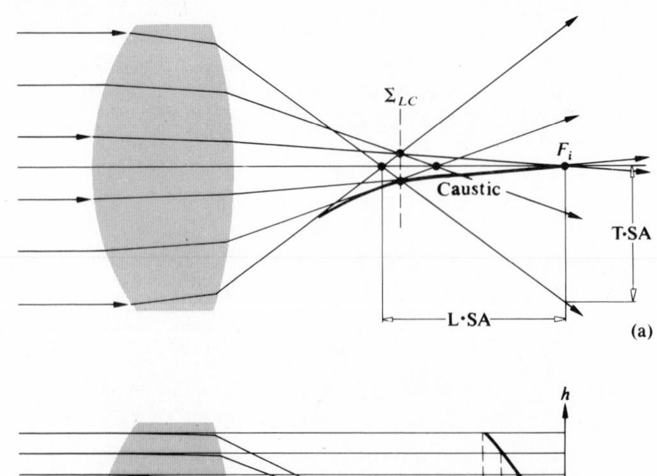

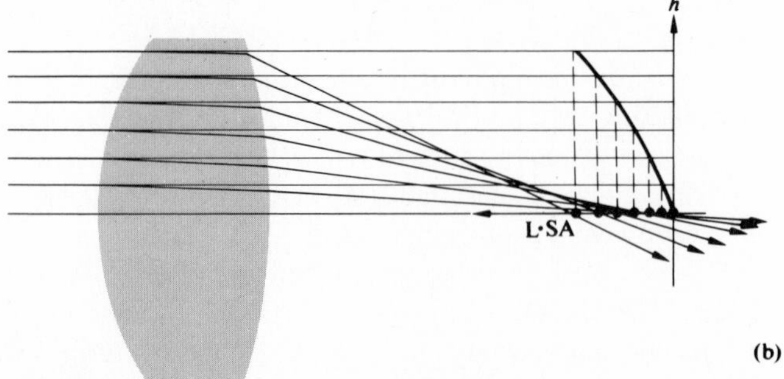

FIGURE 6.15 Spherical aberration for a lens. The envelope of the refracted rays is called a caustic. The intersection of the marginal rays and the caustic locates Σ_{LC}.

paraxial rays. Keep in mind that spherical aberration pertains only to object points that are on the optical axis. The distance between the axial intersection of a ray and the paraxial focus, F_i, is known as the **longitudinal spherical aberration**, or L · SA, of that ray. In this case, the SA is *positive*. In contrast, the marginal rays for a diverging lens will generally intersect the axis behind the paraxial focus, and its spherical aberration is therefore *negative*.

To better appreciate aberrations in terms of their effects on the wavefronts, consider the light from a point source traversing an optical system. Ideally, if the transmitted wavefront at the exit pupil is a sphere centered on the Gaussian image point, then the image is perfect; if not, it's aberrated. The *wave aberrations* are the deviations in optical path length between the actual and ideal wavefronts, often specified by the maximum values given in microns, nanometers, or wavelengths.

Under ideal conditions, a lens will form an image of a point source, which is a tiny bright circular disk surrounded by rings that are so faint they're hardly noticeable (Fig. 10.28); that's the Airy pattern. Spherical aberration essentially shifts light out of the central disk into the surrounding rings, which become far more prominent. For example, Rayleigh established that one quarter-wave of spherical aberration diminishes the irradiance of the image disk by about 20%.

Accordingly, if a screen is placed at F_i in Fig. 6.15, the image of a star will appear as a bright central spot on the axis surrounded by a symmetrical halo delineated by the cone of marginal rays. For an extended image, SA would reduce the contrast and degrade the details.

The height above the axis where a given ray strikes this screen is called the **traverse** (or **lateral**) **spherical aberration**, or T · SA for short. Evidently, SA can be reduced by stopping down the aperture—but that reduces the amount of light entering the system as well. Notice that if the screen is moved to the position labeled Σ_{LC}, the image blur will have its smallest diameter. This is known as the ***circle of least confusion***, and Σ_{LC} is generally the best place to observe the image. If a lens exhibits appreciable SA, it will have to be refocused after it is stopped down, because the position of Σ_{LC} will approach F_i as the aperture decreases.

The amount of spherical aberration, when the aperture and focal length are fixed, varies with both the object distance and the lens shape. For a converging lens, the nonparaxial rays are too strongly bent. Yet if we imagine the lens as roughly resembling two prisms joined at their bases, it is evident that *the incident ray will undergo a minimum deviation when it makes, more or less, the same angle as does the emerging ray* (Section 5.5.1). A striking example is illustrated in Fig. 6.16, where simply turning the lens around markedly reduces the SA. When the object is at "infinity," a simple concave or convex lens that has an almost, but not quite, flat rear side will suffer a minimum amount of spherical aberration. In the same way, if the object and image distances are to be equal ($s_o = s_i = 2f$), the lens should be equiconvex to minimize SA. A combination of a converging and a diverging lens (as in an achromatic doublet) can also be utilized to diminish spherical aberration.

Recall that the aspherical lenses of Section 5.2.1 were completely free of spherical aberration for a specific pair of conjugate points. Moreover, Huygens seems to have been the first to discover that two such axial points exist for spherical surfaces as well. These are shown in Fig. 6.17*a*, which depicts rays issuing from P and leaving the surface as if they came from P'. It is left as a problem to show that the appropriate locations of P and P' are those indicated in the figure. Just as with the aspherical lenses, spherical lenses can be formed that have this same zero SA for the pair of points P and P'. One simply grinds another surface of radius $\overline{PA}$ centered on P to form either a positive- or negative-meniscus lens. The oil-immersion microscope objective uses this principle to great advan-

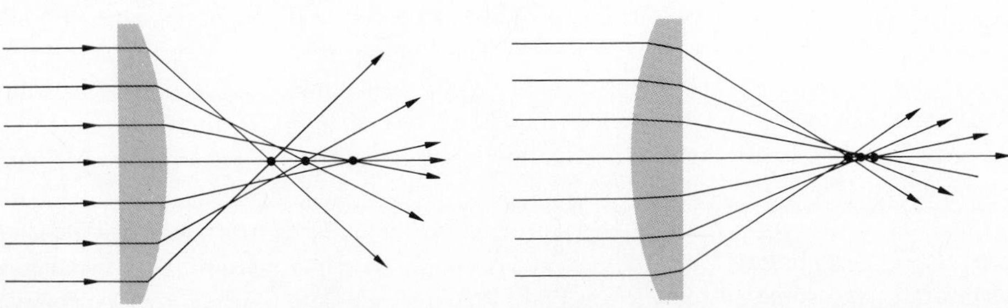

FIGURE 6.16 SA for a planar-convex lens.

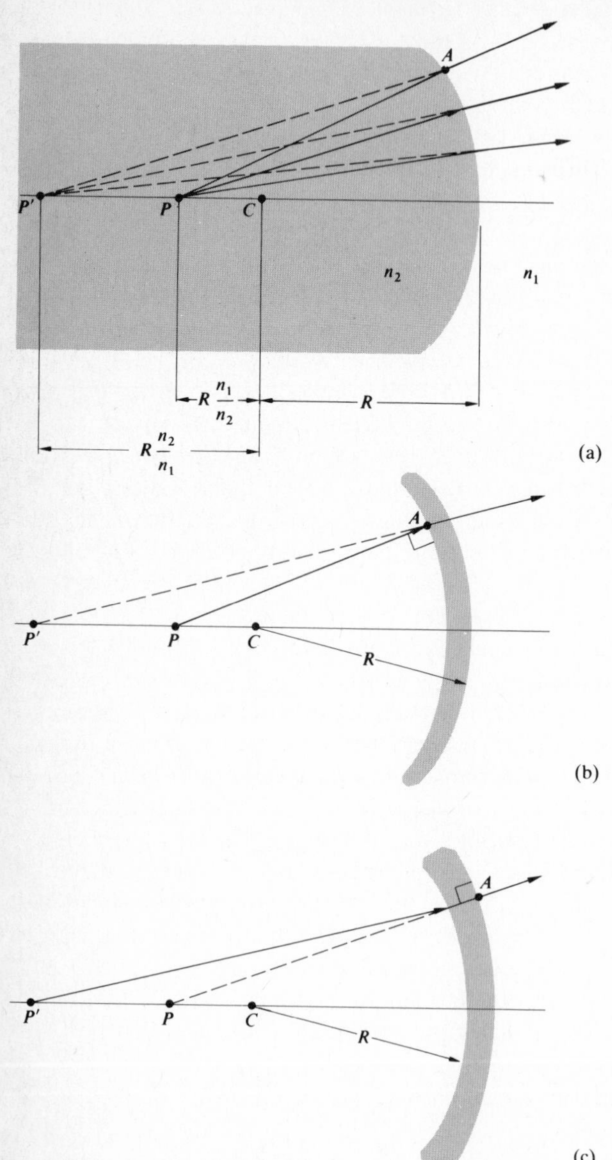

FIGURE 6.17 Corresponding axial points for which SA is zero.

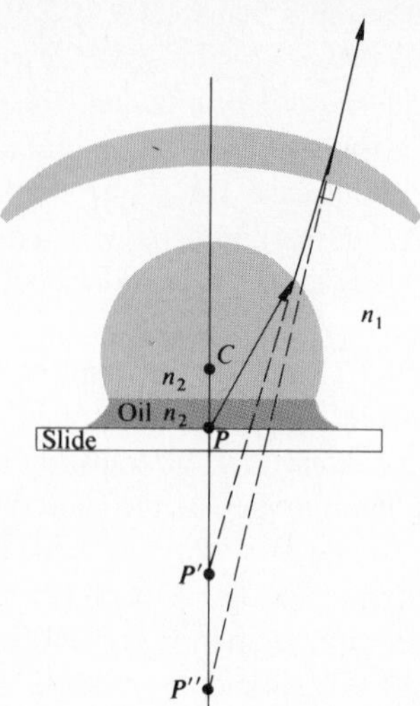

FIGURE 6.18 An oil-immersion microscope objective.

tage. The object under study is positioned at P and surrounded by oil of index n_2, as in Fig. 6.18. P and P' are the proper conjugate points for zero SA for the first element, and P' and P'' are those for the meniscus lens.

Soon after the Hubble Space Telescope (HST) was placed in orbit in April 1990, it became obvious that there was something terribly wrong. The pictures it was returning remained blurred, despite all attempts to adjust the orientation and location of the secondary mirror (p. 228). For a distant star, which was essentially a point source, the size of the image disk was close to the expected diffraction-limited value (about 0.1 arc-second in diameter), but only about 12% of the radiant energy was there, instead of the expected 70% (roughly 84% is the ideal limit). The disk was surrounded by a halo extending out in diameter to about 1.5 arc-seconds containing some 70% of the light. The remaining radiant energy was unavoidably distributed beyond the halo in a radial tendril pattern as a result of a combination of mirror micro-roughness and diffraction from the struts holding the secondary (Fig. 6.19*b*). The situation was a classic example of spherical aberration.

As scientists later determined, the primary mirror had been polished incorrectly; it was too flat at its periphery by about half a wavelength. Rays from its central region were focusing on the optical axis in front of those from the edges. The people at Perkin-Elmer, the company that fashioned the 2.4-m hyperboloid, had polished it superbly well, but to the wrong figure, or curvature. A series of blunders, starting with a 1.3-mm error in the position of a component in the shape-testing device, ultimately led to the flaw. The \$1.6 billion telescope ended up with a debilitating longitudinal spherical aberration of 38 mm (Fig. 6.19*a*).

In 1993 astronauts from the *Endeavor* Space Shuttle successfully executed a dramatic repair mission. They installed a new Wide-Field Planetary Camera (with its own corrective optics that adds about half a wavelength to the edges) and the

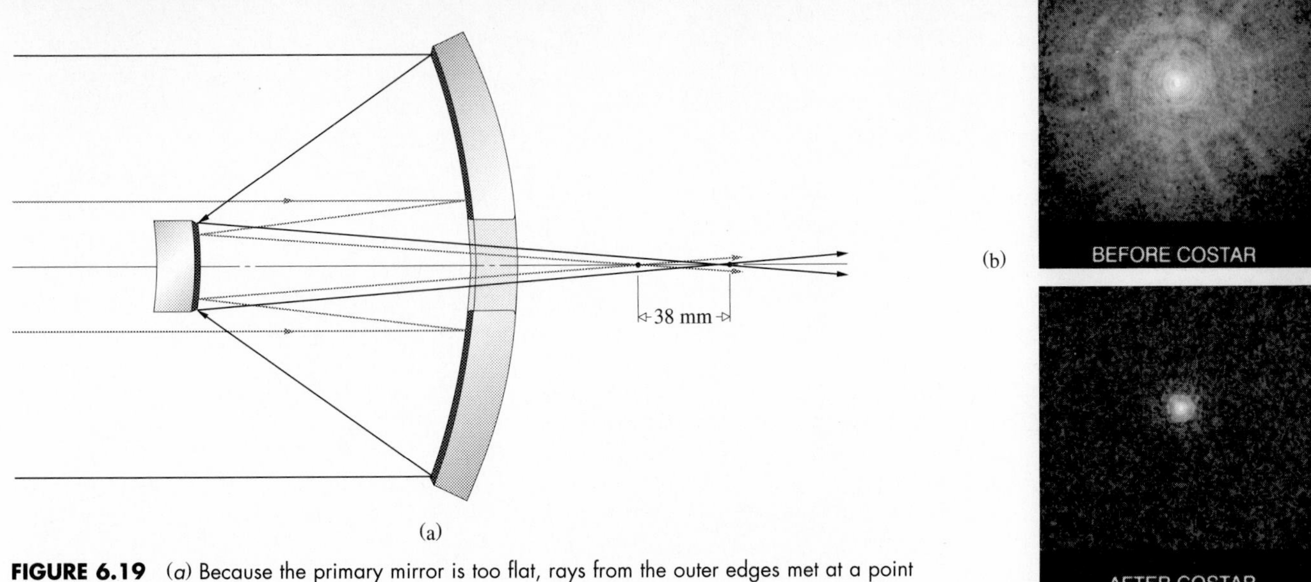

FIGURE 6.19 (*a*) Because the primary mirror is too flat, rays from the outer edges met at a point 38 mm beyond the point where inner rays converge. (*b*) The image of a distant star formed by the HST. (Photo courtesy of NASA.)

Corrective Optics Space Telescope Axial Replacement (COSTAR) module. The job of COSTAR is to reshape the aberrated wavefronts entering the three remaining scientific instruments. It inserts a pair of small mirrors (10 mm and 30 mm) into the beam heading toward each instrument aperture. One of these mirrors simply redirects the light to the other, which is a complex asymmetrical aspheric. That off-axis correcting mirror is configured with the inverse of the spherical aberration of the primary, so that upon reflection the wavefront is reshaped into a perfect wave directed toward the intended aperture. Now better than 70% of the light energy resides in the central image disk, and celestial objects are about 6.5 times brighter than before. People at NASA like to point out that with its vision clearer than ever (Fig. 6.20), and its light-gathering ability improved, the HST could now spot a firefly over a distance equivalent to roughly halfway around the world. (Of course, the bug would have to stay stuck at maximum emission for about 90 minutes.) Moreover, the HST could distinguish two such persistent fireflies provided they were at least 3 m apart.

Coma

Coma, or *comatic aberration*, is an image-degrading, monochromatic, primary aberration associated with an object point even a short distance from the axis. Its origins lie in the fact that the principal "planes" can actually be treated as planes only in the paraxial region. They are, in fact, principal curved surfaces (Fig. 6.1). In the absence of SA, a parallel bundle of rays will focus at the axial point F_i, a distance b.f.l. from the rear vertex. Yet the effective focal lengths and therefore the transverse magnifications will differ for rays traversing off-axis regions of the lens. When the image point is on the optical axis, this situation is of little consequence, but when the ray bundle is oblique and the image point is off-axis, coma will be evident.

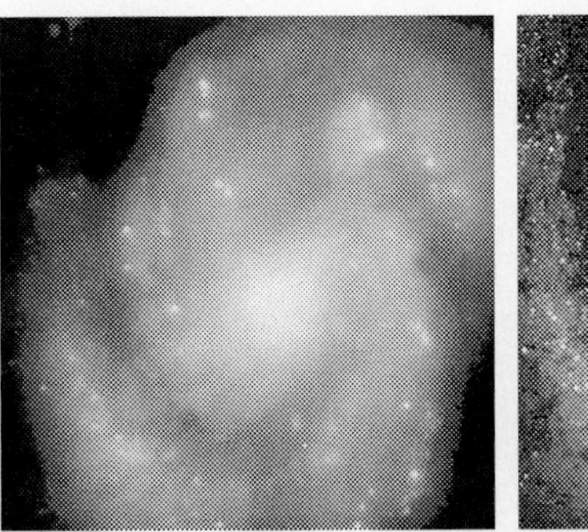
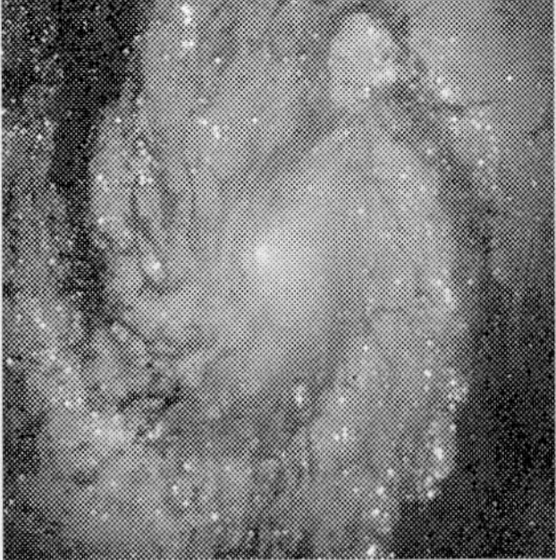

FIGURE 6.20 HST images of the M-100 galaxy with (before repair) and without (after repair) spherical aberration. (Photos courtesy of NASA.)

The dependence of M_T on h, the ray height at the lens, is shown in Fig. 6.21*a*. Here meridional rays traversing the extremities of the lens arrive at the image plane closer to the axis than do the rays in the vicinity of the *principal ray* (i.e., the ray that passes through the principal points). In this instance, the least magnification is associated with the marginal rays that would form the smallest image—the coma is negative. By comparison, the coma in Figs. 6.21*b* and *c* is positive, because the marginal rays focus farther from the axis.

Several skew rays are drawn from an extra-axial object point S in Fig. 6.22 to illustrate the formation of the geometrical comatic image of a point. Observe that each circular cone of rays whose endpoints (1-2-3-4-1-2-3-4) form a ring on the lens is imaged in what H. Dennis Taylor called a *comatic cir-*

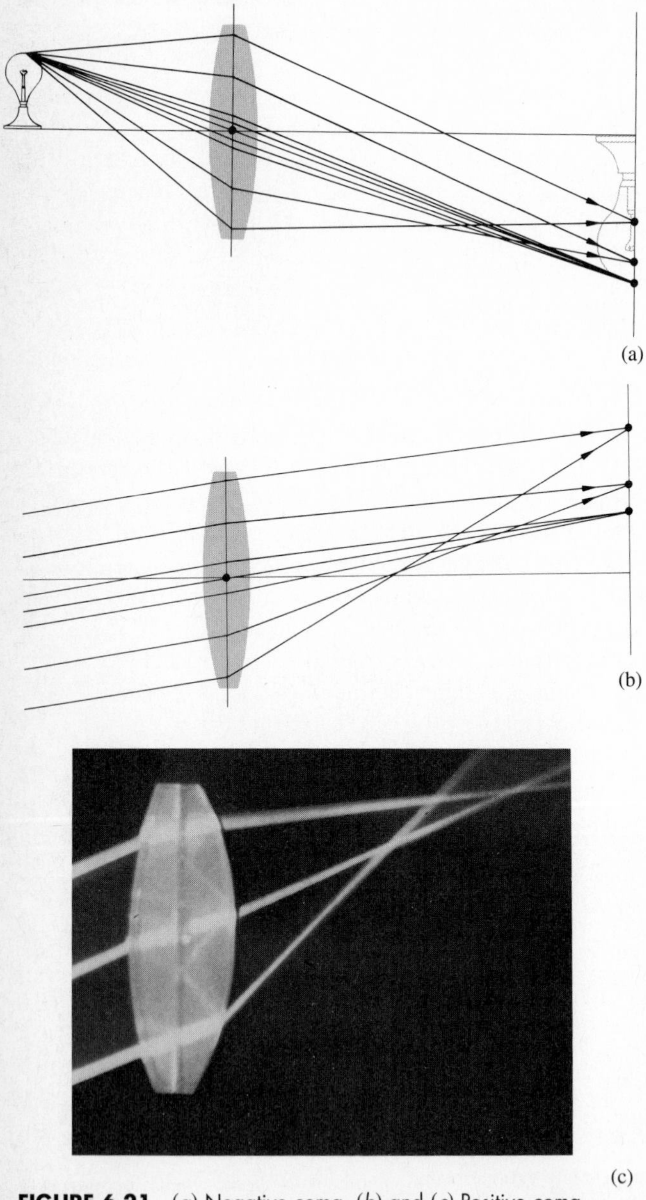

FIGURE 6.21 (*a*) Negative coma. (*b*) and (*c*) Positive coma. (Photo by E.H.)

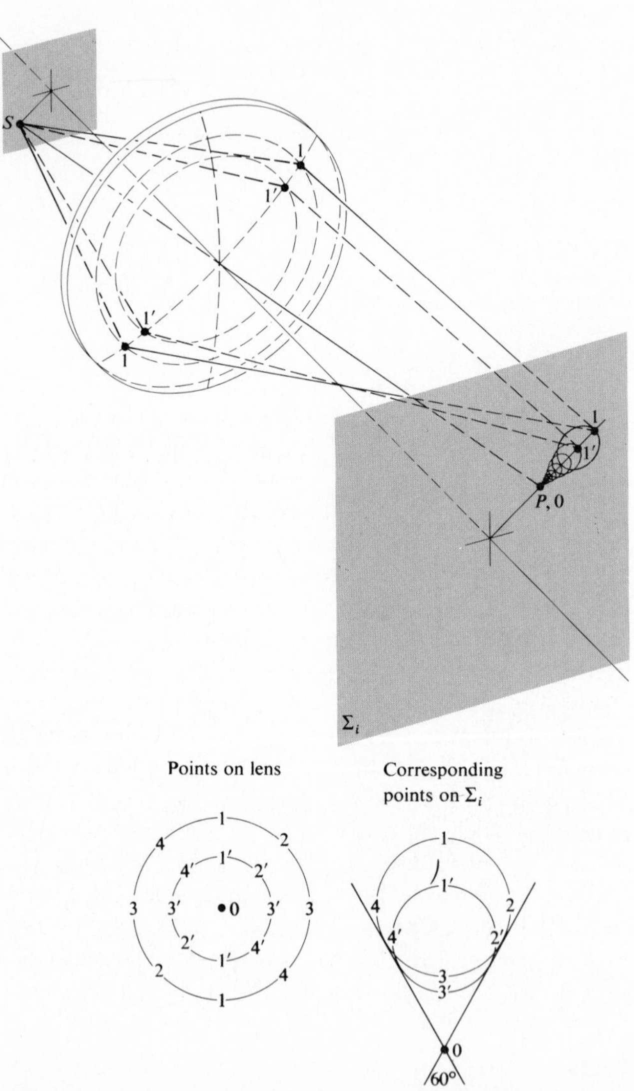

FIGURE 6.22 The geometrical coma image of a monochromatic point source. The central region of the lens forms a point image at the vertex of the cone.

cle on Σ_i. This case corresponds to positive coma, so the larger the ring on the lens, the more distant its comatic circle from the axis. When the outer ring is the intersection of marginal rays, the distance from 0 to 1 in the image is the ***tangential coma***, and the length from 0 to 3 on Σ_i is termed the ***sagittal coma***. A little more than half of the energy in the image appears in the roughly triangular region between 0 and 3. The coma flare, which owes its name to its cometlike tail, is often thought to be the worst of all aberrations, primarily because of its asymmetric configuration.

It's not the purview of Geometrical Optics to be concerned with interference, but when light reaches the screen in Fig. 6.22, it's certainly to be expected. The coma cone, just like the Gaussian image point, is an oversimplification. The image point is really an image disk-ring system, and the coma cone is actually a complicated asymmetrical diffraction pattern. The more coma there is, the more the cone departs from the Airy pattern into an elongated structure of blotches and arcs that only vaguely suggests the disk-ring structure from which it evolved (Fig. 6.23).

Like SA, coma is dependent on the shape of the lens. Thus a strongly concave positive-meniscus lens) with the object at infinity will have a large negative coma. Bending the lens so that it becomes planar-convex |), then equiconvex (), convex-planar (|, and finally convex-meniscus (will change the coma from negative, to zero, to positive. The fact that it can be made exactly zero for a single lens with a given object distance is quite significant. The particular shape it then has ($s_o = \infty$) is almost convex-planar and nearly the configuration for minimum SA.

It is important to realize that *a lens that is well corrected for the case in which one conjugate point is at infinity* ($s_o = \infty$) *may not perform satisfactorily when the object is nearby*. One would therefore do well, when using off-the-shelf lenses in a system operating at finite conjugates, to combine two infinite conjugate corrected lenses, as in Fig. 6.24. In other words, since it is unlikely that a lens with the desired focal length, which is also corrected for the particular set of finite conjugates, can be obtained ready-made, this back-to-back lens approach is an appealing alternative.

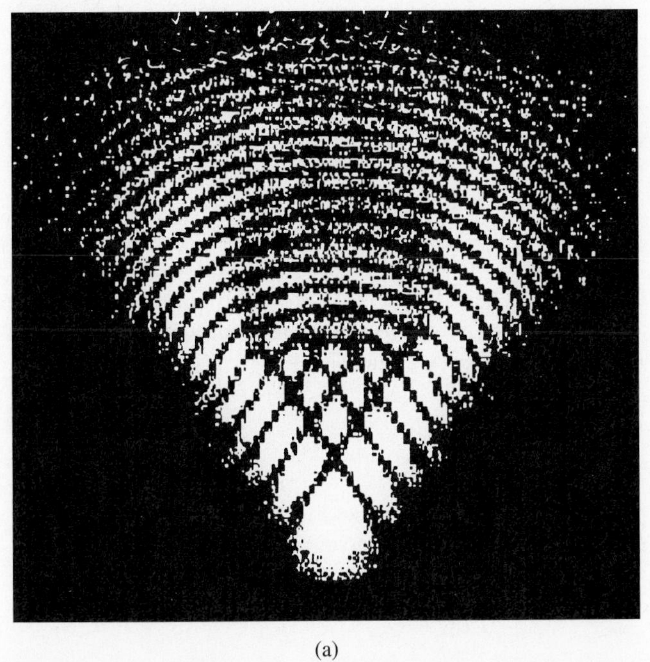

(a)

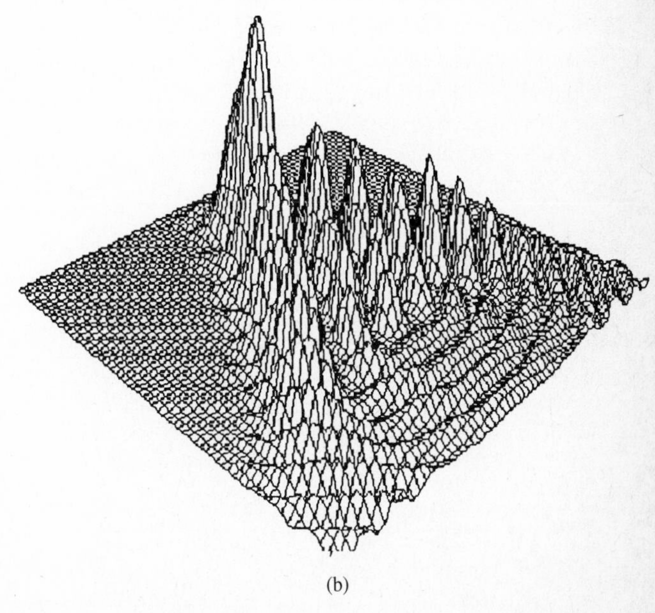

(b)

FIGURE 6.23 Third-order coma. (*a*) A computer-generated diagram of the image of a point source formed by a heavily astigmatic optical system. (*b*) A plot of the corresponding irradiance distribution. (Pictures courtesy of OPAL Group, St. Petersburg, Russia.)

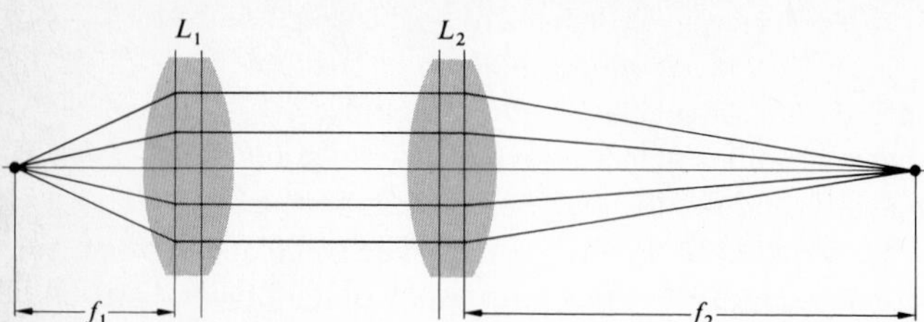

FIGURE 6.24 A combination of two infinite conjugate lenses yielding a system operating at finite conjugates.

Coma can also be negated by using a stop at the proper location, as William Hyde Wollaston (1766–1828) discovered in 1812. The order of the list of primary aberrations (SA, coma, astigmatism, Petzval field curvature, and distortion) is significant, because any one of them, except SA and Petzval curvature, will be affected by the position of a stop, but only if one of the preceding aberrations is also present in the system. Thus, while SA is independent of the location along the axis of a stop, coma will not be, as long as SA is present. This can be appreciated by examining the representation in Fig. 6.25. With the stop at Σ_1, ray 3 is the chief ray and there is SA but no coma; that is, the ray pairs meet on 3. If the stop is moved to Σ_2, the symmetry is upset, ray 4 becomes the chief ray, and the rays on either side of it, such as 3 and 5, meet above, not on it—there is positive coma. With the stop at Σ_3, rays 1 and 3 intersect below the chief ray, 2, and there is negative coma. In this way, controlled amounts of the aberration can be introduced into a compound lens in order to cancel coma in the system as a whole.

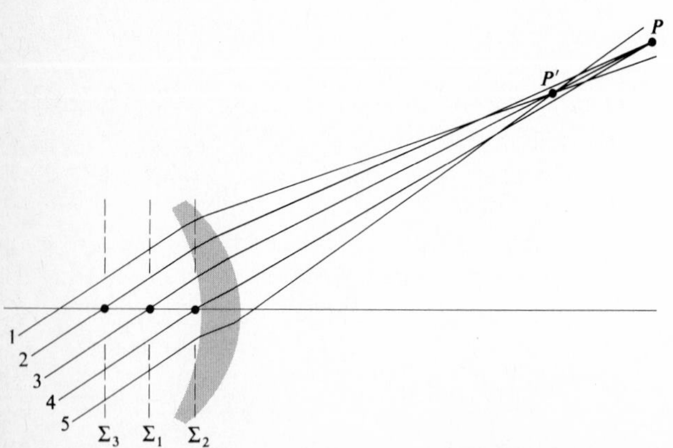

FIGURE 6.25 The effect of stop location on coma.

The **optical sine theorem** is an important relationship that must be introduced here even if space precludes its formal proof. It was discovered independently in 1873 by Abbe and Helmholtz, although a different form of it was given 10 years earlier by R. Clausius (of thermodynamics fame). In any event, it states that

$$n_o y_o \sin \alpha_o = n_i y_i \sin \alpha_i \tag{6.41}$$

where n_o, y_o, α_o and n_i, y_i, α_i are the index, height, and slope angle of a ray in object and image space, respectively, at any aperture size* (Fig. 6.9). If coma is to be zero,

$$M_T = \frac{y_i}{y_o} \qquad [5.24]$$

must be constant for all rays. Suppose then that we send a marginal and a paraxial ray through the system. The former will comply with Eq. (6.41), the latter with its paraxial version (in which $\sin \alpha_o = \alpha_{op}$, $\sin \alpha_i = \alpha_{ip}$). Since M_T is to be constant over the entire lens, we equate the magnification for both marginal and paraxial rays to get

$$\frac{\sin \alpha_o}{\sin \alpha_i} = \frac{\alpha_{op}}{\alpha_{ip}} = \text{constant} \tag{6.42}$$

which is known as the **Sine Condition**. A necessary criterion for the absence of coma is that the system meet the Sine Condition. If there is no SA, compliancy with the Sine Condition will be both necessary and sufficient for zero coma.

It's an easy matter to observe coma. In fact, anyone who has focused sunlight with a simple positive lens has no doubt seen the effects of this aberration. A slight tilt of the lens, so that the nearly collimated rays from the Sun make an angle with the optical axis, will cause the focused spot to flare out into the characteristic comet shape.

Astigmatism

When an object point lies an appreciable distance from the optical axis, the incident cone of rays will strike the lens asymmetrically, giving rise to a third primary aberration known as

*To be precise, the sine theorem is valid for all values of α_o only in the sagittal plane (from the Latin *sagitta*, meaning arrow), which is discussed in the next section.

astigmatism. The word derives from the Greek *a*-, meaning not, and *stigma*, meaning spot or point. To facilitate its description, envision the meridional plane (also called the *tangential plane*) containing both the chief ray (i.e., the one passing through the center of the aperture) and the optical axis. The *sagittal plane* is then defined as the plane containing the chief ray, which, in addition, is perpendicular to the meridional plane (Fig. 6.26). Unlike the latter, which is unbroken from one end of a complicated lens system to the other, the sagittal plane generally changes slope as the chief ray is deviated at the various elements. Hence to be accurate we should say that there are actually several sagittal planes, one attendant with each region within the system. Nevertheless, all skew rays from the object point lying in a sagittal plane are termed ***sagittal rays***.

In the case of an axial object point, the cone of rays is symmetrical with respect to the spherical surfaces of a lens. There is no need to make a distinction between meridional and sagittal planes. The ray configurations in all planes containing the optical axis are identical. In the absence of spherical aberration, all the focal lengths are the same, and consequently all rays arrive at a single focus. In contrast, the configuration of an oblique, parallel ray bundle will be different in the meridional and sagittal planes. As a result, the focal lengths in these planes will be different as well. In effect, here the meridional rays are tilted more with respect to the lens than are the sagittal rays, and they have a shorter focal length. It can be shown,* using Fermat's Principle, that the *focal length difference* depends effectively on the power of the lens (as opposed to the shape or index) and the angle at which the rays are inclined. This *astigmatic difference*, as it is often called, increases rapidly as the rays become more oblique, that is, as the object point moves farther off the axis, and is, of course, zero on axis.

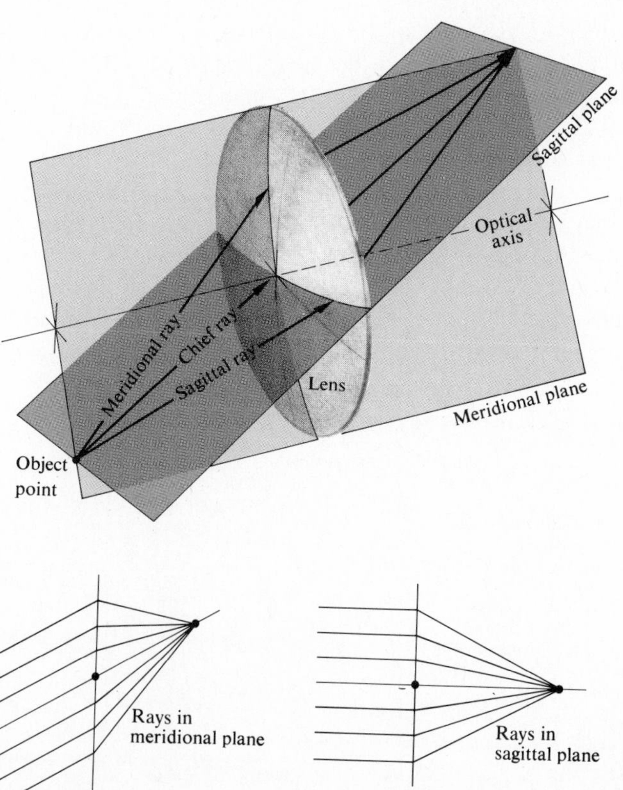

FIGURE 6.26 The sagittal and meridional planes.

Having two distinct focal lengths, the incident conical bundle of rays takes on a considerably altered form after refraction (Fig. 6.27). The cross section of the beam as it leaves the lens is initially circular, but it gradually becomes elliptical with the major axis in the sagittal plane, until at the *tangential* or *meridional focus* F_T, the ellipse degenerates into a "*line*" (at least in third-order theory). Actually, it's a complicated elongated diffraction pattern that looks more linelike the more astigmatism is present. All rays from the object point traverse this "line," which is known as the *primary image*. Beyond this point the beam's cross section rapidly opens out until it is again circular. At that location the image is a circular blur known as the *circle of least confusion*. Moving farther from the lens the beam's cross section again deforms into a "line," called the *secondary image*. This time it's in the meridional plane at the *sagittal focus*, F_S.

The image of a point source formed by a slightly astigmatic optical system ($\lessapprox 0.2\lambda$), in the vicinity of the circle of least confusion, looks very much like the Airy disk-ring pattern, but it's somewhat asymmetrical. As the amount of astigmatism increases (upwards of roughly 0.5λ), the biaxial asymmetry becomes more apparent. The image transforms into a complex distribution of bright and dark regions (resembling the Fresnel diffraction patterns for rectangular openings, p. 489) and only very subtly retains a curved structure arising from the circular aperture. Remember that in all of this we are assuming the absence of SA and coma.

*See A. W. Barton, *A Text Book on Light*, p. 124.

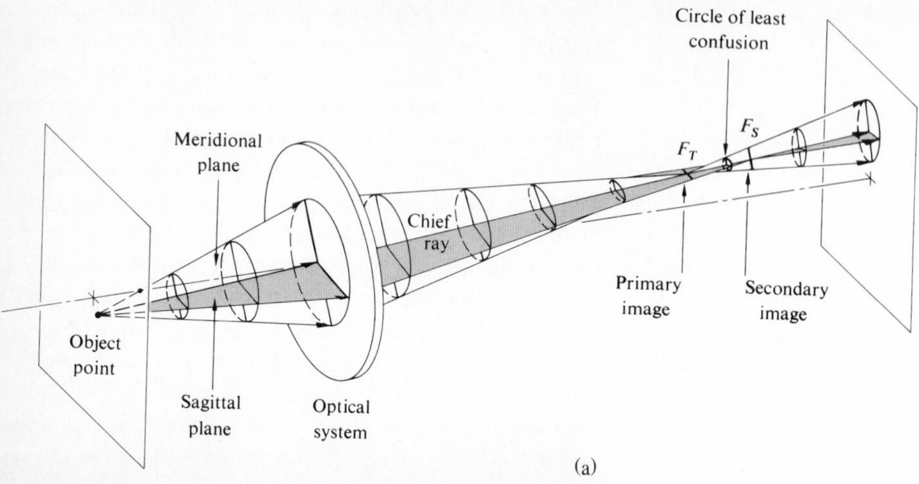

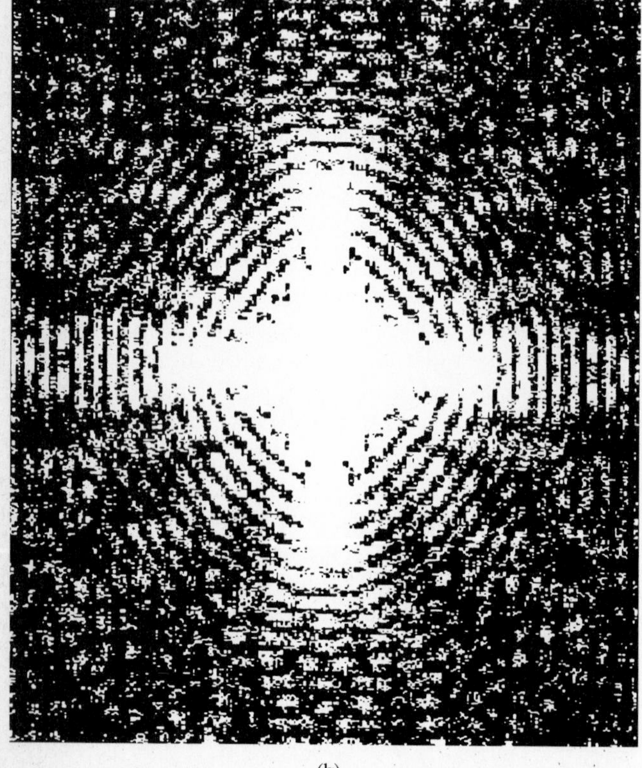

FIGURE 6.27 Astigmatism. (*a*) The light from a monochromatic point source is elongated by an astigmatic lens. (*b*) A computer-generated diagram showing the distribution of light, that is, the diffraction pattern, near the circle of least confusion, corresponding to 0.8λ of astigmatism. (Computer picture courtesy OPAL Group, St. Petersburg Russia.)

Since the circle of least confusion increases in diameter as the astigmatic difference increases (i.e., as the object moves farther off-axis), the image will deteriorate, losing definition around its edges. Observe that the secondary "line" image will change in orientation with changes in the object position, but it will always point toward the optical axis; that is, it will be radial. Similarly, the primary "line" image will vary in orientation, but it will remain normal to the secondary image. This arrangement causes the interesting effect shown in Fig. 6.28 when the object is made up of radial and tangential elements. The primary and secondary images are, in effect, formed of transverse and radial dashes, which increase in size with distance from the axis. In the latter case, the dashes point

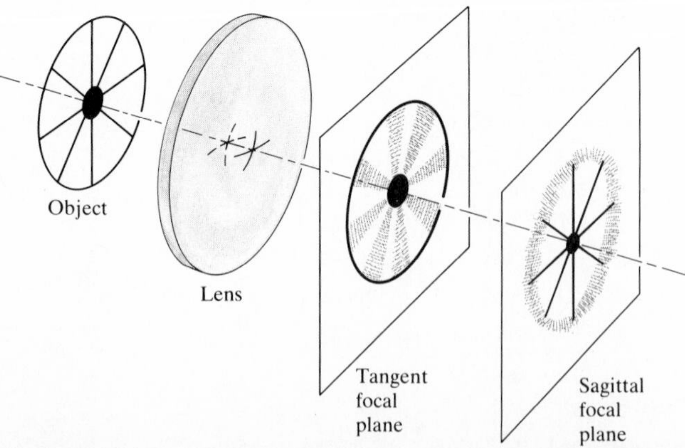

FIGURE 6.28 Images in the tangent and sagittal focal planes.

like arrows toward the center of the image—ergo, the name *sagitta*.

The existence of the sagittal and tangential foci can be verified directly with a fairly simple arrangement. Place a positive lens with a short focal length (about 10 or 20 mm) in the beam of a He–Ne laser. Position another positive test lens with a somewhat longer focal length far enough away so that the now diverging beam fills that lens. A convenient object, to be located between the two lenses, is a piece of ordinary wire screening (or a transparency). Align it so the wires are horizontal (x) and vertical (y). If the test lens is rotated roughly 45° about the vertical (with the x-, y-, and z-axes fixed in the lens), astigmatism should be observable. The meridional is the xz-plane (z being the lens axis, now at about 45° to the laser axis), and the sagittal plane corresponds to the plane of y and the laser axis. As the wire mesh is moved toward the test lens, a point will be reached where the horizontal wires are in focus on a screen beyond the lens, whereas the vertical wires are not. This is the location of the sagittal focus. Each point on the object is imaged as a short line in the meridional (horizontal) plane, which accounts for the fact that only the horizontal wires are in focus. Moving the mesh slightly closer to the lens will bring the vertical lines into clarity while the horizontal ones are blurred. This is the tangential focus. Try rotating the mesh about the central laser axis while at either focus.

Note that unlike visual astigmatism (p. 211), which arose from an actual asymmetry in the surfaces of the optical system, the third-order aberration by that same name applies to spherically symmetrical lenses.

Mirrors, with the singular exception of the plane mirror, suffer much of the same monochromatic aberrations as do lenses. Thus, although a paraboloidal mirror is free of SA for an infinitely distant axial object point, its off-axis imagery is quite poor due to astigmatism and coma. This strongly restricts its use to narrow field devices, such as searchlights and astronomical telescopes. A concave spherical mirror shows SA, coma, and astigmatism. Indeed, one could draw a diagram just like Fig. 6.27 with the lens replaced by an obliquely illuminated spherical mirror. Incidentally, such a mirror displays appreciably less SA than would a simple convex lens of the same focal length.

Field Curvature

Suppose we had an optical system that was free of all the aberrations thus far considered. There would then be a one-to-one correspondence between points on the object and image surfaces (i.e., stigmatic imagery). We mentioned earlier (Section 5.2.3) that a planar object normal to the axis will be imaged approximately as a plane only in the paraxial region. At finite apertures the resulting curved stigmatic image surface is a manifestation of the primary aberration known as **Petzval field curvature**, after the Hungarian mathematician Josef Max Petzval (1807–1891). The effect can readily be appreciated by examining Figs. 5.22 (p. 162) and 6.29. A spherical object segment σ_o is imaged by the lens as a spherical segment σ_i, both centered at O. Flattening out σ_o into the plane σ_o' will cause each object point to move toward the lens along the concomitant chief ray, thus forming a paraboloidal *Petzval surface* Σ_P. Whereas the Petzval surface for a positive lens curves *inward* toward the object plane, for a negative lens it curves

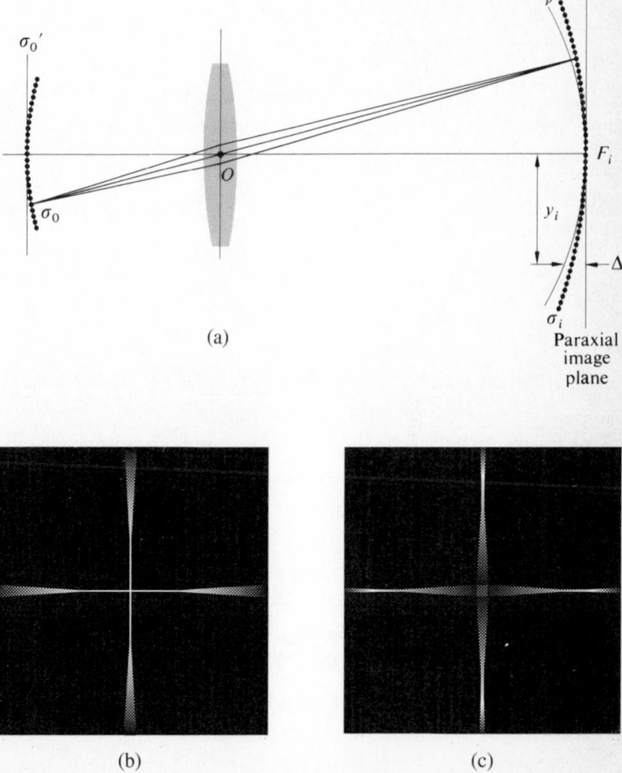

FIGURE 6.29 Field curvature. (*a*) When the object corresponds to σ_o', the image will correspond to surface Σ_P. (*b*) The image formed on a flat screen near the paraxial image plane will be in focus only at its center. (*c*) Moving the screen closer to the lens will bring the edges into focus.

outward away from that plane. Evidently, a suitable combination of positive and negative lenses will negate field curvature. Indeed, the displacement Δx of an image point at height y_i on the Petzval surface from the paraxial image plane is given by

$$\Delta x = \frac{y_i^2}{2} \sum_{j=1}^{m} \frac{1}{n_j f_j} \tag{6.43}$$

where n_j and f_j are the indices and focal lengths of the m thin lenses forming the system. This implies that the Petzval surface will be unaltered by changes in the positions or shapes of the lenses or in the location of the stop, as long as the values of n_j and f_j are fixed. Notice that for the simple case of two thin lenses ($m = 2$) having any spacing, Δx *can be made zero* provided that

$$\frac{1}{n_1 f_1} + \frac{1}{n_2 f_2} = 0$$

or, equivalently,

$$n_1 f_1 + n_2 f_2 = 0 \tag{6.44}$$

This is the so-called **Petzval condition**. As an example of its use, suppose we combine two thin lenses, one positive, the other negative, such that $f_1 = -f_2$ and $n_1 = n_2$. Since

$$\frac{1}{f} = \frac{1}{f_1} + \frac{1}{f_2} - \frac{d}{f_1 f_2} \qquad [6.8]$$

$$f = \frac{f_1^2}{d}$$

the system can satisfy the Petzval condition, have a flat field, and still have a finite positive focal length.

In visual instruments a certain amount of curvature can be tolerated, because the eye can accommodate for it. Clearly, in photographic lenses field curvature is most undesirable, since it has the effect of rapidly blurring the off-axis image when the film plane is at F_i. An effective means of nullifying the inward curvature of a positive lens is to place a negative *field flattener* lens near the focal plane. This is often done in projection and photographic objectives when it is not otherwise practicable to meet the Petzval condition (Fig. 6.30). In this position the flattener will have little effect on other aberrations.

Astigmatism is intimately related to field curvature. In the presence of the former aberration, there will be *two* paraboloidal image surfaces, the tangential, Σ_T, and the sagittal, Σ_S (as in Fig. 6.31). These are the loci of all the primary and secondary images, respectively, as the object point roams over the object plane. At a given height (y_i), a point on Σ_T always lies three times as far from Σ_P as does the corresponding point on Σ_S, and both are on the same side of the Petzval surface (Fig. 6.31). When there is no astigmatism, Σ_S and Σ_T coalesce on Σ_P. It is possible to alter the shapes of Σ_S and Σ_T by bending

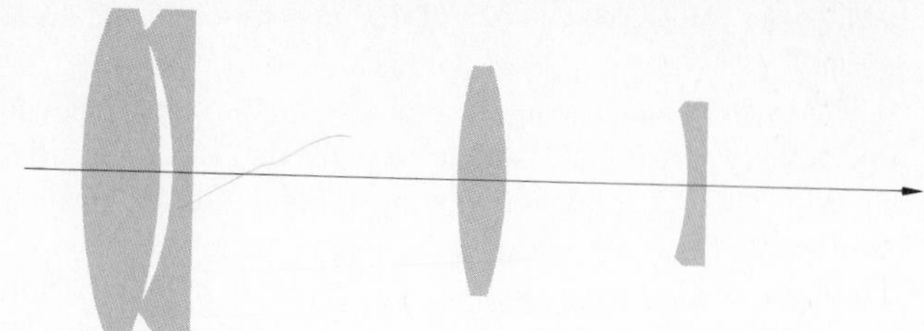

(a) Petzval lens with field flattener

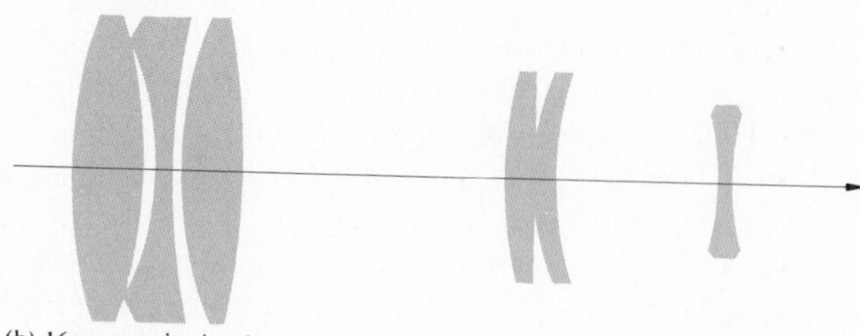

(b) 16 mm projection lens

FIGURE 6.30 The field flattener.

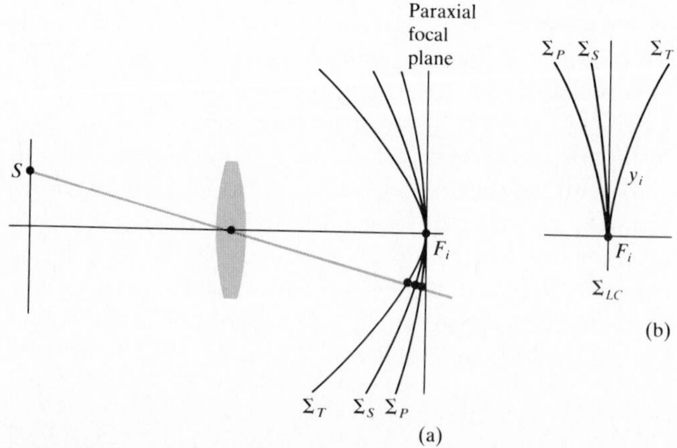

FIGURE 6.31 The tangential, sagittal and Petzval image surfaces.

or relocating the lenses or by moving the stop. The configuration of Fig. 6.31*b* is known as an *artificially flattened* field. A stop in front of an inexpensive meniscus box camera lens is usually arranged to produce just this effect. The surface of least confusion, Σ_{LC}, is planar, and the image there is tolerable, losing definition at the margins because of the astigmatism. That is to say, although their loci form Σ_{LC}, the circles of least confusion increase in diameter with distance off the axis. Modern good-quality photographic objectives are generally ***anastigmats***; that is, they are designed so that Σ_S and Σ_T cross each other, yielding an additional off-axis angle of zero astigmatism. The Cooke Triplet, Tessar, Orthometer, and Biotar (Fig. 5.117) are all anastigmats, as is the relatively fast Zeiss Sonnar, whose residual astigmatism is illustrated graphically in Fig. 6.32. Note the relatively flat field and small amount of astigmatism over most of the film plane.

Let's return briefly to the Schmidt camera shown in Fig. 5.126 (p. 230), since we are now in a better position to appreciate how it functions. With a stop at the center of curvature of the spherical mirror, all chief rays, which by definition pass through *C*, are incident normally on the mirror. Moreover, each pencil of rays from a distant object point is symmetrical about its chief ray. In effect, each chief ray serves as an optical axis, so there are no off-axis points and, in principle, no coma or astigmatism. Instead of attempting to flatten the image surface, the designer has coped with curvature by simply shaping the film plate to conform with it.

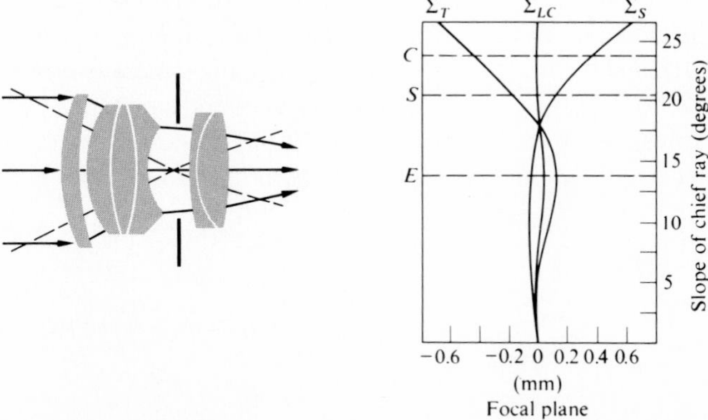

FIGURE 6.32 A typical Sonnar. The markings *C*, *S*, and *E* denote the limits of the 35-mm film format (field stop), that is, corners, sides, and edges. The Sonnar family lies between the double Gauss and the triplet.

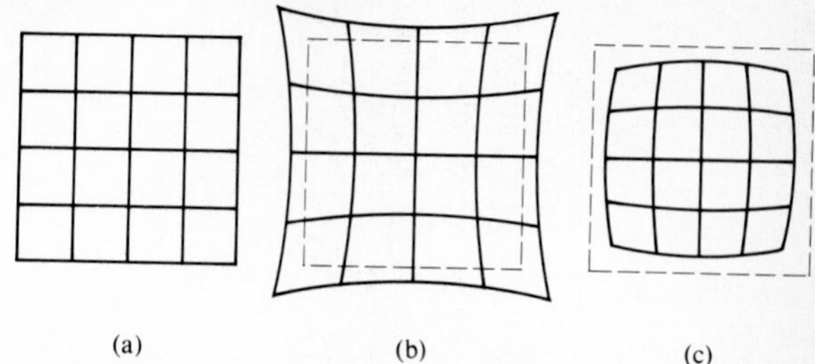

FIGURE 6.33 (*a*) Undistorted object. (*b*) When the magnification on the optical axis is less than the off-axis magnification pincushion distortion results. (*c*) When it is greater on axis than off barrel distortion results.

Distortion

The last of the five primary, monochromatic aberrations is **distortion**. Its origin lies in the fact that the transverse magnification, M_T, may be a function of the off-axis image distance, y_i. Thus, that distance may differ from the one predicted by paraxial theory in which M_T is constant. In other words, distortion arises because different areas of the lens have different focal lengths and different magnifications. In the absence of any of the other aberrations, distortion is manifest in a misshaping of the image as a whole, even though each point is sharply focused. Consequently, when processed by an optical system suffering ***positive*** or ***pincushion distortion***, a square array deforms, as in Fig. 6.33*b*. In that instance, each image point is displaced radially outward from the center, with the most distant points moving the greatest amount (i.e., M_T increases with y_i). Similarly, ***negative*** or ***barrel distortion*** corresponds to the situation in which M_T decreases with the axial distance, and in effect, each point on the image moves radially inward toward the center (Fig. 6.33*c*). Distortion can easily be seen by just looking through an aberrant lens at a piece of lined or graph paper. Fairly thin lenses will show essentially no distortion, whereas ordinary positive or negative, thick, simple lenses will generally suffer positive or negative distortion, respectively. The introduction of a stop into a system of thin lenses is invariably accompanied by distortion, as indicated in Fig. 6.34. One exception is the case in which the aperture stop is at the lens, so that the chief ray is, in effect, the principal ray (i.e., it passes through the principal points, here coalesced at *O*). If the stop is in front of a positive lens, as in Fig. 6.34*b*, the object distance measured along the chief ray will be

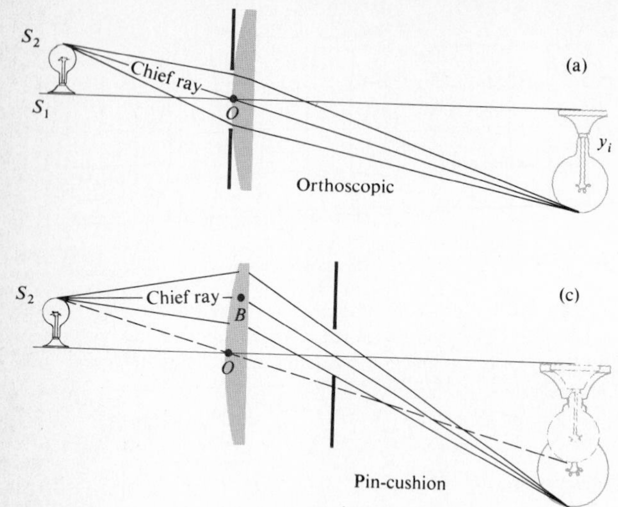

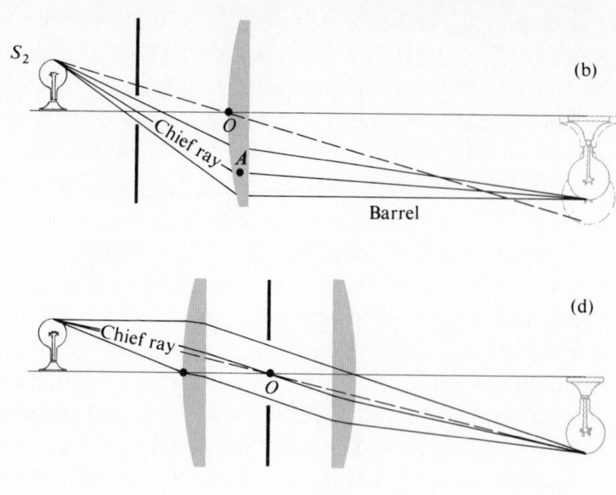

FIGURE 6.34 The effect of stop location on distortion.

greater than it was with the stop at the lens ($S_2A > S_2O$). Thus x_o will be greater and [Eq. (5.26)] M_T will be smaller—ergo, barrel distortion. In other words, M_T for an off-axis point will be less with a front stop in position than it would be without it. The difference is a measure of the aberration, which, by the way, exists regardless of the size of the aperture. In the same way, a rear stop (Fig. 6.34*c*) decreases x_o along the chief ray (i.e., $S_2O > S_2B$), thereby increasing M_T and introducing pin-cushion distortion. *Interchanging the object and image thus has the effect of changing the sign of the distortion* for a given lens and stop. The aforementioned stop positions will produce the opposite effect when the lens is negative.

All of this suggests the use of a stop midway between identical lens elements. The distortion from the first lens will precisely cancel the contribution from the second. This approach has been used to advantage in the design of a number of photographic lenses (Fig. 5.117). To be sure, if the lens is perfectly symmetrical and operating as in Fig. 6.34*d*, the object and image distances will be equal, hence $M_T = 1$. (Incidentally, coma and lateral color will then be identically zero as well.) This applies to (finite conjugate) copy lenses used, for example, to record data. Nonetheless, even when M_T is not 1, making the system approximately symmetrical about a stop is a very common practice, since it markedly reduces these several aberrations.

Distortion can arise in compound lens systems, as for example in the telephoto arrangement shown in Fig. 6.35. For a distant object point, the margin of the positive achromat serves as the aperture stop. In effect, the arrangement is like a negative lens with a front stop, so it displays positive or pin-cushion distortion.

Suppose a chief ray enters and emerges from an optical system in the same direction as, for example, in Fig. 6.34*d*. The point at which the ray crosses the axis is the optical center of the system, but since this is a chief ray, it is also the center of the aperture stop. This is the situation approached in Fig. 6.34*a*, with the stop up against the thin lens. In both instances the incoming and outgoing segments of the chief ray are parallel, and there is zero distortion; that is, the system is *orthoscopic*. This also implies that the entrance and exit pupils will correspond to the principal planes (if the system is immersed in a single medium—see Fig. 6.2). Bear in mind that the chief

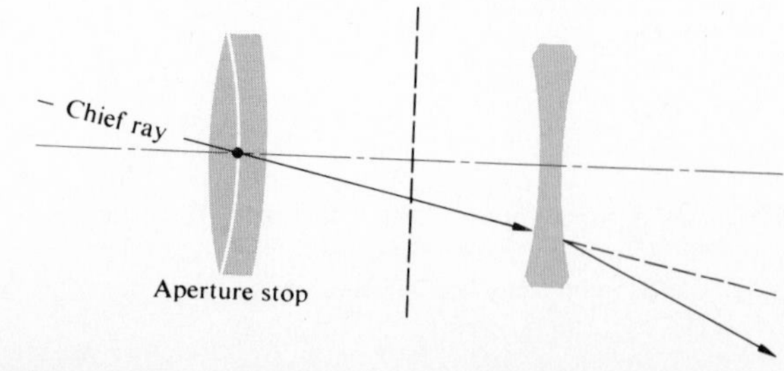

FIGURE 6.35 Distortion in a compound lens.

ray is now a principal ray. *A thin-lens system will have zero distortion if its optical center is coincident with the center of the aperture stop.* By the way, in a pinhole camera, the rays connecting conjugate object and image points are straight and pass through the center of the aperture stop. The entering and emerging rays are obviously parallel (being one and the same), and there is no distortion.

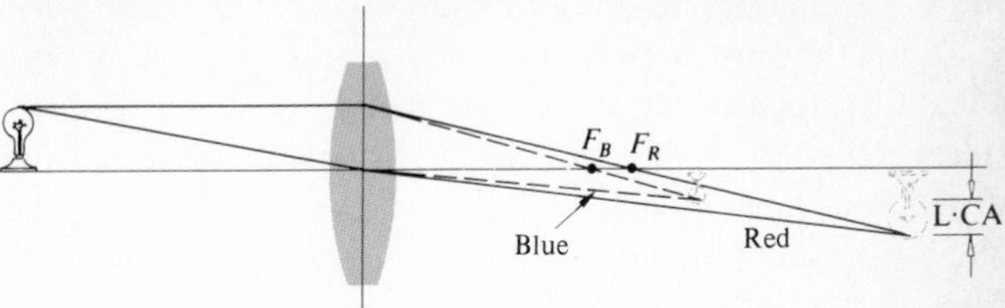

FIGURE 6.37 Lateral chromatic aberration.

6.3.2 Chromatic Aberrations

The five primary or Seidel aberrations have been considered in terms of monochromatic light. To be sure, if the source has a broad spectral bandwidth, these aberrations are influenced accordingly; but the effects are inconsequential, unless the system is quite well corrected. There are, however, **chromatic aberrations** that arise specifically in polychromatic light, which are far more significant. The ray-tracing equation [Eq. (6.12)] is a function of the indices of refraction, which in turn vary with wavelength. Different "colored" rays will traverse a system along different paths, and this is the quintessential feature of chromatic aberration.

Since the thin-lens equation

$$\frac{1}{f} = (n_l - 1)\left(\frac{1}{R_1} - \frac{1}{R_2}\right) \qquad [5.16]$$

is wavelength-dependent via $n_l(\lambda)$, the focal length must also vary with λ. In general (Fig. 3.37, p. 72), $n_l(\lambda)$ decreases with wavelength over the visible region, and thus $f(\lambda)$ increases with λ. The result is illustrated in Fig. 6.36, where the constituent colors in a collimated beam of white light are focused at different points on the axis. The axial distance between two such focal points spanning a given frequency range (e.g., blue to red) is termed the ***axial*** (or *longitudinal*) ***chromatic aberration***, A · CA for short.

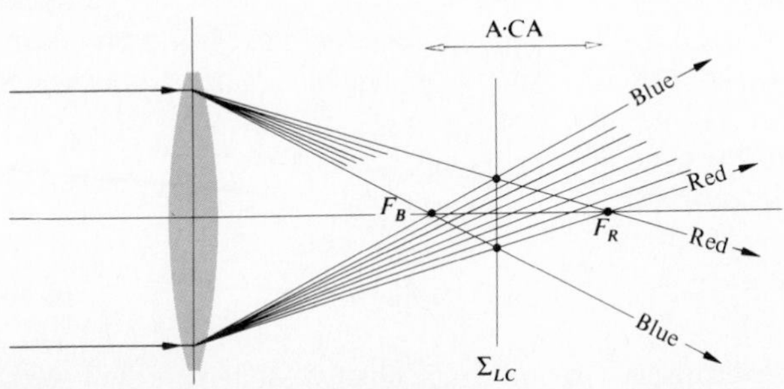

FIGURE 6.36 Axial chromatic aberration.

It's an easy matter to observe chromatic aberrations, or CA, with a thick, simple converging lens. When illuminated by a polychromatic point source (a candle flame will do), the lens will cast a real image surrounded by a halo. If the plane of observation is then moved nearer the lens, the periphery of the blurred image will become tinged in orange-red. Moving it back away from the lens, beyond the best image, will cause the outlines to become tinted in blue-violet. The location of the circle of least confusion (i.e., the plane Σ_{LC}) corresponds to the position where the best image will appear. Try looking directly through the lens at a source—the coloration will be far more striking.

The image of an off-axis point will be formed of the constituent frequency components, each arriving at a different height above the axis (Fig. 6.37). In essence, the frequency dependence of f causes a frequency dependence of the transverse magnification as well. The vertical distance between two such image points (most often taken to be blue and red) is a measure of the ***lateral chromatic aberration***, L · CA, or ***lateral color***. Consequently, a chromatically aberrant lens illuminated by white light will fill a volume of space with a continuum of more or less overlapping images, varying in size and color. Because the eye is most sensitive to the yellow-green portion of the spectrum, the tendency is to focus the lens for that region. With such a configuration one would see all the other colored images superimposed and slightly out of focus, producing a whitish blur or hazed overlay.

When the blue focus, F_B, is to the left of the red focus, F_R, the A · CA is said to be positive, as it is in Fig. 6.36. Conversely, a negative lens would generate negative A · CA, with the more strongly deviated blue rays appearing to originate at the right of the red focus. Physically, what is happening is that the lens, whether convex or concave, is prismatic in shape; that is, it becomes either thinner or thicker as the radial distance from the axis increases. As you well know, rays are therefore deviated either toward or away from the axis, respectively. In both cases the rays are bent toward the thicker "base" of the prismatic cross section. But the angular deviation is an increasing function of n, and therefore it decreases with λ. Hence blue light is deviated the most and is focused nearest the lens. In other words, for a convex lens the red focus is farthest and to the right; for a concave lens it is farthest and to the left.

The human eye has a substantial amount of chromatic aberration which is compensated for by several psychophysical mechanisms. Still, it's possible to see the effect with a small purple dot: held close to the eye, it will appear blue at the center surrounded by red; farther away it will appear red surrounded by blue.

Thin Achromatic Doublets

All of this suggests that a combination of two thin lenses, one positive and one negative, could conceivably result in the precise overlapping of F_R and F_B (Fig. 6.38). Such an arrangement is said to be *achromatized* for those two specific wavelengths. Notice that what we would like to do is effectively eliminate the total dispersion (i.e., the fact that each color is deviated by a different amount) and not the total deviation itself. With the two lenses separated by a distance d,

$$\frac{1}{f} = \frac{1}{f_1} + \frac{1}{f_2} - \frac{d}{f_1 f_2} \qquad [6.8]$$

Rather than writing out the second term in the Thin-Lens Equation [Eq. (5.16)] let's abbreviate the notation and use $1/f_1 = (n_1 - 1)\rho_1$ and $1/f_2 = (n_2 - 1)\rho_2$ for the two elements. Then

$$\frac{1}{f} = (n_1 - 1)\rho_1 + (n_2 - 1)\rho_2 - d(n_1 - 1)\rho_1(n_2 - 1)\rho_2 \qquad (6.45)$$

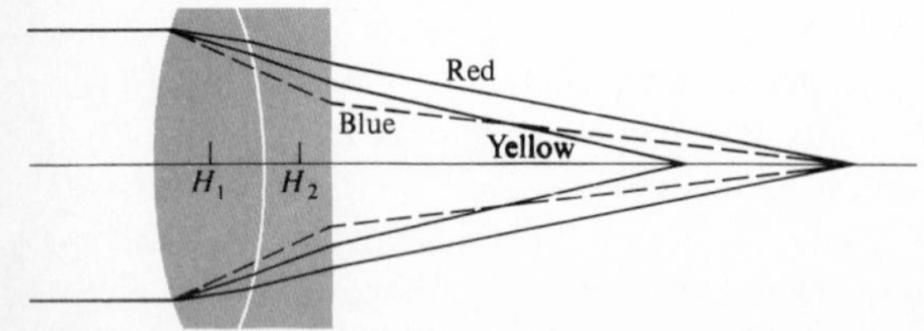

FIGURE 6.38 An achromatic doublet. The paths of the rays are much exaggerated.

This expression will yield the focal length of the doublet for red (f_R) and blue (f_B) light when the appropriate indices are introduced, namely, n_{1R}, n_{2R}, n_{1B}, and n_{2B}. But if f_R is to equal f_B, then

$$\frac{1}{f_R} = \frac{1}{f_B}$$

and, using Eq. (6.45),

$$(n_{1R} - 1)\rho_1 + (n_{2R} - 1)\rho_2 - d(n_{1R} - 1)\rho_1(n_{2R} - 1)\rho_2 = (n_{1B} - 1)\rho_1 + (n_{2B} - 1)\rho_2 - d(n_{1B} - 1)\rho_1(n_{2B} - 1)\rho_2 \qquad (6.46)$$

One case of particular importance corresponds to $d = 0$; that is, the two lenses are in contact. Expanding out Eq. (6.46) with $d = 0$ then leads to

$$\frac{\rho_1}{\rho_2} = -\frac{n_{2B} - n_{2R}}{n_{1B} - n_{1R}} \qquad (6.47)$$

The focal length of the compound lens (f_Y) can conveniently be specified as that associated with yellow light, roughly midway between the blue and red extremes. For the component lenses in yellow light, $1/f_{1Y} = (n_{1Y} - 1)\rho_1$ and $1/f_{2Y} = (n_{2Y} - 1)\rho_2$. Hence

$$\frac{\rho_1}{\rho_2} = \frac{(n_{2Y} - 1)\, f_{2Y}}{(n_{1Y} - 1)\, f_{1Y}} \qquad (6.48)$$

Equating Eqs. (6.47) and (6.48) leads to

$$\frac{f_{2Y}}{f_{1Y}} = -\frac{(n_{2B} - n_{2R})/(n_{2Y} - 1)}{(n_{1B} - n_{1R})/(n_{1Y} - 1)} \qquad (6.49)$$

The quantities

$$\frac{n_{2B} - n_{2R}}{n_{2Y} - 1} \quad \text{and} \quad \frac{n_{1B} - n_{1R}}{n_{1Y} - 1}$$

are known as the **dispersive powers** of the two materials forming the lenses. Their reciprocals, V_2 and V_1, are variously known as the *dispersive indices*, *V-numbers*, or **Abbe numbers**. The lower the Abbe numbers, the greater the dispersive power. Thus

$$\frac{f_{2Y}}{f_{1Y}} = -\frac{V_1}{V_2}$$

or

$$f_{1Y}V_1 + f_{2Y}V_2 = 0 \qquad (6.50)$$

Since the dispersive powers are positive, so too are the V-numbers. This implies, as we anticipated, that one of the two component lenses must be negative, and the other positive, if Eq. (6.50) is to obtain; that is, if f_R is to equal f_B.

At this point we could presumably design an *achromatic doublet*, and indeed we presently shall, but a few additional points must be made first. The designation of wavelengths as red, yellow, and blue is far too imprecise for practical application. Instead it is customary to refer to specific spectral lines whose wavelengths are known with great precision. The **Fraunhofer lines**, as they are called, serve as the needed reference markers across the spectrum. Several of these for the visible region are listed in Table 6.1. The lines F, C, and d (i.e., D_3) are most often used (for blue, red, and yellow), and one generally traces paraxial rays in d-light. Glass manufacturers will usually list their wares in terms of the Abbe number, as in Fig. 6.39, which is a plot of the refractive index versus

$$V_d = \frac{n_d - 1}{n_F - n_C} \tag{6.51}$$

(Take a look at Table 6.2 as well.) Thus Eq. (6.50) might better be written as

$$f_{1d}V_{1d} + f_{2d}V_{2d} = 0 \tag{6.52}$$

where the numerical subscripts pertain to the two glasses used in the doublet, and the letter relates to the d-line.

TABLE 6.1 Several Strong Fraunhofer Lines

Designation	Wavelength (Å)*	Source
C	6562.816 Red	H
D_1	5895.923 Yellow	Na
D	Center of doublet 5892.9	Na
D_2	5889.953 Yellow	Na
D_3 or d	5875.618 Yellow	He
b_1	5183.618 Green	Mg
b_2	5172.699 Green	Mg
c	4957.609 Green	Fe
F	4861.327 Blue	H
f	4340.465 Violet	H
g	4226.728 Violet	Ca
K	3933.666 Violet	Ca

*1 Å = 0.1 nm.

Incidentally, Newton erroneously concluded, on the basis of experiments with the very limited range of materials available at the time, that the dispersive power was constant for all glasses. This is tantamount to saying [Eq. (6.52)] that $f_{1d} = -f_{2d}$, in which case the doublet would have zero power. Newton, accordingly, shifted his efforts from the refracting to the reflecting telescope, and this fortunately turned out to be a good move in the long run. The achromat was invented around 1733 by Chester Moor Hall, Esq., but it lay in limbo until it was seemingly reinvented and patented in 1758 by the London optician John Dollond.

Several forms of the achromatic doublet are shown in Fig. 6.40. Their configurations depend on the glass types selected, as well as on the choice of the other aberrations to be controlled. By the way, when purchasing off-the-shelf doublets of unknown origin, be careful not to buy a lens that has been deliberately designed to include certain aberrations in order to compensate for errors in the original system from which it came. Perhaps the most commonly encountered doublet is the cemented Fraunhofer achromat. It's formed of a crown* double-convex lens in contact with a concave-planar (or nearly planar) flint lens. The use of a crown front element is quite popular because of its resistance to wear. Since the overall shape is roughly convex-planar, by selecting the proper glasses, both spherical aberration and coma can be corrected as well. Suppose that we wish to design a Fraunhofer achromat of focal length 50 cm. We can get some idea of how to select glasses by solving Eq. (6.52) simultaneously with the compound-lens equation

$$\frac{1}{f_{1d}} + \frac{1}{f_{2d}} = \frac{1}{f_d}$$

to get

$$\frac{1}{f_{1d}} = \frac{V_{1d}}{f_d(V_{1d} - V_{2d})} \tag{6.53}$$

and

$$\frac{1}{f_{2d}} = \frac{V_{2d}}{f_d(V_{2d} - V_{1d})} \tag{6.54}$$

Thus, in order to avoid small values of f_{1d} and f_{2d}, which

*Traditionally the glasses roughly in the range $n_d > 1.60$, $V_d > 50$, as well as $n_d < 1.60$, $V_d > 55$ are known as *crowns*, and the others are *flints*. Note the letter designations in Fig. 6.39.

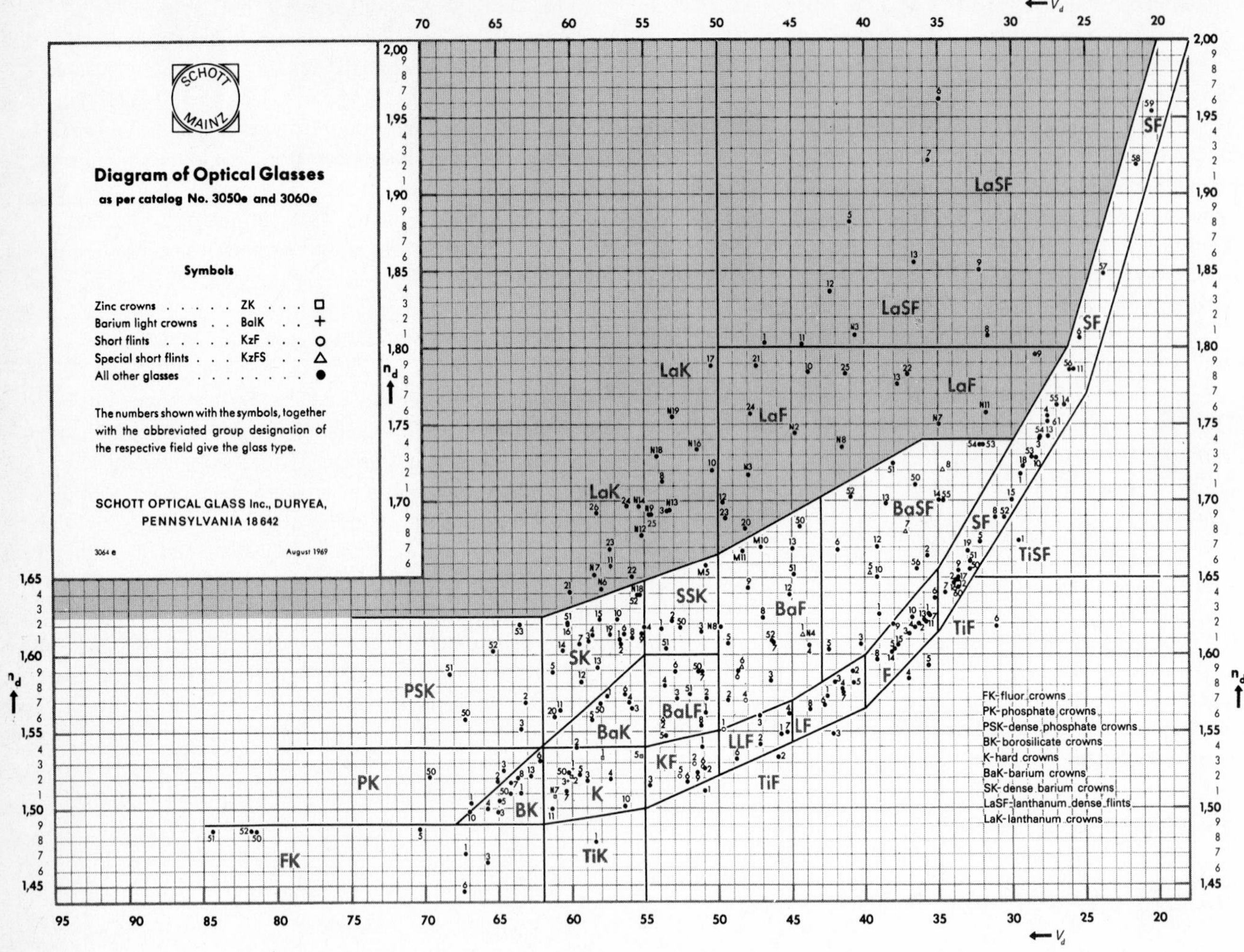

FIGURE 6.39 Refractive index versus Abbe number for various glasses. The specimens in the upper shaded area are the rare-earth glasses, which have high indices of refraction and low dispersions.

would necessitate strongly curved surfaces on the component lenses, the difference $V_{1d} - V_{2d}$ should be made large (roughly 20 or more is convenient). From Fig. 6.39 (or its equivalent) we select, say, BK1 and F2. These have catalogued indices of $n_C = 1.507\,63$, $n_d = 1.510\,09$, $n_F = 1.515\,66$ and $n_C = 1.615\,03$, $n_d = 1.620\,04$, $n_F = 1.632\,08$, respectively. Likewise, their V-numbers are generally given rather accurately, and we needn't compute them. In this instance they are $V_{1d} = 63.46$ and $V_{2d} = 36.37$, respectively. The focal lengths, or if you will, the powers of the two lenses, are given by Eqs. (6.53) and (6.54):

$$\mathfrak{D}_{1d} = \frac{1}{f_{1d}} = \frac{63.46}{0.50(27.09)}$$

TABLE 6.2 Optical Glass

Type Number	Name	n_D	V_D
511:635	Borosilicate crown—BSC-1	1.5110	63.5
517:645	Borosilicate crown—BSC-2	1.5170	64.5
513:605	Crown—C	1.5125	60.5
518:596	Crown	1.5180	59.6
523:586	Crown—C-1	1.5230	58.6
529:516	Crown flint—CF-1	1.5286	51.6
541:599	Light barium crown—LBC-1	1.5411	59.9
573:574	Barium crown—LBC-2	1.5725	57.4
574:577	Barium crown	1.5744	57.7
611:588	Dense barium crown—DBC-1	1.6110	58.8
617:550	Dense barium crown—DBC-2	1.6170	55.0
611:572	Dense barium crown—DBC-3	1.6109	57.2
562:510	Light barium flint—LBF-2	1.5616	51.0
588:534	Light barium flint—LBF-1	1.5880	53.4
584:460	Barium flint—BF-1	1.5838	46.0
605:436	Barium flint—BF-2	1.6053	43.6
559:452	Extra light flint—ELF-1	1.5585	45.2
573:425	Light flint—LF-1	1.5725	42.5
580:410	Light flint—LF-2	1.5795	41.0
605:380	Dense flint—DF-1	1.6050	38.0
617:366	Dense flint—DF-2	1.6170	36.6
621:362	Dense flint—DF-3	1.6210	36.2
649:338	Extra dense flint—EDF-1	1.6490	33.8
666:324	Extra dense flint—EDF-5	1.6660	32.4
673:322	Extra dense flint—EDF-2	1.6725	32.2
689:309	Extra dense flint—EDF	1.6890	30.9
720:293	Extra dense flint—EDF-3	1.7200	29.3

From T. Calvert, "Optical Components," *Electromechanical Design*, May 1971. Type number is given by $(n_D - 1)$:$(10\ V_D)$, where n_D is rounded off to three decimal places. For more data, see Smith, *Modern Optical Engineering*, Fig. 7.5.

and
$$\mathfrak{D}_{2d} = \frac{1}{f_{2d}} = \frac{36.37}{0.50(-27.09)}$$

Hence $\mathfrak{D}_{1d} = 4.685$ D and $\mathfrak{D}_{2d} = -2.685$ D, the sum being 2 D, which is $1/0.5$, as it should be. For ease of fabrication let the first or positive lens be equiconvex. Consequently, its radii

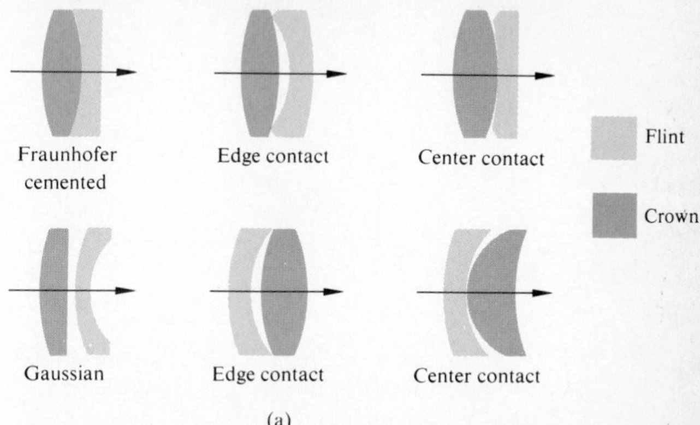

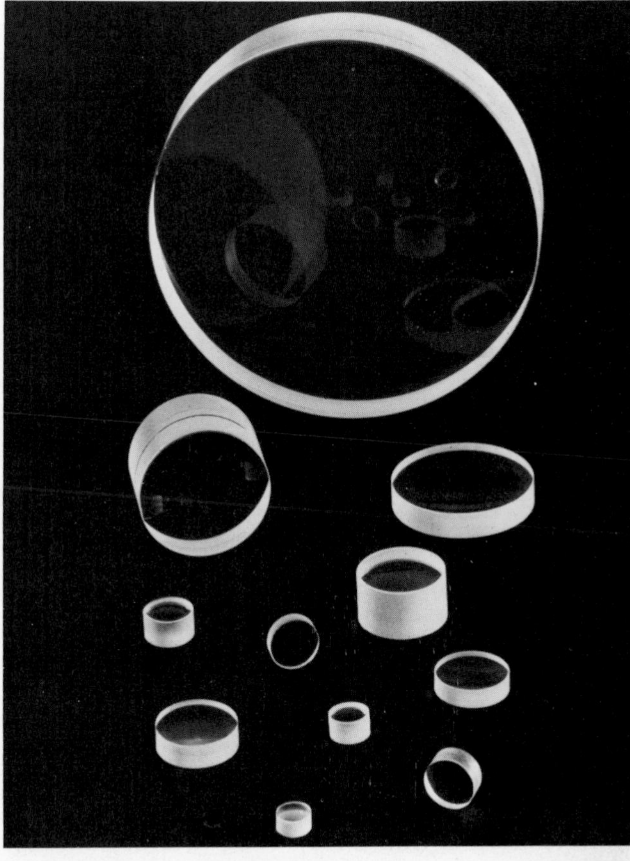

FIGURE 6.40 *(a)* Achromatic doublets. *(b)* Doublets and triplets. (Photo courtesy Melles Griot.)

R_{11} and R_{12} are equal in magnitude. Hence

$$\rho_1 = \frac{1}{R_{11}} - \frac{1}{R_{12}} = \frac{2}{R_{11}}$$

or, equivalently,

$$\frac{2}{R_{11}} = \frac{\mathscr{D}_{1d}}{n_{1d} - 1} = \frac{4.685}{0.51009} = 9.185$$

Thus $R_{11} = -R_{12} = 0.2177$ m. Furthermore, having specified that the lenses be in intimate contact, we have $R_{12} = R_{21}$; that is, the second surface of the first lens matches the first surface of the second lens. For the second lens

$$\rho_2 = \frac{1}{R_{21}} - \frac{1}{R_{22}} = \frac{\mathscr{D}_{2d}}{n_{2d} - 1}$$

or

$$\frac{1}{-0.2177} - \frac{1}{R_{22}} = \frac{-2.685}{0.62004}$$

and $R_{22} = -3.819$ m. In summary, the radii of the crown element are $R_{11} = 21.8$ cm and $R_{12} = -21.8$ cm while the flint has radii of $R_{21} = -21.8$ cm and $R_{22} = -381.9$ cm.

Note that for a thin-lens combination the principal planes coalesce, so that achromatizing the focal length corrects both A · CA and L · CA. In a thick doublet, however, even though the focal lengths for red and blue are made identical, the different wavelengths may have different principal planes. Consequently, although the magnification is the same for all wavelengths, the focal points may not coincide; in other words, correction is made for L · CA but not for A · CA.

In the above analysis, only the *C*- and *F*-rays were brought to a common focus, and the *d*-line was introduced to establish a focal length for the doublet as a whole. It is not possible for *all* wavelengths traversing a doublet achromat to meet at a common focus. The resulting residual chromatism is known as *secondary spectrum*. The elimination of secondary spectrum is particularly troublesome when the design is limited to the glasses currently available. Nevertheless, a fluorite (CaF_2) element combined with an appropriate glass element can form a doublet achromatized at three wavelengths and having very little secondary spectrum. More often triplets are used for color correction at three or even four wavelengths. The secondary spectrum of a binocular can easily be observed by looking at a distant white object. Its borders will be slightly haloed in magenta and green—try shifting the focus forward and backward.

Separated Achromatic Doublets

It is also possible to achromatize the focal length of a doublet composed of two widely separated elements of the same glass. Return to Eq. (6.46) and set $n_{1R} = n_{2R} = n_R$ and $n_{1B} = n_{2B} = n_B$. After a bit of straightforward algebraic manipulation, it becomes

$$(n_R - n_B)[(\rho_1 + \rho_2) - \rho_1\rho_2 d(n_B + n_R - 2)] = 0$$

or

$$d = \frac{1}{(n_B + n_R - 2)}\left(\frac{1}{\rho_1} + \frac{1}{\rho_2}\right)$$

Again introducing the yellow reference frequency, as we did before, namely, $1/f_{1Y} = (n_{1Y} - 1)\rho_1$ and $1/f_{2Y} = (n_{2Y} - 1)\rho_2$, we can replace ρ_1 and ρ_2. Hence

$$d = \frac{(f_{1Y} + f_{2Y})(n_Y - 1)}{n_B + n_R - 2}$$

where $n_{1Y} = n_{2Y} = n_Y$. Assuming $n_Y = (n_B + n_R)/2$, we have

$$d = \frac{f_{1Y} + f_{2Y}}{2}$$

or in *d*-light

$$d = \frac{f_{1d} + f_{2d}}{2} \tag{6.55}$$

This is precisely the form taken by the Huygens ocular (Section 5.7.4). Since the red and blue focal lengths are the same, but the corresponding principal planes for the doublet need not be, the two rays will generally not meet at the same focal point. Thus the ocular's lateral chromatic aberration is well corrected, but axial chromatic aberration is not.

In order for a system to be free of both chromatic aberrations, the red and blue rays must emerge parallel to each other (no L · CA) and must intersect the axis at the same point (no A · CA), which means they must overlap. Since this is effectively the case with a thin achromat, it implies that multi-element systems, as a rule, should consist of achromatic components in order to keep the red and blue rays from separating (Fig. 6.41). As with all such invocations there are exceptions. The Taylor triplet (Section 5.7.7) is one. The two

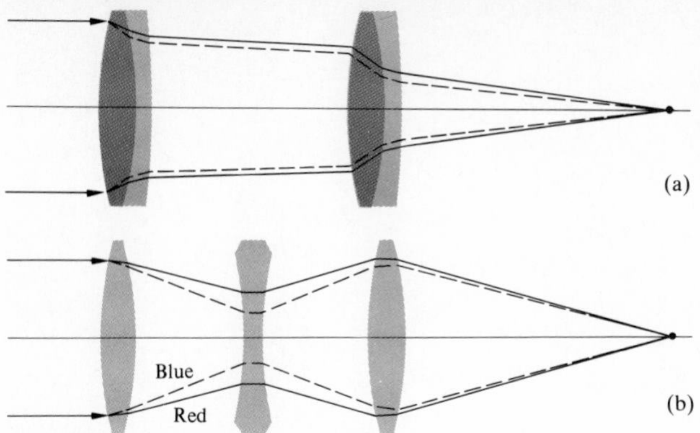

FIGURE 6.41 Achromatized lenses.

colored rays for which it is achromatized separate within the lens but are recombined and emerge together.

6.4 Grin Systems

An ordinary homogeneous lens has two physical features that contribute to the manner in which it reconfigures a wavefront: the difference between its index of refraction and that of the surrounding medium, and the curvature of its interfaces. But as we have already seen, when light propagates through an inhomogeneous medium, wavefronts essentially slow down in optically dense regions and speed up in less dense regions, and bending again occurs. In principle, then, it should be possible to make a lens from some inhomogeneous material, one where there is a GRadient in the INdex of refraction; such a device is known as a **GRIN lens**. A powerful incentive for developing such systems is that they provide the optical designer with an additional set of new parameters with which to control aberrations.

To get a rough sense of how a GRIN lens might work, consider the device pictured in Fig. 6.42 where, for simplicity, we assume $f > r$. This is a flat disk of glass that has been treated so that it has an index $n(r)$ that drops off radially in some as yet undetermined fashion from a maximum value of $n_{\max}$ on the optical axis. Accordingly, it's called a **radial-GRIN** device. A ray that traverses the disk on the optical axis passes along an optical path length of $(OPL)_o = n_{\max}d$, whereas for a ray tra-

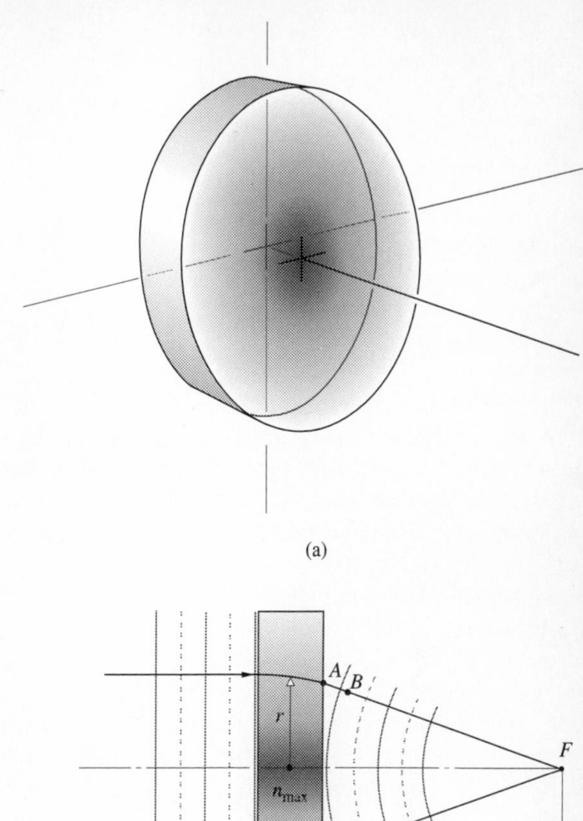

FIGURE 6.42 A disk of transparent glass whose index of refraction decreases radially out from the central axis. (*b*) The geometry corresponding to the focusing of parallel rays by a GRIN lens.

versing at a height r, overlooking the slight bending of its path, $(OPL)_r \approx n(r)d$. Since a planar wavefront must bend into a spherical wavefront, the *OPL*s from one to the other, along any route must be equal (p. 150):

$$(OPL)_r + \overline{AB} = (OPL)_o$$

and $$n(r)d + \overline{AB} = n_{\max}d$$

But $\overline{AF} \approx \sqrt{r^2 + f^2}$; moreover, $\overline{AB} = \overline{AF} - f$ and so

$$n(r) = n_{\max} - \frac{\sqrt{r^2 + f^2} - f}{d}$$

Rewriting the square root via the Binomial Theorem, $n(r)$ becomes

$$n(r) = n_{\max} - \frac{r^2}{2fd}$$

This tells us that if the index of refraction drops off parabolically from its high along the central axis, the GRIN plate will focus a collimated beam at F and serve as a positive lens. Although this is a rather simplistic treatment, it does make the point: a parabolic refractive index profile will focus parallel light.

Today a variety of radial-GRIN lenses are commercially available, and tens of millions of them are already in service in laser printers, photocopiers, and fax machines. The most common device is a GRIN cylinder a few millimeters in diameter, similar in kind to the optical fiber shown in Fig. 5.82*b*. Monochromatically, they provide nearly diffraction-limited performance on axis. Polychromatically, they offer substantial benefits over aspherics.

These small-diameter GRIN rods are usually fabricated via ionic diffusion. A homogeneous base glass is immersed in a molten salt bath for many hours during which ion diffusion/exchange slowly occurs. One type of ion migrates out of the glass, and another from the bath takes its place, changing the index of refraction. The process works its way inward radially toward the optical axis, and the time it takes is roughly proportional to the rod's diameter squared. For a parabolic profile, that sets the practical limit on the aperture size. The focal length is determined by the index change, Δn, and the faster the lens the larger must be Δn. Even so, Δn is usually constrained to be less than about 0.10 for production reasons. Most GRIN cylinders have a parabolic index profile typically expressed as

$$n(r) = n_{\max}(1 - ar^2/2)$$

Figure 6.43 shows one such radial-GRIN rod of length L, under monochromatic illumination. Meridional rays travel sinusoidal paths within the plane-of-incidence, which is perpendicular to the circumference. These sinusoids have a period in space of $2\pi/\sqrt{a}$, where the **gradient constant** $\sqrt{a}$, is a

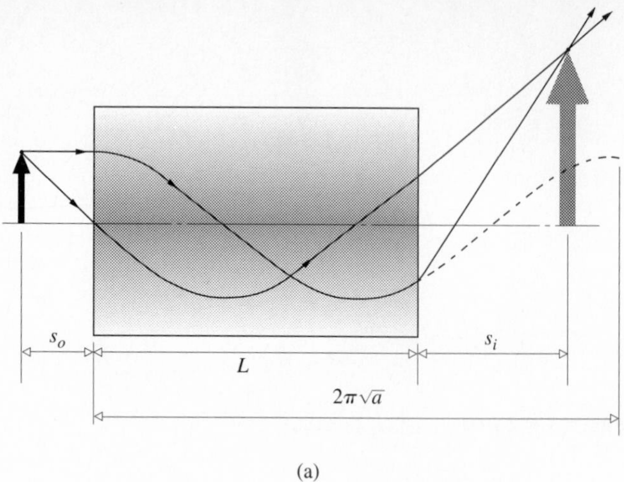

(a)

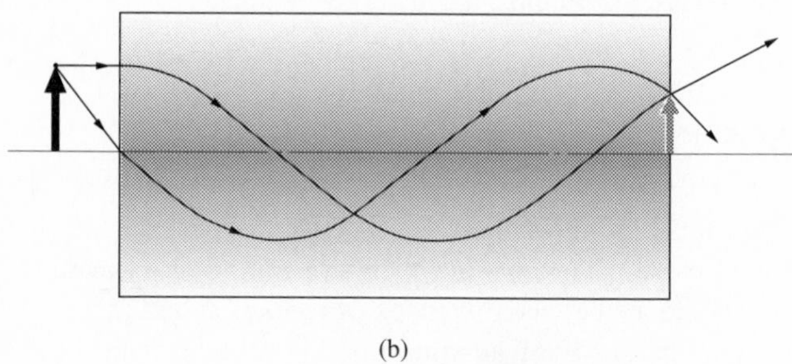

(b)

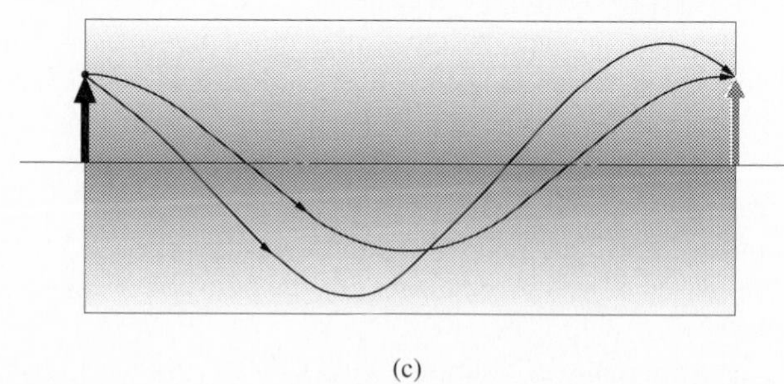

(c)

FIGURE 6.43 (*a*) A radial-GRIN rod producing a real, magnified, erect image. (*b*) Here the image is formed on the face of the rod. (*c*) This is a convenient setup for use in a copy machine.

function of λ and depends on the specific GRIN material. The cross-sectional view in Fig. 6.43*a* shows how a radial-GRIN lens can form an erect, real, magnified image. By changing the object distance or the length of the lens L, a wide range of images can be produced. It's even possible to have the object and image planes on the face of the rod (Figs. 6.43*b* and *c*).

Radial-GRIN lenses are often specified in terms of their length or, equivalently, their **pitch** (Fig. 6.44). A radial-GRIN rod with a pitch of 1.0 is one sine-wave long: $L = 2\pi/\sqrt{a}$. A rod with a pitch of 0.25 has a length of a quarter of a sine-wave $(\pi/2\sqrt{a})$.

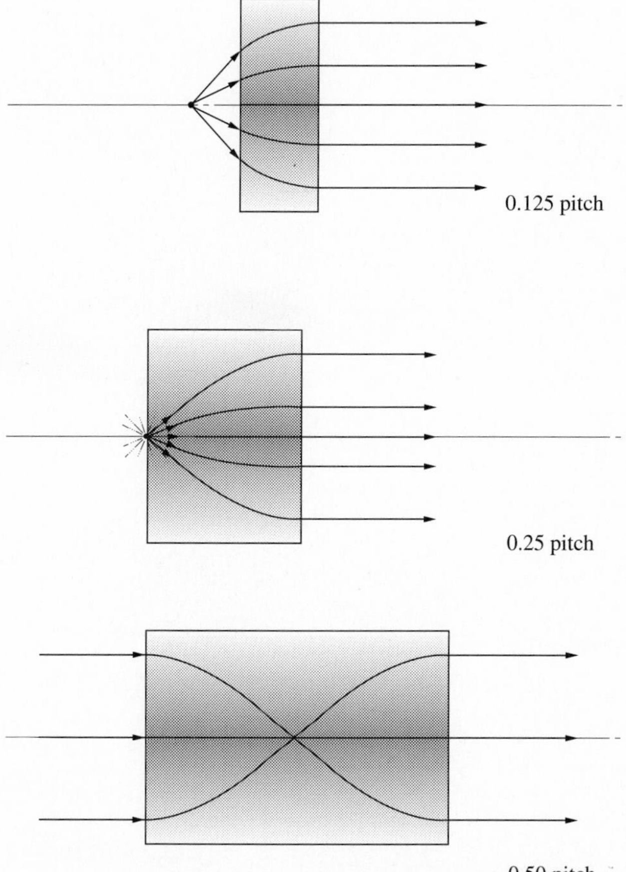

FIGURE 6.44 Radial GRIN lenses with several pitches used in a few typical ways.

An alternative approach to the flat-faced radial-GRIN rod is the axial-GRIN lens which is generally polished with spherical surfaces. As such, it's similar to a bi-aspheric, but without the difficulty of creating the complex surfaces. Usually, a stack of glass plates with appropriate indices are fused together. At high temperatures the glasses meld, diffusing into each other, creating a block of glass with a continuous index profile that can be made linear, quadratic, or even cubic (Fig. 6.45*a*). When such a block is ground into a lens, the process cuts back on the glass and exposes a range of indices. Every annulus (concentric with the optical axis) on the lens's face has a gradually changing index. Rays impinging at different heights above the optical axis encounter glass with different indices and bend appropriately. The spherical aberration evident in Fig. 6.45*c* arises because the edges of the spherical lens refract too strongly. Gradually lowering the index of refraction out toward the edges allows the axial-GRIN lens to correct for spherical aberration.

Generally, introducing a GRIN element into the design of a compound lens greatly simplifies the system, reducing the number of elements by as much as a third while maintaining overall performance.

6.5 Concluding Remarks

For the practical reason of manufacturing ease, the vast majority of optical systems are limited to lenses having spherical surfaces. There are, to be sure, toric and cylindrical lenses as well as many other aspherics. Indeed, very fine, and as a rule very expensive devices, such as high-altitude reconnaisance cameras and tracking systems, may have several aspherical elements. Even so, spherical lenses are here to stay and with them are their inherent aberrations which must satisfactorily be dealt with. As we have seen, the designer (and his faithful electronic companion) must manipulate the system variables (indices, shapes, spacings, stops, etc.) in order to balance out offensive aberrations. This is done to whatever degree and in whatever order is appropriate for the specific optical system. Thus one might tolerate far more distortion and curvature in an

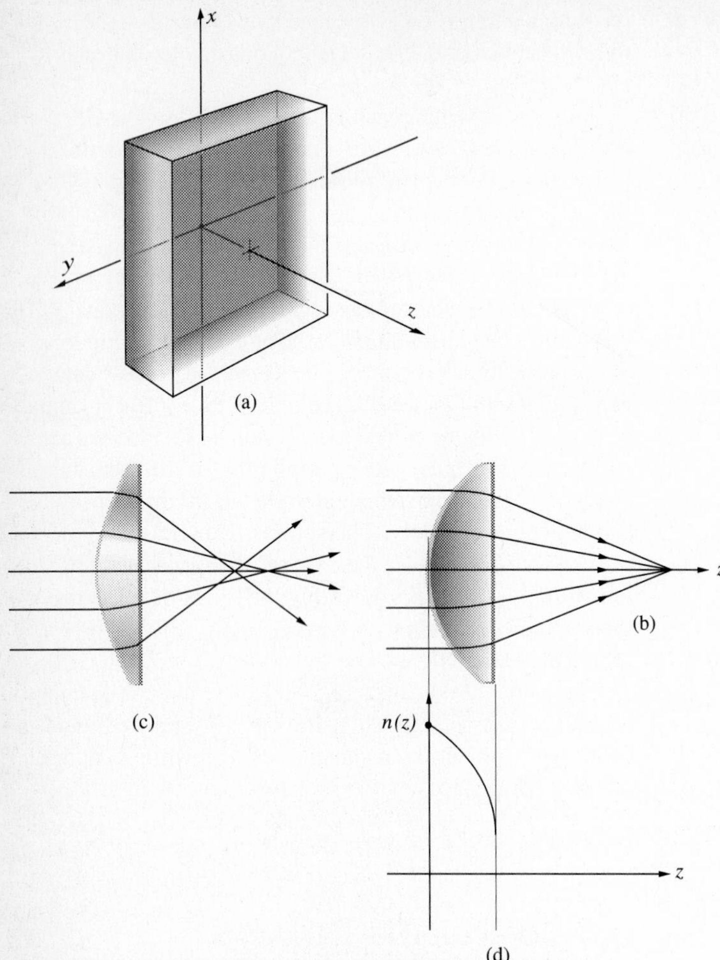

FIGURE 6.45 (*a*) A slab of axial-GRIN material for which the index of refraction is $n(z)$. (*b*) An axial-GRIN lens for which there is no spherical aberration. (*c*) An ordinary lens having SA. (*d*) The index profile.

ordinary telescope than in a good photographic objective. Similarly, there is little need to worry about chromatic aberration if you want to work exclusively with laser light of almost a single frequency. In any event, this chapter has only touched on the problems (more to appreciate than solve them). That they are most certainly amenable to solution is evidenced, for example, by the remarkable aerial photographs in Fig. 6.46, which speak rather eloquently for themselves.

FIGURE 6.46 (*a*) New Orleans and the Mississippi River photographed from 12,500 m (41,000 ft) with Itek's Metritek-21 camera (f = 21 cm). Ground resolution, 1 m; scale, 1:59,492. (*b*) Photo scale, 1:10,000. (*c*) Photo scale, 1:2500. (Photos courtesy of Litton/Itek Optical Systems.)

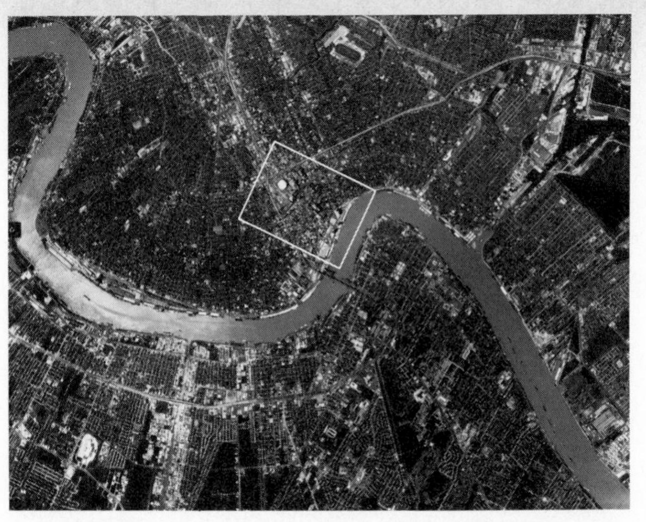

(a)

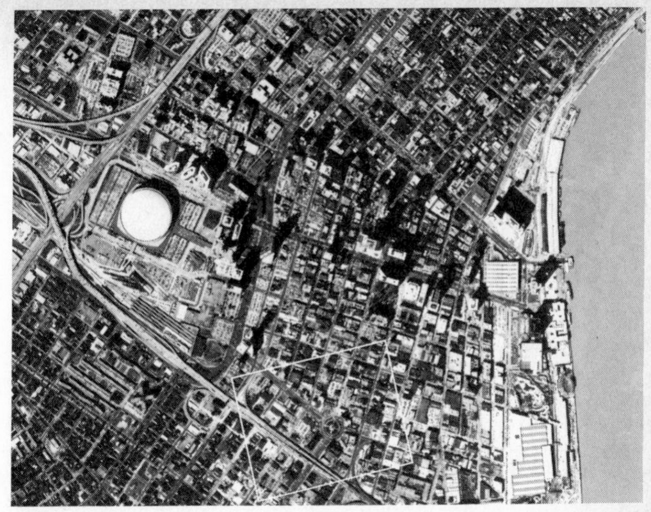

(b)

(c)

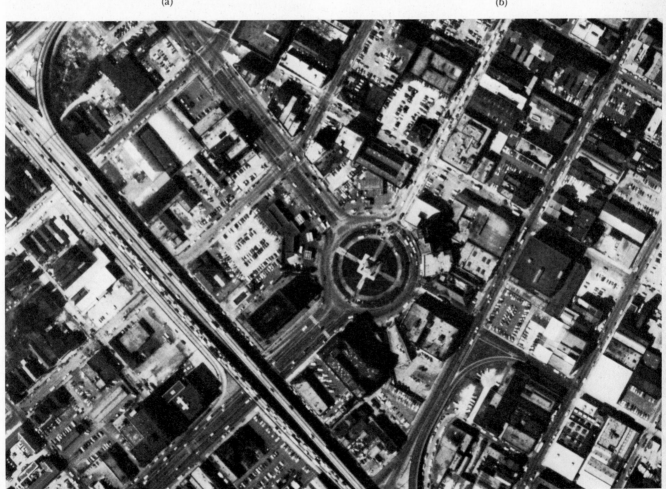

PROBLEMS

6.1* Work out the details leading to Eq. (6.8).

6.2 According to the military handbook MIL-HDBK-141 (23.3.5.3), the Ramsden eyepiece (Fig. 5.106) is made up of two planar-convex lenses of equal focal length f' separated by a distance $2f'/3$. Determine the overall focal length f of the thin-lens combination and locate the principal planes and the position of the field stop.

6.3 Write an expression for the thickness d_l of a double-convex lens such that its focal length is infinite.

6.4 Suppose we have a positive meniscus lens of radii 6 and 10 and a thickness of 3 (any units, as long as you're consistent), with an index of 1.5. Determine its focal length and the locations of its principal points (compare with Fig. 6.3).

6.5* Prove that if the principal points of a biconvex lens of thickness d_l overlap midway between the vertices, the lens is a sphere. Assume the lens is in air.

6.6 Using Eq. (6.2), derive an expression for the focal length of a homogeneous transparent sphere of radius R. Locate its principal points.

6.7* A spherical glass bottle 20 cm in diameter with walls that are negligibly thin is filled with water. The bottle is sitting on the back seat of a car on a nice sunny day. What's its focal length?

6.8* With the previous two problems in mind, compute the magnification that results when the image of a flower 4.0 m from the center of a solid, clear-plastic sphere with a 0.20-m diameter (and a refractive index of 1.4) is cast on a nearby wall. Describe the image in detail.

6.9* A thick glass lens of index 1.50 has radii of +23 cm and +20 cm, so that both vertices are to the left of the corresponding centers of curvature. Given that the thickness is 9.0 cm, find the focal length of the lens. Show that in general $R_1 - R_2 = d/3$ for such afocal zero-power lenses. Draw a diagram showing what happens to an axial incident parallel bundle of rays as it passes through the system.

6.10 It is found that sunlight is focused to a spot 29.6 cm from the back face of a thick lens, which has its principal points H_1 at +0.2 cm and H_2 at −0.4 cm. Determine the location of the image of a candle that is placed 49.8 cm in front of the lens.

6.11* Please establish that the separation between principal planes for a thick glass lens is roughly one-third its thickness. The simplest geometry occurs with a planar-convex lens tracing a ray from the object focus. What can you say about the relationship between the focal length and the thickness for this lens type?

6.12 A crown glass double-convex lens, 4.0 cm thick and operating at a wavelength of 900 nm, has an index of refraction of 3/2. Given that its radii are 4.0 cm and 15 cm, locate its principal points and compute its focal length. If a television screen is placed 1.0 m from the front of the lens, where will the real image of the picture appear?

6.13* Imagine two identical double-convex thick lenses separated by a distance of 20 cm between their adjacent vertices. Given that all the radii of curvature are 50, the refractive indices are 1.5, and the thickness of each lens is 5.0 cm, calculate the combined focal length.

6.14* A compound lens is composed of two thin lenses separated by 10 cm. The first of these has a focal length of +20 cm, and the second a focal length of −20 cm. Determine the focal length of the combination and locate the corresponding principal points. Draw a diagram of the system.

6.15* A convex-planar lens of index 3/2 has a thickness of 1.2 cm and a radius of curvature of 2.5 cm. Determine the system matrix when light is incident on the curved surface.

6.16* A positive meniscus lens with an index of refraction of 2.4 is immersed in a medium of index 1.9. The lens has an axial thickness of 9.6 mm and radii of curvature of 50.0 mm and 100 mm. Compute the system matrix when light is incident on the convex face and show that its determinant is equal to 1.

6.17* Prove that the determinant of the system matrix in Eq. (6.31) is equal to 1.

6.18 Establish that Eqs. (6.36) and (6.37) are equivalent to Eqs. (6.3) and (6.4), respectively.

6.19 Show that the planar surface of a concave-planar or convex-planar lens doesn't contribute to the system matrix.

6.20 Compute the system matrix for a thick biconvex lens of index 1.5 having radii of 0.5 and 0.25 and a thickness of 0.3 (in any units you like). Check that $|\mathcal{A}| = 1$.

6.21* The system matrix for a thick biconvex lens in air is given by

$$\begin{bmatrix} 0.6 & -2.6 \\ 0.2 & 0.8 \end{bmatrix}$$

Knowing that the first radius is 0.5 cm, that the thickness is 0.3 cm, and that the index of the lens is 1.5, find the other radius.

6.22* A concave-planar glass ($n = 1.50$) lens in air has a radius of 10.0 cm and a thickness of 1.00 cm. Determine the system matrix and check that its determinant is 1. At what positive angle (in radians measured above the axis) should a ray strike the lens at a height of 2.0 cm, if it is to emerge from the lens at the same height but parallel to the optical axis?

6.23* Considering the lens in Problem 6.20, determine its focal length and the location of the focal points with respect to its vertices V_1 and V_2.

6.24* Figure P.6.24 shows two identical concave spherical mirrors forming a so-called confocal cavity. Show, without first specifying the value of d, that after traversing the cavity two times the system matrix is

$$\begin{bmatrix} \left(\frac{2d}{r} - 1\right)^2 - \frac{2d}{r} & \frac{4}{r}\left(\frac{d}{r} - 1\right) \\ 2d\left(1 - \frac{d}{r}\right) & 1 - 2\frac{d}{r} \end{bmatrix}$$

Then for the specific case of $d = r$ show that after four reflections the system is back where it started and the light will retrace its original path.

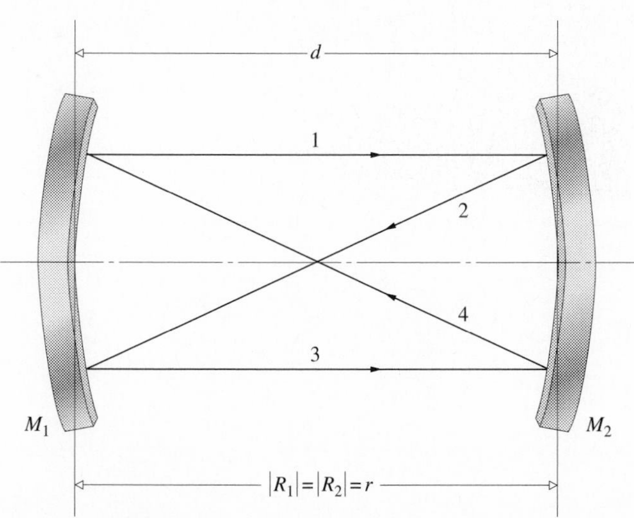

FIGURE P.6.24

6.25 Referring back to Fig. 6.17, show that when $\overline{P'P} = Rn_2/n_1$ and $\overline{PC} = Rn_1/n_2$ all rays originating at P appear to come from P'.

6.26 Starting with the exact expression given by Eq. (5.5), show that Eq. (6.40) results, rather than Eq. (5.8), when the approximations for ℓ_o and ℓ_i are improved a bit.

6.27 Supposing that Fig. P.6.27 is to be imaged by a lens system suffering spherical aberration only, make a sketch of the image.

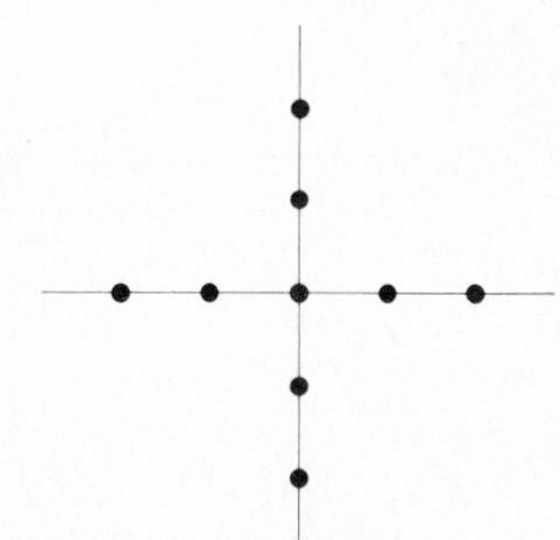

FIGURE P.6.27

6.28* Figure P.6.28 shows the image irradiance distributions arising when a monochromatic point source illuminates three different optical systems, each having only one type of aberration. From the graphs identify that aberration in each case and justify your answer.

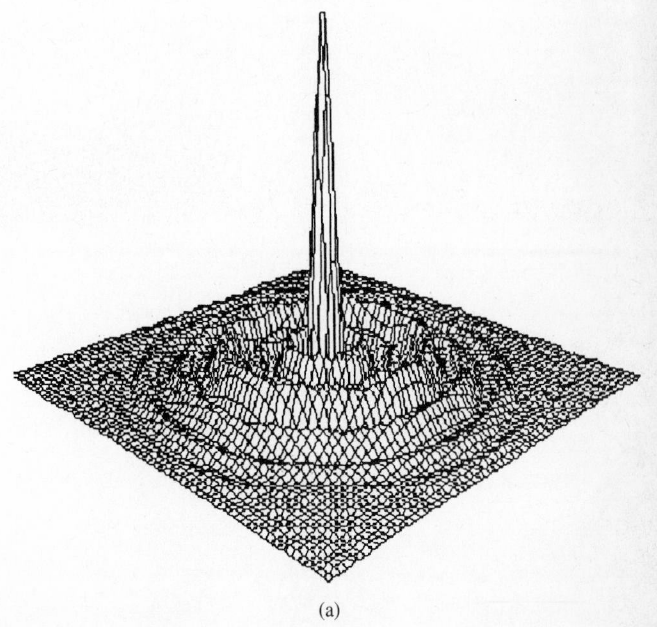

(a)

FIGURE P.6.28a

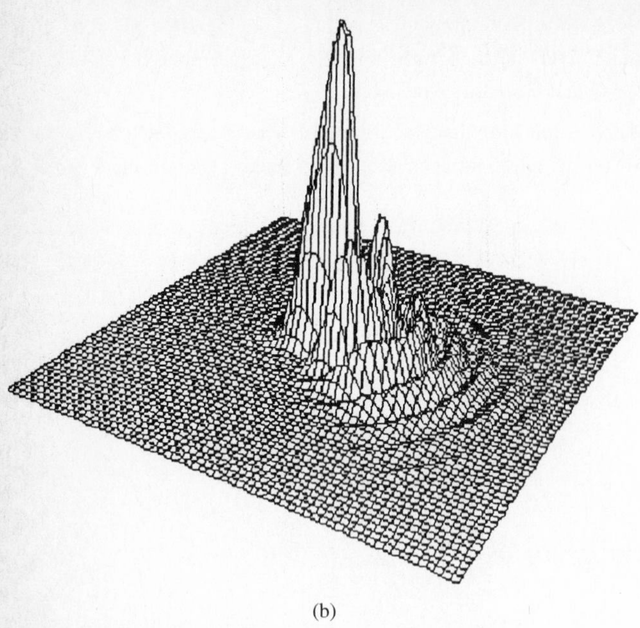

(b)

FIGURE P.6.28b

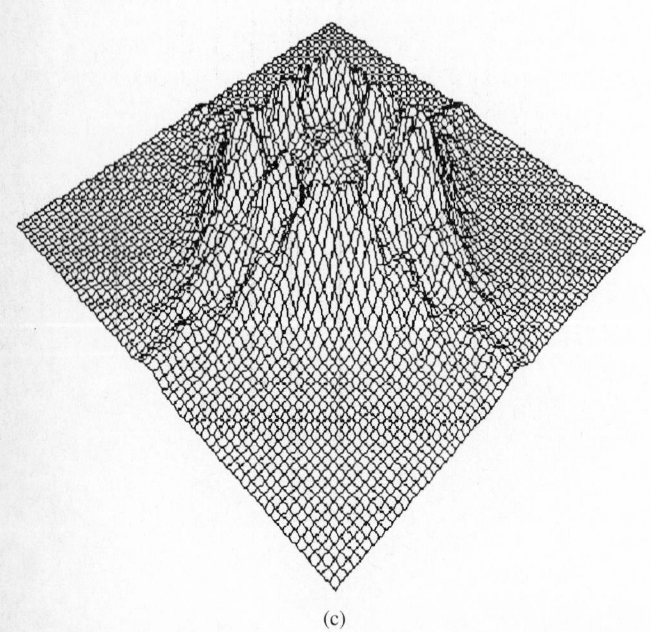

(c)

FIGURE P.6.28c

6.29* Figure P.6.29 shows the distribution of light corresponding to the image arising when a monochromatic point source illuminates two different optical systems each having only one type of aberration. Identify the aberration in each case and justify your answer.

(a)

(b)

FIGURE P.6.29

7 The Superposition of Waves

In succeeding chapters we shall study the phenomena of polarization, interference, and diffraction. These all share a common conceptual basis in that they deal, for the most part, with various aspects of the same process. Stating this in the simplest terms, we are really concerned with what happens when two or more lightwaves overlap in some region of space. The precise circumstances governing this superposition determine the final optical disturbance. Among other things we are interested in learning how the specific properties of each constituent wave (amplitude, phase, frequency, etc.) influence the ultimate form of the composite disturbance.

Recall that each field component of an electromagnetic wave (E_x, E_y, E_z, B_x, B_y, and B_z) satisfies the scalar three-dimensional differential wave equation,

$$\frac{\partial^2\psi}{\partial x^2}+\frac{\partial^2\psi}{\partial y^2}+\frac{\partial^2\psi}{\partial z^2}=\frac{1}{v^2}\frac{\partial^2\psi}{\partial t^2} \qquad [2.60]$$

A significant feature of this expression is that it is *linear*; $\psi(\mathbf{r}, t)$ and its derivatives appear only to the first power. Consequently, if $\psi_1(\mathbf{r}, t)$, $\psi_2(\mathbf{r}, t)$, ... , $\psi_n(\mathbf{r}, t)$ are individual solutions of Eq. (2.59), *any linear combination* of them will, in turn, be a solution. Thus

$$\psi(\mathbf{r}, t)=\sum_{i=1}^{n} C_i\psi_i(\mathbf{r}, t) \qquad (7.1)$$

satisfies the wave equation, where the coefficients C_i are simply arbitrary constants. Known as the ***Principle of Superposition***, this property suggests that the resultant disturbance at any point in a medium is the algebraic sum of the separate constituent waves (Fig. 7.1). At this time we are interested

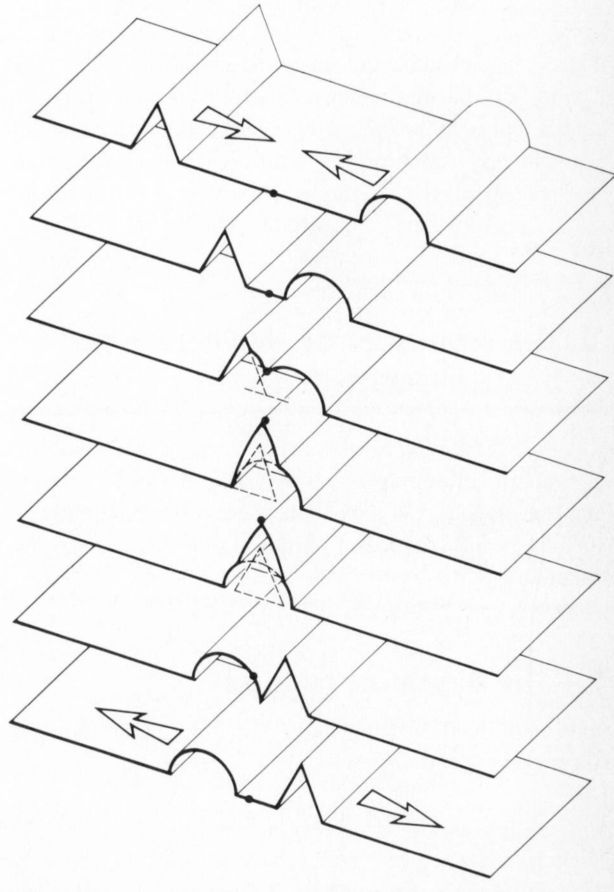

FIGURE 7.1 The superposition of two disturbances.

only in linear systems where the superposition principle is applicable. Do keep in mind, however, that large-amplitude waves, whether sound waves or waves on a string, can generate a nonlinear response. The focused beam of a high-intensity laser (where the electric field might be as high as 10^{10} V/cm) is easily capable of eliciting nonlinear effects (see Chapter 13). By comparison, the electric field associated with sunlight here on Earth has an amplitude of only about 10 V/cm.

In many instances we need not be concerned with the vector nature of light, and for the present we will restrict ourselves to such cases. For example, if the lightwaves all propagate along the same line and share a common constant plane of vibration, they can each be described in terms of one electric-field component. These would all be either parallel or antiparallel at any instant and could thus be treated as scalars. A good deal more will be said about this point as we progress; for now, let's represent the optical disturbance as a scalar function $E(\mathbf{r}, t)$, which is a solution of the differential wave equation. This approach leads to a simple scalar theory that is highly useful as long as we are careful about applying it.

7.1 The Addition of Waves of the Same Frequency

There are several equivalent ways of mathematically adding two or more overlapping waves that have the same frequency and wavelength. Let's examine these different approaches so that, in any particular situation, we can use the one most suitable.

7.1.1 The Algebraic Method

A solution of the differential wave equation can be written in the form

$$E(x, t) = E_0 \sin [\omega t - (kx + \varepsilon)] \tag{7.2}$$

in which E_0 is the amplitude of the harmonic disturbance propagating along the positive x-axis. To separate the space and time parts of the phase, let

$$\alpha(x, \varepsilon) = -(kx + \varepsilon) \tag{7.3}$$

so that

$$E(x, t) = E_0 \sin [\omega t + \alpha(x, \varepsilon)] \tag{7.4}$$

Suppose then that there are two such waves

$$E_1 = E_{01} \sin (\omega t + \alpha_1) \tag{7.5a}$$

and

$$E_2 = E_{02} \sin (\omega t + \alpha_2) \tag{7.5b}$$

each with the same frequency and speed, coexisting in space. The resultant disturbance is the linear superposition of these waves:

$$E = E_1 + E_2$$

or, on expanding Eqs. (7.5a) and (7.5b)

$$E = E_{01} (\sin \omega t \cos \alpha_1 + \cos \omega t \sin \alpha_1)$$
$$+ E_{02} (\sin \omega t \cos \alpha_2 + \cos \omega t \sin \alpha_2)$$

When we separate out the time-dependent terms, this becomes

$$E = (E_{01} \cos \alpha_1 + E_{02} \cos \alpha_2) \sin \omega t$$
$$+ (E_{01} \sin \alpha_1 + E_{02} \sin \alpha_2) \cos \omega t \tag{7.6}$$

Since the parenthetical quantities are constant in time, let

$$E_0 \cos \alpha = E_{01} \cos \alpha_1 + E_{02} \cos \alpha_2 \tag{7.7}$$

and

$$E_0 \sin \alpha = E_{01} \sin \alpha_1 + E_{02} \sin \alpha_2 \tag{7.8}$$

This is not an obvious substitution, but it will be legitimate as long as we can solve for E_0 and α. To that end, square and add Eqs. (7.7) and (7.8) to get

$$E_0^2 = E_{01}^2 + E_{02}^2 + 2E_{01}E_{02} \cos (\alpha_2 - \alpha_1) \tag{7.9}$$

and divide Eq. (7.8) by (7.7) to get

$$\tan \alpha = \frac{E_{01} \sin \alpha_1 + E_{02} \sin \alpha_2}{E_{01} \cos \alpha_1 + E_{02} \cos \alpha_2} \tag{7.10}$$

Provided these last two expressions are satisfied for E_0 and α, the situation of Eqs. (7.7) and (7.8) is valid. The total disturbance [Eq. (7.6)] then becomes

$$E = E_0 \cos \alpha \sin \omega t + E_0 \sin \alpha \cos \omega t$$

or

$$E = E_0 \sin (\omega t + \alpha) \tag{7.11}$$

A single disturbance results from the superposition of the sinusoidal waves E_1 and E_2. *The composite wave [Eq. (7.11)] is harmonic and of the same frequency as the constituents, although its amplitude and phase are different.*

Note that when $E_{01} >> E_{02}$ in Eq. (7.10), $\alpha \approx \alpha_1$ and when $E_{02} >> E_{01}$, $\alpha \approx \alpha_2$; the resultant is in-phase with the dominant component wave (take another look at Fig. 4.11). The flux density of a lightwave is proportional to its amplitude squared, by way of Eq. (3.44). It follows from Eq. (7.9) that the resultant flux density is not simply the sum of the component flux densities; there is an additional contribution $2E_{01}E_{02} \cos (\alpha_2 - \alpha_1)$, known as the **interference term**. The crucial factor is the difference in phase between the two *interfering* waves E_1 and E_2, $\delta \equiv (\alpha_2 - \alpha_1)$. When $\delta = 0, \pm 2\pi, \pm 4\pi, \ldots$ the resultant amplitude is a maximum, whereas $\delta = \pm\pi, \pm 3\pi, \ldots$ yields a minimum (Problem 7.3). In the former case, the waves are said to be in-phase; crest overlaps crest. In the latter instance, the waves are 180° out-of-phase and trough overlaps crest, as shown in Fig. 7.2. Realize that the *phase difference* may arise from a difference in path length traversed by the two waves, as well as a difference in the initial phase angle; that is,

$$\delta = (kx_1 + \varepsilon_1) - (kx_2 + \varepsilon_2) \tag{7.12}$$

or

$$\delta = \frac{2\pi}{\lambda}(x_1 - x_2) + (\varepsilon_1 - \varepsilon_2) \tag{7.13}$$

Here x_1 and x_2 are the distances from the sources of the two waves to the point of observation, and λ is the wavelength in the pervading medium. If the waves are initially in-phase at their respective emitters, then $\varepsilon_1 = \varepsilon_2$, and

$$\delta = \frac{2\pi}{\lambda}(x_1 - x_2) \tag{7.14}$$

This would also apply to the case in which two disturbances from the same source traveled different routes before arriving at the point of observation. Since $n = c/v = \lambda_0/\lambda$,

$$\delta = \frac{2\pi}{\lambda_0} n(x_1 - x_2) \tag{7.15}$$

The quantity $n(x_1 - x_2)$ is known as the **optical path difference** and will be represented by the abbreviation OPD or by the symbol Λ. It's the difference in the two optical path lengths [Eq. (4.9)]. It is possible, in more complicated situations, for each wave to travel through a number of different thicknesses of different media (Problem 7.6). Notice too that $\Lambda/\lambda_0 = (x_1 - x_2)/\lambda$ is the number of waves in the medium corresponding to the path difference; one route is that many wavelengths longer than the other. Since each wavelength is

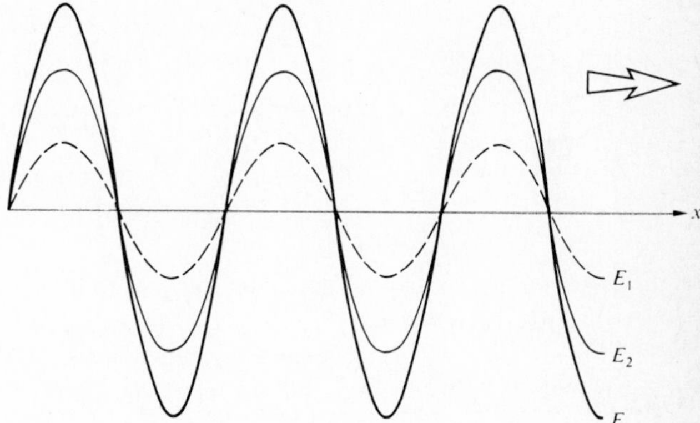

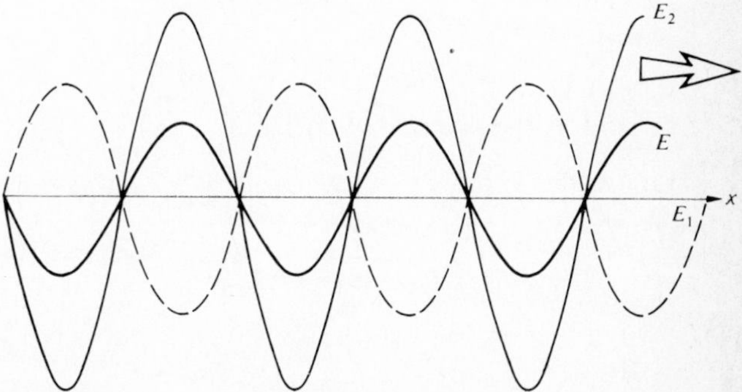

FIGURE 7.2 The superposition of two harmonic waves in-phase and out-of-phase.

associated with a 2π radian phase change, $\delta = 2\pi(x_1 - x_2)/\lambda$, or

$$\delta = k_0\Lambda \tag{7.16}$$

k_0 being the propagation number in vacuum; that is, $2\pi/\lambda_0$. One route is essentially δ radians longer than the other.

Waves for which $\varepsilon_1 - \varepsilon_2$ is *constant*, regardless of its value, are said to be **coherent**, a situation we shall assume obtains throughout most of this discussion.

One special case of some interest is the superposition of the waves

$$E_1 = E_{01} \sin [\omega t - k(x + \Delta x)]$$

and

$$E_2 = E_{02} \sin (\omega t - kx)$$

where in particular $E_{01} = E_{02}$ and $\alpha_2 - \alpha_1 = k\,\Delta x$. It is left to Problem 7.7 to show that in this case Eqs. (7.9), (7.10), and (7.11) lead to a resultant wave of

$$E = 2E_{01} \cos\left(\frac{k\,\Delta x}{2}\right) \sin\left[\omega t - k\left(x + \frac{\Delta x}{2}\right)\right] \tag{7.17}$$

This brings out rather clearly the dominant role played by the path length difference, Δx, especially when the waves are emitted in-phase ($\varepsilon_1 = \varepsilon_2$). There are many practical instances in which one arranges just these conditions, as will be seen later. If $\Delta x << \lambda$, the resultant has an amplitude that is nearly $2E_{01}$, whereas if $\Delta x = \lambda/2$, it is zero. Recall that the former situation (p. 21) is referred to as **constructive interference**, and the latter as **destructive interference** (see Fig. 7.3).

By repeated applications of the procedure used to arrive at Eq. (7.11), we can show that the ***superposition of any number of coherent harmonic waves having a given frequency and traveling in the same direction leads to a harmonic wave of that same frequency*** (Fig. 7.4). We happen to have chosen to represent the two waves above in terms of sine functions, but the same results would prevail if cosine functions were used. In general, then, the sum of N such waves,

$$E = \sum_{i=1}^{n} E_{0i} \cos (\alpha_i \pm \omega t)$$

is given by

$$E = E_0 \cos (\alpha \pm \omega t) \tag{7.18}$$

where

$$E_0^2 = \sum_{i=1}^{N} E_{0i}^2 + 2\sum_{j>i}^{N}\sum_{i=1}^{N} E_{0i}E_{0j} \cos (\alpha_i - \alpha_j) \tag{7.19}$$

and

$$\tan \alpha = \frac{\sum_{i=1}^{N} E_{0i} \sin \alpha_i}{\sum_{i=1}^{N} E_{0i} \cos \alpha_i} \tag{7.20}$$

Pause for a moment and satisfy yourself that these relations are indeed true.

Consider a number (N) of atomic emitters constituting an ordinary source (an incandescent bulb, candle flame, or discharge lamp). A flood of light is emitted that presumably corresponds to a torrent of photons, which manifest themselves *en masse* as an electromagnetic wave. To keep things in a wave perspective, it's useful to imagine the photon as some-

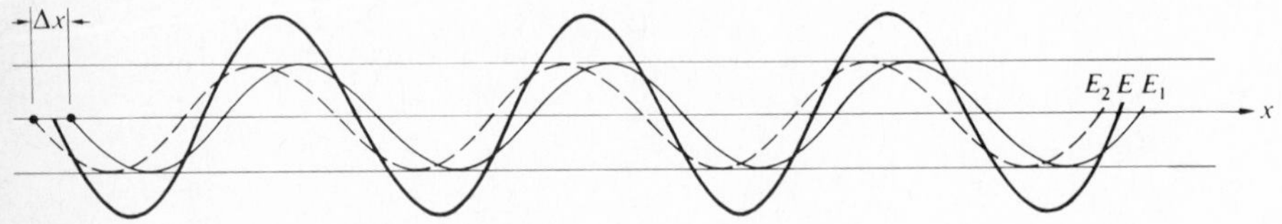

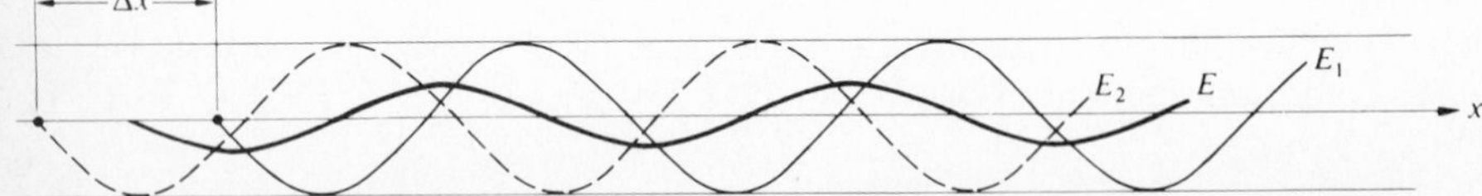

FIGURE 7.3 Waves out-of-phase by $k\Delta x$ radians.

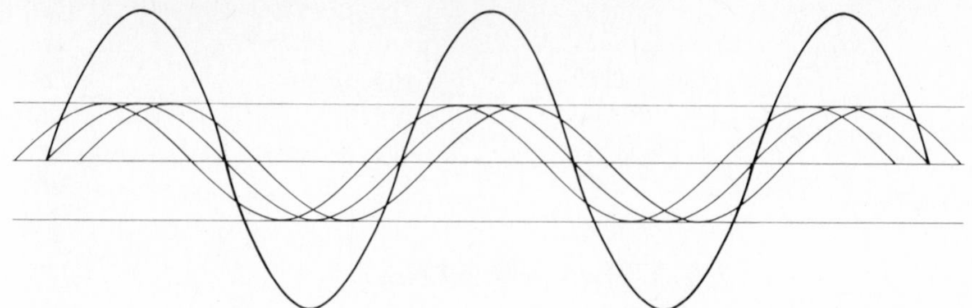

FIGURE 7.4 The superposition of three harmonic waves yields a harmonic wave of the same frequency.

how associated with a short-duration oscillatory wave pulse. Each atom is effectively an independent source of photon wavetrains (Section 3.4.4), and these, in turn, extend in time for roughly 1 to 10 ns. In other words, the atoms can be thought of as emitting wavetrains that have a sustained phase for only up to about 10 ns. After that a new wavetrain may be emitted with a totally random phase, and it too will be sustained for less than approximately 10 ns, and so forth. On the whole each atom emits a disturbance (composed of a stream of photons) that varies in its phase rapidly and randomly.

In any event, the phase of the light from one atom, $\alpha_i(t)$, will remain constant with respect to the phase from another atom $\alpha_j(t)$, for only a time of at most 10 ns before it changes randomly: the atoms are coherent for up to about 10^{-8} s. Since flux density is proportional to the time average of E_0^2, generally taken over a comparatively long interval of time, it follows that the second summation in Eq. (7.19) will contribute terms proportional to $\langle\cos[\alpha_i(t) - \alpha_j(t)]\rangle$, each of which will average out to zero because of the random rapid nature of the phase changes. Only the first summation remains in the time average, and its terms are constants. If each atom is emitting wavetrains of the same amplitude E_{01}, then

$$E_0^2 = NE_{01}^2 \tag{7.21}$$

The resultant flux density arising from N sources having random, rapidly varying phases is given by N times the flux density of any one source. In other words, *it is determined by the sum of the individual flux densities.*

A flashlight bulb, whose atoms are all emitting a random tumult, puts out light which (as the superposition of these essentially "incoherent" wavetrains) is itself rapidly and randomly varying in phase. Thus two or more such bulbs will emit light that is essentially incoherent (i.e., for durations longer than about 10 ns), light whose total combined irradiance will simply equal the sum of the irradiances contributed by each individual bulb. This is also true for candle flames, flashbulbs, and all thermal (as distinct from laser) sources. ***We cannot expect to see interference when the lightwaves from two reading lamps overlap.***

At the other extreme, if the sources are coherent and in-phase at the point of observation (i.e., $\alpha_i = \alpha_j$), Eq. (7.19) will become

$$E_0^2 = \sum_{i=1}^{N} E_{0i}^2 + 2\sum_{j>i}^{N}\sum_{i=1}^{N} E_{0i}E_{0j}$$

or, equivalently,

$$E_0^2 = \left(\sum_{i=1}^{N} E_{0i}\right)^2 \tag{7.22}$$

Again, supposing that each amplitude is E_{01}, we get

$$E_0^2 = (NE_{01})^2 = N^2E_{01}^2 \tag{7.23}$$

In this case of in-phase coherent sources, we have a situation in which the amplitudes are added first and then squared to determine the resulting flux density. The superposition of coherent waves generally has the effect of altering the spatial distribution of the energy but not the total amount present. If there are regions where the flux density is greater than the sum of the individual flux densities, there will be regions where it is less than that sum.

7.1.2 The Complex Method

It is often mathematically convenient to make use of the complex representation when dealing with the superposition of harmonic disturbances. The wave

$$E_1 = E_{01}\cos(kx \pm \omega t + \varepsilon_1)$$

or

$$E_1 = E_{01}\cos(\alpha_1 \mp \omega t)$$

can then be written as

$$\tilde{E}_1 = E_{01}e^{i(\alpha_1 \mp \omega t)} \tag{7.24}$$

if we remember that we are interested only in the real part (see Section 2.5). Suppose that there are N such overlapping waves having the same frequency and traveling in the *positive x-direction*. The resultant wave is given by

$$\tilde{E} = E_0 e^{i(\alpha + \omega t)}$$

which is equivalent to Eq. (7.18) or, upon summation of the component waves,

$$\tilde{E} = \left[\sum_{j=1}^{N} E_{0j}e^{i\alpha_j}\right]e^{+i\omega t} \tag{7.25}$$

The quantity

$$E_0 e^{i\alpha} = \sum_{j=1}^{N} E_{0j}e^{i\alpha_j} \tag{7.26}$$

is known as the *complex amplitude* of the composite wave and is simply the sum of the complex amplitudes of the constituents. Since

$$E_0^2 = (E_0 e^{i\alpha})(E_0 e^{i\alpha})^* \tag{7.27}$$

we can always compute the resultant irradiance from Eqs. (7.26) and (7.27). For example, if $N = 2$,

$$E_0^2 = (E_{01}e^{i\alpha_1} + E_{02}e^{i\alpha_2})(E_{01}e^{-i\alpha_1} + E_{02}e^{-i\alpha_2})$$

whence

$$E_0^2 = E_{01}^2 + E_{02}^2 + E_{01}E_{02}[e^{i(\alpha_1 - \alpha_2)} + e^{-i(\alpha_1 - \alpha_2)}]$$

or

$$E_0^2 = E_{01}^2 + E_{02}^2 + 2E_{01}E_{02}\cos(\alpha_1 - \alpha_2)$$

which is identical to Eq. (7.9).

7.1.3 Phasor Addition

The summation described in Eq. (7.26) can be represented graphically as an addition of vectors in the complex plane (recall the discussion on p. 23). In the parlance of electrical engineering, the complex amplitude is known as a **phasor**, and it's specified by its magnitude and phase, often written simply as $E_0\angle\alpha$. Imagine, then, that we have a disturbance described by

$$E_1 = E_{01}\sin(\omega t + \alpha_1)$$

In Fig. 7.5*a* the wave is represented by a vector of length E_{01} rotating counterclockwise at a rate ω such that its projection on the vertical axis is $E_{01}\sin(\omega t + \alpha_1)$. If we were concerned with cosine waves, we would take the projection on the horizontal axis. Incidentally, the rotating vector is, of course, a phasor $E_{01}\angle\alpha_1$, and the R and I designations signify the real and imaginary axes. Similarly, a second wave

$$E_2 = E_{02}\sin(\omega t + \alpha_2)$$

is depicted along with E_1 in Fig. 7.5*b*. Their algebraic sum, $E = E_1 + E_2$, is the projection on the I-axis of the resultant

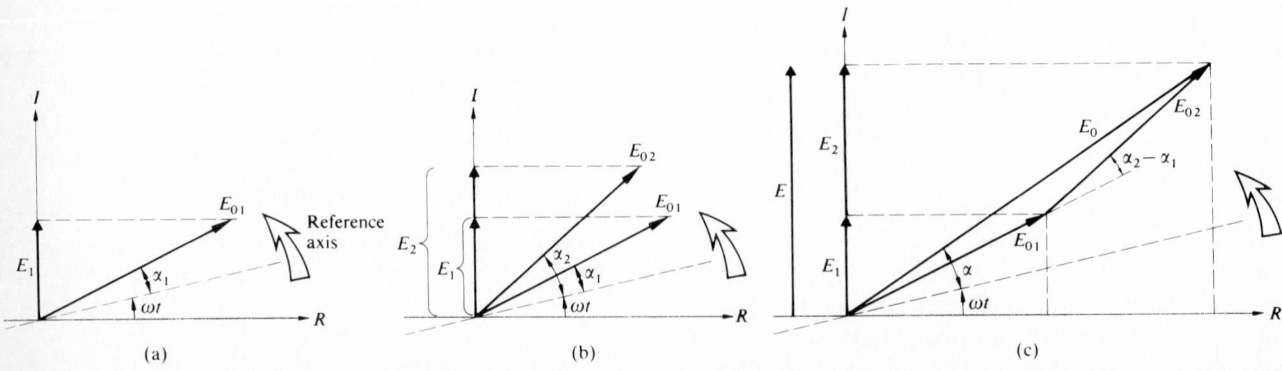

FIGURE 7.5 Phasor addition.

phasor determined by the vector addition of the component phasors, as in Fig. 7.5*c*. The law of cosines applied to the triangle of sides E_{01}, E_{02}, and E_0 yields

$$E_0^2 = E_{01}^2 + E_{02}^2 + 2E_{01}E_{02}\cos(\alpha_2 - \alpha_1)$$

where use was made of the fact that $\cos[\pi - (\alpha_2 - \alpha_1)] = -\cos(\alpha_2 - \alpha_1)$. This is identical to Eq. (7.9), as it must be. Using the same diagram, observe that $\tan\alpha$ is given by Eq. (7.10) as well. We are usually concerned with finding E_0 rather than $E(t)$, and since E_0 is unaffected by the constant revolving of all the phasors, it will often be convenient to set $t = 0$ and eliminate that rotation.

Some rather elegant schemes, such as the *vibration curve* and the *Cornu spiral* (Chapter 10), will be predicated on the technique of phasor addition. As a final example, let's briefly examine the wave resulting from the addition of

$$E_1 = 5\sin\omega t$$

$$E_2 = 10\sin(\omega t + 45°)$$

$$E_3 = \sin(\omega t - 15°)$$

$$E_4 = 10\sin(\omega t + 120°)$$

and

$$E_5 = 8\sin(\omega t + 180°)$$

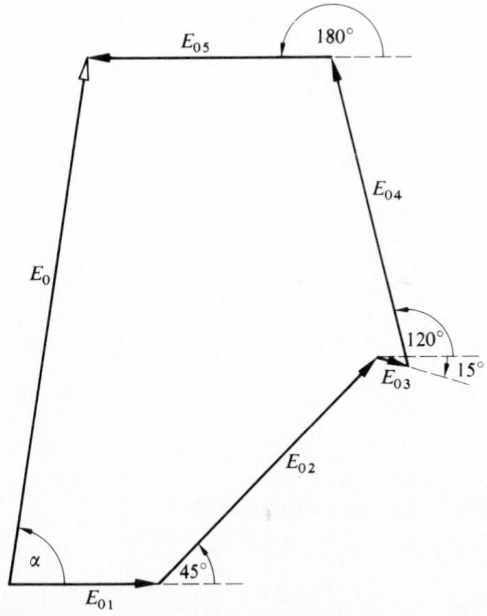

FIGURE 7.6 The sum of E_1, E_2, E_3, E_4, and E_5.

where ω is in degrees per second. The appropriate phasors $5\angle 0°$, $10\angle 45°$, $1\angle -15°$, $10\angle 120°$, and $8\angle 180°$ are plotted in Fig. 7.6. Notice that each phase angle, whether positive or negative, is referenced to the horizontal. One need only read off $E_0\angle\alpha$ with a scale and protractor to get $E = E_0\sin(\omega t + \alpha)$. It is evident that this technique offers a tremendous advantage in speed and simplicity, if not in accuracy.

7.1.4 Standing Waves

We saw earlier (p. 21) that the sum of solutions to the differential wave equation is itself a solution. Thus, in general,

$$\psi(x, t) = C_1 f(x - vt) + C_2 g(x + vt)$$

satisfies the differential wave equation. In particular let's examine *two harmonic waves of the same frequency propagating in opposite directions*. A situation of practical concern arises when the incident wave is reflected backward off some sort of mirror; a rigid wall will do for sound waves or a conducting sheet for electromagnetic waves. Imagine that an incoming wave traveling to the left,

$$E_I = E_{0I}\sin(kx + \omega t + \varepsilon_I) \tag{7.28}$$

strikes a mirror at $x = 0$ and is reflected to the right in the form

$$E_R = E_{0R}\sin(kx - \omega t + \varepsilon_R) \tag{7.29}$$

The composite wave in the region to the right of the mirror is $E = E_I + E_R$. In other words, the two waves (one traveling to the right, the other to the left) exist simultaneously in the region between the source and the mirror.

We could perform the indicated summation and arrive at a general solution* much like that of Section 7.1. However, some valuable physical insights can be gained by taking a slightly more restricted approach.

The initial phase ε_I may be set to zero by merely starting our clock at a time when $E_I = E_{0I}\sin kx$. Certain qualifications determined by the physical setup must be met by the mathematical solution, and these are known formally as ***boundary conditions***. For example, if we were talking about a

*See, for example, J. M. Pearson, *A Theory of Waves*.

rope with one end tied to a wall at $x = 0$, that point must always have a zero displacement. The two overlapping waves, one incident and the other reflected, would have to add in such a way as to yield a zero resultant wave at $x = 0$. Similarly, at the boundary of a perfectly conducting sheet the resultant electromagnetic wave must have a zero electric-field component parallel to the surface. Assuming $E_{0I} = E_{0R}$, the boundary conditions require that at $x = 0$, $E = 0$, and since $\varepsilon_I = 0$, it follows from Eqs. (7.28) and (7.29) that $\varepsilon_R = 0$. The composite disturbance is then

$$E = E_{0I}\,[\sin (kx + \omega t) + \sin (kx - \omega t)]$$

Applying the identity

$$\sin \alpha + \sin \beta = 2 \sin \tfrac{1}{2}(\alpha + \beta) \cos \tfrac{1}{2}(\alpha - \beta)$$

yields

$$E(x, t) = 2E_{0I} \sin kx \cos \omega t \tag{7.30}$$

This is the equation for a **standing** or **stationary wave**, as opposed to a traveling wave (Fig. 7.7). Its profile does not move through space; it is clearly not of the form $f(x \pm vt)$. At any point $x = x'$, the amplitude is a constant equal to $2E_{0I} \sin kx'$, and $E(x', t)$ varies harmonically as $\cos \omega t$. At certain points, namely, $x = 0, \lambda/2, \lambda, 3\lambda/2, \ldots$, the disturbance will be zero at all times. These are known as **nodes** or *nodal points* (Fig. 7.8). Halfway between each adjacent node, that is, at $x = \lambda/4, 3\lambda/4, 5\lambda/4, \ldots$, the amplitude has a maximum value of $\pm 2E_{0I}$, and these points are known as the **antinodes**. The disturbance $E(x, t)$ will be zero at all values of x whenever $\cos \omega t = 0$, that is, when $t = (2m + 1)\tau/4$, where $m = 0, 1, 2, 3, \ldots$ and τ is the period of the component waves.

If the reflection off the mirror is not perfect, as is often the case, the composite wave will contain a traveling component along with the stationary wave. Under such conditions there will be a net transfer of energy, whereas for the pure standing wave there is none.

Although the analysis carried out above is essentially one-dimensional, standing waves exist in two and three dimensions as well. The phenomenon is extremely commonplace: standing waves occur in one dimension on guitar strings and diving boards, in two dimensions on the surface of a drum or in a jiggled pail of water (Fig. 7.9), and in three dimensions when you sing in a shower stall. In fact, standing waves are created within the cavities inside your head whenever you sing, no matter where you are.

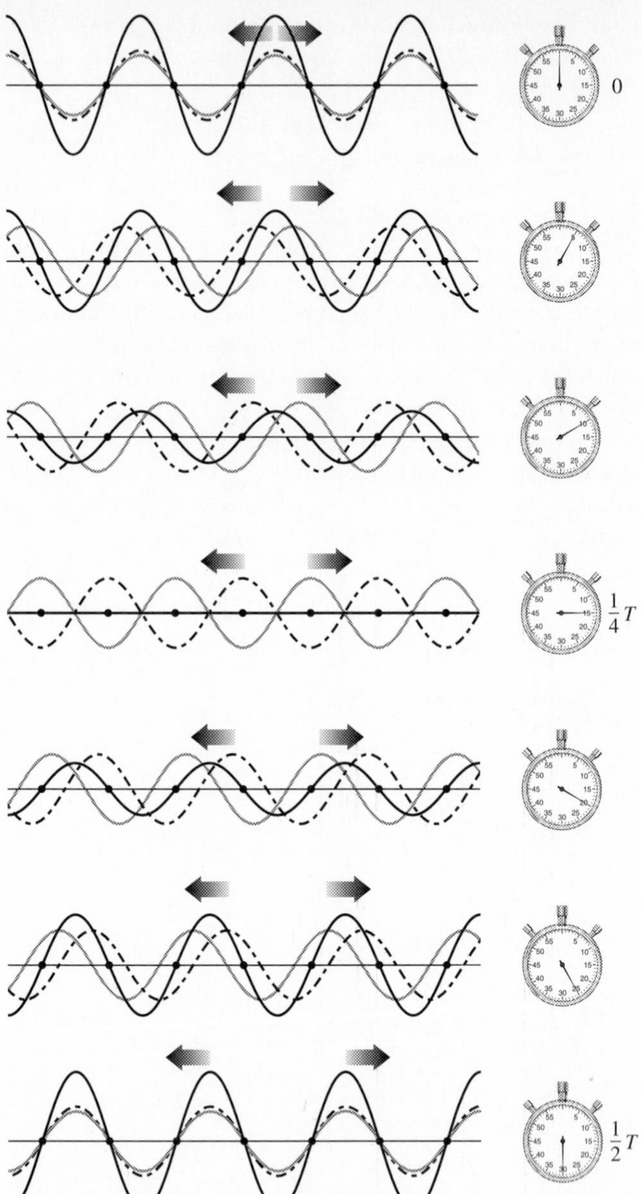

FIGURE 7.7 The creation of standing waves. Two waves of the same amplitude and wavelength traveling in opposite directions form a stationary disturbance that oscillates in place.

If a standing-wave system is driven by an oscillating source, it will efficiently absorb energy provided that the vibrations match one of its standing-wave modes. That process is known as resonance, and it happens every time your

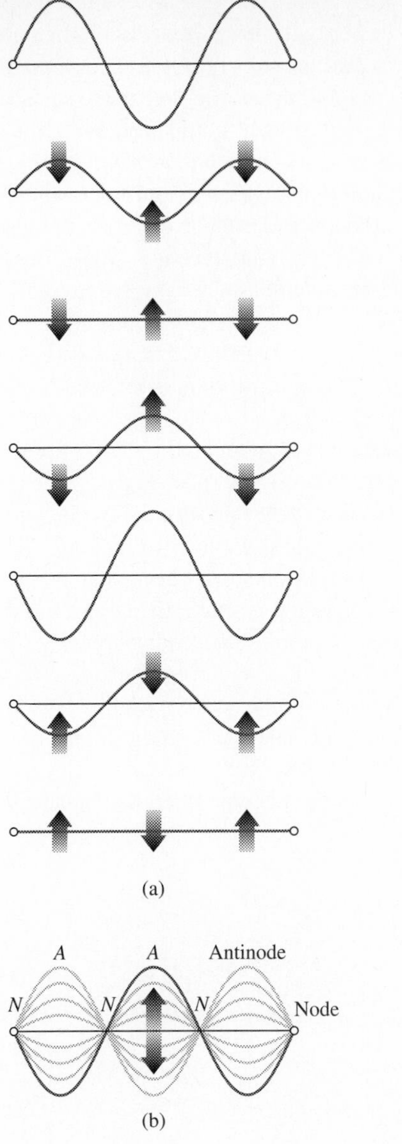

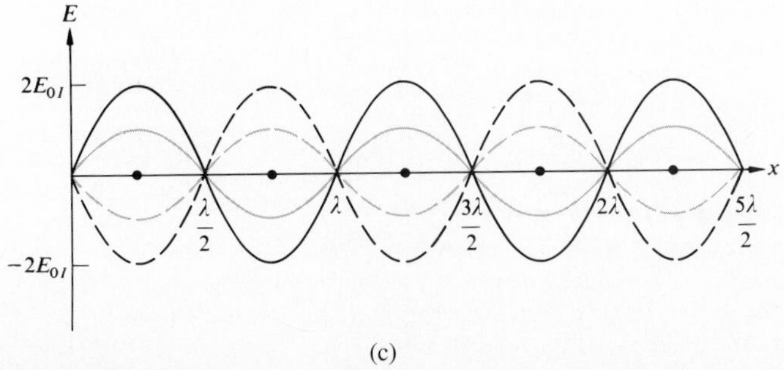

FIGURE 7.8 A standing wave at various times.

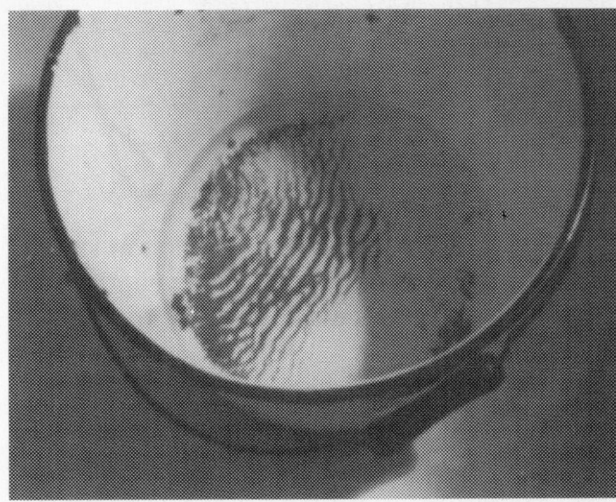

FIGURE 7.9 A pail used to wash a floor contained a suspension of fine dirt particles in water. When placed in a curved sink, the pail gently rocked along a fixed axis, setting up standing waves and distributing the particles in ridges as they settled.

house buzzes when an airplane flies low overhead or when a heavy truck passes by. If the source continues to supply energy, the wave will continue to build until the system's inherent losses equal the energy input and equilibrium is reached. This ability to sustain and simplify an input is an extremely important feature of standing-wave systems. The ear's auditory canal is just such a resonant cavity. It amplifies (by about 100%) sounds in the range from $\approx$3 kHz to $\approx$4 kHz. Similarly, the laser builds its powerful emission within a standing-wave cavity (p. 585). Figure 7.10 shows the standing-wave pattern produced when a reflecting rod is placed in front of an antenna emitting $\approx$3 GHz electromagnetic waves.

It was by measuring the distances between the nodes of standing waves that Hertz was able to determine the wavelength of the radiation in his historic experiments (see Section 3.6.1). A few years later, in 1890, Otto Wiener first demonstrated the existence of standing lightwaves. The arrangement he used is depicted in Fig. 7.11. It shows a normally incident parallel beam of quasimonochromatic light reflecting off a front-silvered mirror. A transparent photographic film, less than $\lambda/20$ thick, deposited on a glass plate, was inclined to the mirror at an angle of about 10^{-3} radians. In that way the film plate cut across the pattern of standing plane waves. After developing the emulsion, it was found to be blackened along a series of equidistant parallel bands. These corresponded to the regions where the photographic layer had intersected the antinodal planes. Significantly, there was no blackening of the emulsion at the mirror's surface. It can be shown that the

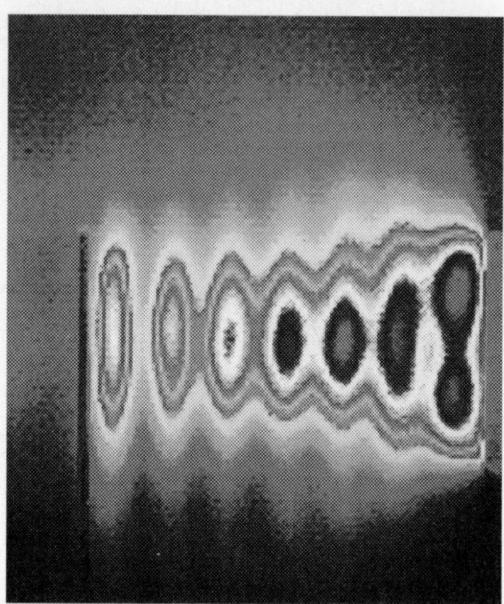

FIGURE 7.10 A two-dimensional standing-wave pattern formed between a source and a reflector. EM-waves from a 3.9-GHz antenna enter from the right. They reflect off a metal rod and travel back to the antenna. The pattern is made visible by absorbing the microwave radiation and recording the resulting temperature distribution with an IR camera. (Photo courtesy H. H. Pohle, Phillips Laboratory, Kirtland Air Force Base.)

nodes and antinodes of the magnetic field component of an electromagnetic standing wave alternate with those of the electric field (Problem 7.10). We might suspect as much from the fact that at $t = (2m + 1)\tau/4$, $E = 0$ for all values of x, so to conserve energy it follows that $B \neq 0$. In agreement with theory, Hertz had previously (1888) determined the existence of a nodal point of the electric field at the surface of his reflector. Accordingly, Wiener could conclude that the blackened regions were associated with antinodes of the **E**-field. ***It is the electric field that triggers the photochemical process.***

In a similar way Drude and Nernst showed that the **E**-field is responsible for fluorescence. These observations are all quite understandable, since the force exerted on an electron by the **B**-field component of an electromagnetic wave is generally negligible in comparison to that of the **E**-field. It is for these reasons that the electric field is referred to as the *optic disturbance* or *light field.*

Standing waves generated by two oppositely propagating disturbances represent a special case of the broader subject of double-beam interference (p. 377). Consider the two point sources sending out waves in Fig. 7.12. When point P, the point of observation somewhere near the middle is far from the sources, angle ϕ is small, the two waves superimpose, and there results a complicated interference pattern (that will be treated in detail in Chapter 9). Suffice it to say here that the space surrounding the sources will be filled with a system of bright and dark bands where the interference is alternately constructive and destructive. As P comes closer and ϕ gets larger, the fringes become finer, that is, narrower, until P is on

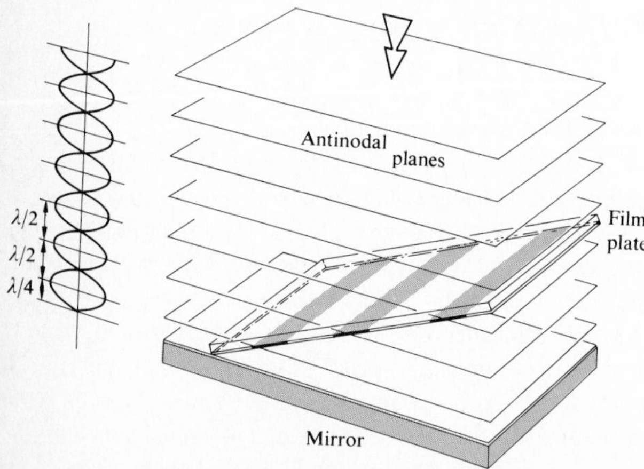

FIGURE 7.11 Wiener's experiment.

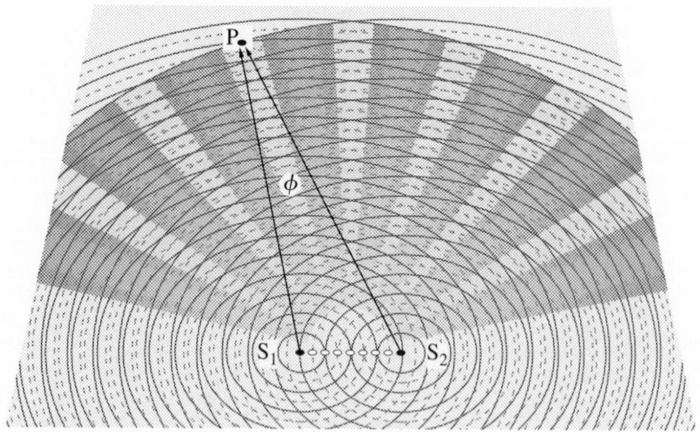

FIGURE 7.12 Two monochromatic point sources. At any point P the resultant wave is maximum where peak (—) overlaps peak (—) or trough (– –) overlaps trough (– –). It's minimum where peak overlaps trough. The maxima that form along the $\overline{S_1S_2}$ line correspond to standing waves.

the line joining the sources and $\phi = 180°$. Then standing waves are set up, and the "fringes" are the finest they'll get, namely, half a wavelength peak-to-peak.

7.2 The Addition of Waves of Different Frequency

Thus far the analysis has been restricted to the superposition of waves all having the same frequency. Yet one never actually has disturbances, of any kind, that are strictly monochromatic. It will be far more realistic, as we shall see, to speak of quasi-monochromatic light, which is composed of a narrow range of frequencies. The study of such light will lead us to the important concepts of bandwidth and coherence time.

The ability to modulate light effectively (Section 8.11.3) makes it possible to couple electronic and optical systems in a way that has had and will certainly continue to have far-reaching effects on the entire technology. Moreover, with the advent of electro-optical techniques, light has taken on a significant role as a carrier of information. This section is devoted to developing some of the mathematical ideas needed to appreciate this new emphasis.

7.2.1 Beats

Consider the composite disturbance arising from a combination of the waves

$$E_1 = E_{01} \cos (k_1 x - \omega_1 t)$$

and

$$E_2 = E_{01} \cos (k_2 x - \omega_2 t)$$

which have equal amplitudes and zero initial phase angles. The net wave

$$E = E_{01}[\cos (k_1 x - \omega_1 t) + \cos (k_2 x - \omega_2 t)]$$

can be reformulated as

$$E = 2E_{01} \cos \tfrac{1}{2}[(k_1 + k_2)x - (\omega_1 + \omega_2)t] \times \cos \tfrac{1}{2}[(k_1 - k_2)x - (\omega_1 - \omega_2)t]$$

using the identity

$$\cos \alpha + \cos \beta = 2 \cos \tfrac{1}{2}(\alpha + \beta) \cos \tfrac{1}{2}(\alpha - \beta)$$

Now define the quantities $\bar{\omega}$ and $\bar{k}$, which are the ***average angular frequency*** and ***average propagation number***, respectively. Similarly, the quantities ω_m and k_m are designated the ***modulation frequency*** and ***modulation propagation number***, respectively. Let

$$\bar{\omega} \equiv \tfrac{1}{2}(\omega_1 + \omega_2) \qquad \omega_m \equiv \tfrac{1}{2}(\omega_1 - \omega_2) \tag{7.31}$$

and

$$\bar{k} \equiv \tfrac{1}{2}(k_1 + k_2) \qquad k_m \equiv \tfrac{1}{2}(k_1 - k_2) \tag{7.32}$$

thus

$$E = 2E_{01} \cos (k_m x - \omega_m t) \cos (\bar{k}x - \bar{\omega}t) \tag{7.33}$$

The total disturbance may be regarded as a traveling wave of frequency $\bar{\omega}$ having a time-varying or modulated amplitude $E_0(x, t)$ such that

$$E(x, t) = E_0(x, t) \cos (\bar{k}x - \bar{\omega}t) \tag{7.34}$$

where

$$E_0(x, t) = 2E_{01} \cos (k_m x - \omega_m t) \tag{7.35}$$

In applications of interest here, ω_1 and ω_2 will always be rather large. In addition, if they are comparable to each other, $\omega_1 \approx \omega_2$, then $\bar{\omega} >> \omega_m$ and $E_0(x, t)$ will change slowly, whereas $E(x, t)$ will vary quite rapidly (Fig. 7.13). The irradiance is proportional to

$$E_0^2(x, t) = 4E_{01}^2 \cos^2 (k_m x - \omega_m t)$$

or

$$E_0^2(x, t) = 2E_{01}^2[1 + \cos (2k_m x - 2\omega_m t)]$$

Notice that $E_0^2(x, t)$ oscillates about a value of $2E_{01}^2$ with an angular frequency of $2\omega_m$ or simply $(\omega_1 - \omega_2)$, which is known as the **beat frequency**. That is, E_0 varies at the modulation frequency, whereas E_0^2 varies at twice that, namely, the beat frequency.

Beats were first observed with the use of light in 1955 by Forrester, Gudmundsen, and Johnson.* To obtain two waves

*A. T. Forrester, R. A. Gudmundsen, and P. O. Johnson, "Photoelectric Mixing of Incoherent Light." *Phys. Rev.* **99**, 1691 (1955).

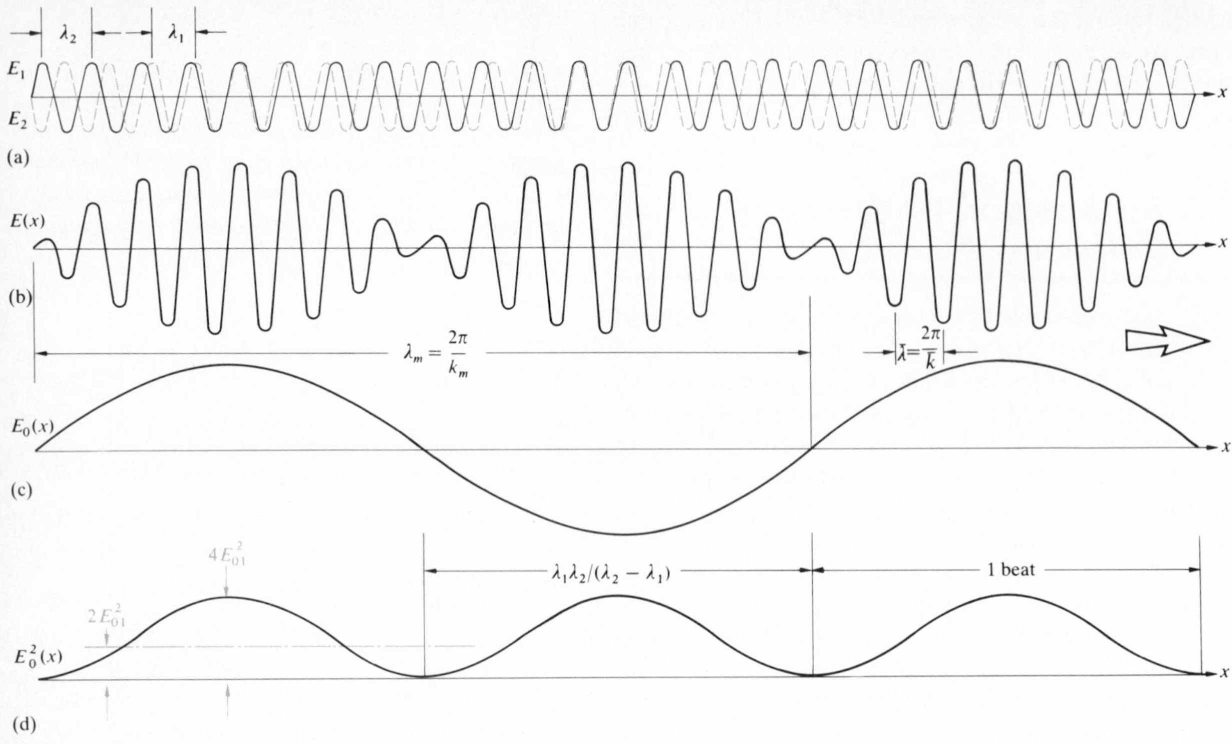

FIGURE 7.13 The superposition of two harmonic waves of different frequency.

of slightly different frequency they used the Zeeman Effect. When the atoms of a discharge lamp, in this case mercury, are subjected to a magnetic field, their energy levels split. As a result, the emitted light contains two frequency components, ν_1 and ν_2, which differ in proportion to the magnitude of the applied field. When these components are recombined at the surface of a photoelectric mixing tube, the beat frequency, $\nu_1 - \nu_2$, is generated. Specifically, the field was adjusted so that $\nu_1 - \nu_2 = 10^{10}$ Hz, which conveniently corresponds to a 3-cm microwave signal. The recorded photoelectric current had the same form as the $E_0^2(x)$ curve in Fig. 7.13*d*.

The advent of the laser has since made the observation of beats using light considerably easier. Even a beat frequency of a few Hz out of 10^{14} Hz can be seen as a variation in phototube current. The observation of beats now represents a particularly sensitive and fairly simple means of detecting small frequency differences. The ring laser (Section 9.8.3), functioning as a gyroscope, utilizes beats to measure frequency differences induced as a result of the rotation of the system. The Doppler Effect, which accounts for the frequency shift when light is reflected off a moving surface, provides another series of applications of beats. By scattering light off a target, whether solid, liquid, or even gaseous, and then beating the original and reflected waves, we get a precise measure of the target speed. In much the same way on an atomic scale, laser light will shift in phase upon interacting with sound waves moving in a material. (This phenomenon is called Brillouin Scattering.) Thus $2\omega_m$ becomes a measure of the speed of sound in the medium.

7.2.2 Group Velocity

The specific relationship between ω and k determines v, the phase velocity of a wave. In a nondispersive medium such as vacuum [Eq. (2.33)] $v = \omega/k$ and **a plot of ω versus k is a**

straight line; the frequency and wavelength change so as to keep v constant. All waves of a particular type (e.g., all EM-waves) travel with the same phase speed in a nondispersive medium. By contrast, in a dispersive medium every wave propagates at a speed that depends on its frequency. When a number of waves combine to form a composite disturbance, the modulation envelope will travel at a speed different from that of the constituent waves. This raises the important notion of the *group velocity* and its relationship to the phase velocity.

The disturbance examined in the previous section,

$$E(x, t) = E_0(x, t) \cos (\bar{k}x - \bar{\omega}t) \qquad [7.34]$$

consists of a high-frequency ($\bar{\omega}$) ***carrier wave***, *amplitude-modulated* by a cosine function. Suppose, for a moment, that the wave in Fig. 7.13*b* was not modulated, that is, E_0 = constant. Each small peak in the carrier would travel to the right with the usual phase velocity. In other words,

$$v = -\frac{(\partial\varphi/\partial t)_x}{(\partial\varphi/\partial x)_t} \qquad [2.32]$$

From Eq. (7.34) the phase is given by $\varphi = (\bar{k}x - \bar{\omega}t)$, hence

$$v = \bar{\omega}/\bar{k} \qquad (7.36)$$

Clearly, this is the phase velocity of the carrier whether it's modulated or not. In the former case, the peaks simply change amplitude periodically as they stream along.

Evidently, there is another motion to be concerned with, and that is the propagation of the modulation envelope. Return to Fig. 7.13*a* and suppose that the constituent waves, $E_1(x, t)$ and $E_2(x, t)$, advance with the same speed, $v_1 = v_2$. Imagine, if you will, the two harmonic functions having different wavelengths and frequencies drawn on separate sheets of clear plastic. When these are overlaid in some way [as in Fig. 7.13*a*] the resultant is a stationary beat pattern. If the sheets are both moved to the right at the same speed so as to resemble traveling waves, the beats will obviously move with that same speed. *The rate at which the modulation envelope advances* is known as the **group velocity**, or v_g. In this instance the group velocity equals the phase velocity of the carrier (the average speed, $\bar{\omega}/\bar{k}$). In other words, $v_g = v = v_1 = v_2$. This applies specifically to nondispersive media in which the phase velocity is independent of wavelength so that the two waves could have the same speed.

For a more generally applicable solution, examine the expression for the modulation envelope:

$$E_0(x, t) = 2E_{01} \cos (k_m x - \omega_m t) \qquad [7.35]$$

The speed with which that wave moves is again given by Eq. (2.32), but now we can forget the carrier wave. The modulation advances at a rate dependent on the phase of the envelope $(k_m x - \omega_m t)$, and

$$v_g = \frac{\omega_m}{k_m}$$

or

$$v_g = \frac{\omega_1 - \omega_2}{k_1 - k_2} = \frac{\Delta\omega}{\Delta k}$$

Realize, however, that ω may be dependent on λ or equivalently on k. The particular function $\omega = \omega(k)$ is called a *dispersion relation*. When the frequency range $\Delta\omega$, centered about $\bar{\omega}$, is small, $\Delta\omega/\Delta k$ is approximately equal to the derivative of the dispersion relation evaluated at $\bar{\omega}$, that is,

$$v_g = \left(\frac{d\omega}{dk}\right)_{\bar{\omega}} \qquad (7.37)$$

The modulation or signal propagates at a speed v_g that may be greater than, equal to, or less than v, the phase velocity of the carrier. Equation (7.37) is quite general and will be true, as well, for any group of overlapping waves, as long as their frequency range is narrow.

A plot of the dispersion relation (Fig. 7.14) produces a curve passing through the origin that is convex upward for normal dispersion and concave downward for anomalous dispersion (p. 67). In either case, the slope of a line drawn from the origin to any point (ω, k) on the curve is the phase velocity at that frequency. Similarly, the slope of the curve at the point $(\bar{\omega}, \bar{k})$ is $(d\omega/dk)_{\bar{\omega}}$, and that's the group velocity for the set of component waves centered at $\bar{\omega}$. In normal dispersion, sinusoidal waves of high frequency (e.g., blue) have larger indices and travel slower than low-frequency waves (e.g., red). Moreover, the slope of the dispersion curve (v_g) is always less than the slope of the line (v); that is, $v_g < v$. Whereas in anomalous dispersion $v_g > v$.

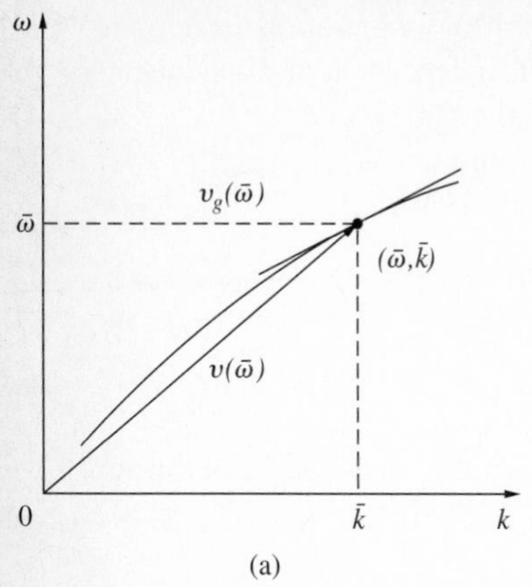

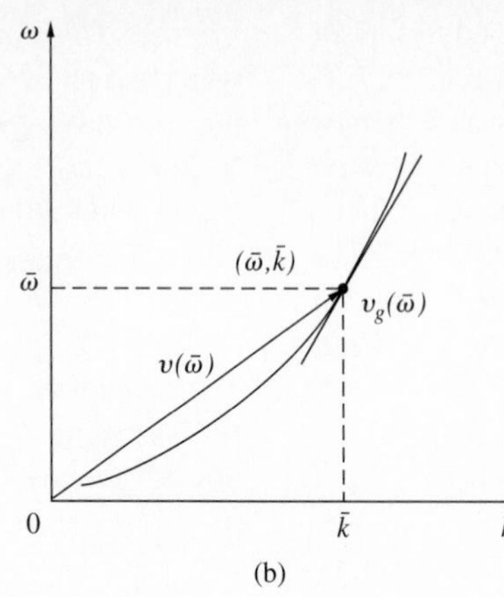

FIGURE 7.14 A plot of the dispersion relation. (*a*) In normal dispersion $v(\bar{\omega}) > v_g(\bar{\omega})$ whereas in (*b*) anomalous dispersion $v_g(\bar{\omega}) > v(\bar{\omega})$. The phase velocity v, of a wave of any frequency ω, is the slope of the line drawn from the origin to that point. The group velocity is the slope of the tangent to the curve at $(\bar{\omega}, \bar{k})$ where $\bar{\omega}$ is the mean frequency of the waves composing the group.

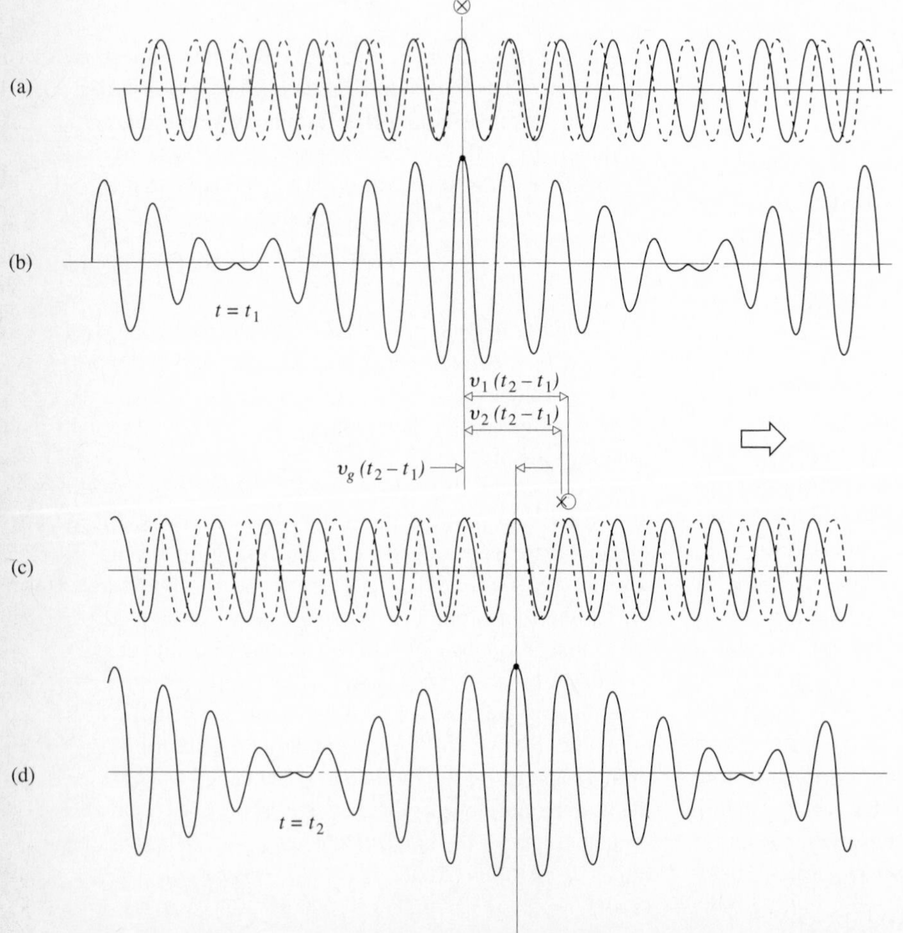

FIGURE 7.15 Group and phase velocities. In (*a*) the two waves coincide at the point indicated by an ⊗. And in (*b*) the peak of the modulated wave occurs at that point. But the waves travel at different speeds in (*c*) and the two original peaks (marked by x and ○) separate. A different pair now coincide in (*d*) to form the high point of the modulated wave, which therefore travels at yet a different speed. Here $v_1 > v_2 > v_g$ and since $\lambda_1 > \lambda_2$ this is a case of normal dispersion.

Since $\omega = kv$, Eq. (7.37) yields

$$v_g = v + k\frac{dv}{dk} \tag{7.38}$$

As a consequence, in nondispersive media in which v is independent of λ, $dv/dk = 0$ and $v_g = v$. Specifically, in vacuum $\omega = kc$, $v = c$, and $v_g = c$. In dispersive media ($v_1 \neq v_2$, as in Fig. 7.15) in which $n(k)$ is known, $\omega = kc/n$, and it is useful to reformulate v_g as

$$v_g = \frac{c}{n} - \frac{kc}{n^2}\frac{dn}{dk}$$

or

$$v_g = v\left(1 - \frac{k}{n}\frac{dn}{dk}\right) \tag{7.39}$$

For optical media, in regions of normal dispersion, the refractive index increases with frequency ($dn/dk > 0$), and as a result $v_g < v$. Clearly, one should also define a ***group index of refraction***

$$n_g \equiv c/v_g \tag{7.40}$$

which must be carefully distinguished from n. In 1885 A. A. Michelson measured n_g in carbon disulfide using pulses of white light and obtained 1.758 in comparison to $n = 1.635$.

As we'll see presently, it is possible to create a wave pulse of the type shown in Fig. 7.16 by properly adding together a large number of sinusoids of the right frequency, amplitude, and phase. Here the medium is normally dispersive and the phase velocity will be taken as the velocity of the carrier, that is, of the roughly sinusoidal wave of frequency $\bar{\omega}$. Because the peaks of the carrier travel faster than does the pulse as a whole, they appear to enter it at the left, sweep through it, and vanish off at the right. Although each peak of the carrier changes height as it progresses across the pulse, $v(\bar{\omega})$ is the speed of any such peak and it's therefore properly the speed of the condition of constant phase. By contrast, the modulation envelope travels at a speed $v_g(\bar{\omega}) = (d\omega/dk)_{\bar{\omega}}$, which in this particular instance equals one quarter of $v(\bar{\omega})$. Any point on the envelope (e.g., the maximum at the center of the pulse) moves at a speed $v_g(\bar{\omega})$, which is the speed of the condition of constant magnitude.

The Special Theory of Relativity makes it clear that there are no circumstances under which a signal can propagate at a speed greater than c. Yet we have already seen that under certain circumstances (Section 3.5.1) the phase velocity can exceed c. The contradiction is only an apparent one, arising from the fact that although a monochromatic wave can indeed have a speed in excess of c, it cannot convey information. In contrast, a signal in the form of any modulated wave will propagate at the group velocity, which is always less than c in normally dispersive media.*

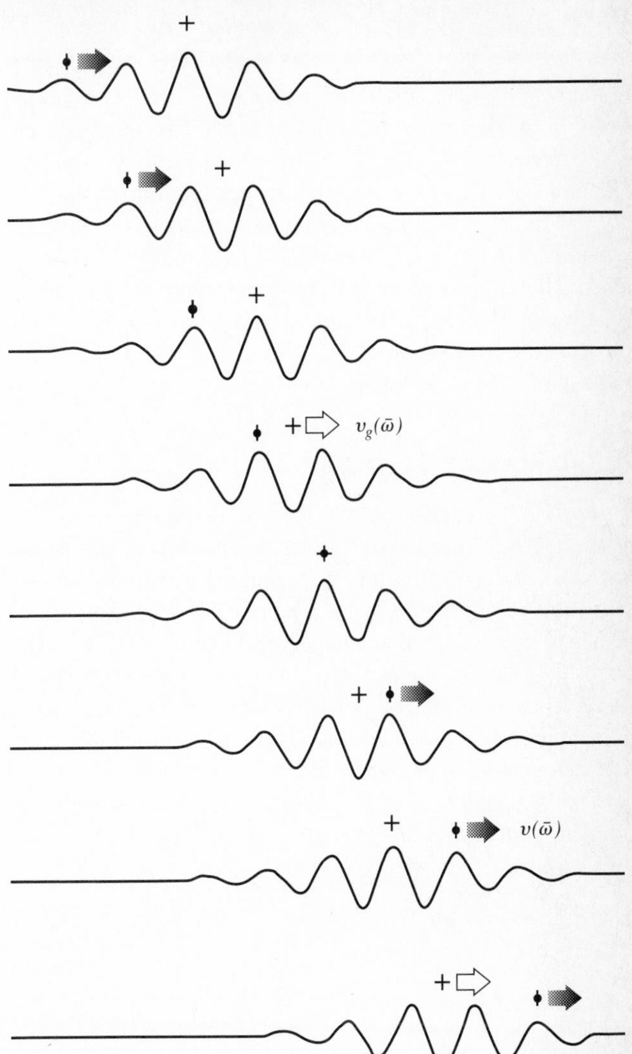

FIGURE 7.16 A wave pulse in a dispersive medium.

*In regions of anomalous dispersion (Section 3.5.1) where $dn/dk < 0$, v_g may be greater than c. Here, however, the signal propagates at yet a different speed, known as the *signal velocity*, v_s. Thus $v_s = v_g$ except in a resonance absorption band. In all cases v_s corresponds to the velocity of energy transfer and never exceeds c.

7.3 Anharmonic Periodic Waves

Figure 7.17 depicts a disturbance that arises from the superposition of two harmonic functions having different amplitudes and wavelengths. Notice that something rather curious has taken place—the composite disturbance is **anharmonic**; it's not sinusoidal. As we have already said, and will certainly say again, *purely sinusoidal* waves have no actual physical existence. This fact emphasizes the practical significance of anharmonic disturbances and is the motivation for much of our present concern with them.

7.3.1 Fourier Series

Figure 7.17 suggests that by using a number of sinusoidal functions whose amplitudes, wavelengths, and relative phases have been judiciously selected, it would be possible to synthesize some rather interesting wave profiles. An exceptionally beautiful mathematical technique for doing precisely this was devised by the French physicist Jean Baptiste Joseph, Baron de Fourier (1768–1830). This theory is predicated on what has come to be known as *Fourier's Theorem*, which states that *a function f(x), having a spatial period λ, can be synthesized by a sum of harmonic functions whose wavelengths are integral submultiples of λ (that is, λ, λ/2, λ/3, etc.).* This Fourier-series representation has the mathematical form

$$f(x) = C_0 + C_1 \cos\left(\frac{2\pi}{\lambda}x + \varepsilon_1\right) + C_2 \cos\left(\frac{2\pi}{\lambda/2}x + \varepsilon_2\right) + \cdots \qquad (7.41)$$

where the C-values are constants, and of course the profile $f(x)$ may correspond to a traveling wave $f(x - vt)$. To get some sense of how this scheme works, observe that although C_0 by itself is obviously a poor substitute for the original function, it will be appropriate at those few points where it crosses the $f(x)$ curve. In the same way, adding on the next term improves things a bit, since the function

$$[C_0 + C_1 \cos(2\pi x/\lambda + \varepsilon_1)]$$

will be chosen so as to cross the $f(x)$ curve even more frequently. If the synthesized function [the right-hand side of Eq. (7.41)] comprises an infinite number of terms, selected to intersect the anharmonic function at an infinite number of points, the series will presumably be identical to $f(x)$.

It is usually more convenient to reformulate Eq. (7.41) by making use of the trigonometric identity

$$C_m \cos(mkx + \varepsilon_m) = A_m \cos mkx + B_m \sin mkx$$

where $k = 2\pi/\lambda$, λ being the wavelength of $f(x)$, $A_m = C_m \cos \varepsilon_m$, and $B_m = -C_m \sin \varepsilon_m$. Thus

$$f(x) = \frac{A_0}{2} + \sum_{m=1}^{\infty} A_m \cos mkx + \sum_{m=1}^{\infty} B_m \sin mkx \qquad (7.42)$$

The first term is written as $A_0/2$ because of the mathematical simplification it will lead to later on. The process of determining the coefficients A_0, A_m, and B_m for a specific periodic function $f(x)$ is referred to as **Fourier analysis**. We'll spend a moment now deriving a set of equations for these coefficients that can be used henceforth. To that end, integrate both sides of Eq. (7.42) over any spatial interval equal to λ, for example, from 0 to λ or from $-\lambda/2$ to $+\lambda/2$ or, more generally, from x' to $x' + \lambda$. Since over any such interval

$$\int_0^\lambda \sin mkx\, dx = \int_0^\lambda \cos mkx\, dx = 0$$

there is only one nonzero term to be evaluated, namely,

$$\int_0^\lambda f(x)\, dx = \int_0^\lambda \frac{A_0}{2}\, dx = A_0 \frac{\lambda}{2}$$

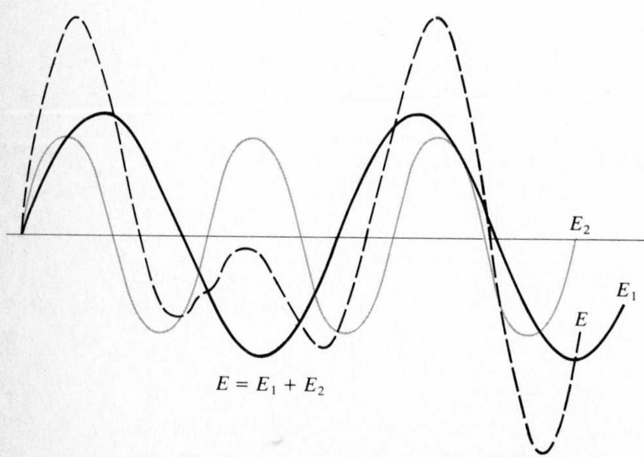

FIGURE 7.17 The superposition of two harmonic waves of different frequency. The resultant wave is periodic but anharmonic.

and thus

$$A_0 = \frac{\lambda}{2} \int_0^\lambda f(x)\,dx \tag{7.43}$$

To find A_m and B_m we will make use of the *orthogonality of sinusoidal functions* (Problem 7.28), that is, the fact that

$$\int_0^\lambda \sin akx \cos bkx\,dx = 0 \tag{7.44}$$

$$\int_0^\lambda \cos akx \cos bkx\,dx = \frac{\lambda}{2}\,\delta_{ab} \tag{7.45}$$

$$\int_0^\lambda \sin akx \sin bkx\,dx = \frac{\lambda}{2}\,\delta_{ab} \tag{7.46}$$

where a and b are nonzero positive integers and δ_{ab}, known as the *Kronecker delta*, is a shorthand notation equal to zero when $a \neq b$ and equal to 1 when $a = b$. To find A_m we now multiply both sides of Eq. (7.42) by cos ℓkx, ℓ being a positive integer, and then integrate over a spatial period. Only one term is nonvanishing, and that is the single contribution in the second sum, which corresponds to $\ell = m$, in which case

$$\int_0^\lambda f(x) \cos mkx\,dx = \int_0^\lambda A_m \cos^2 mkx\,dx = \frac{\lambda}{2} A_m$$

Thus

$$A_m = \frac{2}{\lambda} \int_0^\lambda f(x) \cos mkx\,dx \tag{7.47}$$

This expression can be used to evaluate A_m *for all values of m, including m* $= 0$, as is evident from a comparison of Eqs. (7.43) and (7.47). Similarly, multiplying Eq. (7.42) by sin ℓkx and integrating, leads to

$$B_m = \frac{2}{\lambda} \int_0^\lambda f(x) \sin mkx\,dx \tag{7.48}$$

In summary, a periodic function $f(x)$ can be represented as a Fourier series

$$f(x) = \frac{A_0}{2} + \sum_{m=1}^{\infty} A_m \cos mkx + \sum_{m=1}^{\infty} B_m \sin mkx \tag*{[7.42]}$$

where, knowing $f(x)$, the coefficients are computed using

$$A_m = \frac{2}{\lambda} \int_0^\lambda f(x) \cos mkx\,dx \tag*{[7.47]}$$

and

$$B_m = \frac{2}{\lambda} \int_0^\lambda f(x) \sin mkx\,dx \tag*{[7.48]}$$

Be aware that there are some mathematical subtleties related to the convergence of the series and the number of singularities in $f(x)$, but we need not be concerned with these matters here.

Certain symmetry conditions are well worth recognizing, because they lead to some computational shortcuts. Thus if a function $f(x)$ is *even*, that is, if $f(-x) = f(x)$, or equivalently, if it is symmetric about $x = 0$, its Fourier series will contain only cosine terms ($B_m = 0$ for all m) that are themselves even functions. Likewise, *odd* functions that are antisymmetric about $x = 0$, that is, $f(-x) = -f(x)$, will have series expansions containing only sine functions ($A_m = 0$ for all m). In either case, one need not bother to calculate both sets of coefficients. This is particularly helpful when the location of the origin ($x = 0$) is arbitrary, and we can choose it so as to make life as simple as possible. Nonetheless, keep in mind that many common functions are neither odd nor even (e.g., e^x).

As an example of the technique, let's compute the Fourier series that corresponds to a square wave. Select the location of the origin as shown in Fig. 7.18, and so

$$f(x) = \begin{cases} +1 & \text{when } 0 < x < \lambda/2 \\ -1 & \text{when } \lambda/2 < x < \lambda \end{cases}$$

Since $f(x)$ is odd, $A_m = 0$, and

$$B_m = \frac{2}{\lambda} \int_0^{\lambda/2} (+1) \sin mkx\,dx + \frac{2}{\lambda} \int_{\lambda/2}^{\lambda} (-1) \sin mkx\,dx$$

thus

$$B_m = \frac{1}{m\pi} [-\cos mkx]_0^{\lambda/2} + \frac{1}{m\pi} [\cos mkx]_{\lambda/2}^{\lambda}$$

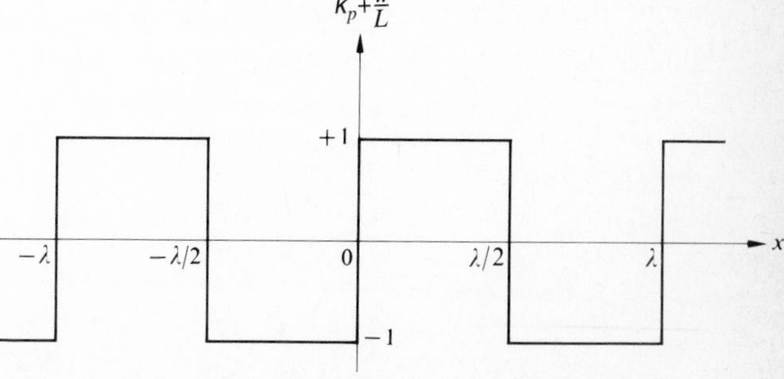

FIGURE 7.18 The profile of a periodic square wave.

Remembering that $k = 2\pi/\lambda$, we obtain

$$B_m = \frac{2}{m\pi}(1 - \cos m\pi)$$

The Fourier coefficients are therefore

$$B_1 = \frac{4}{\pi}, \qquad B_2 = 0, \qquad B_3 = \frac{4}{3\pi},$$

$$B_4 = 0, \qquad B_5 = \frac{4}{5\pi}, \ldots,$$

and the required series is simply

$$f(x) = \frac{4}{\pi}\left(\sin kx + \tfrac{1}{3}\sin 3kx + \tfrac{1}{5}\sin 5kx + \cdots\right) \tag{7.49}$$

Figure 7.19 is a plot of a few partial sums of the series as the number of terms increases. We could pass over to the time domain to find $f(t)$ by just changing kx to ωt. Suppose that we have three ordinary electronic oscillators whose output voltages vary sinusoidally and are controllable in both frequency and amplitude. If these are connected in series with their frequencies set at ω, 3ω, and 5ω and the total signal is examined on an oscilloscope, we can synthesize any of these curves, as in Fig. 7.19*d*. Similarly, we might simultaneously strike three keys on an appropriately tuned piano with just the correct force on each to create a chord, or composite sound wave, having the curve in Fig. 7.19*c* as its profile. Curiously enough, the human ear–brain audio system is capable of Fourier analysis of a simple composite wave into its harmonic constituents—presumably there are people who could even name each note in the chord.

Earlier we postponed any detailed consideration of anharmonic periodic functions, such as that in Fig. 2.6, and restricted the analysis to purely sinusoidal waves. We now have a cogent rationale for having done so. From here on we can envision this kind of disturbance as a superposition of harmonic constituents of different frequencies whose individual behavior can be studied separately. Accordingly, we can write

$$f(x \pm vt) = \frac{A_0}{2} + \sum_{m=1}^{\infty} A_m \cos mk(x \pm vt) + \sum_{m=1}^{\infty} B_m \sin mk(x \pm vt) \tag{7.50}$$

or equivalently

$$f(x \pm vt) = \sum_{m=0}^{\infty} C_m \cos\,[mk(x \pm vt) + \varepsilon_m] \tag{7.51}$$

for any such *anharmonic periodic wave*.

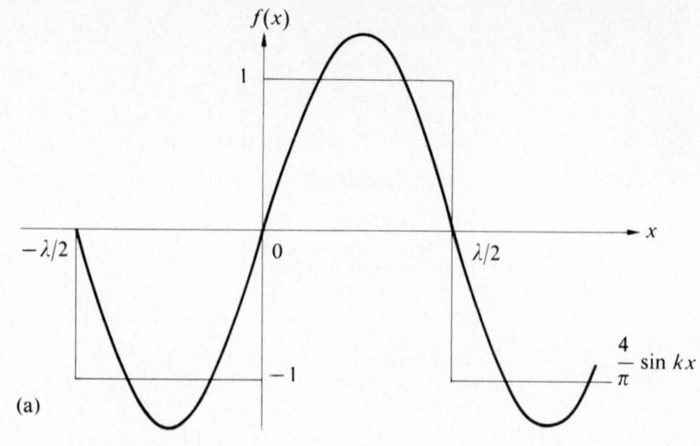

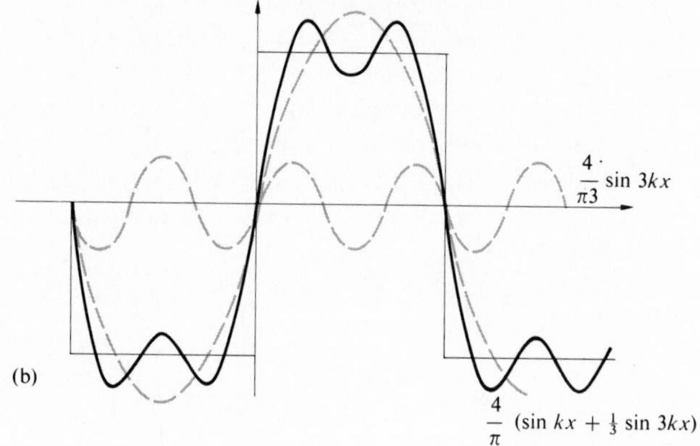

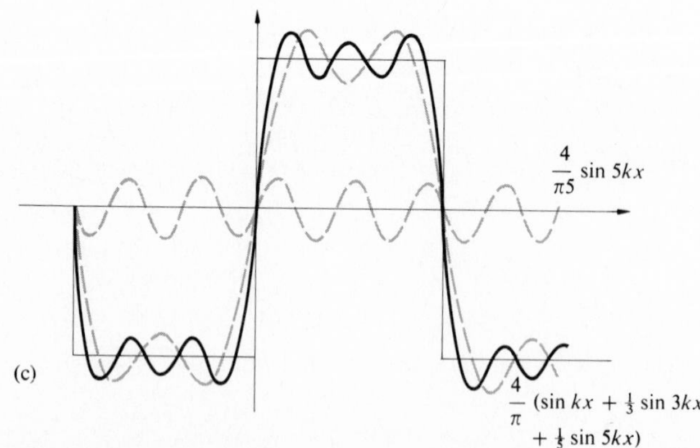

FIGURE 7.19 (Continued)

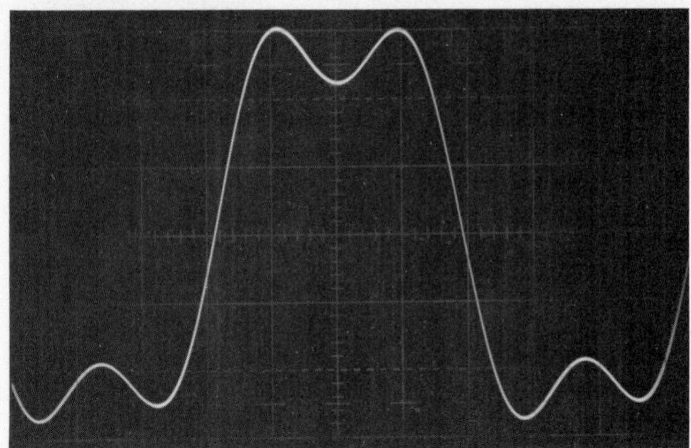

FIGURE 7.19 Synthesis of the profile of a periodic square wave. Notice that all of the constituent waves are in-phase and zero wherever the square wave is zero. Since all the sine waves are in-phase at $x = 0$, all the B_m coefficients are positive.

As a last example, let's analyze the square wave of Fig. 7.20 into its Fourier components. Notice that with the origin chosen as shown, the function is even, and all the B_m terms are zero. The appropriate Fourier coefficients (Problem 7.29) are then

$$A_0 = \frac{4}{a} \quad \text{and} \quad A_m = \frac{4}{a}\left(\frac{\sin m2\pi/a}{m2\pi/a}\right) \tag{7.52}$$

Unlike the previous function, this one has a nonzero value of A_0. You might have already noticed that $A_0/2$ is actually the *mean value* of $f(x)$, and since the curve lies completely above the axis, it will clearly not be zero.

The expression $(\sin u)/u$, which we studied earlier (p. 48), was given the name **sinc *u***, and its values are listed in Table 1 in the appendix. Since the limit of sinc u as u goes to zero is 1, A_m can represent all the coefficients, if we let $m = 0, 1, 2, \ldots$. Notice, too, that because the sinc function has negative values, some of the A_m coefficients will be negative.

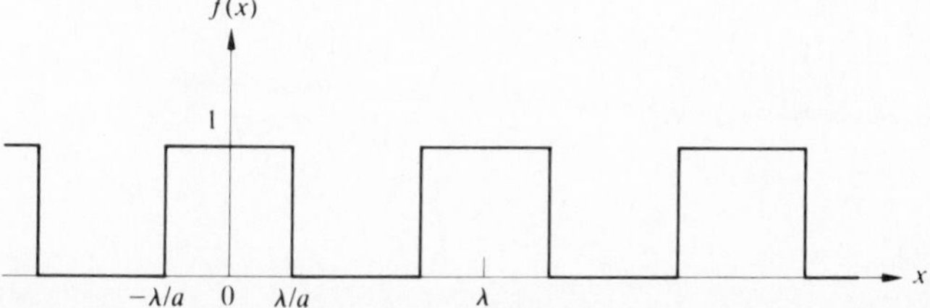

FIGURE 7.20 A periodic anharmonic function.

That means that some of the higher order cosines will be 180° out-of-phase with the $m = 1$ cosine term.

Three things distinguish the functions in Figs. 7.18 and 7.20, which otherwise have the same shape: the location of the $x = 0$ axis, the location of the $f(x) = 0$ axis, and the height of the steps. Consequently, beyond the constant A_0, the constituent harmonic terms must have the same relation to either $f(x)$ when they are plotted. In other words, moving the $x = 0$ axis from where it is in Fig. 7.18 to where it is in Fig. 7.20 will change sines into cosines in the analysis, but otherwise leave the constituent harmonic functions in Fig. 7.19 unaltered; the sinusoids that make up the square pulse in Fig. 7.19 will be the cosinusoids that make it up in Fig. 7.20. With the vertical axis in the middle of the square peak, it's clear from Fig. 7.19 that alternate cosines will have to be negative at $x = 0$.

The width of the square peak, $2(\lambda/a)$, can be any fraction of the total wavelength, depending on a. The Fourier series is then

$$f(x) = \frac{2}{a} + \sum_{m=1}^{\infty} \frac{4}{a} \operatorname{sinc} m2\pi/a \cos mkx \tag{7.53}$$

If we were synthesizing the corresponding function of time, $f(t)$, having a square peak of width $2(\tau/a)$, the same expression, Eq. (7.53), would apply where kx was simply replaced by ωt. Here ω is the *angular temporal frequency* of the periodic function $f(t)$ and is known as the **fundamental**. It is the lowest frequency of the cosine term and arises when $m = 1$. Frequencies of $2\omega, 3\omega, 4\omega, \ldots$, are known as ***harmonics*** of the fundamental and are associated with $m = 2, 3, 4, \ldots$. In much the same way, since λ is the *spatial period*, $\kappa \equiv 1/\lambda$ is the **spatial frequency**, and $k = 2\pi\kappa$ might be called the **angular spatial frequency**. Again one speaks of the harmonics, of frequency $2k, 3k, 4k, \ldots$, where these are spatial alternations. Evidently, the dimensions of κ are cycles per unit length (e.g., cycles per mm or possibly just cm^{-1}), and those of k are radians per unit length. Thus a fence could have a spatial frequency of 1 picket per foot, while a comb might have a k of 10 teeth per inch.

Let's clarify a few points so as to avoid a common confusion concerning the use of the terms *spatial frequency* and *spatial period* (or wavelength). Figure 7.20 shows a one-dimensional periodic square-wave function spread out in space along the x-axis. This might be a pattern seen on the face of an oscilloscope or the profile of a rather extraordinary disturbance moving along a taut rope. In either case, it repeats itself in space over a distance known as the wavelength, and one over that is the spatial frequency.

Now suppose instead that the pattern corresponds to an irradiance distribution, a series of bright and dark stripes, for instance, the kind of thing you might see looking through a narrow horizontal slit against a picket fence or, even better, while scanning on a line across a group of alternately clear and opaque bands (Fig. 13.26) illuminated by monochromatic light. Again the pattern will have some spatial period and frequency determined by the rate at which it repeats in space. As ever the light field will also have an angular spatial frequency (κ) and period ($\boldsymbol{\lambda}$), as well as a temporal frequency and period, quite apart from the other. The pattern might have a wavelength ($\boldsymbol{\lambda}$) of 20 cm, and the light generating it a wavelength (λ) of 500 nm. Herein lies the area of potential confusion. ***Henceforth, we will reserve the symbols k and λ for the lightwave itself and use* k and λ to describe spatial optical patterns.**

Now return to the square function of Fig. 7.20 where now $a = 4$, or in other words the peak has a width of $\boldsymbol{\lambda}/2$. In that instance

$$f(x) = \frac{1}{2} + \frac{2}{\pi}(\cos \kappa x - \tfrac{1}{3}\cos 3\kappa x + \tfrac{1}{5}\cos 5\kappa x - \cdots) \tag{7.54}$$

As a matter of fact, *if the graph of the function f(x) is such that a horizontal line could divide it into equally shaped segments, above and below that line, the Fourier series will consist of only odd harmonics.*

Were we to plot the curve representing the partial sum of the terms through $m = 9$, it would closely resemble the square wave. In contrast, if the width of the peak is reduced, the number of terms in the series needed to produce the same general resemblance to $f(x)$ will increase. This can be appreciated by examining the ratio

$$\frac{A_m}{A_1} = \frac{\sin m2\pi/a}{m \sin 2\pi/a} \tag{7.55}$$

Observe that for $a = 4$, the ninth term (i.e., $m = 9$) is fairly small, $A_9 \approx 10\% A_1$. In comparison, for a peak 100 times narrower (that is, $a = 400$), $A_9 \approx 99\% A_1$. Similarly, whereas it takes terms through $m = 4$ to duplicate the curve of Fig. 7.19*b* when $a = 4$, it will take up to $m = 8$ to produce roughly the equivalent profile when $a = 8$. Making the peak narrower has the effect of introducing higher-order harmonics, which in turn have smaller wavelengths. We might guess, then, that it is not the total number of terms in the series that is of prime importance but rather the relative dimensions of the smallest features being reproduced and the corresponding wavelengths available.* If there are fine details in the profile, the series must contain comparatively short-wavelength (or in the time domain, short-period) contributions.

The negative values of A_m in Eq. (7.53) and in Fig. (7.21) should simply be thought of as the amplitudes of those harmonic contributions that are to be added into the synthesis with their phases shifted by 180°, as compared with the posi-

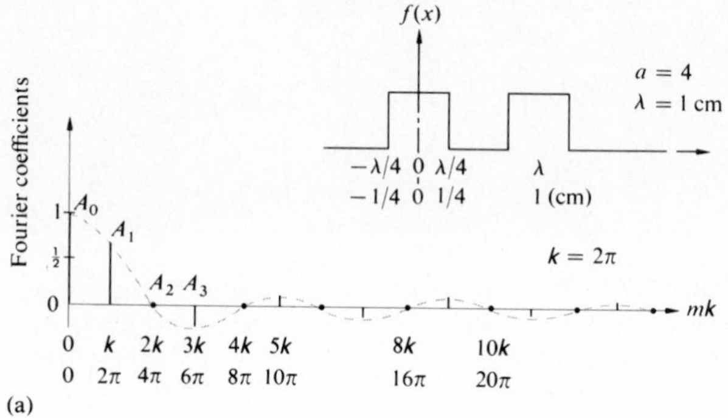

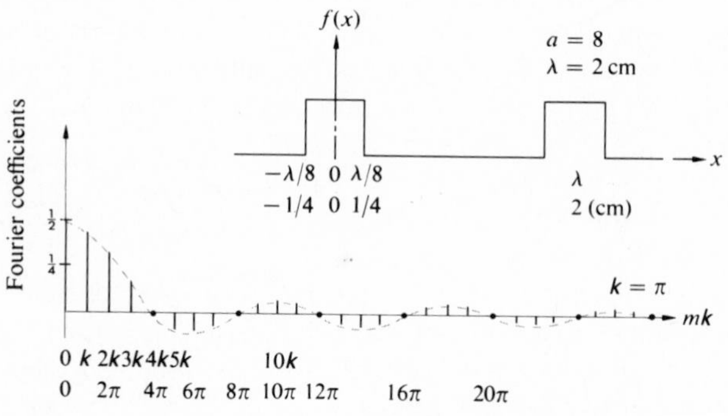

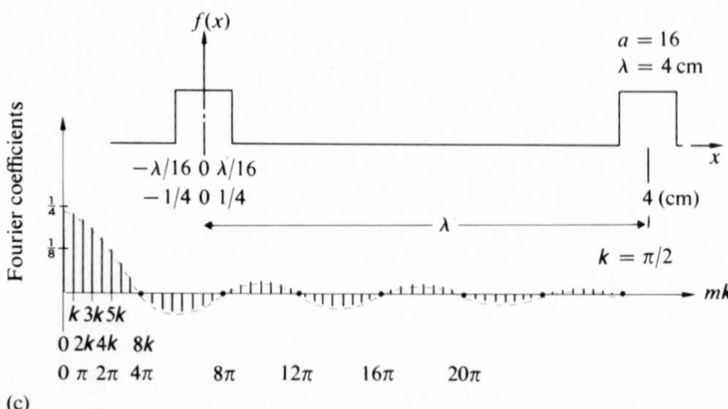

FIGURE 7.21 The square pulse as a limiting case. The negative coefficients correspond to a phase shift of π radians.

*Evidently, one is not going to be able to build a castle of blocks unless the blocks are a good deal smaller than the castle.

tive terms. The equivalence between a negative amplitude and a π-rad phase shift is clear from the fact that $A_m \cos(kx + \pi) = -A_m \cos kx$.

7.4 Nonperiodic Waves

Return to Fig. 7.20 and suppose that we keep the width of the square peak constant while λ is made to increase without limit. As λ approaches infinity, the resulting function will no longer appear periodic. We then have one single square pulse, the adjacent peaks having moved off to infinity. This suggests a possible way of generalizing the method of Fourier series to include nonperiodic functions. Such functions are of great practical interest in physics, particularly in Optics and Quantum Mechanics.

7.4.1 Fourier Integrals

To see how this can be accomplished, let's initially set $a = 4$ and choose some value of λ; anything will do, say, $\lambda = 1$ cm. The peak then has a width of $\frac{1}{2}$ cm, that is, $2(\lambda/a)$, centered at $x = 0$, as illustrated in Fig. 7.21*a*. The importance of each particular frequency, mk, can be appreciated by examining the value of the corresponding Fourier coefficient, in this case A_m. The coefficients may be thought of as weighting factors that appropriately emphasize the various harmonics. Figure 7.21*a* contains a plot of a number of values of A_m (where $m = 0, 1, 2,\ldots$) versus mk for the foregoing square wave—such a curve is known as the ***spatial frequency spectrum***.

We can regard A_m as a function, $A(mk)$, of mk, which may be nonzero only at values of $m = 0, 1, 2, \ldots$. If the quantity a is now made equal to 8 while λ is increased to 2 cm, the peak width will be completely unaffected. The only alteration is a doubling of the space between peaks. Yet a very interesting change in the spatial frequency spectrum is evident in Fig. 7.21*b*. Note that the density of components along the mk-axis has increased markedly. Nonetheless, $A(mk)$ is still zero when $mk = 4\pi, 8\pi, 12\pi, \ldots$, but since k is now π rather than 2π, there will be more terms between these zero points. Finally, let $a = 16$ and increase λ to 4 cm. Again the individual peaks are unaltered in shape, but the terms in the frequency spectrum are now even more densely packed. In effect, the pulse, as compared with λ, is getting smaller and smaller, thereby requiring higher frequencies to synthesize it.

Observe that the envelope of the curve, which was barely discernible in Fig. 7.21*a*, is quite evident in Fig. 7.21*c*. In fact, the envelope is identical in each case, except for a scale factor. It is determined only by the shape of the original signal and will be quite different for other configurations. We conclude that as λ increases and the function takes on the appearance of a single square pulse, the space between each of the $A(mk)$ contributions in the spectrum will decrease. The discrete spectral lines, while decreasing in amplitude, will gradually merge, becoming individually unresolvable. In the limit as λ approaches ∞, the spectral lines will become infinitely close to each other. As k becomes extremely small, m must consequently become exceedingly large, if mk is to be at all appreciable. Changing notation, replace mk, the angular frequency of the harmonics, by k_m. Although it comprises discrete terms, in the limit k_m will be transformed into k (i.e., a continuous frequency distribution). The function $A(k_m)$ in the limit will become the envelope shown in Fig. 7.21. It is obviously no longer meaningful to talk about the fundamental frequency and its harmonics. The pulse being synthesized, $f(x)$, has no apparent fundamental frequency.

Recall that an integral is actually the limit of a sum as the number of elements goes to infinity and their size approaches zero. Thus it should not be surprising that the *Fourier series* must be replaced by the so-called **Fourier integral** as λ goes to infinity. That integral, which is stated here without proof, is

$$f(x) = \frac{1}{\pi}\left[\int_0^{\infty} A(k)\cos kx\,dk + \int_0^{\infty} B(k)\sin kx\,dk\right] \tag{7.56}$$

provided that

$$A(k) = \int_{-\infty}^{\infty} f(x)\cos kx\,dx$$

and

$$B(k) = \int_{-\infty}^{\infty} f(x)\sin kx\,dx \tag{7.57}$$

The similarity with the series representation should be obvious. The quantities $A(k)$ and $B(k)$ are interpreted as the amplitudes of the sine and cosine contributions in the range of angular spatial frequency between k and $k + dk$. They are the **Fourier cosine** and **sine transforms**, respectively. In the foregoing example of a square pulse, it is the cosine transform, $A(k)$, that will be found to correspond to the envelope in Fig. 7.21.

A careful examination of Fig. 7.21 and Eq. (7.53) reveals that except for the zero-frequency term, the amplitudes of the contributions to the synthesis vary as $(4/a)$ sinc $m2\pi/a$: the

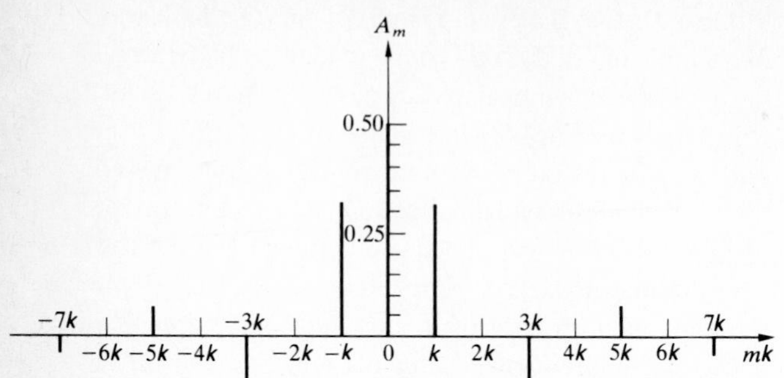

FIGURE 7.22 A symmetrical frequency spectrum for the waveform in Figure 7.21*a*. Note that the zeroth term is actually $A_0/2$, which is indeed the amplitude of the $m = 0$ contribution to the series.

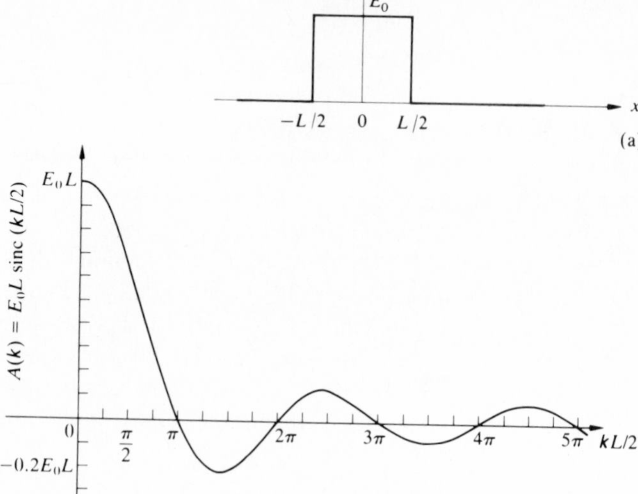

FIGURE 7.23 The square pulse and its transform.

envelope of the curve is a sinc function. Remember that the first term in the series is $\frac{1}{2}A_0$, which suggests another way to represent the frequency spectrum. Inasmuch as $\cos(mkx) = \cos(-mkx)$, we can divide the amplitude of every contribution beyond $m = 0$ in half and plot it twice, once with a positive value of k and again with a negative one (Fig. 7.22). This mathematical contrivance provides a nice symmetrical curve; it's introduced here because it is common practice to represent frequency spectra in that fashion.

As we will see in Chapter 11, the most powerful Fourier transform methods involve a complex representation that automatically gives rise to a symmetrical distribution of positive and negative spatial frequency terms. Certain optical phenomena (such as diffraction) also occur symmetrically in space, and a marvelous relationship can be constructed with the spatial frequency spectrum, provided that it encompasses positive and negative frequencies. Thus the negative frequency is a useful mathematical device that allows us to describe physical systems that are symmetrical (going off in opposite directions from a central point).

7.4.2 Pulses and Wave Packets

Let's now determine the Fourier-integral representation of the square pulse in Fig. 7.23, which is described by the function

$$f(x) = \begin{cases} E_0 & \text{when } |x| < L/2 \\ 0 & \text{when } |x| > L/2 \end{cases}$$

For the moment we'll limit the analysis to positive values of k.

Since $f(x)$ is an even function, the sine transform, $B(k)$, will be found to be zero. Pressing on,

$$A(k) = \int_{-\infty}^{\infty} f(x) \cos kx \, dx = \int_{-L/2}^{+L/2} E_0 \cos kx \, dx$$

Hence

$$A(k) = \frac{E_0}{k} \sin kx \Big|_{-L/2}^{+L/2} = \frac{2E_0}{k} \sin kL/2$$

Multiplying numerator and denominator by L and rearranging terms, we have

$$A(k) = E_0 L \frac{\sin kL/2}{kL/2}$$

or equivalently

$$A(k) = E_0 L \operatorname{sinc}(kL/2) \tag{7.58}$$

The Fourier transform of the square pulse is plotted in Fig. 7.23*b* and should be compared with the envelope in Fig. 7.21. As L increases, the spacing between successive zeros of $A(k)$ decreases and vice versa. Moreover, when $k = 0$, it follows from Eq. (7.58) that $A(0) = E_0 L$.

It is a simple matter to write out the integral representation of $f(x)$ using Eq. (7.56):

$$f(x) = \frac{1}{\pi}\int_0^\infty E_0 L \operatorname{sinc}(kL/2)\cos kx\, dx \tag{7.59}$$

An evaluation of this integral is left for Problem 7.33.

The Cosine Wavetrain

Earlier, when we talked about monochromatic waves, we pointed out that they were in fact fictitious, at least physically. There will always have been some point in time when the generator, however perfect, was turned on. Figure 7.24 depicts a somewhat idealized harmonic pulse corresponding to the function

$$E(x) = \begin{cases} E_0 \cos k_p x & \text{when } -L \leq x \leq L \\ 0 & \text{when } |x| > L \end{cases}$$

We chose to work in the space domain but could certainly have envisioned the disturbance as a function of time. We are effectively examining the spatial profile of the wave $E(x - vt)$ at $t = 0$ rather than the temporal profile at $x = 0$. The spatial frequency k_p is that of the harmonic region of the pulse itself. Proceeding with the analysis, note that $E(x)$ is an even function; consequently, $B(k) = 0$ and

$$A(k) = \int_{-L}^{+L} E_0 \cos k_p x \cos kx\, dx$$

This is identical to

$$A(k) = \int_{-L}^{+L} E_0 \tfrac{1}{2}[\cos(k_p + k)x + \cos(k_p - k)x]\, dx$$

which integrates to

$$A(k) = E_0 L\left[\frac{\sin(k_p + k)L}{(k_p + k)L} + \frac{\sin(k_p - k)L}{(k_p - k)L}\right]$$

or, if you like,

$$A(k) = E_0 L[\operatorname{sinc}(k_p + k)L + \operatorname{sinc}(k_p - k)L] \tag{7.60}$$

When there are many waves in the train ($\lambda_p << L$), $k_p L >> 2\pi$. Thus $(k_p + k)L >> 2\pi$, and therefore $\operatorname{sinc}(k_p + k)L$ is down to fairly small values. In contrast, when $k_p = k$, the second sinc function in the brackets has a maximum value of 1. In other words, the function given by Eq. (7.60) can be thought of as having a peak at $k = -k_p$ as shown in part (*b*) of the drawing. If we limit the treatment to only positive values of k, only the tail of that left-side peak that crosses into the positive k region will contribute. As we have just seen, such contribu-

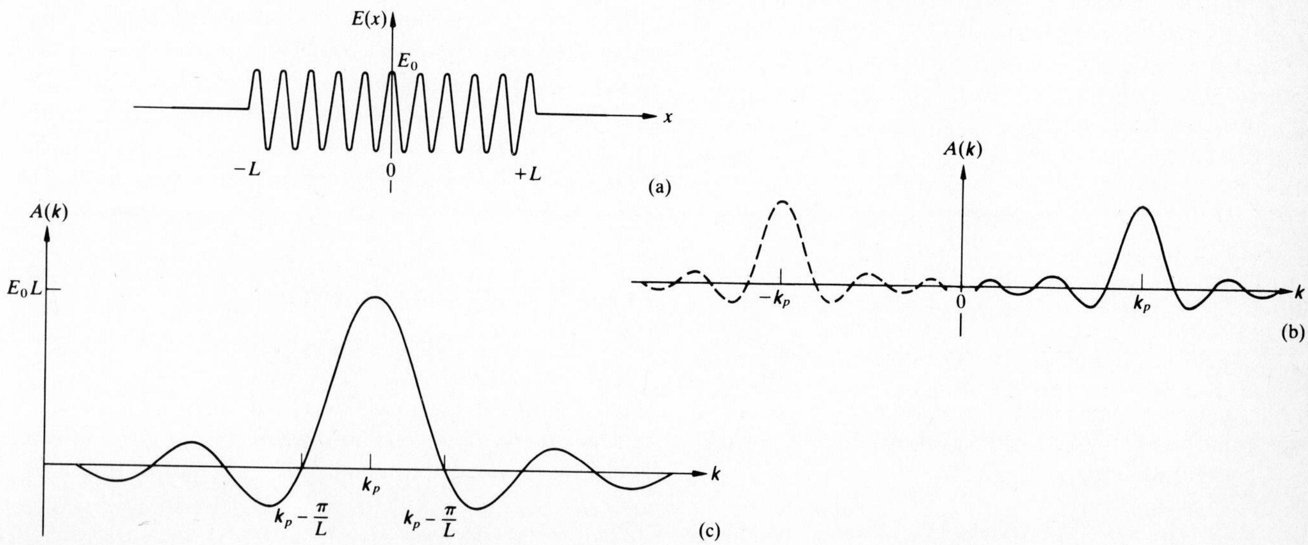

FIGURE 7.24 The profile of a finite cosine wavetrain and its transform.

tions will be negligible far from $k = -k_p$, especially when $L >> \lambda_p$ and the peaks are both narrow and widely spaced. The positive tail of the left-side peak then falls off rapidly beyond $k = -k_p$. Consequently, we can neglect the first sinc in this particular case and write the transform as

$$A(k) = E_0 L \operatorname{sinc}(k_p - k)L \tag{7.61}$$

(Fig. 7.24*c*). Even though the wavetrain is very long, since it is not infinitely long it must be synthesized from a continuous range of spatial frequencies. Thus it can be thought of as the composite of an infinite ensemble of harmonic waves. One speaks of such pulses as ***wave packets*** or ***wave groups***. As we might have expected, the dominant contribution is associated with $k = k_p$. Had the analysis been carried out in the time domain, the same results would have obtained where the transform was centered about the temporal angular frequency ω_p. Clearly, as the wavetrain becomes infinitely long (i.e., $L \to \infty$), its frequency spectrum shrinks, and the curve of Fig. 7.24*c* closes down to a single tall spike at k_p (or ω_p). This is the limiting case of the idealized monochromatic wave.

Since we can think of $A(k)$ as the amplitude of the contributions to $E(x)$ in the range k to $k + dk$, $A^2(k)$ must be related to the energy of the wave in that range (Problem 7.34). We'll come back to this point in Chapter 11 when we consider the *power spectrum*. For the moment, merely observe (Fig. 7.24*c*) that most of the energy is carried in the spatial frequency range from $k_p - \pi/L$ to $k_p + \pi/L$, extending between the minima on either side of the central peak. An increase in the length of the wavetrain causes the energy of the wave to become concentrated in an ever narrowing range of k about k_p.

The wave packet in the time domain, that is,

$$E(t) = \begin{cases} E_0 \cos \omega_p t & \text{when } -T \leq t \leq T \\ 0 & \text{when } |t| > T \end{cases}$$

has the transform

$$A(\omega) = E_0 T \operatorname{sinc}(\omega_p - \omega)T \tag{7.62}$$

where ω and k are related by the phase velocity. The frequency spectrum, except for the notational change from k to ω and L to T, is identical to that of Fig. 7.24*c*.

Frequency Bandwidth

For the particular wave packet being studied the range of angular frequencies (ω or k) that the transform comprises is certainly not finite. Yet if we were to speak of the *width* of the transform ($\Delta\omega$ or Δk), Fig. 7.24*c* suggests that we use $\Delta k = 2\pi/L$ or $\Delta\omega = 2\pi/T$. In contrast, the spatial or temporal extent of the pulse is unambiguous at $\Delta x = 2L$ or $\Delta t = 2T$, respectively. The product of the width of the packet in what might be called *k-space* and its width in *x-space* is $\Delta k\, \Delta x = 4\pi$ or analogously $\Delta\omega\, \Delta t = 4\pi$. The quantities Δk and $\Delta\omega$ are the **frequency bandwidths**. Had we used a differently shaped pulse, the product of the bandwidth and the pulse length might certainly have been somewhat different. The ambiguity arises because we have not yet chosen one of the alternative possibilities for specifying $\Delta\omega$ and Δk. For example, rather than using the first minima of $A(k)$ (there are transforms that have no such minima, such as the Gaussian function of Section 11.2), we could have let Δk be the width of $A^2(k)$ at a point where the curve had dropped to $\frac{1}{2}$ or possibly $1/e$ of its maximum value. In any event, it will suffice for the time being to observe that since $\Delta\omega = 2\pi\Delta\nu$,

$$\Delta\nu \approx 1/\Delta t \tag{7.63}$$

that is, the frequency bandwidth is the same order of magnitude as the reciprocal of the temporal extent of the pulse (Problem 7.35). If the wave packet has a narrow bandwidth, it will extend over a large region of space and time. Accordingly, a radio tuned to receive a bandwidth of $\Delta\nu$ will be capable of detecting pulses of duration no shorter than $\Delta t \approx 1/\Delta\nu$.

These considerations are of profound importance in Quantum Mechanics where wave packets describe particles, and Eq. (7.63) is akin to the Heisenberg Uncertainty Principle.

7.4.3 Coherence Length

Let's now consider the light emitted by what is loosely termed a monochromatic source, for example, a sodium discharge lamp. When the beam is passed through some sort of spectrum analyzer, all its various frequency components are observed. Typically, we find that a number of fairly narrow frequency ranges contain most of the energy and that these are separated by much larger regions of darkness. Each such brightly colored band is known as a **spectral line**. There are devices in which the light enters by way of a slit, and each line is actually a colored image of that slit. Other analyzers represent the frequency distribution on the screen of an oscilloscope. In any event, the individual spectral lines are never infinitely sharp. They always consist of a band of frequencies, however small (Fig. 7.25).

The electron transitions responsible for the generation of light have a duration on the order of 10^{-8} s to 10^{-9} s. Because the emitted wavetrains are finite, there will be a spread in the frequencies present, known as the **natural linewidth** (see Section 11.3.4). Moreover, since the atoms are in random thermal motion, the frequency spectrum will be altered by the Doppler Effect. In addition, the atoms suffer collisions that interrupt the wavetrains and again tend to broaden the frequency distribution. The total effect of all these mechanisms is that each spectral line has a bandwidth $\Delta\nu$ rather than one single frequency. The time that satisfies Eq. (7.63) is referred to as the **coherence time** (henceforth to be written Δt_c), and the length

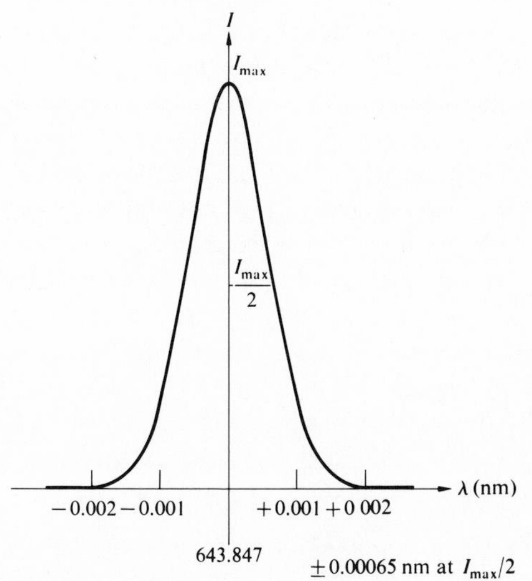

FIGURE 7.25 The cadmium red ($\bar{\lambda}$ = 643.847 nm) spectral line from low-pressure lamp.

Δl_c given by

$$\Delta l_c = c\,\Delta t_c \tag{7.64}$$

is the **coherence length**. As will become evident presently, the coherence length is the extent in space over which the wave is nicely sinusoidal so that its phase can be predicted reliably. The corresponding temporal duration is the coherence time. These concepts are extremely important in considering the interaction of waves, and we will come back to them later in the discussion of interference.

Although the concept of the photon wavetrain is already familiar, we are now in a position, armed with a little Fourier analysis, to deduce something about its configuration. This can be done by essentially working backward from the experimental observation that the frequency distribution of a spectral line from a quasimonochromatic (nonlaser) source can be represented by a bell-shaped Gaussian function (Section 2.1). That is, the irradiance versus frequency is found to be Gaussian. But irradiance is proportional to the electric-field amplitude squared, and since the square of a Gaussian function is a Gaussian function, it follows that the net amplitude of the light field is also bell-shaped.

Now suppose a single photon wavetrain, one of N identical such packets making up the beam, resembles Fig. 7.26*a* in that it is a harmonic function modulated by a Gaussian envelope. Its Fourier transform, $A(\omega)$, is also Gaussian. Imagine that we look at only one and the same harmonic frequency component that goes into making up each photon wavetrain, for example, the one corresponding to ω'. Remember that this component is an infinitely long, constant-amplitude sinusoid. If every packet is indeed identical, the amplitude of the Fourier component associated with ω' will be the same in each. At any point in a stream of photons these ω'-component monochromatic waves, one from each wavetrain, will have a random relative phase distribution that rapidly changes in time with the arrival of each photon. Thus all such contributions taken together [Eq. (7.21)] will correspond on average to a harmonic wave of frequency ω' having an amplitude proportional to $N^{1/2}$, and this is the ω' part of the net observed field. The same will be true for every other frequency constituting the packets. This means that the same amount of energy is present at each frequency in the net light field of the beam as there is in the totality of the

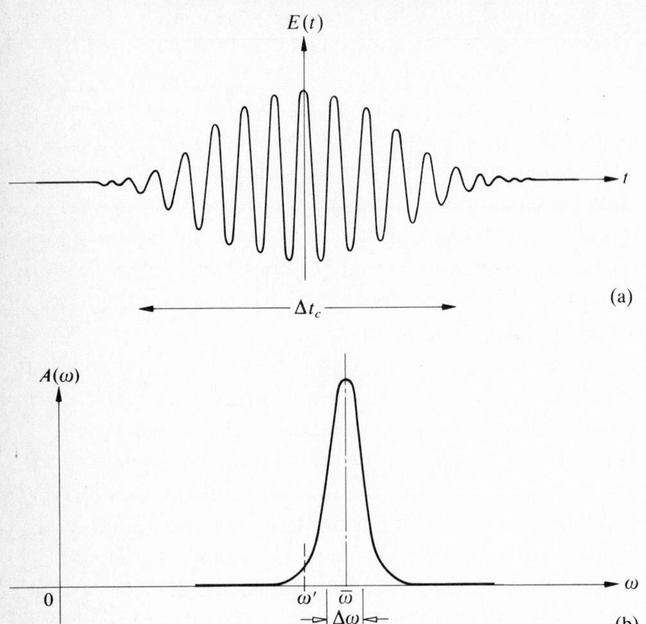

FIGURE 7.26 A cosinusoidal wavetrain modulated by a Gaussian envelope along with its transform, which is also Gaussian.

separate constituent wavetrains. Moreover, we know all about this energy-frequency distribution; it's Gaussian, so the transform of the photon wavetrain must be Gaussian too. In other words, the observed spectral line corresponds to the power spectrum of the beam, but it also corresponds to the power spectrum of an individual photon packet. If the irradiance is Gaussian, the photon wavetrain is Gaussian.

As a result of the randomness of the wavetrains, the individual harmonic components of the resultant wave will not have the same relative phases as they did in each packet. Thus the profile of the resultant will differ from that of the separate wave packets, even though the amplitude of each frequency component present in the resultant is simply $N^{1/2}$ times its amplitude in any one packet. The observed spectral line corresponds to the power spectrum of the resultant beam, to be sure, but it also corresponds to the power spectrum of an individual packet. Ordinarily there will be a tremendous number of arbitrarily overlapping wave groups, so that the envelope of the resultant will rarely, if ever, be zero. If the source is quasimonochromatic (i.e., if the bandwidth is small compared with the mean frequency $\bar{\nu}$), we can envision the resultant as being "almost" sinusoidal.

In summary, the composite lightwave can be pictured as in Fig. 7.27. We might imagine the frequency and amplitude to be randomly varying, the former over a range $\Delta\nu$ centered at $\bar{\nu}$. Accordingly, the **frequency stability**, defined as $\Delta\nu/\bar{\nu}$, is a useful measure of spectral purity. Even a coherence time as short as 10^{-9} s corresponds to roughly a few million wavelengths of the rapidly oscillating carrier ($\bar{\nu}$), so that any amplitude or frequency variations will occur quite slowly in comparison. Equivalently, we can introduce a time-varying phase factor such that the disturbance can be written as

$$E(t) = E_0(t) \cos [\varepsilon(t) - 2\pi\bar{\nu}t] \tag{7.65}$$

where the separation between wave crests changes in time.

The average duration of a wave packet is Δt_c, so two points on the wave in Fig. 7.27 separated by more than Δt_c must lie on different contributing wavetrains. These points would thus be completely uncorrelated in phase. In other words, if we determined the electric field of the composite wave as it passed by an idealized detector, we could predict its phase fairly accurately for times much less than Δt_c later, but not at all for times greater than Δt_c. In Chapter 12 we will consider the *degree of coherence* that applies over the region between these extremes as well.

White light has a frequency range from 0.4×10^{15} Hz to about 0.7×10^{15} Hz, that is, a bandwidth of about 0.3×10^{15}

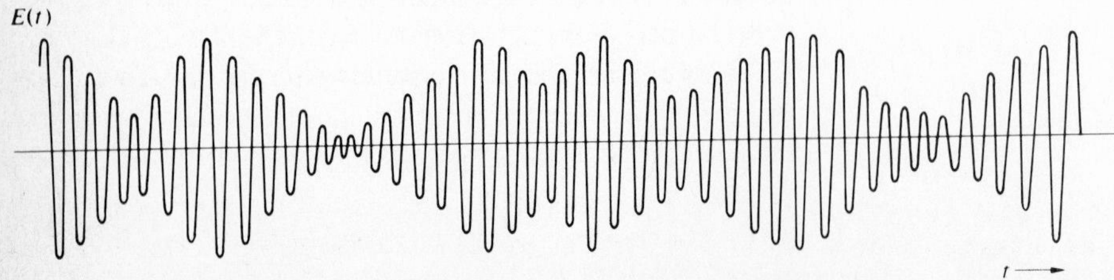

FIGURE 7.27 A quasimonochromatic lightwave.

Hz. The coherence time is then roughly 3×10^{-15} s, which corresponds [Eq. (7.64)] to wavetrains having a spatial extent only a few wavelengths long (Table 7.1). Accordingly, *white light may be envisaged as a random succession of very short pulses.* Were we to synthesize white light, we would have to superimpose a broad, continuous range of harmonic constituents in order to produce the very short wave packets. Inversely, we can pass white light through a Fourier analyzer, such as a diffraction grating or a prism, and in so doing actually generate those components.

The available bandwidth in the visible spectrum ($\approx$300 THz) is so broad that it represents something of a wonderland for the communications engineer. For example, a typical television channel occupies a range of about 4 MHz in the electromagnetic spectrum ($\Delta\nu$ is determined by the duration of the pulses needed to control the scanning electron beam). Thus the visible region could carry roughly 75 million television channels. Needless to say, this is an area of active research (see Section 8.11).

Ordinary discharge lamps have relatively large bandwidths leading to coherence lengths only on the order of several millimeters. In contrast, the spectral lines emitted by low-pressure isotope lamps such as Hg^{198} ($\lambda_{air} = 546.078$ nm) or the international standard Kr^{86} ($\lambda_{air} = 605.616$ nm) have bandwidths of roughly 1000 MHz. The corresponding coherence lengths are approximately 0.3 m, and the coherence times are about 1 ns. The frequency stability is about one part per million—these sources are certainly quasimonochromatic.

TABLE 7.1 Approximate coherence lengths of several sources

Source	Mean Wavelength $\bar{\lambda}$(nm)	Linewidth* $\Delta\lambda$ (nm)	Coherence Length Δl_c
Thermal IR (8000–12 000 nm)	10 000	$\approx$4000	$\approx$25 000 nm = $2.5\bar{\lambda}$
Mid-IR (3000–5000 nm)	4000	$\approx$2000	$\approx$8000 nm = $2\bar{\lambda}$
White light	550	$\approx$300	$\approx$900 nm = $1.6\bar{\lambda}$
Mercury arc	546.1	$\approx$1.0	$\lessapprox$ 0.03 cm
Kr^{86} discharge lamp	605.6	1.2×10^{-3}	0.3 m
Stabilized He-Ne laser	632.8	$\approx 10^{-6}$	$\lessapprox$ 400 m
Special He-Ne laser	1153	8.9×10^{-11}	15×10^6 m

*To find the corresponding frequency bandwidth use, $\Delta\nu/\Delta\lambda_0 = \nu/\bar{\lambda}_0$.

The most spectacular of all present-day sources is the laser. Under optimum conditions, with temperature variations and vibrations meticulously suppressed, a laser was actually operated at quite close to its theoretical limit of frequency constancy. A short-term frequency stability of about 8 parts per 10^{14} was attained* with a He–Ne continuous gas laser at $\lambda_0 = 1153$ nm. That corresponds to a remarkably narrow bandwidth of about 20 Hz. More common and not very difficult to obtain are frequency stabilities of several parts per 10^9. There are commercially available CO_2 lasers that provide a short-term ($\approx 10^{-1}$ s) $\Delta\nu/\bar{\nu}$ ratio of 10^{-9} and a long-term ($\approx 10^3$ s) value of 10^{-8}.

7.4.4 The Discrete Fourier Transform

A function that describes some physical process can be Fourier analyzed, and its transform can be determined analytically. We've already been introduced to the basics of how that's done, and we'll return to elaborate the effort in Chapter 11. But before leaving the subject, it's important to extend the ideas of Fourier analysis to situations where there are no functional representations of the data. Often one has a collection of data points or perhaps a curve created on a plotter or computer screen. In any event, the information can be digitized; that is, numbers can be associated with points on the curve at convenient intervals. To determine the frequency content of such a limited collection of data, a numerical technique known as the **discrete Fourier transform** is used. Since the treatment is computer based, it will suffice for our purposes just to understand the general scheme and be able to appreciate the results.

Until now we dealt with functions such as $f(x)$—representing something interesting like an electric field—that provided values for all x. Instead, suppose we have a finite number of points, N, located at $0, x_1, x_2, \ldots, x_{N-1}$ and the corresponding specific values of whatever quantity is being studied: f_0, f_{x_1}, f_{x_2}, and so on. When the sample points are equally spaced by an interval x_0, they can be represented by the sequence f_0, f_{x_0}, f_{2x_0}, and so forth. In essence, each Fourier integral transform [Eq. (7.57)] is approximated by a summation that is carried out suc-

*T. S. Jaseja, A. Javan, and C. H. Townes, "Frequency Stability of Helium–Neon Lasers and Measurements of Length," *Phys. Rev. Letters* **10**, 165 (1963).

cessively, point by point, over the range of the available data: $f_0, f_{x_0}, f_{2x_0}, \ldots$. Figure 7.28 depicts a hand-drawn pulse and the corresponding computer-calculated discrete Fourier transform (displayed with positive and negative frequency values, as in Fig. 7.22).

It's a straightforward business (Section 11.2.2) to extend Fourier analysis to two-dimensional functions, $f(x, y)$. For example, whereas Fig. 7.29*b* is the transform of the one-dimensional unit-square pulse in terms of the angular spatial frequency k, Fig. 7.29*d* is the transform of the two-dimensional unit-square pulse in terms of the angular spatial frequencies k_x and k_y.

It's natural for physicists to think about processes in relation to energy, especially if any measurements are to be made. The energy associated with a harmonic wave is proportional to the amplitude squared, and since the transform tells us the amplitudes of all the constituent sinusoids that make up the input signal, the square of the transform provides a measure of the distribution of energy, or power, at each and every compo-

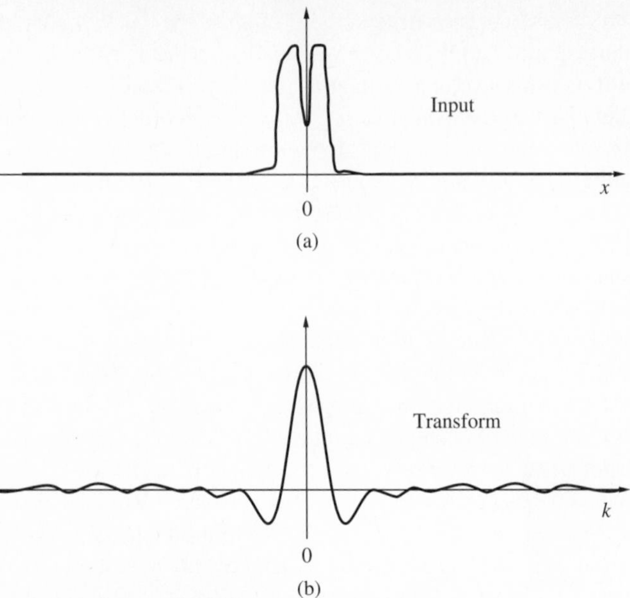

FIGURE 7.28 An input signal and its discrete Fourier transform.

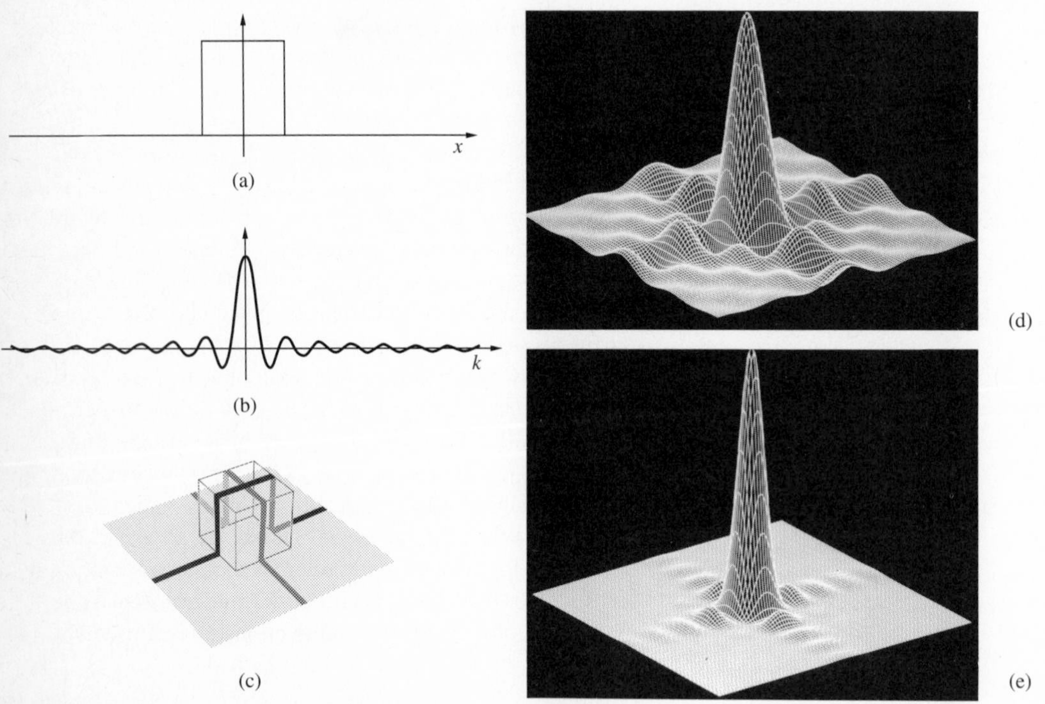

FIGURE 7.29 (*a*) A one-dimensional square pulse and (*b*) its transform. (*c*) A two-dimensional square pulse and (*d*) its transform. (*e*) The power spectrum of the transform in (*d*) plotted in two-dimensional *k*-space. (Photos courtesy R. G. Wilson, Illinois Wesleyan University.)

nent frequency. Consequently, the square of the transform is a function of spatial frequency called the **power spectrum**. Since the transform will most often be written as a complex quantity, the power spectrum can be defined as *the product of the transform and its complex conjugate*, given in units of W/m^{-2} or $W \cdot m^2$.

Figure 7.29*e* is a plot (in *k*-space) of the power spectrum for the two-dimensional square pulse. Notice that it is everywhere positive, which is not the case with the transform. It's clear from the power spectrum that most of the energy in the signal is associated with relatively low frequencies—the frequency increases radially out from the center of the pattern. Because the power spectrum is always positive, it is useful to plot it as a kind of spot diagram in a two-dimensional format; each point then corresponds to the contribution at a particular frequency. Later (p. 532) we'll write the transform in terms of the coordinates (Y, Z) on a distant observing screen and establish that the transform squared is then identical to the irradiance distribution in the diffraction pattern on that screen. Expressed in this way, the transform squared (in units of W/m^2) can be called the *irradiance spectrum*. Although there is a mathematical distinction between the power and irradiance spectra, if you were shown an unlabeled representation of each (the former plotted in *k*-space and the latter in ordinary coordinate space), you'd be hard pressed to tell the difference.

When analytic functions are not available, similar results can be accomplished with the discrete Fourier transform. A two-dimensional field of data (e.g., the picture of the Mona Lisa in Fig. 7.30*a*) can be scanned, digitized, and the discrete transform computed. The graph of the transform of so complicated a signal is itself rather complicated, and so the power spectrum (Fig. 7.30*b*) is pictured instead. Because of the way negative frequencies were introduced, the pattern is symmetrical along any diagonal. The bright narrow central cross arises from the sharp boundary edges of the picture. (As we'll see later, the horizontal edge produces the vertical line and the vertical edge produces the horizontal line—take a look at Fig. 13.37. If the higher spatial frequency terms that carry the fine details (the ones far from center) are filtered out and the picture is reconstructed from what remains, a soft blur results (Fig. 7.30*c*). On the other hand, if the low spatial frequency terms are removed by blocking out the center of the transform, the high frequencies that remain will result in a sharp-edged reconstruction (Fig. 7.30*d*).

The form of the elements within a given image determine its transform and therefore its power spectrum. The pictures in

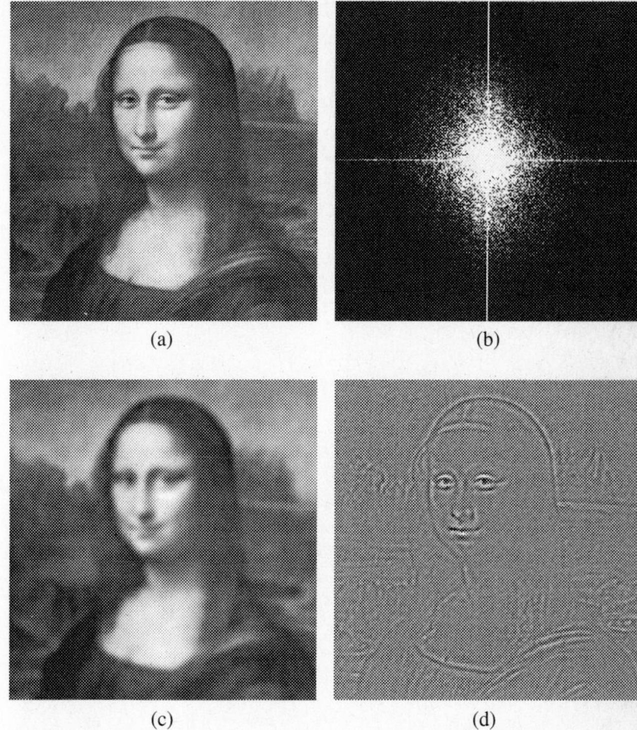

FIGURE 7.30 (*a*) The Mona Lisa (*b*) and the central portion of its power spectrum. (*c*) Mona, with her high spatial frequencies removed. (*d*) Mona with her low spatial frequencies removed. (Photos courtesy Synoptics Image Processing Systems, Cambridge, UK.)

Fig. 7.31 were computer-created, with a vertical sinusoidal pattern superimposed in order to illustrate the point. The idea was to successively isolate several subregions of the picture, to study their transforms, and to filter them. The vertical periodic modulation forms a sinusoidal grid or **grating** that has a single fixed spatial frequency (κ_0). Its presence shows up in the computed power spectrum of any portion of the picture, essentially as two bright spots on the horizontal axis at $\pm\kappa_0$. ***Ideally, the power spectrum of a signal in the form of a sinusoidal grating is remarkably simple. It consists of just two spikes, one at plus and the other at minus the grating frequency.***

The filter (represented by the white square with two black spots) was used to create the inserted small images within each photo. It removed frequencies $+\kappa_0$ and $-\kappa_0$ from each power spectrum (the filtered versions of which are shown in the upper right). The subregion images were then reconstructed using the filtered power spectra. Each of the "cleaned up" images, sans sinusoid, was then returned to its place in the

(a)

original. Notice how different those two spectra are—the facets on the cut-glass cup dominate the spectrum in Fig. 7.31*b*. Clearly, the frequency content of a picture, as spread out before us in the form of either the Fourier transform or the power spectrum, provides a wonderful new way to think about the image.

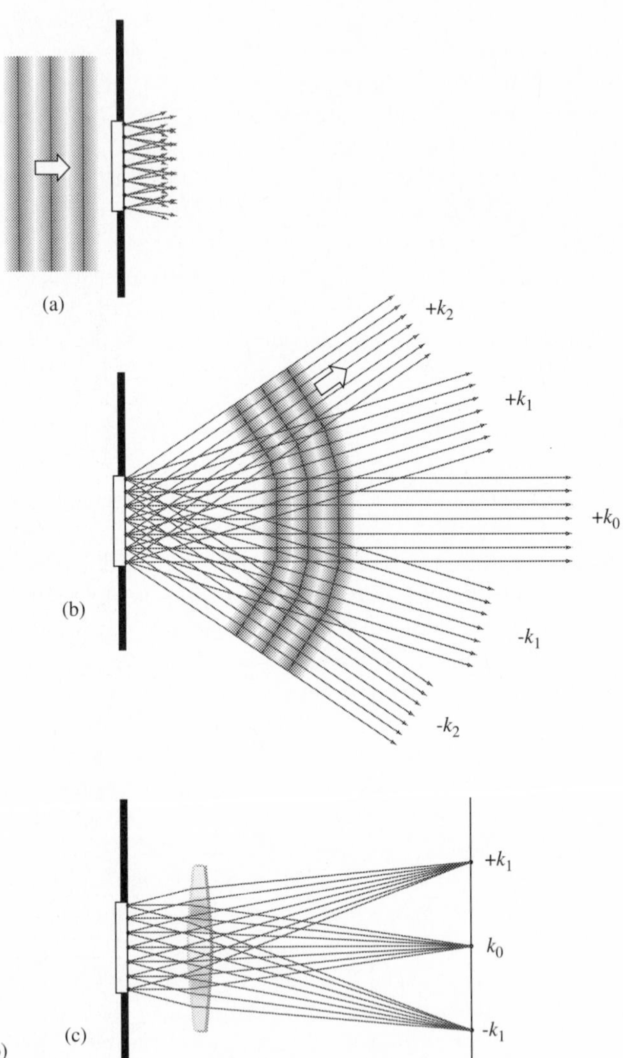

(b)

FIGURE 7.31 Two computer-processed images. The small inserts on the left were created by filtering out the sinusoidal modulation. The white insert represents the filter, and the black one is the filtered power spectrum in each case. (Photos courtesy MountainGate, Reno, Nevada.)

FIGURE 7.32 An illuminated transparency. (*a*) Incident monochromatic plane waves. (*b*) Scattered parallel bundles of rays (plane wave). (*c*) The projection of the power spectrum onto an observing screen.

Fourier Analysis and Diffraction

A discussion of computer image analysis, which is a kind of virtual Optics, can be fascinating in its own right, but it also presages a far more fundamental aspect of diffraction, which can only be touched on in this chapter. The photographic transparency (let it be a slide of the Mona Lisa) shown in Fig. 7.32*a* is a two-dimensional record of the distribution of light that once was an image of the painting. The information so stored can be read out as a signal by illuminating the slide, and that's done here with monochromatic plane waves. Every point on the surface of the slide is a scatterer, and rays emerge from it in a wide range of directions (Fig. 7.32*b*). For every plane wave going off at some angle above the axis, there is one streaming away at the same angle below the axis. Each plane wave (or parallel ray bundle) traveling in a particular κ_i-direction is a Fourier spatial frequency component. A directory of all of these component plane waves constitutes the transform of the transmitted optical field at the transparency. ***The Fourier transform of the electric field at the slide is a weighting function that gives the relative strength of each frequency component composing that field, and therefore each plane-wave stream leaving the aperture.*** The sum total of all the plane waves *is* all the transmitted light, and must be equivalent to the complicated Mona Lisa wavefront leaving the slide, which is also all the transmitted light.

Another nice way to envision what's happening is to suppose that every picture element with a spatial frequency along any direction in the photo plane acts like a sinusoidal grating. And every such grating essentially diffracts light into two symmetrical streams of plane waves traveling at angles proportional to the grating frequency (p. 466).

The region to the right of the slide is filled with waves that increasingly overlap as the distance from the slide increases. Nearby, the light arriving on a viewing screen would show the Mona Lisa fairly clearly, but as the screen was moved away the image would blur and change until it soon became totally unrecognizable. The region beyond the slide contains an intricate distribution of light, the diffraction pattern of the transparency. Mathematically, there are two regimes: *Fresnel diffraction*, which appears close to the aperture (i.e., the slide) and extends out to the region of *Fraunhoffer diffraction*, which comes into being very far from the aperture and goes on from there (p. 437).

If a lens is placed one focal length from the slide, as in Fig. 7.32*c*, it will cause the parallel ray bundles (which produce Fraunhoffer diffraction beyond a distance so great that it's effectively infinite) to conveniently focus on a nearby screen. There, each point of light in the resulting diagonally symmetrical, two-dimensional irradiance distribution corresponds to a specific value of spatial frequency. ***The amplitude of the electric field everywhere in the Fraunhoffer diffraction pattern corresponds to the Fourier transform of the input signal, that is, the electric-field distribution over the aperture***, although neither is measurable directly.

The observable phenomenon is the two-dimensional irradiance distribution which is identical to the square of the Fourier transform of the input field (p. 532). It's also a map of the spatial frequency content of the Mona Lisa, and it "matches" the power spectrum pictured in Fig. 7.30*b*. As we'll see (p. 606), it's possible to spatially filter the optical transform, thereby altering the reconstructed image, just as was done via computer to produce Figs. 7.30*c* and *d*.

Problems

7.1 Determine the resultant of the superposition of the parallel waves $E_1 = E_{01} \sin(\omega t + \varepsilon_1)$ and $E_2 = E_{02} \sin(\omega t + \varepsilon_2)$ when $\omega = 120\pi$, $E_{01} = 6$, $E_{02} = 8$, $\varepsilon_1 = 0$, and $\varepsilon_2 = \pi/2$. Plot each function and the resultant.

7.2* Considering Section 7.1, suppose we began the analysis to find $E = E_1 + E_2$ with two cosine functions $E_1 = E_{01} \cos(\omega t + \alpha_1)$ and $E_2 = E_{02} \cos(\omega t + \alpha_2)$. To make things a little less complicated, let $E_{01} = E_{02}$ and $\alpha_1 = 0$. Add the two waves algebraically and make use of the familiar trigonometric identity $\cos\theta + \cos\Phi = 2\cos\frac{1}{2}(\theta + \Phi)\cos\frac{1}{2}(\theta - \Phi)$ in order to show that $E = E_0 \cos(\omega t + \alpha)$, where $E_0 = 2E_{01} \cos\alpha_2/2$ and $\alpha = \alpha_2/2$. Now show that these same results follow from Eqs. (7.9) and (7.10).

7.3* Show that when the two waves of Eq. (7.5) are in-phase, the resulting amplitude squared is a maximum equal to $(E_{01} + E_{02})^2$, and when they are out-of-phase it is a minimum equal to $(E_{01} - E_{02})^2$.

7.4* Show that the *optical path length*, defined as the sum of the products of the various indices times the thicknesses of media traversed by a beam, that is, $\Sigma_i n_i x_i$, is equivalent to the length of the path in vacuum that would take the same time for that beam to negotiate.

7.5 Answer the following:

(a) How many wavelengths of $\lambda_0 = 500$ nm light will span a 1-m gap in vacuum?
(b) How many waves span the gap when a glass plate 5 cm thick ($n = 1.5$) is inserted in the path?
(c) Determine the *OPD* between the two situations.
(d) Verify that Λ/λ_0 corresponds to the difference between the solutions to (a) and (b) above.

7.6* Determine the optical path difference for the two waves A and B, both having vacuum wavelengths of 500 nm, depicted in Fig. P.7.6; the glass ($n = 1.52$) tank is filled with water ($n = 1.33$). If the waves start out in-phase and all the above numbers are exact, find their relative phase difference at the finishing line.

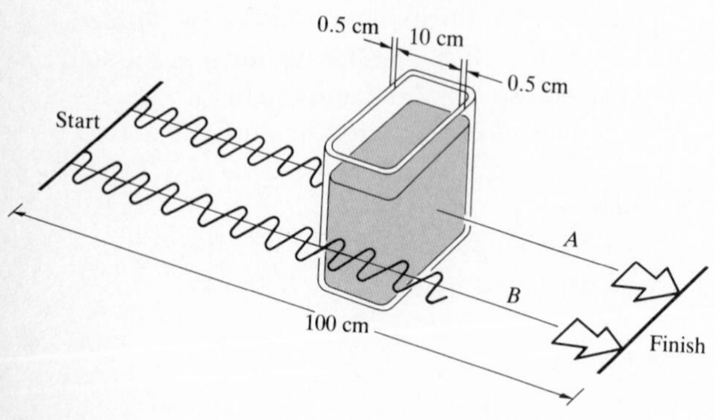

FIGURE P.7.6

7.7* Using Eqs. (7.9), (7.10), and (7.11), show that the resultant of the two waves

$$E_1 = E_{01} \sin [\omega t - k(x + \Delta x)]$$

and

$$E_2 = E_{01} \sin (\omega t - kx)$$

is

$$E = 2E_{01} \cos \left(\frac{k \Delta x}{2}\right) \sin \left[\omega t - k\left(x + \frac{\Delta x}{2}\right)\right] \qquad [7.17]$$

7.8 Add the two waves of Problem 7.7 directly to find Eq. (7.17).

7.9 Use the complex representation to find the resultant $E = E_1 + E_2$, where

$$E_1 = E_0 \cos (kx + \omega t) \quad \text{and} \quad E_2 = - E_0 \cos (kx - \omega t)$$

Describe the composite wave.

7.10 The electric field of a standing electromagnetic plane wave is given by

$$E(x, t) = 2E_0 \sin kx \cos \omega t \qquad [7.30]$$

Derive an expression for $B(x, t)$. (You might want to take another look at Section 3.2.) Make a sketch of the standing wave.

7.11* Considering Wiener's experiment (Fig. 7.11) in monochromatic light of wavelength 550 nm, if the film plane is angled at 1.0° to the reflecting surface, determine the number of bright bands per centimeter that will appear on it.

7.12* Microwaves of frequency 10^{10} Hz are beamed directly at a metal reflector. Neglecting the refractive index of air, determine the spacing between successive nodes in the resulting standing-wave pattern.

7.13* A standing wave is given by

$$E = 100 \sin \tfrac{2}{3}\pi x \cos 5\pi t$$

Determine two waves that can be superimposed to generate it.

7.14* Imagine that we strike two tuning forks, one with a frequency of 340 Hz, the other 342 Hz. What will we hear?

7.15 Figure P.7.15 shows a carrier of frequency ω_c being amplitude-modulated by a sine wave of frequency ω_m, that is,

$$E = E_0(1 + a \cos \omega_m t) \cos \omega_c t$$

Show that this is equivalent to the superposition of three waves of frequencies ω_c, $\omega_c + \omega_m$, and $\omega_c - \omega_m$. When a number of modulating frequencies are present, we write E as a Fourier series and sum over all values of ω_m. The terms $\omega_c + \omega_m$ constitute what is called the *upper sideband*, and all the $\omega_c - \omega_m$ terms form the *lower sideband*.

What bandwidth would you need in order to transmit the complete audible range?

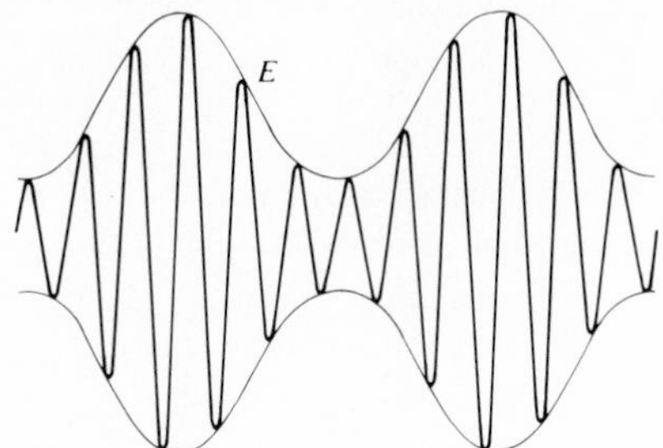

FIGURE P.7.15

7.16 Given the dispersion relation $\omega = ak^2$, compute both the phase and group velocities.

7.17* Using the relation $1/v_g = d\kappa/d\nu$, prove that

$$\frac{1}{v_g} = \frac{1}{v} - \frac{\nu}{v^2}\frac{dv}{d\nu}$$

7.18* In the case of lightwaves, show that

$$\frac{1}{v_g} = \frac{n}{c} + \frac{\nu}{c}\frac{dn}{d\nu}$$

7.19 The speed of propagation of a surface wave in a liquid of depth much greater than λ is given by

$$v = \sqrt{\frac{g\lambda}{2\pi} + \frac{2\pi\Upsilon}{\rho\lambda}}$$

where g = acceleration of gravity, λ = wavelength, ρ = density, Υ = surface tension. Compute the group velocity of a pulse in the long wavelength limit (these are called *gravity waves*).

7.20* Show that the group velocity can be written as

$$v_g = v - \lambda\frac{dv}{d\lambda}$$

7.21 Show that the group velocity can be written as

$$v_g = \frac{c}{n + \omega(dn/d\omega)}$$

7.22* Determine the group velocity of waves when the phase velocity varies inversely with wavelength.

7.23* Show that the group velocity can be written as

$$v_g = \frac{c}{n} + \frac{\lambda c}{n^2}\frac{dn}{d\lambda}$$

7.24* For a wave propagating in a periodic structure for which $\omega(k) = 2\omega_0 \sin(k\ell/2)$, determine both the phase and group velocities. Write the former as a sinc function.

7.25* An ionized gas or plasma is a dispersive medium for EM-waves. Given that the dispersion equation is

$$\omega^2 = \omega_p{}^2 + c^2k^2$$

where ω_p is the constant plasma frequency, determine expressions for both the phase and group velocities and show that $vv_g = c^2$.

7.26 Using the dispersion equation,

$$n^2(\omega) = 1 + \frac{Nq_e^2}{\epsilon_0 m_e}\sum_j\left(\frac{f_j}{\omega_{0j}^2 - \omega^2}\right) \qquad [3.71]$$

show that the group velocity is given by

$$v_g = \frac{c}{1 + Nq_e^2/\epsilon_0 m_e \omega^2 2}$$

for high-frequency electromagnetic waves (e.g., X-rays). Keep in mind that since f_j are the weighting factors, $\Sigma_j f_j = 1$. What is the phase velocity? Show that $vv_g \approx c^2$.

7.27* Analytically determine the resultant when the two functions $E_1 = 2E_0 \cos \omega t$ and $E_2 = \frac{1}{2}E_0 \sin 2\omega t$ are superimposed. Draw E_1, E_2, and $E = E_1 + E_2$. Is the resultant periodic; if so, what is its period in terms of ω?

7.28 Show that

$$\int_0^\lambda \sin akx \cos bkx\, dx = 0 \qquad [7.44]$$

$$\int_0^\lambda \cos akx \cos bkx\, dx = \frac{\lambda}{2}\delta_{ab} \qquad [7.45]$$

$$\int_0^\lambda \sin akx \sin bkx\, dx = \frac{\lambda}{2}\delta_{ab} \qquad [7.46]$$

where $a \neq 0$, $b \neq 0$, and a and b are positive integers.

7.29 Compute the Fourier series components for the periodic function shown in Fig. 7.20.

7.30* Given the function $f(x) = A \cos(\pi x/L)$, determine its Fourier series.

7.31* Take the function $f(\theta) = \theta^2$ in the interval $0 < \theta < 2\pi$ and assume it repeats itself with a period of 2π. Now show that the Fourier expansion of that function is

$$f(x) = \frac{4\pi^2}{3} + \sum_{m=1}^{\infty}\left(\frac{4}{m^2}\cos m\theta - \frac{4\pi}{m}\sin m\theta\right)$$

7.32* Show that the Fourier series representation of the function $f(\theta) = |\sin\theta|$ is

$$f(\theta) = \frac{2}{\pi} - \frac{4}{\pi}\sum_{m=1}^{\infty}\frac{\cos 2m\theta}{4m^2 - 1}$$

7.33 Change the upper limit of Eq. (7.59) from ∞ to a and evaluate the integral. Leave the answer in terms of the so-called *sine integral*:

$$\text{Si}(z) = \int_0^z \text{sinc}\, w\, dw$$

which is a function whose values are commonly tabulated.

7.34 Write an expression for the transform $A(\omega)$ of the harmonic pulse of Fig. P.7.34. Check that sinc u is 50% or greater for values of u *roughly* less than $\pi/2$. With that in mind, show that $\Delta\nu\,\Delta t \approx 1$, where $\Delta\nu$ is the bandwidth of the transform at half its maximum amplitude. Verify that $\Delta\nu\,\Delta t \approx 1$ at half the maximum value of the power spectrum, as well. The purpose here is to get some sense of the kind of approximations used in the discussion.

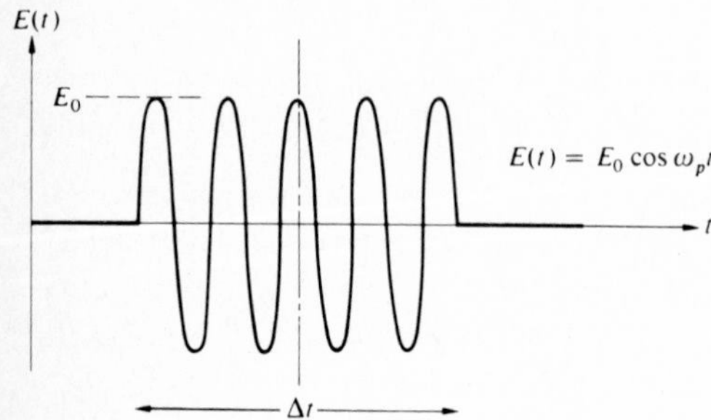

FIGURE P.7.34

7.35 Derive an expression for the coherence length (in vacuum) of a wavetrain that has a frequency bandwidth $\Delta\nu$; express your answer in terms of the *linewidth* $\Delta\lambda_0$ and the mean wavelength $\bar{\lambda}_0$ of the train.

7.36 Consider a photon in the visible region of the spectrum emitted during an atomic transition of about 10^{-8} s. How long is the wave packet? Keeping in mind the results of the previous problem (if you've done it), estimate the linewidth of the packet ($\bar{\lambda}_0 = 500$ nm). What can you say about its monochromaticity, as indicated by the frequency stability?

7.37 The first* experiment directly measuring the bandwidth of a laser (in this case a continuous-wave $Pb_{0.88}Sn_{0.12}$ Te diode laser) was carried out in 1969. The laser, operating at $\lambda_0 = 10\,600$ nm, was heterodyned with a CO_2 laser, and bandwidths as narrow as 54 kHz were observed. Compute the corresponding frequency stability and coherence length for the lead-tin-telluride laser.

7.38* A magnetic-field technique for stabilizing a He–Ne laser to 2 parts in 10^{10} has been patented. At 632.8 nm, what would be the coherence length of a laser with such a frequency stability?

7.39 Imagine that we chop a continuous laserbeam (assumed to be monochromatic at $\lambda_0 = 632.8$ nm) into 0.1-ns pulses, using some sort of shutter. Compute the resultant linewidth $\Delta\lambda$, bandwidth, and coherence length. Find the bandwidth and linewidth that would result if we could chop at 10^{15} Hz.

7.40* Suppose that we have a filter with a pass band of 1.0 Å centered at 600 nm, and we illuminate it with sunlight. Compute the coherence length of the emerging wave.

7.41* A filter passes light with a mean wavelength of $\bar{\lambda}_0 = 500$ nm. If the emerging wavetrains are roughly $20\bar{\lambda}_0$ long, what is the frequency bandwidth of the exiting light?

7.42* Suppose we spread white light out into a fan of wavelengths by means of a diffraction grating and then pass a small select region of that spectrum out through a slit. Because of the width of the slit, a band of wavelengths 1.2 nm wide centered on 500 nm emerges. Determine the frequency bandwidth and the coherence length of this light.

*D. Hinkley and C. Freed, *Phys. Rev. Letters* **23**, 277 (1969).

8 Polarization

8.1 The Nature of Polarized Light

It has already been established that light may be treated as a transverse electromagnetic wave. Thus far we have considered only **linearly polarized** or **plane-polarized** light, that is, light for which the orientation of the electric field is constant, although its magnitude and sign vary in time (Fig. 3.9). In that case, the electric field or optical disturbance resides in what is known as the **plane-of-vibration**. That fixed plane contains both **E** and **k**, the electric field vector and the propagation vector in the direction of motion.

Imagine two harmonic, linearly polarized lightwaves of the same frequency, moving through the same region of space, in the same direction. If their electric field vectors are colinear, the superimposing disturbances will simply combine to form a resultant linearly polarized wave. Its amplitude and phase will be examined in detail, under a diversity of conditions, in the next chapter, when we consider the phenomenon of interference. On the other hand, if the two lightwaves are such that their respective electric-field directions are mutually perpendicular, the resultant wave may or may not be linearly polarized. The exact form the light takes (i.e., its *state of polarization*) and how we can observe it, produce it, change it, and make use of it is the concern of this chapter.

8.1.1 Linear Polarization

The two orthogonal optical disturbances that were considered above can be represented as

$$\mathbf{E}_x(z, t) = \hat{\mathbf{i}}\, E_{0x} \cos (kz - \omega t) \tag{8.1}$$

and

$$\mathbf{E}_y(z, t) = \hat{\mathbf{j}}\, E_{0y} \cos (kz - \omega t + \varepsilon) \tag{8.2}$$

where ε is the relative phase difference between the waves, both of which are traveling in the z-direction. Keep in mind from the start that because the phase is in the form $(kz - \omega t)$, the addition of a *positive* ε means that the cosine function in Eq. (8.2) will not attain the same value as the cosine in Eq. (8.1) until a later time (ε/ω). Accordingly, E_y lags E_x by $\varepsilon > 0$. Of course, if ε is a negative quantity, E_y leads E_x by $\varepsilon < 0$. The resultant optical disturbance is the vector sum of these two perpendicular waves:

$$\mathbf{E}(z, t) = \mathbf{E}_x(z, t) + \mathbf{E}_y(z, t) \tag{8.3}$$

If ε is zero or an integral multiple of $\pm 2\pi$, the waves are said to be in-phase. In that case Eq. (8.3) becomes

$$\mathbf{E} = (\hat{\mathbf{i}} E_{0x} + \hat{\mathbf{j}} E_{0y}) \cos (kz - \omega t) \tag{8.4}$$

The resultant wave has a fixed amplitude equal to $(\hat{\mathbf{i}} E_{0x} + \hat{\mathbf{j}} E_{0y})$; in other words, it too is linearly polarized (Fig. 8.1). The waves advance toward a plane of observation where the fields are to be measured. There one sees a single resultant **E** oscillating, along a tilted line, cosinusoidally in time (Fig. 8.1*b*). The **E**-field progresses through one complete oscillatory cycle as the wave advances along the z-axis through one wavelength. This process can be carried out equally well in reverse; that is, we can resolve any plane-polarized wave into two orthogonal components.

Suppose now that ε is an odd integer multiple of $\pm\pi$. The two waves are 180° out-of-phase, and

$$\mathbf{E} = (\hat{\mathbf{i}} E_{0x} - \hat{\mathbf{j}} E_{0y}) \cos (kz - \omega t) \tag{8.5}$$

This wave is again linearly polarized, but the plane of vibra-

Figure 8.1 Linear light.

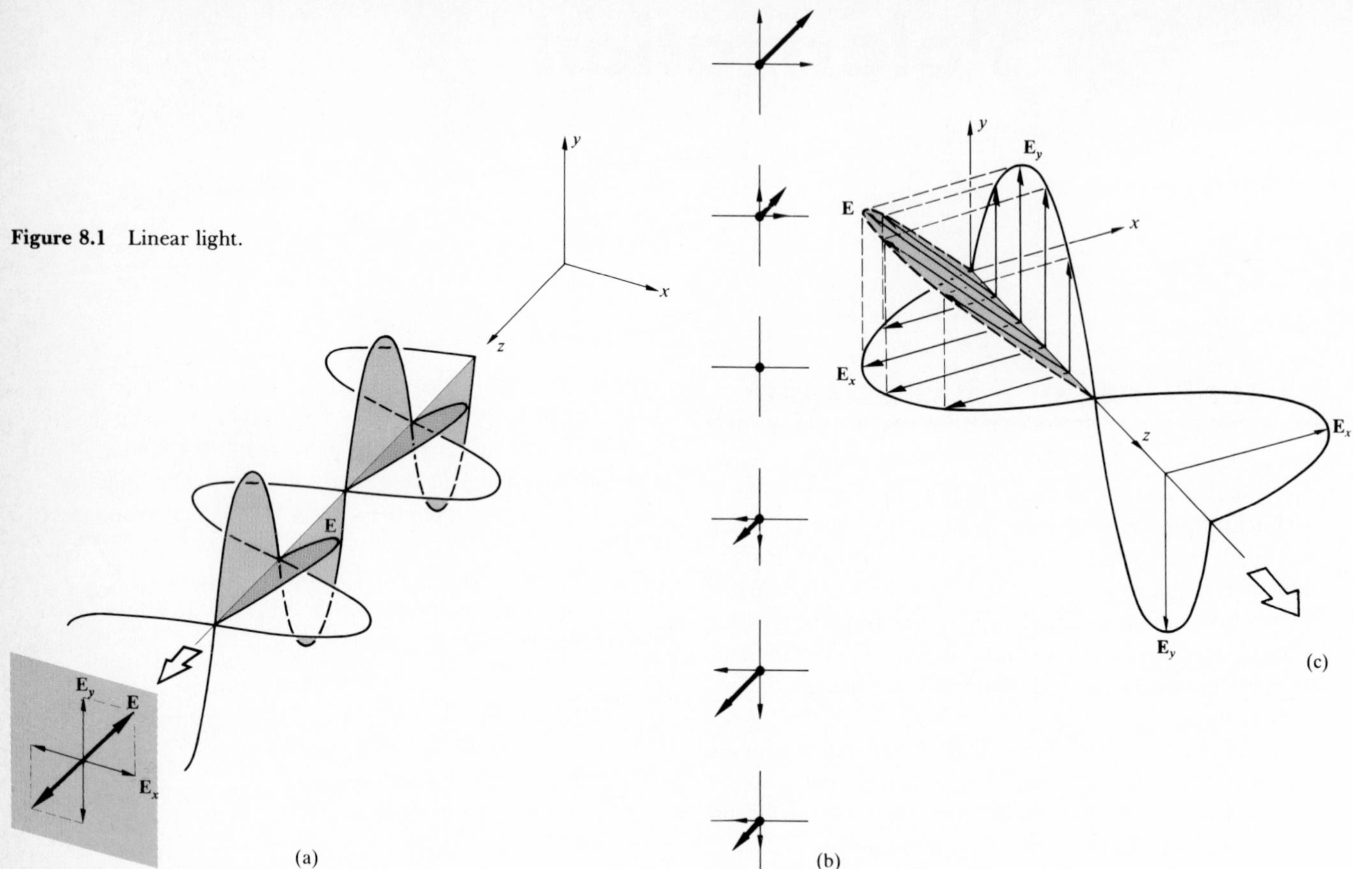

FIGURE 8.1 Linear light. (*a*) The *E*-field linearly polarized in the first and third quadrants. (*b*) That same oscillating field seen head on. (*c*) Light linearly polarized in the second and fourth quadrants.

tion has been rotated (and not necessarily by 90°) from that of the previous condition, as indicated in Fig. 8.2.

8.1.2 Circular Polarization

Another case of particular interest arises when both constituent waves have equal amplitudes (i.e., $E_{0x} = E_{0y} = E_0$), and in addition, their relative phase difference $\varepsilon = -\pi/2 + 2m\pi$, where $m = 0, \pm 1, \pm 2, \ldots$. In other words, $\varepsilon = -\pi/2$ or any value increased or decreased from $-\pi/2$ by whole-number multiples of 2π. Accordingly

$$\mathbf{E}_x(z, t) = \hat{\mathbf{i}} E_0 \cos (kz - \omega t) \tag{8.6}$$

and

$$\mathbf{E}_y(z, t) = \hat{\mathbf{j}} E_0 \sin (kz - \omega t) \tag{8.7}$$

The consequent wave is

$$\mathbf{E} = E_0[\hat{\mathbf{i}} \cos (kz - \omega t) + \hat{\mathbf{j}} \sin (kz - \omega t)] \tag{8.8}$$

(Fig. 8.3). Notice that now the scalar amplitude of **E**, that is, $(\mathbf{E} \cdot \mathbf{E})^{1/2} = E_0$, is a constant. But the direction of **E** is time-varying, and it's not restricted, as before, to a single plane. Figure 8.4 depicts what is happening at some arbitrary point z_0 on the axis. At $t = 0$, **E** lies along the reference axis in Fig. 8.4*a*, and so

$$\mathbf{E}_x = \hat{\mathbf{i}} E_0 \cos kz_0 \quad \text{and} \quad \mathbf{E}_y = \hat{\mathbf{j}} E_0 \sin kz_0$$

At a later time, $t = kz_0/\omega$, $\mathbf{E}_x = \hat{\mathbf{i}} E_0$, $\mathbf{E}_y = 0$, and **E** is along the x-axis. The resultant electric-field vector **E** is rotating *clockwise* at an angular frequency of ω, as seen by an observ-

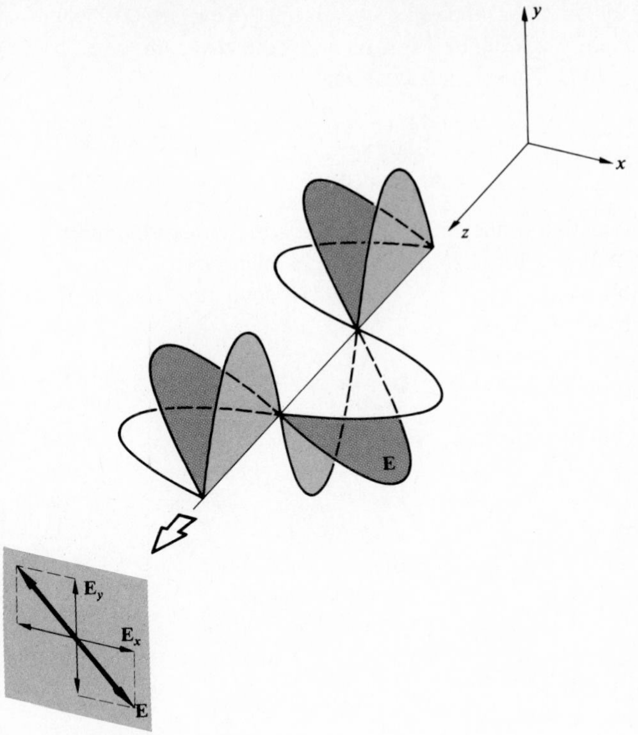

FIGURE 8.2 Linear light.

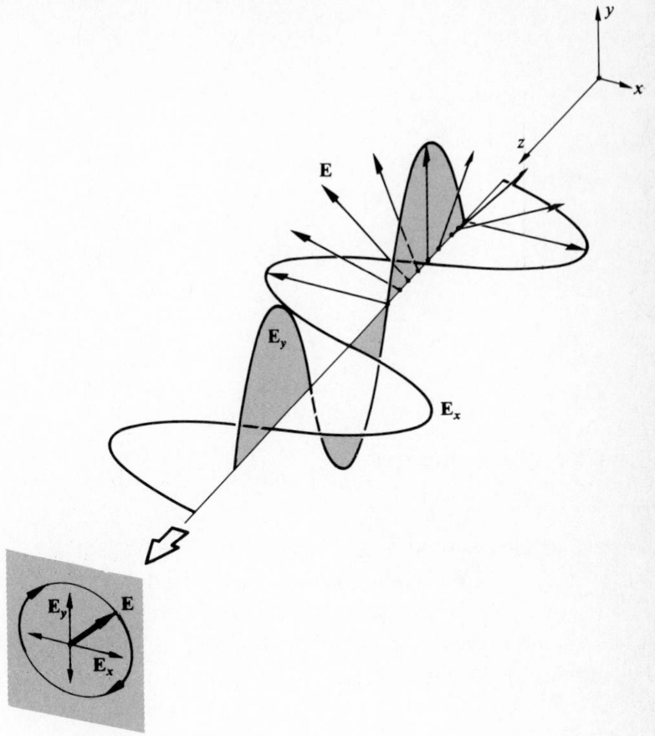

FIGURE 8.3 Right-circular light.

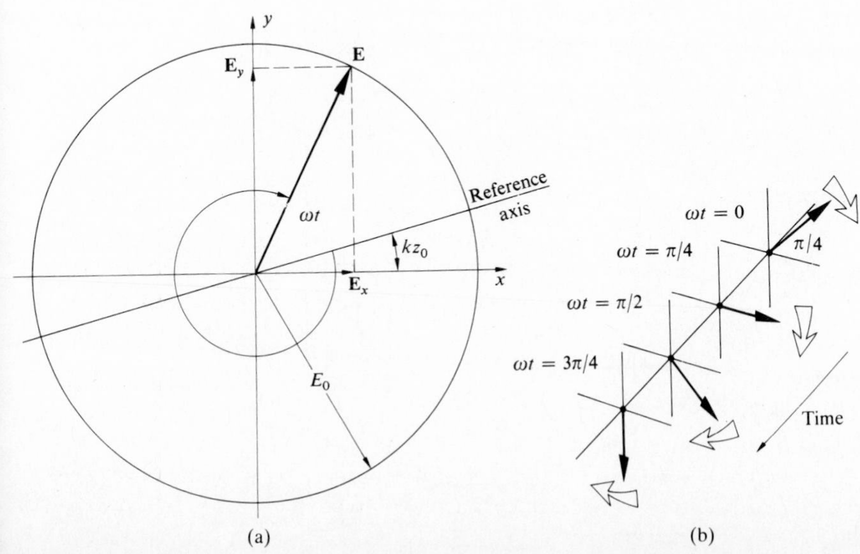

FIGURE 8.4 Rotation of the electric vector in a right-circular wave. Note that the rotation rate is ω and $kz = \pi/4$.

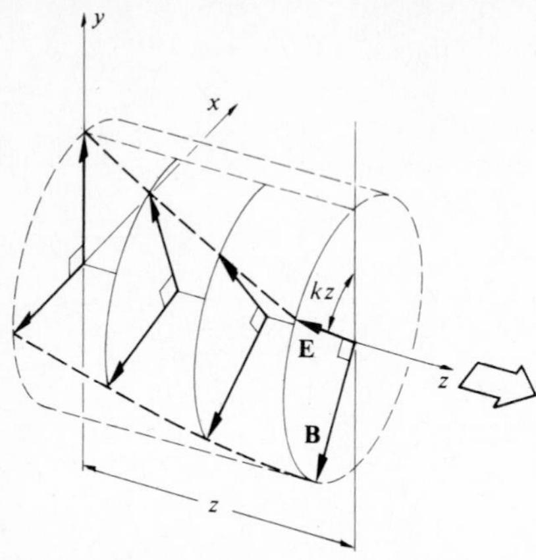

FIGURE 8.5 Right-circular light.

er toward whom the wave is moving (i.e., looking back at the source). Such a wave is **right-circularly polarized** (Fig. 8.5), and one generally simply refers to it as *right-circular light*. The **E**-vector makes one complete rotation as the wave advances through one wavelength. In comparison, if $\varepsilon = \pi/2$, $5\pi/2$, $9\pi/2$, and so on (i.e., $\varepsilon = \pi/2 + 2m\pi$, where $m = 0$, $\pm 1, \pm 2, \pm 3, \ldots$), then

$$\mathbf{E} = E_0[\hat{\mathbf{i}} \cos(kz - \omega t) - \hat{\mathbf{j}} \sin(kz - \omega t)] \tag{8.9}$$

The amplitude is unaffected, but **E** now rotates *counterclockwise*, and the wave is **left-circularly polarized**.

A linearly polarized wave can be synthesized from two oppositely polarized circular waves of equal amplitude. In particular, if we add the right-circular wave of Eq. (8.8) to the left-circular wave of Eq. (8.9), we get

$$\mathbf{E} = 2E_0\hat{\mathbf{i}} \cos(kz - \omega t) \tag{8.10}$$

which has a constant amplitude vector of $2E_0\hat{\mathbf{i}}$ and is therefore linearly polarized.

8.1.3 Elliptical Polarization

As far as the mathematical description is concerned, both linear and circular light may be considered to be special cases of **elliptically polarized** light, or more simply *elliptical light*. This means that, in general, the resultant electric-field vector **E** will rotate, and change its magnitude, as well. In such cases the endpoint of **E** will trace out an ellipse, in a fixed-space perpendicular to **k**, as the wave sweeps by. We can see this better by actually writing an expression for the curve traversed by the tip of **E**. To that end, recall that

$$E_x = E_{0x} \cos(kz - \omega t) \tag{8.11}$$

and

$$E_y = E_{0y} \cos(kz - \omega t + \varepsilon) \tag{8.12}$$

The equation of the curve we are looking for should not be a function of either position or time; in other words, we should be able to get rid of the $(kz - \omega t)$ dependence. Expand the expression for E_y into

$$E_y/E_{0y} = \cos(kz - \omega t)\cos\varepsilon - \sin(kz - \omega t)\sin\varepsilon$$

and combine it with E_x/E_{0x} to yield

$$\frac{E_y}{E_{0y}} - \frac{E_x}{E_{0x}}\cos\varepsilon = -\sin(kz - \omega t)\sin\varepsilon \tag{8.13}$$

It follows from Eq. (8.11) that

$$\sin(kz - \omega t) = [1 - (E_x/E_{0x})^2]^{1/2}$$

so Eq. (8.13) leads to

$$\left(\frac{E_y}{E_{0y}} - \frac{E_x}{E_{0x}}\cos\varepsilon\right)^2 = \left[1 - \left(\frac{E_x}{E_{0x}}\right)^2\right]\sin^2\varepsilon$$

Finally, on rearranging terms, we have

$$\left(\frac{E_y}{E_{0y}}\right)^2 + \left(\frac{E_x}{E_{0x}}\right)^2 - 2\left(\frac{E_x}{E_{0x}}\right)\left(\frac{E_y}{E_{0y}}\right)\cos\varepsilon = \sin^2\varepsilon \tag{8.14}$$

This is the equation of an ellipse making an angle α with the (E_x, E_y)-coordinate system (Fig. 8.6) such that

$$\tan 2\alpha = \frac{2E_{0x}E_{0y}\cos\varepsilon}{E_{0x}^2 - E_{0y}^2} \tag{8.15}$$

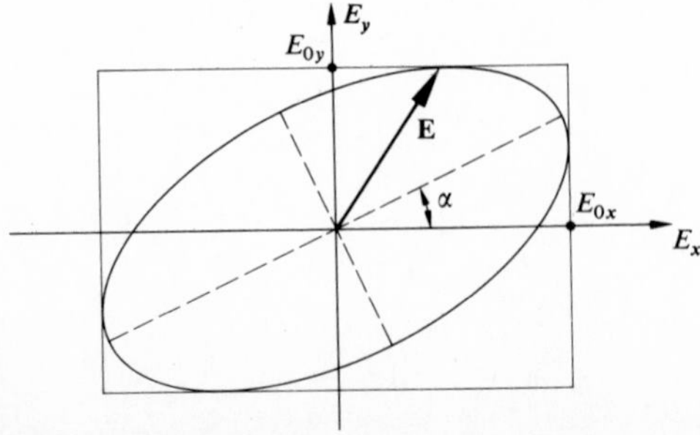

FIGURE 8.6 Elliptical light.

Equation (8.14) might be a bit more recognizable if the principal axes of the ellipse were aligned with the coordinate axes, that is, $\alpha = 0$ or equivalently $\varepsilon = \pm\pi/2, \pm3\pi/2, \pm5\pi/2, \ldots$, in which case we have the familiar form

$$\frac{E_y^2}{E_{0y}^2} + \frac{E_x^2}{E_{0x}^2} = 1 \tag{8.16}$$

Furthermore, if $E_{0y} = E_{0x} = E_0$, this can be reduced to

$$E_y^2 + E_x^2 = E_0^2 \tag{8.17}$$

which, in agreement with our previous results, is a circle. If ε is an even multiple of π, Eq. (8.14) yields

$$E_y = \frac{E_{0y}}{E_{0x}} E_x \tag{8.18}$$

and similarly for odd multiples of π,

$$E_y = -\frac{E_{0y}}{E_{0x}} E_x \tag{8.19}$$

These are both straight lines having slopes of $\pm E_{0y}/E_{0x}$; in other words, we have linear light.

Figure 8.7 diagrammatically summarizes most of these conclusions. This very important diagram is labeled across the bottom "E_x leads E_y by: 0, $\pi/4$, $\pi/2$, $3\pi/4$,...," where these are the positive values of ε to be used in Eq. (8.2). The same set of curves will occur if "E_y leads E_x by: 2π, $7\pi/4$, $3\pi/2$, $5\pi/4$,...," and that happens when ε equals -2π, $-7\pi/4$, $-3\pi/2$, $-5\pi/4$, and so forth. Figure 8.7*b* illustrates how E_x leading E_y by $\pi/2$ is equivalent to E_y leading E_x by $3\pi/2$ (where the sum of these two angles equals 2π). This will be of continuing concern as we go on to shift the relative phases of the two orthogonal components making up the wave.

We are now in a position to refer to a particular lightwave in terms of its specific **state of polarization**. We shall say that linearly polarized or plane-polarized light is in a $\mathscr{P}$-state, and right- or left-circular light is in an $\mathscr{R}$- or $\mathscr{L}$-*state*, respectively. Similarly, the condition of elliptical polarization corresponds to an $\mathscr{E}$-*state*. We've already seen that a $\mathscr{P}$-state can be represented as a superposition of $\mathscr{R}$- and $\mathscr{L}$-states [Eg. (8.10)], and the same is true for an $\mathscr{E}$-state. In this case, as shown in Fig. 8.8, the amplitudes of the two circular waves are different. (An analytical treatment is left for Problem 8.3.)

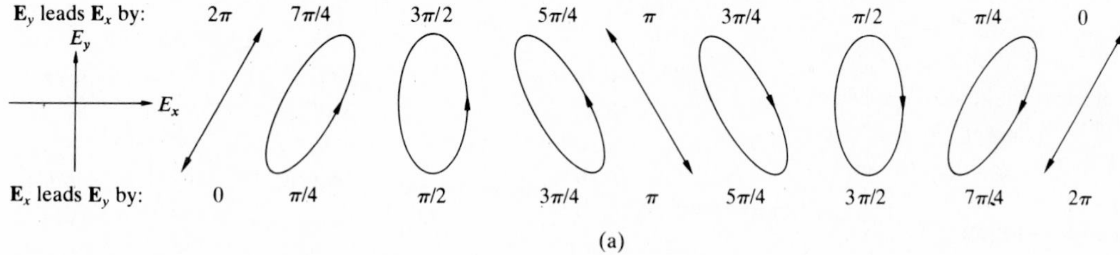

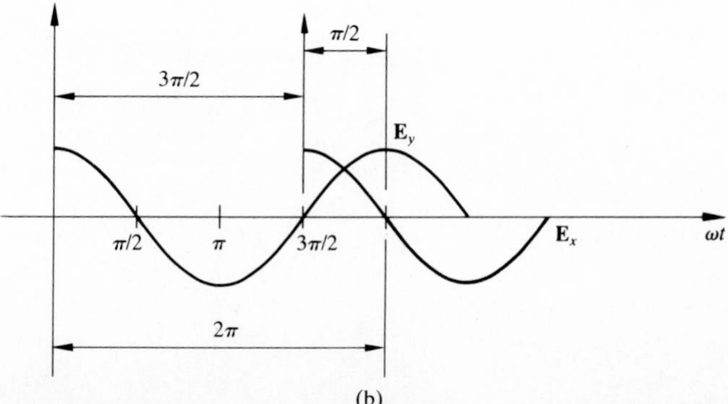

FIGURE 8.7 (*a*) Various polarization configurations. The light would be circular with $\varepsilon = \pi/2$ or $3\pi/2$ if $E_{0x} = E_{0y}$, but here for the sake of generality E_{0y} was taken to be larger than E_{0x}. (*b*) E_x leads E_y (or E_y lags E_x) by $\pi/2$, or alternatively, E_y leads E_x (or E_x lags E_y) by $3\pi/2$.

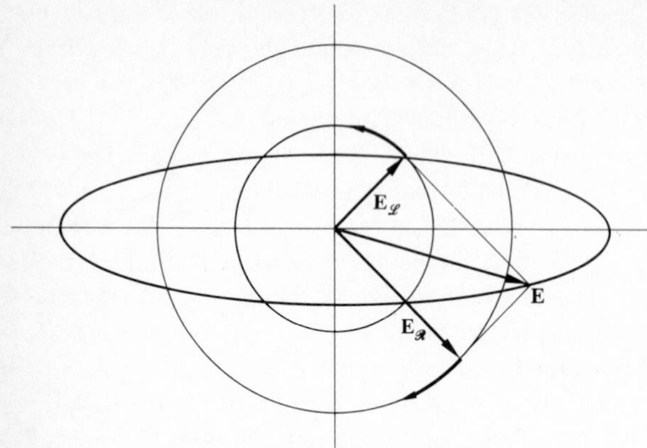

FIGURE 8.8 Elliptical light as the superposition of an $\mathscr{R}$- and $\mathscr{L}$-state.

8.1.4 Natural Light

An ordinary light source consists of a very large number of randomly oriented atomic emitters. Each excited atom radiates a polarized wavetrain for roughly 10^{-8} s. All emissions having the same frequency will combine to form a single resultant polarized wave, which persists for no longer than 10^{-8} s. New wavetrains are constantly emitted, and the overall polarization changes in a completely unpredictable fashion. If these changes take place at so rapid a rate as to render any single resultant polarization state indiscernible, the wave is referred to as **natural light**. It is also known as *unpolarized light*, but this is a misnomer, since in actuality the light is composed of a rapidly varying succession of the different polarization states. *Randomly polarized* is probably a better way to speak of it.

We can mathematically represent natural light in terms of two arbitrary, *incoherent*, orthogonal, linearly polarized waves of equal amplitude (i.e., waves for which the relative phase difference varies rapidly and randomly).

Keep in mind that an idealized monochromatic plane wave must be depicted as an infinite wavetrain. If this disturbance is resolved into two orthogonal components perpendicular to the direction of propagation, they, in turn, must have the same frequency, be infinite in extent, and therefore be mutually coherent (i.e., $\varepsilon =$ constant). In other words, ***a perfectly monochromatic plane wave is always polarized***. In fact, Eqs. (8.1) and (8.2) are just the Cartesian components of a transverse ($E_z = 0$) harmonic plane wave.

Whether natural in origin or artificial, light is generally neither completely polarized nor completely unpolarized; both cases are extremes. More often, the electric-field vector varies in a way that is neither totally regular nor totally irregular, and such an optical disturbance is **partially polarized**. One useful way of describing this behavior is to envision it as the result of the superposition of specific amounts of natural and polarized light.

8.1.5 Angular Momentum and the Photon Picture

We have already seen that an electromagnetic wave impinging on an object can impart both energy and linear momentum to that body. Moreover, if the incident plane wave is circularly polarized, we can expect electrons within the material to be set into circular motion in response to the force generated by the rotating **E**-field. Alternatively, we might picture the field as being composed of two orthogonal $\mathscr{P}$-states that are 90° out-of-phase. These simultaneously drive the electron in two perpendicular directions with a $\pi/2$ phase difference. The resulting motion is again circular. In effect, the torque exerted by the **B**-field averages to zero over an orbit, and the **E**-field drives the electron with an angular velocity ω equal to the frequency of the electromagnetic wave. Angular momentum will thus be imparted by the wave to the substance in which the electrons are imbedded and to which they are bound. We can treat the problem rather simply without actually going into the details of the dynamics. The power delivered to the system is the energy transferred per unit time, $d\mathscr{E}/dt$. Furthermore, the power generated by a torque Γ acting on a rotating body is just $\omega\Gamma$ (which is analogous to vF for linear motion), so

$$\frac{d\mathscr{E}}{dt} = \omega\Gamma \tag{8.20}$$

Since the torque is equal to the time rate-of-change of the angular momentum L, it follows that on the average

$$\frac{d\mathscr{E}}{dt} = \omega \frac{dL}{dt} \tag{8.21}$$

A charge that absorbs a quantity of energy $\mathscr{E}$ from the incident circular wave will simultaneously absorb an amount of angular momentum L such that

$$L = \frac{\mathscr{E}}{\omega} \tag{8.22}$$

If the incident wave is in an $\mathscr{R}$-state, its **E**-vector rotates clockwise, looking toward the source. This is the direction in which a positive charge in the absorbing medium would rotate, and the angular momentum vector is therefore taken to point in the direction opposite to the propagation direction,* as shown in Fig. 8.9.

According to the quantum-mechanical description, an electromagnetic wave transfers energy in quantized packets or photons such that $\mathscr{E} = h\nu$. Thus $\mathscr{E} = \hbar\omega$ (where $\hbar \equiv h/2\pi$), and the *intrinsic* or *spin* angular momentum of a photon is either $-\hbar$ or $+\hbar$, where the signs indicate right- or left-handedness, respectively. Notice that *the angular momentum of a photon is completely independent of its energy*. Whenever a charged particle emits or absorbs electromagnetic radiation, along with changes in its energy and linear momentum, it will undergo a change of $\pm\hbar$ in its angular momentum.†

The energy transferred to a target by an incident monochromatic electromagnetic wave can be envisaged as being transported in the form of a stream of identical photons. We can anticipate a corresponding quantized transport of angular momentum. A purely left-circularly polarized plane wave will impart angular momentum to the target as if all the constituent photons in the beam had their spins aligned in the direction of propagation. Changing the light to right circular reverses the spin orientation of the photons, as well as the torque exerted by them on the target. In 1935, using an extremely sensitive torsion pendulum, Richard A. Beth was actually able to perform such measurements.*

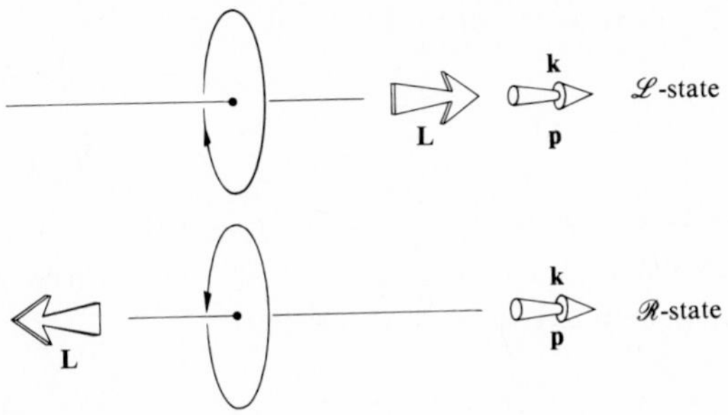

FIGURE 8.9 Angular momentum of a photon.

Thus far we've had no difficulty in describing purely right- and left-circular light in the photon picture; but what is linearly or elliptically polarized light? Classically, light in a $\mathscr{P}$-state can be synthesized by the coherent superposition of equal amounts of light in $\mathscr{R}$- and $\mathscr{L}$-states (with an appropriate phase difference). Any single photon whose angular momentum is somehow measured will be found to have its spin either totally parallel or antiparallel to **k**. A beam of linear light will interact with matter as if it were composed, at that instant, of equal numbers of right- and left-handed photons. There is a subtle point that has to be made here. We cannot say that the beam is actually made up of precisely equal amounts of well-defined right- and left-handed photons; the photons are all identical. Rather, each individual photon exists in either spin state with equal likelihood. If we measured the angular momentum of the constituent photons, $-\hbar$ would result as often as $+\hbar$. This is all we can observe. We are not privy to what the photon is doing before the measurement (if indeed it exists before the measurement). As a whole, a linearly polarized light beam will impart no total angular momentum to a target.

In contrast, if each photon does not occupy both spin states with the same probability, one angular momentum, say $+\hbar$, will be found to occur somewhat more often than the other, $-\hbar$. In this instance, a net positive angular momentum will therefore be imparted to the target. The result *en masse* is elliptically polarized light, that is, a superposition of unequal amounts of $\mathscr{R}$- and $\mathscr{L}$-light bearing a particular phase relationship.

*This choice of terminology is admittedly a bit awkward. Yet its use in Optics is fairly well established, even though it is completely antithetic to the more reasonable convention adopted in elementary particle physics.

†As a rather important yet simple example, consider the hydrogen atom. It is composed of a proton and an electron, each having a spin of $\hbar/2$. The atom has slightly more energy when the spins of both particles are in the same direction. It is possible, however, that once in a very long time, roughly 10^7 years, one of the spins will flip over and be antiparallel to the other. The change in angular momentum of the atom is then $\hbar$, and this is imparted to an emitted photon which carries off the slight excess in energy as well. This is the origin of the 21-cm microwave emission, which is so significant in radio astronomy.

* Richard A. Beth, "Mechanical Detection and Measurement of the Angular Momentum of Light," *Phys. Rev.* **50**, 115 (1936).

8.2 POLARIZERS

Now that we have some idea of what polarized light is, the next logical step is to develop an understanding of the techniques used to generate, change, and manipulate it to fit our needs. An optical device whose input is natural light and whose output is some form of polarized light is a **polarizer**. For example, recall that one possible representation of unpolarized light is the superposition of two equal-amplitude, incoherent, orthogonal $\mathcal{P}$-states. An instrument that separates these two components, discarding one and passing on the other, is known as a *linear polarizer*. Depending on the form of the output, we could also have *circular* or *elliptical polarizers*. All these devices vary in effectiveness down to what might be called leaky or *partial* polarizers.

Polarizers come in many different configurations, but they are all based on one of four fundamental physical mechanisms: *dichroism*, or selective absorption; *reflection*; *scattering*; and *birefringence*, or double refraction. There is, however, one underlying property that they all share: *there must be some form of asymmetry associated with the process.* This is certainly understandable, since the polarizer must somehow select a particular polarization state and discard all others. In truth, the asymmetry may be a subtle one related to the incident or viewing angle, but usually it is an obvious anisotropy in the material of the polarizer itself.

8.2.1 Malus's Law

One matter needs to be settled before we go on: how do we determine experimentally whether or not a device is actually a linear polarizer?

By definition, if natural light is incident on an ideal linear polarizer, as in Fig. 8.10, only light in a $\mathcal{P}$-state will be transmitted. That $\mathcal{P}$-state will have an orientation parallel to a specific direction called the **transmission axis** of the polarizer. Only the component of the optical field parallel to the transmission axis will pass through the device essentially unaffected. If the polarizer in Fig 8.10 is rotated about the z-axis, the reading of the detector (e.g., a photocell) will be unchanged because of the complete symmetry of unpolarized light. Keep in mind that we are dealing with waves, but because of the very high frequency of light, our detector will measure only the incident irradiance. Since the irradiance is proportional to the square of the amplitude of the electric field [Eq. (3.44)], we need only concern ourselves with that amplitude.

Now suppose that we introduce a second identical ideal polarizer, or **analyzer**, whose transmission axis is vertical (Fig. 8.11). If the amplitude of the electric field transmitted by the polarizer is E_0, only its component, $E_0 \cos\theta$, parallel to the transmission axis of the analyzer will be passed on to the detector (assuming no absorption). According to Eq. (3.44), the irradiance reaching the detector is then given by

$$I(\theta) = \frac{c\epsilon_0}{2} E_0^2 \cos^2\theta \tag{8.23}$$

The maximum irradiance, $I(0) = c\epsilon_0 E_0^2/2$, occurs when the angle θ between the transmission axes of the analyzer and polarizer is zero. Equation (8.23) can be rewritten as

$$I(\theta) = I(0) \cos^2\theta \tag{8.24}$$

This is known as **Malus's Law**, having first been published in

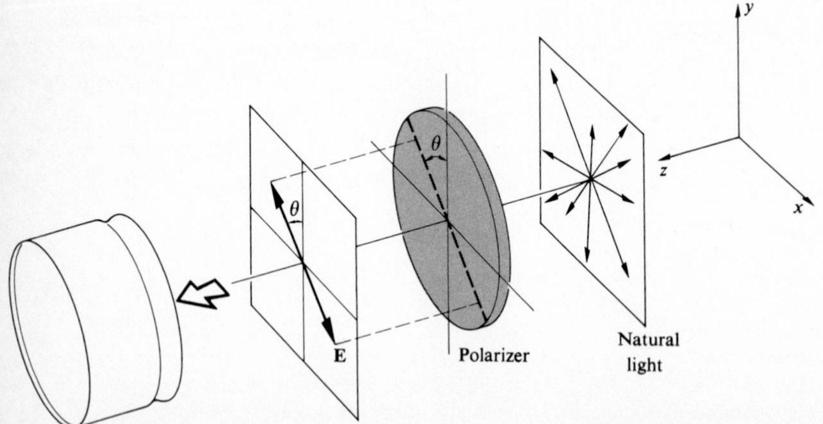

FIGURE 8.10 A linear polarizer.

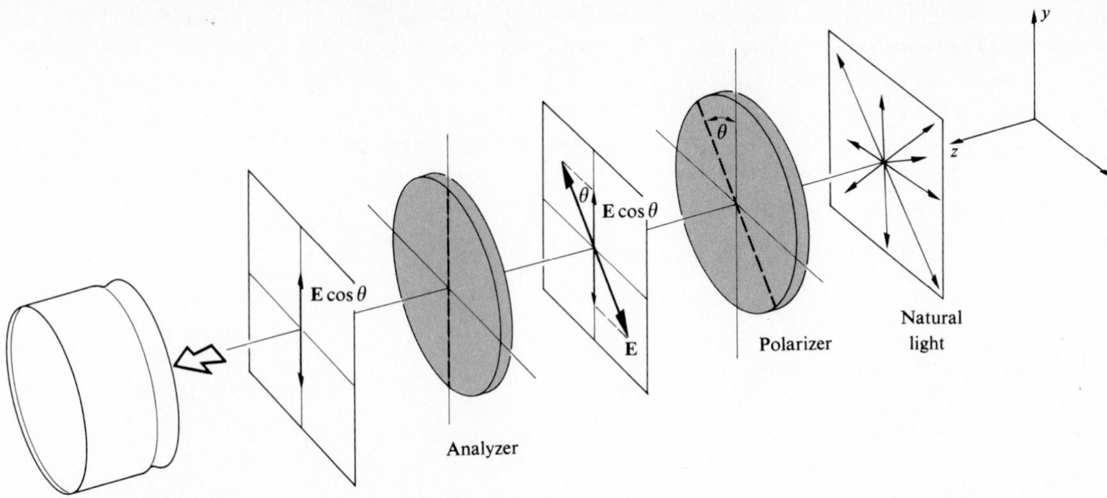

FIGURE 8.11 A linear polarizer and analyzer—Malus's Law.

1809 by Étienne Malus, military engineer and captain in the army of Napoleon.

Observe that $I(90°) = 0$. This arises from the fact that the electric field that passed through the polarizer is perpendicular to the transmission axis of the analyzer (the two devices so arranged are said to be *crossed*). The field is therefore parallel to what is called the *extinction axis* of the analyzer and has no component along the transmission axis. We can use the setup of Fig. 8.11 along with Malus's Law to determine whether a particular device is a linear polarizer.

8.3 Dichroism

In its broadest sense the term **dichroism** refers to the selective absorption of one of the two orthogonal $\mathscr{P}$-state components of an incident beam. The dichroic polarizer itself is physically anisotropic, producing a strong asymmetric or preferential absorption of one field component while being essentially transparent to the other.

8.3.1 The Wire-Grid Polarizer

The simplest device of this sort is a grid of parallel conducting wires, as shown in Fig. 8.12. Imagine that an unpolarized electromagnetic wave impinges on the grid from the right. The electric field can be resolved into the usual two orthogonal components, in this case, one chosen to be parallel to the wires and the other perpendicular to them. The y-component of the field drives the conduction electrons along the length of each wire, thus generating a current. The electrons in turn collide with lattice atoms, imparting energy to them and thereby heating the wires (joule heat). In this manner energy is transferred from the field to the grid. In addition, electrons accelerating along the y-axis radiate in both the forward and backward directions. As should be expected, the incident wave tends to be canceled by the wave reradiated in the forward direction, resulting in little or no transmission of the y-component of the field. The radiation propagating in the backward direction simply appears as a reflected wave. In contrast, the electrons are not free to move very far in the x-direction, and the corre-

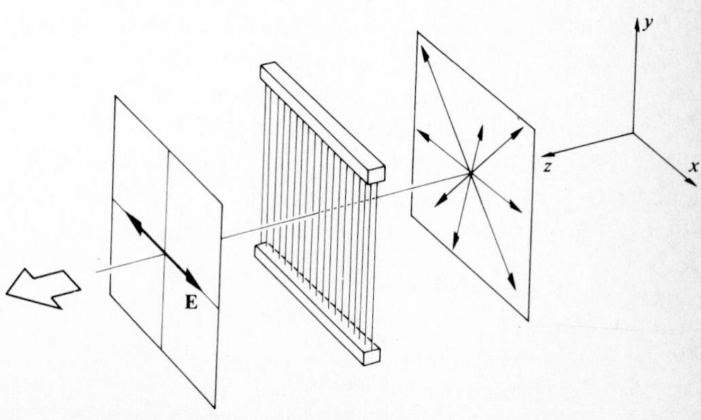

FIGURE 8.12 A wire-grid polarizer.

sponding field component of the wave is essentially unaltered as it propagates through the grid. ***The transmission axis of the grid is perpendicular to the wires***. It is a common error to assume naively that the y-component of the field somehow slips through the spaces between the wires.

One can easily confirm our conclusions using microwaves and a grid made of ordinary electrical wire. It is not so easy a matter, however, to fabricate a grid that will polarize light, but it has been done! In 1960 George R. Bird and Maxfield Parrish, Jr., constructed a grid having an incredible 2160 wires per mm.* Their feat was accomplished by evaporating a stream of gold (or at other times aluminum) atoms at nearly grazing incidence onto a plastic diffraction grating replica (see Section 10.2.7). The metal accumulated along the edges of each step in the grating to form thin microscopic "wires" whose width and spacing were less than one wavelength across.

Although the wire grid is useful, particularly in the infrared, it is mentioned here more for pedagogical than practical reasons. The underlying principle is shared by other, more common, dichroic polarizers.

8.3.2 Dichroic Crystals

Certain materials are inherently dichroic because of an anisotropy in their respective crystalline structures. Probably the best known of these is the naturally occurring mineral *tourmaline*, a semiprecious stone often used in jewelry. Actually there are several tourmalines, which are boron silicates of differing chemical composition [e.g., $NaFe_3B_3Al_6Si_6O_{27}(OH)_4$]. For this substance there is a specific direction within the crystal known as the principal or *optic* axis, which is determined by its atomic configuration. The electric-field component of an incident lightwave that is perpendicular to the principal axis is strongly absorbed by the sample. The thicker the crystal, the more complete the absorption (Fig. 8.13). A plate cut from a tourmaline crystal parallel to its principal axis and several millimeters thick will serve as a linear polarizer. In this instance the crystal's principal axis becomes the polarizer's transmission axis. But the usefulness of tourmaline is rather limited by the fact that its crystals are comparatively small. Moreover, even the transmitted light suffers a certain amount of absorption. To complicate matters, this undesirable absorption is strongly wavelength dependent and the specimen will therefore be colored. A tourmaline crystal held up to natural white light might appear green (they come in other colors as well) when viewed normal to the principal axis and nearly black when

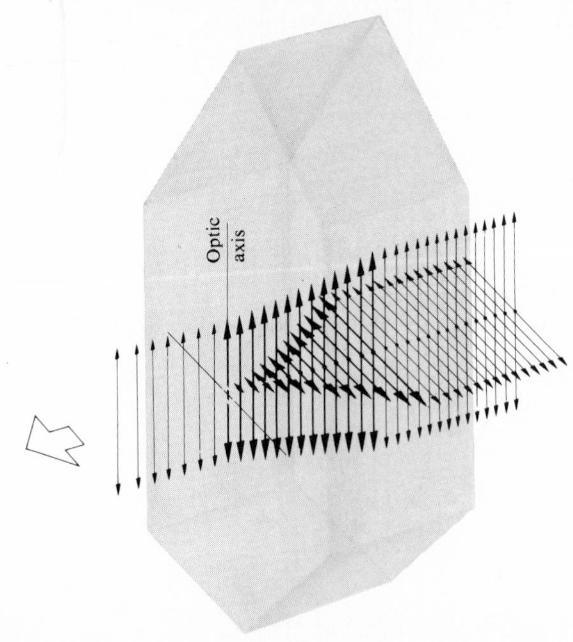

FIGURE 8.13 A dichroic crystal. The naturally occurring ridges evident in the photograph of the tourmaline crystals correspond to the optic axis. (E.H.)

*G. R. Bird and M. Parrish, Jr., "The Wire Grid as a Near-Infrared Polarizer," *J. Opt. Soc. Am.* **50**, 886 (1960).

viewed along that axis, where all the **E**-fields are perpendicular to it (ergo the term dichroic, meaning *two colors*).

There are several other substances that display similar characteristics. A crystal of the mineral hypersthene, a ferromagnesian silicate, might look green under white light polarized in one direction and pink for a different polarization direction.

We can get a qualitative picture of the mechanism that gives rise to crystal dichroism by considering the microscopic structure of the sample. (You might want to take another look at Section 3.5.) Recall that the atoms within a crystal are strongly bound together by short-range forces to form a periodic lattice. The electrons, which are responsible for the optical properties, can be envisioned as elastically tied to their respective equilibrium positions. Electrons associated with a given atom are also under the influence of the surrounding nearby atoms, which themselves may not be symmetrically distributed. As a result, the elastic binding forces on the electrons will be different in different directions. Consequently, their response to the harmonic electric field of an incident electromagnetic wave will vary with the direction of **E**. If in addition to being anisotropic the material is absorbing, a detailed analysis would have to include an orientation-dependent conductivity. Currents will exist, and energy from the wave will be converted into joule heat. The attenuation, in addition to varying in direction, may be dependent on frequency as well. This means that if the incoming white light is in a $\mathscr{P}$-state, the crystal will appear colored, and the color will depend on the orientation of **E**. Substances that display two or even three different colors are said to be dichroic or trichroic, respectively.*

8.3.3 Polaroid

In 1928 Edwin Herbert Land, then a 19-year-old undergraduate at Harvard College, invented the first dichroic sheet polarizer, known commercially as *Polaroid J-sheet*. It incorporated a synthetic dichroic substance called *herapathite*, or *quinine sulfate periodide*.* Land's own retrospective account of his early work is rather informative and makes fascinating reading. It is particularly interesting to follow the sometimes whimsical origins of what is now, no doubt, the most widely used group of polarizers. The following is an excerpt from Land's remarks:

> In the literature there are a few pertinent high spots in the development of polarizers, particularly the work of William Bird Herapath, a physician in Bristol, England, whose pupil, a Mr. Phelps, had found that when he dropped iodine into the urine of a dog that had been fed quinine, little scintillating green crystals formed in the reaction liquid. Phelps went to his teacher, and Herapath then did something which I [Land] think was curious under the circumstances; he looked at the crystals under a microscope and noticed that in some places they were light where they overlapped and in some places they were dark. He was shrewd enough to recognize that here was a remarkable phenomenon, a new polarizing material [now known as herapathite]....
>
> Herapath's work caught the attention of Sir David Brewster, who was working in those happy days on the kaleidoscope.... Brewster, who invented the kaleidoscope, wrote a book about it, and in that book he mentioned that he would like to use herapathite crystals for the eyepiece. When I was reading this book, back in 1926 and 1927, I came across his reference to these remarkable crystals, and that started my interest in herapathite.

Land's initial approach to creating a new form of linear polarizer was to grind herapathite into millions of submicroscopic crystals, which were naturally needle-shaped. Their small size lessened the problem of the scattering of light. In his earliest experiments the crystals were aligned nearly parallel to each other by means of magnetic or electric fields. Later Land found that they would be mechanically aligned when a viscous colloidal suspension of the herapathite needles was extruded through a long narrow slit. The resulting *J*-sheet was effectively a large flat dichroic crystal. The individual submicroscopic crystals still scattered light a bit, and as a result, *J*-sheet was somewhat hazy.

In 1938 Land invented *H-sheet*, which is now probably the most widely used linear polarizer. It does not contain dichroic

*More will be said about these processes later on when we consider birefringence. Suffice it to say now that for crystals classified as *uniaxial* there are two distinct directions, and therefore two colors may be displayed by *absorbing* specimens. In *biaxial* crystals there are three distinct directions and the possibility of three colors.

*E. H. Land, "Some Aspects of the Development of Sheet Polarizers," *J. Opt. Soc. Am.* **41**, 957 (1951).

crystals but is instead a molecular analogue of the wire grid. A sheet of clear polyvinyl alcohol is heated and stretched in a given direction, its long hydrocarbon molecules becoming aligned in the process. The sheet is then dipped into an ink solution rich in iodine. The iodine impregnates the plastic and attaches to the straight long-chain polymeric molecules, effectively forming a chain of its own. The conduction electrons associated with the iodine can move along the chains as if they were long thin wires. The component of **E** in an incident wave that is parallel to the molecules drives the electrons, does work on them, and is strongly absorbed. The transmission axis of the polarizer is therefore perpendicular to the direction in which the film was stretched.

Each separate miniscule dichroic entity is known as a *dichromophore*. In *H*-sheet the dichromophores are of molecular dimensions, so scattering represents no problem. *H*-sheet is a very effective polarizer across the entire visible spectrum, but is somewhat less so at the blue end. When a bright white light is viewed through a pair of crossed *H*-sheet Polaroids, as in Fig. 8.14, the *extinction* color will be a deep blue as a result of this leakage. *HN*-50 would be the designation of a hypothetical, ideal *H*-sheet having a *neutral color* (*N*) and transmitting 50% of the incident natural light while absorbing the other 50%, which is the undesired polarization component. In practice, however, about 4% of the incoming light will be reflected back at each surface (antireflection coatings are not generally used), leaving about 92%. Half of this is presumably absorbed, and thus we might contemplate an *HN*-46 Polaroid. Actually, large quantities of *HN*-38, *HN*-32, and *HN*-22, each differing by the amount of iodine present, are produced commercially and are readily available (Problem 8.9).

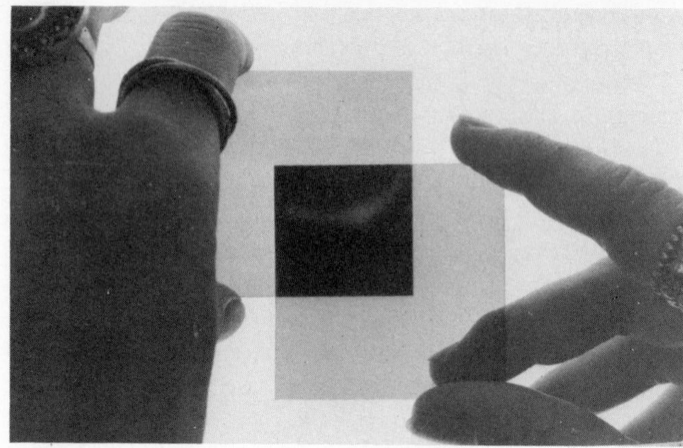

FIGURE 8.14 A pair of crossed polaroids. Each polaroid appears gray because it absorbs roughly half the incident light. (E.H.)

Many other forms of Polaroid have been developed.* *K-sheet,* which is humidity- and heat-resistant, has as its dichromophore the straight-chain hydrocarbon polyvinylene. A combination of the ingredients of *H*- and *K*-sheets leads to *HR-sheet,* a near-infrared polarizer.

Polaroid vectograph is a commercial material at one time designed to be incorporated in a process for making three-dimensional photographs. The stuff never was successful at its intended purpose, but it can be used to produce some rather thought-provoking, if not mystifying, demonstrations. Vectograph film is a water-clear plastic laminate of two sheets of polyvinyl alcohol arranged so that their stretch directions are at right angles to each other. In this form there are no conduction electrons available, and the film is not a polarizer. Using an iodine solution, imagine that we draw an *X* on one side of the film and a *Y* overlapping it on the other. Under natural illumination the light passing through the *X* will be in a $\mathcal{P}$-state perpendicular to the $\mathcal{P}$-state light coming from the *Y*. In other words, the painted regions form two crossed polarizers. They will be seen superimposed on each other. Now, if the vectograph is viewed through a linear polarizer that can be rotated, either the *X*, the *Y*, or both will be seen. Obviously, more imaginative drawings can be made. (One need only remember to make the one on the far side backward.)

*See *Polarized Light: Production and Use*, by Shurcliff, or its more readable little brother, *Polarized Light*, by Shurcliff and Ballard.

8.4 Birefringence

Many crystalline substances (i.e., solids whose atoms are arranged in some sort of regular repetitive array) are *optically anisotropic.* Their optical properties are not the same in all directions within any given sample. The dichroic crystals of the previous section are but one special subgroup. We saw there that if the crystal's lattice atoms were not completely symmetrically arrayed, the binding forces on the electrons would be anisotropic. Earlier, in Fig. 3.35*b* we represented the isotropic oscillator using the simple mechanical model of a spherical charged shell bound by identical springs to a fixed point. This was fine for *optically isotropic* substances (amorphous solids, such as glass and plastic, are usually, but not always, isotropic). Figure 8.15 shows another charged shell, this one bound by springs of differing stiffness (i.e., having different spring constants). An electron that is displaced from equilibrium along a direction parallel to one set of "springs" will evidently oscillate with a different characteristic frequency than it would were it displaced in some other direction.

As was pointed out previously, light propagates through a transparent substance by exciting the atoms within the medium. The electrons are driven by the **E**-field, and they reradiate; these secondary wavelets recombine, and the resultant refracted wave moves on. The speed of the wave, and therefore the index of refraction, is determined by the difference between the frequency of the **E**-field and the natural frequency of the atoms. *An anisotropy in the binding force will be manifest in an anisotropy in the refractive index.* For example, if $\mathscr{P}$-state light was to move through some hypothetical crystal so that it encountered electrons that could be represented by Fig. 8.15, its speed would be governed by the orientation of **E**. If **E** was parallel to the stiff springs, that is, in a direction of strong binding, here along the x-axis, the electron's natural frequency would be high (proportional to the square root of the spring constant). In contrast, with **E** along the y-axis, where the binding force is weaker, the natural frequency would be somewhat lower. Keeping in mind our earlier discussion of dispersion and the $n(\omega)$ curve of Fig. 3.37, the appropriate indices of refraction might look like those in Fig. 8.16. A material of this sort, which displays two different indices of refraction, is said to be **birefringent**.*

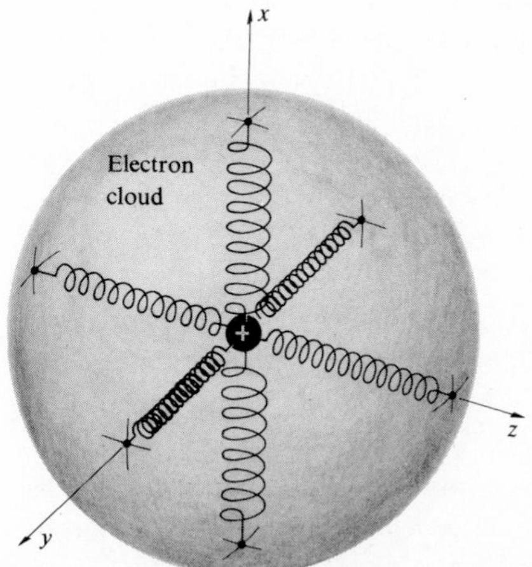

FIGURE 8.15 Mechanical model depicting a negatively charged shell bound to a positive nucleus by pairs of springs having different stiffness.

If the crystal is such that the frequency of the incident light appears in the vicinity of ω_d, in Fig. 8.16, it resides in the absorption band of $n_y(\omega)$. A crystal so illuminated will be strongly absorbing for one polarization direction (y) and transparent for the other (x). A birefringent material that absorbs one of the orthogonal $\mathscr{P}$-states, passing on the other, is *dichroic*. Furthermore, suppose that the crystal symmetry is such that the binding forces in the y- and z-directions are identical; in other words, each of these springs has the same natural frequency and they are equally lossy. The x-axis now defines the direction of the **optic axis**. Inasmuch as a crystal can be represented by an array of these oriented anisotropic charged oscillators, ***the optic axis is actually a direction and not merely a single line***. The model works rather nicely for dichroic crystals, since if light was to propagate along the optic axis (**E** in the yz-plane), it would be strongly absorbed, and if it moved normal to that axis, it would emerge linearly polarized.

Often the natural frequencies of birefringent crystals are above the optical range, and they appear colorless. This is rep-

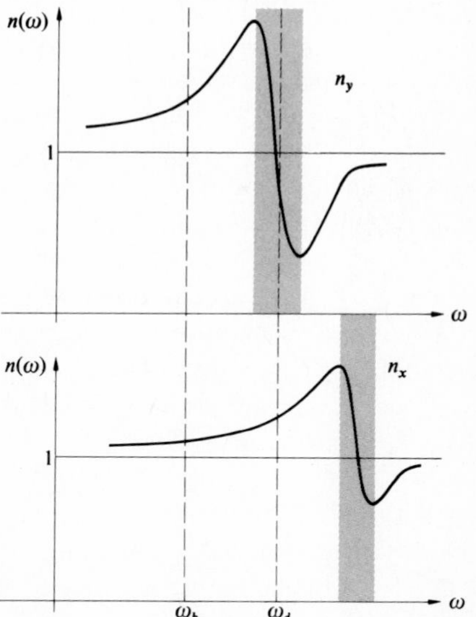

FIGURE 8.16 Refractive index versus frequency along two axes in a crystal. Regions where $dn/d\omega < 0$ correspond to absorption bands.

*The word *refringence* used to be used instead of our present-day term *refraction*. It comes from the Latin *refractus* by way of an etymological route beginning with *frangere*, meaning to break.

resented by Fig. 8.16 where the incident light is now considered to have frequencies in the region of ω_b. Two different indices are apparent, but absorption for either polarization is negligible. Equation (3.71) shows that $n(\omega)$ varies inversely with the natural frequency. This means that a large effective spring constant (i.e., strong binding) corresponds to a low polarizability, a low dielectric constant, and a low refractive index.

We will construct, if only pictorially, a linear polarizer utilizing birefringence by causing the two orthogonal $\mathcal{P}$-states to follow different paths and separate. Even more fascinating things can be done with birefringent crystals, as we shall see later.

8.4.1 Calcite

Let's spend a moment relating the above ideas to a typical birefringent crystal, calcite. Calcite or calcium carbonate ($CaCO_3$) is a common naturally occurring substance. Both marble and limestone are made up of many small calcite crystals bonded together. Of particular interest are the beautiful large single crystals, which, although they are becoming rare, can still be found, particularly in India, Mexico, and South Africa. Calcite is the most common material for making linear polarizers for use with high-power lasers.

Figure 8.17 shows the distribution of carbon, calcium, and oxygen within the calcite structure; Fig. 8.18 is a view from above, looking down along what has, in anticipation, been labeled the optic axis in Fig. 8.17. Each CO_3 group forms a triangular cluster whose plane is perpendicular to the optic axis. If Fig. 8.18 is rotated about a line normal to and passing through the center of any one of the carbonate groups, the same exact configuration of atoms would appear three times during each revolution. The direction designated as the optic axis corresponds to a special crystallographic orientation, in that it is an axis of 3-*fold symmetry*. The large birefringence displayed by calcite arises from the fact that the carbonate groups are all in planes normal to the optic axis. The behavior of their electrons, or rather the mutual interaction of the induced oxygen dipoles, is markedly different when **E** is either in or normal to those planes (Problem 8.24). In any event the asymmetry is clear enough.

Calcite samples can readily be split, forming smooth surfaces known as ***cleavage planes***. The crystal is essentially made to come apart between specific planes of atoms where the interatomic bonding is relatively weak. All cleavage planes in calcite (Fig. 8.18) are normal to three different directions. As a crystal grows, atoms are added layer upon layer, following the same pattern. But more raw material may be available to the growth process on one side than on another, resulting in a crystal with an externally complicated shape. Even so, the cleavage planes are dependent on the atomic configuration, and if one cuts a sample so that each surface is a cleavage plane, its form will be related to the basic arrangement of its atoms. Such a specimen is referred to as a **cleavage form**. In the case of calcite it is a rhombohedron, with each face a parallelogram whose angles are 78° 5′ and 101° 55′ (Fig. 8.19).

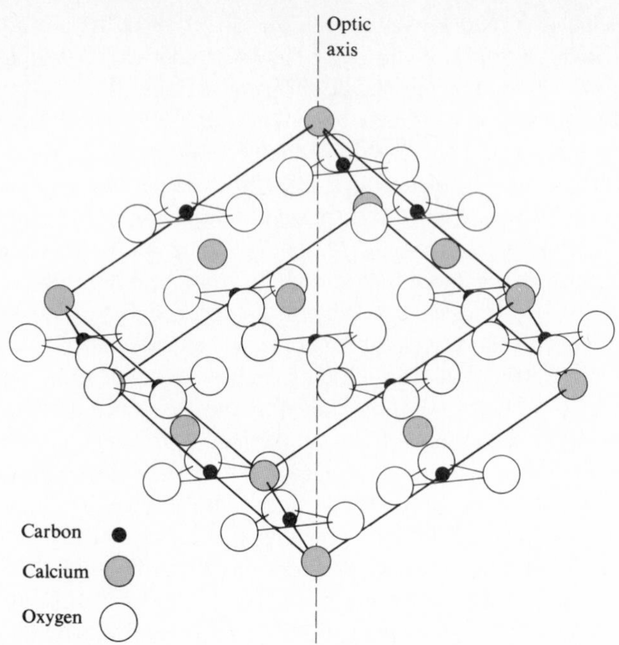

FIGURE 8.17 Arrangement of atoms in calcite.

There are only two *blunt corners* where the surface planes meet to form three obtuse angles. A line passing through the vertex of either of the blunt corners, oriented so that it makes equal angles with each face (45.5°) and each edge (63.8°), is clearly an axis of 3-fold symmetry. (This would be a bit more obvious if we cut the rhomb to have edges of equal length.) Evidently, such a line must correspond to the optic axis. Whatever the natural shape of a particular calcite specimen, you need only find a blunt corner and you have the optic axis.

In 1669 Erasmus Bartholinus (1625–1692), doctor of medicine and professor of mathematics at the University of Copenhagen (and incidentally, Römer's father-in-law), came upon a new and remarkable optical phenomenon in calcite, which he called double refraction. Calcite had been discovered not long before, near Eskifjordur in Iceland, and was then known as ***Iceland spar***. In the words of Bartholinus:*

> Greatly prized by all men is the diamond, and many are the joys which similar treasures bring, such as precious stones and pearls... but he, who, on the other hand, prefers the knowledge of unusual phenomena to these delights, he will, I hope, have no less joy in a new sort of body, namely, a transparent crystal, recently brought to us from Iceland, which perhaps is one of the greatest wonders that nature has produced....
>
> As my investigation of this crystal proceeded there showed itself a wonderful and extraordinary phenomenon: objects which are looked at through the crystal do not show, as in the case of other transparent bodies, a single refracted image, but they appear double.

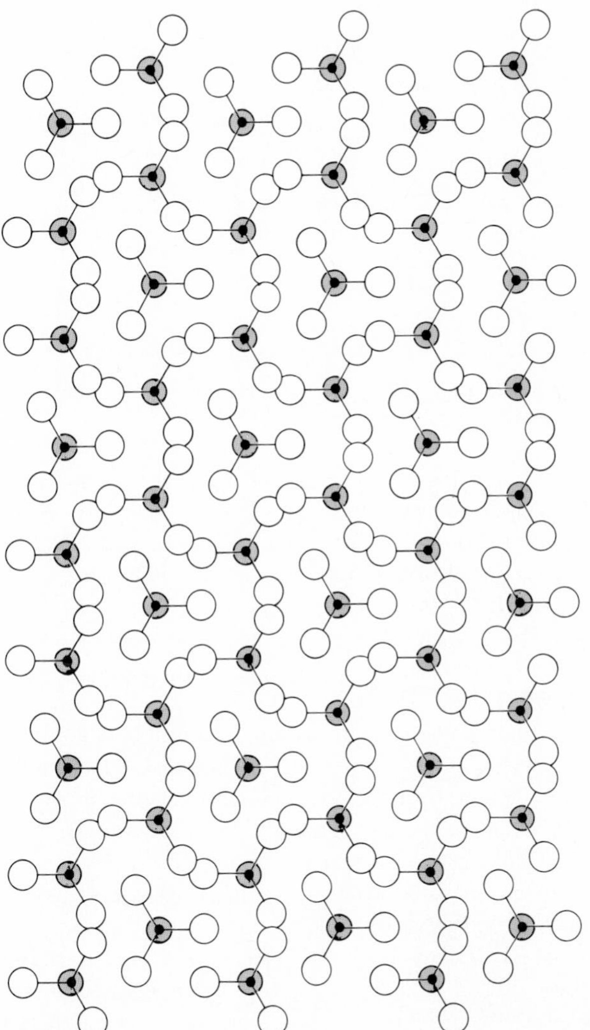

FIGURE 8.18 Atomic arrangement for calcite looking down the optical axis.

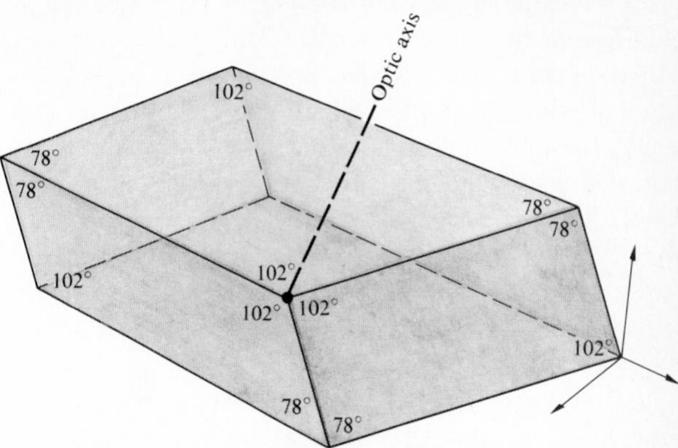

FIGURE 8.19 Calcite cleavage form.

The double image referred to by Bartholinus is quite evident in the photograph in Fig. 8.20. If we send a narrow beam of natural light into a calcite crystal normal to a cleavage

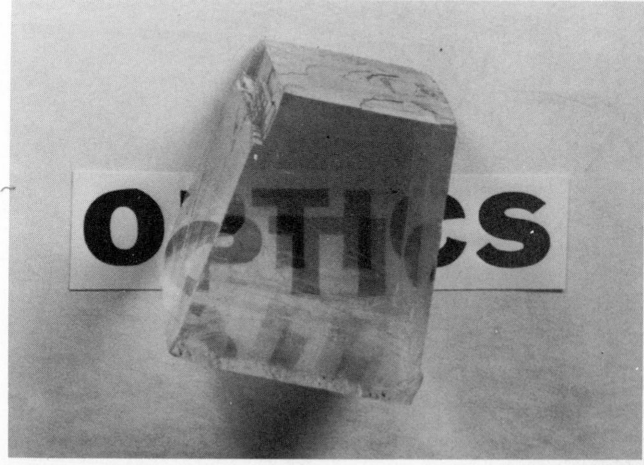

FIGURE 8.20 Double image formed by a calcite crystal (not cleavage form). (E.H.)

*W. F. Magie, *A Source Book in Physics*.

plane, it will split and emerge as two parallel beams. To see the same effect quite simply, we need only place a black dot on a piece of paper and then cover it with a calcite rhomb. The image will now consist of two gray dots (black where they overlap). Rotating the crystal will cause one of the dots to remain stationary while the other appears to move in a circle about it, following the motion of the crystal. The rays forming the fixed dot, which is the one invariably closer to the upper blunt corner, behave as if they had merely passed through a plate of glass. In accord with a suggestion made by Bartholinus, they are known as the **ordinary rays**, or *o-rays*. The rays coming from the other dot, which behave in such an unusual fashion, are known as the **extraordinary rays**, or *e-rays*. If the crystal is examined through an analyzer, it will be found that the ordinary and extraordinary images are linearly polarized (Fig. 8.21). Moreover, the two emerging $\mathscr{P}$-states are orthogonal.

Any number of planes can be drawn through the rhomb so as to contain the optic axis, and these are all called **principal planes**. More specifically, if the principal plane is also normal to a pair of opposite surfaces of the cleavage form, it slices the crystal across a **principal section**. Evidently, three of these pass through any one point; each is a parallelogram having angles of 109° and 71°. Figure 8.22 is a diagrammatic representation of an initially unpolarized beam traversing a principal section of a calcite rhomb. The filled-in circles and arrows drawn along the rays indicate that the *o*-ray has its electric-field vector normal to the principal section, and the field of the *e*-ray is parallel to the principal section.

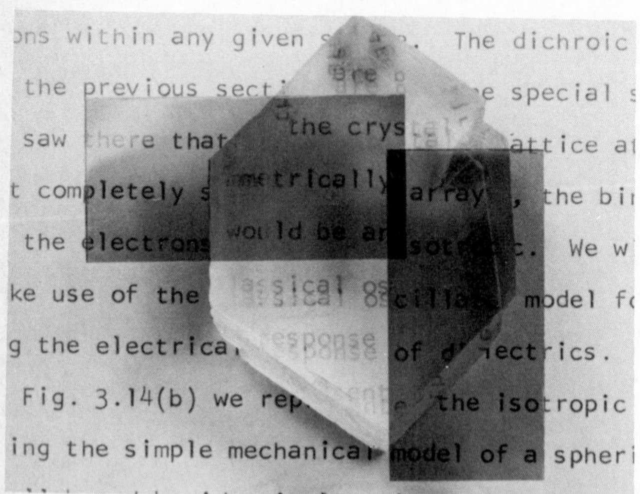

FIGURE 8.21 A calcite crystal (blunt corner on the bottom). The transmission axes of the two polarizers are parallel to their short edges. Where the image is doubled the lower, undeflected one is the ordinary image. Take a long look: there's a lot in this one. (E.H.)

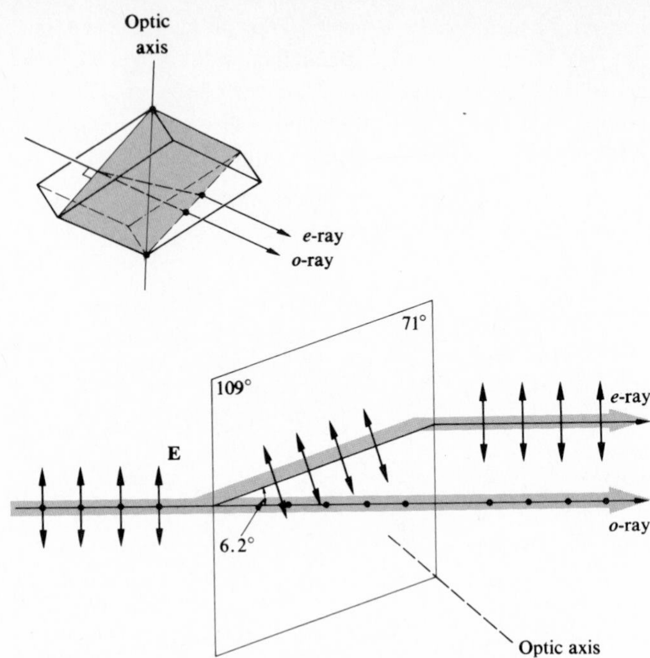

FIGURE 8.22 A light beam with two orthogonal field components traversing a calcite principal section.

To simplify matters a bit, let **E** in the incident plane wave be linearly polarized perpendicular to the optic axis, as shown in Fig. 8.23. The wave strikes the surface of the crystal, thereupon driving electrons into oscillation, and they in turn reradiate secondary wavelets. The wavelets superimpose and recombine to form the refracted wave, and the process is repeated over and over again until the wave emerges from the crystal. This represents a cogent physical argument for applying the ideas of scattering via Huygens's Principle. Huygens himself, though without benefit of electromagnetic theory, used his construction to explain many aspects of double refraction in calcite as long ago as 1690. It should be made clear from the outset, however, that his treatment is incomplete,* in which form it is appealingly, though deceptively, simple.

*A. Sommerfeld, *Optics*, p. 148.

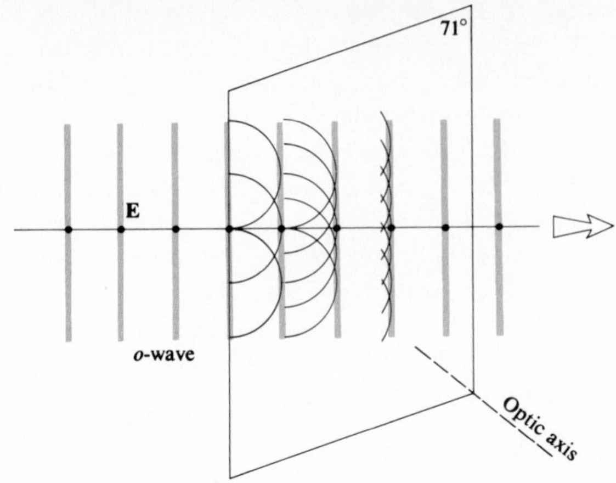

FIGURE 8.23 An incident plane wave polarized perpendicular to the principal section.

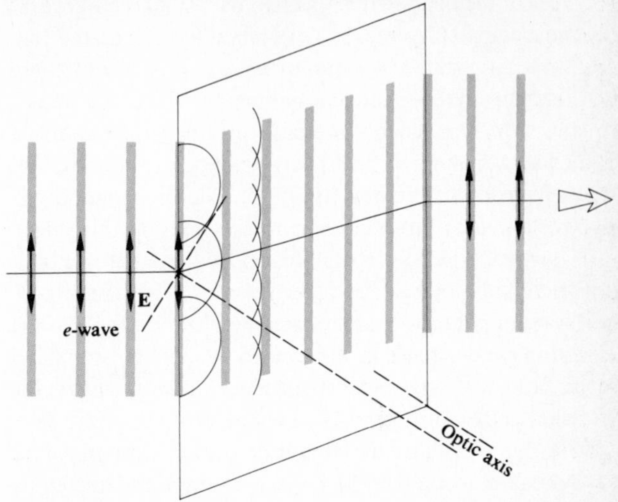

FIGURE 8.24 An incident plane wave polarized parallel to the principal section.

Inasmuch as the **E**-field is perpendicular to the optic axis, one assumes that the wavefront stimulates countless atoms on the surface which then act as sources of spherical wavelets, all of which are in phase. Presumably, as long as the *field of the wavelets is everywhere normal to the optic axis*, they will expand into the crystal in all directions with a speed $v_{\perp}$, as they would in an isotropic medium. (Keep in mind that the speed is a function of frequency.) Since the *o*-wave displays no anomalous behavior, this assumption seems reasonable. The envelope of the wavelets is essentially a portion of a plane wave, which in turn stimulates a distribution of secondary atomic point sources. The process continues, and the wave moves straight across the crystal.

In contrast, consider the incident wave in Fig. 8.24 whose **E**-field is parallel to the principal section. Notice that **E** now has a component normal to the optic axis, as well as a component parallel to it. Since the medium is birefringent, light of a given frequency polarized parallel to the optic axis propagates with a speed $v_{\parallel}$, where $v_{\parallel} \neq v_{\perp}$. In particular for calcite and sodium yellow light ($\lambda = 589$ nm), $1.486v_{\parallel} = 1.658v_{\perp} = c$. What kind of Huygens's wavelets can we expect now? At the risk of oversimplifying matters, we represent each *e*-wavelet, for the moment at least, as a small sphere (Fig. 8.25). But $v_{\parallel} > v_{\perp}$, so that the wavelet will elongate in all directions normal to the optic axis. We therefore speculate, as Huygens did, that the secondary wavelets associated with the *e*-wave are ellipsoids of revolution about the optic axis. The envelope of all the ellipsoidal wavelets is essentially a portion of a plane wave parallel to the incident wave. This plane wave, however, will evidently undergo a sidewise displacement in traversing the crystal. The beam moves in a direction parallel to the lines connecting the origin of each wavelet and the point of tangency with the planar envelope. This is known as the ***ray direction*** *and corresponds to the direction in which energy propagates. Clearly, in an anisotropic crystal the direction of the ray is not normal to the wavefront.*

If the incident beam is natural light, the two situations depicted in Figs. 8.23 and 8.24 will exist simultaneously, with

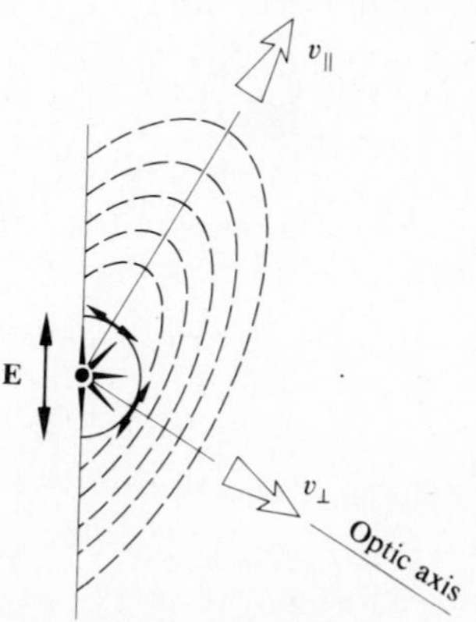

FIGURE 8.25 Wavelets within calcite.

the result that the beam will split into two orthogonal linearly polarized beams (Fig. 8.22). You can actually see the two diverging beams within a crystal by using a properly oriented narrow laserbeam (**E** neither normal nor parallel to the principal plane, which is usually the case). Light will scatter off internal flaws, making its path fairly visible.

The electromagnetic description of what is happening is rather complicated but well worth examining at this point, even if only superficially. Recall from Chapter 3 that the incident **E**-field will polarize the dielectric; that is, it will shift the distribution of charges, thereby creating electric dipoles. The field within the dielectric is thus altered by the inclusion of an induced field, and one is led to introduce a new quantity, the *displacement* **D** (see Appendix 1). In isotropic media **D** is related to **E** by a scalar quantity, and the two are therefore always parallel. In anisotropic crystals **D** and **E** are related by a tensor and are not always parallel. If we now apply Maxwell's Equations to the problem of a wave moving through such a medium, we find that the fields vibrating within the wavefront are **D** and **B** and not, as before, **E** and **B**. The propagation vector **k**, which is normal to the surfaces of constant phase, is now perpendicular to **D** rather than **E**. In fact, **D**, **E**, and **k** are all coplanar. Clearly then, the *ray direction* corresponds to the direction of the Poynting vector $\mathbf{S} = v^2\epsilon\mathbf{E}\times\mathbf{B}$, which is generally different from that of **k**. Because of the manner in which the atoms are distributed, **E** and **D** will, however, be colinear when they are both either parallel or perpendicular to the optic axis.* This means that the *o*-wavelet will encounter an effectively isotropic medium and thus be spherical, having **S** and **k** colinear. In contrast, the *e*-wavelets will have **S** and **k**, or equivalently **E** and **D**, parallel only in directions along or normal to the optic axis. At all other points on the wavelet it is **D** that is tangent to the ellipsoid, and therefore it is always **D** that ends up in the envelope or composite planar wavefront within the crystal (Fig. 8.26).

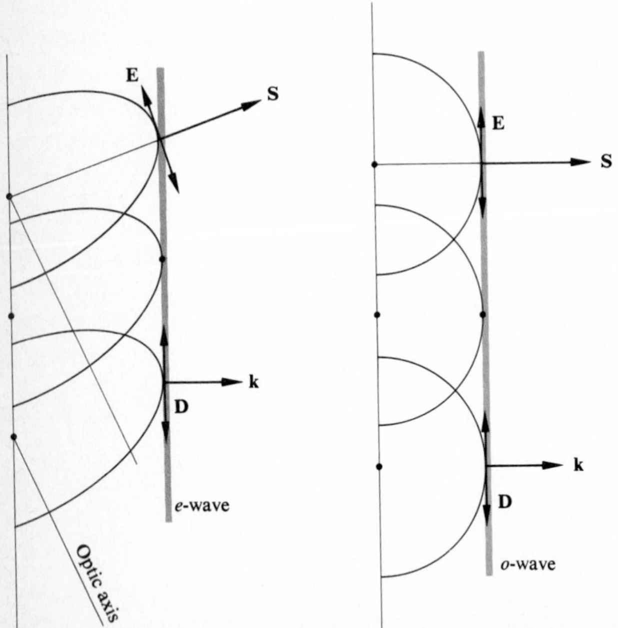

FIGURE 8.26 Orientations of the **E**-, **D**-, **S**-, and **k**-vectors.

8.4.2 Birefringent Crystals

Cubic crystals, such as sodium chloride (i.e., common salt), have their atoms arranged in a relatively simple and highly symmetric form. (There are *four* 3-fold symmetry axes, each running from one corner to an opposite corner, unlike calcite, which has one such axis.) Light emanating from a point source within such a crystal will propagate uniformly in all directions as a spherical wave. As with amorphous solids, there will be no preferred directions in the material. It will have a single index of refraction and be *optically isotropic* (Fig. 8.27). In that case all the springs in the oscillator model will evidently be identical.

Crystals belonging to the *hexagonal*, *tetragonal*, and *trigonal* systems have their atoms arranged so that light propagating in some general direction will encounter an asymmetric structure. Such substances are optically anisotropic and birefringent. The optic axis corresponds to a direction about which the atoms are arranged symmetrically. Crystals like these, for which there is only one such direction, are known as ***uniaxial***.

A point source of natural light embedded within one of these specimens gives rise to spherical *o*-wavelets and ellipsoidal *e*-wavelets. It is the orientation of the field with respect to the optic axis that determines the speeds with which these wavelets expand. The **E**-*field of the o-wave is everywhere nor-*

*In the oscillator model, the general case corresponds to the situation in which **E** is not parallel to any of the spring directions. The field will drive the charge, but its resultant motion will not be in the direction of **E** because of the anisotropy of the binding forces. The charge will be displaced most, for a given force component, in the direction of weakest restraint. The induced field will thus not have the same orientation as **E**.

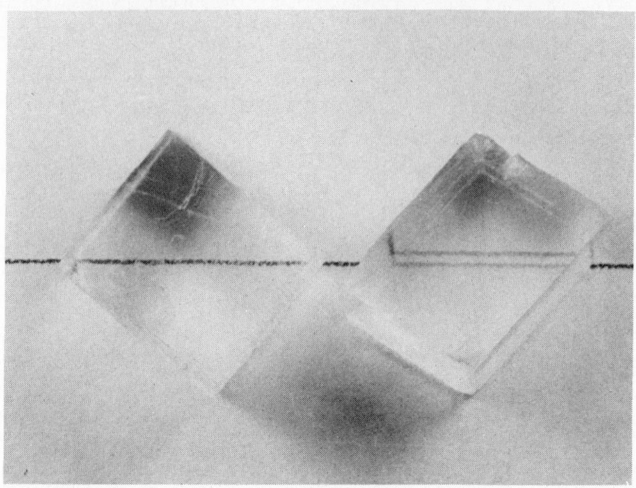

FIGURE 8.27 Images in sodium chloride and calcite single crystals. (E.H.)

mal to the optic axis, so it moves at a speed $v_\perp$ in all directions. Similarly, the *e*-wave has a speed $v_\perp$ only in the direction of the optic axis (Fig. 8.25), along which it is always tangent to the *o*-wave. Normal to this direction, **E** *is parallel to the optic axis*, and that portion of the wavelet expands at a speed $v_\parallel$ (Fig. 8.28). Uniaxial materials have two principal indices of refraction, $n_o \equiv c/v_\perp$ and $n_e \equiv c/v_\parallel$ (Problem 8.35) as indicated in Table 8.1.

TABLE 8.1 Refractive indices of some uniaxial birefringent crystals (λ_0 = 589.3 nm).

Crystal	n_o	n_e
Tourmaline	1.669	1.638
Calcite	1.6584	1.4864
Quartz	1.5443	1.5534
Sodium nitrate	1.5854	1.3369
Ice	1.309	1.313
Rutile (TiO_2)	2.616	2.903

The difference $\Delta n = (n_e - n_o)$ is a measure of the birefringence. In calcite $v_\parallel > v_\perp$, $(n_e - n_o)$ is -0.172, and it is *negative uniaxial*. In comparison, there are other crystals, such as quartz (crystallized silicon dioxide) and ice, for which $v_\perp > v_\parallel$. Consequently, the ellipsoidal *e*-wavelets are enclosed within the spherical *o*-wavelets, as shown in Fig. 8.29. (Quartz is optically active and therefore actually a bit more complicated.) In that case, $(n_e - n_o)$ is positive, and the crystal is *positive uniaxial*.

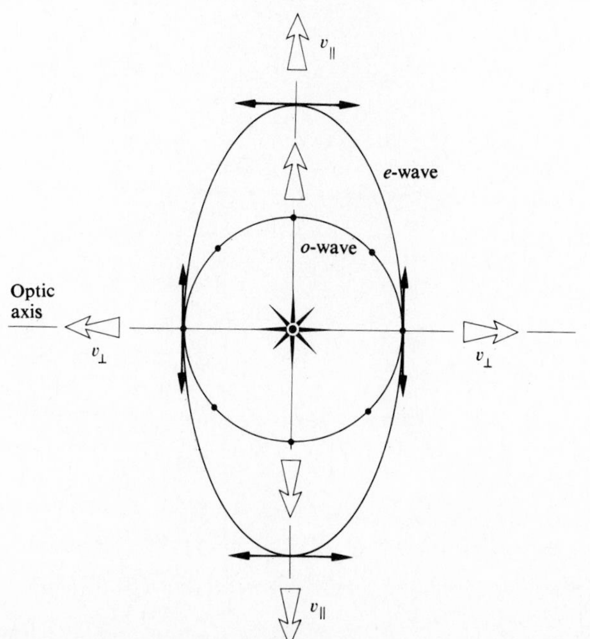

FIGURE 8.28 Wavelets in a negative uniaxial crystal.

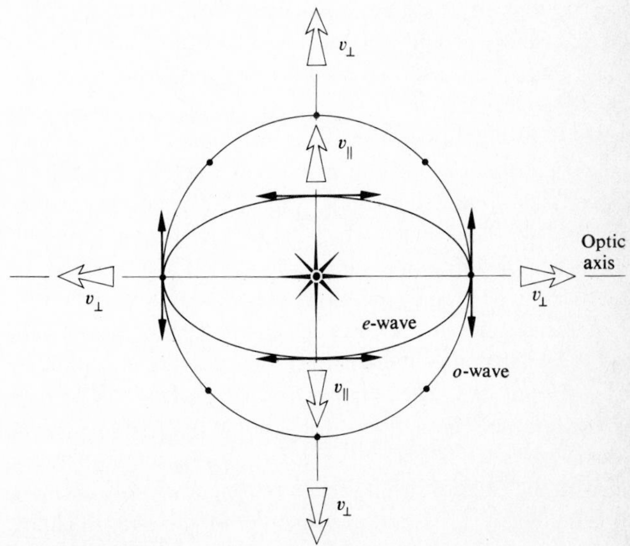

FIGURE 8.29 Wavelets in a positive uniaxial crystal.

The remaining crystallographic systems, namely *orthorhombic*, *monoclinic*, and *triclinic*, have two optic axes and are ***biaxial***. Such substances, for example, mica $[KH_2Al_3(SiO_4)_3]$, have three different principal indices of refraction. Each set of springs in the oscillator model would then be different. The birefringence of biaxial crystals is measured as the numerical difference between the largest and smallest of these indices.

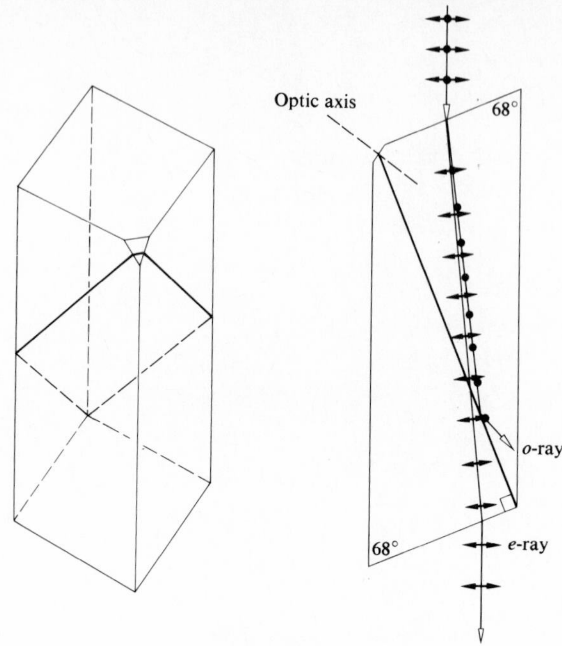

8.4.3 Birefringent Polarizers

It will now be an easy matter, at least conceptually, to make some sort of linear birefringent polarizer. Any number of schemes for separating the *o*- and *e*-waves have been employed, all of them relying on the fact that $n_e \neq n_o$.

The most renowned birefringent polarizer was introduced in 1828 by the Scottish physicist William Nicol (1768–1851). The *Nicol prism* is now mainly of historical interest, having long been superseded by other, more effective polarizers. Putting it rather succinctly, the device is made by first grinding and polishing the ends (from 71° to 68°; see Fig. 8.23) of a suitably long, narrow calcite rhombohedron; after cutting the rhomb diagonally, the two pieces are polished and cemented back together with Canada balsam (Fig. 8.30). The balsam cement is transparent and has an index of 1.55 almost midway between n_e and n_o. The incident beam enters the "prism," the *o*- and *e*-rays are refracted; they separate and strike the balsam layer. The critical angle at the calcite–balsam interface for the *o*-ray is about 69° (Problem 8.37). The *o*-ray (entering within a narrow cone of roughly 28°) will be totally internally reflected and thereafter absorbed by a layer of black paint on the sides of the rhomb. The *e*-ray emerges laterally displaced but otherwise essentially unscathed, at least in the optical region of the spectrum. (Canada balsam absorbs in the ultraviolet.)

FIGURE 8.30 The Nicol prism. The little flat on the blunt corner locates the optic axis. (E.H.)

The *Glan–Foucault polarizer* (Fig. 8.31) is constructed of nothing other than calcite, which is transparent from roughly 5000 nm in the infrared to about 230 nm in the ultraviolet. It therefore can be used over a broad spectral range. The incoming ray strikes the surface normally, and **E** can be resolved into components that are either completely parallel or perpendicular to the optic axis. The two rays traverse the first calcite section without any deviation. (We'll come back to this point later on when we talk about retarders.) If the angle-of-incidence on the calcite–air interface is θ, one need only arrange things so that $n_e < 1/\sin\theta < n_o$ in order for the *o*-ray, and not the *e*-ray, to be totally internally reflected. If the two prisms are now cemented together (glycerine or mineral oil are used

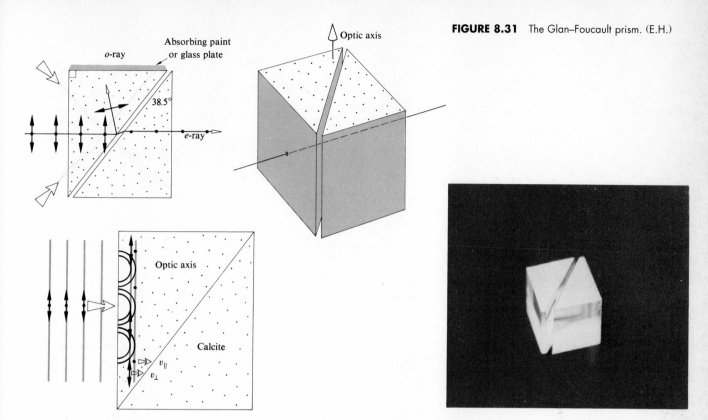

FIGURE 8.31 The Glan–Foucault prism. (E.H.)

in the ultraviolet) and the interface angle is changed appropriately, the device is known as a *Glan–Thompson polarizer*. Its field of view is roughly 30°, in comparison to about 10° for the Glan–Foucault, or *Glan–Air*, as it is often called. The latter, however, has the advantage of being able to handle the considerably higher power levels often encountered with lasers. For example, whereas the maximum irradiance for a Glan–Thompson could be about 1 W/cm^2 (continuous wave as opposed to pulsed), a typical Glan–Air might have an upper limit of 100 W/cm^2 (continuous wave). The difference is due to deterioration of the interface cement (and the absorbing paint, if it's used).

The *Wollaston* prism is a polarizing beamsplitter, because it passes both orthogonally polarized components. It can be made of calcite or quartz in the form indicated in Fig. 8.32. The two component rays separate at the diagonal interface. There, the *e*-ray becomes an *o*-ray, changing its index accordingly. In calcite $n_e < n_o$, and the emerging *o*-ray is bent toward the normal. Similarly, the *o*-ray, whose field is initially perpendicular to the optic axis, becomes an *e*-ray in the right-hand section. This time, in calcite the *e*-ray is bent away from the

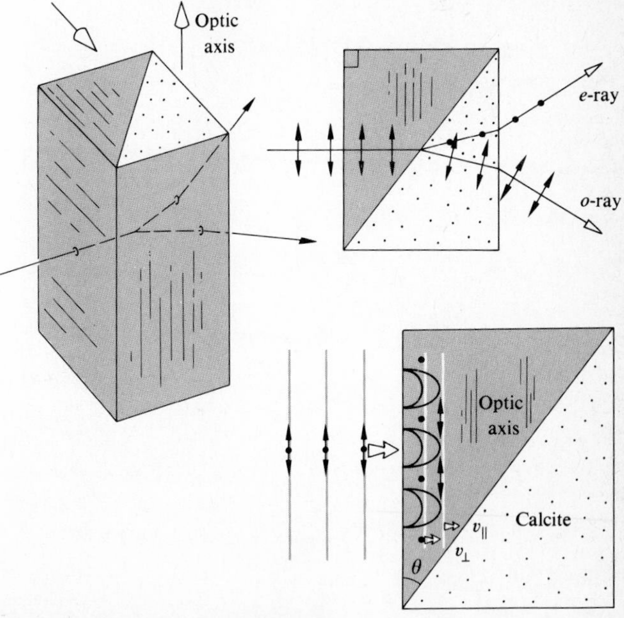

FIGURE 8.32 The Wollaston prism.

normal to the interface (see Problem 8.38). The deviation angle between the two emerging beams is determined by the prism's wedge angle, θ. Prisms providing deviations ranging from about 15° to roughly 45° are available commercially. They can be purchased cemented (e.g., with castor oil or glycerine) or not cemented at all (i.e., optically contacted), depending on the frequency and power requirements.

FIGURE 8.34 Scattering of sky light.

8.5 Scattering and Polarization

Sunlight streaming into the atmosphere from one direction is scattered in all directions by the air molecules (see Section 4.2). Without an atmosphere, the daytime sky would be as black as the void of space, a point well made in the Apollo lunar photographs (Fig. 8.33). You would then see only light that shone directly at you. With an atmosphere, the red end of the spectrum is, for the most part, undeviated, whereas the blue or high-frequency end is substantially scattered. This high-frequency scattered light reaches the observer from many directions, making the entire sky appear bright and blue (Fig. 8.34).

FIGURE 8.33 A half-Earth hanging in the black Moon sky. (Photo courtesy of NASA)

The smoke rising from the end of a lighted cigarette is made up of particles that are smaller than the wavelength of light, making it appear blue when seen against a dark background. In contrast, exhaled smoke contains relatively large water droplets and appears white. Each droplet is larger than the constituent wavelengths of light and thus contains so many oscillators that it is able to sustain the ordinary processes of reflection and refraction. These effects are not preferential to any one frequency component in the incident white light.

The light reflected and refracted several times by a droplet and then finally returned to the observer is therefore also white. This accounts for the whiteness of small grains of salt and sugar, fog, clouds, paper, powders, ground glass, and, more ominously, the typical pallid, polluted city sky.

Particles that are approximately the size of a wavelength (remember that atoms are roughly a fraction of a nanometer across) scatter light in a very distinctive way. A large distribution of such equally sized particles can give rise to a whole range of transmitted colors. In 1883 the volcanic island Krakatoa, located in the Sunda Strait west of Java, blew apart in a fantastic conflagration. Great quantities of fine volcanic dust were spewed high into the atmosphere and drifted over vast regions of the Earth. For a few years afterward the Sun and Moon repeatedly appeared green or blue, and sunrises and sunsets were abnormally colored.

8.5.1 Polarization by Scattering

Imagine a linearly polarized plane wave incident on an air molecule, as pictured in Fig. 8.35. The orientation of the electric field of the scattered radiation (i.e., $\mathbf{E}_s$) follows the dipole pattern such that $\mathbf{E}_s$, the Poynting vector $\mathbf{S}$, and the oscillating dipole are all coplanar (Fig. 3.31). The vibrations induced in the atom are parallel to the **E**-field of the incoming lightwave and so are perpendicular to the propagation direction. Observe once again that the dipole does not radiate in the direction of its axis. Now if the incident wave is unpolarized, it can be represented by two orthogonal, incoherent $\mathscr{P}$-states, in which case the scattered light (Fig. 8.36) is equivalent to a superposition of the conditions shown in Fig. 8.35, *a* and *b*. Evidently, the scattered light in the forward direction is completely unpolarized; off that axis it is partially polarized, becoming increasingly more polarized as the angle increases. When the direction of observation is normal to the primary beam, the light is completely linearly polarized.

You can easily verify these conclusions with a piece of Polaroid. Locate the Sun and then examine a region of the sky at roughly 90° to the solar rays. That portion of the sky will be partially polarized normal to the rays (see Fig. 8.37). It's not completely polarized mainly because of molecular anisotropies, the presence of large particles in the air, and the depolarizing effects of multiple scattering. The latter condition can be illustrated by placing a piece of waxed paper between crossed Polaroids (Fig. 8.38). Because the light undergoes a good deal of scattering and multiple reflections within the waxed paper, a given oscillator may "see" the superposition of many essentially unrelated **E**-fields. The resulting emission is almost completely depolarized.

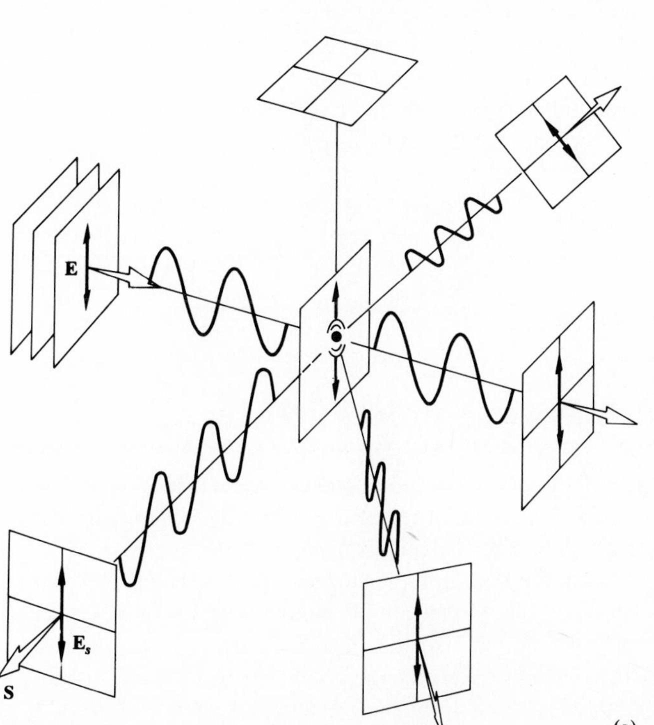

FIGURE 8.35a Scattering of polarized light by a molecule.

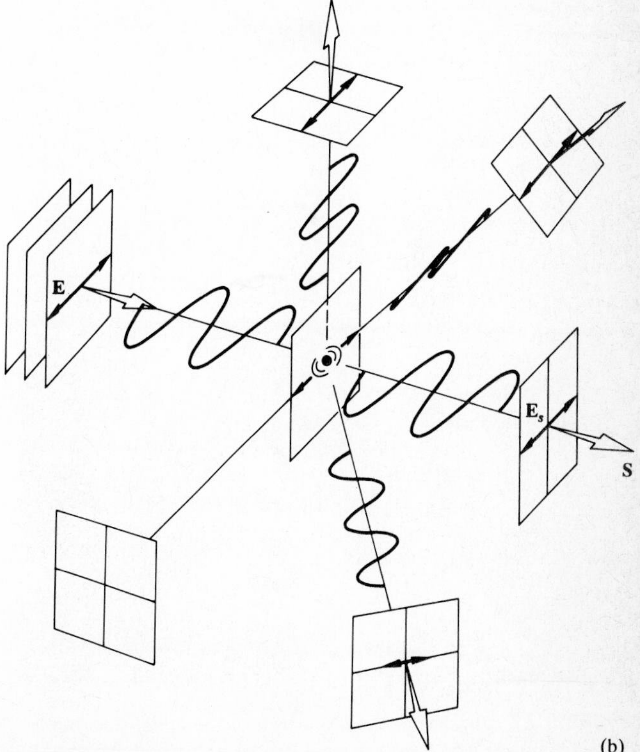

FIGURE 8.35b

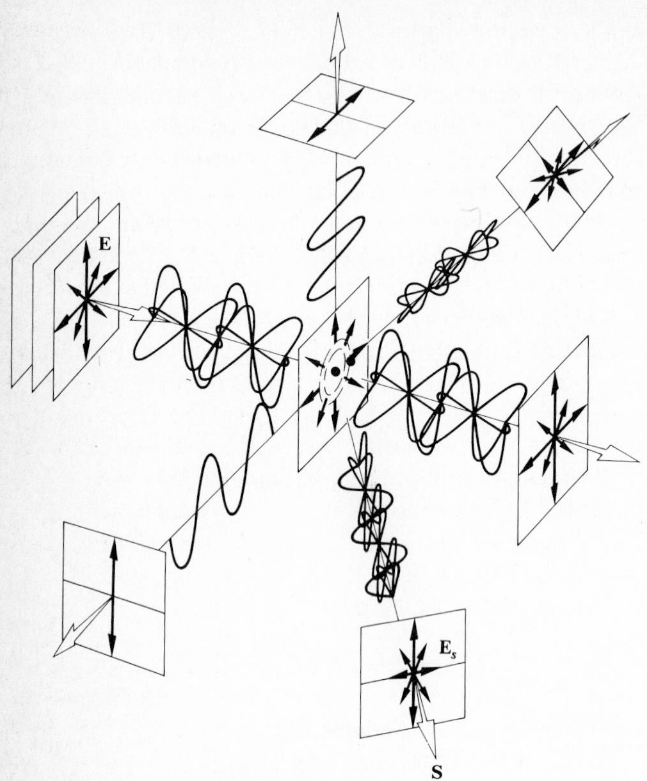

FIGURE 8.36 Scattering of unpolarized light by a molecule.

FIGURE 8.37 A pair of crossed polarizers. The upper polaroid is noticeably darker than the lower one, indicating the partial polarization of sky light. (E.H.)

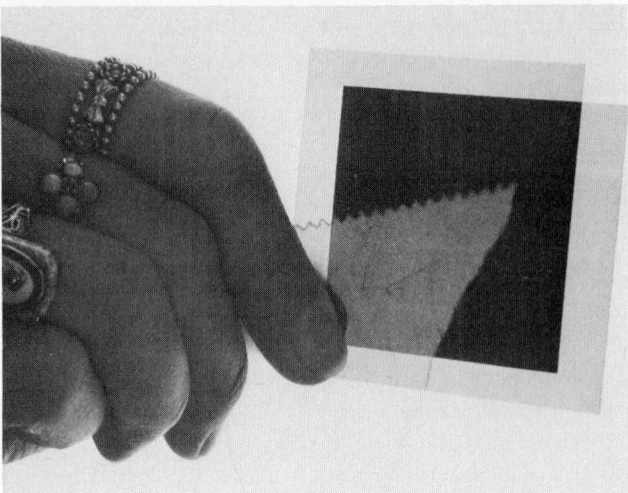

FIGURE 8.38 A piece of waxed paper between crossed polarizers. (E.H.)

As a final experiment, put a few drops of milk in a glass of water and illuminate it (perpendicular to its axis) using a bright flashlight. The solution will appear bluish white in scattered light and orange in direct light, indicating that the operative mechanism is Rayleigh Scattering. The scattered light will also be partially polarized.

Using very much the same ideas, Charles Glover Barkla (1877–1944) in 1906 established the transverse wave nature of X-ray radiation by showing that it could be polarized in certain directions as a result of scattering off matter.

8.6 Polarization by Reflection

One of the most common sources of polarized light is the ubiquitous process of reflection from dielectric media. The glare spread across a window pane, a sheet of paper, or a balding head, the sheen on the surface of a telephone, a billiard ball, or a book jacket are all generally partially polarized.

The effect was first studied by Étienne Malus in 1808. The Paris Academy had offered a prize for a mathematical theory of double refraction, and Malus undertook a study of the problem. He was standing at the window of his house in the Rue d'Enfer one evening, examining a calcite crystal. The Sun was setting, and its image reflected toward him from the windows of the Luxembourg Palace not far away. He held up the crystal and looked through it at the Sun's reflection. To his astonishment, he saw one of the double images disappear as he rotated the calcite. After the Sun had set, he continued to verify his observations into the night, using candlelight reflected from the surfaces of water and glass.* The significance of birefringence and the actual nature of polarized light were first becoming clear. At that time no satisfactory explanation of polarization existed within the context of the wave theory. During the next 13 years the work of many people, principally Thomas Young and Augustin Fresnel, finally led to the representation of light as some sort of transverse vibration. (Keep in mind that all this predates the electromagnetic theory of light by roughly 40 years.)

The electron-oscillator model provides a remarkably simple picture of what happens when light is polarized on reflection. Unfortunately, it's not a complete description, since it does not account for the behavior of magnetic nonconducting materials.† Nonetheless, consider an incoming plane wave linearly polarized so that its **E**-field is perpendicular to the plane of incidence (Fig. 8.39). The wave is refracted at the interface, entering the medium at some transmission angle θ_t. Its electric field drives the bound electrons, in this case normal to the plane-of-incidence, and they in turn reradiate. A portion of that reemitted energy appears in the form of a reflected wave. It should be clear then from the geometry and the dipole radiation pattern that both the reflected and refracted waves must also be in $\mathscr{P}$-states normal to the incident plane.‡ In contradistinction, if the incoming **E**-field is in the incident plane, the electron-oscillators near the surface will vibrate under the influence of the refracted wave, as shown in Fig. 8.39*b*. Observe that a rather interesting thing is happening to the reflected wave. Its flux density is now relatively low, because the reflected ray direction makes a small angle θ with the dipole axis. If we could arrange things so that $\theta = 0$, or equivalently $\theta_r + \theta_t = 90°$, the reflected wave would vanish entirely. *Under those circumstances, for an incoming unpolarized wave made up of two incoherent orthogonal $\mathscr{P}$-states, only the component polarized normal to the incident plane and therefore parallel to the surface will be reflected.* The particular angle of incidence for which this situation occurs is designated by θ_p and referred to as the **polarization angle** or **Brewster's angle**, whereupon $\theta_p + \theta_t = 90°$. Hence, from Snell's Law

$$n_i \sin \theta_p = n_t \sin \theta_t$$

and the fact that $\theta_t = 90° - \theta_p$, it follows that

$$n_i \sin \theta_p = n_t \cos \theta_p$$

and

$$\tan \theta_p = n_t/n_i \tag{8.25}$$

This is known as **Brewster's Law** after the man who discovered it empirically, Sir David Brewster (1781–1868), professor of physics at St. Andrews University and, of course, inventor of the kaleidoscope.

When the incident beam is in air $n_i = 1$, and if the transmitting medium is glass, in which case $n_t \approx 1.5$, the polarization angle is $\approx 56°$. Similarly, if an unpolarized beam strikes the surface of a pond ($n_t \approx 1.33$ for H_2O) at an angle of 53°, the reflected beam will be completely polarized with its **E**-field perpendicular to the plane-of-incidence or, if you like, parallel to the water's surface (Fig. 8.40). This suggests a rather handy way to locate the transmission axis of an unmarked polarizer; one just needs a piece of glass or a pond.

The problem immediately encountered in utilizing this phenomenon to construct an effective polarizer lies in the fact that the reflected beam, although completely polarized, is weak, and the transmitted beam, although strong, is only partially polarized. One scheme, illustrated in Fig. 8.41, is often referred to as a ***pile-of-plates polarizer***. It was invented by Dominique F. J. Arago in 1812. Devices of this kind can be fabricated with glass plates in the visible, silver chloride plates in the infrared, and quartz or vycor in the ultraviolet. It's an easy matter to construct a crude arrangement of this sort with a dozen or so microscope slides. (The beautiful colors that may appear when the slides are in contact are discussed in the next chapter.)

*Try it with a candle flame and a piece of glass. Hold the glass at $\theta_p \approx 56°$ for the most pronounced effect. At near glancing incidence both of the images will be bright, and neither will vanish as you rotate the crystal—Malus apparently lucked out at a good angle to the palace window.

† W. T. Doyle, "Scattering Approach to Fresnel's Equations and Brewster's Law," *Am. J. Phys.* **53**, 463 (1985).

‡The angle of reflection is determined by the scattering array, as discussed in Section 10.2.7. The scattered wavelets in general combine constructively in only one direction, yielding a reflected ray at an angle equal to that of the incident ray.

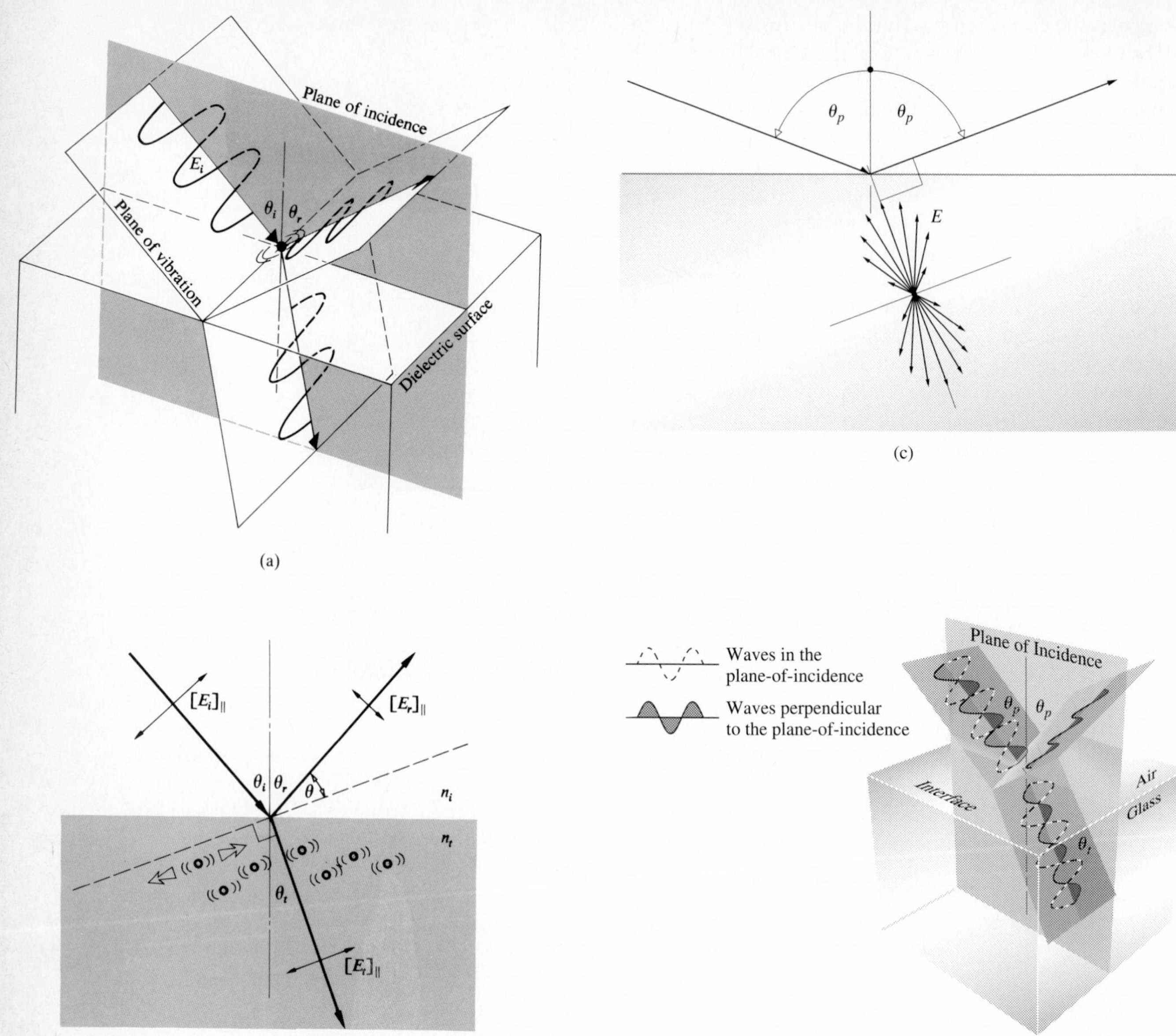

FIGURE 8.39 (*a*) A wave reflecting and refracting at an interface. (*b*) Electron-oscillators and Brewster's Law. (*c*) The dipole radiation pattern. (*d*) The polarization of light that occurs on reflection from a dielectric, such as glass, water, or plastic. At θ_p, the reflected beam is a $\mathcal{P}$-state perpendicular to the plane-of-incidence. The transmitted beam is strong in $\mathcal{P}$-state light parallel to the plane-of-incidence and weak in $\mathcal{P}$-state light perpendicular to the plane-of-incidence—it's partially polarized.

FIGURE 8.40 Light reflecting off a puddle is partially polarized. (*a*) When viewed through a Polaroid filter whose transmission axis is parallel to the ground, the glare is passed and visible. (*b*) When the Polaroid's transmission axis is perpendicular to the water's surface, most of the glare vanishes. (Photo courtesy Martin Seymour.)

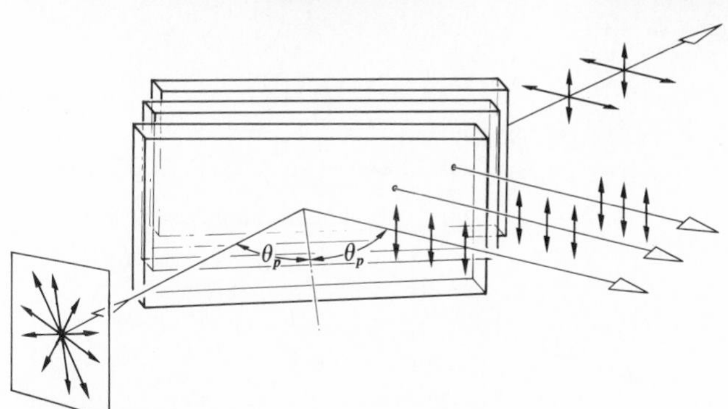

FIGURE 8.41 The pile-of-plates polarizer.

8.6.1 An Application of the Fresnel Equations

In Section 4.6.2 we obtained a set of formulas known as the Fresnel Equations, which describe the effects of an incoming electromagnetic plane wave falling on the interface between two different dielectric media. These equations relate the reflected and transmitted field amplitudes to the incident amplitude by way of the angles-of-incidence θ_i and transmission θ_t. For linear light having its **E**-field parallel to the plane-of-incidence, we defined the *amplitude reflection coefficient* as $r_\parallel \equiv [E_{0r}/E_{0i}]_\parallel$, that is, the ratio of the reflected to incident electric-field *amplitudes*. Similarly, when the electric field is normal to the incident plane, we have $r_\perp \equiv [E_{0r}/E_{0i}]_\perp$. The corresponding irradiance ratio (the incident and reflected beams have the same cross-sectional area) is known as the *reflectance*, and since irradiance is proportional to the square of the amplitude of the field,

$$R_\parallel = r_\parallel^2 = [E_{0r}/E_{0i}]_\parallel^2 \quad \text{and} \quad R_\perp = r_\perp^2 = [E_{0r}/E_{0i}]_\perp^2$$

Squaring the appropriate Fresnel Equations yields

$$R_\parallel = \frac{\tan^2(\theta_i - \theta_t)}{\tan^2(\theta_i + \theta_t)} \tag{8.26}$$

and
$$R_\perp = \frac{\sin^2(\theta_i - \theta_t)}{\sin^2(\theta_i + \theta_t)} \tag{8.27}$$

Whereas $R_\perp$ can never be zero, $R_\parallel$ is indeed zero when the denominator is infinite, that is, when $\theta_i + \theta_t = 90°$. The reflectance, for linear light with **E** parallel to the plane-of-incidence, thereupon vanishes; $E_{r\parallel} = 0$ and the beam is completely transmitted. This is the essence of Brewster's Law.

If the incoming light is unpolarized, we can represent it by two now familiar orthogonal, incoherent, equal-amplitude $\mathcal{P}$-states. Incidentally, the fact that they are equal in amplitude means that the amount of energy in one of these two polarization states is the same as that in the other (i.e., $I_{i\parallel} = I_{i\perp} = I_i/2$), which is quite reasonable. Thus

$$I_{r\parallel} = I_{r\parallel}I_i/2I_{i\parallel} = R_\parallel I_i/2$$

and in the same way $I_{r\perp} = R_\perp I_i/2$. The reflectance in natural light, $R = I_r/I_i$, is therefore given by

$$R = \frac{I_{r\parallel} + I_{r\perp}}{I_i} = \tfrac{1}{2}(R_\parallel + R_\perp) \tag{8.28}$$

Figure 8.42 is a plot of Eqs. (8.26), (8.27), and (8.28) for the particular case when $n_i = 1$ and $n_t = 1.5$. The middle curve, which corresponds to incident natural light, shows that only about 7.5% of the incoming light is reflected when $\theta_i = \theta_p$.

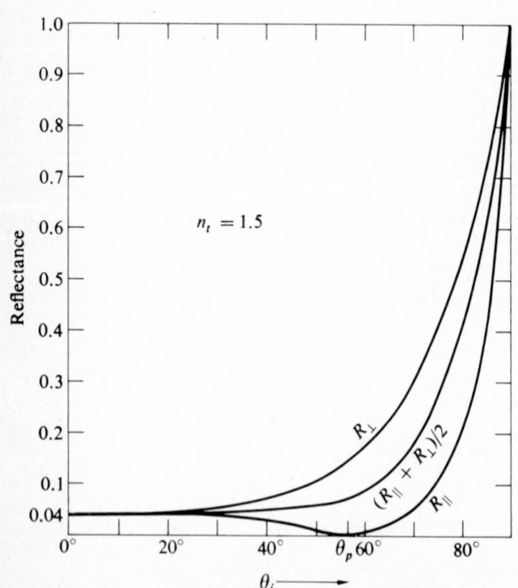

FIGURE 8.42 Reflectance versus incident angle.

The transmitted light is then evidently partially polarized. When $\theta_i \neq \theta_p$ both the transmitted and reflected waves are partially polarized.

It is often desirable to make use of the concept of the **degree of polarization** V, defined as

$$V = \frac{I_p}{I_p + I_n} \tag{8.29}$$

in which I_p and I_n are the constituent flux densities of polarized and "unpolarized" or natural light. For example, if $I_p = 4$ W/m^2 and $I_n = 6$ W/m^2, then $V = 40\%$ and the beam is partially polarized. With "unpolarized" light $I_p = 0$ and obviously $V = 0$, whereas at the opposite extreme, if $I_n = 0$, $V = 1$ and the light is completely polarized; thus $0 \le V \le 1$. One frequently deals with partially polarized, linear, quasimonochromatic light. In that case, if we rotate an analyzer in the beam, there will be an orientation at which the transmitted irradiance is maximum (I_{max}), and perpendicular to this, a direction where it is minimum (I_{min}). Clearly $I_p = I_{max} - I_{min}$, and so

$$V = \frac{I_{max} - I_{min}}{I_{max} + I_{min}} \tag{8.30}$$

Note that V is actually a property of the beam, which may be partially or even completely polarized before encountering any sort of polarizer.

8.7 Retarders

We now consider a class of optical elements known as **retarders**, which serve to change the polarization of an incident wave. In principle, the operation of a retarder is quite simple. One of the two constituent coherent $\mathcal{P}$-states is somehow caused to lag in-phase behind the other by a predetermined amount. Upon emerging from the retarder, the relative phase of the two components is different than it was initially, and thus the polarization state is different as well. Once we have developed the concept of the retarder it will be possible to convert any given polarization state into any other and in so doing create circular and elliptic polarizers as well.

8.7.1 Wave Plates and Rhombs

Recall that a plane monochromatic wave incident on a uniaxial crystal, such as calcite, is generally divided in two, emerging as an ordinary and an extraordinary beam. In contrast, we can cut and polish a calcite crystal so that its optic axis will be normal to both the front and back surfaces (Fig. 8.43). A normally incident plane wave can only have its **E**-field perpendicular to the optic axis. The secondary spherical and ellipsoidal wavelets will be tangent to each other in the direction of the optic axis. The *o*- and *e*-waves, which are envelopes of these wavelets, will be coincident, and a single undeflected plane wave will pass through the crystal; there are no relative phase shifts and no double images.*

Now suppose that the direction of the optic axis is arranged to be parallel to the front and back surfaces, as shown in Fig. 8.44. If the **E**-field of an incident monochromatic plane wave has components parallel and perpendicular to the optic axis, two separate plane waves will propagate through the crystal. Since $v_\parallel > v_\perp$, $n_o > n_e$, and the *e*-wave will move across the specimen more rapidly than the *o*-wave. After traversing a plate of thickness d the resultant electromagnetic wave is the superposition of the *e*- and *o*-waves, which now have a relative phase difference of $\Delta\varphi$. Keep in mind that these are harmonic waves of the same frequency whose **E**-fields are orthogonal. The relative optical path-length difference is given by

$$\Lambda = d(|n_o - n_e|) \tag{8.31}$$

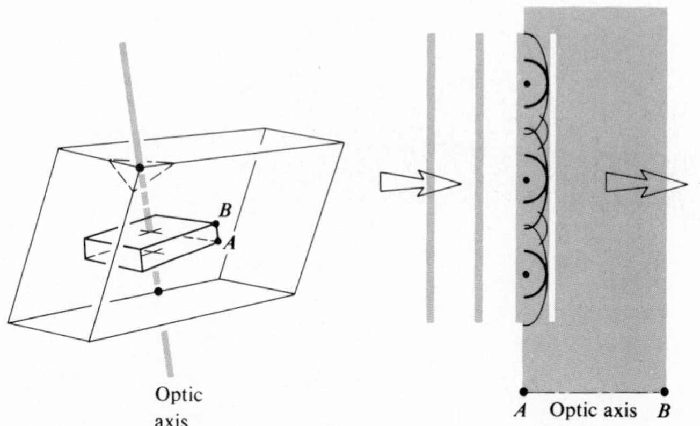

FIGURE 8.43 A calcite plate cut perpendicular to the optic axis.

*If you have a calcite rhomb, find the blunt corner and orient the crystal until you are looking along the direction of the optic axis through one of the faces. The two images will converge until they completely overlap.

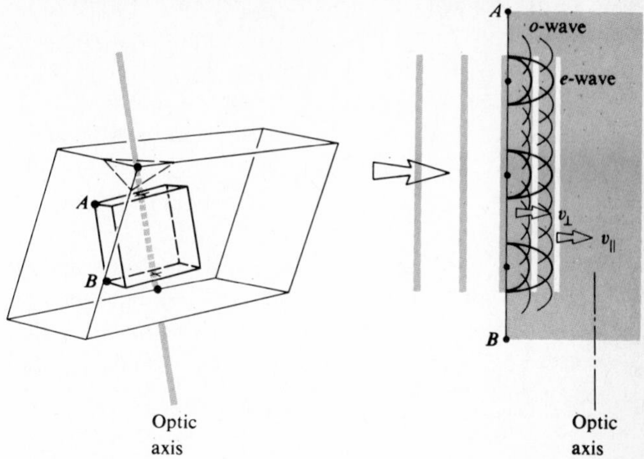

FIGURE 8.44 A calcite plate cut parallel to the optic axis.

and since $\Delta\varphi = k_0\Lambda$,

$$\Delta\varphi = \frac{2\pi}{\lambda_0} d(|n_o - n_e|) \tag{8.32}$$

where λ_0, as always, is the wavelength in vacuum. (The form containing the absolute value of the index difference is the most general statement.) The state of polarization of the emergent light evidently depends on the amplitudes of the incoming orthogonal field components and of course on $\Delta\varphi$.

The Full-Wave Plate

If $\Delta\varphi$ is equal to 2π, the *relative retardation* is one wavelength; the *e*- and *o*-waves are back in-phase, and there is no observable effect on the polarization of the incident monochromatic beam. When the *relative retardation* $\Delta\varphi$, which is also known as the *retardance*, is 360° the device is called a ***full-wave plate***. (This does not mean that $d = \lambda$.) In general, the quantity $|n_o - n_e|$ in Eq. (8.32) changes little over the optical range, so that $\Delta\varphi$ varies effectively as $1/\lambda_0$. Evidently, a full-wave plate can function only in the manner discussed for a particular wavelength, and retarders of this sort are thus said to be *chromatic*. If such a device is placed at some arbitrary orientation between crossed linear polarizers, all the light entering it (in this case let it be white light) will be linear. Only the one wavelength that satisfies Eq. (8.32) will pass through the retarder unaffected, thereafter to be absorbed in the analyzer. All other wavelengths will undergo some retardance and will accordingly emerge from the wave plate as various forms of elliptical light. Some portion of this light will proceed through the analyzer, finally emerging as the complementary color to that which was extinguished. It is a common error to assume that a full-wave plate behaves as if it were isotropic at all frequencies; it obviously doesn't.

Recall that in calcite, the wave whose **E**-field vibrations are parallel to the optic axis travels fastest, that is, $v_{\parallel} > v_{\perp}$. The direction of the optic axis in a *negative* uniaxial retarder is therefore often referred to as the **fast axis**, and the direction perpendicular to it is the **slow axis**. For *positive* uniaxial crystals, such as quartz, these principal axes are reversed, with the slow axis corresponding to the optic axis.

The Half-Wave Plate

A retardation plate that introduces a relative phase difference of π radians or 180° between the o- and e-waves is known as a **half-wave plate**. Suppose that the plane-of-vibration of an incoming beam of linear light makes some arbitrary angle θ with the fast axis, as shown in Fig. 8.45. In a negative material the e-wave will have a higher speed (same ν) and a longer wavelength than the o-wave. When the waves emerge from the plate, there will be a relative phase shift of $\lambda_0/2$ (that is, $2\pi/2$ radians), with the effect that **E** will have rotated through 2θ. Going back to Fig. 8.7, it should be evident that a half-wave plate will similarly flip elliptical light. In addition, it will invert the handedness of circular or elliptical light, changing right to left and vice versa.

As the e- and o-waves progress through any retardation plate, their relative phase difference $\Delta\varphi$ increases, and the state of polarization of the wave therefore gradually changes from one point in the plate to the next. Figure 8.7 can be envisioned as a sampling of a few of these states at one instant in time taken at different locations. Evidently, if the thickness of the material is such that

$$d(|n_o - n_e|) = (2m + 1)\lambda_0/2$$

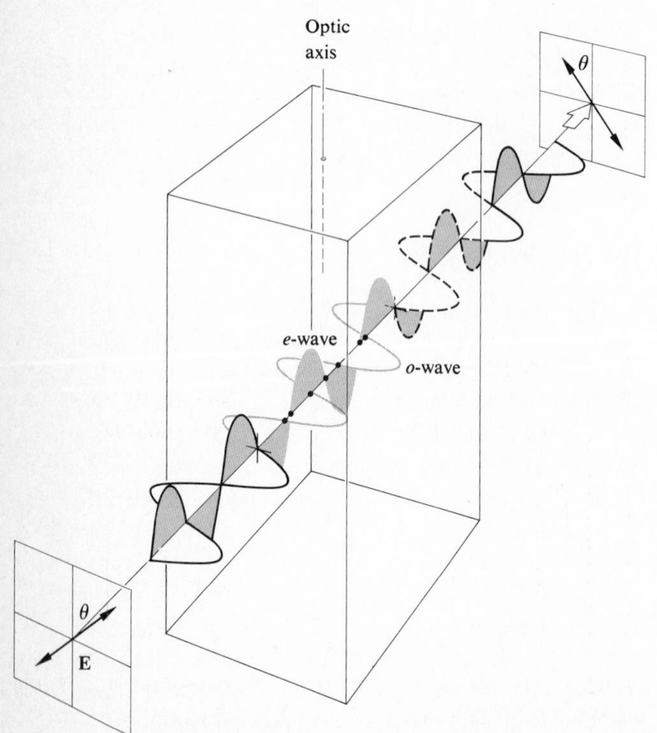

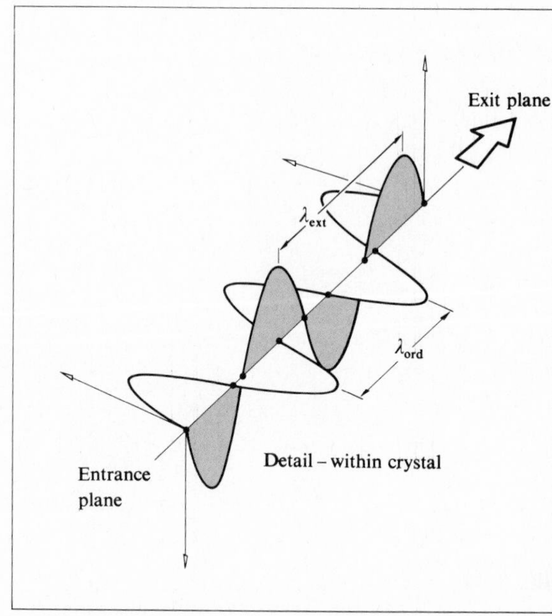

FIGURE 8.45 A half-wave plate.

where $m = 0, 1, 2, \ldots$, it will function as a half-wave plate ($\Delta\varphi = \pi, 3\pi, 5\pi$, etc.).

Although its behavior is simple to visualize, calcite is not often used to make retardation plates. It is brittle and difficult to handle in thin slices, but more than that, its birefringence, the difference between n_e and n_o, is a bit too large for convenience. On the other hand, quartz with its much smaller birefringence is frequently used, but it has no natural cleavage planes and must be cut, ground, and polished, making it rather expensive. The biaxial crystal mica is used most often. Several forms of mica serve the purpose admirably, for example, fluorophlogopite, biotite, or muscovite. The most commonly occurring variety is the pale brown muscovite. It is very easily cleaved into strong, flexible, and exceedingly thin large-area sections. Moreover, its two principal axes are almost exactly parallel to the cleavage planes. Along those axes the indices are about 1.599 and 1.594 for sodium light, and although these numbers vary slightly from one sample to the next, their difference is fairly constant. The minimum thickness of a mica half-wave plate is about 60 microns. Crystalline quartz, single crystal magnesium fluoride (for the IR range from 3000 nm to about 6000 nm), and cadmium sulfide (for the IR range from 6000 nm to about 12,000 nm) are also widely used for wave plates.

Retarders are also made from sheets of polyvinyl alcohol that have been stretched so as to align their long-chain organic molecules. Because of the evident anisotropy, electrons in the material do not experience the same binding forces along and perpendicular to the direction of these molecules. Substances of this sort are therefore permanently birefringent, even though they are not crystalline.

A rather nice half-wave plate can be made by just attaching a strip of old-fashioned glossy cellophane tape over the surface of a microscope slide. (Not all varieties work—the best is LePage's "Transparent Tape.") The fast axis, that is, the vibration direction of the faster of the two waves, corresponds to the transverse direction across the tape's width, and the slow axis is along its length. During its manufacture, cellophane (which is made from regenerated cellulose extracted from cotton or wood pulp) is formed into sheets, and in the process its molecules become aligned, leaving it birefringent. If you put your half-wave plate between crossed linear polarizers, it will show no effect when its principal axes coincide with those of the polarizers. If, however, it is set at 45° with respect to the polarizer, the **E**-field emerging from the tape will be flipped 90° and will be parallel to the transmission axis of the analyzer.

FIGURE 8.46 A hand holding a piece of Scotch tape stuck to a microscope slide between two crossed polaroids. (E.H.)

Light will pass through the region covered by the tape as if it were a hole cut in the black background of the crossed polarizers (Fig. 8.46). A piece of cellophane wrapping will generally also function as a half-wave plate. See if you can determine the orientation of each of its principal axes using the tape retarder and crossed Polaroids. (Notice the fine parallel ridges on the sheet cellophane.)

The Quarter-Wave Plate

The **quarter-wave plate** is an optical element that introduces a relative phase shift of $\Delta\varphi = \pi/2$ between the constituent orthogonal o- and e-components of a wave. It follows once again from Fig. 8.7 that a phase shift of 90° will convert linear to elliptical light and vice versa. It should be apparent that linear light incident parallel to either principal axis will be unaffected by any sort of retardation plate. You can't have a *relative* phase difference without having two components. With incident *natural* light, the two constituent $\mathcal{P}$-states are incoherent; that is, their relative phase difference changes randomly and rapidly. The introduction of an additional constant phase shift by any form of retarder will still result in a random phase difference and thus have no noticeable effect. When linear light at 45° to either principal axis is incident on a quarter-wave plate, its o- and e-components have equal amplitudes. Under these special circumstances, a 90° phase shift converts the wave into circular light. Similarly, an incoming circular beam will emerge linearly polarized.

Quarter-wave plates are also usually made of quartz, mica, or organic polymeric plastic. In any case, the thickness of the birefringent material must satisfy the expression

$$d(|n_o - n_e|) = (4m + 1)\lambda_0/4$$

You can make a crude quarter-wave plate using household plastic food wrap, the thin stretchy stuff that comes on rolls. Like cellophane, it has ridges running in the long direction, which coincides with a principal axis. Overlap about a half dozen layers, being careful to keep the ridges parallel. Position the plastic at 45° to the axes of a polarizer and examine it through a rotating analyzer. Keep adding one layer at a time until the irradiance stays roughly constant as the analyzer turns; at that point you will have circular light and a quarter-wave plate. This is easier said than done in white light, but it's well worth trying.

Commercial wave plates are generally designated by their ***linear retardation***, which might be, for example, 140 nm for a quarter-wave plate. This simply means that the device has a 90° retardance only for green light of wavelength 560 nm (i.e., 4 × 140). The linear retardation is usually not given quite that precisely; 140 ± 20 nm is more realistic. The retardation of a wave plate can be increased or decreased from its specified value by tilting it somewhat. If the plate is rotated about its fast axis, the retardation will increase, whereas a rotation about the slow axis has the opposite effect. In this way a wave plate can be tuned to a specific frequency in a region about its nominal value.

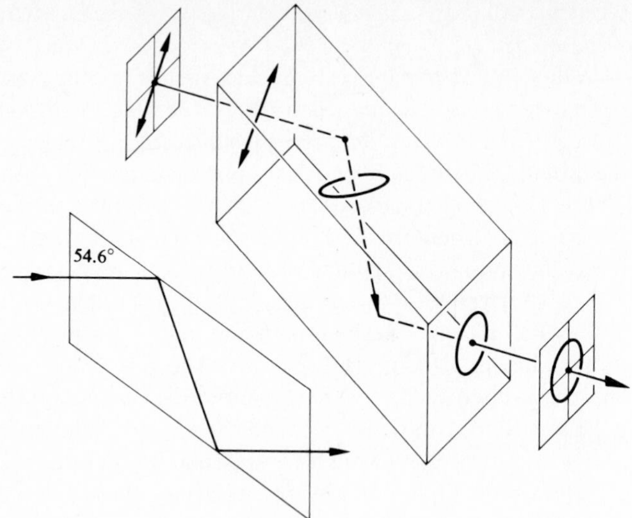

FIGURE 8.47 The Fresnel rhomb.

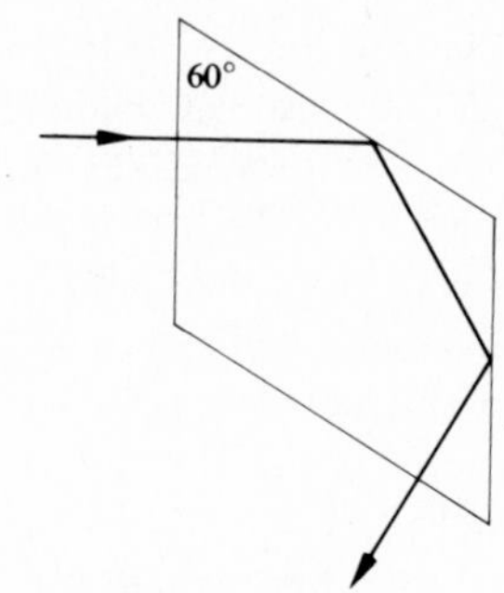

FIGURE 8.48 The Mooney rhomb.

The Fresnel Rhomb

We saw in Chapter 4 that the process of total internal reflection introduced a relative phase difference between the two orthogonal field components. The components parallel and perpendicular to the plane-of-incidence were shifted in phase with respect to each other. In glass (n = 1.51) a shift of 45° accompanies internal reflection at the particular incident angle of 54.6° (Fig. 4.43*e*). The Fresnel rhomb shown in Fig. 8.47 utilizes this effect by causing the beam to be internally reflected twice, thereby imparting a 90° relative phase shift to its components. If the incoming plane wave is linearly polarized at 45° to the plane-of-incidence, the field components $[E_i]_\parallel$ and $[E_i]_\perp$ will initially be equal. After the first reflection the wave within the glass will be elliptically polarized. After the second reflection it will be circular. Since the retardance is almost independent of frequency over a large range, the rhomb is essentially an *achromatic* 90° retarder. The Mooney rhomb (n = 1.65) shown in Fig. 8.48 is similar in principle, although its operating characteristics are different in some respects.

8.7.2 Compensators

A **compensator** *is an optical device that is capable of impressing a controllable retardance on a wave.* Unlike a wave plate where $\Delta\varphi$ is fixed, the relative phase difference arising from a

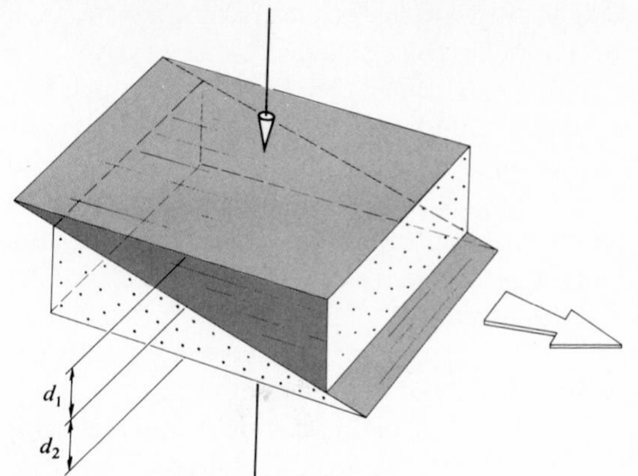

FIGURE 8.49 The Babinet compensator.

compensator can be varied continuously. Of the many different kinds of compensators, we shall consider only two of those that are used most widely. The Babinet compensator, depicted in Fig. 8.49, consists of two independent calcite, or more commonly quartz, wedges whose optic axes are indicated by the lines and dots in the figure. A ray passing vertically downward through the device at some arbitrary point will traverse a thickness of d_1 in the upper wedge and d_2 in the lower one. The relative phase difference imparted to the wave by the first crystal is $2\pi d_1(|n_o - n_e|)/\lambda_0$, and that of the second crystal is $-2\pi d_2(|n_o - n_e|)/\lambda_0$. As in the Wollaston prism, which this system closely resembles but which has larger angles and is much thicker, the *o*- and *e*-rays in the upper wedge become the *e*- and *o*-rays, respectively, in the bottom wedge.

The compensator is thin (the wedge angle is typically about 2.5°), and thus the separation of the rays is negligible. The total phase difference is then

$$\Delta\varphi = \frac{2\pi}{\lambda_0}(d_1 - d_2)(|n_o - n_e|) \tag{8.33}$$

If the compensator is made of calcite, the *e*-wave leads the *o*-wave in the upper wedge, and therefore if $d_1 > d_2$, $\Delta\varphi$ corresponds to the total angle by which the *e*-component leads the *o*-component. The converse is true for a quartz compensator; in other words, if $d_1 > d_2$, $\Delta\varphi$ is the angle by which the *o*-wave leads the *e*-wave. At the center, where $d_1 = d_2$, the effect of one wedge is exactly canceled by the other, and $\Delta\varphi = 0$ for all wavelengths. The retardation will vary from point to point over the surface, being constant in narrow regions running the width of the compensator along which the wedge thicknesses are themselves constant. If light enters by way of a slit parallel to one of these regions and if we then move either wedge horizontally with a micrometer screw, we can get any desired $\Delta\varphi$ to emerge.

When the Babinet is positioned at 45° between crossed polarizers, a series of parallel, equally spaced, dark extinction fringes will appear across the width of the compensator. These mark the positions where the device acts as if it was a full-wave plate. In white light the fringes will be colored, with the exception of the black central band ($\Delta\varphi = 0$). The retardance of an unknown plate can be found by placing it on the compensator and examining the fringe shift it produces.

The Babinet can be modified to produce a uniform retardation over its surface by merely rotating the top wedge 180° about the vertical, so that its thin edge rests on the thin edge of the lower wedge. This configuration will, however, slightly deviate the beam. Another variation of the Babinet, which has the advantage of producing a uniform retardance over its surface and no beam deviation, is the *Soleil compensator* shown in Fig. 8.50. Generally made of quartz (although MgF_2 and CdS are used in the infrared), it consists of two wedges and one plane-parallel slab whose optic axes are oriented as indicated. The quantity d_1 corresponds to the total thickness of both wedges, which is constant for any setting of the positioning micrometer screw.

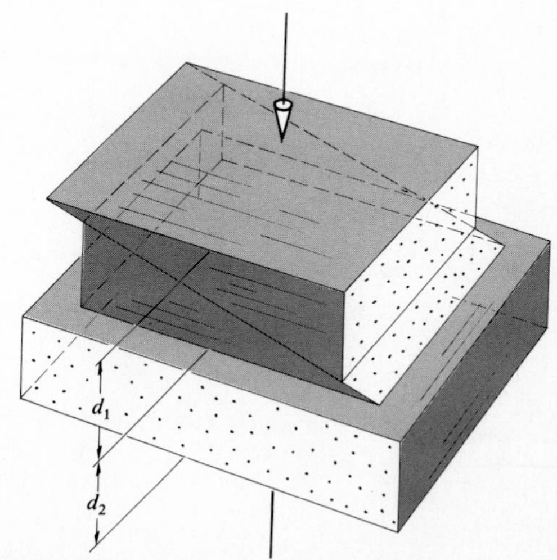

FIGURE 8.50 The Soleil compensator.

8.8 Circular Polarizers

Earlier we concluded that linear light whose **E**-field is at 45° to the principal axes of a quarter-wave plate will emerge from that plate circularly polarized. Any series combination of an appropriately oriented linear polarizer and a 90° retarder will therefore perform as a **circular polarizer**. The two elements function completely independently, and whereas one might be birefringent, the other could be of the reflection type. The handedness of the emergent circular light depends on whether the transmission axis of the linear polarizer is at +45° or −45° to the fast axis of the retarder. Either circular state, $\mathscr{L}$ or $\mathscr{R}$, can be generated quite easily. In fact, if the linear polarizer is situated between two retarders, one oriented at +45° and the other at −45°, the combination will be "ambidextrous." In short, it will yield an $\mathscr{R}$-state for light entering from one side and an $\mathscr{L}$-state when the input is on the other side.

CP-HN is the commercial designation for a popular one-piece circular polarizer. It is a laminate of an *HN* Polaroid and a stretched polyvinyl alcohol 90° retarder. The *input side* of such an arrangement is evidently the face of the linear polarizer. If the beam is incident on the *output side* (i.e., on the retarder), it will thereafter pass through the *H*-sheet and can only emerge linearly polarized.

A circular polarizer can be used as an analyzer to determine the handedness of a wave that is already known to be circular. To see how this might be done, imagine that we have the four elements labeled *A*, *B*, *C*, and *D* in Fig. 8.51. The first two, *A* and *B*, taken together form a circular polarizer, as do *C* and *D*. The precise handedness of these polarizers is unimportant now, as long as they are both the same, which is tantamount to saying that the fast axes of the retarders are parallel. Linear light coming from *A* receives a 90° retardance from B, at which point it is circular. As it passes through *C*, another 90° retardance is added on, resulting once more in a linearly polarized wave. In effect, *B* and *C* together form a half-wave plate, which merely flips the linear light from A through a spatial angle of 2θ, in this case 90°. Since the linear wave from *C* is parallel to the transmission axis of *D*, it passes through it and out of the system.

In this simple process we've actually proved something that is rather subtle. If the circular polarizers $A + B$ and $C + D$ are both left-handed, we've shown that *left-circular light entering a left-circular polarizer from the output side will be transmitted.* Furthermore, it should be apparent, at least after some thought, that right-circular light will produce a $\mathscr{P}$-state perpendicular to the transmission axis of *D* and so will be absorbed. The converse is true as well; that is, *of the two circular forms, only light in an $\mathscr{R}$-state will pass through a right-circular polarizer having entered from the output side.*

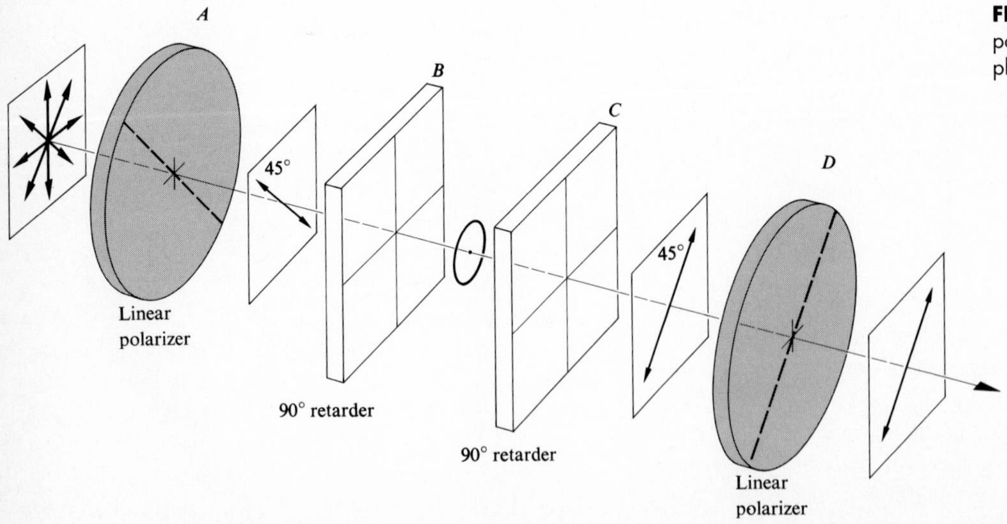

FIGURE 8.51 Two linear polarizers and two quarter-wave plates.

8.9 POLARIZATION OF POLYCHROMATIC LIGHT

8.9.1 Bandwidth and Coherence Time of a Polychromatic Wave

By its very nature purely monochromatic light, which is of course not a physical reality, must be polarized. The two orthogonal components of such a wave have the same frequency, and each has a constant amplitude. If the amplitude of either sinusoidal component varied, it would be equivalent to the presence of other additional frequencies in the Fourier-analyzed spectrum. Moreover, the two components have a constant relative phase difference, that is, they are coherent. A monochromatic disturbance is an infinite wavetrain whose properties have been fixed for all time; whether it is in an $\mathscr{R}$-, $\mathscr{L}$-, $\mathscr{P}$-, or $\mathscr{E}$-state, the wave is completely polarized.

Actual light sources are polychromatic; they emit radiant energy having a range of frequencies. Let's now examine what happens on a submicroscopic scale, paying particular attention to the polarization state of the emitted wave. Envision an electron-oscillator that has been excited into vibration (possibly by a collision) and thereupon radiates. Depending on its precise motion, the oscillator will emit some form of polarized light.

As in Section 7.4.3, we picture the radiant energy from a single atom as a wavetrain having a finite spatial extent Δl_c. Assume for the moment that its polarization state is essentially constant for a duration of the order of the coherence time Δt_c (which, as you recall, corresponds to the temporal extent of the wavetrain, i.e., $\Delta l_c/c$). A typical source generally consists of a large collection of such radiating atoms, which can be envisioned as oscillating with different phases at some dominant frequency $\bar{\nu}$. Suppose then that we examine the light coming from a very small region of the source, such that the emitted rays arriving at a point of observation are essentially parallel. During a time that is short in comparison with the average coherence time, the amplitudes and phases of the wavetrains from the individual atoms will be essentially constant. This means that if we were to look toward the source in some direction, we would, at least for an instant, "see" a coherent superposition of the waves emitted in that direction. We would "see" a resultant wave having a given polarization state. That state would only last for an interval less than the coherence time before it changed, but even so it would correspond to a great many oscillations at the frequency $\bar{\nu}$. Clearly, if the bandwidth $\Delta\nu$ is broad, the coherence time ($\Delta t_c \approx 1/\Delta\nu$) will be small, and any polarization state will be short-lived. Evidently *the concepts of polarization and coherence are related in a fundamental way*.

Now consider a wave whose bandwidth is very small in comparison with its mean frequency, a quasimonochromatic wave. It can be represented by two orthogonal harmonic $\mathscr{P}$-states, as in Eqs. (8.1) and (8.2), but here the amplitudes and initial phase angles are functions of time. Furthermore, the frequency and propagation number correspond to the mean values of the spectrum present in the wave, namely, $\bar{\omega}$ and $\bar{k}$. Thus

$$\mathbf{E}_x(t) = \hat{\mathbf{i}}E_{0x}(t)\cos[\bar{k}z - \bar{\omega}t + \varepsilon_x(t)] \tag{8.34a}$$

and

$$\mathbf{E}_y(t) = \hat{\mathbf{j}}E_{0y}(t)\cos[\bar{k}z - \bar{\omega}t + \varepsilon_y(t)] \tag{8.34b}$$

The polarization state, and accordingly $E_{0x}(t)$, $E_{0y}(t)$, $\varepsilon_x(t)$, and $\varepsilon_y(t)$, will vary slowly, remaining essentially constant over a large number of oscillations. Keep in mind that the narrow bandwidth implies a relatively large coherence time. If we watch the wave during a much longer interval, the amplitudes and phase angles will vary somehow, either independently or in some correlated fashion. If the variations are completely uncorrelated, the polarization state will remain constant only for an interval small compared to the coherence time. In other words, the ellipse describing the polarization state may change shape, orientation, and handedness. Since, speaking practically, no existing detector could discern any one particular state lasting for so short a time, we would conclude that the wave was unpolarized.

Antithetically, if the ratio $E_{0x}(t)/E_{0y}(t)$ was constant even though both terms varied, and if $\varepsilon = \varepsilon_y(t) - \varepsilon_x(t)$ was constant as well, the wave would be polarized. Here the necessity for correlation among these different functions is obvious. Yet we can actually impress these conditions on the wave by merely passing it through a polarizer, thereby removing any undesired constituents. The time interval over which the wave thereafter maintains its polarization state is no longer dependent on the bandwidth, because the wave's components have been appropriately correlated. The light could be polychromatic (even white), yet completely polarized. It will behave very much like the idealized monochromatic waves treated in Section 8.1.

Between the two extremes of completely polarized and unpolarized light is the condition of partial polarization. In fact, it can be shown that any quasimonochromatic wave can be represented as the sum of a polarized and an unpolarized wave, where the two are independent and either may be zero.

8.9.2 Interference Colors

Insert a crumpled sheet of cellophane between two Polaroids illuminated by white light. Alternatively, take an ordinary plastic bag (polyethylene), which shows nothing special between crossed Polaroids, and stretch it. That will align its molecules, making it birefringent. Now crumple it up and examine it again. The resulting pattern will be a profusion of multicolored regions, which vary in hue as either Polaroid rotates. These **interference colors** arise from the wavelength dependence of the retardation. The usual variegated nature of the patterns is due to local variations in thickness, birefringence, or both.

The appearance of interference colors is commonplace and can easily be observed in any number of substances. For example, the effect can be seen with a piece of multilayered mica, a chip of ice, a stretched plastic bag, or finely crushed particles of an ordinary white (quartz) pebble. To appreciate how the phenomenon occurs, examine Fig. 8.52. A narrow beam of monochromatic linear light is schematically shown passing through some small region of a birefringent plate Σ. Over that area the birefringence and thickness are both assumed to be constant. The transmitted light is generally elliptical. Equivalently, envision the light emerging from Σ as composed of two orthogonal linear waves (i.e., the x- and y-components of the total **E**-field), which have a relative phase difference $\Delta\varphi$, determined by Eq. (8.32). Only the components of these two disturbances, which are in the direction of the transmission axis of the analyzer, will pass through it and on to the observer.

Now these components, which also have a phase difference of $\Delta\varphi$, are coplanar and can thus interfere. When $\Delta\varphi = \pi, 3\pi, 5\pi, \ldots$, they are completely out-of-phase and tend to cancel each other. When $\Delta\varphi = 0, 2\pi, 4\pi, \ldots$, the waves are in-phase and reinforce each other. Suppose then that the retardance arising at some point P_1 on Σ for blue light ($\lambda_0 = 435$ nm) is 4π. In that case blue will be strongly transmitted. It follows from Eq. (8.32) that $\lambda_0 \Delta\varphi = 2\pi d(|n_o - n_e|)$ is essentially a constant determined by the thickness and the birefringence. At the point in question, therefore, $\lambda_0 \Delta\varphi = 1740\pi$ for all wavelengths. If we now change to incident yellow light ($\lambda_0 = 580$ nm), $\Delta\varphi \approx 3\pi$ and the light from P_1 is completely canceled. Under white-light illumination that particular point on Σ will seem as if it had removed yellow completely, passing on all the other colors, but none as strongly as blue. Another way of saying this is that the blue light emerging from the region about P_1 is linear ($\Delta\varphi = 4\pi$) and parallel

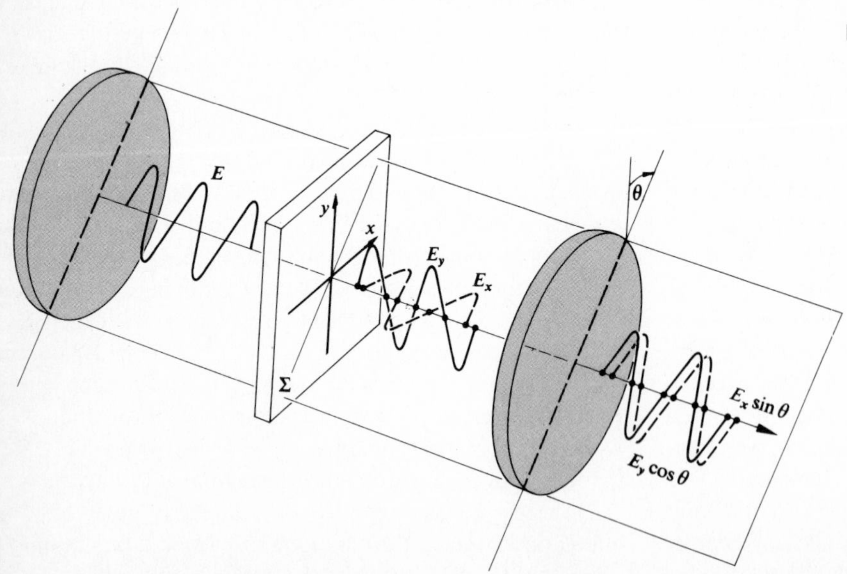

FIGURE 8.52 The origin of interference colors.

to the analyzer's transmission axis. In contrast, the yellow light is linear ($\Delta\varphi = 3\pi$) and along the extinction axis; the other colors are elliptical. The region about P_1 behaves like a half-wave plate for yellow and full-wave plate for blue. If the analyzer were rotated 90°, the yellow would be transmitted, and the blue extinguished.

By definition two colors are said to be complementary when their combination yields white light. Thus when the analyzer is rotated through 90° it will alternately transmit or absorb complementary colors. In much the same way there might be a point P_2 somewhere else on Σ where $\Delta\varphi = 4\pi$ for red ($\lambda_0 = 650$ nm). Then, $\lambda_0 \Delta\varphi = 2600\pi$, whereupon bluish-green light ($\lambda_0 = 520$ nm) will have a retardance of 5π and be extinguished. Clearly, if the retardance varies from one region to the next over the specimen, so too will the color of the light transmitted by the analyzer.

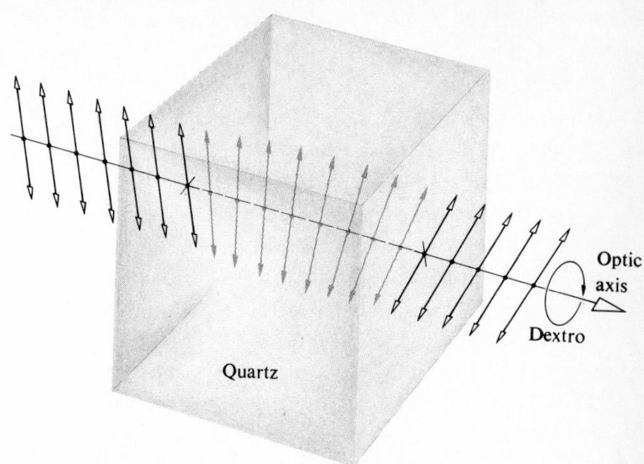

FIGURE 8.53 Optical activity displayed by quartz.

8.10 Optical Activity

The manner in which light interacts with material substances can yield a great deal of valuable information about their molecular structures. The process to be examined next, although of specific interest in the study of Optics, has had and is continuing to have far-reaching effects in the sciences of chemistry and biology.

In 1811 the French physicist Dominique F. J. Arago first observed the rather fascinating phenomenon now known as **optical activity**. It was then that he discovered that the plane of vibration of a beam of linear light underwent a continuous rotation as it propagated along the optic axis of a quartz plate (Fig. 8.53). At about the same time Jean Baptiste Biot (1774–1862) saw this same effect while using both the vaporous and liquid forms of various natural substances like turpentine. Any material that causes the **E**-field of an incident linear plane wave to appear to rotate is said to be *optically active*. Moreover, as Biot found, one must distinguish between right- and left-handed rotation. If while looking in the direction of the source, the plane of vibration appears to have revolved clockwise, the substance is referred to as *dextrorotatory*, or *d-rotatory* (from the Latin *dextro*, meaning right). Alternatively, if **E** appears to have been displaced counterclockwise, the material is *levorotatory*, or *l-rotatory* (from the Latin *levo*, meaning left).

In 1822 the English astronomer Sir John F. W. Herschel (1792–1871) recognized that *d*-rotatory and *l*-rotatory behavior in quartz actually corresponded to two different crystallographic structures. Although the molecules are identical (SiO_2), crystal quartz can be either right- or left-handed, depending on the arrangement of those molecules. As shown in Fig. 8.54, the external appearances of these two forms are the same in all respects, except that one is the mirror image of the other; they are said to be *enantiomorphs* of each other. All transparent enantiomorphic substances are optically active. Furthermore, molten quartz and *fused* quartz, neither of which is crystalline, are not optically active. Evidently, in quartz optical activity is associated with the structural distribution of

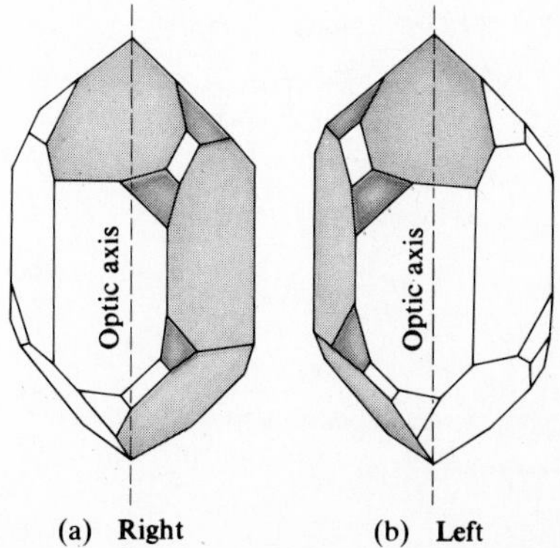

FIGURE 8.54 Right- and left-handed quartz crystals.

the molecules as a whole. There are many substances, both organic and inorganic (e.g., benzil and $NaBrO_3$, respectively), which, like quartz, exhibit optical activity only in crystal form. In contrast, many naturally occurring organic compounds, such as sugar, tartaric acid, and turpentine, are optically active in solution or in the liquid state. Here the *rotatory power*, as it is often referred to, is evidently an attribute of the individual molecules. There are also more complicated substances for which optical activity is associated with both the molecules themselves and their arrangement within the various crystals. An example is rubidium tartrate. A *d*-rotatory solution of that compound will change to *l-rotatory* when crystallized.

In 1825 Fresnel, without addressing the actual mechanism involved, proposed a simple phenomenological description of optical activity. Since the incident linear wave can be represented as a superposition of $\mathscr{R}$- and $\mathscr{L}$-states, he suggested that these two forms of circular light propagate at different speeds. An active material shows *circular birefringence*; that is, it possesses two indices of refraction, one for $\mathscr{R}$-states ($n_{\mathscr{R}}$) and one for $\mathscr{L}$-states ($n_{\mathscr{L}}$). In traversing an optically active specimen, the two circular waves would get out-of-phase, and the resultant linear wave would appear to have rotated. We can see how this is possible analytically by returning to Eqs. (8.8) and (8.9), which described monochromatic right- and left-circular light propagating in the z-direction. It was seen in Eq. (8.10) that the sum of these two waves is indeed linearly polarized. We now alter these expressions slightly in order to remove the factor of two in the amplitude of Eq. (8.10), in which case

$$\mathbf{E}_{\mathscr{R}} = \frac{E_0}{2}[\hat{\mathbf{i}} \cos(k_{\mathscr{R}}z - \omega t) + \hat{\mathbf{j}} \sin(k_{\mathscr{R}}z - \omega t)] \quad (8.35a)$$

and

$$\mathbf{E}_{\mathscr{L}} = \frac{E_0}{2}[\hat{\mathbf{i}} \cos(k_{\mathscr{L}}z - \omega t) - \hat{\mathbf{j}} \sin(k_{\mathscr{L}}z - \omega t)] \quad (8.35b)$$

represent the right- and left-handed constituent waves. Since ω is constant, $k_{\mathscr{R}} = k_0 n_{\mathscr{R}}$ and $k_{\mathscr{L}} = k_0 n_{\mathscr{L}}$. The resultant disturbance is given by $\mathbf{E} = \mathbf{E}_{\mathscr{R}} + \mathbf{E}_{\mathscr{L}}$, and after a bit of trigonometric manipulation, it becomes

$$\mathbf{E} = E_0 \cos[(k_{\mathscr{R}} + k_{\mathscr{L}})z/2 - \omega t][\hat{\mathbf{i}} \cos(k_{\mathscr{R}} - k_{\mathscr{L}})z/2 + \hat{\mathbf{j}} \sin(k_{\mathscr{R}} - k_{\mathscr{L}})z/2] \quad (8.36)$$

At the position where the wave enters the medium ($z = 0$) it is linearly polarized along the x-axis, as shown in Fig. 8.55, that is,

$$\mathbf{E} = E_0 \hat{\mathbf{i}} \cos \omega t \quad (8.37)$$

Notice that at any point along the path, the two components have the same time dependence and are therefore in-phase. This just means that anywhere along the z-axis the resultant is linearly polarized (Fig. 8.56), although its orientation is certainly a function of z. Moreover, if $n_{\mathscr{R}} > n_{\mathscr{L}}$ or equivalently $k_{\mathscr{R}} > k_{\mathscr{L}}$, **E** will rotate counterclockwise, whereas if $k_{\mathscr{L}} > k_{\mathscr{R}}$, the rotation is clockwise (looking toward the source). Traditionally, the angle β through which **E** rotates is defined as positive when it is clockwise. Keeping this sign convention in mind, it should be clear from Eq. (8.36) that the field at point z makes an angle of $\beta = -(k_{\mathscr{R}} - k_{\mathscr{L}})z/2$ with respect to its original orientation. If the medium has a thickness d, the angle through which the plane of vibration rotates is then

$$\beta = \frac{\pi d}{\lambda_0}(n_{\mathscr{L}} - n_{\mathscr{R}}) \quad (8.38)$$

where $n_{\mathscr{L}} > n_{\mathscr{R}}$ is *d*-rotatory and $n_{\mathscr{R}} > n_{\mathscr{L}}$ is *l-rotatory* (Fig. 8.57).

Fresnel was actually able to separate the constituent $\mathscr{R}$- and $\mathscr{L}$-states of a linear beam using the composite prism of Fig. 8.58. It consists of a number of right- and left-handed quartz segments cut with their optic axes as shown. The $\mathscr{R}$-state

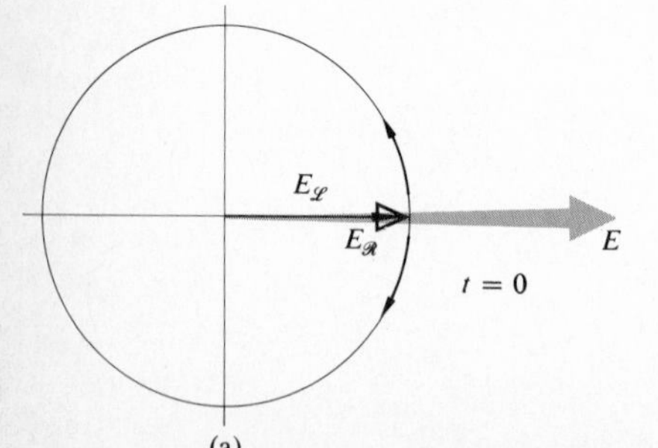

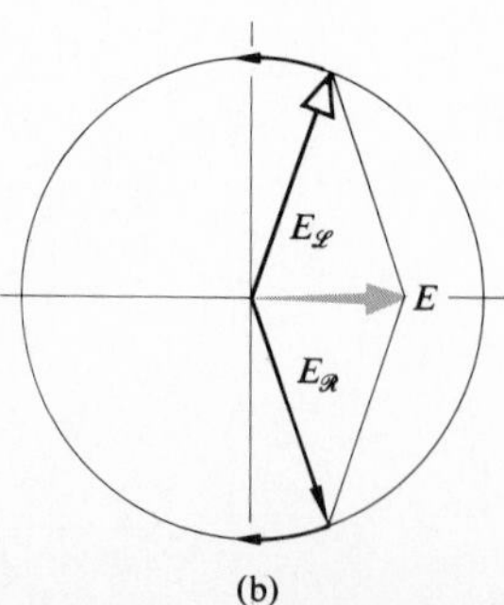

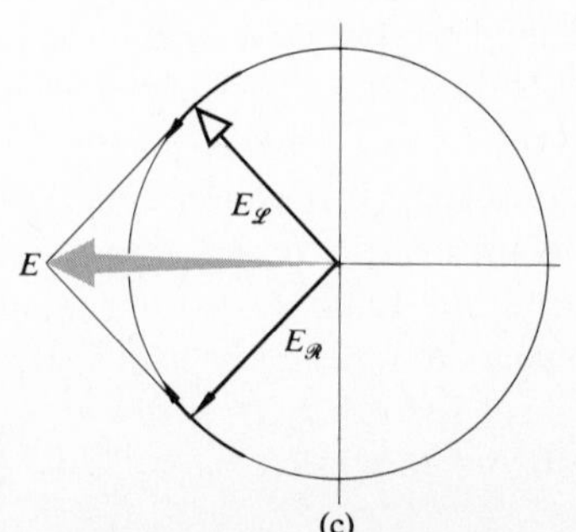

FIGURE 8.55 The superposition of an $\mathscr{R}$- and an $\mathscr{L}$-state at $z = 0$.

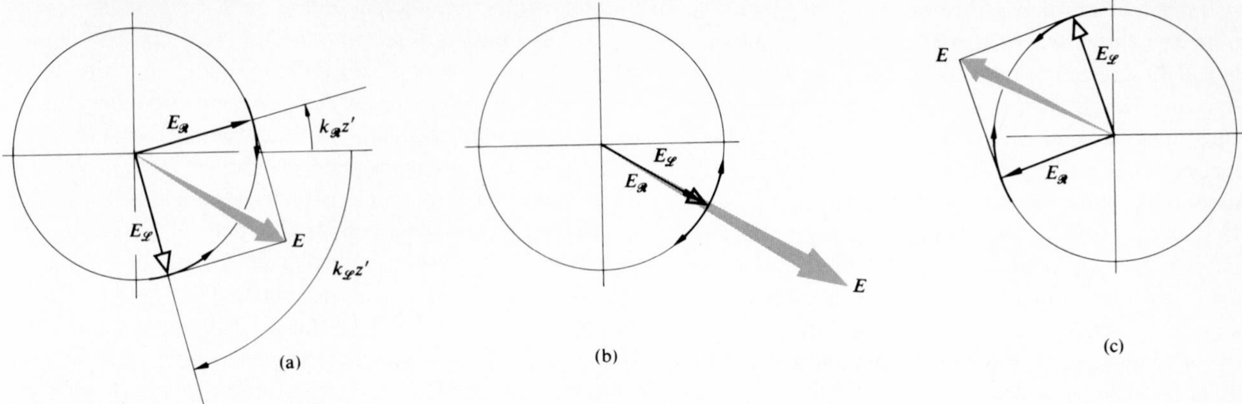

FIGURE 8.56 The superposition of an $\mathscr{R}$- and an $\mathscr{L}$-state at $z = z'$ $(k_{\mathscr{L}} > k_{\mathscr{R}})$.

propagates more rapidly in the first prism than in the second and is thus refracted toward the normal to the oblique boundary. The opposite is true for the $\mathscr{L}$-state, and the two circular waves increase in angular separation at each interface.

In sodium light the *specific rotatory power*, which is defined as β/d, is found to be 21.7°/mm for quartz. It follows that $|n_{\mathscr{L}} - n_{\mathscr{R}}| = 7.1 \times 10^{-5}$ for light propagating along the optic axis. In that particular direction ordinary double refraction vanishes. However, with the incident light propagating normal to the optic axis (as is frequently the case in polarizing prisms, wave plates, and compensators), quartz behaves like any optically inactive, positive, uniaxial crystal. There are other birefringent, optically active crystals, both uniaxial and biaxial, such as cinnabar, HgS ($n_o = 2.854$, $n_e = 3.201$), which has a rotatory power of 32.5°/mm. In contrast, the substance $NaClO_3$ is optically active (3.1°/mm) but not birefringent. The rotatory power of liquids, in comparison, is so relatively small that it is usually specified in terms of 10-cm path lengths; for example, in the case of turpentine ($C_{10}H_6$) it is only $-37°/10$ cm (10°C with $\lambda_0 = 589.3$ nm). The rotatory power of solutions varies with the concentration. This fact is particularly helpful in determining, for example, the amount of sugar present in a urine sample or a commercial sugar syrup.

You can observe optical activity rather easily using colorless corn syrup, the kind available in any grocery store. You

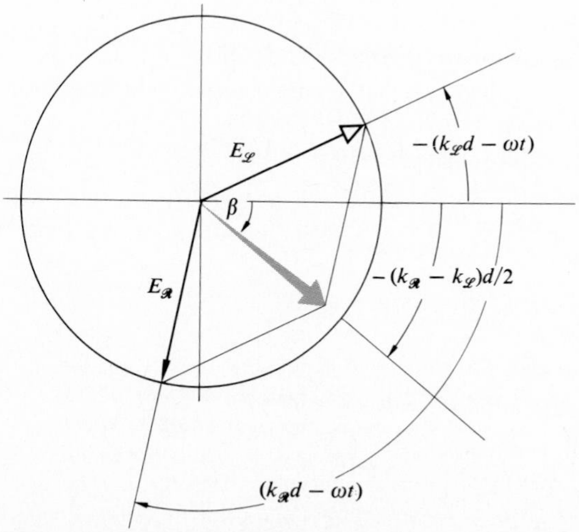

FIGURE 8.57 The superposition of an $\mathscr{R}$- and an $\mathscr{L}$-state at $z = d$ $(k_{\mathscr{L}} > k_{\mathscr{R}}$, $k_{\mathscr{L}} > k_{\mathscr{R}}$, $\lambda_{\mathscr{L}} < \lambda_{\mathscr{R}}$, and $v_{\mathscr{L}} < v_{\mathscr{R}})$.

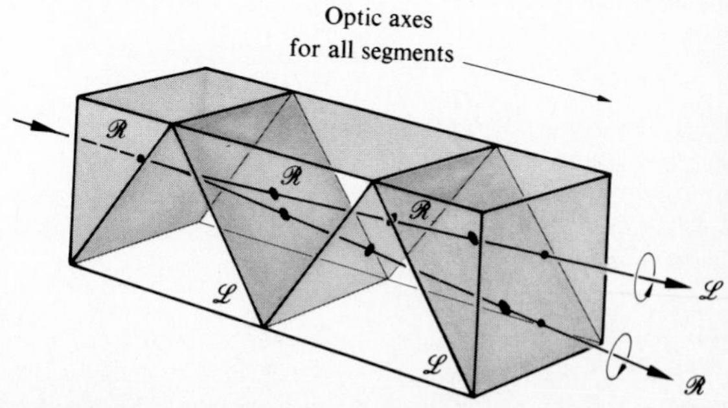

FIGURE 8.58 The Fresnel composite prism.

won't need much of it, since β/d is roughly $+30°$/inch. Put about an inch of syrup in a glass container between crossed Polaroids and illuminate it with a flashlight. The beautiful colors that appear as the analyzer is rotated arise from the fact that β is a function of λ_0, an effect known as *rotatory dispersion*. Using a filter to get roughly monochromatic light, you can readily determine the rotatory power of the syrup.*

The first great scientific contribution made by Louis Pasteur (1822–1895) came in 1848 and was associated with his doctoral research. He showed that racemic acid, which is an optically inactive form of tartaric acid, is actually composed of a mixture containing equal quantities of right- and left-handed constituents. Substances of this sort, which have the same molecular formulas but differ somehow in structure, are called *isomers*. He was able to crystallize racemic acid and then separate the two different types of mirror-image crystals (enantiomorphs) that resulted. When dissolved separately in water, they formed *d*-rotatory and *l*-rotatory solutions. This implied the existence of molecules that, although chemically the same, were themselves mirror images of each other; such molecules are now known as optical *stereoisomers*. These ideas were the basis for the development of the stereochemistry of organic and inorganic compounds, where one is concerned with the three-dimensional spatial distribution of atoms within a given molecule.

8.10.1 A Useful Model

The phenomenon of optical activity is extremely complicated, and although it can be treated in terms of classical Electromagnetic Theory, it actually requires a quantum-mechanical solution.† Despite this, we will consider a simplified model, which will yield a qualitative, yet plausible, description of the process. Recall that we represented an optically isotropic medium by a homogeneous distribution of isotropic electron-oscillators that vibrated parallel to the **E**-field of an incident wave. An optically anisotropic medium was similarly depicted as a distribution of anisotropic oscillators that vibrated at some angle to the driving **E**-field. We now imagine that the electrons in optically active substances are constrained to move along twisting paths that, for simplicity, are assumed to be helical. Such a molecule is pictured much as if it were a conducting helix. The silicon and oxygen atoms in a quartz crystal are known to be arranged in either right- or left-handed spirals about the optic axis, as indicated in Fig. 8.59. In the present representation this crystal would correspond to a parallel array of helices. In comparison, an active sugar solution would be analogous to a distribution of randomly oriented helices, each having the same handedness.*

In quartz we might anticipate that the incoming wave would interact differently with the specimen, depending on whether it "saw" right- or left-handed helices. Thus we could expect different indices for the $\mathscr{R}$- and $\mathscr{L}$-components of the wave. The detailed treatment of the process that leads to circular birefringence in crystals is by no means simple, but at least the necessary asymmetry is evident. How, then, can a random array of helices, corresponding to a solution, produce optical activity? Let us examine one such molecule in this simplified representation, for example, one whose axis happens to be parallel to the harmonic **E**-field of the electromagnetic wave. That field will drive charges up and down along the length of the molecule, effectively producing a time-varying electric dipole moment $\mathscr{p}(t)$, parallel to the axis. In addition, we now have a current associated with the spiraling motion of the electrons. This in turn generates an oscillating magnetic dipole moment $\mathscr{m}(t)$, which is also along the helix axis (Fig. 8.60). In contrast, if the molecule was parallel to the **B**-field of the wave, there would be a time-varying flux and thus an induced electron current circulating around the molecule. This would again yield oscillating axial electric and magnetic dipole moments. In either case $\mathscr{p}(t)$ and $\mathscr{m}(t)$ *will be parallel*

*A gelatin filter works well, but a piece of colored cellophane will also do nicely. Just remember that the cellophane will act as a wave plate (see Section 8.7.1), so don't put it between the polaroids unless you align its principal axes appropriately.

†The review article "Optical Activity and Molecular Dissymmetry," by S. F. Mason, *Contemp. Phys.* **9**, 239 (1968), contains a fairly extensive list of references for further reading.

*In addition to these solid and liquid states, there is a third classification of substances, which is useful because of its remarkable optical properties. It is known as the *mesomorphic* or *liquid crystal* state. Liquid crystals are organic compounds that can flow and yet maintain their characteristic molecular orientations. In particular, *cholesteric* liquid crystals have a helical structure and therefore exhibit extremely large rotatory powers, of the order of 40000°/mm. The pitch of the screw-like molecular arrangement is considerably smaller than that of quartz.

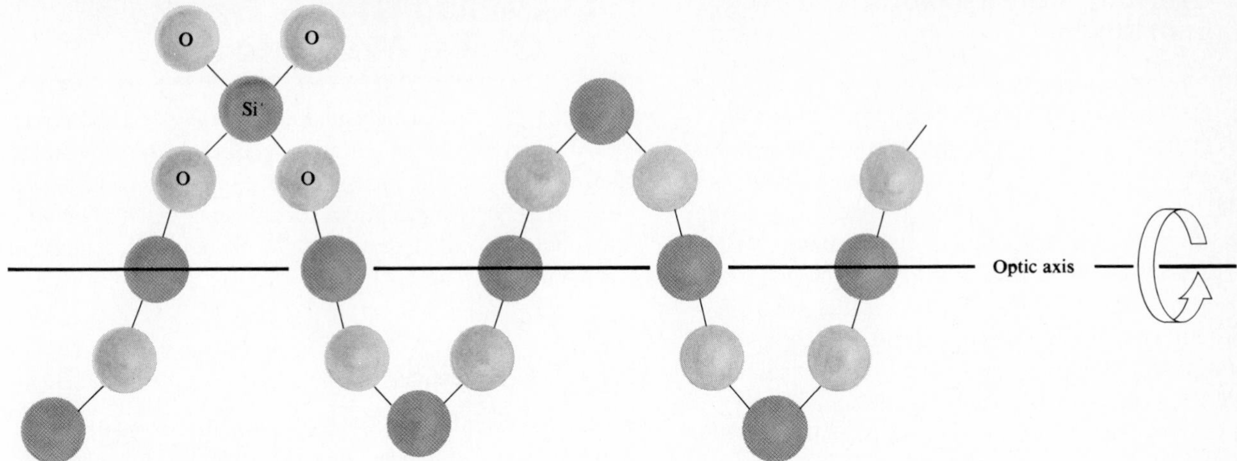

FIGURE 8.59 Right-handed quartz.

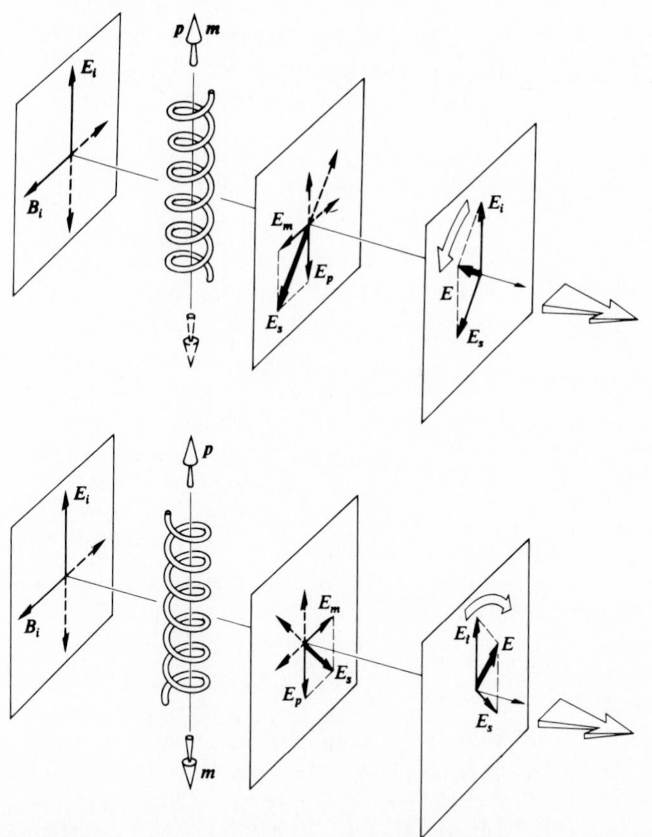

FIGURE 8.60 The radiation from helical molecules.

or antiparallel to each other depending on the sense of the particular molecular helix. Clearly, energy has been removed from the field, and both oscillating dipoles will scatter (i.e., reradiate) electromagnetic waves. The electric field $\mathbf{E}_p$ emitted in a given direction by an electric dipole is perpendicular to the electric field $\mathbf{E}_m$ emitted by a magnetic dipole. The sum of these, which is the resultant field $\mathbf{E}_s$ scattered by a helix, will not be parallel to the incident field $\mathbf{E}_i$ along the direction of propagation. (The same is of course true for the magnetic fields.) The plane of vibration of the resultant transmitted light ($\mathbf{E}_s + \mathbf{E}_i$) will thus be rotated in a direction determined by the sense of the helix. The amount of the rotation will vary with the orientation of each molecule, but it will always be in the same direction for helices of the same sense.

Although this discussion of optically active molecules as helical conductors is admittedly superficial, the analogy is well worth keeping in mind. In fact, if we direct a linear 3-cm microwave beam onto a box filled with a large number of identical copper helices (e.g., 1 cm long by 0.5 cm in diameter and insulated from each other), the transmitted wave will undergo a rotation of its plane of vibration.*

*I. Tinoco and M. P. Freeman, "The Optical Activity of Oriented Copper Helices," *J. Phys. Chem.* **61**, 1196 (1957).

8.10.2 Optically Active Biological Substances

Among the most fascinating observations associated with optical activity are those in biology. Whenever organic molecules are synthesized in the laboratory, an equal number of *d*- and *l*-isomers are produced, with the effect that the compound is optically inactive. One might then expect that if they exist at all, equal amounts of *d*- and *l*-optical stereoisomers will be found in natural organic substances. This is by no means the case. Natural sugar (sucrose, $C_{12}H_{22}O_{11}$), no matter where it is grown, whether extracted from sugar cane or sugar beets, is always *d*-rotatory. Moreover, the simple sugar dextrose or *d*-glucose ($C_6H_{12}O_6$), which as its name implies is *d*-rotatory, is the most important carbohydrate in human metabolism. Evidently, living things can somehow distinguish between optical isomers.

All proteins are fabricated of compounds known as *amino acids*. These in turn are combinations of carbon, hydrogen, oxygen, and nitrogen. There are twenty-odd amino acids, and all of them (with the exception of the simplest one, glycine, which is not enantiomorphic) are generally *l*-rotatory. This means that if we break up a protein molecule, whether it comes from an egg or an eggplant, a beetle or a Beatle, the constituent amino acids will be *l*-rotatory. One important exception is the group of antibiotics, such as penicillin, which do contain some dextro amino acids. In fact, this may well account for the toxic effect penicillin has on bacteria.

It is intriguing to speculate about the possible origins of life on this and other planets. For example, did life on Earth originally consist of both mirror-image forms? Five amino acids were found in a meteorite that fell in Victoria, Australia, on September 28, 1969, and analysis has revealed the existence of roughly equal amounts of the optically right- and left-handed forms. This is in marked contrast to the overwhelming predominance of the left-handed form found in terrestrial rocks. The implications are many and marvelous.*

*See *Physics Today*, Feb. 1971, p. 17, for additional discussion and references for further reading.

8.11 Induced Optical Effects—Optical Modulators

A number of different physical effects involving polarized light all share the single common feature of somehow being externally induced. In these instances, one exerts an external influence (e.g., a mechanical force, a magnetic or electric field) on the optical medium, thereby changing the manner in which it transmits light.

8.11.1 Photoelasticity

In 1816 Sir David Brewster discovered that normally transparent isotropic substances could be made optically anisotropic by the application of mechanical stress. The phenomenon is known as *mechanical birefringence*, **photoelasticity**, or *stress birefringence*. Under compression or tension, the material takes on the properties of a negative or positive uniaxial crystal, respectively. In either case, the effective optic axis is in the direction of the stress, and the induced birefringence is proportional to the stress. If the stress is not uniform over the sample, neither is the birefringence or the retardance imposed on a transmitted wave [Eq. (8.32)].

Photoelasticity serves as the basis of a technique for studying the stresses in both transparent and opaque mechanical structures (Fig. 8.61). Improperly annealed or carelessly mounted glass, whether serving as an automobile windshield or a telescope lens, will develop internal stresses that can easily be detected. Information concerning the surface strain on opaque objects can be obtained by bonding photoelastic coatings to the parts under study. More commonly, a transparent scale model of the part is made out of a material *optically sensitive to stress*, such as epoxy, glyptol, or modified polyester resins. The model is then subjected to the forces that the actual component would experience in use. Since the birefringence varies from point to point over the surface of the model, when it is placed between crossed polarizers, a complicated variegated fringe pattern will reveal the internal stresses. Examine almost any piece of clear plastic or even a block of unflavored gelatin between two Polaroids; try stressing it further and watch the pattern change accordingly (Fig. 8.62).

The retardance at any point on the sample is proportional to the *principal stress difference*; that is, $(\sigma_1 - \sigma_2)$, where the sigmas are the orthogonal principal stresses. For example, if

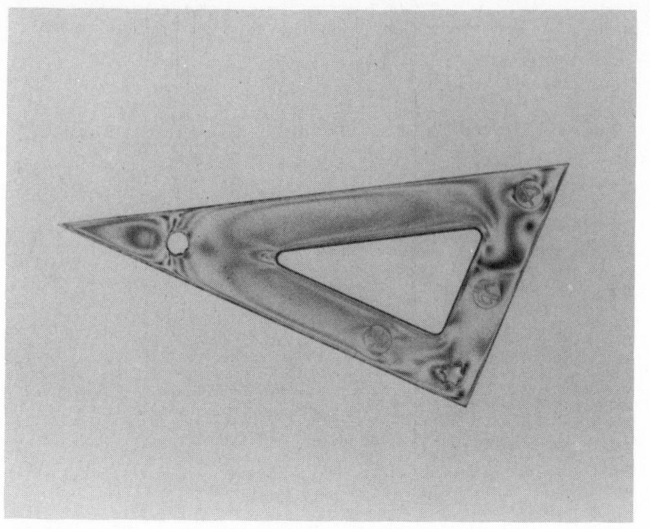
(a)

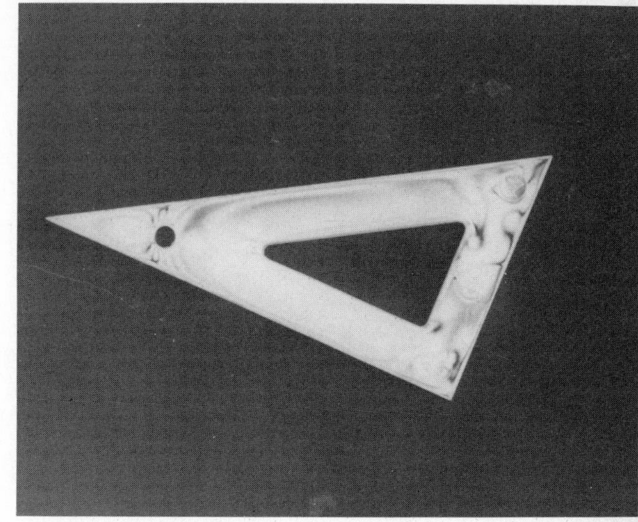
(b)

FIGURE 8.61 A clear plastic triangle between polaroids. (E.H.)

the sample were a plate under vertical tension, σ_1 would be the maximum principal stress in the vertical direction and σ_2 would be the minimum principal stress, in this case zero, horizontally. In more complicated situations, the principal stresses, as well as their differences, will vary from one region to the next. Under white-light illumination, the loci of all points on the specimen for which $(\sigma_1 - \sigma_2)$ is constant are known as *isochromatic regions*, and each such region corresponds to a particular color. Superimposed on these colored fringes will be a separate system of black bands. At any point where the **E**-field of the incident linear light is parallel to either local principal stress axis, the wave will pass through the sample

(a)

(b)

FIGURE 8.62 A stressed piece of clear plastic between polaroids. (E.H.)

unaffected, regardless of wavelength. With crossed polarizers, that light will be absorbed by the analyzer, yielding a black region known as an *isoclinic* band (Problem 8.48). In addition to being beautiful to look at, the fringes also provide both a qualitative map of the stress pattern and a basis for quantitative calculations.

8.11.2 The Faraday Effect

Michael Faraday in 1845 discovered that the manner in which light propagated through a material medium could be influenced by the application of an external magnetic field. In particular, he found that the plane-of-vibration of linear light incident on a piece of glass rotated when a strong magnetic field was applied in the propagation direction. The **Faraday Effect** was one of the earliest indications of the interrelationship between electromagnetism and light. Although it is reminiscent of optical activity, there is an important distinction.

The angle β (measured in minutes of arc) through which the plane-of-vibration rotates is given by the empirically determined expression

$$\beta = \mathcal{V}Bd \tag{8.39}$$

where B is the static magnetic flux density (usually in gauss), d is the length of medium traversed (in cm), and $\mathcal{V}$ is a factor of proportionality known as the ***Verdet constant***. The Verdet constant for a particular medium varies with both frequency (dropping off rapidly as ν decreases) and temperature. It is roughly of the order of 10^{-5} min of arc gauss^{-1} cm^{-1} for gases and 10^{-2} min of arc gauss^{-1} cm^{-1} for solids and liquids (see Table 8.2). You can get a better feeling for the meaning of these numbers by imagining, for example, a 1-cm-long sample of H_2O in the moderately large field of 10^4 gauss. (The Earth's field is about one half gauss.) In that particular case, a rotation of $2° \; 11'$ would result since $\mathcal{V} = 0.0131$.

By convention, *a positive Verdet constant corresponds to a (diamagnetic) material for which the Faraday Effect is l-rotatory when the light moves parallel to the applied* **B***-field and d-rotatory when it propagates antiparallel to* **B**. No such reversal of handedness occurs in the case of natural optical activity. For a convenient mnemonic, imagine the **B**-field to be generated by a solenoidal coil wound about the sample. The plane-of-vibration, when $\mathcal{V}$ is positive, rotates in the same direction as the current in the coil, regardless of the beam's propagation direction along its axis. Consequently, the effect can be amplified by reflecting the light back and forth a few times through the sample.

TABLE 8.2 Verdet constants for some selected substances.

Material	Temperature (°C)	$\mathcal{V}$ (min of arc gauss^{-1} cm^{-1})
Light flint glass	18	0.0317
Water	20	0.0131
NaCl	16	0.0359
Quartz	20	0.0166
$NH_4Fe(SO_4)_2.12H_2O$	26	−0.00058
Air*	0	6.27×10^{-6}
CO_2*	0	9.39×10^{-6}

*λ = 578 nm and 760 mm Hg.
More extensive listings are given in the usual handbooks.

The theoretical treatment of the Faraday Effect involves the quantum-mechanical theory of dispersion, including the effects of **B** on the atomic or molecular energy levels. It will suffice here merely to outline the limited classical argument for nonmagnetic materials.

Suppose the incident light to be circular and monochromatic. An elastically bound electron will take on a steady-state circular orbit being driven by the rotating **E**-field of the wave. (The effect of the wave's **B**-field is negligible.) The introduction of a large constant applied magnetic field perpendicular to the plane of the orbit will result in a radial force F_M on the electron. That force can point either toward or away from the circle's center, depending on the handedness of the light and the direction of the constant **B**-field. The total radial force (F_M plus the elastic restoring force) can therefore have two different values and so too can the radius of the orbit. Consequently, for a given magnetic field there will be two possible values of the electric dipole moment, the polarization, and the permittivity, as well as two values of the index of refraction, $n_{\mathcal{R}}$ and $n_{\mathcal{L}}$. The discussion can then proceed in precisely the same fashion as that of Fresnel's treatment of optical activity. As before, one speaks of two normal modes of propagation of electromagnetic waves through the medium, the $\mathcal{R}$- and $\mathcal{L}$-states.

For ferromagnetic substances things are somewhat more complicated. In the case of a magnetized material β is proportional to the component of the magnetization in the direction of propagation rather than the component of the applied dc field.

There are a number of practical applications of the Faraday Effect. It can be used to analyze mixtures of hydrocarbons, since each constituent has a characteristic magnetic rotation. When utilized in spectroscopic studies, it yields information about the properties of energy states above the ground level. Interestingly, the Faraday Effect has been used to make optical modulators. An infrared version, constructed by R. C. LeCraw, utilized the synthetic magnetic crystal yttrium–iron garnet (YIG), to which has been added a quantity of gallium. YIG has a structure similar to that of natural gem garnets. The device is depicted schematically in Fig. 8.63. A linear infrared laser beam enters the crystal from the left. A transverse dc magnetic field saturates the magnetization of the YIG crystal in that direction. The total magnetization vector (arising from the constant field and the field of the coil) can vary in direction, being tilted toward the axis of the crystal by an amount proportional to the modulating current in the coil. Since the Faraday rotation depends on the axial component of the magnetization, the coil current controls β. The analyzer then converts this polarization modulation to amplitude modulation by way of Malus's Law [Eq. (8.24)]. In short, the signal to be transmitted is introduced across the coil as a modulating voltage, and the emerging laser beam carries that information in the form of amplitude variations.

There are actually several other magneto-optic effects. We shall consider only two of these, and rather succinctly at that. The *Voigt* and *Cotton–Mouton Effects* both arise when a constant magnetic field is applied to a transparent medium perpendicular to the direction of propagation of the incident light beam. The former occurs in vapors, whereas the latter, which is considerably stronger, occurs in liquids. In either case the medium displays birefringence similar to that of a uniaxial crystal whose optic axis is in the direction of the dc magnetic field, that is, normal to the light beam [Eq. (8.32)]. The two indices of refraction now correspond to the situations in which the plane-of-vibration of the wave is either normal or parallel to the constant magnetic field. Their difference Δn (i.e., the birefringence) is proportional to the square of the applied magnetic field. It arises in liquids from an aligning of the optically and magnetically anisotropic molecules of the medium with that field. If the incoming light propagates at some angle to the static field other than 0 or $\pi/2$, the Faraday and Cotton–Mouton Effects occur concurrently, with the former generally being much the larger of the two. The Cotton–Mouton is the magnetic analogue of the Kerr (electro-optic) Effect, to be considered next.

8.11.3 The Kerr and Pockels Effects

The first electro-optic effect was discovered by the Scottish physicist John Kerr (1824–1907) in 1875. He found that an isotropic transparent substance becomes birefringent when placed in an electric field **E**. The medium takes on the charac-

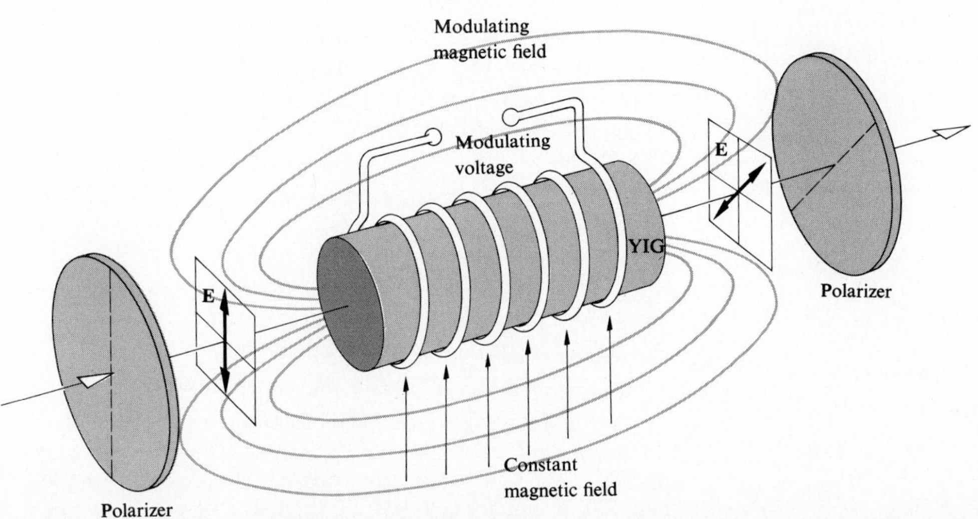

FIGURE 8.63 A Faraday Effect modulator.

teristics of a uniaxial crystal whose optic axis corresponds to the direction of the applied field. The two indices, $n_{\parallel}$ and $n_{\perp}$, are associated with the two orientations of the plane-of-vibration of the wave, namely, parallel and perpendicular to the applied electric field, respectively. Their difference, Δn, is the birefringence, and it is found to be

$$\Delta n = \lambda_0 K E^2 \tag{8.40}$$

where K is the ***Kerr constant***. When K is positive, as it most often is, Δn, which can be thought of as $n_e - n_o$, is positive, and the substance behaves like a positive uniaxial crystal. Values of the Kerr constant (Table 8.3) are often listed in electrostatic units, so that one must remember to enter E in Eq. (8.40) in statvolts per cm (one statvolt $\approx$ 300 V). Observe that, as with the Cotton–Mouton Effect, *the Kerr Effect is proportional to the square of the field and is often referred to as the quadratic electro-optic effect*. The phenomenon in liquids is attributed to a partial alignment of anisotropic molecules by the **E**-field. In solids the situation is considerably more complicated.

Figure 8.64 depicts an arrangement known as a Kerr shutter or optical modulator. It consists of a glass cell containing two electrodes, which is filled with a polar liquid. This *Kerr cell*, as it is called, is positioned between crossed linear polarizers whose transmission axes are at $\pm 45°$ to the applied **E**-field. With zero voltage across the plates no light will be transmitted; the shutter is closed. The application of a modulating voltage generates a field, causing the cell to function as a variable wave plate and thus opening the shutter proportionately. The great value of such a device lies in the fact that it can respond effectively to frequencies roughly as high as 10^{10} Hz. Kerr cells, usually containing nitrobenzene or carbon disulfide, have been used for a number of years in a variety of applications. They serve as shutters in high-speed photography and as light-beam choppers to replace rotating toothed wheels. As such, they have been utilized in measurements of the speed of light. Kerr cells are also used as Q-switches in pulsed laser systems.

TABLE 8.3 Kerr constants for some selected liquids (20°C, λ_0 = 589.3 nm).

	Substance	K (in units of 10^{-7} cm statvolt^{-2})
Benzene	C_6H_6	0.6
Carbon disulfide	CS_2	3.2
Chloroform	$CHCl_3$	−3.5
Water	H_2O	4.7
Nitrotoluene	$C_5H_7NO_2$	123
Nitrobenzene	$C_6H_5NO_2$	220

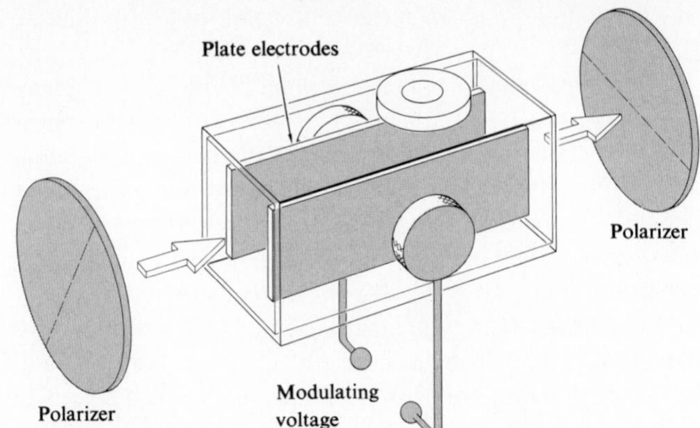

FIGURE 8.64 A Kerr cell.

If the plates functioning as the electrodes have an effective length of ℓ cm and are separated by a distance d, the retardation is given by

$$\Delta\varphi = 2\pi K \ell V^2/d^2 \tag{8.41}$$

where V is the applied voltage. Thus a nitrobenzene cell in which d is one cm and ℓ is several cm will require a rather large voltage, roughly 3×10^4 V, in order to respond as a half-wave plate. This is a characteristic quantity known as the *half-wave voltage*, $V_{\lambda/2}$. Another drawback is that nitrobenzene is both poisonous and explosive. Transparent solid substances, such as the mixed crystal potassium tantalate niobate ($KTa_{0.65}Nb_{0.35}O_3$), KTN for short, or barium titanate ($BaTiO_3$), which show a Kerr Effect, are therefore of interest as electro-optical modulators.

There is another very important electro-optical effect known as the *Pockels Effect*, after the German physicist Friedrich Carl Alwin Pockels (1865–1913), who studied it extensively in 1893. It is a linear electro-optical effect, inasmuch as the induced birefringence is proportional to the first power of the applied **E**-field and therefore the applied voltage. The Pockels Effect exists only in certain crystals that lack a center of symmetry—in other words, crystals having no central point through which every atom can be reflected into an identical atom. There are 32 crystal symmetry classes, 20 of which may show the Pockels Effect. Incidentally, these same 20 classes are also piezoelectric. Thus, many crystals and all liquids are excluded from displaying a linear electro-optic effect.

The first practical Pockels cell, which could perform as a shutter or modulator, was not made until the 1940s, when suitable crystals were finally developed. The operating principle for such a device is one we've already discussed. In brief, the birefringence is varied electronically by means of a controlled applied electric field. The retardance can be altered as desired, thereby changing the state of polarization of the incident linear wave. In this way, the system functions as a polarization modulator. Early devices were made of ammonium dihydrogen phosphate ($NH_4H_2PO_4$), or ADP, and potassium dihydrogen phosphate (KH_2PO_4), known as KDP; both are still in use. A great improvement was provided by the introduction of single crystals of potassium dideuterium phosphate (KD_2PO_4), or KD*P, which yields the same retardation with voltages less than half of those needed for KDP. This process of infusing crystals with deuterium is accomplished by growing them in a solution of heavy water. Cells made with KD*P or CD*A (cesium dideuterium arsenate) have been produced commercially for some time.

A *Pockels cell* is simply an appropriate noncentrosymmetric, oriented, single crystal immersed in a controllable electric field. Such devices can usually be operated at fairly low voltages (roughly 5 to 10 times less than that of an equivalent Kerr cell); they are linear, and of course there is no problem with toxic liquids. The response time of KDP is quite short, typically less than 10 ns, and it can modulate a light beam at up to about 25 GHz (i.e., 25×10^9 Hz).

There are two common cell configurations, referred to as *transverse* and *longitudinal*, depending on whether the applied **E**-field is perpendicular or parallel to the direction of propagation, respectively. The longitudinal type is illustrated, in its most basic form, in Fig. 8.65. Since the beam traverses the electrodes, these are usually made of transparent metal-oxide coatings (e.g., SnO, InO, or CdO), thin metal films, grids, or rings. The crystal itself is generally uniaxial in the absence of an applied field, and it is aligned such that its optic axis is along the beam's propagation direction. For such an arrangement the retardance is given by

$$\Delta\varphi = 2\pi n_o^3 r_{63} V/\lambda_0 \tag{8.42}$$

where r_{63} is the *electro-optic constant* in m/V, n_o is the *ordinary* index of refraction, V is the potential difference in volts, and λ_0 is the vacuum wavelength in meters.* Since the crystals are anisotropic, their properties vary in different directions, and they must be described by a group of terms referred to collectively as the second-rank electro-optic tensor r_{ij}. Fortunately, we need only concern ourselves here with one of its components, namely, r_{63}, values of which are given in Table 8.4. The half-wave voltage corresponds to a value of $\Delta\varphi = \pi$, in which case

$$\Delta\varphi = \pi \frac{V}{V_{\lambda/2}} \tag{8.43}$$

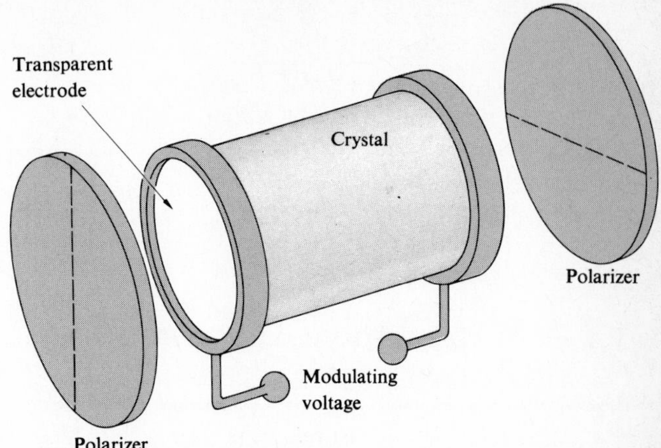

FIGURE 8.65 A Pockels cell.

TABLE 8.4 Electro-optic constants (room temperature, λ_0 = 546.1 nm).

Material	r_{63} (units of 10^{-12} m/V)	n_o (approx.)	$V_{\lambda/2}$ (in kV)
ADP ($NH_4H_2PO_4$)	8.5	1.52	9.2
KDP (KH_2PO_4)	10.6	1.51	7.6
KDA (KH_2AsO_4)	~13.0	1.57	~6.2
KD*P (KD_2PO_4)	~23.3	1.52	~3.4

*This expression, along with the appropriate one for the transverse mode, is derived rather nicely in A. Yariv, *Quantum Electronics*. Even so, the treatment is sophisticated and not recommended for casual reading.

and from Eq. (8.42)

$$V_{\lambda/2} = \frac{\lambda_0}{2n_o^3 r_{63}} \tag{8.44}$$

As an example, for KDP, $r_{63} = 10.6 \times 10^{-12}$ m/V, $n_o = 1.51$, and we obtain $V_{\lambda/2} \approx 7.6 \times 10^3$ V at $\lambda_0 = 546.1$ nm.

Pockels cells have been used as ultra-fast shutters, *Q*-switches for lasers, and dc to 30-GHz light modulators.*

8.12 A Mathematical Description of Polarization

Until now we've considered polarized light in terms of the electric field component of the wave. The most general representation was, of course, that of elliptical light. The endpoint of the vector **E** was envisioned continuously sweeping along the path of an ellipse having a particular shape—the circle and line being special cases. The period over which the ellipse was traversed equaled that of the lightwave (i.e., roughly 10^{-15} s) and was far too short to be detected. In contrast, measurements made in practice are generally averages over comparatively long time intervals.

Clearly, it would be advantageous to formulate an alternative description of polarization in terms of convenient observables, namely, irradiances. Our motives are far more than the ever-present combination of aesthetics and pedagogy. The formalism to be considered has far-reaching significance in other areas of study, for example, particle physics (the photon is, after all an elementary particle) and Quantum Mechanics. It serves in some respects to link the classical and quantum-mechanical pictures. But even more demanding of our present attention are the considerable practical advantages to be gleaned from this alternative description.

We shall evolve an elegant procedure for predicting the effects of complex systems of polarizing elements on the ultimate state of an emergent wave. The mathematics, written in the compressed form of matrices, will require only the simplest manipulation of those matrices. The complicated logic associated with phase retardations, relative orientations, and so forth, for a tandem series of wave plates and polarizers is almost all built in. One need only select appropriate matrices from a chart and drop them into the mathematical mill.

*The reader interested in light modulation in general should consult D. F. Nelson, "The Modulation of Laser Light," *Scientific American* (June 1968). Also see Chapter 14, Vol. II of *Handbook of Optics* (1995).

8.12.1 The Stokes Parameters

The modern representation of polarized light actually had its origins in 1852 in the work of G. G. Stokes. He introduced four quantities that are functions only of observables of the electromagnetic wave and are now known as the **Stokes parameters**.* The polarization state of a beam of light (either natural or totally or partially polarized) can be described in terms of these quantities. We will first define the parameters operationally and then relate them to electromagnetic theory.

Imagine that we have a set of four filters, each of which, under *natural* illumination, will transmit half the incident light, the other half being discarded. The choice is not a unique one, and a number of equivalent possibilities exist. Suppose then that the first filter is simply isotropic, passing all states equally, whereas the second and third are linear polarizers whose transmission axes are horizontal and at +45° (diagonal along the first and third quadrants), respectively. The last filter is a circular polarizer opaque to $\mathscr{L}$-states. Each of these four filters is positioned alone in the path of the beam under investigation, and the transmitted irradiances I_0, I_1, I_2, I_3 are measured with a type of meter that is insensitive to polarization (not all of them are). The operational definition of the Stokes parameters is then given by the relations

$$\mathcal{S}_0 = 2I_0 \tag{8.45a}$$

$$\mathcal{S}_1 = 2I_1 - 2I_0 \tag{8.45b}$$

$$\mathcal{S}_2 = 2I_2 - 2I_0 \tag{8.45c}$$

$$\mathcal{S}_3 = 2I_3 - 2I_0 \tag{8.45d}$$

Notice that $\mathcal{S}_0$ is simply the incident irradiance, and $\mathcal{S}_1$, $\mathcal{S}_2$, and $\mathcal{S}_3$ specify the state of polarization. Thus $\mathcal{S}_1$ reflects a ten-

*Much of the material in this section is treated more extensively in Shurcliff's *Polarized Light: Production and Use*, which is something of a classic on the subject. You might also look at M. J. Walker, "Matrix Calculus and the Stokes Parameters of Polarized Radiation," *Am. J. Phys.* **22**, 170 (1954), and W. Bickel and W. Bailey, "Stokes Vectors, Mueller Matrices, and Polarized Scattered Light," *Am. J. Phys.* **53**, 468 (1985).

dency for the polarization to resemble either a horizontal $\mathscr{P}$-state (whereupon $\mathscr{S}_1 > 0$) or a vertical one (in which case $\mathscr{S}_1 < 0$). When the beam displays no preferential orientation with respect to these axes ($\mathscr{S}_1 = 0$) it may be elliptical at $\pm 45°$, circular, or unpolarized. Similarly, $\mathscr{S}_2$ implies a tendency for the light to resemble a $\mathscr{P}$-state oriented in the direction of $+45°$ (when $\mathscr{S}_2 > 0$) or in the direction of $-45°$ (when $\mathscr{S}_2 < 0$) or neither ($\mathscr{S}_2 = 0$). In the same way $\mathscr{S}_3$ reveals a tendency of the beam toward right-handedness ($\mathscr{S}_3 > 0$), left-handedness ($\mathscr{S}_3 < 0$), or neither ($\mathscr{S}_3 = 0$).

Now recall the expressions for quasimonochromatic light,

$$\mathbf{E}_x(t) = \hat{\mathbf{i}} E_{0x}(t) \cos [(\bar{k}z - \bar{\omega}t) + \varepsilon_x(t)] \qquad [8.34a]$$

and

$$\mathbf{E}_y(t) = \hat{\mathbf{j}} E_{0y}(t) \cos [(\bar{k}z - \bar{\omega}t) + \varepsilon_y(t)] \qquad [8.34b]$$

where $\mathbf{E}(t) = \mathbf{E}_x(t) + \mathbf{E}_y(t)$. Using these in a fairly straightforward way, we can recast the Stokes parameters* as

$$\mathscr{S}_0 = \langle E_{0x}^2 \rangle_T + \langle E_{0y}^2 \rangle_T \qquad (8.46a)$$

$$\mathscr{S}_1 = \langle E_{0x}^2 \rangle_T - \langle E_{0y}^2 \rangle_T \qquad (8.46b)$$

$$\mathscr{S}_2 = \langle 2E_{0x}E_{0y} \cos \varepsilon \rangle_T \qquad (8.46c)$$

$$\mathscr{S}_3 = \langle 2E_{0x}E_{0y} \sin \varepsilon \rangle_T \qquad (8.46d)$$

Here $\varepsilon = \varepsilon_y - \varepsilon_x$ and we've dropped the constant $\epsilon_0 c/2$, so that the parameters are now *proportional* to irradiances. For the hypothetical case of perfectly monochromatic light, $E_{0x}(t)$, $E_{0y}(t)$, and $\varepsilon(t)$ are time-independent, and one need only drop the $\langle\ \rangle$ brackets in Eq. (8.46) to get the applicable Stokes parameters. Interestingly enough, these same results can be obtained by time averaging Eq. (8.14), which is the general equation for elliptical light.†

If the beam is unpolarized, $\langle E_{0x}^2 \rangle_T = \langle E_{0y}^2 \rangle_T$; neither averages to zero, because the amplitude squared is always positive. In that case $\mathscr{S}_0 = \langle E_{0x}^2 \rangle_T + \langle E_{0y}^2 \rangle_T$, but $\mathscr{S}_1 = \mathscr{S}_2 = \mathscr{S}_3 = 0$. The latter two parameters go to zero, since both cos ε and sin ε average to zero independently of the amplitudes. It is often convenient to *normalize* the Stokes parameters by dividing each one by the value of $\mathscr{S}_0$. This has the effect of using an incident beam of unit irradiance. The set of parameters ($\mathscr{S}_0$, $\mathscr{S}_1$, $\mathscr{S}_2$, $\mathscr{S}_3$) for *natural light* in the normalized representation is then (1, 0, 0, 0). If the light is horizontally polarized, it has no vertical component, and the normalized parameters are (1, 1, 0, 0). Similarly, for vertically polarized light we have (1, −1, 0, 0). Representations of a few other polarization states are listed in Table 8.5. (The parameters are displayed vertically for reasons to be discussed later.) Notice that for completely polarized light it follows from Eq. (8.46) that

$$\mathscr{S}_0^2 = \mathscr{S}_1^2 + \mathscr{S}_2^2 + \mathscr{S}_3^2 \qquad (8.47)$$

TABLE 8.5 Stokes and Jones vectors for some polarization states.

State of polarization	Stokes vectors	Jones vectors
Horizontal $\mathscr{P}$-state	$\begin{bmatrix}1\\1\\0\\0\end{bmatrix}$	$\begin{bmatrix}1\\0\end{bmatrix}$
Vertical $\mathscr{P}$-state	$\begin{bmatrix}1\\-1\\0\\0\end{bmatrix}$	$\begin{bmatrix}0\\1\end{bmatrix}$
$\mathscr{P}$-state at $+45°$	$\begin{bmatrix}1\\0\\1\\0\end{bmatrix}$	$\frac{1}{\sqrt{2}}\begin{bmatrix}1\\1\end{bmatrix}$
$\mathscr{P}$-state at $-45°$	$\begin{bmatrix}1\\0\\-1\\0\end{bmatrix}$	$\frac{1}{\sqrt{2}}\begin{bmatrix}1\\-1\end{bmatrix}$
$\mathscr{R}$-state	$\begin{bmatrix}1\\0\\0\\1\end{bmatrix}$	$\frac{1}{\sqrt{2}}\begin{bmatrix}1\\-i\end{bmatrix}$
$\mathscr{L}$-state	$\begin{bmatrix}1\\0\\0\\-1\end{bmatrix}$	$\frac{1}{\sqrt{2}}\begin{bmatrix}1\\i\end{bmatrix}$

*For the details, see E. Hecht, "Note on an Operational Definition of the Stokes Parameters," *Am. J. Phys.* **38**, 1156 (1970).

†E. Collett, "The Description of Polarization in Classical Physics," *Am. J. Phys.* **36**, 713 (1968).

Moreover, for partially polarized light it can be shown that the degree of polarization [Eq. (8.29)] is given by

$$V = (\mathcal{S}_1^2 + \mathcal{S}_2^2 + \mathcal{S}_3^2)^{1/2}/\mathcal{S}_0 \tag{8.48}$$

Imagine now that we have two quasimonochromatic waves described by $(\mathcal{S}_0', \mathcal{S}_1', \mathcal{S}_2', \mathcal{S}_3')$ and $(\mathcal{S}_0'', \mathcal{S}_1'', \mathcal{S}_2'', \mathcal{S}_3'')$, which are superimposed in some region of space. As long as the waves are *incoherent*, any one of the Stokes parameters of the resultant will be the sum of the corresponding parameters of the constituents (all of which are proportional to irradiance). In other words, the set of parameters describing the resultant is $(\mathcal{S}_0' + \mathcal{S}_0'', \mathcal{S}_1' + \mathcal{S}_1'', \mathcal{S}_2' + \mathcal{S}_2'', \mathcal{S}_3' + \mathcal{S}_3'')$. For example, if a unit-flux density vertical $\mathcal{P}$-state $(1, -1, 0, 0)$ is added to an *incoherent* $\mathcal{L}$-state (see Table 8.5) of flux density 2, $(2, 0, 0, -2)$, the composite wave has parameters $(3, -1, 0, -2)$. It is an ellipse of flux density 3, more nearly vertical than horizontal $(\mathcal{S}_1 < 0)$, left-handed $(\mathcal{S}_3 < 0)$, and having a degree of polarization of $\sqrt{5}/3$.

The set of Stokes parameters for a given wave can be envisaged as a *vector*; we have already seen how two such (incoherent) vectors add.* Indeed, it will not be the usual kind of three-dimensional vector, but this sort of representation is widely used in physics to great advantage. More specifically, the parameters $(\mathcal{S}_0, \mathcal{S}_1, \mathcal{S}_2, \mathcal{S}_3)$ are arranged in the form of what is called a *column vector*,

$$\mathcal{S} = \begin{bmatrix} \mathcal{S}_0 \\ \mathcal{S}_1 \\ \mathcal{S}_2 \\ \mathcal{S}_3 \end{bmatrix} \tag{8.49}$$

8.12.2 The Jones Vectors

Another representation of polarized light, which complements that of the Stokes parameters, was invented in 1941 by the American physicist R. Clark Jones. The technique he evolved has the advantages of being applicable to coherent beams and at the same time being extremely concise. Yet unlike the previous formalism, it is *only applicable to polarized waves*. In that case it would seem that the most natural way to represent the beam would be in terms of the electric vector itself. Written in column form, this *Jones vector* is

$$\tilde{\mathbf{E}} = \begin{bmatrix} E_x(t) \\ E_y(t) \end{bmatrix} \tag{8.50}$$

where $E_x(t)$ and $E_y(t)$ are the instantaneous scalar components of $\tilde{\mathbf{E}}$. Obviously, knowing $\tilde{\mathbf{E}}$, we know everything about the polarization state. And if we preserve the phase information, we will be able to handle coherent waves. With this in mind, rewrite Eq. (8.50) as

$$\tilde{\mathbf{E}} = \begin{bmatrix} E_{0x}e^{i\varphi_x} \\ E_{0y}e^{i\varphi_y} \end{bmatrix} \tag{8.51}$$

where φ_x and φ_y are the appropriate phases. Horizontal and vertical $\mathcal{P}$-states are thus given by

$$\tilde{\mathbf{E}}_h = \begin{bmatrix} E_{0x}e^{i\varphi_x} \\ 0 \end{bmatrix} \quad \text{and} \quad \tilde{\mathbf{E}}_v = \begin{bmatrix} 0 \\ E_{0y}e^{i\varphi_y} \end{bmatrix} \tag{8.52}$$

respectively. The sum of two coherent beams, as with the Stokes vectors, is formed by a sum of the corresponding components. Since $\tilde{\mathbf{E}} = \tilde{\mathbf{E}}_h + \tilde{\mathbf{E}}_v$, when, for example $E_{0x} = E_{0y}$ and $\varphi_x = \varphi_y$, $\mathbf{E}$ is given by

$$\tilde{\mathbf{E}} = \begin{bmatrix} E_{0x}e^{i\varphi_x} \\ E_{0x}e^{i\varphi_x} \end{bmatrix} \tag{8.53}$$

or, after factoring, by

$$\tilde{\mathbf{E}} = E_{0x}e^{i\varphi_x}\begin{bmatrix} 1 \\ 1 \end{bmatrix} \tag{8.54}$$

which is a $\mathcal{P}$-state at $+45°$. This is the case since the amplitudes are equal and the phase difference is zero.

In many applications it is not necessary to know the exact amplitudes and phases. In such instances we can *normalize* the irradiance to unity, thereby forfeiting some information but gaining much simpler expressions. This is done by dividing both elements in the vector by the same scalar (real or complex) quantity, such that the sum of the squares of the components is one. For example, dividing both terms of Eq. (8.53) by $\sqrt{2}\,E_{0x}e^{i\varphi_x}$ leads to

$$\mathbf{E}_{45} = \frac{1}{\sqrt{2}}\begin{bmatrix} 1 \\ 1 \end{bmatrix} \tag{8.55}$$

Similarly, in normalized form

$$\mathbf{E}_h = \begin{bmatrix} 1 \\ 0 \end{bmatrix} \quad \text{and} \quad \mathbf{E}_v = \begin{bmatrix} 0 \\ 1 \end{bmatrix} \tag{8.56}$$

Right-circular light has $E_{0x} = E_{0y}$, and the y-component leads

*The detailed requirements for a collection of objects to form a vector space and themselves be vectors in such a space are discussed in, for example, Davis, *Introduction to Vector Analysis*.

the x-component by 90°. Since we are using the form $(kz - \omega t)$, we will have to add $-\pi/2$ to φ_y, thus

$$\tilde{\mathbf{E}}_{\mathscr{R}} = \begin{bmatrix} E_{0x}e^{i\varphi_x} \\ E_{0x}e^{i(\varphi_x - \pi/2)} \end{bmatrix}$$

Dividing both components by $E_{0x}e^{i\varphi_x}$ yields

$$\begin{bmatrix} 1 \\ e^{-i\pi/2} \end{bmatrix} = \begin{bmatrix} 1 \\ -i \end{bmatrix}$$

Hence the normalized Jones vector is*

$$\tilde{\mathbf{E}}_{\mathscr{R}} = \frac{1}{\sqrt{2}}\begin{bmatrix} 1 \\ -i \end{bmatrix} \quad \text{and similarly} \quad \tilde{\mathbf{E}}_{\mathscr{L}} = \frac{1}{\sqrt{2}}\begin{bmatrix} 1 \\ i \end{bmatrix} \tag{8.57}$$

The sum $\tilde{\mathbf{E}}_{\mathscr{R}} + \tilde{\mathbf{E}}_{\mathscr{L}}$ is

$$\frac{1}{\sqrt{2}}\begin{bmatrix} 1+1 \\ -i+i \end{bmatrix} = \frac{2}{\sqrt{2}}\begin{bmatrix} 1 \\ 0 \end{bmatrix}$$

This is a horizontal $\mathscr{P}$-state having an amplitude twice that of either component, a result in agreement with our earlier calculation of Eq. (8.10). The Jones vector for elliptical light can be obtained by the same procedure used to arrive at $\tilde{\mathbf{E}}_{\mathscr{R}}$ and $\tilde{\mathbf{E}}_{\mathscr{L}}$, where now E_{0x} may not be equal to E_{0y}, and the phase difference need not be 90°. In essence, for vertical and horizontal $\mathscr{E}$-states all we need to do is stretch out the circular form into an ellipse by multiplying either component by a scalar. Thus

$$\frac{1}{\sqrt{5}}\begin{bmatrix} 2 \\ -i \end{bmatrix} \tag{8.58}$$

describes one possible form of horizontal, right-handed, elliptical light.

Two vectors $\mathbf{A}$ and $\mathbf{B}$ are said to be orthogonal when $\mathbf{A}\cdot\mathbf{B} = 0$; similarly, two complex vectors are orthogonal when $\tilde{\mathbf{A}}\cdot\tilde{\mathbf{B}}^* = 0$. One refers to two polarization states as being *orthogonal* when their Jones vectors are orthogonal. For example,

$$\tilde{\mathbf{E}}_{\mathscr{R}}\cdot\tilde{\mathbf{E}}_{\mathscr{L}}^* = \tfrac{1}{2}[(1)(1)^* + (-i)(i)^*] = 0$$

or

$$\tilde{\mathbf{E}}_h\cdot\tilde{\mathbf{E}}_v^* = [(1)(0)^* + (0)(1)^*] = 0$$

where taking the complex conjugates of real numbers obviously leaves them unaltered. Any polarization state will have a corresponding orthogonal state. Notice that

$$\tilde{\mathbf{E}}_{\mathscr{R}}\cdot\tilde{\mathbf{E}}_{\mathscr{R}} = \tilde{\mathbf{E}}_{\mathscr{L}}\cdot\tilde{\mathbf{E}}_{\mathscr{L}}^* = 1$$

and

$$\tilde{\mathbf{E}}_{\mathscr{R}}\cdot\tilde{\mathbf{E}}_{\mathscr{L}}^* = \tilde{\mathbf{E}}_{\mathscr{L}}\cdot\tilde{\mathbf{E}}_{\mathscr{R}}^* = 0$$

Such vectors form an *orthogonal set*, as do $\tilde{\mathbf{E}}_h$ and $\tilde{\mathbf{E}}_v$. As we have seen, any polarization state can be described by a linear combination of the vectors in either one of the orthonormal sets. These same ideas are of considerable importance in Quantum Mechanics, where one deals with orthonormal wavefunctions.

8.12.3 The Jones and Mueller Matrices

Suppose that we have a polarized incident beam represented by its Jones vector $\tilde{\mathbf{E}}_i$, which passes through an optical element, emerging as a new vector $\tilde{\mathbf{E}}_t$ corresponding to the transmitted wave. The optical element has transformed $\tilde{\mathbf{E}}_i$ into $\tilde{\mathbf{E}}_t$, a process that can be described mathematically using a 2×2 matrix. Recall that a matrix is just an array of numbers that has prescribed addition and multiplication operations. Let $\mathscr{A}$ represent the transformation matrix of the optical element in question. Then

$$\tilde{\mathbf{E}}_t = \mathscr{A}\tilde{\mathbf{E}}_i \tag{8.59}$$

where

$$\mathscr{A} = \begin{bmatrix} a_{11} & a_{12} \\ a_{21} & a_{22} \end{bmatrix} \tag{8.60}$$

and the column vectors are to be treated like any other matrices. As a reminder, write Eq. (8.59) as

$$\begin{bmatrix} \tilde{E}_{tx} \\ \tilde{E}_{ty} \end{bmatrix} = \begin{bmatrix} a_{11} & a_{12} \\ a_{21} & a_{22} \end{bmatrix}\begin{bmatrix} \tilde{E}_{ix} \\ \tilde{E}_{iy} \end{bmatrix} \tag{8.61}$$

and, upon expanding,

$$\tilde{E}_{tx} = a_{11}\tilde{E}_{ix} + a_{12}\tilde{E}_{iy}$$

$$\tilde{E}_{ty} = a_{21}\tilde{E}_{ix} + a_{22}\tilde{E}_{iy}$$

Table 8.6 contains a brief listing of Jones matrices for various optical elements. To appreciate how these are used let's examine a few applications. Suppose that $\tilde{\mathbf{E}}_i$ represents a $\mathscr{P}$-state at

*Had we used $(\omega t - kz)$ for the phase, the terms in $\tilde{\mathbf{E}}_{\mathscr{R}}$ would have been interchanged. The present notation, although possibly a bit more difficult to keep straight (e.g., $-\pi/2$ for a phase lead), is more often used in modern works. Be wary when consulting references (e.g., Shurcliff).

TABLE 8.6 Jones and Mueller matrices.

Linear optical element		Jones matrix	Mueller matrix
Horizontal linear polarizer	↔	$\begin{bmatrix} 1 & 0 \\ 0 & 0 \end{bmatrix}$	$\frac{1}{2}\begin{bmatrix} 1 & 1 & 0 & 0 \\ 1 & 1 & 0 & 0 \\ 0 & 0 & 0 & 0 \\ 0 & 0 & 0 & 0 \end{bmatrix}$
Vertical linear polarizer	↕	$\begin{bmatrix} 0 & 0 \\ 0 & 1 \end{bmatrix}$	$\frac{1}{2}\begin{bmatrix} 1 & -1 & 0 & 0 \\ -1 & 1 & 0 & 0 \\ 0 & 0 & 0 & 0 \\ 0 & 0 & 0 & 0 \end{bmatrix}$
Linear polarizer at +45°	⤢	$\frac{1}{2}\begin{bmatrix} 1 & 1 \\ 1 & 1 \end{bmatrix}$	$\frac{1}{2}\begin{bmatrix} 1 & 0 & 1 & 0 \\ 0 & 0 & 0 & 0 \\ 1 & 0 & 1 & 0 \\ 0 & 0 & 0 & 0 \end{bmatrix}$
Linear polarizer at −45°	⤡	$\frac{1}{2}\begin{bmatrix} 1 & -1 \\ -1 & 1 \end{bmatrix}$	$\frac{1}{2}\begin{bmatrix} 1 & 0 & -1 & 0 \\ 0 & 0 & 0 & 0 \\ -1 & 0 & 1 & 0 \\ 0 & 0 & 0 & 0 \end{bmatrix}$
Quarter-wave plate, fast axis vertical		$e^{i\pi/4}\begin{bmatrix} 1 & 0 \\ 0 & -i \end{bmatrix}$	$\begin{bmatrix} 1 & 0 & 0 & 0 \\ 0 & 1 & 0 & 0 \\ 0 & 0 & 0 & -1 \\ 0 & 0 & 1 & 0 \end{bmatrix}$
Quarter-wave plate, fast axis horizontal		$e^{i\pi/4}\begin{bmatrix} 1 & 0 \\ 0 & i \end{bmatrix}$	$\begin{bmatrix} 1 & 0 & 0 & 0 \\ 0 & 1 & 0 & 0 \\ 0 & 0 & 0 & 1 \\ 0 & 0 & -1 & 0 \end{bmatrix}$
Homogeneous circular polarizer right	↻	$\frac{1}{2}\begin{bmatrix} 1 & i \\ -i & 1 \end{bmatrix}$	$\frac{1}{2}\begin{bmatrix} 1 & 0 & 0 & 1 \\ 0 & 0 & 0 & 0 \\ 0 & 0 & 0 & 0 \\ 1 & 0 & 0 & 1 \end{bmatrix}$
Homogeneous circular polarizer left	↺	$\frac{1}{2}\begin{bmatrix} 1 & -i \\ i & 1 \end{bmatrix}$	$\frac{1}{2}\begin{bmatrix} 1 & 0 & 0 & -1 \\ 0 & 0 & 0 & 0 \\ 0 & 0 & 0 & 0 \\ -1 & 0 & 0 & 1 \end{bmatrix}$

+45°, which passes through a quarter-wave plate whose fast axis is vertical (i.e., in the y-direction). The polarization state of the emergent wave is found as follows, where we drop the constant-amplitude factors for convenience:

$$\begin{bmatrix} 1 & 0 \\ 0 & -i \end{bmatrix}\begin{bmatrix} 1 \\ 1 \end{bmatrix} = \begin{bmatrix} \tilde{E}_{tx} \\ \tilde{E}_{ty} \end{bmatrix}$$

and thus

$$\tilde{\mathbf{E}}_t = \begin{bmatrix} 1 \\ -i \end{bmatrix}$$

The beam, as you well know, is right-circular. If the wave passes through a series of optical elements represented by the matrices $\mathscr{A}_1, \mathscr{A}_2, \ldots, \mathscr{A}_n$, then

$$\tilde{\mathbf{E}}_t = \mathscr{A}_n \cdots \mathscr{A}_2 \mathscr{A}_1 \tilde{\mathbf{E}}_i$$

The matrices do not commute; they must be applied in the proper order. The wave leaving the first optical element in the series is $\mathscr{A}_1 \tilde{\mathbf{E}}_i$; after passing through the second element, it becomes $\mathscr{A}_2\mathscr{A}_1 \tilde{\mathbf{E}}_i$, and so on. To illustrate the process, return to the wave considered above (i.e., a $\mathscr{P}$-state at +45°), but now have it pass through two quarter-wave plates, both with their fast axes vertical. Thus, again discarding the amplitude factors, we have

$$\tilde{\mathbf{E}}_t = \begin{bmatrix} 1 & 0 \\ 0 & -i \end{bmatrix}\begin{bmatrix} 1 & 0 \\ 0 & -i \end{bmatrix}\begin{bmatrix} 1 \\ 1 \end{bmatrix}$$

whereupon

$$\tilde{\mathbf{E}}_t = \begin{bmatrix} 1 & 0 \\ 0 & -i \end{bmatrix}\begin{bmatrix} 1 \\ -i \end{bmatrix}$$

and finally

$$\tilde{\mathbf{E}}_t = \begin{bmatrix} 1 \\ -1 \end{bmatrix}$$

The transmitted beam is a $\mathscr{P}$-state at −45°, having essentially been flipped through 90° by a half-wave plate. When the same series of optical elements is used to examine various states, it becomes desirable to replace the product $\mathscr{A}_n \cdots \mathscr{A}_2\mathscr{A}_1$ by the single 2 × 2 *system matrix* obtained by carrying out the multiplication. (The order in which it is calculated should be $\mathscr{A}_2\mathscr{A}_1$, then $\mathscr{A}_3\mathscr{A}_2\mathscr{A}_1$, etc.)

In 1943 Hans Mueller, then a professor of physics at the Massachusetts Institute of Technology, devised a matrix method for dealing with the Stokes vectors. Recall that the Stokes vectors have the attribute of being applicable to both polarized and partially polarized light. The Mueller method shares this quality and thus serves to complement the Jones method. The latter, however, can easily deal with coherent waves, whereas the former cannot. The Mueller, 4 × 4, matrices are applied in much the same way as are the Jones matrices. There is therefore little need to discuss the method at length; a few simple examples, augmented by Table 8.6, should suffice. Imagine that we pass a unit-irradiance unpolarized wave through a linear horizontal polarizer. The Stokes vector of the emerging wave $\mathcal{S}_t$ is

$$\mathcal{S}_t = \frac{1}{2}\begin{bmatrix} 1 & 1 & 0 & 0 \\ 1 & 1 & 0 & 0 \\ 0 & 0 & 0 & 0 \\ 0 & 0 & 0 & 0 \end{bmatrix}\begin{bmatrix} 1 \\ 0 \\ 0 \\ 0 \end{bmatrix} = \begin{bmatrix} \frac{1}{2} \\ \frac{1}{2} \\ 0 \\ 0 \end{bmatrix}$$

The transmitted wave has an irradiance of $\frac{1}{2}$(i.e., $\mathcal{S}_0 = \frac{1}{2}$) and is linearly polarized horizontally ($\mathcal{S}_1 > 0$). As another example, suppose we have a partially polarized elliptical wave whose Stokes parameters have been determined to be, say, (4, 2, 0, 3). Its irradiance is 4; it is more nearly horizontal than vertical ($\mathcal{S}_1 > 0$), it is right-handed ($\mathcal{S}_3 > 0$), and it has a degree of polarization of 90%. Since none of the parameters can be larger than $\mathcal{S}_0$, a value of $\mathcal{S}_3 = 3$ is fairly large, indicating that the ellipse resembles a circle. If the wave is now made to traverse a quarter-wave plate with a vertical fast axis, then

$$\mathcal{S}_t = \begin{bmatrix} 1 & 0 & 0 & 0 \\ 0 & 1 & 0 & 0 \\ 0 & 0 & 0 & -1 \\ 0 & 0 & 1 & 0 \end{bmatrix}\begin{bmatrix} 4 \\ 2 \\ 0 \\ 3 \end{bmatrix}$$

and thus

$$\mathcal{S}_t = \begin{bmatrix} 4 \\ 2 \\ -3 \\ 0 \end{bmatrix}$$

The emergent wave has the same irradiance and degree of polarization but is now partially linearly polarized.

We have only touched on a few of the more important aspects of the matrix methods. The full extent of the subject goes far beyond these introductory remarks.*

*One can weave a more elaborate and mathematically satisfying development in terms of something called the coherence matrix. For further, but more advanced, reading, see O'Neill, *Introduction to Statistical Optics*.

PROBLEMS

8.1 Described completely the state of polarization of each of the following waves:

(a) $\mathbf{E} = \hat{\mathbf{i}}E_0 \cos(kz - \omega t) - \hat{\mathbf{j}}E_0 \cos(kz - \omega t)$

(b) $\mathbf{E} = \hat{\mathbf{i}}E_0 \sin 2\pi(z/\lambda - \nu t) - \hat{\mathbf{j}}E_0 \sin 2\pi(z/\lambda - \nu t)$

(c) $\mathbf{E} = \hat{\mathbf{i}}E_0 \sin(\omega t - kz) + \hat{\mathbf{j}}E_0 \sin(\omega t - kz - \pi/4)$

(d) $\mathbf{E} = \hat{\mathbf{i}}E_0 \cos(\omega t - kz) + \hat{\mathbf{j}}E_0 \cos(\omega t - kz + \pi/2)$.

8.2 Consider the disturbance given by the expression $\mathbf{E}(z, t) = [\hat{\mathbf{i}} \cos \omega t + \hat{\mathbf{j}} \cos(\omega t - \pi/2)]E_0 \sin kz$. What kind of wave is it? Draw a rough sketch showing its main features.

8.3 Analytically, show that the superposition of an $\mathcal{R}$- and an $\mathcal{L}$-state having different amplitudes will yield an $\mathcal{E}$-state, as shown in Fig. 8.8. What must ε be to duplicate that figure?

8.4 Write an expression for a $\mathcal{P}$-state lightwave of angular frequency ω and amplitude E_0 propagating along the x-axis with its plane of vibration at an angle of 25° to the xy-plane. The disturbance is zero at $t = 0$ and $x = 0$.

8.5* Write an expression for a $\mathcal{P}$-state lightwave of angular frequency ω and amplitude E_0 propagating along a line in the xy-plane at 45° to the x-axis and having its plane of vibration corresponding to the xy-plane. At $t = 0$, $y = 0$, and $x = 0$ the field is zero.

8.6 Write an expression for an $\mathcal{R}$-state lightwave of frequency ω propagating in the positive x-direction such that at $t = 0$ and $x = 0$ the **E**-field points in the negative z-direction.

8.7* A beam of linearly polarized light with its electric field vertical impinges perpendicularly on an ideal linear polarizer with a vertical transmission axis. If the incoming beam has an irradiance of 200-W/m^2, what is the irradiance of the transmitted beam?

8.8* Given that 300 W/m^2 of light from an ordinary tungsten bulb arrives at an ideal linear polarizer. What is its radiant flux density on emerging?

8.9 If light that is initially natural and of flux density I_i passes through two sheets of *HN*-32 whose transmission axes are parallel, what will be the flux density of the emerging beam?

8.10* What will be the irradiance of the emerging beam if the analyzer of the previous problem is rotated 30°?

8.11* The irradiance of a beam of natural light is 400 W/m^2. It impinges on the first of two consecutive ideal linear polarizers whose transmission axes are 40.0° apart. How much light emerges from the two?

8.12* As we saw in Section 8.10, substances such as sugar and insulin are *optically active*; they rotate the plane of polarization in proportion to both the path length and the concentration of the solution. A glass vessel is placed between a pair of crossed *HN*-50 linear polarizers, and 50% of the natural light incident on the first polarizer is transmitted through the second polarizer. By how much did the sugar solution in the cell rotate the light passed by the first polarizer?

8.13* The light from an ordinary flashlight is passed through a linear polarizer with its transmission axis vertical. The resulting beam, having an irradiance of 200 W/m^2, is incident normally on a vertical *HN*-50 linear polarizer whose transmission axis is tilted at 30° above the horizontal. How much light is transmitted?

8.14* Linearly polarized light (with an irradiance of 200 W/m^2) aligned with its electric-field vector at +55° from the vertical impinges perpendicularly on an ideal sheet polarizer whose transmission axis is at +10° from the vertical. What fraction of the incoming light emerges?

8.15* Two ideal linear sheet polarizers are arranged with respect to the vertical with their transmission axis at 10° and 60°, respectively. If a linearly polarized beam of light with its electric field at 40° enters the first polarizer, what fraction of its irradiance will emerge?

8.16* Given a pair of crossed polarizers with transmission axes vertical and horizontal. The beam emerging from the first polarizer has flux density I_1, and of course no light passes through the analyzer (i.e., $I_2 = 0$). Now insert a perfect linear polarizer (*HN*-50) with its transmission axis at 45° to the vertical between the two elements—compute I_2. Think about the motion of the electrons that are radiating in each polarizer.

8.17* Imagine that you have two identical perfect linear polarizers and a source of natural light. Place them one behind the other and position their transmission axes at 0° and 50°, respectively. Now insert between them a third linear polarizer with its transmission axes at 25°. If 1000 W/m^2 of light is incident, how much will emerge with and without the middle polarizer in place?

8.18* Given that 200 W/m^2 of randomly polarized light is incident normally on a stack of ideal linear polarizers that are positioned one behind the other with the transmission axis of the first vertical, the second at 30°, the third at 60°, and the fourth at 90°. How much light emerges?

8.19* Two *HN*-50 linear polarizers are positioned one behind the other. What angle should their transmission axes make if an incident unpolarized 100-W/m^2 beam is to be reduced to 30.0 W/m^2 on emerging from the pair?

8.20 An ideal polarizer is rotated at a rate ω between a similar pair of stationary crossed polarizers. Show that the emergent flux density will be modulated at four times the rotational frequency. In other words, show that

$$I = \frac{I_1}{8}(1 - \cos 4\omega t)$$

where I_1 is the flux density emerging from the first polarizer and I is the final flux density.

8.21 Figure P.8.21 shows a ray traversing a calcite crystal at nearly normal incidence, bouncing off a mirror, and then going through the crystal again. Will the observer see a double image of the spot on Σ?

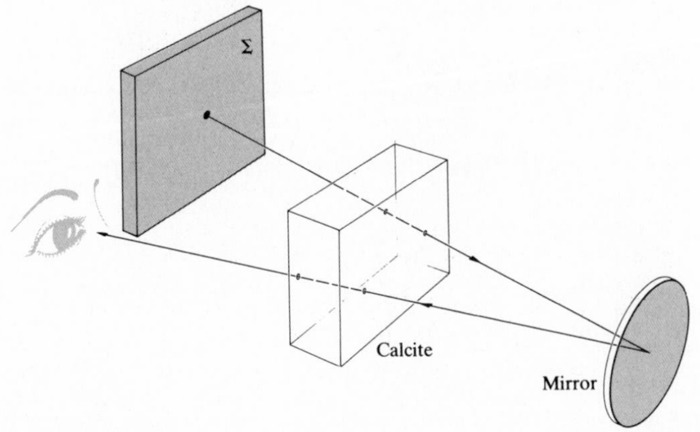

FIGURE P.8.21

8.22* A pencil mark on a sheet of paper is covered by a calcite crystal. With illumination from above, isn't the light impinging on the paper already polarized, having passed through the crystal? Why then do we see two images? Test your solution by polarizing the light from a flashlight and then reflecting it off a sheet of paper. Try specular reflection off glass; is the reflected light polarized?

8.23 Discuss in detail what you see in Fig. P.8.23. The crystal in the photograph is calcite, and it has a blunt corner at the upper left. The two Polaroids have their transmission axes parallel to their *short* edges.

FIGURE P.8.23

8.24 The calcite crystal in Fig. P.8.24 is shown in three different orientations. Its blunt corner is on the left in (*a*), the lower left in (*b*), and the bottom in (*c*). The Polaroid's transmission axis is horizontal. Explain each photograph, particularly (*b*).

8.25 In discussing calcite, we pointed out that its large birefringence arises from the fact that the carbonate groups lie in parallel planes (normal to the optic axis). Show in a sketch and explain why the polarization of the group will be less when **E** is perpendicular to the CO_3 plane than when **E** is parallel to it. What does this mean with respect to $v_\perp$ and $v_\parallel$, that is, the wave's speeds when **E** is linearly polarized perpendicular or parallel to the optic axis?

8.26* Imagine that we have a transmitter of microwaves that radiates a linearly polarized wave whose **E**-field is known to be parallel to the dipole direction. We wish to reflect as much energy as possible off the surface of a pond (having an index of refraction of 9.0). Find the necessary incident angle and comment on the orientation of the beam.

8.27* At what angle will the reflection of the sky coming off the surface of a pond ($n = 1.33$) completely vanish when seen through a Polaroid filter?

8.28* What is Brewster's angle for reflection of light from the surface of a piece of glass ($n_g = 1.65$) immersed in water ($n_w = 1.33$)?

(a)

(b)

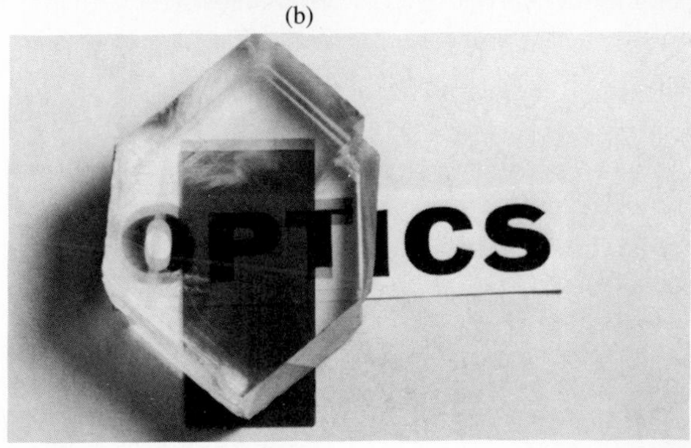

(c)

FIGURE P.8.24

8.29* A beam of light is reflected off the surface of some unknown liquid, and the light is examined with a linear sheet polarizer. It is found that when the central axis of the polarizer (that is, the perpendicular to the plane of the sheet) is tilted down from the vertical at an angle of 54.30°, the reflected light is completely passed, provided the transmission axis is parallel to the plane of the interface. From this information, compute the index of refraction of the liquid.

8.30* Light reflected from a glass ($n_g = 1.65$) plate immersed in ethyl alcohol ($n_e = 1.36$) is found to be completely linearly polarized. At what angle will the partially polarized beam be transmitted into the plate?

8.31* A beam of natural light is incident on an air–glass interface ($n_{ti} = 1.5$) at 40°. Compute the degree of polarization of the reflected light.

8.32* A beam of natural light incident in air on a glass ($n = 1.5$) interface at 70° is partially reflected. Compute the overall reflectance. How would this compare with the case of incidence at, say, 56.3°? Explain.

8.33 A ray of yellow light is incident on a calcite plate at 50°. The plate is cut so that the optic axis is parallel to the front face and perpendicular to the plane-of-incidence. Find the angular separation between the two emerging rays.

8.34* A beam of light is incident normally on a quartz plate whose optic axis is perpendicular to the beam. If $\lambda_0 = 589.3$ nm, compute the wavelengths of both the ordinary and extraordinary waves. What are their frequencies?

8.35 A beam of light enters a calcite prism from the left, as shown in Fig. P.8.35. There are three possible orientations of the optic axis of particular interest, and these correspond to the x-, y-, and z-directions. Imagine that we have three such prisms. In each case sketch the entering and emerging beams, showing the state of polarization. How can any one of these be used to determine n_o and n_e?

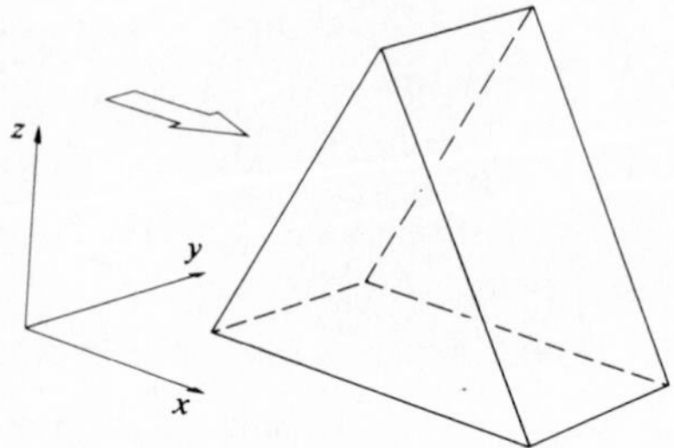

FIGURE P.8.35

8.36 The electric-field vector of an incident $\mathscr{P}$-state makes an angle of +30° with the horizontal fast axis of a quarter-wave plate. Describe, in detail, the state of polarization of the emergent wave.

8.37 Compute the critical angle for the ordinary ray, that is, the angle for total internal reflection at the calcite–balsam layer of a Nicol prism.

8.38* Draw a quartz Wollaston prism, showing all pertinent rays and their polarization states.

8.39 The prism shown in Fig. P.8.39 is known as a *Rochon polarizer*. Sketch all the pertinent rays, assuming

(a) that it is made of calcite.

(b) that it is made of quartz.

(c) Why might such a device be more useful than a dichroic polarizer when functioning with high–flux density laser light?

(d) What valuable feature of the Rochon is lacking in the Wollaston polarizer?

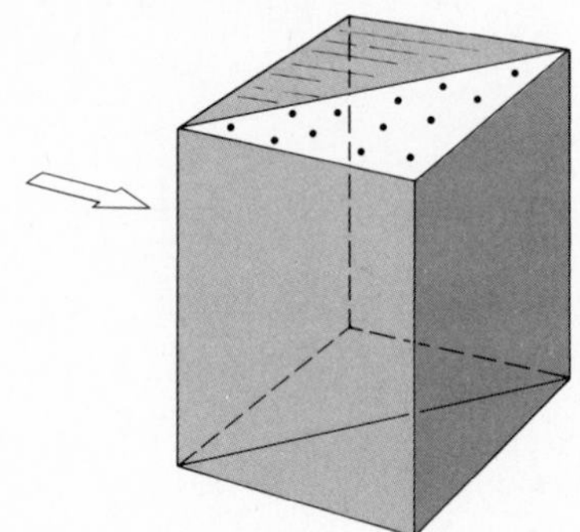

FIGURE P.8.39

8.40* Take two ideal Polaroids (the first with its axis vertical and the second, horizontal) and insert between them a stack of 10 half-wave plates, the first with its fast axis rotated $\pi/40$ rad from the vertical, and each subsequent one rotated $\pi/40$ rad from the previous one. Determine the ratio of the emerging to incident irradiance, showing your logic clearly.

8.41* Suppose you were given a linear polarizer and a quarter-wave plate. How could you determine which was which, assuming you also had a source of natural light?

8.42* An $\mathscr{L}$-state traverses an eighth-wave plate having a horizontal fast axis. What is its polarization state on emerging?

8.43* Figure P.8.43 shows two Polaroid linear polarizers and between them a microscope slide to which is attached a piece of cellophane tape. Explain what you see.

8.44 A Babinet compensator is positioned at 45° between crossed linear polarizers and is being illuminated with sodium light. When a thin sheet of mica (indices 1.599 and 1.594) is placed on the com-

FIGURE P.8.43

pensator, the black bands all shift by $\frac{1}{4}$ of the space separating them. Compute the retardance of the sheet and its thickness.

8.45 Imagine that we have randomly polarized room light incident almost normally on the glass surface of a radar screen. A portion of it would be specularly reflected back toward the viewer and would thus tend to obscure the display. Suppose now that we cover the screen with a right-circular polarizer, as shown in Fig. P.8.45. Trace the incident and reflected beams, indicating their polarization states. What happens to the reflected beam?

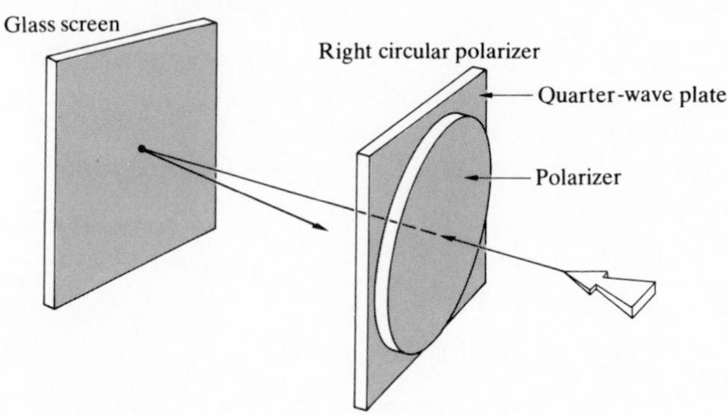

FIGURE P.8.45

8.46 Is it possible for a beam to consist of two orthogonal incoherent $\mathscr{P}$-states and not be natural light? Explain. How might you arrange to have such a beam?

8.47* The specific rotatory power for sucrose dissolved in water at 20°C ($\lambda_0 = 589.3$ nm) is +66.45° per 10 cm of path traversed through a solution containing 1 g of active substance (sugar) per cm^3 of solution. A vertical $\mathscr{P}$-state (sodium light) enters at one end of a 1-m tube containing 1000 cm^3 of solution, of which 10 g is sucrose. At what orientation will the $\mathscr{P}$-state emerge?

8.48 On examining a piece of stressed photoelastic material between crossed linear polarizers, we would see a set of colored bands (isochromatics) and, superimposed on these, a set of dark bands (isoclinics). How might we remove the isoclinics, leaving only the isochromatics? Explain your solution. Incidentally, the proper arrangement is independent of the orientation of the photoelastic sample.

8.49* Consider a Kerr cell whose plates are separated by a distance d. Let ℓ be the effective length of those plates (slightly different from the actual length because of fringing of the field). Show that

$$\Delta\varphi = 2\pi K\ell V^2/d^2 \qquad [8.41]$$

8.50 Compute the half-wave voltage for a longitudinal Pockels cell made of ADA (ammonium dihydrogen arsenate) at $\lambda_0 \approx 550$ nm, where $r_{63} = 5.5 \times 10^{-12}$ and $n_o = 1.58$.

8.51 Find a Jones vector $\tilde{\mathbf{E}}_2$ representing a polarization state orthogonal to

$$\tilde{\mathbf{E}}_1 = \begin{bmatrix} 1 \\ -2i \end{bmatrix}$$

Sketch both of these.

8.52* Two incoherent light beams represented by (1, 1, 0, 0) and (3, 0, 0, 3) are superimposed.

(a) Describe in detail the polarization states of each of these.

(b) Determine the resulting Stokes parameters of the combined beam and describe its polarization state.

(c) What is its degree of polarization?

(d) What is the resulting light produced by overlapping the incoherent beams (1, 1, 0, 0) and (1, −1, 0, 0)? Explain.

8.53* Show by direct calculation, using Mueller matrices, that a unit-irradiance beam of natural light passing through a vertical linear polarizer is converted into a vertical $\mathscr{P}$-state. Determine its relative irradiance and degree of polarization.

8.54* Show by direct calculation, using Mueller matrices, that a unit-irradiance beam of natural light passing through a linear polarizer with its transmission axis at $+45°$ is converted into a $\mathscr{P}$-state at $+45°$. Determine its relative irradiance and degree of polarization.

8.55* Show by direct calculation, using Mueller matrices, that a beam of horizontal $\mathscr{P}$-state light passing through a $\frac{1}{4}\lambda$-plate with its fast axis horizontal emerges unchanged.

8.56* Confirm that the matrix

$$\begin{bmatrix} 1 & 0 & 0 & 0 \\ 0 & 0 & 0 & -1 \\ 0 & 0 & 1 & 0 \\ 0 & 1 & 0 & 0 \end{bmatrix}$$

will serve as a Mueller matrix for a quarter-wave plate with its fast axis at $+45°$. Shine linear light polarized at 45° through it. What happens? What emerges when a horizontal $\mathscr{P}$-state enters the device?

8.57 Derive the Mueller matrix for a quarter-wave plate with its fast axis at $-45°$. Check that this matrix effectively cancels the previous one, so that a beam passing through the two wave plates successively remains unaltered.

8.58* Pass a beam of horizontally polarized linear light through each one of the $\frac{1}{4}\lambda$-plates in the two previous questions and describe the states of the emerging light. Explain which field component is leading which and how Fig. 8.7 compares with these results.

8.59 Use Table 8.6 to derive a Mueller matrix for a half-wave plate having a vertical fast axis. Utilize your result to convert an $\mathscr{R}$-state into an $\mathscr{L}$-state. Verify that the same wave plate will convert an $\mathscr{L}$- to an $\mathscr{R}$-state. Advancing or retarding the relative phase by $\pi/2$ should have the same effect. Check this by deriving the matrix for a half-wave plate with a horizontal fast axis.

8.60 Construct one possible Mueller matrix for a right-circular polarizer made out of a linear polarizer and a quarter-wave plate. Such a device is obviously an inhomogeneous two-element train and will differ from the *homogeneous* circular polarizer of Table 8.6. Test your matrix to determine that it will convert natural light to an $\mathscr{R}$-state. Show that it will pass $\mathscr{R}$-states, as will the homogeneous matrix. Your matrix should convert $\mathscr{L}$-states incident on the input side to $\mathscr{R}$-states, whereas the homogeneous polarizer will totally absorb them. Verify this.

8.61* If the Pockels cell modulator shown in Fig. 8.65 is illuminated by light of irradiance I_i, it will transmit a beam of irradiance I_t such that

$$I_t = I_i \sin^2(\Delta\varphi/2)$$

Make a plot of I_t/I_i versus applied voltage. What is the significance of the voltage that corresponds to maximum transmission? What is the lowest voltage above zero that will cause I_t to be zero for ADP ($\lambda_0 = 546.1$ nm)? How can things be rearranged to yield a maximum value of I_t/I_i for zero voltage? In this new configuration what irradiance results when $V = V_{\lambda/2}$?

8.62 Construct a Jones matrix for an isotropic plate of absorbing material having an amplitude transmission coefficient of t. It might sometimes be desirable to keep track of the phase, since even if $t = 1$, such a plate is still an isotropic phase retarder. What is the Jones matrix for a region of vacuum? What is it for a perfect absorber?

8.63 Construct a Mueller matrix for an isotropic plate of absorbing material having an amplitude transmission coefficient of t. What Mueller matrix will completely depolarize any wave without affecting its irradiance? (It has no physical counterpart.)

8.64 Keeping Eq. (8.29) in mind, write an expression for the randomly polarized flux density component (I_n) of a partially polarized beam in terms of the Stokes parameters. To check your result, add a randomly polarized Stokes vector of flux density 4 to an $\mathscr{R}$-state of flux density 1. Then see if you get $I_n = 4$ for the resultant wave.

9 Interference

The intricate color patterns shimmering across an oil slick on a wet asphalt pavement result from one of the more common manifestations of the phenomenon of interference.* On a macroscopic scale we might consider the related problem of the interaction of surface ripples on a pool of water. Our everyday experience with this kind of situation allows us to envision a complex distribution of disturbances (as shown, e.g., in Fig. 9.1). There might be regions where two (or more) waves have overlapped, partially or even completely canceling each other. Still other regions might exist in the pattern, where the resultant troughs and crests are even more pronounced than those of any of the constituent waves. After being superimposed, the individual waves separate and continue on, completely unaffected by their previous encounter.

Although the subject could be treated from the perspective of QED (p. 138), we'll take a much simpler approach. The wave theory of the electromagnetic nature of light provides a natural basis from which to proceed. Recall that the expression describing the optical disturbance is a second-order, homogeneous, linear, partial, differential equation [Eq. (3.22)]. As we have seen, it therefore obeys the important *Superposition Principle*. Accordingly, the resultant electric-field intensity **E**, at a point in space where two or more lightwaves overlap, is equal to the *vector sum* of the individual constituent disturbances. Briefly then, ***optical interference corresponds to the interaction of two or more lightwaves yielding a resultant irradiance that deviates from the sum of the component irradiances***.

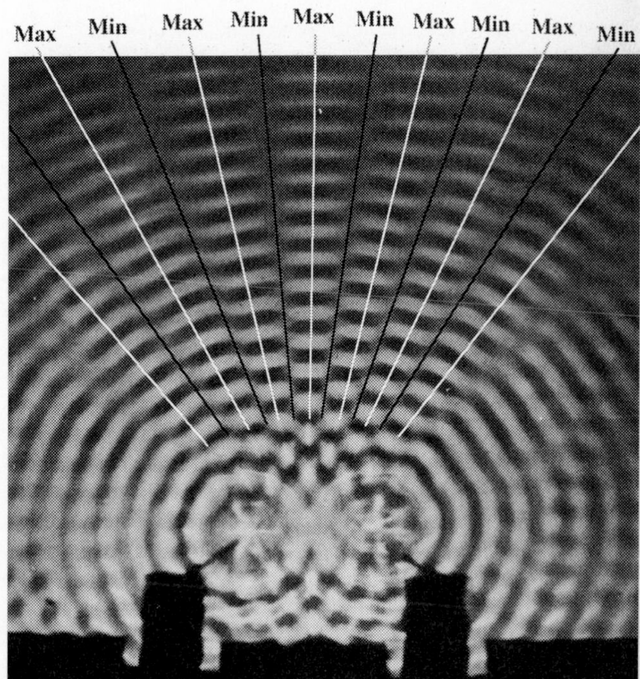

FIGURE 9.1 Water waves from two in-phase point sources in a ripple tank. In the middle of the pattern the wave peaks (thin bright bands) and troughs (thin black bands) lie within long wedge-shaped areas (maxima) separated by narrow dark regions of calm (minima). The optical equivalent is the electric field distribution depicted in Fig. 9.3c. (Photos courtesy PSSC *College Physics*, 1968, @ 1965 Educational Development Center, Inc.)

Out of the multitude of optical systems that produce interference, we will choose a few of the more important to examine. Interferometric devices will be divided, for the sake of

*The layer of water on the asphalt allows the oil film to assume the shape of a smooth planar surface. The black asphalt absorbs the transmitted light, preventing back reflection, which would tend to obscure the fringes.

discussion, into two groups: *wavefront splitting* and *amplitude splitting*. In the first instance, portions of the primary wavefront are used either directly as sources to emit secondary waves or in conjunction with optical devices to produce virtual sources of secondary waves. These secondary waves are then brought together, thereupon to interfere. In the case of amplitude splitting, the primary wave itself is divided into two segments, which travel different paths before recombining and interfering.

9.1 General Considerations

We have already examined the problem of the superposition of two scalar waves (Section 7.1), and in many respects those results will again be applicable. But light is, of course, a vector phenomenon; the electric and magnetic fields are vector fields. An appreciation of this fact is fundamental to any kind of intuitive understanding of interference. Still, there are many situations in which the particular optical system can be so configured that the vector nature of light is of little practical significance. We will derive the basic interference equations within the context of the vector model, thereafter delineating the conditions under which the scalar treatment is applicable.

In accordance with the Principle of Superposition, the electric field intensity $\mathbf{E}$, at a point in space, arising from the separate fields $\mathbf{E}_1, \mathbf{E}_2, \ldots$ of various contributing sources is given by

$$\mathbf{E} = \mathbf{E}_1 + \mathbf{E}_2 + \cdots \tag{9.1}$$

The optical disturbance, or light field $\mathbf{E}$, varies in time at an exceedingly rapid rate, roughly

$$4.3 \times 10^{14} \text{ Hz} \quad \text{to} \quad 7.5 \times 10^{14} \text{ Hz}$$

making the actual field an impractical quantity to detect. On the other hand, the irradiance I can be measured directly with a wide variety of sensors (e.g., photocells, bolometers, photographic emulsions, or eyes). The study of interference is therefore best approached by way of the irradiance.

Much of the analysis to follow can be performed without specifying the particular shape of the wavefronts, and the results are therefore quite general (Problem 9.1). For the sake of simplicity, however, consider two point sources, S_1 and S_2, emitting monochromatic waves of the same frequency in a homogeneous medium. Let their separation a be much greater than λ. Locate the point of observation P far enough away from the sources so that at P the wavefronts will be planes (Fig. 9.2). For the moment, consider only linearly polarized waves of the form

$$\mathbf{E}_1(\mathbf{r}, t) = \mathbf{E}_{01} \cos (\mathbf{k}_1 \cdot \mathbf{r} - \omega t + \varepsilon_1) \tag{9.2a}$$

and

$$\mathbf{E}_2(\mathbf{r}, t) = \mathbf{E}_{02} \cos (\mathbf{k}_2 \cdot \mathbf{r} - \omega t + \varepsilon_2) \tag{9.2b}$$

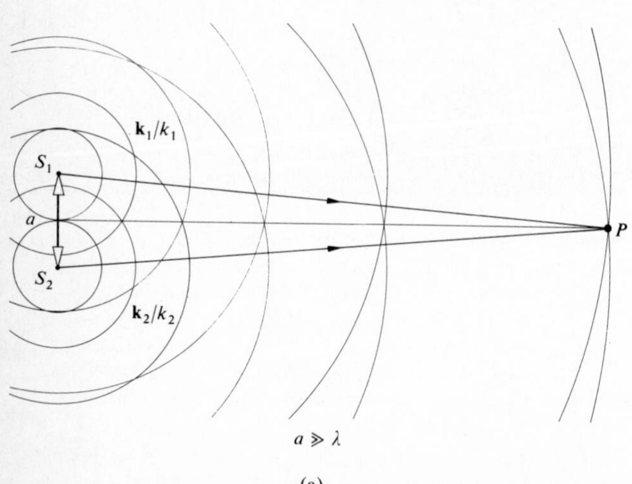

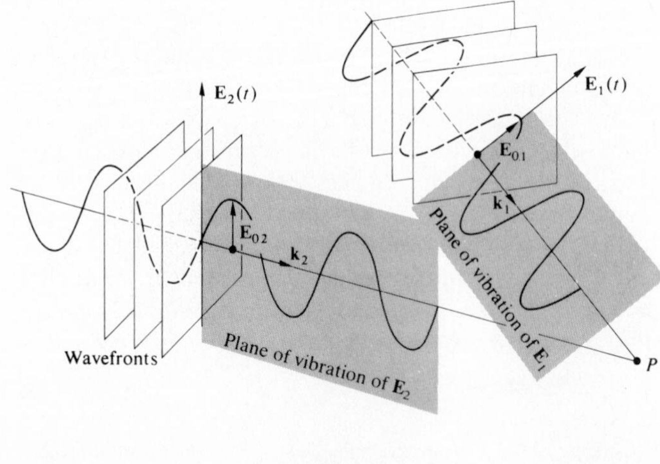

FIGURE 9.2 Waves from two point sources overlapping in space.

We saw in Chapter 3 that the irradiance at P is given by

$$I = \epsilon v \langle \mathbf{E}^2 \rangle_T$$

Inasmuch as we will be concerned only with relative irradiances within the same medium, we will, for the time being at least, simply neglect the constants and set

$$I = \langle \mathbf{E}^2 \rangle_T$$

What is meant by $\langle \mathbf{E}^2 \rangle_T$ is of course the time average of the magnitude of the electric-field intensity squared, or $\langle \mathbf{E} \cdot \mathbf{E} \rangle_T$. Accordingly

$$\mathbf{E}^2 = \mathbf{E} \cdot \mathbf{E}$$

where now

$$\mathbf{E}^2 = (\mathbf{E}_1 + \mathbf{E}_2) \cdot (\mathbf{E}_1 + \mathbf{E}_2)$$

and thus

$$\mathbf{E}^2 = \mathbf{E}_1^2 + \mathbf{E}_2^2 + 2\mathbf{E}_1 \cdot \mathbf{E}_2 \tag{9.3}$$

Taking the time average of both sides, we find that the irradiance becomes

$$I = I_1 + I_2 + I_{12} \tag{9.4}$$

provided that

$$I_1 = \langle \mathbf{E}_1^2 \rangle_T \tag{9.5}$$

$$I_2 = \langle \mathbf{E}_2^2 \rangle_T \tag{9.6}$$

and

$$I_{12} = 2\langle \mathbf{E}_1 \cdot \mathbf{E}_2 \rangle_T \tag{9.7}$$

The latter expression is known as the *interference term.* To evaluate it in this specific instance, we form

$$\mathbf{E}_1 \cdot \mathbf{E}_2 = \mathbf{E}_{01} \cdot \mathbf{E}_{02} \cos(\mathbf{k}_1 \cdot \mathbf{r} - \omega t + \varepsilon_1) \times \cos(\mathbf{k}_2 \cdot \mathbf{r} - \omega t + \varepsilon_2) \tag{9.8}$$

or equivalently

$$\begin{aligned}\mathbf{E}_1 \cdot \mathbf{E}_2 = \mathbf{E}_{01} \cdot \mathbf{E}_{02} &[\cos(\mathbf{k}_1 \cdot \mathbf{r} + \varepsilon_1) \\ &\times \cos \omega t + \sin(\mathbf{k}_1 \cdot \mathbf{r} + \varepsilon_1) \sin \omega t] \\ &\times [\cos(\mathbf{k}_2 \cdot \mathbf{r} + \varepsilon_2) \cos \omega t \\ &+ \sin(\mathbf{k}_2 \cdot \mathbf{r} + \varepsilon_2) \sin \omega t]\end{aligned} \tag{9.9}$$

Recall that the time average of some function $f(t)$, taken over an interval T, is

$$\langle f(t) \rangle_T = \frac{1}{T}\int_t^{t+T} f(t')\, dt' \tag{9.10}$$

The period τ of the harmonic functions is $2\pi/\omega$, and for our present concern $T >> \tau$. In that case the $1/T$ coefficient in front of the integral has a dominant effect. After multiplying out and averaging Eq. (9.9) we have

$$\langle \mathbf{E}_1 \cdot \mathbf{E}_2 \rangle_T = \tfrac{1}{2}\mathbf{E}_{01} \cdot \mathbf{E}_{02} \cos(\mathbf{k}_1 \cdot \mathbf{r} + \varepsilon_1 - \mathbf{k}_2 \cdot \mathbf{r} - \varepsilon_2)$$

where use was made of the fact (p. 48) that $\langle \cos^2 \omega t \rangle_T = \frac{1}{2}$, $\langle \sin^2 \omega t \rangle_T = \frac{1}{2}$, and $\langle \cos \omega t \sin \omega t \rangle_T = 0$. The interference term is then

$$I_{12} = \mathbf{E}_{01} \cdot \mathbf{E}_{02} \cos \delta \tag{9.11}$$

and δ, equal to $(\mathbf{k}_1 \cdot \mathbf{r} - \mathbf{k}_2 \cdot \mathbf{r} + \varepsilon_1 - \varepsilon_2)$, is the *phase difference* arising from a combined path length and initial phase-angle difference. Notice that if $\mathbf{E}_{01}$ and $\mathbf{E}_{02}$ (and therefore $\mathbf{E}_1$ and $\mathbf{E}_2$) are perpendicular, $I_{12} = 0$ and $I = I_1 + I_2$. Two such orthogonal $\mathcal{P}$-states will combine to yield an $\mathcal{R}$-, $\mathcal{L}$-, $\mathcal{P}$-, or $\mathcal{E}$-state, but the flux-density distribution will be unaltered.

The most common situation in the work to follow corresponds to $\mathbf{E}_{01}$ parallel to $\mathbf{E}_{02}$. In that case, the irradiance reduces to the value found in the scalar treatment of Section 7.1. Under those conditions

$$I_{12} = E_{01}E_{02} \cos \delta$$

This can be written in a more convenient way by noticing that

$$I_1 = \langle \mathbf{E}_1^2 \rangle_T = \frac{E_{01}^2}{2} \tag{9.12}$$

and

$$I_2 = \langle \mathbf{E}_2^2 \rangle = \frac{E_{02}^2}{2} \tag{9.13}$$

The interference term becomes

$$I_{12} = 2\sqrt{I_1 I_2} \cos \delta$$

whereupon the total irradiance is

$$I = I_1 + I_2 + 2\sqrt{I_1 I_2} \cos \delta \tag{9.14}$$

At various points in space, the resultant irradiance can be greater, less than, or equal to $I_1 + I_2$, depending on the value of I_{12}, that is, depending on δ. A maximum irradiance is obtained when $\cos\delta = 1$, so that

$$I_{max} = I_1 + I_2 + 2\sqrt{I_1 I_2} \qquad (9.15)$$

when $\delta = 0, \pm 2\pi, \pm 4\pi, \ldots$

In this case of ***total constructive interference*** the phase difference between the two waves is an integer multiple of 2π, and the disturbances are *in-phase*. When $0 < \cos\delta < 1$ the waves are *out-of-phase*, $I_1 + I_2 < I < I_{max}$, and the result is *constructive interference*. At $\delta = \pi/2$, $\cos\delta = 0$, the optical disturbances are 90° out-of-phase, and $I = I_1 + I_2$. For $0 > \cos\delta > -1$ we have the condition of *destructive interference*, $I_1 + I_2 > I > I_{min}$. A minimum irradiance results when the waves are 180° out-of-phase, troughs overlap crests, $\cos\delta = -1$, and

$$I_{min} = I_1 + I_2 - 2\sqrt{I_1 I_2} \qquad (9.16)$$

This occurs when $\delta = \pm\pi, \pm 3\pi, \pm 5\pi, \ldots$, and it is referred to as ***total destructive interference***.

Another somewhat special yet very important case arises when the amplitudes of both waves reaching P in Fig. 9.2 are equal (i.e., $\mathbf{E}_{01} = \mathbf{E}_{02}$). Since the irradiance contributions from both sources are then equal, let $I_1 = I_2 = I_0$. Equation (9.14) can now be written as

$$I = 2I_0(1 + \cos\delta) = 4I_0 \cos^2\frac{\delta}{2} \qquad (9.17)$$

from which it follows that $I_{min} = 0$ and $I_{max} = 4I_0$. For an analysis in terms of the angle between the two beams, see Problem 9.3.

Equation (9.14) holds equally well for the spherical waves emitted by S_1 and S_2. Such waves can be expressed as

$$\mathbf{E}_1(r_1, t) = \mathbf{E}_{01}(r_1) \exp[i(kr_1 - \omega t + \varepsilon_1)] \qquad (9.18a)$$

and

$$\mathbf{E}_2(r_2, t) = \mathbf{E}_{02}(r_2) \exp[i(kr_2 - \omega t + \varepsilon_2)] \qquad (9.18b)$$

The terms r_1 and r_2 are the radii of the spherical wavefronts overlapping at P; they specify the distances from the sources to P. In this case

$$\delta = k(r_1 - r_2) + (\varepsilon_1 - \varepsilon_2) \qquad (9.19)$$

The flux density in the region surrounding S_1 and S_2 will certainly vary from point to point as $(r_1 - r_2)$ varies. Nonetheless, from the principle of conservation of energy, we expect the spatial average of I to remain constant and equal to the average of $I_1 + I_2$. The space average of I_{12} must therefore be zero, a property verified by Eq. (9.11), since the average of the cosine term is, in fact, zero. (For further discussion of this point, see Problem 9.2.)

Equation (9.17) will be applicable when the separation between S_1 and S_2 is small in comparison with r_1 and r_2 and when the interference region is also small in the same sense. Under these circumstances $\mathbf{E}_{01}$ and $\mathbf{E}_{02}$ may be considered independent of position, that is, constant over the small region examined. If the emitting sources are of equal strength, $\mathbf{E}_{01} = \mathbf{E}_{02}$, $I_1 = I_2 = I_0$ and we have

$$I = 4I_0 \cos^2 \tfrac{1}{2}[k(r_1 - r_2) + (\varepsilon_1 - \varepsilon_2)]$$

Irradiance maxima occur when

$$\delta = 2\pi m$$

provided that $m = 0, \pm 1, \pm 2, \ldots$ Similarly, minima, for which $I = 0$, arise when

$$\delta = \pi m'$$

where $m' = \pm 1, \pm 3, \pm 5, \ldots$, or if you like, $m' = 2m + 1$. Using Eq. (9.19) these two expressions for δ can be rewritten such that maximum irradiance occurs when

$$(r_1 - r_2) = [2\pi m + (\varepsilon_2 - \varepsilon_1)]/k \qquad (9.20a)$$

and minimum when

$$(r_1 - r_2) = [\pi m' + (\varepsilon_2 - \varepsilon_1)]/k \qquad (9.20b)$$

Either one of these equations defines a family of surfaces, each of which is a hyperboloid of revolution. The vertices of the hyperboloids are separated by distances equal to the right-hand sides of Eqs. (9.20a) and (9.20b). The foci are located at S_1 and S_2. If the waves are in-phase at the emitter, $\varepsilon_1 - \varepsilon_2 = 0$, and Eqs. (9.20a) and (9.20b) can be simplified to

[maxima] $$(r_1 - r_2) = 2\pi m/k = m\lambda \qquad (9.21a)$$

[minima] $$(r_1 - r_2) = \pi m'/k = \tfrac{1}{2}m'\lambda \qquad (9.21b)$$

for maximum and minimum irradiance, respectively. Figure 9.3a shows a few of the surfaces over which there are irradi-

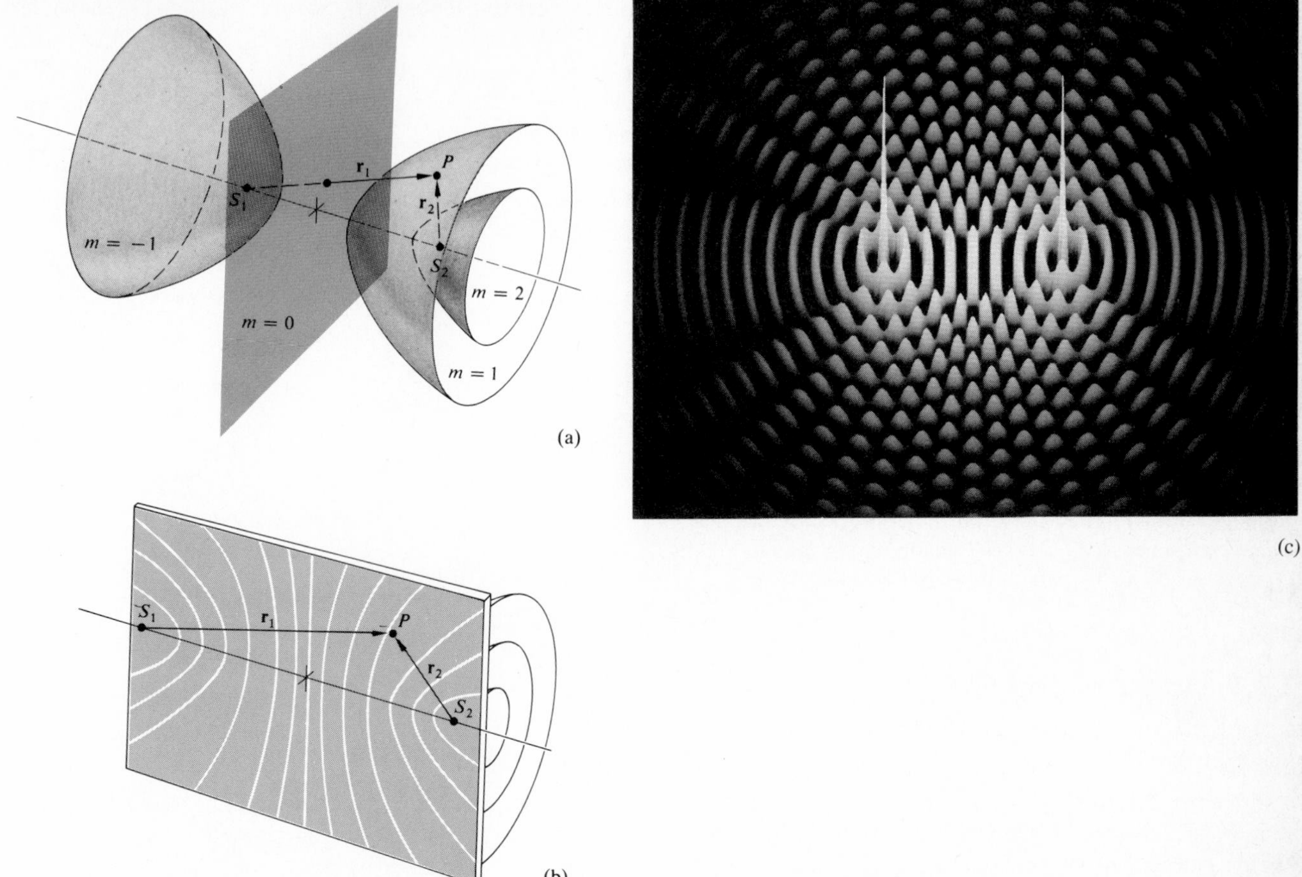

FIGURE 9.3 (*a*) Hyperboloidal surfaces of maximum irradiance for two point sources. The quantity m is positive where $r_1 > r_2$. (*b*) Here we see how the irradiance maxima are distributed on a plane containing S_1 and S_2. (*c*) The electric-field distribution in the plane shown in part (*b*). The tall peaks, are the point sources S_1 and S_2. Note that the spacing of the sources is different in (*b*) and (*c*). (Photo courtesy, The Optics Project, Mississippi State University.)

ance maxima. The dark and light zones that would be seen on a screen placed in region of interference are known as **interference fringes** (Fig. 9.3b). Notice that the central bright band, equidistant from the two sources, is the so-called zeroth-order fringe ($m = 0$), which is straddled by the $m' = \pm 1$ minima, and these, in turn, are bounded by the first-order ($m = \pm 1$) maxima, which are straddled by the $m' = \pm 3$ minima, and so forth.

Since the wavelength λ for light is very small, a large number of surfaces corresponding to the lower values of m will exist close to, and on either side of, the plane $m = 0$. A number of fairly straight parallel fringes will therefore appear on a screen placed perpendicular to that ($m = 0$) plane and in the vicinity of it, and for this case the approximation $r_1 \approx r_2$ will hold. If S_1 and S_2 are then displaced normal to the $\overline{S_1S_2}$ line, the fringes will merely be displaced parallel to themselves. *Two narrow slits will generate a large number of exactly overlapping fringes, thereby increasing the irradiance, leaving the central region of the two-point source pattern otherwise essentially unchanged.*

9.2 Conditions for Interference

If two beams are to interfere to produce a stable pattern, they must have very nearly the same frequency. A significant frequency difference would result in a rapidly varying, time-dependent phase difference, which in turn would cause I_{12} to average to zero during the detection interval (see Section 7.1). Still, if the sources both emit white light, the component reds will interfere with reds, and the blues with blues. A great many fairly similar, slightly displaced, overlapping monochromatic patterns will produce one total white-light pattern. It will not be as sharp or as extensive as a quasimonochromatic pattern, but *white light will produce observable interference*.

The clearest patterns exist when the interfering waves have equal or nearly equal amplitudes. The central regions of the dark and light fringes then correspond to complete destructive and constructive interference, respectively, yielding maximum contrast.

For a fringe pattern to be observed, the two sources need not be in-phase with each other. A somewhat shifted but otherwise identical interference pattern will occur if there is some initial phase difference between the sources, as long as it remains constant. Such sources (which may or may not be in step, but are always marching together) are **coherent**.*

9.2.1 Temporal and Spatial Coherence

Remember that because of the granular nature of the emission process, conventional quasimonochromatic sources produce light that is a mix of photon wavetrains. At each illuminated point in space there is a net field that oscillates nicely (through roughly a million cycles) for less than 10 ns or so before it randomly changes phase. This interval over which the lightwave resembles a sinusoid is a measure of its **temporal coherence**. The average time interval during which the lightwave oscillates in a predictable way we have already designated as the coherence time of the radiation. The longer the coherence time, the greater the temporal coherence of the source.

As observed from a fixed point in space, the passing lightwave appears fairly sinusoidal for some number of oscillations between abrupt changes of phase. The corresponding spatial extent over which the lightwave oscillates in a regular, predictable way is the coherence length [Eq. (7.64)]. Once again, it will be convenient to picture the light beam as a progression of well-defined, more or less sinusoidal, wavegroups of average length Δl_c, whose phases are uncorrelated to one another. Bear in mind that ***temporal coherence is a manifestation of spectral purity***. If the light were ideally monochromatic, the wave would be a perfect sinusoid with an infinite coherence length. All real sources fall short of this, and all actually emit a range of frequencies, albeit sometimes quite narrow. For instance, an ordinary laboratory discharge lamp has a coherence length of several millimeters, whereas certain kinds of lasers routinely provide coherence lengths of tens of kilometers.

Figure 9.4 summarizes some of these ideas. In (*a*) the wave, which arises from a point source, is monochromatic and

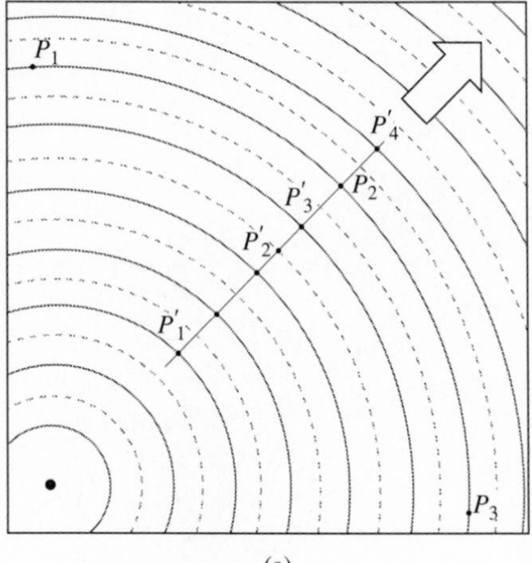

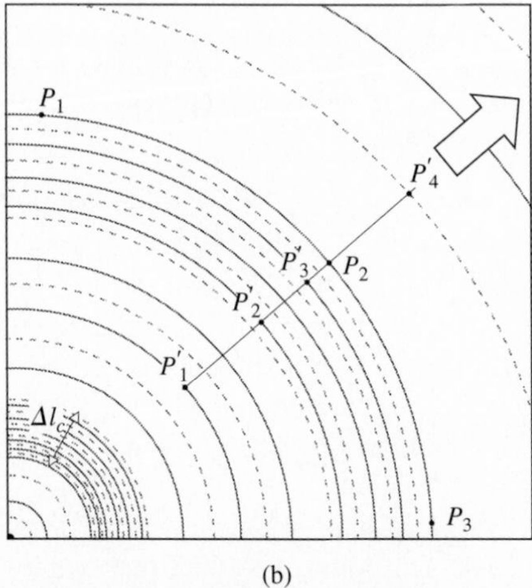

FIGURE 9.4 Temporal and spatial coherence. (*a*) Here the waves display both forms of coherence perfectly. (*b*) Here there is complete spatial coherence, but only partial temporal coherence.

*Chapter 12 is devoted to the study of coherence, so here we'll merely touch on those aspects that are immediately pertinent.

has complete temporal coherence. What happens at P_1' will, a moment later, happen at P_2' and still later at P_3'—all totally predictably. In fact, by watching P_4' we can determine what the wave will be doing at P_1' at any time. Every point on the wave is correlated; its coherence time is unlimited. By contrast, Fig. 9.4*b* shows a point source that changes frequency from moment to moment. Now there's no correlation of the wave at points that are far apart like P_1' and P_4'. The waves lack the total temporal coherence displayed in (*a*), but they're not completely unpredictable; the behavior at points that are close together such as P_2' and P_3' are somewhat correlated. This is an instance of *partial temporal coherence*, a measure of which is the coherence length—the shortest distance over which the disturbance is sinusoidal, that is, the distance over which the phase is predictable.

Notice, in both parts of Fig. 9.4, that the behavior of the waves at points P_1, P_2, and P_3 is completely correlated. Each of the two wave streams arises from a single point source and P_1, P_2, and P_3 lie on the same wavefront in both cases; the disturbances at each of these laterally separated points is in-phase and stays in-phase. Both waves therefore exhibit complete **spatial coherence**. By contrast, suppose the source is broad, that is, composed of many widely spaced point sources (monochromatic ones of period τ), as is Fig. 9.5. If we could take a picture of the wave pattern in Fig. 9.5 every τ seconds, it would be the same; each wavefront would be replaced by an identical one, one wavelength behind it. The disturbances at P_1', P_2', and P_3' are correlated, and the wave is temporally coherent.

Now to insert a little realism; suppose each point source changes phase rapidly and randomly, emitting 10-ns long sinusoidal wavetrains. The waves in Fig. 9.5 would randomly change phase, shifting, combining, and recombining in a frenzied tumult. The disturbances at P_1', P_2', and P_3' would only be correlated for a time less than 10 ns. And the wave field at two, even modestly spaced, lateral points such as P_1 and P_2 would be almost completely uncorrelated depending on the size of the source. The beam from a candle flame or a shaft of sunlight is a multi-frequency mayhem much like this.

Two ordinary sources, two lightbulbs, can be expected to maintain a constant relative phase for a time no greater than Δt_c, so the interference pattern they produce will randomly shift around in space at an exceedingly rapid rate, averaging out and making it quite impractical to observe. Until the advent of the laser, it was a working principle that no two individual sources could ever produce an observable interference

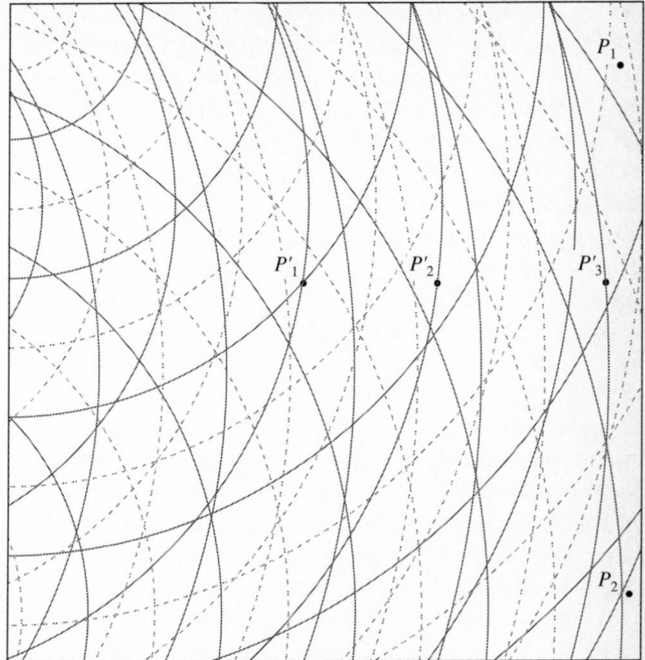

FIGURE 9.5 With multiple (here four) widely spaced point sources, the resultant wave is still coherent. But if those sources change phase rapidly and randomly, both the spatial and temporal coherence diminish accordingly.

pattern. The coherence time of lasers, however, can be appreciable, and interference via independent lasers has been observed and photographed. The most common means of overcoming this problem with ordinary thermal sources is to make one source serve to produce two coherent secondary sources.

9.2.2 The Fresnel–Arago Laws

In Section 9.1 it was assumed that the two overlapping optical disturbance vectors were linearly polarized and parallel. Nonetheless, the formulae apply as well to more complicated situations; indeed, the treatment is applicable regardless of the polarization state of the waves. To appreciate this, recall that any polarization state can be synthesized out of two orthogonal $\mathcal{P}$-states. For natural light these $\mathcal{P}$-states are mutually incoherent, but that represents no particular difficulty.

Suppose that every wave has its propagation vector in the same plane, so that we can label the constituent orthogonal $\mathcal{P}$-

states, with respect to that plane, for example, $\mathbf{E}_{\parallel}$ and $\mathbf{E}_{\perp}$, which are parallel and perpendicular to the plane, respectively (Fig. 9.6*a*). Thus any plane wave, whether polarized or not, can be written in the form $(\mathbf{E}_{\parallel} + \mathbf{E}_{\perp})$. Imagine that the waves $(\mathbf{E}_{\parallel 1} + \mathbf{E}_{\perp 1})$ and $(\mathbf{E}_{\parallel 2} + \mathbf{E}_{\perp 2})$ emitted from two identical coherent sources superimpose in some region of space. The resulting flux-density distribution will consist of two independent, precisely, overlapping interference patterns $\langle(\mathbf{E}_{\parallel 1} + \mathbf{E}_{\parallel 2})^2\rangle_T$ and $\langle(\mathbf{E}_{\perp 1} + \mathbf{E}_{\perp 2})^2\rangle_T$. Therefore, *although we derived the equations of the previous section specifically for linear light, they are applicable to any polarization state, including natural light.*

Notice that even though $\mathbf{E}_{\perp 1}$ and $\mathbf{E}_{\perp 2}$ are always parallel to each other, $\mathbf{E}_{\parallel 1}$ and $\mathbf{E}_{\parallel 2}$, which are in the reference plane, need not be. They will be parallel only when the two beams are themselves parallel (i.e., $\mathbf{k}_1 = \mathbf{k}_2$). The inherent vector nature of the interference process as manifest in the dot-product representation [Eq. (9.11)] of I_{12} cannot be ignored. There are many practical situations in which the beams approach being parallel, and in these cases the scalar theory will do nicely. Even so, (*b*) and (*c*) in Fig. 9.6 are included as an urge to caution. They depict the imminent overlapping of two coherent linearly polarized waves. In Fig. 9.6*b* the optical vectors are parallel, even though the beams aren't, and interference would nonetheless result. In Fig. 9.6*c* the optical vectors are perpendicular, and $I_{12} = 0$, which would be the case here even if the beams were parallel.

Fresnel and Arago made an extensive study of the conditions under which the interference of polarized light occurs, and their conclusions summarize some of the above considerations. The **Fresnel–Arago Laws** are as follows:

1. Two orthogonal, coherent $\mathcal{P}$-states cannot interfere in the sense that $I_{12} = 0$ and no fringes result.
2. Two parallel, coherent $\mathcal{P}$-states will interfere in the same way as will natural light.
3. The two constituent orthogonal $\mathcal{P}$-states of natural light cannot interfere to form a readily observable fringe pattern even if rotated into alignment. This last point is understandable, since these $\mathcal{P}$-states are incoherent.

9.3 Wavefront-Splitting Interferometers

The main problem in producing interference is the sources: they must be *coherent*. And yet separate, independent, adequately coherent sources, other than the modern laser, don't exist! That dilemma was first solved two hundred years ago by

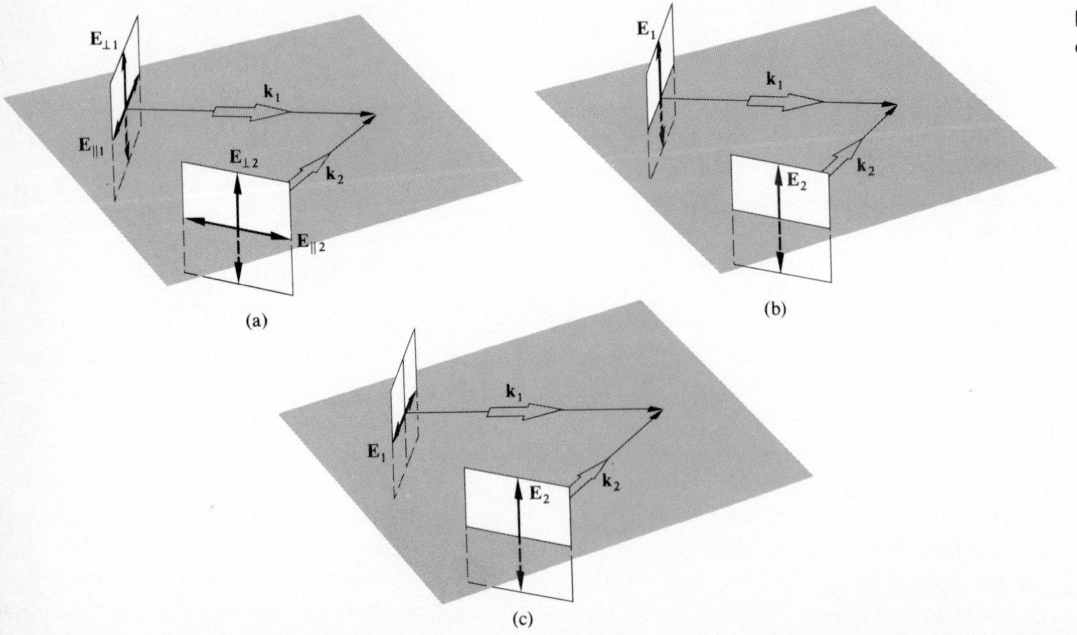

FIGURE 9.6 Interference of polarized light.

Thomas Young in his classic double-beam experiment. He brilliantly took a single wavefront, split off from it two coherent portions, and had them interfere.

9.3.1 Young's Experiment

In 1665 Grimaldi described an experiment he had performed to examine the interaction between two beams of light. He admitted sunlight into a dark room through two close-together pinholes in an opaque screen. Like a camera obscura (p. 219), each pinhole cast an image of the Sun on a distant white surface. The idea was to show that where the circles of light overlapped, darkness could result. Although at the time he couldn't possibly understand why, the experiment failed because the primary source, the Sun's disk (which subtends about 32 minutes of arc), was too large and therefore the incident light didn't have the necessary spatial coherence to properly simultaneously illuminate the two pinholes; to do that, the Sun would have had to subtend only a few seconds of arc.

A hundred and forty years later, Dr. Thomas Young (guided by the phenomenon of beats, which was understood to be produced by two overlapping sound waves) began his efforts to establish the wave nature of light. He redid Grimaldi's experiment, but this time the sunlight passed through an initial pinhole, which became the primary source (Fig. 9.7). This had

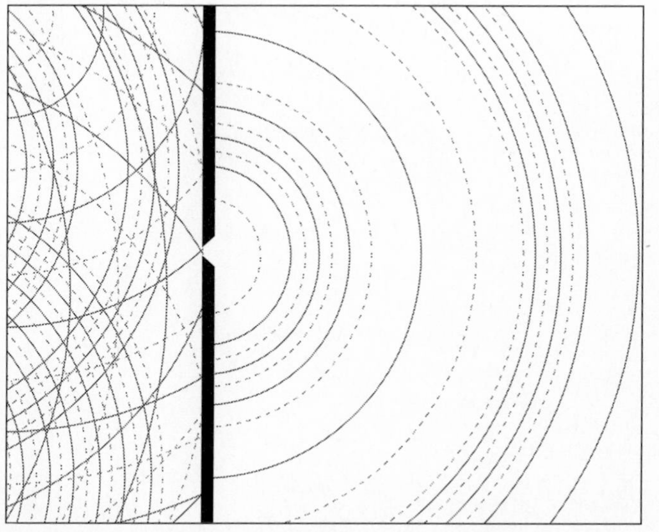

FIGURE 9.7 The pinhole scatters a wave that is spatially coherent, even though it's not temporally coherent.

the effect of creating a spatially coherent beam that could identically illuminate the two apertures. In this way Young succeeded in producing a system of alternating bight and dark bands—interference fringes. Today, aware of the physics involved, we generally replace the pinholes with narrow slits that let through much more light (Fig. 9.8*a*).

Consider a hypothetical monochromatic plane wave illuminating a long narrow slit. From that primary slit light will be diffracted out at all angles in the forward direction and a cylindrical wave will emerge. Suppose that this wave, in turn, falls on two parallel, narrow, closed spaced slits, S_1 and S_2. This is shown in a three-dimensional view in Fig. 9.8*a*. When symmetry exists, the segments of the primary wavefront arriving at the two slits will be exactly in-phase, and the slits will constitute two coherent secondary sources. We expect that wherever the two waves coming from S_1 and S_2 overlap, interference will occur (provided that the optical path difference is less than the coherence length, $c\,\Delta t_c$).

Figures 9.8*a*, *b*, and *c* correspond to the classic arrangement of **Young's Experiment**, although there are other variations. Nowadays the first screen is usually dispensed with, and plane waves from a laser directly illuminate the aperture screen (Fig. 9.8*f*). In a realistic physical situation, the distance between each of the screens (Σ_a and Σ_o) in Fig. 9.8*c* would be very large in comparison with the distance a between the two slits, several thousand times as much, and all the fringes would be fairly close to the center O of the screen. The optical path difference between the rays along $\overline{S_1P}$ and $\overline{S_2P}$ can be determined, to a good approximation, by dropping a perpendicular from S_2 onto $\overline{S_1P}$. This path difference is given by

$$(\overline{S_1B}) = (\overline{S_1P}) - (\overline{S_2P}) \tag{9.22}$$

or

$$(\overline{S_1B}) = r_1 - r_2$$

Continuing with this approximation (see Problem 9.15), $(r_1 - r_2) = a \sin\theta$ and so

$$r_1 - r_2 \approx a\theta \tag{9.23}$$

since $\theta \approx \sin\theta$. Notice that

$$\theta \approx \frac{y}{s} \tag{9.24}$$

and so

$$r_1 - r_2 \approx \frac{a}{s} y \tag{9.25}$$

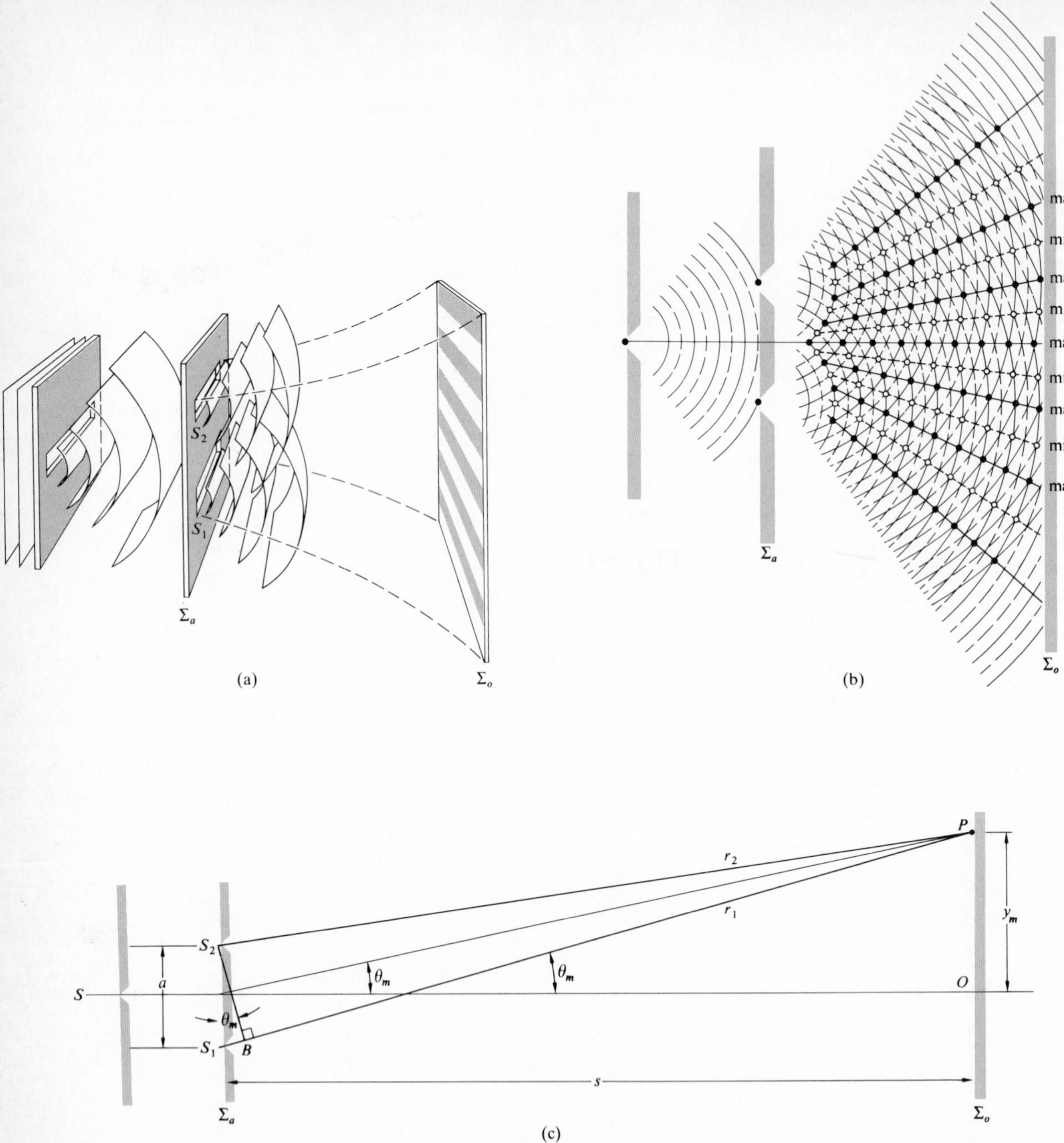

FIGURE 9.8 Young's Experiment. (*a*) Cylindrical waves superimposed in the region beyond the aperture screen. (*b*) Overlapping waves showing peaks and troughs. The maxima and minima lie along nearly straight hyperbolas. (*c*) The geometry of Young's Experiment. (*d*) A path length difference of one wavelength corresponds to $m = \pm 1$ and the first-order maximum. (*e*) (Photo courtesy M. Cagnet, M. Francon, and J. C. Thierr: *Atlas optischer Erscheinungen*, Berlin–Heidelberg–New York: Springer, 1962.) (*f*) A modern version of Young's Experiment using a photodetector (e.g., a photovoltaic cell or photodiode like the RS 305-462) and an $X - Y$ recorder. The detector rides on a motor-driven slide and scans the interference pattern.

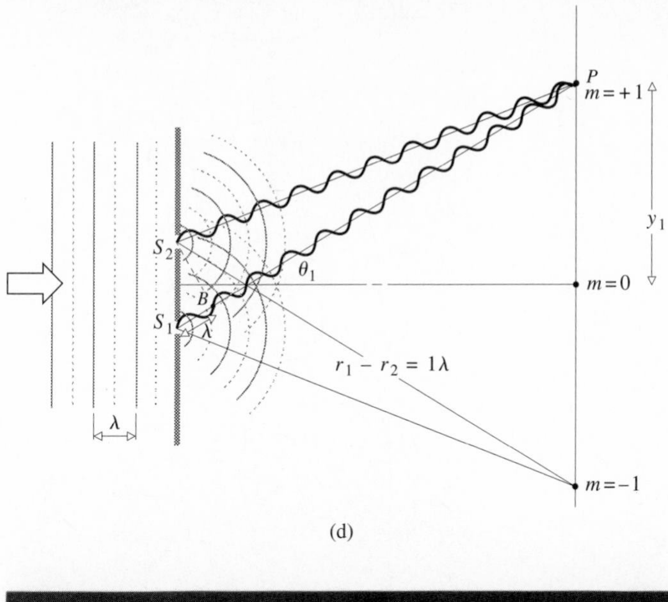

(d)

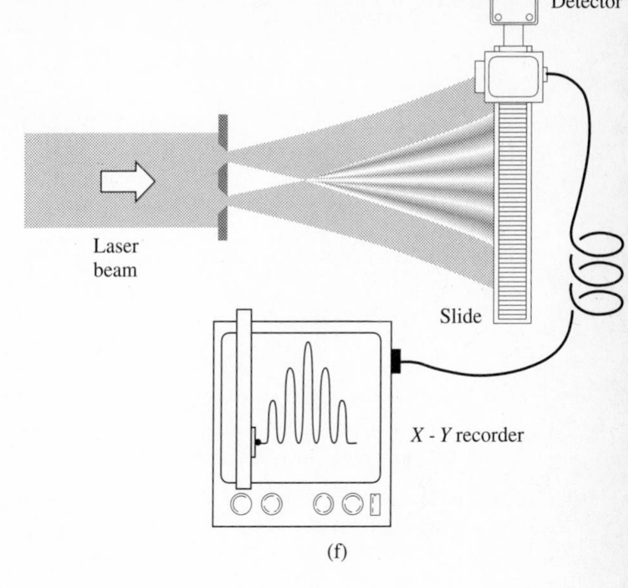

(f)

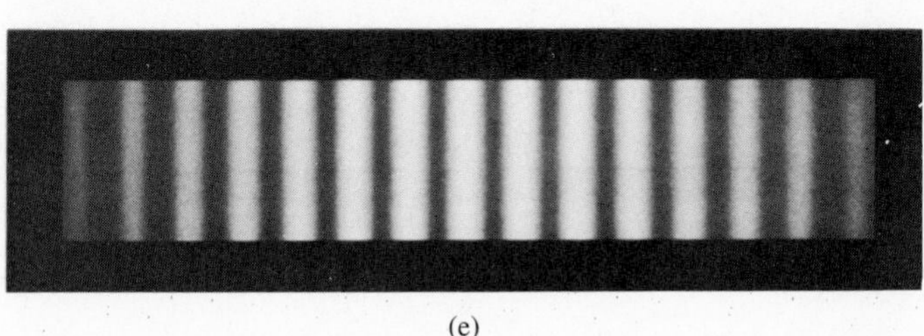
(e)

In accordance with Section 9.1, *constructive* interference will occur when

$$r_1 - r_2 = m\lambda \tag{9.26}$$

Thus, from the last two relations we obtain

$$y_m \approx \frac{s}{a} m\lambda \tag{9.27}$$

This gives the position of the mth bright fringe on the screen, if we count the maximum at 0 as the zeroth fringe. The angular position of the fringe is obtained by substituting the last expression into Eq. (9.24); thus

$$\theta_m = \frac{m\lambda}{a} \tag{9.28}$$

This relationship can be obtained directly by inspecting Fig. 9.8*c*. For the mth-order interference maximum, m whole wavelengths should fit within the distance $r_1 - r_2$. Therefore, from the triangle S_1S_2B,

$$a \sin \theta_m = m\lambda \tag{9.29}$$

or

$$\theta_m \approx m\lambda/a$$

The spacing of the fringes on the screen can be gotten readily from Eq. (9.27). The difference in the positions of two consecutive maxima is

$$y_{m+1} - y_m \approx \frac{s}{a}(m + 1)\lambda - \frac{s}{a}m\lambda$$

or

$$\Delta y \approx \frac{s}{a}\lambda \tag{9.30}$$

Evidently, red fringes are broader than blue ones.

Since this pattern is equivalent to that obtained for two overlapping spherical waves (at least in the $r_1 \approx r_2$ region), we can apply Eq. (9.17). Using the phase difference

$$\delta = k(r_1 - r_2)$$

Eq. (9.17) can be rewritten as

$$I = 4I_0 \cos^2 \frac{k(r_1 - r_2)}{2}$$

provided, of course, that the two beams are coherent and have equal irradiances I_0. With

$$r_1 - r_2 \approx ya/s$$

the resultant irradiance becomes

$$I = 4I_0 \cos^2 \frac{ya\pi}{s\lambda} \tag{9.31}$$

As shown in Fig. 9.9, consecutive maxima are separated by the Δy given in Eq. (9.30). Remember that we effectively assumed that each slit was infinitesimally wide, and so the cosine-squared fringes of Fig. 9.9 are really an unattainable idealization.* The actual pattern, Fig. 9.8*e*, drops off with distance on either side of O because of diffraction.

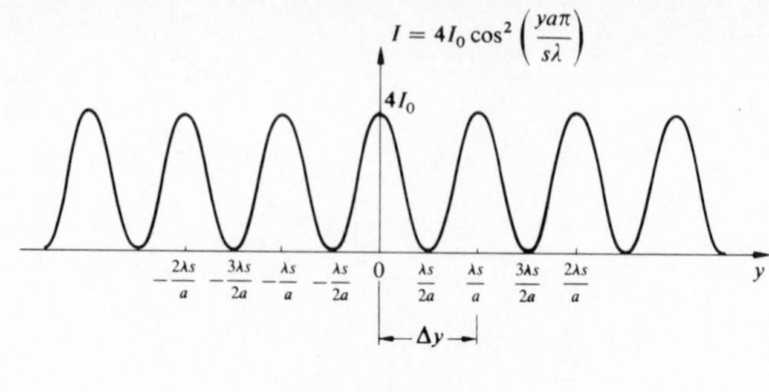

(a)

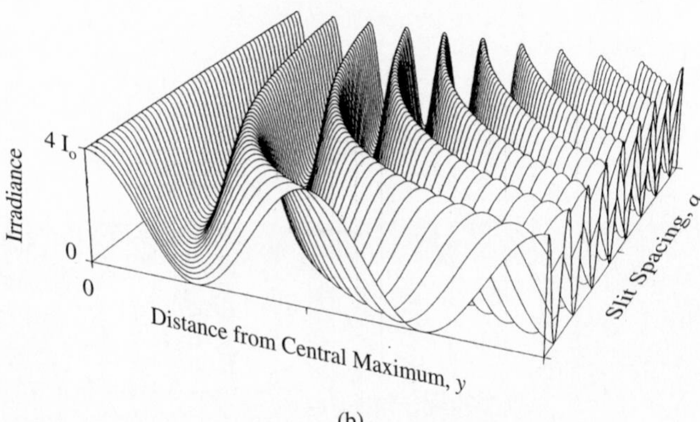

(b)

FIGURE 9.9 (*a*) Idealized irradiance versus distance curve. (*b*) The fringe separation Δy varies inversely with the slit separation, as one might expect from Fourier considerations; remember the inverse nature of spatial intervals and spatial frequency intervals. (Reprinted from"Graphical Representations of Fraunhofer Interference and Diffraction" *Am. J. Phys* **62**, 6 (1994), with permission of A.B. Bartlett, University of Colorado and B. Mechtly, Northeast Missouri State University and the American Association of Physics Teachers.)

The Effects of Finite Coherence Length

As P in Fig. 9.8*c* is taken farther above or below the axis, $\overline{S_1B}$ (which is less than or equal to $\overline{S_1S_2}$) increases. If the primary source has a short coherence length, as the optical path difference increases, identically paired wavegroups will no longer be able to arrive at P exactly together—there will be an increasing amount of overlap in portions of uncorrelated wavegroups, and the contrast of the fringes will degrade. It is possible for Δl_c to be less than $\overline{S_1B}$. In that case, instead of two correlated portions of the same wavegroup arriving at P, only segments of different wavegroups will overlap, and the fringes will vanish.

As depicted in Fig. 9.10*a*, when the path length difference exceeds the coherence length, wavegroup-E_1 from source S_1 arrives at P with wavegroup-D_2 from S_2. There is interference, but it lasts only for a short time before the pattern shifts as wavegroup-D_1 begins to overlap wavegroup-C_2, since the relative phases are different. If the coherence length was larger or the path difference smaller, wavegroup-D_1 would more or less interact with its clone wavegroup-D_2, and so on for each pair. The phases would then be correlated, and the interference pattern stable (Fig. 9.10*b*). Since a whitelight source will have a coherence length of less than about three wavelengths, it follows from Eq. (9.27) that only about three fringes will be seen on either side of the central maximum.

Under white light (or with broad bandwidth illumination), all the constituent colors will arrive at $y = 0$ in-phase, having traveled equal distances from each aperture. The zeroth-order fringe will be essentially white, but all other higher order maxima will show a spread of wavelengths, since y_m is a function of λ, according to Eq. (9.27). Thus in white light we can think of the mth maximum as the mth-order band of wavelengths—

*Modifications of this pattern arising as a result of increasing the width of either the primary S or secondary-source slits will be considered in later chapters (10 and 12). In the latter case, fringe contrast will be used as a measure of the degree of coherence (Section 12.1). In the latter, diffraction effects become significant.

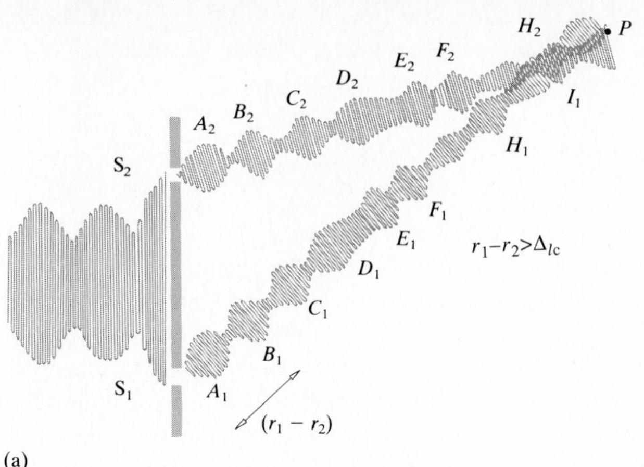

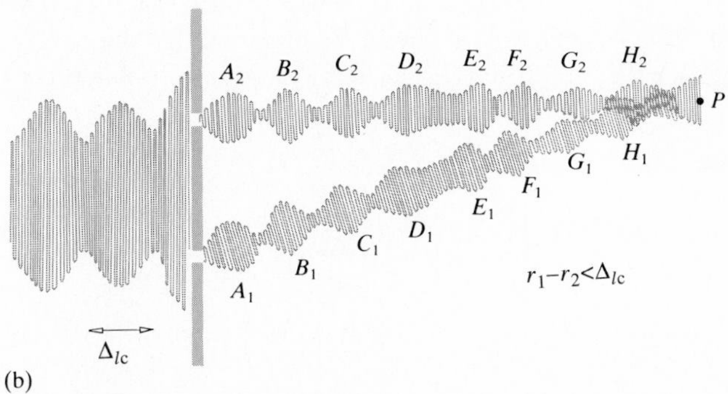

FIGURE 9.10 A schematic representation of how light, composed of a progression of wavegroups with a coherence length Δl_c, produces interference when (*a*) the path length difference exceeds Δl_c and (*b*) the path length difference is less than Δl_c.

a notion that will lead directly to the diffraction grating in the next chapter.

The fringe pattern can be observed by punching two small pinholes in a thin card. The holes should be approximately the size of the type symbol for a period on this page, and the separation between their centers about three radii. A street lamp, car headlight, or traffic signal at night, located a few hundred feet away, will serve as a plane wave source. The card should be positioned directly in front of and *very close to the eye*. The fringes will appear perpendicular to the line of centers. The pattern is much more readily seen with slits, as discussed in Section 10.2.2, but you should give the pinholes a try.

Microwaves, because of their long wavelength, also offer an easy way to observe double-slit interference. Two slits (e.g, $\lambda/2$ wide by λ long, separated by 2λ) cut in a piece of sheet metal or foil will serve quite well as secondary sources (Fig. 9.11).

The Fourier Perspective

When the plane waves in Fig. 9.8*b* illuminated the first narrow slit, light spilled out (i.e., diffracted) beyond the opaque screen in a form resembling a cylindrical wave; the narrower the slit, the more nearly cylindrical the wave. Beyond the screen the light spread over a very wide range of angles, or equivalently a wide range of spatial frequencies. From a Fourier perspective, this happened because an infinitesimally narrow source (i.e., narrow in space) generates a light field that is infinitely broad (i.e., broad in spatial frequency). The transform of a point source, an ideal one-dimensional signal spike (known as a Dirac delta function, p. 517), is a continuous constant spectrum containing all spatial frequencies, a spherical wave. In the same way, an ideal line source results in a cylindrical wave.

In practice, Young's Experiment usually consists of two in-phase slit sources arranged such that $s >> a$. As a rule, s is so large that the resulting fringe system corresponds to a Fraunhoffer diffraction pattern (p. 449). The two very thin slits resemble two line sources, two ideally narrow signal spikes, and the transform of two delta functions is a cosine function—we saw that in Fig. 7.31. To the extent that the slits can be considered infinitesimally narrow, the amplitude of the electric field in the diffraction pattern will be cosinusoidal, and the irradiance distribution will vary as the cosine squared, as in Fig. 9.6.

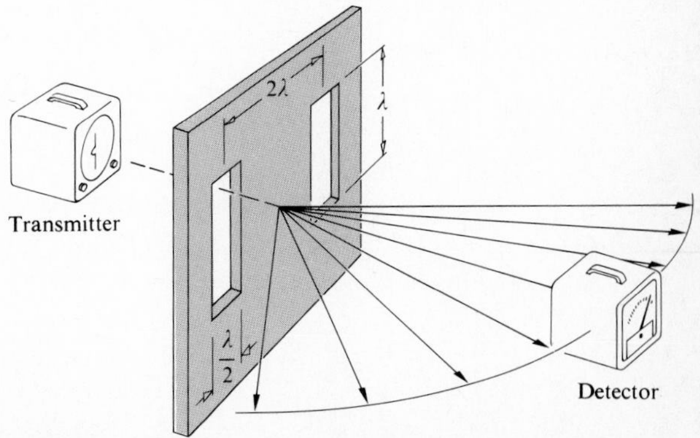

FIGURE 9.11 A microwave interferometer.

Several Other Interferometers

The same physical and mathematical considerations applied to Young's Experiment relate directly to a number of other wavefront-splitting interferometers. Most common among these are Fresnel's double mirror, Fresnel's double prism, and Lloyd's mirror.

Fresnel's double mirror consists of two plane front-silvered mirrors inclined to each other at a very small angle, as shown in Fig. 9.12. One portion of the cylindrical wavefront coming from slit S is reflected from the first mirror, and another portion of the wavefront is reflected from the second mirror. An interference field exists in space in the region where the two reflected waves are superimposed. The images (S_1 and S_2) of the slit S in the two mirrors can be considered as separate coherent sources, placed at a distance a apart. It follows from the Laws of Reflection, as illustrated in Fig. 9.12*a*, that $\overline{SA} = \overline{S_1A}$ and $\overline{SB} = \overline{S_2B}$ so that $\overline{SA} + \overline{AP} = r_1$ and $\overline{SB} + \overline{BP} = r_2$. The optical path length difference between the two rays is then $r_1 - r_2$. The various maxima occur at $r_1 - r_2 = m\lambda$, as they do with Young's Interferometer. Again, the separation of the fringes is given by

$$\Delta y \approx \frac{s}{a}\lambda$$

where s is the distance between the plane of the two virtual sources (S_1, S_2) and the screen. The arrangement in Fig. 9.12 has been deliberately exaggerated to make the geometry somewhat clearer. The angle θ between the mirrors must be quite small if the electric-field vectors for each of the two beams are to be parallel, or nearly so. Let $\mathbf{E}_1$ and $\mathbf{E}_2$ represent the lightwaves emitted from the coherent virtual sources S_1 and S_2. At any instant in time at the point P in space, each of these vectors can be resolved into components, parallel and perpendicular to the plane of the figure. With $\mathbf{k}_1$ and $\mathbf{k}_2$ parallel to $\overline{AP}$ and $\overline{BP}$, respectively, it should be apparent that the components of $\mathbf{E}_1$ and $\mathbf{E}_2$ in the plane of the figure will approach

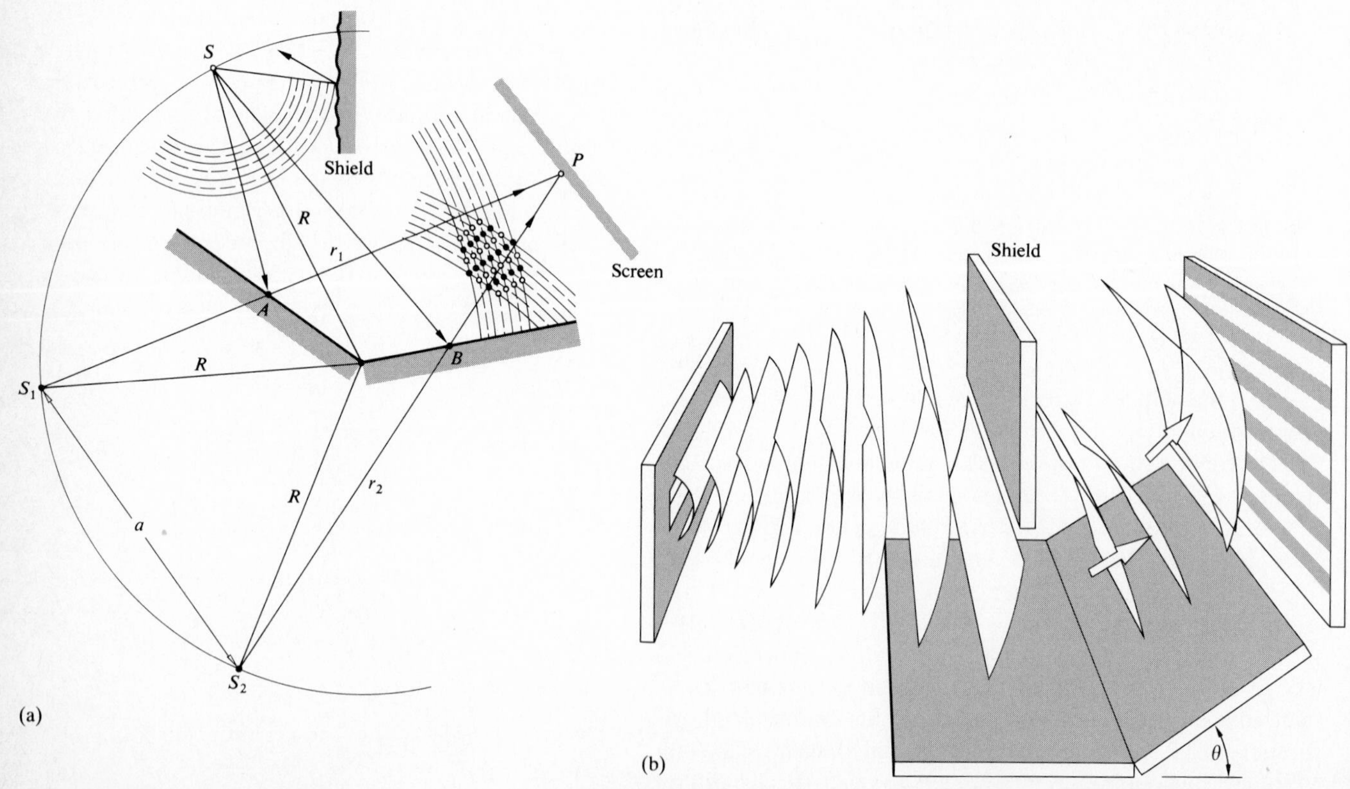

FIGURE 9.12 Fresnel's double mirror.

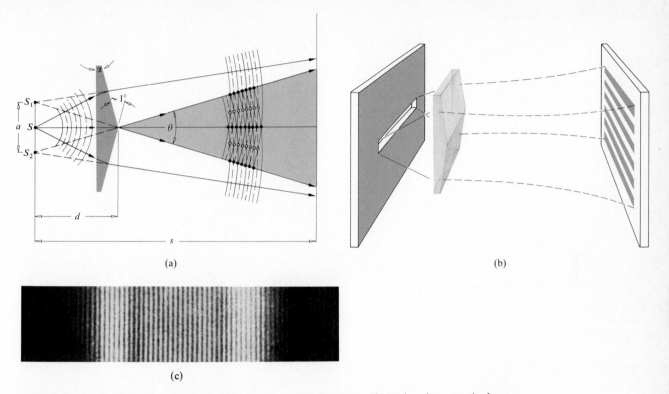

FIGURE 9.13 Fresnel's biprism. (*a*) The biprism creates two image sources. (*b*) With a slit source the fringes are bright bands. (*c*) Interference fringes observed with an electron biprism arrangement by G. Möllenstedt. Once again electrons behave like photons. (Photo from *Handbuch der Physik*, edited by S. Flügge, Springer-Verlag, Heidelberg.)

being parallel only for small θ. As θ decreases a decreases and the fringes broaden.

The **Fresnel double prism** or biprism consists of two thin prisms joined at their bases, as shown in Fig. 9.13. A single cylindrical wavefront impinges on both prisms. The top portion of the wavefront is refracted downward, and the lower segment is refracted upward. In the region of superposition, interference occurs. Here, again, two virtual sources S_1 and S_2 exist, separated by a distance a, which can be expressed in terms of the prism angle α (Problem 9.17), where $s >> a$. The expression for the separation of the fringes is the same as before.

The last wavefront-splitting interferometer that we will consider is **Lloyd's mirror**, shown in Fig. 9.14. It consists of a flat piece of either dielectric or metal that serves as a mirror, from which is reflected a portion of the cylindrical wavefront coming from slit S. Another portion of the wavefront proceeds

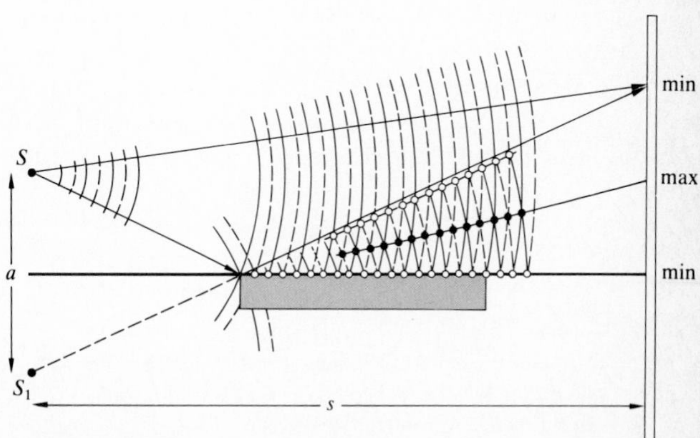

FIGURE 9.14 Lloyd's mirror.

directly from the slit to the screen. For the separation a, between the two coherent sources, we take the distance between the actual slit and its image S_1 in the mirror. The spacing of the fringes is once again given by $(s/a)\lambda$. The distinguishing feature of this device is that at glancing incidence ($\theta_i = \pi/2$) the reflected beam undergoes a 180° phase shift. (Recall that the amplitude-reflection coefficients are then both equal to −1). With an additional phase shift of $\pm\pi$,

$$\delta = k(r_1 - r_2) \pm \pi$$

and the irradiance becomes

$$I = 4I_0 \sin^2\left(\frac{\pi a y}{s\lambda}\right)$$

The fringe pattern for Lloyd's mirror is complementary to that of Young's Interferometer; the maxima of one pattern exist at values of y that correspond to minima in the other pattern. The top edge of the mirror is equivalent to $y = 0$ and will be the center of a dark fringe rather than a bright one, as in Young's device. The lower half of the pattern will be obstructed by the presence of the mirror itself. Consider what would happen if a thin sheet of transparent material were placed in the path of the rays traveling directly to the screen. The transparent sheet would have the effect of increasing the number of wavelengths in each direct ray. The entire pattern would accordingly move upward, where the reflected rays would travel a bit farther before interfering. Because of the obvious inherent simplicity of this device, it has been used over a very wide region of the electromagnetic spectrum. The actual reflecting surfaces have ranged from crystals for X-rays, ordinary glass for light, and wire screening for microwaves to a lake or even the Earth's ionosphere for radiowaves.*

All the above interferometers can be demonstrated quite readily either using a laser or, for white light, something a bit more old-fashioned like a discharge lamp or a carbon arc (Fig. 9.15).

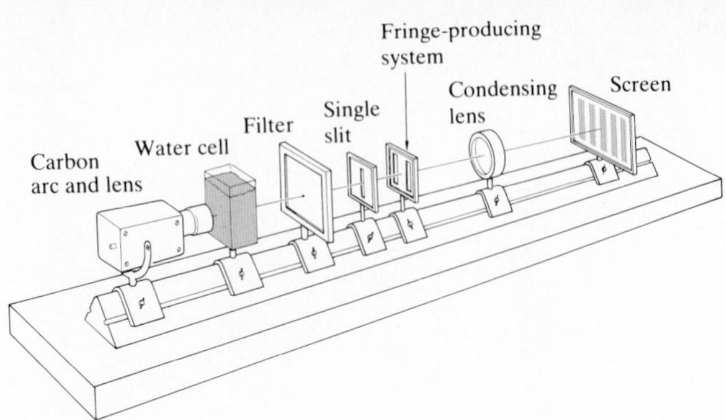

FIGURE 9.15 Bench setup to study wavefront-splitting arrangements with a carbon arc source.The water cell is needed to keep things cool.

9.4 Amplitude-Splitting Interferometers

Suppose that a lightwave is incident on a half-silvered mirror†, or simply on a sheet of glass. Part of the wave is transmitted and part reflected. Both the transmitted and reflected waves have lower amplitudes than the original one. It can be said figuratively that the amplitude has been "split."

If the two separate waves could be brought together again at a detector, interference would result, as long as the original coherence between the two had not been destroyed. If the path lengths differed by a distance greater than that of the wavegroup (i.e., the coherence length), the portions reunited at the detector would correspond to different wavegroups. No unique phase relationship would exist between them in that case, and the fringe pattern would be unstable to the point of being unobservable. We will get back to these ideas when we consider coherence theory in more detail. For the moment the discussion is restricted, for the most part, to those cases for which the path difference is less than the coherence length.

*For a discussion of the effects of a finite slit width and a finite frequency bandwidth, see R. N. Wolfe and F. C. Eisen, "Irradiance Distribution in a Lloyd Mirror Interference Pattern," *J. Opt. Soc. Am.* **38**, 706 (1948).

†A *half-silvered mirror* is one that is semitransparent, because the metallic coating is too thin to be opaque. You can look through it, and at the same time you can see your reflection in it. *Beamsplitters*, as devices of this kind are called, can also be made of thin stretched plastic films, known as *pellicles*, or even uncoated glass plate.

9.4.1 Dielectric Films—Double-Beam Interference

Interference effects are observable in sheet transparent materials, the thicknesses of which vary over a very broad range, from films less than the length of a lightwave (e.g., for green light λ_0 equals about $\frac{1}{150}$ the thickness of this printed page) to plates several centimeters thick. A layer of material is referred to as a *thin film* for a given wavelength of electromagnetic radiation when its thickness is of the order of that wavelength. Before the early 1940s interference phenomena associated with thin dielectric films, although well known, had fairly limited practical value. The rather spectacular color displays arising from oil slicks and soap films, however pleasing aesthetically and theoretically, were mainly curiosities.

With the advent of suitable vacuum deposition techniques in the 1930s, precisely controlled coatings could be produced on a commercial scale, and that, in turn, led to a rebirth of interest in dielectric films. During the Second World War, both sides were finding the enemy with a variety of coated optical devices, and by the 1960s multilayered coatings were in widespread use.

FRINGES OF EQUAL INCLINATION

Consider the simple case of a transparent parallel plate of dielectric material having a thickness d (Fig. 9.16). Suppose that the film is nonabsorbing and that the amplitude-reflection coefficients at the interfaces are so low that only the first two reflected beams E_{1r} and E_{2r} (both having undergone only one reflection) need be considered (Fig. 9.17). In practice, the amplitudes of the higher-order reflected beams (E_{3r}, etc.) generally decrease very rapidly, as can be shown for the air–water and air–glass interfaces (Problem 9.23). For the moment, consider S to be a monochromatic point source.

The film serves as an amplitude-splitting device, so that E_{1r} and E_{2r} may be considered as arising from two coherent virtual sources lying behind the film; that is, the two images of S formed by reflection at the first and second interfaces. The reflected rays are parallel on leaving the film and can be brought together at a point P on the focal plane of a telescope objective or on the retina of the eye when focused at infinity. From Fig. 9.17, the optical path length difference for the first two reflected beams is given by

$$\Lambda = n_f[(\overline{AB}) + (\overline{BC})] - n_1(\overline{AD})$$

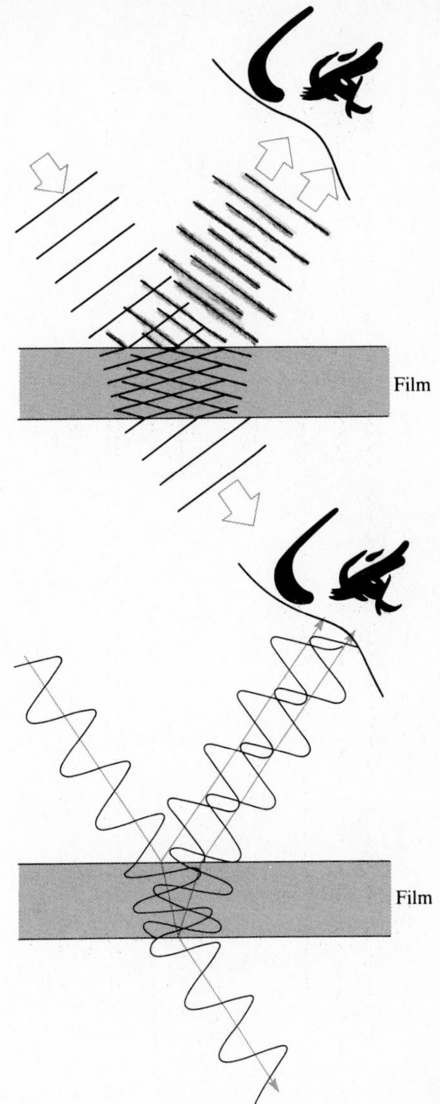

FIGURE 9.16 The wave and ray representations of thin-film interference. Light reflected from the top and bottom of the film interferes to create a fringe pattern.

and since $(\overline{AB}) = (\overline{BC}) = d/\cos\theta_t$,

$$\Lambda = \frac{2n_f d}{\cos\theta_t} - n_1(\overline{AD})$$

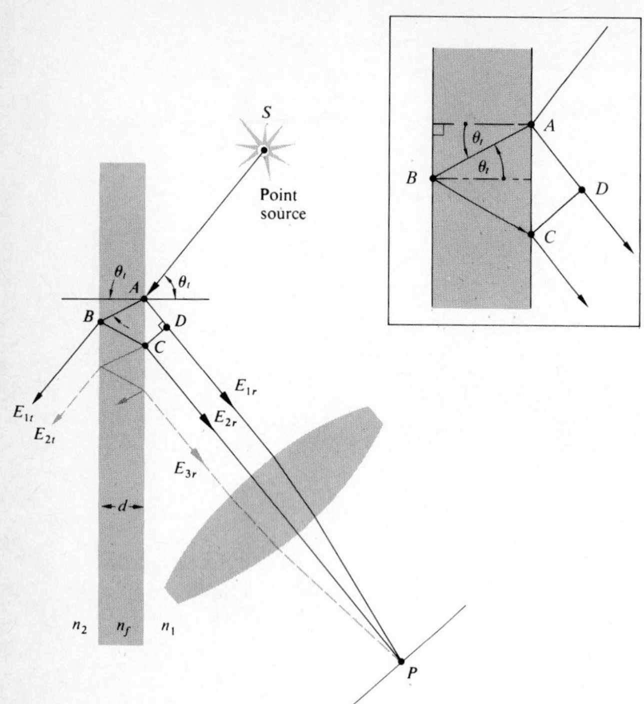

FIGURE 9.17 Fringes of equal inclination.

Now, to find an expression for $(\overline{AD})$, write

$$(\overline{AD}) = (\overline{AC})\sin\theta_i$$

Using Snell's Law, this becomes

$$(\overline{AD}) = (\overline{AC})\frac{n_f}{n_1}\sin\theta_t$$

where

$$(\overline{AC}) = 2d\tan\theta_t \tag{9.32}$$

The expression for Λ now becomes

$$\Lambda = \frac{2n_f d}{\cos\theta_t}(1 - \sin^2\theta_t)$$

or finally

$$\Lambda = 2n_f d\cos\theta_t \tag{9.33}$$

The corresponding phase difference associated with the optical path length difference is then just the product of the free-space propagation number and Λ, that is, $k_0\Lambda$. If the film is immersed in a single medium, the index of refraction can simply be written as $n_1 = n_2 = n$. Realize that n may be less than n_f, as in the case of a soap film in air, or greater than n_f, as with an air film between two sheets of glass. In either case *there will be an additional phase shift arising from the reflections themselves*. Recall that for incident angles up to about 30°, regardless of the polarization of the incoming light, the two beams, one internally and one externally reflected, will experience a *relative phase shift* of π radians (Fig. 4.43 and Section 4.3). Accordingly,

$$\delta = k_0\Lambda \pm \pi$$

and more explicitly

$$\delta = \frac{4\pi n_f}{\lambda_0}d\cos\theta_t \pm \pi \tag{9.34}$$

or

$$\delta = \frac{4\pi d}{\lambda_0}(n_f^2 - n^2\sin^2\theta_i)^{1/2} \pm \pi \tag{9.35}$$

The sign of the phase shift is immaterial, so we will choose the negative sign to make the equations a bit simpler. In reflected

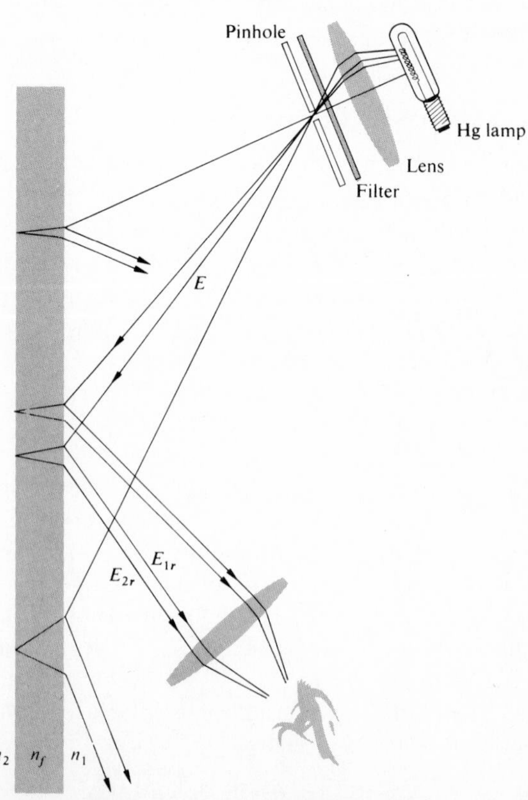

FIGURE 9.18 Fringes seen on a small portion of the film.

light an interference maximum, a bright spot, appears at P when $\delta = 2m\pi$—in other words, an even multiple of π. In that case Eq. (9.34) can be rearranged to yield

$$d \cos \theta_t = (2m + 1)\frac{\lambda_f}{4} \qquad m = 0,1,2,\ldots \qquad \text{[maxima]} \qquad (9.36)$$

where use has been made of the fact that $\lambda_f = \lambda_0/n_f$. This also corresponds to minima in the transmitted light. Interference minima in reflected light (maxima in transmitted light) result when $\delta = (2m \pm 1)\pi$, that is, odd multiples of π. For such cases Eq. (9.34) yields

$$d \cos \theta_t = 2m\frac{\lambda_f}{4} \qquad \text{[minima]} \qquad (9.37)$$

The appearance of odd and even multiples of $\lambda_f/4$ in Eqs. (9.36) and (9.37) is significant , as we will see presently. We could, of course, have a situation in which $n_1 > n_f > n_2$ or $n_1 < n_f < n_2$, as with a fluoride film deposited on an optical element of glass immersed in air. The π phase shift would then not be present, and the above equations would simply be modified appropriately.

If the lens used to focus the rays has a small aperture, interference fringes will appear on a small portion of the film. Only the rays leaving the point source that are reflected directly into the lens will be seen (Fig. 9.18). For an extended source, light will reach the lens from various directions, and the fringe pattern will spread out over a large area of the film (Fig. 9.19).

The angle θ_i or equivalently θ_t, determined by the position of P, will in turn control δ. The fringes appearing at points P_1 and P_2 in Fig. 9.20 are known as **fringes of equal inclination**. (Problem 9.28 discusses some easy ways to see these fringes.) Keep in mind that each source point on the extended source is incoherent with respect to the others. When the image of the extended source is reflected in the surface, it will be seen to be banded with bright and dark fringes. Each of these is an arc of a circle centered on the intersection of a perpendicular dropped from the eye to the film.

As the film becomes thicker, the separation $\overline{AC}$ between E_{1r} and E_{2r} also increases, since

$$\overline{AC} = 2d \tan \theta_t \qquad [9.32]$$

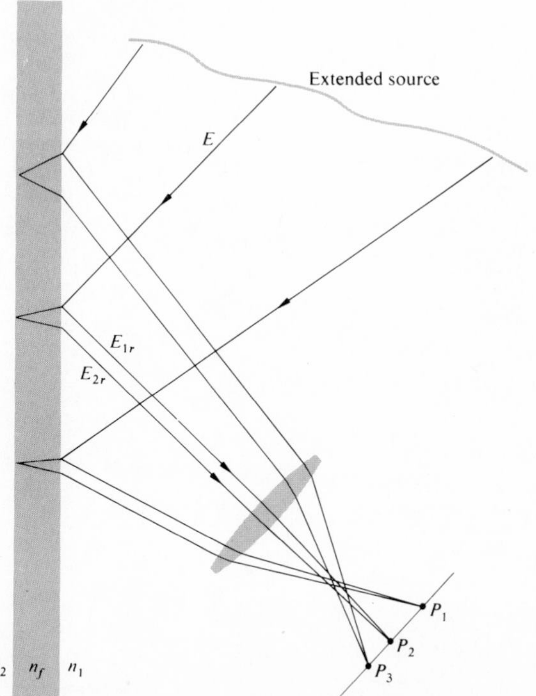

FIGURE 9.19 Fringes seen on a large region of the film.

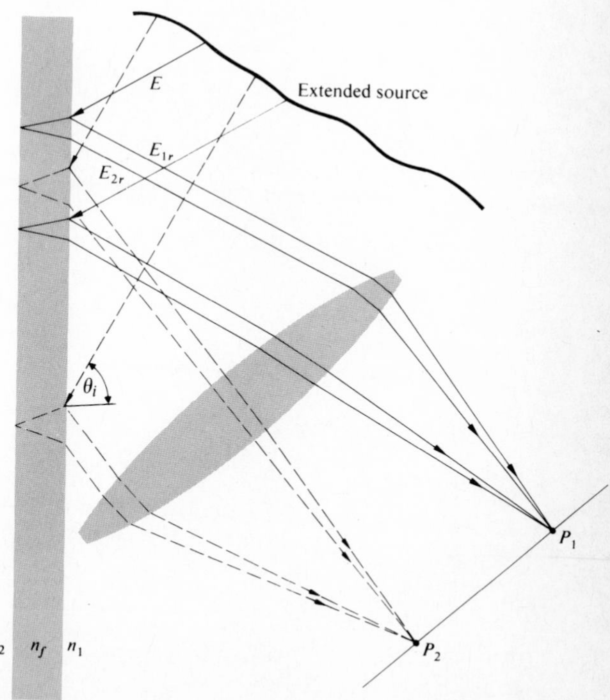

FIGURE 9.20 All rays inclined at the same angle arrive at the same point.

When only one of the two rays is able to enter the pupil of the eye, the interference pattern will disappear. The larger lens of a telescope can then be used to gather in both rays, once again making the pattern visible. The separation can also be reduced by reducing θ_t and therefore θ_i, that is, by viewing the film at nearly normal incidence. The equal-inclination fringes that are seen in this manner for thick plates are known as **Haidinger fringes**, after the Austrian physicist Wilhelm Karl Haidinger (1795–1871). With an extended source, the symmetry of the setup requires that the interference pattern consists of a series of concentric circular bands centered on the perpendicular drawn from the eye to the film (Fig. 9.21). As the observer moves, the interference pattern follows along.

Haidinger fringes can be seen in the ordinary window glass of a store front. Find one with a neon sign in the window and look out at the street, at night, very close to the glowing tube. You'll see circular fringes centered on your eye floating off in the distance.

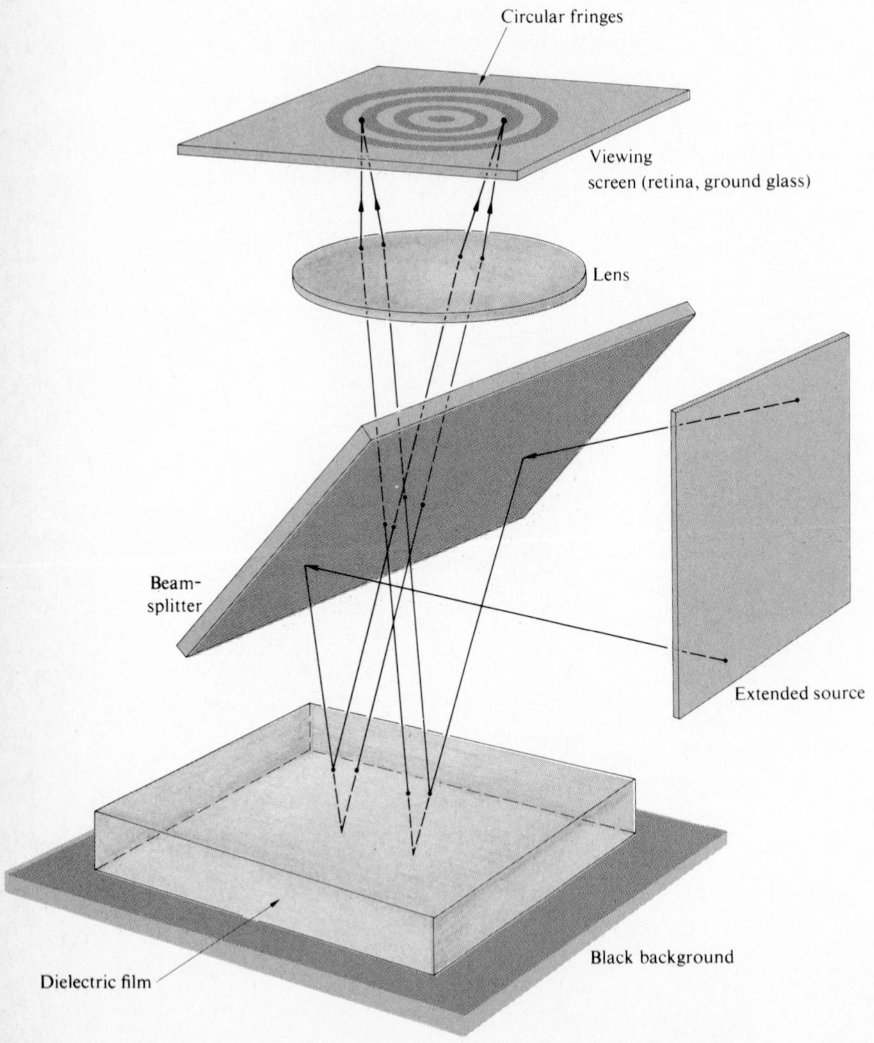

FIGURE 9.21 Circular Haidinger fringes centered on the lens axis.

Fringes of Equal Thickness

A whole class of interference fringes exists for which the optical thickness, $n_f d$, is the dominant parameter rather than θ_i. These are referred to as **fringes of equal thickness**. Under white-light illumination the iridescence of soap bubbles, oil slicks (a few wavelengths thick), and even oxidized metal surfaces is the result of variations in film thickness. Interference bands of this kind are analogous to the constant-height contour lines of a topographical map. Each fringe is the locus of all points in the film for which the optical thickness is a constant. In general, n_f does not vary, so that the fringes correspond to regions of constant film thickness. As such, they can be quite useful in determining the surface features of optical elements (lenses, prisms, etc.). For example, a surface to be examined may be put into contact with an *optical flat*.* The air in the space between the two generates a thin-film interference pattern. If the test surface is flat, a series of straight, equally spaced bands indicates a wedge-shaped air film, usually resulting from dust between the flats. Two pieces of plate glass separated at one end by a strip of paper will form a satisfactory wedge with which to observe these bands.

When viewed at nearly normal incidence in the manner illustrated in Fig. 9.22, the contours arising from a nonuniform film are called **Fizeau fringes**. For a thin wedge of small angle α, the optical path length difference between two reflected rays may be approximated by Eq. (9.33), where d is the thickness at a particular point, that is,

$$d = x\alpha \tag{9.38}$$

For small values of θ_i the condition for an interference maximum becomes

$$(m + \tfrac{1}{2})\lambda_0 = 2n_f d_m$$

or

$$(m + \tfrac{1}{2})\lambda_0 = 2\alpha x_m n_f$$

*A surface is said to be optically flat when it deviates by not more than about $\lambda/4$ from a perfect plane. In the past, the best flats were made of clear fused quartz. Now glass-ceramic materials (e.g., CERVIT) having extremely small thermal coefficients of expansion (about one-sixth that of quartz) are available. Individual flats of $\lambda/200$ or a bit better can be made.

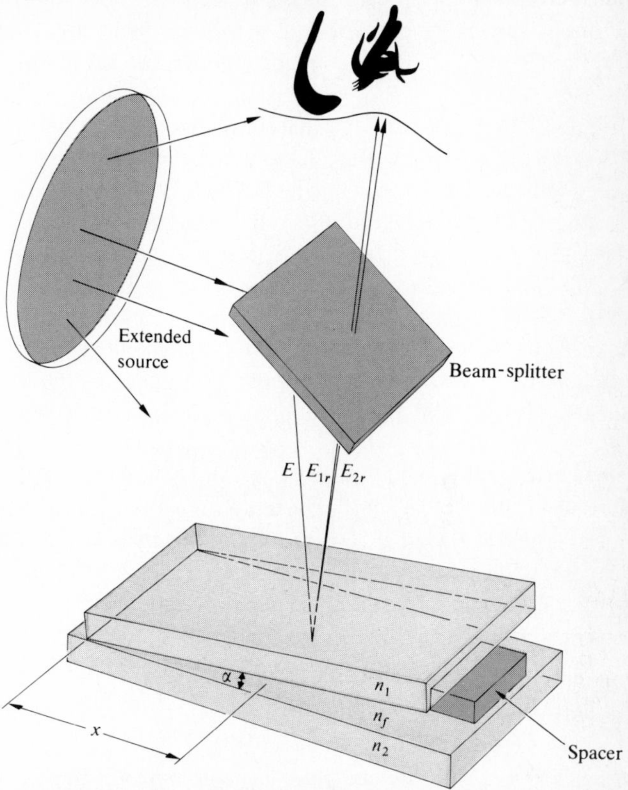

FIGURE 9.22 Fringes from a wedge-shaped film.

Since $n_f = \lambda_0/\lambda_f$, x_m may be written as

$$x_m = \left(\frac{m + 1/2}{2\alpha}\right)\lambda_f \tag{9.39}$$

Maxima occur at distances from the apex given by $\lambda_f/4\alpha$, $3\lambda_f/4\alpha$, and so on, and consecutive fringes are separated by a distance Δx, given by

$$\Delta x = \lambda_f/2\alpha \tag{9.40}$$

Notice that the difference in film thickness between adjacent maxima is simply $\lambda_f/2$. Since the beam reflected from the lower surface traverses the film twice ($\theta_i \approx \theta_t \approx 0$), adjacent maxima differ in optical path length by λ_f. Note, too, that the film thickness at the various maxima is given by

$$d_m = (m + \tfrac{1}{2})\frac{\lambda_f}{2} \tag{9.41}$$

which is an odd multiple of a quarter wavelength. Traversing the film twice yields a phase shift of π, which when added to the shift of π resulting from reflection, puts the two rays back in-phase.

Figure 9.23 is a photograph of a soap film held vertically so that it settles into a wedge shape under the influence of gravity. When illuminated with white light, the bands are various colors. The black region at the top is a portion where the film is less than $\lambda_f/4$ thick. Twice this, plus an additional shift of $\lambda_f/2$ due to the reflection, is less than a whole wavelength. The reflected rays are therefore out-of-phase. As the thickness decreases still further, the total phase difference approaches π. The irradiance at the observer goes to a minimum (Eq. 9.16), and the film appears black in reflected light.*

Press two well-cleaned microscope slides together. The enclosed air film will usually not be uniform. In ordinary room light a series of irregular, colored bands (fringes of equal thickness) will be clearly visible across the surface (Fig. 9.24). The thin glass slides distort under pressure, and the fringes move and change accordingly. Tape two slides together with transparent (matt-surfaced) tape. It will scatter light and make the reflected fringes more easily seen.

FIGURE 9.23 A wedge-shaped film made of liquid dishwashing soap. (E. H.)

*The relative phase shift of π between internal and external reflection is required if the reflected flux density is to go to zero smoothly, as the film gets thinner and finally disappears.

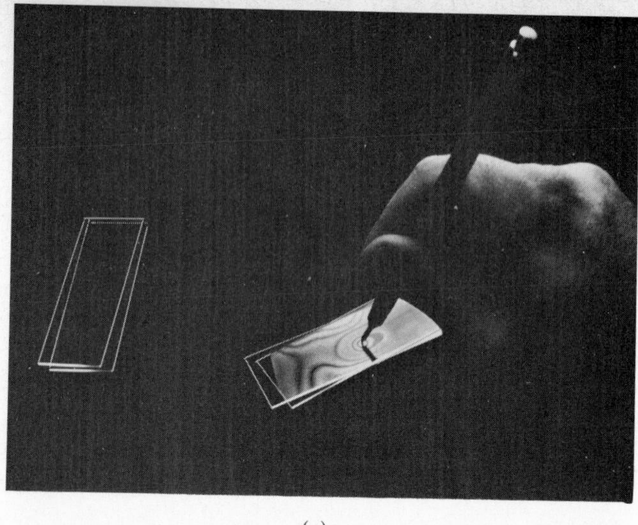

(a)

FIGURE 9.24 Fringes created by an air film between two microscope slides. (Photo by E. H.)

If the two pieces of glass are forced together at a point, as might be done by pressing on them with a sharp pencil, a series of concentric, nearly circular, fringes is formed about that point. Known as **Newton's rings*** (Fig. 9.25), this pattern is more precisely examined with the arrangement of Fig. 9.26. Here a lens is placed on an optical flat and illuminated at normal incidence with quasimonochromatic light. The amount of uniformity in the concentric circular pattern is a measure of the degree of perfection in the shape of the lens. With R as the radius of curvature of the convex lens, the relation between the distance x and the film thickness d is given by

$$x^2 = R^2 - (R - d)^2$$

or more simply by

$$x^2 = 2Rd - d^2$$

*Robert Hooke (1635–1703) and Isaac Newton independently studied a whole range of thin-film phenomena, from soap bubbles to the air film between lenses. Quoting from Newton's *Opticks*:

> I took two Object-glasses, the one a Planoconvex for a fourteen Foot Telescope, and the other a large double Convex for one of about fifty Foot; and upon this, laying the other with its plane side downwards, I pressed them slowly together to make the Colours successfully emerge in the middle of the Circles.

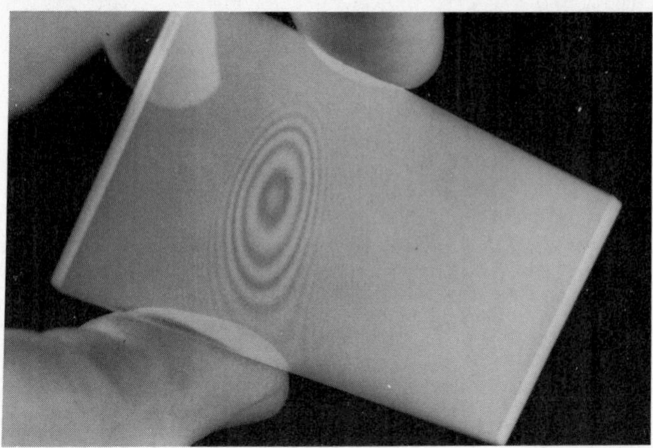

FIGURE 9.25 Newton's rings with two microscope slides. (E. H.)

Since $R >> d$, this becomes

$$x^2 = 2Rd$$

Assume that we need only examine the first two reflected beams E_{1r} and E_{2r}. The mth-order interference maximum will occur in the thin film when its thickness is in accord with the relationship

$$2n_f d_m = (m + \tfrac{1}{2})\lambda_0$$

The radius of the mth bright ring is therefore found by combining the last two expressions to yield

[bright ring] $$x_m = [(m + \tfrac{1}{2})\lambda_f R]^{1/2} \qquad (9.42)$$

Similarly, the radius of the mth dark ring is

[dark ring] $$x_m = (m\lambda_f R)^{1/2} \qquad (9.43)$$

If the two pieces of glass are in good contact (no dust), the central fringe at that point ($x_0 = 0$) will clearly be a minimum in irradiance, an understandable result since d goes to zero at that point. In transmitted light, the observed pattern will be the complement of the reflected one discussed above, so that the center will now appear bright.

Newton's rings, which are Fizeau fringes, can be distinguished from the circular pattern of Haidinger's fringes by the manner in which the diameters of the rings vary with the order m. The central region in the Haidinger pattern corresponds to the maximum value of m (Problem 9.27), whereas just the opposite applies to Newton's rings.

An optical shop, in the business of making lenses, will have a set of precision spherical test plates or gauges. A designer can specify the surface accuracy of a new lens in terms of the number and regularity of the Newton rings that will be seen with a particular test gauge. The use of test plates in the manufacture of high-quality lenses, however, is giving way to far more sophisticated techniques involving laser interferometers (Section 9.8.2).

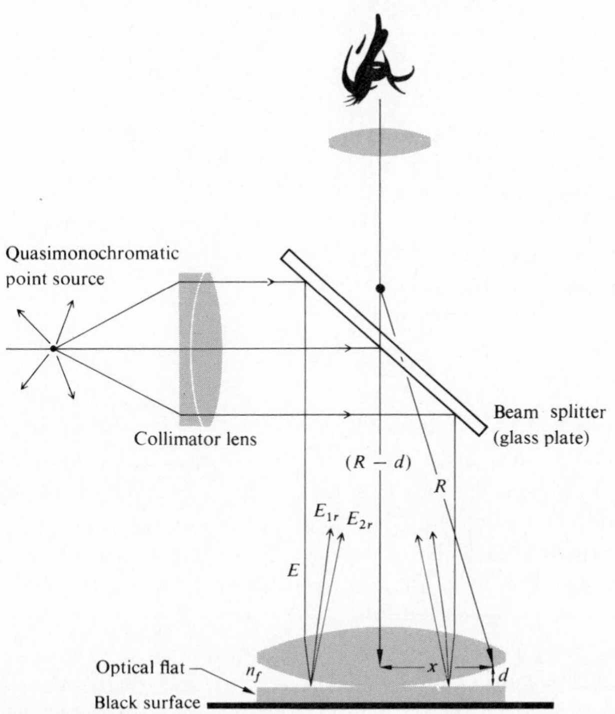

FIGURE 9.26 A standard setup to observe Newton's rings.

9.4.2 Mirrored Interferometers

There are a good number of amplitude-splitting interferometers that utilize arrangements of mirrors and beamsplitters. By far the best known and historically the most important of these is the **Michelson Interferometer**. Its configuration is illustrated in Fig. 9.27. An extended source (e.g., a diffusing ground-glass plate illuminated by a discharge lamp) emits a wave, part of which travels to the right. The beamsplitter at O divides the wave into two, one segment traveling to the right

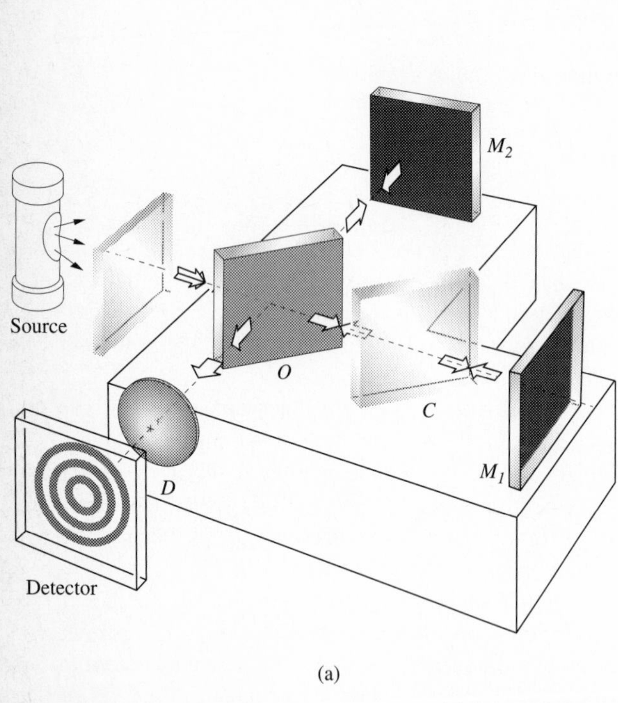

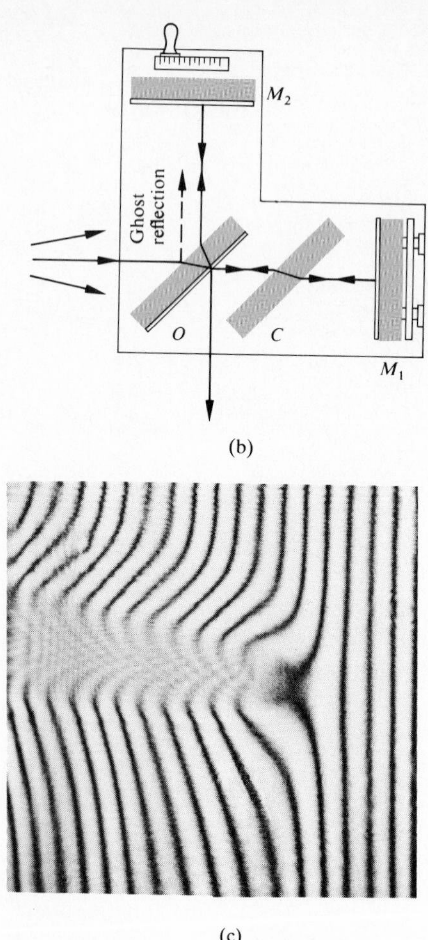

FIGURE 9.27 The Michelson Interferometer. (*a*) Circular fringes are centered on the lens. (*b*) Top view of the interferometer showing the path of the light. (*c*) A wedge fringe pattern was distorted when the tip of a hot soldering iron was placed in one arm. Observe the interesting perceptual phenomenon whereby the region corresponding to the iron's tip appears faintly yellow. (E. H.)

and one up into the background. The two waves are reflected by mirrors M_1 and M_2 and return to the beamsplitter. Part of the wave coming from M_2 passes through the beamsplitter going downward, and part of the wave coming from M_1 is deflected by the beamsplitter toward the detector. The two waves are united, and interference can be expected.

Notice that one beam passes through O three times, whereas the other traverses it only once. Consequently, each beam will pass through equal thicknesses of glass only when a *compensator plate* C is inserted in the arm OM_1. The compensator is an exact duplicate of the beamsplitter, with the exception of any possible silvering or thin film coating on the beamsplitter. It is positioned at an angle of 45°, so that O and C are parallel to each other. With the compensator in place, any optical path difference arises from the actual path difference. In addition, because of the dispersion of the beamsplitter, the optical path is a function of λ. Accordingly, for quantitative work, the interferometer without the compensator plate can be used only with a quasimonochromatic source. The inclusion of a compensator negates the effect of dispersion, so that even a source with a very broad bandwidth will generate discernible fringes.

To understand how fringes are formed, refer to the construction shown in Fig. 9.28, where the physical components are represented more as mathematical surfaces. An observer at the position of the detector will simultaneously see both mirrors M_1 and M_2 along with the source Σ in the beamsplitter. We can redraw the interferometer as if all the elements were in a straight line. Here M_1' corresponds to the image of mirror M_1 in the beamsplitter, and Σ has been swung over in line with O and M_2. The positions of these elements in the diagram depend on their relative distances from O (e.g., M_1' can be in front of, behind, or coincident with M_2 and can even pass through it). The surfaces Σ_1 and Σ_2 are the images of the source Σ in mirrors M_1 and M_2, respectively. Now consider a single point S on the source emitting light in all directions; let's follow the course of one emerging ray. In actuality a wave from S will be split at O, and its segments will thereafter be reflected by M_1 and M_2. In our schematic diagram we represent this by reflecting the ray off both M_2 and M_1'. To an observer at D, the two reflected rays will appear to have come from the image points S_1 and S_2 [note that all rays shown in (*a*) and (*b*) of Fig. 9.28 share a common plane-of-incidence]. For all practical purposes, S_1 and S_2 are coherent point sources, and we can anticipate a flux-density distribution obeying Eq. (9.14).

As the figure shows, the optical path difference for these rays is nearly $2d \cos \theta$, which represents a phase difference of $k_0 2d \cos \theta$. There is an additional phase term arising from the fact that the wave traversing the arm OM_2 is internally reflected in the beamsplitter, whereas the OM_1-wave is externally reflected at O. If the beamsplitter is simply an uncoated glass plate, the relative phase shift resulting from the two reflections

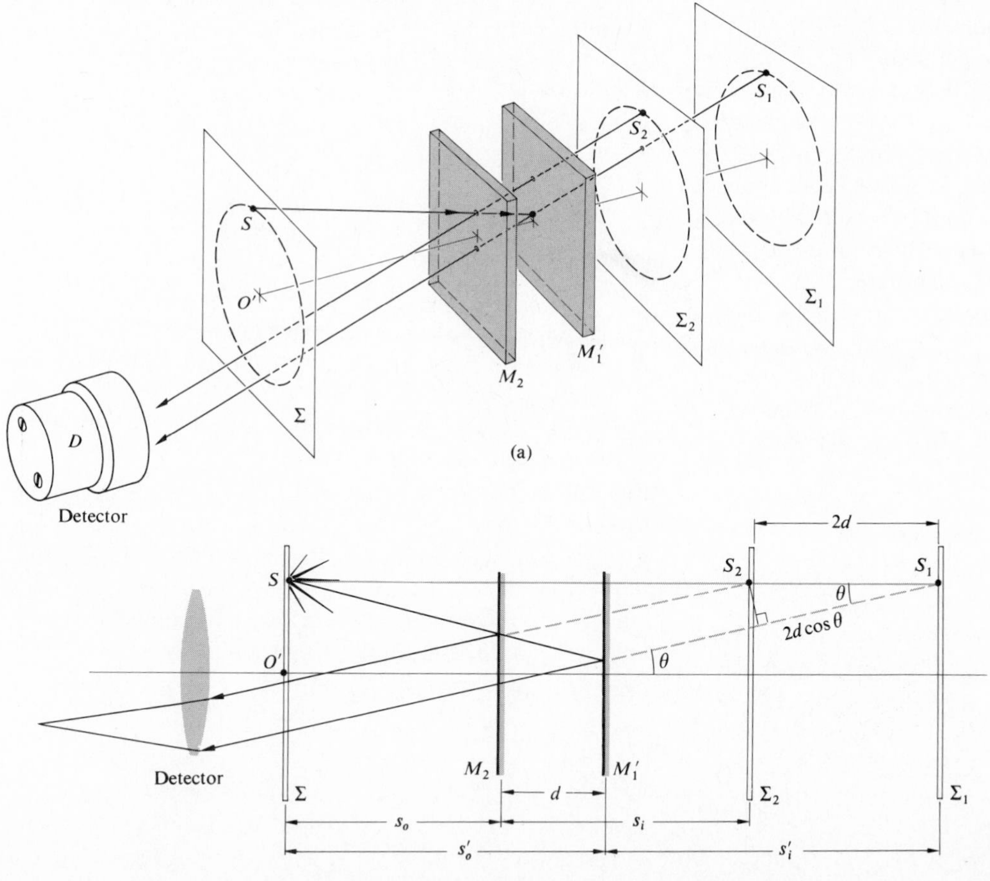

FIGURE 9.28 A conceptual rearrangement of the Michelson Interferometer.

will be π radians. *Destructive*, rather than constructive, interference will then exist when

$$2d \cos \theta_m = m\lambda_0 \qquad (9.44)$$

where m is an integer. If this condition is fulfilled for the point S, then it will be equally well fulfilled for any point on Σ that lies on the circle of radius $O'S$, where O' is located on the axis of the detector. As illustrated in Fig. 9.29, an observer will see a circular fringe system concentric with the central axis of her eye's lens. Because of the small aperture of the eye, the observer will not be able to see the entire pattern without the use of a large lens near the beamsplitter to collect most of the emergent light.

If we use a source containing a number of frequency components (e.g., a mercury discharge lamp), the dependence of θ_m on λ_0 in Eq. (9.44) requires that each such component generate a fringe system of its own. Note, too, that since $2d \cos \theta_m$ must be less than the coherence length of the source, it follows that laser light will be particularly easy to use in demonstrating the interferometer (see Section 9.5). This point would be made strikingly evident were we to compare the fringes produced by laser light with those generated by "white" light from an ordinary tungsten bulb or a candle. In the latter case, the path difference must be very nearly zero, if we are to see any fringes at all, whereas in the former instance a difference of 10 cm has little noticeable effect.

An interference pattern in quasimonochromatic light typically consists of a large number of alternatively bright and dark rings. A particular ring corresponds to a fixed *order m*. As M_2 is moved toward M_1', d decreases, and according to Eq. (9.44), $\cos \theta_m$ increases while θ_m therefore decreases. The rings shrink toward the center, with the highest-order one disappearing whenever d decreases by $\lambda_0/2$. Each remaining ring broadens as more and more fringes vanish at the center, until only a few fill the whole screen. By the time $d = 0$ has been reached, the central fringe will have spread out, filling the entire field of view. With a phase shift of π resulting from reflection off the beamsplitter, the whole screen will then be an interference minimum. (Lack of perfection in the optical elements can render this unobservable.) Moving M_2 still farther causes the fringes to reappear at the center and move outward.

Notice that a central dark fringe for which $\theta_m = 0$ in Eq. (9.44) can be represented by

$$2d = m_0\lambda_0 \qquad (9.45)$$

(Keep in mind that this is a special case. The central region might correspond to neither a maximum nor a minimum.) Even if d is 10 cm, which is fairly modest in laser light, and $\lambda_0 = 500$ nm, m_0 will be quite large, namely 400 000. At a fixed value of d, successive dark rings will satisfy the expressions

$$\begin{aligned} 2d \cos \theta_1 &= (m_0 - 1)\lambda_0 \\ 2d \cos \theta_2 &= (m_0 - 2)\lambda_0 \\ &\vdots \\ 2d \cos \theta_p &= (m_0 - p)\lambda_0 \end{aligned} \qquad (9.46)$$

The angular position of any ring, for example, the pth ring, is determined by combining Eqs. (9.45) and (9.46) to yield

$$2d(1 - \cos \theta_p) = p\lambda_0 \qquad (9.47)$$

Since $\theta_m \equiv \theta_p$, both are just the half-angle subtended at the detector by the particular ring, and since $m = m_0 - p$, Eq. (9.47) is equivalent to Eq. (9.44). The new form is somewhat more convenient, since (using the same example as above) with $d = 10$ cm, the sixth dark ring can be specified by stating that $p = 6$, or in terms of the *order* of the pth ring, that $m =$ 399 994. If θ_p is small,

$$\cos \theta_p = 1 - \frac{\theta_p^2}{2}$$

and Eq. (9.47) yields

$$\theta_p = \left(\frac{p\lambda_0}{d}\right)^{1/2} \qquad (9.48)$$

for the angular radius of the pth fringe.

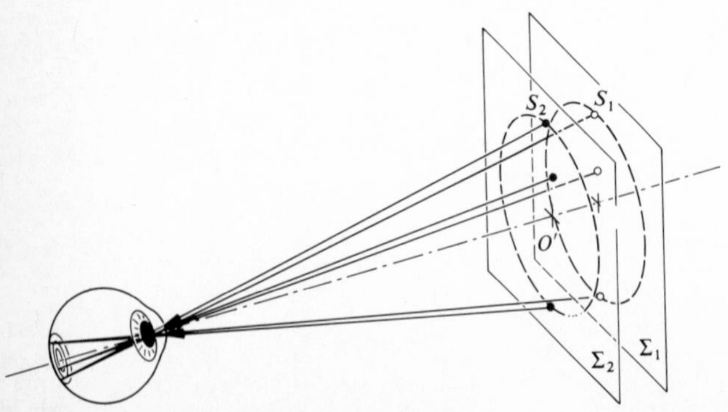

FIGURE 9.29 Formation of circular fringes.

The construction of Fig. 9.28 represents one possible configuration, the one in which we consider only pairs of parallel emerging rays. Since these rays do not actually meet, they cannot form an image without a condensing lens of some sort. Indeed, that lens is most often provided by the observer's eye focused at infinity. The resulting *fringes of equal inclination* (θ_m = constant) located at infinity are also *Haidinger fringes*. A comparison of Figs. 9.28*b* and 9.3*a*, both showing two coherent point sources, suggests that in addition to these (virtual) fringes at infinity, there might also be (real) fringes formed by converging rays. These fringes do in fact exist. Hence, if you illuminate the interferometer with a *broad source* and shield out all extraneous light, you can easily see the projected pattern on a screen in a darkened room (see Section 9.5). The fringes will appear in the space in front of the interferometer (i.e., where the detector is shown), and their size will increase with increasing distance from the beamsplitter. We will consider the (real) fringes arising from point-source illumination a little later on.

When the mirrors of the interferometer are inclined with respect to each other, making a small angle (i.e., when M_1 and M_2 are not quite perpendicular), *Fizeau fringes* are observed. The resultant wedge-shaped air film between M_2 and M_1' creates a pattern of straight parallel fringes. The interfering rays appear to diverge from a point behind the mirrors. The eye would have to focus on this point in order to make these *localized fringes* observable. It can be shown analytically* that by appropriate adjustment of the orientation of the mirrors M_1 and M_2, fringes can be produced that are straight, circular, elliptical, parabolic, or hyperbolic—this holds as well for the real and virtual fringes.

The Michelson Interferometer can be used to make extremely accurate length measurements. As the moveable mirror is displaced by $\lambda_0/2$, each fringe will move to the position previously occupied by an adjacent fringe. Using a microscope arrangement, one need only count the number of fringes N, or portions thereof, that have moved past a reference point to determine the distance traveled by the mirror Δd, that is,

$$\Delta d = N(\lambda_0/2)$$

Nowadays this can be done fairly easily by electronic means. Michelson used the method to measure the number of wavelengths of the red cadmium line corresponding to the standard meter in Sévres near Paris.†

The Michelson Interferometer can be used along with a few polaroid filters to verify the Fresnel–Arago Laws. A polarizer inserted in each arm will allow the optical path length difference to remain fairly constant, while the vector field directions of the two beams are easily changed.

A microwave Michelson Interferometer can be constructed with sheet-metal mirrors and a chicken-wire beamsplitter. With the detector located at the central fringe, it can easily measure shifts from maxima to minima as one of the mirrors is moved, thereby determining λ. A few sheets of plywood, plastic, or glass inserted in one arm will change the central fringe. Counting the number of fringe shifts yields a value for the index of refraction, and from that we can compute the dielectric constant of the material.

The **Mach–Zehnder Interferometer** is another amplitude-splitting device. As shown in Fig. 9.30, it consists of two beamsplitters and two totally reflecting mirrors. The two waves within the apparatus travel along separate paths. A difference between the optical paths can be introduced by a slight tilt of one of the beamsplitters. Since the two paths are separated, the interferometer is relatively difficult to align. For the

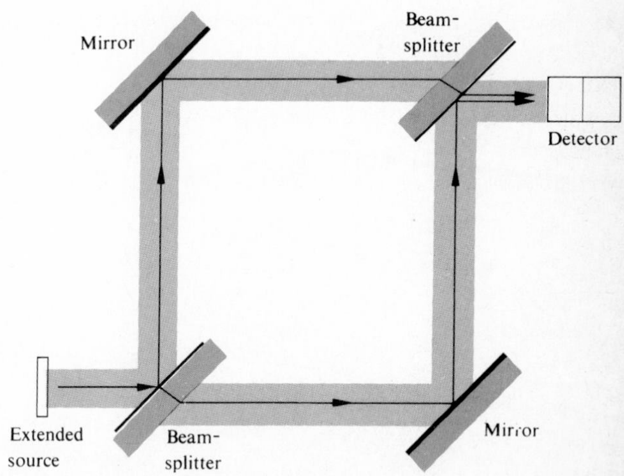

FIGURE 9.30 The Mach–Zehnder Interferometer.

*See, for example, Valasek, *Optics*, p. 135.

†A discussion of the procedure he used to avoid counting the 3 106 327 fringes directly can be found in Strong, *Concepts of Classical Optics*, p. 238, or Williams, *Applications of Interferometry*, p. 51.

same reason, however, the interferometer finds myriad applications. It has even been used, in a somewhat altered yet conceptually similar form, to obtain electron interference fringes.*

An object interposed in one beam will alter the optical path length difference, thereby changing the fringe pattern. A common application of the device is to observe the density variations in gas-flow patterns within research chambers (wind tunnels, shock tubes, etc.). One beam passes through the optically flat windows of the test chamber, while the other beam traverses appropriate compensator plates. The beam within the chamber will propagate through regions having a spatially varying index of refraction. The resulting distortions in the wavefront generate the fringe contours. A particularly nice application is shown in Fig. 9.31, which is a photograph of the magnetic compression device known as Scylla IV. It was used to study controlled thermonuclear reactions at the Los Alamos Scientific Laboratory. In this application, the Mach–Zehnder Interferometer appears in the form of a parallelogram, as illustrated in Fig. 9.32. The two ruby laser *interferograms*, as these photographs are called, show (Fig. 9.33) the background pattern without a plasma in the tube and the density contours within the plasma during a reaction (Fig. 9.34).

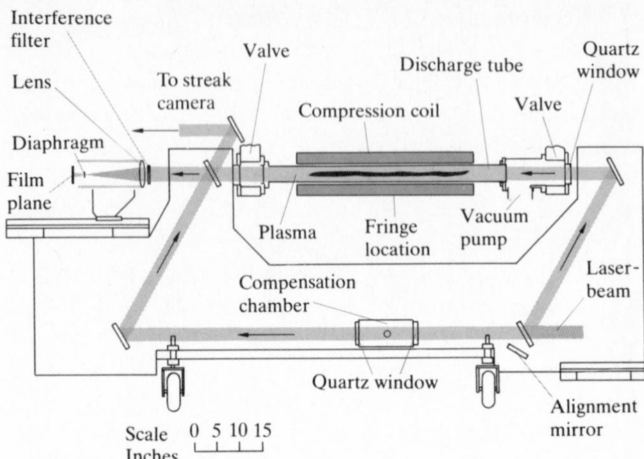

FIGURE 9.32 Schematic of Scylla IV.

Another amplitude-splitting device, which differs from the previous instrument in many respects, is the **Sagnac Interferometer**. It is very easy to align and quite stable. An interesting application of the device is discussed in the last section of this chapter, where we consider its use as a gyroscope. One form

FIGURE 9.31 Scylla IV (Courtesy of University of California, Lawrence Livermore National Laboratory, and the Department of Energy)

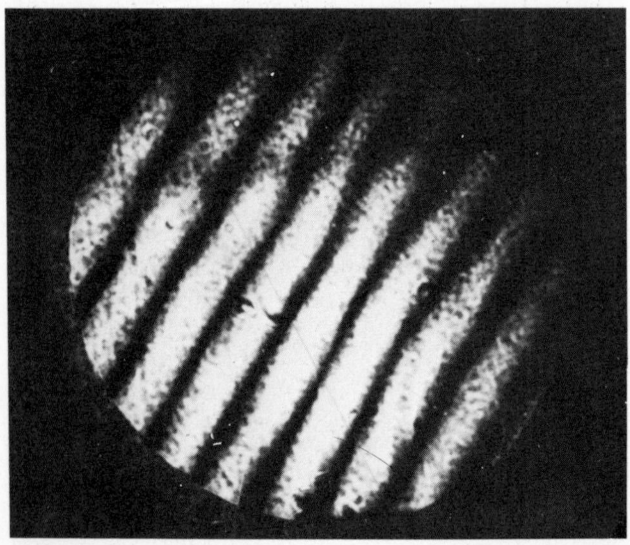

FIGURE 9.33 Interferogram without plasma. (Photo courtesy Los Alamos National Laboratory.)

*L. Marton, J. Arol Simpson, and J. A. Suddeth, *Rev. Sci. Instr.* **25**, 1099 (1954), and *Phys. Rev.* **90**, 490 (1953).

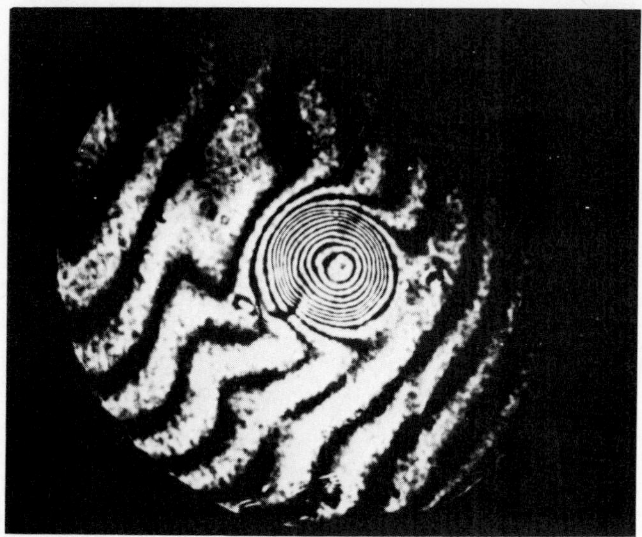

FIGURE 9.34 Interferogram with plasma. (Photo courtesy Los Alamos National Laboratory.)

of the Sagnac Interferometer is shown in Fig. 9.35*a* and another in Fig. 9.35*b*; still others are possible. Notice that the main feature of the device is that there are two identical but oppositely directed paths taken by the beams and that both form closed loops before they are united to produce interference. A deliberate slight shift in the orientation of one of the mirrors will produce a path length difference and a resulting fringe pattern. Since the beams are superimposed and therefore inseparable, the interferometer cannot be put to any of the conventional uses. These in general depend on the possibility of imposing variations on only one of the constituent beams.

Real Fringes

Before we examine the creation of real, as opposed to virtual, fringes, let's first consider another amplitude-splitting interferometric device, the **Pohl fringe-producing system**, illustrated in Fig. 9.36. It is simply a thin transparent film illuminated by the light coming from a point source. In this case, the fringes are real and can accordingly be intercepted on a screen placed anywhere in the vicinity of the interferometer without a condensing-lens system. A convenient light source to use is a mercury lamp covered with a shield having a small hole ($\approx\frac{1}{4}$ inch diameter) in it. As a thin film, use a piece of ordinary mica taped to a dark-colored book cover, which serves as an opaque backing. If you have a laser, its remarkable coherence length and high flux density will allow you to perform this same experiment with almost anything smooth and transparent. Expand the beam to about an inch or two in diameter by passing it through a lens (a focal length of 50 to 100 mm will do). Then just reflect the beam off the surface of a glass plate (e.g., a microscope slide), and the fringes will be

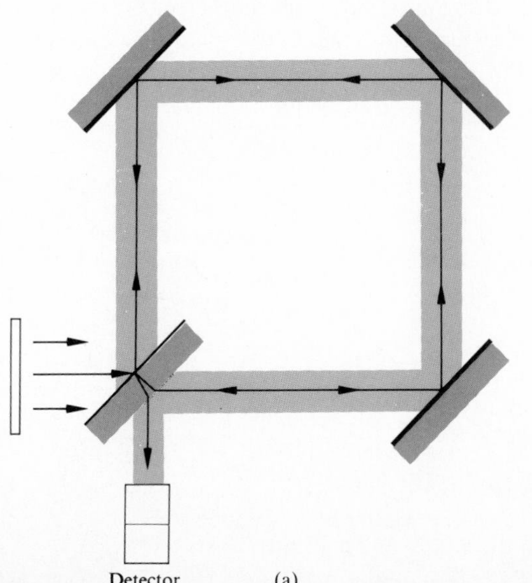

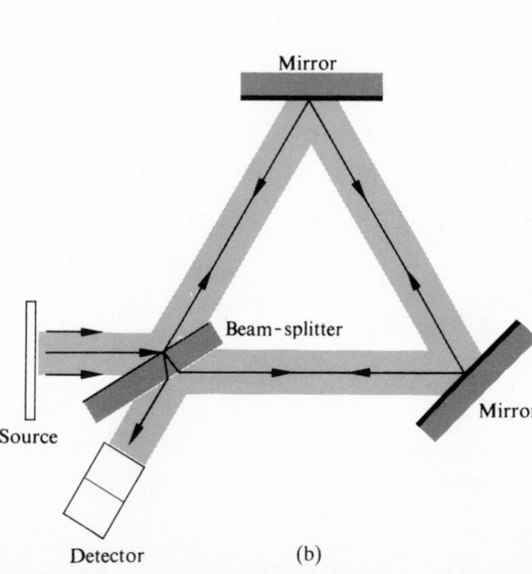

FIGURE 9.35 (*a*) A Sagnac Interferometer. (*b*) Another variation of the Sagnac Interferometer.

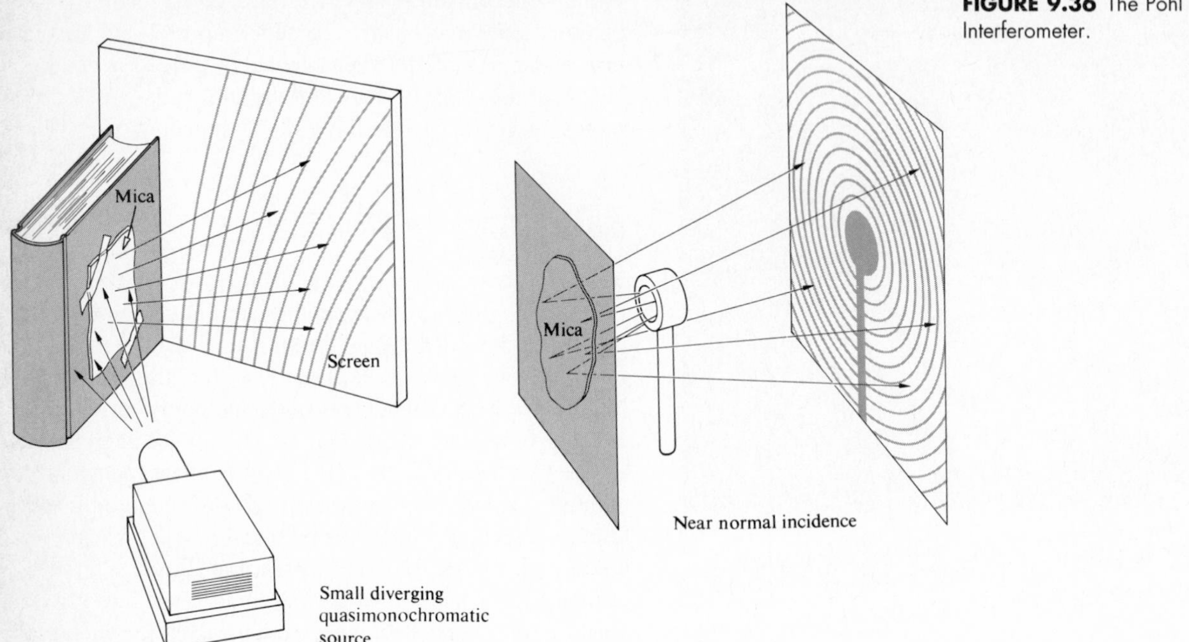

FIGURE 9.36 The Pohl Interferometer.

evident within the illuminated disk wherever it strikes a screen.

The underlying physical principle involved with point-source illumination for all four of the interferometric devices considered above can be appreciated with the help of a construction, variations of which are shown in Figs. 9.37 and 9.38.* The two vertical lines in Fig. 9.37, or the inclined ones in Fig. 9.38, represent either the positions of the mirrors or the two sides of the thin sheet in the Pohl Interferometer. Let's assume that point P in the surrounding medium is a point at which there is constructive interference. A screen placed at that point would intercept this maximum, as well as a whole fringe pattern, without any condensing system. The coherent virtual sources emitting the interfering beams are mirror images S_1 and S_2 of the actual point source S. It should be noted that this kind of real fringe pattern can be observed with both the Michelson and Sagnac Interferometers (Fig. 9.39). If either

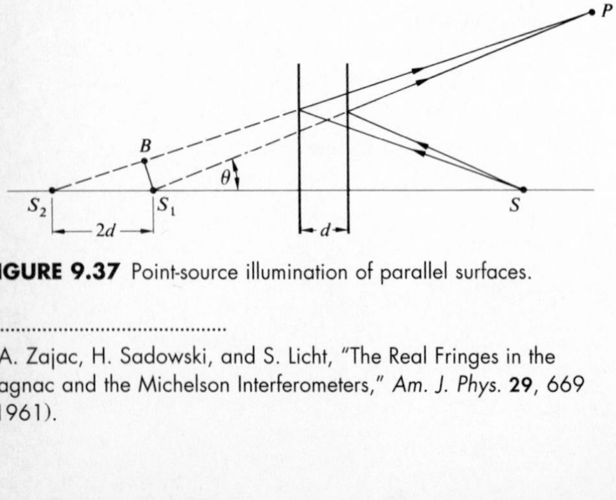

FIGURE 9.37 Point-source illumination of parallel surfaces.

*A. Zajac, H. Sadowski, and S. Licht, "The Real Fringes in the Sagnac and the Michelson Interferometers," *Am. J. Phys.* **29**, 669 (1961).

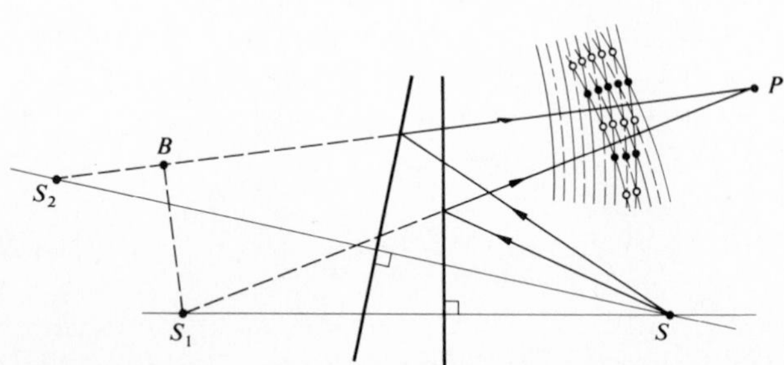

FIGURE 9.38 Point-source illumination of inclined surfaces.

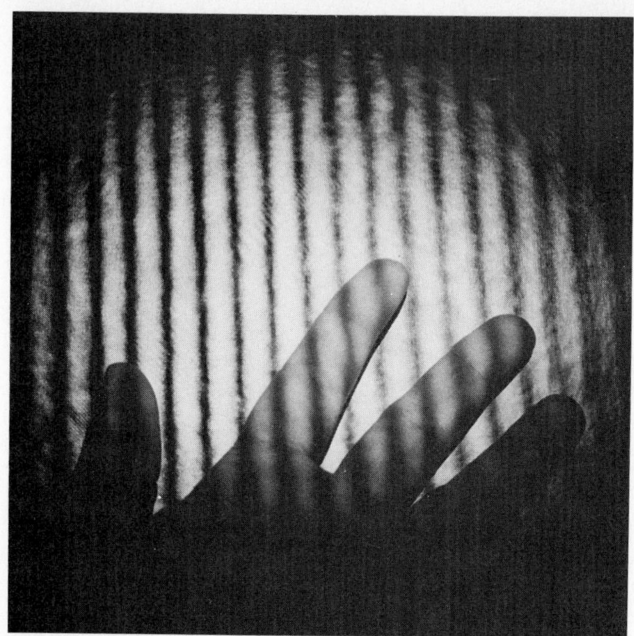

FIGURE 9.39 Real Michelson fringes using He–Ne laser light. (E. H.)

device is illuminated with an expanded laserbeam, a real fringe pattern will be generated directly by the emerging waves. This is an extremely simple and beautiful demonstration.

9.5 Types and Localization of Interference Fringes

Often it is important to know where the fringes produced in a given interferometric system will be located, since that is the region where we need to focus our detector (eye, camera, telescope). In general, the problem of locating fringes is characteristic of a given interferometer; that is, it has to be solved for each individual device.

Fringes can be classified, first, as either *real* or *virtual* and, second, as either *nonlocalized* or *localized*. Real fringes are those that can be seen on a screen without the use of an additional focusing system. The rays forming these fringes converge to the point of observation, all by themselves. Virtual fringes cannot be projected onto a screen without a focusing system. In this case the rays obviously do not converge.

Nonlocalized fringes are real and exist everywhere within an extended (three-dimensional) region of space. The pattern is literally nonlocalized, in that it is not restricted to some small region. Young's Experiment, as illustrated in Fig. 9.8, fills the space beyond the secondary sources with a whole array of real fringes. Nonlocalized fringes of this sort are generally produced by small sources, that is, point or line sources, be they real or virtual. In contrast, localized fringes are clearly observable only over a particular surface. The pattern is literally localized, whether near a thin film or at infinity. This type of fringe will always result from the use of extended sources but can be generated with a point source as well.

The Pohl Interferometer (Fig. 9.36) is particularly useful in illustrating these principles, since with a point source it will produce both real nonlocalized and virtual localized fringes. The real nonlocalized fringes (Fig. 9.40, upper half) can be intercepted on a screen almost anywhere in front of the mica film.

For the nonconverging rays, realize that since the aperture of the eye is quite small, it will intercept only those rays that are directed almost exactly at it. For this small pencil of rays, the eye, at a particular position, sees either a bright or dark spot but not much more. To perceive an extended fringe pattern formed by parallel rays of the type shown in the bottom half of Fig. 9.40, a large lens will have to be used to gather in light entering at other orientations. In practice, however, the source is usually somewhat extended, and fringes can generally be seen by looking into the film with the eye focused at infinity. These virtual fringes are localized at infinity and are equivalent to the *equal-inclination fringes* of Section 9.4. Similarly, if the mirrors M_1 and M_2 in the Michelson Interferometer are parallel, the usual circular, virtual, equal-inclination fringes localized at infinity will be seen. We can imagine a thin air film between the surfaces of the mirrors M_2 and M_1' acting to generate these fringes. As with the configuration of Fig. 9.40 for the Pohl device, real nonlocalized fringes will also be present.

The geometry of the fringe pattern seen in reflected light from a transparent wedge of small angle α is shown in Fig. 9.41. The fringe location P will be determined by the direction of incidence of the incoming light. Newton's rings have this same kind of localization, as do the Michelson, Sagnac, and other interferometers for which the equivalent interference

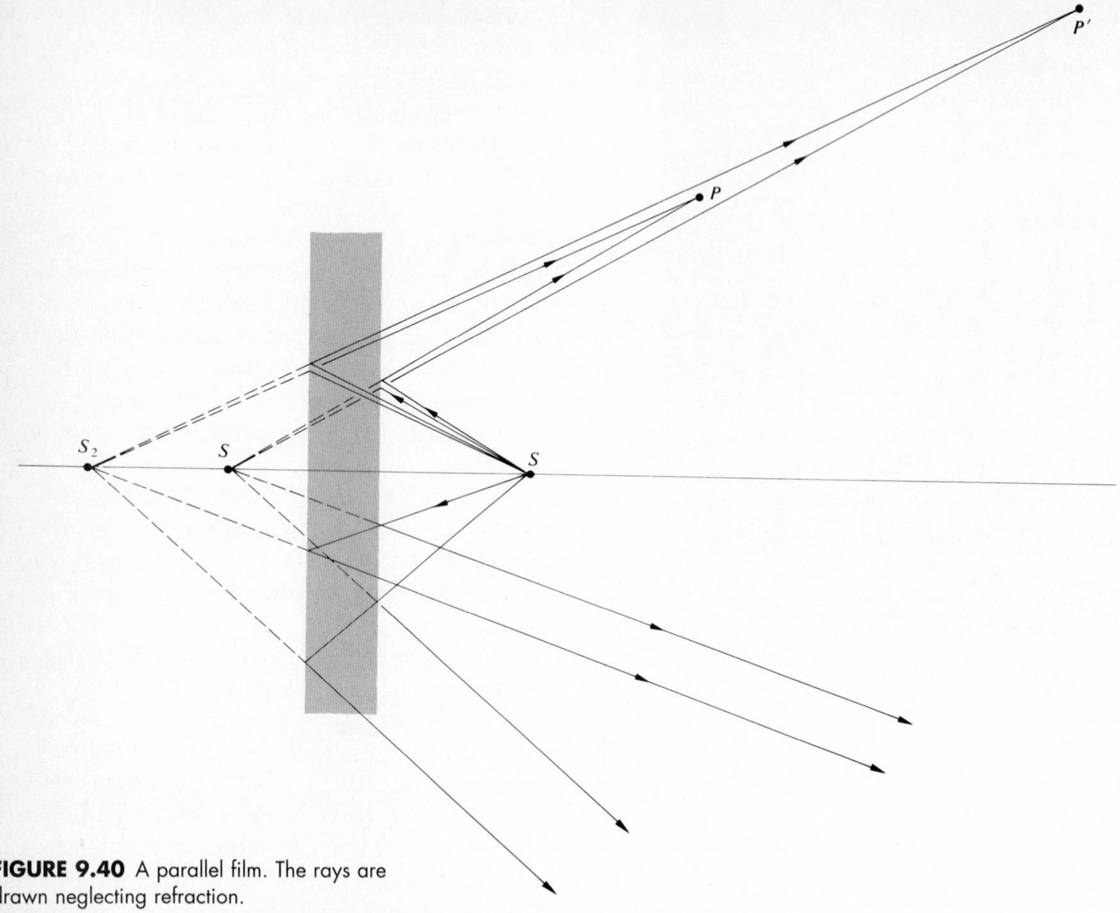

FIGURE 9.40 A parallel film. The rays are drawn neglecting refraction.

system consists of two reflecting planes inclined slightly to each other. The wedge setup of the Mach–Zehnder Interferometer is distinctive in that by rotating the mirrors, one can localize the resulting virtual fringes on any plane within the region generally occupied by the test chamber (Fig. 9.42).

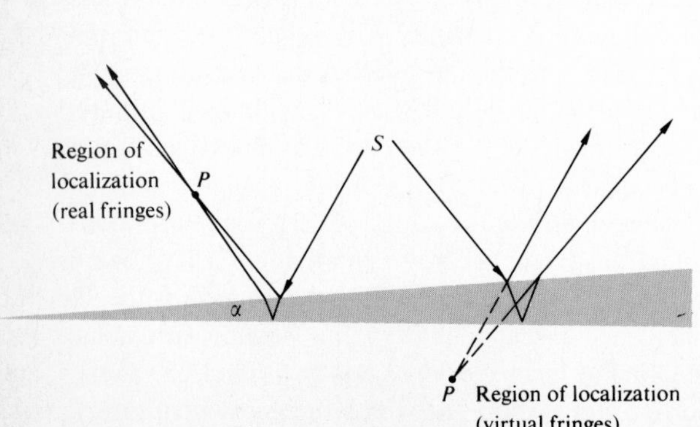

FIGURE 9.41 Fringes formed by a wedge-shaped film.

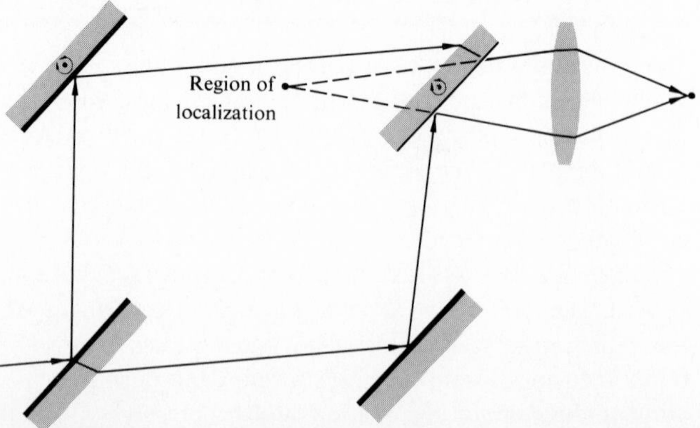

FIGURE 9.42 Fringes in the Mach–Zehnder Interferometer.

9.6 Multiple Beam Interference

Thus far we have examined a number of situations in which two coherent beams are combined under diverse conditions to produce interference patterns. There are, however, circumstances under which a much larger number of mutually coherent waves are made to interfere. In fact, whenever the amplitude-reflection coefficients, the r's, for the parallel plate illustrated in Fig. 9.17 are not small, as was previously the case, the higher-order reflected waves $\mathbf{E}_{3r}$, $\mathbf{E}_{4r}$,... become quite significant. A glass plate, slightly silvered on both sides so that the r's approach unity, will generate a large number of multiply internally reflected rays. For the moment, we will consider only situations in which the film, substrate, and surrounding medium are transparent dielectrics. This avoids the more complicated phase changes resulting from metal-coated surfaces.

To begin the analysis as simply as possible, let the film be nonabsorbing and let $n_1 = n_2$. The notation will be in accord with that of Section 4.10; the amplitude-transmission coefficients are represented by t, the fraction of the amplitude of a wave transmitted on entering into the film, and t', the fraction transmitted when a wave leaves the film. The rays are actually lines drawn perpendicular to the wavefronts and therefore are also perpendicular to the optical fields $\mathbf{E}_{1r}$, $\mathbf{E}_{2r}$, and so forth. Since the rays will remain nearly parallel, the scalar theory will suffice as long as we are careful to account for any possible phase shifts.

As shown in Fig. 9.43, the scalar amplitudes of the reflected waves $\mathbf{E}_{1r}$, $\mathbf{E}_{2r}$, $\mathbf{E}_{3r}$,..., are respectively E_0r, $E_0tr't'$, $E_0tr'^3t'$,..., where E_0 is the amplitude of the initial incoming wave and $r = -r'$ via Eq. (4.89). The minus sign indicates a phase shift, which we will consider later. Similarly, the transmitted waves $\mathbf{E}_{1t}$, $\mathbf{E}_{2t}$, $\mathbf{E}_{3t}$,... will have amplitudes E_0tt', $E_0tr'^2t'$, $E_0tr'^4t'$,.... Consider the set of parallel reflected rays. Each ray bears a fixed phase relationship to all the other reflected rays. The phase differences arise from a combination of optical path length differences and phase shifts occurring at the various reflections. Nonetheless, the waves are mutually coherent, and if they are collected and brought to focus at a point P by a lens, they will all interfere. The resultant irradiance expression has a particularly simple form for two special cases.

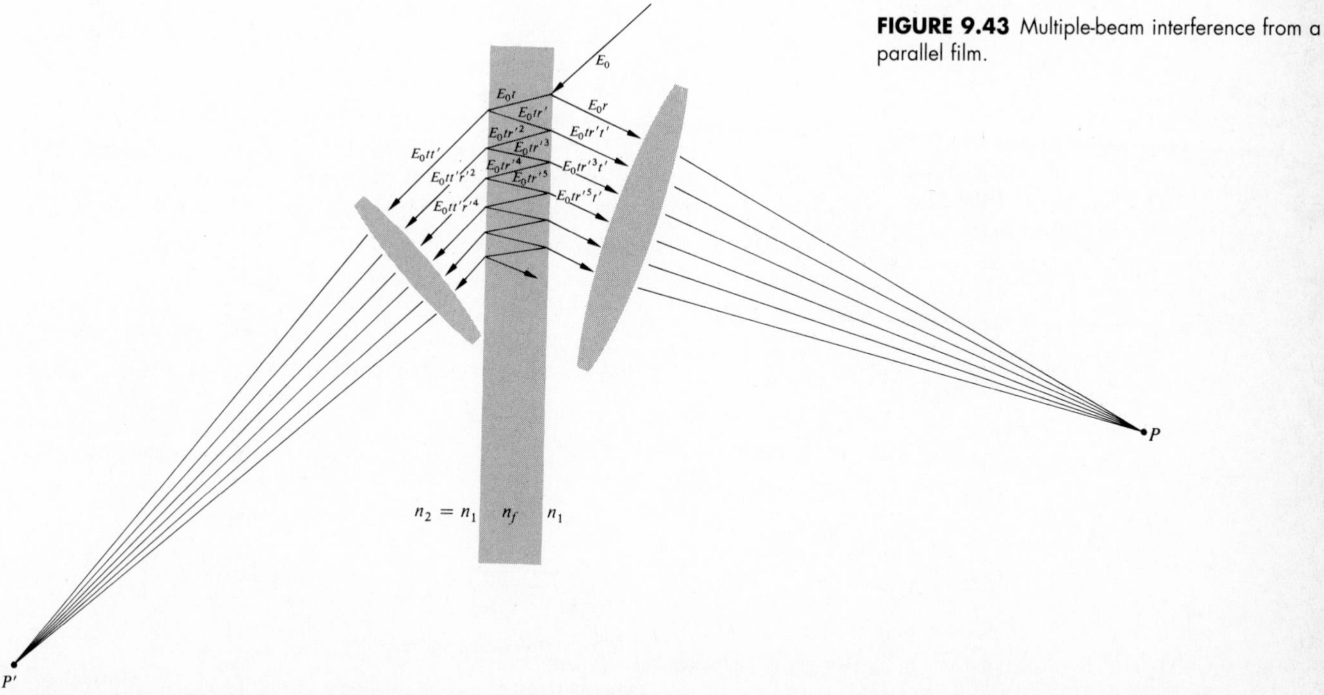

FIGURE 9.43 Multiple-beam interference from a parallel film.

The difference in optical path length between adjacent rays is given by

$$\Lambda = 2n_f d \cos\theta_t \qquad [9.33]$$

All the waves except for the first, $\mathbf{E}_{1r}$, undergo an odd number of reflections *within* the film. It follows from Fig. 4.43 that at each internal reflection the component of the field parallel to the plane-of-incidence changes phase by either 0 or π, depending on the internal incident angle $\theta_i < \theta_c$. The component of the field perpendicular to the plane-of-incidence suffers no change in-phase on internal reflection when $\theta_i < \theta_c$. Clearly then, no relative change in-phase among these waves results from an odd number of such reflections (Fig. 9.44). As the *first special case*, if $\Lambda = m\lambda$, the second, third, fourth, and successive waves will all be in-phase at P. The wave $\mathbf{E}_{1r}$, however, because of its reflection at the top surface of the film, will be out-of-phase by 180° with respect to all the other waves. The phase shift is embodied in the fact that $r = -r'$ and r' occurs only in odd powers. the sum of the scalar amplitudes, that is, the total *reflected amplitude* at point P, is then

$$E_{0r} = E_0 r - (E_0 t r' t' + E_0 t r'^3 t' + E_0 t r'^5 t' + \cdots)$$

or

$$E_{0r} = E_0 r - E_0 t r t'(1 + r^2 + r^4 + \cdots)$$

where since $\Lambda = m\lambda$, we've just replaced r' by $-r$. The geometric series in parentheses converges to the finite sum $1/(1 - r^2)$ as long as $r^2 < 1$, so that

$$E_{0r} = E_0 r - \frac{E_0 t r t'}{(1 - r^2)} \qquad (9.49)$$

It was shown in Section 4.10, when we considered Stokes's treatment of the principle of reversibility (Eq. 4.86), that $tt' = 1 - r^2$, and it follows that

$$E_{0r} = 0$$

Thus when $\Lambda = m\lambda$ the second, third, fourth, and successive waves exactly cancel the first reflected wave, as shown in Fig. 9.45. In this case no light is reflected; all the incoming energy is transmitted. The *second special case* arises when $\Lambda = (m + \frac{1}{2})\lambda$. Now the first and second rays are in-phase, and all other adjacent waves are $\lambda/2$ out-of-phase; that is, the second is out-of-phase with the third, the third is out-of-phase with the fourth, and so on. The resultant *scalar amplitude* is then

$$E_{0r} = E_0 r + E_0 t r t' - E_0 t r^3 t' + E_0 t r^5 t' - \cdots$$

or

$$E_{0r} = E_0 r + E_0 r t t'(1 - r^2 + r^4 - \cdots)$$

The series in parentheses is equal to $1/(1 + r^2)$, in which case

$$E_{0r} = E_0 r \left[1 + \frac{tt'}{(1 + r^2)}\right]$$

Again, $tt' = 1 - r^2$; therefore, as illustrated in Fig. 9.46,

$$E_{0r} = \frac{2r}{(1 + r^2)} E_0$$

Since this particular arrangement results in the addition of the first and second waves, which have relatively large amplitudes, it should yield a large reflected flux density. The irradiance is proportional to $E_{0r}^2/2$, so from Eq. (3.44)

$$I_r = \frac{4r^2}{(1 + r^2)^2}\left(\frac{E_0^2}{2}\right) \qquad (9.50)$$

That this is in fact the maximum, $(I_r)_{\max}$, will be shown later.

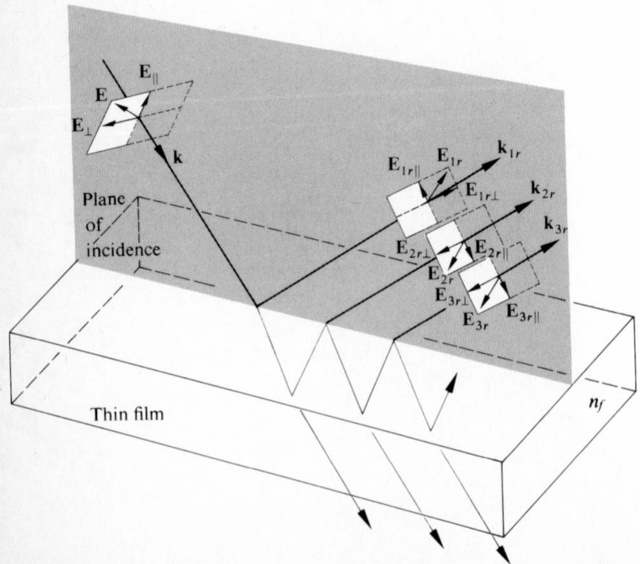

FIGURE 9.44 Phase shifts arising purely from the reflections (internal $\theta_i < \theta_p'$).

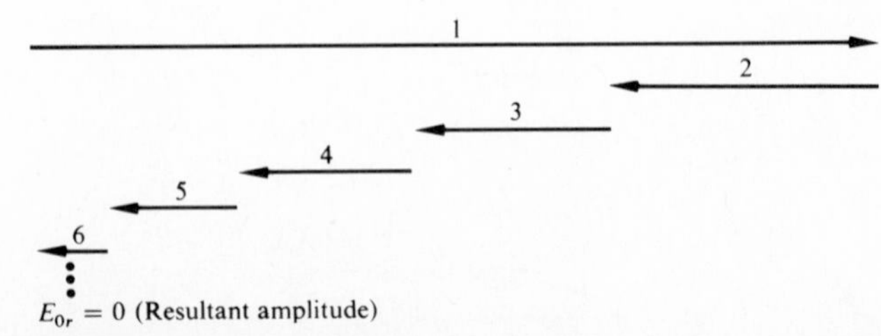

FIGURE 9.45 Phasor diagram.

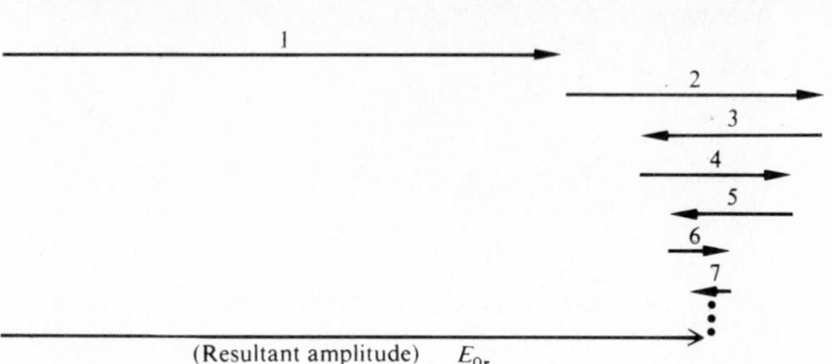

FIGURE 9.46 Phasor diagram.

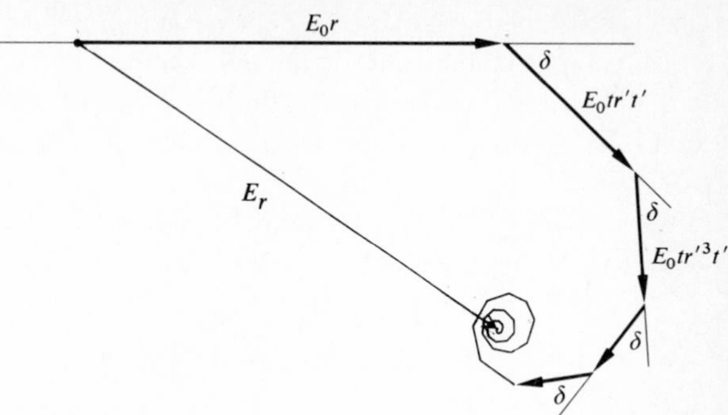

FIGURE 9.47 Phasor diagram.

We will now consider the problem of multiple-beam interference in a more general fashion, making use of the complex representation. Again let $n_1 = n_2$, thereby avoiding the need to introduce different reflection and transmission coefficients at each interface. The optical fields at point P are given by

$$\tilde{E}_{1r} = E_0 re^{i\omega t}$$
$$\tilde{E}_{2r} = E_0 tr't'e^{i(\omega t-\delta)}$$
$$\tilde{E}_{3r} = E_0 tr'^3t'e^{i(\omega t-2\delta)}$$
$$\vdots$$
$$\tilde{E}_{Nr} = E_0 tr'^{(2N-3)}t'e^{i[\omega t-(N-1)\delta]}$$

where $E_0 e^{i\omega t}$ is the incident wave.

The terms $\delta, 2\delta, \ldots, (N-1)\delta$ are the contributions to the phase arising from an optical path length difference between adjacent rays ($\delta = k_0\Lambda$). There is an additional phase contribution arising from the optical distance traversed in reaching point P, but this is common to each ray and has been omitted. The relative phase shift undergone by the first ray as a result of the reflection is embodied in the quantity r'. The resultant *reflected scalar wave* is then

$$\tilde{E}_r = \tilde{E}_{1r} + \tilde{E}_{2r} + \tilde{E}_{3r} + \cdots + \tilde{E}_{Nr}$$

or upon substitution (Fig. 9.47)

$$\tilde{E}_r = E_0 re^{i\omega t} + E_0 tr't'e^{i(\omega t-\delta)} + \cdots + E_0 tr'^{(2N-3)}t' \times e^{i[\omega t-(N-1)\delta]}$$

This can be rewritten as

$$\tilde{E}_r = E_0 e^{i\omega t}\{r + r'tt'e^{-i\delta}[1 + (r'^2e^{-i\delta}) + (r'^2e^{-i\delta})^2 + \cdots + (r'^2e^{-i\delta})^{N-2}]\}$$

If $|r'^2e^{-i\delta}| < 1$, and if the number of terms in the series approaches infinity, the series converges. The resultant wave becomes

$$\tilde{E}_r = E_0 e^{i\omega t}\left[r + \frac{r'tt'e^{-i\delta}}{1 - r'^2e^{-i\delta}}\right] \tag{9.51}$$

In the case of zero absorption, no energy being taken out of the waves, we can use the relations $r = -r'$ and $tt' = 1 - r^2$ to rewrite Eq. (9.51) as

$$\tilde{E}_r = E_0 e^{i\omega t}\left[\frac{r(1 - e^{-i\delta})}{1 - r^2e^{-i\delta}}\right]$$

The reflected flux density at P is then $I_r = \tilde{E}_r\tilde{E}_r^*/2$, that is,

$$I_r = \frac{E_0^2 r^2(1 - e^{-i\delta})(1 - e^{+i\delta})}{2(1 - r^2e^{-i\delta})(1 - r^2e^{+i\delta})}$$

which can be transformed into

$$I_r = I_i \frac{2r^2(1 - \cos\delta)}{(1 + r^4) - 2r^2\cos\delta} \tag{9.52}$$

The symbol $I_i = E_0^2/2$ represents the incident flux density, since, of course, E_0 was the amplitude of the incident wave. Similarly, the amplitudes of the transmitted waves given by

$$\tilde{E}_{1t} = E_0 tt'e^{i\omega t}$$
$$\tilde{E}_{2t} = E_0 tt'r'^2e^{i(\omega t-\delta)}$$
$$\tilde{E}_{3t} = E_0 tt'r'^4e^{i(\omega t-2\delta)}$$
$$\vdots$$
$$\tilde{E}_{Nt} = E_0 tt'r'^{2(N-1)}e^{i[\omega-(N-1)\delta]}$$

can be added to yield

$$\tilde{E}_t = E_0 e^{i\omega t}\left[\frac{tt'}{1 - r^2e^{-i\delta}}\right] \tag{9.53}$$

(Because we are interested in the irradiance a common factor of $e^{-i\delta/2}$, arising from the transmission through the film, was omitted. It contributes to the fact that there is a phase difference of $\pi/2$ between the reflected and transmitted waves, but is of no concern here.)

Multiplying Eq. (9.53) by its complex conjugate yields (Problem 9.37) the irradiance of the transmitted beam

$$I_t = \frac{I_i(tt')^2}{(1 + r^4) - 2r^2 \cos \delta} \tag{9.54}$$

Using the trigonometric identity $\cos \delta = 1 - 2 \sin^2 (\delta/2)$, Eqs. (9.52) and (9.54) become

$$I_r = I_i \frac{[2r/(1 - r^2)]^2 \sin^2(\delta/2)}{1 + [2r/(1 - r^2)]^2 \sin^2(\delta/2)} \tag{9.55}$$

and

$$I_t = I_i \frac{1}{1 + [2r/(1 - r^2)]^2 \sin^2(\delta/2)} \tag{9.56}$$

where energy is not absorbed, that is, $tt' + r^2 = 1$. If indeed none of the incident energy is absorbed, the flux density of the incoming wave should exactly equal the sum of the flux density reflected off the film and the total transmitted flux density emerging from the film. It follows from Eqs. (9.55) and (9.56) that this is indeed the case, namely,

$$I_i = I_r + I_t \tag{9.57}$$

This will not be true, however, if the dielectric film is coated with a thin layer of semitransparent metal. Surface currents induced in the metal will dissipate a portion of the incident electromagnetic energy.

Consider the transmitted waves as described by Eq. (9.54). A maximum will exist when the denominator is as small as possible, that is, when $\cos \delta = 1$, in which case $\delta = 2\pi m$ and

$$(I_t)_{\max} = I_i$$

Under these conditions, Eq. (9.52) indicates that

$$(I_r)_{\min} = 0$$

as we would expect from Eq. (9.57). Again, from Eq. (9.54) it is clear that a minimum transmitted flux density will exist when the denominator is a maximum, that is, when $\cos \delta = -1$. In that case $\delta = (2m + 1)\pi$ and

$$(I_t)_{\min} = I_i \frac{(1 - r^2)^2}{(1 + r^2)^2} \tag{9.58}$$

The corresponding maximum in the reflected flux density is

$$(I_r)_{\max} = I_i \frac{4r^2}{(1 + r^2)^2} \tag{9.59}$$

Notice that the constant-inclination fringe pattern has its maxima when $\delta = (2m + 1)\pi$ or

$$\frac{4\pi n_f}{\lambda_0} d \cos \theta_t = (2m + 1)\pi$$

which is the same as the result we arrived at previously, in Eq. (9.36), by using only the first two reflected waves. Note, too, that Eq. (9.59) verifies that Eq. (9.50) was indeed a maximum.

The form of Eqs. (9.55) and (9.56) suggests that we introduce a new quantity, the ***coefficient of finesse*** F, such that

$$F \equiv \left(\frac{2r}{1 - r^2}\right)^2 \tag{9.60}$$

whereupon these equations can be written as

$$\frac{I_r}{I_i} = \frac{F \sin^2 (\delta/2)}{1 + F \sin^2(\delta/2)} \tag{9.61}$$

and

$$\frac{I_t}{I_i} = \frac{1}{1 + F \sin^2(\delta/2)} \tag{9.62}$$

The term $[1 + F \sin^2 (\delta/2)]^{-1} \equiv \mathscr{A}(\theta)$ is known as the **Airy function**. It represents the transmitted flux-density distribution and is plotted in Fig. 9.48. The complementary function $[1 - \mathscr{A}(\theta)]$, that is, Eq. (9.61), is plotted as well, in Fig. 9.49. When $\delta/2 = m\pi$ the Airy function is equal to unity for all values of F and therefore r. When r approaches 1, the transmitted

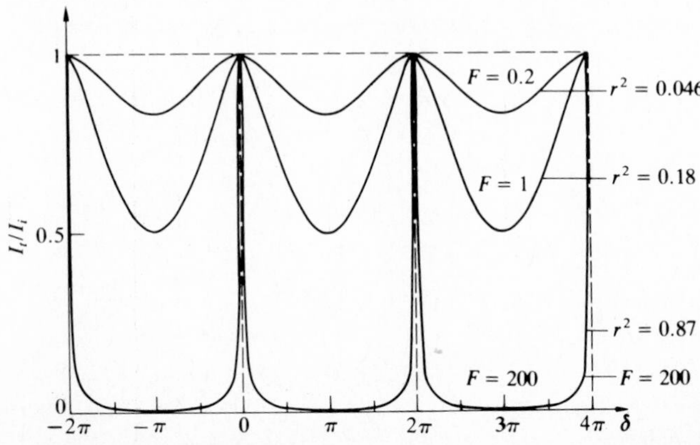

FIGURE 9.48 Airy function.

As might be expected from the considerations of Section 9.5, it is possible to produce real nonlocalized fringes using a bright point source.

The partially transparent metal films that are often used to increase the reflectance ($R = r^2$) will absorb a fraction A of the flux density; this fraction is referred to as the ***absorptance***.

The expression

$$tt' + r^2 = 1$$

or

$$T + R = 1 \qquad [4.60]$$

where T is the transmittance, must now be rewritten as

$$T + R + A = 1 \qquad (9.63)$$

One further complication introduced by the metallic films is an additional phase shift $\phi(\theta_i)$, which can differ from either zero or π. The phase difference between two successively transmitted waves is then

$$\delta = \frac{4\pi n_f}{\lambda_0} d \cos \theta_t + 2\phi \qquad (9.64)$$

For the present conditions, θ_i is small and ϕ may be considered to be constant. In general, d is so large, and λ_0 so small, that ϕ can be neglected. We can now express Eq. (9.54) as

$$\frac{I_t}{I_i} = \frac{T^2}{1 + R^2 - 2R \cos \delta}$$

or equivalently

$$\frac{I_t}{I_i} = \left(\frac{T}{1 - R}\right)^2 \frac{1}{1 + [4R/(1 - R)^2] \sin^2(\delta/2)} \qquad (9.65)$$

Making use of Eq. (9.63) and the definition of the Airy function, we obtain

$$\frac{I_t}{I_i} = \left[1 - \frac{A}{(1 - R)}\right]^2 \mathscr{A}(\theta) \qquad (9.66)$$

as compared with the equation for zero absorption

$$\frac{I_t}{I_i} = \mathscr{A}(\theta) \qquad [9.62]$$

Inasmuch as the absorbed portion A is never zero, the transmitted flux-density maxima $(I_t)_{max}$ will always be somewhat less than I_i. [Recall that for $(I_t)_{max}$, $\mathscr{A}(\theta) = 1$.]

Accordingly, the *peak transmission* is defined as $(I_t/I_i)_{max}$:

$$\frac{(I_t)_{max}}{I_i} = \left[1 - \frac{A}{(1 - R)}\right]^2 \qquad (9.67)$$

A silver film 50 nm thick would be approaching its maximum value of R (e.g., about 0.94), while T and A might be, respectively, 0.01 and 0.05. In this case, the peak transmission will be down to $\frac{1}{36}$. The relative irradiance of the fringe pattern will still be determined by the Airy function, since

$$\frac{I_t}{(I_t)_{max}} = \mathscr{A}(\theta) \qquad (9.68)$$

A measure of the sharpness of the fringes, that is, how rapidly the irradiance drops off on either side of the maximum, is given by the half-width γ. Shown in Fig. 9.52, γ is the width of the peak, in radians, when $I_t = (I_t)_{max}/2$.

Peaks in the transmission occur at specific values of the phase difference $\delta_{max} = 2\pi m$. Accordingly, the irradiance will drop to half its maximum value [i.e., $\mathscr{A}(\theta) = \frac{1}{2}$] whenever $\delta = \delta_{max} \pm \delta_{1/2}$. Inasmuch as

$$\mathscr{A}(\theta) = [1 + F \sin^2(\delta/2)]^{-1}$$

then when

$$[1 + F \sin^2 (\delta_{1/2}/2)]^{-1} = \tfrac{1}{2}$$

it follows that

$$\delta_{1/2} = 2 \sin^{-1} (1/\sqrt{F})$$

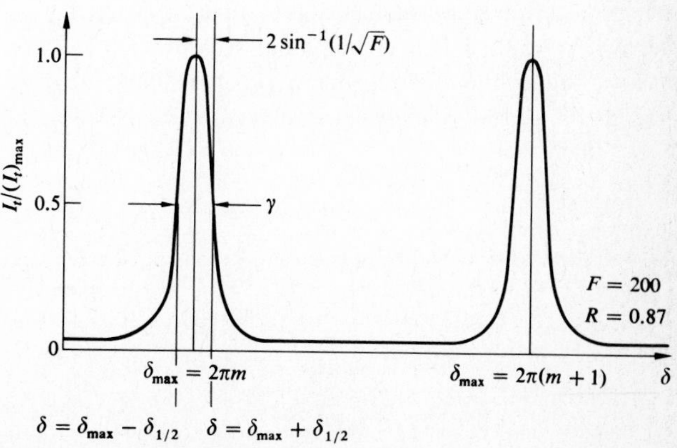

FIGURE 9.52 Fabry–Perot fringes.

Since F is generally rather large, $\sin^{-1}(1/\sqrt{F}) \approx 1/\sqrt{F}$, and therefore the half-width, $\gamma = 2\delta_{1/2}$, becomes

$$\gamma = 4/\sqrt{F} \tag{9.69}$$

Recall that $F = 4R/(1 - R)^2$, so that the larger R is, the sharper the transmission peaks will be.

Another quantity of particular interest is the ratio of the separation of adjacent maxima to the half-width. Known as the **finesse**, $\mathcal{F} \equiv 2\pi/\gamma$ or, from Eq. (9.69),

$$\mathcal{F} = \frac{\pi\sqrt{F}}{2} \tag{9.70}$$

Over the visible spectrum, the finesse of most ordinary Fabry–Perot instruments is about 30. The physical limitation on $\mathcal{F}$ is set by deviations in the mirrors from plane parallelism. Keep in mind that as the finesse increases, the half-width decreases, but so too does the peak transmission. Incidentally, a finesse of about 1000 is attainable with curved-mirror systems using dielectric thin-film coatings.*

Fabry–Perot Spectroscopy

The Fabry–Perot Interferometer is frequently used to examine the detailed structure of spectral lines. We will not attempt a complete treatment of interference spectroscopy, but rather will define the relevant terminology, briefly outlining appropriate derivations.†

As we have seen, a hypothetical, purely monochromatic lightwave generates a particular circular fringe system. But δ is a function of λ_0, so that if the source were made up of two such monochromatic components, two superimposed ring systems would result. When the individual fringes partially overlap, a certain amount of ambiguity exists in deciding when the two systems are individually discernible, that is, when they are said to be *resolved*. Lord Rayleigh's‡ criterion for resolving two equal-irradiance overlapping slit images is well accepted, even if somewhat arbitrarily in the present application. Its use, however, will allow a comparison with prism or grating instruments. The essential feature of this criterion is that the fringes are *just resolvable* when the combined irradiance of both fringes at the center, or saddle point, of the resultant broad fringe is $8/\pi^2$ times the maximum irradiance. This simply means that one would see a broad bright fringe with a grey central region. To be a bit more analytic about it, examine Fig. 9.53, keeping in mind the previous derivation of the half-width. Consider the case in which the two constituent fringes have equal irradiances, $(I_a)_{max} = (I_b)_{max}$. The peaks in the resultant, occurring at $\delta = \delta_a$ and $\delta = \delta_b$, will have equal irradiances,

$$(I_t)_{max} = (I_a)_{max} + I' \tag{9.71}$$

At the saddle point, the irradiance $(8/\pi^2)(I_t)_{max}$, is the sum of the two constituent irradiances, so that, recalling Eq. (9.68),

$$(8/\pi^2)\frac{(I_t)_{max}}{(I_a)_{max}} = [\mathcal{A}(\theta)]_{\delta=\delta_a+\Delta\delta/2} + [\mathcal{A}(\theta)]_{\delta=\delta_b+\Delta\delta/2} \tag{9.72}$$

Using $(I_t)_{max}$ given by Eq. (9.71), along with the fact that

$$\frac{I'}{(I_a)_{max}} = [\mathcal{A}(\theta)]_{\delta=\delta_a+\Delta\delta}$$

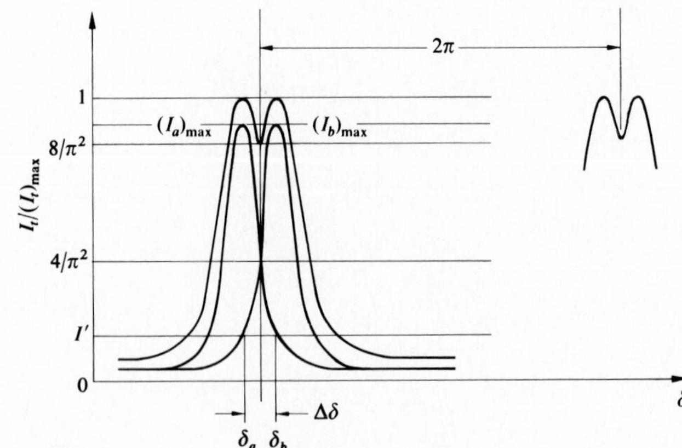

FIGURE 9.53 Overlapping fringes.

*The paper "Multiple Beam Interferometry," by H. D. Polster, *Appl. Opt.* **8**, 522 (1969), should be of interest. Also look at "The Optical Computer," E. Abraham, C. Seaton, and S. Smith, *Sci. Am.* (Feb. 1983), p. 85, for a discussion of the use of the Fabry–Perot Interferometer as an optical transistor.

†A more complete treatment can be found in Born and Wolf, *Principles of Optics*, and in W. E. Williams, *Applications of Interferometry*, to name only two.

‡The criterion will be reconsidered with respect to diffraction in the next chapter (see Fig. 10.40).

we can solve Eq. (9.72) for $\Delta\delta$. For large values of F,

$$(\Delta\delta) \approx \frac{4.2}{\sqrt{F}} \tag{9.73}$$

This then represents the smallest phase increment, $(\Delta\delta)_{\min}$, separating two resolvable fringes. It can be related to equivalent minimum increments in wavelength $(\Delta\lambda_0)_{\min}$, frequency $(\Delta\nu)_{\min}$, and wave number $(\Delta k)_{\min}$. From Eq. (9.64), for $\delta = 2\pi m$, we have

$$m\lambda_0 = 2n_f d \cos\theta_t + \frac{\phi\lambda_0}{\pi} \tag{9.74}$$

Dropping the term $\phi\lambda_0/\pi$, which is clearly negligible, and then differentiating, yields

$$m(\Delta\lambda_0) + \lambda_0(\Delta m) = 0$$

or

$$\frac{\lambda_0}{(\Delta\lambda_0)} = -\frac{m}{(\Delta m)}$$

The minus will be omitted, since it means only that the order increases when λ_0 decreases. When δ changes by 2π, m changes by 1,

$$\frac{2\pi}{(\Delta\delta)} = \frac{1}{(\Delta m)}$$

and thus

$$\frac{\lambda_0}{(\Delta\lambda_0)} = \frac{2\pi m}{(\Delta\delta)} \tag{9.75}$$

The ratio of λ_0 to the least resolvable wavelength difference, $(\Delta\lambda_0)_{\min}$, is known as the ***chromatic resolving power*** $\mathcal{R}$ of any spectroscope. At nearly normal incidence

$$\mathcal{R} \equiv \frac{\lambda_0}{(\Delta\lambda_0)_{\min}} \approx \mathcal{F}\frac{2n_f d}{\lambda_0} \tag{9.76}$$

or

$$\mathcal{R} \approx \mathcal{F}m$$

For a wavelength of 500 nm, $n_f d = 10$ mm, and $R = 90\%$, the resolving power is well over a million, a range achieved by the finest diffraction gratings. It follows as well, in this example, that $(\Delta\lambda_0)_{\min}$ is less than a millionth of λ_0. In terms of frequency, the ***minimum resolvable bandwidth*** is

$$(\Delta\nu)_{\min} = \frac{c}{\mathcal{F}2n_f d} \tag{9.77}$$

inasmuch as $|\Delta\nu| = |c\Delta\lambda_0/\lambda_0^2|$.

As the two components present in the source become increasingly different in wavelength, the peaks shown overlapping in Fig. 9.53 separate. As the wavelength difference increases, the mth-order fringe for one wavelength λ_0 will approach the $(m + 1)$th-order for the other wavelength $(\lambda_0 - \Delta\lambda_0)$. The particular wavelength difference at which overlapping takes place, $(\Delta\lambda_0)_{\text{fsr}}$, is known as the ***free spectral range***. From Eq. (9.75), a change in δ of 2π corresponds to $(\Delta\lambda_0)_{\text{fsr}} = \lambda_0/m$, or at near normal incidence,

$$(\Delta\lambda_0)_{\text{fsr}} \approx \lambda_0^2/2n_f d \tag{9.78}$$

and similarly

$$(\Delta\nu)_{\text{fsr}} \approx \text{c}/2n_f\text{d} \tag{9.79}$$

Continuing with the above example (i.e., $\lambda_0 = 500$ nm and $n_f d = 10$ mm), $(\Delta\lambda_0)_{\text{fsr}} = 0.0125$ nm. If we attempt to increase the resolving power by merely increasing d, the free spectral range will decrease, bringing with it the resulting confusion from the overlapping of orders. What is needed is that $(\Delta\lambda_0)_{\min}$ *be as small as possible and* $(\Delta\lambda_0)_{\text{fsr}}$ *be as large as possible*. But lo and behold,

$$\frac{(\Delta\lambda_0)_{\text{fsr}}}{(\Delta\lambda_0)_{\min}} = \mathcal{F} \tag{9.80}$$

This result should not be too surprising in view of the original definition of $\mathcal{F}$.

Both the applications and configurations of the Fabry–Perot Interferometer are numerous indeed. Etalons have been arranged in series with other etalons, as well as with grating and prism spectroscopes, and multilayer dielectric films have been used to replace the metallic mirror coatings.

Scanning techniques are now widely in use. These take advantage of the superior linearity of photoelectric detectors over photographic plates, to obtain more reliable flux-density measurements. The basic setup for *central-spot scanning* is illustrated in Fig. 9.54. Scanning is accomplished by varying δ, by changing n_f or d rather than $\cos\theta_t$. In some arrangements, n_f is smoothly varied by altering the air pressure within the etalon. Alternatively, mechanical vibration of one

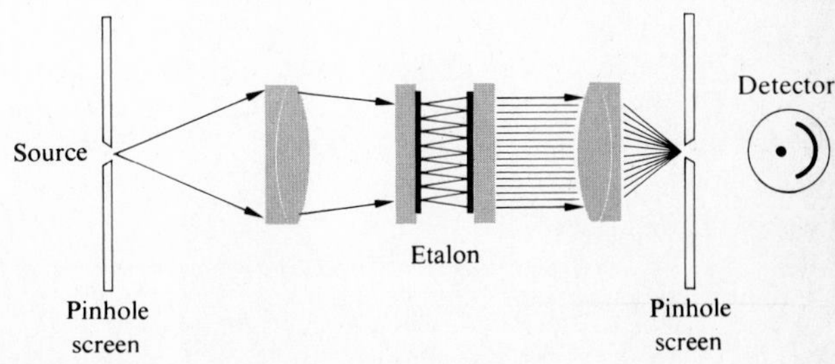

FIGURE 9.54 Central spot scanning.

mirror with a displacement of $\lambda_0/2$ will be enough to scan the free spectral range, corresponding as it does to $\Delta\delta = 2\pi$. A popular technique for accomplishing this utilizes a piezoelectric mirror mount. This kind of material will change its length, and therefore d, as a voltage is applied to it. The voltage profile determines the mirror motion.

Instead of photographically recording irradiance over a large region in space, at a single point in time, this method records irradiance over a large region in time, at a single point in space.

The actual configuration of the etalon itself has also undergone some significant variations. Pierre Connes in 1956 first described the *spherical-mirror Fabry–Perot Interferometer*. Since then, curved-mirror systems have become prominent as laser cavities and are also finding increasing use as spectrum analyzers.

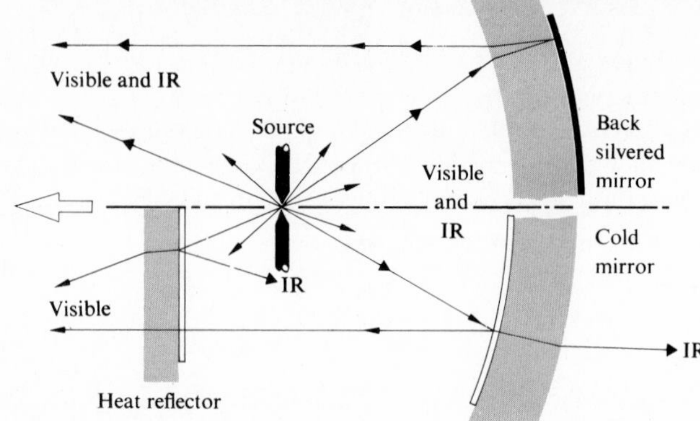

FIGURE 9.56 A composite drawing showing an ordinary system in the top half and a coated one in the bottom.

9.7 Applications of Single and Multilayer Films

The optical uses to which coatings of thin dielectric films have been put in recent times are many indeed. Coatings to eliminate unwanted reflections off a diversity of surfaces, from showcase glass to high-quality camera lenses, are now commonplace (Fig. 9.55). Multilayer, nonabsorbing beamsplitters and *dichroic* mirrors (color-selective beamsplitters that transmit and reflect particular wavelengths) can be purchased commercially.

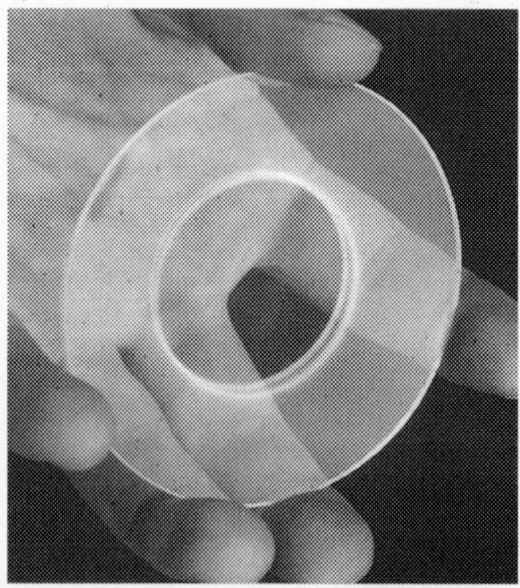

FIGURE 9.55 This glass disk has an antireflection coating in the shape of a circle applied to both its sides. (E. H.)

Figure 9.56 is a segmented diagram illustrating the use of a *cold mirror* in combination with a *heat reflector* to channel infrared radiation to the rear of a motion-picture projector. The intense unwanted infrared radiation emitted by the source is removed from the beam to avoid heating problems at the photographic film. The top half of Fig. 9.56 is an ordinary back-silvered mirror shown for comparison. Solar cells, which are one of the prime power-supply systems for space vehicles, and even the astronauts' helmets and visors, are shielded with similar heat control coverings.

Multilayer broad and narrow band-pass filters that transmit only over a specific spectral range can be made to span the region from infrared to ultraviolet. In the visible, for example, they play an important part in splitting up the image in color television cameras, and in the infrared they're used in missile guidance systems, CO_2 lasers, and satellite horizon sensors. The applications of thin-film devices are manifold, as are their structures, which extend from the simplest single coatings to intricate arrangements of one hundred or more layers.

The treatment of multilayer film theory used here will deal with the *total* electric and magnetic fields and their boundary conditions in the various regions. This is a far more practical approach for many-layered systems than is the multiple-wave technique used earlier.*

*For a very readable nonmathematical discussion, see P. Baumeister and G. Pincus, "Optical Interference Coatings," *Sci. Amer.* **223**, 59 (December 1970).

9.7.1 Mathematical Treatment

Consider the linearly polarized wave shown in Fig. 9.57, impinging on a thin dielectric film between two semi-infinite transparent media. In practice, this might correspond to a dielectric layer a fraction of a wavelength thick, deposited on the surface of a lens, a mirror, or a prism. One point must be made clear at the outset: each wave $E_{r\text{I}}$, $E'_{r\text{II}}$, $E_{t\text{II}}$, and so forth, represents the resultant of all possible waves traveling in that direction, at that point in the medium. The summation process is therefore built in. As discussed in Section 4.6.2, the boundary conditions require that the tangential components of both the electric (**E**) and magnetic ($\mathbf{H} = \mathbf{B}/\mu$) fields be continuous across the boundaries (i.e., equal on both sides). At boundary I

$$E_\text{I} = E_{i\text{I}} + E_{r\text{I}} = E_{t\text{I}} + E'_{r\text{II}} \tag{9.81}$$

and
$$H_\text{I} = \sqrt{\frac{\epsilon_0}{\mu_0}}(E_{i\text{I}} - E_{r\text{I}})n_0 \cos\theta_{i\text{I}}$$
$$= \sqrt{\frac{\epsilon_0}{\mu_0}}(E_{t\text{I}} - E'_{r\text{II}})n_1 \cos\theta_{i\text{II}} \tag{9.82}$$

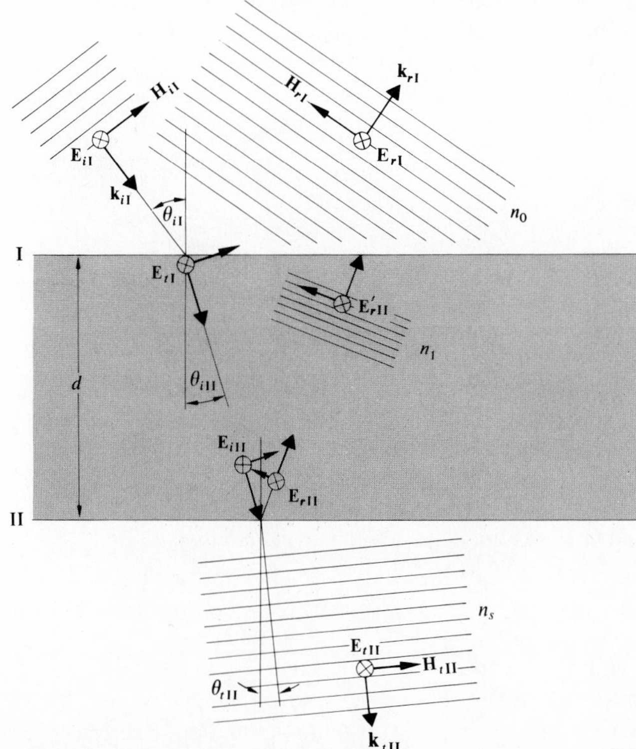

FIGURE 9.57 Fields at the boundaries.

where use is made of the fact that **E** and **H** in nonmagnetic media are related through the index of refraction and the unit propagation vector:

$$\mathbf{H} = \sqrt{\frac{\epsilon_0}{\mu_0}}\, n\hat{\mathbf{k}} \times \mathbf{E}$$

At boundary II

$$E_\text{II} = E_{i\text{II}} + E_{r\text{II}} = E_{t\text{II}} \tag{9.83}$$

and

$$H_\text{II} = \sqrt{\frac{\epsilon_0}{\mu_0}}(E_{i\text{II}} - E_{r\text{II}})n_1 \cos\theta_{i\text{II}}$$
$$= \sqrt{\frac{\epsilon_0}{\mu_0}} E_{t\text{II}} n_s \cos\theta_{t\text{II}} \tag{9.84}$$

the substrate having an index n_s. In accord with Eq. (9.33), a wave that traverses the film once undergoes a shift in-phase of $k_0(2n_1 d \cos\theta_{i\text{II}})/2$, which will be denoted by $k_0 h$, so that

$$E_{i\text{II}} = E_{t\text{I}} e^{-ik_0 h} \tag{9.85}$$

and
$$E_{r\text{II}} = E'_{r\text{II}} e^{+ik_0 h} \tag{9.86}$$

Equations (9.83) and (9.84) can now be written as

$$E_\text{II} = E_{t\text{I}} e^{-ik_0 h} + E'_{r\text{II}} e^{+ik_0 h} \tag{9.87}$$

and
$$H_\text{II} = (E_{t\text{I}} e^{-ik_0 h} - E'_{r\text{II}} e^{+ik_0 h})\sqrt{\frac{\epsilon_0}{\mu_0}}\, n_1 \cos\theta_{i\text{II}} \tag{9.88}$$

These last two equations can be solved for $E_{t\text{I}}$ and $E'_{r\text{II}}$, which when substituted into Eqs. (9.81) and (9.82), yield

$$E_\text{I} = E_\text{II} \cos k_0 h + H_\text{II}(i \sin k_0 h)/\Upsilon_1 \tag{9.89}$$

and
$$H_\text{I} = E_\text{II} \Upsilon_1 i \sin k_0 h + H_\text{II} \cos k_0 h \tag{9.90}$$

where

$$\Upsilon_1 \equiv \sqrt{\frac{\epsilon_0}{\mu_0}}\, n_1 \cos\theta_{i\text{II}}$$

When **E** is in the plane-of-incidence, the above calculations result in similar equations, provided that now

$$\Upsilon_1 \equiv \sqrt{\frac{\epsilon_0}{\mu_0}}\, n_1/\cos\theta_{i\text{II}}$$

In matrix notation, the above linear relations take the form

$$\begin{bmatrix} E_\text{I} \\ H_\text{I} \end{bmatrix} = \begin{bmatrix} \cos k_0 h & (i \sin k_0 h)/\Upsilon_1 \\ \Upsilon_1 i \sin k_0 h & \cos k_0 h \end{bmatrix} \begin{bmatrix} E_\text{II} \\ H_\text{II} \end{bmatrix} \tag{9.91}$$

or
$$\begin{bmatrix} E_\text{I} \\ H_\text{I} \end{bmatrix} = \mathcal{M}_\text{I} \begin{bmatrix} E_\text{II} \\ H_\text{II} \end{bmatrix} \tag{9.92}$$

The *characteristic matrix* $\mathcal{M}_{\text{I}}$ relates the fields at the two adjacent boundaries. It follows, therefore, that if two overlaying films are deposited on the substrate, there will be three boundaries or interfaces, and now

$$\begin{bmatrix} E_{\text{II}} \\ H_{\text{II}} \end{bmatrix} = \mathcal{M}_{\text{II}} \begin{bmatrix} E_{\text{III}} \\ H_{\text{III}} \end{bmatrix} \tag{9.93}$$

Multiplying both sides of this expression by $\mathcal{M}_{\text{I}}$, we obtain

$$\begin{bmatrix} E_{\text{I}} \\ H_{\text{I}} \end{bmatrix} = \mathcal{M}_{\text{I}}\mathcal{M}_{\text{II}} \begin{bmatrix} E_{\text{III}} \\ H_{\text{III}} \end{bmatrix} \tag{9.94}$$

In general, if p is the number of layers, each with a particular value of n and h, then the first and the last boundaries are related by

$$\begin{bmatrix} E_{\text{I}} \\ H_{\text{I}} \end{bmatrix} = \mathcal{M}_{\text{I}}\mathcal{M}_{\text{II}} \cdots \mathcal{M}_p \begin{bmatrix} E_{(p+1)} \\ H_{(p+1)} \end{bmatrix} \tag{9.95}$$

The characteristic matrix of the entire system is the resultant of the product (in the proper sequence) of the individual 2×2 matrices, that is,

$$\mathcal{M} = \mathcal{M}_{\text{I}}\mathcal{M}_{\text{II}} \cdots \mathcal{M}_p = \begin{bmatrix} m_{11} & m_{12} \\ m_{21} & m_{22} \end{bmatrix} \tag{9.96}$$

To see how all this fits together, we will derive expressions for the amplitude coefficients of reflection and transmission using the above scheme. By reformulating Eq. (9.92) in terms of the boundary conditions [(9.81), (9.82), and (9.84)] and setting

$$\text{Y}_0 = \sqrt{\frac{\epsilon_0}{\mu_0}}\, n_0 \cos \theta_{i\text{I}}$$

and

$$\text{Y}_s = \sqrt{\frac{\epsilon_0}{\mu_0}}\, n_s \cos \theta_{t\text{II}}$$

we obtain

$$\begin{bmatrix} (E_{i\text{I}} + E_{r\text{I}}) \\ (E_{i\text{I}} - E_{r\text{I}})\text{Y}_0 \end{bmatrix} = \mathcal{M}_1 \begin{bmatrix} E_{t\text{II}} \\ E_{t\text{II}}\text{Y}_s \end{bmatrix}$$

When the matrices are expanded, the last relation becomes

$$1 + r = m_{11}t + m_{12}\text{Y}_s t$$

and

$$(1 - r)\text{Y}_0 = m_{21}t + m_{22}\text{Y}_s t$$

inasmuch as

$$r = E_{r\text{I}}/E_{i\text{I}} \quad \text{and} \quad t = E_{t\text{II}}/E_{i\text{I}}$$

Consequently,

$$r = \frac{\text{Y}_0 m_{11} + \text{Y}_0\text{Y}_s m_{12} - m_{21} - \text{Y}_s m_{22}}{\text{Y}_0 m_{11} + \text{Y}_0\text{Y}_s m_{12} + m_{21} + \text{Y}_s m_{22}} \tag{9.97}$$

and

$$t = \frac{2\text{Y}_0}{\text{Y}_0 m_{11} + \text{Y}_0\text{Y}_s m_{12} + m_{21} + \text{Y}_s m_{22}} \tag{9.98}$$

To find either r or t for any configuration of films, we need only compute the characteristic matrices for each film, multiply them, and then substitute the resulting matrix elements into the above equations.

9.7.2 Antireflection Coatings

Now consider the extremely important case of normal incidence, that is,

$$\theta_{i\text{I}} = \theta_{i\text{II}} = \theta_{t\text{II}} = 0$$

which in addition to being the simplest, is also quite frequently approximated in practical situations. If we put a subscript on r to indicate the number of layers present, the reflection coefficient for a single film becomes

$$r_1 = \frac{n_1(n_0 - n_s)\cos k_0 h + i(n_0 n_s - n_1^2)\sin k_0 h}{n_1(n_0 + n_s)\cos k_0 h + i(n_0 n_s + n_1^2)\sin k_0 h} \tag{9.99}$$

Multiplying r_1 by its complex conjugate leads to the reflectance

$$R_1 = \frac{n_1^2(n_0 - n_s)^2 \cos^2 k_0 h + (n_0 n_s - n_1^2)^2 \sin^2 k_0 h}{n_1^2(n_0 + n_s)^2 \cos^2 k_0 h + (n_0 n_s + n_1^2)^2 \sin^2 k_0 h} \tag{9.100}$$

This formula becomes particularly simple when $k_0 h = \frac{1}{2}\pi$, which is equivalent to saying that the optical thickness h of the film is an odd multiple of $\frac{1}{4}\lambda_0$. In this case $d = \frac{1}{4}\lambda_f$, and

$$R_1 = \frac{(n_0 n_s - n_1^2)^2}{(n_0 n_s + n_1^2)^2} \tag{9.101}$$

which, quite remarkably, will equal zero when

$$n_1^2 = n_0 n_s \tag{9.102}$$

Generally, d is chosen so that h equals $\frac{1}{4}\lambda_0$ in the yellow-green portion of the visible spectrum, where the eye is most sensitive. Cryolite ($n = 1.35$), a sodium aluminum fluoride compound, and magnesium fluoride ($n = 1.38$) are common low-index films. Since MgF_2 is by far the more durable, it is used more frequently. On a glass substrate, ($n_s \approx 1.5$), both these films have indices that are still somewhat too large to satisfy Eq. (9.102). Nonetheless, a single $\frac{1}{4}\lambda_0$ layer of MgF_2 will reduce the reflectance of glass from about 4% to a bit more than 1%, over the visible spectrum. It is now common practice to apply antireflection coatings to the elements of optical instruments. On camera lenses, such coatings produce a decrease in the haziness caused by stray internally scattered light, as well as a marked increase in image brightness. At wavelengths on either side of the central yellow-green region, R increases and the lens surface will appear blue-red in reflected light.

For a double-layer, quarter-wavelength antireflection coating,

$$\mathcal{M} = \mathcal{M}_{\text{I}}\mathcal{M}_{\text{II}}$$

or more specifically

$$\mathcal{M} = \begin{bmatrix} 0 & i/\Upsilon_1 \\ i\Upsilon_1 & 0 \end{bmatrix} \begin{bmatrix} 0 & i/\Upsilon_2 \\ i\Upsilon_2 & 0 \end{bmatrix} \tag{9.103}$$

At normal incidence this becomes

$$\mathcal{M} = \begin{bmatrix} -n_2/n_1 & 0 \\ 0 & -n_1/n_2 \end{bmatrix} \tag{9.104}$$

Substituting the appropriate matrix elements into Eq. (9.97) yields r_2, which, when squared, leads to the reflectance

$$R_2 = \left[\frac{n_2^2 n_0 - n_s n_1^2}{n_2^2 n_0 + n_s n_1^2}\right]^2 \tag{9.105}$$

For R_2 to be exactly zero at a particular wavelength, we need

$$\left(\frac{n_2}{n_1}\right)^2 = \frac{n_s}{n_0} \tag{9.106}$$

This kind of film is referred to as a *double-quarter, single-minimum* coating. When n_1 and n_2 are as small as possible, the reflectance will have its single broadest minimum equal to zero at the chosen frequency. It should be clear from Eq. (9.106) that $n_2 > n_1$; accordingly, it is now common practice to designate a (glass)–(high index)–(low index)–(air) system as $gHLa$. Zirconium dioxide ($n = 2.1$), titanium dioxide ($n = 2.40$), and zinc sulfide ($n = 2.32$) are commonly used for H-layers, and magnesium fluoride ($n = 1.38$) and cerium fluoride ($n = 1.63$) often serve as L-layers.

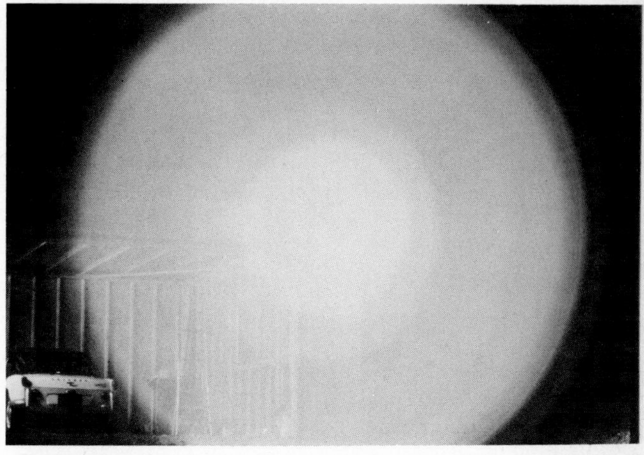

FIGURE 9.58 Lens elements coated with a single layer of MgF_2. (Photo courtesy Optical Coating Laboratory, Inc. Santa Rosa, CA.)

Other double- and triple-layer schemes can be designed to satisfy specific requirements for spectral response, incident angle, cost, and so on. Fig. 9.58 is a scene photographed through a 15-element zoom lens, with a 150-W lamp pointing directly into the camera. The lens elements were covered with a single layer of MgF_2. For Fig. 9.59 a triple-layer antireflection coating was used. The improved contrast and glare reduction are apparent.

FIGURE 9.59 Lens elements coated with a multilayer film structure. (Photo courtesy Optical Coating Laboratory, Inc., Santa Rosa, CA.)

9.7.3 Multilayer Periodic Systems

The simplest kind of periodic system is the *quarter-wave stack*, which is made up of a number of quarter-wave layers. The periodic structure of alternately high- and low-index materials, illustrated in Fig. 9.60, is designated by

$$g(HL)^3a$$

Figure 9.61 illustrates the general form of a portion of the spectral reflectance for a few multilayer filters. The width of the high-reflectance central zone increases with increasing values of the index ratio n_H/n_L, and its height increases with the number of layers. Note that the maximum reflectance of a periodic structure such as $g(HL)^m a$ can be increased further by adding another H-layer, so that it has the form $g(HL)^m Ha$. Mirror surfaces with very high reflectance can be produced using this arrangement.

The small peak on the short-wavelength side of the central zone can be decreased by adding an eighth-wave low-index film to both ends of the stack, in which case the whole arrangement will be denoted by

$$g(0.5L)(HL)^m H(0.5L)a$$

This has the effect of increasing the short-wavelength high-frequency transmittance and is therefore known as a *high-pass filter*. Similarly, the structure

$$g(0.5H)L(HL)^m(0.5H)a$$

merely corresponds to the case in which the end H-layers are $\lambda_0/8$ thick. It has a higher transmittance at the long-wavelength, low-frequency range and serves as a *low-pass filter*.

At nonnormal incidence, up to about 30°, there is quite frequently little degradation in the response of thin-film coatings. In general, the effect of increasing the incident angle is a shift in the whole reflectance curve down to slightly shorter wavelengths. This kind of behavior is evidenced by several naturally occurring periodic structures, for example, peacock and hummingbird feathers, butterfly wings, and the backs of several varieties of beetles.

The last multilayer system to be considered is the *interference*, or more precisely the *Fabry–Perot, filter*. If the separation between the plates of an etalon is of the order of λ, the transmission peaks will be widely separated in wavelength. It will then be possible to block all the peaks but one by using absorbing filters of colored glass or gelatin. The transmitted light corresponds to a single sharp peak, and the etalon serves as a narrow band-pass filter. Such devices can be fabricated by depositing a semitransparent metal film onto a glass support, followed by a MgF_2 spacer and another metal coating.

All-dielectric, essentially nonabsorbing Fabry–Perot filters have an analogous structure, two possible examples of which are

$$g\ HLH\ LL\ HLH\ a$$

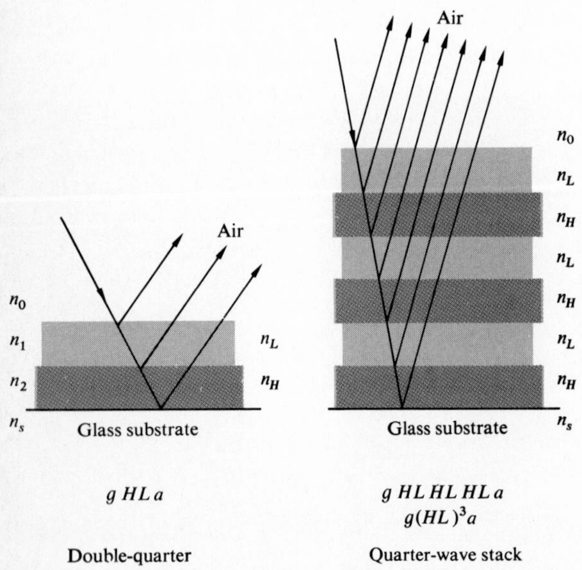

FIGURE 9.60 A periodic structure. Refraction has been omitted for simplicity.

FIGURE 9.61 Reflectance and transmittance for several periodic structures.

and $$g\ HLHL\ HH\ LHLH\ a$$

The characteristic matrix for the first of these is

$$\mathcal{M} = \mathcal{M}_H \mathcal{M}_L \mathcal{M}_H \mathcal{M}_L \mathcal{M}_L \mathcal{M}_H \mathcal{M}_L \mathcal{M}_H$$

but from Eq. (9.104)

$$\mathcal{M}_L \mathcal{M}_L = \begin{bmatrix} -1 & 0 \\ 0 & -1 \end{bmatrix}$$

or $$\mathcal{M}_L \mathcal{M}_L = -\mathcal{I}$$

where $\mathcal{I}$ is the unity or identity matrix. The central double layer, corresponding to the Fabry–Perot cavity, is a half-wavelength thick ($d = \frac{1}{2}\lambda_f$). It therefore has no effect on the reflectance *at the particular wavelength under consideration.* Thus, it said to be an absentee layer, and as a consequence,

$$\mathcal{M} = -\mathcal{M}_H \mathcal{M}_L \mathcal{M}_H \mathcal{M}_H \mathcal{M}_L \mathcal{M}_H$$

The same conditions prevail over and over again at the center and will finally result in

$$\mathcal{M} = \begin{bmatrix} 1 & 0 \\ 0 & 1 \end{bmatrix}$$

At the special frequency for which the filter was designed, r at normal incidence, according to Eq. (9.97), reduces to

$$r = \frac{n_0 - n_s}{n_0 + n_s}$$

the value for the uncoated substrate. In particular, for glass ($n_s = 1.5$), in air ($n_0 = 1$) the theoretical peak transmission is 96% (neglecting reflections from the back surface of the substrate, as well as losses in both the blocking filter and the films themselves).

9.8 Applications of Interferometry

There have been many physical applications of the principles of interferometry. Some of these are only of historical or pedagogical significance, whereas others are now being used extensively. The advent of the laser and the resultant availability of highly coherent quasimonochromatic light have made it particularly easy to create new interferometer configurations.

9.8.1 Scattered-Light Interference

Probably the earliest recorded study of interference fringes arising from scattered light is to be found in Sir Isaac Newton's *Optiks* (1704, Book Two, Part IV). Our present interest in this phenomenon is twofold. First, it provides an extremely easy way to see some rather beautiful colored interference fringes. Second, it is the basis for a remarkably simple and highly useful interferometer.

To see the fringes, lightly rub a thin layer of ordinary talcum powder onto the surface of any common back-silvered mirror (dew will do as well). Neither the thickness nor the uniformity of the coating is particularly important. The use of a bright point source, however, is crucial. A satisfactory source can be made by taping a heavy piece of cardboard having a hole about $\frac{1}{4}$ inch in diameter over a good flashlight. Initially, stand back from the mirror about 3 or 4 feet; the fringes will be too fine and closely spaced to see if you stand much nearer. Hold the flashlight alongside your cheek and illuminate the mirror so that you can see the brightest reflection of the bulb in it. The fringes will then be clearly seen as a number of alternately bright and dark bands.

In Fig. 9.62 two coherent rays leaving the point source are shown arriving at point P after traveling different routes. One ray is reflected from the mirror and then scattered by a single transparent talcum grain toward P. The second ray is first scattered downward by the grain, after which it crosses the mirror and is reflected back toward P. The resulting optical path length difference determines the interference at P. At normal incidence, the pattern is a series of concentric rings of radius*

$$\rho \approx \left[\frac{nm\lambda a^2 b^2}{d(a^2 - b^2)}\right]^{1/2}$$

Now consider a related device, which is very useful in testing optical systems. Known as a **scatter plate**, it generally

*For more of the details, see A. J. deWitte, "Interference in Scattered Light," *Am. J. Phys.* **35**, 301 (1967).

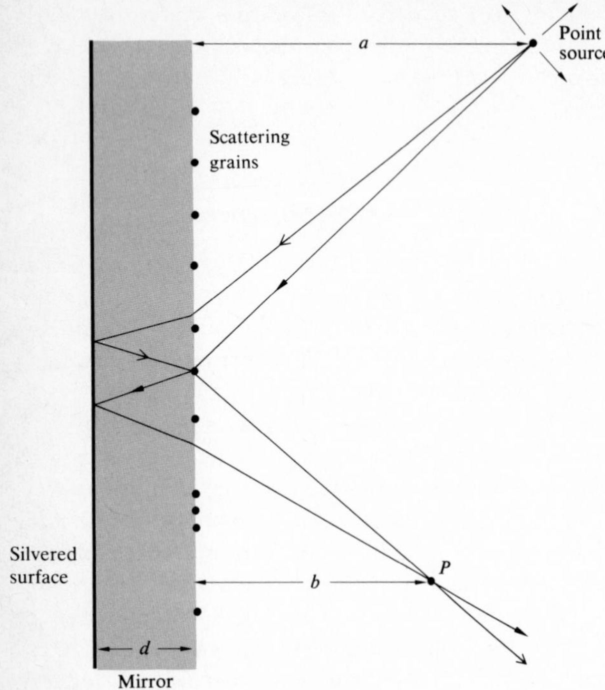

FIGURE 9.62 Interference of scattered light. (Source unknown)

consists of a slightly rough-surfaced, transparent sheet. In an arrangement such as the one shown in Fig. 9.63, it serves as an amplitude-splitting element. In this application it must have a center of symmetry; that is, each scattering site is required to have a duplicate, symmetrically located about a central point.

In the system under consideration, a point source of quasi-monochromatic light S is imaged, by means of lens L_1 on the surface, at point A of the mirror being tested. A portion of the light coming from the source is scattered by the scatter plate and thereafter illuminates the entire surface of the mirror. The mirror, in turn, reflects light back to the scatter plate. This wave, as well as the light forming the image of the pinhole at point A, passes through the scatter plate again and finally reaches the image plane (either on a screen or in a camera). Fringes are formed on this latter plane. The interference process, which is manifest in the formation of these fringes, occurs because each point in the final image plane is illuminated by light arriving via two dissimilar routes, one originating at A and the other at some point B, which reflects scattered light. Indeed, as strange as they may look at first sight, well-defined fringes do result, as shown in Fig. 9.64.

Examining the passage of light through the system in a bit more detail, consider the light initially incident on the scatter

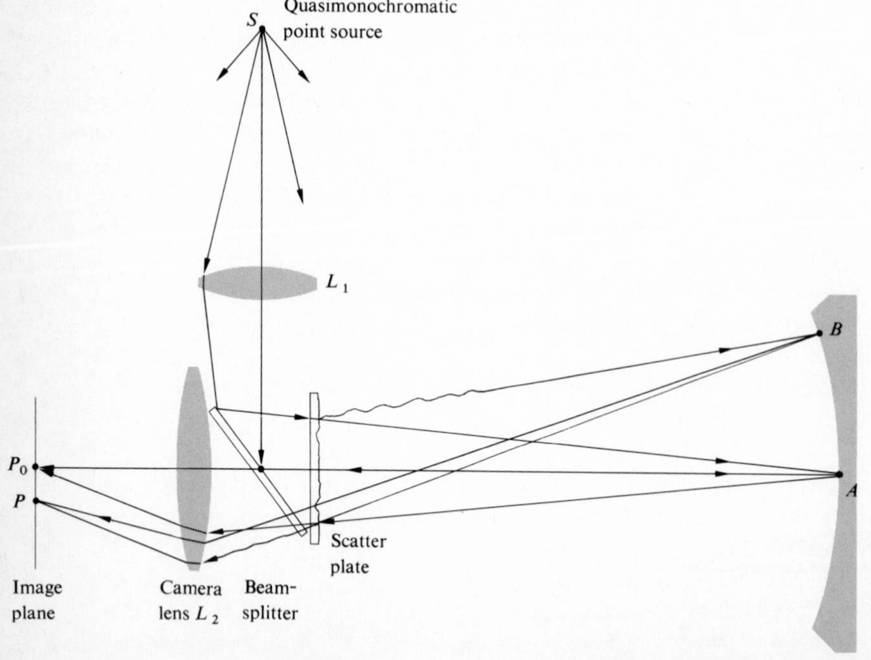

FIGURE 9.63 Scatter plate setup. Adapted from R. M. Scott, *Appl. Opt.* **8**, 531 (1969).

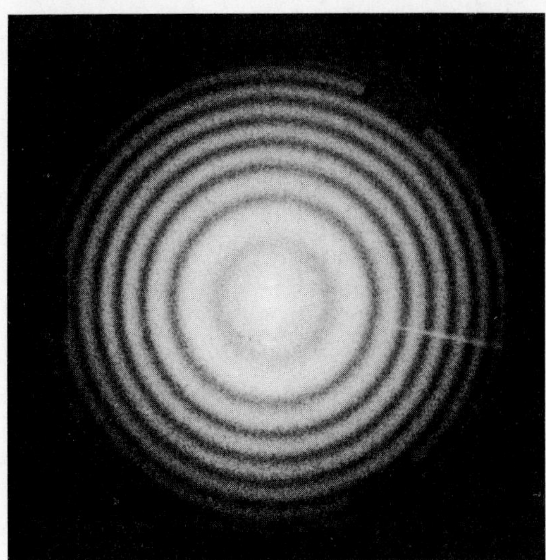

FIGURE 9.64 Fringes in scattered light.

plate and assume that the wave is planar, as shown in Fig. 9.65. After it passes through the scatter plate, the incident plane wavefront $\mathbf{E}_i$ will be distorted into a transmitted wavefront $\mathbf{E}_T$. We envision this wave, in turn, split into a series of Fourier components consisting of plane waves, that is,

$$\mathbf{E}_T = \mathbf{E}_1 + \mathbf{E}_2 + \cdots \tag{9.107}$$

Two of these constituents are shown in Fig. 9.65*a*. Now suppose we attach a specific meaning to these components; namely, $\mathbf{E}_1$ is taken to represent the light traveling to the point A in Fig. 9.63, and $\mathbf{E}_2$ that traveling toward B. The analysis of the stages that follow could be continued in the same way. Let the portion of the wavefront returning from A be represented by the wavefront $\mathbf{E}_A$ in Fig. 9.65*b*. The scatter plate will transform it into an irregular transmitted wave, denoted by $\mathbf{E}_{AT}$ in the same figure. This again corresponds to a complicated configuration, but it can be split into Fourier components consisting of plane waves, as in the above case. In Fig. 9.65*b*, two of these component wavefronts have been drawn, one traveling to the left, and the other inclined at an angle θ. The latter wavefront, which is denoted by $\mathbf{E}_{A\theta}$, is focused by lens L_2 at the point P on the screen (Fig. 9.63).

The wavefront returning from B to the scatter plate is denoted by $\mathbf{E}_B$ in Fig. 9.65*c*. Upon traversing the scatter plate, it will be reshaped into the wave $\mathbf{E}_{BT}$. One of the Fourier components of this wavefront, denoted by $\mathbf{E}_{B\theta}$, is inclined at the angle θ and will therefore be focused at the same point P on the screen.

Some of the waves arriving at P will be coherent in the sense that interference occurs. To obtain the resultant irradiance I_P, first add the amplitudes of all the waves arriving at P, that is, $\mathbf{E}_P$, and then square and time average $\mathbf{E}_P$.

In the discussion above, only two point sources at the mirror were considered. Actually, of course, the whole surface of the mirror is illuminated by the ongoing light, and every point of it will serve as a secondary source of returning waves. All the waves will be deformed by the scatter plate, and these, in turn, can be split into plane wave components. In each series of component waves, there will be one inclined at an angle θ,

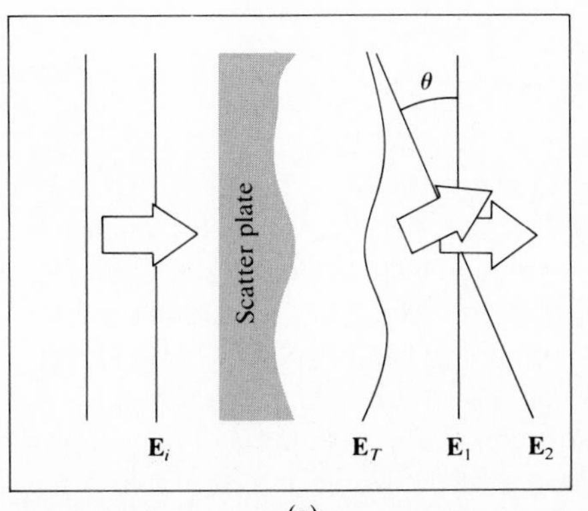

(a)

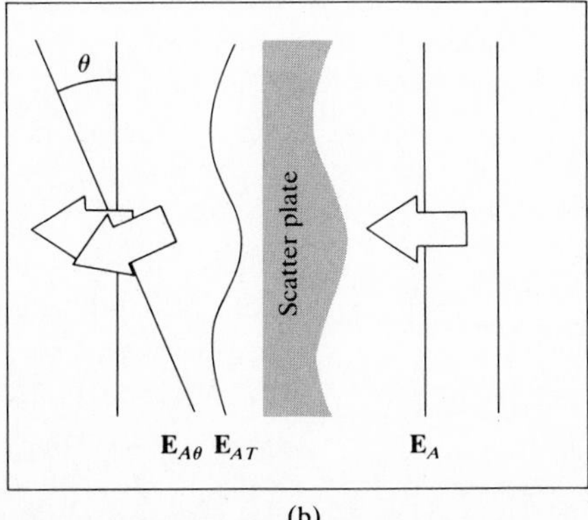

(b)

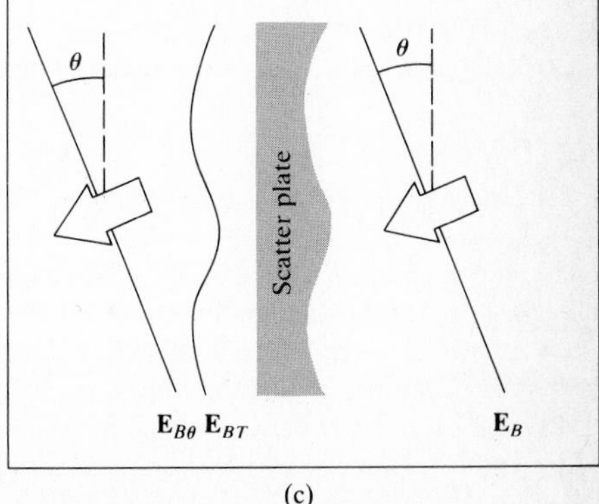

(c)

FIGURE 9.65 Wavefronts passing through the scatter plate.

and all of these will be focused at the same point P on the screen. The resultant amplitude will then have the form

$$\mathbf{E}_P = \mathbf{E}_{A\theta} + \mathbf{E}_{B\theta} + \cdots$$

The light reaching the image plane can be envisioned as made up in part of two optical fields of special interest. One of these results from light that was scattered only on its passage through the plate toward the mirror, and the other results from light that was scattered only on the way toward the image plane. The former broadly illuminates the test mirror and ultimately results in an image of it on the screen. The latter, which was initially focused to the region about A, scatters a diffuse blur across the screen. The point A is chosen so that the small area in the vicinity of it is free of aberrations. In that case, the wave reflected from it serves as a reference with which to compare the wavefront corresponding to the entire mirror surface. The interference pattern will show, as a series of contour fringes, any deviations from perfection in the mirror surface.*

*For further discussion of the scatter plate, the reader might consult the rather succinct papers by J. M. Burch, *Nature* **171**, 889 (1953), and *J. Opt. Soc. Am.* **52**, 600 (1962). Reference should be made to J. Strong, *Concepts of Classical Optics*, p. 383. Also see R. M. Scott, "Scatter Plate Interferometry," *Appl. Opt.* **8**, 531 (1969), and J. B. Houston, Jr., "How to Make and Use a Scatterplate Interferometer," *Optical Spectra* (June 1970), p. 32.

9.8.2 The Twyman-Green Interferometer

The Twyman–Green is essentially a variation of the Michelson Interferometer. It's an instrument of great importance in the domain of modern optical testing. Among its distinguishing physical characteristics (illustrated in Fig. 9.66) are a quasimonochromatic point source and lens L_1, to provide a source of incoming *plane waves*, and a lens L_2, which permits all the light from the aperture to enter the eye so that the entire field can be seen, that is, any portion of M_1 and M_2. A continuous laser serves as a superior source in that it provides the convenience of long path length differences and, in addition, short photographic exposure times. These tend to minimize unwanted vibration effects. Laser versions of the Twyman–Green are among the most effective testing tools in Optics. As shown in the figure, the device is set up to examine a lens. The spherical mirror M_2 has its center of curvature coincident with the focal point of the lens. If the lens being tested is free of aberrations, the emerging reflected light returning to the beamsplitter will again be a plane wave. If, however, astigmatism, coma, or spherical aberration deforms the wavefront, a fringe pattern clearly manifesting these distortions can be seen and photographed. When M_2 is replaced by a plane mirror, a number of other elements (prisms, optical flats, etc.) can be tested equally well. The optician interpreting the fringe pattern can then mark the surface for further polishing to correct high or low spots. In the fabrication of the finest optical systems, telescopes, high-altitude cameras, and so forth, the interferograms may even be scanned electronically, and the resulting data analyzed by computer. Computer-controlled plotters can then automatically produce surface contour maps or perspective "three-dimensional" drawings of the distorted wavefront generated by the element being tested. These procedures can be used throughout the fabrication process to ensure the highest quality optical instruments. Complex systems with wavefront aberrations in the fractional-wavelength range are the result of what might be called the new technology.

9.8.3 The Rotating Sagnac Interferometer

The Sagnac Interferometer is widely used to measure rotational speed. In particular, the ***ring laser***, which is essentially a Sagnac Interferometer containing a laser in one or more of its arms, was designed specifically for that purpose. The first ring laser gyroscope was introduced in 1963, and work is continuing on various devices of this sort (Fig. 9.67). The initial experiments that gave impetus to these efforts were performed by Sagnac in 1911. At that time he rotated the entire interferometer, mirrors, source, and detector, about a perpendicular axis passing through its center (Fig. 9.68). Recall, from Section 9.4.2, that two overlapping beams traverse the interferometer, one clockwise, the other counterclockwise. The rotation effectively shortens the path taken by one beam in comparison to that of the other. In the interferometer, the result is a fringe shift proportional to the angular speed of rota-

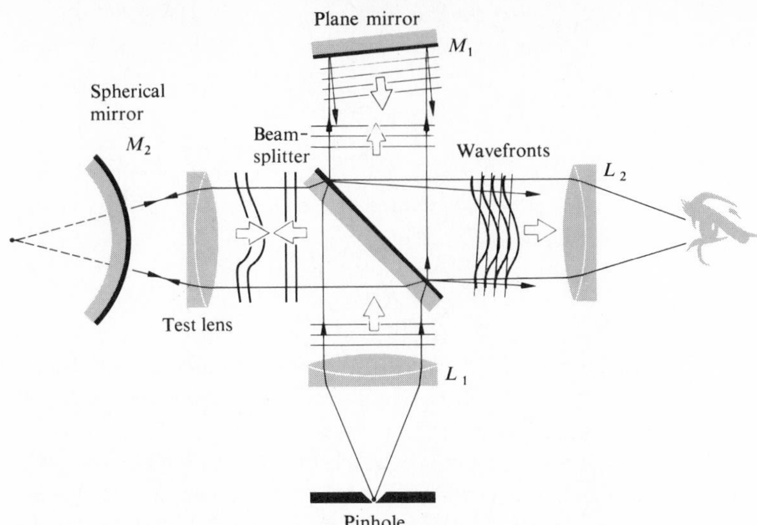

FIGURE 9.66 The Twyman–Green Interferometer. (E. H.)

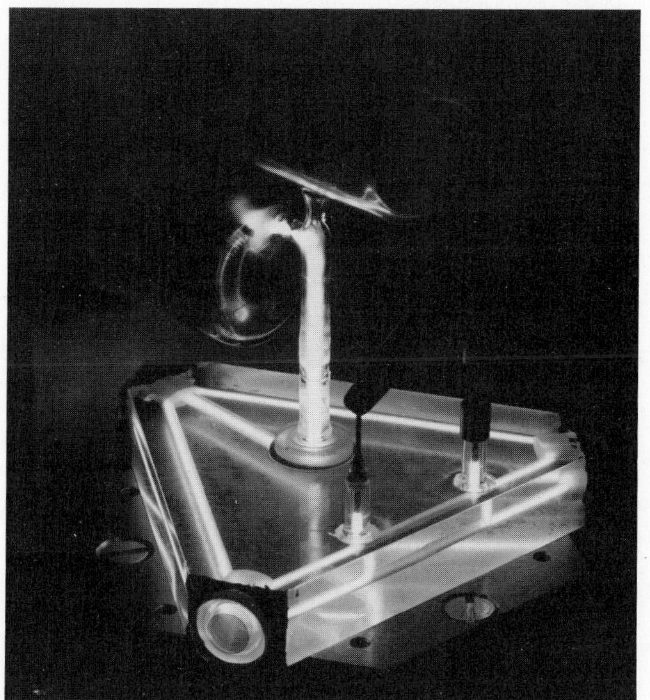

FIGURE 9.67 A ring laser gyro. (Photo courtesy Autonetics, a Division of Boeing North America, Inc., a wholly owned subsidiary of The Boeing Company)

tion ω. In the ring laser, it is a frequency difference between the two beams that is proportional to ω.

Consider the arrangement depicted in Fig. 9.68. The corner A (and every other corner) moves with a linear speed $v = R\omega$,

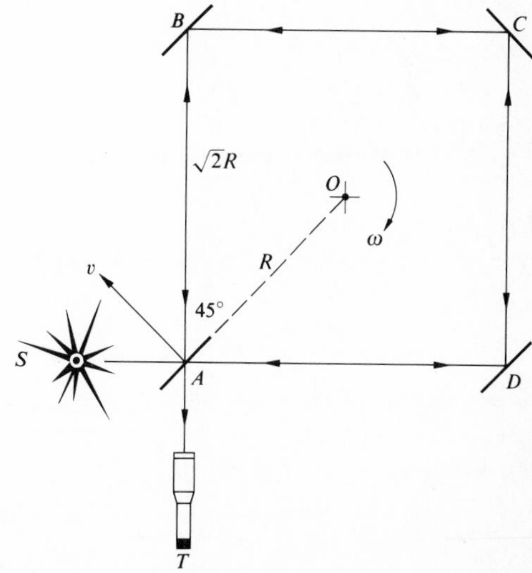

FIGURE 9.68 The rotating Sagnac Interferometer. Originally it was 1 m × 1 m with $\omega = 120$ rev/min.

where R is half the diagonal of the square. Using classical reasoning, we find that the time of travel of light along AB is

$$t_{AB} = \frac{R\sqrt{2}}{c - v/\sqrt{2}}$$

or

$$t_{AB} = \frac{2R}{\sqrt{2}c - \omega R}$$

The time of travel of the light from A to D is

$$t_{AD} = \frac{2R}{\sqrt{2}c + \omega R}$$

The total time for counterclockwise and clockwise travel is given respectively by

$$t_{\circlearrowleft} = \frac{8R}{\sqrt{2}c + \omega R}$$

and

$$t_{\circlearrowright} = \frac{8R}{\sqrt{2}c - \omega R}$$

For $\omega R << c$ the difference between these two intervals is

$$\Delta t = t_{\circlearrowright} - t_{\circlearrowleft}$$

or, using the Binomial Series,

$$\Delta t = \frac{8R^2\omega}{c^2}$$

This can be expressed in terms of area $A = 2R^2$ of the square formed by the beams of light as

$$\Delta t = \frac{4A\omega}{c^2}$$

Let the period of the monochromatic light used be $\tau = \lambda/c$; then the fractional displacement of the fringes, given by $\Delta N = \Delta t/\tau$, is

$$\Delta N = \frac{4A\omega}{c\lambda}$$

a result that has been verified experimentally. In particular, Michelson and Gale* used this method to determine the angular velocity of the Earth.

The preceding classical treatment is obviously lacking, inasmuch as it assumes speeds in excess of c, an assumption that is contrary to the dictates of Special Relativity. Furthermore, it would appear that since the system is accelerating, General Relativity would prevail. In fact, these formalisms yield the same results.

*Michelson and Gale, *Astrophys. J.* **61**, 140 (1925).

PROBLEMS

9.1 Returning to Section 9.1, let

$$\tilde{\mathbf{E}}_1(r,t) = \tilde{\mathbf{E}}_1(\mathbf{r})e^{-i\omega t}$$

and

$$\tilde{\mathbf{E}}_2(r,t) = \tilde{\mathbf{E}}_2(\mathbf{r})e^{-i\omega t}$$

where the wavefront shapes are not explicitly specified, and $\tilde{\mathbf{E}}_1$ and $\tilde{\mathbf{E}}_2$ are complex vectors depending on space and initial phase angle. Show that the interference term is then given by

$$I_{12} = \tfrac{1}{2}(\tilde{\mathbf{E}}_1 \cdot \tilde{\mathbf{E}}_2^* + \tilde{\mathbf{E}}_1^* \cdot \tilde{\mathbf{E}}_2) \qquad (9.108)$$

You will have to evaluate terms of the form

$$\langle \tilde{\mathbf{E}}_1 \cdot \tilde{\mathbf{E}}_2 e^{-2i\omega t} \rangle_T = \tilde{\mathbf{E}}_1 \cdot \tilde{\mathbf{E}}_2 / T \int_t^{t+T} e^{-2i\omega t'} dt'$$

for $T >> \tau$ (take another look at Problem 3.7). Show that Eq. (9.108) leads to Eq. (9.11) for plane waves.

9.2 In Section 9.1 we considered the spatial distribution of energy for two point sources. We mentioned that for the case in which the separation $a >> \lambda$, I_{12} spatially averages to zero. Why is this true? What happens when a is much less than λ?

9.3* Return to Fig. 2.20 and prove that if two electromagnetic plane waves making an angle θ have the same amplitude, E_0, the resulting interference pattern on the yx-plane is a cosine-squared irradiance distribution given by

$$I(y) = 4E_0^2 \cos^2\left(\frac{\pi}{\lambda} y \sin\theta\right)$$

Locate the zeros of irradiance. What is the value of the fringe separation? What happens to the separation as θ increases? Compare your analysis with that leading to Eq. (9.17). [*Hint*: Begin with the wave expressions given in Section 2.7, which have the proper phases already worked out, and write them as exponentials.]

9.4 Will we get an interference pattern in Young's Experiment (Fig. 9.8) if we replace the source slit S by a single long-filament light-bulb? What would occur if we replaced the slits S_1 and S_2 by these same bulbs?

9.5* Figure P.9.5 shows an output pattern that was measured by a tiny microphone when two small piezo-loudspeakers separated by 15 cm were pointed toward the microphone at a distance of 1.5 m away. Given that the speed of sound at 20°C is 343 m/s, determine the approximate frequency at which the speakers were driven. Discuss the nature of the pattern and explain why it has a central minimum.

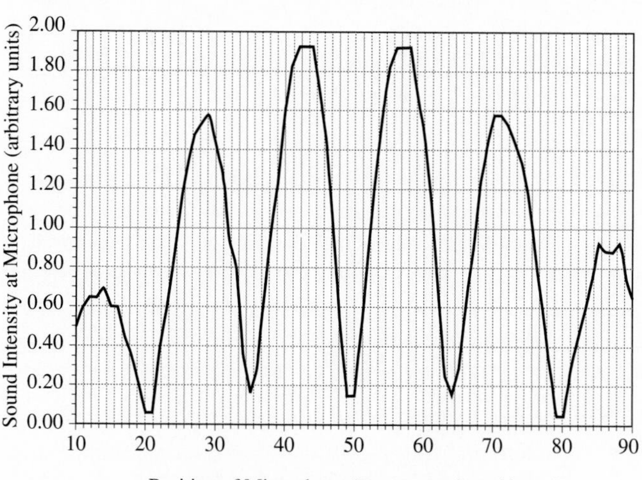

FIGURE P.9.5 (Data courtesy of CENCO.)

9.6* Two 1.0-MHz radio antennas emitting in-phase are separated by 600 m along a north–south line. A radio receiver placed 2.0 km east is equidistant from both transmitting antennas and picks up a fairly strong signal. How far north should that receiver be moved if it is again to detect a signal nearly as strong?

9.7 An expanded beam of red light from a He–Ne laser ($\lambda_0 = 632.8$ nm) is incident on a screen containing two very narrow horizontal slits separated by 0.200 mm. A fringe pattern appears on a white screen held 1.00 m away.

(a) How far (in radians and millimeters) above and below the central axis are the first zeros of irradiance?

(b) How far (in mm) from the axis is the fifth bright band?

(c) Compare these two results.

9.8* Red plane waves from a ruby laser ($\lambda_0 = 694.3$ nm) in air impinge on two parallel slits in an opaque screen. A fringe pattern forms on a distant wall, and we see the fourth bright band 1.0° above the central axis. Kindly calculate the separation between the slits.

9.9* A 3 × 5 card containing two pinholes, 0.08 mm in diameter and separated center to center by 0.10 mm, is illuminated by parallel rays of blue light from an argon ion laser ($\lambda_0 = 487.99$ nm). If the fringes on an observing screen are to be 10 mm apart, how far away should the screen be?

9.10* White light falling on two long narrow slits emerges and is observed on a distant screen. If red light ($\lambda_0 = 780$ nm) in the first-order fringe overlaps violet in the second-order fringe, what is the latter's wavelength?

9.11* Considering the double-slit experiment, derive an equation for the distance $y_{m'}$ from the central axis to the m'th irradiance *minimum*, such that the first dark bands on either side of the central maximum correspond to $m' = \pm 1$. Identify and justify all your approximations.

9.12* With regard to Young's Experiment, derive a general expression for the shift in the vertical position of the mth *maximum* as a result of placing a thin parallel sheet of glass of index n and thickness d directly over one of the slits. Identify your assumptions.

9.13* Plane waves of monochromatic light impinge at an angle θ_i on a screen containing two narrow slits separated by a distance a. Derive an equation for the angle measured from the central axis which locates the mth maximum.

9.14* Sunlight incident on a screen containing two long narrow slits 0.20 mm apart casts a pattern on a white sheet of paper 2.0 m beyond. What is the distance separating the violet ($\lambda_0 = 400$ nm) in the first-order band from the red ($\lambda_0 = 600$ nm) in the second-order band?

9.15 To examine the conditions under which the approximations of Eq. (9.23) are valid:

(a) Apply the law of cosines to triangle S_1S_2P in Fig. 9.8*c* to get

$$\frac{r_2}{r_1} = \left[1 - 2\left(\frac{a}{r_1}\right)\sin\theta + \left(\frac{a}{r_1}\right)^2\right]^{1/2}$$

(b) Expand this in a Maclaurin series yielding

$$r_2 = r_1 - a \sin\theta + \frac{a^2}{2r_1}\cos^2\theta + \cdots$$

(c) In light of Eq. (9.17), show that if $(r_1 - r_2)$ is to equal $a \sin\theta$, it is required that $r_1 >> a^2/\lambda$.

9.16 A stream of electrons, each having an energy of 0.5 eV, impinges on a pair of extremely thin slits separated by 10^{-2} mm. What is the distance between adjacent minima on a screen 20 m behind the slits? ($m_e = 9.108 \times 10^{-31}$ kg, 1 eV $= 1.602 \times 10^{-19}$ J.)

9.17* Show that a for the Fresnel biprism of Fig. 9.13 is given by $a = 2d(n-1)\alpha$.

9.18* In the Fresnel double mirror $s = 2$ m, $\lambda_0 = 589$ nm, and the separation of the fringes was found to be 0.5 mm. What is the angle of inclination of the mirrors, if the perpendicular distance of the actual point source to the intersection of the two mirrors is 1 m?

9.19* The Fresnel biprism is used to obtain fringes from a point source that is placed 2 m from the screen, and the prism is midway between the source and the screen. Let the wavelength of the light be $\lambda_0 = 500$ nm and the index of refraction of the glass be $n = 1.5$. What is the prism angle, if the separation of the fringes is 0.5 mm?

9.20 What is the general expression for the separation of the fringes of a Fresnel biprism of index n immersed in a medium having an index of refraction n'?

9.21 Using Lloyd's mirror, X-ray fringes were observed, the spacing of which was found to be 0.002 5 cm. The wavelength used was 8.33 Å. If the source–screen distance was 3 m, how high above the mirror plane was the point source of X-rays placed?

9.22 Imagine that we have an antenna at the edge of a lake picking up a signal from a distant radio star (Fig. P.9.22), which is just coming up above the horizon. Write expressions for δ and for the angular position of the star when the antenna detects its first maximum.

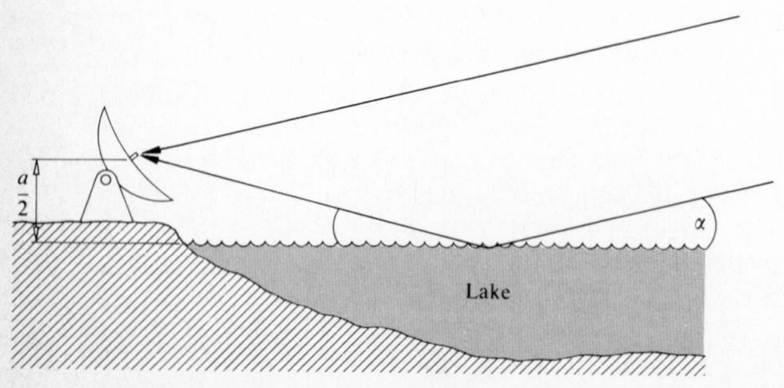

FIGURE P.9.22

9.23* If the plate in Fig. 9.17 is glass in air, show that the amplitudes of E_{1r}, E_{2r}, and E_{3r} are respectively $0.2E_{0i}$, $0.192E_{0i}$, and $0.008E_{0i}$, where E_{0i} is the incident amplitude. Make use of the Fresnel coefficients at normal incidence, assuming no absorption. You might repeat the calculation for a water film in air.

9.24 A soap film surrounded by air has an index of refraction of 1.34. If a region of the film appears bright red ($\lambda_0 = 633$ nm) in normally reflected light, what is its minimum thickness there?

9.25* A thin film of ethyl alcohol ($n = 1.36$) spread on a flat glass plate and illuminated with white light shows a color pattern in reflection. If a region of the film reflects only green light (500 nm) strongly, how thick is it?

9.26* A soap film of index 1.34 has a region where it is 550.0 nm thick. Determine the vacuum wavelengths of the radiation that is not reflected when the film is illuminated from above with sunlight.

9.27 Consider the circular pattern of Haidinger's fringes resulting from a film with a thickness of 2 mm and an index of refraction of 1.5. For monochromatic illumination of $\lambda_0 = 600$ nm, find the value of m for the central fringe ($\theta_t = 0$). Will it be bright or dark?

9.28 Illuminate a microscope slide (or even better, a thin coverglass slide). Colored fringes can easily be seen with an ordinary fluorescent lamp (although some of the newer versions don't work well at all) serving as a broad source or a mercury street light as a point source. Describe the fringes. Now rotate the glass. Does the pattern change? Duplicate the conditions shown in Figs. 9.18 and 9.19. Try it again with a sheet of plastic food wrap stretched across the top of a cup.

9.29 Figure P.9.29 illustrates a setup used for testing lenses. Show that

$$d = x^2(R_2 - R_1)/2R_1R_2$$

when d_1 and d_2 are negligible in comparison with $2R_1$ and $2R_2$, respectively. (Recall the theorem from plane geometry that relates the products of the segments of intersecting chords.) Prove that the radius of the mth dark fringe is then

$$x_m = [R_1R_2m\lambda_f/(R_2 - R_1)]^{1/2}$$

How does this relate to Eq. (9.43)?

9.30* Newton rings are observed on a film with quasimonochromatic light that has a wavelength of 500 nm. If the 20th bright ring has a radius of 1 cm, what is the radius of curvature of the lens forming one part of the interfering system?

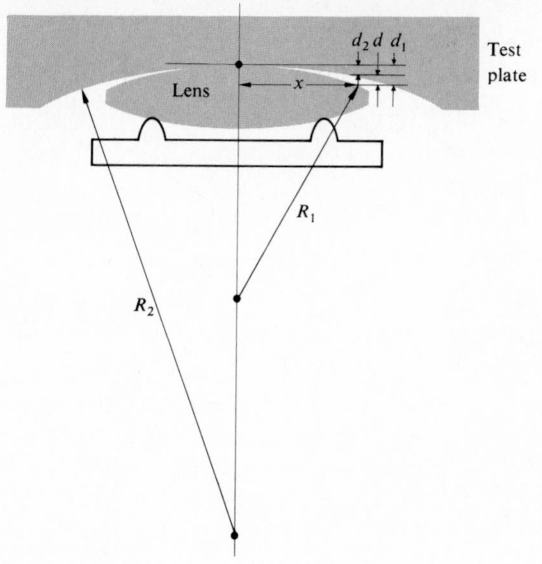

FIGURE P.9.29

9.31 Fringes are observed when a parallel beam of light of wavelength 500 nm is incident perpendicularly onto a wedge-shaped film with an index of refraction of 1.5. What is the angle of the wedge if the fringe separation is $\frac{1}{3}$ cm?

9.32* Suppose a wedge-shaped air film is made between two sheets of glass, with a piece of paper 7.618×10^{-5} m thick used as the spacer at their very ends. If light of wavelength 500 nm comes down from directly above, determine the number of bright fringes that will be seen across the wedge.

9.33 A Michelson Interferometer is illuminated with monochromatic light. One of its mirrors is then moved 2.53×10^{-5} m, and it is observed that 92 fringe-pairs, bright and dark, pass by in the process. Determine the wavelength of the incident beam.

9.34* One of the mirrors of a Michelson Interferometer is moved, and 1000 fringe-pairs shift past the hairline in a viewing telescope during the process. If the device is illuminated with 500-nm light, how far was the mirror moved?

9.35* Suppose we place a chamber 10.0 cm long with flat parallel windows in one arm of a Michelson Interferometer that is being illuminated by 600-nm light. If the refractive index of air is 1.000 29 and all the air is pumped out of the cell, how many fringe-pairs will shift by in the process?

9.36* A form of the Jamin Interferometer is illustrated in Fig. P.9.36. How does it work? To what use might it be put?

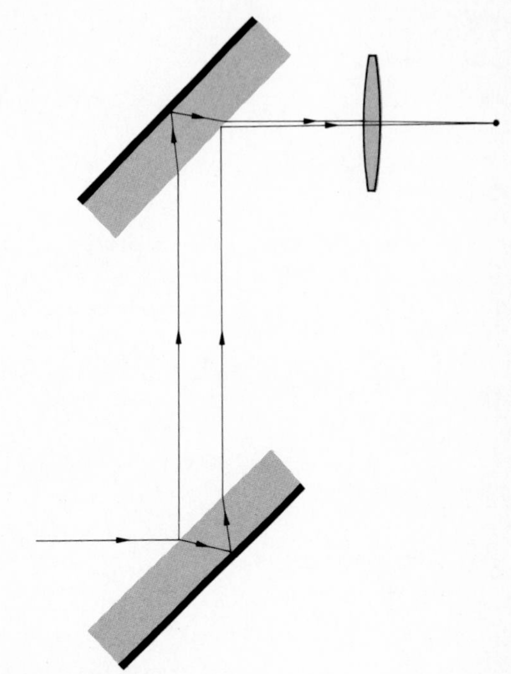

FIGURE P.9.36

9.37 Starting with Eq. (9.53) for the transmitted wave, compute the flux density, that is, Eq. (9.54).

9.38 Given that the mirrors of a Fabry–Perot Interferometer have an amplitude reflection coefficient of $r = 0.8944$, find

(a) the coefficient of finesse,

(b) the half-width,

(c) the finesse, and,

(d) the *contrast factor* defined by

$$C \equiv \frac{(I_t/I_i)_{\max}}{(I_t/I_i)_{\min}}$$

9.39 To fill in some of the details in the derivation of the smallest phase increment separating two resolvable Fabry–Perot fringes, that is,

$$(\Delta\delta) \approx 4.2/\sqrt{F} \qquad [9.73]$$

satisfy yourself that

$$[\mathscr{A}(\theta)]_{\delta = \delta_a \pm \Delta\delta/2} = [\mathscr{A}(\theta)]_{\delta = \Delta\delta/2}$$

Show that Eq. (9.72) can be rewritten as

$$2[\mathscr{A}(\theta)]_{\delta=\Delta\delta/2} = 0.81\{1 + [\mathscr{A}(\theta)]_{\delta=\Delta\delta}\}$$

When F is large γ is small, and $\sin(\Delta\delta) = \Delta\delta$. Prove that Eq. (9.73) then follows.

9.40 Consider the interference pattern of the Michelson Interferometer as arising from two beams of equal flux density. Using Eq. (9.17), compute the half-width. What is the separation, in δ, between adjacent maxima? What then is the finesse?

9.41* Satisfy yourself of the fact that a film of thickness $\lambda_f/4$ and index n_1 will always reduce the reflectance of the substrate on which it is deposited, as long as $n_s > n_1 > n_0$. Consider the simplest case of normal incidence and $n_0 = 1$. Show that this is equivalent to saying that the waves reflected back from the two interfaces cancel one another.

9.42 Verify that the reflectance of a substrate can be increased by coating it with a $\lambda_f/4$, high-index layer, that is, $n_1 > n_s$. Show that the reflected waves interfere constructively. The quarter-wave stack $g(HL)^m Ha$ can be thought of as a series of such structures.

9.43 Determine the refractive index and thickness of a film to be deposited on a glass surface ($n_g = 1.54$) such that no normally incident light of wavelength 540 nm is reflected.

9.44 A glass microscope lens having an index of 1.55 is to be coated with a magnesium fluoride film to increase the transmission of normally incident yellow light ($\lambda_0 = 550$ nm). What minimum thickness should be deposited on the lens?

9.45* A glass camera lens with an index of 1.55 is to be coated with a cryolite film ($n \approx 1.30$) to decrease the reflection of normally incident green light ($\lambda_0 = 500$ nm). What thickness should be deposited on the lens?

10 Diffraction

10.1 Preliminary Considerations

An opaque body placed midway between a screen and a point source casts an intricate shadow made up of bright and dark regions quite unlike anything one might expect from the tenets of Geometrical Optics (Fig. 10.1).* The work of Francesco Grimaldi in the 1600s was the first published detailed study of this ***deviation of light from rectilinear propagation***, something he called "*diffractio*." *The effect is a general characteristic of wave phenomena occurring whenever a portion of a wavefront, be it sound, a matter wave, or light, is obstructed in some way.* If in the course of encountering an obstacle, either transparent or opaque, a region of the wavefront is altered in amplitude or phase, diffraction will occur.† The various segments of the wavefront that propagate beyond the obstacle interfere, causing the particular energy-density distribution referred to as the diffraction pattern. There is no significant physical distinction between *interference* and *diffraction.* It has, however, become somewhat customary, if not always appropriate, to speak of interference when considering the superposition of only a few waves and diffraction when treat-

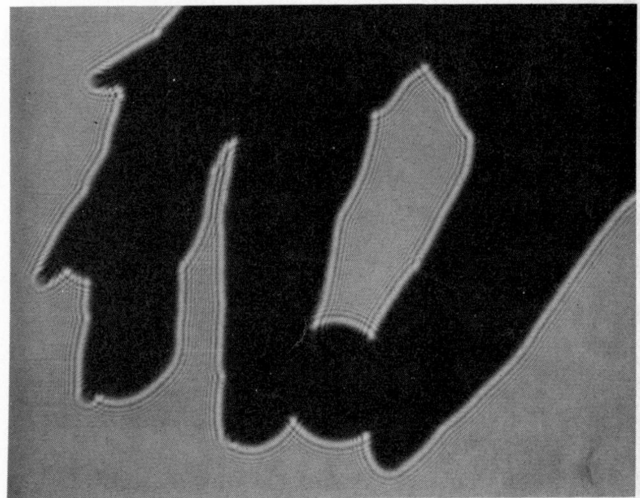

(a)

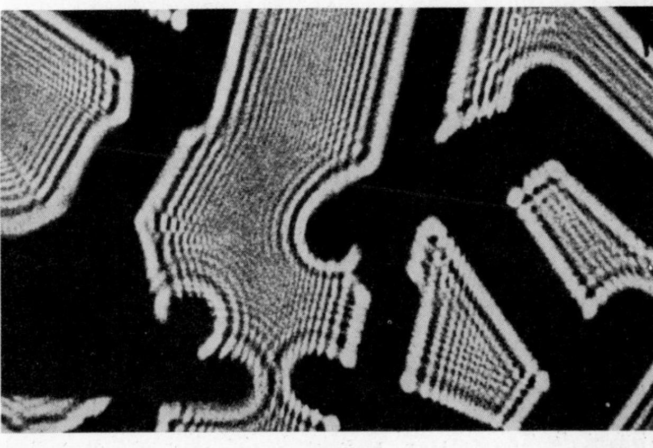

(b)

FIGURE 10.1 (*a*) The shadow of Mary's hand holding a dime, cast directly on 4 × 5 Polaroid A.S.A. 3000 film using a He–Ne beam and no lenses. (E. H.) (*b*) Fresnel diffraction of electrons by zinc oxide crystals. (After H. Boersch from *Handbuck der Physik*, edited by S.Flügge, Springer-Verlag, Heidelberg.)

*The effect is easily seen, but you need a fairly strong source. A high-intensity lamp shining through a small hole works well. If you look at the shadow pattern arising from a pencil under point-source illumination, you will see an unusual bright region bordering the edge and even a faintly illuminated band down the middle of the shadow. Take a close look at the shadow cast by your hand in direct sunlight.

†Diffraction associated with transparent obstacles is not usually considered, although if you have ever driven an automobile at night with a few rain droplets on your eyeglasses, you are no doubt quite familiar with the effect. If you have not, put a droplet of water or saliva on a glass plate, hold it close to your eye, and look directly through it at a point source. You'll see bright and dark fringes.

ing a large number of waves. Even so, one refers to multiple-beam interference in one context and diffraction from a grating in another.

It would be nice to treat diffraction from the perspective of the most powerful contemporary theory of light, Quantum Electrodynamics (QED), but that's terribly impractical; the analysis is far too complicated and wouldn't add much at that. What we *can* do is show qualitatively how QED applies to a few basic situations. For our purposes, however, the classical wave theory, which provides the simplest effective formalism, will more than suffice. Still, wherever it's appropriate, the discussion will be illuminated with insights from Fourier analysis, even though the detailed treatment of that subject is postponed to the next chapter.

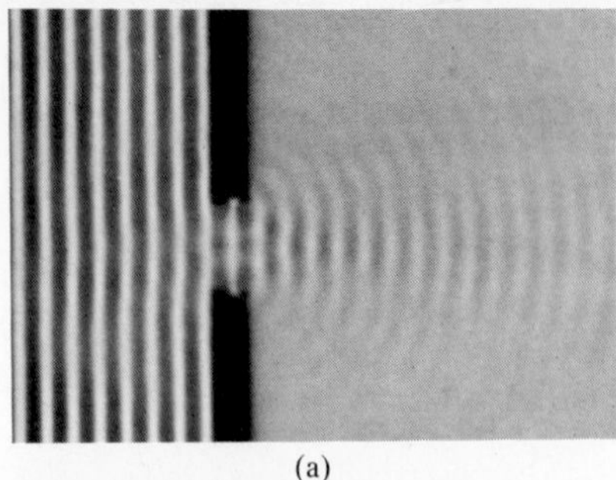

(a)

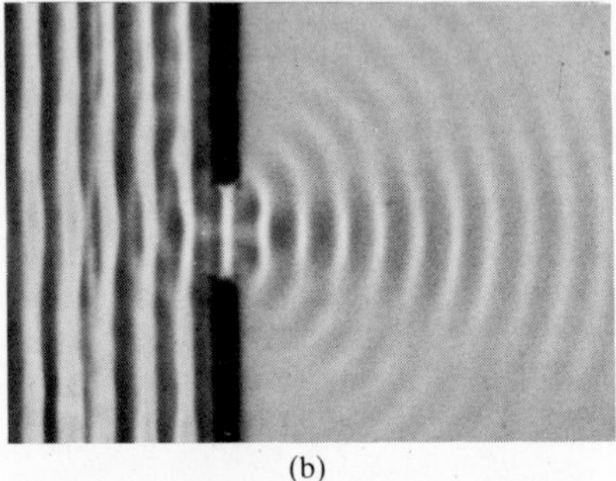

(b)

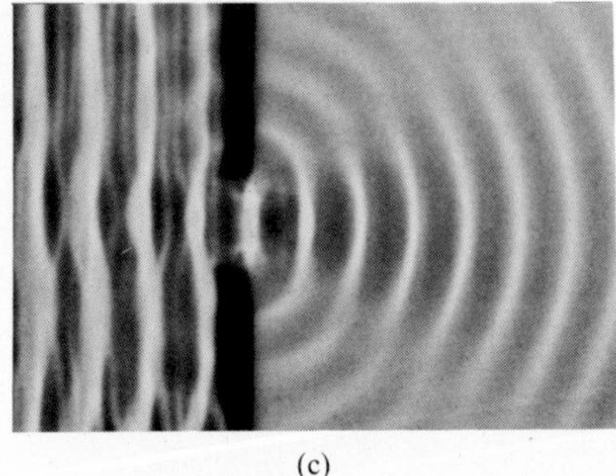

(c)

FIGURE 10.2 Diffraction through an aperture with varying λ as seen in a ripple tank. (Photo courtesy PSSC *Physics*, D. C. Heath, Boston, 1960.)

The Huygens-Fresnel Principle

As an initial approach to the problem, let's reconsider Huygens's Principle (Section 4.4.2). Each point on a wavefront can be envisaged as a source of secondary spherical wavelets. The progress through space of the wavefront, or any portion thereof can then presumably be determined. At any particular time, the shape of the wavefront is supposed to be the envelope of the secondary wavelets (Fig. 4.27). The technique, however, ignores most of each secondary wavelet, retaining only that portion common to the envelope. As a result of this inadequacy, Huygens's Principle by itself is unable to account for the details of the diffraction process. That this is indeed the case is born out by everyday experience. Sound waves (e.g., ν = 500 Hz, $\lambda \approx 68$ cm) easily "bend" around large objects like telephone poles and trees, yet these objects cast fairly distinct shadows when illuminated by light. Huygens's Principle is independent of any wavelength considerations, however, and would predict the same wavefront configurations in both situations.

The difficulty was resolved by Fresnel with his addition of the concept of interference. The corresponding **Huygens–Fresnel Principle** states that ***every unobstructed point of a wavefront, at a given instant, serves as a source of spherical secondary wavelets (with the same frequency as that of the primary wave). The amplitude of the optical field at any point beyond is the superposition of all these wavelets (considering their amplitudes and relative phases).***

Applying these ideas on the very simplest qualitative level, refer to the ripple tank photographs in Fig. 10.2 and the illustration in Fig. 10.3. If each unobstructed point on the incoming plane wave acts as a coherent secondary source, the maximum optical path length difference among them will be $\Lambda_{max} = |\overline{AP} - \overline{BP}|$, corresponding to a source point at each edge of the aperture. But Λ_{max} is less than or equal to $\overline{AB}$, the latter being the case when P is on the screen. When $\lambda > \overline{AB}$, as in Fig. 10.3, it follows that $\lambda > \Lambda_{max}$, and since the waves were initially in phase, they all interfere constructively (to varying degrees) wherever P happens to be (see Fig. 10.2*c*). Thus, ***if***

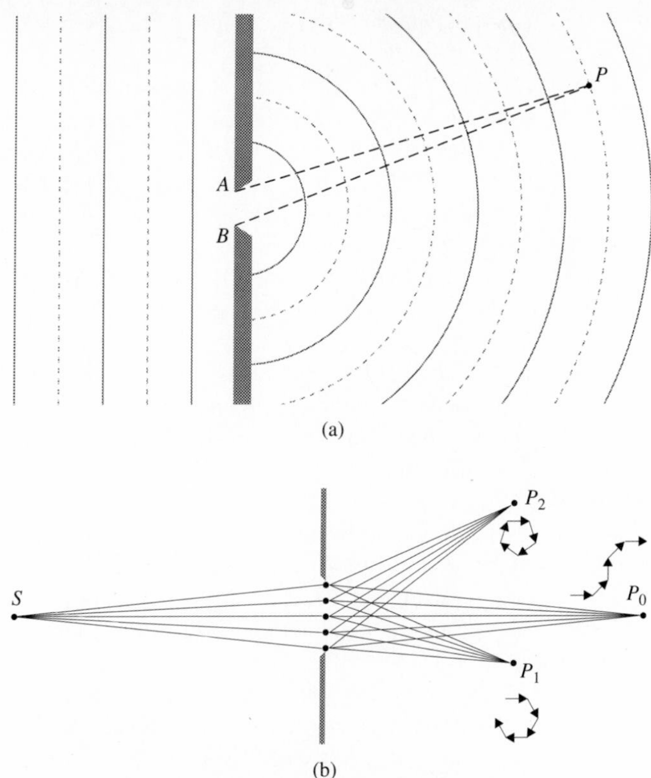

FIGURE 10.3 Diffraction at a small aperture. (*a*) The classical wave picture. (*b*) The view via QED and probability amplitudes.

the wavelength is large compared to the aperture, the waves will spread out at large angles into the region beyond the obstruction. And the smaller the aperture gets, the more nearly circular the diffracted waves become (recall the discussion of this point from a Fourier context, p. 389).

The antithetic situation occurs when $\lambda < \overline{AB}$, as in Fig. 10.2*a*. The area where $\lambda > \Lambda_{max}$ is limited to a small region extending out directly in front of the aperture, and it is only there that all the wavelets will interfere constructively. Beyond this zone some of the wavelets can interfere destructively, and the "shadow" begins. Keep in mind that the idealized *geometric shadow* corresponds to $\lambda \rightarrow 0$.

Classically, the reason light goes where it does beyond the screen is that the multitude of wavelets emitted from the aperture "interfere"; that is, they combine (as phasors) at every point in the region, some places enhancing, some canceling, depending on the *OPL*.

Quantum mechanically (Section 4.11.1), the reason light goes where it does beyond the screen is that the multitude of probability amplitudes for photons from the aperture "interfere"; that is, they combine (as phasors) at every point in the region, some places enhancing, some canceling, depending on the *OPL*. When the hole is several wavelengths wide, as in Fig. 10.2*a*, the many paths to any point P correspond to a broad range of phasor phases. Consider all the paths to a point in the forward direction such as P_0. The straight-line route from S to P_0 corresponds to a minimum in *OPL*. Any other paths through the aperture to P_0 are somewhat longer (depending on the size of the hole) and have phasors (all of which we will take to be the same size) that are grouped around that stationary *OPL* value, much as those in Fig. 4.71. They have small mutual phase-angle differences (half +, half −) and so added tip-to-tail they turn one way, then the other, to produce a substantial resultant probability amplitude. A photon counter at P_0 will see lots of light. Off the forward direction (where the *OPL* is not stationary), the phasors each have relatively large phase-angle differences for every path and all are of the same sign. Placed tip-to-tail they spiral around, adding up to little or nothing. A detector at P_1 will record few counts, and one at P_2 fewer still.

If the aperture is now made much smaller, the number of counts at P_1 and P_2 increases, even as the number at P_0 drops off. With a narrow hole, all the paths to either P_1 or P_2 are much closer together and have nearly the same *OPL*. The phase-angle differences are therefore much smaller, the phasor spirals no longer close on themselves, and the resultant probability amplitudes, though small, are appreciable everywhere.

Qualitatively, both QED and the classical Huygens–Fresnel Principle lead to much the same general conclusion: *light diffracts and interference is at the heart of the process.*

The Huygens–Fresnel Principle has some shortcomings (which we will examine later), in addition to the fact that the whole thing at this point is rather hypothetical. Gustav Kirchhoff developed a more rigorous theory based directly on the solution of the differential wave equation. Kirchhoff, though a contemporary of Maxwell, did his work before Hertz's demonstration (and the resulting popularization) of the propagation of electromagnetic waves in 1887. Accordingly, Kirchhoff employed the older elastic-solid theory of light. His refined analysis lent credence to the assumptions of Fresnel and led to an even more precise formulation of Huygens's Principle as an exact consequence of the wave equation. Even so, the Kirchhoff theory is itself an approximation that is valid for sufficiently small wavelengths—that is, when the diffracting apertures have dimensions that are large in comparison to λ. The difficulty arises from the fact that what's required is the solution of a partial differential equation that meets the boundary conditions imposed by the obstruction. This kind of rigorous solution is obtainable only in a few special cases. Kirchhoff's theory works fairly well, even though it deals only with scalar waves and is insensitive to the fact that light is a transverse vector field.*

It should be stressed that the problem of determining an exact solution for a particular diffracting configuration is among the most troublesome to be dealt with in Optics. The first such solution, utilizing the electromagnetic theory of light, was published by Arnold Johannes Wilhelm Sommerfeld (1868–1951) in 1896. Although the problem was physically somewhat unrealistic, in that it involved an infinitely thin yet opaque, perfectly conducting plane screen, the result was nonetheless extremely valuable, providing a good deal of insight into the fundamental processes involved.

Rigorous solutions of this sort do not exist even today for many of the configurations of practical interest. We will therefore, out of necessity, rely on the approximate treatments of Huygens–Fresnel and Kirchhoff. In recent times, microwave techniques have been employed to conveniently study features of the diffraction field that might otherwise be almost impossible to examine optically. The Kirchhoff theory has held up remarkably well under this kind of scrutiny.† In many cases, the simpler Huygens–Fresnel treatment will prove adequate to our needs.

10.1.1 Opaque Obstructions

Diffraction may be envisioned as arising from the interaction of electromagnetic waves with some sort of physical obstruction. We would therefore do well to reexamine briefly the processes involved; what actually takes place within the material of the opaque object?

One possible description is that a screen may be considered to be a continuum; that is, its microscopic structure may be neglected. For a nonabsorbing metal sheet (no joule heating, therefore infinite conductivity) we can write Maxwell's Equations for the metal and for the surrounding medium, and then match the two at the boundaries. Precise solutions can be obtained for very simple configurations. The reflected and diffracted waves then result from the current distribution within the sheet.

Examining the screen on a submicroscopic scale, imagine the electron cloud of each atom set into vibration by the electric field of the incident radiation. The classical model, which speaks of electron-oscillators vibrating and reemitting at the source frequency, serves quite well so that we need not be concerned with the quantum-mechanical description. The amplitude and phase of a particular oscillator within the screen are determined by the local electric field surrounding it. This in turn is a superposition of the incident field and the fields of all the other vibrating electrons. A large opaque screen with no apertures, be it made of black paper or aluminum foil, has one obvious effect: there is no optical field in the region beyond it. Electrons near the illuminated surface are driven into oscillation by the impinging light. They emit radiant energy, which is ultimately "reflected" backward, absorbed by the material, or both. In any case, the incident wave and the electron-oscillator fields superimpose in such a way as to yield zero light at any point beyond the screen. This might seem a remarkably special balance, but it actually is not. If the incident wave were not canceled completely, it would propagate deeper into the material of the screen, exciting more electrons to radiate. This in turn would further weaken the wave until it ultimately vanished (if the screen were thick enough). Even an ordinarily opaque material such as silver, in the form of a sufficiently thin sheet, is partially transparent (recall the half-silvered mirror).

Now, remove a small disk-shaped segment from the center of the screen, so that light streams through the aperture. The oscillators that uniformly cover the disk are removed along with it, so the remaining electrons within the screen are no longer affected by them. As a first and certainly approximate approach, *assume that the mutual interaction of the oscillators is essentially negligible*; that is, the electrons in the screen are completely unaffected by the removal of the electrons in the disk. The field in the region beyond the aperture will then be

*A vectorial formulation of the scalar Kirchhoff theory is discussed in J. D. Jackson, *Classical Electrodynamics*, p. 283. Also see Sommerfeld, *Optics*, p. 325. You might as well take a look at B. B. Baker and E. T. Copson, *The Mathematical Theory of Huygens's Principle*, as a general reference to diffraction. None of these texts is easy reading.

†C. L. Andrews, *Am. J. Phys.* **19**, 250 (1951); S. Silver, *J. Opt. Soc. Am.* **52**, 131 (1962).

that which existed before the removal of the disk, namely zero, minus the contribution from the disk alone. Except for the sign, it is as if the source and screen had been taken away, leaving only the oscillators on the disk, rather than vice versa. In other words, the diffraction field can be pictured as arising exclusively from a set of fictitious noninteracting oscillators distributed uniformly over the region of the aperture. This of course, is the essence of the Huygens–Fresnel Principle.

We can expect, however, that instead of no interaction at all between electron-oscillators, there is a short-range effect, since the oscillator fields drop off with distance. In this physically more realistic view, the electrons within the vicinity of the aperture's edge are affected when the disk is removed. For larger apertures, the number of oscillators in the disk is much greater than the number along the edge. In such cases, if the point of observation is far away and in the forward direction, the Huygens–Fresnel Principle should, and does, work well (Fig. 10.4). For very small apertures, or at points of observation in the vicinity of the aperture, edge effects become important, and we can anticipate difficulties. Indeed, at a point within the aperture itself, the electron-oscillators on the edge are of the greatest significance because of their proximity. Yet these electrons were certainly not unaffected by the removal of the adjacent oscillators of the disk. Thus, the deviation from the Huygens–Fresnel Principle should be appreciable.

10.1.2 Fraunhofer and Fresnel Diffraction

Imagine that we have an opaque screen, Σ, containing a single small aperture, which is being illuminated by plane waves from a very distant point source, S. The plane of observation σ is a screen parallel with, and very close to, Σ. Under these conditions an image of the aperture is projected onto the screen, which is clearly recognizable despite some slight fringing around its periphery (Fig. 10.5). If the plane of observation is moved farther away from Σ, the image of the aperture, though still easily recognizable, becomes increasingly more structured as the fringes become more prominent. This phenomenon is known as **Fresnel** or **near-field** diffraction. If the plane of observation is slowly moved out still farther, a continuous change in the fringes results. At a very great distance from Σ the projected pattern will have spread out considerably, bearing little or no resemblance to the actual aperture. Thereafter moving σ essentially changes only the size of the pattern and not its shape. This is **Fraunhofer** or **far-field** diffraction. If at that point we could sufficiently reduce the wavelength of the incoming radiation, the pattern would revert to the Fresnel case. If λ were decreased even more, so that it approached zero, the fringes would disappear, and the image would take on the limiting shape of the aperture, as predicted by Geometrical Optics. Returning to the original setup, if the point source

FIGURE 10.4 Ripple-tank photos. In one case, the waves are simply diffracted by a slit; in the other, a series of equally spaced point sources span the aperture and generate a similar pattern. (Photos courtesy PSSC *Physics*, D. C. Heath, Boston, 1960.)

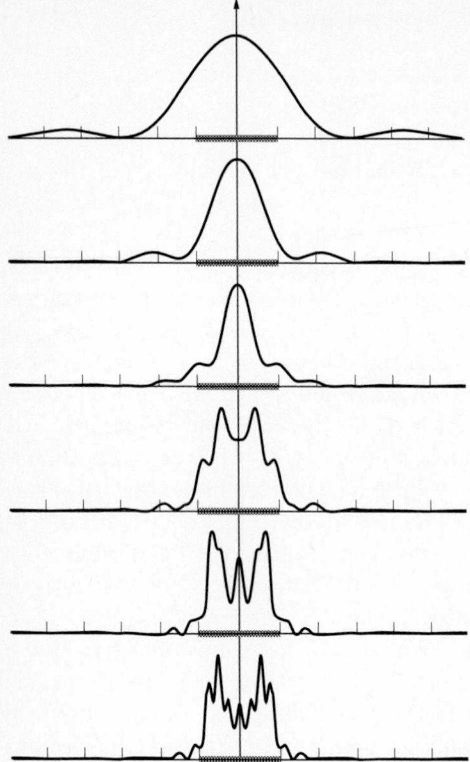

FIGURE 10.5 A succession of diffraction patterns at increasing distance form a single slit; Fresnel at the bottom (nearby), going toward Fraunhofer at the top (faraway). Adapted from *Fundamentals of Waves and Oscillations* by K. U. Ingard.

was now moved toward Σ, spherical waves would impinge on the aperture, and a Fresnel pattern would exist, even on a distant plane of observation.

Consider a point source S and a point of observation P, where both are very far from Σ and no lenses are present (Problem 10.1). ***As long as both the incoming and outgoing waves approach being planar (differing therefrom by a small fraction of a wavelength) over the extent of the diffracting apertures (or obstacles), Fraunhofer diffraction obtains.*** Another way to appreciate this is to realize that the *phase* of each contribution at P, due to differences in the path traversed, is crucial to the determination of the resultant field. Moreover, if the wavefronts impinging on, and emerging from, the aperture are planar, then these path differences will be describable by a linear function of the two aperture variables. ***This linearity in the aperture variables is the definitive mathematical criterion of Fraunhofer diffraction.*** On the other hand, when S or P or both are too near Σ for the curvature of the incoming and outgoing wavefronts to be negligible, Fresnel diffraction prevails.

Each point on the aperture is to be visualized as a source of Huygens wavelets, and we should be a little concerned about their relative strengths. When S is nearby, compared with the size of the aperture, a spherical wavefront will illuminate the hole. The distances from S to each point on the aperture will be different, and the strength of the incident electric field (which drops off inversely with distance) will vary from point to point over the diffracting screen. That would not be the case for incoming homogeneous plane waves. Much the same thing is true for the diffracted waves going from the aperture to P. Even if they are all emitted with the same amplitude, if P is nearby, the waves converging on it are spherical and vary in amplitude, because of the different distances from various parts of the aperture to P. Ideally, for P at infinity (whatever that means) the waves arriving there will be planar, and we need not worry about differences in field strength. That too contributes to the simplicity of the limiting Fraunhofer case.

As a practical rule of thumb, Fraunhofer diffraction will occur at an aperture (or obstacle) of greatest width a when

$$R > a^2/\lambda$$

where R is the smaller of the two distances from S to Σ and Σ to P (Problem 10.1). Of course, when $R = \infty$ the finite size of the aperture is of little concern. Moreover, an increase in λ clearly shifts the phenomenon toward the Fraunhofer extreme.

A practical realization of the Fraunhofer condition, where both S and P are effectively at infinity, is achieved by using an arrangement equivalent to that of Fig. 10.6. The point source S is located at F_1, the principal focus of lens L_1, and the plane of observation is the second focal plane of L_2. In the terminology of Geometrical Optics, the source plane and σ are conjugate planes.

These same ideas can be generalized to any lens system forming an image of an extended source or object (Problem

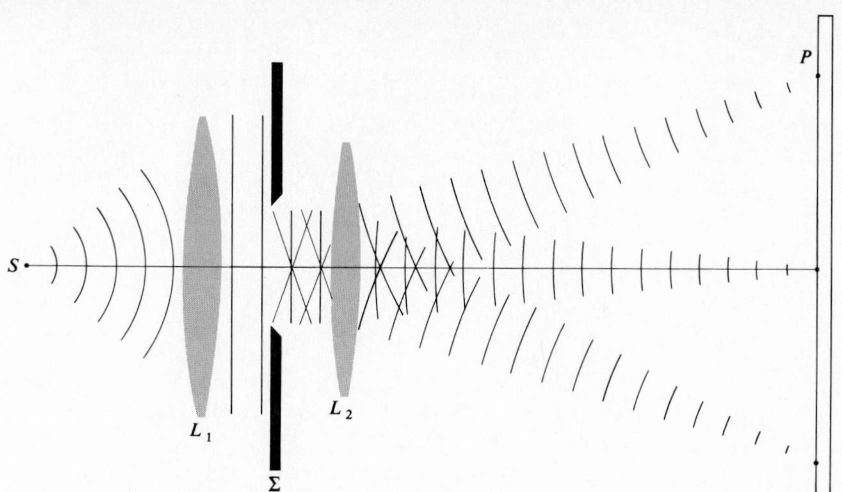

FIGURE 10.6 Fraunhofer diffraction.

10.5).* Indeed, the image would be a Fraunhofer diffraction pattern. It is because of these important practical considerations, as well as the inherent simplicity of Fraunhofer diffraction, that we will examine it before Fresnel diffraction, even though it is a special case of the latter.

10.1.3 Several Coherent Oscillators

As a simple yet logical bridge between the studies of interference and diffraction, consider the arrangement in Fig. 10.7. The illustration depicts a linear array of N coherent point oscillators (or radiating antennas), which are all identical, even to their polarization. For the moment, assume that the oscillators have no intrinsic phase difference; that is, they each have the same initial phase angle. The rays shown are all almost parallel, meeting at some very distant point P. If the spatial extent of the array is comparatively small, the separate wave amplitudes arriving at P will be essentially equal, having traveled nearly equal distances, that is,

$$E_0(r_1) = E_0(r_2) = \cdots = E_0(r_N) = E_0(r)$$

*A He–Ne laser can be set up to generate magnificent patterns without any auxiliary lenses, but this requires plenty of space.

The sum of the interfering spherical wavelets yields an electric field at P, given by the real part of

$$\tilde{E} = E_0(r)e^{i(kr_1-\omega t)} + E_0(r)e^{i(kr_2-\omega t)} + \cdots + E_0(r)e^{i(kr_N-\omega t)} \tag{10.1}$$

It should be clear, from Section 9.1, that we need not be concerned with the vector nature of the electric field for this configuration. Now then

$$\tilde{E} = E_0(r)e^{-i\omega t}e^{ikr_1} \times [1 + e^{ik(r_2-r_1)} + e^{ik(r_3-r_1)} + \cdots + e^{ik(r_N-r_1)}]$$

The phase difference between adjacent sources is obtained from the expression $\delta = k_0\Lambda$, and since $\Lambda = nd \sin\theta$, in a medium of index n, $\delta = kd \sin\theta$. Making use of Fig. 10.7, it follows that $\delta = k(r_2 - r_1)$, $2\delta = k(r_3 - r_1)$, and so on. Thus the field at P may be written as

$$\tilde{E} = E_0(r)e^{-i\omega t}e^{ikr_1} \times [1 + (e^{i\delta}) + (e^{i\delta})^2 + (e^{i\delta})^3 + \cdots + (e^{i\delta})^{N-1}] \tag{10.2}$$

The bracketed geometric series has the value

$$(e^{i\delta N}-1)/(e^{i\delta} - 1)$$

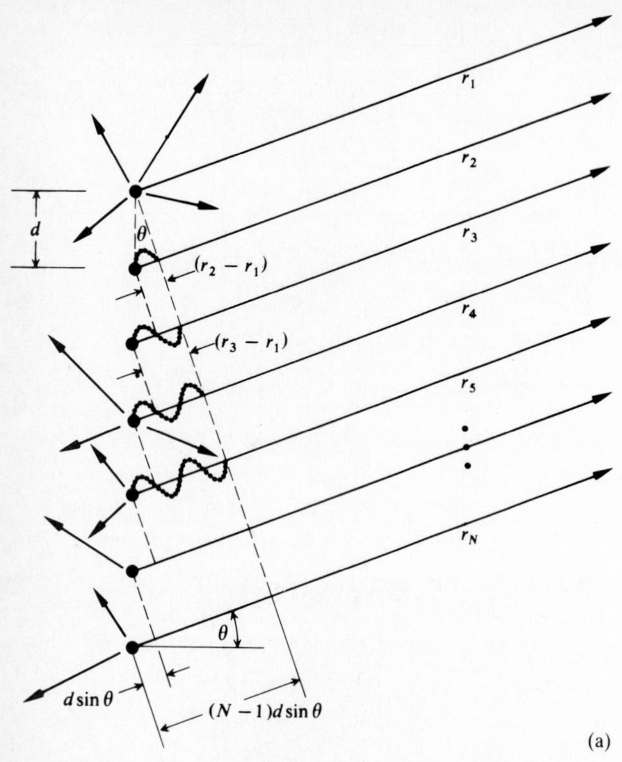

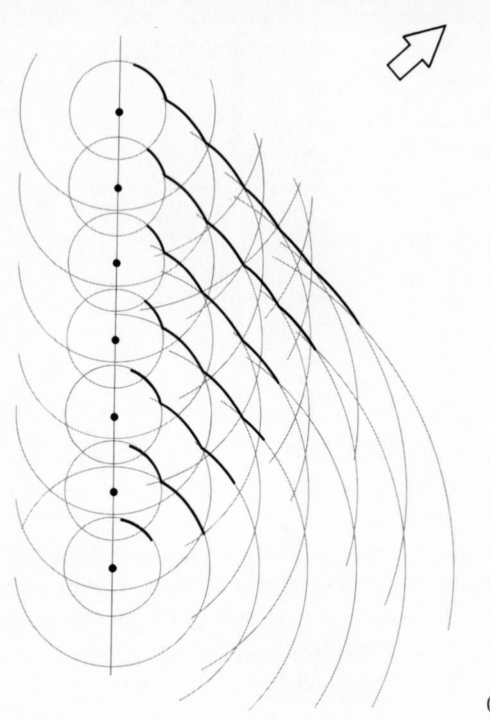

FIGURE 10.7 A linear array of in-phase coherent oscillators. (*a*) Note that at the angle shown $\delta = \pi$ while at $\theta = 0$, δ would be zero. (*b*) One of many sets of wavefronts emitted from a line of coherent point sources.

which can be rearranged into the form

$$\frac{e^{iN\delta/2}[e^{iN\delta/2} - e^{-iN\delta/2}]}{e^{i\delta/2}[e^{i\delta/2} - e^{-i\delta/2}]}$$

or equivalently

$$e^{i(N-1)\delta/2}\left(\frac{\sin N\delta/2}{\sin \delta/2}\right)$$

The field then becomes

$$\tilde{E} = E_0(r)e^{-i\omega t}\, e^{i[kr_1 + (N-1)\delta/2]}\left(\frac{\sin N\delta/2}{\sin\delta/2}\right) \quad (10.3)$$

Notice that if we define R as the distance from the center of the line of oscillators to the point P, that is,

$$R = \tfrac{1}{2}(N-1)d\sin\theta + r_1$$

then Eq. (10.3) takes on the form

$$\tilde{E} = E_0(r)e^{i(kR-\omega t)}\left(\frac{\sin N\delta/2}{\sin \delta/2}\right) \quad (10.4)$$

Finally, then, the flux-density distribution within the diffraction pattern due to N coherent, identical, distant point sources in a linear array is proportional to $EE^*/2$ for complex E or

$$I = I_0\frac{\sin^2(N\delta/2)}{\sin^2(\delta/2)} \quad (10.5)$$

where I_0 is the flux density from any single source arriving at P. (See Problem 10.2 for a graphic derivation of the irradiance.) For $N = 0$, $I = 0$, for $N = 1$, $I = I_0$, and for $N = 2$, $I = 4I_0\cos^2(\delta/2)$, in accord with Eq. (9.17). The functional dependence of I on θ is more apparent in the form

$$I = I_0 \frac{\sin^2 [N(kd/2)\sin \theta]}{\sin^2 [(kd/2)\sin \theta]} \tag{10.6}$$

The $\sin^2 [N(kd/2) \sin \theta]$ term undergoes rapid fluctuations, whereas the function that modulates it, $\{\sin [(kd/2) \sin \theta]\}^{-2}$, varies relatively slowly. The combined expression gives rise to a series of sharp principal peaks separated by small subsidiary maxima. The principal maxima occur in directions θ_m, such that $\delta = 2m\pi$, where $m = 0, \pm 1, \pm 2, \ldots$. Because $\delta = kd \sin \theta$,

$$d \sin \theta_m = m\lambda \tag{10.7}$$

Since $[\sin^2 N\delta/2]/[\sin^2 \delta/2] = N^2$ for $\delta = 2m\pi$ (from L'Hospital's rule), the principal maxima have values of $N^2 I_0$. This is to be expected, inasmuch as all the oscillators are in-phase at that orientation. The system will radiate a maximum in a direction perpendicular to the array ($m = 0$, $\theta_0 = 0$ and π). As θ increases, δ increases and I falls off to zero at $N\delta/2 = \pi$, its first minimum. Note that if $d < \lambda$ in Eq. (10.7), only the $m = 0$ or zero-order principal maximum exists. *If we were looking at an idealized line source of electron-oscillators separated by atomic distances, we could expect only that one principal maximum in the light field.*

The antenna array in Fig. 10.8 can transmit radiation in the narrow beam or lobe corresponding to a principal maximum. (The parabolic dishes shown reflect in the forward direction, and the radiation pattern is no longer symmetrical around the common axis.) Suppose that we have a system in which we can introduce an intrinsic phase shift of ε between adjacent oscillators. In that case

$$\delta = kd \sin \theta + \varepsilon$$

The various principal maxima will occur at new angles

$$d \sin \theta_m = m\lambda - \varepsilon/k$$

Concentrating on the central maximum $m = 0$, we can vary its orientation θ_0 at will by merely adjusting the value of ε.

The Principle of Reversibility, which states that without absorption, wave motion is reversible, leads to the same field pattern for an antenna used as either a transmitter or a receiver. The array, functioning as a radio telescope, can therefore be "pointed" by combining the output from the individual antennas with an appropriate phase shift, ε, introduced between each of them. For a given ε the output of the system corresponds to the signal impinging on the array from a specific direction in space.

Figure 10.8 is a photograph of the first multiple radio interferometer, designed by W. N. Christiansen and built in Australia in 1951. It consists of 32 parabolic antennas, each 2 m in diameter, designed to function in phase at the wavelength of the 21-cm hydrogen emission line. The antennas are arranged along an east–west base line with 7 m separating each one. This particular array utilized the Earth's rotation as the scanning mechanism.*

Examine Fig. 10.9, which depicts an idealized line source of electron-oscillators (e.g., the secondary sources of the Huygens–Fresnel Principle for a long slit whose width is much less than λ, illuminated by plane waves). Each point emits a spherical wavelet, which we write as

$$E = \left(\frac{\mathcal{E}_0}{r}\right) \sin (\omega t - kr)$$

explicitly indicating the inverse r-dependence of the amplitude. The quantity $\mathcal{E}_0$ is said to be the **source strength**. The present situation is distinct from that of Fig. 10.7, since now

FIGURE 10.8 Interferometric radio telescope at the University of Sydney, Australia ($N = 32$, $\lambda = 21$ cm, $d = 7$ m, 2 m diameter, 700 ft. east–west base line). (Photo courtesy of W. N. Christiansen.)

*See E. Brookner, "Phased-Array Radars," *Sci. Am.* (Feb. 1985), p. 94.

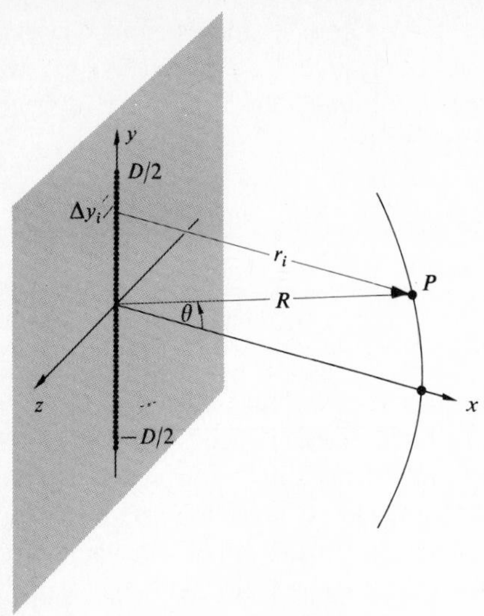

FIGURE 10.9 A coherent line source.

the sources are very weak, their number, N, is tremendously large, and the separation between them is vanishingly small. A minute but finite segment of the array Δy_i will contain $\Delta y_i(N/D)$ sources, where D is the entire length of the array. Imagine that the array is divided up into M such segments (i.e., i goes from 1 to M). The contribution to the electric-field intensity at P from the ith segment is accordingly

$$E_i = \left(\frac{\mathcal{E}_0}{r_i}\right) \sin(\omega t - kr_i)\left(\frac{N\Delta y_i}{D}\right)$$

provided that Δy_i is so small that the oscillators within it have a negligible relative phase difference (r_i = constant), and their fields simply add constructively. We can cause the array to become a continuous (coherent) line source by letting N approach infinity. This description, besides being fairly realistic on a macroscopic scale, also allows the use of the calculus for more complicated geometries. Certainly as N approaches infinity, the source strengths of the individual oscillators must diminish to nearly zero, if the total output is to be finite. We can therefore define a constant $\mathcal{E}_L$ as the *source strength per unit length* of the array, that is,

$$\mathcal{E}_L \equiv \frac{1}{D} \lim_{N \to \infty} (\mathcal{E}_0 N) \tag{10.8}$$

The net field at P from all M segments is

$$E = \sum_{i=1}^{M} \frac{\mathcal{E}_L}{r_i} \sin(\omega t - kr_i)\Delta y_i$$

For a continuous line source the Δy_i must become infinitesimal ($M \to \infty$), and the summation is then transformed into a definite integral

$$E = \mathcal{E}_L \int_{-D/2}^{+D/2} \frac{\sin(\omega t - kr)}{r} \, dy \tag{10.9}$$

where $r = r(y)$. The approximation used to evaluate Eq. (10.9) must depend on the position of P with respect to the array and will therefore make the distinction between Fraunhofer and Fresnel diffraction. The coherent optical line source does not exist as a physical entity, but we will make good use of it as a mathematical device.

10.2 Fraunhofer Diffraction

10.2.1 The Single Slit

Return to Fig. 10.9, where now the point of observation is very distant from the coherent line source and $R >> D$. Under these circumstances $r(y)$ never deviates appreciably from its midpoint value R, so that the quantity $(\mathcal{E}_L/R)$ at P is essentially constant for all elements dy. It follows from Eq. (10.9) that the field at P due to the differential segment of the source dy is

$$dE = \frac{\mathcal{E}_L}{R} \sin(\omega t - kr) dy \tag{10.10}$$

where $(\mathcal{E}_L/R)\, dy$ is the amplitude of the wave. Notice that the phase is much more sensitive to variations in $r(y)$ than is the amplitude, so that we will have to be more careful about introducing approximations into it. We can expand $r(y)$, in precisely the same manner as was done in Problem (9.15), to make it an explicit function of y; thus

$$r = R - y \sin\theta + (y^2/2R)\cos^2\theta + \cdots \tag{10.11}$$

where θ is measured from the xz-plane. The third term can be ignored as long as its contribution to the phase is insignificant even when $y = \pm D/2$; that is, $(\pi D^2/4\lambda R)\cos^2\theta$ must be negligible. This will be true for all values of θ when R is adequately large. We now have the **Fraunhofer condition**, where the distance r is linear in y: the distance to the point of observation and therefore the phase can be written as a linear function of the aperture variables. Substituting into Eq. (10.10) and integrating leads to

$$E = \frac{\mathcal{E}_L}{R}\int_{-D/2}^{+D/2} \sin\,[\omega t - k(R - y\sin\theta)]\,dy \qquad (10.12)$$

and finally

$$E = \frac{\mathcal{E}_L D}{R}\,\frac{\sin[(kD/2)\sin\theta]}{(kD/2)\sin\theta}\,\sin\,(\omega t - kR) \qquad (10.13)$$

To simplify the appearance of things, let

$$\beta \equiv (kD/2)\sin\theta \qquad (10.14)$$

so that

$$E = \frac{\mathcal{E}_L D}{R}\left(\frac{\sin\beta}{\beta}\right)\sin\,(\omega t - kR) \qquad (10.15)$$

The quantity most readily measured is the irradiance (forgetting the constants) $I(\theta) = \langle E^2\rangle_T$ or

$$I(\theta) = \frac{1}{2}\left(\frac{\mathcal{E}_L D}{R}\right)^2\left(\frac{\sin\beta}{\beta}\right)^2 \qquad (10.16)$$

where $\langle\sin^2(\omega t - kR)\rangle_T = \frac{1}{2}$. When $\theta = 0$, $\sin\beta/\beta = 1$ and $I(\theta) = I(0)$, which corresponds to the *principal maximum. The irradiance resulting from an idealized coherent line source in the Fraunhofer approximation* is then

$$I(\theta) = I(0)\left(\frac{\sin\beta}{\beta}\right)^2 \qquad (10.17)$$

or, using the *sinc function* (see Table 1 of the Appendix),

$$I(\theta) = I(0)\,\mathrm{sinc}^2\,\beta$$

There is symmetry about the y-axis, and this expression holds for θ measured in any plane containing that axis. Notice that since $\beta = (\pi D/\lambda)\sin\theta$, when $D >> \lambda$, the irradiance drops extremely rapidly as θ deviates from zero. This arises from the fact that β becomes very large for large values of length D (a centimeter or so when using light). The phase of the line source is equivalent, by way of Eq. (10.15), to that of a point source located at the center of the array, a distance R from P. Finally, a relatively long coherent line source ($D >> \lambda$) can be envisioned as a single-point emitter radiating predominantly in the forward, $\theta = 0$, direction; in other words, its emission resembles a circular wave in the xz-plane. In contrast, notice that if $\lambda >> D$, β is small, $\sin\beta \approx \beta$, and $I(\theta) \approx I(0)$. The irradiance is then constant for *all* θ, and the line source resembles a point source emitting spherical waves.

We can now turn our attention to the problem of Fraunhofer diffraction by a slit or elongated narrow rectangular hole (Fig. 10.10). An aperture of this sort might typically have a width of several hundred λ and a length of a few centimeters. The usual procedure to follow in the analysis is to divide the slit into a series of long differential strips (dz by ℓ) parallel to the y-axis, as shown in Fig. 10.11. We immediately recognize, however, that each strip is a long coherent line source and can therefore be replaced by a point emitter on the z-axis. In effect, each such emitter radiates a circular wave in the ($y = 0$ or) xz-plane. This is certainly reasonable, since the slit is long and the merging wavefronts are practically unobstructed in the slit direction. There will thus be very little diffraction parallel to the edges of the slit. The problem has been reduced to that of finding the field in the xz-plane due to an infinite number of point sources extending across the width of the slit along the z-axis. We then need only evaluate the integral of the contribution dE from each element dz in the Fraunhofer approximation. But once again, this is equivalent to a coherent line source, so that the complete solution for the slit is, as we have seen,

$$I(\theta) = I(0)\left(\frac{\sin\beta}{\beta}\right)^2 \qquad [10.17]$$

provided that

$$\beta = (kb/2)\sin\theta \qquad (10.18)$$

and θ is measured from the xy-plane (see Problem 10.3). Note that here the line source is short, $D = b$, β is not large, and although the irradiance falls off rapidly, higher-order subsidiary maxima will be observable. The extrema of $I(\theta)$ occur at values of β that cause $dI/d\beta$ to be zero, that is,

$$\frac{dI}{d\beta} = I(0)\frac{2\sin\beta(\beta\cos\beta - \sin\beta)}{\beta^3} = 0 \qquad (10.19)$$

The irradiance has minima, equal to zero, when $\sin\beta = 0$, whereupon

$$\beta = \pm\pi, \pm 2\pi, \pm 3\pi, \ldots \qquad (10.20)$$

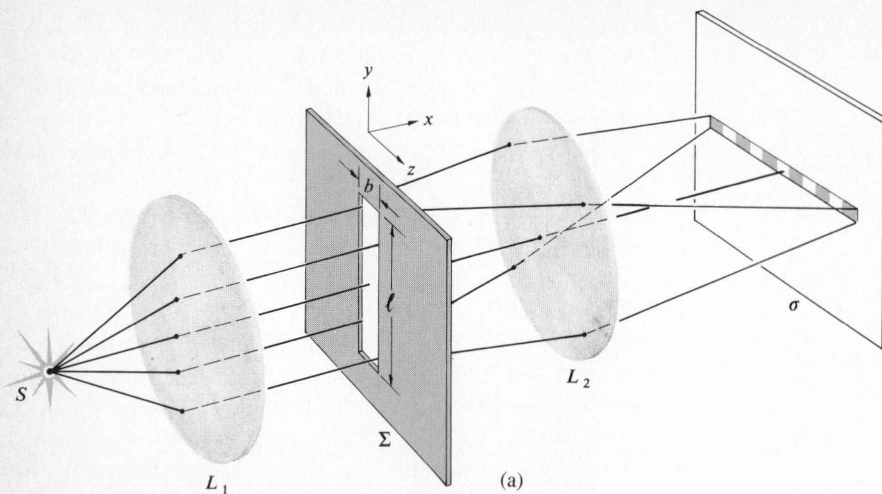

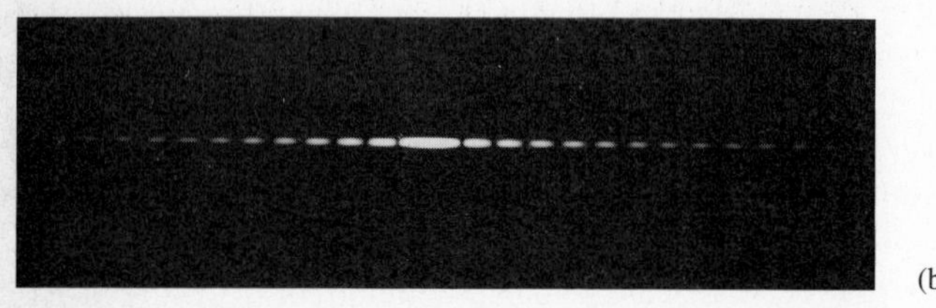

FIGURE 10.10 (*a*) Single-slit Fraunhofer diffraction. (*b*) Diffraction pattern of a single vertical slit under point-source illumination.

It also follows from Eq. (10.19) that when

$$\beta \cos \beta - \sin \beta = 0$$

$$\tan \beta = \beta \tag{10.21}$$

The solutions to this transcendental equation can be determined graphically, as shown in Fig. 10.12. The points of intersection of the curves $f_1(\beta) = \tan \beta$ with the straight line $f_2(\beta) = \beta$ are common to both and so satisfy Eq. (10.21). Only one such extremum exists between adjacent minima [Eq. (10.20)], so that $I(\theta)$ must have subsidiary maxima at these values of β (viz, $\pm 1.4303\pi$, $\pm 2.4590\pi$, $\pm 3.4707\pi, \ldots$).

There is a particularly easy way to appreciate what's happening here with the aid of Fig. 10.13. We envision every point in the aperture emitting rays in all directions in the *xz*-plane. The light that continues to propagate directly forward in Fig. 10.13*a* is the undiffracted beam, all the rays arrive on the viewing screen in-phase, and a central bright spot will be formed by them. If the screen is not actually at infinity, the rays that converge to it are not quite parallel but with it at infinity, or better still, with a lens in place, the rays are as drawn. Figure 10.13*b* shows the specific bundle of rays coming off at an angle θ_1 where the path length difference between the rays from the very top and bottom, $b \sin \theta_1$, is made equal to one wavelength. A ray from the middle of the slit will then lag $\frac{1}{2}\lambda$ behind a ray from the top and exactly cancel it. Similarly, a ray from just below center will cancel a ray from just below the top, and so on; all across the aperture ray-pairs will cancel, yielding a minimum. The irradiance has dropped from its high central maximum to the first zero on either side at $\sin \theta_1 = \pm\lambda/b$.

As the angle increases further, some small fraction of the rays will again interfere constructively, and the irradiance will rise to form a subsidiary peak. A further increase in the angle produces another minimum, as shown in Fig. 10.13*c*, when $b \sin \theta_2 = 2\lambda$. Now imagine the aperture divided into quarters. Ray by ray, the top quarter will cancel the one beneath it, and the next, the third, will cancel the last quarter. Ray-pairs at the same locations in adjacent segments are $\lambda/2$ out-of-phase and destructively interfere. In general then, zeros of irradiance will occur when

$$b \sin \theta_m = m\lambda$$

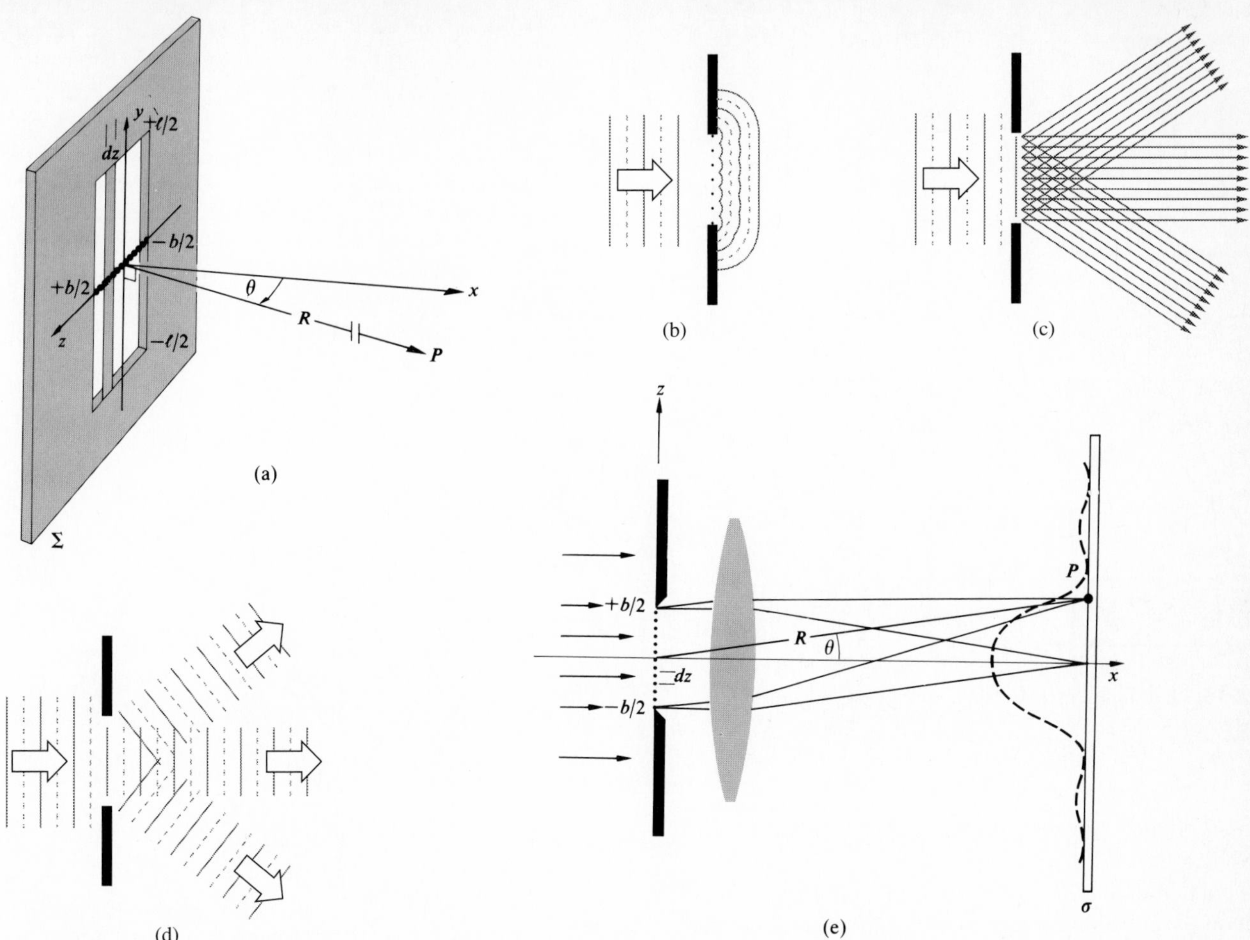

FIGURE 10.11 (*a*) Point P on σ is essentially infinitely far from Σ. (*b*) Huygens wavelets emitted across the aperture. (*c*) The equivalent representation in terms of rays. Each point emits rays in all directions. The parallel rays in various directions are seen. (*d*) These ray bundles correspond to plane waves, which can be thought of as the three-dimensional Fourier components. (*e*) A single slit illuminated by monochromatic plane waves.

where $m = \pm 1, \pm 2, \pm 3, \ldots$, which is equivalent to Eq. (10.20), since $\beta = m\pi = (kb/2) \sin \theta_m$.

We should inject a note of caution at this point: one of the frailties of the Huygens–Fresnel Principle is that it does not take proper regard of the variations in amplitude, with angle, over the surface of each secondary wavelet. We will come back to this when we consider the *obliquity factor* in Fresnel diffraction, where the effect is significant. In Fraunhofer diffraction the distance from the aperture to the plane of observation is so large that we need not be concerned about it, provided that θ remains small.

Figure 10.14 is a plot of the flux density, as expressed by Eq. (10.17). Envision some point on the curve, for example, the third subsidiary maximum at $\beta = 3.4707\pi$; since $\beta = (\pi b/\lambda) \sin \theta$, an increase in the slit width b requires a decrease in θ, if β is to be constant. Under these conditions the pattern shrinks in toward the principal maximum, as it would if λ were decreased. If the source emits white light, the higher-order maxima show a succession of colors trailing off into red with increasing θ. Each different colored light component has its minima and subsidiary maxima at angular positions characteristic of that wavelength (Problem 10.6). Indeed, only in

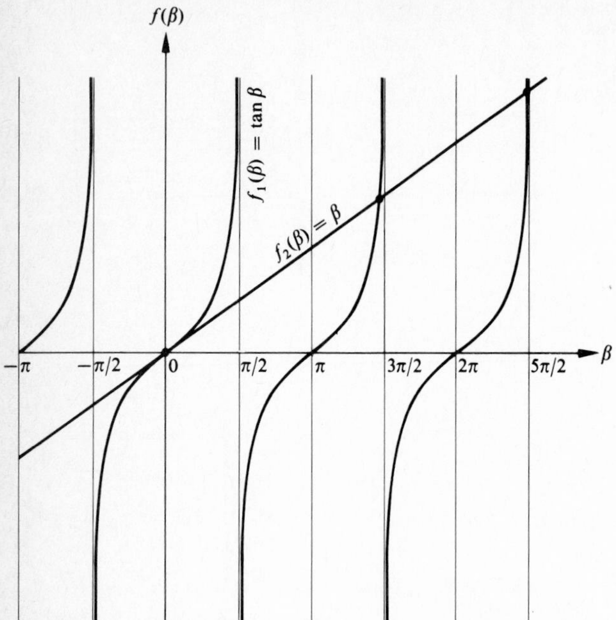

FIGURE 10.12 The points of intersection of the two curves are the solutions of Eq. (10.21).

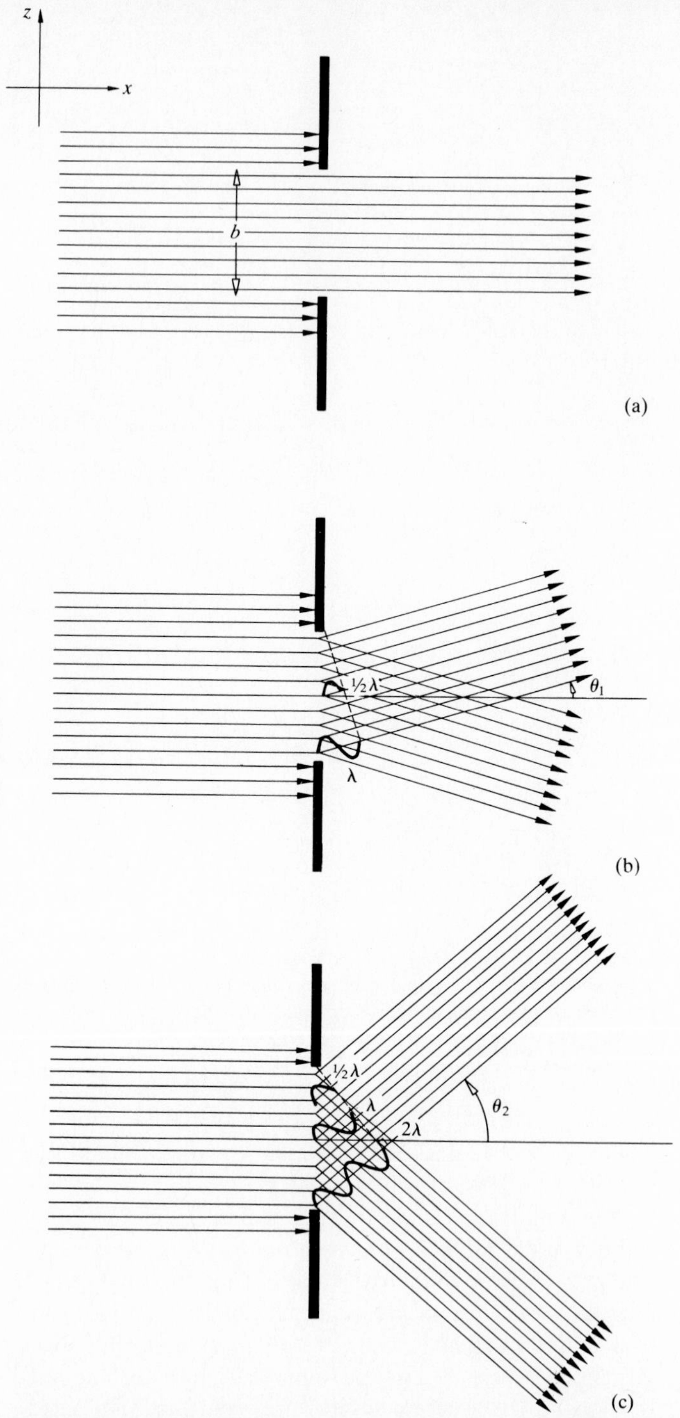

FIGURE 10.13 The diffraction of light in various directions. Here the aperture is a single slit, as in Fig. 10.11.

the region about $\theta = 0$ will all the constituent colors overlap to yield white light.

The point source S in Fig. 10.10 would be imaged at the position of the center of the pattern, if the diffracting screen Σ were removed. Under this sort of illumination, the pattern produced with the slit in place is a series of dashes in the yz-plane of the screen σ, much like a spread-out image of S (Fig. 10.10*b*). An incoherent line source (in place of S) positioned parallel to the slit, in the focal plane of the collimator L_1, will broaden the pattern out into a series of bands. Any point on the line source generates an independent diffraction pattern, which is displaced, with respect to the others, along the y-direction. With no diffracting screen present, the image of the line source would be a line parallel to the original slit. With the screen in place the line is spread out, as was the point image of S (Fig. 10.15). Keep in mind that it's the small dimension of the slit that does the spreading out.

The single-slit pattern is easily observed without the use of special equipment. Any number of sources will do (e.g., a distant street light at night, a small incandescent lamp, sunlight streaming through a narrow space in a window shade); almost anything that resembles a point or line source will serve. Prob-

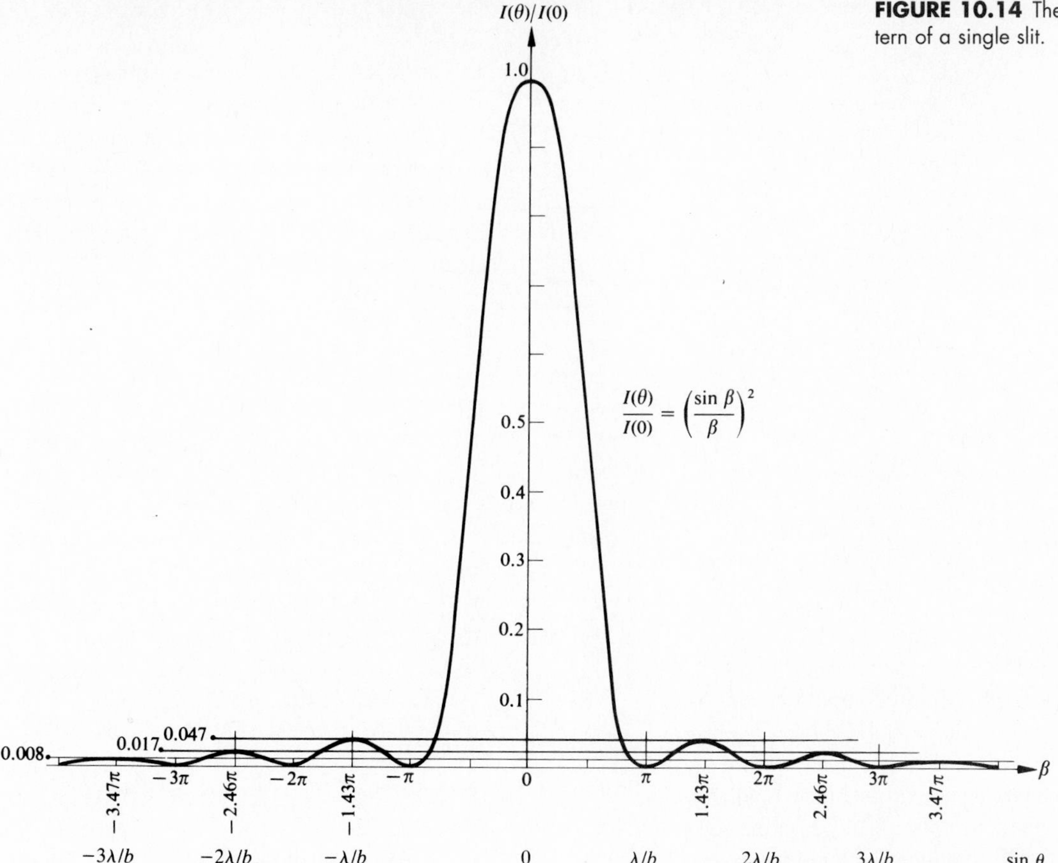

FIGURE 10.14 The Fraunhofer diffraction pattern of a single slit.

FIGURE 10.15 The single-slit pattern with a line source. See first photograph of Fig. 10.18.

ably the best source for our purposes is an ordinary clear, *straight-filament* display bulb (the kind in which the filament is vertical and about 3 inches long). You can use your imagination to generate all sorts of single-slit arrangements (e.g., a comb or fork rotated to decrease the projected space between the tines, or a scratch across a layer of india ink on a microscope slide). An inexpensive vernier caliper makes a remarkably good variable slit. Hold the caliper close to your eye with the slit, a few thousandths of an inch wide, parallel to the filament of the lamp. Focus your eye beyond the slit at infinity, so that its lens serves as L_2.

10.2.2 The Double Slit

It might at first seem from Fig. 10.11 that the location of the principal maximum is always to be in line with the center of the diffracting aperture; this, however, is not generally true.

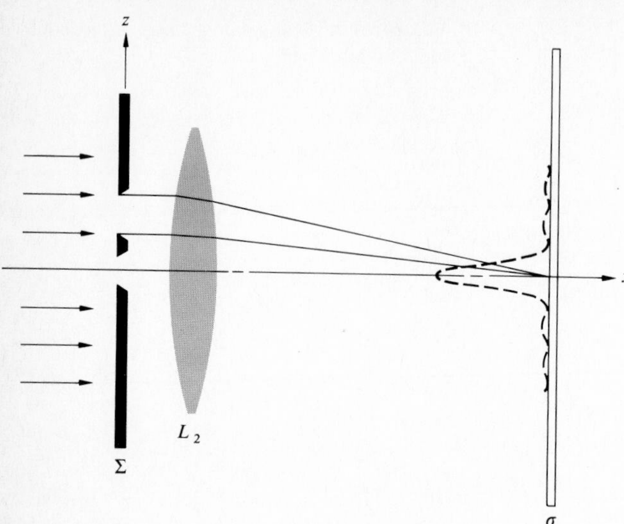

FIGURE 10.16 The double-slit setup.

The diffraction pattern is actually centered about the axis of the lens and has exactly the same shape and location, regardless of the slit's position, as long as its orientation is unchanged and the approximations are valid (Fig. 10.16). All waves traveling parallel to the lens axis converge on the second focal point of L_2; this then is the image of S and the center of the diffraction pattern. Suppose now that we have two long slits of width b and center-to-center separation a (Fig. 10.17). Each aperture, by itself, would generate the same single-slit diffraction pattern on the viewing screen σ. At any point on σ, the contributions from the two slits overlap, and even though each must be essentially equal in amplitude, they may well differ significantly in-phase. Since the same primary wave excites the secondary sources at each slit, the resulting wavelets will be coherent, and interference must occur. If the primary plane wave is incident on Σ at some angle θ_i (see Problem 10.3), there will be a constant relative phase difference between the secondary sources. At normal incidence, the wavelets are all emitted in-phase. The interference fringe at a particular point of observation is determined by the differences in the optical path lengths traversed by the overlapping wavelets from the two slits. As we will see, the flux-density distribution (Fig. 10.18) is the result of a rapidly varying double-slit interference system modulated by a single-slit diffraction pattern.

To obtain an expression for the optical disturbance at a point on σ, we need only slightly reformulate the single-slit analysis. Each of the two apertures is divided into differential strips (dz by ℓ), which in turn behave like an infinite number of point sources aligned along the z-axis. The total contribution to the electric field, in the Fraunhofer approximation [Eq. (10.12)], is then

$$E = C\int_{-b/2}^{b/2} F(z)\,dz + C\int_{a-b/2}^{a+b/2} F(z)\,dz \qquad (10.22)$$

where $F(z) = \sin[\omega t - k(R - z\sin\theta)]$. The constant-amplitude factor C is the secondary source strength per unit length along the z-axis (assumed to be independent of z over each aperture) divided by R, which is measured from the origin to P and is taken as constant. We will be concerned only with relative flux densities on σ, so that the actual value of C is of little interest to us now. Integration of Eq. (10.22) yields

$$E = bC\left(\frac{\sin\beta}{\beta}\right)[\sin(\omega t - kR) + \sin(\omega t - kR + 2\alpha)] \qquad (10.23)$$

with $\alpha \equiv (ka/2)\sin\theta$ and, as before, $\beta \equiv (kb/2)\sin\theta$. This is just the sum of the two fields at P, one from each slit, as given by Eq. (10.15). The distance from the first slit to P is R, giving a phase contribution of $-kR$. The distance from the second slit to P is $(R - a\sin\theta)$ or $(R - 2\alpha/k)$, yielding a phase term equal to $(-kR + 2\alpha)$, as in the second sine function. The quantity 2β is the phase difference $(k\Lambda)$ between two nearly parallel rays, arriving at a point P on σ, from the edges of one of the slits. The quantity 2α is the phase difference between two waves arriving at P, one having originated at any point in the first slit, the other coming from the corresponding point in the second slit. Simplifying Eq. (10.23) a bit further, it becomes

$$E = 2bC\left(\frac{\sin\beta}{\beta}\right)\cos\alpha\,\sin(\omega t - kR + \alpha)$$

which when squared and averaged over a relatively long inter-

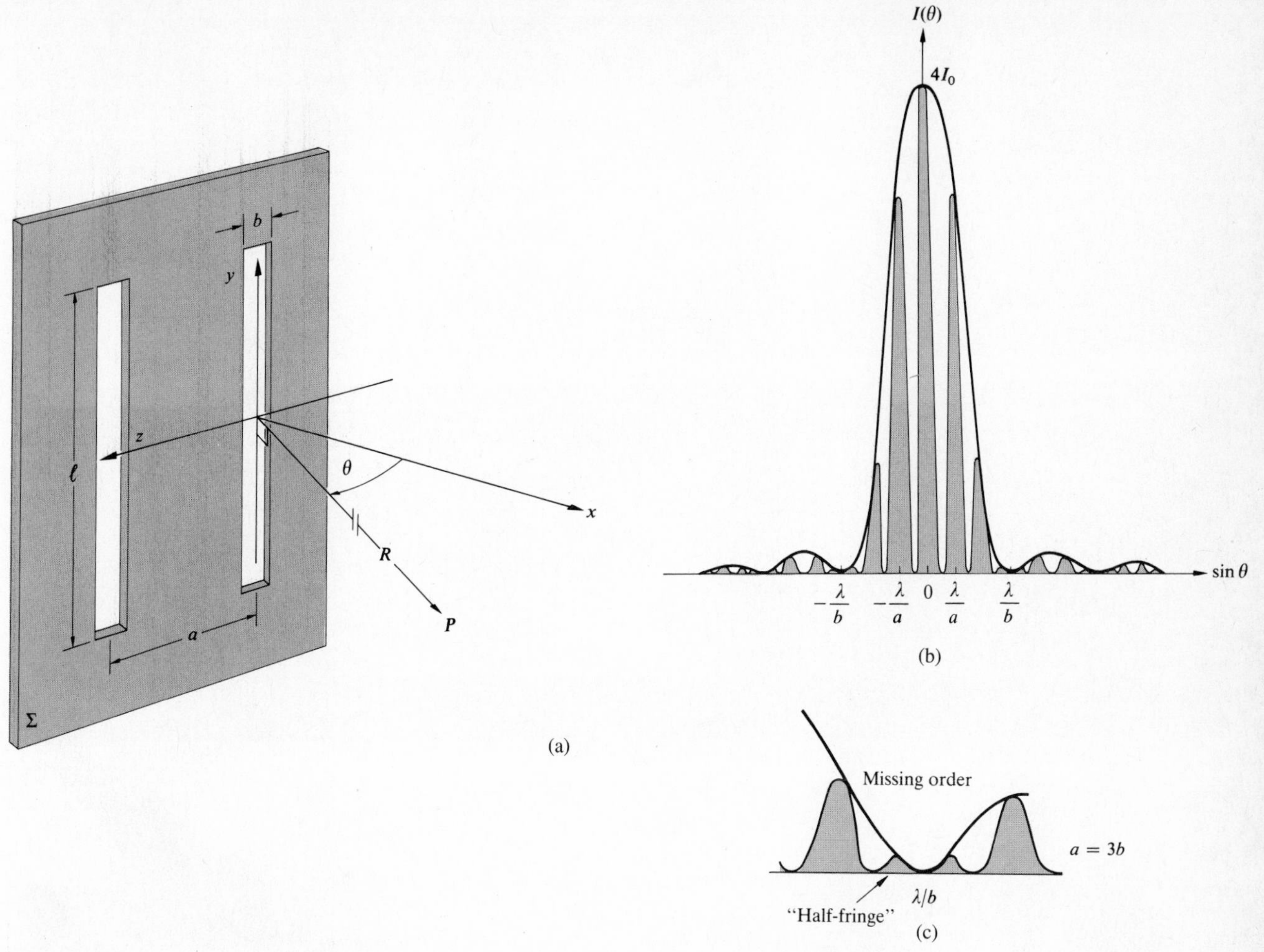

FIGURE 10.17 (*a*) Double-slit geometry. Point *P* on σ is essentially infinitely far away. (*b*) A double-slit pattern ($a = 3b$).

val in time is the irradiance

$$I(\theta) = 4I_0\left(\frac{\sin^2\beta}{\beta^2}\right)\cos^2\alpha \qquad (10.24)$$

In the $\theta = 0$ direction (i.e., when $\beta = \alpha = 0$), I_0 is the flux-density contribution from either slit, and $I(0) = 4I_0$ is the total flux density. The factor of 4 comes from the fact that the amplitude of the electric field is twice what it would be at that point with one slit covered.

If in Eq. (10.24)b becomes vanishingly small ($kb << 1$), then $(\sin\beta)/\beta \approx 1$, and the equation reduces to the flux-density expression for a pair of long line sources, that is, Young's Experiment, Eq. (9.17). If on the other hand $a = 0$, the two slits coalesce into one, $\alpha = 0$, and Eq. (10.24) becomes $I(0) = 4I_0(\sin^2\beta)/\beta^2$. This is the equivalent of Eq. (10.17) for single-slit diffraction with the source strength doubled. We might then envision the total expression as being generated by a $\cos^2\alpha$ interference term modulated by a $(\sin^2\beta)/\beta^2$ diffraction term. If the slits are finite in width but very narrow, the dif-

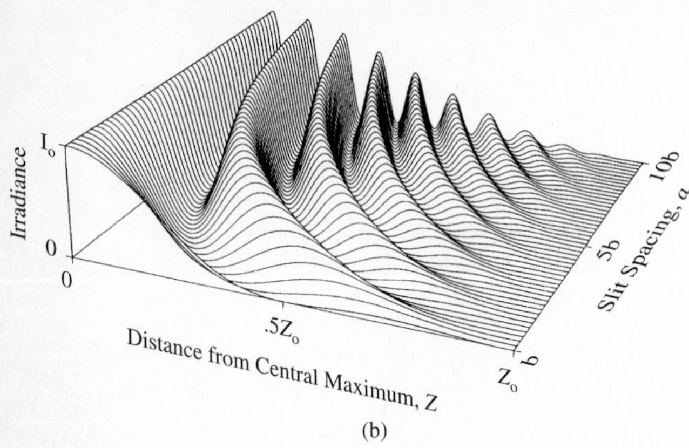

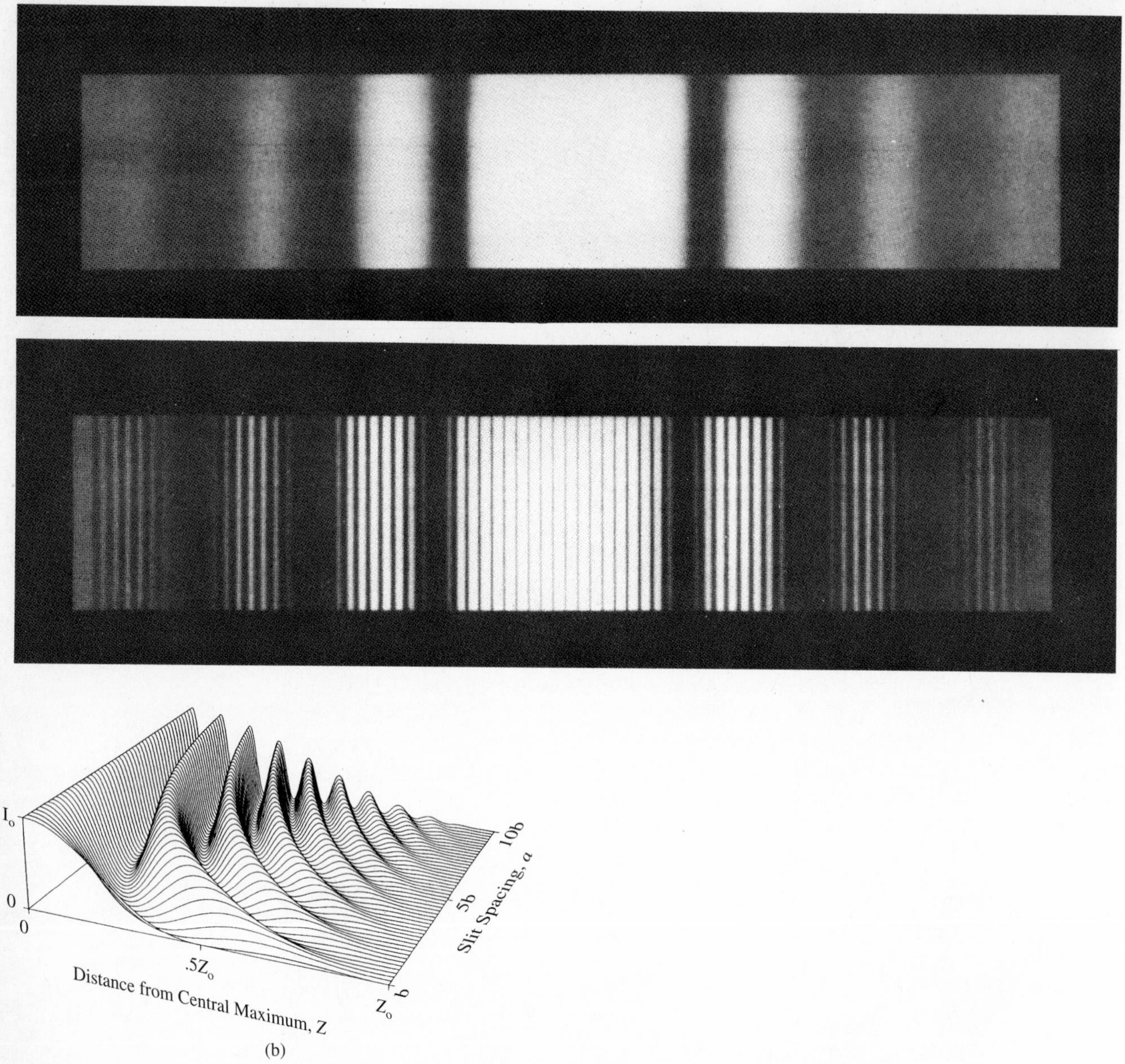

(b)

FIGURE 10.18 Single- and double-slit Fraunhofer patterns. (*a*) Photographs taken with monochromatic light. The faint cross-hatching arises entirely in the printing process. (Photos courtesy M. Cagnet, M. Francon, and J. C. Thrierr: *Atlas optischer Erscheinungen*, Berlin–Heidelberg–New York: Springer, 1962.) (*b*) When the slit spacing equals b, the two slits coalesce into one (of width $2b$) and the single-slit pattern appears—that's the first curve closest to you. The farthest curve corresponds to the two slits separated by $a = 10b$. Notice that the two-slit patterns all have their first diffraction minimum at a distance from the central maximum of Z_0. Note how the curves gradually match Fig. 10.17*b* as b gets smaller in comparison to a. (Reprinted from "Graphical Representations of Fraunhofer Interference and Diffraction" *Am. J. Phys.*, **62,** 6, (1994), with permission of A. B. Bartlett, University of Colorado and B. Mechtly, Northeast Missouri State University and the American Association of Physics Teachers.)

fraction pattern from either slit will be uniform over a broad central region, and bands resembling the idealized Young's fringes will appear within that region. At angular positions (θ-values) where

$$\beta = \pm\pi, \pm 2\pi, \pm 3\pi, \ldots$$

diffraction effects are such that no light reaches σ, and clearly none is available for interference. At points on σ where

$$\alpha = \pm\pi/2, \pm 3\pi/2, \pm 5\pi/2, \ldots$$

the various contributions to the electric field will be completely out-of-phase and will cancel, regardless of the actual amount of light made available from the diffraction process.

The irradiance distribution for a double-slit Fraunhofer pattern is illustrated in Fig. 10.17*b*. Notice that it is a combination of Figs. 9.9 and 10.14. The curve is for the particular case in which $a = 3b$ (i.e., $\alpha = 3\beta$). You can get a rough idea of what the pattern will look like, since if $a = mb$, where m is any number, there will be $2m$ bright fringes (counting "fractional fringes" as well)* within the central diffraction peak (Problem 10.10). An interference maximum and a diffraction minimum (zero) may correspond to the same θ-value. In that case no light is available at that precise position to partake in the interference process, and the suppressed peak is said to be a *missing order*.

The double-slit pattern is also rather easily observed, and the seeing is well worth the effort. A straight-filament, tubular bulb is again the best line source. For slits, coat a microscope slide with India ink; if you happen to have some, a colloidal suspension of graphite in alcohol works even better (it's more opaque). Scratch a pair of slits across the dry ink with a razor blade and stand about 10 feet from the source. Hold the slits parallel to the filament and close to your eye, which, when focused at infinity, will serve as the needed lens. Interpose red or blue cellophane and observe the change in the width of the fringes. Find out what happens when you cover one and then both of the slits with a microscope slide. Move the slits slowly in the z-direction; then holding them stationary, move your eye in the z-direction. Verify that the position of the center of the pattern is indeed determined by the lens and not the aperture.

*Notice that m need not be an integer. Moreover, if m is an integer, there will be "half-fringes," as shown in Fig. 10.17*c*.

10.2.3 Diffraction by Many Slits

The procedure for obtaining the irradiance function for a monochromatic wave diffracted by many slits is essentially the same as that used when considering two slits. Here again, the limits of integration must be appropriately altered. Consider the case of N long, parallel, narrow slits, each of width b and center-to-center separation a, as illustrated in Fig. 10.19. With the origin of the coordinate system once more at the center of the first slit, the total optical disturbance at a point on the screen σ is given by

$$E = C\int_{-b/2}^{b/2} F(z)\,dz + C\int_{a-b/2}^{a+b/2} F(z)\,dz + C\int_{2a-b/2}^{2a+b/2} F(z)\,dz + \cdots + C\int_{(N-1)a-b/2}^{(N-1)a+b/2} F(z)\,dz \tag{10.25}$$

where as before, $F(z) = \sin[\omega t - k(R - z\sin\theta)]$. This applies to the Fraunhofer condition, so that the aperture configuration must be such that all the slits are close to the origin, and the approximation [Eq. (10.11)]

$$r = R - z\sin\theta \tag{10.26}$$

applies over the entire array. The contribution from the jth slit (where the first one is numbered zero), obtained by evaluating only that one integral in Eq. (10.25), is then

$$E_j = \frac{C}{k\sin\theta}[\sin(\omega t - kR)\sin(kz\sin\theta) - \cos(\omega t - kR)\cos(kz\sin\theta)]_{ja-b/2}^{ja+b/2}$$

provided that we require $\theta_j \approx \theta$. After some manipulation this becomes

$$E_j = bC\left(\frac{\sin\beta}{\beta}\right)\sin(\omega t - kR + 2\alpha j) \tag{10.27}$$

recalling that $\beta = (kb/2)\sin\theta$ and $\alpha = (ka/2)\sin\theta$. Notice that this is equivalent to the expression for a line source [Eq. (10.15)] or, of course, a single slit, where in accord with Eq. (10.26) and Fig. 10.19, $R_j = R - ja\sin\theta$, so that $-kR + 2\alpha j = -kR_j$. The total optical disturbance, as given by Eq.

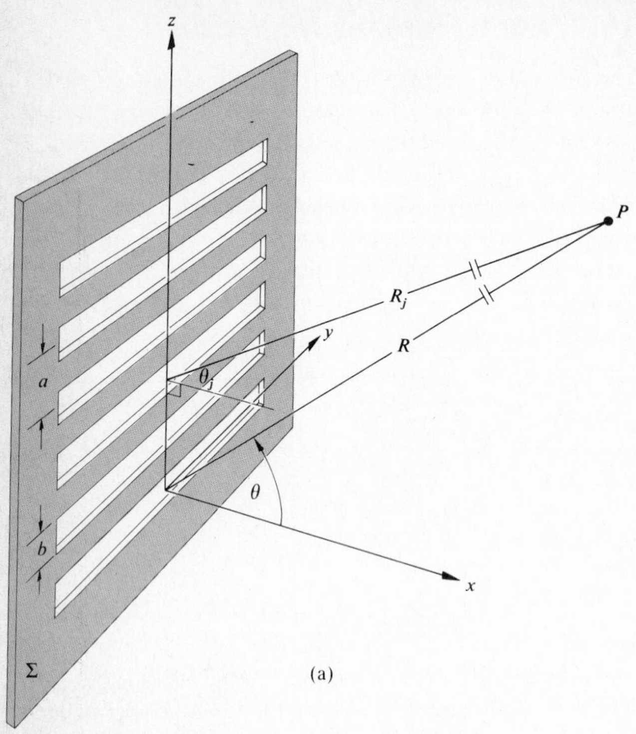

FIGURE 10.19*a* Multi-slit geometry. Again point P is on σ essentially infinitely far from Σ.

FIGURE 10.19*b, c, d*

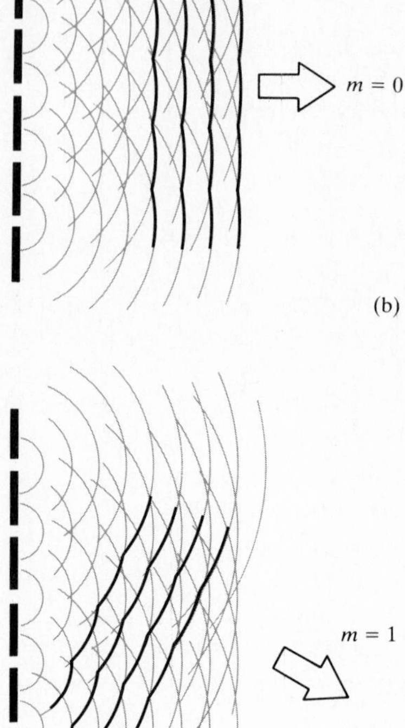

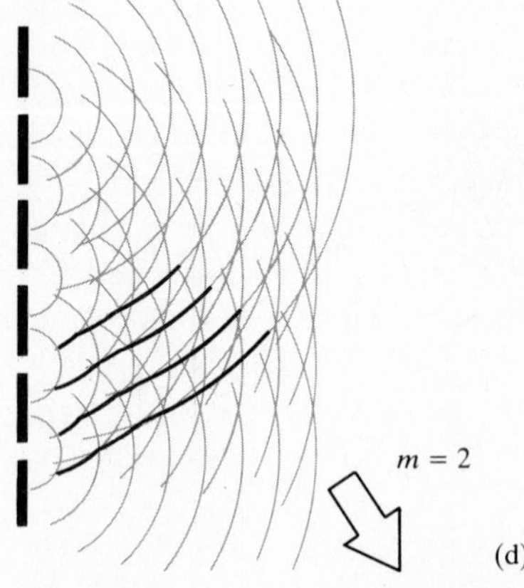

(10.25), is simply the sum of the contributions from each of the slits; that is,

$$E = \sum_{j=0}^{N-1} E_j$$

or

$$E = \sum_{j=0}^{N-1} bC\left(\frac{\sin\beta}{\beta}\right)\sin(\omega t - kR + 2\alpha j) \quad (10.28)$$

This in turn can be written as the imaginary part of a complex exponential:

$$E = \text{Im}\left[bC\left(\frac{\sin\beta}{\beta}\right)e^{i(\omega t - kR)}\sum_{j=0}^{N-1}(e^{i2\alpha})^j\right] \quad (10.29)$$

But we have already evaluated this same geometric series in the process of simplifying Eq. (10.2). Equation (10.29) therefore reduces to the form

$$E = bC\left(\frac{\sin\beta}{\beta}\right)\left(\frac{\sin N\alpha}{\sin\alpha}\right)\sin[\omega t - kR + (N-1)\alpha] \quad (10.30)$$

The distance from the center of the array to the point P is equal to $[R - (N-1)(a/2)\sin\theta]$, and therefore the phase of E at P corresponds to that of a wave emitted from the midpoint of the source. The flux-density distribution function is

$$I(\theta) = I_0\left(\frac{\sin\beta}{\beta}\right)^2\left(\frac{\sin N\alpha}{\sin\alpha}\right)^2 \tag{10.31}$$

Note that I_0 is the flux density in the $\theta = 0$ direction emitted by any one of the slits and that $I(0) = N^2 I_0$. In other words, the waves arriving at P in the forward direction are all in-phase, and their fields add constructively. Each slit by itself would generate precisely the same flux-density distribution. Superimposed, the various contributions yield a multiple-wave interference system modulated by the single-slit diffraction envelope. If the width of each aperture were shrunk to zero, Eq. (10.31) would become the flux-density expression [Eq. (10.6)] for a linear coherent array of oscillators. As in that earlier treatment [Eq. (10.17)], **principal maxima** occur when $(\sin N\alpha/\sin\alpha) = N$, that is, when

$$\alpha = 0, \pm\pi, \pm 2\pi, \ldots$$

or equivalently, since $\alpha = (ka/2)\sin\theta$,

$$a \sin\theta_m = m\lambda \tag{10.32}$$

with $m = 0, \pm 1, \pm 2, \ldots$. This is quite general and gives rise to the same θ-locations for these maxima, regardless of the value of $N \geq 2$. Minima, of zero flux density, exist whenever $(\sin N\alpha/\sin\alpha)^2 = 0$ or when

$$\alpha = \pm\frac{\pi}{N}, \pm\frac{2\pi}{N}, \pm\frac{3\pi}{N}, \ldots, \pm\frac{(N-1)\pi}{N}, \pm\frac{(N+1)\pi}{N}, \ldots \tag{10.33}$$

Between consecutive principal maxima (i.e., over the range in α of π) there will therefore be $N - 1$ minima. And, of course, between each pair of minima there will have to be a **subsidiary maximum**. The term $(\sin N\alpha/\sin\alpha)^2$, which we can think of as embodying the interference effects, has a rapidly varying numerator and a slowly varying denominator. The subsidiary maxima are therefore located approximately at points where $\sin N\alpha$ has its greatest value, namely,

$$\alpha = \pm\frac{3\pi}{2N}, \pm\frac{5\pi}{2N}, \ldots \tag{10.34}$$

The $N - 2$ *subsidiary maxima* between consecutive principal maxima are clearly visible in Fig. 10.20. We can get some idea of the flux density at these peaks by rewriting Eq. (10.31) as

$$I(\theta) = \frac{I(0)}{N^2}\left(\frac{\sin\beta}{\beta}\right)^2\left(\frac{\sin N\alpha}{\sin\alpha}\right)^2 \tag{10.35}$$

where at the points of interest $|\sin N\alpha| = 1$. For large N, α is small and $\sin^2\alpha \approx \alpha^2$. At the first subsidiary peak $\alpha = 3\pi/2N$, in which case

$$I \approx I(0)\left(\frac{\sin\beta}{\beta}\right)^2\left(\frac{2}{3\pi}\right)^2 \tag{10.36}$$

and the flux density has dropped to about $\frac{1}{22}$ of that of the adjacent principal maximum (see Problem 10.12). Since $(\sin\beta)/\beta$ for small β varies slowly, it will not differ from 1 appreciably, close to the zeroth-order principal maximum, so that $I/I(0) \approx \frac{1}{22}$. This flux-density ratio for the next secondary peak is down to $\frac{1}{62}$, and it continues to decrease as α approaches a value halfway between the principal maxima. At that symmetry point, $\alpha \approx \pi/2$, $\sin\alpha \approx 1$, and the flux-density ratio has its lowest value, approximately $1/N^2$. Thereafter $\alpha > \pi/2$, and the flux densities of the subsidiary maxima begin to increase.

Try duplicating Fig. 10.20 using a tubular bulb and homemade slits. You'll probably have difficulty seeing the subsidiary maxima clearly, with the effect that the only perceptible difference between the double- and multiple-slit patterns may be an apparent broadening in the dark regions between principal maxima. As in Fig. 10.20, the dark regions will become wider than the bright bands as N increases and the secondary peaks fade out. If we consider each principal maximum to be bounded in width by two adjacent zeros, then each will extend over a length in θ, ($\sin\theta \approx \theta$) of approximately $2\lambda/Na$. As N increases, the principal maxima maintain their relative spacing (λ/a) while becoming increasingly narrow. Figure 10.21 shows the case of six slits, with $a = 4b$.

The multiple-slit interference term in Eq. 10.35 has the form $(\sin^2 N\alpha)/N^2\sin^2\alpha$; thus for large N, $(N^2\sin^2\alpha)^{-1}$ may be envisioned as the curve beneath which $\sin^2 N\alpha$ rapidly varies. Notice that for small α this interference term looks like $\text{sinc}^2 N\alpha$.

10.2.4 The Rectangular Aperture

Consider the configuration depicted in Fig. 10.22. A monochromatic plane wave propagating in the x-direction is incident on the opaque diffracting screen Σ. We wish to find the consequent (far-field) flux-density distribution in space or equivalently at some arbitrary distant point P. According to

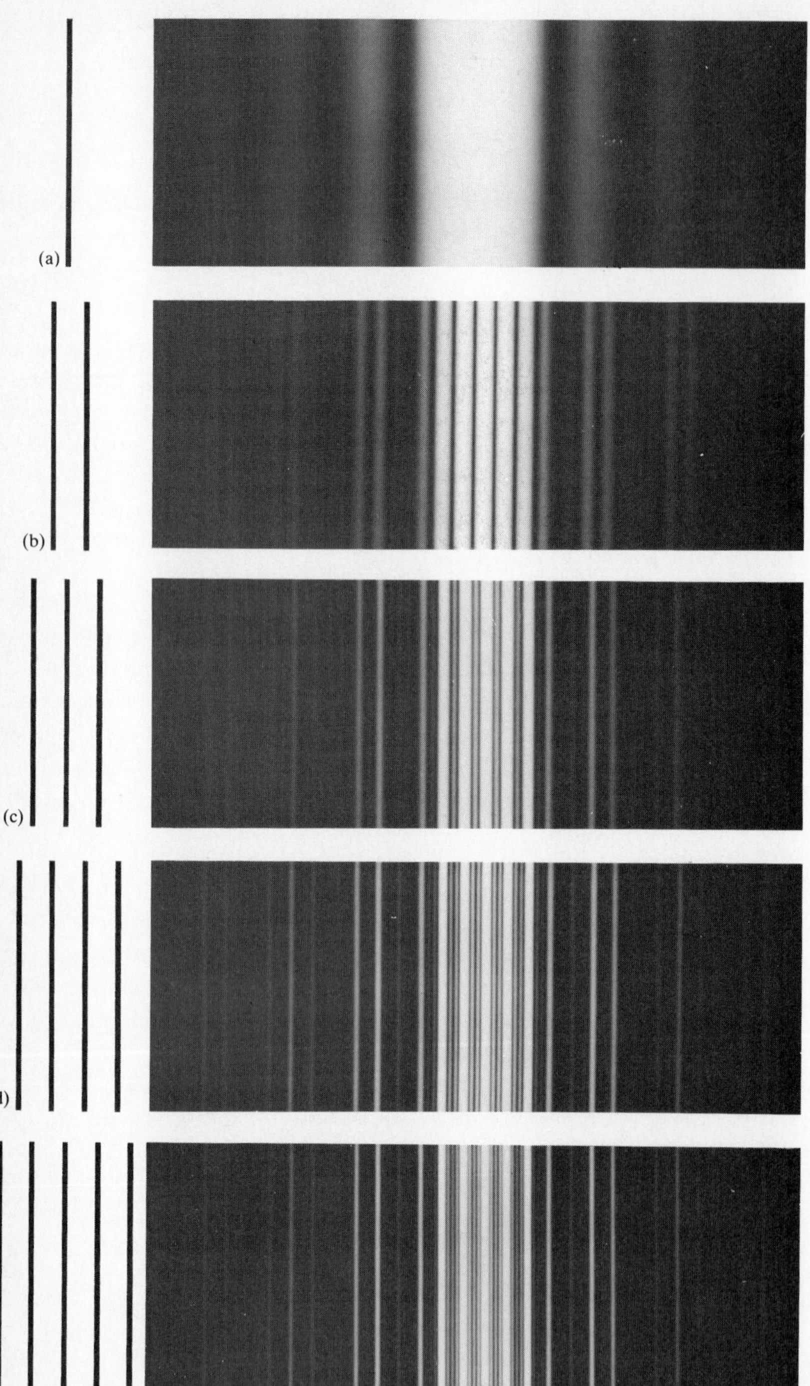

FIGURE 10.20 Diffraction patterns for slit systems shown at left. (Francis Weston Sears, Optics, ©1949, Addison-Wesley, Inc. Reading, MA. Reprinted with permission of Addison Wesley Longman, Inc.)

the Huygens–Fresnel Principle, a differential area dS, within the aperture, may be envisioned as being covered with coherent secondary point sources. But dS is much smaller in extent than is λ, so that all the contributions at P remain in-phase and interfere constructively. This is true regardless of θ; that is, dS emits a spherical wave (Problem 10.13). If ε_A is the source

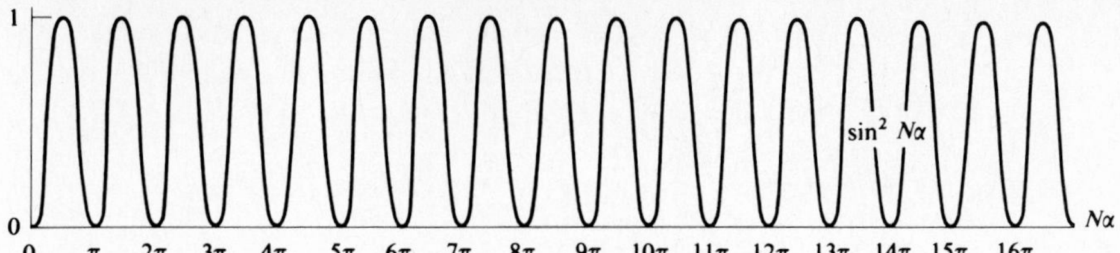

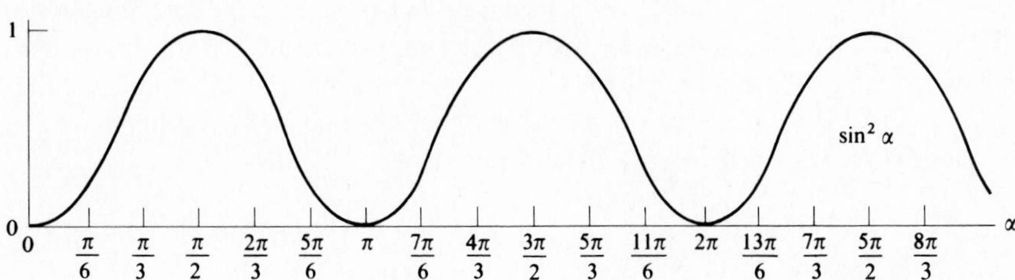

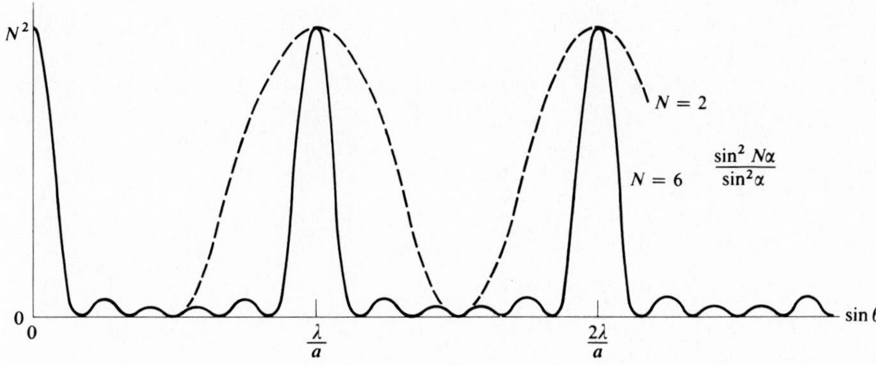

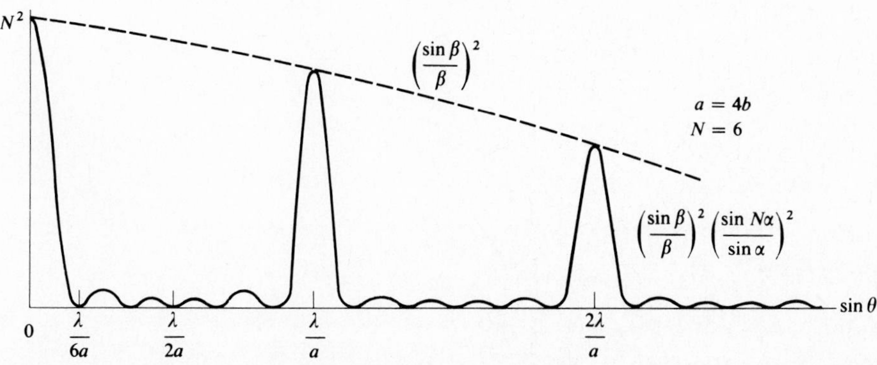

FIGURE 10.21 Multiple-slit pattern ($a = 4b$, $N = 6$).

strength per unit area, *assumed to be constant over the entire aperture*, then the optical disturbance at P due to dS is either the real or imaginary part of

$$dE = \left(\frac{\mathcal{E}_A}{r}\right) e^{i(\omega t - kr)}\, dS \tag{10.37}$$

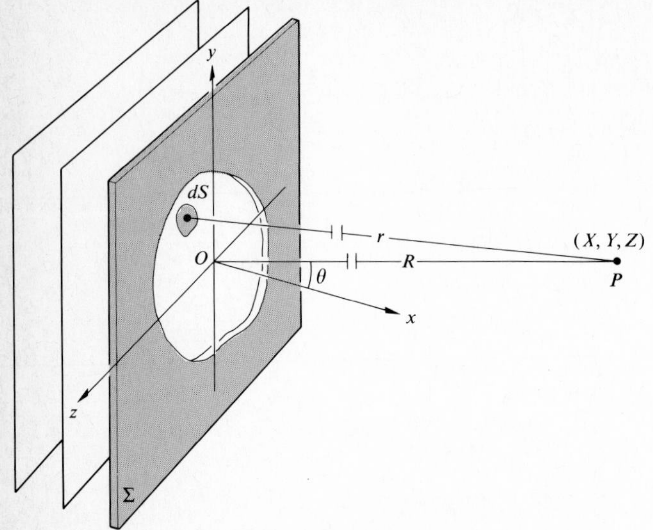

FIGURE 10.22 Fraunhofer diffraction from an arbitrary aperture, where r and R are very large compared to the size of the hole.

The choice is yours and depends only on whether you like sine or cosine waves, there being no difference except for a phase shift. The distance from dS to P is

$$r = [X^2 + (Y - y)^2 + (Z - z)^2]^{1/2} \tag{10.38}$$

and as we have seen, the Fraunhofer condition occurs when this distance approaches infinity. As before, it will suffice to replace r by the distance $\overline{OP}$, that is, R, in the amplitude term, as long as the aperture is relatively small. But the approximation for r in the phase needs to be treated a bit more carefully; $k = 2\pi/\lambda$ is a large number. To that end we expand out Eq. (10.38) and, by making use of

$$R = [X^2 + Y^2 + Z^2]^{1/2} \tag{10.39}$$

obtain

$$r = R[1 + (y^2 + z^2)/R^2 - 2(Yy + Zz)/R^2]^{1/2} \tag{10.40}$$

In the far-field case R is very large in comparison to the

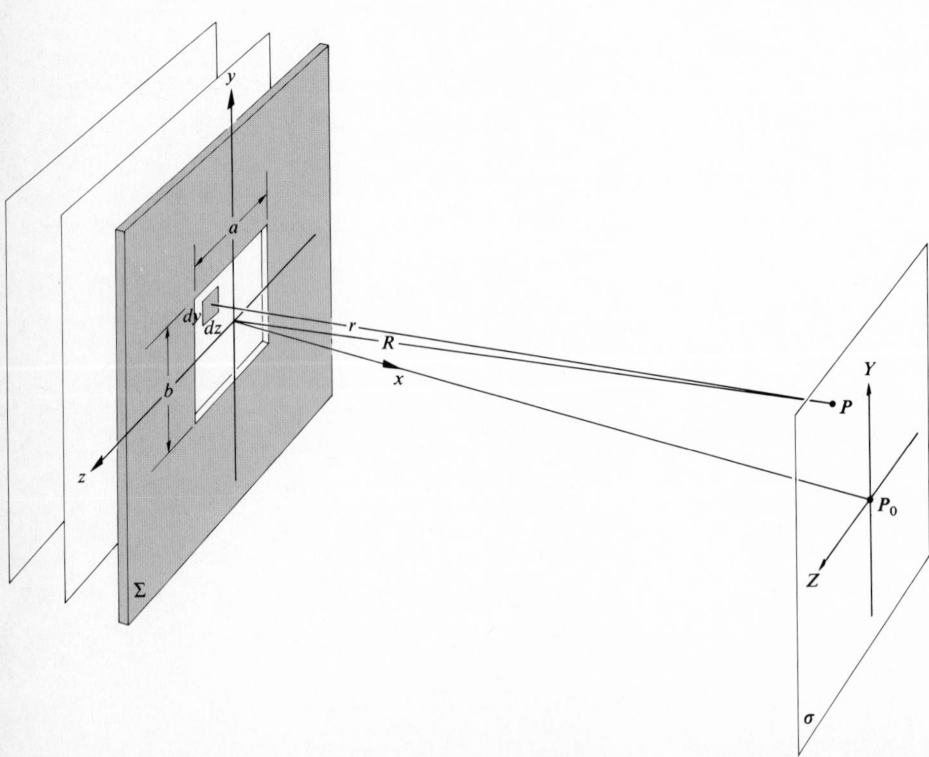

FIGURE 10.23 A rectangular aperture.

dimensions of the aperture, and the $(y^2 + z^2)/R^2$ term is certainly negligible. Since P is very far from Σ, θ can still be kept small, even though Y and Z are fairly large, and this mitigates any concern about the directionality of the emitters (the obliquity factor). Now

$$r = R[1 - 2(Yy + Zz)/R^2]^{1/2}$$

and dropping all but the first two terms in the binomial expansion, we have

$$r = R[1 - (Yy + Zz)/R^2]$$

The total disturbance arriving at P is

$$\tilde{E} = \frac{\mathcal{E}_A e^{i(\omega t - kR)}}{R} \iint_{\text{Aperture}} e^{ik(Yy+Zz)/R}\, dS \qquad (10.41)$$

Consider the specific configuration shown in Fig. 10.23. Equation (10.41) can now be written as

$$\tilde{E} = \frac{\mathcal{E}_A e^{i(\omega t - kR)}}{R} \int_{-b/2}^{b/2} e^{ikYy/R}\, dy \int_{-a/2}^{a/2} e^{ikZz/R}\, dz$$

where $dS = dy\, dz$. With $\beta' \equiv kbY/2R$ and $\alpha' \equiv kaZ/2R$, we have

$$\int_{-b/2}^{+b/2} e^{ikYy/R}\, dy = b\left(\frac{e^{i\beta'} - e^{-i\beta'}}{2i\beta'}\right) = b\left(\frac{\sin \beta'}{\beta'}\right)$$

and similarly

$$\int_{-a/2}^{+a/2} e^{ikZz/R}\, dz = a\left(\frac{e^{i\alpha'} - e^{-i\alpha'}}{2i\alpha'}\right) = a\left(\frac{\sin \alpha'}{\alpha'}\right)$$

so that

$$\tilde{E} = \frac{A\mathcal{E}_A e^{i(\omega t - kR)}}{R}\left(\frac{\sin \alpha'}{\alpha'}\right)\left(\frac{\sin \beta'}{\beta'}\right) \qquad (10.42)$$

where A is the area of the aperture. Since $I = \langle(\text{Re}\, \tilde{E})^2\rangle_T$,

$$I(Y, Z) = I(0)\left(\frac{\sin \alpha'}{\alpha'}\right)^2\left(\frac{\sin \beta'}{\beta'}\right)^2 \qquad (10.43)$$

where $I(0)$ is the irradiance at P_0; that is, at $Y = 0$, $Z = 0$ (see Fig. 10.24). At values of Y and Z such that $\alpha' = 0$ or $\beta' = 0$, $I(Y, Z)$ assumes the familiar shape of Fig. 10.14. When β' or

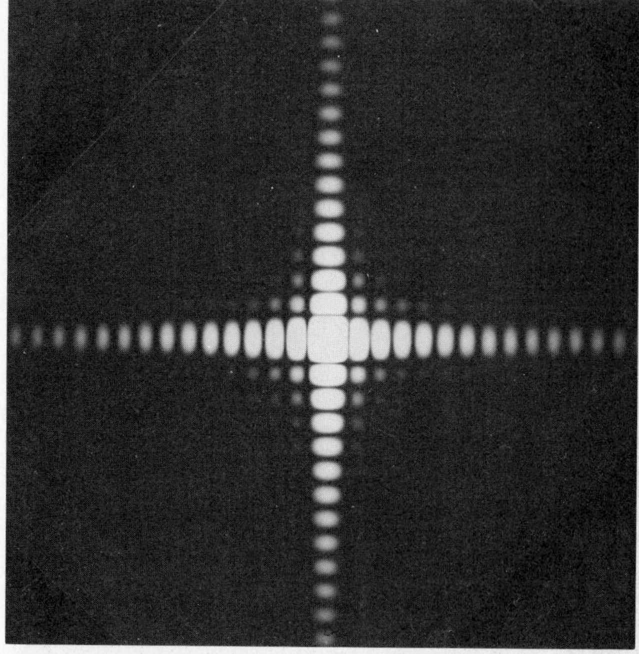

(a)

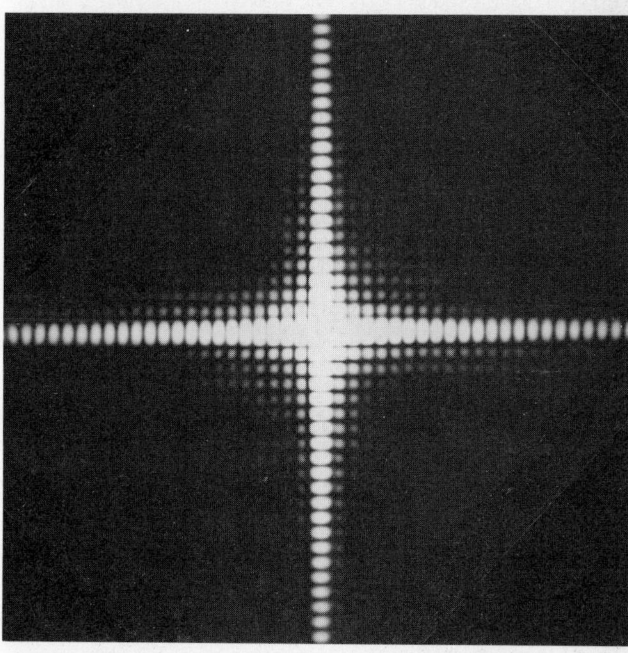

(b)

FIGURE 10.24 (*a*) Fraunhofer pattern of a square aperture. (*b*) The same pattern further exposed to bring out some of the faint terms. (E. H.)

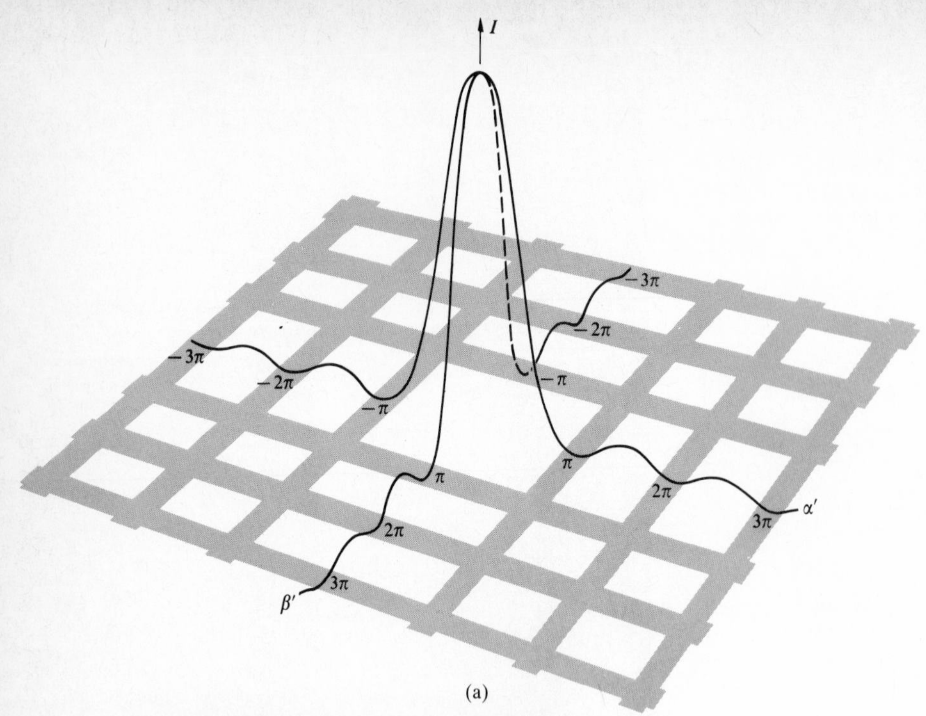

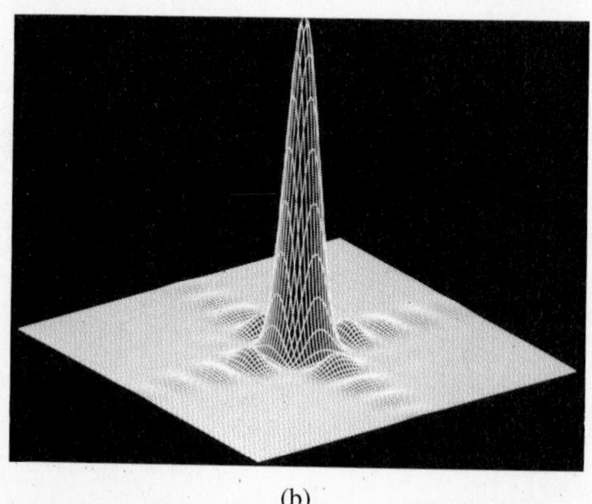

(b)

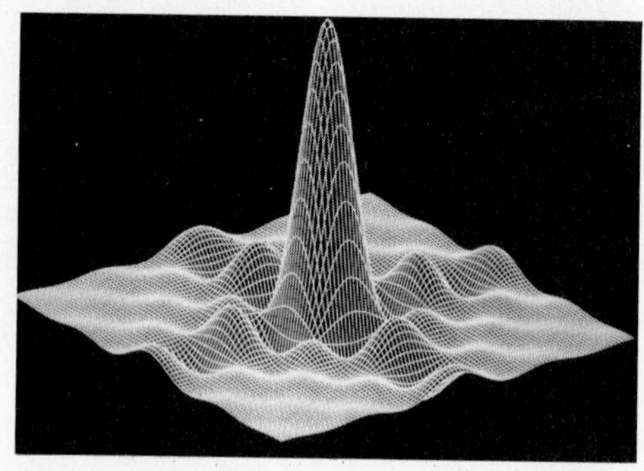

(c)

FIGURE 10.25 *(a)* The irradiance distribution for a square aperture. *(b)* The irradiance produced by Fraunhofer diffraction at a square aperture. *(c)* The electric-field distribution produced by Fraunhofer diffraction via a square aperture. (Photos courtesy R. G. Wilson, Illinois Wesleyan University.)

α' are nonzero integer multiples of π or equivalently when Y and Z are nonzero integer multiples of $\lambda R/b$ and $\lambda R/a$, respectively, $I(Y, Z) = 0$, and we have a rectangular grid of nodal lines, as indicated in Fig. 10.25. Notice that the pattern in the Y-, Z-directions varies *inversely* with the y-, z-aperture dimensions. A horizontal, rectangular opening will produce a pattern with a vertical rectangle at its center.

Along the β'-axis, $\alpha' = 0$ and the subsidiary maxima are located approximately halfway between zeros, that is, at $\beta'_m = \pm 3\pi/2, \pm 5\pi/2, \pm 7\pi/2, \ldots$. At each subsidiary maximum $\sin \beta'_m = 1$, and of course along the β'-axis, since $\alpha' = 0$, $(\sin \alpha')/\alpha' = 1$, so that the relative irradiances are approximated simply by

$$\frac{I}{I(0)} = \frac{1}{\beta_m'^2} \tag{10.44}$$

Similarly, along the α'-axis

$$\frac{I}{I(0)} = \frac{1}{\alpha_m'^2} \tag{10.45}$$

The flux-density ratio* drops off rather rapidly from 1 to $\frac{1}{22}$ to $\frac{1}{62}$ to $\frac{1}{122}$, and so on. Even so, the off-axis secondary peaks are still smaller; for example, the four corner peaks (whose coordinates correspond to appropriate combinations of $\beta' = \pm 3\pi/2$ and $\alpha' = \pm 3\pi/2$) nearest to the central maximum each have relative irradiances of $(\frac{1}{22})^2$.

10.2.5 The Circular Aperture

Fraunhofer diffraction at a circular aperture is an effect of great practical significance in the study of optical instrumentation. Envision a typical arrangement: plane waves impinging on a screen Σ containing a circular aperture and the consequent far-field diffraction pattern spread across a distant observing screen σ. By using a focusing lens L_2, we can bring σ in close to the aperture without changing the pattern. Now, if L_2 is positioned within and exactly fills the diffracting opening in Σ, the form of the pattern is essentially unaltered. The lightwave reaching Σ is cropped, so that only a circular segment propagates through L_2 to form an image in the focal plane. This is obviously the same process that takes place in an eye, telescope, microscope, or camera lens. The image of a distant point source, as formed by a perfectly aberration-free converging lens, is never a point but rather some sort of diffraction pattern. We are essentially collecting only a fraction of the incident wavefront and therefore cannot hope to form a perfect image. As shown in the last section, the expression for the optical disturbance at P, arising from an arbitrary aperture in the far-field case, is

$$\tilde{E} = \frac{\mathcal{E}_A e^{i(\omega t - kR)}}{R} \iint_{\text{Aperture}} e^{ik(Yy+Zz)/R}\, dS \qquad [10.41]$$

For a circular opening, symmetry would suggest introducing spherical coordinates in both the plane of the aperture and the plane of observation, as shown in Fig. 10.26. Therefore, let

$$z = \rho \cos\phi \qquad y = \rho \sin\phi$$

$$Z = q \cos\Phi \qquad Y = q \sin\Phi$$

The differential element of area is now

$$dS = \rho\, d\rho\, d\phi$$

Substituting these expressions into Eq. (10.41), it becomes

$$\tilde{E} = \frac{\mathcal{E}_A e^{i(\omega t - kR)}}{R} \int_{\rho=0}^{a} \int_{\phi=0}^{2\pi} e^{i(k\rho q/R)\cos(\phi - \Phi)} \rho\, d\rho\, d\phi \qquad (10.46)$$

Because of the complete axial symmetry, the solution must be independent of Φ. We might just as well solve Eq. (10.46) with $\Phi = 0$ as with any other value, thereby simplifying things slightly.

The portion of the double integral associated with the variable ϕ,

$$\int_0^{2\pi} e^{i(k\rho q/R)\cos\phi}\, d\phi$$

is one that arises quite frequently in the mathematics of physics. It is a unique function in that it cannot be reduced to any of the more common forms, such as the various hyperbolic, exponential, or trigonometric functions, and indeed with the exception of these, it is perhaps the most often encountered. The quantity

$$J_0(u) = \frac{1}{2\pi} \int_0^{2\pi} e^{iu\cos v}\, dv \qquad (10.47)$$

is known as the *Bessel function* (of the first kind) of order zero. More generally,

$$J_m(u) = \frac{i^{-m}}{2\pi} \int_0^{2\pi} e^{i(mv + u\cos v)}\, dv \qquad (10.48)$$

represents the Bessel function of order m. Numerical values of $J_0(u)$ and $J_1(u)$ are tabulated for a large range of u in most mathematical handbooks. Just like sine and cosine, the Bessel functions have series expansions and are certainly no more esoteric than these familiar childhood acquaintances. As seen in Fig. 10.27, $J_0(u)$ and $J_1(u)$ are slowly decreasing oscillatory functions that do nothing particularly dramatic.

*These particular photographs were taken during an undergraduate laboratory session. A 1.5-mW He–Ne laser was used as a plane-wave source. The apparatus was set up in a long darkened room, and the pattern was cast directly on 4 × 5 Polaroid (ASA 3000) film. The film was located about 30 feet from a small aperture, so that no focusing lens was needed. The shutter, placed directly in front of the laser, was a student-contrived cardboard guillotine arrangement, and therefore no exposure times are available. Any camera shutter (a single-lens reflex with the lens removed and the back open) will serve, but the cardboard one was more fun.

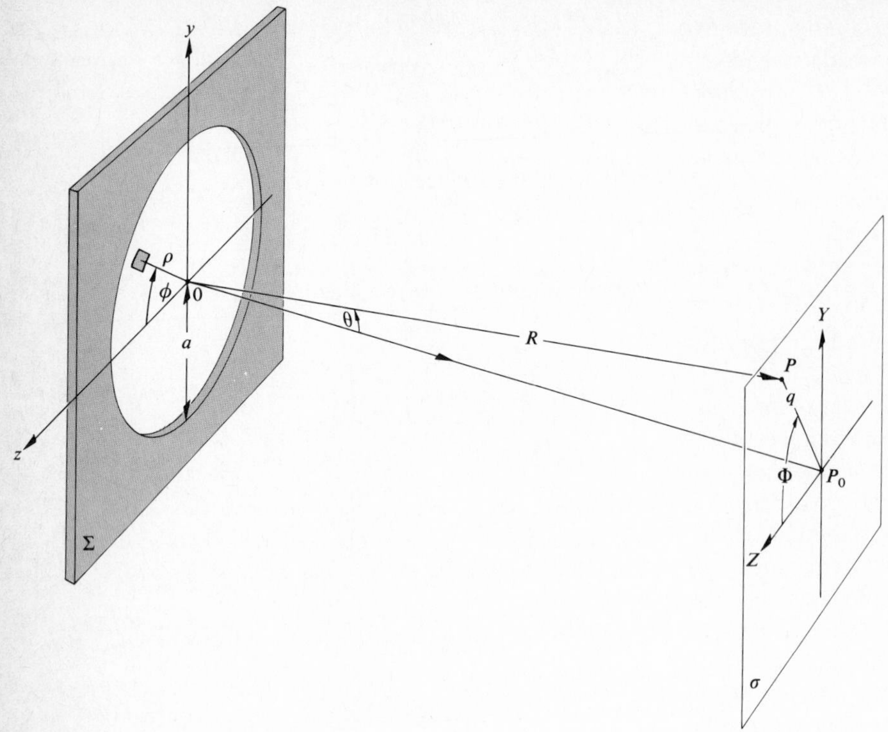

FIGURE 10.26 Circular aperture geometry.

Equation (10.46) can be rewritten as

$$\tilde{E} = \frac{\mathcal{E}_A e^{i(\omega t - kR)}}{R} 2\pi \int_0^a J_0(k\rho q/R)\rho \, d\rho \qquad (10.49)$$

Another general property of Bessel functions, referred to as a recurrence relation, is

$$\frac{d}{du}[u^m J_m(u)] = u^m J_{m-1}(u)$$

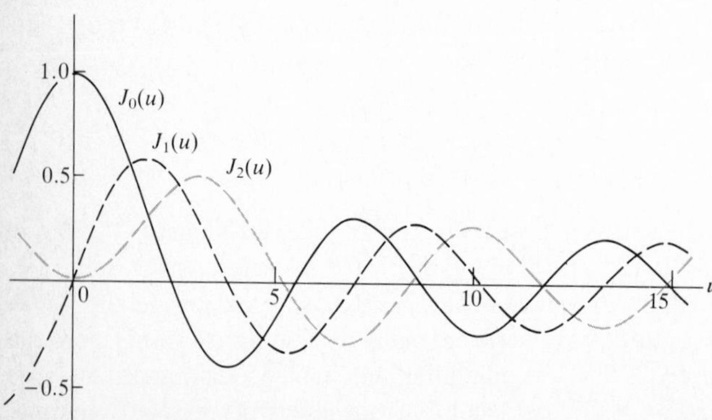

FIGURE 10.27 Bessel functions.

When $m = 1$, this clearly leads to

$$\int_0^u u' J_0(u') \, du' = u J_1(u) \qquad (10.50)$$

with u' just serving as a dummy variable. If we now return to the integral in Eq. (10.49) and change the variable such that $w = k\rho q/R$, then $d\rho = (R/kq) \, dw$ and

$$\int_{\rho=0}^{\rho=a} J_0(k\rho q/R)\rho \, d\rho = (R/kq)^2 \int_{w=0}^{w=kaq/R} J_0(w) w \, dw$$

Making use of Eq. (10.50), we get

$$\tilde{E}(t) = \frac{\mathcal{E}_A e^{i(\omega t - kR)}}{R} 2\pi a^2 (R/kaq) J_1(kaq/R) \qquad (10.51)$$

The irradiance at point P is $\langle(\text{Re}\,\tilde{E})^2\rangle$ or $\frac{1}{2}\tilde{E}\tilde{E}^*$, that is,

$$I = \frac{2\mathcal{E}_A^2 A^2}{R^2} \left[\frac{J_1(kaq/R)}{kaq/R} \right]^2 \qquad (10.52)$$

where A is the area of the circular opening. To find the irradiance at the center of the pattern (i.e., at P_0), set $q = 0$. It follows from the above recurrence relation ($m = 1$) that

$$J_0(u) = \frac{d}{du} J_1(u) + \frac{J_1(u)}{u} \qquad (10.53)$$

From Eq. (10.47) we see that $J_0(0) = 1$, and from Eq. (10.48), $J_1(0) = 0$. The ratio of $J_1(u)/u$ as u approaches zero has the same limit (L'Hospital's rule) as the ratio of the separate derivatives of its numerator and denominator, namely, $dJ_1(u)/du$ over 1. But this means that the right-hand side of Eq. (10.53) is twice that limiting value, so that $J_1(u)/u = \frac{1}{2}$ at $u = 0$. The irradiance at P_0 is therefore

$$I(0) = \frac{\varepsilon_A^2 A^2}{2R^2} \qquad (10.54)$$

which is the same result obtained for the rectangular opening [Eq. (10.43)]. If R is assumed to be essentially constant over the pattern, we can write

$$I = I(0)\left[\frac{2J_1(kaq/R)}{kaq/R}\right]^2 \qquad (10.55)$$

Since $\sin\theta = q/R$, the irradiance can be written as a function of θ,

$$I(\theta) = I(0)\left[\frac{2J_1(ka\sin\theta)}{ka\sin\theta}\right]^2 \qquad (10.56)$$

and as such is plotted in Fig. 10.28. Because of the axial symmetry, the towering central maximum corresponds to a high-irradiance circular spot known as the **Airy disk**. It was Sir George Biddell Airy (1801–1892), Astronomer Royal of England, who first derived Eq. (10.56). The central disk is surrounded by a dark ring that corresponds to the first zero of the function $J_1(u)$. From Table 10.1 $J_1(u) = 0$ when $u = 3.83$, that is, $kaq/R = 3.83$. The radius q_1 drawn to the center of this first dark ring can be thought of as the extent of the Airy disk. It is given by

$$q_1 = 1.22\frac{R\lambda}{2a} \qquad (10.57)$$

For a lens focused on the screen σ, the focal length $f \approx R$, so

$$q_1 \approx 1.22\frac{f\lambda}{D} \qquad (10.58)$$

where D is the aperture diameter, in other words, $D = 2a$. (The *diameter* of the Airy disk in the visible spectrum is *very roughly* equal to the $f/\#$ of the lens in millionths of a meter.)

As shown in Figs. 10.29 to 10.31, q_1 varies inversely with the hole's diameter. As D approaches λ, the Airy disk can be very large indeed, and the circular aperture begins to resemble a point source of spherical waves.

The higher-order zeros occur at values of kaq/R equal to 7.02, 10.17, and so forth. The secondary maxima are located where u satisfies the condition

$$\frac{d}{du}\left[\frac{J_1(u)}{u}\right] = 0$$

which is equivalent to $J_2(u) = 0$. From the tables then, these secondary peaks occur when kaq/R equals 5.14, 8.42, 11.6, and so on, whereupon $I/I(0)$ drops from 1 to 0.017 5, 0.004 2, and 0.001 6, respectively (Problem 10.22).

Circular apertures are preferable to rectangular ones, as far as lens shapes go, since the circle's irradiance curve is broader around the central peak and drops off more rapidly thereafter. Exactly what fraction of the total light energy incident on σ is confined to within the various maxima is a question of interest, but one somewhat too involved to solve here.* On integrating the irradiance over a particular region of the pattern, one finds that 84% of the light arrives within the Airy disk, and 91% within the bounds of the second dark ring.

10.2.6 Resolution of Imaging Systems

Imagine that we have some sort of lens system that forms an image of an extended object. If the object is self-luminous, it is likely that we can regard it as made up of an array of incoherent sources. On the other hand, an object seen in reflected light will surely display some phase correlation between its various scattering points. When the point sources are in fact incoherent, the lens system will form an image of the object that consists of a distribution of partially overlapping, yet independent, Airy patterns. In the finest lenses, which have negligible aberrations, the spreading out of each image point due to diffraction represents the ultimate limit on image quality.

Suppose that we simplify matters somewhat and examine only two equal-irradiance, incoherent, distant point sources. For example, consider two stars seen through the objective

*See Born and Wolf, *Principles of Optics*, p. 398, or the very fine elementary text by Towne, *Wave Phenomena*, p. 464.

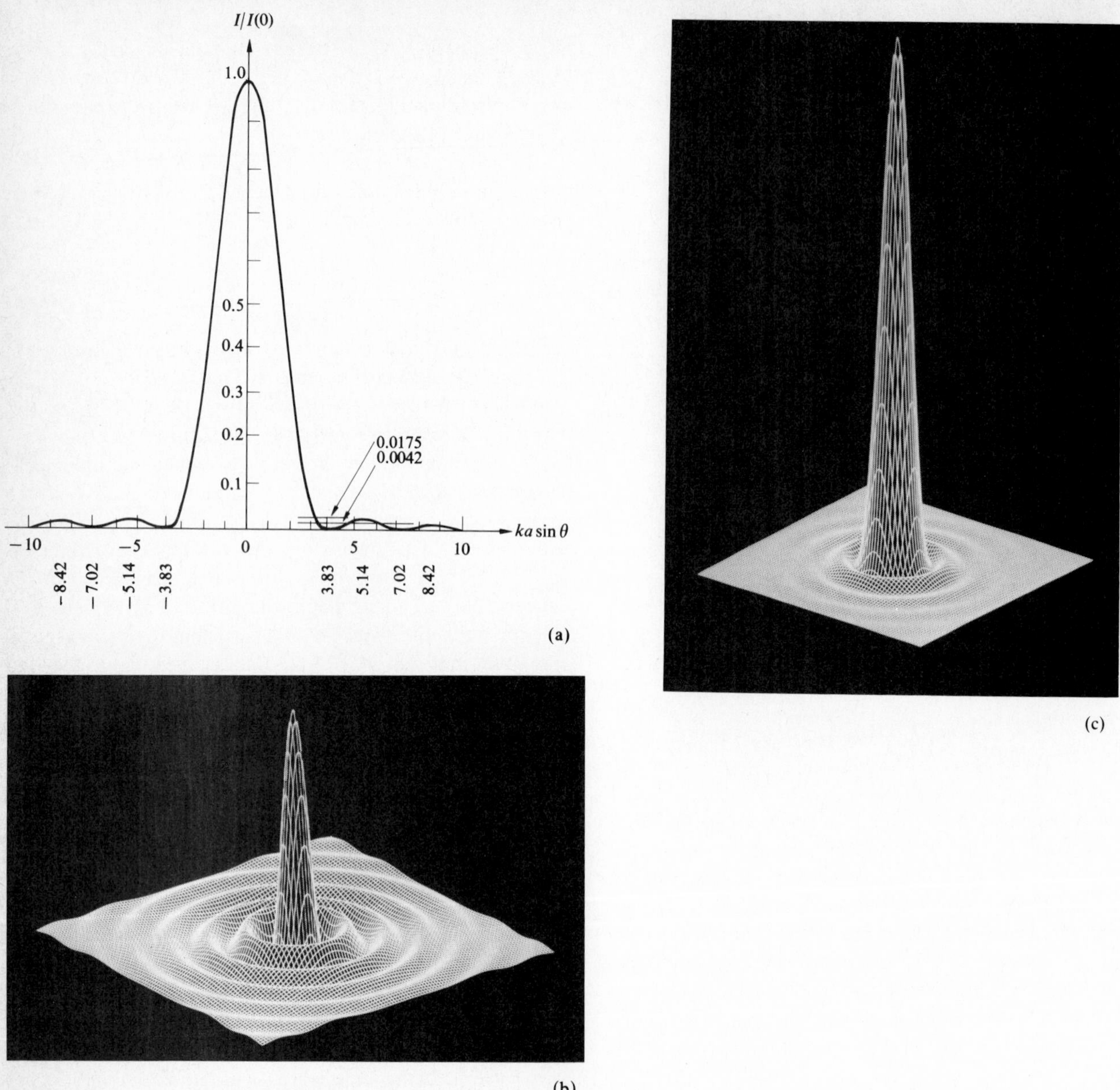

FIGURE 10.28 (*a*) The Airy pattern. (*b*) Electric field created by Fraunhofer diffraction at a circular aperture. (*c*) Irradiance resulting from Fraunhofer diffraction at a circular aperture. (Photos courtesy R. G. Wilson, Illinois Wesleyan University.)

lens of a telescope, where the entrance pupil corresponds to the diffracting aperture. In the previous section we saw that the radius of the Airy disk was given by $q_1 = 1.22f\lambda/D$. If $\Delta\theta$ is the corresponding angular measure, then $\Delta\theta = 1.22\lambda/D$, inasmuch as $q_1/f = \sin\Delta\theta \approx \Delta\theta$. The Airy disk for each star will be spread out over an angular half-width $\Delta\theta$ about its geomet-

TABLE 10.1 Bessel Functions*

x	$J_1(x)$*	x	$J_1(x)$	x	$J_1(x)$
0.0	0.0000	3.0	0.3391	6.0	−0.2767
0.1	0.0499	3.1	0.3009	6.1	−0.2559
0.2	0.0995	3.2	0.2613	6.2	−0.2329
0.3	0.1483	3.3	0.2207	6.3	−0.2081
0.4	0.1960	3.4	0.1792	6.4	−0.1816
0.5	0.2423	3.5	0.1374	6.5	−0.1538
0.6	0.2867	3.6	0.0955	6.6	−0.1250
0.7	0.3290	3.7	0.0538	6.7	−0.0953
0.8	0.3688	3.8	0.0128	6.8	−0.0652
0.9	0.4059	3.9	−0.0272	6.9	−0.0349
1.0	0.4401	4.0	−0.0660	7.0	−0.0047
1.1	0.4709	4.1	−0.1033	7.1	0.0252
1.2	0.4983	4.2	−0.1386	7.2	0.0543
1.3	0.5220	4.3	−0.1719	7.3	0.0826
1.4	0.5419	4.4	−0.2028	7.4	0.1096
1.5	0.5579	4.5	−0.2311	7.5	0.1352
1.6	0.5699	4.6	−0.2566	7.6	0.1592
1.7	0.5778	4.7	−0.2791	7.7	0.1813
1.8	0.5815	4.8	−0.2985	7.8	0.2014
1.9	0.5812	4.9	−0.3147	7.9	0.2192
2.0	0.5767	5.0	−0.3276	8.0	0.2346
2.1	0.5683	5.1	−0.3371	8.1	0.2476
2.2	0.5560	5.2	−0.3432	8.2	0.2580
2.3	0.5399	5.3	−0.3460	8.3	0.2657
2.4	0.5202	5.4	−0.3453	8.4	0.2708
2.5	0.4971	5.5	−0.3414	8.5	0.2731
2.6	0.4708	5.6	−0.3343	8.6	0.2728
2.7	0.4416	5.7	−0.3241	8.7	0.2697
2.8	0.4097	5.8	−0.3110	8.8	0.2641
2.9	0.3754	5.9	−0.2951	8.9	0.2559

*$J_1(x) = 0$ for x = 0, 3.832, 7.016, 10.173, 13.324, . . .
Adapted from E. Kreyszig, *Advanced Engineering Mathematics*, reprinted by permission of John Wiley & Sons, Inc.

FIGURE 10.29 Airy rings (0.5-mm hole diameter). (E. H.)

ric image point, as shown in Fig. 10.32. If the angular separation of the stars is $\Delta\varphi$ and if $\Delta\varphi >> \Delta\theta$, the images will be distinct and easily resolved. As the stars approach each other, their respective images come together, overlap, and commingle into a single blend of fringes. If Lord Rayleigh's criterion is applied, the stars are said to be *just resolved* when the center of one Airy disk falls on the first minimum of the Airy pattern of the other star. (We can certainly do a bit better than this, but Rayleigh's criterion, however arbitrary, has the virtue of being particularly uncomplicated.*) The *minimum resolvable angular separation* or *angular limit of resolution* is

$$(\Delta\varphi)_{\min} = \Delta\theta = 1.22\lambda/D \qquad (10.59)$$

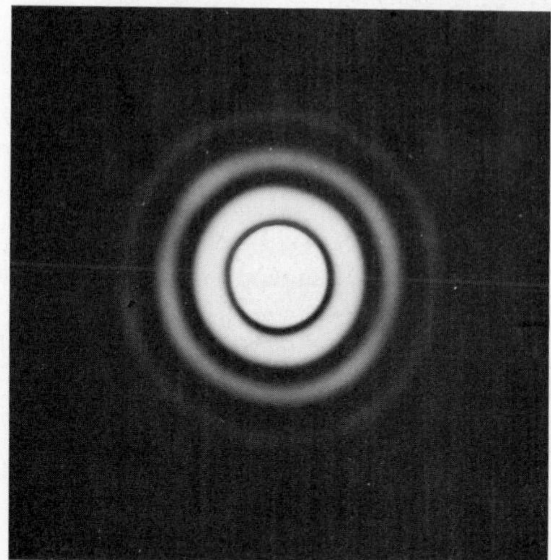

FIGURE 10.30 Airy rings (1.0-mm hole diameter). (E. H.)

*In Rayleigh's own words: "This rule is convenient on account of its simplicity and it is sufficiently accurate in view of the necessary uncertainty as to what exactly is meant by resolution." See Section 9.6.1. for further discussion.

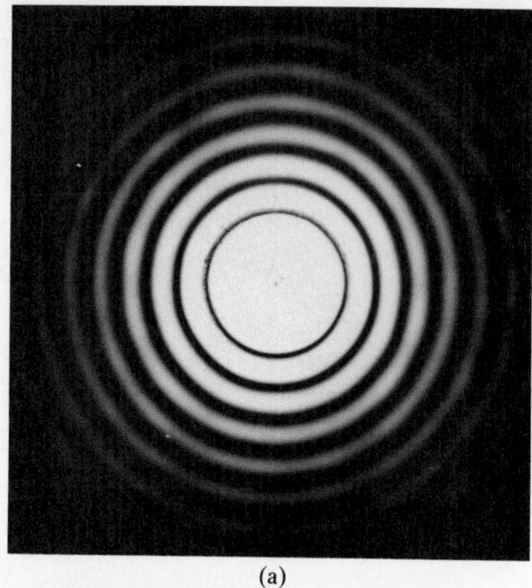
(a)

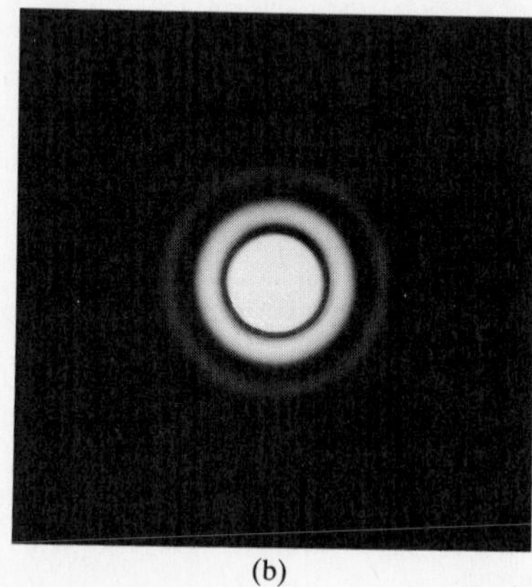
(b)

FIGURE 10.31 (*a*) Airy rings—long exposure (1.5-mm hole diameter). (*b*) Central Airy disk—short exposure with the same aperture. (E. H.)

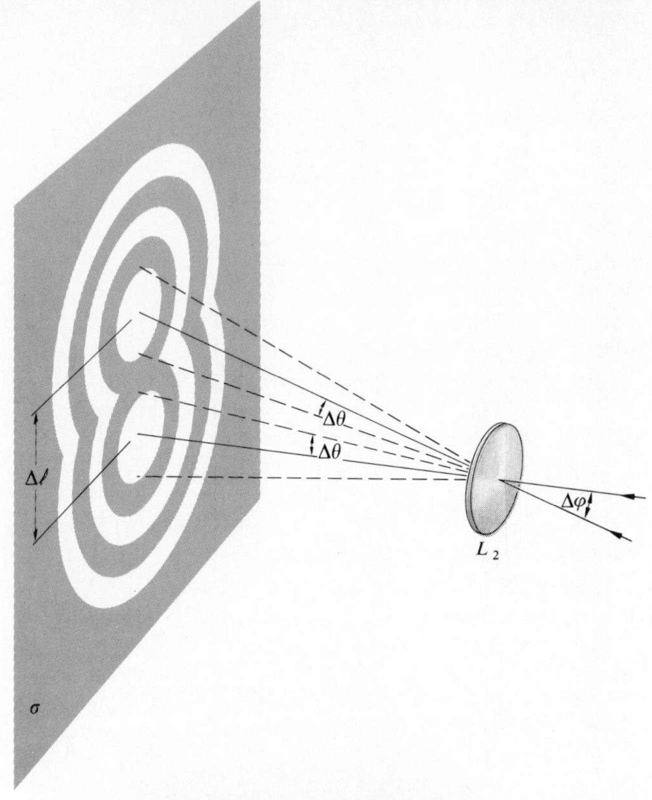

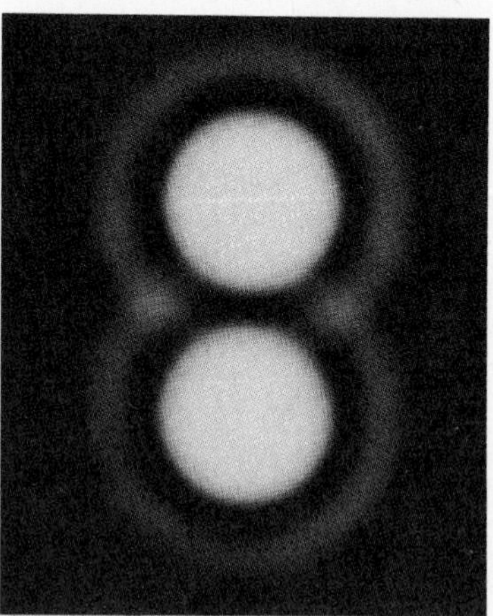

FIGURE 10.32 Overlapping images.

as depicted in Fig. 10.33. If $\Delta\ell$ is the center-to-center separation of the images, the **limit of resolution** is

$$(\Delta\ell)_{\min} = 1.22f\lambda/D \qquad (10.60)$$

The **resolving power** for an image-forming system is generally defined as either $1/(\Delta\varphi)_{\min}$ or $1/(\Delta\ell)_{\min}$.

If the smallest resolvable separation between images is to be reduced (i.e., if the resolving power is to be increased), the wavelength, for instance, might be made smaller. Using ultraviolet rather than visible light in microscopy allows for the perception of finer detail. The electron microscope utilizes equivalent wavelengths of about 10^{-4} to 10^{-5} that of light. This makes it possible to examine objects that would otherwise be completely obscured by diffraction effects in the visi-

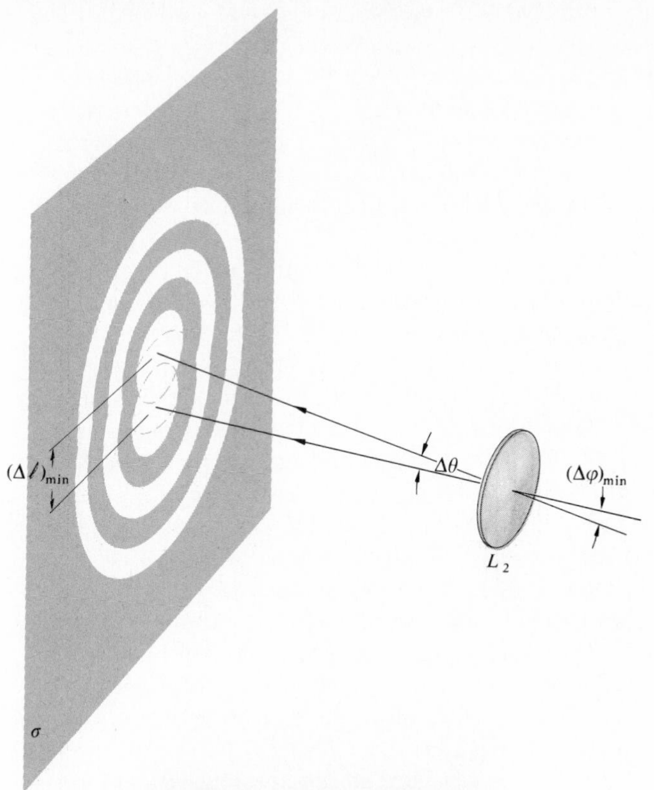

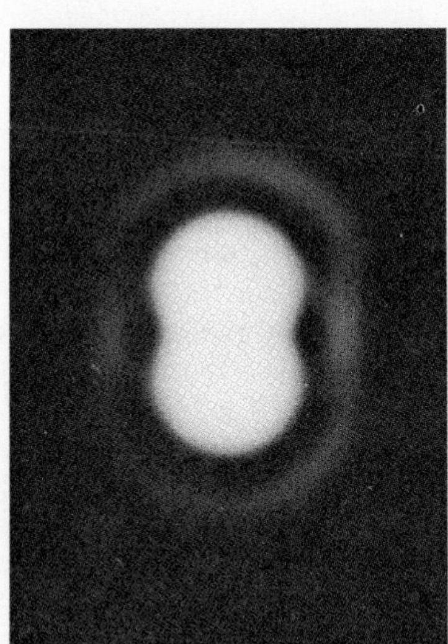

FIGURE 10.33 Overlapping images.

ble spectrum. On the other hand, the resolving power of a telescope can be increased by increasing the diameter of the objective lens or mirror. Besides collecting more of the incident radiation, this will also result in a smaller Airy disk and therefore a sharper, brighter image. The Mount Palomar 200-in telescope has a mirror 5 m in diameter (neglecting the obstruction of a small region at its center). At 550 nm it has an angular limit of resolution of 2.7×10^{-2} s of arc. In contrast, the Jodrell Bank radio telescope, with a 250-ft diameter, operates at a rather long, 21-cm wavelength. It therefore has a limit of resolution of only about 700 s of arc. The human eye has a pupil diameter that of course varies. Taking it, under bright conditions, to be about 2 mm, with $\lambda = 550$ nm, $(\Delta\varphi)_{min}$ turns out to be roughly 1 min of arc. With a focal length of about 20 mm, $(\Delta\ell)_{min}$ on the retina is 6700 nm. This is roughly twice the mean spacing between receptors. The human eye should therefore be able to resolve two points, an inch apart, at a distance of some 100 yards. You will probably not be able to do quite that well; one part in one thousand is more likely.

A more appropriate criterion for resolving power has been proposed by C. Sparrow. Recall that at the Rayleigh limit there is a central minimum or saddle point between adjacent peaks. A further decrease in the distance between the two point sources will cause the central dip to grow shallower and ultimately disappear. The angular separation corresponding to that configuration is Sparrow's limit. The resultant maximum has a broad flat top. In other words, at the origin, which is the center of the peak, the second derivative of the irradiance function is zero; there is no change in slope (Fig. 10.40).

Unlike the Rayleigh rule, which rather tacitly assumes incoherence, the Sparrow condition can readily be generalized to coherent sources. In addition, astronomical studies of equal-brightness stars have shown that Sparrow's criterion is by far the more realistic.

10.2.7 The Diffraction Grating

A repetitive array of diffracting elements, either apertures or obstacles, that has the effect of producing periodic alterations in the phase, amplitude, or both of an emergent wave is said to be a **diffraction grating**. One of the simplest such arrangements is the multiple-slit configuration of Section 10.2.3. It seems to have been invented by the American astronomer David Rittenhouse in about 1785. Some years later Joseph von Fraunhofer independently rediscovered the principle and went on to make a number of important contributions to both the theory and technology of gratings. The earliest devices were indeed multiple-slit assemblies, usually consisting of a grid of fine wire or thread wound about and extending between two parallel screws, which served as spacers. A wavefront, in passing through such a system, is confronted by alternate opaque and transparent regions, so that it undergoes a modulation in *amplitude*. Accordingly, a multiple-slit configuration is said to be a ***transmission amplitude grating***. Another, more common form of transmission grating is made by ruling or scratching parallel notches into the surface of a flat, clear glass plate (Fig. 10.34*a*). Each of the scratches serves as a source of scattered light, and together they form a regular array of parallel line sources. When the grating is totally transparent, so that there is negligible amplitude modulation, the regular variations in the optical thickness across the grating yield a modulation in-*phase*, and we have what is known as a ***transmission phase grating*** (Fig. 10.35*a*). In the Huygens–Fresnel representation you can envision the wavelets as radiated with different phases over the grating surface. An emerging wavefront therefore contains periodic variations in its shape rather than its amplitude. This in turn is equivalent to an angular distribution of constituent plane waves.

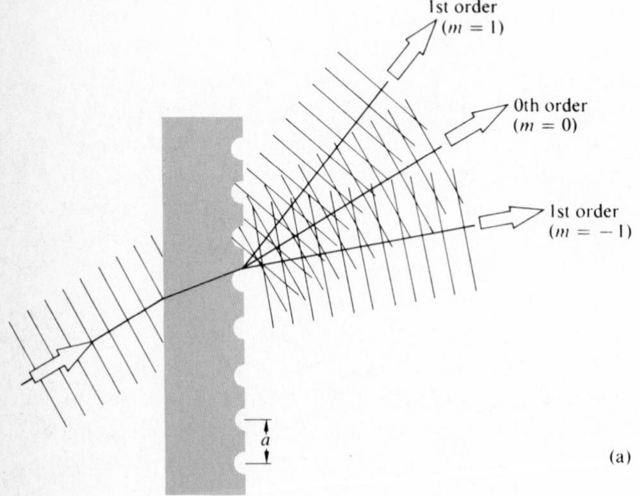

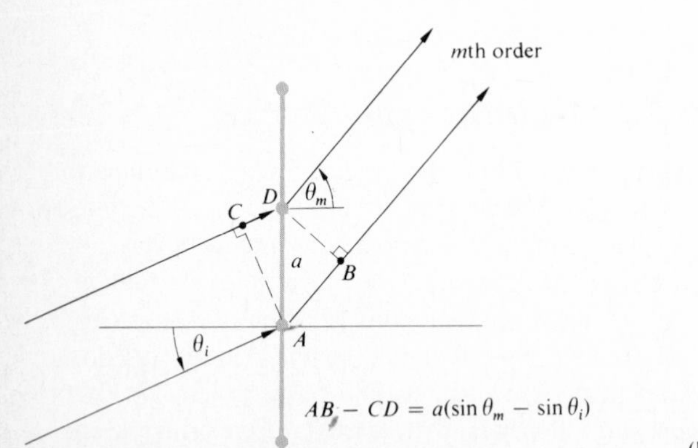

FIGURE 10.34 A transmission grating.

On reflection from this kind of grating, light scattered by the various periodic surface features will arrive at some point P with a definite phase relationship. The consequent interference pattern generated after reflection is quite similar to that arising from transmission. Gratings designed specifically to function in this fashion are known as ***reflection phase gratings*** (Fig. 10.36). Gratings of this sort have traditionally been ruled in thin films of aluminum that have been evaporated onto optically flat glass blanks. The aluminum, being fairly soft, results in less wear on the diamond ruling tool and is also a better reflector in the ultraviolet region.

The manufacture of ruled gratings is extremely difficult, and relatively few are made. In actuality, most gratings are exceedingly good plastic castings or *replicas* of fine, master ruled gratings. Today, large numbers of gratings are made holographically (p. 617).

If you were to look perpendicularly through a transmission grating at a distant parallel line source, your eye would serve as a focusing lens for the diffraction pattern. Recall the analysis of Section 10.2.3 and the expression

$$a \sin \theta_m = m\lambda \qquad [10.32]$$

which is known as the **grating equation** for normal incidence. The values of m specify the *order* of the various principal maxima. For a source having a broad continuous spectrum, such as a tungsten filament, the $m = 0$, or zeroth-order, image corresponds to the undeflected, $\theta_0 = 0$, white-light view of the source (Fig. 10.35*b*). The grating equation is dependent on λ, and so for any value of $m \neq 0$ the various colored images of the source corresponding to slightly different angles (θ_m) spread out into a continuous spectrum. The regions occupied by the faint subsidiary maxima will show up as bands seemingly devoid of any light. The first-order spectrum $m = \pm 1$ appears on either side of $\theta = 0$ and is followed, along with alternate intervals of darkness, by the higher-order spectra, $m = \pm 2, \pm 3, \ldots$. Notice that the smaller a becomes in Eq. (10.32), the fewer will be the number of visible orders.

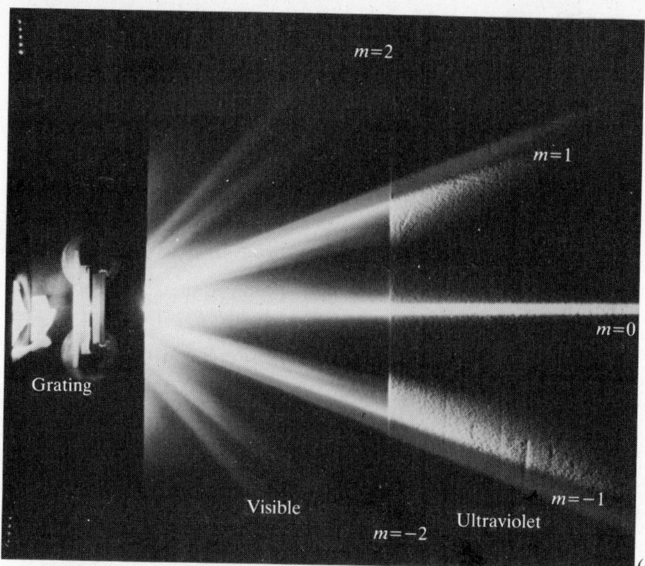

(a)

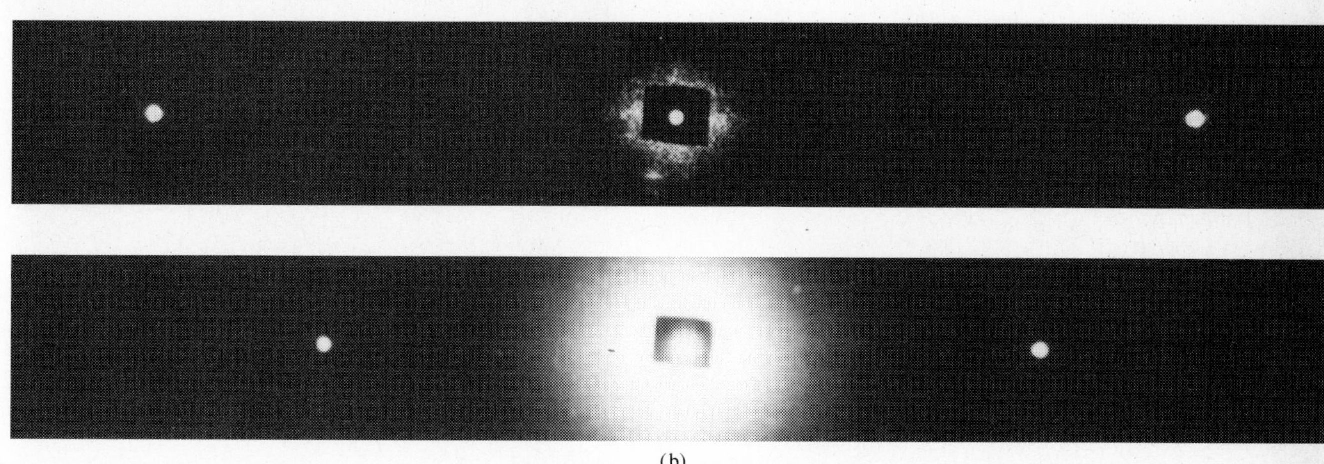

(b)

FIGURE 10.35 Light passing through a grating. (*a*) The region on the left is the visible spectrum, that on the right, the ultraviolet. (Photo courtesy Klinger Educational Prod. Corp., College Point, N. Y.) (*b*) Head-on views of the $m = 0$ and $m = \pm 1$ diffracted beams arising when light from a He–Ne laser passed through a 530 lines/mm grating. In the upper version, the grating was in air ($\lambda = 632.8$ nm). In the lower version, the grating was immersed in water. From the measured value of θ_1 the grating equation yielded $\lambda_w = 471$ nm and therefore $n_w = 1.34$. (Photo courtesy A. F. Leung, The Chinese University of Hong Kong.)

It should be no surprise that the grating equation is in fact Eq. (9.29), which describes the location of the maxima in Young's double-slit setup. The interference maxima, all located at the same angles, are now simply sharper (just as the multiple-beam operation of the Fabry–Perot etalon made its fringes sharper). In the double-slit case when the point of observation is somewhat off the exact center of an irradiance maximum the two waves, one from each slit, will still be more or less in-phase, and the irradiance, though reduced, will still be appreciable. Thus the bright regions are fairly broad. By contrast, with multiple-beam systems though all the waves interfere constructively at the centers of the maxima, even a small displacement will cause certain ones to arrive out-of-phase by $\frac{1}{2}\lambda$ with respect to others. For example, suppose P is slightly off from θ_1 so that $a \sin \theta = 1.010\lambda$ instead of 1.000λ. Each of the waves from successive slits will arrive at P shifted by 0.01λ with respect to the previous one. Then 50 slits down from the first, the path length will have shifted by $\frac{1}{2}\lambda$, and the light from slit 1 and slit 51 will essentially cancel. The same would be true for slit-pairs 2 and 52, 3 and 53, and so forth. The result is a rapid falloff in irradiance beyond the centers of the maxima.

Consider next the somewhat more general situation of oblique incidence, as depicted in Figs. 10.34 and 10.36. The

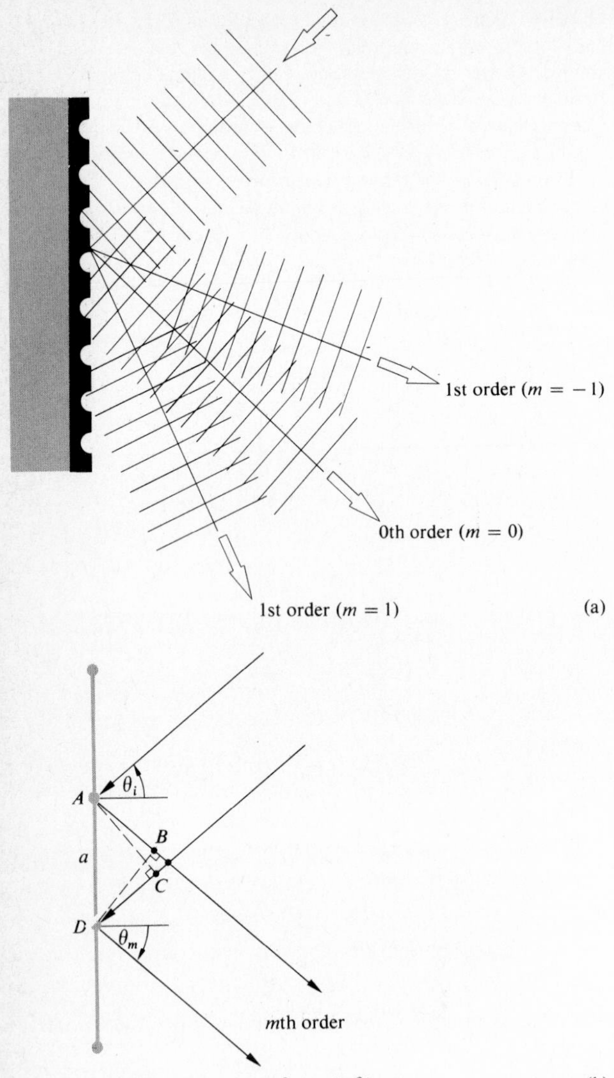

FIGURE 10.36 A reflection grating.

grating equation, for both transmission and reflection, becomes

$$a(\sin\theta_m - \sin\theta_i) = m\lambda \tag{10.61}$$

This expression applies equally well, regardless of the refractive index of the transmission grating itself (Problem 10.37). One of the main disadvantages of the devices examined thus far, and in fact the reason for their obsolescence, is that they spread the available light energy out over a number of low-irradiance spectral orders. For a grating like that shown in Fig. 10.36, most of the incident light undergoes *specular reflection*, as if from a plane mirror. It follows from the grating equation that $\theta_m = \theta_i$ corresponds to the zeroth order, $m = 0$. All of this light is essentially wasted, at least for spectroscopic purposes, since the constituent wavelengths overlap.

In an article in the *Encyclopedia Britannica* of 1888 Lord Rayleigh suggested that it was at least theoretically possible to shift energy out of the useless zeroth order into one of the higher-order spectra. So motivated, Robert Williams Wood (1868–1955) succeeded in 1910 in ruling grooves with a controlled shape, as shown in Fig. 10.37. Most modern gratings are of this shaped or **blazed** variety. The angular positions of the nonzero orders, θ_m-values, are determined by a, λ, and, of more immediate interest, θ_i. But θ_i and θ_m are measured from

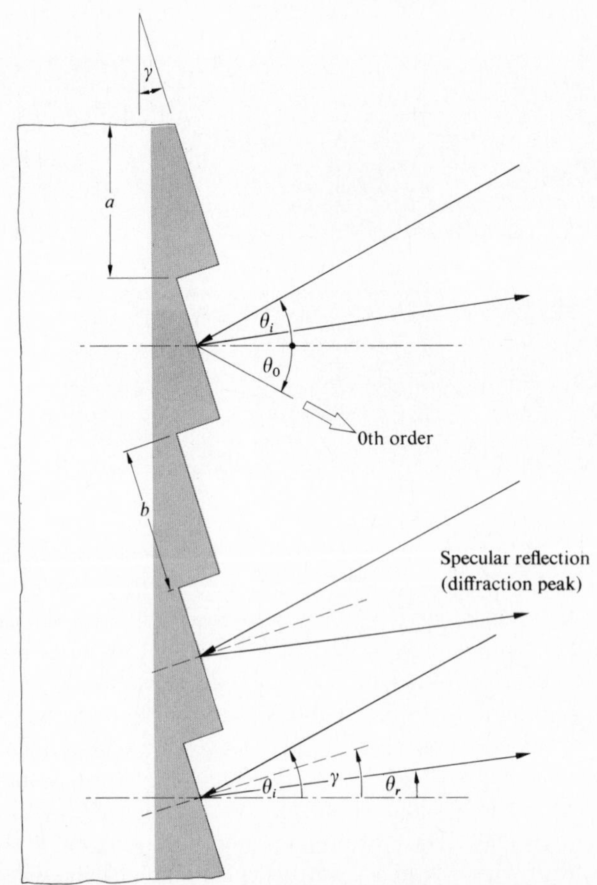

FIGURE 10.37 Section of a blazed reflection phase grating.

the normal to the grating plane and not with respect to the individual groove surfaces. On the other hand, the location of the peak in the single-facet diffraction pattern corresponds to *specular reflection* off that face, for each groove. It is governed by the *blaze angle* γ and can be varied independently of θ_m. This is somewhat analogous to the antenna array of Section 10.1.3, where we were able to control the spatial position of the interference pattern [Eq. (10.6)] by adjusting the relative phase shift between sources without actually changing their orientations.

Consider the situation depicted in Fig. 10.38 when the incident wave is normal to the plane of a blazed reflection grating; that is, $\theta_i = 0$, so for $m = 0$, $\theta_0 = 0$. For *specular reflection* $\theta_i - \theta_r = 2\gamma$ (Fig. 10.37), most of the diffracted radiation is concentrated about $\theta_r = -2\gamma$. (θ_r is negative because the incident and reflected rays are on the same side of the grating normal.) This will correspond to a particular nonzero order, on one side of the central image, when $\theta_m = -2\gamma$; in other words, $a \sin(-2\gamma) = m\lambda$ for the desired λ and m.

Grating Spectroscopy

Quantum Mechanics, which evolved in the early 1920s, had its initial thrust in the area of atomic physics. Predictions were made concerning the detailed structure of the hydrogen atom as manifested by its emitted radiation, and spectroscopy provided the vital proving ground. The need for larger and better gratings became apparent. Grating spectrometers, used over the range from soft X-rays to the far infrared, have enjoyed continued interest. In the hands of the astrophysicist or rocket-borne, they yield information concerning the very origins of the Universe, information as varied as the temperature of a star, the rotation of a galaxy, and the red shift in the spectrum of a quasar. In the mid-1900s George R. Harrison and George W. Stroke remarkably improve the quality of high-resolution gratings. They used a ruling engine* whose operation was controlled by an interferometrically guided servomechanism.

Let us now examine in some detail a few of the major features of the grating spectrum. Assume an infinitesimally narrow incoherent source. The effective width of an emergent spectral line may be defined as the angular distance between the zeros on either side of a principal maximum; in other words, $\Delta\alpha = 2\pi/N$, which follows from Eq. (10.33). At oblique incidence we can redefine α as $(ka/2)(\sin\theta - \sin\theta_i)$, and so a small change in α is given by

$$\Delta\alpha = (ka/2)\cos\theta\,(\Delta\theta) = 2\pi/N \qquad (10.62)$$

where the angle-of-incidence is constant, that is, $\Delta\theta_i = 0$. Thus, even when the incident light is monochromatic,

$$\Delta\theta = 2\lambda/(Na\cos\theta_m) \qquad (10.63)$$

is the *angular width of a line*, due to *instrumental broadening*. Interestingly enough, the angular linewidth varies inversely with the width of the grating itself, *Na*. Another important quantity is the difference in angular position corresponding to a difference in wavelength. The **angular dispersion**, as in the case of a prism, is defined as

$$\mathcal{D} \equiv d\theta/d\lambda \qquad (10.64)$$

Differentiating the grating equation yields

$$\mathcal{D} = m/(a\cos\theta_m) \qquad (10.65)$$

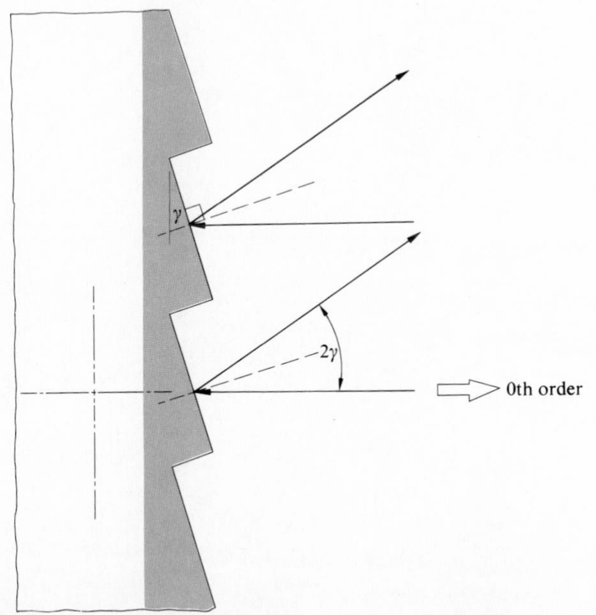

FIGURE 10.38 Blazed grating.

*For more details about these marvelous machines, see A. R. Ingalls, *Sci. Amer.* **186**, 45 (1952), or the article by E. W. Palmer and J. F. Verrill, *Contemp. Phys.* **9**, 257 (1968).

This means that the angular separation between two different frequency lines will increase as the order increases.

Blazed plane gratings with nearly rectangular grooves are most often mounted so that the incident propagation vector is almost normal to either one of the groove faces. This is the condition of *autocollimation*, in which θ_i and θ_m are on the same side of the normal and $\gamma \approx \theta_i \approx -\theta_m$ (see Fig. 10.39), whereupon

$$\mathcal{D}_{\text{auto}} = (2 \tan \theta_i)/\lambda \tag{10.66}$$

which is independent of a.

When the wavelength difference between two lines is small enough so that they overlap, the resultant peak becomes somewhat ambiguous. The chromatic resolving power $\mathcal{R}$ of a spectrometer is defined as

$$\mathcal{R} \equiv \lambda/(\Delta\lambda)_{\min} \qquad [9.76]$$

where $(\Delta\lambda)_{\min}$ is the least resolvable wavelength difference, or **limit of resolution**, and λ is the mean wavelength. Lord Rayleigh's criterion for the resolution of two fringes with equal flux density requires that the principal maximum of one coincide with the first minimum of the other. (Compare this with the equivalent statement used in Section 9.6.1.) As shown in Fig. 10.40, at the limit of resolution the angular separation is half the linewidth, or from Eq. (10.63)

$$(\Delta\theta)_{\min} = \lambda/(Na \cos \theta_m)$$

Applying the expression for the dispersion, we get

$$(\Delta\theta)_{\min} = (\Delta\lambda)_{\min}\, m/(a \cos \theta_m)$$

The combination of these two equations provides us with $\mathcal{R}$, that is,

$$\lambda/(\Delta\lambda)_{\min} = mN \tag{10.67}$$

or

$$\mathcal{R} = \frac{Na(\sin \theta_m - \sin \theta_i)}{\lambda} \tag{10.68}$$

The resolving power is a function of the grating width Na, the angle-of-incidence, and λ. A grating 6 inches wide and containing 15,000 lines per inch will have a total of 9×10^4 lines and a resolving power, in the second order, of 1.8×10^5. In the vicinity of 540 nm, the grating could resolve a wavelength difference of 0.003 nm. Notice that the resolving power cannot exceed $2Na/\lambda$, which occurs when $\theta_i = -\theta_m = 90°$. The

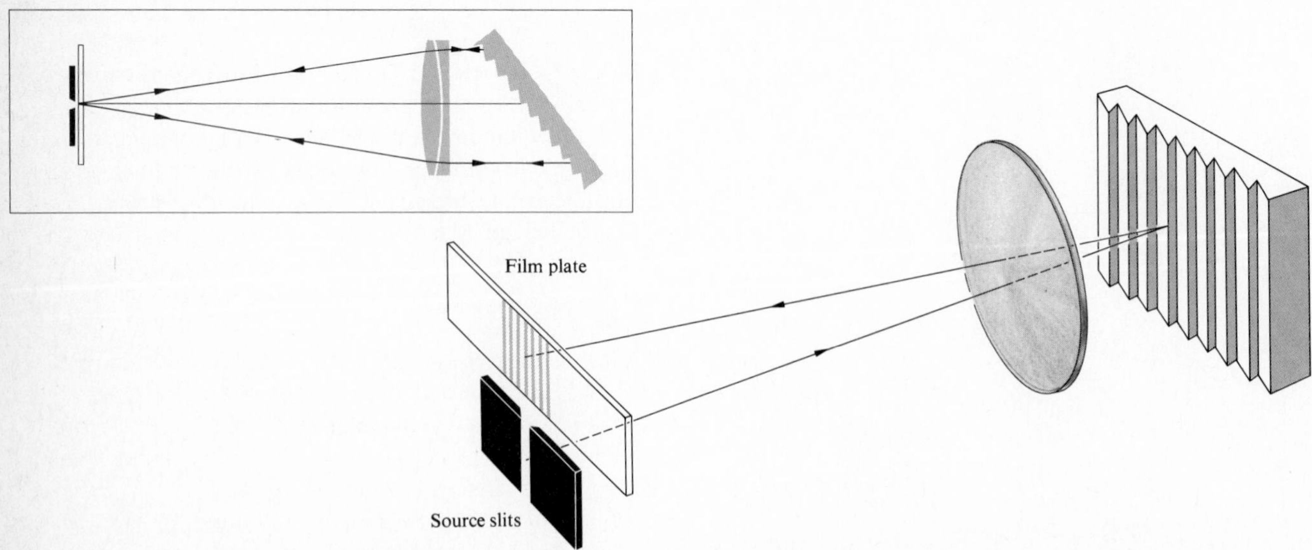

FIGURE 10.39 The Littrow autocollimation mounting.

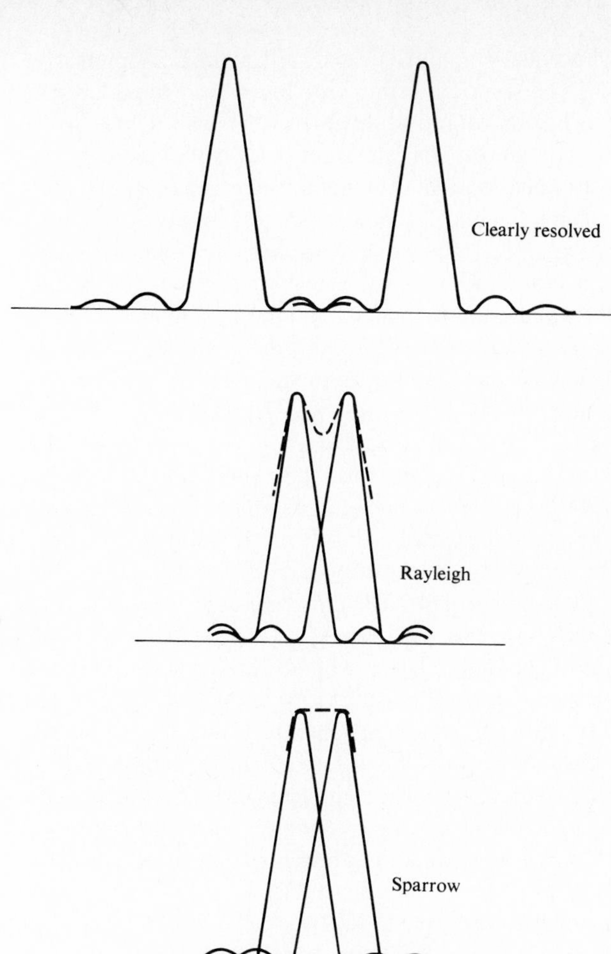

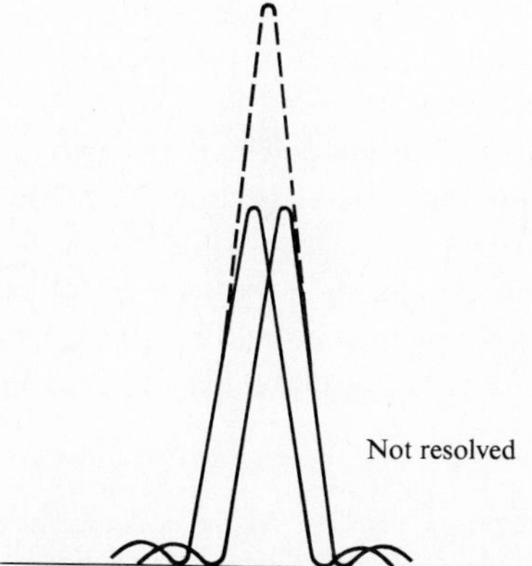

FIGURE 10.40 Overlapping point images.

largest values of $\mathcal{R}$ are obtained when the grating is used in autocollimation, whereupon

$$\mathcal{R}_{\text{auto}} = \frac{2Na \sin \theta_i}{\lambda} \tag{10.69}$$

and again θ_i and θ_m are on the same side of the normal. For one of Harrison's 260-mm-wide blazed gratings at about 75° in a Littrow mount, with $\lambda = 500$ nm, the resolving power just exceeds 10^6.

We now need to consider the problem of overlapping orders. The grating equation makes it quite clear that a line of 600 nm in the first order will have precisely the same position, θ_m, as a 300-nm line in the second order or a 200-nm line when $m = 3$. If two lines of wavelength λ and $(\lambda + \Delta\lambda)$ in successive orders $(m + 1)$ and m just coincide, then

$$a(\sin \theta_m - \sin \theta_i) = (m + 1)\lambda = m(\lambda + \Delta\lambda)$$

The precise wavelength difference is known as the **free spectral range**,

$$(\Delta\lambda)_{\text{fsr}} = \lambda/m \tag{10.70}$$

as it was for the Fabry–Perot Interferometer. In comparison with that device, whose resolving power was

$$\mathcal{R} = \mathcal{F}m \tag*{[9.76]}$$

we might take N to be the finesse of a diffraction grating (Problem 10.38).

A high-resolution grating blazed for the first order, so as to have the greatest free spectral range, will require a high groove density (up to about 1200 lines per millimeter) in order to maintain $\mathcal{R}$. Equation (10.68) shows that $\mathcal{R}$ can be kept constant by ruling fewer lines with increasing spacing, such that the grating width Na is constant. But this requires an increase in m and a subsequent decrease in free spectral range, characterized by overlapping orders. If this time N is held constant while a alone is made larger, $\mathcal{R}$ increases as does m, so that $(\Delta\lambda)_{\text{fsr}}$ again decreases. The angular width of a line is reduced (i.e., the spectral lines become sharper), the coarser the grating is, but the dispersion in a given order diminishes, with the effect that the lines in that spectrum approach each other.

Thus far we have considered a particular type of periodic array, namely the *line grating*. A good deal more information is available in the literature* concerning their shapes, mountings, uses, and so forth.

There are a few unlikely household items that can be used as crude gratings. The grooved surface of a phonograph record works nicely near grazing incidence and CDs are lovely reflection gratings. Surprisingly enough, with $\theta_i \approx 90°$ an ordinary fine-toothed comb will separate out the constituent wavelengths of white light. This occurs in exactly the same fashion as it would with a more orthodox reflection grating. In a letter to a friend dated May 12, 1673, James Gregory pointed out that sunlight passing through a feather would produce a colored pattern, and he asked that his observations be conveyed to Mr. Newton. If you've got one, a feather makes a nice transmission grating.

Two- and Three-Dimensional Gratings

Suppose that the diffracting screen Σ contains a large number, N, of identical diffracting objects (apertures or obstacles). These are to be envisioned as distributed over the surface of Σ in a completely random manner. We also require that each and every one be similarly oriented. Imagine the diffracting screen to be illuminated by plane waves that are focused by a perfect lens L_2, after emerging from Σ (see Fig. 10.16). The individual apertures generate identical Fraunhofer diffraction patterns, all of which overlap on the image plane σ. If there is no regular periodicity in the location of the apertures, we cannot anticipate anything but a random distribution in the relative phases of the waves arriving at an arbitrary point P on σ. We have to be rather careful, however, because there is one exception, which occurs when P is on the central axis, that is, $P = P_0$. All rays, from all apertures, parallel to the central axis will traverse equal optical path lengths before reaching P_0. They will therefore arrive in-phase and interfere constructively.

Now consider a group of arbitrarily directed parallel rays (not in the direction of the central axis), each one emitted from a different aperture. These will be focused at some point on σ, such that each corresponding wave will have an equal probability of arriving with any phase between 0 and 2π. What must be determined is the resultant field arising from the superposition of N equal-amplitude phasors all having random relative phases. The solution to this problem requires an elaborate analysis in terms of probability theory, which is a little too far afield to do here.* The important point is that the sum of a number of phasors taken at random angles is not simply zero, as might be thought. The general analysis begins, for statistical reasons, by assuming that there are a large number of individual aperture screens, each containing N random diffracting apertures and each illuminated, in turn, by a monochromatic wave. We shouldn't be surprised if there is some difference, however small, between the diffraction patterns of two different random distributions of, say, $N = 100$ holes—after all, they are different, and the smaller N is, the more obvious that becomes. Indeed, we can expect their similarities to show up statistically on considering a large number of such masks—ergo the general approach.

If the many individual resulting irradiance distributions are all averaged for a particular *off-axis* point on σ, it will be found that the average irradiance (I_{av}) there equals N times the irradiance (I_0) due to a single aperture: $I_{av} = NI_0$. Still, the irradiance at any point arising from any one aperture screen can differ from this average value by a fairly large amount, regardless of how great N is. These point-to-point fluctuations about the average manifest themselves in each particular pattern as a granularity that tends to show a radial fiberlike structure. If this fine-grained mottling is averaged over a small region of the pattern, which nonetheless contains many fluctuations, it will average out to NI_0.

Of course, in any real experiment the situation will not quite match the ideal—there is no such thing as monochromatic light or a truly random array of (nonoverlapping) diffracting objects. Nonetheless, with a screen containing N "random" apertures illuminated by quasimonochromatic, nearly plane-wave illumination, we can anticipate seeing a mottled flux-density distribution closely resembling that of an individual aperture but N times as strong. Moreover, a bright spot will exist on-axis at its center, which will have a flux density of N^2 times that of a single aperture. If, for example, the screen contains N rectangular holes (Fig. 10.41*a*), the resultant pattern (Fig. 10.41*b*) will resemble Fig. 10.24. Similarly, the

*See F. Kneubühl, "Diffraction Grating Spectroscopy," *Appl. Opt.* **8**, 505 (1969); R. S. Longhurst, *Geometrical and Physical Optics*; and the extensive article by G. W. Stroke in the *Encyclopedia of Physics*, Vol. 29, edited by S. Flügge, p. 426.

*For a statistical treatment, consult J. M. Stone, *Radiation and Optics*, p. 146, and Sommerfeld, *Optics*, p. 194. Also take a look at "Diffraction Plates for Classroom Demonstrations," by R. B. Hoover, *Am. J. Phys.* **37**, 871 (1969), and T. A. Wiggins, "Hole Gratings for Optics Experiments," *Am. J. Phys.* **53**, 227 (1985).

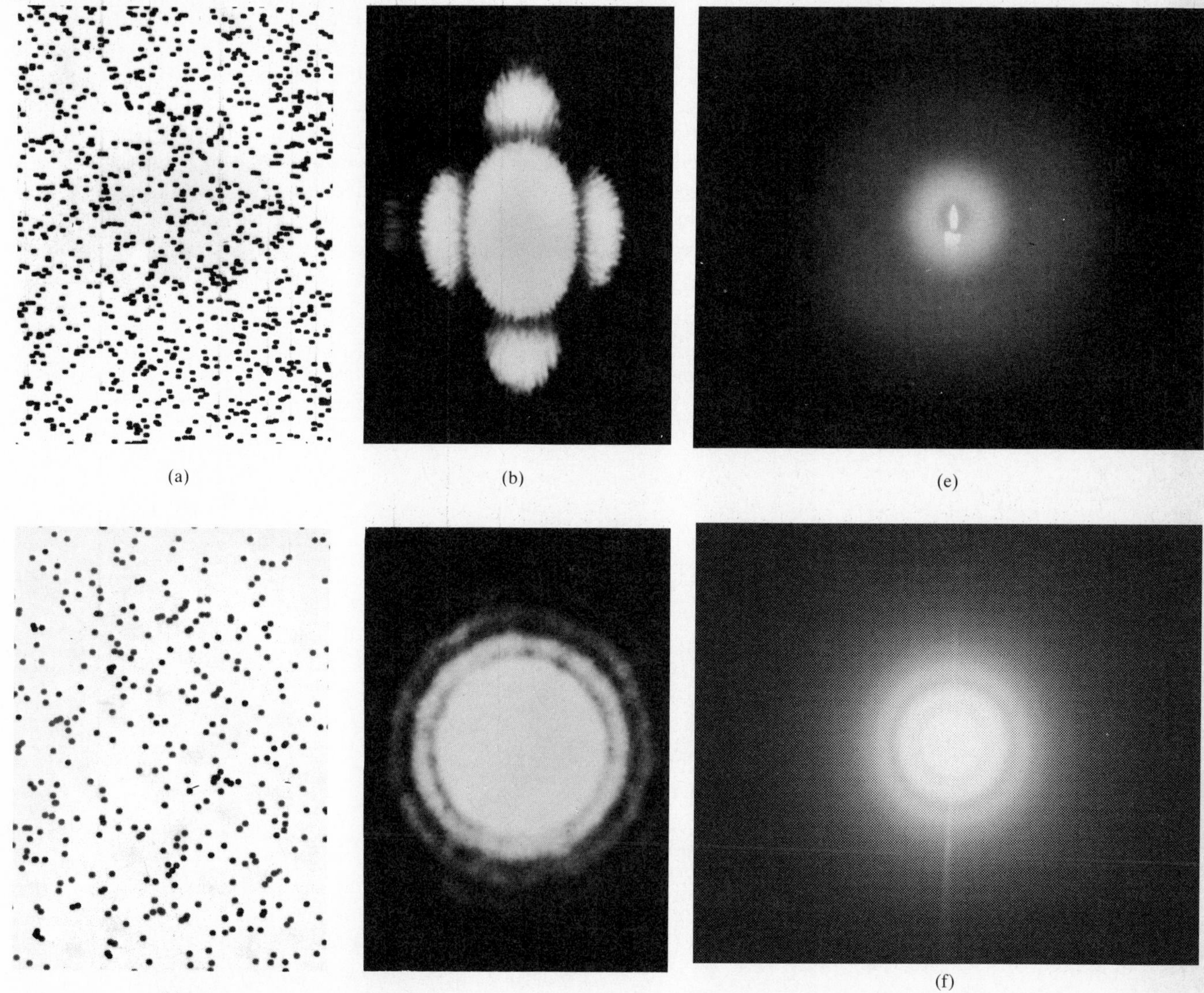

FIGURE 10.41 (*a*) A random array of rectangular apertures. (*b*) The resulting white-light Fraunhofer pattern. (*c*) A random array of circular apertures. (*d*) The resulting white-light Fraunhofer pattern. (Photos courtesy The Ealing Corporation and Richard B. Hoover.) (*e*) A candle flame viewed through a fogged piece of glass. The spectral colors are visible as concentric rings. (E. H.) (*f*) A similar colored ring system created by viewing a white-light point source through a glass plate covered with transparent spherical lycopodium spores. (E. H.)

array of circular holes depicted in Fig. 10.41*c* will produce the diffraction rings of Fig. 10.41*d*.

As the number of apertures increases, the central spot will tend to become so bright as to obscure the rest of the pattern. Note as well that the above considerations apply when all the apertures are illuminated completely coherently. In actuality, the diffracted flux-density distribution will be determined by the degree of coherence (see Chapter 12). The pattern will run the gamut from no interference with completely incoherent light to the case discussed above for completely coherent illumination (Problem 10.40).

The same kind of effects arise from what we might call a two-dimensional *phase grating*. For example, the halo or corona often seen about the Sun or Moon results from diffraction by random droplets of water vapor (i.e., cloud particles). If you would like to duplicate the effect, fog up a microscope slide with your breath, or rub a very thin film of talcum powder on it and then fog it up. Look at a white-light point source. You should see a pattern of clear, concentric, colored rings [Eq. (10.56)] surrounding a white central disk. If you just see a white blur, you don't have a distribution of roughly equal sized droplets; have another try at the talcum. Strikingly beautiful patterns approximating concentric ring systems can be seen through an ordinary *mesh* nylon stocking. If you are fortunate enough to have mercury-vapor street lights, you'll have no trouble seeing all their constituent visible spectral frequencies. (If not, block out most of a fluorescent lamp, leaving something resembling a small source.) Notice the increased

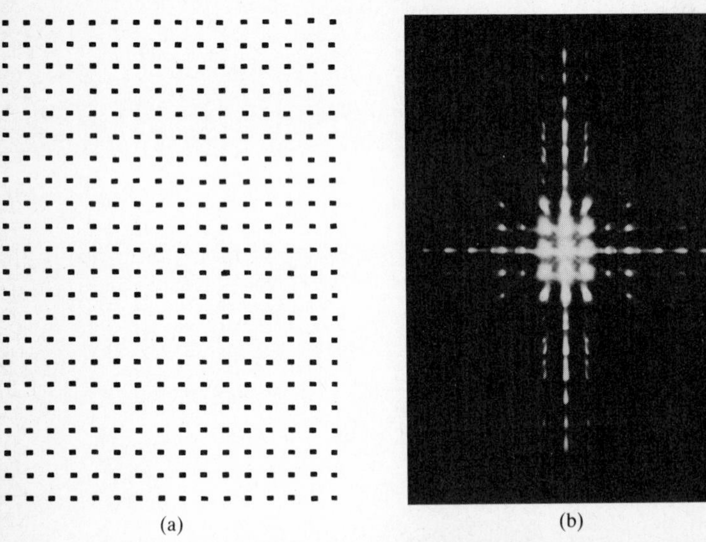

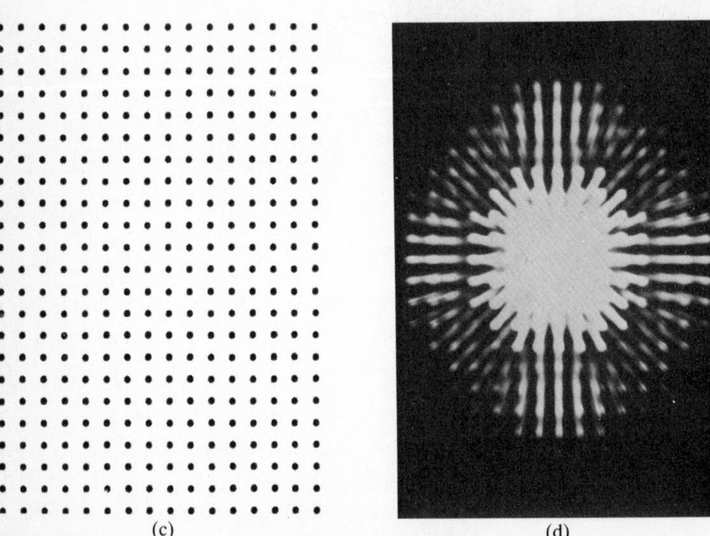

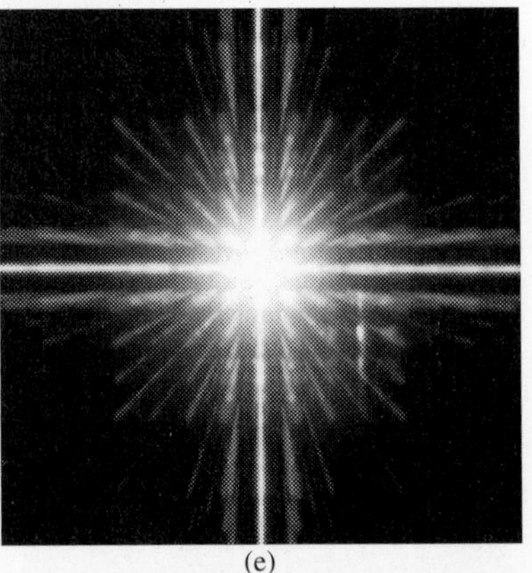

FIGURE 10.42 (*a*) An ordered array of rectangular apertures. (*b*) The resulting white-light Fraunhofer pattern. (*c*) An ordered array of circular apertures. (*d*) The resulting white-light Fraunhofer pattern. (Photos courtesy The Ealing Corporation and Richard B. Hoover.) (*e*) A white-light point source seen through a piece of tightly woven cloth. (E. H.)

symmetry as you increase the number of layers of nylon. Incidentally, this is precisely the way Rittenhouse, the inventor of the grating, became interested in the problem, only he used a silk handkerchief.

Consider the case of a *regular* two-dimensional array of diffracting elements (Fig. 10.42) under normally incident plane-wave illumination. Each small element behaves as a coherent source. And because of the regular periodicity of the lattice of emitters, each emergent wave bears a fixed-phase relation to the others. There will now be certain directions in which constructive interference prevails. Obviously, these occur when the distances from each diffracting element to P are such that the waves are nearly in-phase at arrival. The phenomenon can be observed by looking at a point source through a piece of *square woven*, thin cloth (such as nylon curtain material; see Fig. 10.42*e*) or the fine metal mesh of a tea strainer. The diffracted image is effectively the superposition of two grating patterns at right angles. Examine the center of the pattern carefully to see its gridlike structure.

As for the possibility of a *three-dimensional grating*, there seems to be no particular conceptual difficulty. A regular spatial array of scattering centers would certainly yield interference maxima in preferred directions. In 1912 Max von Laue (1879–1960) conceived the ingenious idea of using the regularly spaced atoms within a crystal as a three-dimensional grating. It is apparent from the grating equation [Eq. (10.61)] that if λ is much greater than the grating spacing, only the zeroth order ($m = 0$) is possible. This is equivalent to $\theta_0 = \theta_i$, that is, specular reflection. Since the spacing between atoms in a crystal is generally several angstroms ($1 \text{ Å} = 10^{-1}$ nm), light can be diffracted only in the zeroth order.

Von Laue's solution to the problem was to probe the lattice, not with light but with X-rays whose wavelengths were comparable to the interatomic distances (Fig. 10.43). A narrow beam of white radiation (the broad continuous frequency range emitted by an X-ray tube) was directed onto a thin single crystal. The film plate (Fig. 10.44) revealed a Fraunhofer pattern consisting of an array of precisely located spots. These sites of constructive interference occurred whenever the angle between the beam and a set of atomic planes within the crystal obeyed Bragg's Law:

$$2d \sin \theta = m\lambda \tag{10.71}$$

Notice that in X-ray work θ is traditionally measured from the plane and not the normal to it. Each set of planes diffracts a particular wavelength into a particular direction. Figure 10.45 rather strikingly shows the analogous behavior in a ripple tank.

Instead of reducing λ to the X-ray range, we could have scaled everything up by a factor of about a billion and made a lattice of metal balls as a grating for microwaves.

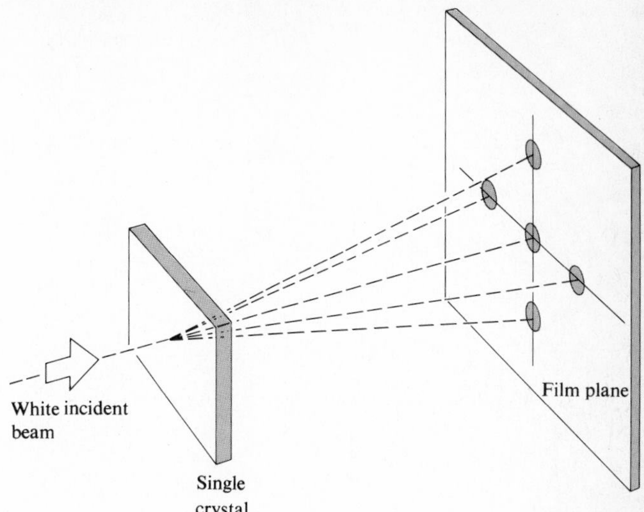

FIGURE 10.43 Transmission Laue pattern.

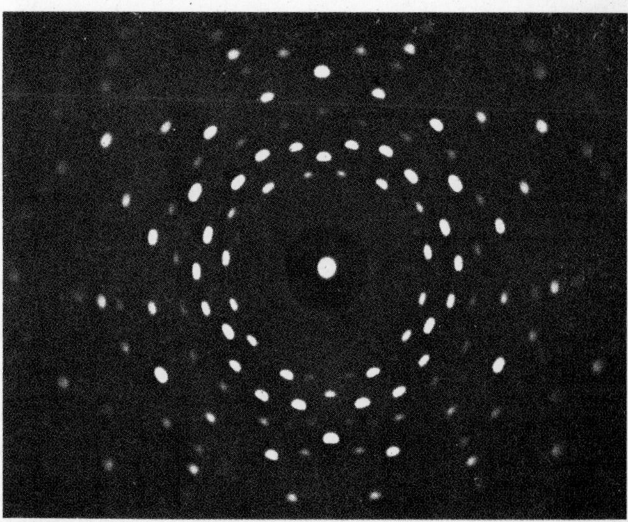

FIGURE 10.44 X-ray diffraction pattern for quartz (SiO_2). (Source unknown)

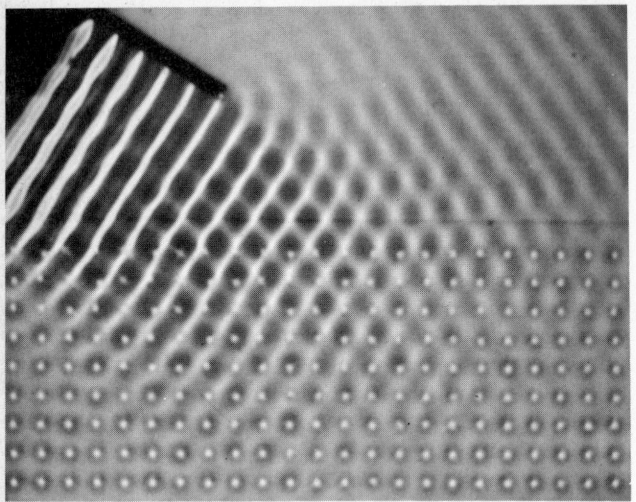

FIGURE 10.45 Water waves in a ripple tank reflecting off an array of pegs acting as point scatterers. (Photo courtesy PSSC *Physics*, D. C. Heath, Boston, 1960.)

10.3 Fresnel Diffraction

10.3.1 The Free Propagation of a Spherical Wave

In the Fraunhofer configuration, the diffracting system was relatively small, and the point of observation was very distant. Under these circumstances a few potentially problematic features of the Huygens–Fresnel Principle could be completely passed over without concern. But we are now going to deal with the near-field region, which extends right up to the diffracting element itself, and any such approximations would be inappropriate. We therefore return to the Huygens–Fresnel Principle in order to reexamine it more closely. At any instant, every point on the primary wavefront is envisioned as a continuous emitter of spherical secondary wavelets. But if each wavelet radiated uniformly in all directions, in addition to generating an ongoing wave, there would also be a reverse wave traveling back toward the source. No such wave is found experimentally, so we must somehow modify the radiation pattern of the secondary emitters. We now introduce the function $K(\theta)$, known as the **obliquity** or **inclination factor**, in order to describe the directionality of the secondary emissions. Fresnel recognized the need to introduce a quantity of this kind, but he did little more than conjecture about its form.* It remained for the more analytic Kirchhoff formulation to provide an actual expression for $K(\theta)$, which, as we will see in Section 10.4, turns out to be

$$K(\theta) = \tfrac{1}{2}(1 + \cos\theta) \tag{10.72}$$

As shown in Fig. 10.46, θ is the angle made with the normal to the primary wavefront, $\mathbf{k}$. This has its maximum value, $K(0) = 1$, in the forward direction and also dispenses with the back wave, since $K(\pi) = 0$.

Let us now examine the free propagation of a spherical monochromatic wave emitted from a *point source S*. If the Huygens–Fresnel Principle is correct, we should be able to add up the secondary wavelets arriving at a point P and thus obtain the unobstructed primary wave. In the process we will gain some insights, recognize a few shortcomings, and develop a very useful technique. Consider the construction shown in Fig. 10.47. The spherical surface corresponds to the primary wavefront at some arbitrary time t' after it has been emitted from S at $t = 0$. The disturbance, having a radius ρ, can be represented by any one of the mathematical expressions describing a harmonic spherical wave, for example,

$$E = \frac{\mathcal{E}_0}{\rho}\cos(\omega t' - k\rho) \tag{10.73}$$

As illustrated, we have divided the wavefront into a number of annular regions. The boundaries of the various regions correspond to the intersections of the wavefront with a series of spheres centered at P of radius $r_0 + \lambda/2$, $r_0 + \lambda$, $r_0 + 3\lambda/2$, and so forth. These are the **Fresnel** or **half-period zones**. Notice that, for a secondary point source in one zone, there will be a point source in the adjacent zone that is further from P by an amount $\lambda/2$. Since each zone, though small, is finite

*It is interesting to read Fresnel's own words on the matter, keeping in mind that he was talking about light as an elastic vibration of the aether.

> Since the impulse communicated to every part of the primitive wave was directed along the normal, the motion which each tends to impress upon the aether ought to be more intense in this direction than in any other; and the rays which would emanate from it, if acting alone, would be less and less intense as they deviated more and more from this direction.
>
> The investigation of the law according to which their intensity varies about each center of disturbance is doubtless a very difficult matter;...

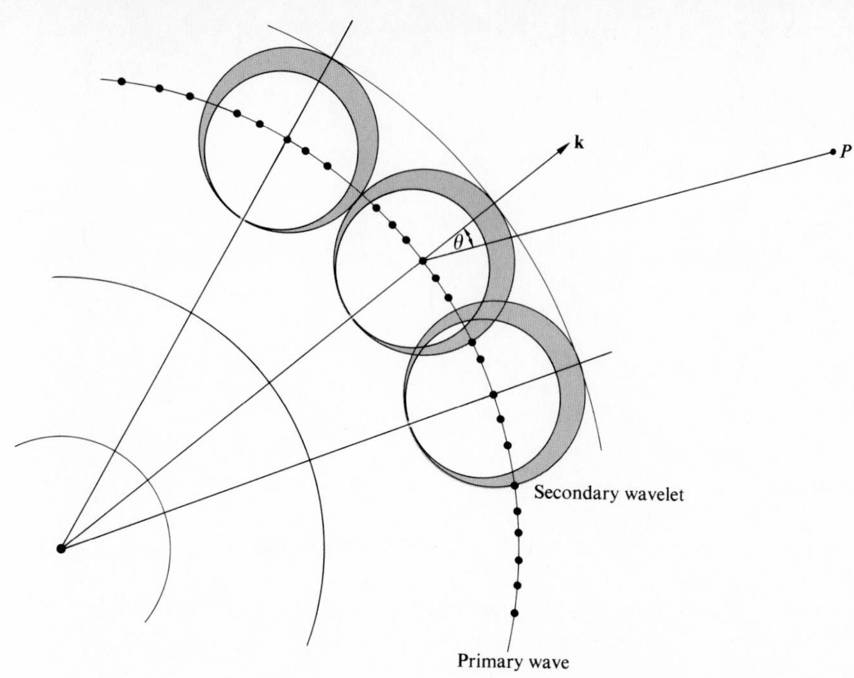

FIGURE 10.46 Secondary wavelets.

in extent, we define a ring-shaped differential area element dS, as indicated in Fig. 10.48. All the point sources within dS are coherent, and *we assume that each radiates in-phase with the primary wave* [Eq. (10.73)]. The secondary wavelets travel a distance r to reach P, at a time t, all arriving there with the same phase, $\omega t - k(\rho + r)$. The amplitude of the primary wave at a distance ρ from S is $\mathcal{E}_0/\rho$. We assume, accordingly, that the source strength per unit area $\mathcal{E}_A$ of the secondary emitters on dS is proportional to $\mathcal{E}_0/\rho$ by way of a constant Q, that is, $\mathcal{E}_A = Q\mathcal{E}_0/\rho$. The contribution of the optical disturbance at P

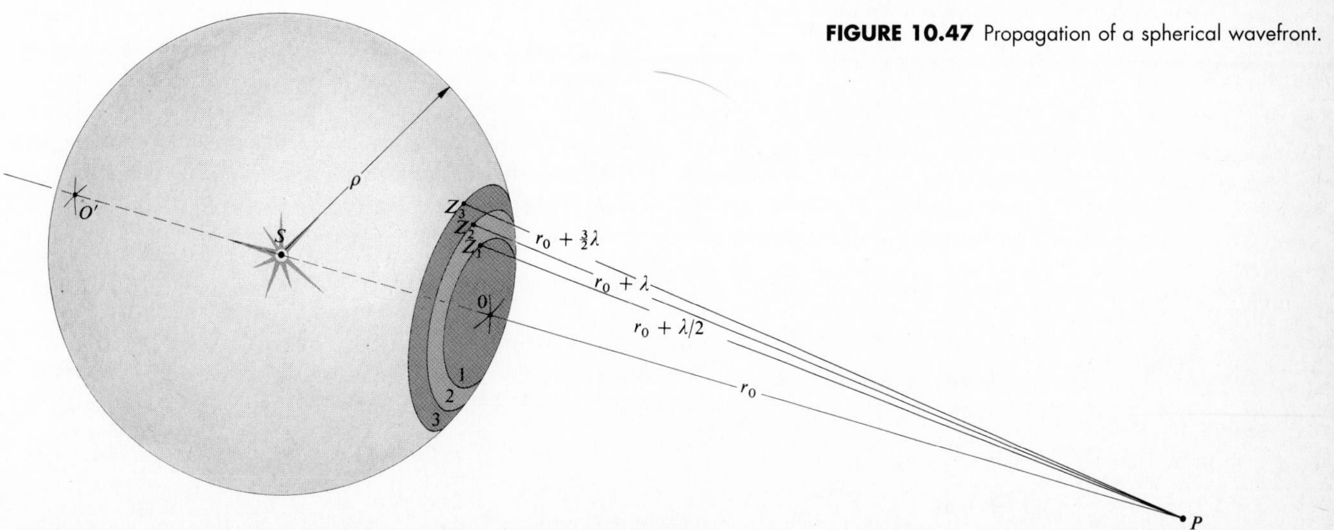

FIGURE 10.47 Propagation of a spherical wavefront.

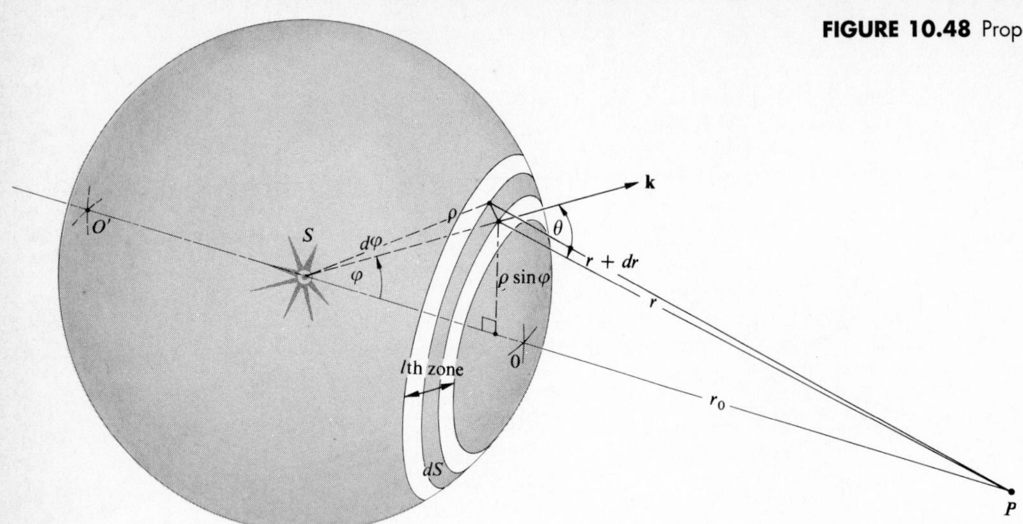

FIGURE 10.48 Propagation of a spherical wavefront.

from the secondary sources on dS is, therefore,

$$dE = K\frac{\mathcal{E}_A}{r}\cos\,[\omega t - k(\rho + r)]\,dS \tag{10.74}$$

The obliquity factor must vary slowly and may be assumed to be constant over a single Fresnel zone. To get dS as a function of r, begin with

$$dS = \rho\,d\varphi\,2\pi(\rho\sin\varphi)$$

Applying the law of cosines, we get

$$r^2 = \rho^2 + (\rho + r_0)^2 - 2\rho(\rho + r_0)\cos\varphi$$

Upon differentiation this yields

$$2r\,dr = 2\rho(\rho + r_0)\sin\varphi\,d\varphi$$

with ρ and r_0 held constant. Making use of the value of $d\varphi$, we find that the area of the element is therefore

$$dS = 2\pi\frac{\rho}{(\rho + r_0)}\,r\,dr \tag{10.75}$$

The disturbance arriving at P from the lth zone is

$$E_l = K_l 2\pi\frac{\mathcal{E}_A\rho}{(\rho + r_0)}\int_{r_{l-1}}^{r_l}\cos\,[\omega t - k(\rho + r)]\,dr$$

Hence

$$E_l = \frac{-K_l\mathcal{E}_A\rho\lambda}{(\rho + r_0)}\,[\sin(\omega t - k\rho - kr)]\,\Big|_{r=r_{l-1}}^{r=r_l}$$

Upon the introduction of $r_{l-1} = r_0 + (l-1)\lambda/2$ and $r_l = r_0 + l\lambda/2$, the expression reduces (Problem 10.42) to

$$E_l = (-1)^{l+1}\frac{2K_l\mathcal{E}_A\rho\lambda}{(\rho + r_0)}\sin\,[\omega t - k(\rho + r_0)] \tag{10.76}$$

Observe that the amplitude of E_l alternates between positive and negative values, depending on whether l is odd or even. This means that the contributions from adjacent zones are out-of-phase and tend to cancel. It is here that the obliquity factor makes a crucial difference. As l increases, θ increases and K decreases, so that successive contributions do not in fact completely cancel each other. It is interesting to note that E_l/K_l is independent of any position variables. Although the areas of each zone are almost equal, they do increase slightly as l increases, which means an increased number of emitters. But the average distance from each zone to P also increases, such that E_l/K_l remains constant (see Problem 10.43).

The sum of the optical disturbances from all m zones at P is

$$E = E_1 + E_2 + E_3 + \cdots + E_m$$

and since these alternate in sign, we can write

$$E = |E_1| - |E_2| + |E_3| - \cdots \pm |E_m| \tag{10.77}$$

If m is odd, the series can be reformulated in two ways, either as

$$E = \frac{|E_1|}{2} + \left(\frac{|E_1|}{2} - |E_2| + \frac{|E_3|}{2}\right) + \left(\frac{|E_3|}{2} - |E_4| + \frac{|E_5|}{2}\right) + \cdots$$
$$+ \left(\frac{|E_{m-2}|}{2} - |E_{m-1}| + \frac{|E_m|}{2}\right) + \frac{|E_m|}{2} \qquad (10.78)$$

or as

$$E = |E_1| - \frac{|E_2|}{2} - \left(\frac{|E_2|}{2} - |E_3| + \frac{|E_4|}{2}\right)$$
$$- \left(\frac{|E_4|}{2} - |E_5| + \frac{|E_6|}{2}\right) + \cdots$$
$$+ \left(\frac{|E_{m-3}|}{2} - |E_{m-2}| + \frac{|E_{m-1}|}{2}\right) - \frac{|E_{m-1}|}{2} + |E_m| \qquad (10.79)$$

There are now two possibilities: either $|E_l|$ is greater than the arithmetic mean of its two neighbors $|E_{l-1}|$ and $|E_{l+1}|$, or it is less than that mean. This is really a question concerning the rate of change of $K(\theta)$. When

$$|E_l| > (|E_{l-1}| + |E_{l+1}|)/2$$

each bracketed term is negative. It follows from Eq. (10.78) that

$$E < \frac{|E_1|}{2} + \frac{|E_m|}{2} \qquad (10.80)$$

and from Eq. (10.79) that

$$E > |E_1| - \frac{|E_2|}{2} - \frac{|E_{m-1}|}{2} + |E_m| \qquad (10.81)$$

Since the obliquity factor goes from 1 to 0 over a great many zones, we can neglect any variation between adjacent zones, that is, $|E_1| \approx |E_2|$ and $|E_{m-1}| \approx |E_m|$. Expression (10.81), to the same degree of approximation, becomes

$$E > \frac{|E_1|}{2} + \frac{|E_m|}{2} \qquad (10.82)$$

We conclude from Eqs. (10.80) and (10.82) that

$$E \approx \frac{|E_1|}{2} + \frac{|E_m|}{2} \qquad (10.83)$$

This same result is obtained when

$$|E_l| < (|E_{l-1}| + |E_{l+1}|)/2$$

If the last term, $|E_m|$, in the series of Eq. (10.77) corresponds to an even m, the same procedure (Problem 10.44) leads to

$$E \approx \frac{|E_1|}{2} - \frac{|E_m|}{2} \qquad (10.84)$$

Fresnel conjectured that the obliquity factor was such that the last contributing zone occurred at $\theta = 90°$, that is,

$$K(\theta) = 0 \text{ for } \pi/2 \leq |\theta| \leq \pi$$

In that case Eqs. (10.83) and (10.84) both reduce to

$$E \approx \frac{|E_1|}{2} \qquad (10.85)$$

when $|E_m|$ goes to zero, because $K_m(\pi/2) = 0$. Alternatively, using Kirchhoff's correct obliquity factor, we divide the *entire* spherical wave into zones with the last or mth zone surrounding O'. Now θ approaches π, $K_m(\pi) = 0$, $|E_m| = 0$, and once again $E \approx |E_1|/2$. *The optical disturbance generated by the entire unobstructed wavefront is approximately equal to one-half the contribution from the first zone.*

If the primary wave were simply to propagate from S to P in a time t, it would have the form

$$E = \frac{\mathcal{E}_0}{(\rho + r_0)} \cos [\omega t - k(\rho + r_0)] \qquad (10.86)$$

Yet the disturbance synthesized from secondary wavelets, Eqs. (10.76) and (10.85), is

$$E = \frac{K_1 \mathcal{E}_A \rho \lambda}{(\rho + r_0)} \sin [\omega t - k(\rho + r_0)] \qquad (10.87)$$

These two equations must, however, be exactly equivalent, and we interpret the constants in Eq. (10.87) to make them so. Note that there is some latitude in how we do this. We prefer to have the obliquity factor equal to 1 in the forward direction, that is, $K_1 = 1$ (rather than $1/\lambda$), from which it follows that Q must be equal to $1/\lambda$. In that case, $\mathcal{E}_A \rho \lambda = \mathcal{E}_0$, which is fine dimensionally. Keep in mind that $\mathcal{E}_A$ is the secondary-wavelet source strength per unit area over the primary wavefront of radius ρ, and $\mathcal{E}_0/\rho$ is the amplitude of that primary wave $E_0(\rho)$. Thus $\mathcal{E}_A = E_0(\rho)/\lambda$. There is one other problem, and that is the $\pi/2$ phase difference between Eqs. (10.86) and

(10.87). This can be accounted for if we are willing to assume that the secondary sources radiate one-quarter of a wavelength out-of-phase with the primary wave (see Section 4.2.3).

We have found it necessary to modify the initial statement of the Huygens–Fresnel Principle, but this should not distract us from our rather pragmatic reasons for using it, which are twofold. First, the Huygens–Fresnel theory can be shown to be an approximation of the Kirchhoff formulation and as such is no longer merely a contrivance. Second, it yields, in a fairly simple way, many predictions that are in fine agreement with experimental observations. Don't forget that it worked quite well in the Fraunhofer approximation.

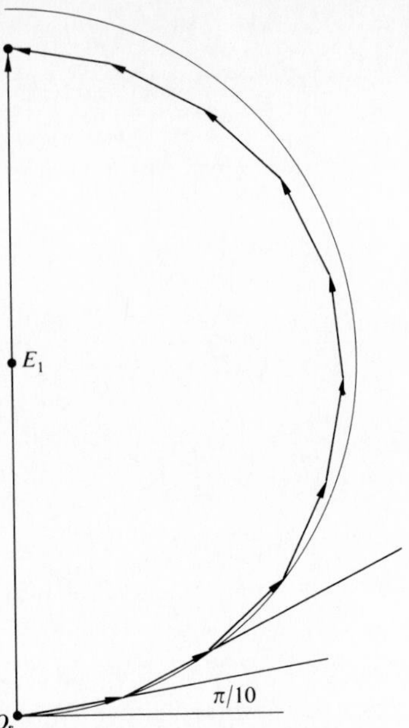

FIGURE 10.49 Phasor addition.

10.3.2 The Vibration Curve

We now develop a graphic method for qualitatively analyzing a number of diffraction problems that arise predominantly from circularly symmetric configurations.

Imagine that the first, or polar, Fresnel zone in Fig. 10.47 is divided into N subzones by the intersection of spheres, centered on P, of radii

$$r_0 + \lambda/2N,\ r_0 + \lambda/N,\ r_0 + 3\lambda/2N, \ldots,\ r_0 + \lambda/2$$

Each subzone contributes to the disturbance at P, the resultant of which is of course just E_1. Since the phase difference across the entire zone, from O to its edge, is π rad (corresponding to $\lambda/2$), each subzone is shifted by π/N rad. Figure 10.49 depicts the vector addition of the subzone phasors, where, for convenience, $N = 10$. The chain of phasors deviates very slightly from the circle, because the obliquity factor shrinks each successive amplitude. When the number of subzones is increased to infinity (i.e., $N \to \infty$), the polygon of vectors blends into a segment of a smooth spiral called a **vibration curve**. For each additional Fresnel zone, the vibration curve swings through *one half-turn* and a phase of π as it spirals inward. As shown in Fig. 10.50, the points O_s, Z_{s1}, Z_{s2}, $Z_{s3}, \ldots$, O'_s on the spiral correspond to points O, Z_1, Z_2, Z_3, ..., O', respectively, on the wavefront in Fig. 10.47. Each point $Z_1, Z_2, \ldots, Z_m$ lies on the periphery of a zone, so each point $Z_{s1}, Z_{s2}, \ldots, Z_{sm}$ is separated by a half-turn. We will see later, in Eq. (10.91), that the radius of each zone is proportional to the square root of its numerical designation, m. The radius of the hundredth zone will be only 10 times that of the first zone. Initially, therefore, the angle θ increases rapidly; thereafter it gradually slows down as m becomes larger. Accordingly, $K(\theta)$ decreases rapidly only for

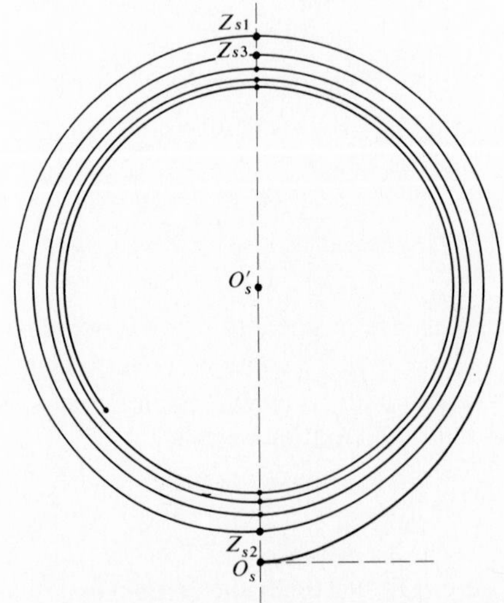

FIGURE 10.50 The vibration curve.

the first few zones. The result is that as the spiral circulates around with increasing m, it becomes tighter and tighter, deviating from a circle by a smaller amount for each revolution.

Keep in mind that the spiral is made up of an infinite number of phasors, each shifted by a small phase angle. The relative phase between any two disturbances at P, coming from two points on the wavefront, say O and A, can be depicted as shown in Fig. 10.51. The angle made by the tangents to the vibration curve, at points O_s and A_s, is β, and this is the desired phase difference. If the point A is considered to lie on the boundary of a cap-shaped region of the wavefront, the resultant at P from the whole region is $\overrightarrow{O_sA_s}$ at an angle δ.

The total disturbance arriving at P from an unimpeded wave is the sum of the contributions from all the zones between O and O'. The length of the vector from O_s to O'_s is therefore precisely that amplitude. Note that as expected, the amplitude $O_sO'_s$ is just about one-half the contribution from the first zone, O_sZ_{s1}. Observe that $\overrightarrow{O_sO'_s}$ has a phase of 90° with respect to the wave arriving at P from O. A wavelet emitted at O in-phase with the primary excitation gets to P still in-phase with the primary wave. This means that $\overrightarrow{O_sO'_s}$ is 90° out-of-phase with the unobstructed primary wave. This, as we have seen, is one of the shortcomings of the Fresnel formulation.

10.3.3 Circular Apertures

Spherical Waves

Fresnel's procedure, applied to a point source, can be used as a semiquantitative method to study diffraction at a circular aperture. Envision a monochromatic spherical wave impinging on a screen containing a small hole, as illustrated in Fig. 10.52. We first record the irradiance arriving at a very small sensor placed at point P on the symmetry axis. Our intention is to move the sensor around in space and so get a point-by-point map of the irradiance of the region beyond Σ.

Let us assume that the sensor at P "sees" an integral number of zones, m, filling the aperture. In actuality, the sensor merely records the irradiance at P, the zones having no reality. If m is even, then since $K_m \neq 0$,

$$E = (|E_1| - |E_2|) + (|E_3| - |E_4|) + \cdots + (|E_{m-1}| - |E_m|)$$

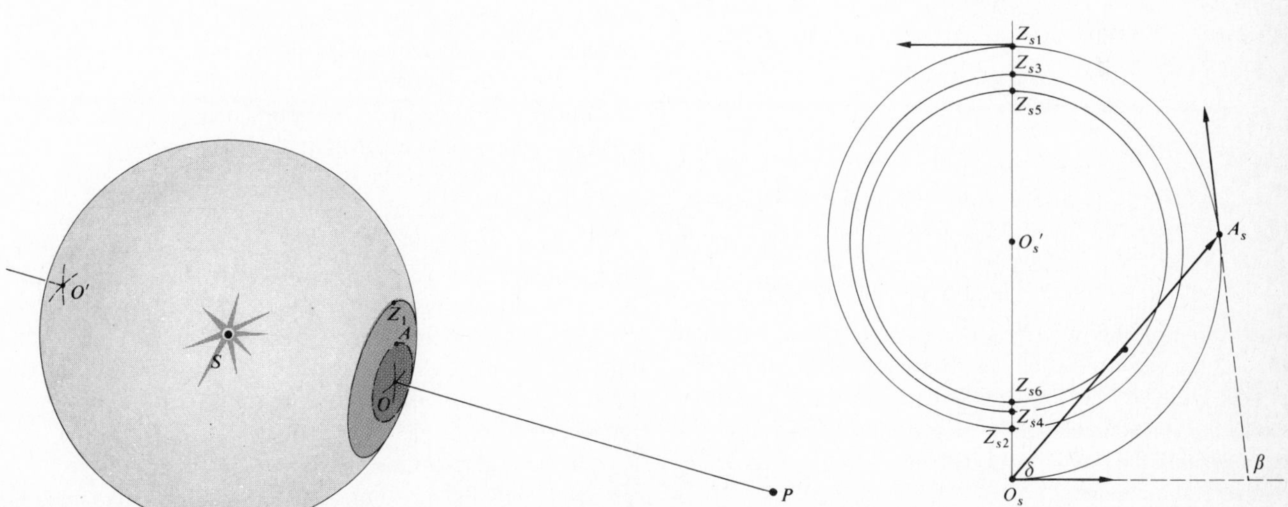

FIGURE 10.51 Wavefront and corresponding vibration curve.

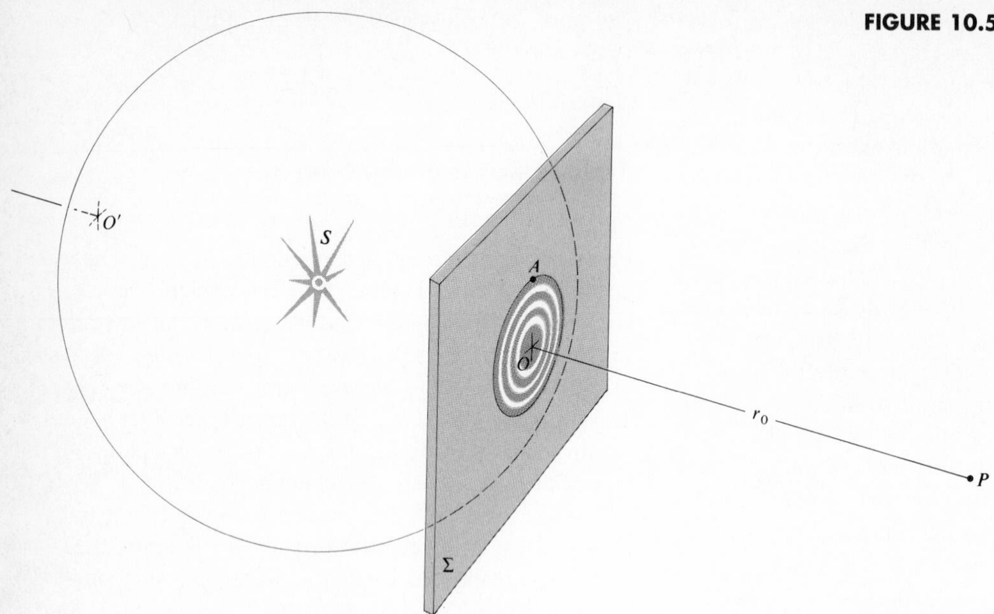

FIGURE 10.52 A circular aperture.

Because each adjacent contribution is nearly equal,

$$E \approx 0$$

and $I \approx 0$. If, on the other hand, m is odd,

$$E = |E_1| - (|E_2| - |E_3|) - (|E_4| - |E_5|) - \cdots - (|E_{m-1}| - |E_m|)$$

and

$$E \approx |E_1|$$

which is roughly twice the amplitude of the unobstructed wave. This is truly an amazing result. By inserting a screen in the path of the wave, thereby blocking out most of the wavefront, we have increased the irradiance at P by a factor of four. Conservation of energy clearly demands that there be other points where the irradiance has decreased. Because of the complete symmetry of the setup, we can expect a circular ring pattern. If m is not an integer (i.e., a fraction of a zone appears in the aperture), the irradiance at P is somewhere between zero and its maximum value.

You might see this all a bit more clearly if you imagine that the aperture is expanding smoothly from an initial value of nearly zero. The amplitude at P can be determined from the vibration curve, where A is any point on the edge of the hole. The phasor magnitude O_sA_s is the desired amplitude of the optical field. Return to Fig. 10.51; as the hole increases, A_s moves counterclockwise around the spiral toward Z_{s1} and a maximum. Allowing the second zone in reduces O_sA_s to O_sZ_{s2}, which is nearly zero, and P becomes a dark spot. As the aperture increases, O_sA_s oscillates in length from nearly zero to a number of successive maxima, which themselves gradually decrease. Finally, when the hole is fairly large, the wave is essentially unobstructed, A_s approaches O'_s, and further changes in O_sA_s are imperceptible.

To map the rest of the pattern, we now move the sensor along any line perpendicular to the axis, as shown in Fig. 10.53. At P we assume that two complete zones fill the aperture and $E \approx 0$. At P_1 the second zone has been partially obscured and the third begins to show; E is no longer zero. At P_2 a good fraction of the second zone is hidden, whereas the third is even more evident. Since the contributions from the first and third zones are in-phase, the sensor, placed anywhere on the dotted circle passing through P_2, records a bright spot. As it moves radially outward and portions of successive zones are uncovered, the sensor detects a series of relative maxima and minima. Figure 10.54 shows the diffraction patterns for a number of holes ranging in diameter from 1 mm to 4 mm as they appear on a screen 1 m away. Starting from the top left and moving right, the first four holes are so small that only a fraction of the first zone is uncovered. The sixth hole uncovers the first and second zones and is therefore black at its center. The ninth hole uncovers the first three zones and is once again bright at its center. Notice that even slightly beyond the geometric shadow at P_3, in Fig. 10.53, the first zone is partially uncovered. Each of the last few contributing segments is only a small fraction of its respective zone and as such is negligible. The sum of all the amplitudes of the fractional zones, though small, is therefore still finite. Farther into the geometric shadow, however, the entire first zone is obscured, the last terms are again negligible, and this time the series does indeed go to zero and darkness.

We can gain a better appreciation of the actual size of the things we are dealing with by computing the number of zones in a given aperture. The area of each zone (from Problem 10.43) is given by

$$A = \frac{\rho}{(\rho + r_0)} \pi r_0 \lambda \tag{10.88}$$

If the aperture has a radius R, a good approximation of the number of zones within it is simply

$$\frac{\pi R^2}{A} = \frac{(\rho + r_0)R^2}{\rho r_0 \lambda} \tag{10.89}$$

For example, with a point source 1 m behind the aperture ($\rho \approx 1$ m), a plane of observation 1 m in front of it ($r_0 = 1$ m), and $\lambda = 500$ nm, there are four zones when $R = 1$ mm, and 400 zones when $R = 1$ cm. When both ρ and r_0 are increased to the point where only a small fraction of a zone appears in

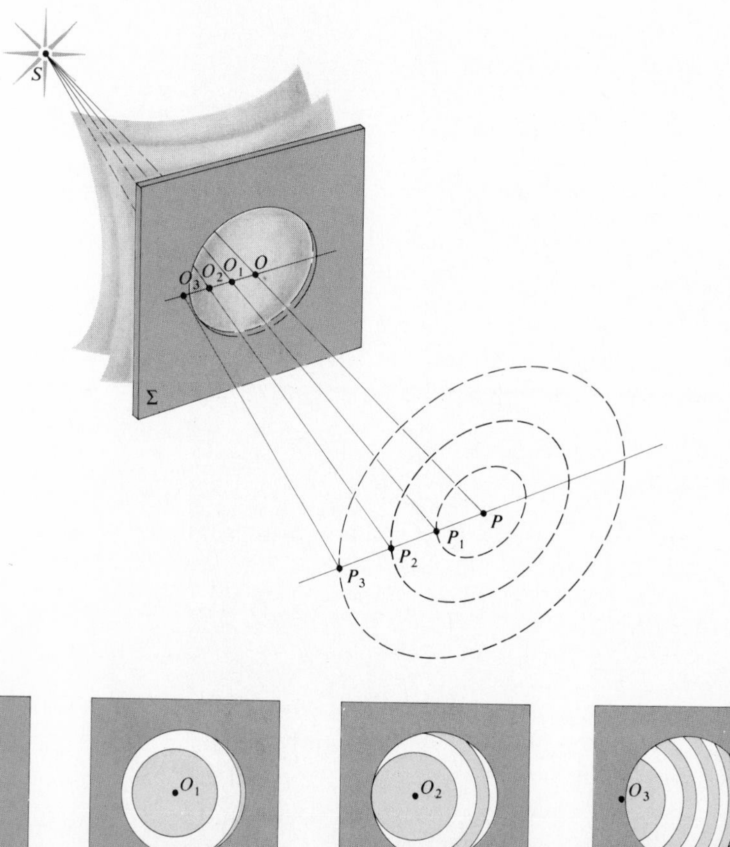

FIGURE 10.53 Zones in a circular aperture.

the aperture, Fraunhofer diffraction occurs. This is essentially a restatement of the Fraunhofer condition of Section 10.1.2; see Problem 10.1 as well.

It follows from Eq. (10.89) that the number of zones filling the aperture depends on the distance r_0 from P to O. As P moves in either direction along the central axis, the number of uncovered zones, whether increasing or decreasing, oscillates between odd and even integers. As a result, the irradiance goes through a series of maxima and minima. Clearly, this does not occur in the Fraunhofer configuration, where by definition, more than one zone cannot appear in the aperture.*

*D. S. Burch, "Fresnel Diffraction by a Circular Aperture," *Am. J. Phys.* **53**, 255 (1985).

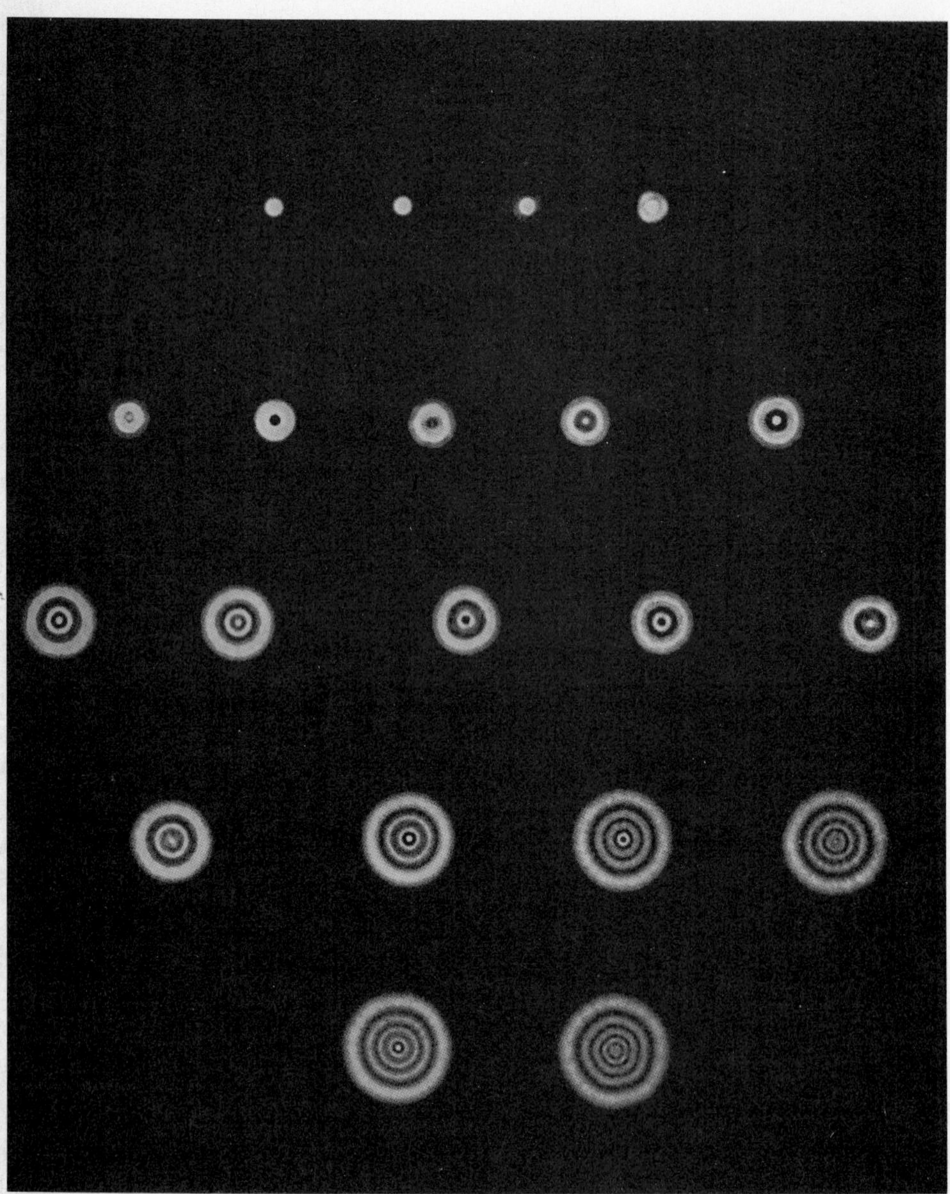

FIGURE 10.54 Diffraction patterns for circular apertures of increasing size. (Francis Weston Sears, *Optics*, ©1949, Addison-Wesley Reading, MA. Reprinted by permission of Addison Wesley Longman, Inc.)

Plane Waves

Suppose now that the point source has been moved so far from the diffracting screen that the incoming light can be regarded as a plane wave ($\rho \to \infty$). Referring to Fig. 10.55, we derive an expression for the radius of the mth zone, R_m. Since $r_m = r_0 + m\lambda/2$,

$$R_m^2 = (r_0 + m\lambda/2)^2 - r_0^2$$

and so

$$R_m^2 = mr_0\lambda + m^2\lambda^2/4 \tag{10.90}$$

Under most circumstances the second term in Eq. (10.90) is negligible as long as m is not extremely large; consequently

$$R_m^2 = mr_0\lambda \tag{10.91}$$

and *the radii are proportional to the square roots of integers.* Using a collimated He–Ne laser ($\lambda_0 = 632.8$ nm), the radius of the first zone is 1 mm when viewed from a distance of 1.58 m. Under these particular conditions Eq. (10.91) is applicable as long as $m << 10^7$, in which case $R_m = \sqrt{m}$ in millimeters.

Figure 10.53 requires a slight modification in that now the lines $\overline{O_1P_1}$, $\overline{O_2P_2}$, and $\overline{O_3P_3}$ are perpendiculars dropped from the points of observation to Σ.

10.3.4 Circular Obstacles

In 1818 Fresnel entered a competition sponsored by the French Academy. His paper on the theory of diffraction ultimately won first prize and the title *Mémoire Courronné*, but not until it had provided the basis for a rather interesting story. The judging committee consisted of Pierre Laplace, Jean B. Biot, Siméon D. Poisson, Dominique F. Arago, and Joseph L. Gay-Lussac—a formidable group indeed. Poisson, who was an ardent critic of the wave description of light, deduced a remarkable and seemingly untenable conclusion from Fresnel's theory. He showed that a bright spot would be visible at the center of the shadow of a circular opaque obstacle, a result that he felt proved the absurdity of Fresnel's treatment. We can come to the same conclusion by considering the following, somewhat oversimplified argument. Recall that an unobstructed wave yields a disturbance [Eq. (10.85)] given by $E \approx |E_1|/2$. If some sort of obstacle precisely covers the first Fres-

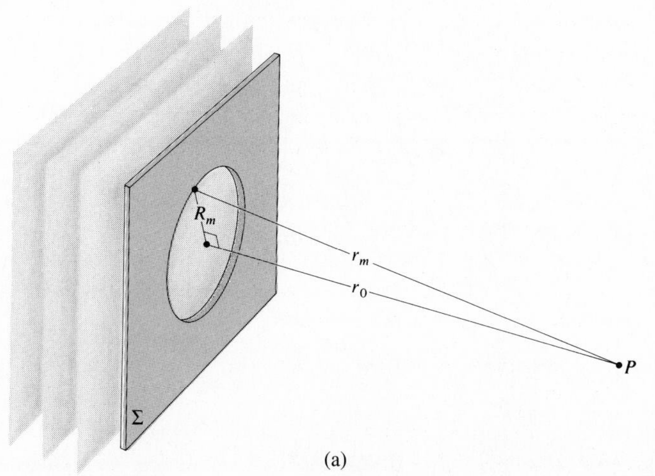

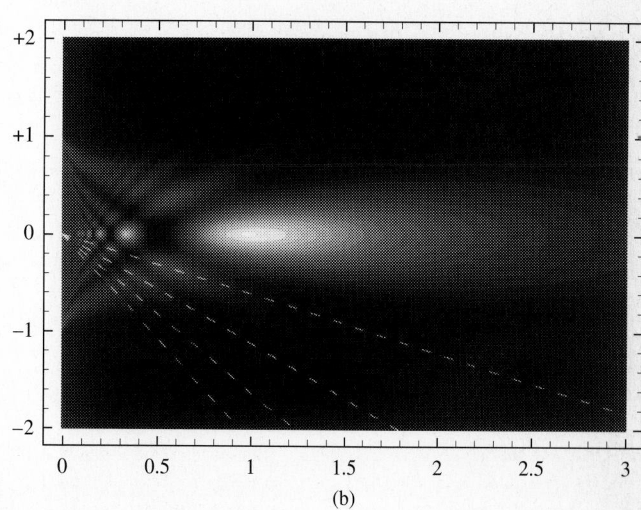

FIGURE 10.55 (*a*) Plane waves incident on a circular hole. (*b*) A cross section of the three-dimensional irradiance distribution. The horizontal axis is scaled in units of R^2/λ and the vertical axis in R, where R is the radius of the hole. Thus the aperture extends from +1 to −1. At a distance of $R^2/\lambda = r_0$ one Fresnel zone fills the aperture and the irradiance has a maximum. Beyond that $I(r)$ falls off monotonically until it reaches the far-field regime. The first four zeros of the Fraunhofer irradiance distribution lie on the dashed lines. (Photo courtesy, G. W. Forbes, The Institute of Optics, University of Rochester.)

nel zone, so that its contribution of $|E_1|$ is subtracted out, then $E \approx -|E_1|/2$. It is therefore possible that at some point P on the axis, the irradiance will be unaltered by the insertion of that obstruction. This surprising prediction, fashioned by Poisson as the death blow to the wave theory, was almost immediately verified experimentally by Arago; the spot actually existed. Amusingly enough, Poisson's spot, as it is now called, had been observed many years earlier (1723) by Maraldi, but this work had long gone unnoticed.*

We now examine the problem a bit more closely, since it is quite evident from Fig. 10.56 that there is a good deal of structure in the actual shadow pattern. If the opaque obstacle, be it a disk or sphere, obscures the first ℓ zones, then

$$E = |E_{\ell+1}| - |E_{\ell+2}| + \cdots + |E_m|$$

(where, as before, there is no absolute significance to the signs other than that alternate terms must subtract). Unlike the analysis for the circular aperture, E_m now approaches zero, because $K_m \to 0$. The series must be evaluated in the same manner as that of the unobstructed wave [Eqs. (10.78) and (10.79)]. Repeating that procedure yields

$$E \approx \frac{|E_{\ell+1}|}{2} \tag{10.92}$$

and the irradiance on the central axis is generally only slightly less than that of the unobstructed wave. *There is a bright spot everywhere along the central axis except immediately behind the circular obstacle.* The wavelets propagating beyond the disk's circumference meet in-phase on the central axis. Notice that as P moves close to the disk, θ increases, $K_{\ell+1} \to 0$, and the irradiance gradually falls off to zero. If the disk is large, the $(\ell + 1)$th zone is very narrow, and any irregularities in the obstacle's surface may seriously obscure that zone. For Poisson's spot to be readily observable, the obstacle must be smooth and circular.

If A is a point on the periphery of the disk or sphere, A_s is the corresponding point on the vibration curve (Fig. 10.57). As the disk increases for a fixed P, A_s spirals in counterclockwise toward O'_s, and the amplitude $A_sO'_s$ gradually decreases. The same thing happens as P moves toward a disk of constant size.

Off the axis, the zones covered in Fig. 10.53 for the circular aperture will now be exposed and vice versa. Accordingly, a whole series of concentric bright and dark rings will surround the central spot.

The opaque disk images S at P and would similarly form a crude image of every point in an extended source. R. W. Pohl

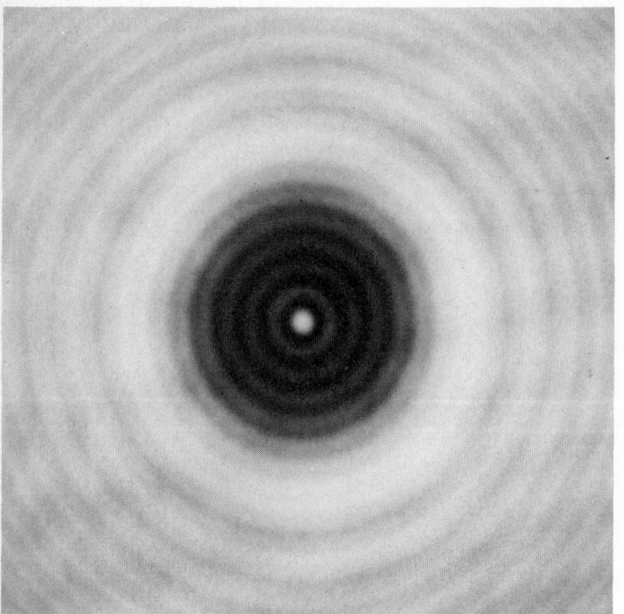

FIGURE 10.56 Shadow of a 1/8-inch diameter ball bearing. The bearing was glued to an ordinary microscope slide and illuminated with a He–Ne laserbeam. There are some faint extraneous nonconcentric fringes arising from both the microscope slide and a lens in the beam. (E. H.)

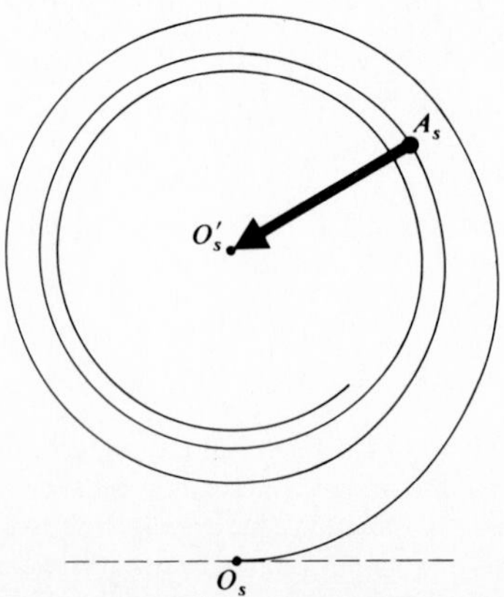

FIGURE 10.57 The vibration curve applied to a circular obstruction.

*See J. E. Harvey and J. L. Forgham, "The Spot of Arago: New Relevance for an Old Phenomenon," *Am. J. Phys.* **52**, 243 (1984).

has shown that a small disk can therefore be used as a crude positive lens.

The diffraction pattern can be seen with little difficulty, but you need a telescope or binoculars. Glue a small ball bearing ($\approx\frac{1}{8}$ or $\frac{1}{4}$ inch in diameter) to a microscope slide, which then serves as a handle. Place the bearing a few meters beyond the point source and observe it from 3 or 4 meters away. Position it so that it is directly in front of and completely obscuring the source. You will need the telescope to magnify the image, since r_0 is so large. If you can hold the telescope steady, the ring system should be quite clear.

10.3.5 The Fresnel Zone Plate

In our previous considerations we utilized the fact that successive Fresnel zones tended to nullify each other. This suggests that we will observe a tremendous increase in irradiance at P, if we remove either all the even or all the odd zones. A screen that alters the light, either in amplitude or phase, coming from every other half-period zone is called a **zone plate**.*

Suppose that we construct a zone plate that passes only the first 20 odd zones and obstructs the even zones,

$$E = E_1 + E_3 + E_5 + \cdots + E_{39}$$

and each of these terms is approximately equal. For an unobstructed wavefront, the disturbance at P would be $E_1/2$, whereas with the zone plate in place, $E \approx 20E_1$. The irradiance has been increased by a factor of 1600. The same result would obviously be true if the even zones were passed instead.

To calculate the radii of the zones shown in Fig. 10.58, refer to Fig. 10.59. The outer edge of the mth zone is marked by the point A_m. By definition, a wave that travels the path S–A_m–P must arrive out-of-phase by $m\lambda/2$ with a wave that traverses the path S–O–P, that is,

$$(\rho_m + r_m) - (\rho_0 + r_0) = m\lambda/2 \qquad (10.93)$$

Clearly $\rho_m = (R_m^2 + \rho_0^2)^{1/2}$ and $r_m = (R_m^2 + r_0^2)^{1/2}$. Expand both these expressions using the binomial series. Since R_m is comparatively small, retaining only the first two terms yields

$$\rho_m = \rho_0 + \frac{R_m^2}{2\rho_0} \quad \text{and} \quad r_m = r_0 + \frac{R_m^2}{2r_0}$$

Finally, substituting into Eq. (10.93), we obtain

$$\left(\frac{1}{\rho_0} + \frac{1}{r_0}\right) = \frac{m\lambda}{R_m^2} \qquad (10.94)$$

Under plane-wave illumination ($\rho_0 \to \infty$), and Eq. (10.94)

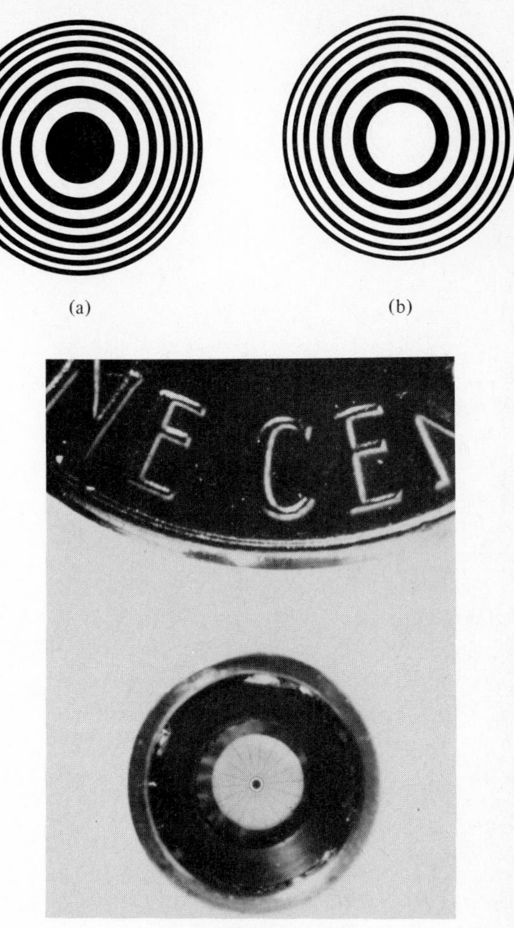

FIGURE 10.58 (*a*) and (*b*) Zone plates. (*c*) A zone plate used to image alpha particles coming from a target 1 cm in front, on photographic film 5 cm behind. The plate is 2.5 mm in diameter and contains 100 zones, the narrowest of which is 5.3 μm wide. (Photo courtesy of the University of California, Lawrence Livermore National Laboratory and the Department of Energy.)

*Lord Rayleigh seems to have invented the zone plate, as witnessed by this entry of April 11, 1871, in his notebook: "The experiment of blocking out the odd Huygens zones so as to increase the light at centre succeeded very well."

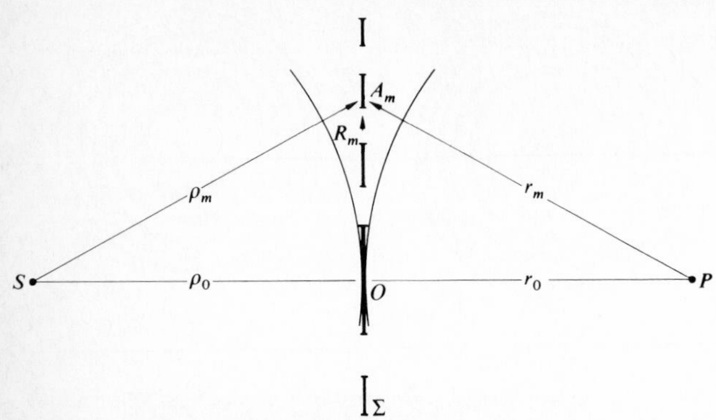

FIGURE 10.59 Zone-plate geometry.

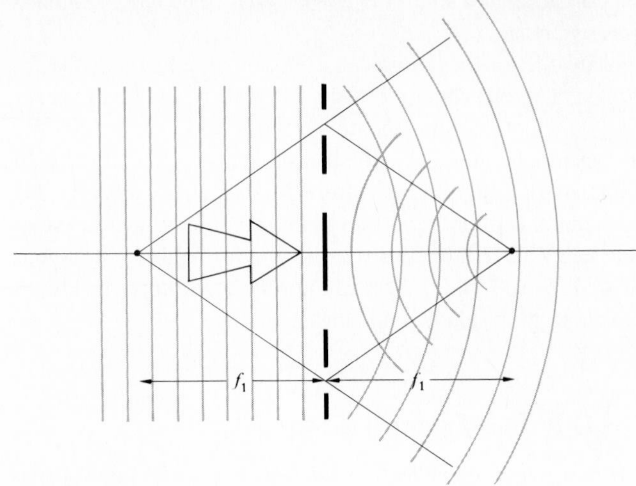

FIGURE 10.60 Zone-plate foci.

reduces to

$$R_m^2 = mr_0\lambda \qquad [10.91]$$

which is an approximation of the exact expression stated by Eq. (10.90). Equation (10.94) has a form identical to that of the thin-lens equation, which is not merely a coincidence, since S is actually imaged in converging diffracted light at P. Accordingly, the *primary focal length* is said to be

$$f_1 = \frac{R_m^2}{m\lambda} \qquad (10.95)$$

(Note that the zone plate will show extensive chromatic aberration.) The points S and P are said to be conjugate foci. With a collimated incident beam (Fig. 10.60) the image distance is the primary or *first-order* focal length, which in turn corresponds to a principal maximum in the irradiance distribution. In addition to this real image, there is also a virtual image formed of diverging light a distance f_1 in front of Σ. At a distance of f_1 from Σ each ring on the plate is filled by exactly one half-period zone on the wavefront. If we move a sensor along the S–P axis toward Σ, it registers a series of very small irradiance maxima and minima until it arrives at a point $f_1/3$ from Σ. At that *third-order focal point*, there is a pronounced irradiance peak. Additional focal points will exist at $f_1/5$, $f_1/7$, and so forth, unlike a lens but even more unlike a simple opaque disk.

Following a suggestion by Lord Rayleigh, R. W. Wood constructed a *phase-reversal zone plate*. Instead of blocking out every other zone, he increased the thickness of alternate zones, thereby retarding their phase by π. Since the entire plate is transparent, the amplitude should double, and the irradiance increase by a factor of four. In actuality, the device does not work quite that well, because the phase is not really constant over each zone. Ideally, the retardation should be made to vary gradually over a zone, jumping back by π at the start of the next zone.*

The usual way to make an optical zone plate is to draw a large-scale version and then photographically reduce it. Plates with hundreds of zones can be made by photographing a Newton's ring pattern, in collimated quasimonochromatic light. Rings of aluminum foil on cardboard work very well for microwaves.

Zone plates can be made of metal with a self-supporting spoked structure, so that the transparent regions are devoid of any material. These will function as lenses in the range from ultraviolet to soft X-rays, where ordinary glass is opaque.

*See Ditchburn, *Light*, 2nd ed., p. 232; M. Sussman, "Elementary Diffraction Theory of Zone Plates," *Am. J. Phys.* **28**, 394 (1960); Ora E. Myers, Jr., "Studies of Transmission Zone Plates," *Am. J. Phys.* **19**, 359 (1951); and J. Higbie, "Fresnel Zone Plate: Anomalous Foci," *Am. J. Phys.* **44**, 929 (1976).

10.3.6 Fresnel Integrals and the Rectangular Aperture

We now consider a class of problems within the domain of Fresnel diffraction, which no longer have the circular symmetry of the previously studied configurations. Consider Fig. 10.61 where dS is an area element situated at some arbitrary point A whose coordinates are (y, z). The location of the origin O is determined by a perpendicular drawn to Σ from the position of the monochromatic point source. The contribution to the optical disturbance at P from the secondary sources on dS has the form given by Eq. (10.74). Making use of what we learned from the freely propagating wave ($\mathcal{E}_A\rho\lambda = \mathcal{E}_0$), we can rewrite that equation as

$$dE_p = \frac{K(\theta)\mathcal{E}_0}{\rho r\lambda} \cos\,[k(\rho + r) - \omega t]\, dS \qquad (10.96)$$

The sign of the phase has changed from that of Eq. (10.74) and is written in this way to conform with traditional treatment. *In the case where the dimensions of the aperture are small* in comparison to ρ_0 and r_0, we can set $K(\theta) = 1$ and let $1/\rho r$ equal $1/\rho_0 r_0$ in the amplitude coefficient. Being more careful about approximations introduced into the phase, apply the Pythagorean theorem to triangles SOA and POA to get

$$\rho = (\rho_0^2 + y^2 + z^2)^{1/2}$$

and

$$r = (r_0^2 + y^2 + z^2)^{1/2}$$

Expand these using the binomial series and form

$$\rho + r \approx \rho_0 + r_0 + (y^2 + z^2)\frac{\rho_0 + r_0}{2\rho_0 r_0} \qquad (10.97)$$

Observe that this is a more sensitive approximation than that used in the Fraunhofer analysis [Eq. (10.40)], where the terms quadratic and higher in the aperture variables were neglected. The disturbance at P in the complex representation is

$$\tilde{E}_p = \frac{\mathcal{E}_0 e^{-i\omega t}}{\rho_0 r_0 \lambda} \int_{y_1}^{y_2}\int_{z_1}^{z_2} e^{ik(\rho + r)}\, dy\, dz \qquad (10.98)$$

Following the usual form of derivation, we introduce the dimensionless variables u and v defined by

$$u \equiv y\left[\frac{2(\rho_0 + r_0)}{\lambda\rho_0 r_0}\right]^{1/2} \qquad v \equiv z\left[\frac{2(\rho_0 + r_0)}{\lambda\rho_0 r_0}\right]^{1/2} \qquad (10.99)$$

Substituting Eq. (10.97) into Eq. (10.98) and utilizing the new variables, we arrive at

$$\tilde{E}_p = \frac{\mathcal{E}_0}{2(\rho_0 + r_0)} e^{i[k(\rho_0 + r_0) - \omega t]} \int_{u_1}^{u_2} e^{i\pi u^2/2}\, du \int_{v_1}^{v_2} e^{i\pi v^2/2}\, dv \qquad (10.100)$$

The term in front of the integral represents the unobstructed disturbance at P divided by 2; let us call it $\tilde{E}_u/2$. The integral itself can be evaluated using two functions, $\mathscr{C}(w)$ and $\mathscr{S}(w)$, where w represents either u or v. These quantities, which are known as the **Fresnel integrals**, are defined by

$$\mathscr{C}(w) \equiv \int_0^w \cos\,(\pi w'^2/2)\, dw'$$

$$\mathscr{S}(w) \equiv \int_0^w \sin\,(\pi w'^2/2)\, dw' \qquad (10.101)$$

Both functions have been extensively studied, and their numerical values are well tabulated. Their interest to us at this

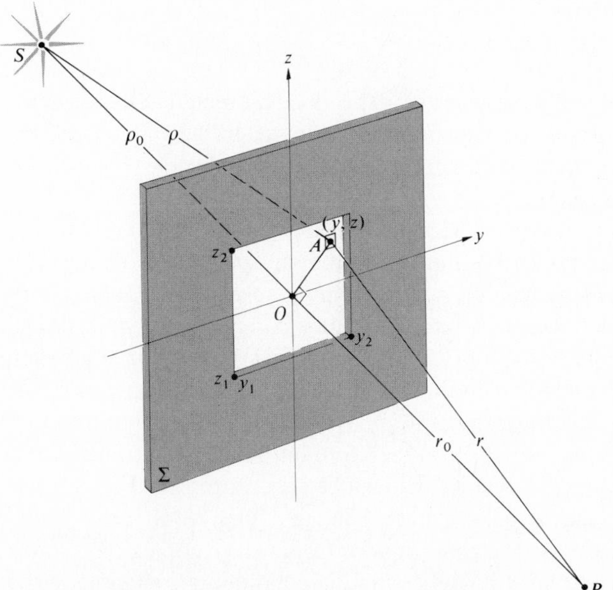

FIGURE 10.61 Fresnel diffraction at a rectangular aperture.

point derives from the fact that

$$\int_0^w e^{i\pi w'^2/2}\, dw' = \mathscr{C}(w) + i\mathscr{S}(w)$$

and this, in turn, has the form of the integrals in Eq. (10.100). The disturbance at P is then

$$\tilde{E}_p = \frac{\tilde{E}_u}{2}\,[\mathscr{C}(u) + i\mathscr{S}(u)]_{u_1}^{u_2}[\mathscr{C}(v) + i\mathscr{S}(v)]_{v_1}^{v_2} \qquad (10.102)$$

which can be evaluated using the tabulated values of $\mathscr{C}(u_1)$, $\mathscr{C}(u_2)$, $\mathscr{S}(u_1)$, and so on. The mathematics becomes rather involved if we compute the disturbance at all points of the plane of observation, leaving the position of the aperture fixed. Instead we will fix the *S–O–P* line and imagine that we move the aperture through small displacements in the Σ-plane. This has the effect of translating the origin O with respect to the fixed aperture, thereby scanning the pattern over the point P. Each new position of O corresponds to a new set of relative boundary locations y_1, y_2, z_1, and z_2. These in turn mean new values of u_1, u_2, v_1, and v_2, which, when substituted into Eq. (10.102), yield a new $\tilde{E}_p$. The error encountered in such a procedure is negligible, as long as the aperture is displaced by distances that are small compared with ρ_0. This approach is therefore even more appropriate to incident plane waves. In that case if E_0 is the amplitude of the incoming plane wave at Σ, Eq. (10.96) becomes simply

$$dE_p = \frac{E_0 K(\theta)}{r\lambda}\cos(kr - \omega t)\, dS$$

where, as before, $\mathcal{E}_A = E_0/\lambda$. This time, with

$$u = y\left(\frac{2}{\lambda r_0}\right)^{1/2} \qquad v = z\left(\frac{2}{\lambda r_0}\right)^{1/2} \qquad (10.103)$$

where we have divided the numerator and denominator in Eq. (10.99) by ρ_0 and then let it go to infinity, $\tilde{E}_p$ takes the same form as Eq. (10.102), where $\tilde{E}_u$ is again the unobstructed disturbance. The irradiance at P is $\tilde{E}_p\tilde{E}_p^*/2$; hence

$$I_p = \frac{I_0}{4}\{[\mathscr{C}(u_2) - \mathscr{C}(u_1)]^2 + [\mathscr{S}(u_2) - \mathscr{S}(u_1)]^2\}$$
$$\times \{[\mathscr{C}(v_2) - \mathscr{C}(v_1)]^2 + [\mathscr{S}(v_2) - \mathscr{S}(v_1)]^2\} \qquad (10.104)$$

where I_0 is the unobstructed irradiance at P.

As a simple example, envision a square hole 2 mm on each side under plane-wave illumination at 500 nm. If P is 4 m away and directly opposite point O at the center of the aperture, $u_2 = 1.0$, $u_1 = -1.0$, $v_2 = 1.0$, and $v_1 = -1.0$. The Fresnel integrals are both odd functions, that is,

$$\mathscr{C}(w) = -\,\mathscr{C}(-w) \quad\text{and}\quad \mathscr{S}(w) = -\mathscr{S}(-w)$$

Consequently

$$I_p = \frac{I_0}{4}\{[2\mathscr{C}(1)]^2 + [2\mathscr{S}(1)]^2\}^2$$

and a numerical value is easily obtained. To find the irradiance somewhere else in the pattern, for example, 0.1 mm to the left of center, move the aperture relative to the *OP*-line accordingly, whereupon $u_2 = 1.1$, $u_1 = -0.9$, $v_2 = 1.0$, and $v_1 = -1.0$. The resultant I_p will also be equal to that found at 0.1 mm to the right of center. Indeed, because the aperture is square, the same value obtains 0.1 mm directly above and below center as well (Fig. 10.62).

We can approach the limiting case of free propagation by allowing the aperture dimensions to increase indefinitely. Making use of the fact that $\mathscr{C}(\infty) = \mathscr{S}(\infty) = \frac{1}{2}$ and $\mathscr{C}(-\infty) = \mathscr{S}(-\infty) = -\frac{1}{2}$ the irradiance at P, opposite the center of the aperture, is

$$I_p = I_0$$

which is exactly correct. This is rather remarkable, considering that when the length $\overline{OA}$ is large, all the approximations made in the derivation are no longer applicable. It should be realized, however, that a relatively small aperture satisfying the approximations can still be large enough to effectively show no diffraction in the region opposite its center. For example, with $\rho_0 = r_0 = 1$ m an aperture that subtends an angle of about 1° or 2° at P may correspond to values of $|u|$ and $|v|$ of roughly 25 to 50. The quantities $\mathscr{C}$ and $\mathscr{S}$ are then very close to their limiting values of $\frac{1}{2}$. Further increases in the aperture dimensions beyond the point where the approximations are violated can therefore introduce only a small error. This implies that we need not be very concerned about restricting the actual aperture size (as long as $r_0 >> \lambda$ and $\rho_0 >> \lambda$). The contributions from wavefront regions remote from O must be quite small, a condition attributable to the obliquity factor and the inverse r-dependence of the amplitude of the secondary wavelets.

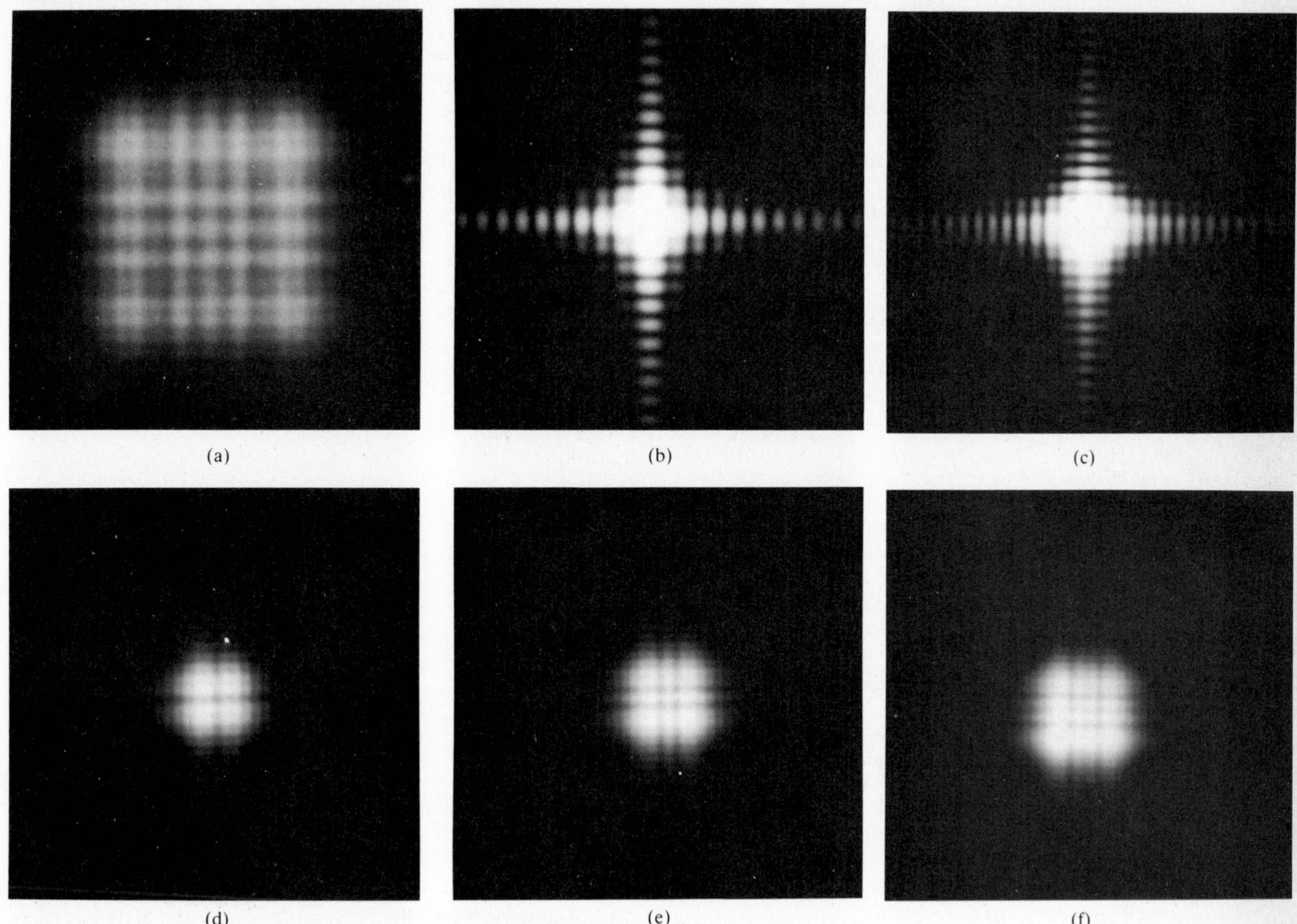

FIGURE 10.62 *(a)* A typical Fresnel pattern for a square aperture. *(b)*–*(f)* A series of Fresnel patterns for increasing square apertures under identical conditions. Note that as the hole gets larger, the pattern changes from a spread-out Fraunhofer-like distribution to a far more localized structure. (E. H.)

10.3.7 The Cornu Spiral

Marie Alfred Cornu (1841–1902), professor at the École Polytechnique in Paris, devised an elegant geometrical depiction of the Fresnel integrals, akin to the vibration curve already considered. Figure 10.63, which is known as the ***Cornu spiral***, is a plot in the complex plane of the points $\tilde{B}(w) \equiv \mathscr{C}(w) + i\mathscr{S}(w)$ as w takes on all possible values from 0 to $\pm\infty$. This just means that we plot $\mathscr{C}(w)$ on the horizontal or real axis and $\mathscr{S}(w)$ on the vertical or imaginary axis. The appropriate numerical values are taken from Table 10.2. If $d\ell$ is an element of arc length measured along the curve, then

$$d\ell^2 = d\mathscr{C}^2 + d\mathscr{S}^2$$

From the definitions (10.101),

$$d\ell^2 = (\cos^2 \pi w^2/2 + \sin^2 \pi w^2/2)dw^2$$

and

$$d\ell = dw$$

Values of w correspond to the arc length and are marked off

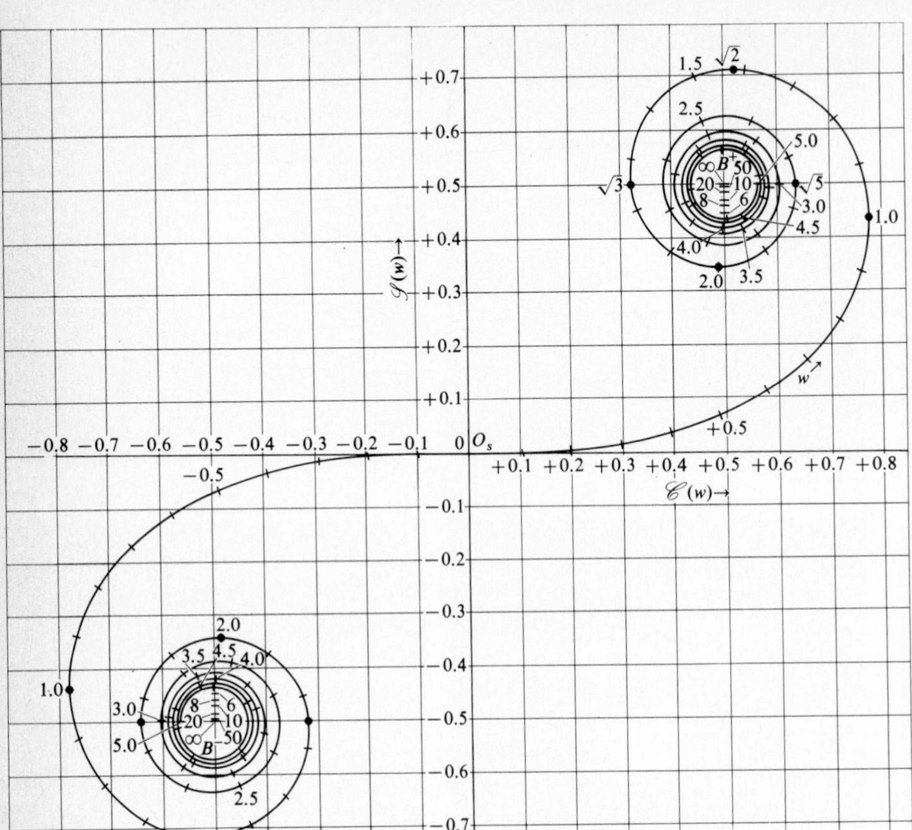

FIGURE 10.63 The Cornu spiral.

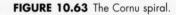

along the spiral in Fig. 10.63. As w approaches $\pm\infty$, the curve spirals into its limiting values at $\tilde{B}^+ = \frac{1}{2} + i\frac{1}{2}$ and $\tilde{B}^- = -\frac{1}{2} - i\frac{1}{2}$. The slope of the spiral is

$$\frac{d\mathscr{S}}{d\mathscr{C}} = \frac{\sin \pi w^2/2}{\cos \pi w^2/2} = \tan \frac{\pi w^2}{2} \tag{10.105}$$

and so the angle between the tangent to the spiral at any point and the $\mathscr{C}$-axis is $\beta = \pi w^2/2$.

The Cornu spiral can be used either as a convenient tool for quantitative determinations or as an aid to gaining a qualitative picture of a diffraction pattern (which was also the case with the vibration curve). As an example of its quantitative uses, reconsider the problem of a 2-mm-square hole, dealt with in the previous section ($\lambda = 500$ nm, $r_0 = 4$ m, and plane-wave illumination). We wish to find the irradiance at P directly opposite the aperture's center, where in this case $u_1 = -1.0$ and $u_2 = 1.0$. The variable u is measured along the arc; that is, w is replaced by u on the spiral. Place two points on the spiral at distances from O_s equal to u_1 and u_2. (These are symmetrical with respect to O_s, because P is now opposite the aperture's center.) Label the two points $\tilde{B}_1(u)$ and $\tilde{B}_2(u)$, respectively, as in Fig. 10.64. The phasor $\tilde{\mathbf{B}}_{12}(u)$ drawn from $\tilde{B}_1(u)$ to $\tilde{B}_2(u)$ is just the complex number $\tilde{B}_2(u) - \tilde{B}_1(u)$,

$$\tilde{\mathbf{B}}_{12}(u) = [\mathscr{C}(u) + i\mathscr{S}(u)]_{u_1}^{u_2}$$

and is the first term in the expression [Eq. (10.102)] for $\tilde{E}_p$. Similarly for $v_1 = -1.0$ and $v_2 = 1.0$, $\tilde{B}_2(v) - \tilde{B}_1(v)$ is

$$\tilde{\mathbf{B}}_{12}(v) = [\mathscr{C}(v) + i\mathscr{S}(v)]_{v_1}^{v_2}$$

which is the latter portion of $\tilde{E}_p$. The magnitudes of these two complex numbers are just the lengths of the appropriate $\tilde{\mathbf{B}}_{12}$-phasors, which can be read off the curve with a ruler, using either axis as a scale. The irradiance is then simply

$$I_p = \frac{I_0}{4} |\tilde{\mathbf{B}}_{12}(u)|^2 |\tilde{\mathbf{B}}_{12}(v)|^2 \tag{10.106}$$

TABLE 10.2 Fresnel Integrals

w	$\mathscr{C}(w)$	$\mathscr{S}(w)$	w	$\mathscr{C}(w)$	$\mathscr{S}(w)$
0.00	0.0000	0.0000	4.50	0.5261	0.4342
0.10	0.1000	0.0005	4.60	0.5673	0.5162
0.20	0.1999	0.0042	4.70	0.4914	0.5672
0.30	0.2994	0.0141	4.80	0.4338	0.4968
0.40	0.3975	0.0334	4.90	0.5002	0.4350
0.50	0.4923	0.0647	5.00	0.5637	0.4992
0.60	0.5811	0.1105	5.05	0.5450	0.5442
0.70	0.6597	0.1721	5.10	0.4998	0.5624
0.80	0.7230	0.2493	5.15	0.4553	0.5427
0.90	0.7648	0.3398	5.20	0.4389	0.4969
1.00	0.7799	0.4383	5.25	0.4610	0.4536
1.10	0.7638	0.5365	5.30	0.5078	0.4405
1.20	0.7154	0.6234	5.35	0.5490	0.4662
1.30	0.6386	0.6863	5.40	0.5573	0.5140
1.40	0.5431	0.7135	5.45	0.5269	0.5519
1.50	0.4453	0.6975	5.50	0.4784	0.5537
1.60	0.3655	0.6389	5.55	0.4456	0.5181
1.70	0.3238	0.5492	5.60	0.4517	0.4700
1.80	0.3336	0.4508	5.65	0.4926	0.4441
1.90	0.3944	0.3734	5.70	0.5385	0.4595
2.00	0.4882	0.3434	5.75	0.5551	0.5049
2.10	0.5815	0.3743	5.80	0.5298	0.5461
2.20	0.6363	0.4557	5.85	0.4819	0.5513
2.30	0.6266	0.5531	5.90	0.4486	0.5163
2.40	0.5550	0.6197	5.95	0.4566	0.4688
2.50	0.4574	0.6192	6.00	0.4995	0.4470
2.60	0.3890	0.5500	6.05	0.5424	0.4689
2.70	0.3925	0.4529	6.10	0.5495	0.5165
2.80	0.4675	0.3915	6.15	0.5146	0.5496
2.90	0.5624	0.4101	6.20	0.4676	0.5398
3.00	0.6058	0.4963	6.25	0.4493	0.4954
3.10	0.5616	0.5818	6.30	0.4760	0.4555
3.20	0.4664	0.5933	6.35	0.5240	0.4560
3.30	0.4058	0.5192	6.40	0.5496	0.4965
3.40	0.4385	0.4296	6.45	0.5292	0.5398
3.50	0.5326	0.4152	6.50	0.4816	0.5454
3.60	0.5880	0.4923	6.55	0.4520	0.5078
3.70	0.5420	0.5750	6.60	0.4690	0.4631
3.80	0.4481	0.5656	6.65	0.5161	0.4549
3.90	0.4223	0.4752	6.70	0.5467	0.4915
4.00	0.4984	0.4204	6.75	0.5302	0.5362
4.10	0.5738	0.4758	6.80	0.4831	0.5436
4.20	0.5418	0.5633	6.85	0.4539	0.5060
4.30	0.4494	0.5540	6.90	0.4732	0.4624
4.40	0.4383	0.4622	6.95	0.5207	0.4591

and the problem is solved. Notice that the arc lengths along the spiral (i.e., $\Delta u = u_2 - u_1$ and $\Delta v = v_2 - v_1$) are proportional to the aperture's overall dimensions in the y- and z-direction, respectively. *The arc lengths are therefore constant, regardless of the position of P in the plane of observation.* On the other hand, the phasors $\tilde{\mathbf{B}}_{12}(u)$ and $\tilde{\mathbf{B}}_{12}(v)$, which span the arc lengths, are not constant, and they do depend on the location of P.

Maintaining the position of P opposite the center of the diffracting hole, now suppose that the aperture size is adjustable. As the square hole is gradually opened, Δv and Δu increase accordingly. The endpoints $\tilde{B}_1$ and $\tilde{B}_2$ of either of these arc lengths spiral around counterclockwise toward their limiting values of $\tilde{B}^-$ and $\tilde{B}^+$, respectively. The phasors $\tilde{\mathbf{B}}_{12}(u)$ and $\tilde{\mathbf{B}}_{12}(v)$, which are identical in this instance because of the symmetry, pass through a series of extrema. The central spot in the pattern therefore gradually shifts from relative brightness to darkness and back. All the while, the entire irradiance distribution varies continually from one beautifully intricate display to the next (Fig. 10.62). For any particular aperture size, the off-center diffraction pattern can be computed by repositioning P. It is helpful to visualize the arc length as a piece of string, whose measure is equal to either Δv or Δu. Imagine it lying on the spiral, with O_s initially at its midpoint. As P is moved, for example, to the left along the y-axis (Fig. 10.61), y_1 and therefore u_1 both become less negative, and y_2 and u_2

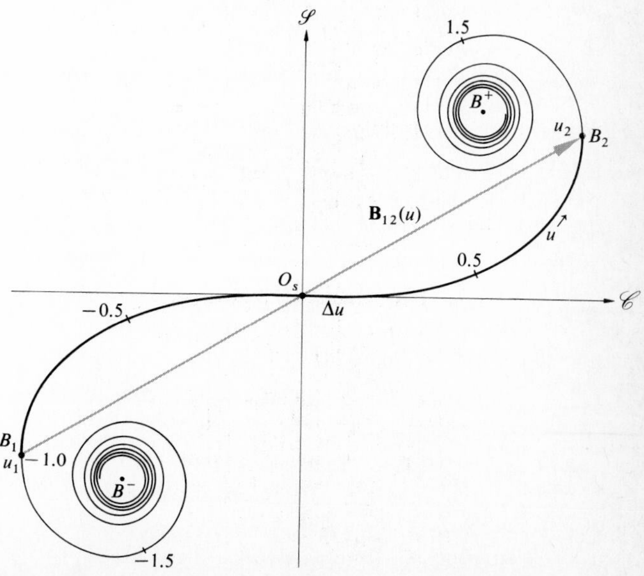

FIGURE 10.64 Cornu spiral.

increase positively. The result is that our Δu-string slides up the spiral. As the distance between the endpoints of the Δu-string changes, $|\tilde{\mathbf{B}}_{12}(u)|$ changes, and the irradiance [Eq. (10.106)] varies accordingly. When P is at the left edge of the geometric shadow, $y_1 = u_1 = 0$. As the point of observation moves into the geometric shadow, u_1 increases *positively*, and the Δu-string is now entirely on the upper half of the Cornu spiral. As u_1 and u_2 continue to increase, the string winds ever more tightly about the $\tilde{B}^+$-limit. Its ends, $\tilde{B}_1$ and $\tilde{B}_2$, become closer to each other, with the result that $|\tilde{\mathbf{B}}_{12}(u)|$ becomes quite small, and I_p decreases within the geometric shadow region. (We will come back to this point in more detail in the next section.) The same process applies when we scan in the z-direction; Δv is constant and $\tilde{\mathbf{B}}_{12}(v)$ varies.

If the aperture is completely opened out, revealing an unobstructed wave, $u_1 = v_1 = -\infty$, which means that $\tilde{B}_1(u) = \tilde{B}_1(v) = \tilde{B}^-$ and $\tilde{B}_2(u) = \tilde{B}_2(v) = \tilde{B}^+$. The $\tilde{B}^-\tilde{B}^+$-line makes a 45° angle with the $\mathscr{C}$-axis and has a length equal to $\sqrt{2}$. Consequently, the phasors $\tilde{\mathbf{B}}_{12}(u)$ and $\tilde{\mathbf{B}}_{12}(v)$ each have magnitude $\sqrt{2}$ and phase $\pi/4$, that is, $\tilde{\mathbf{B}}_{12}(u) = \sqrt{2}\exp(i\pi/4)$ and $\mathbf{B}_{12}(v) = \sqrt{2}\exp(i\pi/4)$. It follows from Eq. (10.102) that

$$\tilde{E}_p = \tilde{E}_u e^{i\pi/2} \tag{10.107}$$

and as in Section 10.3.1, we have the unobstructed amplitude, except for a $\pi/2$ phase discrepancy.* Finally, using (10.106), $I_p = I_0$.

We can construct a more palpable picture of what the Cornu spiral represents by considering Fig. 10.65, which depicts a cylindrical wavefront propagating from a coherent line source. The present procedure is exactly the same as that used in deriving the vibration curve, and the reader is referred back to Section 10.3.2 for a more leisurely discussion. Suffice it to say that the wavefront is divided into *half-period strip zones* by its intersection with a family of cylinders having a common axis and radii of $r_0 + \lambda/2$, $r_0 + \lambda$, $r_0 + 3\lambda/2$, and so on. *The contributions from these strip zones are proportional to their areas, which decrease rapidly.* This is in contrast to the circular zones, whose radii increase, thereby keeping the areas nearly constant. Each strip zone is similarly divided into N subzones, which have a relative phase difference of π/N. The vector sum of all the amplitude contributions from zones above the center line is a spiraling polygon. If N goes to ∞ and the contributions generated by the strip zones below the center line are included, the polygon smooths out into a continuous Cornu spiral. This is not surprising, since the coherent line source generates an infinite number of overlapping point-source patterns.

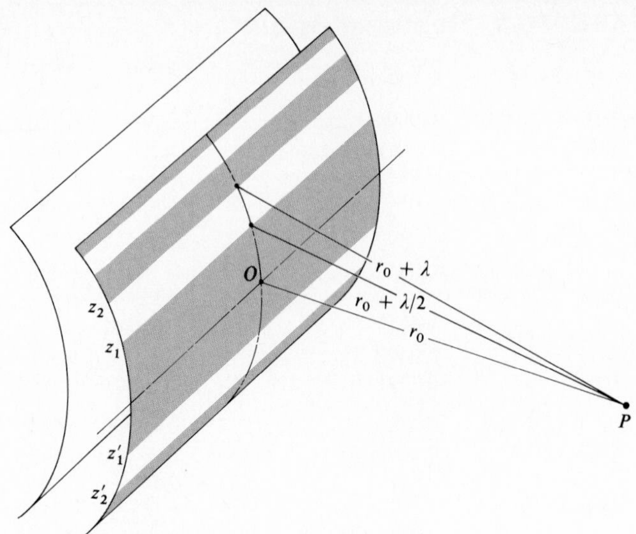

FIGURE 10.65 Cylindrical wavefront zones.

Figure 10.66 shows a number of unit tangent vectors at various positions along the spiral. The vector at O_s corresponds to the contribution from the central axis passing through O on the wavefront. The points associated with the boundaries of each strip zone can be located on the spiral, since at those positions the relative phase, β, is either an even or odd multiple of π. For example, the point Z_{s1} on the spiral (Fig. 10.66), which is related to z_1 (Fig. 10.65) on the wavefront, is by definition 180° out-of-phase with O_s. Therefore Z_{s1} must be located at the top of the spiral, where $w = \sqrt{2}$ inasmuch as there $\beta = \pi w^2/2 = \pi$.

It will be helpful as we go along in the treatment to visualize the blocking out of these strip zones when analyzing the effects of obstructions. Obviously, one could even make an appropriate zone plate, which would accomplish this to some advantage, and such devices are in use.

*The phase discrepancy will be resolved by the Kirchhoff theory in Section 10.4.

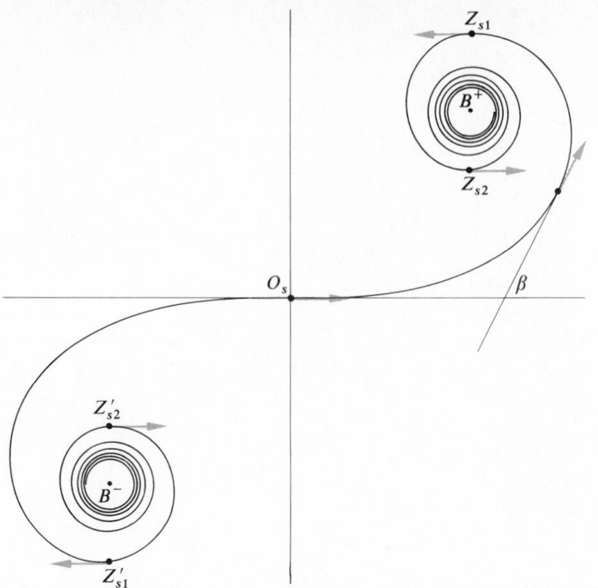

FIGURE 10.66 Cornu spiral related to the cylindrical wavefront.

10.3.8 Fresnel Diffraction by a Slit

We can treat Fresnel diffraction at a long slit as an extension of the rectangular-aperture problem. We need only elongate the rectangle by allowing y_1 and y_2 to move very far from O, as shown in Fig. 10.67. As the point of observation moves along the y-axis, as long as the vertical boundaries at either end of the slit are still essentially at infinity, $u_2 \approx \infty$, $u_1 \approx -\infty$, and $\tilde{\mathbf{B}}_{12}(u) \approx \sqrt{2}e^{i\pi/4}$. From Eq. (10.106), for either point-source or plane-wave illumination,

$$I_p = \frac{I_0}{2} \, |\tilde{\mathbf{B}}_{12}(v)|^2 \tag{10.108}$$

and the pattern is independent of y. The values of z_1 and z_2, which fix the slit width, determine the important parameter $\Delta v = v_2 - v_1$, which in turn governs $\tilde{\mathbf{B}}_{12}(v)$. Imagine once again that we have a string of length Δv lying along the spiral. At P, opposite point O, the aperture is symmetrical, and the string is centered on O_s (Fig. 10.68). The chord $|\tilde{\mathbf{B}}_{12}(v)|$ need only be measured and substituted into Eq. (10.108) to find I_p. At point P_1, z_1 and therefore v_1 are smaller negative numbers, whereas z_2 and v_2 have increased positively. The arc length Δv (the string) moves up the spiral (Fig. 10.68),

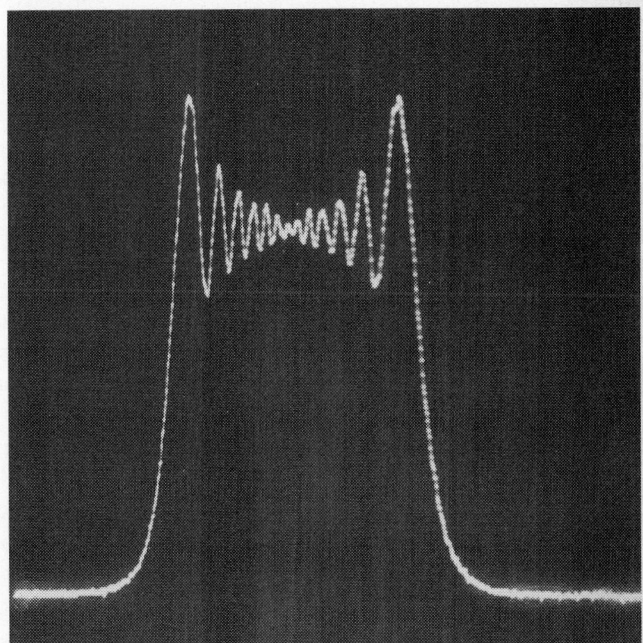

FIGURE 10.67 (*a*) Single-slit geometry. (*b*) A typical near-field irradiance distribution fairly close to a wide slit. The aperture was illuminated by a He–Ne laser and the pattern detected via a photodiode. Here the horizontal is parallel to the *z*-axis in the diagram. (Photo courtesy W. Klein, I. Physikalisches Institut, Germany.)

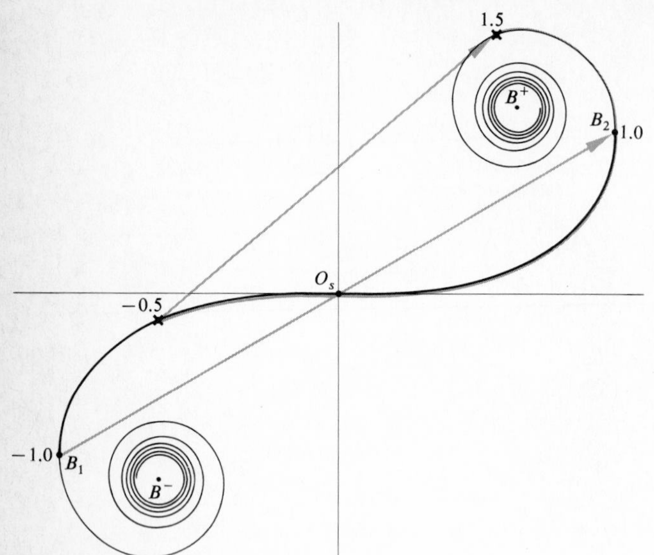

FIGURE 10.68 Cornu spiral for the slit.

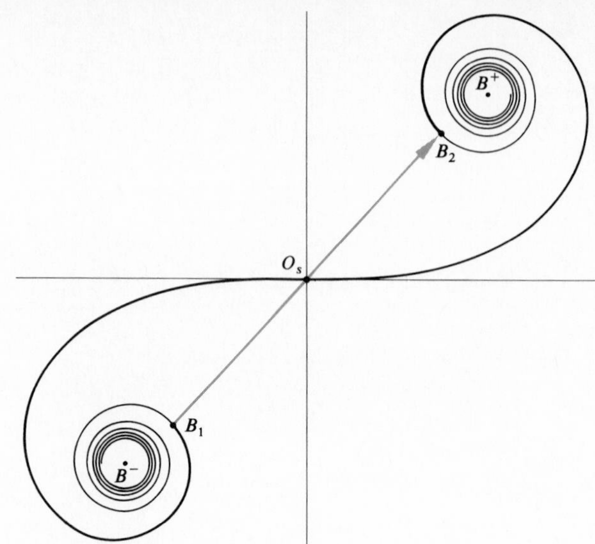

FIGURE 10.69 An irradiance minimum in the slit pattern.

and the chord decreases. As the point of observation moves down into the geometric shadow, the string winds about B^+, and the chord goes through a series of relative extrema. If Δv is very small, our imaginary piece of string is small, and the chord $|\tilde{\mathbf{B}}_{12}(v)|$ decreases appreciably only when the radius of curvature of the spiral itself is small. This occurs in the vicinity of $\tilde{B}^+$ or $\tilde{B}^-$, that is, far out into the geometric shadow. There will therefore be light well beyond the edges of the aperture, as long as the aperture is relatively small. Note too that with small Δv there will be a broad central maximum. In fact, if Δv is much less than 1, $r_0\lambda$ is much greater than the aperture width, and the Fraunhofer condition prevails. This transition of Eq. (10.108) into the form of Eq. (10.17) is more plausible when we realize that for large w the Fresnel integrals have trigonometric representations (see Problem 10.46).

As the slit widens, Δv becomes larger, for a fixed r_0, until a configuration like that in Fig. 10.69 exists for a point opposite the slit's center. If the point of observation is moved vertically either up or down, Δv slides either down or up the spiral. Yet the chord increases in both cases, so that the center of the diffraction pattern must be a relative minimum. Fringes now appear within the geometric image of the slit, unlike the Fraunhofer pattern.

Figure 10.70 shows two curves of $|\tilde{\mathbf{B}}_{12}(w)|^2$ plotted against $(w_1 + w_2)/2$, which is the center point of the arc length Δw. (Recall that the symbol w stands for either u or v.) A family of such curves running the range in Δw from about 1 to 10 would cover the region of interest. The curves are computed by first choosing a particular Δw and then reading the appropriate $|\tilde{\mathbf{B}}_{12}(w)|$ values off the Cornu spiral as Δw slides along it. For a long slit

$$I_p = \frac{I_0}{2}|\tilde{\mathbf{B}}_{12}(v)|^2 \qquad [10.108]$$

and since Δz is the slit width that corresponds to Δv, each curve in Fig. 10.70 *is proportional to the irradiance distribution* for a given slit. For example, Fig. 10.70*a* can be read as $|\tilde{\mathbf{B}}_{12}(v)|^2$ versus $(v_1 + v_2)/2$ for $\Delta v = 2.5$. The abscissa relates to $(z_1 + z_2)/2$, that is, the displacement of the point of observation from the center of the slit. In Fig. 10.70*b* $\Delta w = 3.5$, which means that a slit having a $\Delta v = 3.5$ clearly has fringes appearing within the geometric image as expected (Problem 10.45). The curves could, of course, be plotted in terms of values of Δz or Δy explicitly, but that would unnecessarily limit them to one set of configuration parameters ρ_0, r_0, and λ.

As the slit is widened still further, Δv approaches and then surpasses 10. An increasing number of fringes appear within the geometric image, and the pattern no longer extends appreciably beyond that image.

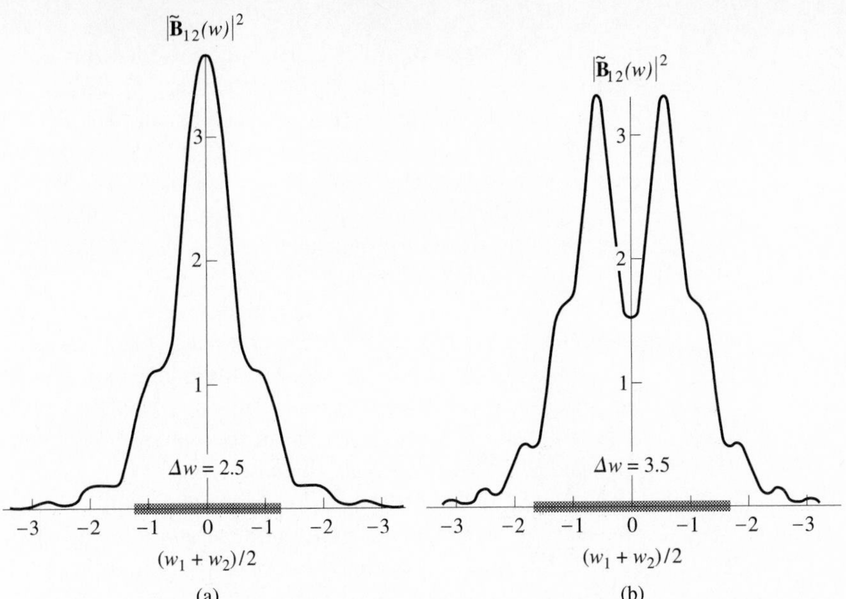

FIGURE 10.70 $|\tilde{\mathbf{B}}_{12}(w)|^2$ versus $(w_1 + w_2)/2$ for (*a*) $\Delta w = 2.5$ and (*b*) $\Delta w = 3.5$.

The same kind of reasoning applies equally well to the analysis of the rectangular aperture, where use can also be made of the curves in Fig. 10.70.

To observe Fresnel slit diffraction, form a long narrow space between two fingers held at arm's length. Make a similar parallel slit close to your eye, using your other hand. With a *bright* source, such as the daytime sky or a large lamp, illuminating the far slit, observe it through the nearby aperture. After inserting the near slit the far slit will appear to widen, and rows of fringes will be evident.

10.3.9 The Semi-Infinite Opaque Screen

We now form a semi-infinite planar opaque screen by removing the upper half of Σ in Fig. 10.67. This is done simply enough, by letting $z_2 = y_1 = y_2 = \infty$. Remembering the original approximations, we limit the geometry so that the point of observation is close to the screen's edge. Since $v_2 = u_2 = \infty$ and $u_1 = -\infty$, Eq. (10.104) or (10.108) leads to

$$I_p = \frac{I_0}{2}\{[\tfrac{1}{2} - \mathscr{C}(v_l)]^2 + [\tfrac{1}{2} - \mathscr{S}(v_l)]^2\} \quad (10.109)$$

When the point P is directly opposite the edge, $v_1 = 0$, $\mathscr{C}(0) = \mathscr{S}(0) = 0$, and $I_p = I_0/4$. This was to be expected, since half the wavefront is obstructed, the amplitude of the disturbance is halved, and the irradiance drops to one quarter. This occurs at point (3) in Figs. 10.71 and 10.72. Moving into the geometric shadow region to point (2) and then on to (1)

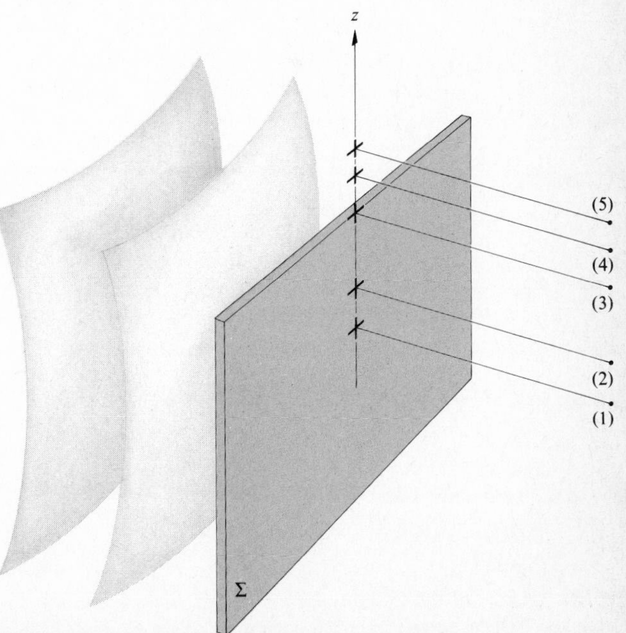

FIGURE 10.71 The semi-infinite opaque screen.

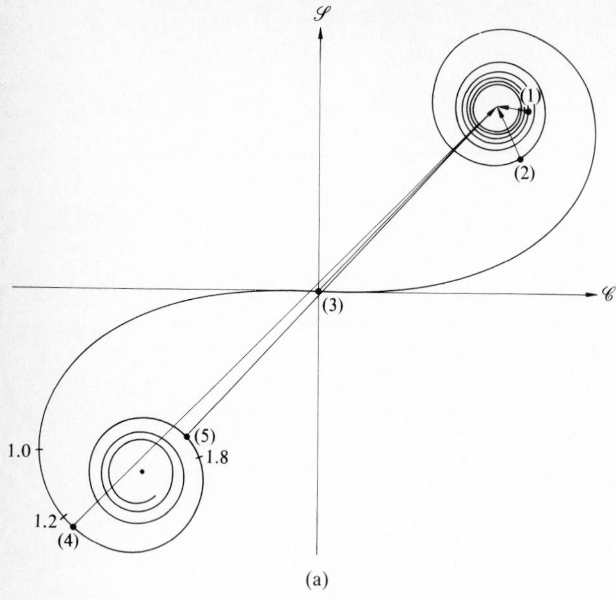

and still further, the successive chords clearly decrease monotonically (Problem 10.46). No irradiance oscillations exist within that region; the irradiance merely drops off rapidly. At any point above (3) the screen's edge will be below it, in other words, $z_1 < 0$ and $v_1 < 0$. At about $v_1 = -1.2$ the chord reaches a maximum, and the irradiance is a maximum. Thereafter, I_p oscillates about I_0, gradually diminishing in magnitude. With sensitive electronic techniques, many hundreds of these fringes can be observed.*

It is evident that the diffraction pattern of Fig. 10.73 would appear in the vicinity of the edges of a wide *slit* (Δv greater than about 10) as a limiting case. The irradiance distribution suggested by geometrical optics is obtained only when λ goes to zero. Indeed as λ decreases, the fringes move closer to the edge and become increasingly fine in extent.

The straight-edge pattern can be observed using any kind of slit, held up in front of a broad lamp at arm's length, as a source. Introduce an opaque obstruction (e.g., a blackened

*J. D. Barnett and F. S. Harris, Jr., *J. Opt. Soc. Amer.* **52**, 637 (1962).

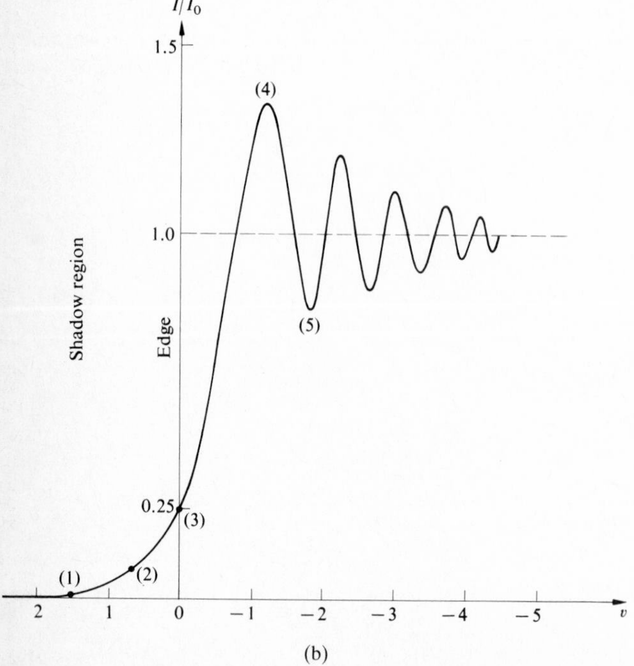

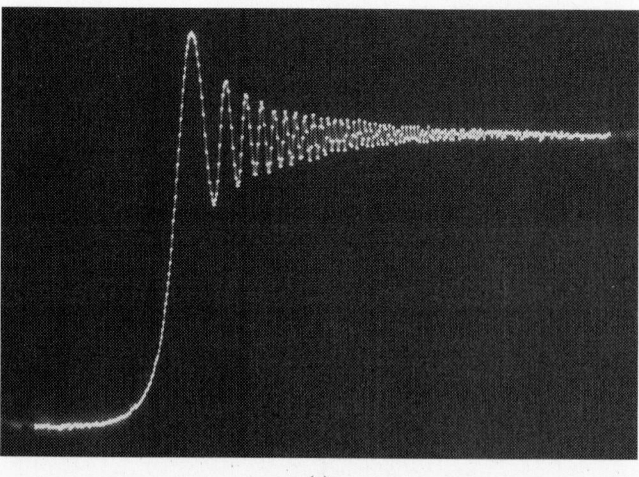

FIGURE 10.72 (*a*) The Cornu spiral for a semi-infinite screen. (*b*) The corresponding calculated irradiance distribution. (*c*) The same irradiance pattern under He–Ne laser illumination measured with a photodiode. (Photo courtesy W. Klein, I. Physikalisches Institut, Koln, Germany.)

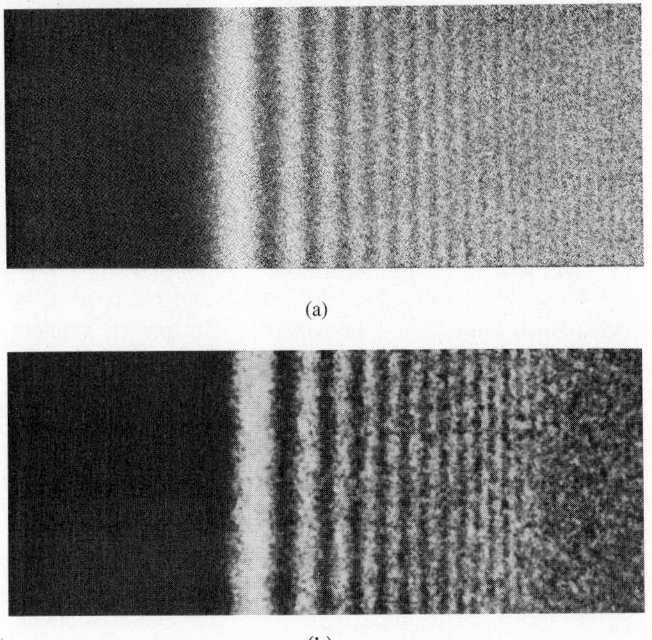

FIGURE 10.73 (*a*) The fringe pattern for a half-screen formed with light. (Francis Weston Sears, *Optics*, ©1949, Addison-Wesley Reading, MA. Reprinted by permission of Addison Wesley Longman, Inc.) (*b*) Fresnel electron diffraction at a half plane (MgO crystal)–electrons behave like photons. (Photo from *Handbuch der Physik*, edited by S. Flügge, Springer-Verlag, Heidelberg.)

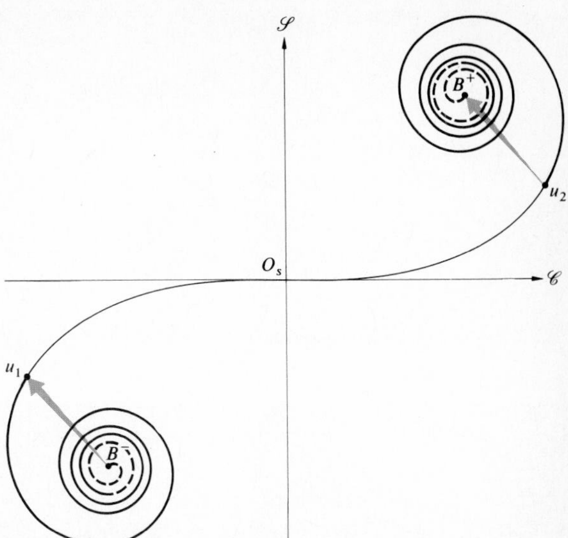

FIGURE 10.74 The Cornu spiral as applied to a narrow obstacle.

microscope slide or a razor blade) very near your eye. As the edge of the obstruction passes in front of the source slit parallel to it, a series of fringes will appear.

10.3.10 Diffraction by a Narrow Obstacle

Refer back to the description of the single narrow slit; consider the complementary case in which the slit is opaque, and the screen transparent. Let's envision, for example, a vertical opaque wire. At a point directly opposite the wire's center, there will be two separate contributing regions extending from y_1 to $-\infty$ and from y_2 to $+\infty$. On the Cornu spiral these correspond to two arc lengths from u_1 to B^- and from u_2 to B^+. The amplitude of the disturbance at a point P on the plane of observation is the magnitude of the *vector* sum of the two phasors $\overrightarrow{B^-u_1}$ and $\overrightarrow{u_2B^+}$, illustrated in Fig. 10.74. As with the opaque disk, the symmetry is such that there will always be an illuminated region along the central axis. This can be seen from the spiral, since when P is on the central axis, $\overrightarrow{B^-u_1} = \overrightarrow{u_2B^+}$ and their sum can never be zero. The arc length Δu represents the obscured region of the spiral, which increases as the diameter of the wire increases. For thick wires, u_1 approaches B^-, u_2 approaches B^+, the phasors decrease in length, and the irradiance on the shadow's axis drops off. This is evident in Fig. 10.75, which shows the patterns actually cast by a thin piece of lead from a mechanical pencil and by a rod with a $\frac{1}{8}$-inch diameter. Imagine that we have a small irradiance sensor at point P on the plane of observation (or the film plate). As P moves off the central axis to the right, y_1 and u_1 increase negatively, whereas y_2 and u_2, which are positive, decrease. The opaque region, Δu, slides down the spiral. When the sensor is at the right edge of the geometric shadow $y_2 = 0$, $u_2 = 0$; in other words, u_2 is at O_s. Notice that if the wire is thin, that is, if Δu is small, the sensor will record a gradual decrease in irradiance as u_2 approaches O_s. On the other hand, if the wire is thick, Δu is large and u_1 and u_2 are large. As Δu slides down the spiral, the two phasors revolve through a number of complete rotations, going in- and out-of-phase in the process. The resulting additional extrema appearing within the geometric shadow are evident in Fig. 10.75*b*. In fact, the separation between internal

(a)

(b)

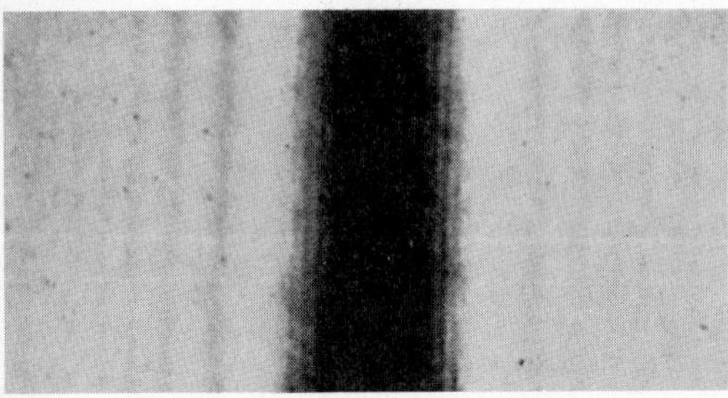

(c)

FIGURE 10.75 (*a*) The shadow pattern cast by the lead from a mechanical pencil. (*b*) The pattern cast by a 1/8-inch diameter rod. (E. H.) (c) Matter-wave diffraction. Fresnel electron-diffraction pattern of a 2-μm diameter metallized quartz filament. [Photo from O. E. Klemperer, *Electron Physics*, Butterworths, London (1972), pgs 188–191.]

fringes varies inversely with the width of the rod, just as if the pattern arose from the interference of two waves (Young's Experiment) reflected at the rod's edges.

10.3.11 Babinet's Principle

Two diffracting screens are said to be *complementary* when the transparent regions on one exactly correspond to the opaque regions on the other and vice versa. When two such screens are overlapped, the combination is obviously completely opaque. Now then, let E_1 or E_2 be the scalar optical disturbance arriving at P when either complementary screen Σ_1 or Σ_2, respectively, is in place. The total contribution from each aperture is determined by integrating over the area bounded by that aperture. If both *apertures* are present at once, there are no opaque regions at all; the limits of integration go to infinity, and we have the unobstructed disturbance E_0, whereupon

$$E_1 + E_2 = E_0 \tag{10.110}$$

which is the statement of **Babinet's Principle**. Take a close look at Figs. 10.69 and 10.74, which depict the Cornu spiral configurations for a transparent slit and a narrow opaque obstacle. If the two arrangements are made complementary, Fig. 10.76 illustrates Babinet's Principle quite clearly. The phasor arising from a narrow obstacle ($\overrightarrow{B^-B_1} + \overrightarrow{B_2B^+}$) added to that from a slit $\overrightarrow{B_2B_1}$ yields the unobstructed phase $\overrightarrow{B^-B^+}$.

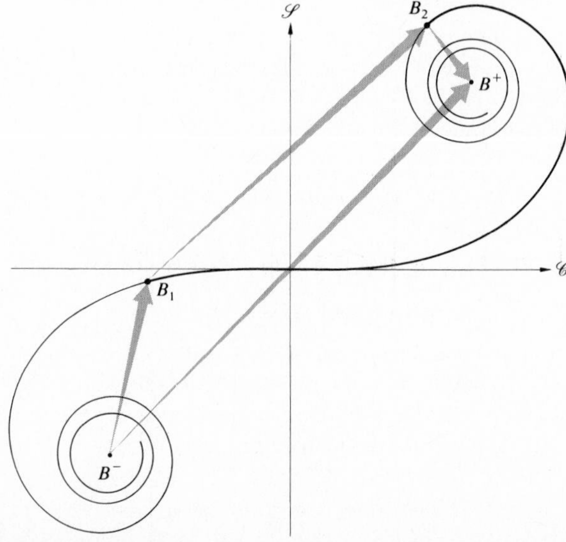

FIGURE 10.76 The Cornu spiral illustrating Babinet's Principle.

The principle implies that when $E_0 = 0$, $E_1 = -E_2$; in other words, these disturbances are precisely equal in magnitude and 180° out-of-phase. One would therefore observe exactly the same irradiance distribution with either Σ_1 or Σ_2 in place, an interesting result indeed. It is evident, however, that the principle cannot be exactly true, since for an unobstructed wave from a point source, there are no zero-amplitude points (i.e., $E_0 \neq 0$ everywhere). Yet if the source is imaged at P_0 by perfect lenses, as in Fig. 10.10 (with neither Σ_1 nor Σ_2 present), there will be a large, essentially zero-amplitude region beyond the immediate vicinity of P_0 (beyond the Airy disk) in which $E_1 + E_2 = E_0 = 0$. It is therefore only for the case of Fraunhofer diffraction that complementary screens will generate equivalent irradiance distributions, that is, $E_1 = -E_2$ (excluding point P_0). Nonetheless, Eq. (10.110) is valid in Fresnel diffraction, even though the irradiances obey no simple relationship. This is exemplified by the slit and narrow obstacle of Fig. 10.76. Moreover, for a circular hole and disk, refer back to Figs. 10.52 and 10.58 and then examine Fig. 10.77. Equation (10.110) is again clearly applicable, even though the diffraction patterns are certainly not equivalent.

The real beauty of Babinet's Principle is most evident when applied to Fraunhofer diffraction, as shown in Fig. 10.78, where the patterns from complementary screens are almost identical.

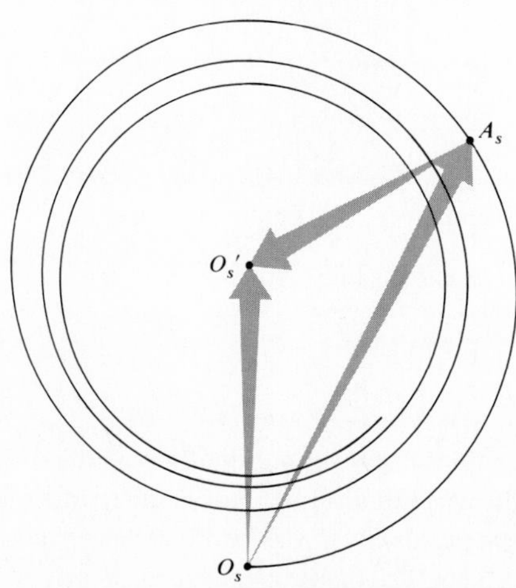

FIGURE 10.77 The vibration curve illustrating Babinet's Principle.

10.4 Kirchhoff's Scalar Diffraction Theory

We have described a number of diffracting configurations, quite satisfactorily, within the context of the relatively simple Huygens–Fresnel theory. Yet the whole imagery of surfaces covered with fictitious point sources, which was the basis of that analysis, was merely postulated rather than derived from fundamental principles. The Kirchhoff treatment shows that these results are actually derivable from the *scalar* differential wave equation.

The discussion to follow is rather formal and involved. Portions of it have therefore been relegated to an appendix, where we can indulge in succinctness and risk sacrificing readability for rigor.

In the past, when dealing with a distribution of monochromatic point sources, we computed the resultant optical disturbance at point P (i.e., E_p) by carrying out a superposition of the individual waves. There is, however, a completely different approach, which is founded in potential theory. Here one is concerned not with the sources themselves but rather with the scalar optical disturbance and its derivatives over an arbitrary closed surface surrounding P. We assume that a Fourier analysis can separate the constituent frequencies, so that we need only deal with one such frequency at a time. The monochromatic optical disturbance E is a solution of the differential wave equation

$$\nabla^2 E = \frac{1}{c^2}\frac{\partial^2 E}{\partial t^2} \tag{10.111}$$

Without specifying the precise spatial nature of the wave, we can write it as

$$\tilde{E} = \tilde{\mathscr{E}} e^{-ikct} \tag{10.112}$$

Here $\tilde{\mathscr{E}}$ represents the complex space part of the disturbance. Substituting this into the wave equation, we obtain

$$\nabla^2 \tilde{\mathscr{E}} + k^2 \tilde{\mathscr{E}} = 0 \tag{10.113}$$

This is known as the *Helmholtz Equation* and is solved, with the aid of Green's Theorem, in Appendix 2. The optical disturbance existing at a point P, expressed in terms of the optical disturbance and its gradient evaluated on an arbitrary

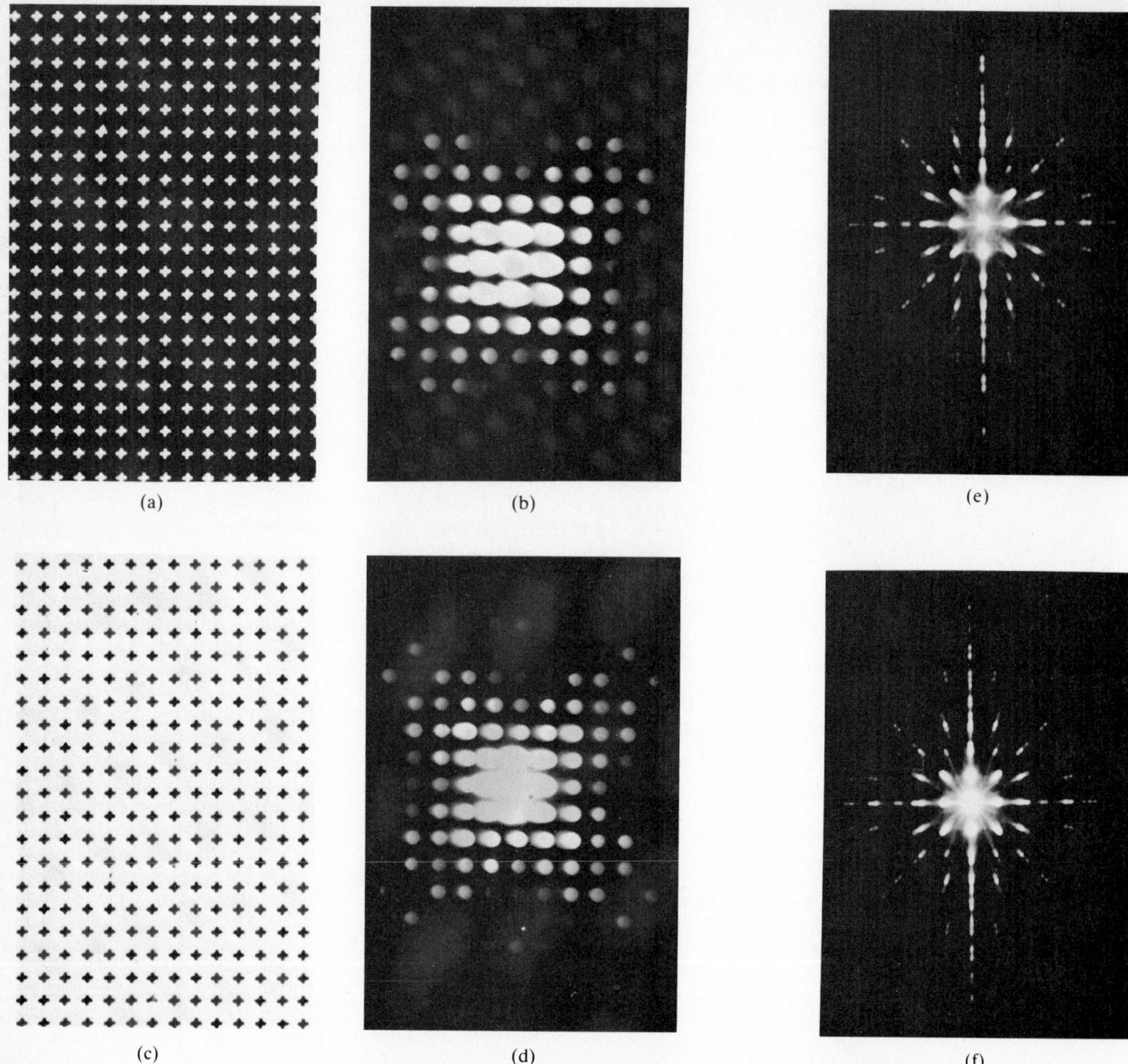

FIGURE 10.78 *(a)–(d)* White-light diffraction patterns for regular arrays of apertures and complementary obstacles in the form of rounded plus signs. *(e)* and *(f)* Diffraction patterns for a regular array of rectangular apertures and obstacles, respectively. (Photos courtesy The Ealing Corporation and Richard B. Hoover.)

closed surface S, enclosing P, is

$$\tilde{\mathscr{E}}_p = \frac{1}{4\pi}\left[\oiint_S \frac{e^{ikr}}{r}\,\nabla\tilde{\mathscr{E}}\cdot d\mathbf{S} - \oiint_S \tilde{\mathscr{E}}\,\nabla\left(\frac{e^{ikr}}{r}\right)\cdot d\mathbf{S}\right] \tag{10.114}$$

Known as the *Kirchhoff Integral Theorem*, Eq. (10.114) relates to the geometric configuration illustrated in Fig. 10.79.

We now apply the theorem to the specific instance of an unobstructed spherical wave originating at a point source s, as shown in Fig. 10.80. The disturbance has the form

$$\tilde{E}(\rho, t) = \frac{\mathcal{E}_0}{\rho}\,e^{i(k\rho-\omega t)} \tag{10.115}$$

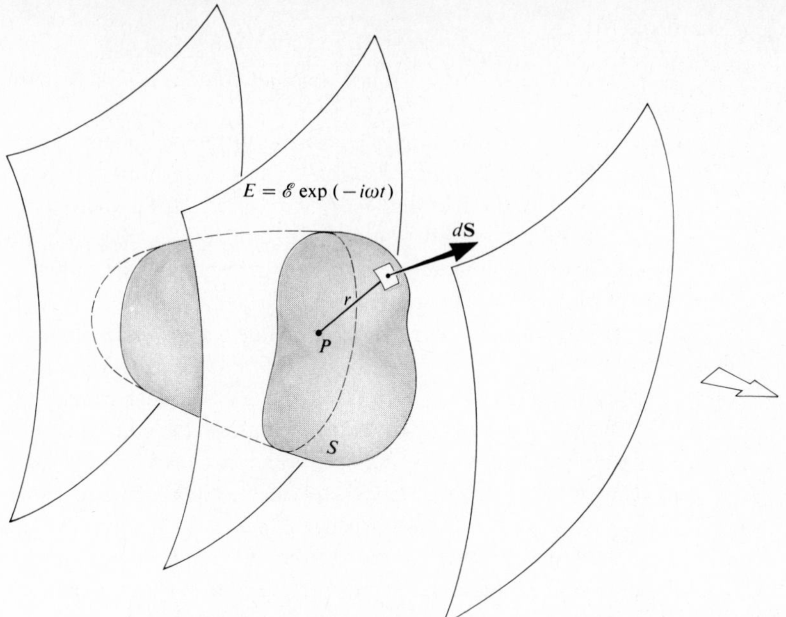

FIGURE 10.79 An arbitrary closed surface *S* enclosing point *P*.

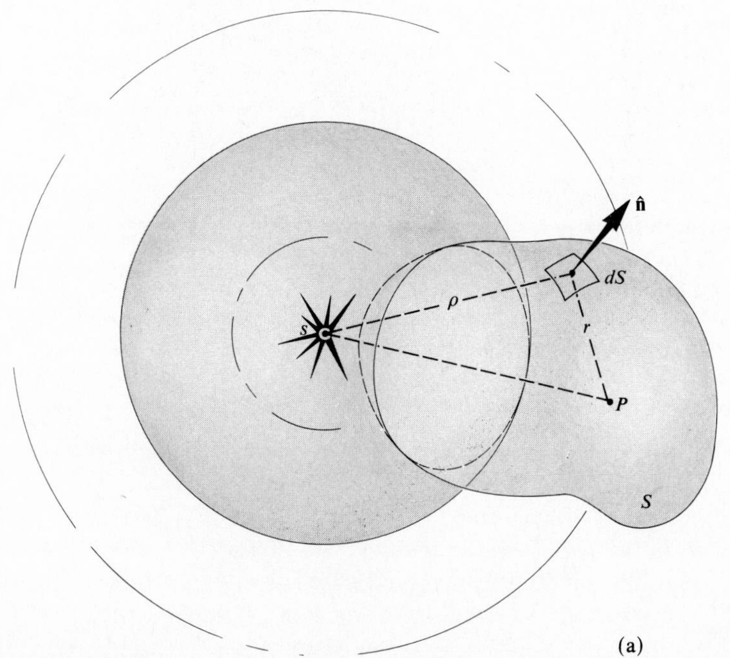

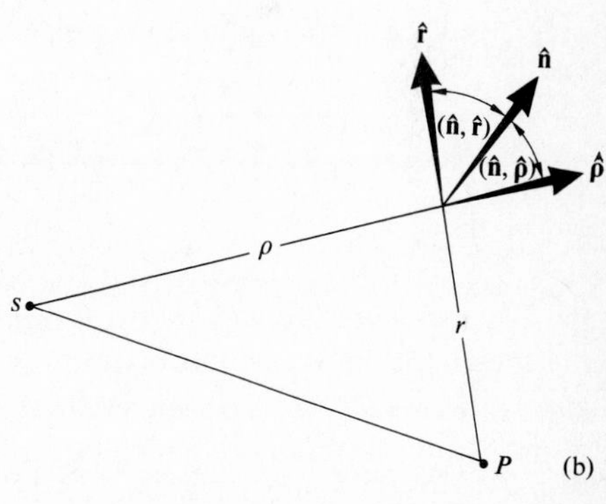

FIGURE 10.80 A spherical wave emitted from point *s*.

in which case

$$\tilde{\mathscr{E}}(\rho) = \frac{\mathcal{E}_0}{\rho} e^{ik\rho} \tag{10.116}$$

If we substitute this into Eq. (10.114), it becomes

$$\tilde{\mathscr{E}}_p = \frac{1}{4\pi}\left[\oiint_S \frac{e^{ikr}}{r}\frac{\partial}{\partial\rho}\left(\frac{\mathcal{E}_0}{\rho}e^{ik\rho}\right)\cos(\hat{\mathbf{n}}, \hat{\boldsymbol{\rho}})\, dS - \oiint_S \frac{\mathcal{E}_0}{\rho}e^{ik\rho}\frac{\partial}{\partial r}\left(\frac{e^{ikr}}{r}\right)\cos(\hat{\mathbf{n}}, \hat{\mathbf{r}})\, dS\right]$$

where $d\mathbf{S} = \hat{\mathbf{n}}\, dS$, $\hat{\mathbf{n}}$, $\hat{\mathbf{r}}$, and $\hat{\boldsymbol{\rho}}$ are unit vectors,

$$\nabla\left(\frac{e^{ikr}}{r}\right) = \hat{\mathbf{r}}\frac{\partial}{\partial r}\left(\frac{e^{ikr}}{r}\right)$$

and

$$\nabla\mathscr{E}(\boldsymbol{\rho}) = \hat{\boldsymbol{\rho}}\,\partial\mathscr{E}/\partial\rho$$

The differentiations under the integral signs are

$$\frac{\partial}{\partial\rho}\left(\frac{e^{ik\rho}}{\rho}\right) = e^{ik\rho}\left(\frac{ik}{\rho} - \frac{1}{\rho^2}\right)$$

and

$$\frac{\partial}{\partial r}\left(\frac{e^{ikr}}{r}\right) = e^{ikr}\left(\frac{ik}{r} - \frac{1}{r^2}\right)$$

When $\rho >> \lambda$ and $r >> \lambda$, the $1/\rho^2$ and $1/r^2$ terms can be neglected. This approximation is fine in the optical spectrum but certainly need not be true for microwaves. Proceeding, we write

$$\tilde{\mathscr{E}}_p = -\frac{\mathcal{E}_0 i}{\lambda}\oiint_S \frac{e^{ik(\rho+r)}}{\rho r}\left[\frac{\cos(\hat{\mathbf{n}}, \hat{\mathbf{r}}) - \cos(\hat{\mathbf{n}}, \hat{\boldsymbol{\rho}})}{2}\right] dS \tag{10.117}$$

This is the ***Fresnel–Kirchhoff diffraction formula***.

Take a long look at Eq. (10.96), which represents the disturbance at P arising from an element dS in the Huygens–Fresnel theory, and compare it with Eq. (10.117). In Eq. (10.117) the angular dependence is contained in the single term $\frac{1}{2}[\cos(\hat{\mathbf{n}}, \hat{\mathbf{r}}) - \cos(\hat{\mathbf{n}}, \hat{\boldsymbol{\rho}})]$, which we shall call the **obliquity factor** $K(\theta)$, showing it to be equivalent to Eq. (10.72) later on. Notice as well that k can be replaced by $-k$ everywhere, since we certainly could have chosen the phase of Eq. (10.115) to have been $(\omega t - k\rho)$. With Eq. (10.112) in mind multiply both sides of Eq. (10.117) by $\exp(-i\omega t)$; the differential element is then

$$dE_p = \frac{K(\theta)\mathcal{E}_0}{\rho r\lambda}\cos[k(\rho + r) - \omega t - \pi/2]\, dS \tag{10.118}$$

This is the contribution to E_p arising from an element of surface area dS a distance r from P. The $\pi/2$ term in the phase results from the fact that $-i = \exp(-i\pi/2)$. The Kirchhoff formulation therefore leads to the same total result, with the exception that it includes the correct $\pi/2$ phase shift, which is lacking in the Huygens–Fresnel treatment [Eq. (10.96)].

We have yet to ensure that the surface S can be made to correspond to the unobstructed portion of the wavefront, as it does in the Huygens–Fresnel theory. For the case of a freely propagating spherical wave emanating from the point source s, we construct the doubly connected region shown in Fig. 10.81. The surface S_2 completely surrounds the small spherical surface S_1. At $\rho = 0$ the disturbance $E(\rho, t)$ has a singularity and is therefore properly excluded from the volume V between S_1 and S_2. The integral must now include both surfaces S_1 and S_2. But we can have S_2 increase outward indefinitely by requiring its radius to go to infinity. In that case, the contribution to the

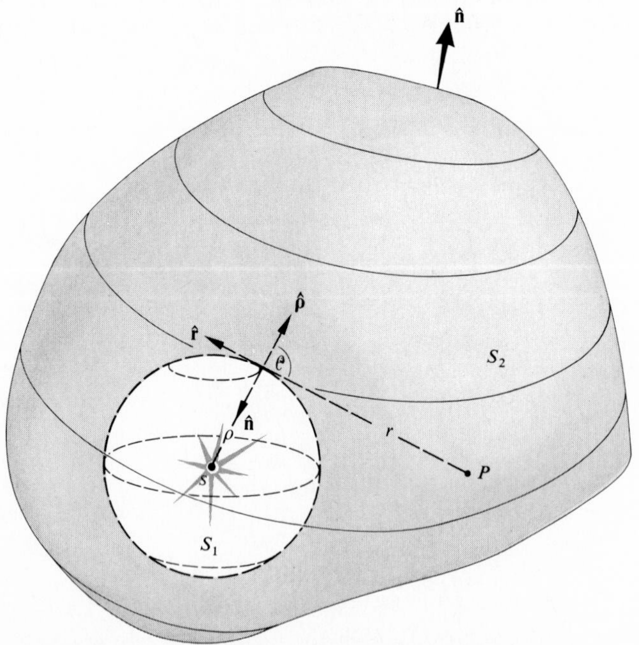

FIGURE 10.81 A doubly connected region surrounding point s.

surface integral vanishes. (This is true whatever the form of the incoming disturbance, as long as it drops off at least as rapidly as a spherical wave.) The remaining surface S_1 is a sphere centered at the point source. Since, over S_1, $\hat{\mathbf{n}}$ and $\hat{\boldsymbol{\rho}}$ are antiparallel, it is evident form Fig. 10.80*b* that the angles ($\hat{\mathbf{n}}$, $\hat{\mathbf{r}}$) and ($\hat{\mathbf{n}}$, $\hat{\boldsymbol{\rho}}$) are θ and 180°, respectively. The obliquity factor then becomes

$$K(\theta) = \frac{\cos\theta + 1}{2}$$

which is Eq. (10.72). Clearly, since the surface of integration S_1 is centered at s, it does indeed correspond to the spherical wavefront at some instant. ***The Huygens–Fresnel Principle is therefore directly traceable to the scalar differential wave equation.***

We shall not pursue the Kirchhoff formulation any further, other than to point out briefly how it is applied to diffracting screens. The single closed surface of integration surrounding the point of observation P is generally taken to be the entire screen Σ capped by an infinite hemisphere. There are then three distinct areas with which to be concerned. The contribution to the integral from the region of the infinite hemisphere is zero. Moreover, it is assumed that there is no disturbance immediately behind the opaque screen, so that this second region contributes nothing. The disturbance at P is therefore determined solely by the contributions arising from the aperture, and one need only integrate Eq. (10.117) over that area.

The fine results obtained by using the Huygens–Fresnel Principle are now justified theoretically, the main limitations being that $\rho >> \lambda$ and $r >> \lambda$.

10.5 Boundary Diffraction Waves

In Section 10.1.1 we said that the diffracted wave could be envisioned as arising from a fictitious distribution of secondary emitters spread across the unobstructed portion of the wavefront, namely, the Huygens–Fresnel Principle. There is, however, another, completely different, and rather appealing possibility. Suppose that an incoming wave sets the electrons on the rear of the diffracting screen Σ into oscillation, and these in turn radiate. We anticipate a twofold effect. First, all the oscillators that are remote from the edge of the aperture radiate back toward the source in such a fashion as to cancel the incoming wave at all points, except within the projection of the aperture itself. In other words, if this were the only contributing mechanism, a perfect geometrical image of the aperture would appear on the plane of observation. There is, however, an additional contribution arising from those oscillators in the vicinity of the aperture's edge. A portion of the energy radiated by these secondary sources propagates in the forward direction. The superposition of this scattered wave (known as the *boundary diffraction wave*) and the unobstructed portion of the primary wave (known as the *geometrical wave*) yield the diffraction pattern. A rather cogent reason for contemplating such a scheme becomes apparent when one examines the following arrangement. Tear a small hole ($\approx\frac{1}{2}$ cm in diameter) of arbitrary shape in a piece of paper, and holding it at arm's length, view an ordinary lightbulb some meters distant. Even with your eye in the shadow region, the edges of the aperture will be brightly illuminated. The ripple-tank photograph in Fig. 10.82 also illustrates the process. Notice how each edge of the slit seems to serve as a center for a circular disturbance, that then propagates beyond the aper-

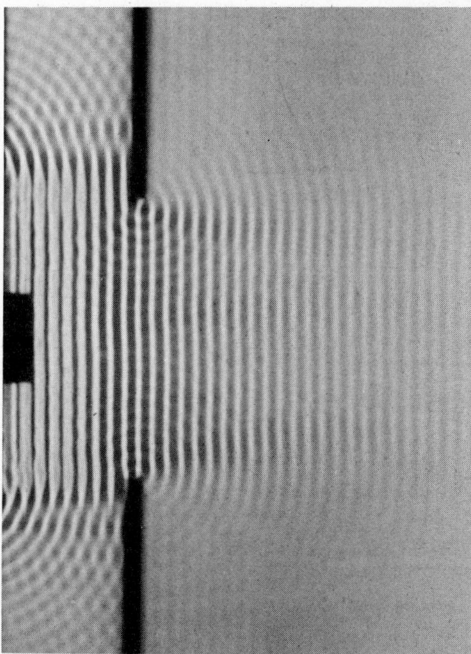

FIGURE 10.82 Ripple-tank waves passing through a slit. (Photo courtesy PSSC *Physics*, D. C. Heath, Boston, 1960.)

ture. There are no electron-oscillators here, which implies that these ideas have a certain generality, being applicable to elastic waves as well.

The formulation of diffraction in terms of the interference of a scattered edge wave and a geometrical wave is perhaps more physically appealing than the fictitious emitters of the Huygens–Fresnel Principle. It is not, however, a new concept. Indeed, it was first propounded by the ubiquitous Thomas Young even before Fresnel's celebrated memoir on diffraction. But in time Fresnel's brilliant successes unfortunately convinced Young to reject his own ideas, and he finally did so in a letter to Fresnel in 1818. Strengthened by Kirchhoff's work, the Fresnel conception of diffraction became generally accepted and has persisted (right up to Section 10.4). The resurrection of Young's theory began in 1888. At that time, Gian Antonio Maggi proved that Kirchhoff's analysis, for a point source at least, was equivalent to two contributing terms. One of these was a geometrical wave, but the other, unhappily, was an integral, which allowed no clear physical interpretation at the time.

In his doctoral thesis (1893) Eugen Maey showed that an edge wave could indeed be extracted from a modified Kirchhoff formulation for a semi-infinite half-plane. Arnold Sommerfeld's rigorous solution of the half-plane problem (see Section 10.1) showed that a cylindrical wave actually does proceed from the screen's edge. It propagates into both the geometrical shadow region and the illuminated region. In the latter, the boundary diffraction wave combines with the geometrical wave, in complete accord with Young's theory. In 1917 Adalbert (Wojciech) Rubinowicz was able to prove that Kirchhoff's formula for a plane or spherical wave can be appropriately decomposed into the two desired waves, thereby revealing the basic correctness of Young's ideas. He also later established that the boundary diffraction wave, to a first approximation, was generated by reflection of the primary wave from the aperture's edge. In 1923 Friedrich Kottler pointed out the equivalence of the solutions of Maggi and Rubinowicz, and one now speaks of the Young-Maggi-Rubinowicz theory. Most recently, Kenro Miyamoto and Emil Wolf (1962) have extended the boundary diffraction theory to the case of arbitrary incident waves.*

A very useful contemporary approach to the problem has been devised by Joseph B. Keller. He has developed a geometric theory of diffraction that is closely related to Young's edge wave picture. Along with the usual rays of geometrical optics, Keller hypothesizes the existence of diffracted rays. Rules governing these diffracted rays, which are analogous to the Laws of Reflection and Refraction, are employed to determine the resultant fields.

*A fairly complete bibliography can be found in the article by A. Rubinowicz in *Progress in Optics*, Vol. 4, p. 199.

PROBLEMS

10.1 A point source S is a perpendicular distance R away from the center of a circular hole of radius a in an opaque screen. If the distance to the periphery is $(R + \ell)$, show that Fraunhofer diffraction will occur on a very distant screen when

$$\lambda R >> a^2/2$$

What is the smallest satisfactory value of R if the hole has a radius of 1 mm, $\ell \leq \lambda/10$, and $\lambda = 500$ nm?

10.2 Using Fig. P.10.2, derive the irradiance equation for N coherent oscillators, Eq. (10.5).

10.3* In Section 10.1.3 we talked about introducing an intrinsic phase shift ε between oscillators in a linear array. With this in mind, show that Eq. (10.18) becomes

$$\beta = (kb/2)(\sin\theta - \sin\theta_i)$$

when the incident plane wave makes an angle θ_i with the plane of the slit.

10.4 Referring back to the multiple antenna system of Fig. 10.8, compute the angular separation between successive lobes or principal maxima and the width of the central maximum.

FIGURE P.10.2

10.5 Examine the setup of Fig. 10.6 in order to determine what is happening in the image space of the lenses; in other words, locate the exit pupil and relate it to the diffraction process. Show that the configurations in Fig. P.10.5 are equivalent to that of Fig. 10.6 and will therefore result in Fraunhofer diffraction. Design at least one more such arrangement.

10.6 The angular distance between the center and the first minimum of a single-slit Fraunhofer diffraction pattern is called the *half-angular breadth*; write an expression for it. Find the corresponding *half-linear width* (a) when no focusing lens is present and the slit-viewing screen distance is L, and (b) when a lens of focal length f_2 is very close to the aperture. Notice that the half-linear width is also the distance between the successive minima.

10.7* A single slit in an opaque screen 0.10 mm wide is illuminated (in air) by plane waves from a krypton ion laser ($\lambda_0 = 461.9$ nm). If the observing screen is 1.0 m away, determine whether or not the

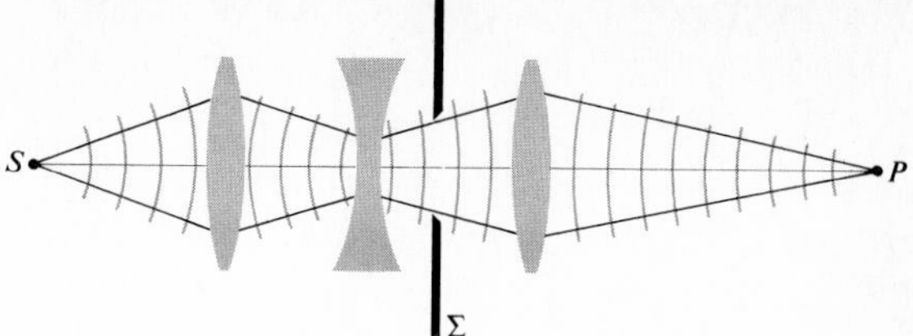

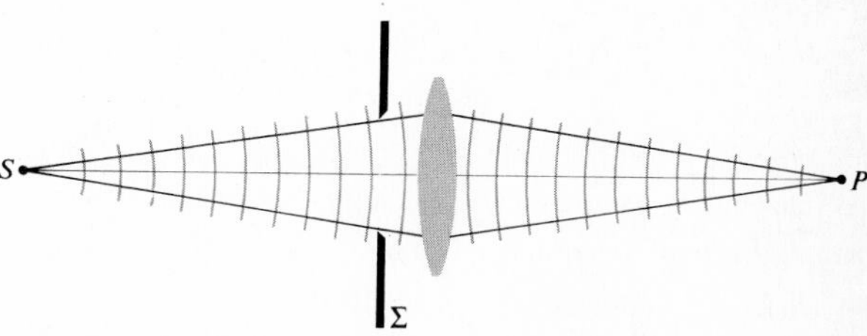

FIGURE P.10.5

resulting diffraction pattern will be of the far-field variety and then compute the angular width of the central maximum.

10.8* A narrow single slit (in air) in an opaque screen is illuminated by infrared from a He–Ne laser at 1152.2 nm, and it is found that the center of the tenth dark band in the Fraunhofer pattern lies at an angle of 6.2° off the central axis. Please determine the width of the slit. At what angle will the tenth minimum appear if the entire arrangement is immersed in water ($n_w = 1.33$) rather than air ($n_a = 1.000\,29$)?

10.9 A collimated beam of microwaves impinges on a metal screen that contains a long horizontal slit that is 20 cm wide. A detector moving parallel to the screen in the far-field region locates the first minimum of irradiance at an angle of 36.87° above the central axis. Determine the wavelength of the radiation.

10.10 Show that for a double-slit Fraunhofer pattern, if $a = mb$, the number of bright fringes (or parts thereof) within the central diffraction maximum will be equal to $2m$.

10.11* Two long slits 0.10 mm wide, separated by 0.20 mm, in an opaque screen are illuminated by light with a wavelength of 500 nm. If the plane of observation is 2.5 m away, will the pattern correspond to Fraunhofer or Fresnel diffraction? How many Young's fringes will be seen within the central bright band?

10.12 What is the relative irradiance of the subsidiary maxima in a three-slit Fraunhofer diffraction pattern? Draw a graph of the irradiance distribution, when $a = 2b$, for two and then three slits.

10.13* Starting with the irradiance expression for a finite slit, shrink the slit down to a minuscule area element and show that it emits equally in all directions.

10.14* Show that Fraunhofer diffraction patterns have a center of symmetry [i.e., $I(Y, Z) = I(-Y, -Z)$], regardless of the configuration of the aperture, as long as there are no phase variations in the field over the region of the hole. Begin with Eq. (10.41). We'll see later (Chapter 11) that this restriction is equivalent to saying that the aperture function is real.

10.15 With the results of Problem 10.14 in mind, discuss the symmetries that would be evident in the Fraunhofer diffraction pattern of an aperture that is itself symmetrical about a line (assuming normally incident quasimonochromatic plane waves).

10.16 From symmetry considerations, create a rough sketch of the Fraunhofer diffraction patterns of an equilateral triangular aperture and an aperture in the form of a plus sign.

10.17 Figure P.10.17 is the irradiance distribution in the far field for a configuration of elongated rectangular apertures. Describe the arrangement of holes that would give rise to such a pattern and give your reasoning in detail.

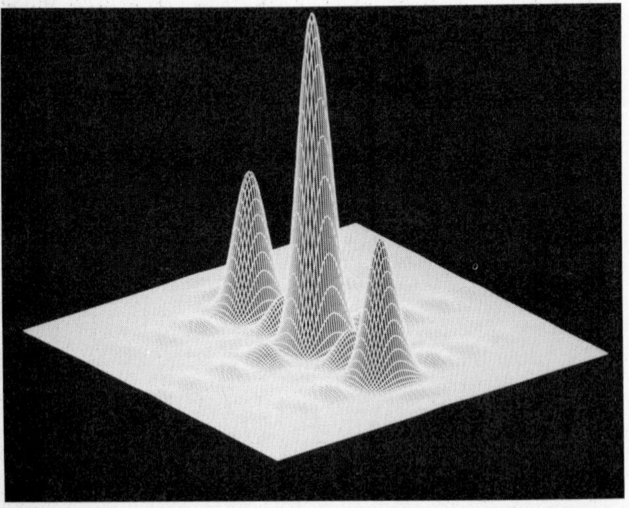

FIGURE P.10.17 Photo courtesy R. G. Wilson, Illinois Wesleyan University.

10.18 In Fig. P.10.18*a* and *b* are the electric field and irradiance distributions, respectively, in the far field for a configuration of elongated rectangular apertures. Describe the arrangement of holes that would give rise to such patterns and discuss your reasoning.

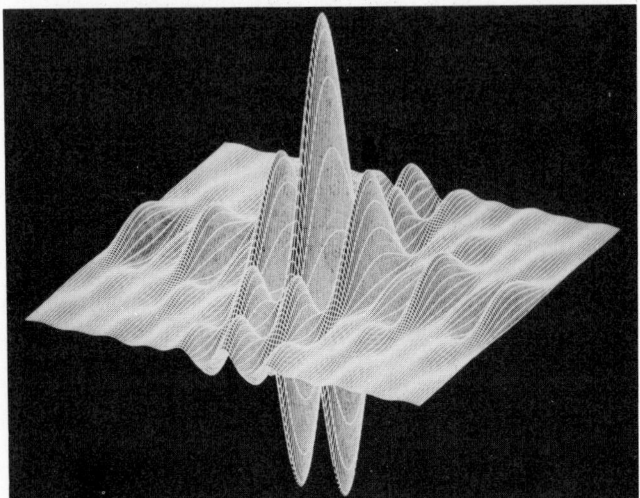

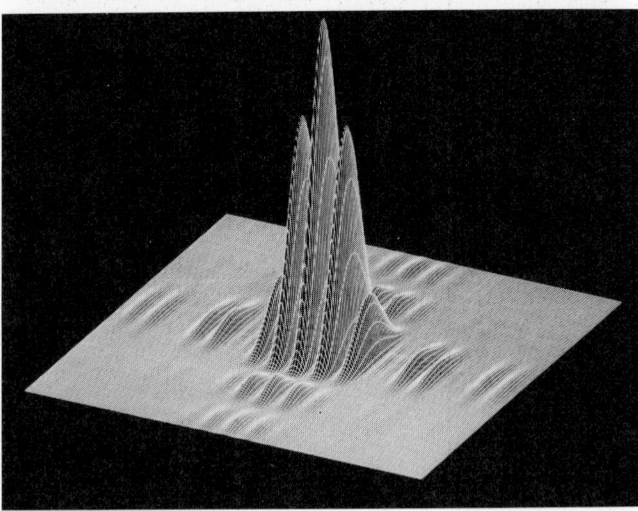

FIGURE P.10.18 Photos courtesy R. G. Wilson, Illinois Wesleyan University.

10.19 Figure P.10.19 is a computer-generated Fraunhofer irradiance distribution. Describe the aperture that would give rise to such a pattern and give your reasoning in detail.

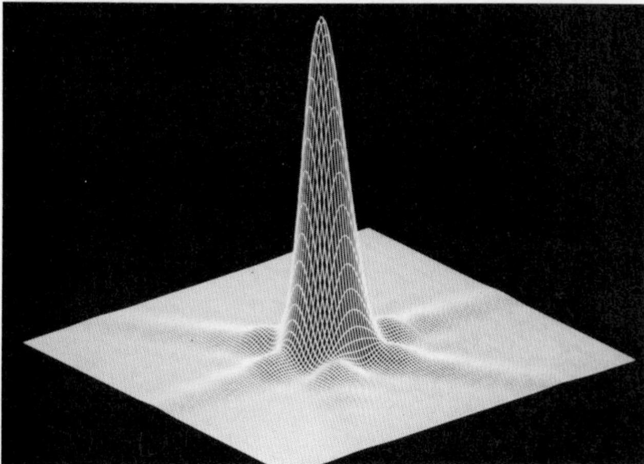

FIGURE P.10.19 Photo courtesy R. G. Wilson, Illinois Wesleyan University.

10.20 Figure P.10.20 is the electric-field distribution in the far field for a hole of some sort in an opaque screen. Describe the aperture that would give rise to such a pattern and give your reasoning in detail.

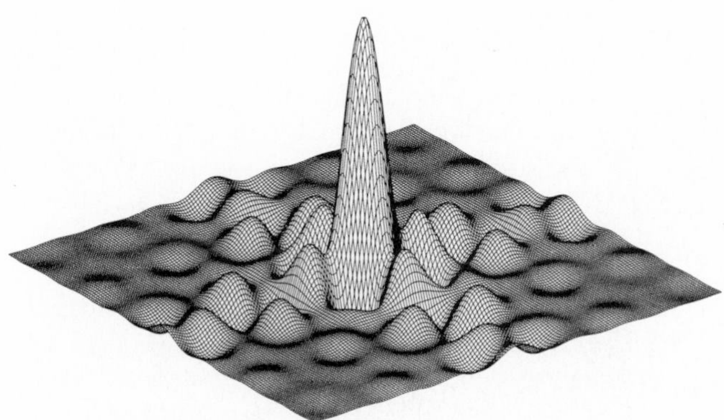

FIGURE P.10.20 Photos courtesy R. G. Wilson, Illinois Wesleyan University.

10.21 In light of the five previous questions, identify Fig. P.10.21, explaining what it is and what aperture gave rise to it.

10.22* Verify that the peak irradiance I_1 of the first "ring" in the Airy pattern for far-field diffraction at a circular aperture is such that $I_1/I(0) = 0.0175$. You might want to use the fact that

$$J_1(u) = \frac{u}{2}\left[1 - \frac{1}{1!2!}\left(\tfrac{1}{2}u\right)^2 + \frac{1}{2!3!}\left(\tfrac{1}{2}u\right)^4 - \frac{1}{3!4!}\left(\tfrac{1}{2}u\right)^6 + \cdots\right]$$

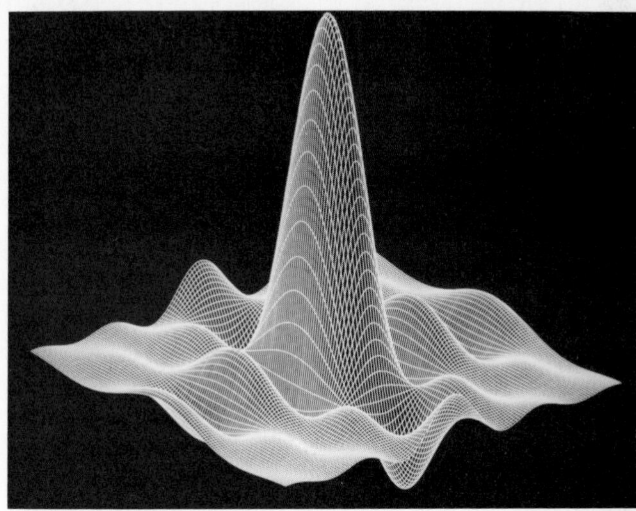

FIGURE P.10.21 Photo courtesy R. G. Wilson, Illinois Wesleyan University.

10.23 No lens can focus light down to a perfect point, because there will always be some diffraction. Estimate the size of the minimum spot of light that can be expected at the focus of a lens. Discuss the relationship among the focal length, the lens diameter, and the spot size. Take the *f-number* of the lens to be roughly 0.8 or 0.9, which is just about what you can expect for a fast lens.

10.24 Figure P.10.24 shows several aperture configurations. Roughly sketch the Fraunhofer patterns for each. Note that the circular regions should generate Airy-like ring systems centered at the origin.

FIGURE P.10.24 L 2 Z S 5

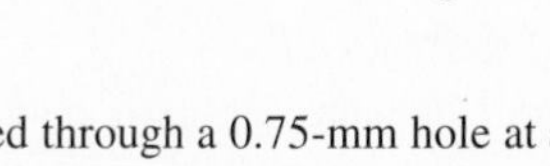

10.25* Suppose that we have a laser emitting a diffraction-limited beam ($\lambda_0 = 632.84$ nm) with a 2-mm diameter. How big a light spot would be produced on the surface of the Moon a distance of 376×10^3 km away from such a device? Neglect any effects of the Earth's atmosphere.

10.26* If you peered through a 0.75-mm hole at an eye chart, you would probably notice a decrease in visual acuity. Compute the angular limit of resolution, assuming that it's determined only by diffraction; take $\lambda_0 = 550$ nm. Compare your results with the value of 1.7×10^{-4} rad, which corresponds to a 4.0-mm pupil.

10.27 The neoimpressionist painter Georges Seurat was a member of the pointillist school. His paintings consist of an enormous number of closely spaced small dots ($\approx\frac{1}{10}$ inch) of pure pigment. The illusion of color mixing is produced only in the eye of the observer. How far from such a painting should one stand in order to achieve the desired blending of color?

10.28* The Mount Palomar telescope has an objective mirror with a 508-cm diameter. Determine its angular limit of resolution at a wavelength of 550 nm, in radians, degrees, and seconds of arc. How far apart must two objects be on the surface of the Moon if they are to be resolvable by the Palomar telescope? The Earth–Moon distance is 3.844×10^8 m; take $\lambda_0 = 550$ nm. How far apart must two objects be on the Moon if they are to be distinguished by the eye? Assume a pupil diameter of 4.00 mm.

10.29* A transmission grating whose lines are separated by 3.0×10^{-6} m is illuminated by a narrow beam of red light ($\lambda_0 = 694.3$ nm) from a ruby laser. Spots of diffracted light, on both sides of the undeflected beam, appear on a screen 2.0 m away. How far from the central axis is each of the two nearest spots?

10.30* A diffraction grating with slits 0.60×10^{-3} cm apart is illuminated by light with a wavelength of 500 nm. At what angle will the third-order maximum appear?

10.31* A diffraction grating produces a second-order spectrum of yellow light ($\lambda_0 = 550$ nm) at 25°. Determine the spacing between the lines on the grating.

10.32 White light falls normally on a transmission grating that contains 1000 lines per centimeter. At what angle will red light ($\lambda_0 = 650$ nm) emerge in the first-order spectrum?

10.33* Light from a laboratory sodium lamp has two strong yellow components at 589.5923 nm and 588.9953 nm. How far apart in the first-order spectrum will these two lines be on a screen 1.00 m from a grating having 10 000 lines per centimeter?

10.34* Sunlight impinges on a transmission grating that is formed with 5000 lines per centimeter. Does the third-order spectrum overlap the second-order spectrum? Take red to be 780 nm and violet to be 390 nm.

10.35 Light having a frequency of 4.0×10^{14} Hz is incident on a grating formed with 10 000 lines per centimeter. What is the highest-order spectrum that can be seen with this device? Explain.

10.36* Suppose that a grating spectrometer while in vacuum on Earth sends 500-nm light off at an angle of 20.0° in the first-order spectrum. By comparison, after landing on the planet Mongo, the same light is diffracted through 18.0°. Determine the index of refraction of the Mongoian atmosphere.

10.37 Prove that the equation

$$a(\sin\theta_m - \sin\theta_i) = m\lambda \qquad [10.61]$$

when applied to a transmission grating, is independent of the refractive index.

10.38 A high-resolution grating 260 mm wide, with 300 lines per millimeter, at about 75° in autocollimation has a resolving power of just about 10^6 for $\lambda = 500$ nm. Find its free spectral range. How do these values of $\mathcal{R}$ and $(\Delta\lambda)_{fsr}$ compare with those of a Fabry–Perot etalon having a 1-cm air gap and a finesse of 25?

10.39 What is the total number of lines a grating must have in order just to separate the sodium doublet ($\lambda_1 = 5895.9$ Å, $\lambda_2 = 5890.0$ Å) in the third order?

10.40* Imagine an opaque screen containing 30 randomly located circular holes. The light source is such that every aperture is coherently illuminated by its own plane wave. Each wave in turn is completely incoherent with respect to all the others. Describe the resulting far-field diffraction pattern.

10.41 Imagine that you are looking through a piece of square woven cloth at a point source ($\lambda_0 = 600$ nm) 20 m away. If you see a square arrangement of bright spots located about the point source (Fig. P.10.41), each separated by an apparent nearest-neighbor distance of 12 cm, how close together are the strands of cloth?

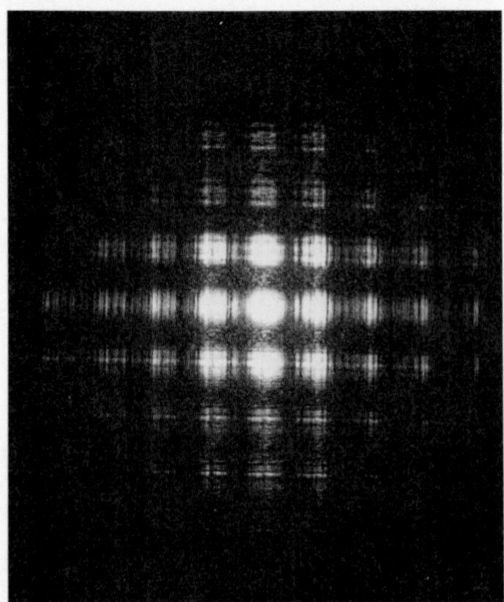

FIGURE P.10.41 Photo by E. H.

10.42* Perform the necessary mathematical operations needed to arrive at Eq. (10.76).

10.43 Referring to Fig. 10.48, integrate the expression $dS = 2\pi\rho^2 \sin\varphi \, d\varphi$ over the lth zone to get the area of that zone,

$$A_l = \frac{\lambda \pi \rho}{\rho + r_0}\left[r_0 + \frac{(2l-1)\lambda}{4}\right]$$

Show that the mean distance to the lth zone is

$$r_l = r_0 + \frac{(2l-1)\lambda}{4}$$

so that the ratio A_l/r_l is constant.

10.44* Derive Eq. (10.84).

10.45 Use the Cornu spiral to make a rough sketch of $|\tilde{\mathbf{B}}_{12}(w)|^2$ versus $(w_1 + w_2)/2$ for $\Delta w = 5.5$. Compare your results with those of Fig. 10.70.

10.46 The Fresnel integrals have the asymptotic forms (corresponding to large values of w) given by

$$\mathscr{C}(w) \approx \tfrac{1}{2} + \left(\frac{1}{\pi w}\right)\sin\left(\frac{\pi w^2}{2}\right)$$

$$\mathscr{S}(w) \approx \tfrac{1}{2} - \left(\frac{1}{\pi w}\right)\cos\left(\frac{\pi w^2}{2}\right)$$

Using this fact, show that the irradiance in the shadow of a semi-infinite opaque screen decreases in proportion to the inverse square of the distance to the edge, as z_1 and therefore v_1 become large.

10.47 What would you expect to see on the plane of observation if the half-plane Σ in Fig. 10.71 were semi-transparent?

10.48 Plane waves from a collimated He–Ne laserbeam ($\lambda_0 = 632.8$ nm) impinge on a steel rod with a 2.5-mm diameter. Draw a rough *graphic* representation of the diffraction pattern that would be seen on a screen 3.16 m from the rod.

10.49 Make a rough sketch of the irradiance function for a Fresnel diffraction pattern arising from a double slit. What would the Cornu spiral picture look like at point P_0?

10.50* Make a rough sketch of a possible Fresnel diffraction pattern arising from each of the indicated apertures (Fig. P.10.50).

FIGURE P.10.50

10.51* Suppose the slit in Fig. 10.67 is made very wide. What will the Fresnel diffraction pattern look like?

10.52* Collimated light from a krypton ion laser at 568.19 nm impinges normally on a circular aperture. When viewed axially from a distance of 1.00 m, the hole uncovers the first half-period Fresnel zone. Determine its diameter.

10.53* Plane waves impinge perpendicularly on a screen with a small circular hole in it. It is found that when viewed from some axial point P the hole uncovers $\frac{1}{2}$ of the first half-period zone. What is the irradiance at P in terms of the irradiance there when the screen is removed?

10.54* A collimated beam from a ruby laser (694.3 nm) having an irradiance of 10 W/m^2 is incident perpendicularly on an opaque screen containing a square hole 5.0 mm on a side. Compute the irradiance at a point on the central axis 250 cm from the aperture.

10.55* A long narrow slit 0.10 mm wide is illuminated by light of wavelength 500 nm coming from a point source 0.90 m away. Determine the irradiance at a point 2.0 m beyond the screen when the slit is centered on, and perpendicular to, the line from the source to the point of observation. Write your answer in terms of the unobstructed irradiance.

11 Fourier Optics

11.1 Introduction

In what is to follow we will extend the discussion of Fourier methods introduced in Chapter 7. It is our intent to provide a strong basic introduction to the subject rather than a complete treatment. Besides its real mathematical power, Fourier analysis leads to a marvelous way of treating optical processes in terms of spatial frequencies.* It is always exciting to discover a new bag of analytic toys, but it's perhaps even more valuable to unfold yet another way of thinking about a broad range of physical problems—we shall do both.†

The primary motivation here is to develop an understanding of the way optical systems process light to form images. In the end we want to know all about the amplitudes and phases of the lightwaves reaching the image plane. Fourier methods are especially suited to that task, so we first extend the treatment of Fourier transforms begun earlier. Several transforms are particularly useful in the analysis, and these will be considered first. Among them is the delta function, which will subsequently be used to represent a point source of light. How an optical system responds to an object comprising a large number of delta-function point sources will be considered in Section 11.3.1. The relationship between Fourier analysis and Fraunhofer diffraction is explored throughout the discussion, but is given special attention in Section 11.3.3. The chapter ends with a return to the problem of image evaluation, this time from a different, though related, perspective: the object is treated not as a collection of point sources but as a scatterer of plane waves.

*See Section 3.2 for a further nonmathematical discussion.

†As general references for this chapter, see R. C. Jennison, *Fourier Transforms and Convolutions for the Experimentalist;* N. F. Barber, *Experimental Correlograms and Fourier Transforms;* A. Papoulis, *Systems and Transforms with Applications in Optics;* J. W. Goodman, *Introduction to Fourier Optics;* J Gaskill, *Linear Systems, Fourier Transforms, and Optics;* R. G. Wilson, *Fourier Series and Optical Transform Techniques in Contemporary Optics;* and the excellent series of booklets *Images and Information,* by B. W. Jones et al.

11.2 Fourier Transforms

11.2.1 One-Dimensional Transforms

It was seen in Section 7.4 that a one-dimensional function of some space variable $f(x)$ could be expressed as a linear combination of an infinite number of harmonic contributions:

$$f(x) = \frac{1}{\pi}\left[\int_0^\infty A(k)\cos kx\,dk + \int_0^\infty B(k)\sin kx\,dk\right] \quad [7.56]$$

The weighting factors that determine the significance of the various angular spatial frequency (k) contributions, that is, $A(k)$ and $B(k)$, are the *Fourier cosine and sine transforms of* $f(x)$ given by

$$A(k) = \int_{-\infty}^{+\infty} f(x')\cos kx'\,dx'$$

and

$$B(k) = \int_{-\infty}^{+\infty} f(x')\sin kx'\,dx' \quad [7.57]$$

respectively. Here the quantity x' is a dummy variable over which the integration is carried out, so that neither $A(k)$ nor $B(k)$ is an explicit function of x', and the choice of symbol used to denote it is irrelevant. The sine and cosine transforms can be consolidated into a single complex exponential expression as follows: substituting [Eq. 7.57] into [Eq. 7.56], we obtain

$$f(x) = \frac{1}{\pi}\int_0^{\infty} \cos kx \int_{-\infty}^{+\infty} f(x') \cos kx'\, dx'\, dk$$

$$+ \frac{1}{\pi}\int_0^{\infty} \sin kx \int_{-\infty}^{+\infty} f(x') \sin kx'\, dx'\, dk$$

But since $\cos k(x' - x) = \cos kx \cos kx' + \sin kx \sin kx'$, this can be rewritten as

$$f(x) = \frac{1}{\pi}\int_0^{\infty}\left[\int_{-\infty}^{+\infty} f(x') \cos k(x' - x)\, dx'\right] dk \quad (11.1)$$

The quantity in the square brackets is an even function of k, and therefore changing the limits on the outer integral leads to

$$f(x) = \frac{1}{2\pi}\int_{-\infty}^{+\infty}\left[\int_{-\infty}^{+\infty} f(x') \cos k(x' - x)\, dx'\right] dk \quad (11.2)$$

Inasmuch as we are looking for an exponential representation, Euler's theorem comes to mind. Consequently, observe that

$$\frac{i}{2\pi}\int_{-\infty}^{+\infty}\left[\int_{-\infty}^{+\infty} f(x') \sin k(x' - x)\, dx'\right] dk = 0$$

because the factor in brackets is an odd function of k. Adding these last two expressions yields the complex* form of the Fourier integral,

$$f(x) = \frac{1}{2\pi}\int_{-\infty}^{+\infty}\left[\int_{-\infty}^{+\infty} f(x') e^{ikx'}\, dx'\right] e^{-ikx}\, dk \quad (11.3)$$

Thus we can write

$$f(x) = \frac{1}{2\pi}\int_{-\infty}^{+\infty} F(k) e^{-ikx}\, dk \quad (11.4)$$

provided that

$$F(k) = \int_{-\infty}^{+\infty} f(x) e^{ikx}\, dx \quad (11.5)$$

*To keep the notation in standard form, and when there's no loss of clarity, we omit the tilde symbol that would otherwise indicate a complex quantity.

having set $x' = x$ in Eq. (11.5). The function $F(k)$ is the **Fourier transform** of $f(x)$, which is symbolically denoted by

$$F(k) = \mathscr{F}\{f(x)\} \quad (11.6)$$

Actually, several equivalent, slightly different ways of defining the transform appear in the literature. For example, the signs in the exponentials could be interchanged, or the factor of $1/2\pi$ could be split symmetrically between $f(x)$ and $F(k)$; each would then have a coefficient of $1/\sqrt{2\pi}$. Note that $A(k)$ is the real part of $F(k)$, while $B(k)$ is its imaginary part, that is,

$$F(k) = A(k) + iB(k) \quad (11.7a)$$

As was seen in Section 2.4, a complex quantity like this can also be written in terms of a real-valued amplitude, $|F(k)|$, the *amplitude spectrum*, and a real-valued phase, $\phi(k)$, the *phase spectrum*:

$$F(k) = |F(k)| e^{i\phi(k)} \quad (11.7b)$$

and sometimes this form can be quite useful [see Eq. (11.96)].

Just as $F(k)$ is the transform of $f(x)$, $f(x)$ itself is said to be the **inverse Fourier transform** of $F(k)$, or symbolically

$$f(x) = \mathscr{F}^{-1}\{F(k)\} = \mathscr{F}^{-1}\{\mathscr{F}\{f(x)\}\} \quad (11.8)$$

and $f(x)$ and $F(k)$ are frequently referred to as a Fourier-transform pair. It's possible to construct the transform and its inverse in an even more symmetrical form in terms of the spatial frequency $\kappa = 1/\lambda = k/2\pi$. Still, in whatever way it's expressed, the transform will not be precisely the same as the inverse transform, because of the minus sign in the exponential. As a result (Problem 11.10), in the present formulation,

$$\mathscr{F}\{F(k)\} = 2\pi f(-x) \quad \text{while} \quad \mathscr{F}^{-1}\{F(k)\} = f(x)$$

This is most often inconsequential, especially for even functions where $f(x) = f(-x)$, so we can expect a good deal of parity between functions and their transforms.

Obviously, if f were a function of time rather than space, we would merely have to replace x by t and then k, the angular spatial frequency, by ω, the angular temporal frequency, in order to get the appropriate transform pair in the time domain, that is,

$$f(t) = \frac{1}{2\pi}\int_{-\infty}^{+\infty} F(\omega) e^{-i\omega t}\, d\omega \quad (11.9)$$

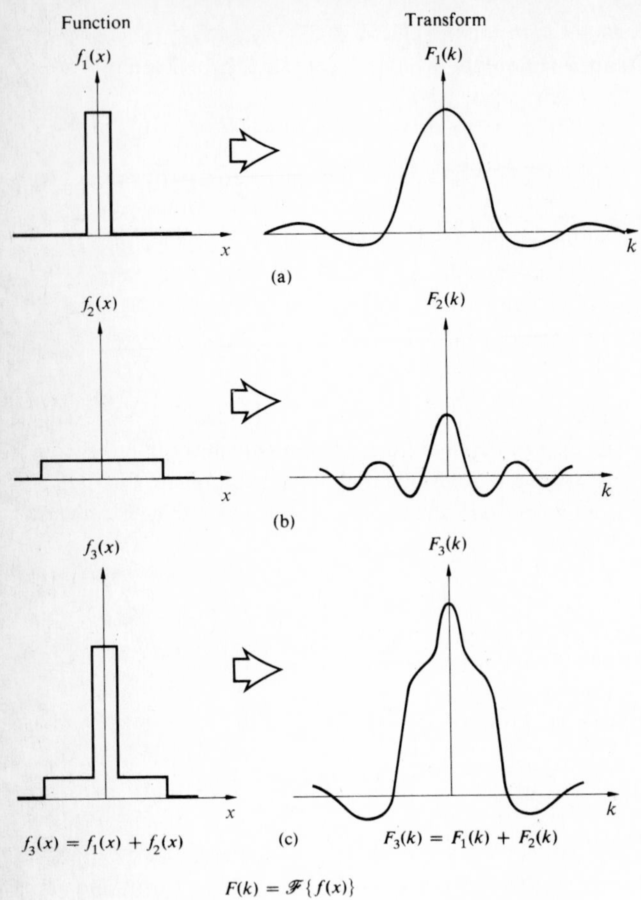

FIGURE 11.1 A composite function and its Fourier transform.

and
$$F(\omega) = \int_{-\infty}^{+\infty} f(t)e^{i\omega t}\,dt \tag{11.10}$$

It should be mentioned that if we write $f(x)$ as a sum of functions, its transform [Eq. (11.5)] will apparently be the sum of the transforms of the individual component functions. This can sometimes be quite a convenient way of establishing the transforms of complicated functions that can be constructed from well-known constituents. Figure 11.1 makes this procedure fairly self-evident.

TRANSFORM OF THE GAUSSIAN FUNCTION

As an example of the method, let's examine the Gaussian probability function,

$$f(x) = Ce^{-ax^2} \tag{11.11}$$

where $C = \sqrt{a/\pi}$ and a is a constant. If you like, you can imagine this to be the profile of a pulse at $t = 0$. The familiar bell-shaped curve (Fig. 11.2*a*) is quite frequently encountered in Optics. It will be germane to a diversity of considerations, such as the wave packet representation of individual photons, the cross-sectional irradiance distribution of a laserbeam in the TEM_{00} mode, and the statistical treatment of thermal light in coherence theory. Its Fourier transform, $\mathcal{F}\{f(x)\}$, is obtained by evaluating

$$F(k) = \int_{-\infty}^{+\infty} (Ce^{-ax^2})e^{ikx}\,dx$$

On completing the square, the exponent, $-ax^2 + ikx$, becomes $-(x\sqrt{a} - ik/2\sqrt{a})^2 - k^2/4a$, and letting $x\sqrt{a} - ik/2\sqrt{a} = \beta$ yields

$$F(k) = \frac{C}{\sqrt{a}}e^{-k^2/4a}\int_{-\infty}^{+\infty} e^{-\beta^2}\,d\beta$$

The definite integral can be found in tables and equals $\sqrt{\pi}$; hence

$$F(k) = e^{-k^2/4a} \tag{11.12}$$

which is again a Gaussian function (Fig. 11.2b), this time with k as the variable. The standard deviation is defined as the range of the variable (x or k) over which the function drops by

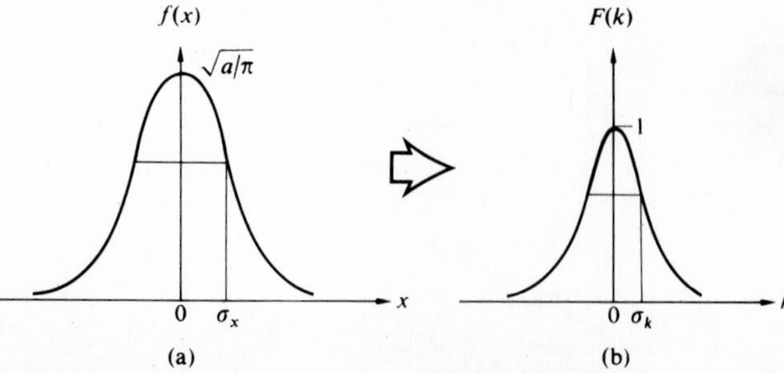

FIGURE 11.2 A Gaussian and its Fourier transform.

a factor of $e^{-1/2} = 0.607$ of its maximum value. Thus the standard deviations for the two curves are $\sigma_x = 1/\sqrt{2a}$ and $\sigma_k = \sqrt{2a}$ and $\sigma_x \sigma_k = 1$. As a increases, $f(x)$ becomes narrower while, in contrast, $F(k)$ broadens. In other words, the shorter the pulse length, the broader the spatial frequency bandwidth.

11.2.2 Two-Dimensional Transforms

Thus far the discussion has been limited to one-dimensional functions, but Optics generally involves two-dimensional signals: for example, the field across an aperture or the flux-density distribution over an image plane. The Fourier-transform pair can readily be generalized to two dimensions, whereupon

$$f(x, y) = \frac{1}{(2\pi)^2} \iint_{-\infty}^{+\infty} F(k_x, k_y) e^{-i(k_x x + k_y y)} dk_x \, dk_y \tag{11.13}$$

and

$$F(k_x, k_y) = \iint_{-\infty}^{+\infty} f(x, y) e^{i(k_x x + k_y y)} \, dx \, dy \tag{11.14}$$

The quantities k_x and k_y are the angular spatial frequencies along the two axes. Suppose we were looking at the image of a tiled floor made up alternately of black and white squares aligned with their edges parallel to the x- and y-directions. If the floor were infinite in extent, the mathematical distribution of reflected light could be regarded in terms of a two-dimensional Fourier series. With each tile having a length ℓ, the spatial period along either axis would be 2ℓ, and the associated fundamental angular spatial frequencies would equal π/ℓ. These and their harmonics would certainly be needed to construct a function describing the scene. If the pattern was finite in extent, the function would no longer be truly periodic, and the Fourier integral would have to replace the series. In effect, Eq. (11.13) says that $f(x, y)$ can be constructed out of a linear combination of elementary functions having the form $\exp[-i(k_x x + k_y y)]$, each appropriately weighted in amplitude and phase by a complex factor $F(k_x, k_y)$. The transform simply tells you how much of and with what phase each elementary component must be added to the recipe. In three dimensions, the elementary functions appear as $\exp[-i(k_x x + k_y y + k_z z)]$ or $\exp(-i\mathbf{k}\cdot\mathbf{r})$, which correspond to planar surfaces. Furthermore, if f is a wave function, that is, some sort of three-dimensional wave $f(\mathbf{r}, t)$, these elementary contributions become plane waves that look like $\exp[-i(\mathbf{k}\cdot\mathbf{r} - \omega t)]$. In other words, *the disturbance can be synthesized out of a linear combination of plane waves having various propagation numbers and moving in various directions*. Similarly, in two dimensions the elementary functions are "oriented" in different directions as well. That is to say, for a given set of values of k_x and k_y, the exponent or phase of the elementary functions will be constant along lines

$$k_x x + k_y y = \text{constant} = A$$

or

$$y = -\frac{k_x}{k_y} x + \frac{A}{k_y} \tag{11.15}$$

The situation is analogous to one in which a set of planes normal to and intersecting the xy-plane does so along the lines given by Eq. (11.15) for differing values of A. A vector perpendicular to the set of lines, call it $\mathbf{k}_\alpha$, would have components k_x and k_y. Figure 11.3 shows several of these lines (for a given k_x and k_y), where $A = 0, \pm 2\pi, \pm 4\pi \ldots$. The slopes are all equal to $-k_x/k_y$ or $-\lambda_y/\lambda_x$ while the y-intercepts equal $A/k_y = A\lambda_y/2\pi$. The orientation of the constant phase lines is

$$\alpha = \tan^{-1}\frac{k_y}{k_x} = \tan^{-1}\frac{\lambda_x}{\lambda_y} \tag{11.16}$$

The wavelength, or spatial period λ_α, measured along $\mathbf{k}_\alpha$, is obtained from the similar triangles in the diagram, where $\lambda_\alpha/\lambda_y = \lambda_x/\sqrt{\lambda_x^2 + \lambda_y^2}$ and

$$\lambda_\alpha = \frac{1}{\sqrt{\lambda_x^{-2} + \lambda_y^{-2}}} \tag{11.17}$$

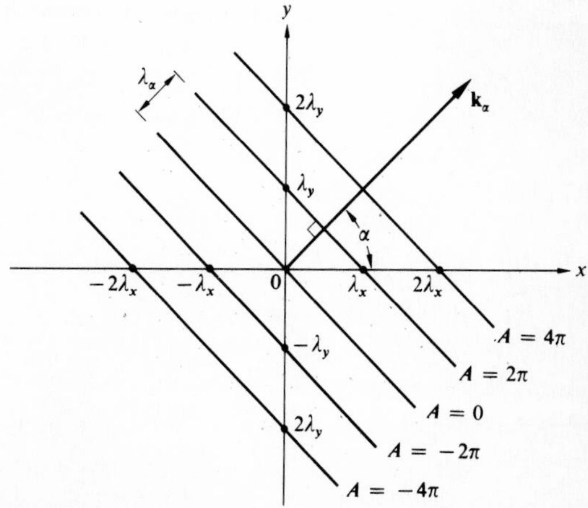

FIGURE 11.3 Geometry for Eq. (11.15).

The angular spatial frequency k_α, being $2\pi/\lambda_\alpha$, is then

$$k_\alpha = \sqrt{k_x^2 + k_y^2} \tag{11.18}$$

as expected. All of this just means that in order to construct a two-dimensional function, harmonic terms in addition to those of spatial frequency k_x and k_y will generally have to be included as well, and these are oriented in directions other than along the x- and y-axes.

Return for a moment to Fig. 10.11, which shows an aperture, with the diffracted wave leaving it represented by several different conceptions. One of these ways to envision the complicated emerging wavefront is as a superposition of plane waves coming off in a whole range of directions. These are the Fourier-transform components, which emerge in specific directions with specific values of angular spatial frequency—the zero spatial frequency term corresponding to the undeviated axial wave, the higher spatial frequency terms coming off at increasingly great angles from the central axis. These Fourier components make up the diffracted field as it emerges from the aperture.

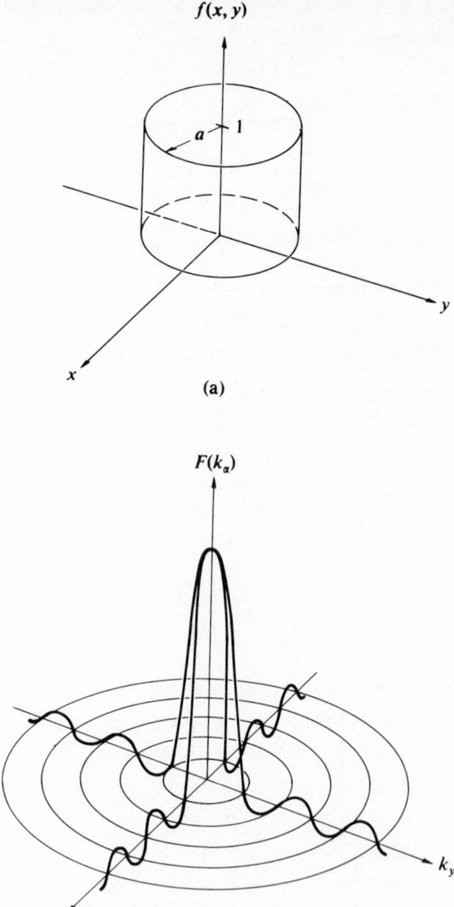

FIGURE 11.4 The cylinder, or top-hat, function and its transform.

Transform of the Cylinder Function

The cylinder function

$$f(x, y) = \begin{cases} 1 & \sqrt{x^2 + y^2} \le a \\ 0 & \sqrt{x^2 + y^2} > a \end{cases} \tag{11.19}$$

(Fig. 11.4*a*) provides an important practical example of the application of Fourier methods to two dimensions. The mathematics will not be particularly simple, but the relevance of the calculation to the theory of diffraction by circular apertures and lenses amply justifies the effort. The evident circular symmetry suggests polar coordinates, and so let

$$\begin{aligned} k_x &= k_\alpha \cos\alpha \\ k_y &= k_\alpha \sin\alpha \\ x &= r\cos\theta \\ y &= r\sin\theta \end{aligned} \tag{11.20}$$

in which case $dx\,dy = r\,dr\,d\theta$. The transform, $\mathscr{F}\{f(x)\}$, then reads

$$F(k_\alpha, \alpha) = \int_{r=0}^{\alpha} \left[\int_{\theta=0}^{2\pi} e^{ik_\alpha r\cos(\theta - a)}\, d\theta \right] r\, dr \tag{11.21}$$

Inasmuch as $f(x, y)$ is circularly symmetric, its transform must be symmetrical as well. This implies that $F(k_\alpha, \alpha)$ is independent of α. The integral can therefore be simplified by letting α equal some constant value, which we choose to be zero, whereupon

$$F(k_\alpha, \alpha) = \int_0^{\alpha} \left[\int_0^{2\pi} e^{ik_\alpha r\cos\theta}\, d\theta \right] r\, dr \tag{11.22}$$

It follows from Eq. (10.47) that

$$F(k_\alpha) = 2\pi \int_0^a J_0(k_\alpha r) r\, dr \tag{11.23}$$

the $J_0(k_\alpha r)$ being a Bessel function of order zero. Introducing a change of variable, namely, $k_\alpha r = w$, we have $dr = k_\alpha^{-1}\, dw$, and the integral becomes

$$\frac{1}{k_\alpha^2}\int_{w=0}^{k_\alpha a} J_0(w)w\, dw \tag{11.24}$$

Using Eq. (10.50), the transform takes the form of a first-order Bessel function (see Fig. 10.27), that is,

$$F(k_\alpha) = \frac{2\pi}{k_\alpha^2}\, k_\alpha a J_1(k_\alpha a)$$

or

$$F(k_\alpha) = 2\pi a^2\left[\frac{J_1(k_\alpha a)}{k_\alpha a}\right] \tag{11.25}$$

The similarity between this expression (Fig. 11.4b) and the formula for the electric field in the Fraunhofer diffraction pattern of a circular aperture [Eq. (10.51)] is, of course, not accidental.

The Lens as a Fourier Transformer

Figure 11.5 shows a transparency, located in the front focal plane of a converging lens, being illuminated by parallel light. This object, in turn, scatters plane waves, which are collected by the lens, and parallel bundles of rays are brought to convergence at its back focal plane. If a screen were placed there, at Σ_t, the so-called **transform plane**, we would see the far-field diffraction pattern of the object spread across it [this is essentially the configuration of Fig. 10.11*e*]. In other words, the electric-field distribution across the object mask, which is known as the **aperture function**, is transformed by the lens into the far-field diffraction pattern. Although this assertion is true enough for most purposes, it's not exactly true. After all, the lens doesn't actually form its image on a plane.

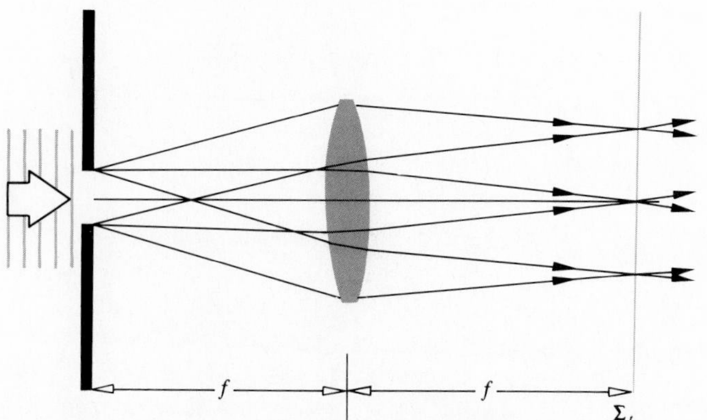

FIGURE 11.5 The light diffracted by a transparency at the front (or object) focal point of a lens converges to form the far-field diffraction pattern at the back (or image) focal point of the lens.

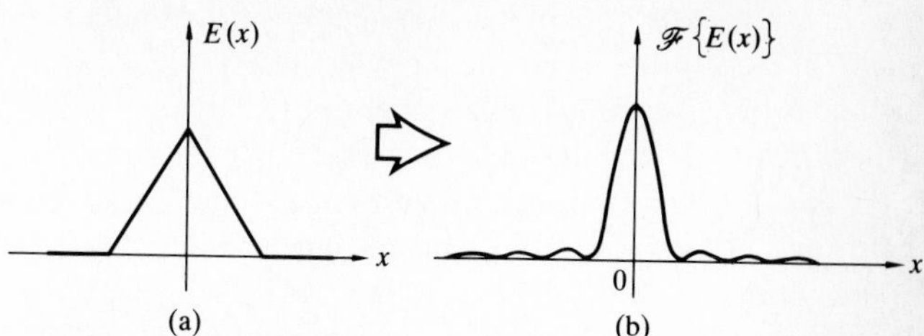

FIGURE 11.6 The transform of the triangle function is the sinc² function.

Remarkably, that Fraunhofer **E**-field pattern corresponds to the exact Fourier transform of the aperture function—a fact we shall confirm more rigorously in Section 11.3.3. Here the object is in the front focal plane, and all the various diffracted waves maintain their phase relationships traveling essentially equal optical path lengths to the transform plane. That doesn't quite happen when the object is displaced from the front focal plane. Then there will be a phase deviation, but that is actually of little consequence, since we are generally interested in the irradiance where the phase information is averaged out and the phase distortion is unobservable.

Thus if an otherwise opaque object mask contains a single circular hole, the **E**-field across it will resemble the top hat of Fig. 11.4*a*, and the diffracted field, the Fourier transform, will be distributed in space as a Bessel function, looking very much like Fig. 11.4*b*. Similarly, if the object transparency varies in density only along one axis, such that its amplitude transmission profile is triangular (Fig. 11.6*a*), then the amplitude of the electric field in the diffraction pattern will correspond to Fig. 11.6*b*—the Fourier transform of the triangle function is the sinc-squared function.

11.2.3 The Dirac Delta Function

Many physical phenomena occur over very short durations in time with great intensity, and one is frequently concerned with the consequent response of some system to such stimuli. For example: How will a mechanical device, like a billiard ball, respond to being slammed with a hammer? Or how will a particular circuit behave if the input is a short burst of current? In much the same way, we can envision some stimulus that is a sharp pulse in the space, rather than the time, domain. A bright minute source of light imbedded in a dark background is essentially a highly localized, two-dimensional, spatial pulse—a spike of irradiance. A convenient idealized mathematical representation of this sort of sharply peaked stimulus is the **Dirac delta function** $\delta(x)$. This is a quantity that is zero everywhere except at the origin, where it goes to infinity in a manner so as to encompass a *unit area*, that is,

$$\delta(x) = \begin{cases} 0 & x \neq 0 \\ \infty & x = 0 \end{cases} \tag{11.26}$$

and

$$\int_{-\infty}^{+\infty} \delta(x)\, dx = 1 \tag{11.27}$$

This is not really a function in the traditional mathematical sense. In fact, because it is so singular in nature, it remained the focus of considerable controversy long after it was reintroduced and brought into prominence by P.A.M. Dirac in 1930. Yet physicists, pragmatic as they sometimes are, found it so highly useful that it soon became an established tool, despite what seemed a lack of rigorous justification. The precise mathematical theory of the delta function evolved roughly 20 years later, in the early 1950s, principally at the hands of Laurent Schwartz.

Perhaps the most basic operation to which $\delta(x)$ can be applied is the evaluation of the integral

$$\int_{-\infty}^{+\infty} \delta(x) f(x)\, dx$$

Here the expression $f(x)$ corresponds to any continuous function. Over a tiny interval running from $x = -\gamma$ to $+\gamma$ centered about the origin, $f(x) \approx f(0) \approx$ constant, since the function is continuous at $x = 0$. From $x = -\infty$ to $x = -\gamma$ and from $x = +\gamma$ to $x = +\infty$, the integral is zero, simply because the δ-function is zero there. Thus the integral equals

$$f(0) \int_{-\gamma}^{+\gamma} \delta(x)\, dx$$

Because $\delta(x) = 0$ for all x other than 0, the interval can be vanishingly small, that is, $\gamma \to 0$, and still

$$\int_{-\gamma}^{+\gamma} \delta(x)\, dx = 1$$

from Eq. (11.27). Hence we have the exact result that

$$\int_{-\infty}^{+\infty} \delta(x) f(x)\, dx = f(0) \tag{11.28}$$

This is often spoken of as the **sifting property** of the δ-function, because it manages to extract only the one value of $f(x)$ taken at $x = 0$ from all its possible values. Similarly, with a shift of origin of an amount x_0,

$$\delta(x - x_0) = \begin{cases} 0 & x \neq x_0 \\ \infty & x = x_0 \end{cases} \tag{11.29}$$

and the spike resides at $x = x_0$ rather than $x = 0$, as shown in Fig. 11.7. The corresponding sifting property can be appreci-

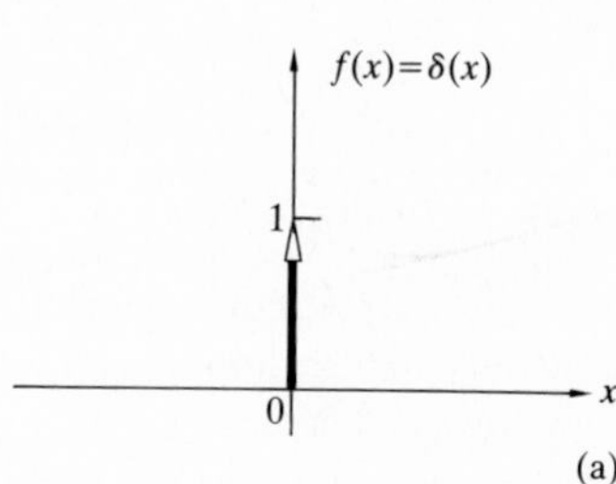

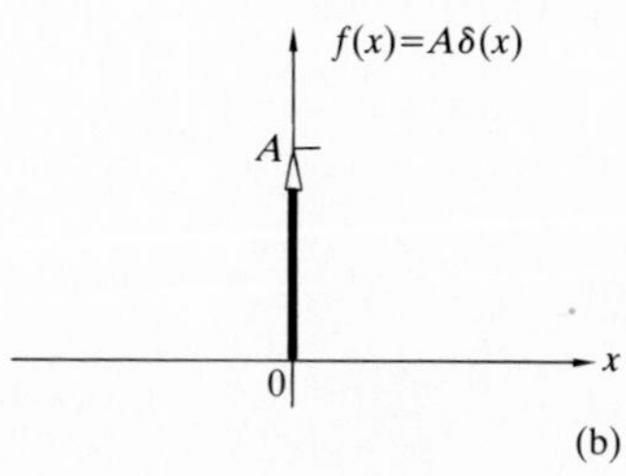

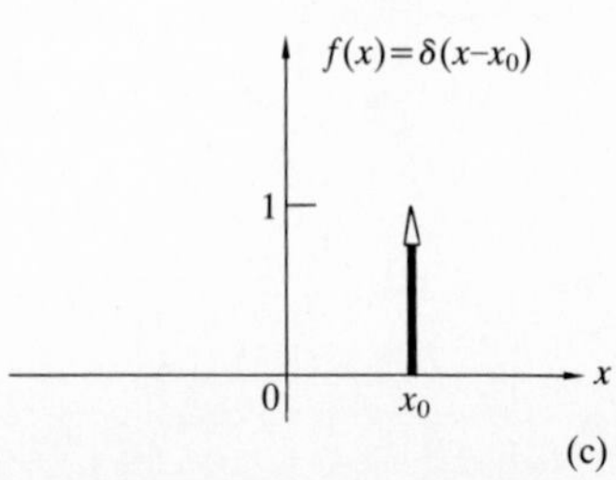

FIGURE 11.7 The height of the arrow representing the delta function corresponds to the area under the function.

ated by letting $x - x_0 = x'$, then with $f(x' + x_0) = g(x')$,

$$\int_{-\infty}^{+\infty} \delta(x - x_0) f(x)\, dx = \int_{-\infty}^{+\infty} \delta(x') g(x')\, dx' = g(0)$$

and since $g(0) = f(x_0)$,

$$\int_{-\infty}^{+\infty} \delta(x - x_0) f(x)\, dx = f(x_0). \tag{11.30}$$

Formally, rather than worrying about a precise definition of $\delta(x)$ for each value of x, it would be more fruitful to continue along the lines of defining the effect of $\delta(x)$ on some other function $f(x)$. Accordingly, Eq. (11.28) is really the definition of an entire operation that assigns a number $f(0)$ to the function $f(x)$. Incidentally, an operation that performs this service is called a *functional*.

It is possible to construct a number of sequences of pulses, each member of which has an ever-decreasing width and a concomitantly increasing height, such that any one pulse encompasses a unit area. A sequence of square pulses of height a/L and width L/a for which $a = 1, 2, 3, \ldots$ would fit the bill; so would a sequence of Gaussians [Eq. (11.11)],

$$\delta_a(x) = \sqrt{\frac{a}{\pi}}\, e^{-ax^2} \tag{11.31}$$

as in Fig. 11.8, or a sequence of sinc functions

$$\delta_a(x) = \frac{a}{\pi} \operatorname{sinc}(ax) \tag{11.32}$$

Such strongly peaked functions that approach the sifting property, that is, for which

$$\lim_{a\to\infty} \int_{-\infty}^{+\infty} \delta_a(x) f(x)\, dx = f(0) \tag{11.33}$$

are known as *delta sequences*. It is often useful, but not actually rigorously correct, to imagine $\delta(x)$ as the convergence limit of such sequences as $a \to \infty$. The extension of these ideas into two dimensions is provided by the definition

$$\delta(x,y) = \begin{cases} \infty & x = y = 0 \\ 0 & \text{otherwise} \end{cases} \tag{11.34}$$

and

$$\iint_{-\infty}^{+\infty} \delta(x, y)\, dx\, dy = 1 \tag{11.35}$$

and the sifting property becomes

$$\iint_{-\infty}^{+\infty} f(x,y)\delta(x - x_0)\delta(y - y_0)\, dx\, dy = f(x_0,y_0) \tag{11.36}$$

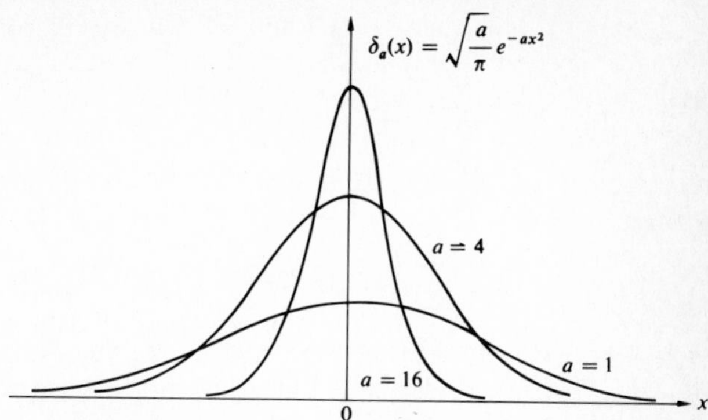

FIGURE 11.8 A sequence of Gaussians.

Another representation of the δ-function follows from Eq. (11.3), the Fourier integral, which can be restated as

$$f(x) = \int_{-\infty}^{+\infty} \left[\frac{1}{2\pi} \int_{-\infty}^{+\infty} e^{-ik(x-x')} dk \right] f(x')\, dx'$$

and hence

$$f(x) = \int_{-\infty}^{+\infty} \delta(x - x') f(x')\, dx' \tag{11.37}$$

provided that

$$\delta(x - x') = \frac{1}{2\pi} \int_{-\infty}^{+\infty} e^{-ik(x-x')} dk \tag{11.38}$$

Equation (11.37) is identical to Eq. (11.30), since by definition from Eq. (11.29) $\delta(x - x') = \delta(x' - x)$. The (divergent) integral of Eq. (11.38) is zero everywhere except at $x = x'$. Evidently, with $x' = 0$, $\delta(x) = \delta(-x)$ and

$$\delta(x) = \frac{1}{2\pi} \int_{-\infty}^{+\infty} e^{-ikx} dk = \frac{1}{2\pi} \int_{-\infty}^{+\infty} e^{ikx} dk \tag{11.39}$$

This implies, via Eq. (11.4), that the delta function can be thought of as the inverse Fourier transform of unity, that is, $\delta(x) = \mathcal{F}^{-1}\{1\}$ and so $\mathcal{F}\{\delta(x)\} = 1$. We can imagine a square pulse becoming narrower and taller as its transform, in turn, grows broader, until finally the pulse is infinitesimal in width, and its transform is infinite in extent, in other words, a constant.

Displacements and Phase Shifts

If the δ-spike is shifted off $x = 0$ to, say, $x = x_0$, its transform will change phase but not amplitude—that remains equal to one. To see this, evaluate

$$\mathscr{F}\{\delta(x - x_0)\} = \int_{-\infty}^{+\infty} \delta(x - x_0)e^{ikx}\,dx$$

From the sifting property [Eq. (11.30)] the expression becomes

$$\mathscr{F}\{\delta(x - x_0)\} = e^{ikx_0} \tag{11.40}$$

What we see is that only the phase is affected, the amplitude being one as it was when $x_0 = 0$. This whole process can be appreciated somewhat more intuitively if we switch to the time domain and think of an infinitesimally narrow pulse (such as a spark) occurring at $t = 0$. This results in the generation of an infinite range of frequency components, which are all initially in-phase at the instant of creation ($t = 0$). On the other hand, suppose the pulse occurs at a time t_0. Again every frequency is produced, but in this situation the harmonic components are all in phase at $t = t_0$. Consequently, if we extrapolate back, the phase of each constituent at $t = 0$ will now have to be different, depending on the particular frequency. Besides, we know that all these components superimpose to yield zero everywhere except at t_0, so that a frequency-dependent phase shift is quite reasonable. This phase shift is evident in Eq. (11.40) for the space domain. Note that it does vary with the angular spatial frequency k.

All of this is quite general in its applicability, and we observe that ***the Fourier transform of a function that is displaced in space (or time) is the transform of the undisplaced function multiplied by an exponential that is linear in phase*** (Problem 11.14). This property of the transform will be of special interest presently, when we consider the image of several point sources that are separated but otherwise identical. The process can be appreciated diagramatically with the help of Figs. 11.9 and 7.19. To shift the square wave by $\pi/4$ to the right, the fundamental must be shifted $\frac{1}{8}$-wavelength (or, say, 1.0 mm), and every component must then be displaced an equal distance (i.e., 1.0 mm). Thus each component must be shifted in phase by an amount specific to it that produces a 1.0-mm displacement. Here each is displaced, in turn, by a phase of $m\pi/4$.

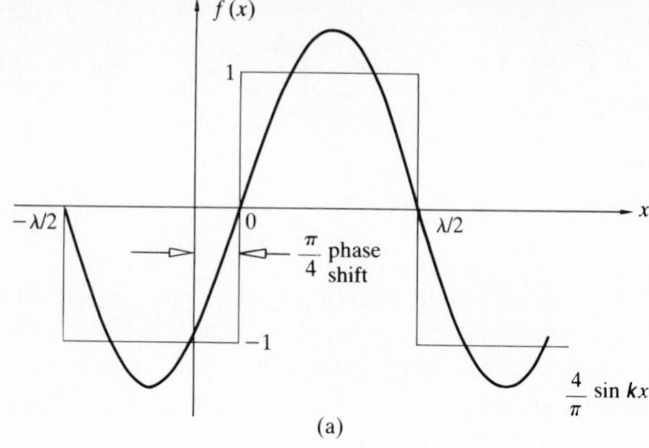

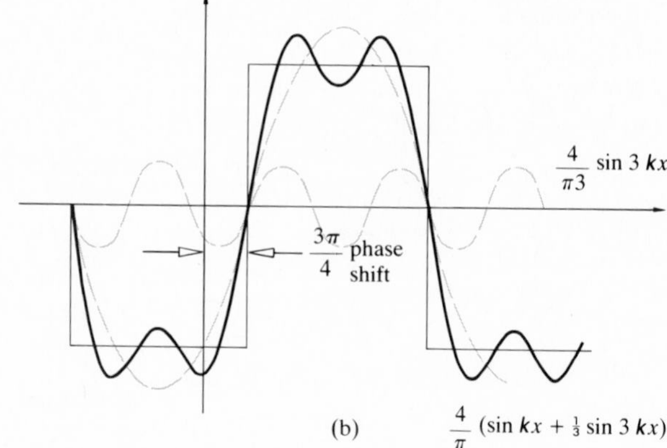

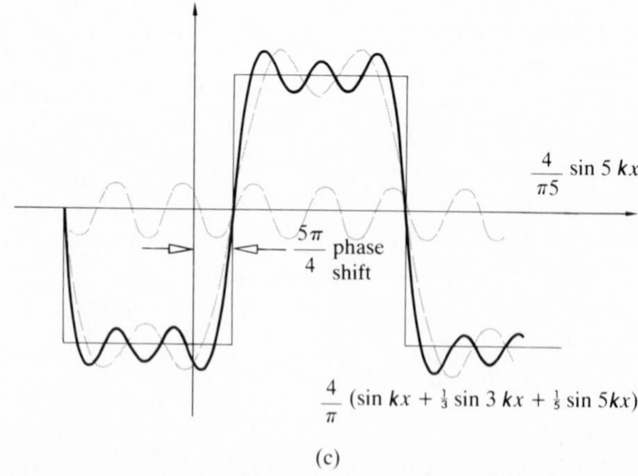

FIGURE 11.9 A shifted square wave showing the corresponding change in phase for each component wave.

SINES AND COSINES

We saw earlier (Fig. 11.1) that if the function at hand can be written as a sum of individual functions, its transform is simply the sum of the transforms of the component functions. Suppose we have a string of delta functions spread out uniformly like the teeth on a comb,

$$f(x) = \sum_j \delta(x - x_j) \tag{11.41}$$

When the number of terms is infinite, this periodic function is often called *comb*(x). In any event, the transform will simply be a sum of terms, such as that of Eq. (11.40):

$$\mathscr{F}\{f(x)\} = \sum_j e^{ikx_j} \tag{11.42}$$

In particular, if there are two δ-functions, one at $x_0 = d/2$ and the other at $x_0 = -d/2$,

$$f(x) = \delta[x - (+d/2)] + \delta[x - (-d/2)]$$

and

$$\mathscr{F}\{f(x)\} = e^{ikd/2} + e^{-ikd/2}$$

which is just

$$\mathscr{F}\{f(x)\} = 2\cos(kd/2) \tag{11.43}$$

as in Fig. 11.10. Thus the transform of the sum of these two symmetrical δ-functions is a cosine function and vice versa. The composite is a real even function, and $F(k) = \mathscr{F}\{f(x)\}$ will also be real and even. This should be reminiscent of Young's Experiment with infinitesimally narrow slits—we'll come back to it later. If the phase of one of the δ-functions is shifted, as in Fig. 11.11, the composite function is asymmetrical, it's odd,

$$f(x) = \delta[x - (+d/2)] - \delta[x - (-d/2)]$$

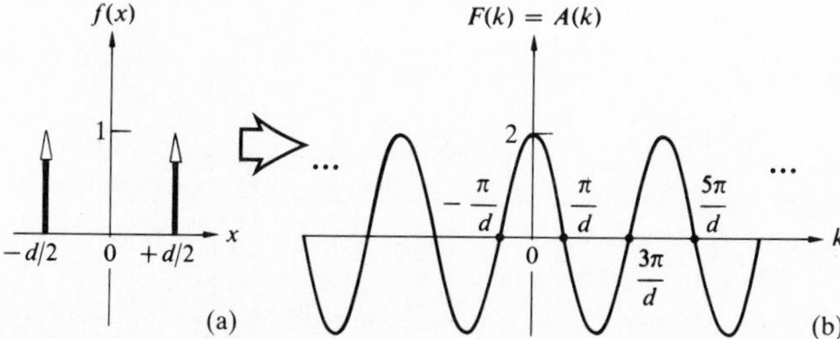

FIGURE 11.10 Two delta functions and their cosine-function transform.

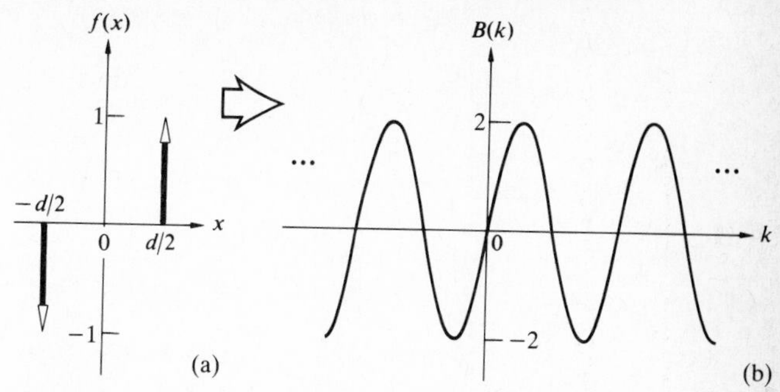

FIGURE 11.11 Two delta functions and their sine-function transform.

and

$$\mathscr{F}\{f(x)\} = e^{ikd/2} - e^{ikd/2} = 2i\sin(kd/2) \tag{11.44}$$

The real sine transform [Eq. (11.7)] is then

$$B(k) = 2\sin(kd/2) \tag{11.45}$$

and it too is an odd function.

This raises an interesting point. Recall that there are two alternative ways to consider the complex transform: either as the sum of a real and an imaginary part, from Eq. (11.7a), or as the product of an amplitude and a phase term, from Eq. (11.7b). It happens that the cosine and sine are rather special functions; the former is associated with a purely real contribution and the latter is associated with a purely imaginary one. Most functions, even harmonic ones, will usually be a blend of real and imaginary parts. For example, once a cosine is displaced a little, the new function, which is typically neither odd nor even, has both a real and an imaginary part. Moreover, it can be expressed as a cosinusoidal amplitude spectrum, which is appropriately phase-shifted (Fig. 11.12). Notice that when the cosine is shifted $\frac{1}{4}\lambda$ into a sine, the relative phase difference between the two component delta functions is again π rad.

Figure 11.13 displays in summary form a number of transforms, mostly of harmonic functions. Observe how the functions and transforms in (*a*) and (*b*) combine to produce the function and its transform in (*d*). As a rule, each member of the pair of δ-pulses in the frequency spectrum of a harmonic function is located along the k-axis at a distance from the origin equal to the fundamental angular spatial frequency of $f(x)$. Since any well-behaved periodic function can be expressed as

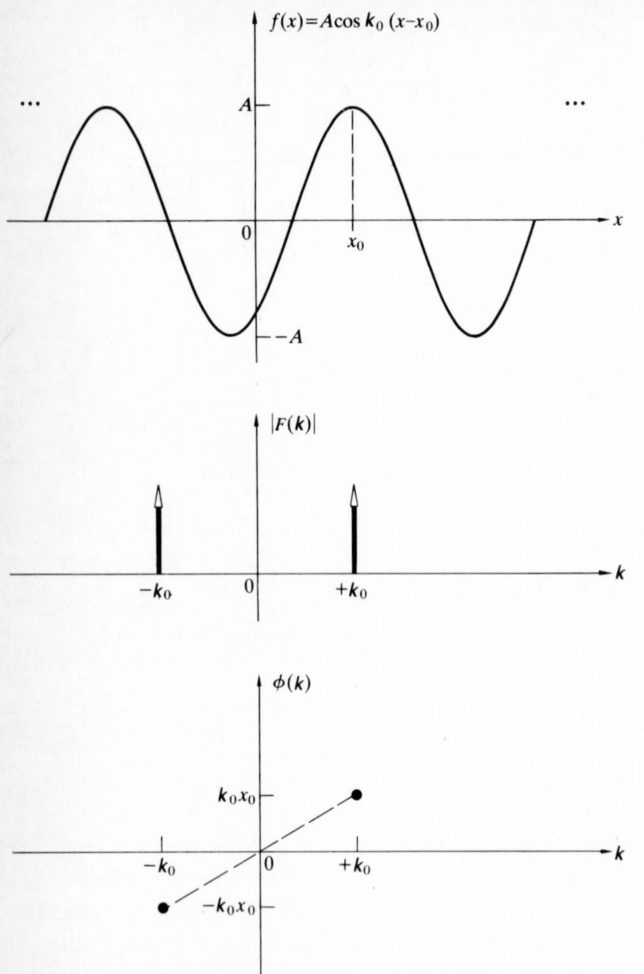

FIGURE 11.12 The spectra of a shifted cosine function.

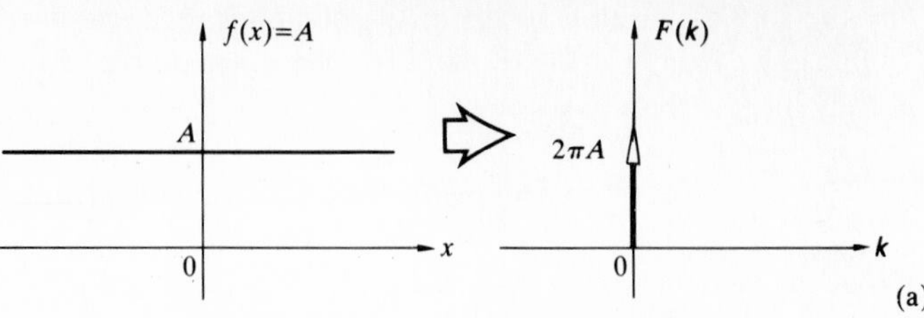

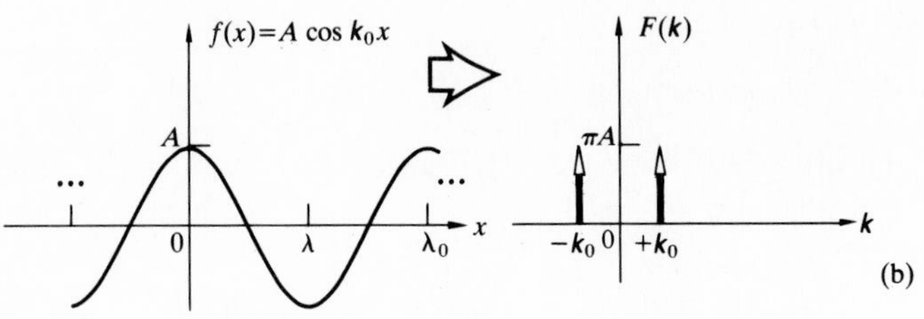

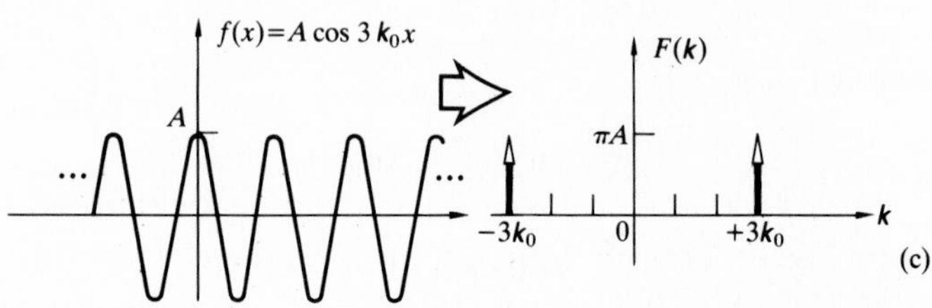

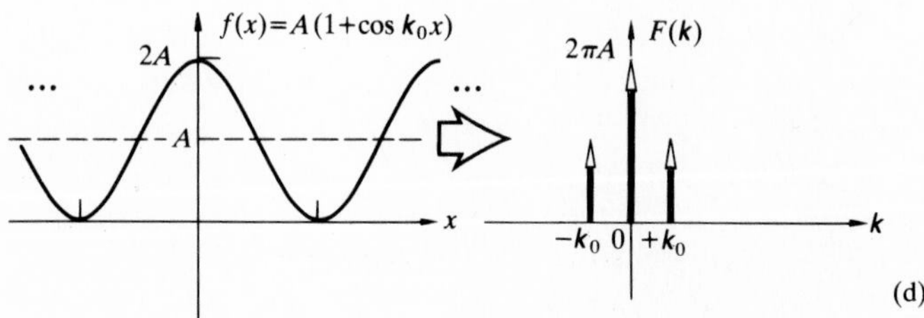

a Fourier series, it can also be represented as an array of pairs of delta functions, each weighted appropriately and each a distance from the k-origin equal to the angular spatial frequency of the particular harmonic contribution—*the frequency spectrum of any periodic function will be discrete*. One of the most remarkable of the periodic functions is *comb*(x): as shown in Fig. 11.14, its transform is also a comb function.

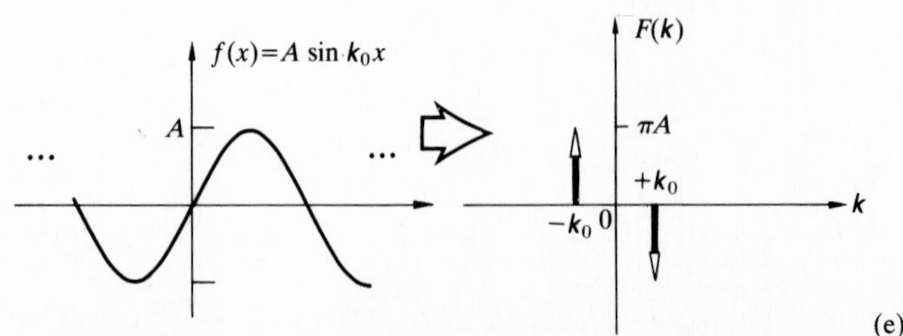

FIGURE 11.13 Some functions and their transforms.

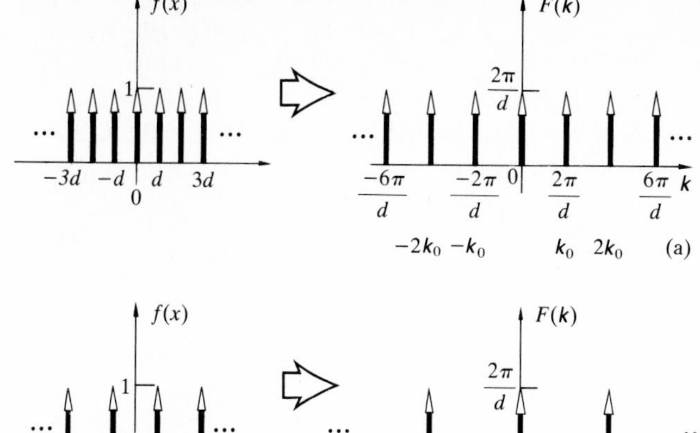

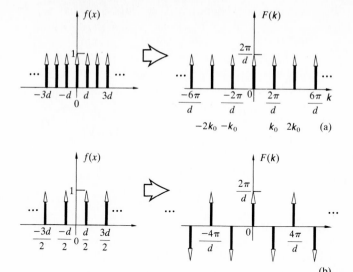

FIGURE 11.14 (*a*) The comb function and its transform. (*b*) A shifted comb function and its transform.

11.3 Optical Applications

11.3.1 Linear Systems

Fourier techniques provide a particularly elegant framework from which to evolve a description of the formation of images. And for the most part, this will be the direction in which we shall be moving, although some side excursions are unavoidable in order to develop the needed mathematics.

A key point in the analysis is the concept of a **linear system**, which in turn is defined in terms of its input–output relations. Suppose then that an input signal $f(y, z)$ passing through some optical system results in an output $g(Y, Z)$. The system is linear if:

1. multiplying $f(y, z)$ by a constant a produces an output $ag(Y, Z)$.
2. when the input is a weighted sum of two (or more) functions, $af_1(y, z) + bf_2(y, z)$, the output will similarly have the form $ag_1(Y, Z) + bg_2(Y, Z)$, where $f_1(y, z)$ and $f_2(y, z)$ generate $g_1(Y, Z)$ and $g_2(Y, Z)$ respectively.

Furthermore, a linear system will be *space invariant* if it possesses the property of *stationarity*; that is, in effect, changing the position of the input merely changes the location of the output without altering its functional form. The idea behind much of this is that the output produced by an optical system can be treated as a linear superposition of the outputs arising from each of the individual points on the object. In fact, if we symbolically represent the operation of the linear system as $\mathscr{L}\{\ \}$, the input and output can be written as

$$g(Y, Z) = \mathscr{L}\{f(y, z)\} \tag{11.46}$$

Using the sifting property of the δ-function [Eq. (11.36)], this becomes

$$g(Y, Z) = \mathscr{L}\left\{\iint_{-\infty}^{+\infty} f(y', z')\delta(y' - y)\delta(z' - z)\,dy'\,dz'\right\}$$

The integral expresses $f(y, z)$ as a linear combination of elementary delta functions, each weighted by a number $f(y', z')$. It follows from the second linearity condition that the system operator can equivalently act on each of the elementary functions; thus

$$g(Y, Z) = \iint_{-\infty}^{+\infty} f(y', z')\mathscr{L}\{\delta(y' - y)\delta(z' - z)\}\,dy'\,dz' \tag{11.47}$$

The quantity $\mathscr{L}\{\delta(y' - y)\delta(z' - z)\}$ is the response of the system [Eq. (11.46)] to a delta function located at the point (y', z') in the input space—it's called the **impulse response**. Apparently, if the impulse response of a system is known, the output can be determined directly from the input by means of Eq. (11.47). If the elementary sources are coherent, the input and output signals will have to be electric fields; if incoherent, they'll be flux densities.

Consider the self-luminous and, therefore, incoherent source depicted in Fig. 11.15. We can imagine that each point on the object plane, Σ_0, emits light that is processed by the optical system. It emerges to form a spot on the focal or image plane, Σ_i. In addition, *assume that the magnification between object and image planes is one*. The image will be life-sized and erect, which makes it a little easier to deal with for the time being. Notice that if the magnification (M_T) was greater than one, the image would be larger than the object. Consequently, all of its structural details would be larger and broader, so the spatial frequencies of the harmonic contributions that go into synthesizing the image would be lower than

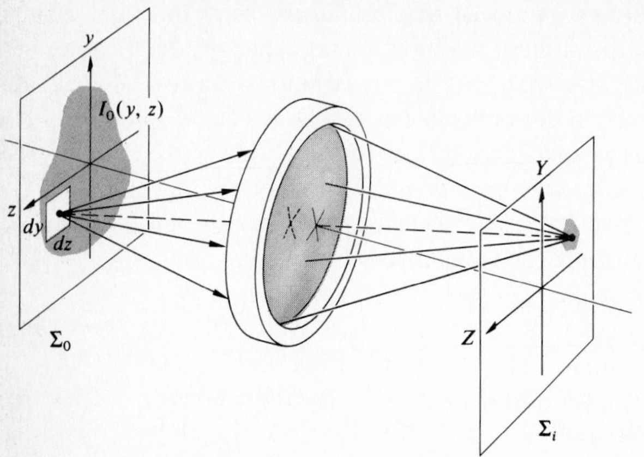

FIGURE 11.15 A lens system forming an image.

those of the object. For example, an object that is a transparency of a sinusoidally varying black and white linear pattern (a sinusoidal amplitude grating) would be imaged having a greater space between maxima and therefore a lower spatial frequency. Besides that, the image irradiance would be decreased by M_T^2, because the image area would be increased by a factor of M_T^2.

If $I_0(y, z)$ is the irradiance distribution on the object plane, an element $dy\,dz$ located at (y, z) will emit a radiant flux of $I_0(y, z)\,dy\,dz$. Because of diffraction (and the possible presence of aberrations), this light is smeared out into some sort of blur spot over a finite area on the image plane rather than focused to a point. The spread of radiant flux is described mathematically by the function $\mathcal{S}(y, z; Y, Z)$, such that the flux density arriving at the image point from $dy\,dz$ is

$$dI_i(Y, Z) = \mathcal{S}(y, z; Y, Z)I_0(y, z)\,dy\,dz \qquad (11.48)$$

This is the patch of light in the image plane at (Y, Z), and $\mathcal{S}(y, z; Y, Z)$ is known as the **point-spread function**. In other words, when the irradiance $I_0(y, z)$ over the source element $dy\,dz$ is 1 W/m^2, $\mathcal{S}(y, z; Y, Z)dy\,dz$ is the profile of the resulting irradiance distribution in the image plane. Because of the incoherence of the source, the flux-density contributions from each of its elements are additive, so

$$I_i(Y, Z) = \iint_{-\infty}^{+\infty} I_0(y, z)\mathcal{S}(y, z; Y, Z)\,dy\,dz \qquad (11.49)$$

In a "perfect," diffraction-limited optical system having no aberrations, $\mathcal{S}(y, z; Y, Z)$ would correspond in shape to the diffraction figure of a point source at (y, z). Evidently, if we set the input equal to a δ-pulse centered at (y_0, z_0), then $I_0(y, z) = A\delta(y - y_0)\delta(z - z_0)$. Here the constant A of magnitude one carries the needed units (i.e., irradiance times area). Thus

$$I_i(Y, Z) = A\iint_{-\infty}^{+\infty} \delta(y - y_0)\delta(z - z_0)\mathcal{S}(y, z; Y, Z)\,dy\,dz$$

and so from the sifting property,

$$I_i(Y, Z) = A\mathcal{S}(y_0, z_0; Y, Z)$$

The point-spread function has a functional form identical to that of the image generated by a δ-pulse input. It's the impulse response of the system [compare Eqs. (11.47) and (11.49)], whether optically perfect or not. In a well-corrected system $\mathcal{S}$, apart from a multiplicative constant, is the Airy irradiance distribution function [Eq. (10.56)] centered on the Gaussian image point (Fig. 11.16).

If the system is space invariant, a point-source input can be moved about over the object plane without any effect other than changing the location of its image. Equivalently, one can say that the spread function is the same for any point (y, z). In practice, however, the spread function will vary, but even so, the image plane can be divided into small regions, over each of which $\mathcal{S}$ doesn't change appreciably. Thus if the object, and

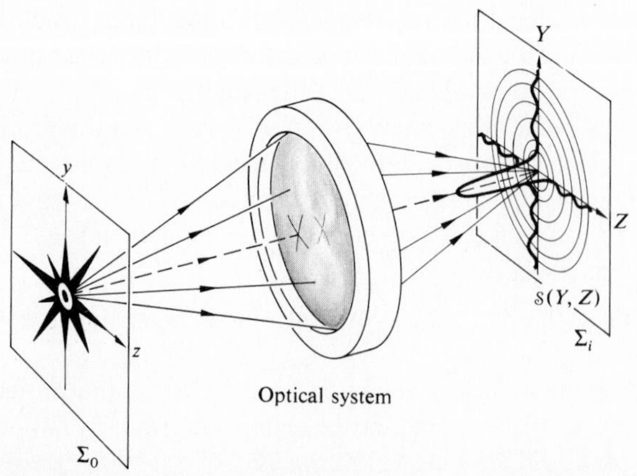

FIGURE 11.16 The point-spread function: the irradiance produced by the optical system with an input point source.

therefore its image, is small enough, the system can be taken to be space invariant. We can imagine a spread function sitting at every Gaussian image point on Σ_i, each multiplied by a different weighting factor $I_0(y, z)$ but all of the same general shape independent of (y, z). Since the magnification was set at one, the coordinates of any object and conjugate image point have the same magnitude.

If we were dealing with coherent light, we would have to consider how the system acted upon an input that was again a δ-pulse, but this time one representing the field amplitude.

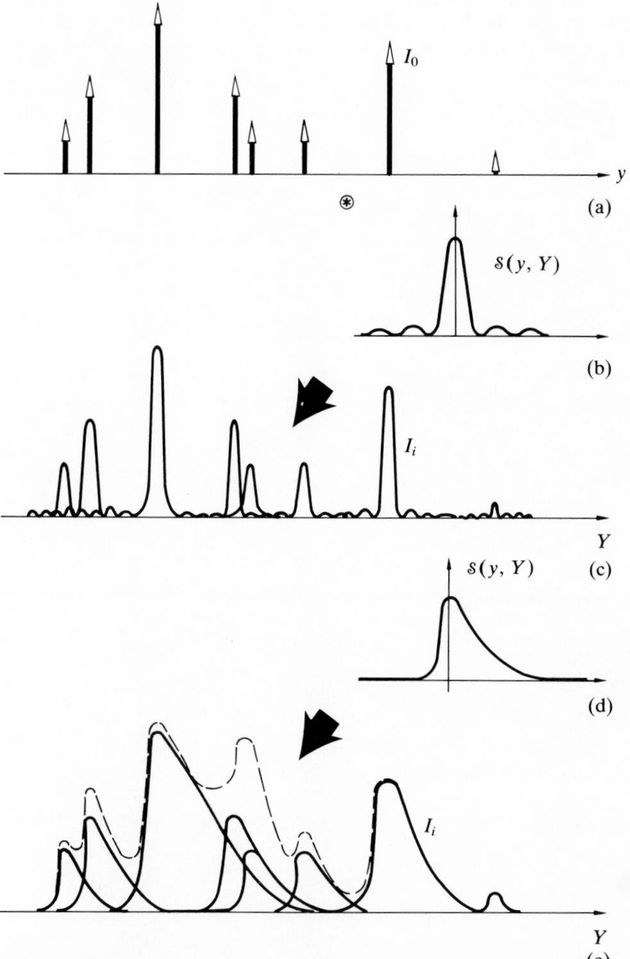

FIGURE 11.17 Here (*a*) is convolved first with (*b*) to produce (*c*) and then with (*d*) to produce (*e*). The resulting pattern is the sum of all the spread-out contributions as indicated by the dashed curve in (*e*).

Once more the resulting image would be described by a spread function, although it would be an *amplitude* spread function. For a diffraction-limited circular aperture, the amplitude spread function looks like Fig. 10.28*b*. And finally, we would have to be concerned about the interference that would take place on the image plane as the coherent fields interacted. By contrast, with incoherent object points the process occurring on the image plane is simply the summation of overlapping irradiances, as depicted in one dimension in Fig. 11.17. Each source point, with its own strength, corresponds to an appropriately scaled δ-pulse, and in the image plane each of these is smeared out, via the spread function. The sum of all the overlapping contributions is the image irradiance.

What kind of dependence on the image and object space variables will $\mathcal{S}(y, z; Y, Z)$ have? The spread function can only depend on (y, z) as far as the location of its center is concerned. Thus the value of $\mathcal{S}(y, z; Y, Z)$ anywhere on Σ_i merely depends on the displacement at that location from the particular Gaussian image point $(Y = y, Z = z)$ on which $\mathcal{S}$ is centered (Fig. 11.18). In other words,

$$\mathcal{S}(y, z; Y, Z) = \mathcal{S}(Y - y, Z - z) \tag{11.50}$$

When the object point is on the central axis $(y = 0, z = 0)$, the Gaussian image point is as well, and the spread function is then just $\mathcal{S}(Y, Z)$, as depicted in Fig. 11.16. Under the circumstances of space invariance and incoherence,

$$I_i(Y, Z) = \iint_{-\infty}^{+\infty} I_0(y, z)\mathcal{S}(Y - y, Z - z)\, dy\, dz \tag{11.51}$$

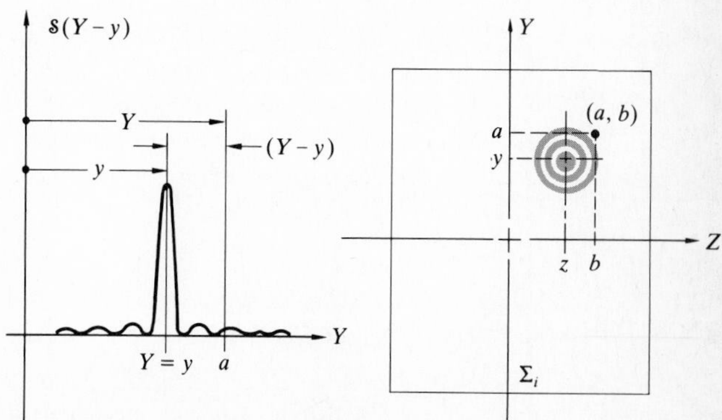

FIGURE 11.18 The point-spread function.

11.3.2 The Convolution Integral

Figure 11.17 shows a one-dimensional representation of the distribution of point-source δ-functions that make up the object. The corresponding image is essentially obtained by "dealing out" an appropriately weighted point-spread function to the location of each image point on Σ_i and then adding up all the contributions at each point along Y. This dealing out of one function to every point of (and weighted by) another function is a process known as **convolution**, and we say that one function $I_0(y)$, is convolved with another, $\mathcal{S}(y, Y)$, or vice versa.

This procedure can be carried out in two dimensions as well, and that's essentially what is being done by Eq. (11.51), the so-called *convolution integral*. The corresponding one-dimensional expression describing the convolution of two functions $f(x)$ and $h(x)$,

$$g(X) = \int_{-\infty}^{+\infty} f(x)h(X - x)\,dx \qquad (11.52)$$

is easier to appreciate. In Fig. 11.17 one of the two functions was a group of δ-pulses, and the convolution operation was particularly easy to visualize. Still, we can imagine any function to be composed of a "densely packed" continuum of δ-pulses and treat it in much the same fashion. Let us now examine in some detail exactly how the integral of Eq. (11.52) mathematically manages to perform the convolution. The essential features of the process are illustrated in Fig. 11.19. The resulting signal $g(X_1)$, at some point X_1 in the output space, is a linear superposition of all the individual overlapping contributions that exist at X_1. In other words, each source element dx yields a signal of a particular strength $f(x)\,dx$, which is then smeared out by the system into a region centered

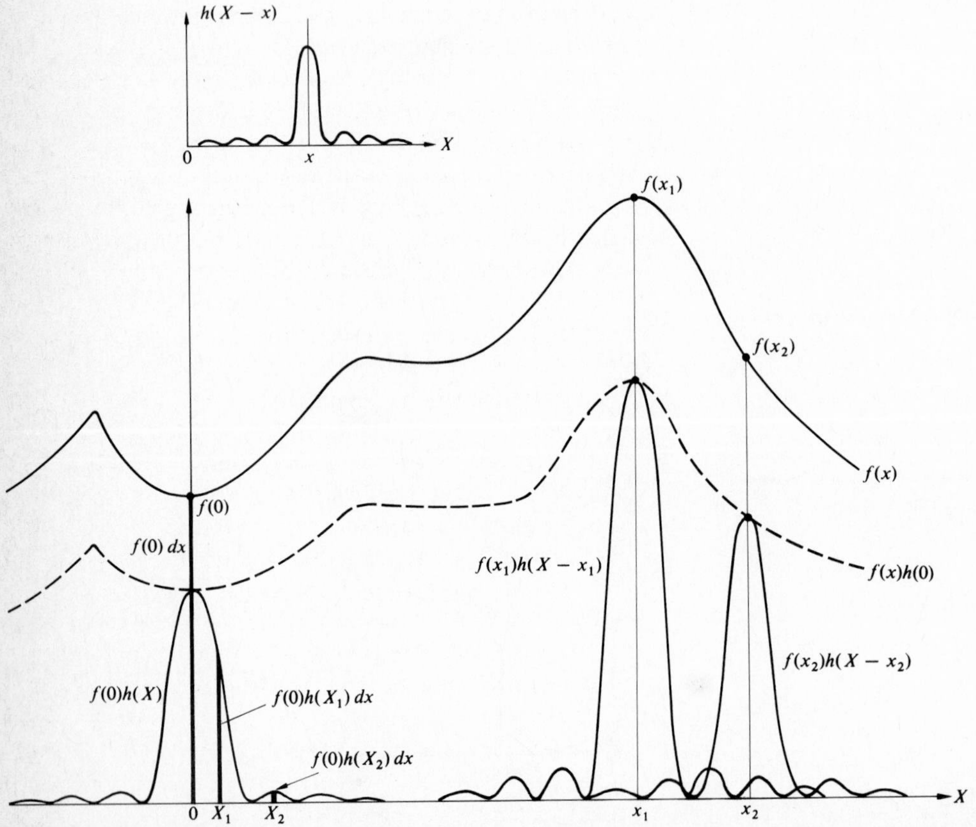

FIGURE 11.19 The overlapping of weighted spread functions.

about the Gaussian image point ($X = x$). The output at X_1 is then $dg(X_1) = f(x)h(X_1 - x)\,dx$. The integral sums up all of these contributions from each source element. Of course the elements more remote from a given point on Σ_i contribute less, because the spread function generally drops off with displacement. Thus we can imagine $f(x)$ to be a one-dimensional irradiance distribution, such as a series of vertical bands, as in Fig. 11.20. If the one-dimensional **line-spread function**, $h(X - x)$, is that of Fig. 11.20*d*, the resulting image will simply be a somewhat blurred version of the input (Fig. 11.20*e*).

Let's now examine the convolution a bit more as a mathematical entity. Actually it's a rather subtle beast, performing a process that might certainly not be obvious at first glance, so let's approach it from a slightly different viewpoint. Accordingly, we will have two ways of thinking about the convolution integral, and we shall show that they are equivalent.

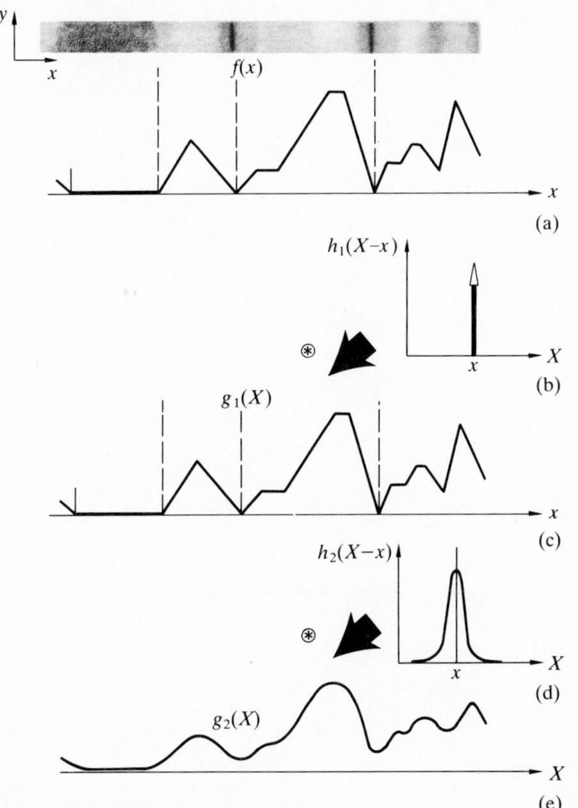

FIGURE 11.20 The irradiance distribution is converted to a function $f(x)$ shown in (*a*). This is convolved with a δ-function (*b*) to yield a duplicate of $f(x)$. By contrast, convolving $f(x)$ with the spread function h_2 in (*d*) yields a smoothed-out curve represented by $g_2(x)$ in (*e*).

Suppose $h(x)$ looks like the asymmetrical function in Fig. 11.21*a*. Then $h(-x)$ appears in Fig. 11.21*b*, and its shifted form $h(X - x)$ is shown in (*c*). The convolution of $f(x)$ [depicted in (*d*)] and $h(x)$ is $g(X)$, as given by Eq. (11.52). This is often written more concisely as $f(x) \circledast h(x)$. The integral simply says that the area under the product function $f(x)h(X - x)$ for all x is $g(X)$. Evidently, the product is nonzero only over the range d wherein $h(X - x)$ is nonzero, that is, where the two curves overlap (Fig. 11.21*e*). At a particular point X_1 in the output space, the area under the product $f(x)h(X_1 - x)$ is $g(X_1)$. This fairly direct interpretation can be related back to the physically more pleasing view of the integral in terms of overlapping point contributions, as depicted previously in Fig. 11.19. Remember that there we said that each source element was smeared out in a blur spot on the image plane having the shape of the spread function. Now suppose we take the direct approach and wish to compute the product area of Fig. 11.21*e* at X_1, that is, $g(X_1)$. A differential element dx centered on any point in the region of overlap (Fig. 11.22*a*), say x_1, will contribute an amount $f(x_1)h(X_1 - x_1)\,dx$ to the area. This same differential element will make an identical contribution when viewed in the overlapping spread-function scheme. To see this, examine (*b*) and (*c*) in Fig. 11.22, which are *now drawn in the output space*. The latter shows the spread function "centered" at $X = x_1$. A source element dx, in this case located on the object at x_1, generates a smeared-out signal proportional to $f(x_1)h(X - x_1)$, as in (*d*), where $f(x_1)$ is just a number. The piece of this signal that exists at X_1 is $f(x_1)h(X_1 - x_1)\,dx$, which indeed is identical to the contribution made by dx at x_1 in (*a*). Similarly, each differential element of the product area (at any $x = x'$) in Fig. 11.22*a* has its counterpart in a curve like that of (*d*) but "centered" on a new point ($X = x'$). Points beyond $x = x_2$ make no contribution, because they are not in the overlap region of (*a*) and, equivalently, because they are too far from X_1 for the smear to reach it, as shown in (*e*).

If the functions being convolved are simple enough, $g(X)$ can be determined roughly without any calculations at all. The convolution of two identical square pulses is illustrated, from both of the viewpoints discussed above, in Figs. 11.23 and 11.24. In Fig. 11.23 each impulse constituting $f(x)$ is spread out into a square pulse and summed. In Fig. 11.24 the overlap-

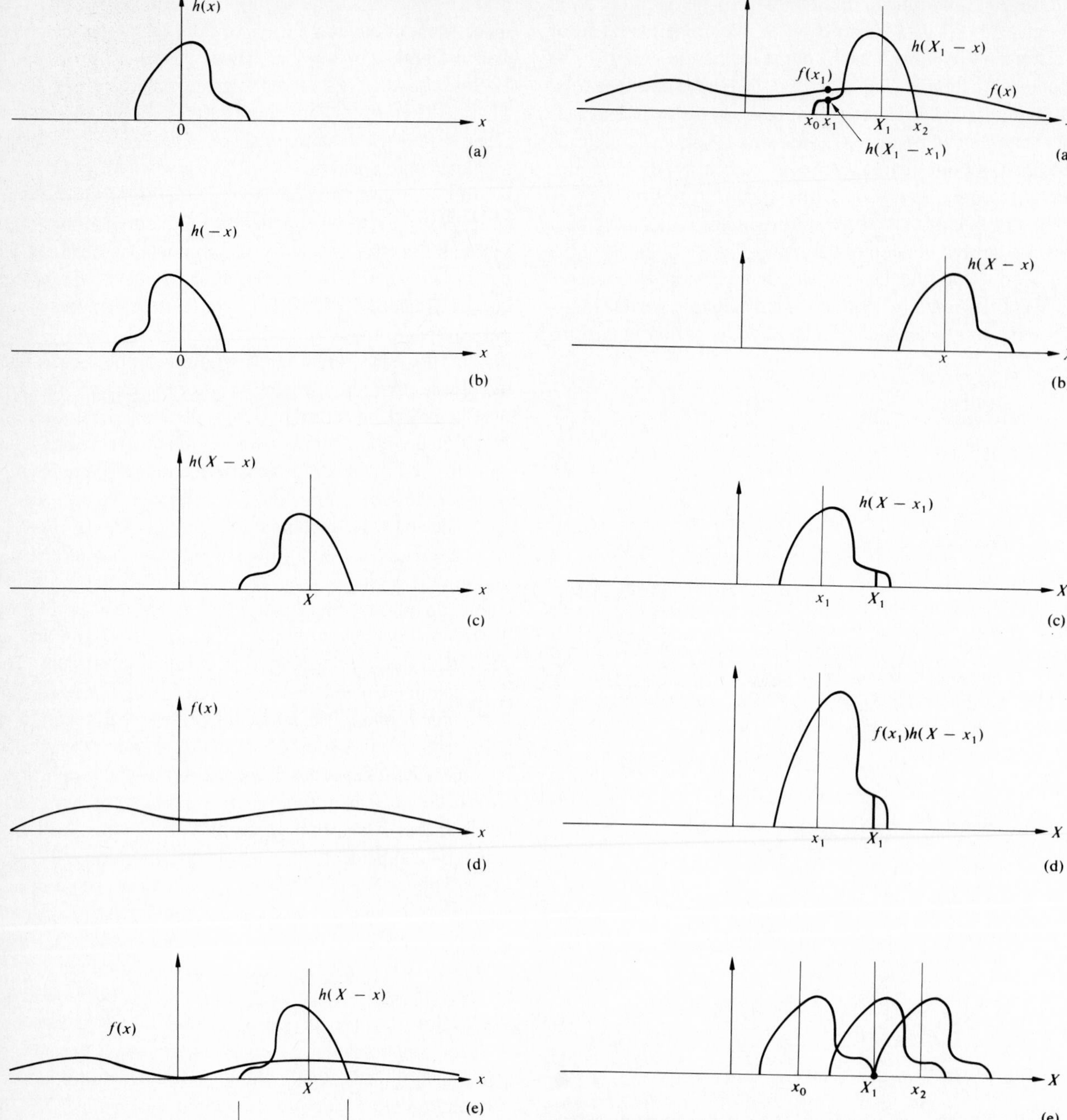

FIGURE 11.21 The geometry of the convolution process in the object coordinates.

FIGURE 11.22 The geometry of the convolution process in the image coordinates.

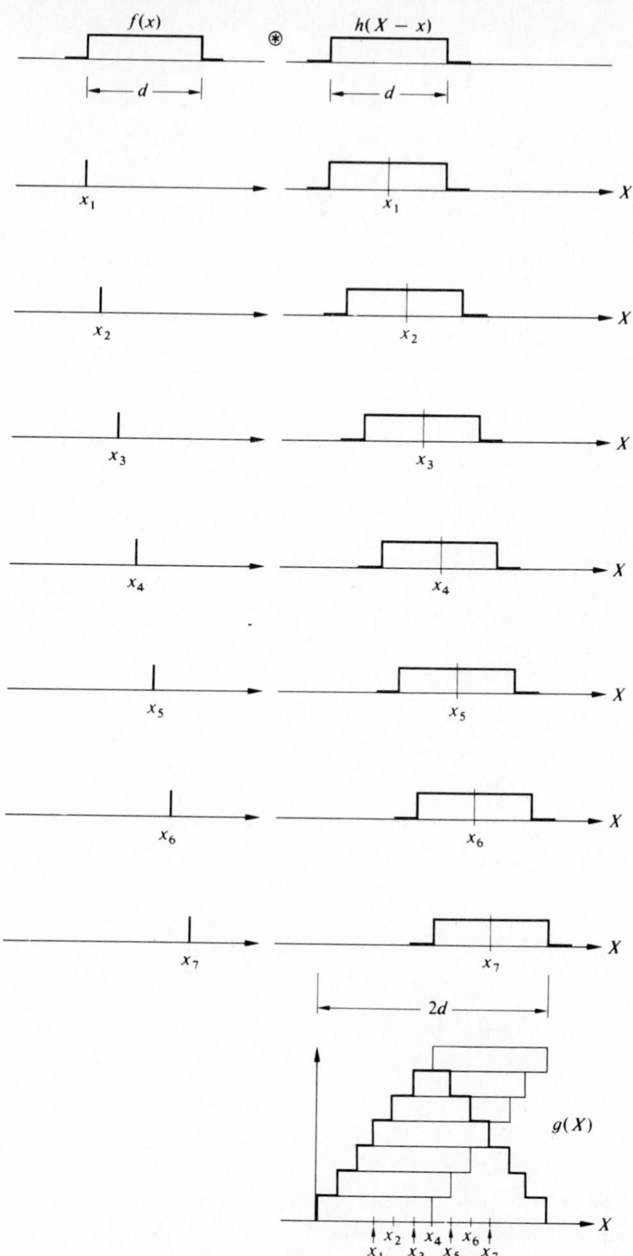

FIGURE 11.23 Convolution of two square pulses. The fact that we represented *f(x)* by a finite number of delta functions (viz., 7) accounts for the steps in *g(X)*.

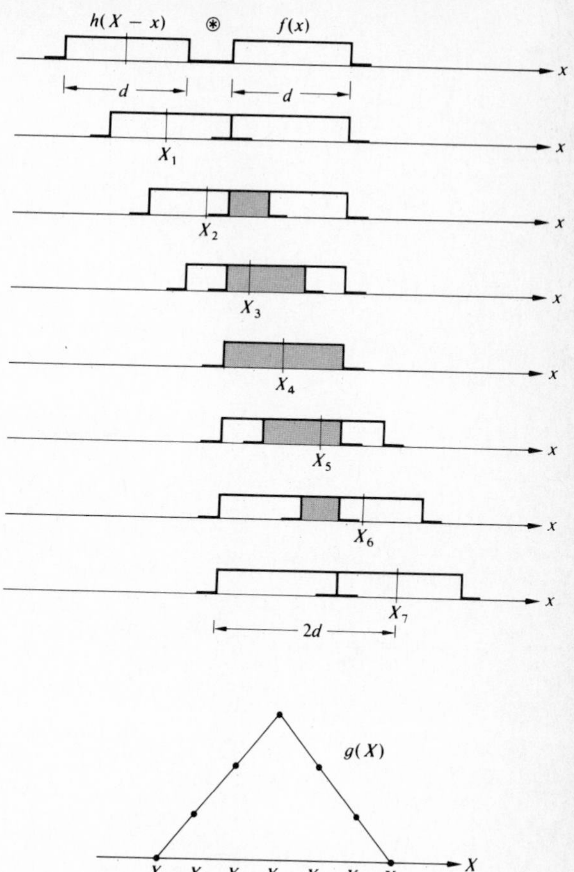

FIGURE 11.24 Convolution of two square pulses.

ping area, as h varies, is plotted against X. In both instances the result is a triangular pulse. Incidentally, observe that $(f \circledast h) = (h \circledast f)$, as can be seen by a change of variable $(x' = X - x)$ in Eq. (11.52), being careful with the limits (see Problem 11.15).

Figure 11.25 illustrates the convolution of two functions $I_0(y, z)$ and $\mathcal{S}(y, z)$ in two dimensions, as given by Eq. (11.51). Here the volume under the product curve $I_0(y, z)\mathcal{S}(Y - y, Z - z)$, that is, the region of overlap, equals $I_i(Y, Z)$ at (Y, Z); see Problem 11.16.

The Convolution Theorem

Suppose we have two functions $f(x)$ and $h(x)$ with Fourier transforms $\mathscr{F}\{f(x)\} = F(k)$ and $\mathscr{F}\{h(x)\} = H(k)$, respectively.

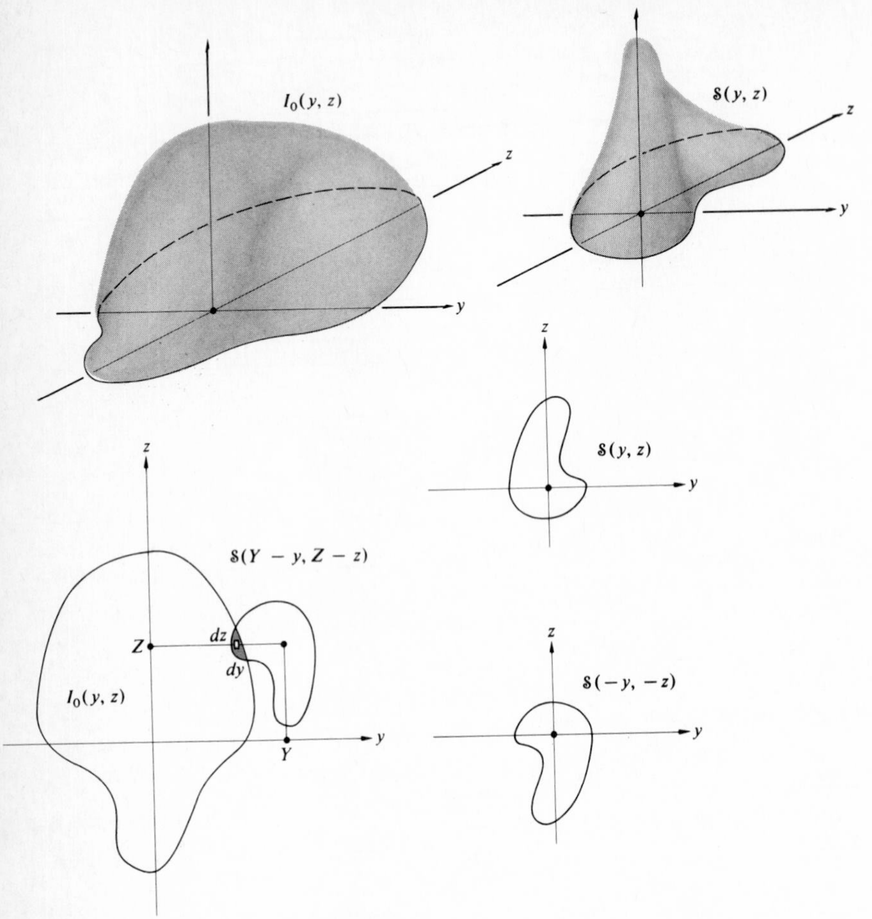

FIGURE 11.25 Convolution in two dimensions.

The **convolution theorem** states that if $g = f \circledast h$,

$$\mathscr{F}\{g\} = \mathscr{F}\{f \circledast h\} = \mathscr{F}\{f\} \cdot \mathscr{F}\{h\} \tag{11.53}$$

or

$$G(k) = F(k)H(k) \tag{11.54}$$

where $\mathscr{F}\{g\} = G(k)$. The proof is quite straightforward:

$$\mathscr{F}\{f \circledast h\} = \int_{-\infty}^{+\infty} g(X)e^{ikX}\,dX$$

$$= \int_{-\infty}^{+\infty} e^{ikX}\left[\int_{-\infty}^{+\infty} f(x)h(X - x)\,dx\right]dX$$

Thus

$$G(k) = \int_{-\infty}^{+\infty}\left[\int_{-\infty}^{+\infty} h(X - x)e^{ikX}\,dX\right]f(x)\,dx$$

If we put $w = X - x$ in the inner integral, then $dX = dw$ and

$$G(k) = \int_{-\infty}^{+\infty} f(x)e^{ikx}\,dx \int_{-\infty}^{+\infty} h(w)e^{ikw}\,dw$$

Hence

$$G(k) = F(k)H(k)$$

which verifies the theorem. As an example of its application, refer to Fig. 11.26. Since the convolution of two identical square pulses ($f \circledast h$) is a triangular pulse (g), the product of their transforms (Fig. 7.17) must be the transform of g, namely,

$$\mathscr{F}\{g\} = [d \operatorname{sinc}(kd/2)]^2 \tag{11.55}$$

As an additional example, convolve a square pulse with the two δ-functions of Fig. 11.11. The transform of the resulting

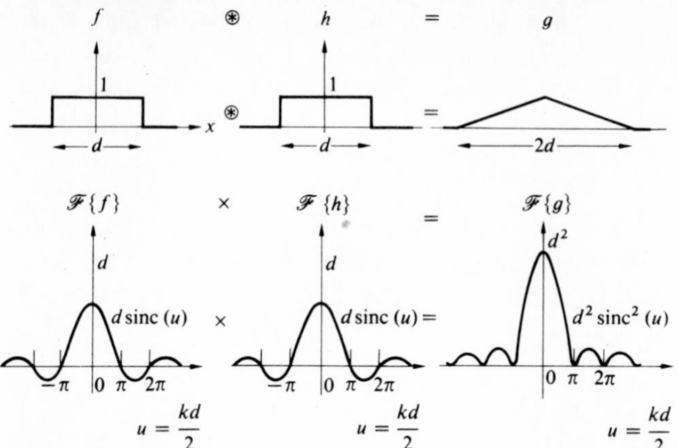

FIGURE 11.26 An illustration of the convolution theorem.

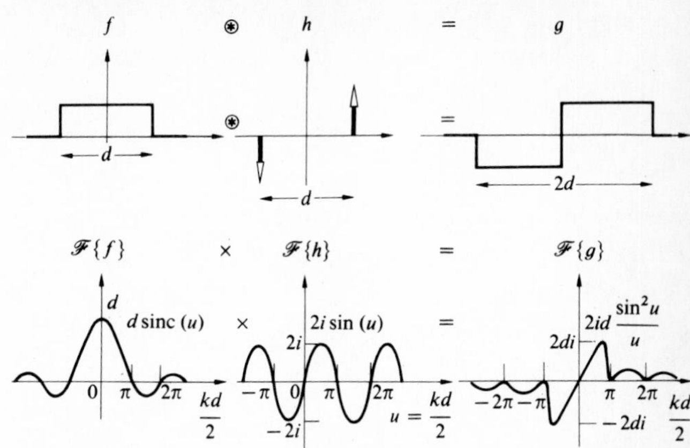

FIGURE 11.27 An illustration of the convolution theorem.

double pulse (Fig. 11.27) is again the product of the individual transforms.

The k-space counterpart of Eq. (11.53), namely, the *frequency convolution theorem*, is given by

$$\mathscr{F}\{f\cdot h\} = \frac{1}{2\pi}\mathscr{F}\{f\} \circledast \mathscr{F}\{h\} \tag{11.56}$$

That is, the transform of the product is the convolution of the transforms.

Figure 11.28 makes the point rather nicely. Here an infinitely long cosine, $f(x)$, is multiplied by a rectangular pulse, $h(x)$, which truncates it into a short oscillatory wavetrain, $g(x)$. The transform of $f(x)$ is a pair of delta functions, the transform of the rectangular pulse is a sinc function, and the convolution of the two is the transform of $g(x)$. Compare this result with that of Eq. (7.60).

Transform of the Gaussian Wave Packet

As a further example of the usefulness of the convolution theorem, let's evaluate the Fourier transform of a pulse of light in

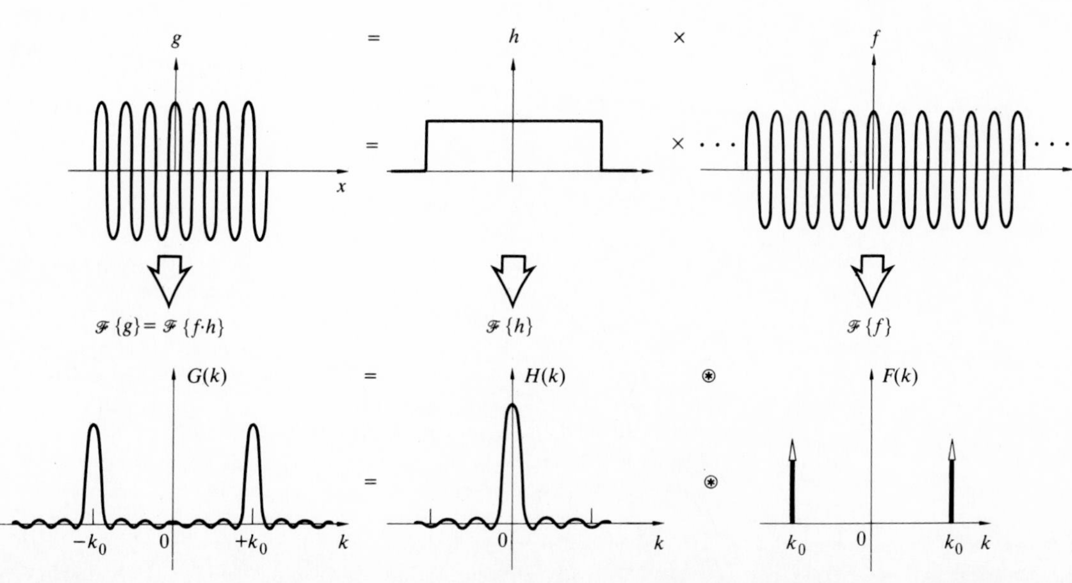

FIGURE 11.28 An example of the frequency convolution theorem.

the configuration of the wave packet of Fig. 11.29. Taking a rather general approach, notice that since a one-dimensional harmonic wave has the form

$$\tilde{E}(x, t) = E_0 e^{-i(k_0 x - \omega t)}$$

one need only modulate the amplitude to get a pulse of the desired structure. Assuming the wave's profile to be independent of time, we can write it as

$$\tilde{E}(x, 0) = f(x) e^{-ik_0 x}$$

Now, to determine $\mathcal{F}\{f(x)e^{-ik_0x}\}$ evaluate

$$\int_{-\infty}^{+\infty} f(x) e^{-ik_0 x} e^{ikx} \, dx \tag{11.57}$$

Letting $k' = k - k_0$, we get

$$F(k') = \int_{-\infty}^{+\infty} f(x) e^{ik'x} \, dx = F(k - k_0) \tag{11.58}$$

In other words, if $F(k) = \mathcal{F}\{f(x)\}$, then $F(k - k_0) = \mathcal{F}\{f(x)e^{-ik_0x}\}$. For the specific case of a Gaussian envelope [Eq. (11.11)], as in the figure, $f(x) = \sqrt{a/\pi}\, e^{-ax^2}$, that is,

$$\tilde{E}(x, 0) = \sqrt{a/\pi}\, e^{-ax^2} e^{-ik_0 x} \tag{11.59}$$

From the foregoing discussion and Eq. (11.12), it follows that

$$\mathcal{F}\{\tilde{E}(x, 0)\} = e^{-(k-k_0)^2/4a} \tag{11.60}$$

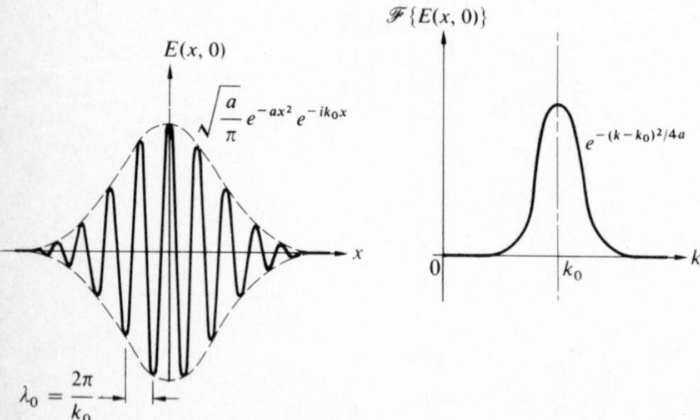

FIGURE 11.29 A Gaussian wave packet and its transform.

In quite a different way, the transform can be determined from Eq. (11.56). The expression $\tilde{E}(x, 0)$ is now viewed as the product of the two functions $f(x) = \sqrt{a/\pi}\, \exp(-ax^2)$ and $h(x) = \exp(-ik_0x)$. One way to evaluate $\mathcal{F}\{h\}$ is to set $f(x) = 1$ in Eq. (11.57). This yields the transform of 1 with k replaced by $k - k_0$. Since $\mathcal{F}\{1\} = 2\pi\delta(k)$ (see Problem 11.4), we have $\mathcal{F}\{e^{-ik_0x}\} = 2\pi\delta(k - k_0)$. Thus $\mathcal{F}\{\tilde{E}(x, 0)\}$ is $1/2\pi$ times the convolution of $2\pi\delta(k - k_0)$, with the Gaussian $e^{-k^2/4a}$ centered on zero. The result* is once again a Gaussian centered on k_0, namely, $e^{-(k-k_0)^2/4a}$.

11.3.3 Fourier Methods in Diffraction Theory

Fraunhofer Diffraction

Fourier-transform theory provides a particularly beautiful insight into the mechanism of Fraunhofer diffraction. Let's go back to Eq. (10.41), rewritten as

$$E(Y, Z) = \frac{\mathcal{E}_A e^{i(\omega t - kR)}}{R} \iint_{\text{Aperture}} e^{ik(Yy+Zz)/R} \, dy \, dz \tag{11.61}$$

This formula refers to Fig. 10.22, which depicts an arbitrary diffracting aperture in the yz-plane upon which is incident a monochromatic plane wave. The quantity R is the distance from the center of the aperture to the output point where the field is $E(Y, Z)$. The source strength per unit area of the aperture is denoted by $\mathcal{E}_A$. We are talking about electric fields that are of course time-varying; the term $\exp i(\omega t - kR)$ just relates

*We should actually have used the real part of $\exp(-ik_0x)$ to start with in this derivation, since the transform of the complex exponential is different from the transform of $\cos k_0x$ and taking the real part afterward is insufficient. This is the same sort of difficulty one always encounters when forming products of complex exponentials. The final answer [Eq. (11.60)] should, in fact, contain an additional $\exp[-(k + k_0)^2/4a]$ term, as well as a multiplicative constant of $\frac{1}{2}$. This second term is usually negligible in comparison, however. Even so, had we used $\exp(+ik_0x)$ to start with [Eq. (11.59)], only the negligible term would have resulted! Using the complex exponential to represent the sine or cosine in this fashion is *rigorously incorrect,* albeit pragmatically common practice. As a short-cut, it should be indulged in only with the greatest caution!

the phase of the net disturbance at the point (Y, Z) to that at the center of the aperture. The $1/R$ corresponds to the dropoff of field amplitude with distance from the aperture. The phase term in front of the integral is of little present concern, since we are interested in the relative amplitude distribution of the field, and it doesn't much matter what the resultant phase is at any particular output point. Thus if we limit ourselves to a small region of output space over which R is essentially constant, everything in front of the integral, with the exception of $\mathcal{E}_A$, can be lumped into a single constant.

The $\mathcal{E}_A$ has thus far been assumed to be invariant over the aperture, but that certainly need not be the case. Indeed, if the aperture were filled with a bumpy piece of dirty glass, the field emanating from each area element $dy\,dz$ could differ in both amplitude and phase. There would be nonuniform absorption, as well as a position-dependent optical path length through the glass, which would certainly affect the diffracted field distribution. The variations in $\mathcal{E}_A$, as well as the multiplicative constant, can be combined into a single complex quantity

$$\mathcal{A}(y, z) = \mathcal{A}_0(y, z)e^{i\phi(y,z)} \tag{11.62}$$

which we call the **aperture function**. The amplitude of the field over the aperture is described by $\mathcal{A}_0(y, z)$, while the point-to-point phase variation is represented by $\exp[i\phi(y, z)]$. Accordingly, $\mathcal{A}(y, z)\,dy\,dz$ is proportional to the diffracted field emanating from the differential source element $dy\,dz$. Consolidating this much, we can reformulate Eq. (11.61) more generally as

$$E(Y, Z) = \iint_{-\infty}^{+\infty} \mathcal{A}(y, z)e^{ik(Yy+Zz)/R}\,dy\,dz \tag{11.63}$$

The limits on the integral can be extended to $\pm\infty$, because the aperture function is nonzero only over the region of the aperture.

It might be helpful to envision $dE(Y, Z)$ at a given point P as if it were a plane wave propagating in the direction of $\mathbf{k}$ as in Fig. 11.30, and having an amplitude determined by $\mathcal{A}(y, z)\,dy\,dz$. To underscore the similarity between Eq. (11.63) and Eq. (11.14), let's define the *spatial frequencies* k_Y *and* k_Z *as*

$$k_Y \equiv kY/R = k\sin\phi = k\cos\beta \tag{11.64}$$

and

$$k_Z \equiv kZ/R = k\sin\theta = k\cos\gamma \tag{11.65}$$

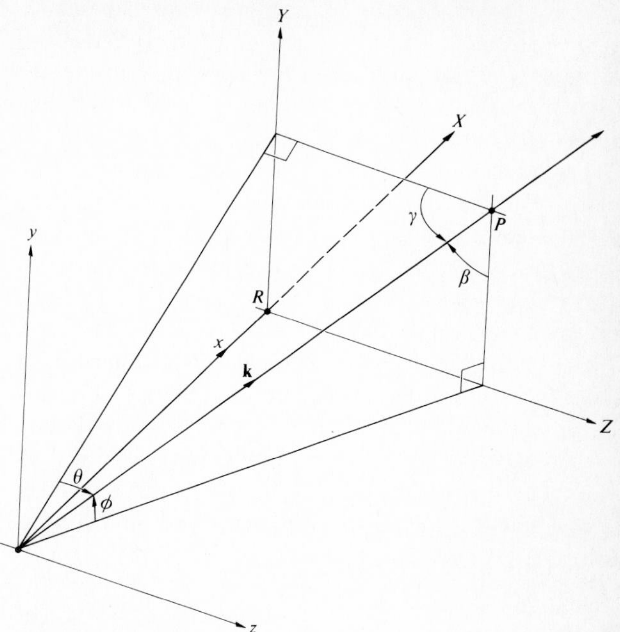

FIGURE 11.30 A bit of geometry.

For each point on the image plane, there is a corresponding spatial frequency. The diffracted field can now be written as

$$E(k_Y, k_Z) = \iint_{-\infty}^{+\infty} \mathcal{A}(y,z)e^{i(k_Y y+k_Z z)}\,dy\,dz \tag{11.66}$$

and we've arrived at the key point: ***the field distribution in the Fraunhofer diffraction pattern is the Fourier transform of the field distribution across the aperture (i.e., the aperture function)***. Symbolically, this is written as

$$E(k_Y, k_Z) = \mathcal{F}\{\mathcal{A}(y, z)\} \tag{11.67}$$

The field distribution in the image plane is the spatial-frequency spectrum of the aperture function. The inverse transform is then

$$\mathcal{A}(y, z) = \frac{1}{(2\pi)^2}\iint_{-\infty}^{+\infty} E(k_Y, k_Z)e^{-i(k_Y y+k_Z z)}\,dk_Y\,dk_Z \tag{11.68}$$

that is,

$$\mathcal{A}(y, z) = \mathcal{F}^{-1}\{E(k_Y, k_Z)\} \tag{11.69}$$

As we have seen time and again, the more localized the signal, the more spread out is its transform—the same is true in two dimensions. The smaller the diffracting aperture, the larger the angular spread of the diffracted beam or, equivalently, the larger the spatial frequency bandwidth.

There is a minor issue that should be mentioned here. If we actually try to observe a Fraunhofer pattern on a distant screen (without a lens), what we get will only be an approximation; the true Fraunhofer pattern is formed in parallel light that doesn't converge at any finite distance. That doesn't generally cause any grief because what we do observe is the irradiance, and that is indistinguishable from the ideal distribution at great distances. Still, at any distant, but finite, location the diffracted electric-field distribution will differ in phase very slightly from the Fourier transform of the aperture function. Since we cannot even measure the electric field, the problem is not likely to be a practical one and we shall henceforth simply overlook it.

The Single Slit

As an illustration of the method, consider the long slit in the y-direction of Fig. 10.11, illuminated by a plane wave. Assuming that there are no phase or amplitude variations across the aperture, $\mathscr{A}(y, z)$ has the form of a square pulse (Fig. 7.23):

$$\mathscr{A}(y, z) = \begin{cases} \mathscr{A}_0 & \text{when } |z| \leq b/2 \\ 0 & \text{when } |z| > b/2 \end{cases}$$

where $\mathscr{A}_0$ is no longer a function of y and z. If we take it as a one-dimensional problem,

$$E(k_Z) = \mathscr{F}\{\mathscr{A}(z)\} = \mathscr{A}_0 \int_{z=-b/2}^{+b/2} e^{ik_z z}\, dz$$

$$E(k_Z) = \mathscr{A}_0 b \text{ sinc } k_Z b/2$$

With $k_Z = k \sin \theta$, this is precisely the form derived in Section 10.2.1. The far-field diffraction pattern of a rectangular aperture (Section 10.2.4) is the two-dimensional counterpart of the slit. With $\mathscr{A}(y, z)$ again equal to $\mathscr{A}_0$ over the aperture (Fig. 10.23),

$$E(k_Y, k_Z) = \mathscr{F}\{\mathscr{A}(y, z)\}$$

$$= \int_{y=-b/2}^{+b/2} \int_{z=-a/2}^{+a/2} \mathscr{A}_0 e^{i(k_y y + k_z z)}\, dy\, dz$$

hence,

$$E(k_Y, k_Z) = \mathscr{A}_0 ba \text{ sinc } \frac{bkY}{2R} \text{ sinc } \frac{akZ}{2R}$$

just as in Eq. (10.42), where ba is the area of the hole.

Young's Experiment: The Double Slit

In our first treatment of Young's Experiment (Section 9.3) we took the slits to be infinitesimally wide. The aperture function was then two symmetrical δ-pulses, and the corresponding idealized field amplitude in the diffraction pattern was the Fourier transform, namely, a cosine function. Squared, this yields the familiar cosine-squared irradiance distribution of Fig. 9.9. More realistically, each aperture actually has some finite shape, and the real diffraction pattern will never be quite so simple. Figure 11.31 shows the case in which the holes are actual slits. The aperture function, $g(x)$, is obtained by convolving the δ-function spikes, $h(x)$, that locate each slit with the rectangular pulse, $f(x)$, that corresponds to the particular opening. From the convolution theorem, the product of the transforms is the modulated cosine amplitude function representing the diffracted field as it appears on the image plane. Squaring that would produce the anticipated double-slit irradiance distribution shown in Fig. 10.18. The one-dimensional transform curves are plotted against k, but that's equivalent to plotting against image-space variables by means of Eq. (11.64). (The same reasoning applied to circular apertures yields the fringe pattern of Fig. 12.2.)

Three Slits

Looking at Fig. 11.13*d* it should be clear that the transform of the array of three δ-functions in the diagram will generate a cosine that is raised by an amount proportional to the zero-frequency term, that is, the δ-function at the origin. When that delta function has twice the amplitude of the other two, the cosine is totally positive. Now suppose we have three ideally narrow parallel slits uniformly illuminated. The aperture function corresponds to Fig. 11.32*a*, where the central δ-function is half its previous size. Accordingly, the cosine transform will drop one quarter of the way down, as indicated in Fig. 11.32*b*. This corresponds to the diffracted electric-field amplitude, and its square, Fig. 11.32*c*, is the three-slit irradiance pattern.

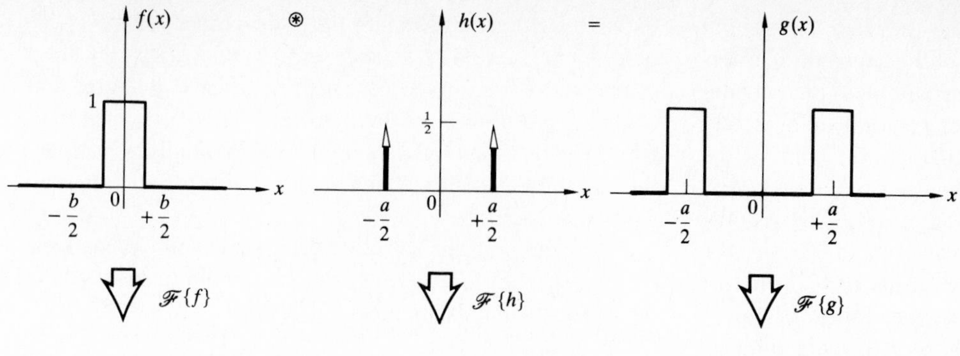

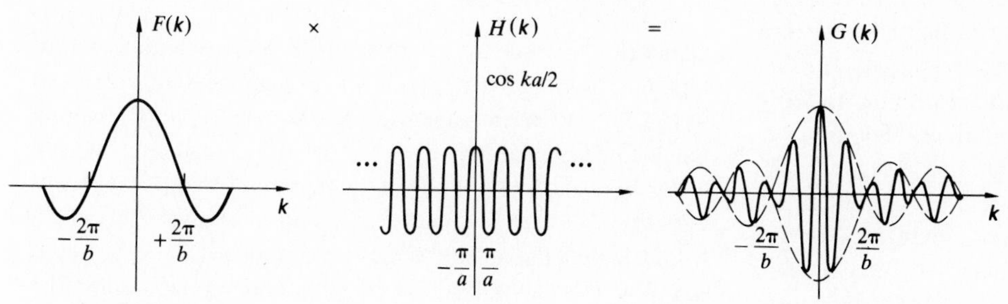

FIGURE 11.31 An illustration of the convolution theorem.

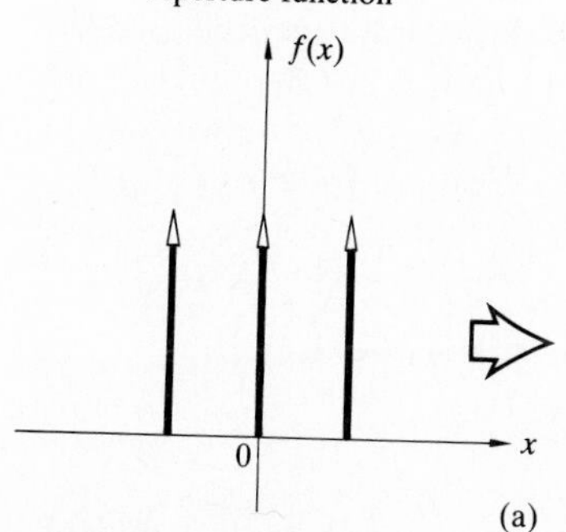

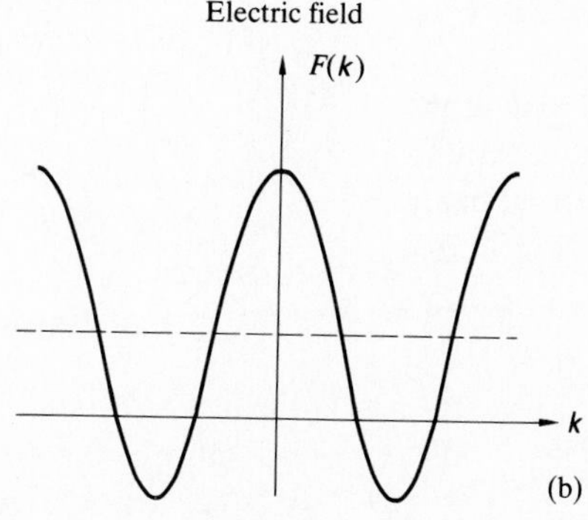

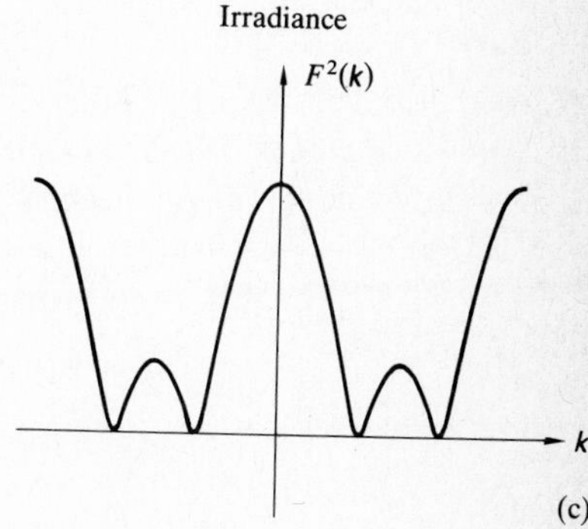

FIGURE 11.32 The Fourier transform of three equal δ-functions representing three slits.

APODIZATION

The term **apodization** derives from the Greek α, to take away, and $\pi o\delta o\sigma$, meaning foot. It refers to the process of suppressing the secondary maxima (side lobes) or feet of a diffraction pattern. In the case of a circular pupil (Section 10.2.5), the diffraction pattern is a central spot surrounded by concentric rings. The first ring has a flux density of 1.75% that of the central peak—it's small but it can be troublesome. About 16% of the light incident on the image plane is distributed in the ring system. The presence of these side lobes can diminish the resolving power of an optical system to a point where apodization is called for, as is often the case in astronomy and spectroscopy. For example, the star Sirius, which appears as the brightest star in the sky (it's in the constellation *Canis Major*—the big dog), is actually one of a binary system. It's accompanied by a faint white dwarf as they both orbit about their mutual center of mass. Because of the tremendous difference in brightness (10^4 to 1), the image of the faint companion, as viewed with a telescope, is generally completely obscured by the side lobes of the diffraction pattern of the main star.

Apodization can be accomplished in several ways, for example, by altering the shape of the aperture or its transmission characteristics.* We already know from Eq. (11.66) that the diffracted field distribution is the transform of $\mathscr{A}(y, z)$. Thus we could effect a change in the side lobes by altering $\mathscr{A}_0(y, z)$ or $\phi(y, z)$. Perhaps the simplest approach is the one in which only $\mathscr{A}_0(y, z)$ is manipulated. This can be accomplished physically by covering the aperture with a suitably coated flat glass plate (or coating the objective lens itself). Suppose that the coating becomes increasingly opaque as it goes radially out from the center (in the yz-plane) toward the edges of a circular pupil. The transmitted field will correspondingly decrease off-axis until it is made to become negligible at the periphery of the aperture. In particular, imagine that this dropoff in amplitude follows a Gaussian curve. Then $\mathscr{A}_0(y, z)$ is a Gaussian function, as is its transforms $E(Y, Z)$, and consequently the ring system vanishes. Even though the central peak is broadened, the side lobes are indeed suppressed (Fig. 11.33).

Another rather heuristic but appealing way to look at the process is to realize that the higher spatial frequency contributions go into sharpening up the details of the function being synthesized. As we saw earlier in one dimension (Fig. 7.19), the high frequencies serve to fill in the corners of the square pulse. In the same way, since $\mathscr{A}(y, z) = \mathscr{F}^{-1}\{E(k_Y, k_Z)\}$, sharp edges on the aperture necessitate the presence of appreciable contributions of high spatial frequency in the diffracted field. It follows that making $\mathscr{A}_0(y, z)$ fall off gradually will reduce these high frequencies, which in turn is manifest in a suppression of the side lobes.

Apodization is one aspect of the more encompassing technique of *spatial filtering*, which is discussed in an extensive yet nonmathematical treatment in Chapter 13.

The Array Theorem

Generalizing some of our previous ideas to two dimensions, imagine that we have a screen containing N identical holes, as in Fig. 11.34. In each aperture, at the same relative position, we locate a point $O_1, O_2, \ldots, O_N$ at $(y_1, z_1), (y_2, z_2), \ldots, (y_N, z_N)$, respectively. Each of these, in turn, fixes the origin of a local coordinate system (y', z'). Thus a point (y', z') in the local frame of the jth aperture has coordinates $(y_j + y', z_j + z')$ in the (y, z)-system. Under coherent monochromatic illumination, the resulting Fraunhofer diffraction field $E(Y, Z)$ at some point P on the image plane will be a superposition of the individual fields at P arising from each separate aperture; in other words,

$$E(Y, Z) = \sum_{j=1}^{N} \iint_{-\infty}^{+\infty} \mathscr{A}_I(y', z')e^{ik[Y(y_j+y')+Z(z_j+z')]/R}\,dy'\,dz' \tag{11.70}$$

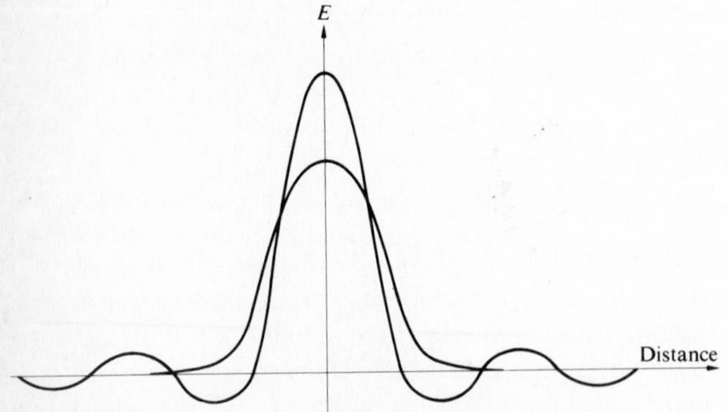

FIGURE 11.33 An Airy pattern compared with a Gaussian.

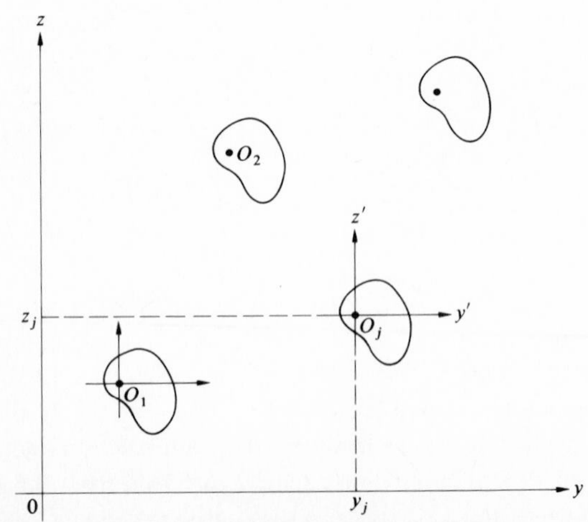

FIGURE 11.34 Multiple-aperture configuration.

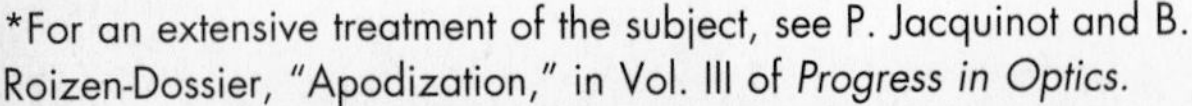
*For an extensive treatment of the subject, see P. Jacquinot and B. Roizen-Dossier, "Apodization," in Vol. III of *Progress in Optics*.

or $$E(Y,Z) = \iint_{-\infty}^{+\infty} \mathcal{A}_I(y',z')e^{ik(Yy'+Zz')/R}dy'\,dz' \times \sum_{j=1}^{N} e^{ik(Yy_j+Zz_j)/R} \tag{11.71}$$

where $\mathcal{A}_I(y', z')$ is the individual aperture function applicable to each hole. This can be recast, using Eqs. (11.64) and (11.65), as

$$E(k_Y, k_Z) = \iint_{-\infty}^{+\infty} \mathcal{A}_I(y',z')e^{i(k_Yy'+k_Zz')}\,dy'\,dz' \times \sum_{j=1}^{N} e^{i(k_Yy_j)}e^{i(k_Zz_j)} \tag{11.72}$$

Notice that the integral is the Fourier transform of the individual aperture function, while the sum is the transform [Eq. (11.42)] of an array of delta functions

$$A_\delta = \sum_j \delta(y - y_j)\delta(z - z_j) \tag{11.73}$$

Inasmuch as $E(k_Y, k_Z)$ itself is the transform $\mathcal{F}\{\mathcal{A}(y, z)\}$ of the total aperture function for the entire array, we have

$$\mathcal{F}\{\mathcal{A}(y,z)\} = \mathcal{F}\{\mathcal{A}_I(y',z')\}\cdot\mathcal{F}\{A_\delta\} \tag{11.74}$$

This equation is a statement of the **array theorem**, which says that ***the field distribution in the Fraunhofer diffraction pattern of an array of similarly oriented identical apertures equals the Fourier transform of an individual aperture function (i.e., its diffracted field distribution) multiplied by the pattern that would result from a set of point sources arrayed in the same configuration (which is the transform of A_δ).***

This can be seen from a slightly different point of view. The total aperture function may be formed by convolving the individual aperture function with an appropriate array of delta functions, each sitting at one of the coordinate origins (y_1, z_1), (y_2, z_2), and so on. Hence

$$\mathcal{A}(y,z) = \mathcal{A}_I(y',z') \circledast A_\delta \tag{11.75}$$

whereupon the array theorem follows directly from the convolution theorem [Eq. (11.53)].

As a simple example, imagine that we again have Young's Experiment with two slits along the y-direction, of width b and separation a. The individual aperture function for each slit is a step function,

$$\mathcal{A}_I(z') = \begin{cases} \mathcal{A}_{I0} & \text{when } |z'| \leq b/2 \\ 0 & \text{when } |z'| > b/2 \end{cases}$$

and so

$$\mathcal{F}\{\mathcal{A}_I(z')\} = \mathcal{A}_{I0}b \operatorname{sinc} k_Zb/2$$

With the slits located at $z = \pm a/2$,

$$A_\delta = \delta(z - a/2) + \delta(z + a/2)$$

and from Eq. (11.43)

$$\mathcal{F}\{A_\delta\} = 2\cos k_Za/2$$

Thus

$$E(k_Z) = 2\mathcal{A}_{I0}b \operatorname{sinc}\left(\frac{k_Zb}{2}\right)\cos\left(\frac{k_Za}{2}\right)$$

which is the same conclusion arrived at earlier (Fig. 11.31). The irradiance pattern is a set of *cosine-squared* interference fringes modulated by a *sinc-squared* diffraction envelope.

11.3.4 Spectra and Correlation

Parseval's Formula

Suppose that $f(x)$ is a pulse of finite extent, and $F(k)$ is its Fourier transform [Eq. (11.5)]. Thinking back to Section 7.8, we recognize the function $F(k)$ as the amplitude of the spatial frequency spectrum of $f(x)$. And $F(k)\,dk$ then connotes the amplitude of the contributions to the pulse within the frequency range from k to $k + dk$. Hence it seems that $|F(k)|$ serves as a spectral amplitude density, and its square, $|F(k)|^2$, should be proportional to the energy per unit spatial frequency interval. Similarly, in the time domain, if $f(t)$ is a radiated electric field, $|f(t)|^2$ is proportional to the radiant flux or power, and the total emitted energy is proportional to $\int_0^\infty |f(t)|^2\,dt$. With $F(\omega) = \mathcal{F}\{f(t)\}$ it appears that $|F(\omega)|^2$ must be a measure of the radiated energy per unit temporal frequency interval. To be a bit more precise, let's evaluate $\int_{-\infty}^{+\infty}|f(t)|^2\,dt$ in terms of the appropriate Fourier transforms. Inasmuch as $|f(t)|^2 = f(t)f^*(t) = f(t)\cdot[\mathcal{F}^{-1}\{\mathcal{F}(\omega)\}]^*$,

$$\int_{-\infty}^{+\infty} |f(t)|^2\,dt = \int_{-\infty}^{+\infty} f(t)\left[\frac{1}{2\pi}\int_{-\infty}^{+\infty} F^*(\omega)e^{+i\omega t}\,d\omega\right]dt$$

Interchanging the order of integration, we obtain

$$\int_{-\infty}^{+\infty} |f(t)|^2\, dt = \frac{1}{2\pi}\int_{-\infty}^{+\infty} F^*(\omega)\left[\int_{-\infty}^{+\infty} f(t)e^{i\omega t}\, dt\right] d\omega$$

and so

$$\int_{-\infty}^{+\infty} |f(t)|^2\, dt = \frac{1}{2\pi}\int_{-\infty}^{+\infty} |F(\omega)|^2\, d\omega \qquad (11.76)$$

where $|F(\omega)|^2 = F^*(\omega)F(\omega)$. This is *Parseval's formula*. As expected, the total energy is proportional to the area under the $|F(\omega)|^2$ curve, and consequently $|F(\omega)|^2$ is sometimes called the **power spectrum** or *spectral energy distribution*. The corresponding formula for the space domain is

$$\int_{-\infty}^{+\infty} |f(x)|^2\, dx = \frac{1}{2\pi}\int_{-\infty}^{+\infty} |F(k)|^2\, dk \qquad (11.77)$$

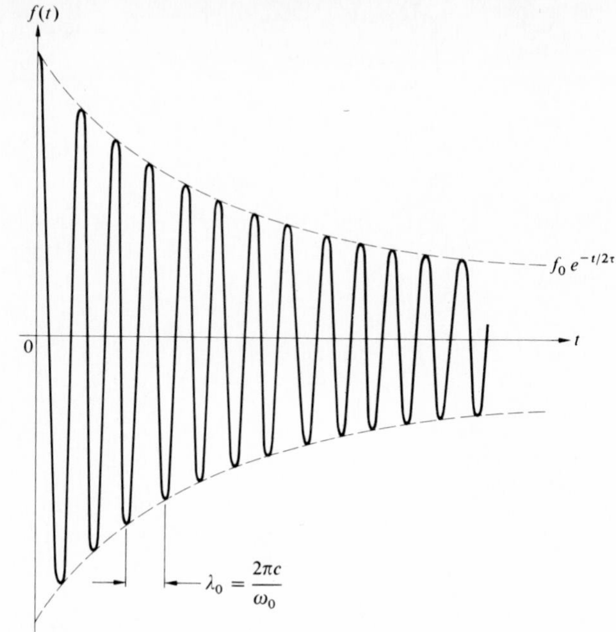

FIGURE 11.35 A damped harmonic wave.

The Lorentzian Profile

As an indication of the manner in which these ideas are applied in practice, consider the damped harmonic wave $f(t)$ at $x = 0$ depicted in Fig. 11.35. Here

$$f(t) = \begin{cases} 0 & \text{from } t = -\infty \text{ to } t = 0 \\ f_0 e^{-t/2\tau}\cos\omega_0 t & \text{from } t = 0 \text{ to } t = +\infty \end{cases}$$

The negative exponential dependence arises, quite generally, whenever the rate-of-change of a quantity depends on its instantaneous value. In this case, we might suppose that the power radiated by an atom varies as $(e^{-t/\tau})^{1/2}$. In any event, τ is known as the time constant of the oscillation, and $\tau^{-1} = \gamma$ is the damping constant. The transform of $f(t)$ is

$$F(\omega) = \int_0^{\infty} (f_0 e^{-t/2\tau}\cos\omega_0 t)e^{i\omega t}\, dt \qquad (11.78)$$

The evaluation of this integral is explored in the problems. One finds on performing the calculation that

$$F(\omega) = \frac{f_0}{2}\left[\frac{1}{2\tau} - i(\omega + \omega_0)\right]^{-1} + \frac{f_0}{2}\left[\frac{1}{2\tau} - i(\omega - \omega_0)\right]^{-1}$$

When $f(t)$ is the radiated field of an atom, τ denotes the *lifetime* of the excited state (from around 1.0 ns to 10 ns). Now if we form the power spectrum $F(\omega)F^*(\omega)$, it will be composed of two peaks centered on $\pm\,\omega_0$ and thus separated by $2\omega_0$. At optical frequencies where $\omega_0 >> \gamma$, these will be both narrow and widely spaced, with essentially no overlap. The shape of these peaks is determined by the transform of the modulation envelope in Fig. 11.35, that is, a negative exponential. The location of the peaks is fixed by the frequency of the modulated cosine wave, and the fact that there are two such peaks is a reflection of the spectrum of the cosine in this symmetrical frequency representation (Section 7.8). To determine the observable spectrum from $F(\omega)F^*(\omega)$, we need only consider the positive frequency term, namely,

$$|F(\omega)|^2 = \frac{f_0^2}{\gamma^2}\,\frac{\gamma^2/4}{(\omega - \omega_0)^2 + \gamma^2/4} \qquad (11.79)$$

This has a maximum value of f_0^2/γ^2 at $\omega = \omega_0$, as shown in Fig. 11.36. At the half-power points $(\omega - \omega_0) = \pm\gamma/2$, $|F(\omega)|^2 = f_0^2/2\gamma^2$, which is half its maximum value. The width of the spectral line between these points is equal to γ.

The curve given by Eq. (11.79) is known as the *resonance* or *Lorentz profile*. The frequency bandwidth arising from the finite duration of the excited state is called the *natural linewidth*.

If the radiating atom suffers a collision, it can lose energy and thereby further shorten the duration of emission. The frequency bandwidth increases in the process, which is known as *Lorentz broadening*. Here again, the spectrum is found to have a Lorentz profile. Furthermore, because of the random

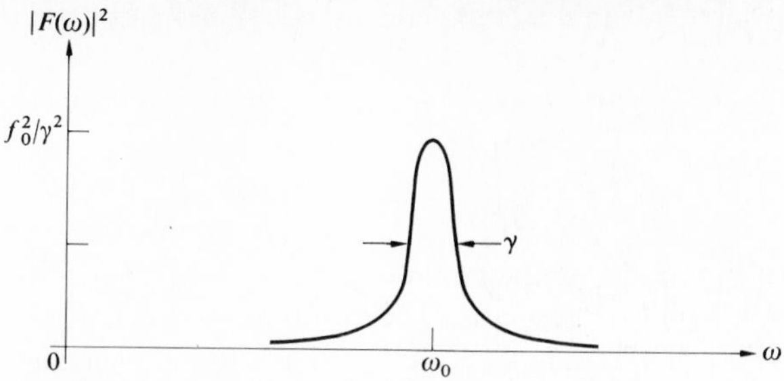

FIGURE 11.36 The resonance or Lorentz profile.

thermal motion of the atoms in a gas, the frequency bandwidth will be increased via the Doppler effect. *Doppler broadening*, as it is called, results in a Gaussian spectrum (Section 7.10). The Gaussian drops more slowly in the immediate vicinity of ω_0 and then more quickly away from it than does the Lorentzian profile. These effects can be combined mathematically to yield a single spectrum by convolving the Gaussian and Lorentzian functions. In a low-pressure gaseous discharge, the Gaussian profile is by far the wider and generally predominates.

AUTOCORRELATION AND CROSS-CORRELATION

Let's now go back to the derivation of Parseval's formula and follow it through again, this time with a slight modification. We wish to evaluate $\int_{-\infty}^{+\infty} f(t+\tau)f^*(t)\,dt$, using much the same approach as before. Thus, if $F(\omega) = \mathscr{F}\{f(t)\}$,

$$\int_{-\infty}^{+\infty} f(t+\tau)f^*(t)\,dt = \int_{-\infty}^{+\infty} f(t+\tau) \times \left[\frac{1}{2\pi}\int_{-\infty}^{+\infty} F^*(\omega)e^{+i\omega t}\,d\omega\right]dt \tag{11.80}$$

Changing the order of integration, we obtain

$$\frac{1}{2\pi}\int_{-\infty}^{+\infty} F^*(\omega)\left[\int_{-\infty}^{+\infty} f(t+\tau)e^{i\omega t}\,dt\right]d\omega$$

$$= \frac{1}{2\pi}\int_{-\infty}^{+\infty} F^*(\omega)\mathscr{F}\{f(t+\tau)\}d\omega$$

To evaluate the transform within the last integral, notice that

$$f(t+\tau) = \frac{1}{2\pi}\int_{-\infty}^{+\infty} F(\omega)e^{-i\omega(t+\tau)}\,d\omega$$

by a change of variable in Eq. (11.9). Hence,

$$f(t+\tau) = \mathscr{F}^{-1}\{F(\omega)e^{-i\omega\tau}\}$$

so as discussed earlier, $\mathscr{F}\{f(t+\tau)\} = F(\omega)e^{-i\omega\tau}$, Eq. (11.80) becomes

$$\int_{-\infty}^{+\infty} f(t+\tau)f^*(t)\,dt = \frac{1}{2\pi}\int_{-\infty}^{+\infty} F^*(\omega)F(\omega)e^{-i\omega\tau}\,d\omega \tag{11.81}$$

and both sides are functions of the parameter τ. The left-hand side of this formula is said to be the **autocorrelation** of $f(t)$, denoted by

$$c_{ff}(\tau) \equiv \int_{-\infty}^{+\infty} f(t+\tau)f^*(t)\,dt \tag{11.82}$$

which is often written symbolically as $f(t) \odot f^*(t)$. If we take the transform of both sides, Eq. (11.81) then becomes

$$\mathscr{F}\{c_{ff}(\tau)\} = |F(\omega)|^2 \tag{11.83}$$

This is a form of the *Wiener–Khintchine theorem*. It allows for determination of the spectrum by way of the autocorrelation of the generating function. The definition of $c_{ff}(\tau)$ applies when the function has finite energy. When it doesn't, things will have to be changed slightly. The integral can also be restated as

$$c_{ff}(\tau) = \int_{-\infty}^{+\infty} f(t)f^*(t-\tau)\,dt \tag{11.84}$$

by a simple change of variable ($t+\tau$ to t). Similarly, the **cross-correlation** of the functions $f(t)$ and $h(t)$ is defined as

$$c_{fh}(\tau) = \int_{-\infty}^{+\infty} f^*(t)h(t+\tau)\,dt \tag{11.85}$$

Correlation analysis is essentially a means for comparing two signals in order to determine the degree of similarity between them. In autocorrelation the original function is displaced in time by an amount τ, the product of the displaced and undisplaced versions is formed, and the area under that product (corresponding to the degree of overlap) is computed by means of the integral. The autocorrelation function, $c_{ff}(\tau)$, provides the result that will be obtained in such a process for all values of τ. One reason for doing such a thing, for example, is to extract a signal from a background of random noise.

To see how the business works step by step, let's take the autocorrelation of a simple function, such as $A \sin(\omega t + \varepsilon)$, shown in Fig. 11.37. In each part of the diagram the function is shifted by a value of τ, the product $f(t) \cdot f(t + \tau)$ is formed, and then the area under that product function is computed and plotted in part (e). Notice that the process is indifferent to the value of ε. The final result is $c_{ff}(\tau) = \frac{1}{2}A^2 \cos \omega\tau$, where this function unfolds through one cycle as τ goes through 2π, so it has the same frequency as $f(t)$. Accordingly, if we had a process for generating the autocorrelation, we could reconstruct from that both the original amplitude A and the angular frequency ω.

Assuming the functions to be real, we can rewrite $c_{fh}(\tau)$ as

$$c_{fh}(\tau) = \int_{-\infty}^{+\infty} f(t)h(t + \tau)\, dt \tag{11.86}$$

which is obviously similar to the expression for the convolution of $f(t)$ and $h(t)$. Equation (11.86) is written symbolically as $c_{fh}(\tau) = f(t) \odot h(t)$. Indeed, if either $f(t)$ or $h(t)$ is even, then $f(t) \circledast h(t) = f(t) \odot h(t)$, as we shall see by example presently. Recall that the convolution flips one of the functions over and then sums up the overlap area (Fig. 11.21), that is, the area

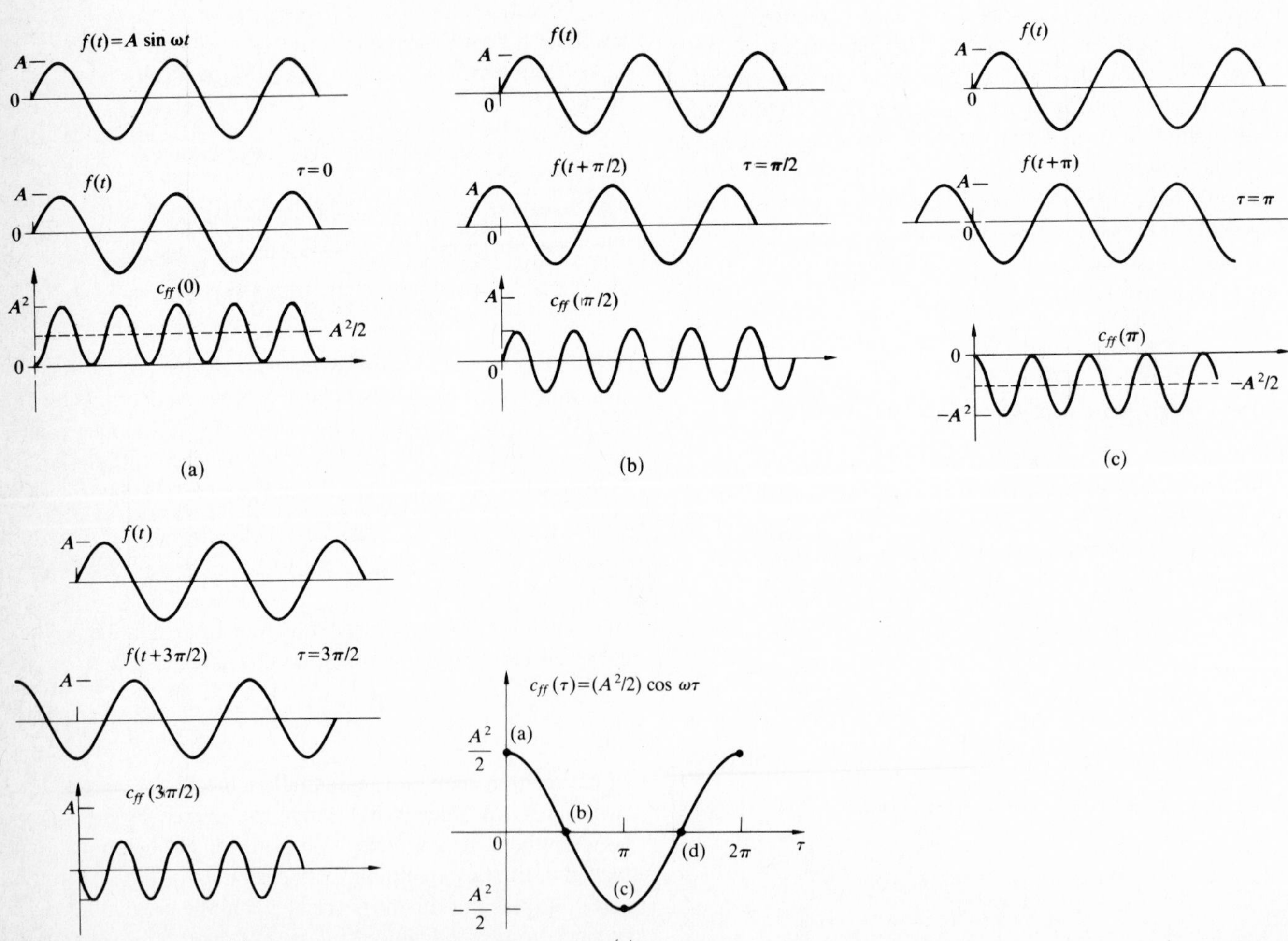

FIGURE 11.37 The autocorrelation of a sine function.

under the product curve. In contrast, the correlation sums up the overlap without flipping the function, and thus if the function is even, $f(t) = f(-t)$, it isn't changed by being flipped (or folded about the symmetry axis), and the two integrands are identical. For this to obtain, either function must be even, since $f(t) \circledast h(t) = h(t) \circledast f(t)$. The autocorrelation of a square pulse is therefore equal to the convolution of the pulse with itself, which yields a triangular signal, as in Fig. 11.24. This same conclusion follows from Eq. (11.83) and Fig. 11.26. The transform of a square pulse is a sinc function, so that the power spectrum varies as $\text{sinc}^2\, u$. The inverse transform of $|F(\omega)|^2$, that is, $\mathcal{F}^{-1}\{\text{sinc}^2\, u\}$, is $c_{ff}(\tau)$, which as we have seen, is again a triangular pulse (Fig. 11.38).

It is clearly possible for a function to have infinite energy [Eq. (11.76)] over an integration ranging from $-\infty$ to $+\infty$ and yet still have a finite *average power*

$$\lim_{T\to\infty} \frac{1}{2T} \int_{-T}^{+T} |f(t)|^2\, dt$$

Accordingly, we will define a correlation that is divided by the integration interval:

$$C_{fh}(\tau) \equiv \lim_{T\to\infty} \frac{1}{2T} \int_{-T}^{+T} f(t)h(t+\tau)\, dt \qquad (11.87)$$

For example, if $f(t) = A$ (i.e., a constant), its autocorrelation

$$C_{ff}(\tau) = \lim_{T\to\infty} \frac{1}{2T} \int_{-T}^{+T} (A)(A)\, dt = A^2$$

and the power spectrum, which is the transform of the autocorrelation, becomes

$$\mathcal{F}\{C_{ff}(\tau)\} = A^2 2\pi\delta(\omega)$$

a single impulse at the origin ($\omega = 0$), which is sometimes referred to as a *dc*-term. Notice that $C_{fh}(\tau)$ can be thought of as the time average of a product of two functions, one of which is shifted by an interval τ. In the next chapter, expressions of the form $\langle f^*(t)h(t+\tau)\rangle$ arise as coherence functions relating electric fields. They are also quite useful in the analysis of noise problems, for example, film grain noise.

We can obviously reconstruct a function from its transform, but once the transform is squared, as in Eq. (11.83), we lose information about the signs of the frequency contributions, that is, their relative phases. In the same way, the autocorrelation of a function contains no phase information and is not unique. To see this more clearly, imagine we have a number of harmonic functions of different amplitude and frequency. If their relative phases are altered, the resultant function

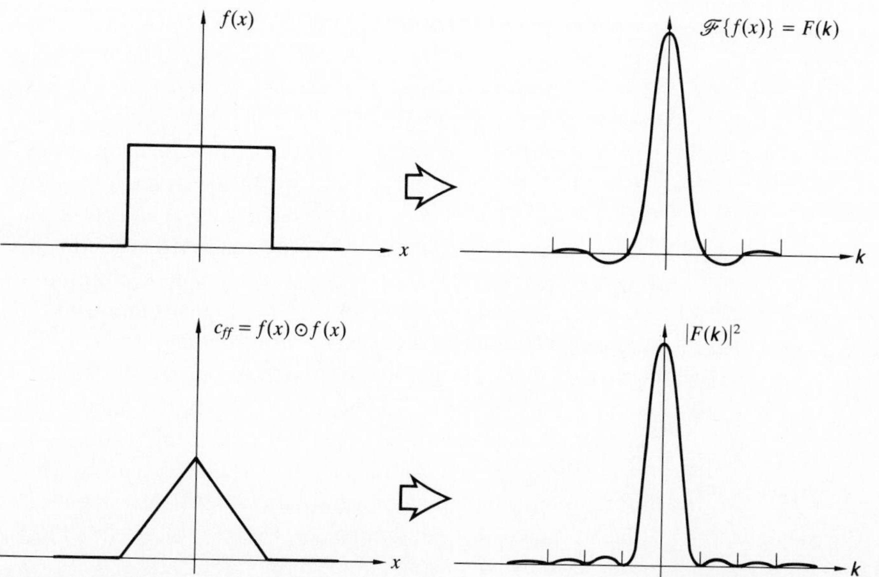

FIGURE 11.38 The square of the Fourier transform of the rectangular pulse $f(x)$ (i.e., $|F(k)|^2$) equals the Fourier transform of the autocorrelation of $f(x)$.

changes, as does its transform, but in all cases the amount of energy available at any frequency must be constant. Thus, whatever the form of the resultant profile, its autocorrelation is unaltered. It is left as a problem to show analytically that when $f(t) = A\sin(\omega t + \varepsilon)$, $C_{ff}(\tau) = (A^2/2)\cos\omega\tau$, which confirms the loss of phase information.

Figure 11.39 shows a means of optically correlating two two-dimensional spatial functions. Each of these signals is represented as a point-by-point variation in the irradiance transmission property of a photographic transparency (T_1 and T_2). For relatively simple signals opaque screens with appropriate apertures could serve instead of transparencies (e.g., for square pulses).* The irradiance at any point P on the image is due to a focused bundle of parallel rays that has traversed both transparencies. The coordinates of P, $(\theta f, \varphi f)$, are fixed by the orientation of the ray bundle, that is, the angles θ and φ. If the transparencies are identical, a ray passing through any point (x, y) on the first film with a transmittance $g(x, y)$ will pass through a corresponding point $(x + X, y + Y)$ on the second film where the transmittance is $g(x + X, y + Y)$. The shifts in coordinate are given by $X = \ell\theta$ and $Y = \ell\varphi$, where ℓ is the separation between the transparencies. The irradiance at P is therefore proportional to the autocorrelation of $g(x, y)$, that is,

$$C_{ff}(X, Y) = \iint_{-\infty}^{+\infty} g(x, y)g(x + X, y + Z)\,dx\,dy \tag{11.88}$$

and the entire flux-density pattern is called a *correlogram*. If the transparencies are different, the image is of course representative of the cross-correlation of the functions. Similarly, if one of the transparencies is rotated by 180° with respect to the other, the convolution can be obtained (see Fig. 11.25).

Before moving on, let's make sure that we actually do have a good physical feeling for the operation performed by the correlation functions. Accordingly, suppose we have a random noise-like signal (e.g., a fluctuating irradiance at a point in space or a time-varying voltage or electric field), as in Fig. 11.40*a*. The autocorrelation of $f(t)$ in effect compares the function with its value at some other time, $f(t + \tau)$. For example, with $\tau = 0$ the integral runs along the signal in time, summing up and averaging the product of $f(t)$ and $f(t + \tau)$; in this case it's simply $f^2(t)$. Since at each value of t, $f^2(t)$ is positive, $C_{ff}(0)$ will be a comparatively large number. On the other hand, when the noise is compared with itself shifted by an amount $+\tau_1$, $C_{ff}(\tau_1)$ will be somewhat reduced. There will be points in time where $f(t)f(t + \tau_1)$ is positive and other points where it will be negative, so that the value of the integral drops off (Fig. 11.40*b*). In other words, by shifting the signal with respect to itself, we have reduced the point-by-point similarity that previously ($\tau = 0$) occurred at any instant. As this shift τ increases, what little correlation existed quickly vanishes, as depicted in Fig. 11.40*c*. We can assume from the fact that the autocorrelation and the power spectrum form a Fourier transform pair [Eq. (11.83)] that the broader the frequency bandwidth of the noise, the narrower the autocorrelation. Thus for wide-bandwidth noise even a slight shift markedly reduces any similarity between $f(t)$ and $f(t + \tau)$. Furthermore, if the signal comprises a random distribution of rectangular pulses, we can see intuitively that the similarity we spoke of earlier persists for a time commensurate with the width of the pulses. The wider (in time) the pulses are, the more slowly the correlation decreases as τ increases. But this is equivalent to saying that reducing the signal bandwidth broadens $C_{ff}(\tau)$. All of this is in keeping with our previous observation that the autocorrelation tosses out any phase information, which in this case would correspond to the locations in time of the random pulses. Clearly, $C_{ff}(\tau)$ shouldn't be affected by the position of the pulses along t.

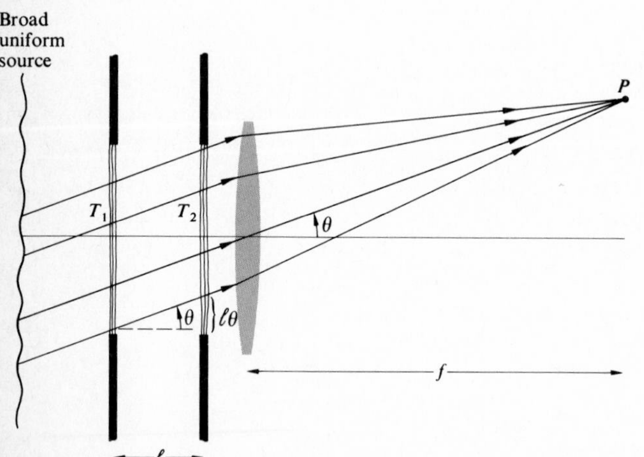

FIGURE 11.39 Optical correlation of two functions.

*See L. S. G. Kovasznay and A. Arman, *Rev. Sci. Instr.* **28**, 793 (1958), and D. McLachlan, Jr., *J. Opt. Soc. Am.* **52**, 454 (1962).

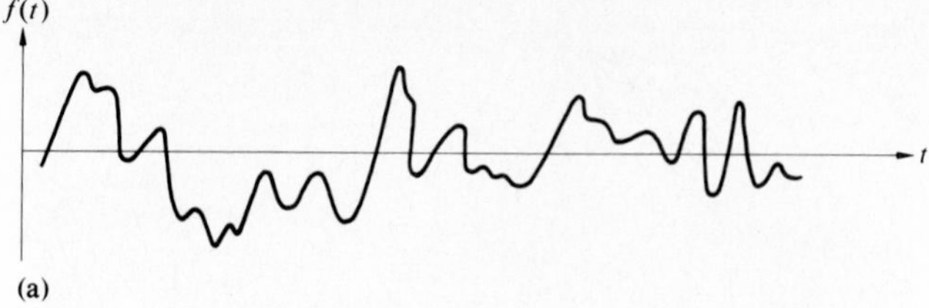

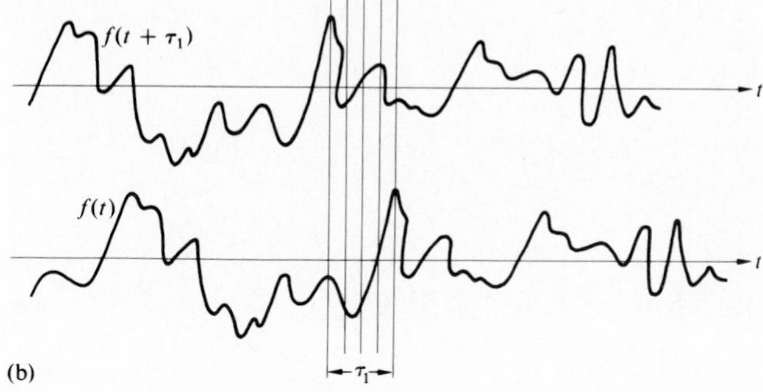

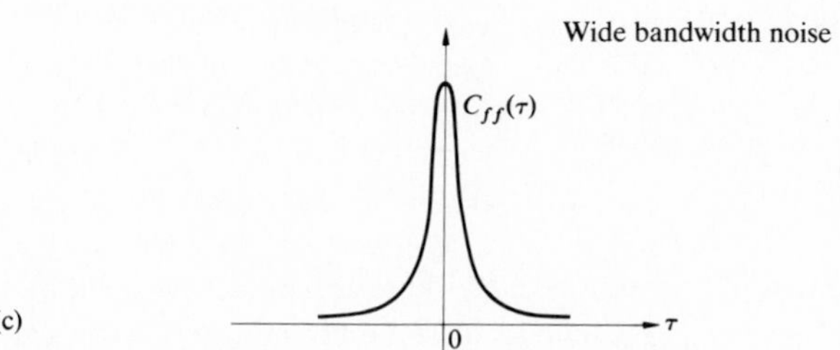

FIGURE 11.40 A signal $f(t)$ and its autocorrelation.

In very much the same way, the cross-correlation is a measure of the similarity between two different waveforms, $f(t)$ and $h(t)$, as a function of the relative time shift τ. Unlike the autocorrelation, there is now nothing special about $\tau = 0$. Once again, for each value of τ we average the product $f(t)h(t + \tau)$ to get $C_{fh}(\tau)$ via Eq. (11.87). For the functions shown in Fig. 11.41, $C_{fh}(\tau)$ would have a positive peak at $\tau = \tau_1$.

Since the 1960s a great deal of effort has gone into the development of optical processors that can rapidly analyze pictorial data. The potential uses range from comparing fingerprints to scanning documents for words or phrases; from screening aerial reconnaissance pictures to creating terrain-following guidance systems for missiles. An example of this kind of *optical pattern recognition*, accomplished using correlation techniques, is shown in Fig. 11.42. The input signal $f(x, y)$ depicted in photograph (*a*) is a broad view of some region that is to be searched for a particular group of structures [photograph (*b*)] isolated as the reference signal $h(x, y)$. Of course, that small frame is easy enough to scan directly by eye, so to make things more realistic, imagine the input to be a few hundred feet of reconnaissance film. The result of optically correlating these two signals is displayed in photograph (*c*), where we immediately see, from the correlation peak (i.e., the spike of light), that indeed the desired group of structures is in the input picture, and moreover its location is marked by the peak.

11.3.5 Transfer Functions

An Introduction to the Concepts

Until recent times, the traditional means of determining the quality of an optical element or system of elements was to evaluate its limit of resolution. The greater the resolution, the better the system was presumed to be. In the spirit of this approach one might train an optical system on a resolution tar-

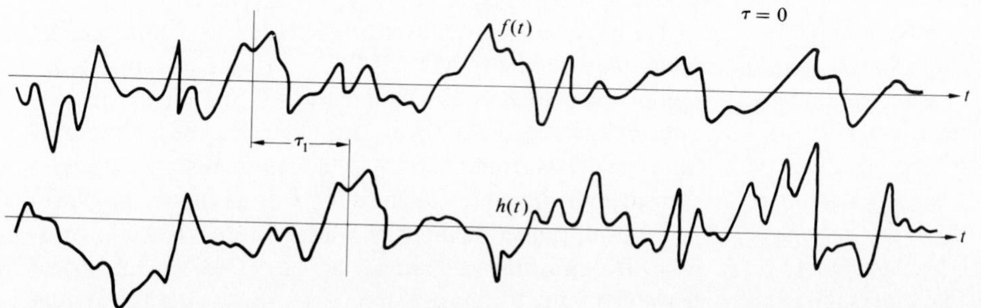

FIGURE 11.41 The cross-correlation of $f(t)$ and $h(t)$.

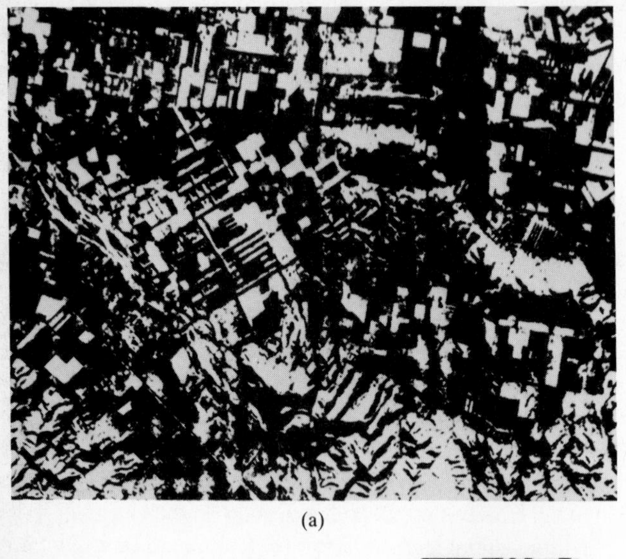

(a)

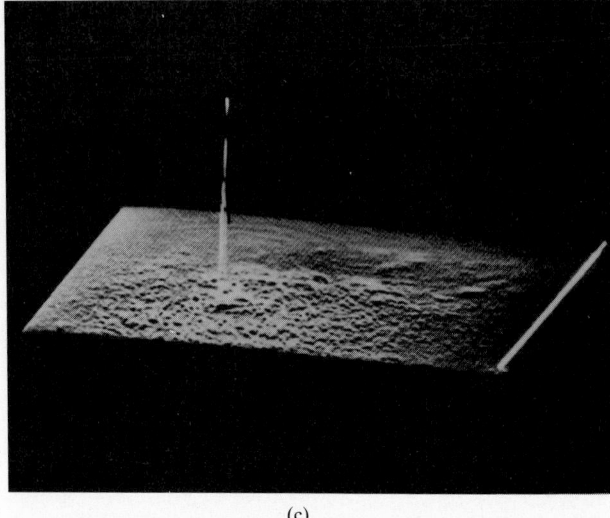

(c)

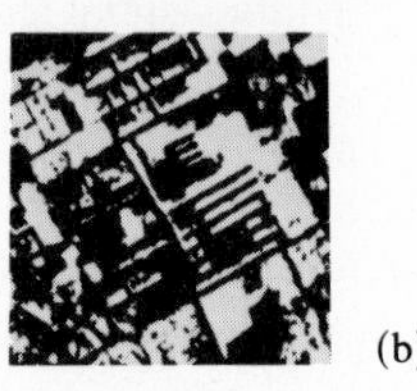

(b)

FIGURE 11.42 An example of optical pattern recognition. (*a*) Input signal, (*b*) reference data, (*c*) correlation pattern. (Reprinted with permission from the November 1980 issue of *Electro-Optical Systems Design*. David Casasent.)

get consisting, for instance, of a series of alternating light and dark parallel rectangular bars. We have already seen that an object point is imaged as a smear of light described by the point-spread function $\mathcal{S}(Y, Z)$, as in Fig. 11.18. Under incoherent illumination, these elementary flux-density patterns overlap and add linearly to create the final image. The one-dimensional counterpart is the *line-spread function* $\mathcal{S}(Z)$, which corresponds to the flux-density distribution across the image of a geometrical line source having infinitesimal width (Fig. 11.43). Because even an ideally perfect system is limited by diffraction effects, the image of a resolution target (Fig. 11.44) will be somewhat blurred (see Fig. 11.20). Thus, as the width of the bars on the target is made narrower, a limit will be reached where the fine-line structure (akin to a *Ronchi ruling*) will no longer be discernible—this then is the resolution limit of the system. We can think of it as a spatial frequency cutoff where each bright and dark bar pair constitutes one cycle on the object (a common measure of which is *line pairs per mm*). An obvious analogy which underscores the shortcomings of this approach would be to evaluate a high-fidelity sound system simply on the basis of its upper-frequency cutoff. The limitations of this scheme became quite apparent with the introduction of detectors such as the plumbicon, image orthicon, and vidicon. These tubes have a relatively coarse scanning raster, which fixes the resolution limit of the lens-tube system at a fairly low spatial frequency. Accordingly, it would seem reasonable to design the optics preceding such detectors so that it provided the most contrast over this limited frequency range. It would clearly be unnecessary and perhaps, as we shall see, even detrimental to select a mating lens system merely because of its own high limit of resolution. Evidently, it would be more helpful to have some figure of merit applicable to the entire operating frequency range.

We have already represented the object as a collection of point sources, each of which is imaged as a point-spread function by the optical system, and that patch of light is then convolved into the image. Now we approach the problem of image analysis from a different, though related, perspective. Consider the object to be the source of an input lightwave, which itself is made up of plane waves. These travel off in specific directions corresponding, via Eqs. (11.64) and (11.65), to particular values of spatial frequency. How does the system

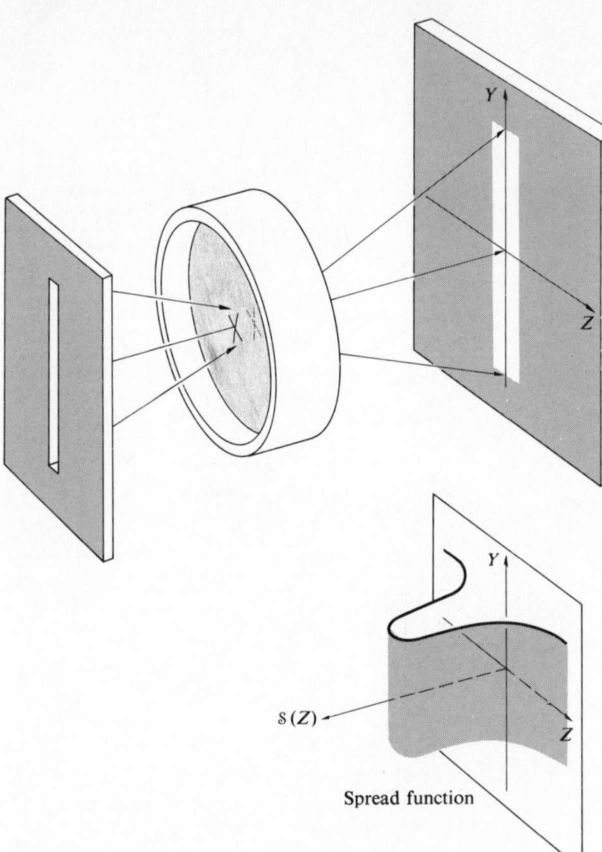

FIGURE 11.43 The line-spread function.

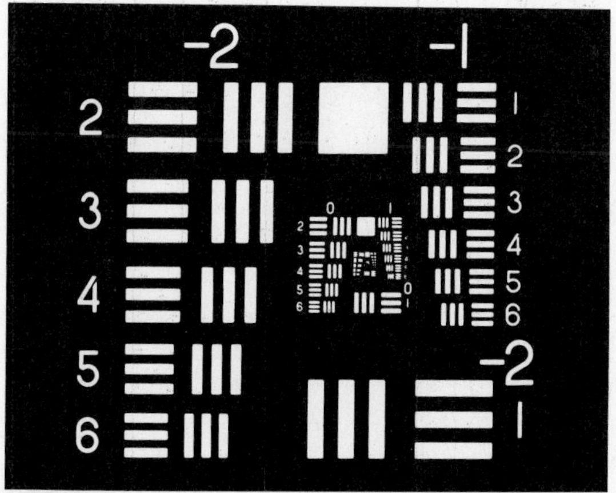

FIGURE 11.44 A bar target resolution chart.

modify the amplitude and phase of each plane wave as it transfers it from object to image?

A highly useful parameter in evaluating the performance of a system is the **contrast** or **modulation**, defined by

$$\text{Modulation} \equiv \frac{I_{\text{max}} - I_{\text{min}}}{I_{\text{max}} + I_{\text{min}}} \tag{11.89}$$

As a simple example, suppose the input is a cosinusoidal irradiance distribution arising from an incoherently illuminated transparency (Fig. 11.45). Here the output is also a cosine, but one that's somewhat altered. The modulation, which corresponds to the amount the function varies about its mean value divided by that mean value, is a measure of how readily the fluctuations will be discernible against the dc background. For the input the modulation is a maximum of 1.0, but the output modulation is only 0.17. This is only the response of our hypothetical system to essentially one spatial frequency input—it would be nice to know what it does at all such frequencies. Moreover, here the input modulation was 1.0, and the com-

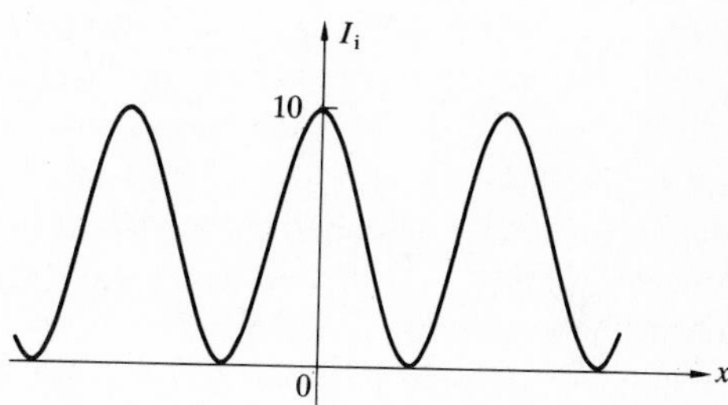

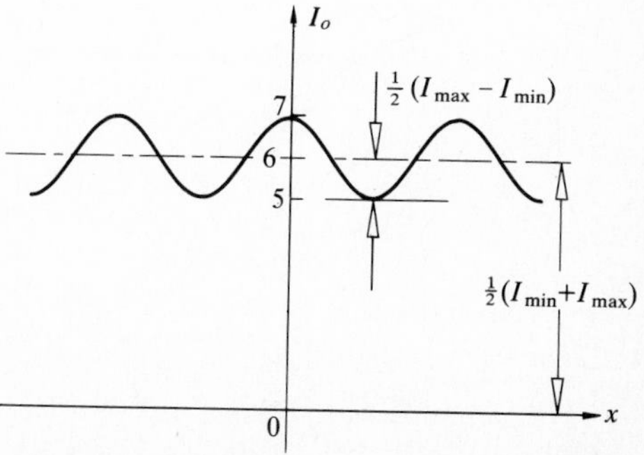

FIGURE 11.45 The irradiance into and out of a system.

parison with the output was easy. In general it will not be 1.0, and so we define *the ratio of the image modulation to the object modulation at all spatial frequencies* as the **modulation transfer function**, or MTF.

Figure 11.46 is a plot of the MTF for two hypothetical lenses. Both start off with a zero-frequency (dc) value of 1.0, and both cross the zero axis somewhere where they can no longer resolve the data at that *cutoff frequency*. Had they both been diffraction-limited lenses, that cutoff would have depended only on diffraction and, hence, on the size of the aperture. In any event, suppose one of these is to be coupled to a detector whose cutoff frequency is indicated in the diagram. Despite the fact that lens-1 has a higher limit of resolution, lens-2 would certainly provide better performance when coupled to the particular detector.

It should be pointed out that a square bar target provides an input signal that is a series of square pulses, and the contrast in image is actually a superposition of contrast variations due to the constituent Fourier components. Indeed, one of the key points in what is to follow is that *optical elements functioning as linear operators transform a sinusoidal input into an undistorted sinusoidal output.* Despite this, the input and output irradiance distributions as a rule will not be identical. For example, the system's magnification affects the spatial frequency of the output (henceforth, the magnification will be taken as one). Diffraction and aberrations reduce the sinusoid's amplitude (contrast). Finally, asymmetrical aberrations (e.g., coma) and poor centering of elements produce a shift in the position of the output sinusoid corresponding to the introduction of a phase shift. This latter point, which was considered in Fig. 11.12, can be appreciated using a diagram like that of Fig. 11.47.

If the spread function is symmetrical, the image irradiance will be an unshifted sinusoid, whereas an asymmetrical spread function will apparently push the output over a bit, as in Fig.

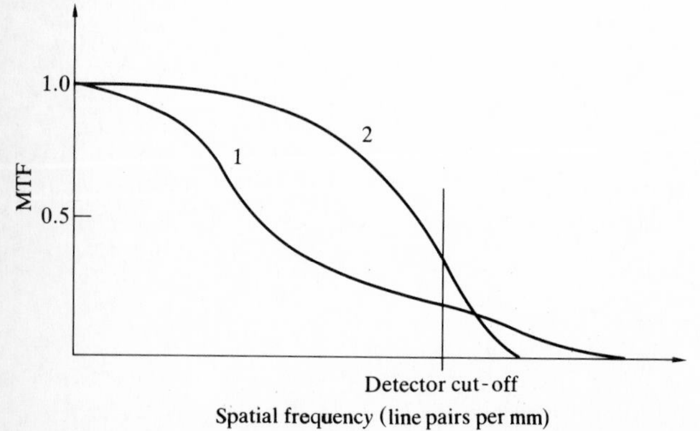

FIGURE 11.46 Modulation versus spatial frequency for two lenses.

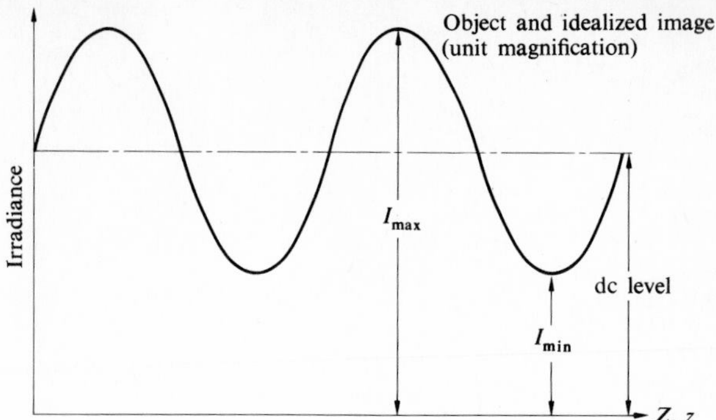

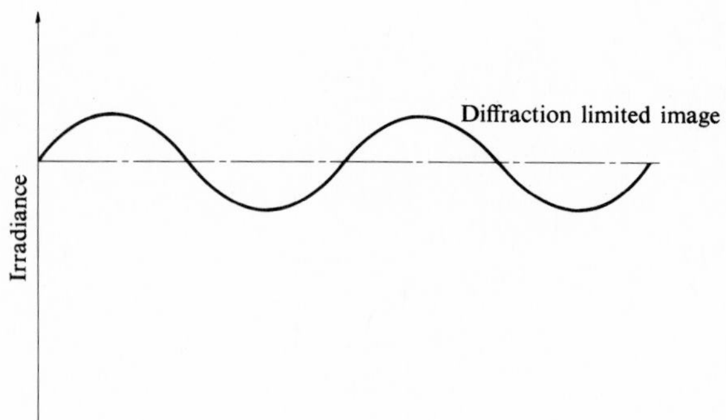

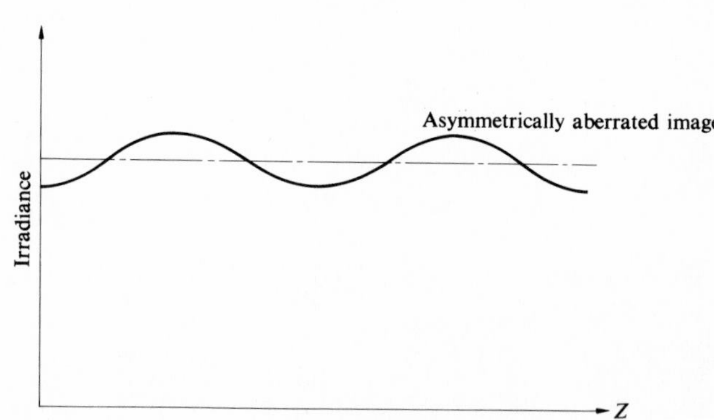

FIGURE 11.47 Harmonic input and resulting output.

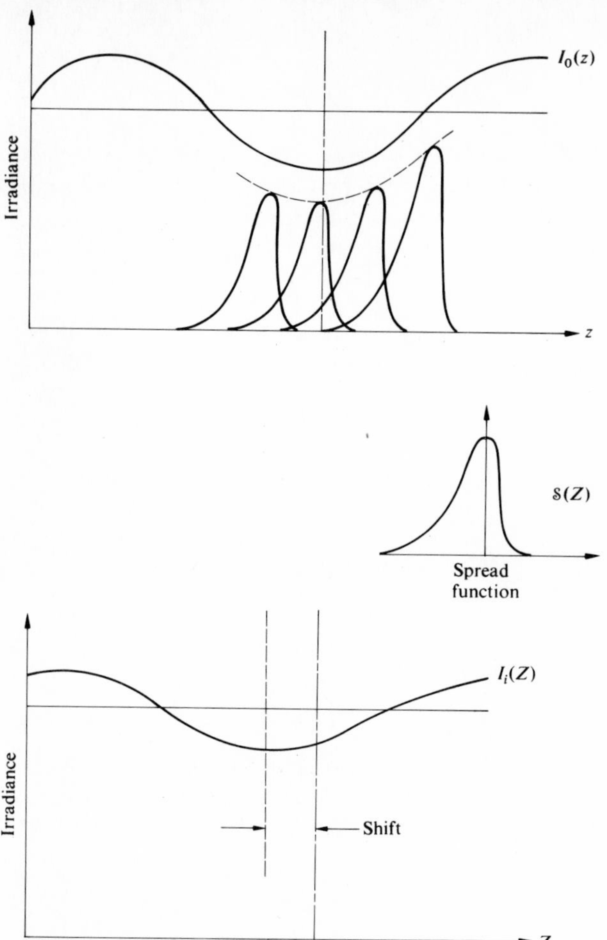

FIGURE 11.48 Harmonic input and output with an asymmetric spread function.

11.48. In either case, *regardless of the form of the spread function, the image is harmonic if the object is harmonic*. Consequently, if we envision an object as being composed of Fourier components, the manner in which these individual harmonic components are transformed by the optical system into the corresponding harmonic constituents of the image is the quintessential feature of the process. The function that performs this service is known as the **optical transfer function**, or OTF. It is a spatial frequency-dependent complex quantity whose modulus is the *modulation transfer function* (MTF) and whose phase, naturally enough, is the **phase transfer function** (PTF). The former is a measure of the reduction in contrast from object to image over the spectrum. The latter represents the commensurate relative phase shift. Phase shifts in centered optical systems occur only off-axis, and often the PTF is of less interest than the MTF. Even so, each application of the transfer function must be studied carefully; there are situations wherein the PTF plays a crucial role. In point of fact, the MTF has become a widely used means of specifying the performance of all sorts of elements and systems, from lenses, magnetic tape, and films to telescopes, the atmosphere, and the eye, to mention but a few. Moreover, it has the advantage that if the MTFs for the individual independent components in a system are known, the total MTF is often simply their product. This is inapplicable to the cascading of lenses, since the aberrations in one lens can compensate for those of another lens in tandem with it, and they are therefore not independent. Thus if we photograph an object having a modulation of 0.3 at 30 cycles per mm, using a camera whose lens at the appropriate setting has an MTF of 0.5 at 30 cycles/mm and a film* such as Tri-X with an MTF of 0.4 at 30 cycles/mm, the image modulation will be $0.3 \times 0.5 \times 0.4 = 0.06$.

A More Formal Discussion

We saw in Eq. (11.51) that the image (under the conditions of space invariance and incoherence) could be expressed as the convolution of the object irradiance and the point-spread function, in other words,

$$I_i(Y, Z) = I_0(y, z) \circledast \mathcal{S}(y, z) \tag{11.90}$$

The corresponding statement in the spatial frequency domain is obtained by a Fourier transform, namely,

$$\mathcal{F}\{I_i(Y, Z)\} = \mathcal{F}\{I_0(y, z)\} \cdot \mathcal{F}\{\mathcal{S}(y, z)\} \tag{11.91}$$

where use was made of the convolution theorem [Eq. (11.53)]. This says that *the frequency spectrum of the image irradiance distribution equals the product of the frequency spectrum of*

*Incidentally, the whole idea of treating film as a noise-free linear system is somewhat suspect. For further reading see J. B. De Velis and G. B. Parrent, Jr., "Transfer Function for Cascaded Optical Systems," *J. Opt. Soc. Am.* **57**, 1486 (1967).

the object irradiance distribution and the transform of the spread function (Fig. 11.49). Thus, it is multiplication by $\mathcal{F}\{\mathcal{S}(y, z)\}$ that produces the alteration in the frequency spectrum of the object, converting it into that of the image spectrum. In other words, it is $\mathcal{F}\{\mathcal{S}(y, z)\}$ that, in effect, transfers the object spectrum into the image spectrum. This is just the service performed by the OTF, and indeed we shall define the **unnormalized OTF** as

$$\mathcal{T}(k_Y, k_Z) \equiv \mathcal{F}\{\mathcal{S}(y, z)\} \tag{11.92}$$

The modulus of $\mathcal{T}(k_Y, k_Z)$ will effect a change in the amplitudes of the various frequency components of the object spectrum, while its phase will, of course, appropriately alter the phase of these components to yield $\mathcal{F}\{I_i(Y, Z)\}$. Bear in mind that in the right-hand side of Eq. (11.90) the only quantity dependent on the actual optical system is $\mathcal{S}(y, z)$, so it's not surprising that the spread function is the spatial counterpart of the OTF.

Let's now verify the statement made earlier that a harmonic input transforms into a somewhat altered harmonic output. To that end, suppose

$$I_0(z) = 1 + a \cos (k_Z z + \varepsilon) \tag{11.93}$$

where for simplicity's sake, we'll again use a one-dimensional distribution. The 1 is a dc bias, which makes sure the irradiance doesn't take on any unphysical negative values. Insofar as $f \circledast h = h \circledast f$, it will be more convenient here to use

$$I_i(Z) = \mathcal{S}(Z) \circledast I_0(z)$$

and so

$$I_i(Z) = \int_{-\infty}^{+\infty} \{1 + a \cos [k_Z(Z - z) + \varepsilon]\}\mathcal{S}(z)\, dz$$

Expanding out the cosine, we obtain

$$I_i(Z) = \int_{-\infty}^{+\infty} \mathcal{S}(z)\,dz + a \cos (k_Z Z + \varepsilon) \int_{-\infty}^{+\infty} \cos k_Z z \mathcal{S}(z)\, dz$$
$$+ a \sin (k_Z Z + \varepsilon) \int_{-\infty}^{+\infty} \sin k_Z z \mathcal{S}(z)\, dz$$

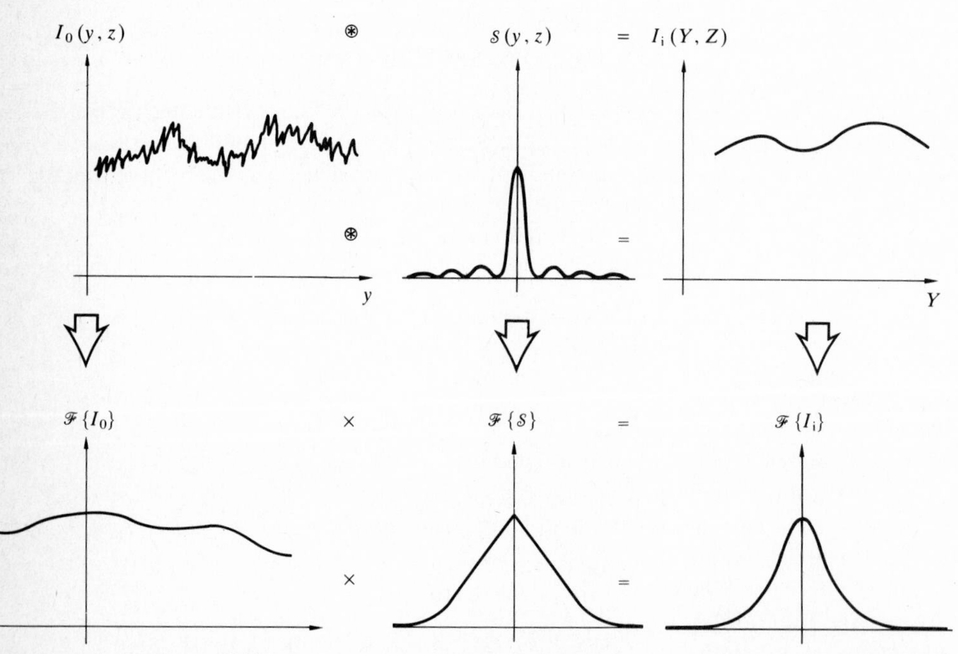

FIGURE 11.49 The relationships between the object and image spectra by way of the OTF, and the object and image irradiances by way of the point-spread function—all in incoherent illumination.

Referring back to Eq. (7.57), we recognize the second and third integrals as the Fourier cosine and sine transforms of $\mathcal{S}(z)$, respectively, that is to say, $\mathcal{F}_c\{\mathcal{S}(z)\}$ and $\mathcal{F}_s\{\mathcal{S}(z)\}$. Hence

$$I_i(z) = \int_{-\infty}^{+\infty} \mathcal{S}(z)\,dz + \mathcal{F}_c\{\mathcal{S}(z)\}a \cos(k_Z Z + \varepsilon) + \mathcal{F}_s\{\mathcal{S}(z)\}a \sin(k_Z Z + \varepsilon) \tag{11.94}$$

Recall that the complex transform we've become so used to working with was defined such that

$$\mathcal{F}\{f(z)\} = \mathcal{F}_c\{f(z)\} + i\mathcal{F}_s\{f(z)\} \tag{11.95}$$

or

$$F(k_Z) = A(k_Z) + iB(k_Z) \qquad [11.7]$$

In addition,

$$\mathcal{F}\{f(z)\} = |F(k_Z)|e^{i\varphi(k_Z)} = |F(k_Z)|[\cos\varphi + i\sin\varphi]$$

where

$$|F(k_Z)| = [A^2(k_Z) + B^2(k_Z)]^{1/2} \tag{11.96}$$

and

$$\varphi(k) = \tan^{-1}\frac{B(k_Z)}{A(k_Z)} \tag{11.97}$$

In precisely the same way, we apply this to the OTF, writing it as

$$\mathcal{F}\{\mathcal{S}(z)\} \equiv \mathcal{T}(k_Z) = \mathcal{M}(k_Z)e^{i\Phi(k_Z)} \tag{11.98}$$

where $\mathcal{M}(k_Z)$ and $\Phi(k_Z)$ are the unnormalized MTF and the PTF, respectively. It is left as a problem to show that Eq. (11.94) can be recast as

$$I_i(Z) = \int_{-\infty}^{+\infty} \mathcal{S}(z)\,dz + a\mathcal{M}(k_Z)\cos[k_Z Z + \varepsilon - \Phi(k_Z)] \tag{11.99}$$

Notice that this is a function of the same form as the input signal [Eq. (11.93)], $I_0(z)$, which is just what we set out to determine. If the line-spread function is symmetrical (i.e., even), $\mathcal{F}_s\{\mathcal{S}(z)\} = 0$, $\mathcal{M}(k_Z) = \mathcal{F}_c\{\mathcal{S}(z)\}$, and $\Phi(k_Z) = 0$; there is no phase shift, as was pointed out in the previous section. For an asymmetric (odd) spread function, $\mathcal{F}_s\{\mathcal{S}(z)\}$ is nonzero, as is the PTF.

It has now become customary practice to define a set of *normalized transfer functions* by dividing $\mathcal{T}(k_Z)$ by its zero spatial frequency value, that is, $\mathcal{T}(0) = \int_{-\infty}^{+\infty} \mathcal{S}(z)\,dz$. The normalized spread function becomes

$$\mathcal{S}_n(z) = \frac{\mathcal{S}(z)}{\displaystyle\int_{-\infty}^{+\infty} \mathcal{S}(z)\,dz} \tag{11.100}$$

while the **normalized OTF** is

$$T(k_Z) \equiv \frac{\mathcal{F}\{\mathcal{S}(z)\}}{\displaystyle\int_{-\infty}^{+\infty} \mathcal{S}(z)\,dz} = \mathcal{F}\{\mathcal{S}_n(z)\} \tag{11.101}$$

or in two dimensions

$$T(k_Y, k_Z) = M(k_Y, k_Z)e^{i\Phi(k_Y, k_Z)} \tag{11.102}$$

where $M(k_Y, k_Z) \equiv \mathcal{M}(k_Y, k_Z)/\mathcal{T}(0, 0)$. Therefore $I_i(Z)$ in Eq. (11.99) would then be proportional to

$$1 + aM(k_Z)\cos[k_Z Z + \varepsilon - \Phi(k_Z)]$$

The image modulation [Eq. (11.89)] becomes $aM(k_Z)$, the object modulation [Eq. (11.93)] is a, and the ratio is, as expected, the normalized MTF $= M(k_Z)$.

This discussion is only an introductory one designed more as a strong foundation than a complete structure. There are many other insights to be explored, such as the relationship between the autocorrelation of the pupil function and the OTF, and from there, the means of computing and measuring transfer functions (Fig. 11.50)—but for this the reader is directed to the literature.*

*See the series of articles "The Evolution of the Transfer Function," by F. Abbott, beginning in March 1970 in *Optical Spectra*; the articles "Physical Optics Notebook," by G. B. Parrent, Jr., and B. J. Thompson, beginning in December 1964, in the *S.P.I.E. Journal*, Vol. 3; or "Image Structure and Transfer," by K. Sayanagi, 1967, available from the Institute of Optics, University of Rochester. A number of books are worth consulting for practical emphasis, e.g., *Modern Optics*, by E. Brown; *Modern Optical Engineering*, by W. Smith; and *Applied Optics*, by L. Levi. In all of these, be careful of the sign convention in the transforms.

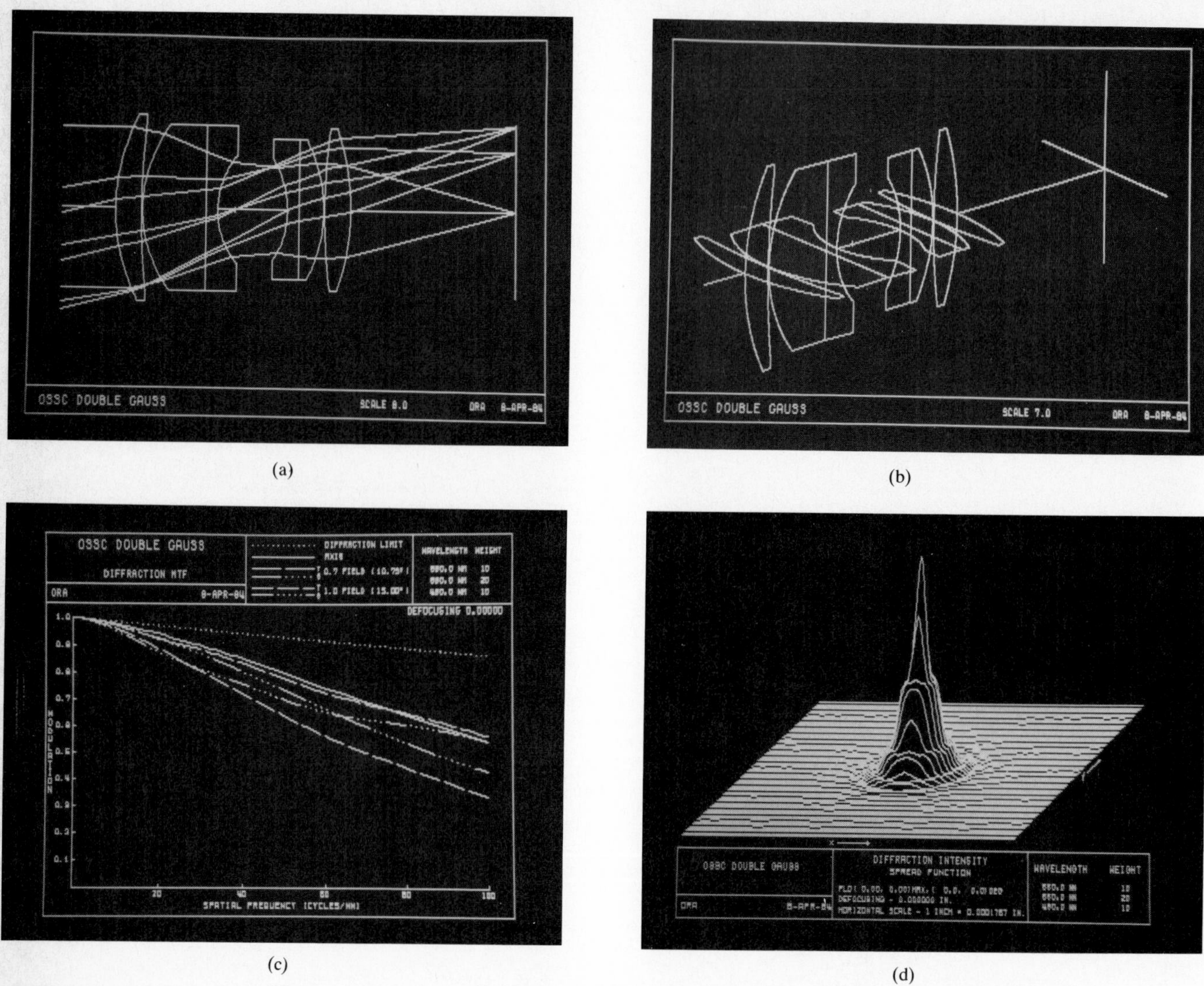

(a) (b) (c) (d)

FIGURE 11.50 An example of the kind of lens design information available via modern computer techniques. (Photos courtesy Optical Research Associates)

PROBLEMS

11.1 Determine the Fourier transform of the function

$$E(x) = \begin{cases} E_0 \sin k_p x & |x| < L \\ 0 & |x| > L \end{cases}$$

Make a sketch of $\mathscr{F}\{E(x)\}$. Discuss its relationship to Fig. 11.11.

11.2* Determine the Fourier transform of

$$f(x) = \begin{cases} \sin^2 k_p x & |x| < L \\ 0 & |x| > L \end{cases}$$

Make a sketch of it.

11.3 Determine the Fourier transform of

$$f(t) = \begin{cases} \cos^2 \omega_p t & |t| < T \\ 0 & |t| > T \end{cases}$$

Make a sketch of $F(\omega)$, then sketch its limiting form as $T \to \pm\infty$.

11.4* Show that $\mathscr{F}\{1\} = 2\pi\delta(k)$.

11.5* Determine the Fourier transform of the function $f(x) = A \cos k_0 x$.

11.6 Given that $\mathscr{F}\{f(x)\} = F(k)$ and $\mathscr{F}\{h(x)\} = H(k)$, if a and b are constants, determine $\mathscr{F}\{af(x) + bh(x)\}$.

11.7* Figure P.11.7 shows two periodic functions, $f(x)$ and $h(x)$, which are to be added to produce $g(x)$. Sketch $g(x)$; then draw diagrams of the real and imaginary frequency spectra, as well as the amplitude spectra for each of the three functions.

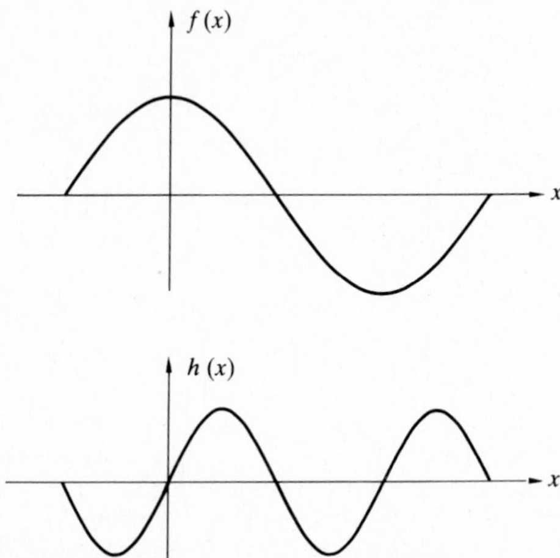

FIGURE P.11.7

11.8 Compute the Fourier transform of the triangular pulse shown in Fig. P.11.8. Make a sketch of your answer, labeling all the pertinent values on the curve.

11.9* Given that $\mathscr{F}\{f(x)\} = F(k)$, introduce a constant scaling factor $1/a$ and determine the Fourier transform of $f(x/a)$. Show that the transform of $f(-x)$ is $F(-k)$.

11.10* Show that the Fourier transform of the transform, $\mathscr{F}\{f(x)\}$, equals $2\pi f(-x)$, and that this is not the inverse transform of the transform, which equals $f(x)$. This problem was suggested by Mr. D. Chapman while a student at the University of Ottawa.

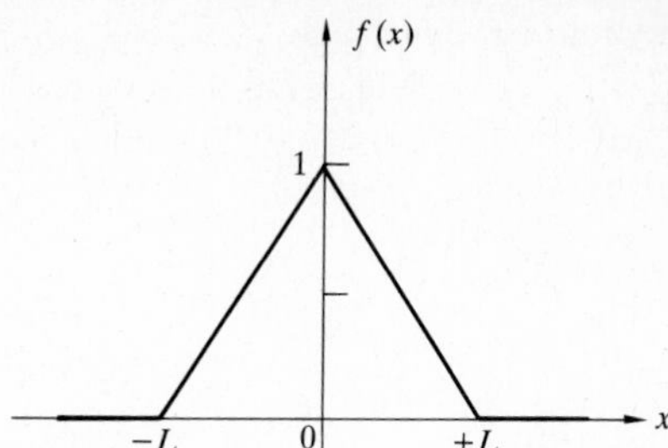

FIGURE P.11.8

11.11* The rectangular function is often defined as

$$\text{rect}\left|\frac{x - x_0}{a}\right| = \begin{cases} 0, & |(x - x_0)/a| > \frac{1}{2} \\ \frac{1}{2}, & |(x - x_0)/a| = \frac{1}{2} \\ 1, & |(x - x_0)/a| < \frac{1}{2} \end{cases}$$

where it is set equal to $\frac{1}{2}$ at the discontinuities (Fig. P.11.11). Determine the Fourier transform of

$$f(x) = \text{rect}\left|\frac{x - x_0}{a}\right|$$

Notice that this is just a rectangular pulse, like that in Fig. 11.1*b*, shifted a distance x_0 from the origin.

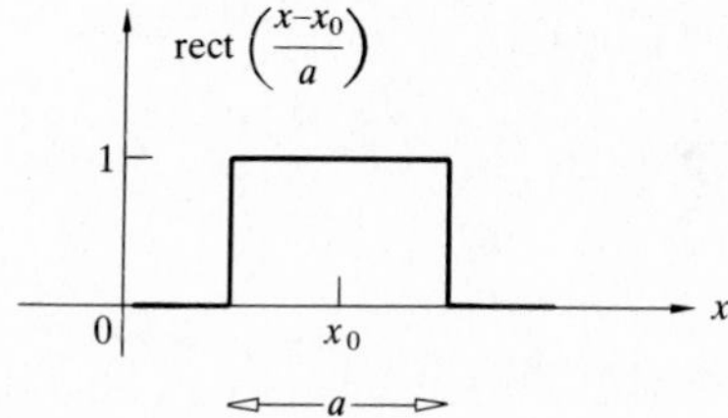

FIGURE P.11.11

11.12* With the last two problems in mind, show that $\mathscr{F}\{(1/2\pi)\,\text{sinc}\,(\frac{1}{2}x)\} = \text{rect}(k)$, starting with the knowledge that $\mathscr{F}\{\text{rect}(x)\} = \text{sinc}\,(\frac{1}{2}k)$, in other words, Eq. (7.58) with $L = a$, where $a = 1$.

11.13* Utilizing Eq. (11.38), show that $\mathscr{F}^{-1}\{\mathscr{F}\{f(x)\}\} = f(x)$.

11.14* Given $\mathscr{F}\{f(x)\}$, show that $\mathscr{F}\{f(x - x_0)\}$ differs from it only by a linear phase factor.

11.15 Prove that $f \circledast h = h \circledast f$ directly. Now do it using the convolution theorem.

11.16* Suppose we have two functions, $f(x, y)$ and $h(x, y)$, where both have a value of 1 over a square region in the xy-plane and are zero everywhere else (Fig. P.11.16). If $g(X, Y)$ is their convolution, make a plot of $g(X, 0)$.

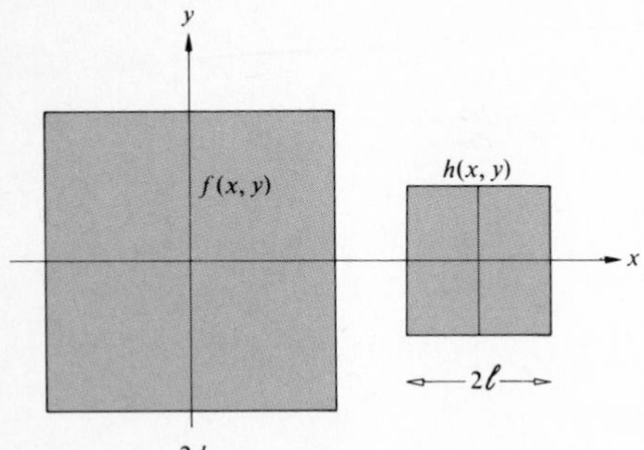

FIGURE P.11.16

11.17 Referring to the previous problem, justify the fact that the convolution is zero for $|X| \geq d + \ell$ when ℓ is viewed as a spread function.

11.18* Use the method illustrated in Fig. 11.23 to convolve the two functions depicted in Fig. P.11.18.

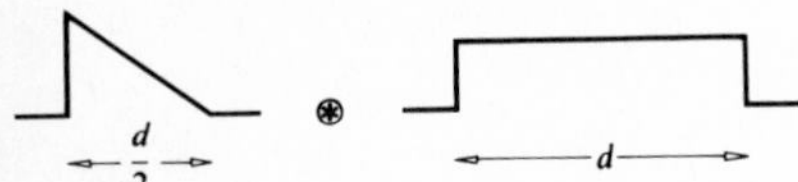

FIGURE P.11.18

11.19 Given that $f(x) \circledast h(x) = g(X)$, show that after shifting one of the functions an amount x_0, we get $f(x - x_0) \circledast h(x) = g(X - x_0)$.

11.20* Prove analytically that the convolution of any function $f(x)$ with a delta function, $\delta(x)$, generates the original function $f(X)$. You might make use of the fact that $\delta(x)$ is even.

11.21 Prove that $\delta(x - x_0) \circledast f(x) = f(X - x_0)$ and discuss the meaning of this result. Make a sketch of two appropriate functions and convolve them. Be sure to use an asymmetrical $f(x)$.

11.22* Show that $\mathscr{F}\{f(x) \cos k_0 x\} = [F(k - k_0) + F(k + k_0)]/2$ and that $\mathscr{F}\{f(x) \sin k_0 x\} = [F(k - k_0) - F(k + k_0)]/2i$.

11.23* Figure P.11.23 shows two functions. Convolve them graphically and draw a plot of the result.

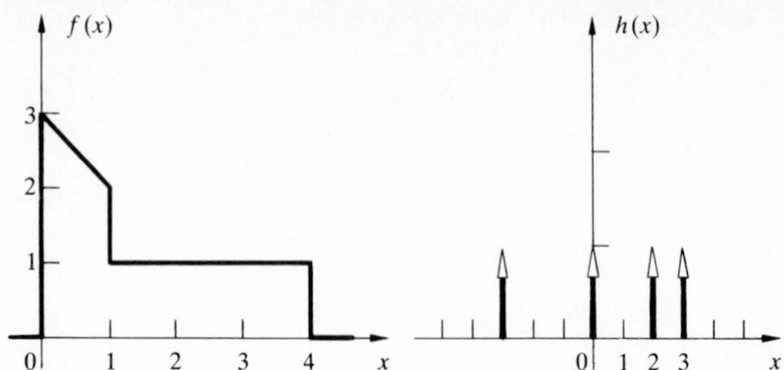

FIGURE P.11.23

11.24 Given the function

$$f(x) = \text{rect}\left|\frac{x - a}{a}\right| + \text{rect}\left|\frac{x + a}{a}\right|$$

determine its Fourier transform. (See Problem 11.11.)

11.25 Given the function $f(x) = \delta(x + 3) + \delta(x - 2) + \delta(x - 5)$, convolve it with the arbitrary function $h(x)$.

11.26* Make a sketch of the function arising from the convolution of the two functions depicted in Fig. P.11.26.

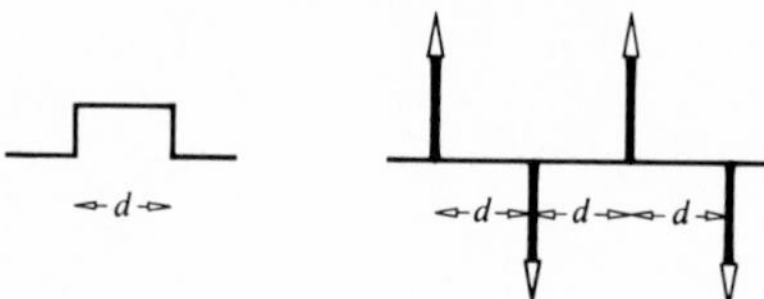

FIGURE P.11.26

11.27* Figure P.11.27 depicts a *rect* function (as defined above) and a periodic *comb* function. Convolve the two to get $g(x)$. Now sketch the transform of each of these functions against spatial frequency $k/2\pi = 1/\lambda$. Check your results with the convolution theorem. Label all the relevant points on the horizontal axes in terms of d—like the zeros of the transform of $f(x)$.

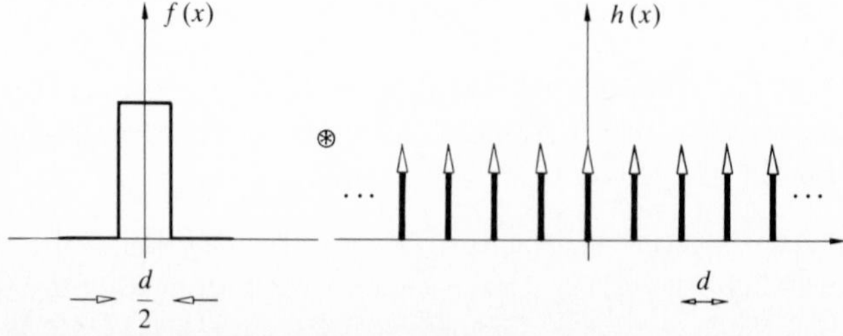

FIGURE P.11.27

11.28 Figure P.11.28 shows, in one dimension, the electric field across an illuminated aperture consisting of several opaque bars forming a grating. Considering it to be created by taking the product of a periodic rectangular wave $h(x)$ and a unit rectangular function $f(x)$, sketch the resulting electric field in the Fraunhofer region.

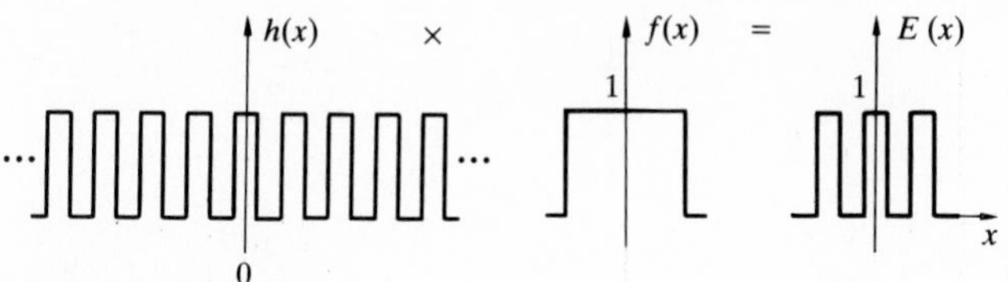

FIGURE P.11.28

11.29 Show (for normally incident plane waves) that if an aperture has a center of symmetry (i.e., if the aperture function is even), then the diffracted field in the Fraunhofer case also possesses a center of symmetry.

11.30 Suppose a given aperture produces a Fraunhofer field pattern $E(Y, Z)$. Show that if the aperture's dimensions are altered such that the aperture function goes from $\mathcal{A}(y, z)$ to $\mathcal{A}(\alpha y, \beta z)$, the newly diffracted field will be given by

$$E'(Y, Z) = \frac{1}{\alpha\beta} E\left(\frac{Y}{\alpha}, \frac{Z}{\beta}\right)$$

11.31 Show that when $f(t) = A \sin(\omega t + \varepsilon)$, $C_{ff}(\tau) = (A^2/2)\cos \omega t$, which confirms the loss of phase information in the autocorrelation.

11.32 Suppose we have a single slit along the y-direction of width b where the aperture function is constant across it at a value of $\mathcal{A}_0$. What is the diffracted field if we now apodize the slit with a cosine function amplitude mask? In other words, we cause the aperture function to go from $\mathcal{A}_0$ at the center to 0 at $\pm b/2$ via a cosinusoidal dropoff.

11.33* Show, from the integral definitions, that $f(x) \odot g(x) = f(x) \circledast g(-x)$.

11.34* Figure P.11.34 shows a transparent ring on an otherwise opaque mask. Make a rough sketch of its autocorrelation function, taking l to be the center-to-center separation against which you plot that function.

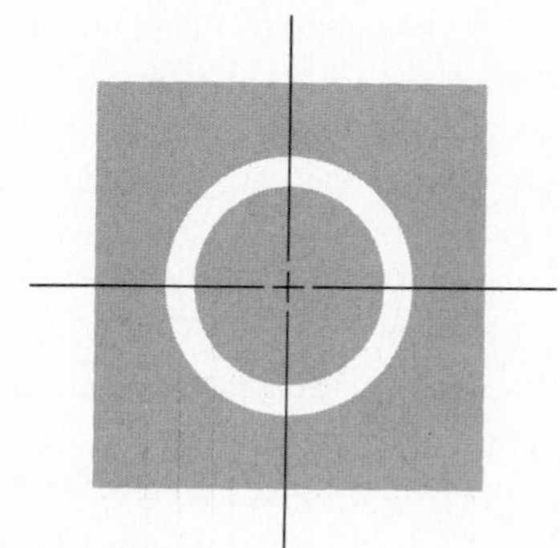

FIGURE P.11.34

11.35* Consider the function in Fig. 11.35 as a cosine carrier multiplied by an exponential envelope. Use the frequency convolution theorem to evaluate its Fourier transform.

12 Basics of Coherence Theory

Thus far in our discussion of phenomena involving the superposition of waves, we've restricted the treatment to that of either completely coherent or completely incoherent disturbances. This was done primarily as a mathematical convenience, since, as is quite often the case, the extremes in a physical situation are the easiest to deal with analytically. In fact, both of these limiting conditions are more conceptual idealizations than actual physical realities. There is a middle ground between these antithetic poles, which is of considerable contemporary concern—the domain of ***partial coherence***. Even so, the need for extending the theoretical structure is not new; it dates back at least to the mid-1860s, when Emile Verdet demonstrated that a primary source commonly considered to be incoherent, such as the Sun, could produce observable fringes when it illuminated the closely spaced pinholes ($\lesssim$0.05 mm) of Young's Experiment (Section 9.3). Theoretical interest in the study of partial coherence lay dormant until it was revived in the 1930s by P. H. van Cittert and later by Fritz Zernike. And as the technology flourished, advancing from traditional light sources, which were essentially optical frequency noise generators, to the laser, a new practical impetus was given the subject. Moreover, the recent advent of individual-photon detectors has made it possible to examine related processes associated with the corpuscular aspects of the optical field.

Optical coherence theory is currently an area of active research. Thus, even though much of the excitement in the field is associated with material beyond the level of this book, we shall nonetheless introduce some of the basic ideas.

12.1 Introduction

Earlier (Section 7.10) we evolved the highly useful picture of quasimonochromatic light as resembling a series of randomly phased finite wavetrains (Fig. 7.27). Such a disturbance is nearly sinusoidal, although the frequency does vary slowly (in comparison to the rate of oscillation, 10^{15} Hz) about some mean value. Moreover, the amplitude fluctuates as well, but this too is a comparatively slow variation. The average constituent wavetrain exists roughly for a time Δt_c, which is the *coherence time* given by the inverse of the frequency bandwidth $\Delta\nu$.

It is often convenient, even if artificial, to divide coherence effects into two classifications, *temporal and spatial* (p. 382). *The former relates directly to the finite bandwidth of the source, the latter to its finite extent in space.*

To be sure, if the light were monochromatic, $\Delta\nu$ would be zero, and Δt_c infinite, but this is, of course, unattainable. However, over an interval much shorter that Δt_c an actual wave behaves essentially as if it were monochromatic. In effect, the coherence time is *the temporal interval over which we can reasonably predict the phase of the lightwave at a given point in space*. This then is what is meant by **temporal coherence**; namely, if Δt_c is large, the wave has a high degree of temporal coherence and vice versa.

The same characteristic can be viewed somewhat differently. To that end, imagine that we have two separate points $P_1'$and P_2' lying on the same radius drawn from a quasimonochromatic point source (see Fig. 9.4). If the coherence length, $c\Delta t_c$, is much larger than the distance (r_{12}) between P_1' *and* P_2', then a single wavetrain can easily extend over the whole separation. The disturbance at P_1' would then be highly correlated with the disturbance occurring at P_2'. On the other hand, if this longitudinal separation were much greater than the coherence length, many wavetrains, each with an unrelated phase, would span the gap r_{12}. In that case, the disturbances at the two points in space would be independent at any given time. The degree to which a correlation exists is sometimes spoken of alternatively as the amount of *longitudinal coherence*. Whether we think in terms of coherence time (Δt_c) or coherence length ($c\Delta t_c$), the effect still arises from the finite bandwidth of the source.

The idea of **spatial coherence** is most often used to describe effects arising from the finite spatial extent of ordinary light sources. Suppose then that we have a classical broad monochromatic source. Two point radiators on it, separated by a lateral distance that is large compared with λ, will presumably behave quite independently. That is to say, there will be a lack of correlation existing between the phases of the two emitted disturbances. Extended sources of this sort are generally referred to as incoherent, but this description is somewhat misleading, as we shall see in a moment. Usually one is interested not so much in what is happening on the source itself but rather in what is occurring within some distant region of the radiation field. The question to be answered is really: How do the nature of the source and the geometrical configuration of the situation relate to the resulting phase correlation between two laterally spaced points in the light field?

This brings to mind Young's Experiment, in which a primary monochromatic source S illuminates two pinholes in an opaque screen. These in turn serve as secondary sources, S_1 and S_2, to generate a fringe pattern on a distant plane of observation, Σ_o (Fig. 9.8). We already know that if S is an idealized point source, the wavelets issuing from any set of apertures S_1 and S_2 on Σ_a will maintain a constant relative phase; they will be precisely correlated and therefore coherent. A well-defined array of stable fringes results, and the field is spatially coherent. At the other extreme, if the pinholes are illuminated by separate thermal sources (even with narrow bandwidths), no correlation exists; no fringes will be observable with existing detectors, and the fields at S_1 and S_2 are said to be incoherent. The generation of interference fringes is then seemingly a very convenient measure of the coherence.

We can gain some important insights into the process by returning to the general considerations of Section 9.1 and Eq. (9.7). Imagine two scalar waves $E_1(t)$ and $E_2(t)$ traveling toward, and overlapping at, point P, as in Fig. 9.2. If the light is monochromatic and both beams have the same frequency, the resulting interference pattern will depend on their relative phase at P. If the waves are in-phase, $E_1(t)E_2(t)$ will be positive for all t as the fields rise and fall in together. Hence, $I_{12} = 2\langle E_1(t)E_2(t)\rangle_T$ will be a nonzero positive number, and the net irradiance I will exceed $I_1 + I_2$. Similarly, if the lightwaves are completely out-of-phase, one will be positive when the other is negative, with the result that the product $E_1(t)E_2(t)$ will always be negative, yielding a negative interference term I_{12}, and the result that I will be less than $I_1 + I_2$. In both cases, the product of the two fields moment by moment is oscillatory, but it is nonetheless either totally positive or negative and so averages in time to a nonzero value.

Now consider the more realistic case in which the two lightwaves are quasimonochromatic, resembling the disturbance in Fig. 7.27, which has a finite coherence length. If we again form the product $E_1(t)E_2(t)$, we see in Fig. 12.1c that it varies in time, drifting from negative to positive values. Accordingly, the interference term $\langle E_1(t)E_2(t)\rangle_T$, which is averaged over a relatively long interval compared with the periods of the waves, will be quite small, if not zero: $I \approx I_1 + I_2$. In other words, insofar as the two lightwaves are uncorrelated in their risings and fallings, they will not preserve a constant phase relationship, they will not be completely coherent, and they will not produce the ideal high-contrast interference pattern considered in Chapter 9. We should be reminded here of Eq. (11.87), which expresses the cross-correlation of two functions—with $\tau = 0$. Indeed, if P is shifted in space (e.g.,

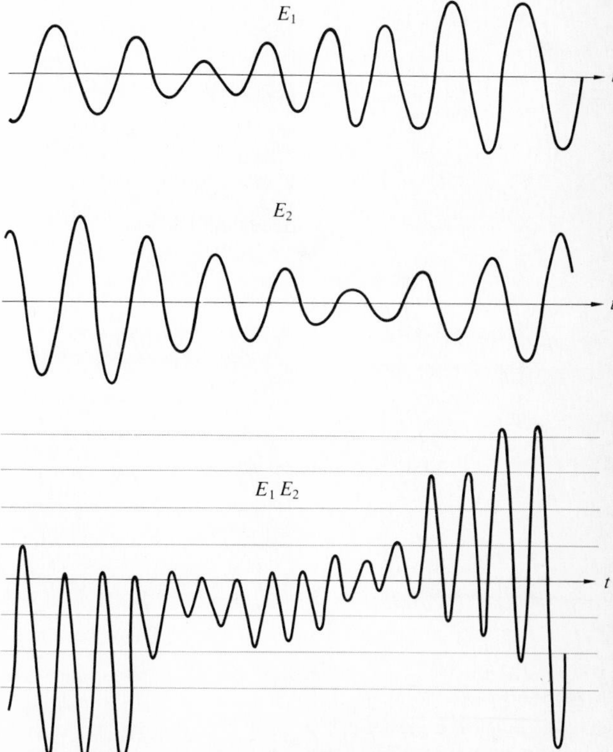

FIGURE 12.1 Two overlapping E-fields and their product as functions of time. The more uncorrelated the fields, the more nearly the product will average to zero.

along the plane of observation in Young's Experiment), thereby introducing a relative time delay of τ between the two lightwaves, then the interference term becomes $\langle E_1(t)E_2(t+\tau)\rangle_T$, which is the cross-correlation. Coherence is correlation, a point that will be made formally in Section 12.3.

Young's Experiment can also be used to demonstrate temporal coherence effects with a finite bandwidth source. Figure 12.2*a* shows the fringe patterns obtained with two small circular apertures illuminated by a He–Ne laser. Before the photograph in Fig. 12.2*b* was taken, an optically flat piece of glass, 0.5 mm thick, was positioned over one of the pinholes (say S_1). No change in the form of the pattern (other than a shift in its location) is evident, because the coherence length of the laser light far exceeds the optical path length difference introduced by the glass. On the other hand, when the same experiment is repeated using the light from a collimated mercury arc [(*c*) and (*d*) in Fig. 12.2], the fringes disappear. Here the coherence length is short enough and the additional optical path length difference of the glass is long enough for uncorrelated wavetrains from the two apertures to arrive at the plane of observation. In other words, of any two coherent wavetrains that leave S_1 and S_2, the one from S_1 is now delayed so long in the glass that it falls completely behind the other and arrives at Σ_o to meet a totally different wavetrain from S_2.

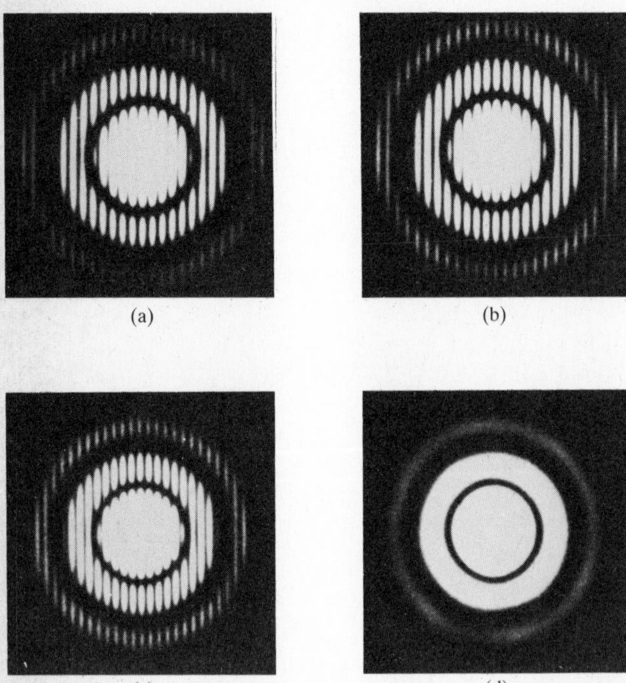

FIGURE 12.2 Double-beam interference from a pair of circular apertures. (*a*) He–Ne laserlight illuminating the holes. (*b*) Laserlight once again but now a glass plate, 0.5 mm thick, is covering one of the holes. (*c*) Fringes with collimated mercury-arc illumination but no glass plate. (*d*) This time the fringes disappear when the plate is inserted using mercury light. [From B. J. Thompson, *J. Soc. Photo. Inst. Engr.* **4**, 7 (1965).]

In both cases of temporal and spatial coherence we are really concerned with one phenomenon, namely, the correlation between optical disturbances. That is, we are generally interested in determining the effects arising from relative fluctuations in the fields at two points in space–time. Admittedly, the term *temporal coherence* seems to imply an effect that is exclusively temporal. However, it relates back to the finite extent of the wavetrain in either space or time, and some people even prefer to refer to it as *longitudinal spatial* rather than temporal coherence. Even so, it does depend intrinsically on the stability of phase in time, and accordingly we will continue to use the term *temporal coherence*. Spatial coherence, or if you will, *lateral spatial coherence*, is perhaps easier to appreciate, because it's so closely related to the concept of the wavefront. Thus if two laterally displaced points reside on the same wavefront at a given time, the fields at those points are spatially coherent (see Section 12.3.1).

12.2 Visibility

The quality of the fringes produced by an interferometric system can be described quantitatively using the **visibility** $\mathscr{V}$, which, as first formulated by Michelson, is given by

$$\mathscr{V}(\mathbf{r}) \equiv \frac{I_{\max} - I_{\min}}{I_{\max} + I_{\min}} \tag{12.1}$$

This is identical to the *modulation* of Eq. (11.89). Here $I_{\max}$ and $I_{\min}$ are the irradiances corresponding to the maximum and adjacent minimum in the fringe system. If we set up Young's Experiment, we could vary the separation of the apertures or the size of the primary incoherent quasimonochromatic source, measure $\mathscr{V}$ as it changes in turn, and then relate all this to the idea of coherence. An analytic expression can be derived for the flux-density distribution with the aid of Fig. 12.3.* Here we use a lens L to localize the fringe pattern

*This treatment in part follows that given by Towne in Chapter 11 of *Wave Phenomena*. See Klein, *Optics*, Section 6.3, or Problem 12.6 for different versions.

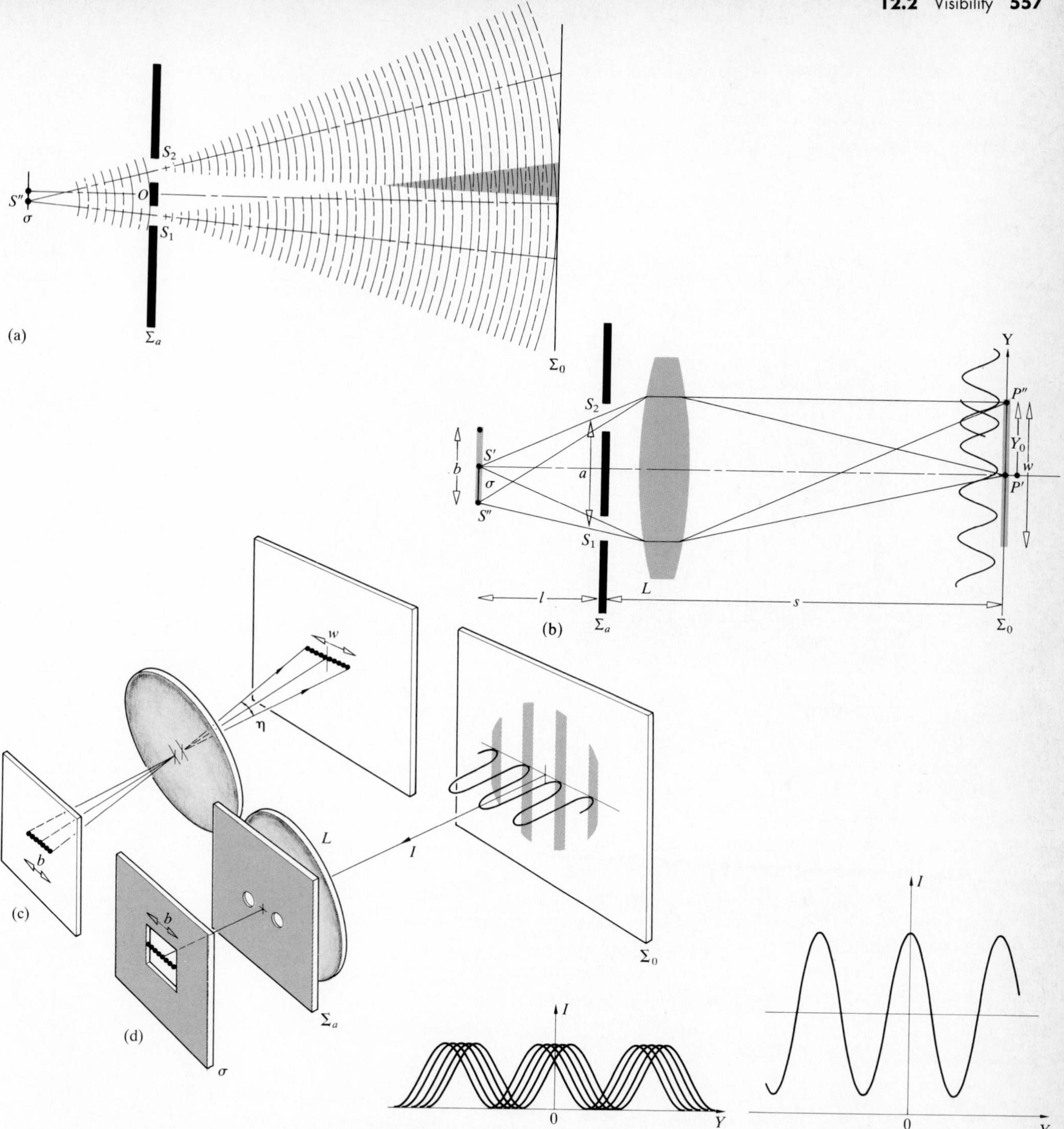

FIGURE 12.3 Young's Experiment with an extended slit source. (*e*) A simple representation of how shifted fringes with the same spatial frequency overlap and combine to form a net disturbance of that same spatial frequency with a reduced visibility (see Fig. 7.4).

more effectively, that is, to make the cones of light diffracted by the finite pinholes more completely overlap on the plane Σ_o. A point source S' located on the central axis would generate the usual pattern given by

$$I = 4I_0 \cos^2\left(\frac{Ya\pi}{s\lambda}\right) \tag{12.2}$$

from Section 9.3. Similarly, a point source above or below S' and lying on a line normal to the line $\overline{S_1S_2}$ would generate the same straight band fringe system slightly displaced in a direction parallel to the fringes. Thus replacing S' by an incoherent line source (normal to the plane of the drawing) effectively just increases the amount of light available. This is something we presumably already knew. In contrast, an off-axis point source, at say S'', will generate a pattern centered about P'', its image point on Σ_o in the absence of the aperture screen. A "spherical" wavelet leaving S'' is focused at P''; thus all rays from S'' to P'' traverse equal optic paths, and the interference must be constructive; in other words, the central maximum appears at P''. The path difference $\overline{S_1P''} - \overline{S_2P''}$ accounts for the displacement $\overline{P'P''}$. Consequently, S'' produces a fringe system identical to that of S' but shifted by an amount $\overline{P'P''}$ with respect to it. Since these source points are incoherent, their irradiances add on Σ_o rather than their field amplitudes (Fig. 12.3*e*).

The pattern arising from a broad source having a rectangular aperture of width b can be determined by finding the irradiance due to an incoherent continuous line source parallel to $\overline{S_1S_2}$. Notice, in Fig. 12.3*b*, that the variable Y_0 describes the location of any point on the image of the source when the aperture screen is absent. With Σ_a in place, each differential element of the line source will contribute a fringe system centered about its own image point, a distance Y_0 from the origin on Σ_o. Moreover, its contribution to the flux-density pattern dI is proportional to the differential line element or, more conveniently, to its image, dY_0, on Σ_o. Thus, using Eq. (9.31), the contribution to the total irradiance arising from dY_0 becomes

$$dI = A\, dY_0 \cos^2\left[\frac{a\pi}{s\lambda}(Y - Y_0)\right] \tag{12.3}$$

where A is an appropriate constant. This, in analogy to Eq. (12.2), is the expression for an entire fringe system of minute irradiance centered at Y_0 contributed by the tiny piece of the source whose image corresponds to dY_0 at Y_0. By integrating over the extent w of the image of the line source, we effectively integrate over the source and get the entire pattern:

$$I(Y) = A\int_{-w/2}^{+w/2} \cos^2\left[\frac{a\pi}{s\lambda}(Y - Y_0)\right] dY_0 \tag{12.4}$$

After a good bit of straightforward trigonometric manipulation, this becomes

$$I(Y) = \frac{Aw}{2} + \frac{A}{2}\frac{s\lambda}{a\pi}\sin\left(\frac{a\pi}{s\lambda}w\right)\cos\left(2\frac{a\pi}{s\lambda}Y\right) \tag{12.5}$$

The irradiance oscillates about an average value of $\bar{I} = Aw/2$,

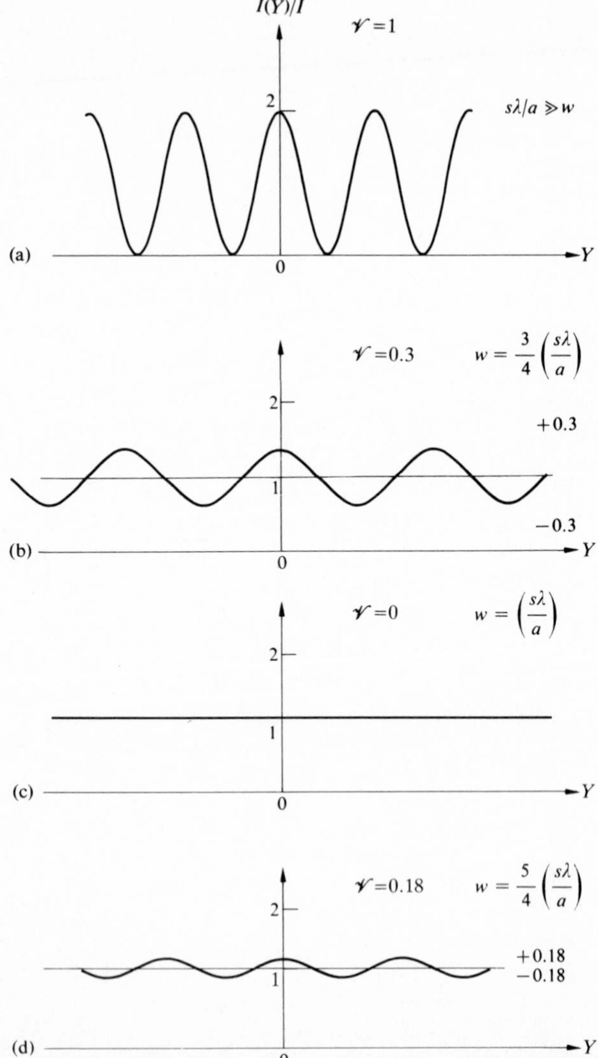

FIGURE 12.4 Fringes with varying source slit size. Here w is the width of the image of the slit, and $s\lambda/a$ is the peak-to-peak width of the fringes.

which increases with w, which in turn increases with the width of the source slit. Accordingly,

$$\frac{I(Y)}{\bar{I}} = 1 + \left(\frac{\sin a\pi w/s\lambda}{a\pi w/s\lambda}\right)\cos\left(2\frac{a\pi}{s\lambda}Y\right) \tag{12.6}$$

or

$$\frac{I(Y)}{\bar{I}} = 1 + \operatorname{sinc}\left(\frac{a\pi w}{s\lambda}\right)\cos\left(2\frac{a\pi}{s\lambda}Y\right) \tag{12.7}$$

It follows that the extreme values of the relative irradiance are given by

$$\frac{I_{max}}{\bar{I}} = 1 + \left|\operatorname{sinc}\left(\frac{a\pi w}{s\lambda}\right)\right| \tag{12.8}$$

and

$$\frac{I_{min}}{\bar{I}} = 1 - \left|\operatorname{sinc}\left(\frac{a\pi w}{s\lambda}\right)\right| \tag{12.9}$$

When w is very small in comparison to the fringe width $(s\lambda/a)$, the sinc function (p. 48) approaches 1 and $I_{max}/\bar{I} = 2$, while $I_{min}/\bar{I} = 0$ (see Fig. 12.4). As w increases, I_{min} begins to differ from zero, and the fringes lose contrast until they finally vanish entirely at $w = s\lambda/a$. Between the arguments of π and 2π (i.e., $w = s\lambda/a$ and $w = 2s\lambda/a$), the sinc is negative. As the primary slit source widens beyond $w = s\lambda/a$, the fringes reappear but shifted in phase; in other words, whereas previously there was a maximum at $Y = 0$, now there will be a minimum.

In actuality, the light diffracted by the apertures is localized (Section 10.2) so that the fringe system does not continue out uniformly indefinitely as Y increases. Instead, the pattern of Fig. 12.4*a* will look more like Fig. 12.5.

As a rule, the extent of the source (b) and the separation of the slits (a) are very small compared with the distances between the screens (l) and (s), and consequently we can make some simplifying approximations. While the above considerations were expressed in terms of w and s, it follows from Fig. 12.3*c*, using the central angle η, that $b \approx l\eta$ and $w \approx s\eta$; hence $w/s \approx b/l$. Accordingly, $(a\pi w/s\lambda) \approx (a\pi\eta/\lambda) \approx (a\pi b/l\lambda)$. The visibility of the fringes follows from Eq. (12.1):

$$\mathscr{V} = \left|\operatorname{sinc}\left(\frac{a\pi w}{s\lambda}\right)\right| = \left|\operatorname{sinc}\left(\frac{a\pi b}{l\lambda}\right)\right| \tag{12.10}$$

which is plotted in Fig. 12.6. Observe that $\mathscr{V}$ is a function of both the source breadth and the aperture separation a. Holding

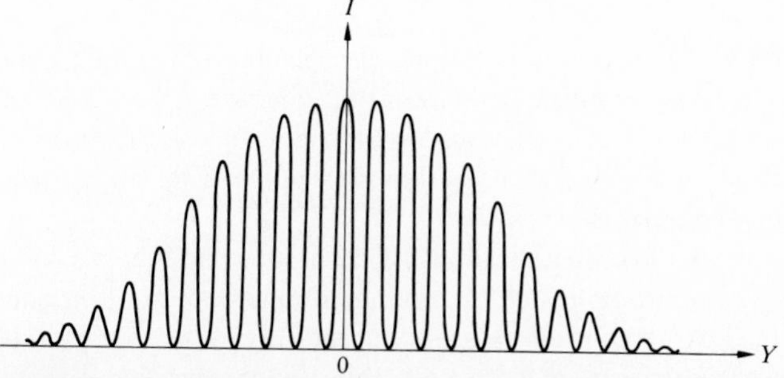

FIGURE 12.5 Double-beam interference fringes showing the effect of diffraction.

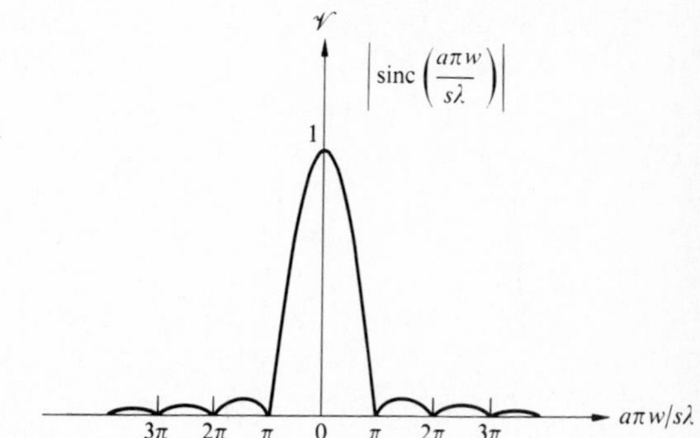

FIGURE 12.6 The visibility as given by Eq. (12.10).

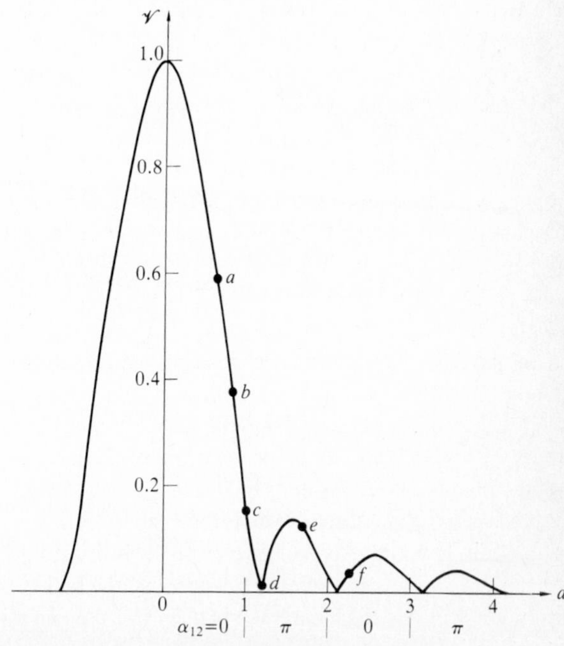

FIGURE 12.7 The visibility for a circular source.

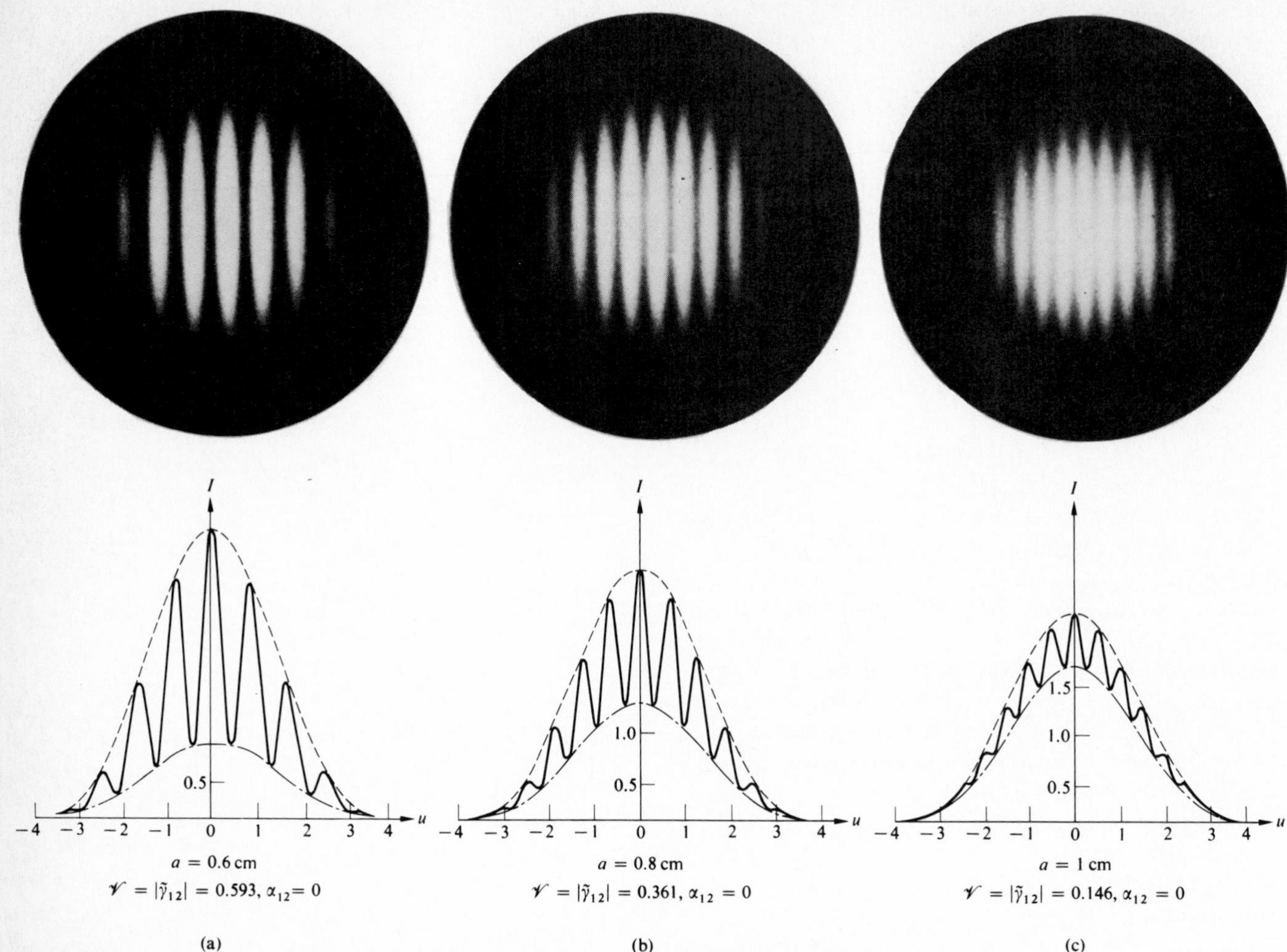

FIGURE 12.8 Double-beam interference patterns using partially coherent light. The photographs correspond to a variation in visibility associated with changes in *a*, the separation between the apertures. In the theoretical curves $I_{max} \propto 1 + |2J_1(u)/u|$ and $I_{min} \propto 1 - |2J_1(u)/u|$. Several of the symbols will be discussed later. [From B. J. Thompson and E. Wolf, *J. Opt. Soc. Am.* **47**, 895 (1957).]

either one of these parameters constant and varying the other will cause $\mathscr{V}$ to change in precisely the same way. Note that the visibilities in both Figs. 12.4*a* and 12.5 are equal to one, because $I_{min} = 0$. Clearly then, the visibility of the fringe system on the plane of observation is linked to the way the light is distributed over the aperture screen. If the primary source were in fact a point, *b* would equal zero, and the visibility would be a perfect 1. Shy of that, the smaller $(a\pi b/l\lambda)$ is, the better; that is, the bigger $\mathscr{V}$ is and the clearer the fringes are. We can think of $\mathscr{V}$ as a measure of the degree of coherence of the light from the primary source as spread over the aperture screen. Keep in mind that we have encountered the sinc function before, in connection with the diffraction pattern resulting from a rectangular aperture.

When the primary source is circular, the visibility is a good deal more complicated to calculate. It turns out to be proportional to a first-order Bessel function (Fig. 12.7). This too is quite reminiscent of diffraction, this time at a circular aperture

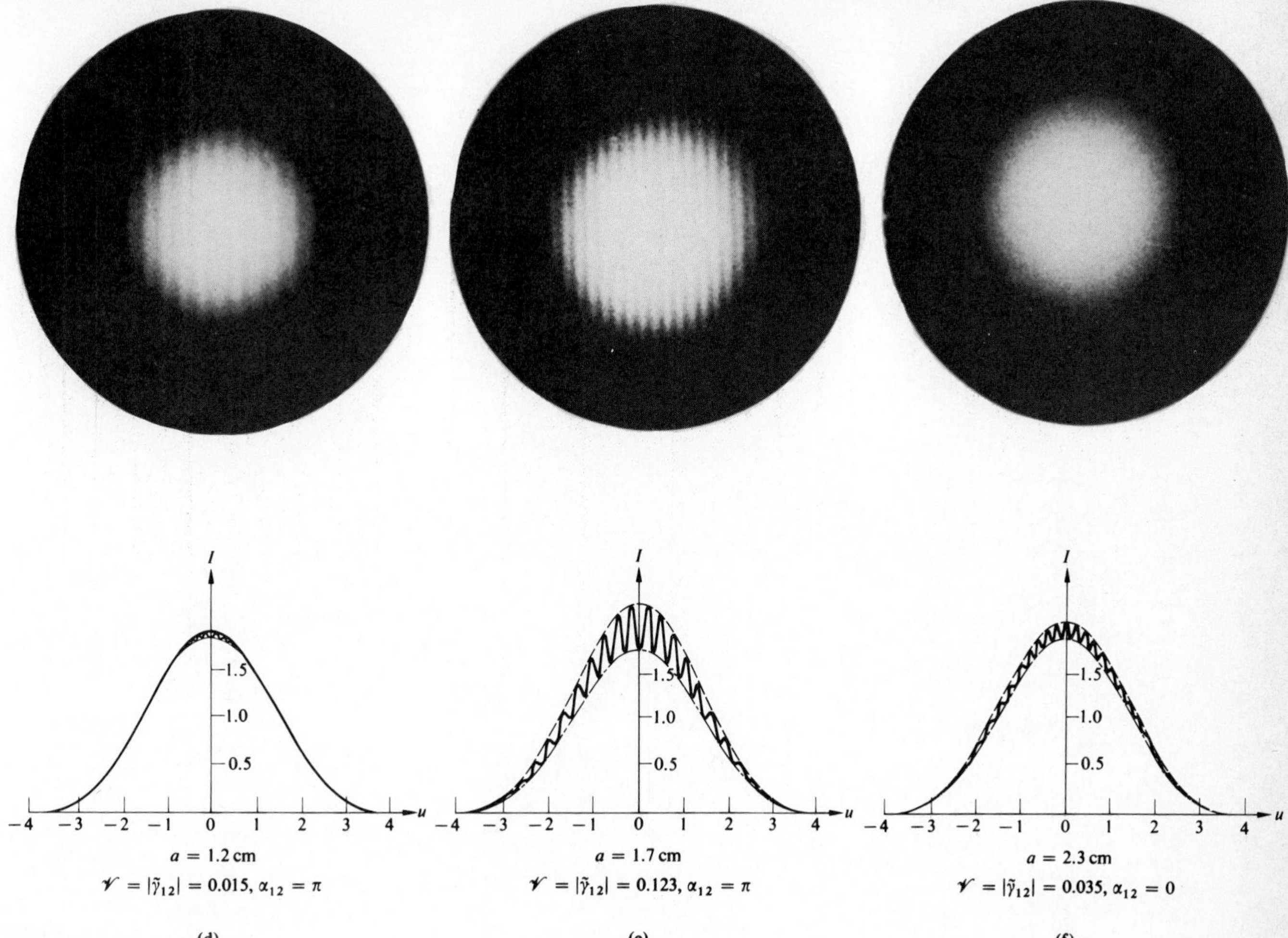

[Eq. (10.56)]. These similarities between expressions for $\mathcal{V}$ and the corresponding diffraction patterns for an aperture of the same shape are not merely fortuitous but rather are a manifestation of something called the van Cittert–Zernike theorem, as we will see presently.

Figure 12.8 shows a sequence of fringe systems in which the circular incoherent primary source is constant in size but the separation a between S_1 and S_2 is increased. The visibility decreases from (a) to (d) in the figure, then increases for (e) and decreases again at (f). All the associated $\mathcal{V}$-values are plotted in Fig. 12.7. Note the shift in the peaks, that is, the change in phase at the center of the pattern for each point on the second lobe of Fig. 12.7 (the Bessel function is negative over that range). In other words, (a), (b), and (c) have a central maximum, while (d) and (e) have a central minimum, and (f) on the third lobe is back to a maximum. In the same way, for a slit source the domain where sinc $(a\pi w/s\lambda)$ in Eq. (12.7) is positive or negative will yield a maximum or minimum, respectively, in $I(0)/\bar{I}$. These in turn correspond to the odd or even lobes of the visibility curve of Fig. 12.6. Bear in mind that we could define a complex visibility of magnitude $\mathcal{V}$, having an argument corresponding to the phase shift—we'll come back to this idea later.

Since the width of the fringes is inversely proportional to a, the spatial frequency of the bright and dark bands increases accordingly from (a) to (f) in Fig. 12.8. Figure 12.9 results when the separation a is held constant while the primary incoherent source diameter is increased.

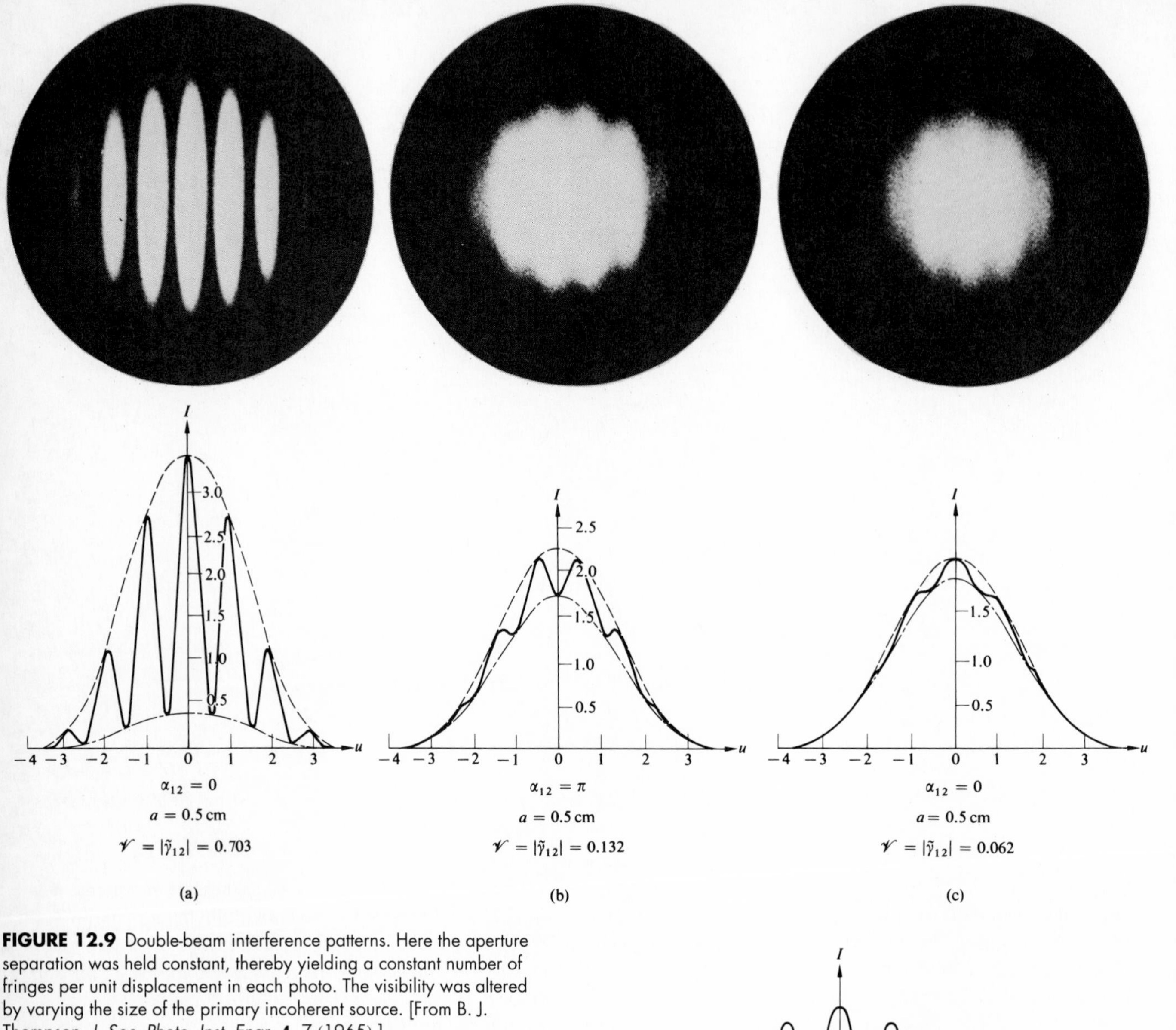

I
3.0
2.5
2.0
1.5
1.0
0.5
−4 −3 −2 −1 0 1 2 3
u
$\alpha_{12} = 0$
$a = 0.5$ cm
$\mathscr{V} = |\tilde{\gamma}_{12}| = 0.703$
(a)
I
2.5
2.0
1.5
1.0
0.5
−4 −3 −2 −1 0 1 2 3
u
$\alpha_{12} = \pi$
$a = 0.5$ cm
$\mathscr{V} = |\tilde{\gamma}_{12}| = 0.132$
(b)
I
1.5
1.0
0.5
−4 −3 −2 −1 0 1 2 3
u
$\alpha_{12} = 0$
$a = 0.5$ cm
$\mathscr{V} = |\tilde{\gamma}_{12}| = 0.062$
(c)

FIGURE 12.9 Double-beam interference patterns. Here the aperture separation was held constant, thereby yielding a constant number of fringes per unit displacement in each photo. The visibility was altered by varying the size of the primary incoherent source. [From B. J. Thompson, *J. Soc. Photo. Inst. Engr.* **4**, 7 (1965).]

We should also mention that the effects of the finite bandwidth will show up in a given fringe pattern as a gradually decreasing value of $\mathscr{V}$ with Y, as in Fig. 12.10 (see Problem 12.3). When the visibility is determined in these cases, using the central region of each of a series of patterns, the dependence of $\mathscr{V}$ on aperture separation will again match Fig. 12.7.

I
0
Y

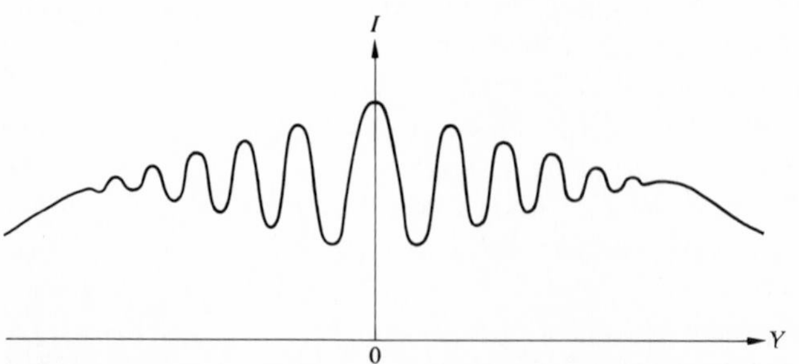

FIGURE 12.10 A finite bandwidth results in a decreasing value of $\mathscr{V}$ with increasing Y.

12.3 The Mutual Coherence Function and the Degree of Coherence

Let's now carry the discussion a bit further in a more formal fashion. Again suppose we have a broad, narrow bandwidth source, which generates a light field whose complex representation is $\tilde{E}(\mathbf{r}, t)$. We'll overlook polarization effects, and therefore a scalar treatment will do. The disturbances at two points in space S_1 and S_2 are then $\tilde{E}(S_1, t)$ and $\tilde{E}(S_2, t)$ or, more succinctly, $\tilde{E}_1(t)$ and $\tilde{E}_2(t)$. If these two points are then isolated using an opaque screen with two circular apertures (Fig. 12.11), we're back to Young's Experiment. The two apertures serve as sources of secondary wavelets, which propagate out to some point P on Σ_o. There the resultant field is

$$\tilde{E}_P(t) = \tilde{K}_1\tilde{E}_1(t - t_1) + \tilde{K}_2\tilde{E}_2(t - t_2) \qquad (12.11)$$

where $t_1 = r_1/c$ and $t_2 = r_2/c$. This says that the field at the space–time point (P, t) can be determined from the fields that existed at S_1 and S_2 at t_1 and t_2, respectively, these being the instants when the light, which is now overlapping, first emerged from the apertures. The quantities $\tilde{K}_1$ and $\tilde{K}_2$, which are known as *propagators*, depend on the size of the apertures and their relative locations with respect to P. They mathematically affect the alterations in the field resulting from its having traversed either of the apertures. For example, the secondary wavelets issuing from the pinholes in this setup are out-of-phase by $\pi/2$ rad, with the primary wave incident on the aperture screen, Σ_a (Section 10.3.1). Clearly someone is going to have to tell $\tilde{E}(\mathbf{r}, t)$ to shift phase beyond Σ_a—that's just what the $\tilde{K}$ factors are for. Moreover, they reflect a reduction in the field that might arise from a number of physical causes: absorption, diffraction, and so forth. Here, since there is a $\pi/2$ phase shift in the field, which can be introduced by multiplying by exp $i\pi/2$, $\tilde{K}_1$ and $\tilde{K}_2$ are purely imaginary numbers.

The resultant irradiance at P measured over some finite time interval, which is long compared with the coherence time, is

$$I = \langle \tilde{E}_P(t)\tilde{E}_P^*(t)\rangle_T \qquad (12.12)$$

It should be remembered that Eq. (12.12) is written sans several multiplicative constants. Hence using Eq. (12.11),

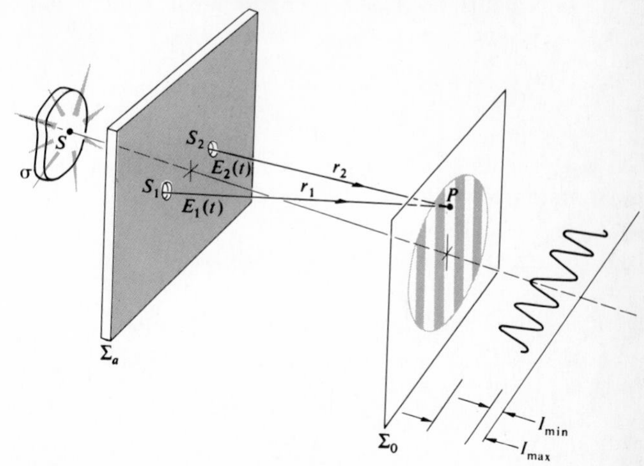

FIGURE 12.11 Young's Experiment.

$$\begin{aligned} I = {} & \tilde{K}_1\tilde{K}_1^*\langle \tilde{E}_1(t - t_1)\tilde{E}_1^*(t - t_1)\rangle_T \\ & + \tilde{K}_2\tilde{K}_2^*\langle \tilde{E}_2(t - t_2)\tilde{E}_2^*(t - t_2)\rangle_T \\ & + \tilde{K}_1\tilde{K}_2^*\langle \tilde{E}_1(t - t_1)\tilde{E}_2^*(t - t_2)\rangle_T \\ & + \tilde{K}_1^*\tilde{K}_2\langle \tilde{E}_1^*(t - t_1)\tilde{E}_2(t - t_2)\rangle_T \end{aligned} \qquad (12.13)$$

It is now assumed that the wave field is *stationary*, as is almost universally the case in classical Optics; in other words, it does not alter its statistical nature with time, so that the time average is independent of whatever origin we select. Thus, even though there are fluctuations in the field variables, the time origin can be shifted, and the averages in Eq. (12.13) will be unaffected. The particular moment over which we decide to measure I shouldn't matter. Accordingly, the first two time averages can be rewritten as

$$I_{S_1} = \langle \tilde{E}_1(t)\tilde{E}_1^*(t)\rangle_T \quad \text{and} \quad I_{S_2} = \langle \tilde{E}_2(t)\tilde{E}_2^*(t)\rangle_T$$

where the origin was displaced by amounts t_1 and t_2, respectively. Here the subscripts underscore the fact that these are the irradiances at points S_1 and S_2. Furthermore, if we let $\tau = t_2 - t_1$, we can shift the time origin by an amount t_2 in the last two terms of Eq. (12.13) and write them as

$$\tilde{K}_1\tilde{K}_2^*\langle \tilde{E}_1(t + \tau)\tilde{E}_2^*(t)\rangle_T + \tilde{K}_1^*\tilde{K}_2\langle \tilde{E}_1^*(t + \tau)\,\tilde{E}_2(t)\rangle_T$$

But this is a quantity plus its own complex conjugate and is therefore just twice its real part; that is, it equals

$$2 \text{ Re } [\tilde{K}_1 \tilde{K}_2^* \langle \tilde{E}_1(t + \tau) \tilde{E}_2^*(t) \rangle_T]$$

The $\tilde{K}$-factors are purely imaginary, and so $\tilde{K}_1 \tilde{K}_2^* = \tilde{K}_1^* \tilde{K}_2 = |\tilde{K}_1||\tilde{K}_2|$. The time-average portion of this term is a cross-correlation function [Section 11.3.4(iii)], which we denote by

$$\tilde{\Gamma}_{12}(\tau) \equiv \langle \tilde{E}_1(t + \tau) \tilde{E}_2^*(t) \rangle_T \tag{12.14}$$

and refer to as the **mutual coherence function** of the light field at S_1 and S_2. If we make use of all this, Eq. (12.13) takes the form

$$I = |\tilde{K}_1|^2 I_{S_1} + |\tilde{K}_2|^2 I_{S_2} + 2|\tilde{K}_1||\tilde{K}_2| \text{ Re} \tilde{\Gamma}_{12}(\tau) \tag{12.15}$$

The terms $|\tilde{K}_1|^2 I_{S_1}$ and $|\tilde{K}_2|^2 I_{S_2}$, if we again overlook multiplicative constants, are the irradiance at P arising when one or the other of the apertures is open alone, in other words, $\tilde{K}_2 = 0$ or $\tilde{K}_1 = 0$, respectively. Denoting these as I_1 and I_2, Eq. (12.15) becomes

$$I = I_1 + I_2 + 2|\tilde{K}_1||\tilde{K}_2| \text{ Re} \tilde{\Gamma}_{12}(\tau) \tag{12.16}$$

Note that when S_1 and S_2 are made to coincide, the mutual coherence function becomes

$$\tilde{\Gamma}_{11}(\tau) = \langle \tilde{E}_1(t + \tau) \tilde{E}_1^*(t) \rangle_T$$

or

$$\tilde{\Gamma}_{22}(\tau) = \langle \tilde{E}_2(t + \tau) \tilde{E}_2^*(t) \rangle_T$$

We can imagine that two wavetrains emerge from this coalesced source point and somehow pick up a relative phase delay proportional to τ. In the present situation τ becomes zero (since the optical path difference goes to zero), and these functions are reduced to the corresponding irradiances $I_{S_1} = \langle \tilde{E}_1(t) \tilde{E}_1^*(t) \rangle_T$ and $I_{S_2} = \langle \tilde{E}_2(t) \tilde{E}_2^*(t) \rangle_T$ on Σ_a. Hence

$$\Gamma_{11}(0) = I_{S_1} \quad \text{and} \quad \Gamma_{22}(0) = I_{S_2}$$

and these are called *self-coherence functions*. Thus

$$I_1 = |\tilde{K}_1|^2 \Gamma_{11}(0) \quad \text{and} \quad I_2 = |\tilde{K}_2|^2 \Gamma_{22}(0)$$

Keeping Eq. (12.16) in mind, observe that

$$|\tilde{K}_1||\tilde{K}_2| = \sqrt{I_1}\sqrt{I_2} \,/\, \sqrt{\Gamma_{11}(0)}\sqrt{\Gamma_{22}(0)}$$

Hence the normalized form of the mutual coherence function is defined as

$$\tilde{\gamma}_{12}(\tau) \equiv \frac{\tilde{\Gamma}_{12}(\tau)}{\sqrt{\Gamma_{11}(0)\Gamma_{22}(0)}} = \frac{\langle \tilde{E}_1(t + \tau) \; \tilde{E}_2^*(t) \rangle_T}{\sqrt{\langle |\tilde{E}_1|^2 \rangle \langle |\tilde{E}_2|^2 \rangle}} \tag{12.17}$$

and it's spoken of as the **complex degree of coherence**, for reasons that will be clear imminently. Equation (12.16) can then be recast as

$$I = I_1 + I_2 + 2\sqrt{I_1 I_2} \text{ Re} \, \tilde{\gamma}_{12}(\tau) \tag{12.18}$$

which is the *general interference law for partially coherent light.*

For quasimonochromatic light the phase-angle difference concomitant with the optical path difference is given by

$$\varphi = \frac{2\pi}{\bar{\lambda}}(r_2 - r_1) = 2\pi\bar{\nu}\tau \tag{12.19}$$

where $\bar{\lambda}$ and $\bar{\nu}$ are the mean wavelength and frequency. Now $\tilde{\gamma}_{12}(\tau)$ is a complex quantity expressible as

$$\tilde{\gamma}_{12}(\tau) = |\tilde{\gamma}_{12}(\tau)| e^{i\Phi_{12}(\tau)} \tag{12.20}$$

The phase angle of $\tilde{\gamma}_{12}(\tau)$ relates back to Eq. (12.14) and the phase angle between the fields. If we set $\Phi_{12}(\tau) = \alpha_{12}(\tau) - \varphi$, then

$$\text{Re } \tilde{\gamma}_{12}(\tau) = |\tilde{\gamma}_{12}(\tau)| \cos [\alpha_{12}(\tau) - \varphi]$$

Equation (12.18) is then expressible as

$$I = I_1 + I_2 + 2\sqrt{I_1 I_2} |\tilde{\gamma}_{12}(\tau)| \cos [\alpha_{12}(\tau) - \varphi] \tag{12.21}$$

It can be shown from Eq. (12.17) and the Schwarz inequality that $0 \leq |\tilde{\gamma}_{12}(\tau)| \leq 1$. In fact, a comparison of Eqs. (12.21) and (9.14), the latter having been derived for the case of complete coherence, makes it evident that if $|\tilde{\gamma}_{12}(\tau)| = 1$, I is the same as that generated by two *coherent* waves out-of-phase at S_1 and S_2 by an amount $\alpha_{12}(\tau)$. If at the other extreme $|\tilde{\gamma}_{12}(\tau)| = 0$, $I = I_1 + I_2$, there is no interference, and the two disturbances are said to be *incoherent*. When $0 < |\tilde{\gamma}_{12}(\tau)| < 1$ we have *partial coherence*, the measure of which is $|\tilde{\gamma}_{12}(\tau)|$ itself; this is known as the **degree of coherence**. In summary then,

$$|\tilde{\gamma}_{12}| = 1 \quad \text{coherent limit}$$

$$|\tilde{\gamma}_{12}| = 0 \quad \text{incoherent limit}$$

$$0 < |\tilde{\gamma}_{12}| < 1 \quad \text{partial coherence}$$

The basic statistical nature of the entire process must be underscored. Clearly $\tilde{\Gamma}_{12}(\tau)$ and, therefore, $\tilde{\gamma}_{12}(\tau)$ are the key quantities in the various expressions for the irradiance distribution; they are the essence of what we previously called the interference term. It should be pointed out that $\tilde{E}_1(t+\tau)$ and $\tilde{E}_2(t)$ are in fact two disturbances occurring at different points in both space and time. We anticipate, as well, that the amplitudes and phases of these disturbances will somehow fluctuate in time. If these fluctuations at S_1 and S_2 are completely independent, then $\tilde{\Gamma}_{12}(\tau) = \langle \tilde{E}_1(t+\tau)\tilde{E}_2^*(t)\rangle_T$ will go to zero, since $\tilde{E}_1$ and $\tilde{E}_2$ can be either positive or negative with equal likelihood, and their product averages to zero. In that case no correlation exists, and $\tilde{\Gamma}_{12}(\tau) = \tilde{\gamma}_{12}(\tau) = 0$. If the field at S_1 at a time $(t+\tau)$ were perfectly correlated with the field at S_2 at a time t, their relative phase would remain unaltered despite individual fluctuations. The time average of the product of the fields would certainly not be zero, just as it would not be zero even if the two were only slightly correlated.

Both $|\tilde{\gamma}_{12}(\tau)|$ and $\alpha_{12}(\tau)$ are slowly varying functions of τ in comparison to $\cos 2\pi\bar{\nu}\tau$ and $\sin 2\pi\bar{\nu}\tau$. In other words, as P is moved across the resultant fringe system, the point-by-point spatial variations in I are predominantly due to the changes in φ as $(r_2 - r_1)$ changes.

The maximum and minimum values of I occur when the cosine term in Eq. (12.21) is $+1$ and -1, respectively. The visibility at P (Problem 12.7) is then

$$\mathcal{V} = \frac{2\sqrt{I_1}\sqrt{I_2}}{I_1 + I_2}|\tilde{\gamma}_{12}(\tau)| \tag{12.22}$$

Perhaps the most common arrangement occurs when things are adjusted so that $I_1 = I_2$, whereupon

$$\mathcal{V} = |\tilde{\gamma}_{12}(\tau)| \tag{12.23}$$

That is, *the modulus of the complex degree of coherence is identical to the visibility of the fringes* (take another look at Fig. 12.8).

It is essential to realize that Eqs. (12.17) and (12.18) clearly suggest the way in which the real parts of $\tilde{\Gamma}_{12}(\tau)$ and $\tilde{\gamma}_{12}(\tau)$ can be determined from direct measurement. When the flux densities of two disturbances are adjusted to be equal, Eq. (12.23) provides an experimental means of obtaining $|\tilde{\gamma}_{12}(\tau)|$ from the resultant fringe pattern. Furthermore, the off-axis shift in the location of the central fringe (from $\varphi = 0$) is a measure of $\alpha_{12}(\tau)$, the apparent relative retardation of the phase of the disturbances at S_1 and S_2. Thus, measurements of the visibility and fringe position yield both the amplitude and phase of the complex degree of coherence.

By the way, it can be shown* that $|\tilde{\gamma}_{12}(\tau)|$ will equal 1 for all values of τ and any pair of spatial points, if and only if the optical field is strictly monochromatic, and therefore such a situation is unattainable. Moreover, a nonzero radiation field for which $|\tilde{\gamma}_{12}(\tau)| = 0$ for all values of τ and any pair of spatial points cannot exist in free space either.

12.3.1 Temporal and Spatial Coherence

Let's now relate the ideas of temporal and spatial coherence to the above formalism.

If the primary source S in Fig. 12.11 shrinks down to a point source on the central axis having a finite frequency bandwidth, temporal coherence effects will predominate. The optical disturbances at S_1 and S_2 will then be identical. In effect, the mutual coherence [Eq. (12.14)] between the two points will be the self-coherence of the field. Hence $\tilde{\Gamma}(S_1, S_2, \tau) = \tilde{\Gamma}_{12}(\tau) = \tilde{\Gamma}_{11}(\tau)$ or $\tilde{\gamma}_{12}(\tau) = \tilde{\gamma}_{11}(\tau)$. The same thing obtains when S_1 and S_2 coalesce, and $\tilde{\gamma}_{11}(\tau)$ is sometimes referred to as the **complex degree of temporal coherence** at that point for two instances of time separated by an interval τ. This would be the case in an amplitude-splitting interferometer, such as Michelson's, in which τ equals the path length difference divided by c. The expression for I, that is, Eq. (12.18), would then contain $\tilde{\gamma}_{11}(\tau)$ rather than $\tilde{\gamma}_{12}(\tau)$.

Suppose a lightwave is divided into two identical disturbances of the form

$$\tilde{E}(t) = E_0 e^{i\phi(t)} \tag{12.24}$$

*The proofs are given in Beran and Parrent, *Theory of Partial Coherence*, Section 4.2.

by an amplitude-splitting interferometer, which later recombines them to generate a fringe pattern. Then

$$\tilde{\gamma}_{11}(\tau) = \frac{\langle \tilde{E}(t+\tau)\tilde{E}^*(t)\rangle_T}{|\tilde{E}|^2} \tag{12.25}$$

or

$$\tilde{\gamma}_{11}(\tau) = \langle e^{i\phi(t+\tau)}e^{-i\phi(t)}\rangle_T$$

Hence

$$\tilde{\gamma}_{11}(\tau) = \lim_{T\to\infty}\frac{1}{T}\int_0^T e^{i[\phi(t+\tau)-\phi(t)]}\,dt \tag{12.26}$$

and

$$\tilde{\gamma}_{11}(\tau) = \lim_{T\to\infty}\frac{1}{T}\int_0^T (\cos\Delta\phi + i\sin\Delta\phi)\,dt$$

where $\Delta\phi = \phi(t+\tau) - \phi(t)$. For a strictly monochromatic plane wave of infinite coherence length, $\phi(t) = \mathbf{k}\cdot\mathbf{r} - \omega t$, $\Delta\phi = -\omega\tau$, and

$$\tilde{\gamma}_{11}(\tau) = \cos\omega\tau - i\sin\omega\tau = e^{-i\omega\tau}$$

Hence $|\tilde{\gamma}_{11}| = 1$; the argument of $\tilde{\gamma}_{11}$ is just $-2\pi\nu\tau$, and we have complete coherence. In contradistinction, for a quasimonochromatic wave where τ is greater than the coherence time, $\Delta\phi$ will be random, varying between 0 and 2π such that the integral averages to zero, $|\tilde{\gamma}_{11}(\tau)| = 0$, corresponding to complete incoherence. A path difference of 60 cm, produced when the two arms of a Michelson Interferometer differ in length by 30 cm, corresponds to a time delay between the recombining beams of $\tau \approx 2$ ns. This is roughly the coherence time of a good isotope discharge lamp, and the visibility of the pattern under this sort of illumination will be quite poor. If white light is used instead, $\Delta\nu$ is large, Δt_c is very small, and the coherence length is less than one wavelength. In order for τ to be less than Δt_c (i.e., in order that the visibility be good), the optical path difference will have to be a small fraction of a wavelength. The other extreme is laserlight, in which Δt_c can be so long that a value of $c\tau$ that will cause an appreciable decrease in visibility would require an impractically large interferometer.

We see that $\tilde{\Gamma}_{11}(\tau)$, being a measure of temporal coherence, must be intimately related to the coherence time and therefore the bandwidth of the source. Indeed, ***the Fourier transform of the self-coherence function,*** $\tilde{\Gamma}_{11}(\tau)$, ***is the power spectrum, which describes the spectral energy distribution of the light*** (Section 11.3.4).

If we go back to Young's Experiment (Fig. 12.11) with a very narrow-bandwidth extended source, spatial coherence effects will predominate. The optical disturbances at S_1 and S_2 will differ, and the fringe pattern will depend on $\tilde{\Gamma}(S_1, S_2, \tau) = \tilde{\Gamma}_{12}(\tau)$. By examining the region about the central fringe where $(r_2 - r_1) = 0$, $\tau = 0$ and $\tilde{\Gamma}_{12}(0)$ and $\tilde{\gamma}_{12}(0)$ can be determined. This latter quantity is the **complex degree of spatial coherence** of the two points at the same instant in time. $\tilde{\Gamma}_{12}(0)$ plays a central role in the description of the Michelson stellar interferometer to be discussed forthwith.

There is a very convenient relationship between the complex degree of coherence in a region of space and the corresponding irradiance distribution across the extended source giving rise to the light fields. We shall make use of that relationship, the **van Cittert–Zernike theorem**, as a calculational aid without going through its formal derivation. Indeed, the analysis of Section 12.2 already suggests some of the essentials. Figure 12.12 represents an extended quasimonochromatic incoherent source, S, located on the plane σ and having an irradiance given by $I(y, z)$. Also shown in an observation screen on which are two points, P_1 and P_2. These are at distances R_1 and R_2, respectively, from a tiny element of S. It is on this plane that we wish to determine $\tilde{\gamma}_{12}(0)$, which describes the correlation of the field vibrations at the two points. Note that although the source is incoherent, the light reaching P_1 and P_2 will generally be correlated to some degree, since each source element contributes to the field at each such point.

Calculation of $\tilde{\gamma}_{12}(0)$ from the fields at P_1 and P_2 results in an integral that has a familiar structure. The integral has the same form and will yield the same results as a well-known diffraction integral, provided we reinterpret each term appropriately. For instance, $I(y, z)$ appears in that coherence integral where an aperture function would be if it were, in fact, a diffraction integral. Thus, suppose that S is not a source but an aperture of identical size and shape, and suppose that $I(y, z)$ is not a description of irradiance, but instead its functional form corresponds to the field distribution across that aperture. In other words, imagine that there is a transparency at the aperture with amplitude transmission characteristics that correspond functionally to $I(y, z)$. Furthermore, imagine that the aperture is illuminated by a spherical wave converging toward the fixed point P_2 (see Fig. 12.12*b*), so that *there will be a diffraction pattern centered on* P_2. This diffracted field distribution, normalized to unity at P_2, is everywhere (i.e., at P_1) equal to the value of $\tilde{\gamma}_{12}(0)$ at that point. This is the van Cittert–Zernike theorem.

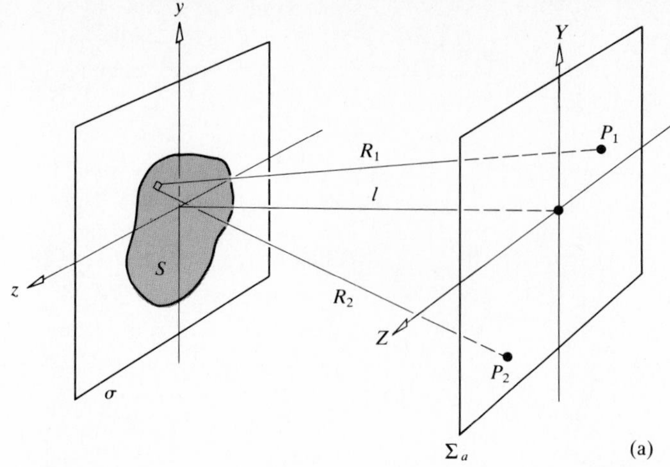

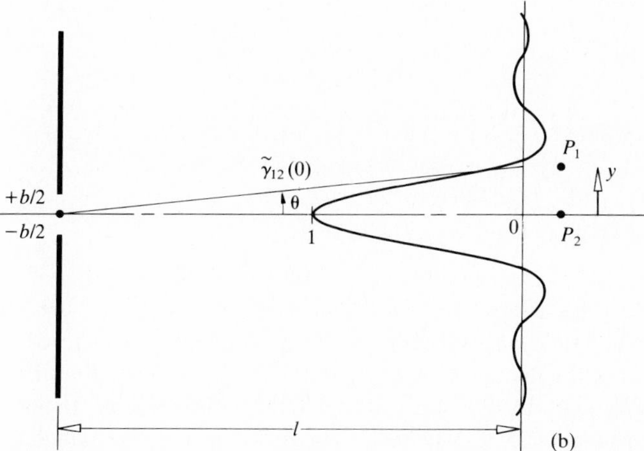

FIGURE 12.12 (*a*) The geometry of the van Cittert–Zernike theorem. (*b*) The normalized diffraction pattern corresponds to the degree of coherence. Here for a rectangular source slit the diffraction pattern is sinc $(\pi by/l\lambda)$.

When P_1 and P_2 are close together and S is small compared with l, the complex degree of coherence equals the normalized Fourier transform of the irradiance distribution across the source. Furthermore, if the source has a uniform irradiance, then $\tilde{\gamma}_{12}(0)$ is simply a sinc function when the source is a slit and a Bessel function when it's circular. Observe that in Fig. 12.12*b* the sinc function corresponds to that of Fig. 10.13, where $\beta = (kb/2)\sin\theta$ and $\theta \approx \sin\theta$. Thus if P_1 is a distance y from P_2, $\beta = kb\theta/2$ and $\theta = y/l$, hence $|\tilde{\gamma}_{12}(0)| = |\text{sinc}\,(\pi by/l\lambda)|$. This result is explored further in the problem set.

12.4 Coherence and Stellar Interferometry

12.4.1 The Michelson Stellar Interferometer

In 1890 A. A. Michelson, following an earlier suggestion by Fizeau, proposed an interferometric device (Fig. 12.13) that is of interest here both because it was the precursor of some important modern techniques, and because it lends itself to an interpretation in terms of coherence theory. The function of the *stellar interferometer*, as it is called, is to measure the small angular dimensions of remote astronomical bodies.

Two widely spaced movable mirrors, M_1 and M_2, collect rays, assumed to be parallel, from a very distant star. The light is then channeled via mirrors M_3 and M_4 through apertures S_1 and S_2 of a mask and thence into the objective of a telescope. The optical paths $M_1M_3S_1$ and $M_2M_4S_2$ are made equal, so that the relative phase-angle difference between a disturbance at M_1 and M_2 is the same as that between S_1 and S_2. The two apertures generate the usual Young's Experiment fringe system in the focal plane of the objective. Actually, the mask and openings are not really necessary; the mirrors alone could serve as apertures.

Suppose we now point the device so that its central axis is directed toward one of the stars in a closely spaced double-star configuration. Because of the tremendous distances involved, the rays reaching the interferometer from either star are well collimated. Furthermore, we assume, at least for the moment, that the light has a narrow linewidth centered about a mean wavelength of $\bar{\lambda}_0$. The disturbances arising at S_1 and S_2 from the axial star are in-phase, and a pattern of bright and dark bands forms, centered on P_0.

Similarly, rays from the other star arrive at some angle θ, but this time the disturbances at M_1 and M_2 (and therefore at S_1 and S_2) are out-of-phase by approximately $\bar{k}_0 h\theta$ or, if you will, retarded by a time $h\theta/c$, as indicated in Fig. 12.13*b*. The resulting fringe system is centered about a point P shifted by an angle θ' from P_0 such that $h\theta/c = a\theta'/c$. Since these stars behave as though they were incoherent point sources, the individual irradiance distributions simply overlap. The separation between the fringes set up by either star is equal and dependent solely on a. Yet the visibility varies with h. Thus, if h is increased from nearly zero until $\bar{k}_0 h\theta = \pi$, that is, until

$$h = \frac{\bar{\lambda}_0}{2\theta} \tag{12.27}$$

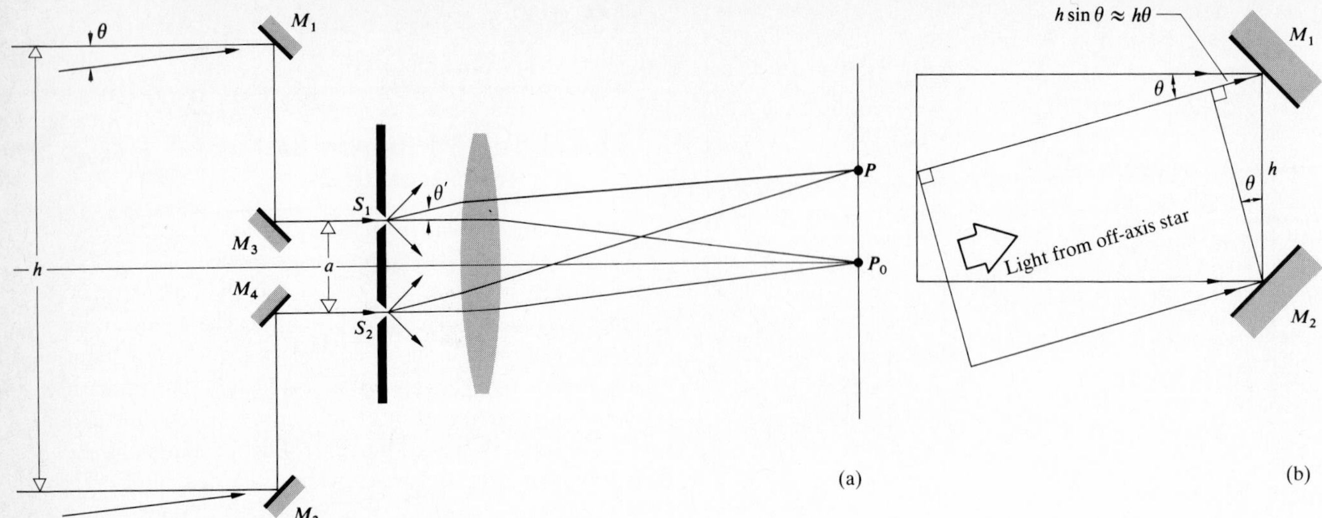

FIGURE 12.13 Michelson stellar interferometer.

the two fringe systems take on an increasing relative displacement, until finally the maxima from one star overlap the minima from the other, at which point, if their irradiances are equal, $\mathcal{V} = 0$. Hence, when the fringes vanish, one need only measure h to determine the angular separation between the stars, θ. Notice that the appropriate value of h varies inversely with θ.

Note that even though the source points, the two stars, are assumed to be completely uncorrelated, the resulting optical fields at any two points (M_1 and M_2) are not necessarily incoherent. For that matter, as h becomes very small, the light from each point source arrives with essentially zero relative phase at M_1 and M_2; $\mathcal{V}$ approaches 1, and the fields at those locations are highly coherent.

In much the same way as with a double star system, the angular diameter (θ) of certain single stars can be measured. Once again the fringe visibility corresponds to the degree of coherence of the optical field at M_1 and M_2. If the star is assumed to be a circular distribution of incoherent point sources such that it has a uniform brilliance, its visibility is equivalent to that already plotted in Fig. 12.7. Earlier, we alluded to the fact that $\mathcal{V}$ for this sort of source was given by a first-order Bessel function, and in fact it is expressible as

$$\mathcal{V} = |\tilde{\gamma}_{12}(0)| = 2\left|\frac{J_1(\pi h\theta/\bar{\lambda}_0)}{\pi h\theta/\bar{\lambda}_0}\right| \tag{12.28}$$

Recall that $J_1(u)/u = \frac{1}{2}$ at $u = 0$, and the maximum value of $\mathcal{V}$ is 1. The first zero of $\mathcal{V}$ occurs when $\pi h\theta/\bar{\lambda}_0 = 3.83$, as in Fig. 10.28. Equivalently, the fringes disappear when

$$h = 1.22\,\frac{\bar{\lambda}_0}{\theta} \tag{12.29}$$

and as before, one simply measures h to find θ.

In Michelson's arrangement, the two outrigged mirrors were movable on a long girder, which was mounted on the 100-inch reflector of the Mount Wilson Observatory. Betelgeuse (α Orionis) was the first star whose angular diameter was measured with the device. It's the orange-looking star in the upper left of the constellation Orion. In fact, its name is a contraction for the Arabic phrase meaning *the armpit of the central one* (i.e., Orion). The fringes formed by the interferometer, one cold December night in 1920, were made to vanish at $h = 121$ inches, and with $\bar{\lambda}_0 = 570$ nm, $\theta = 1.22(570 \times 10^{-9})/121(2.54 \times 10^{-2}) = 22.6 \times 10^{-8}$ rad, or 0.047 second of arc. Using its known distance, determined from parallax measurements, the star's diameter turned out to be about 240 million miles, or roughly 280 times that of the Sun. Actually, Betelgeuse is an irregular variable star whose maximum diameter is so tremendous that it's larger than the orbit of Mars about the Sun. The main limitation on the use of the stellar interferometer is due to the inconveniently long mirror separations required for all but the largest stars. This is true as well in radio astronomy, where an analogous setup has been widely used to measure the extent of celestial sources of radiofrequency emissions.

Incidentally, we assume, as is often done, that "good" coherence means a visibility of 0.88 or better. For a disk source this occurs when $\pi h\theta/\bar{\lambda}_0$ in Eq. (12.28) equals one, that is, when

$$h = 0.32 \frac{\bar{\lambda}_0}{\theta} \tag{12.30}$$

For a narrow-bandwidth source of diameter D a distance R away, there is an **area of coherence** equal to $\pi(h/2)^2$ over which $|\tilde{\gamma}_{12}| \geq 0.88$. Since $D/R = \theta$,

$$h = 0.32 \frac{R\bar{\lambda}_0}{D} \tag{12.31}$$

These expressions are very handy for estimating the required physical parameters in an interference or diffraction experiment. For example, if we put a red filter over a 1-mm-diameter disk-shaped flashlight source and stand back 20 m from it, then

$$h = 0.32(20)(600 \times 10^{-9})/10^{-3} = 3.8 \text{ mm}$$

where the mean wavelength is taken as 600 nm. This means that a set of apertures spaced at about h or less should produce nice fringes. Evidently, the area of coherence increases with R, and this is why you can always find a distant bright streetlight to use as a convenient source.

12.4.2 Correlation Interferometry

Let's return for a moment to the representation of a disturbance emanating from a thermal source, as discussed in Section 7.4.3. Here the word *thermal* connotes a light field arising predominantly from the superposition of spontaneously emitted waves issuing from a great many independent atomic sources.* A quasimonochromatic optical field can be represented by

$$E(t) = E_0(t) \cos\,[\varepsilon(t) - 2\pi\bar{\nu}t] \qquad [7.65]$$

The amplitude is a relatively slowly varying function of time, as is the phase. For that matter, the wave might undergo tens of thousands of oscillations before either the amplitude (i.e., the envelope of the field vibrations) or the phase would change appreciably. Thus, just as the coherence time is a measure of the fluctuation interval of the phase, it is also a measure of the interval over which $E_0(t)$ is fairly predictable. Large fluctuations in ε are generally accompanied by correspondingly large fluctuations of E_0. Presumably, knowledge of these amplitude fluctuations of the field could be related to the phase fluctuations and therefore to the correlation (i.e., coherence) functions. Accordingly, at two points in space–time where the phases of the field are correlated, we could expect the amplitudes to be related as well.

When a fringe pattern exists for the Michelson stellar interferometer, it is because the fields at M_1 and M_2, the apertures, are somehow correlated, that is, $\tilde{\Gamma}_{12}(0) = \langle \tilde{E}_1(t)\tilde{E}_2^*(t)\rangle_T \neq 0$. If we could measure the field amplitudes at these points, their fluctuations would likewise show an interrelationship. Since this isn't practicable because of the high frequencies involved, we might instead measure and compare the fluctuations in irradiance at the locations of M_1 and M_2 and from this, in some as yet unknown way, infer $|\tilde{\gamma}_{12}(0)|$. In other words, if there are values of τ for which $\tilde{\gamma}_{12}(\tau)$ is nonzero, the field at the two points is partially coherent, and a correlation between the irradiance fluctuations at these locations is implied. This is the essential idea behind a series of remarkable experiments conducted in the years 1952 to 1956 by R. Hanbury–Brown in collaboration with R. Q. Twiss and others. The culmination of their work was the so-called *correlation interferometer*.

Thus far we have evolved only an intuitive justification for the phenomenon rather than a firm theoretical treatment. Such an analysis, however, is beyond the scope of this discussion, and we shall have to content ourselves with merely outlining its salient features.* Just as in Eq. (12.14), we are interested in determining the cross-correlation function, this time, of the irradiances at two points in a partially coherent field, $\langle I_1(t + \tau)I_2(t)\rangle_T$. The contributing wavetrains, which are again represented by complex fields, are assumed to have been randomly emitted in accord with Gaussian statistics, with the final result

*Thermal light is sometimes spoken of as *Gaussian light*, because the amplitude of the field follows a Gaussian probability distribution (p.12).

*For a complete discussion, see, for example, L. Mandel, "Fluctuations of Light Beams," *Progress in Optics*, Vol. II, p. 193, or Françon, *Optical Interferometry*, p. 182.

that

$$\langle I_1(t+\tau)I_2(t)\rangle_T = \langle I_1\rangle_T\langle I_2\rangle_T + |\tilde{\Gamma}_{12}(\tau)|^2 \quad (12.32)$$

or

$$\langle I_1(t+\tau)I_2(t)\rangle_T = \langle I_1\rangle_T\langle I_2\rangle_T[1 + |\tilde{\gamma}_{12}(\tau)|^2] \quad (12.33)$$

The instantaneous irradiance fluctuations $\Delta I_1(t)$ and $\Delta I_2(t)$ are given by the variations of the instantaneous irradiances $I_1(t)$ and $I_2(t)$ about their mean values $\langle I_1(t)\rangle_T$ and $\langle I_2(t)\rangle_T$, as in Fig. 12.14. Consequently, if we use

$$\Delta I_1(t) = I_1(t) - \langle I_1\rangle_T \quad \text{and} \quad \Delta I_2(t) = I_2(t) - \langle I_2\rangle_T$$

and the fact that

$$\langle \Delta I_1(t)\rangle_T = 0 \quad \text{and} \quad \langle \Delta I_2(t)\rangle_T = 0$$

Eqs. (12.32) and (12.33) become

$$\langle \Delta I_1(t+\tau)\Delta I_2(t)\rangle_T = |\tilde{\Gamma}_{12}(\tau)|^2 \quad (12.34)$$

or

$$\langle \Delta I_1(t+\tau)\Delta I_2(t)\rangle_T = \langle I_1\rangle_T\langle I_2\rangle_T|\tilde{\gamma}_{12}(\tau)|^2 \quad (12.35)$$

(Problem 12.11). These are the desired cross-correlations of the irradiance fluctuations. They exist as long as the field is partially coherent at the two points in question. Incidentally, these expressions correspond to linearly polarized light. When the wave is unpolarized. a multiplicative factor of $\frac{1}{2}$ must be introduced on the right-hand side.

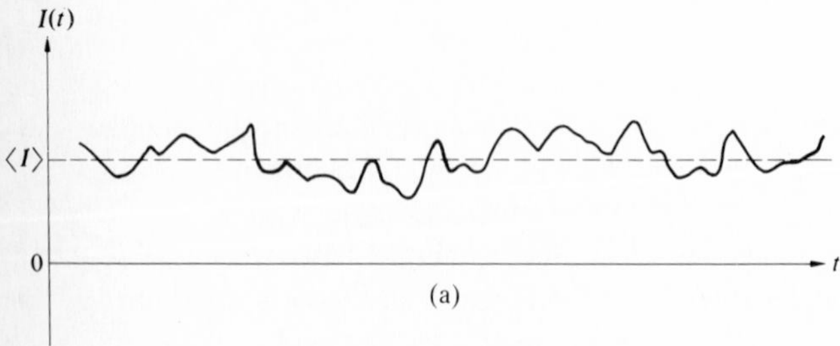

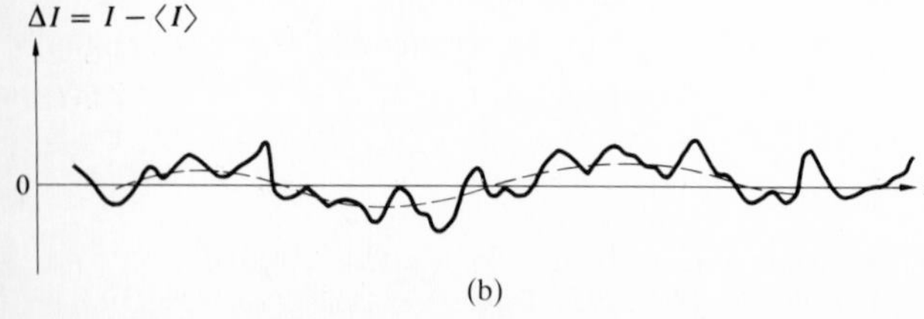

FIGURE 12.14 Irradiance variations.

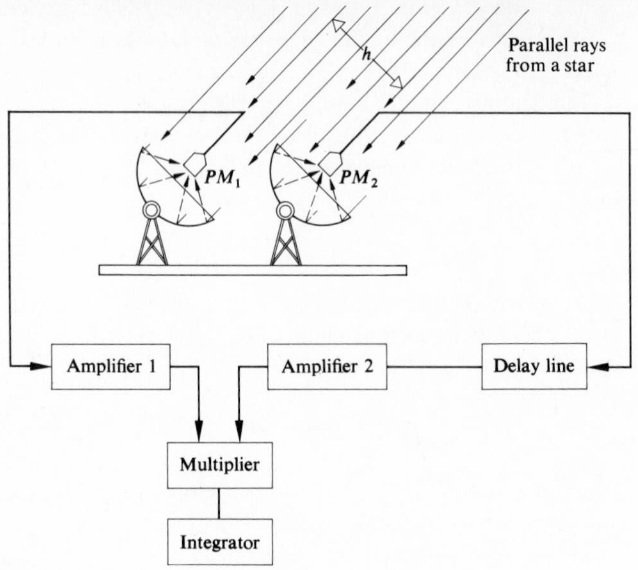

FIGURE 12.15 Stellar correlation interferometer.

The validity of the principle of correlation interferometry was first established in the radiofrequency region of the spectrum, where signal detection was a fairly straightforward matter. Soon afterward, in 1956, Hanbury–Brown and Twiss proposed the optical stellar interferometer illustrated in Fig. 12.15. But the only suitable detectors that could be used at optical frequencies were photoelectric devices whose very operation is keyed to the quantized nature of the light field. Thus

> ...it was by no means certain that the correlation would be fully preserved in the process of photo-electric emission. For these reasons a laboratory experiment was carried out as described below.*

That experiment is shown in Fig. 12.16. Filtered light from a Hg arc was passed through a rectangular aperture, and different portions of the emerging wavefront were sampled by two

*Taken from R. Hanbury–Brown and R. Q. Twiss, "Correlation Between Photons in Two Coherent Beams of Light," *Nature* **127**, 27 (1956).

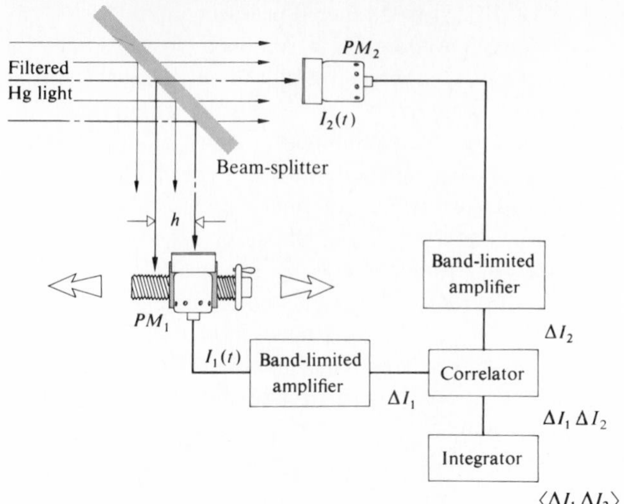

FIGURE 12.16 Hanbury–Brown and Twiss experiment.

photomultipliers, PM_1 and PM_2. The degree of coherence was altered by moving PM_1, that is, by varying h. The signals from the two photomultipliers were presumably proportional to the incident irradiances $I_1(t)$ and $I_2(t)$. These were then filtered and amplified, such that the steady, or dc, component of each of the signals (being proportional to $\langle I_1\rangle_T$ and $\langle I_2\rangle_T$) was removed, leaving only the fluctuations, in other words, $\Delta I_1(t) = I_1(t) - \langle I_1\rangle_T$ and $\Delta I_2(t) = I_2(t) - \langle I_2\rangle_T$. The two signals were then multiplied together in the correlator, and the time average of the product, which was proportional to $\langle \Delta I_1(t)\Delta I_2(t)\rangle_T$, was finally recorded. The values of $|\tilde{\gamma}_{12}(0)|^2$ for various separations, h, as deduced experimentally via Eq. (12.35), were in fine agreement with those calculated from theory. For the given geometry, the correlation definitely existed; moreover, it was preserved through photoelectric detection.

The irradiance fluctuations have a frequency bandwidth roughly equivalent to the bandwidth ($\Delta\nu$) of the incident light, in other words, $(\Delta t_c)^{-1}$, which is about 100 MHz or more. This is much better than trying to follow the field alternations at 10^{15} Hz. Even so, fast circuitry with roughly a 100-MHz pass bandwidth is required. In actuality the detectors have a finite resolving time T, so that the signal currents $\mathcal{I}_1$ and $\mathcal{I}_2$ are actually proportional to averages of $I_1(t)$ and $I_2(t)$ over T and not their instantaneous values. In effect, the measured fluctuations are smoothed out, as illustrated by the dashed curve of Fig. 12.14*b*. For $T > \Delta t_c$, which is normally the case, this just leads to a reduction, by a factor of $\Delta t_c/T$, in the correlation actually observed:

$$\langle \Delta\mathcal{I}_1(t)\Delta\mathcal{I}_2(t)\rangle = \langle\mathcal{I}_1\rangle\langle\mathcal{I}_2\rangle\,\frac{\Delta t_c}{T}\,|\tilde{\gamma}_{12}(0)|^2 \qquad (12.36)$$

For example, in the preceding laboratory arrangement, the filtered mercury light had a coherence time of about 1 ns, while the electronics had a reciprocal pass bandwidth or effective integration time of $\approx$40 ns. Note that Eq. (12.36) isn't any different conceptually from Eq. (12.35)—it's just been made a bit more realistic.

Shortly after their successful laboratory experiment, Hanbury–Brown and Twiss constructed the stellar interferometer shown in Fig. 12.15. Searchlight mirrors were used to collect starlight and focus it onto two photomultipliers. One arm contained a delay line, so that the mirrors could physically be located at the same height, with compensation for any differences in the arrival times of the light. The measurement of $\langle \Delta\mathcal{I}_1(t)\Delta\mathcal{I}_2(t)\rangle_T$ at various separations of the detectors allowed the square of the modulus of the degree of coherence, $|\tilde{\gamma}_{12}(0)|^2$, to be deduced, and this in turn yielded the angular diameter of the source, just as it did with the Michelson stellar interferometer. This time, however, the separation h could be very large, because one no longer had to worry about messing up the phase of the waves, as was the case in the Michelson device. There, a slight shift in a mirror of a fraction of a wavelength was fatal. Here, in contrast, the phase was discarded, so that the mirrors didn't even have to be of high optical quality. The star Sirius was the first to be examined, and it was found to have an angular diameter of 0.006 9 second of arc. More recently, a correlation interferometer with a baseline of 618 feet has been constructed at Narrabri, Australia. For certain stars, angular diameters of as little as 0.000 5 second of arc can be measured with this instrument—that's a long way from the angular diameter of Betelgeuse (0.047 second of arc).*

The electronics involved in irradiance correlation could be greatly simplified if the incident light were very nearly monochromatic and of considerably higher flux density. Laserlight isn't thermal and doesn't display the same statistical fluctuations, but it can nonetheless be used to generate *pseudother-*

*For a discussion of the photon aspects of irradiance correlation, see Garbuny, *Optical Physics*, Section 6.2.5.2, or Klein, *Optics*, Section 6.4.

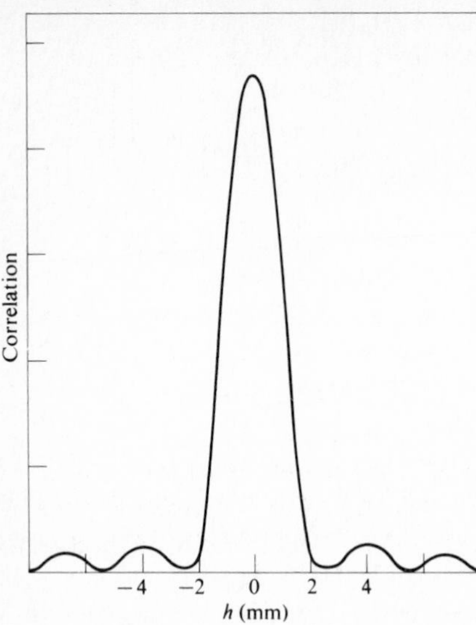

FIGURE 12.17 A correlation function for a pseudothermal source. [From A. B. Haner and N. R. Isenor, *Am J. Phys.* **38**, 748 (1970).]

*mal** light. A pseudothermal source is composed of an ordinary bright source (a laser is most convenient) and a moving medium of *nonuniform* optical thickness, such as a rotating ground glass disk. If the scattered beam emerging from a stationary piece of ground glass is examined with a *sufficiently slow detector*, the inherent irradiance fluctuations will be smoothed out completely. By setting the ground glass in motion, irradiance fluctuations appear with a simulated coherence time commensurate with the disk's speed. In effect, one has an extremely brilliant thermal source of variable Δt_c (from, say, 1 s to 10^{-5} s), which can be used to examine a whole range of coherence effects. For example, Fig. 12.17 shows the correlation function, which is proportional to $[2J_1(u)/(u)]^2$, for a pseudothermal circular aperture source determined from irradiance fluctuations. The experiment setup resembles that of Fig. 12.16, although the electronics is considerably simpler.*

*See W. Martienssen and E. Spiller, "Coherence and Fluctuations in Light Beams," *Am J. Phys.* **32**, 919 (1964), and A. B. Haner and N. R. Isenor, "Intensity Correlations from Pseudothermal Light Sources," *Am. J. Phys.* **38**, 748 (1970). Both of these articles are well worth studying.

*A good overall reference for this chapter is the review article by L. Mandel and E. Wolf, "Coherence Properties of Optical Fields," *Revs. Modern Phys.* **37**, 231 (1965); this is rather heavy reading. Take a look at K. I. Kellermann, "Intercontinental Radio Astronomy," *Sci. Am.* **226**, 72 (February 1972).

PROBLEMS

12.1 Suppose we set up a fringe pattern using a Michelson Interferometer with a mercury vapor lamp as the source. Switch on the lamp in your mind's eye and discuss what will happen to the fringes as the mercury vapor pressure builds to its steady-state value.

12.2* We wish to examine the irradiance produced on the plane of observation in Young's Experiment when the slits are illuminated simultaneously by two monochromatic plane waves of somewhat different frequency, E_1 and E_2. Sketch these against time, taking $\lambda_1 = 0.8\lambda_2$. Now draw the product E_1E_2 (at a point P) against time. What can you say about its average over a relatively long interval? What does $(E_1 + E_2)^2$ look like? Compare it with $E_1^2 + E_2^2$. Over a time that is long compared with the periods of the waves, approximate $\langle(E_1 + E_2)^2\rangle_T$.

12.3* With the previous problem in mind, now consider things spread across space at a given moment in time. Each wave separately would result in an irradiance distribution I_1 and I_2. Plot both on the same space axis and then draw their sum $I_1 + I_2$. Discuss the meaning of your results. Compare your work with Fig. 7.13. What happens to the net irradiance as more waves of different frequency are added in? Explain in terms of the coherence length. Hypothetically, what would happen to the pattern as the frequency bandwidth approached infinity?

12.4 With the previous problem in mind, return to the autocorrelation of a sine function, shown in Fig. 11.37. Now suppose we have a signal composed of a great many sinusoidal components. Imagine that you take the autocorrelation of this complicated signal and plot the result (use three or four components to start with), as in part (*e*) of Fig. 11.37. What will the autocorrelation function look like when the number of waves is very large and the signal resembles random noise? What is the significance of the $\tau = 0$ value? How does this compare with the previous problem?

12.5* Imagine that we have the arrangement depicted in Fig. 12.3. If the separation between fringes (max. to max.) is 1 mm and if the projected width of the source slit on the screen is 0.5 mm, compute the visibility.

12.6 Referring to the slit source and pinhole screen arrangement of Fig. P.12.6, show by integration over the source that

$$I(Y) \propto b + \frac{\sin (\pi a/\lambda l)b}{\pi a/\lambda l} \cos (2\pi aY/\lambda s)$$

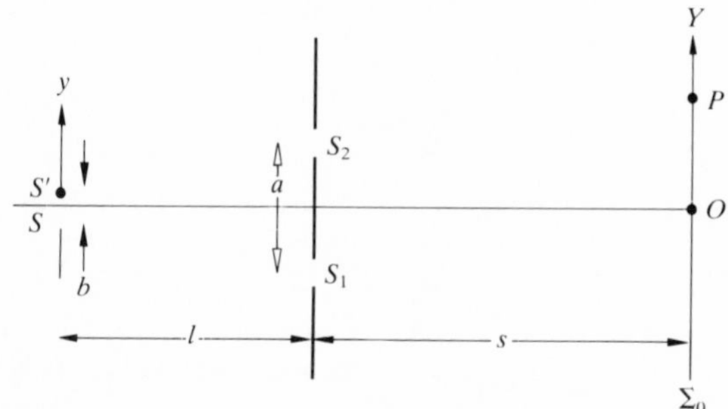

FIGURE P.12.6

12.7 Carry out the details leading to the expression for the visibility given by Eq. (12.22).

12.8 Under what circumstances will the irradiance on Σ_o in Fig. P.12.8 be equal to $4I_0$, where I_0 is the irradiance due to either incoherent point source alone?

12.9* Suppose we set up Young's Experiment with a small circular hole of diameter 0.1 mm in front of a sodium lamp ($\bar{\lambda}_0 = 589.3$ nm) as the source. If the distance from the source to the slits is 1 m, how far apart will the slits be when the fringe pattern disappears?

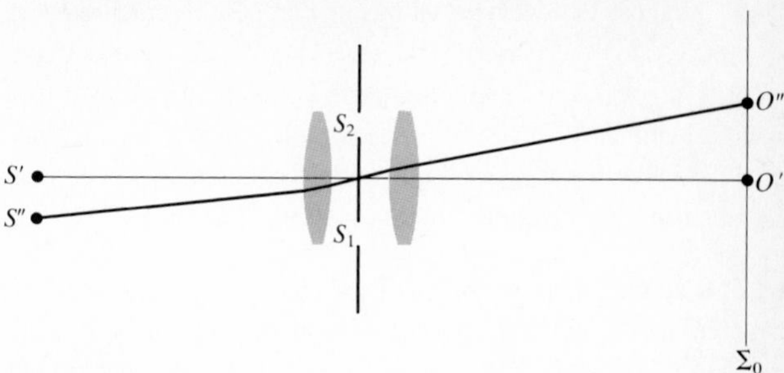

FIGURE P.12.8

12.10 Taking the angular diameter of the Sun viewed from the Earth to be about $1/2°$, determine the diameter of the corresponding area of coherence, neglecting any variations in brightness across the surface.

12.11 Show that Eqs. (12.34) and (12.35) follow from Eqs. (12.32) and (12.33).

12.12* Return to Eq. (12.21) and separate it into two terms representing a coherent and an incoherent contribution, the first arising from the superposition of two coherent waves with irradiances of $|\tilde{\gamma}_{12}(\tau)|I_1$ and $|\tilde{\gamma}_{12}(\tau)|\, I_2$ having relative phase of $\alpha_{12}(\tau) - \varphi$, and the second from the superposition of incoherent waves of irradiance $[1 - |\tilde{\gamma}_{12}(\tau)|]I_1$ and $[1 - |\tilde{\gamma}_{12}(\tau)|]I_2$. Now derive expressions for I_{coh}/I_{incoh} and for I_{incoh}/I_{total}. Discuss the physical significance of this alternative formulation and how we might view the visibility of fringes in terms of it.

12.13 Imagine that we have Young's Experiment, where one of the two pinholes is now covered by a neutral-density filter that cuts the irradiance by a factor of 10, and the other hole is covered by a transparent sheet of glass, so there is no relative phase shift introduced. Compute the visibility in the hypothetical case of completely coherent illumination.

12.14* Suppose that Young's double-slit apparatus is illuminated by sunlight with a mean wavelength of 550 nm. Determine the separation of the slits that would cause the fringes to vanish.

12.15 We wish to construct a double-pinhole setup illuminated by a uniform, quasimonochromatic, incoherent slit source of mean wavelength 500 nm and width b, a distance of 1.5 m from the aperture screen. If the pinholes are 0.50 mm apart, how wide can the source be if the visibility of the fringes on the plane of observation is not to be less than 85%?

12.16* Suppose that we have an incoherent, quasimonochromatic, uniform slit source, such as a discharge lamp with a mask and filter in front of it. We wish to illuminate a region on an aperture screen 10.0 m away, such that the modulus of the complex degree of coherence everywhere within a region 1.0 mm wide is equal to or greater than 90% when the wavelength is 500 nm. How wide can the slit be?

12.17* Figure P.12.17 shows two incoherent quasimonochromatic point sources illuminating two pinholes in a mask. Show that the fringes formed on the plane of observation have minimum visibility when

$$a(\alpha_2 - \alpha_1) = \tfrac{1}{2}m$$

where $m = \pm 1, \pm 3, \pm 5 \ldots.$

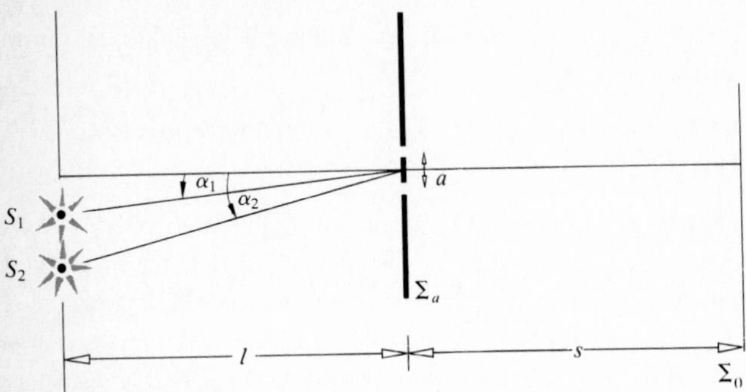

FIGURE P.12.17

12.18 Imagine that we have a wide quasimonochromatic source ($\lambda = 500$ nm) consisting of a series of vertical, incoherent, infinitesimally narrow line sources, each separated by 500 μm. This is used to illuminate a pair of exceedingly narrow vertical slits in an aperture screen 2.0 m away. How far apart should the apertures be to create a fringe system of maximum visibility?

13 Modern Optics: Lasers and Other Topics

13.1 LASERS AND LASERLIGHT

During the early 1950s a remarkable device known as the *maser* came into being through the efforts of a number of scientists. Principal among these people were Charles Hard Townes of the United States and Alexandr Mikhailovich Prokhorov and Nikolai Gennadievich Basov of the USSR, all of whom shared the 1964 Nobel Prize in Physics for their work. The maser, which is an acronym for Microwave Amplification by Stimulated Emission of Radiation, is, as the name implies, an extremely low-noise, microwave amplifier.* It functioned in what was then a rather unconventional way, making direct use of the quantum-mechanical interaction of matter and radiant energy. Almost immediately after its inception, speculation arose as to whether or not the same technique could be extended into the optical region of the spectrum. In 1958 Townes and Arthur L. Schawlow prophetically set forth the general physical conditions that would have to be met in order to achieve Light Amplification by Stimulated Emission of Radiation. And then in July of 1960 Theodore H. Maiman announced the first successful operation of an optical maser or **laser**—certainly one of the great milestones in the history of Optics, and indeed in the history of science, had been achieved.

The laser is a quantum-mechanical device that manages to produce its "marvelous light" by taking advantage of the subtle ways in which atoms interact with electromagnetic radiation. To gain a solid, if only introductory, understanding of how the laser works and what makes its emissions so special, we'll first lay out some basic theory about ordinary thermal sources, such as lightbulbs and stars. That will require an introduction to blackbody radiation, but those insights are also basic to any treatment of the interaction of EM-radiation and matter. To that will be added a discussion of the Boltzmann distribution (p. 580) as applied to atomic energy levels. With this to stand on, we can appreciate the central notion of stimulated emission via the Einstein A and B coefficients (p. 580); the rest, more or less, follows.

13.1.1 Radiant Energy and Matter in Equilibrium

It shouldn't surprise anyone that if physics was to be turned on its head, it would be done while trying to figure out what light (i.e., radiant energy) was all about. Quantum theory had its earliest beginnings back in 1859 with the study of a seemingly obscure phenomenon known as **blackbody radiation**. That year, Charles Darwin published *The Origin of Species*, and Gustav Robert Kirchhoff proffered an intellectual challenge that would lead to a revolution in physics.

Kirchhoff was involved in analyzing the way bodies in thermal equilibrium behave in the process of exchanging radiant energy. This *thermal radiation* is electromagnetic energy emitted by all objects, the source of which is the random motion of their constituent atoms. He characterized the abilities of a body to emit and absorb electromagnetic energy by an *emission coefficient* ε_λ and an *absorption coefficient* α_λ. Epsilon is the energy per unit area per unit time emitted in a tiny wavelength range around λ (in units of $\mathrm{W/m^2/m}$): thermal radiation comprises a wide range of frequencies, and an energy-measuring device by necessity admits a band of wavelengths. *Alpha is the fraction of the incident radiant energy absorbed per unit area per unit time in that wavelength range*; it's unitless. The emission and absorption coefficients depend on both the nature of the surface of the body (color, texture,

*See James P. Gorden, "The Maser," *Sci. Am.* **199**, 42 (December 1958).

etc.) and the wavelength—a body that emits or absorbs well at one wavelength may emit or absorb poorly at another.

Consider an isolated chamber of some sort in thermal equilibrium at a fixed temperature T. Presumably, it would be filled with radiant energy at a myriad of different wavelengths—think of a glowing furnace. Kirchhoff assumed there was some formula, or *distribution function* $I_\lambda(\lambda)$, which depends on T and which provides values of the ***energy per unit area per unit time at each wavelength***; call it the ***spectral flux density*** within the cavity (or ***spectral exitance*** when it leaves it). He concluded that the *total* amount of energy at all wavelengths being absorbed by the walls versus the amount emitted by them must be the same, or else T would change, and it doesn't. Furthermore, Kirchhoff argued that if the walls were made of different materials (which behave differently with T), that same balance would have to apply for *each* wavelength range individually. The energy absorbed at λ, namely, $\alpha_\lambda I_\lambda$, must equal the energy radiated, ε_λ, *and this is true for all materials no matter how different.* **Kirchhoff's Radiation Law** is therefore

$$\frac{\varepsilon_\lambda}{\alpha_\lambda} = I_\lambda \tag{13.1}$$

wherein the distribution I_λ, in units of $J/m^3 \cdot s$ or W/m^3, is a universal function the same for every type of cavity wall regardless of material, color, size, and shape and is only dependent on T and λ. That's quite extraordinary! Still, the British ceramist Thomas Wedgwood had commented long before (1792) that objects in a fired kiln all turned glowing red together along with the furnace walls, regardless of their size, shape, or material constitution.

Although Kirchhoff could not provide the energy distribution function in general, he did observe that a perfectly absorbing body, one for which $\alpha_\lambda = 1$, will appear black and, in that special case, $I_\lambda = \varepsilon_\lambda$. Moreover, the distribution function for a perfectly black object is the same as for an isolated chamber at that same temperature (visualize such a blackbody at equilibrium inside a hot oven). The radiant energy distribution at equilibrium within an isolated cavity is in every regard the same, "as if it came from a completely blackbody of the same temperature." Therefore ***the energy that would emerge from a small hole in the chamber should be identical to the radiation coming from a perfectly black object at the same temperature.***

The scientific community accepted the challenge of experimentally determining I_λ, but the technical difficulties were great and progress came slowly. The basic setup (Fig. 13.1*a*) is simple enough, although coming up with a reliable source was a daunting problem for a long time. Data must be extracted that is independent of the construction of the specific detector, and so the best thing to plot is the radiant energy per unit

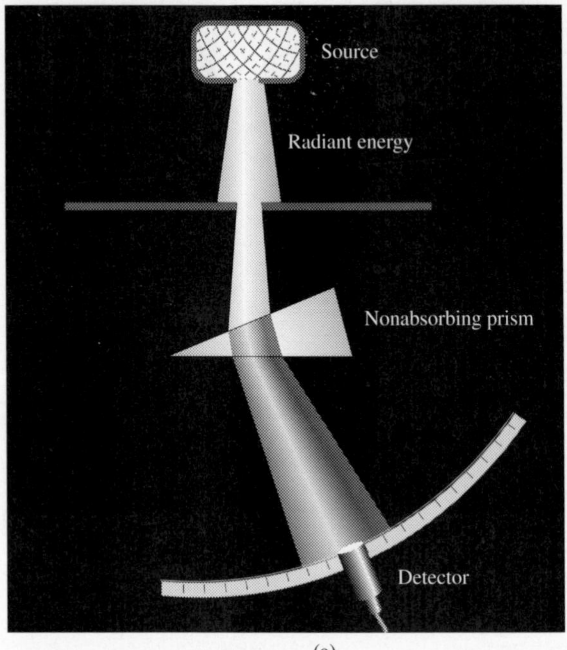

(a)

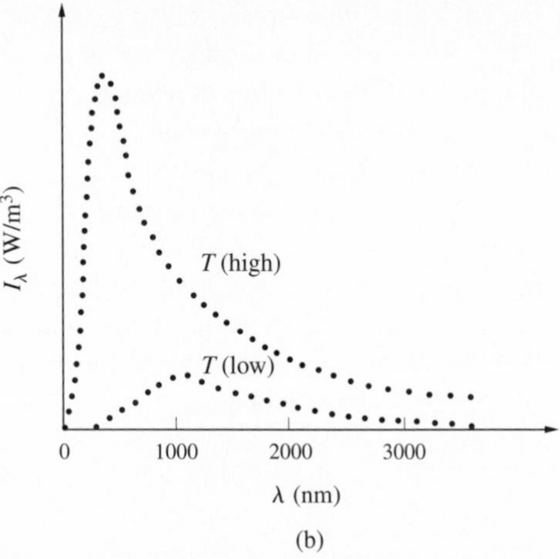

(b)

FIGURE 13.1 (*a*) A basic experimental setup for measuring blackbody radiation. (*b*) Values of I_λ at successive wavelengths as measured by a detector. Each curve corresponds to a specific source temperature.

time, which enters the detector per unit area (of the entrance window) per unit wavelength range (admitted by the detector). The kind of curves that were ultimately recorded are shown in Fig. 13.1*b*, and each is a plot of I_λ at a specific temperature.

Stefan-Boltzmann Law

In 1865 John Tyndall published some experimental results, including the determination that the total energy emitted by a heated platinum wire was 11.7 times greater when operating at 1200°C (1473 K) than it was at 525°C (798 K). Rather amazingly, Josef Stefan (1879) noticed that the ratio of $(1473\ \text{K})^4$ to $(798\ \text{K})^4$ was 11.6, nearly 11.7, and he surmised that the rate at which energy is radiated is proportional to T^4. In this observation Stefan was quite right (and quite lucky); Tyndall's results were actually far from those of a blackbody. Still, the conclusion was subsequently given a theoretical foundation by L. Boltzmann (1884). His was a traditional treatment of the radiation pressure exerted on a piston in a cylinder using the laws of thermodynamics and Kirchhoff's Law. The analysis progressed in much the same way one would treat a gas in a cylinder, but instead of atoms, the active agency was electromagnetic waves. The resulting Stefan–Boltzmann Law for blackbodies (which is correct, though nowadays we would derive it differently) is

$$P = \sigma A T^4 \tag{13.2}$$

where P is the total radiant power at all wavelengths, A is the area of the radiating surface, T is the absolute temperature in kelvins, and σ is a universal constant now given as

$$\sigma = 5.670\,33 \times 10^{-8}\ \text{W/m}^2\cdot\text{K}^4$$

The total area under any one of the blackbody-radiation curves of Fig. 13.1*b* for a specific T is the power per unit area, and from Eq. (13.2) that's just $P/A = \sigma T^4$.

Real objects are not perfect blackbodies; carbon black has an absorptivity of nearly one, but only at certain frequencies (obviously including the visible). Its absorptivity is much lower in the far infrared. Nonetheless, most objects resemble a blackbody (at least at certain temperatures and wavelengths)—you, for instance, are nearly a blackbody for infrared. Because of that, it's useful to write a similar expression for ordinary objects. This can be done by introducing a multiplicative factor called the total emissivity (ε), which relates the radiated power to that of a blackbody for which $\varepsilon = 1$, at the same temperature, thus

$$P = \varepsilon\sigma A T^4$$

Table 13.1 provides a few values of ε (at room temperature), where $0 < \varepsilon < 1$. Note that emissivity is unitless.

If an object with a *total absorptivity* of α is placed in an enclosure such as a cavity or a room having an emissivity ε_e and a temperature T_e, the body will radiate at a rate $\varepsilon\sigma A T^4$ and absorb energy inside the enclosure at a rate $\alpha(\varepsilon_e \sigma A T_e^4)$. Yet at any temperature at which the body and enclosure are in equilibrium (i.e., $T = T_e$), these rates must be equal; hence, $\alpha\varepsilon_e = \varepsilon$ and that has to be true for all temperatures. The net power radiated (when $T > T_e$) or absorbed (when $T < T_e$) by the body is then

$$P = \varepsilon\sigma A(T^4 - T_e^4)$$

All bodies not at zero kelvin radiate, and the fact that T is raised to the fourth power makes the radiation highly sensitive to temperature changes. When a body at 0°C (273 K) is brought up to 100°C (373 K), it radiates about 3.5 times the previous power. Increasing the temperature increases the net power radiated; that's why it gets more and more difficult to increase the temperature of an object. (Try heating a steel spoon to 1300°C.) Increasing the temperature of an object also shifts the emitted distribution of energy among the various wavelengths present. At the moment when the filament of a lightbulb "blows," the resistance, current, and temperature

TABLE 13.1 Some Representative Values of Total Emissivity*

Material	ε
Aluminum foil	0.02
Copper, polished	0.03
Copper, oxidized	0.5
Carbon	0.8
White paint, flat	0.87
Red brick	0.9
Concrete	0.94
Black paint, flat	0.94
Soot	0.95

*$T = 300$ K, room temperature.

rise; it goes from its normal operating reddish-white color to a bright flash of blue-white.

Wien Displacement Law

Perhaps the last notable success in applying classical theory to the problem of blackbody radiation came in 1893 at the hands of the German physicist and Nobel laureate Wilhelm Otto Fritz Franz Wien (1864–1928), known to his friends as Willy. He derived what is today called the **Displacement Law**. Each blackbody curve reaches a maximum height at a value of wavelength (λ_{max}) that is particular to it and therefore to the absolute temperature T. At that wavelength, the blackbody radiates the most energy. Wien was able to show that

$$\lambda_{max} T = \text{constant} \tag{13.3}$$

where the constant was found experimentally to be 0.002 898 m·K. The peak wavelength is inversely proportional to the temperature. *Raise the temperature, and the bulk of the radiation shifts to shorter wavelengths and higher frequencies* (see the dashed curve in Fig. 13.2). As a glowing coal or a blazing star gets hotter, it goes from IR warm to red-hot to blue-white. A person or a piece of wood, both only roughly blackbodies, radiates for the most part in the infrared and would only begin to glow faintly in the visible at round 600°C or 700°C, long after either had decomposed. The bright cherry red of a chunk of "red" hot iron sets in at around 1300°C.

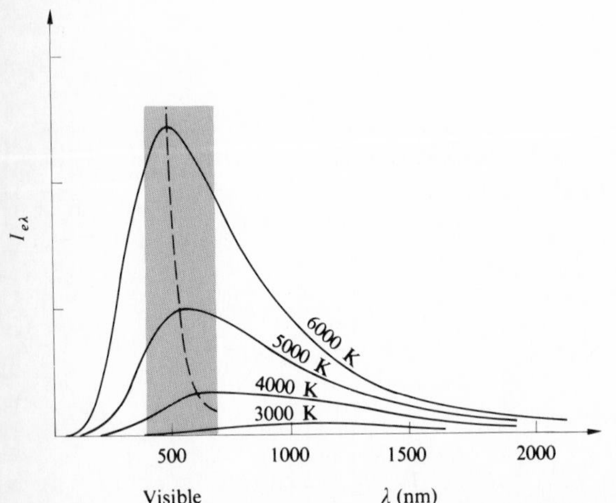

FIGURE 13.2 Blackbody radiation curves. The hyperbola passing through peak points corresponds to Wien's Law.

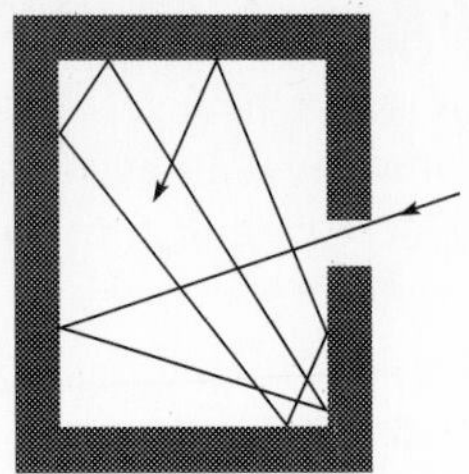

FIGURE 13.3 Radiant energy entering a tiny hole in a chamber will rattle around with little chance of ever emerging through the aperture and so the hole looks black. In reverse, the aperture of a heated chamber appears as a blackbody source.

In 1899 researchers greatly advanced the state of experimentation by using, as a source of blackbody radiation, a small hole in a heated cavity (Fig. 13.3). Energy entering such an aperture reflects around inside until it's absorbed. (The pupil of the eye appears black for precisely the same reason.) A near-perfect absorber is a near-perfect emitter, and the region of a small hole in the face of an oven is a wonderful source of *blackbody radiation*.

It was at this point in time that classical theory began to falter. All attempts to fit the entire radiation curve (Fig. 13.2) with some theoretical expression based on electromagnetism led only to the most limited successes. Wien produced a formula that agreed with the observed data fairly well in the short wavelength region but deviated from it substantially at large λ. Lord Rayleigh and later Sir James Jeans (1877–1946) developed a description in terms of the standing-wave modes of the field within the enclosure. But the resulting *Rayleigh–Jeans formula* matched the experimental curves only in the very long wavelength region. The failure of classical theory was totally inexplicable; a turning point in the history of physics had arrived.

Planck Radiation Law

Max Karl Ernst Ludwig Planck at 42 was the somewhat reluctant father of quantum theory. Like so many other theoreticians at the turn of the century, he, too, was working on blackbody radiation. But Planck would succeed not only in producing Kirchhoff's distribution function, but also in turning physics upside-down in the process. We cannot follow the details of his derivation here; besides, the original version was wrong. (Bose and Einstein corrected it years later.) Still, it had such a powerful impact that it's worth looking at some of the features that are right.

Planck knew that if an arbitrary distribution of energetic molecules was injected into a constant-temperature chamber, it would ultimately rearrange itself into the Maxwell–Boltzmann distribution of speeds as it inevitably reached equilibrium. Presumably, if an arbitrary distribution of radiant energy is injected into a constant-temperature cavity, it, too, will ultimately rearrange itself into the Kirchhoff distribution of energies as it inevitably reaches equilibrium.

In October 1900, Planck produced a distribution formula that was based on the latest experimental results. This mathematical contrivance, concocted "by happy guesswork," fit all the data available. It contained two fundamental constants, one of which (h) would come to be known as **Planck's Constant**. That much by itself was quite a success, even if it didn't explain anything. Although Planck had no idea of it at the time, he was about to take a step that would inadvertently revolutionize our perception of the physical Universe.

Naturally enough, Planck set out to construct a theoretical scheme that would logically lead to the equation he had already devised. He assumed that the radiation in a chamber interacted with simple microscopic oscillators of some unspecified type. These vibrated on the surfaces of the cavity walls, absorbing and reemitting radiant energy independent of the material. (In fact, the atoms of the walls do exactly that. Because of their tightly packed configuration in the solid walls, the atoms interact with a huge number of their neighbors. That completely blurs their usual characteristic sharp resonance vibrations, allowing them to oscillate over a broad range of frequencies and emit a continuous spectrum.) Try as he might, Planck was unsuccessful. At that time, he was a devotee of E. Mach, who had little regard for the reality of atoms, and yet the obstinate insolubility of the problem ultimately led Planck to "an act of desperation." He hesitantly turned to Boltzmann's "distasteful" statistical method, which had been designed to deal with the clouds of atoms that constitute a gas.

Boltzmann, the great proponent of the atom, and Planck were intellectual adversaries for a while. And now Planck was forced to use his rival's statistical analysis, which—ironically—he would misapply. If Boltzmann's scheme for counting atoms was to be applied to something continuous, such as energy, some adjustments would have to be made in the procedure. Thus, according to Planck, the total energy of the oscillators had to be thought of, at least temporarily, as apportioned into "energy elements" so that they could be counted. These energy elements were given a value proportional to the frequency ν of the resonators. Remember that he already had the formula he was after and in it there appeared the term $h\nu$. Planck's Constant,

$$6.626\,075\,5 \times 10^{-34}\ \mathrm{J\cdot s} \quad \text{or} \quad 4.135\,669\,2 \times 10^{-15}\ \mathrm{eV\cdot s}$$

is a very small number and so $h\nu$, which has the units of energy, is itself a very small quantity. Accordingly, he set the value of the energy element equal to it: $\varepsilon = h\nu$.

This was a statistical analysis, and counting was central. Still, when the method was applied as Boltzmann intended, it naturally smoothed out energy, making it continuous as usual. Again, we needn't worry about the details. The amazing thing was that Planck had stumbled on a hidden mystery of nature: **energy is quantized**—it comes in tiny bursts, but he didn't realize it then.

Planck derived the following formula for the spectral exitance (or spectral irradiance)—which he had already arrived at by fitting curves to the data—and it's the answer to Kirchhoff's challenge:

$$I_\lambda = \frac{2\pi hc^2}{\lambda^5}\left[\frac{1}{e^{\frac{hc}{\lambda k_B T}} - 1}\right] \tag{13.4}$$

where k_B is Boltzmann's Constant. This is **Planck's Radiation Law**, and, of course, it fit blackbody data splendidly (Fig. 13.4). Notice how the expression contains the speed of light,

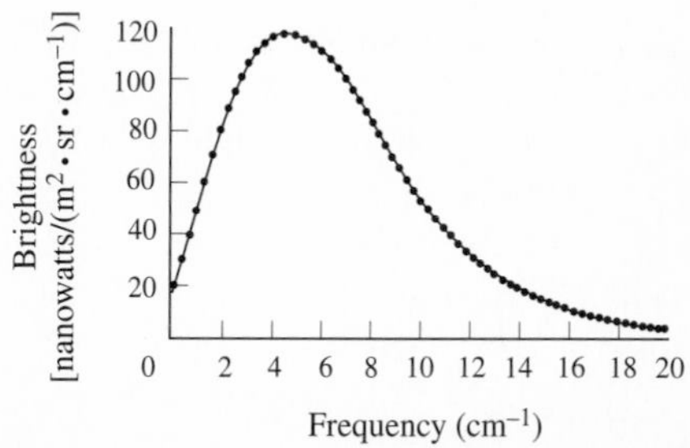

FIGURE 13.4 The cosmic background radiation of the Universe. Since the creation of the Universe with the Big Bang, it has expanded and cooled. The data points (measured in the microwave spectrum) were detected by the Cosmic Background Explorer (COBE) satellite. The solid line is the Planck blackbody curve for a temperature of 2.735 ± 0.06 K.

Boltzmann's Constant, and Planck's Constant (h). It bridges Electromagnetic Theory to the domain of the atom.

Although Eq. (13.4) represents a great departure from previous ideas, Planck did not mean to break with classical theory. It would have been unthinkable for him even to suggest that radiant energy was anything but continuous. "That energy is forced, at the outset, to remain together in certain quanta ...," Planck later remarked, "was purely a formal assumption and I really did not give it much thought." It was only around 1905, at the hands of a much bolder thinker, Albert Einstein, that we learned that the atomic oscillators were real and that their energies were quantized. Each oscillator could only exist with an energy that was a whole-number multiple of $h\nu$ (a little like the *gravitational*-PE of someone walking up a flight of stairs). Moreover, **radiant energy itself is quantized**, existing in localized blasts of an amount $\mathscr{E} = h\nu$.

13.1.2 Stimulated Emission

The LAser accomplishes "light amplification" by making use of energetic atoms in a medium to reinforce the light field. Let's therefore examine the manner in which the energy states of a system of atoms at some arbitrary temperature is normally distributed. The problem is part of the broader discipline of Statistical Mechanics and is addressed specifically in terms of the Maxwell–Boltzmann distribution.

Population of Energy Levels

Imagine a chamber filled with a gas in equilibrium at some temperature T. If T is relatively low, as it is in a typical room, most of the atoms will be in their ground states, but a few will momentarily pick up enough energy to "rise" into an excited state. The classical Maxwell–Boltzmann distribution maintains that, on average, a number of atoms per unit volume, N_i, will be in any excited state of energy $\mathscr{E}_i$ such that

$$N_i = N_0 e^{-\mathscr{E}_i/k_B T}$$

where N_0 is a constant for a given temperature. *The higher the energy state, that is, the greater the value of* $\mathscr{E}$ *(the smaller is the exponential) and the fewer atoms there will be in that state.*

Since we will be interested in atomic transition between arbitrary states, consider the jth energy level where $\mathscr{E}_j > \mathscr{E}_i$. Then for it $N_j = N_0 e^{-\mathscr{E}_j/k_B T}$, and the ratio of the populations occupying these two states is

$$\frac{N_j}{N_i} = \frac{e^{-\mathscr{E}_j/k_B T}}{e^{-\mathscr{E}_i/k_B T}} \tag{13.5}$$

This is the *relative population*, and it follows that

$$N_j = N_i e^{-(\mathscr{E}_j - \mathscr{E}_i)/k_B T} = N_i e^{-h\nu_{ji}/k_B T} \tag{13.6}$$

where use was made of the fact that a transition for the jth-state to the ith-state corresponds to an energy change of $(\mathscr{E}_j - \mathscr{E}_i)$ and since such transitions are accompanied by the emission of a photon of frequency ν_{ji}, we can substitute $(\mathscr{E}_j - \mathscr{E}_i) = h\nu_{ji}$.

The Einstein *A* and *B* Coefficients

In 1916 Einstein devised an elegant and rather simple theoretical treatment of the dynamic equilibrium existing for a material medium bathed in electromagnetic radiation, absorbing and reemitting. The analysis was used to affirm Planck's Radiation Law, but more importantly it also created the theoretical foundation for the laser. The reader should already be familiar with the basic mechanism of *absorption* (see Fig. 3.33, p. 65). Suppose the atom is in its lowest energy or ground state configuration. A photon having an adequate amount of energy interacts with the atom, imparting that energy to the atom, thereby causing the electron cloud to take on a new configuration. The atom jumps into a higher-energy excited state (Fig. 13.5). In a dense medium, the atom is likely to interact with its jiggling neighbors and pass off its bounty of energy via collisions.

Such an excess-energy configuration is usually (though not always) exceedingly short-lived, and in 10 ns or so, without the intercession of any external influence, the atom will emit its overload of energy as a photon. As it does, it reverts to a stable state in a process called *spontaneous emission* (Fig. 13.5*b*).

The remarkable thing is that there is a third alternative process, one first appreciated by Einstein and crucial to the operation of the laser—which wasn't invented until almost a half century later. For a medium inundated with EM-radiation, it's possible for a photon to interact with an excited atom while that atom is still in its higher-energy configuration. The atom can then dump its excess energy in-step with the incoming photon, in a process now called **stimulated emission** (Fig. 13.5).

In the case of absorption, the rate-of-change of the number of atoms in some initial state, as they leave to some higher state, must depend on the strength of the photon field inundating those atoms. In other words, it must depend on the energy density u, given by Eq. (3.34), but more specifically it must depend on the energy density in the frequency range driving the transition, that is, the spectral energy density u_ν, which has units of joules per meter-squared per inverse second ($\mathrm{J \cdot s/m^2}$). (Note that if we consider the radiation field as a photon gas, the spectral energy density can be thought of as the photon density per unit frequency range.) The rate-of-change of the number of atoms, the **transition rate**, will also be proportional to the population, that is, the number density of atoms in

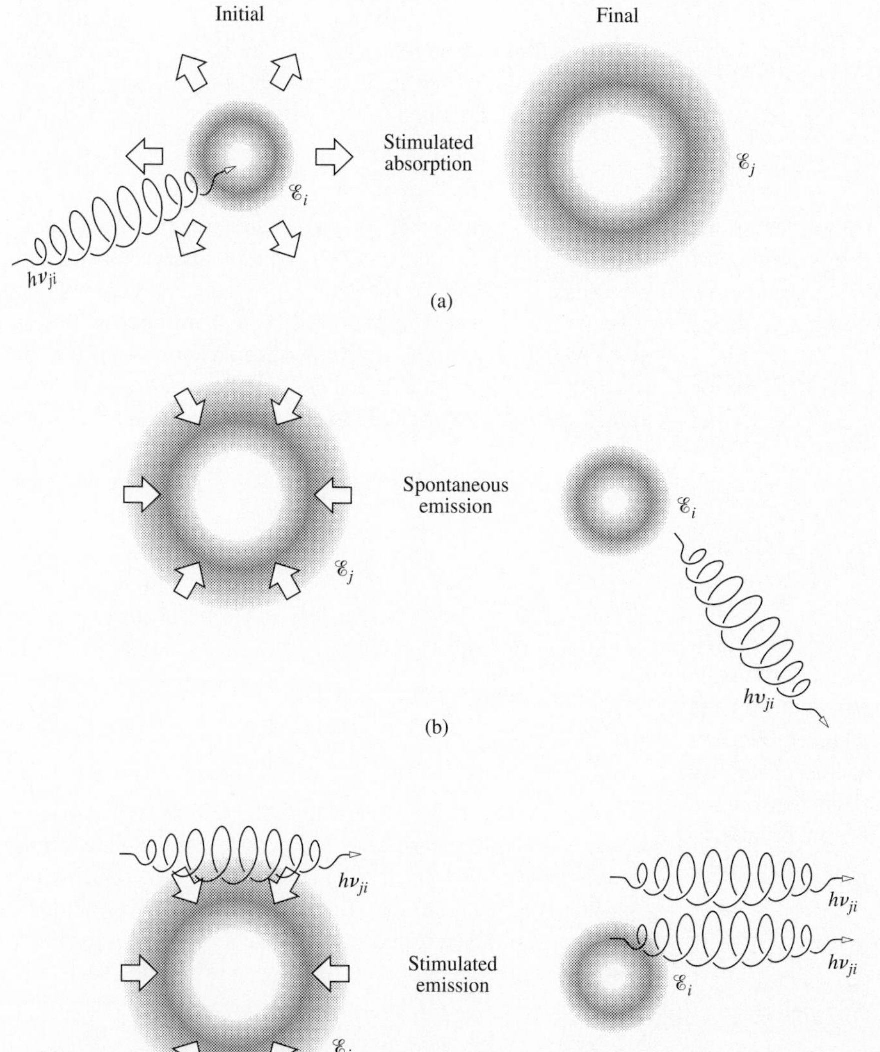

FIGURE 13.5 A schematic representation of (*a*) stimulated absorption, (*b*) spontaneous emission, and (*c*) stimulated emission.

that state (N_i); the more there are, the more can leave (via absorption) per second. Because the process is driven by the photon field, let's call it **stimulated absorption**, whereupon the transition rate is

[stimulated absorption] $$\left(\frac{dN_i}{dt}\right)_{\text{ab}} = -B_{ij}N_i u_\nu \tag{13.7}$$

Here B_{ij} is a constant of proportionality, and the minus arises because N_i is decreasing. Similarly, for stimulated emission

[stimulated emission] $$\left(\frac{dN_j}{dt}\right)_{\text{st}} = -B_{ji}N_j u_\nu \tag{13.8}$$

In the case of spontaneous emission, the process is independent of the field environment and

[spontaneous emission] $$\left(\frac{dN_j}{dt}\right)_{\text{sp}} = -A_{ji}N_j \tag{13.9}$$

Keep in mind that the transition rate, the number of atoms making transitions per second, divided by the number of atoms, is the probability of a transition occurring per second, $\mathcal{P}$. Consequently, the probability per second of spontaneous emission is $\mathcal{P}_{\text{sp}} = A_{ji}$.

For a single excited atom making a spontaneous transition to a lower state, the inverse of the transition probability per second is the **mean life** or **lifetime** of the excited state τ. Thus (operating under conditions that exclude any other mechanism but spontaneous emission), if N atoms are in that excited state, the total rate of transitions, that is, the number of emitted photons per second, is $N\mathcal{P}_{\text{sp}} = NA_{ji} = N/\tau$. A low-transition probability means a long lifetime.

The three constants A_{ji}, B_{ji}, and B_{ij} are Einstein's coefficients. Following his lead we assume (1) that thermodynamic equilibrium exists between the radiation field and the atoms in it at any T; (2) that the energy density has the characteristics of a blackbody at T; and (3) that the number densities of the two states are in accord with the Maxwell–Boltzmann distribution.

Given that the system is in equilibrium, the rate of upward ($i \rightarrow j$) transitions must equal the rate of downward transitions ($j \rightarrow i$):

$$B_{ij}N_i u_\nu = B_{ji}N_j u_\nu + A_{ji}N_j$$

Dividing both sides by N_i and rearranging terms yields

$$\frac{N_j}{N_i} = \frac{B_{ij}u_\nu}{A_{ji} + B_{ji}u_\nu}$$

Making use of Eq. (13.6), that is, what we found from the application of the Maxwell–Boltzmann distribution, this becomes

$$e^{-h\nu_{ji}/k_BT} = \frac{B_{ij}u_\nu}{A_{ji} + B_{ji}u_\nu}$$

and solving for u_ν leads to

$$u_\nu = \frac{A_{ji}/B_{ji}}{(B_{ij}/B_{ji})e^{h\nu_{ji}/k_BT} - 1} \tag{13.10}$$

Here Einstein pointed out that as $T \rightarrow \infty$, the spectral energy density, that is, the spectral photon density, approaches infinity. Figure 13.3 shows that I_λ increases with T, and that implies that u_ν will behave in a like fashion. In fact, $I_\nu = \frac{c}{4}u_\nu$, a point we will address presently. In any event, since $e^0 = 1$, the only way u_ν will be large is if

$$B_{ij} = B_{ji} = B$$

for large T, but since these constants are temperature independent, they must be equal at all T. The probabilities of stimulated emission and absorption are $\mathcal{P}_{\text{st}} = B_{ji}u_\nu$ and $\mathcal{P}_{\text{ab}} = B_{ij}u_\nu$, respectively. Hence, **the probability of stimulated emission is identical to the probability of stimulated absorption**; an atom in the lower state is just as likely to make a stimulated transition up, as an excited atom is to make a stimulated transition down.

Simplifying the notation (let $A = A_{ji}$), Eq. (13.10) becomes

$$u_\nu = \frac{A}{B}\left[\frac{1}{e^{h\nu_{ji}/k_BT} - 1}\right] \tag{13.11}$$

The ratio A/B can be expressed via basic quantities by comparing this equation with

$$I_\lambda = \frac{2\pi hc^2}{\lambda^5}\left[\frac{1}{e^{\frac{hc}{\lambda k_BT}} - 1}\right] \tag*{[13.4]}$$

But first transform I_λ into I_ν where these are expressions for exitance (which is irradiance going outward) per interval $d\lambda$ and $d\nu$, respectively. Using the fact that $\lambda = c/\nu$, differentiating yields $d\lambda = -cd\nu/\nu^2$. Because $I_\lambda\, d\lambda = I_\nu d\nu$, and dropping the sign (since it just says that one differential increases while the other decreases), we get $I_\lambda c/\nu^2 = I_\nu$; and so

$$I_\nu = \frac{2\pi h\nu^3}{c^2}\left[\frac{1}{e^{\frac{h\nu}{k_BT}} - 1}\right] \tag{13.12}$$

Now as a last step we need only to compare the spectral

energy density u_ν in the chamber with the spectral exitance,

$$I_\nu = \frac{c}{4} u_\nu \tag{13.13}$$

emerging from it. Rather than burden the reader with a complete derivation of this relationship, let it suffice merely to justify it. Keep in mind that I_ν corresponds to a flow of energy across a unit normal area, in one side and out the other—a beam leaving the chamber. In Section 3.3.1 we saw that the instantaneous flow of power per unit normal area, the Poynting vector, was given by $S = cu$ and so on average $I = cu$ for a beam. Inside a chamber, however, with light traveling in every direction, not all the photons that contribute to u will contribute to the exitance in a particular direction. Presumably, inside the chamber a unit area held horizontally would have as much energy flowing up through it as down. Moreover, only the components perpendicular to the area contribute to S, so a factor of $1/4$ is not unreasonable.

From Eqs. (13.11), (13.12), and (13.14) it follows that

$$\frac{A}{B} = \frac{8\pi h\nu^3}{c^3} \tag{13.14}$$

The probability of spontaneous emission is proportional to the probability of stimulated emission; an atom susceptible to one mechanism is proportionately susceptible to the other. Lasers work by stimulated emission, and anything that enhances spontaneous emission (i.e., A) at the price of stimulated emission (i.e., B) can be expected to work to the detriment of the process. Because the ratio of A/B varies as ν^3, it would seem that X-ray lasers ought to be very difficult to build—they are!

Imagine a system of atoms in thermal equilibrium having only two possible states. Furthermore, require that the atoms have a long mean life so that we can ignore spontaneous emission. When the system is inundated by photons of the proper energy, stimulated absorption depopulates the lower i-level, while stimulated emission depopulates the upper j-level. The number of photons vanishing from the system per second via stimulated absorption is proportional to $\mathcal{P}_{ab}N_i$, and the number entering it via stimulated emission is proportional to $\mathcal{P}_{st}N_j$, but from the equality of the B-coefficients it follows that $\mathcal{P}_{st} = \mathcal{P}_{ab}$. Therefore $\mathcal{P}_{ab}N_j = \mathcal{P}_{st}N_j$. However, if the system is in thermal equilibrium, $N_i > N_j$, which means that the number of photons vanishing per second exceeds the number entering per second; there's a net absorption of photons by the lower state because there are more atoms in the lower state an any given temperature. The reverse would be true if we could create a situation—a *population inversion*—in which $N_i < N_j$; then stimulated emission would dominate over stimulated absorption.

13.1.3 The Laser

Consider an ordinary medium in which a few atoms are in some excited state; call it $|j\rangle$ to conform with quantum-mechanical notation. If a photon in an incident beam is to trigger one of these excited atoms into stimulated emission, it must have the frequency ν_{ji}, as per Fig. 13.5*c*. A remarkable feature of this process is that ***the emitted photon is in-phase with, has the polarization of, and propagates in the same direction as, the stimulating radiation***. The emitted photon is said to be in the same radiation mode as the incident wave and tends to add to it, increasing its flux density. However, since most atoms are ordinarily in the ground state, absorption is usually far more likely than stimulated emission.

This raises an intriguing point: What would happen if a substantial percentage of the atoms could somehow be excited into an upper state, leaving the lower state all but empty? For obvious reasons this is known as **population inversion**. An incident photon of the proper frequency could then trigger an avalanche of stimulated photons—*all in-phase*. The initial wave would continue to build, so long as there were no dominant competitive processes (such as scattering) and provided the population inversion could be maintained. In effect, energy (electrical, chemical, optical, etc.) would be pumped in to sustain the inversion, and a beam of light would be extracted after sweeping across the *active medium*.

The First (Pulsed Ruby) Laser

To see how all of this is accomplished in practice, let's take a look at Maiman's original device (Fig. 13.6). The first operative laser had as its active medium a small, cylindrical, synthetic, pale pink ruby, that is, an Al_2O_3 crystal containing about 0.05 percent (by weight) of Cr_2O_3. Ruby, which is still one of the most common of the crystalline laser media, had been used earlier in maser applications and was suggested for use in the laser by Schawlow. The rod's end faces were polished flat, parallel and normal to the axis. Then both were silvered (one only partially) to form a **resonant cavity**.

It was surrounded by a helical gaseous discharge flashtube, which provided broadband **optical pumping**. Ruby appears red because the chromium atoms have absorption bands in the

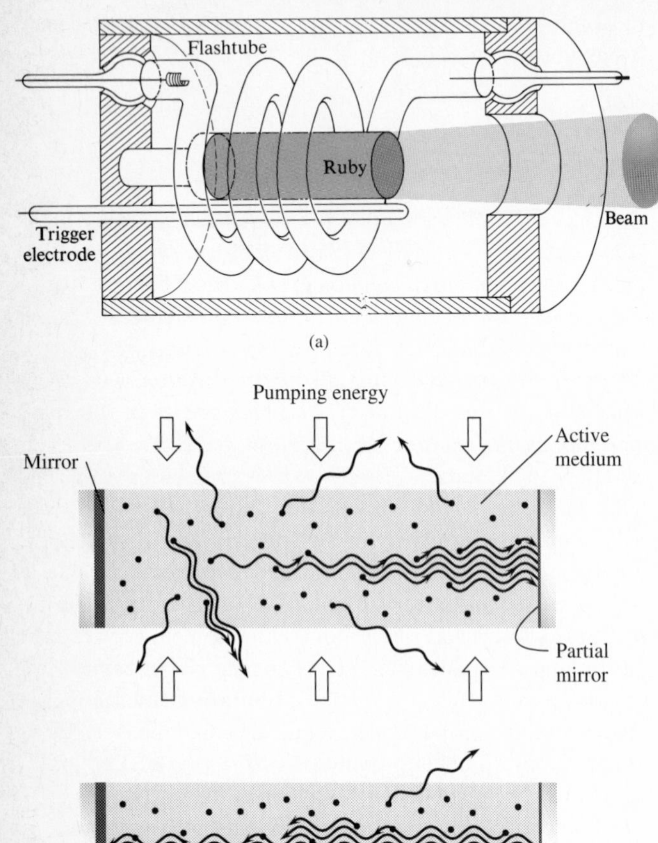

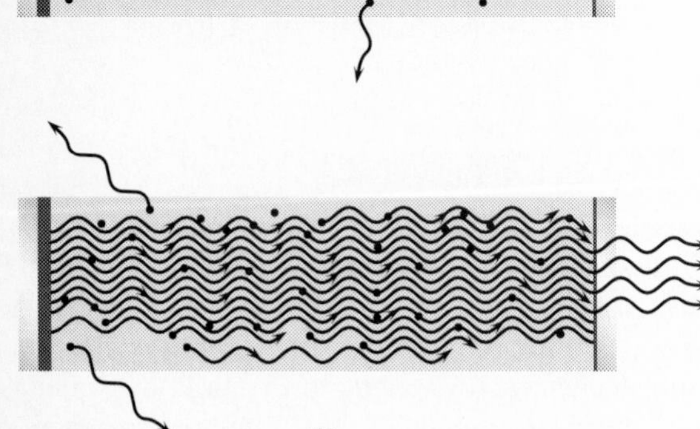

FIGURE 13.6 The first ruby-laser configuration, just about life-sized.

blue and green regions of the spectrum (Fig. 13.7*a*). Firing the flashtube generates an intense burst of light lasting a few milliseconds. Much of this energy is lost in heat, but many of the Cr^{3+} ions are excited into the absorption bands. A simplified energy-level diagram appears in Fig. 13.7*b*. The excited ions rapidly relax (in about 100 ns), giving up energy to the crystal lattice and making nonradiative transitions. They preferentially drop "down" to a pair of closely spaced, especially long-lived, interim states. They remain in these so-called **metastable states** for up to several milliseconds ($\approx$3 ms at room temperature) before randomly, and in most cases spontaneously, dropping down to the ground state. This is accompanied by the emission of the characteristic red fluorescent radiation of ruby. The lower-level transition dominates, and the resulting emission occurs in a relatively broad spectral range centered about 694.3 nm; it emerges in all directions and is incoherent.

When the pumping rate is increased somewhat, a popula-

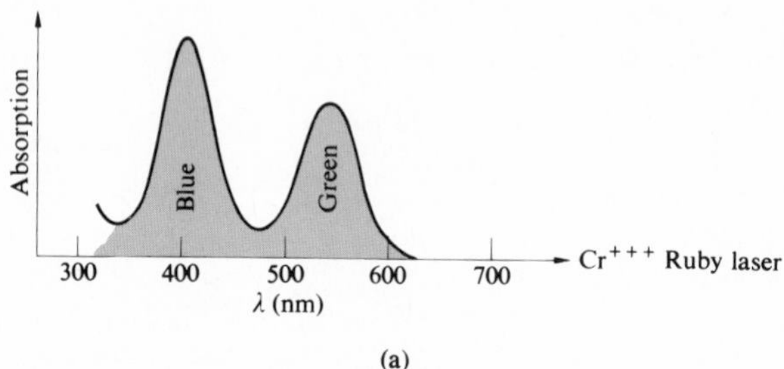

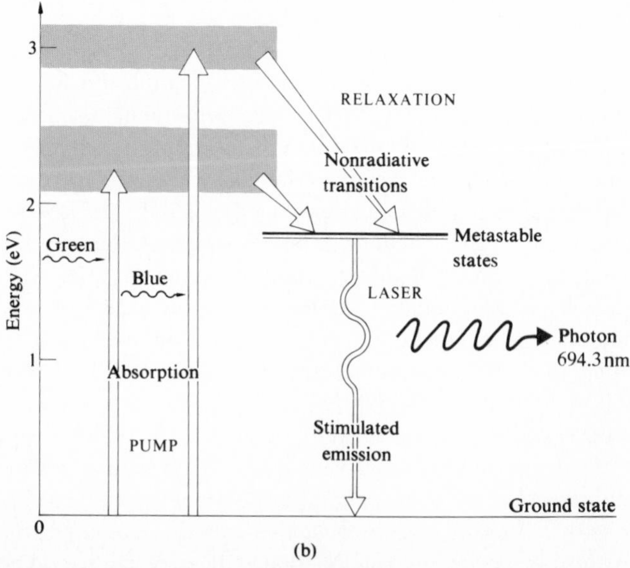

FIGURE 13.7 Ruby-laser energy levels.

tion inversion occurs, and the first few spontaneously emitted photons stimulate a chain reaction. One quantum triggers the rapid, in-phase emission of another, dumping energy from the metastable atoms into the evolving lightwave (Fig. 13.6*b*). The wave continues to grow as it sweeps back and forth across the active medium (provided enough energy is available to overcome losses at the mirrored ends). Since one of those reflecting surfaces was partially silvered, an intense pulse of red laser light (lasting about 0.5 ms and having a line-width of about 0.01 nm) emerges from that end of the ruby rod.

Notice how neatly everything works out. The broad absorption bands make the initial excitation rather easy, while the long lifetime of the metastable state facilitates the population inversion. The atomic system in effect consists of (1) the absorption bands, (2) the metastable state, and (3) the ground state. Accordingly, it is spoken of as a *three-level* laser.

Today's ruby laser is generally a high-power source of pulsed coherent radiation used extensively in work on interferometry, plasma diagnostics, holography, and so forth. Such devices operate with coherence lengths of from 0.1 m to 10 m. Modern configurations usually have flat external mirrors, one totally and the other partially reflecting. As an oscillator, the ruby laser generates millisecond pulses in the energy range from around 50 J to 100 J, but by using a tandem oscillator-amplifier setup, energies well in excess of 100 J can be produced. The commercial ruby laser typically operates at a modest overall efficiency of less than 1%, producing a beam that has a diameter ranging from 1 mm to about 25 mm, with a divergence of from 0.25 mrad to about 7 mrad.

Optical Resonant Cavities

The resonant cavity, which in this case is of course a Fabry–Perot etalon, plays a significant role in the operation of the laser. In the early stages of the laser process, spontaneous photons are emitted in every direction, as are the stimulated photons. But all of these, with the singular exception of those propagating very nearly along the cavity axis, quickly pass out of the sides of the ruby. In contrast, the axial beam continues to build as it bounces back and forth across the active medium. This accounts for the amazing degree of collimation of the issuing laserbeam, which is then effectively a coherent plane wave. Although the medium acts to amplify the wave, the *optical feedback* provided by the cavity converts the system into an oscillator and hence into a light generator—the acronym is thus somewhat of a misnomer.

In addition, the disturbance propagating within the cavity takes on a standing-wave configuration determined by the separation (L) of the mirrors. The cavity resonates (i.e., standing waves exist within it) when there is an integer number (m) of half wavelengths spanning the region between the mirrors. The idea is simply that there must be a node at each mirror, and this can only happen when L equals a whole number multiple of $\lambda/2$ (where $\lambda = \lambda_0/n$). Thus

$$m = \frac{L}{\lambda/2}$$

and

$$\nu_m = \frac{mv}{2L} \tag{13.15}$$

There are therefore an infinite number of possible oscillatory **longitudinal cavity modes**, each with a distinctive frequency ν_m. Consecutive modes are separated by a constant difference,

$$\nu_{m+1} - \nu_m = \Delta\nu = \frac{v}{2L} \tag{13.16}$$

which is the free spectral range of the etalon [Eq. (9.79)] and, incidentally, the inverse of the round-trip time. For a gas laser 1 m long, $\Delta\nu \approx 150$ MHz.

The resonant modes of the cavity are considerably narrower in frequency than the bandwidth of the normal spontaneous atomic transition. These modes, whether the device is constructed so that there is one or more, will be the ones that are sustained in the cavity, and hence the emerging beam is restricted to a region close to those frequencies (Fig. 13.8). In other words, the radiative transition makes available a relatively broad range of frequencies out of which the cavity will select and amplify only certain narrow bands and, if desired, even only one such band. This is the origin of the laser's extreme quasimonochromaticity. Thus while the bandwidth of the ruby transition to the ground state is roughly a rather broad 0.53 nm (330 GHz)—because of interactions of the chromium ions with the lattice—the corresponding laser cavity bandwidth, the frequency spread of the radiation of a single resonant mode, is a much narrower 0.000 05 nm (30 MHz). This situation is depicted in Fig. 13.8*b*, which shows a typical transition lineshape and a series of corresponding cavity spikes—in this case each is separated by $v/2L$, and each is 30 MHz wide.

A possible way to generate only a single mode in the cavity would be to have the mode separation, as given by Eq. (13.16), exceed the transition bandwidth. Then only one mode

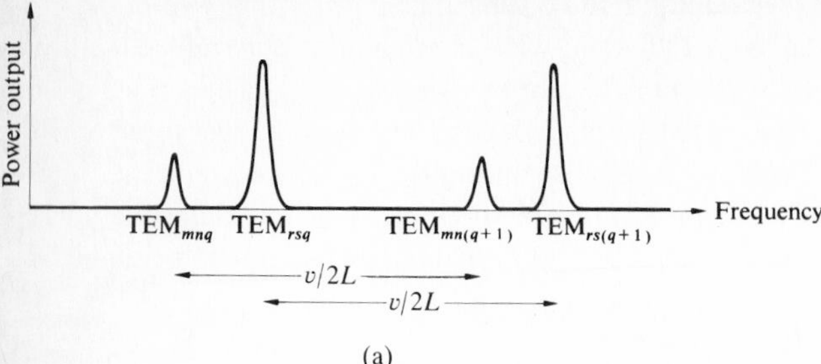

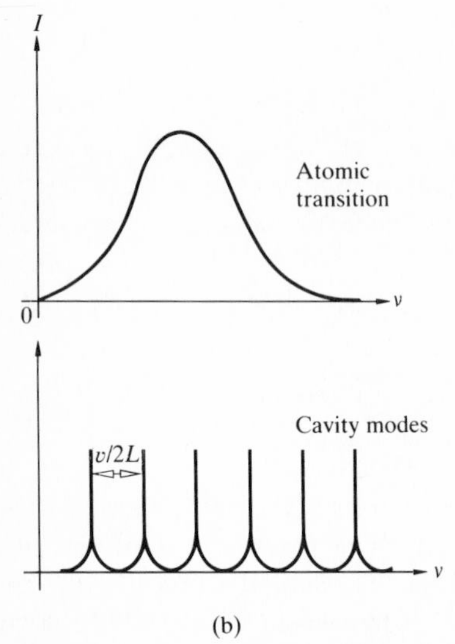

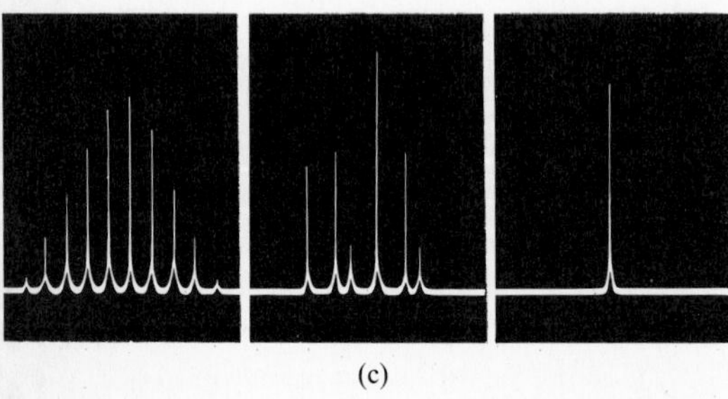

FIGURE 13.8 Laser modes: (*a*) illustrates the nomenclature; (*b*) compares the broad atomic emission with the narrow cavity modes; (*c*) depicts three operation configurations for a c-w gas laser, first showing several longitudinal modes under a roughly Gaussian envelope, then several longitudinal and transverse modes, and finally a single longitudinal mode.

would fit within the range of available frequencies provided by the transition. For a ruby laser (with an index of refraction of 1.76) a cavity length of a few centimeters will easily ensure single longitudinal mode operation. The drawback of this particular approach is that it limits the length of the active region contributing energy to the beam and so limits the output power of the laser.

In addition to the longitudinal or axial modes of oscillation, which correspond to standing waves set up along the cavity or z-axis, **transverse modes** can be sustained as well. Since the

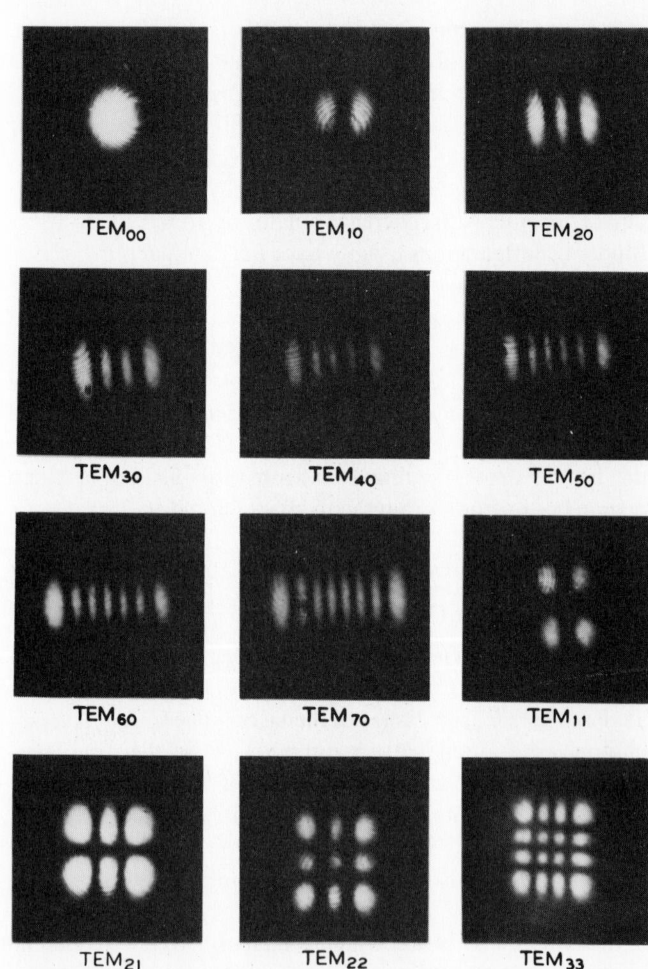

FIGURE 13.9 Mode patterns (without the faint interference fringes this is what the beam looks like in cross section). (Photos courtesy Bell Telephone Laboratories.)

fields are very nearly normal to z, these are known as TEM_{mn} modes (transverse electric and magnetic). The m and n subscripts are the integer number of transverse nodal lines in the x- and y-directions across the emerging beam. That is to say, the beam is segmented in its cross section into one or more regions. Each such array is associated with a given TEM mode, as shown in Figs. 13.9 and 13.10. The lowest order or TEM_{00} transverse mode is perhaps the most widely used, and this for several compelling reasons: the flux density is ideally Gaussian over the beam's cross section (Fig. 13.11); there are no phase shifts in the electric field across the beam, as there are in other modes, and so it is completely spatially coherent; the beam's angular divergence is the smallest; and it can be focused down to the smallest-sized spot. Note that the amplitude in this mode is actually not constant over the wavefront, and it is consequently an inhomogeneous wave.

A complete specification of each mode has the form TEM_{mnq}, where q is the longitudinal mode number. For each transverse mode (m, n) there can be many longitudinal modes (i.e., values of q). Often, however, it's unnecessary to work with a particular longitudinal mode, and the q subscript is usually simply dropped.*

Several additional cavity arrangements are of considerably more practical significance than is the original plane-parallel setup (Fig. 13.12). For example, if the planar mirrors are replaced by identical concave spherical mirrors separated by a distance very nearly equal to their radius of curvature, we have the *confocal* resonator. Thus the focal points are almost coincident on the axis midway between mirrors—ergo the name confocal.

If one of the spherical mirrors is made planar, the cavity is termed a *hemispherical* or *hemiconcentric* resonator. Both of these configurations are considerably easier to align than is the plane-parallel form. Laser cavities are either *stable* or *unstable* to the degree that the beam tends to retrace itself and so remain relatively close to the optical axis (Fig. 13.13). A beam in an unstable cavity will "walk out," going farther from the axis on each reflection until it quickly leaves the cavity altogether. By contrast, in a stable configuration (with mirrors that are, say, 100% and 98% reflective) the beam might traverse the resonator 50 times or more. Unstable resonators are commonly used in high-power lasers, where the fact that the beam traces across a wide region of the active medium enhances the amplification and allows for more energy to be extracted. This approach will be especially useful for media (like carbon dioxide or argon) wherein the beam gains a good deal of energy on each sweep of the cavity. The needed number of sweeps is determined by the so-called *small-signal gain* of the active medium. The actual selection of a resonator configuration is governed by the specific requirements of the system—there is no universally best arrangement.

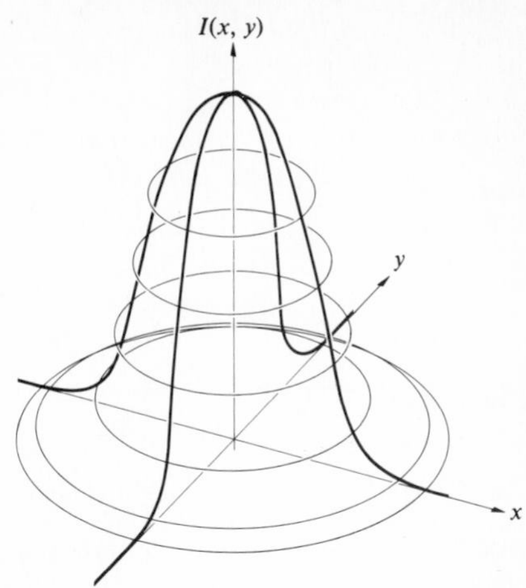

FIGURE 13.11 Gaussian irradiance distribution.

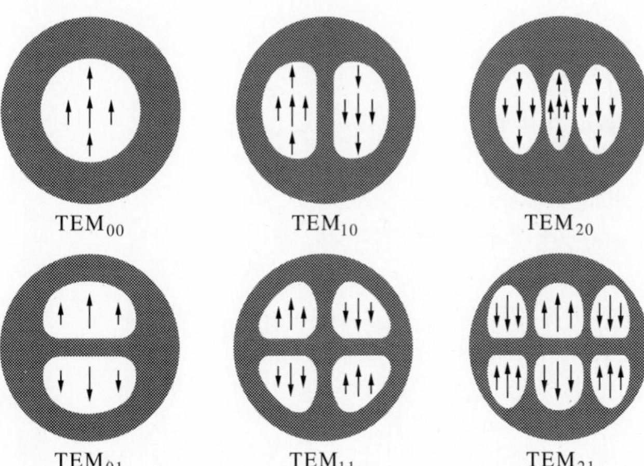

FIGURE 13.10 Mode configurations (rectangular symmetry). Circularly symmetric modes are also observable, but any slight asymmetry (such as Brewster windows) destroys them.

*Take a look at R. A. Phillips and R. D. Gehrz, "Laser Mode Structure Experiments for Undergraduate Laboratories," *Am. J. Phys.* **38**, 429 (1970).

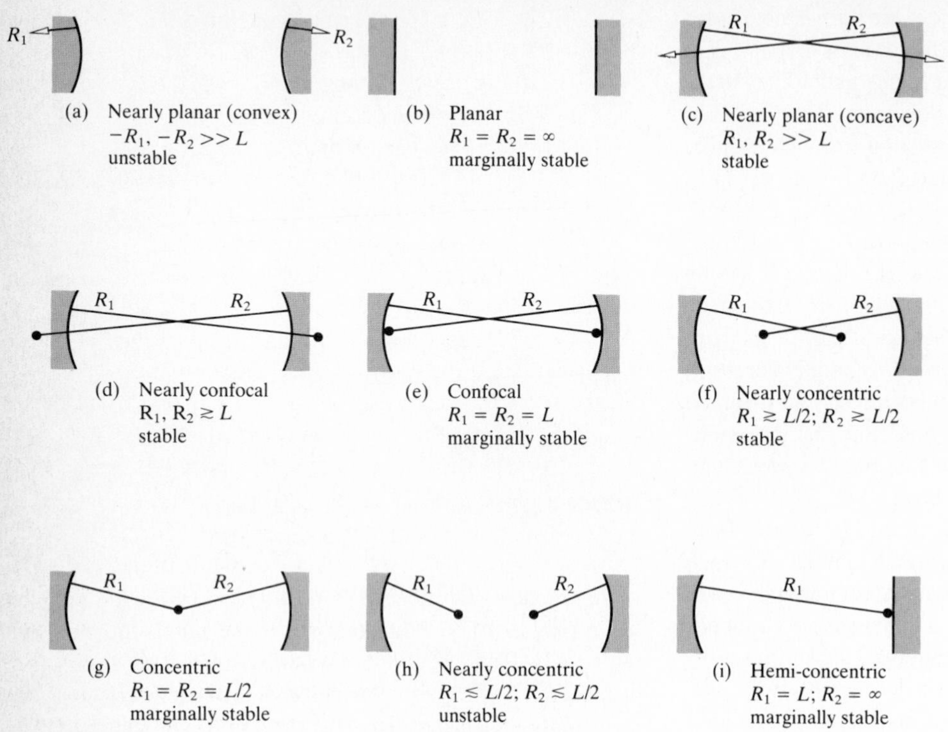

R_1 R_2

(a) Nearly planar (convex)
$-R_1, -R_2 >> L$
unstable

(b) Planar
$R_1 = R_2 = \infty$
marginally stable

R_1 R_2

(c) Nearly planar (concave)
$R_1, R_2 >> L$
stable

R_1 R_2

(d) Nearly confocal
$R_1, R_2 \gtrsim L$
stable

R_1 R_2

(e) Confocal
$R_1 = R_2 = L$
marginally stable

R_1 R_2

(f) Nearly concentric
$R_1 \gtrsim L/2$; $R_2 \gtrsim L/2$
stable

R_1 R_2

(g) Concentric
$R_1 = R_2 = L/2$
marginally stable

R_1 R_2

(h) Nearly concentric
$R_1 \lesssim L/2$; $R_2 \lesssim L/2$
unstable

R_1

(i) Hemi-concentric
$R_1 = L$; $R_2 = \infty$
marginally stable

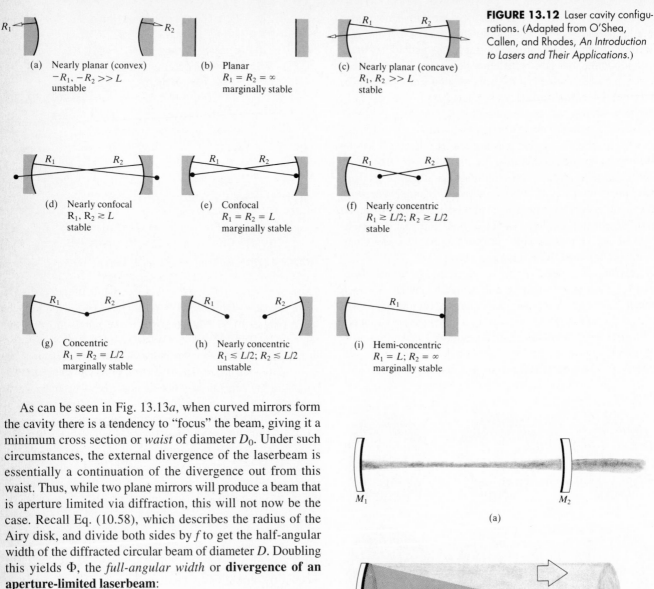

FIGURE 13.12 Laser cavity configurations. (Adapted from O'Shea, Callen, and Rhodes, *An Introduction to Lasers and Their Applications.*)

As can be seen in Fig. 13.13*a*, when curved mirrors form the cavity there is a tendency to "focus" the beam, giving it a minimum cross section or *waist* of diameter D_0. Under such circumstances, the external divergence of the laserbeam is essentially a continuation of the divergence out from this waist. Thus, while two plane mirrors will produce a beam that is aperture limited via diffraction, this will not now be the case. Recall Eq. (10.58), which describes the radius of the Airy disk, and divide both sides by f to get the half-angular width of the diffracted circular beam of diameter D. Doubling this yields Φ, the *full-angular width* or **divergence of an aperture-limited laserbeam**:

$$\Phi \approx 2.44\lambda/D$$

By comparison, far from the region of minimum cross section, the full-angular width of a waisted laserbeam is

$$\Phi \approx 1.27\lambda/D_0 \qquad (13.17)$$

where D_0, the beam-waist diameter, can be calculated from the particular cavity configuration.

M_1 M_2

(a)

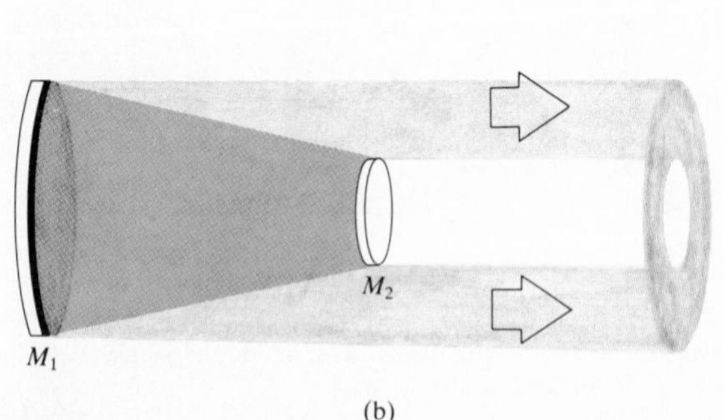

(b)

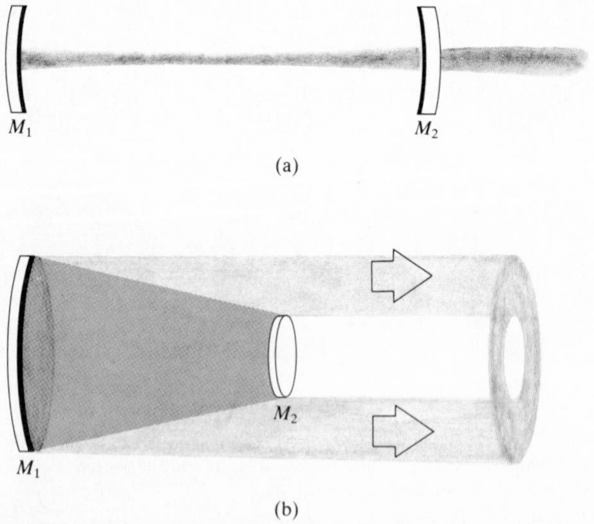

FIGURE 13.13 Stable and unstable laser resonators. (Adapted from O'Shea, Callen, and Rhodes, *An Introduction to Lasers and Their Applications.*)

The decay of energy in a cavity is expressed in terms of the Q or **quality factor** of the resonator. The origin of the expression dates back to the early days of radio engineering, when it was used to describe the performance of an oscillating (tuning) circuit. A high-Q, low-loss circuit meant a narrow bandpass and a sharply tuned radio. If an optical cavity is somehow disrupted, as for example by the displacement or removal of one of the mirrors, the laser action generally ceases. When this is done deliberately in order to delay the onset of oscillation in the laser cavity, it's known as *Q-spoiling* or **Q-switching**. The power output of a laser is self-limited in the sense that the population inversion is continuously depleted through stimulated emission by the radiation field within the cavity. However, if oscillation is prevented, the number of atoms pumped into the (long-lived) metastable state can be considerably increased, thereby creating a very extensive population inversion. When the cavity is switched on at the proper moment, a tremendously powerful *giant pulse* (perhaps up to several hundred megawatts) will emerge as the atoms drop down to the lower state almost in unison. A great many *Q-switching* arrangements utilizing various control schemes, for example, bleachable absorbers that become transparent under illumination, rotating prisms and mirrors, mechanical choppers, ultrasonic cells, or electro-optic shutters such as Kerr or Pockels cells, have all been used.

The Helium–Neon Laser

Maiman's announcement of the first operative laser came at a New York news conference on July 7, 1960.* By February of 1961 Ali Javan and his associates W. R. Bennett, Jr., and D. R. Herriott had reported the successful operation of a *continuous-wave* (c-w) helium–neon, gas laser at 1152.3 nm. The He–Ne laser (Fig. 13.14) is still among the most popular devices of its kind, most often providing a few milliwatts of continuous power in the visible (632.8 nm). Its appeal arises primarily because it's easy to construct, relatively inexpensive, and fairly reliable and in most cases can be operated by a flick of a single switch. Pumping is usually accomplished by electrical discharge (via either dc, ac, or electrodeless rf excitation). Free electrons and ions are accelerated by an applied field and, as a result of collisions, cause further ionization and excitation of the gaseous medium (typically, a mixture of about 0.8 torr of He and about 0.1 torr of Ne). Many helium atoms, after dropping down from several upper levels, accumulate in the long-lived 2^1S- and 2^3S-states. These are metastable states (Fig. 13.15) from which there are no allowed radiative transitions. The excited He atoms inelastically collide with and transfer energy to ground-state Ne atoms, raising them in turn to the 5s- and 4s-states. These are the upper laser levels, and there then exists a population inversion with respect to the lower 4p- and 3p-states. Transitions between the 5s- and 4s-states are forbidden. Spontaneous photons initiate stimulated emission, and the chain reaction begins. The dominant laser transitions correspond to 1152.3 nm and 3391.2 nm in the infrared and, of course, the ever-popular 632.8 nm in the visible (bright red). The p-states drain off into the 3s-state, thus themselves remaining uncrowded and thereby continuously sustaining the inversion. The 3s-level is metastable, so that 3s-atoms return to the ground state after losing energy to the walls of the enclosure. This is why the plasma tube's diameter inversely affects the gain and is, accordingly, a significant design para-

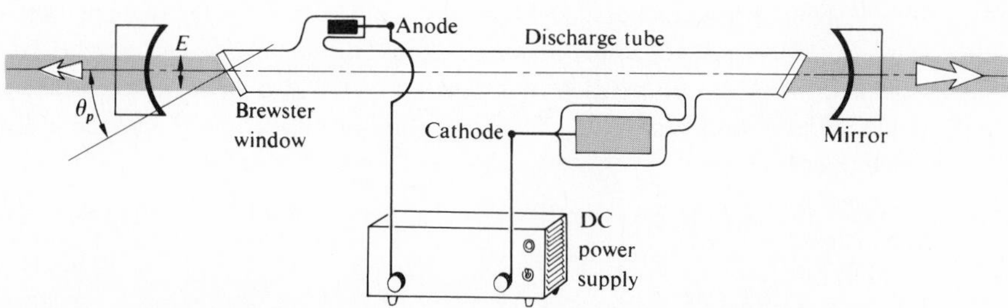

FIGURE 13.14 A simple, early He–Ne laser configuration.

*His initial paper, which would have made his findings known in a more traditional fashion, was rejected for publication by the editors of *Physical Review Letters*—this to their everlasting chagrin.

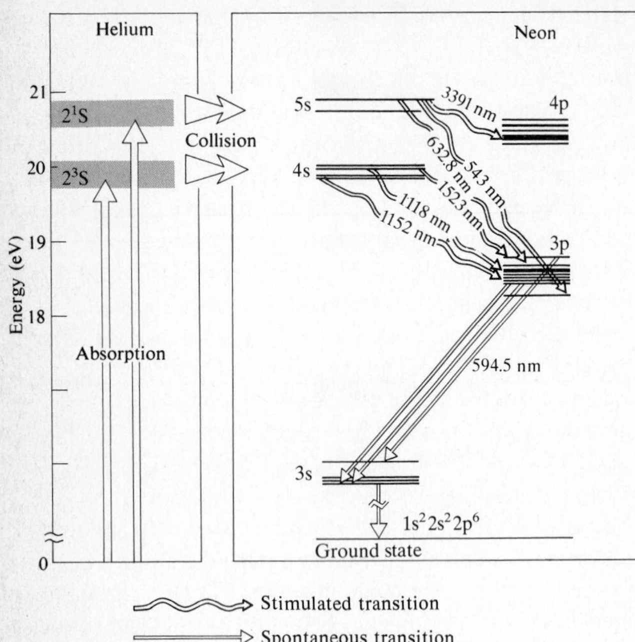

FIGURE 13.15 He–Ne laser energy levels.

meter. In contrast to the ruby, where the laser transition is down to the ground state, stimulated emission in the He–Ne laser occurs between two upper levels. The significance of this, for example, is that since the 3p-state is ordinarily only sparsely occupied, a population inversion is very easily obtained, and this without having to half empty the ground state.

Return to Fig. 13.14, which pictures the relevant features of a basic early He–Ne laser. The mirrors are coated with a multilayered dielectric film having a reflectance of over 99%. The laser output is made linearly polarized by the inclusion of Brewster end windows (i.e., plates tilted at the polarization angle) terminating the discharge tube. If these end faces were instead normal to the axis, reflection losses (4% at each interface) would become unbearable. By tilting them at the polarization angle, the windows presumably have 100% transmission for light whose electric-field component is parallel to the plane-of-incidence (the plane of the drawing). This polarization state rapidly becomes dominant, since the normal component is partially reflected off-axis at each transit of the windows. Linearly polarized light in the plane-of-incidence soon becomes the preponderant stimulating mechanism in the cavity, to the ultimate exclusion of the orthogonal polarization.*

Epoxying the windows to the ends of the laser tube and mounting the mirrors externally was a typical though dreadful approach used commercially until the mid-1970s. Inevitably, the epoxy leaked, allowing water vapor in and helium out. Today, such lasers are *hard sealed*; the glass is bonded directly to metal (Kovar) mounts, which support the mirrors within the tube. The mirrors (one of which is generally ≈100% reflective) have modern resistive coatings so they can tolerate the discharge environments within the tube. Operating lifetimes of 20 000 hours and more are now the rule (up from only a few hundred hours in the 1960s). Brewster windows are usually optional, and most commercial He–Ne lasers generate more or less "unpolarized" beams. The typical mass-produced He–Ne laser (with an output of from 0.5 mW to 5 mW) operates in the TEM_{00} mode, has a coherence length of around 25 cm, a beam diameter of approximately 1 mm, and a low overall efficiency of only 0.01% to about 0.1%. Although there are infrared He–Ne lasers, and even a new green (543.5 nm) He–Ne laser, the bright red 632.8-nm version remains the most popular.

A Survey of Laser Developments

Laser technology is so dynamic a field that what was a laboratory breakthrough a year or two ago may be a commonplace off-the-shelf item today. The whirlwind will certainly not pause to allow descriptive terms like "the smallest," "the largest," "the most powerful," and so on to be applicable for very long. With this in mind, we briefly survey the existing scene without trying to anticipate the wonders that will surely come after this type is set. Laserbeams have already been bounced off the Moon; they have spot welded detached retinas, generated fusion neutrons, stimulated seed growth, served as communications links, read CD discs, guided milling machines, missiles, ships, and grating engines, carried color

*Half of the output power of the laser is *not* lost in reflections at the Brewster windows when the transverse $\mathcal{P}$-state light is scattered. Energy simply isn't continuously channeled into that polarization component by the cavity. If it's reflected out of the plasma tube, it's not present to stimulate further emission.

television pictures, drilled holes in diamonds, levitated tiny objects,* and intrigued countless among the curious.

Along with ruby there are a great many other **solid-state lasers** whose outputs range in wavelength from roughly 170 nm to 3900 nm. For example, the trivalent rare earths Nd^{3+}, Ho^{3+}, Gd^{3+}, Tm^{3+}, Er^{3+}, Pr^{3+}, and Eu^{3+} undergo laser action in a host of hosts, such as $CaWO_4$, Y_2O_3, $SrMoO_4$, LaF_3, yttrium aluminum garnet (YAG for short), and glass, to name only a few. Of these, neodymium-doped glass and neodymium-doped YAG are of particular importance. Both constitute high-powered laser media operating at approximately 1060 nm. Nd: YAG lasers generating in excess of a kilowatt of continuous power have been constructed. Tremendous power outputs in pulsed systems have been obtained by operating several lasers in tandem. The first laser in the train serves as a Q-switched oscillator that fires into the next stage, which functions as an amplifier; and there may be one or more such amplifiers in the system. By reducing the feedback of the cavity, a laser will no longer be self-oscillatory, but it will amplify an incident wave that has triggered stimulated emission. Thus the amplifier is, in effect, an active medium, which is pumped, but for which the end faces are only partially reflecting or even nonreflecting. Ruby systems of this kind, delivering a few GW (gigawatts, i.e., 10^9 W) in the form of pulses lasting several nanoseconds, are available commercially. On December 19, 1984, the largest laser then in existence, the Nova at the Lawrence Livermore National Laboratory in California, fired all 10 of its beams at once for the first time, producing a warm-up shot of a mere 18 kJ of 350-nm radiation in a 1-ns pulse (Fig. 13.16). This immense neodymium-doped glass laser can focus up to 120 TW onto a fusion pellet—that's roughly 500 times more power than all the electrical generating stations in the United States—albeit only for about 10^{-9} s. In 1996, using just one beamline of the Nova, LLNL researchers were able to produce 1.25-PW pulses, each lasting 490 fs and carrying 580 J. This petawatt laser is designed to produce a beam with an irradiance of 10^{21} W/cm^2.

A large group of **gas lasers** operate across the spectrum from the far IR to the UV (1 mm to 150 nm). Primary among these are helium–neon, argon, and krypton, as well as several molecular gas systems, such as carbon dioxide, hydrogen fluoride, and molecular nitrogen (N_2). Argon lases mainly in the green, blue-green, and violet (predominantly at 488.0 and 514.5 nm) in either pulsed or continuous operation. Although its output is usually several watts c-w, it has gone as high as 150 W c-w. The argon ion laser is similar in some respects to the He–Ne laser, although it evidently differs in its usually greater power, shorter wavelength, broader linewidth, and higher price. All of the noble gases (He, Ne, Ar, Kr, Xe) have been made to lase individually, as have the gaseous ions of many other elements, but the former grouping has been studied most extensively.

FIGURE 13.16 The Nova laser. (Photo courtesy Lawrence Livermore National Laboratory.)

The CO_2 molecule, which lases between vibrational modes, emits in the IR at 10.6 μm, with typical c-w power levels of from watts to several kilowatts. Its efficiency can be an unusually high 15% when aided by additions of N_2 and He. While it once took a discharge tube nearly 200 m long to generate 10 kW c-w, considerably smaller "table models" are now available commercially. For a while in the 1970s, the record output belonged to an experimental gas-dynamic laser utilizing ther-

*See M. Lubin and A. Fraas, "Fusion by Laser," *Sci. Am.* **224**, 21 (June 1971); R. S. Craxton, R. L. McCrory, and J. M. Soures, "Progress in Laser Fusion," *Sci. Am.* **255**, 69 (August 1986); and A. Ashkin, "The Pressure of Laser Light," *Sci. Am.* **226**, 63 (February 1972).

mal pumping on a mixture of CO_2, N_2, and H_2O to generate 60 kW c-w at 10.6 μm in multimode operation.

The pulsed nitrogen laser operates at 337.1 nm in the UV, as does the c-w helium–cadmium laser. A number of metal vapors (e.g., Zn, Hg, Sn, Pb) have displayed laser transitions in the visible, but problems such as maintaining uniformity of the vapor in the discharge region have handicapped their exploitation. The He–Cd laser emits at 325.0 nm and 441.6 nm. These are transitions of the cadmium ion arising after excitation resulting from collisions with metastable helium atoms.

The **semiconductor laser**—alternatively known as the junction or diode laser—was invented in 1962, soon after the development of the light-emitting diode (LED). Today it serves a central role in electro-optics, primarily because of its spectral purity, high efficiency ($\approx$100%), ruggedness, ability to be modulated at extremely rapid rates, long lifetimes, and moderate power (as much as 200 mW) despite its pinhead size. Junction lasers have already been used in the millions in fiberoptic communications, laser disk audio systems, and so forth.

The first such lasers were made of one material, gallium arsenide, appropriately doped to form a *p–n* junction. The associated high lasing threshold of these so-called homostructures limited them to pulsed mode operation and cryogenic temperatures; otherwise the heat developed in their small structures would destroy them. The first tunable lead–salt diode laser was developed in 1964, but it was not until almost a dozen years later that it became commercially available. It operates at liquid nitrogen temperatures, which is certainly inconvenient, but it can scan from 2 μm to 30 μm.

Later advances have since allowed a reduction in the threshold and resulted in the advent of the continuous-wave (c-w), room temperature diode laser. Transitions occur between the conduction and valence bands, and stimulated emission results in the immediate vicinity of the *p–n* junction (Fig. 13.17). Quite generally, as a current flows in the forward direction through a semiconductor diode, electrons from the *n*-layer conduction band will recombine with *p*-layer holes, thereupon emitting energy in the form of photons. This radiative process, which competes for energy with the existing absorption mechanisms (such as phonon production), comes to predominate when the recombination layer is small and the current is large. To make the system lase, the light emitted from the diode is retained within a resonant cavity, and that's usually accomplished by simply polishing the end faces perpendicular to the junction channel.

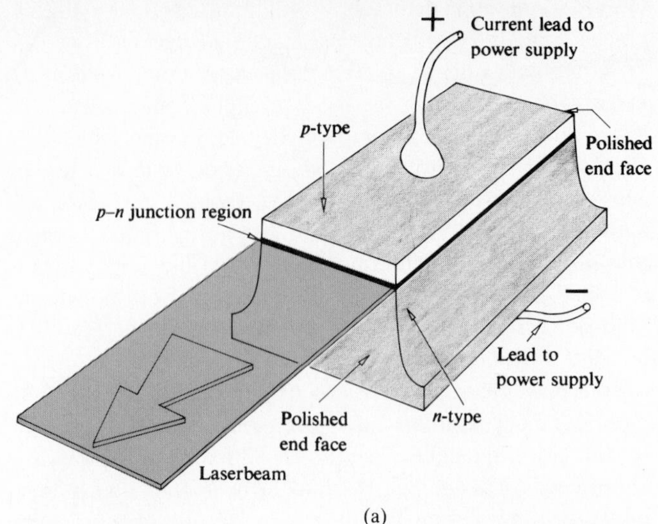

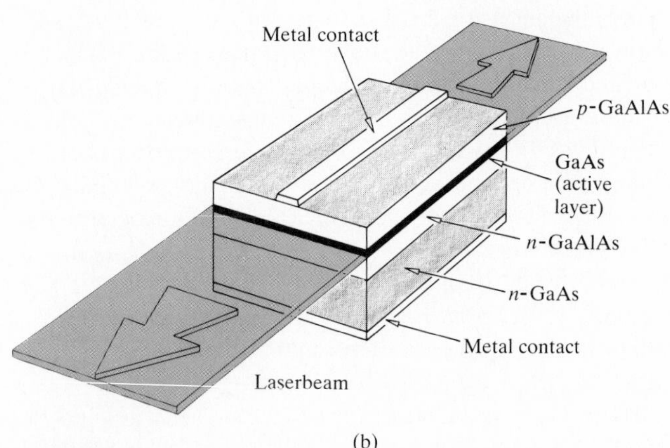

FIGURE 13.17 (*a*) An early GaAs *p–n* junction laser. (*b*) A more modern diode laser.

Nowadays semiconductor lasers are created to meet specific needs, and there are many designs producing wavelengths ranging from around 700 nm to about 30 μm. The early 1970s saw the introduction of the c-w GaAs/GaAlAs laser. Operating at room temperature in the 750-nm to 900-nm region (depending on the relative amounts of aluminum and gallium), the tiny diode chip is usually about a sixteenth of a cubic centimeter in volume. Figure 13.17*b* shows a typical heterostructure (a device formed of different materials) diode laser of this kind. Here the beam emerges in two directions from the

0.2-μm thick active layer of GaAs. These little lasers usually produce upward of 20 mW of continuous wave power. To take advantage of the low-loss region ($\lambda \approx 1.3\ \mu$m) in fiberoptic glass the GaInAsP/InP laser was devised in the mid-1970s with an output of 1.2 μm to 1.6 μm.

The cleaved–coupled-cavity laser is shown in Fig. 13.18. In it the number of axial modes is controlled in order to produce very-narrow-bandwidth tunable radiation. Two cavities coupled together across a small gap restrict the radiation to the extremely narrow bandwidth that can be sustained in both resonant chambers.*

The first **liquid laser** was operated in January of 1963.† All of the early devices of this sort were exclusively *chelates* (i.e., metallo-organic compounds formed of a metal ion with organic radicals). That original liquid laser contained an alcohol solution of europium benzoylacetonate emitting at 613.1 nm. The discovery of laser action in nonchelate organic liquids was made in 1966. It came with the fortuitous lasing (at 755.5 nm) of a chloroaluminum phthalocyanine solution during a search for stimulated Raman emission in that substance.‡

A great many fluorescent dye solutions of such families as the fluoresceins, coumarins, and rhodamines have since been made to lase at frequencies from the IR into the UV. These have usually been pulsed, although c-w operation has been obtained. There are so many organic dyes that it would seem possible to build such a laser at any frequency in the visible. Moreover, these devices are distinctive in that they inherently can be tuned continuously over a range of wavelengths (of perhaps 70 nm or so, although a pulsed system tunable over 170 nm exists). Indeed, there are other arrangements that will vary the frequency of a primary laserbeam (i.e., the beam enters with one color and emerges with another, Section 13.4), but in the case of the dye laser, the primary beam itself is tuned internally. This is accomplished, for example, by changing the concentration or the length of the dye cell or by adjusting a diffraction grating reflector at the end of the cavity. Multicolor dye laser systems, which can easily be switched from one dye to another and thereby operate over a very broad frequency range, are available commercially.

FIGURE 13.18 The cleaved–coupled-cavity laser. (Photos courtesy of Bell Laboratories.)

A **chemical laser** is one that is pumped with energy released via a chemical reaction. The first of this kind was operated in 1964, but it was not until 1969 that a continuous-wave chemical laser was developed. One of the most promising of these is the deuterium fluoride–carbon dioxide ($DF–CO_2$) laser. It is self-sustaining in that it requires no

*See Y. Suematsu, "Advances in Semiconductor Lasers," *Phys. Today*, 32 (May 1985). For a discussion of heterostructure diode lasers refer to M. B. Panish and I. Hayashi, "A New Class of Diode Lasers," *Sci. Am.* **225**, 32 (July 1971).

†See Adam Heller, "Laser Action in Liquids," *Phys. Today* (November 1967), p. 35, for a more detailed account.

‡P. Sorokin, "Organic Lasers," *Sci. Amer.* **220**, 30 (February 1969).

external power source. In brief, the reaction $F_2 + D_2 \rightarrow 2DF$, which occurs on the mixing of these two fairly common gases, generates enough energy to pump a CO_2 laser.

There are solid-state, gaseous, liquid, and vapor (e.g., H_2O) lasers; there are semiconductor lasers, free electron (600 nm to 3 mm) lasers, X-ray lasers, and lasers with very special properties, such as those that generate extremely short pulses, or those that have extraordinary frequency stability. These latter devices are very useful in the field of high-resolution spectroscopy, but there is a growing need for them in other research areas as well (e.g., in the interferometers used to attempt to detect gravity waves). In any event, these lasers must have precisely controlled cavity configurations despite the disturbing influences of temperature variations, vibrations, and even sound waves. To date, the record is held by a laser at the Joint Institute for Laboratory Astrophysics in Boulder, Colorado, which maintains a frequency stability (p. 310) of nearly one part in 10^{14}.

13.1.4 The Light Fantastic

Laserbeams differ somewhat from one type of laser to another; yet there are several remarkable features that are displayed, to varying degrees, by all of them. Quite apparent is the fact that most laserbeams are exceedingly directional, or if you will, highly collimated. One need only blow some smoke into the otherwise invisible, visible laserbeam to see (via scattering) a fantastic thread of light stretched across a room. A He–Ne beam in the TEM_{00} mode generally has a divergence of only about one minute of arc or less. Recall that in that mode the emission closely approximates a Gaussian irradiance distribution; that is, the flux density drops off from a maximum at the central axis of the beam and has no side lobes. The typical laserbeam is quite narrow, usually issuing at no more than a few millimeters in diameter. Since the beam resembles a truncated plane wave, it is of course highly *spatially coherent*. In fact, its directionality may be thought of as a manifestation of that coherence. Laserlight is quasimonochromatic, generally having an exceedingly narrow frequency bandwidth (p. 308). In other words, it is highly *temporally coherent*.

Another attribute is the large flux or *radiant power* that can be delivered in that narrow frequency band. As we've seen, the laser is distinctive in that it emits all its energy in the form of a narrow beam. In contrast, a 100-W incandescent lightbulb may pour out considerably more radiant energy in toto than a lower-power c-w laser, but the emission is incoherent, spread over a large solid angle, and it has a broad bandwidth as well. A good lens* can totally intercept a laserbeam and focus essentially all of its energy into a minute spot (whose diameter varies directly with λ and the focal length and inversely with the beam diameter). Spot diameters of just a few thousandths of an inch can readily be attained with lenses that have a conveniently short focal length. And a spot diameter of a few hundred-millionths of an inch is possible in principle. Thus flux densities can readily be generated in a focused laserbeam of over 10^{17} W/cm^2, in contrast to, say, an oxyacetylene flame having roughly 10^3 W/cm^2. To get a better feel for these power levels, note that a focused CO_2 laserbeam of a few kilowatts c-w can burn a hole through a quarter-inch stainless steel plate in about 10 seconds. By comparison, a pinhole and filter positioned in front of an ordinary source will certainly produce spatially and temporally coherent light, but only at a minute fraction of the total power output.

Femtosecond Optical Pulses

The advent of the mode-locked dye laser in the early part of the 1970s gave a great boost to the efforts then being made at generating extremely short pulses of light.† Indeed, by 1974 subpicosecond (1 ps $= 10^{-12}$ s) optical pulses were already being produced, although the remainder of the decade saw little significant progress. In 1981 two separate advances resulted in the creation of femtosecond laser pulses (i.e., <0.1 ps or <100 fs)—a group at Bell Labs developed a colliding-pulse ring dye laser, and a team at IBM devised a new pulse-compression scheme.

Above and beyond the implications in the practical domain of electro-optical communications, these accomplishments have firmly established a new field of research known as *ultrafast phenomena*. The most effective way to study the progression of a process that occurs exceedingly rapidly (e.g., carrier dynamics in semiconductors, fluorescence, photochemical biological processes, and molecular configuration changes) is to examine it on a time scale that is comparatively short with

*Spherical aberration is usually the main problem, since laserbeams are, as a rule, both quasimonochromatic and incident along the axis of the lens.

†See Chandrashekhar Joshi and Paul Corkum, "Interactions of Ultra-Intense Laser Light with Matter," *Phys. Today* **36** (January 1995).

respect to what's happening. Pulses lasting $\approx$10 fs allow an entirely new access into previously obscure areas in the study of matter.

Pulses lasting a mere 8 fs (10^{-15} s), which corresponds to wavetrains only about 4 wavelengths of red light in length, can now be produced routinely. One of the techniques that makes these femtosecond wavegroups possible is based on an idea used in radar work in the 1950s called *pulse compression*. Here an initial laser pulse has its frequency spectrum broadened, thereby allowing the inverse or temporal pulse width to be shortened—remember that $\Delta\nu$ and Δt are conjugate Fourier quantities [Eq. (7.63)]. The input pulse (several picoseconds long) is passed into a nonlinear dispersive medium, namely, a single-mode optical fiber. When the light intensity is high enough, the index of refraction has an appreciable nonlinear term (Section 13.4), and the carrier frequency of the pulse experiences a time-dependent shift. On traversing perhaps 30 m of fiber, the frequency of the pulse is drawn out or "chirped." That is, a spread occurs in the spectrum of the pulse, with the low frequencies leading and the high frequencies trailing. Next the spectrally broadened pulse is passed through another dispersive system (a delay line), such as a pair of diffraction gratings. By traveling different paths, the blue-shifted trailing edge of the pulse is made to catch up to the red-shifted leading edge, creating a time-compressed output pulse.

The Speckle Effect

A rather striking and easily observable manifestation of the spatial coherence of laserlight is its granular appearance on reflection from a diffuse surface. Using a He–Ne laser (632.8 nm), expand the beam a bit by passing it through a simple lens and project it onto a wall or a piece of paper. The illuminated disk appears speckled with bright and dark regions that sparkle and shimmer in a dazzling psychedelic dance. Squint and the grains grow in size; step toward the screen and they shrink; take off your eyeglasses and the pattern stays in perfect focus. In fact, if you are nearsighted, the diffraction fringes caused by dust on the lens blur out and disappear, but the speckles do not. Hold a pencil at varying distances from your eye so that the disk appears just above it. At each position, focus on the pencil; wherever you focus, the granular display is crystal clear. Indeed, look at the pattern through a telescope; as you adjust the scope from one extreme to the other, the ubiquitous granules remain perfectly distinct, even though the wall is completely blurred.

The spatially coherent light scattered from a diffuse surface fills the surrounding region with a *stationary* interference pattern (just as in the case of the wavefront-splitting arrangements of Section 9.3). At the surface the granules are exceedingly small, and they increase in size with distance. At any location in space the resultant field is the superposition of many contributing scattered wavelets. These must have a constant relative phase determined by the optical path length from each scatterer to the point in question, if the interference pattern is to be sustained. Figure 13.19 illustrates this point rather nicely. It shows a cement block illuminated in one case by laserlight and in the other by collimated light from a Hg arc lamp, both of about the same spatial coherence. While the laser's coherence length is much greater than the height of the surface features, the coherence length of the Hg light is not. In the former case, the speckles in the photograph are large, and they obscure the surface structure; in the latter, despite its spatial coherence, the speckle pattern is not observable in the photograph, and the surface features predominate. Because of the rough texture the optical path length difference between two wavelets arriving at a point in space, scattered from different surface bumps, is generally greater than the coherence length of the mercury light. This means that the relative phases of the overlapping wavetrains change rapidly and randomly in time, washing out the large-scale interference pattern.

A real system of fringes is formed of the scattered waves that converge in front of the screen. The fringes can be viewed by intersecting the interference pattern with a sheet of paper at a convenient location. After forming the real image in space, the rays proceed to diverge, and any region of the

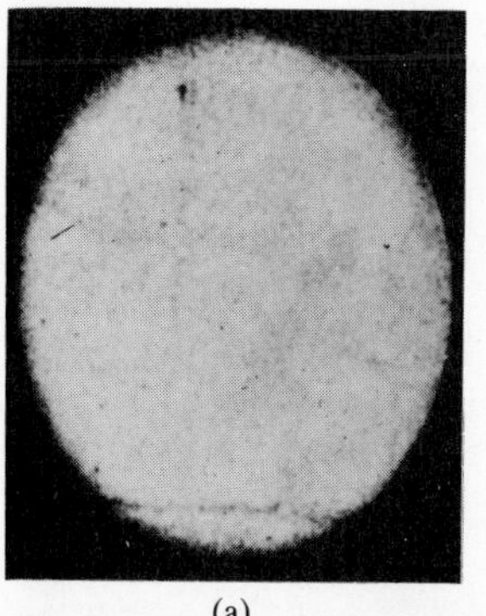
(a)

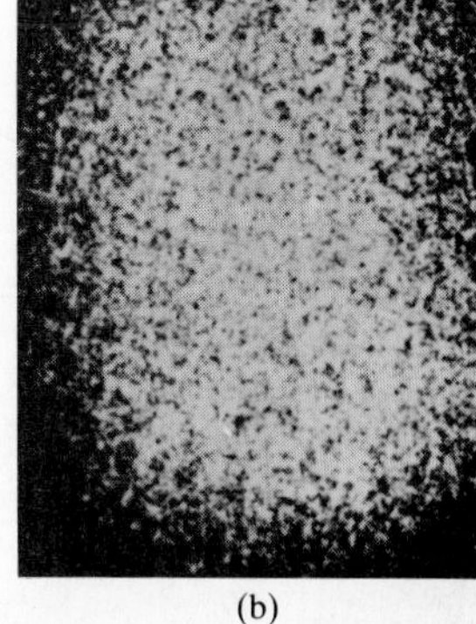
(b)

FIGURE 13.19 Speckle patterns. (*a*) A cement block illuminated by a mercury arc and (*b*) a He–Ne laser. [From B. J. Thompson, *J. Soc. Phot. Inst. Engr.* **4**, 7 (1965).]

image can therefore be viewed directly with the eye appropriately focused. In contrast, rays that initially diverge appear to the eye as if they had originated behind the scattering screen and thus form a virtual image.

It seems that as a result of chromatic aberration, normal and farsighted eyes tend to focus red light behind the screen. Contrarily, a nearsighted person observes the real field in front of the screen (regardless of wavelength). Thus if the viewer moves her head to the right, the pattern will move to the right in the first instance (where the focus is beyond the screen) and to the left in the second (focus in front). The pattern will follow the motion of your head, if you're viewing it very close to the surface. The same apparent parallax motion can be seen by looking through a window; outside objects will seem to move with your head, inside ones opposite to it. The brilliant, narrow-bandwidth, spatially coherent laserbeam is ideally suited for observing the granular effect, although other means are certainly possible.* In unfiltered sunlight the grains are minute, on the surface, and multicolored. The effect is easy to observe on a smooth, flat-black material (e.g., poster-painted paper), but you can see it on a fingernail or a worn coin as well.

Although it provides a marvelous demonstration, both aesthetically and pedagogically, the granular effect can be a real practical nuisance in coherently illuminated systems. For example, in holographic imagery the speckle pattern corresponds to troublesome background noise. Incidentally, very much the same kind of thing is observable when listening to a mobile radio where the signal strength fluctuates from one location to the next, depending on the environment and the resulting interference pattern.

The Spontaneous Raman Effect

It is possible that an excited atom will not return to its initial state after the emission of a photon. This kind of behavior had been observed and studied extensively by George Stokes prior to the advent of quantum theory. Since the atom drops down to an interim state, it emits a photon of lower energy than the incident primary photon, in what is usually referred to as a ***Stokes transition***. If the process takes place rapidly (roughly 10^{-7} s), it is called **fluorescence**, whereas if there is an appreciable delay (in some cases seconds, minutes, or even many hours), it is known as **phosphorescence**. Using ultraviolet quanta to generate a fluorescent emission of visible light has become an accepted occurrence in our everyday lives. Any number of commonplace materials (e.g., detergents, organic dyes, and tooth enamel) will emit characteristic visible photons so that they appear to glow under ultraviolet illumination; ergo the widespread use of the phenomenon for commercial display purposes and for "whitening" cloths.

If quasimonochromatic light is scattered from a substance, it will thereafter consist mainly of light of the same frequency. Yet it is possible to observe very weak additional components having higher and lower frequencies (side bands). Moreover, the difference between the side bands and the incident frequency ν_i is found to be characteristic of the material and therefore suggests an application to spectroscopy. The **Spontaneous Raman Effect**, as it is now called, was predicted in 1923 by Adolf Smekal and observed experimentally in 1928 by Sir Chandrasekhara Vankata Raman (1888–1970), then professor of physics at the University of Calcutta. The effect was difficult to put to actual use, because one needed strong sources (usually Hg discharges were used) and large samples. Often the ultraviolet from the source would further complicate matters by decomposing the specimen. And so it is not surprising that little sustained interest was aroused by the promising practical aspects of the Raman Effect. The situation was changed dramatically when the laser became a reality. ***Raman spectroscopy*** is now a unique and powerful analytical tool.

To appreciate how the phenomenon operates, let's review the germane features of molecular spectra. A molecule can absorb radiant energy in the far-infrared and microwave regions, converting it to rotational kinetic energy. Furthermore, it can absorb infrared photons (i.e., ones within a wavelength range from roughly 10^{-2} mm down to about 700 nm), transforming that energy into vibrational motion of the molecule. Finally, a molecule can absorb energy in the visible and ultraviolet regions through the mechanism of electron transitions, much like those of an atom. Suppose then that we have a molecule in some vibrational state, which, using quantum-mechanical notation, we call $|b\rangle$, as indicated diagrammatically in Fig. 13.20*a*. This need not necessarily be an excited state. An incident photon of energy $h\nu_i$ is absorbed, raising the system to some intermediate or virtual state, whereupon it immediately makes a Stokes transition, emitting a (scattered) photon of energy $h\nu_s < h\nu_i$. In conserving energy, the difference $h\nu_i - h\nu_s = h\nu_{cb}$ goes into exciting the molecule to a

*For further reading on this effect, see L. I. Goldfischer, *J. Opt. Soc. Am.* **55**, 247 (1965); D. C. Sinclair, *J. Opt. Soc. Am.* **55**, 575 (1965); J. D. Rigden and E. I. Gordon, *Proc. IRE* **50**, 2367 (1962); B. M. Oliver, *Proc. IEEE* **51**, 220 (1963).

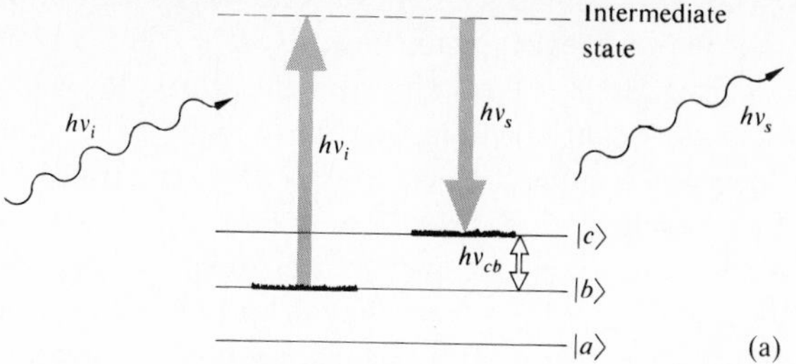

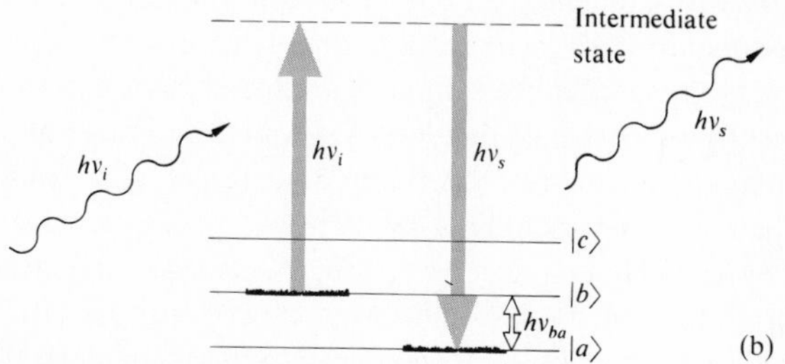

FIGURE 13.20 Spontaneous Raman Scattering.

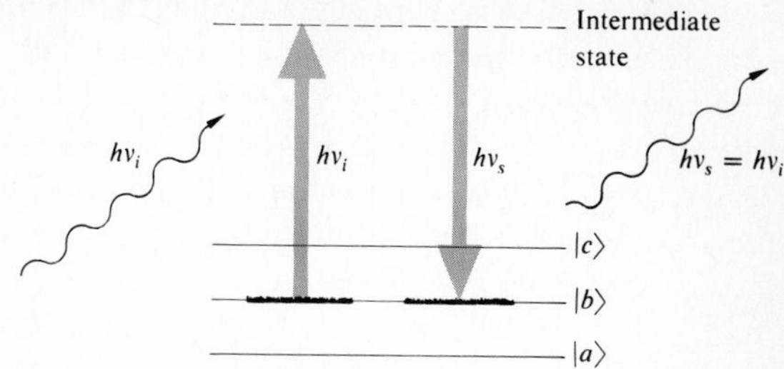

FIGURE 13.21 Rayleigh Scattering.

higher vibrational energy level $|c\rangle$. It is possible that electronic or rotational excitation results as well.

Alternatively, if the initial state is an excited one (just heat the sample), the molecule, after absorbing and emitting a photon, may drop back to an even lower state (Fig. 13.20*b*), thereby making an ***anti-Stokes transition***. In this instance $h\nu_s > h\nu_i$, which means that some vibrational energy of the molecule ($h\nu_{ba} = h\nu_s - h\nu_i$) has been converted into radiant energy. In either case, the resulting differences between ν_s and ν_i correspond to specific energy-level differences for the substance under study and as such yield insights into its molecular structure. Figure 13.21, for comparison's sake, depicts Rayleigh Scattering where $\nu_s = \nu_i$.

The laser is an ideal source for spontaneous Raman Scattering. It is bright, quasimonochromatic, and available in a wide range of frequencies. Figure 13.22 illustrates a typical laser-Raman system. Complete research instruments of this sort are commercially available, including the laser (usually helium–neon, argon, or krypton), focusing lens systems, and

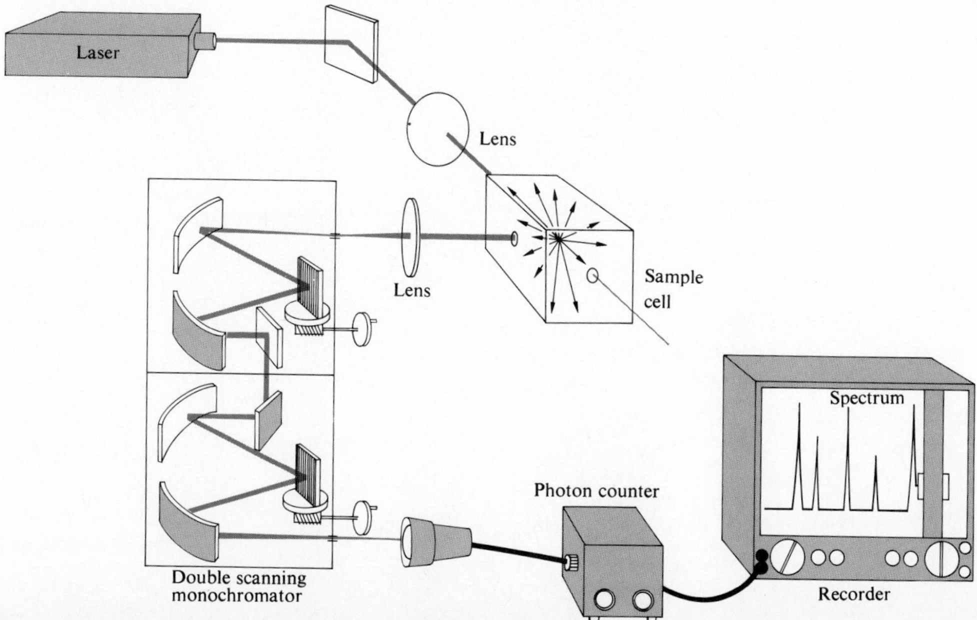

FIGURE 13.22 A laser–Raman system.

photon-counting electronics. The double scanning monochromator provides the needed discrimination between ν_i and ν_s, since unshifted laserlight (ν_i) is scattered along with the Raman spectra (ν_s). Although Raman Scattering associated with molecular rotation was observed prior to the use of the laser, the increased sensitivity now available makes the process easier and allows even the effects of electron motion to be examined.

THE STIMULATED RAMAN EFFECT

In 1962 Eric J. Woodbury and Won K. Ng rather fortuitously discovered a remarkable related effect known as *Stimulated Raman Scattering*. They had been working with a million-watt pulsed ruby laser incorporating a nitrobenzene Kerr cell shutter (see Section 8.11.3). They found that about 10% of the incident energy at 694.3 nm was shifted in wavelength and appeared as a *coherent* scattered beam at 766.0 nm. It was subsequently determined that the corresponding frequency shift of about 40 THz was characteristic of one of the vibrational modes of the nitrobenzene molecule, as were other new frequencies also present in the scattered beam. Stimulated Raman Scattering can occur in solids, liquids, or dense gases under the influence of focused high-energy laser pulses (Fig. 13.23). The effect is schematically depicted in Fig. 13.24. Here two photon beams are simultaneously incident on a molecule, one corresponding to the laser frequency ν_i, the other having the scattered frequency ν_s. In the original setup, the scattered beam was reflected back and forth through the specimen, but the effect can occur without a resonator. The laserbeam loses a photon $h\nu_i$, while the scattered beam gains a photon $h\nu_s$ and is subsequently *amplified*. The remaining energy ($h\nu_i - h\nu_s = h\nu_{ba}$) is transmitted to the sample. The chain reaction in which a large portion of the incident beam is converted into stimulated Raman light can only occur above a certain high-threshold flux density of the exciting laserbeam.

Stimulated Raman Scattering provides a whole new range of high flux-density coherent sources extending from the infrared to the ultraviolet. It should be mentioned that in prin-

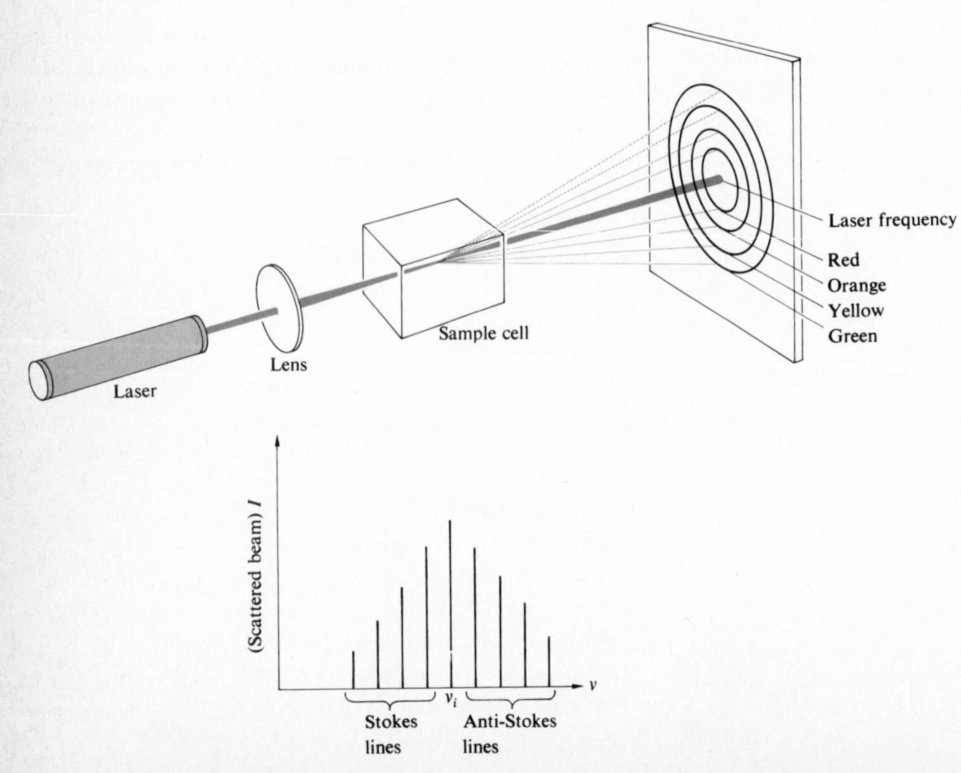

FIGURE 13.23 Stimulated Raman Scattering. [See R. W. Minck, R. W. Terhune, and C. C. Wang, *Proc. IEEE* **54**, 1357 (1966).]

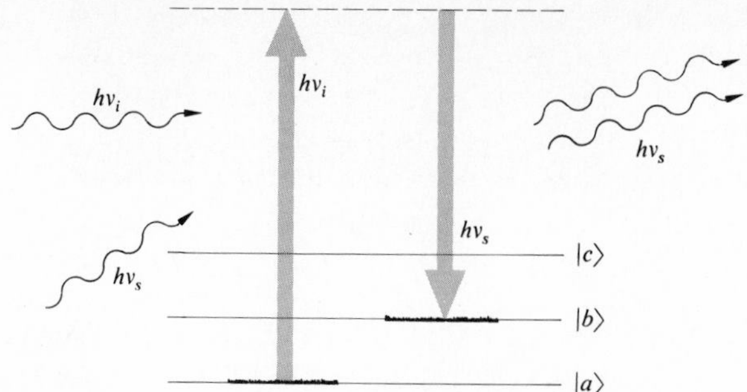

FIGURE 13.24 Energy-level diagram of Stimulated Raman Scattering.

ciple each spontaneous scattering mechanism (e.g., Rayleigh and Brillouin Scattering) has its stimulated counterpart.*

13.2 Imagery — The Spatial Distribution of Optical Information

The manipulation of all sorts of data via optical techniques has already become a technological *fait accompli*. The literature since the 1960s reflects, in a diversity of areas, this far-reaching interest in the methodology of optical data processing. Practical applications have been made in the fields of television and photographic image enhancement, radar and sonar signal processing (phased and synthetic array antenna analysis), as well as in pattern recognition (e.g., aerial photointerpretation and fingerprint studies), to list only a very few.

Our concern here is to develop the nomenclature and some of the ideas necessary for an appreciation of this contemporary thrust in Optics.

*For further reading on these subjects you might try the review-tutorial paper by Nicolaas Bloembergen, "The Stimulated Raman Effect," *Am. J. Phys.* **35**, 989 (1967). It contains a fairly good bibliography as well as a historical appendix. Many of the papers in *Lasers and Light* also deal with this material and are highly recommended reading.

13.2.1 Spatial Frequencies

In electrical processes one is most frequently concerned with signal variations in time, that is, the moment-by-moment alteration in voltage that might appear across a pair of terminals at some fixed location in space. By comparison, in Optics we are most often concerned with information spread across a region of space at a fixed location in time. For example, we can think of the scene depicted in Fig. 13.25*a* as a two-dimensional flux-density distribution. It might be an illuminated transparency, a television picture, or an image projected on a screen; in any event there is presumably some function $I(y, z)$, which assigns a value of I to each point in the picture. To simplify matters a bit, suppose we scan across the screen on a horizontal line ($z = 0$) and plot point-by-point variations in irradiance with distance, as in Fig. 13.25*b*. The function $I(y, 0)$ can be synthesized out of harmonic functions, using the techniques of Fourier analysis treated in Chapters 7 and 11. In this instance, the function is rather complicated, and it would take many terms to represent it adequately. Yet if the functional form of $I(y, 0)$ is known, the procedure is straightforward enough. Scanning across another line, for example, $z = a$, we get $I(y, a)$, which is drawn in Fig. 13.25*c* and which just happens to turn out to be a series of equally spaced square pulses. This function is one that was considered at length in Section 7.3, and a few of its constituent Fourier components are roughly sketched in Fig. 13.25*d*. If the peaks in (*c*) are separated, center to center, by say, 1-cm intervals, the spatial period equals 1 cm per cycle, and its reciprocal, which is the spatial frequency, equals 1 cycle per cm.

Quite generally, we can transform the information associated with any scan line into a series of sinusoidal functions of appropriate amplitude and spatial frequency. In the case of either of the simple sine- or square-wave targets of Fig. 13.26, each such horizontal scan line is identical, and the patterns are effectively one-dimensional. The spatial frequency spectrum of Fourier components needed to synthesize the square wave is shown in Fig. 7.21. On the other hand, $I(y, z)$ for the wine bottle candelabra scene is two-dimensional, and we have to think in terms of two-dimensional Fourier transforms (Sections 7.4.4 and 11.2.2). We might mention as well that, at least in principle, we could have recorded the amplitude of the electric field at each point of the scene and then performed a similar decomposition of that signal into its Fourier components.

Recall (Section 11.3.3) that the far-field or Fraunhofer diffraction pattern is essentially identical to the Fourier transform of the aperture function $\mathcal{A}(y, z)$. The aperture function is pro-

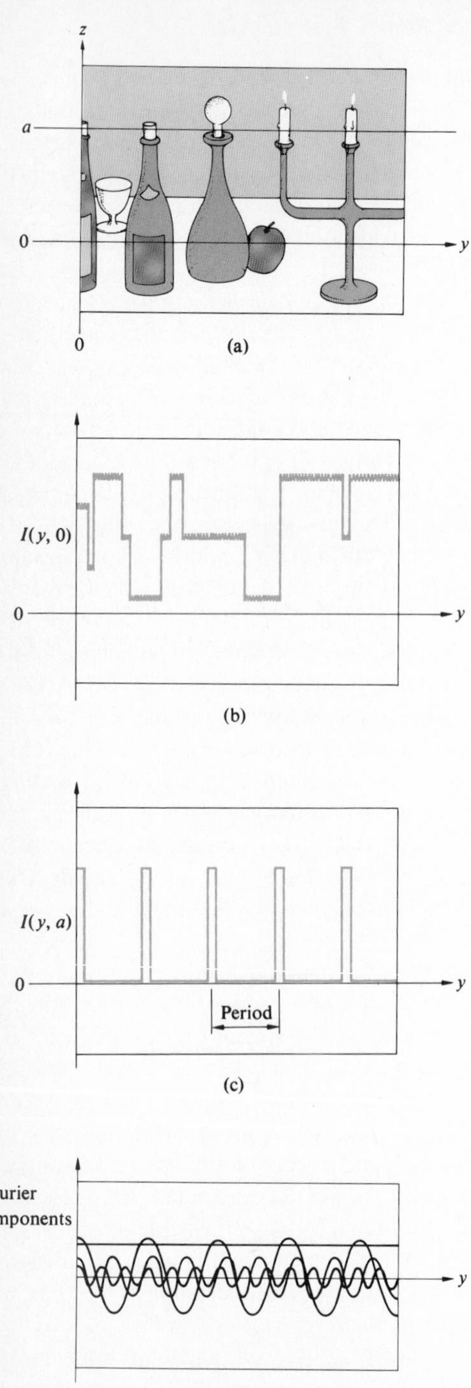

FIGURE 13.25 A two-dimensional irradiance distribution.

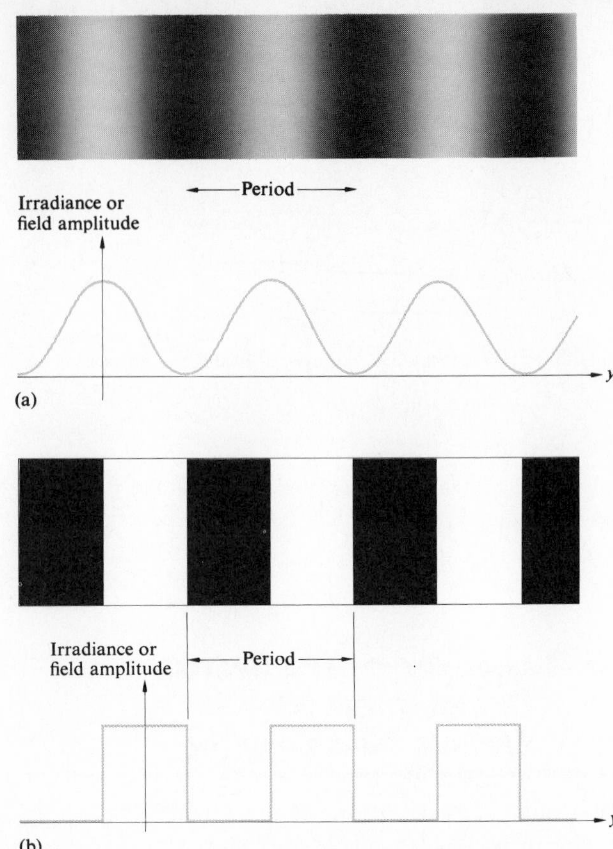

FIGURE 13.26 (*a*) Sine-wave target and (*b*) square-wave target.

portional to $\varepsilon_A(y, z)$, the source strength per unit area [Eq. (10.37)] over the input or object plane. In other words, if the field distribution on the object plane is given by $\mathscr{A}(y, z)$, its two-dimensional Fourier transform will appear as the field distribution $E(Y, Z)$ on a very distant screen. As in Figs. 7.32 and 10.11, we can introduce a lens (L_t) after the object in order to shorten the distance to the image plane. That objective lens is commonly referred to as the ***transform lens***, since we can imagine it as if it were an *optical computer* capable of generating instant Fourier transforms. Now, suppose we illuminate a somewhat idealized transmission grating with a spatially coherent, quasimonochromatic wave, such as the plane wave emanating from a laser or a collimated, filtered Hg arc source (Fig. 13.27). In either case, the amplitude of the field is assumed to be fairly constant over the incident wavefront. The aperture function is then a periodic step function (Fig. 13.28);

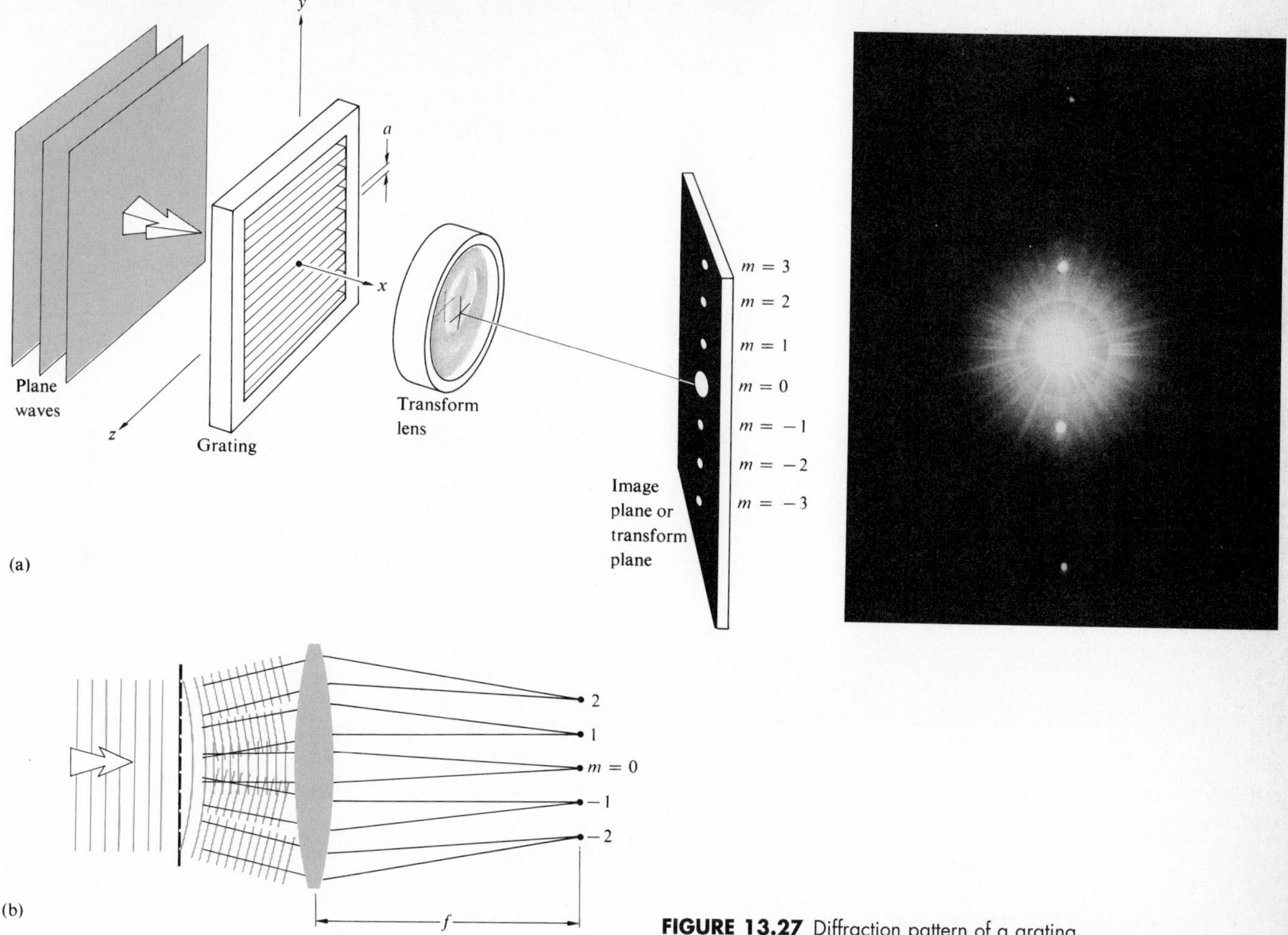

FIGURE 13.27 Diffraction pattern of a grating.

in other words, as we move from point to point on the object plane, the amplitude of the field is either zero or a constant. If a is the grating spacing, it is also the spatial period of the step function, and its reciprocal is the fundamental spatial frequency of the grating. The central spot ($m = 0$) in the diffraction pattern is the dc term corresponding to a zero spatial frequency—it's the bias level that arises from the fact that the input $\mathscr{A}(y)$ is everywhere positive. This bias level can be shifted by constructing the step-function pattern on a uniform gray background. As the spots in the image (or in this case the transform) plane get farther from the central axis, their associated spatial frequencies (m/a) increase in accord with the grating equation $\sin \theta_m = \lambda(m/a)$. A coarser grating would have a larger value of a, so that a given order (m) would be concomitant with a lower frequency, (m/a), and the spots would all be closer to the central or optical axis.

Had we used as an object a transparency resembling the sine target (Fig. 13.26*a*), such that the aperture function varied sinusoidally, there would ideally have only been three spots on the transform plane, these being the zero-frequency central peak and the first order or fundamental ($m = \pm 1$) on either side of the center. Extending things into two dimensions, a crossed grating (or mesh) yields the diffraction pattern shown in Fig. 13.29. Note that in addition to the obvious periodicity horizontally and vertically across the mesh, it is also repetitive, for example, along diagonals. A more involved object, such as a transparency of the surface of the Moon, would generate an extremely complex diffraction pattern. Because of the

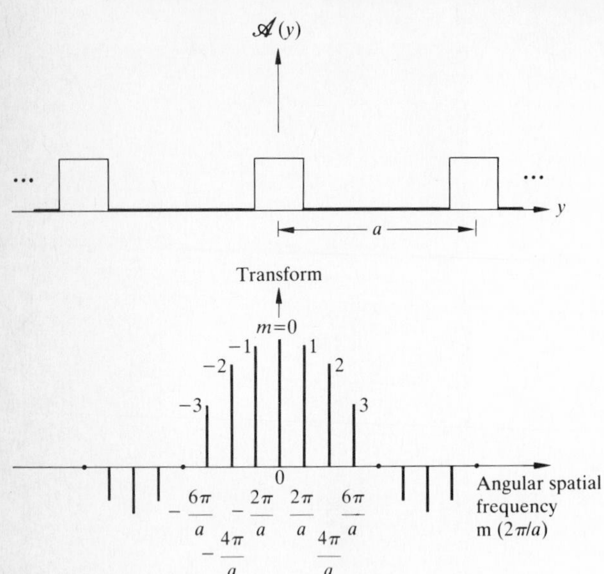

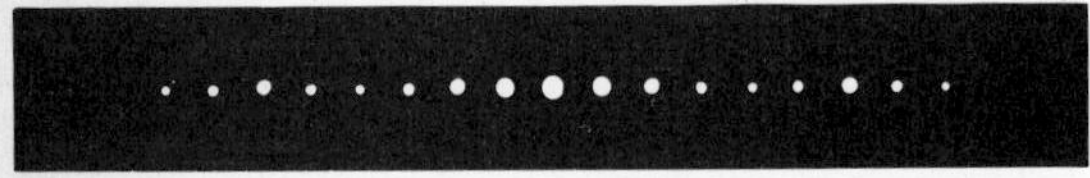

Diffraction pattern

FIGURE 13.28 Square wave and its transform.

FIGURE 13.29 Diffraction pattern of a crossed grating.

simple periodic nature of the grating, we could think of its Fourier-series components, but now we will certainly have to think in terms of Fourier transforms. In any case, *each spot of light in the diffraction pattern denotes the presence of a specific spatial frequency, which is proportional to its distance from the optical axis (zero-frequency location).* Frequency components of positive and negative sign appear diametrically opposite each other about the central axis. If we could measure the electric field at each point in the transform plane, we would indeed observe the transform of the aperture function, but this is not practicable. Instead, what will be detected is the flux-density distribution, where at each point the irradiance is proportional to the time average of the electric field squared or equivalently to the square of the amplitude of the particular spatial frequency contribution at that point.

13.2.2 Abbe's Theory of Image Formation

Consider the system depicted in Fig. 13.30*a*, which is just an elaborated version of Fig. 13.30*b*. Plane monochromatic wavefronts emanating from the collimating lens (L_c) are diffracted by a grating. The result is a distorted wavefront, which we resolve into a new set of plane waves, each corresponding to a given order $m = 0, \pm 1, \pm 2, \ldots$ or spatial frequency and each traveling in a specific direction (Fig. 13.30*b*). The objective lens (L_t) serves as a *transform lens*, forming the Fraunhofer diffraction pattern of the grating on the transform plane Σ_t (which is also the back focal plane of L_t). The waves, of course, propagate beyond Σ_t and arrive at the image plane Σ_i. There they overlap and interfere to form an inverted image of the grating. Accordingly, points G_1 and G_2 are imaged at P_1 and P_2, respectively. The objective lens forms two distinct patterns of interest. One is the Fourier transform on the focal plane conjugate to the plane of the source, and the other is the image of the object, formed on the plane conjugate to the object plane. Figure 13.31 shows the same setup for a long, narrow, horizontal slit coherently illuminated.

We can envision the points S_0, S_1, S_2, and so forth in Fig. 13.30*a* as if they were point emitters of Huygens wavelets, and the resulting diffraction pattern on Σ_t is then the grating's image. In other words, *the image arises from a double diffraction process.* Alternatively, we can imagine that the incoming wave is diffracted by the object, and the resulting diffracted wave is then diffracted once again by the objective lens. If that

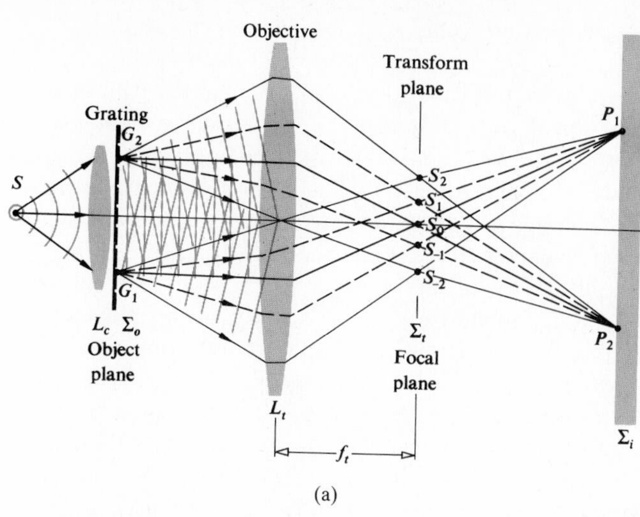

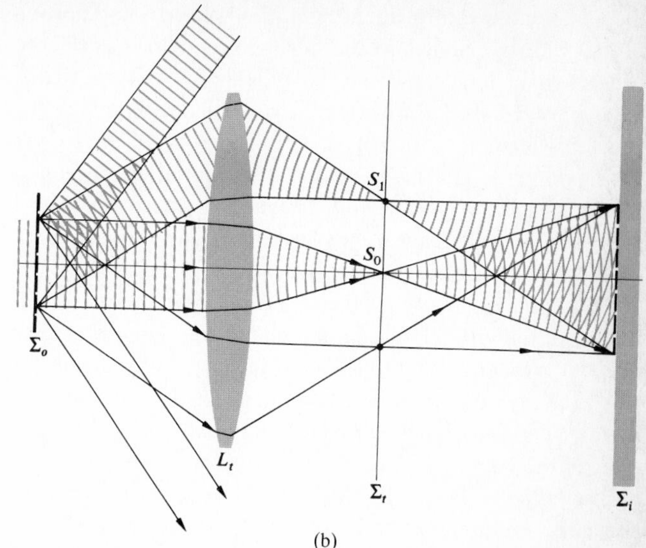

FIGURE 13.30 Image formation.

lens were not there, a diffraction pattern of the object would appear on Σ_i in place of the image.

These ideas were first propounded by Professor Ernst Abbe (1840–1905) in 1873.* His interest at the time concerned the theory of microscopy, whose relationship to the above discussion is clear if we consider L_t as a microscope objective. Moreover, if the grating is replaced by a piece of some thin translucent material (i.e., the specimen being examined), which is illuminated by light from a small source and condenser, the system certainly resembles a microscope.

Carl Zeiss (1816–1888), who in the mid-1800s was running a small microscope factory in Jena, realized the shortcomings of the trial-and-error development techniques of that era. In 1866 he enlisted the services of Ernst Abbe, then lecturer at the University of Jena, to establish a more scientific approach to microscope design. Abbe soon found by experimentation that a larger aperture resulted in higher resolution, even though

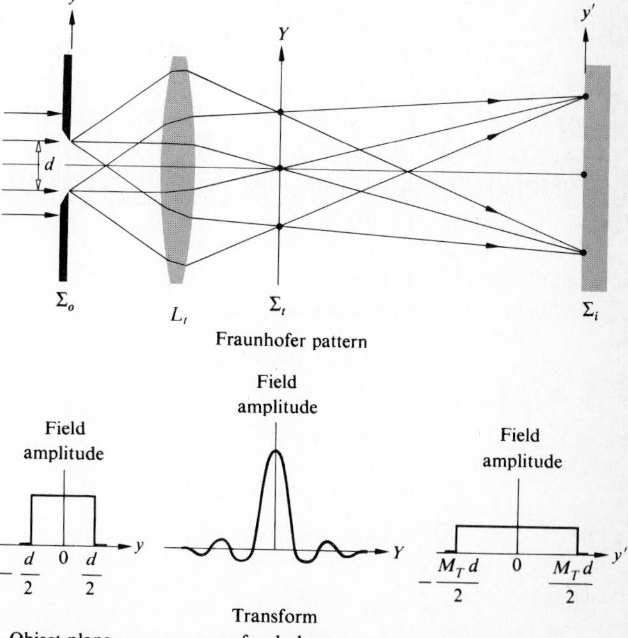

FIGURE 13.31 The image of a slit.

*An alternative and yet ultimately equivalent approach was put forth in 1896 by Lord Rayleigh. He envisaged each point on the object as a coherent source whose emitted wave was diffracted by the lens into an Airy pattern. Each of these in turn was centered on the ideal image point (on Σ_i) of the corresponding point source. Thus Σ_i was covered with a distribution of somewhat overlapping and interfering Airy patterns.

the apparent cone of incident light filled only a small portion of the objective. Somehow the surrounding "dark space" contributed to the image. Consequently, he took the approach that the then well-known diffraction process that occurs at the edge of a lens (leading to the Airy pattern for a point source) was not operative in the same sense as it was for an incoherently illuminated telescope objective. Specimens, whose size was of the order of λ, were apparently scattering light into the "dark space" of the microscope objective. Observe that if, as in Fig. 13.30*b*, the aperture of the objective is not large enough to collect all of the diffracted light, the image does not correspond exactly to that object. Rather, it relates to a fictitious object whose complete diffraction pattern matches the one collected by L_t. We know from the previous section that these lost portions of the outer region of the Fraunhofer pattern are associated with the higher spatial frequencies. And, as we shall see presently, their removal will result in a loss in image sharpness and resolution.

Practically speaking, unless the grating considered earlier has an infinite width, it cannot be strictly periodic. This means that it has a continuous Fourier spectrum dominated by the usual discrete Fourier-series terms, the other being much smaller in amplitude. Complicated, irregular objects clearly display the continuous nature of their Fourier transforms. In any event, it should be emphasized that *unless the objective lens has an infinite aperture, it functions as a low-pass filter rejecting spatial frequencies above a given value and passing all those below* (the former being those that extend beyond the physical boundary of the lens). Consequently, all practical lens systems will be limited in their ability to reproduce the high spatial frequency content of an actual object under coherent illumination.* It might be mentioned as well that there is a basic nonlinearity associated with optical imaging systems operating at high spatial frequencies.†

*Refer to H. Volkmann, "Ernst Abbe and His Work," *Appl. Opt.* **5** 1720 (1966), for a more detailed account of Abbe's many accomplishments in Optics.

†R. J. Becherer and G. B. Parrent, Jr., "Nonlinearity in Optical Imaging Systems," *J. Opt. Soc. Am.* **57**, 1479 (1967).

13.2.3 Spatial Filtering

Suppose we actually set up the system shown in Fig. 13.30*a*, using a laser as a plane-wave source. If the points S_0, S_1, S_2, and so on are to be the sources of a Fraunhofer pattern, the image screen must presumably be located at $x = \infty$ (although 30 or 40 ft will often do). At the risk of being repetitious, recall that the reason for using L_t originally was to bring the diffraction pattern of the object in from infinity. We now introduce an *imaging lens* L_i (Figs. 13.32 and 13.33) in order to bring in from infinity the diffraction pattern of the set of source points $S_0, S_1, S_2, \ldots$, thereby relocating Σ_i at a convenient distance. The transform lens causes the light from the object to converge in the form of a diffraction pattern on the plane Σ_t; that is, it produces on Σ_t a two-dimensional Fourier transform of the object. To wit, the spatial frequency spectrum of the object is spread across the transform plane. Thereafter, L_i (the "*inverse*" *transform lens*) projects the diffraction pattern of the light distributed over Σ_t onto the image plane. In other words, it diffracts the diffracted beam, which effectively means that it generates an (inverted) inverse transform. Thus essentially an inverse transform of the data on Σ_t appears as the final image.

Quite frequently, in practice L_t and L_i are identical ($f_t = f_i$) well-corrected multi-element lenses [for quality work these might have resolutions of about 150 line pairs/mm—one line pair being a period in Fig. 13.26*b*. For less demanding applications, two projector objectives of large aperture (about 100 mm) having convenient focal lengths of roughly 30 or 40 cm serve quite nicely. One of these lenses is then merely turned around so that both their back focal planes coincide with Σ_t. Incidentally, the input or object plane need not be located a focal length away from L_t; the transform still appears on Σ_t. Moving Σ_0 affects only the phase of the amplitude distribution, and that is generally of little interest. The device shown

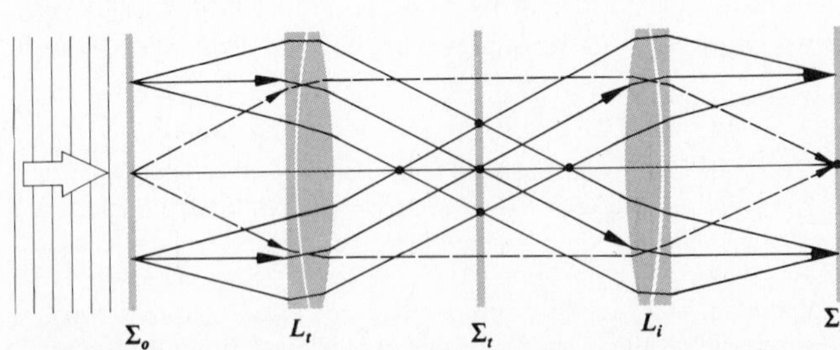

FIGURE 13.32 Object, transform, and image planes.

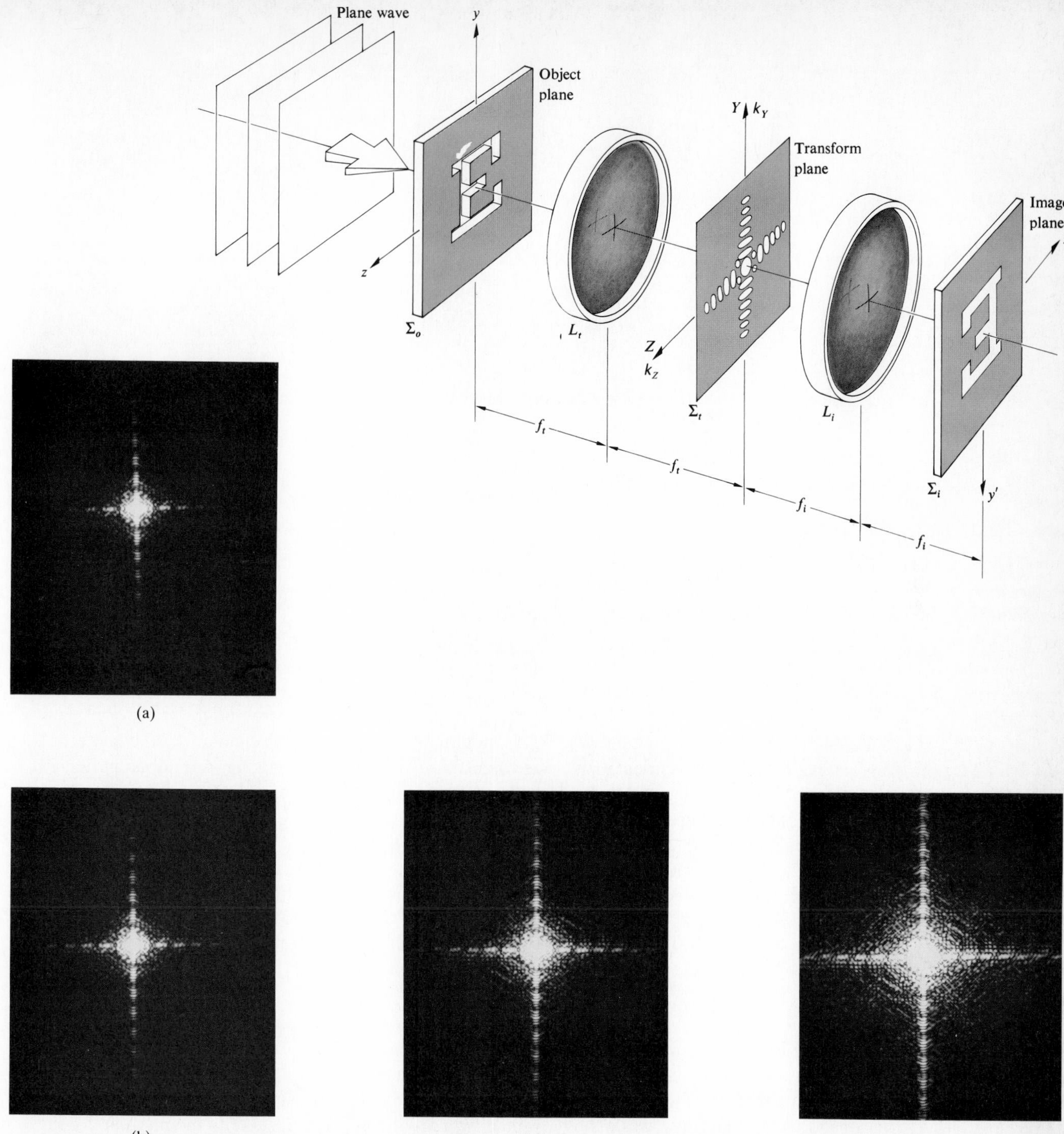

FIGURE 13.33 The Fourier transform of the letter E via an optical computer. Parts (*a*) through (*g*) show more and more of the detail of the transform as the exposure time is increased. (Photos by E. H.) (Continued...)

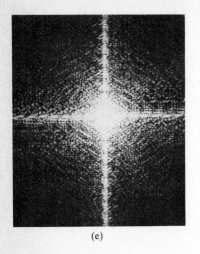

(e)

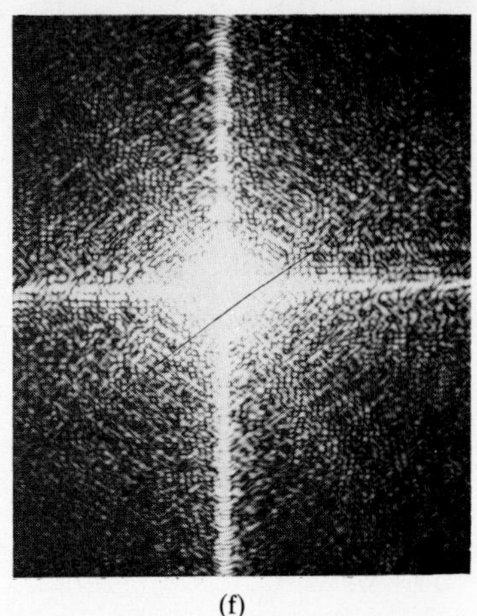

(f)

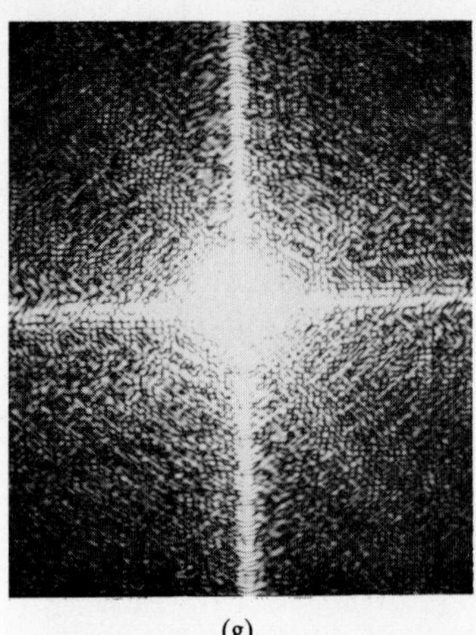

(g)

FIGURE 13.33 (Continued.)

in Figs. 13.32 and 13.33 is often referred to as a ***coherent optical computer***. It allows us to insert obstructions (i.e., masks or filters) into the transform plane and in so doing partially or completely block out certain spatial frequencies, stopping them from reaching the image plane. *This process of altering the frequency spectrum of the image is known as* **spatial filtering** (see Section 7.4.4).

From our earlier discussion of Fraunhofer diffraction we know that a long narrow slit at Σ_0, regardless of its orientation and location, generates a transform at Σ_t consisting of a series of dashes of light lying along a straight line perpendicular to the slit (Fig. 10.10) and *passing through the origin*. Consequently, if the straight-line object is described by $y = mz + b$, the diffraction pattern lies along the line $Y = -Z/m$ or equivalently, from Eqs. (11.64) and (11.65), $k_Y = -k_Z/m$. With this and the Airy pattern in mind, we should be able to anticipate some of the gross structure of the transforms of various objects. Be aware as well that these transforms are centered about the zero-frequency optical axis of the system. For example, a transparent plus sign whose horizontal line is thicker than its vertical one has a two-dimensional transform again shaped more or less like a plus sign. The thick horizontal line generates a series of short vertical dashes, while the thin vertical element produces a line of long horizontal dashes. Remember that object elements with small dimensions diffract through relatively large angles. Along with Abbe, one could think of this entire subject in these terms rather than using the concepts of spatial frequency filtering and transforms, which represent the more modern influence of communication theory.

The vertical portions of the symbol E in Fig. 13.33 generate the broad frequency spectrum appearing as the horizontal pattern. Note that all parallel line sources on a given object correspond to a single linear array on the transform plane. This, in turn, passes through the origin on Σ_t (the intercept is zero), just as in the case of the grating. A transparent figure 5 will generate a pattern consisting of both a horizontal and vertical distribution of spots extending over a relatively large frequency range. There will also be a comparatively low-frequency, concentric ring-like structure. The transforms of disks and rings and the like will obviously be circularly symmetric. Similarly, a horizontal elliptical aperture will generate vertically oriented concentric elliptical bands. Most often, far-field patterns possess a center of symmetry (see Problems 10.14 and 11.29).

We are now in a better position to appreciate the process of spatial filtering and to that end will consider an experiment similar to one published in 1906 by A. B. Porter. Figure 13.34*a* shows a fine wire mesh whose periodic pattern is disrupted by a few particles of dust. With the mesh at Σ_0, Fig. 13.34*b* shows the transform as it would appear on Σ_t. Now the fun starts—since the transform information relating to the dust is located in an irregular cloud-like distribution about the center point, we can easily eliminate it by inserting an opaque mask at Σ_t. If the mask has holes at each of the principal maxima, thus passing on only those frequencies, the image appears dustless (Fig. 13.35*a*). At the other extreme, if we just pass the cloud-like pattern near center, very little of the periodic structure appears, leaving an image consisting of essentially just the dust particles (Fig. 13.35*b*). Passing only the zero-order central spot generates a uniformly illuminated (dc) field, just as if the mesh were no longer in position. Observe that as more and more of the higher frequencies are eliminated, the detail of the image deteriorates markedly [(*d*), (*e*), and (*f*) in Fig. 13.35]. This can be understood quite simply by remembering how a function, with what we might call "sharp edges," was synthesized out of harmonic components. The square wave of Fig. 7.19 serves to illustrate the point. It is evident that the addition of higher harmonics serves predominantly to square up the corners and flatten out the peaks and troughs of the profile. In this way, ***the high spatial frequencies contribute to the sharp edge detail between light and dark regions of the image***. The removal of the high-frequency terms causes a rounding out of the step function and a consequent loss of resolution in the two-dimensional case.

What would happen if we took out the dc component (Fig. 13.35*c*) by passing everything but the central spot? A point on the original image that appears black in the photo denotes a near-zero irradiance and perforce a near-zero field amplitude. Presumably, all of the various optical field components completely cancel each other at that point—ergo, no light. Yet with the removal of the dc term, the point in question must certainly then have a nonzero field amplitude. When squared ($I \propto E_0^2/2$) this will generate a nonzero irradiance. It follows that regions that were originally black in the photo will now appear whitish, while regions that were white will become grayish, as in Fig. 13.36.

Let's now examine some of the possible applications of this technique. Figure 13.37*a* shows a composite photograph of the Moon consisting of film strips pieced together to form a single mosaic. The video data were telemetered to Earth by *Lunar Orbiter 1*. Clearly, the grating-like regular discontinuities between adjacent strips in the object photograph generate the broad-bandwidth, vertical-frequency distribution evident in Fig. 13.37*c*. When these frequency components are blocked, the enhanced image shows no sign of having been a mosaic. In very much the same way, one can suppress extraneous data in bubble chamber photographs of subatomic particle tracks.* These photographs are made difficult to analyze

*D. G. Falconer, "Optical Processing of Bubble Chamber Photographs," *Appl. Opt.* **5**, 1365 (1966), includes some additional uses for the coherent optical computer.

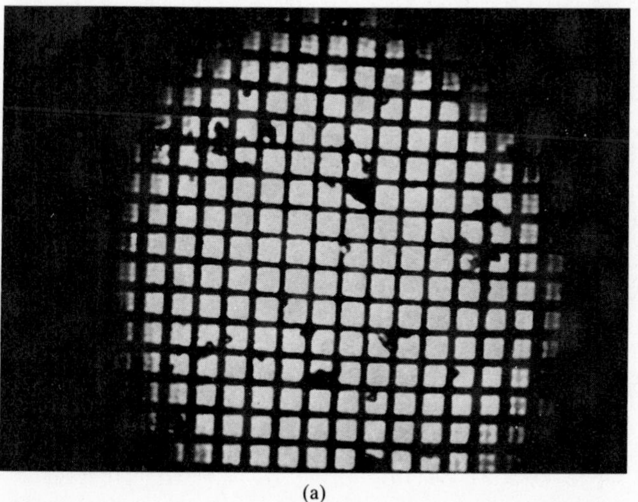

(a)

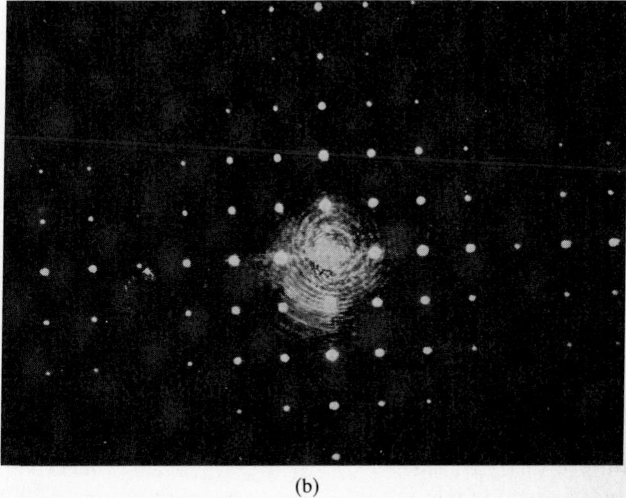

(b)

FIGURE 13.34 A fine, slightly dusty mesh and its transform. (Photos from D. Dutton, M. P. Givens, and R. E. Hopkins, *Spectra-Physics Laser Technical Bulletin Number 3*.)

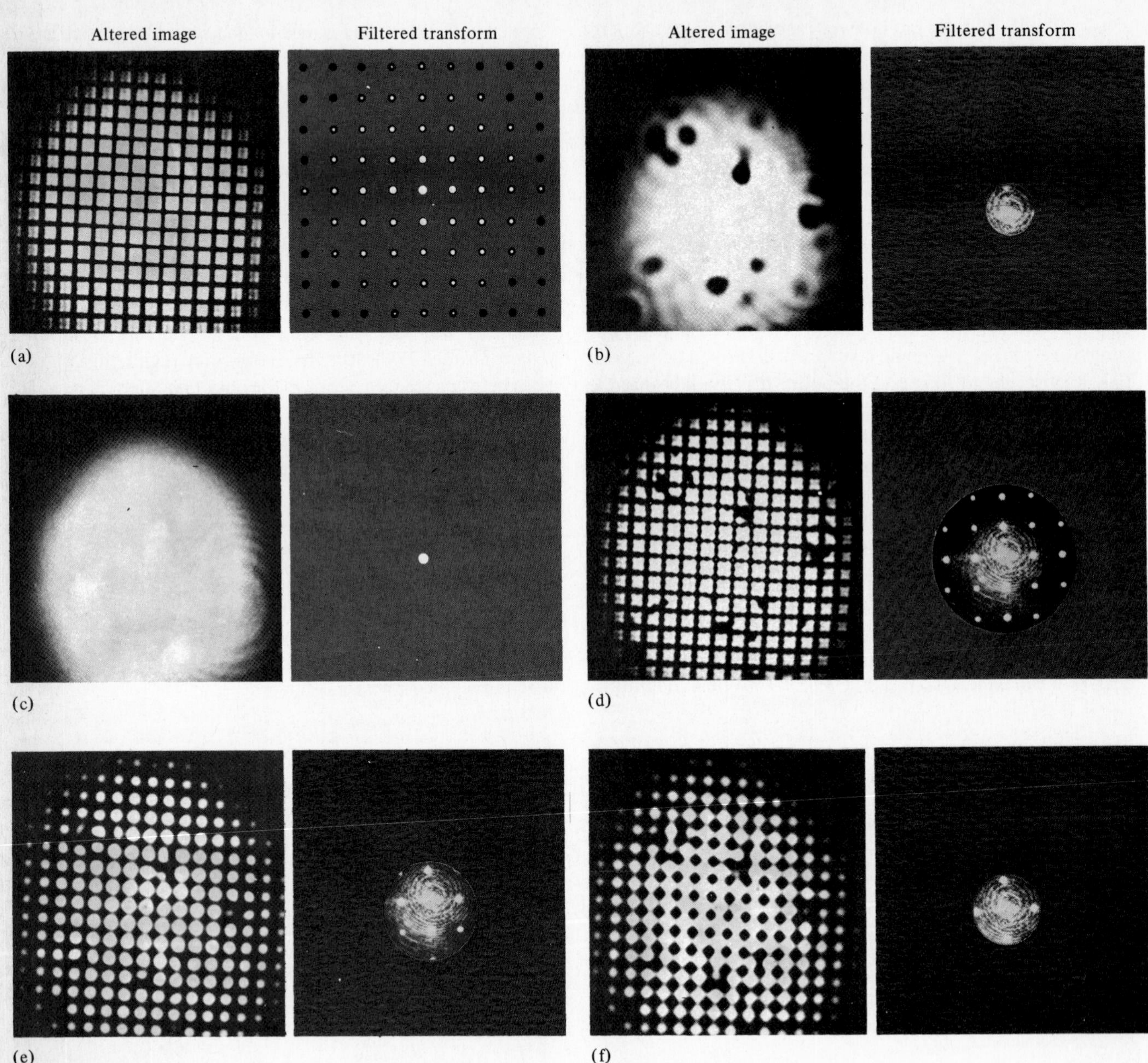

FIGURE 13.35 Images resulting when various portions of the diffraction pattern of Fig. 13.34*b* are obscured by the accompanying masks or spatial filters. (Photos from D. Dutton, M. P. Givens, and R. E. Hopkins, *Spectra-Physics Laser Technical Bulletin Number 3*.)

because of the presence of the unscattered beam tracks (Fig. 13.38), which, since they are all parallel, are easily removed by spatial filtering.

Consider the familiar half-tone or facsimile process by which a printer can create the illusion of various tones of gray while using only black ink and white paper (take a close look

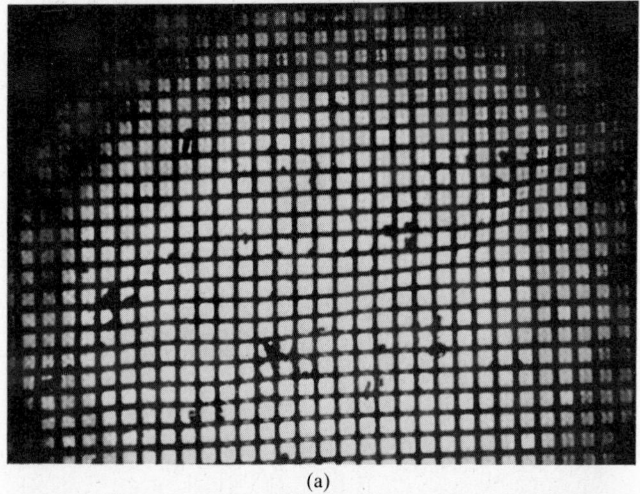

(a)

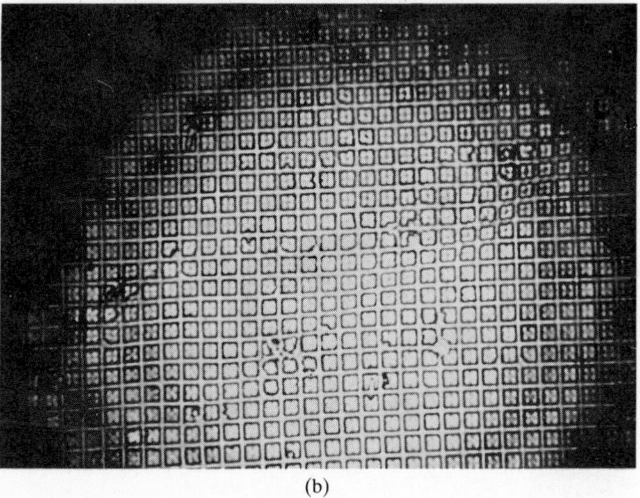

(b)

FIGURE 13.36 Part (*b*) is a filtered version of (*a*) where the zeroth order was removed. (Photos from D. Dutton, M. P. Givens, and R. E. Hopkins, *Spectra-Physics Laser Technical Bulletin Number 3.*)

at a newspaper photograph). If a transparency* of such a facsimile is inserted at Σ_0 in Fig. 13.32, its frequency spectrum will appear on Σ_t. Once again the relatively high-frequency components arising from the half-tone mesh can easily be eliminated. This yields an image in shades of gray (Fig. 13.39) showing none of the discontinuous nature of the original. One could construct a precise filter to obstruct only the square mesh frequencies by actually using a negative transparency of the transform of the basic checkerboard array. Alternatively, it usually suffices to use a low-pass circular aperture filter, and in so doing inadvertently discard some of the high-frequency detail of the original scene, at least as long as the mesh frequency is comparatively high.

The same procedure can be used to remove the graininess of highly enlarged photographs, which is of value, for example, in aerial photo reconnaissance. In contrast, we could sharpen up the details in a slightly blurred photograph by emphasizing its high-frequency components. This could be done with a filter that preferentially absorbed the low-frequency portion of the spectrum. A great deal of effort, beginning in the 1950s, has gone into the study of photographic image enhancement, and the ensuing successes have been notable indeed. Prominent among these contributors was A. Maréchal of the Institut d'Optique, Université de Paris, who combined absorbing and phase-shifting filters to reconstitute the detail in badly blurred photographs. These filters are transparent coatings deposited on optical flats so as to retard the phase of various portions of the spectrum (Section 13.2.4).

As this work in optical data processing continues into the coming decades, we will surely see the replacement of the photographic stages, in increasingly many applications, by real-time electro-optical devices (e.g., arrays of ultrasonic light modulators forming a multichannel input are already in use).* The coherent optical computer will reach a certain maturity, becoming an even more powerful tool when the input, filtering, and output functions are performed electro-optically. A continuous stream of real-time data could flow into and out of such a device.

*Polaroid 55 P/N film is satisfactory for medium resolution work, while Kodak 649 plates are good where higher resolution is required of the transparency.

*We have only touched on the subject of optical data processing; a more extensive discussion of these matters is given, for example, by Goodman in *Introduction to Fourier Optics*, Chapter 7. That text also includes a good reference list for further reading in the journal literature. Also see P. F. Mueller, "Linear Multiple Image Storage," *Appl. Opt.* **8**, 267 (1969). Here, as in much of modern Optics, the frontiers are fast moving, and obsolescence is a hard rider.

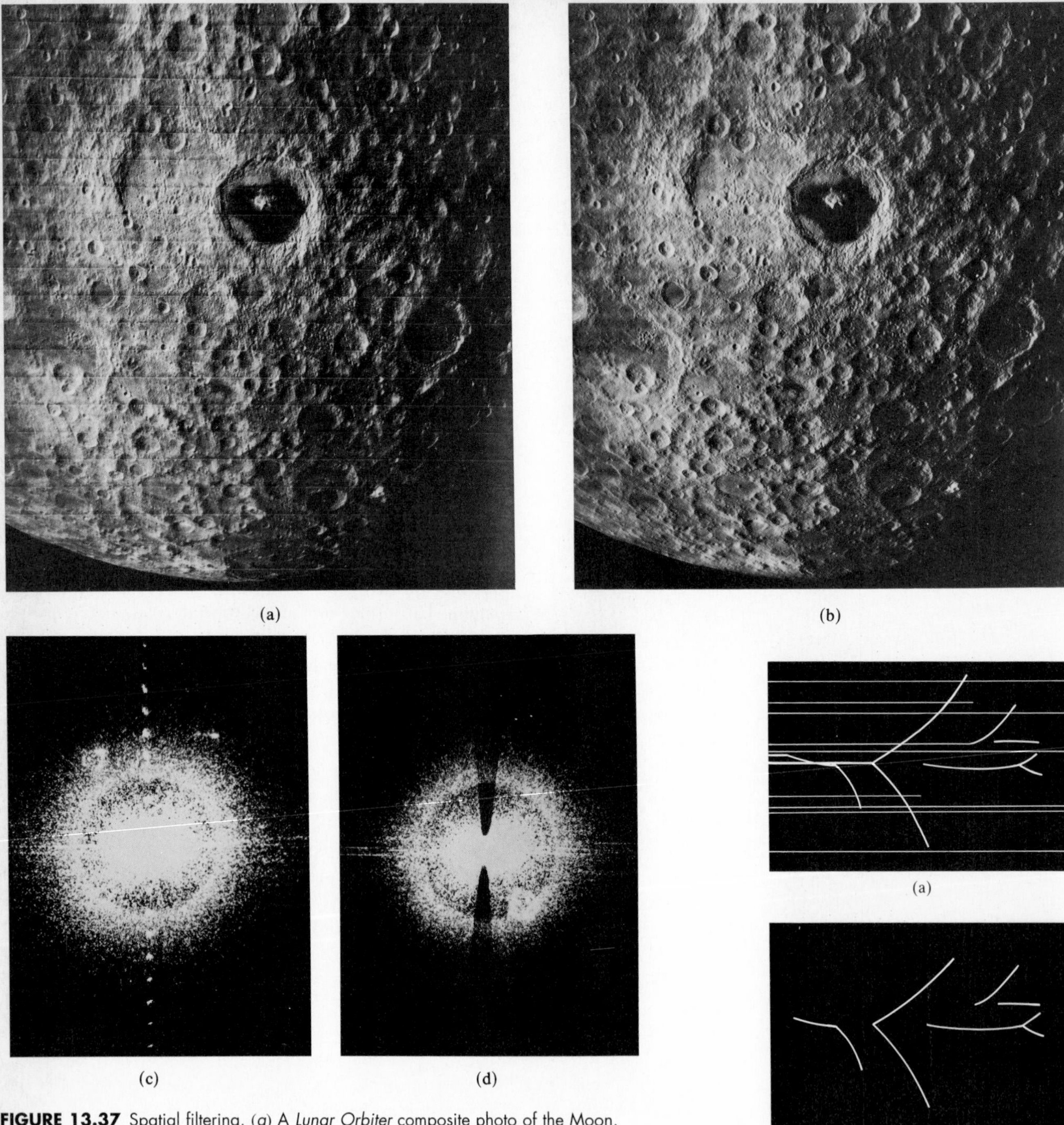

(a) (b) (c) (d)

FIGURE 13.37 Spatial filtering. (*a*) A *Lunar Orbiter* composite photo of the Moon. (*b*) Filtered version of the photo sans horizontal lines. (*c*) A typical unfiltered transform (power spectrum) of a moonscape. (*d*) Diffraction pattern with the vertical dot pattern filtered out. (Photos courtesy D. A. Ansley, W. A. Blikken, The Conductron Corporation, and NASA.)

(a) (b)

FIGURE 13.38 Unfiltered and filtered bubble-chamber tracks.

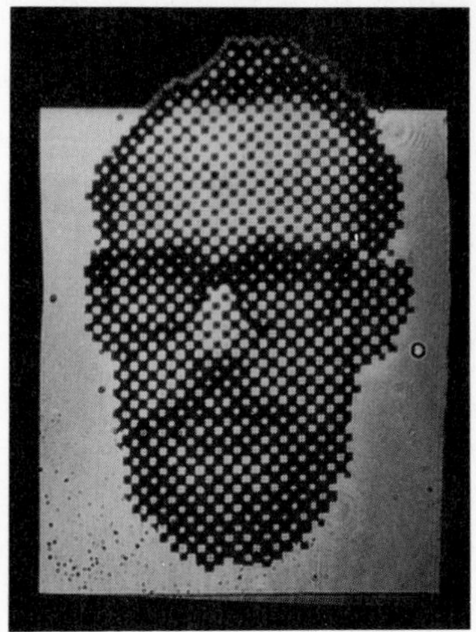

FIGURE 13.39 A self-portrait of K. E. Bethke consisting of only black and white regions as in a halftone. When the high frequencies are filtered out, shades of gray appear and the sharp boundaries vanish. [From R. A. Phillips, *Am. J. Phys.* **37**, 536 (1969).]

13.2.4 Phase Contrast

It was mentioned briefly in the last section that the reconstructed image could be altered by introducing a phase-shifting filter. Probably the best-known example of this technique dates back to 1934 and the work of the Dutch physicist Fritz Zernike, who invented the method of **phase contrast** and applied it in the *phase-contrast microscope*.

An object can be "seen" because it stands out from its surroundings—it has a color, tone, or lack of color, which provides contrast with the background. This kind of structure is known as an *amplitude object*, because it is observable by dint of variations that it causes in the amplitude of the lightwave. The wave that is either reflected or transmitted by such an object becomes *amplitude modulated* in the process. In contradistinction, it is often desirable to "see" *phase objects*, that is, ones that are transparent, thereby providing practically no contrast with their environs and altering only the phase of the detected wave. The optical thickness of such objects generally varies from point to point as either the refractive index or the actual thickness, or both, vary. Obviously, since the eye cannot detect phase variations, such objects are invisible. This is the problem that led biologists to develop techniques for staining transparent microscope specimens and in so doing to convert phase objects into amplitude objects. But this approach is unsatisfactory in many respects, for example, when the stain kills the specimen whose life processes are under study, as is all too often the case.

Recall that diffraction occurs when a portion of the surface of constant phase is obstructed in some way, that is, when a region of the wavefront is altered (either in amplitude or phase, i.e., shape). Suppose then that a plane wave passes through a transparent particle, which retards the phase of a region of the front. The emerging wave is no longer perfectly planar but contains a small indentation corresponding to the area retarded by the specimen; the wave is *phase modulated*.

Taking a rather simplistic view of things, we can imagine the phase-modulated wave $E_{PM}(\mathbf{r}, t)$ (Fig. 13.40) to consist of the original incident plane wave $E_i(x, t)$ plus a localized disturbance $E_d(\mathbf{r}, t)$. (The symbol $\mathbf{r}$ means that E_{PM} and E_d depend on x, y, and z; i.e., they vary over the yz-plane, whereas E_i is uniform and does not.) Indeed, if the phase retardation is very small, the localized disturbance is a wave of very small amplitude, E_{0d}, lagging by just about $\lambda_0/4$, as in Fig. 13.41. There the difference between $E_{PM}(\mathbf{r}, t)$ and $E_i(x, t)$ is shown to be $E_d(\mathbf{r}, t)$. The disturbance $E_i(x, t)$ is called the direct or *zeroth-order wave*, while $E_d(\mathbf{r}, t)$ is the *diffracted wave*. The former produces a uniformly illuminated field at Σ_i, which is unaffected by the object, while the latter carries all of the information about the optical structure of the particle. After broadly diverging from the object, these higher-order spatial frequency terms (see Section 13.2.2) are caused to converge

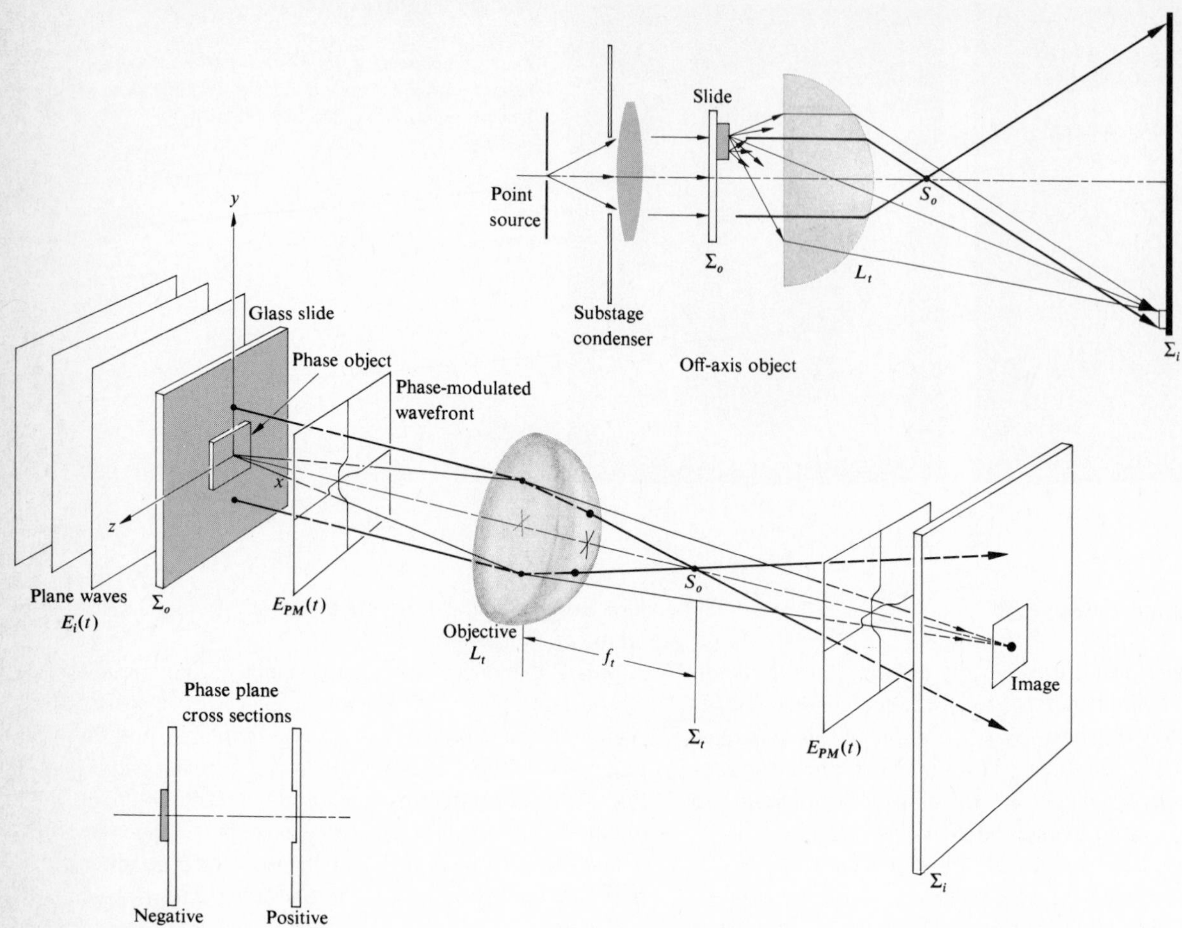

FIGURE 13.40 Phase-contrast setup.

on the image plane. The direct and diffracted waves recombine out-of-phase by $\pi/2$, again forming the phase-modulated wave. Since the amplitude of the reconstructed wave $E_{PM}(\mathbf{r}, t)$ is everywhere the same on Σ_i, even though the phase varies from point to point, the flux density is uniform, and no image is perceptible. Likewise, the zeroth-order spectrum of a phase grating will be $\pi/2$ out-of-phase with the higher-order spectra.

If we could somehow shift the relative phase between the diffracted and direct beams by an additional $\pi/2$ prior to their recombination, they would still be coherent and could then interfere either constructively or destructively (Fig. 13.42). In either case, the reconstructed wavefront over the region of the image would then be amplitude modulated—the image would be visible.

We can see this in a very simple analytical way where

$$E_i(x, t)|_{x=0} = E_0 \sin \omega t \tag{13.18}$$

is the incoming monochromatic lightwave at Σ_o without the specimen in place. The particle will induce a position-dependent phase variation $\phi(y, z)$ such that the wave just leaving it is

$$E_{PM}(\mathbf{r}, t)|_{x=0} = E_0 \sin [\omega t + \phi(y, z)] \tag{13.19}$$

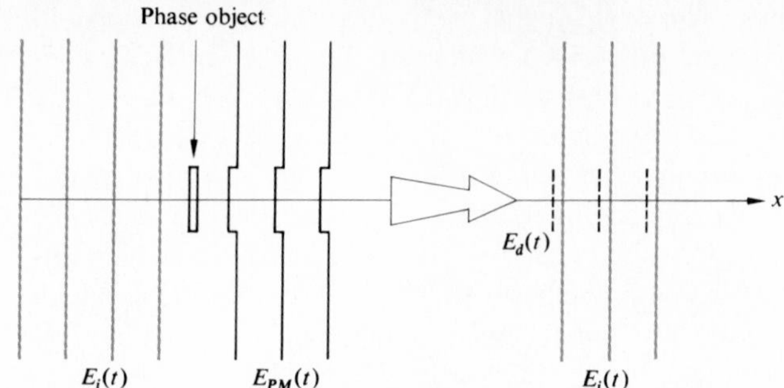

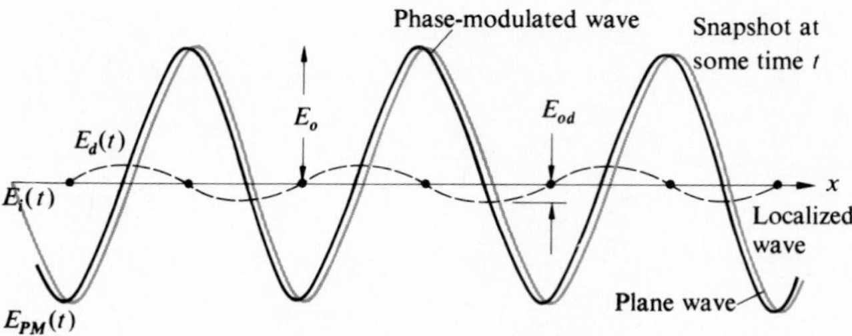

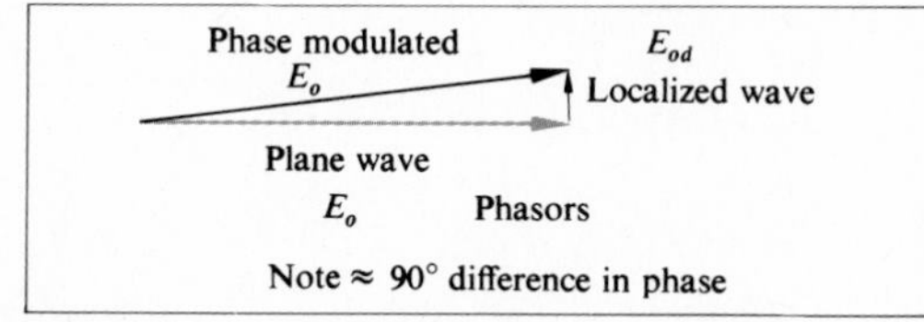

FIGURE 13.41 Wavefronts in the phase-contrast process.

This is a constant-amplitude wave, which is essentially the same on the conjugate image plane. That is, there are some losses, but if the lens is large and aberration-free and we neglect the orientation and size of the image, Eq. (13.19) will suffice to represent the PM wave on either Σ_o or Σ_i. Reformulating that disturbance as

$$E_{PM}(y, z, t) = E_0 \sin \omega t \cos \phi + E_0 \cos \omega t \sin \phi \tag{13.20}$$

and limiting ourselves to *very small values* of ϕ, we obtain

$$E_{PM}(y, z, t) = E_0 \sin \omega t + E_0\phi(y, z) \cos \omega t$$

The first term is independent of the object, while the second term obviously isn't. Thus, as above, if we change their relative phase by $\pi/2$, that is, either change the cosine to sine or vice versa, we get

$$E_{AM}(y, z, t) = E_0[1 + \phi(y, z)] \sin \omega t \tag{13.21}$$

which is an amplitude-modulated wave. Observe that $\phi(y, z)$ can be expressed in terms of a Fourier expansion, thereby introducing the spatial frequencies associated with the object. Incidentally, this discussion is precisely analogous to the one proposed in 1936 by E. H. Armstrong for converting AM radio waves to FM [$\phi(t)$ could be thought of as a frequency modulation wherein the zeroth-order term is the carrier]. An electrical bandpass filter was used to separate the carrier from the remaining information spectrum so that the $\pi/2$ phase shift could be accomplished. Zernike's method of doing essentially the same thing is as follows. He inserted a spatial filter in the transform plane Σ_t of the objective, which was capable of inducing the $\pi/2$ phase shift. Observe that the direct light actually forms a small image of the source on the optical axis at the location of Σ_t. The filter could then be a small circular

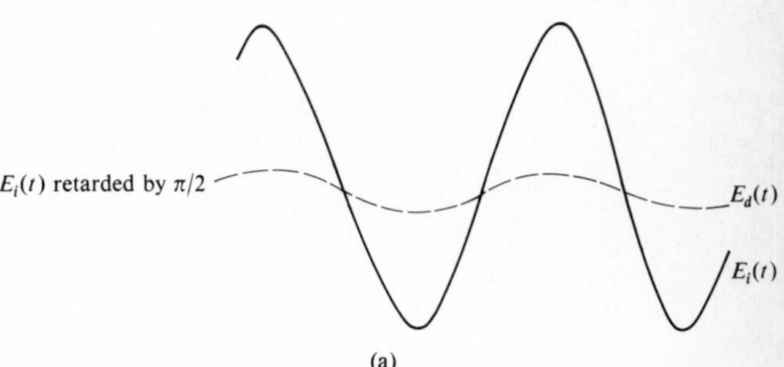

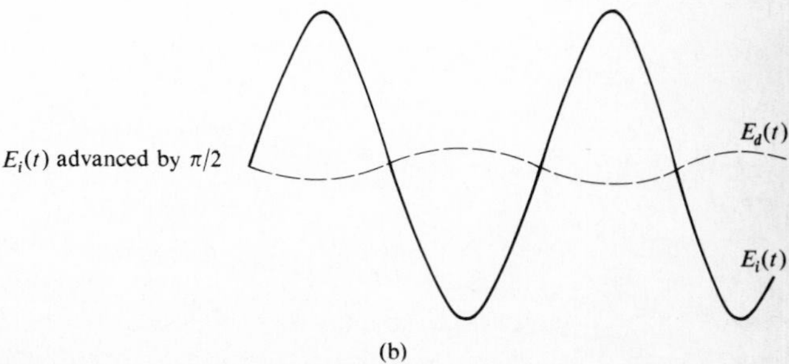

FIGURE 13.42 Effect of phase shifts.

indentation of depth d etched in a transparent glass plate of index n_g. Ideally, only the direct beam would pass through the indentation, and in so doing it would take on a *phase advance* with respect to the diffracted wave of $(n_g - 1)d$, which is made to equal $\lambda_0/4$. A filter of this sort is known as a ***phase plate***, and since its effect corresponds to Fig. 13.42*b*, that is, destructive interference, phase objects that are thicker or have higher indices appear dark against a bright background. If, instead, the phase plate had a small raised disk at its center, the opposite would be true. The former case is called *positive-phase contrast*; the latter, *negative-phase contrast.*

In actual practice, a brighter image is obtained by using a broad, rather than a point, source along with a substage condenser. The emerging plane waves illuminate an annular diaphragm (Fig. 13.43), which, since it is the source plane, is conjugate to the transform plane of the objective. The zeroth-order waves, shown in the figure, pass through the object according to the tenets of geometrical optics. They then traverse the thin annular region of the phase plate located at Σ_t. That region of the plate is quite small, and so the cone of diffracted rays, for the most part, misses it. By making the annular region absorbing as well (a thin metal film will do), the very large uniform zeroth-order term (Fig. 13.44) is reduced with respect to the higher orders, and the contrast improves. Or, if you like, E_0 is reduced to a value comparable with that of the diffracted wave E_{0d}. Generally, a microscope will come

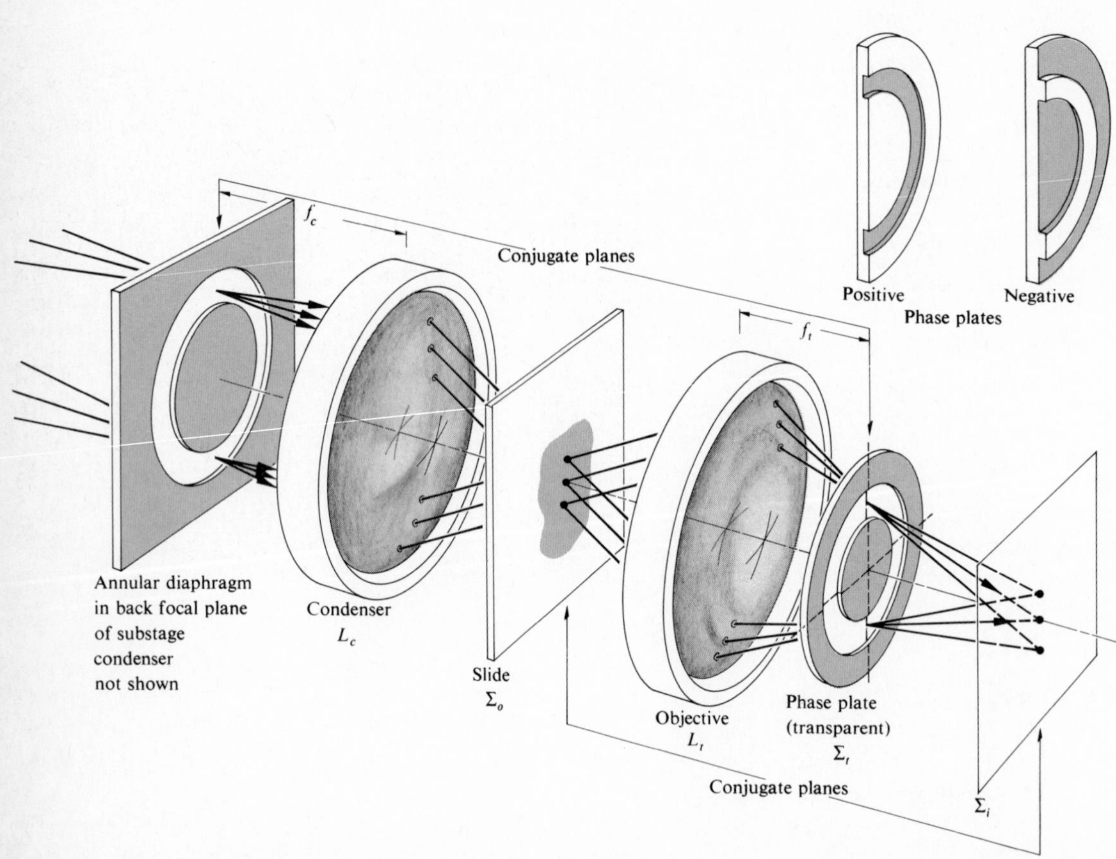

FIGURE 13.43 Phase contrast (only zeroth order shown).

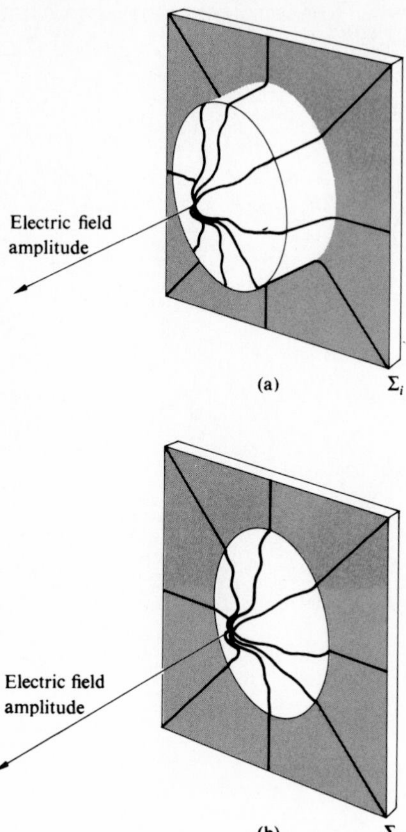

FIGURE 13.44 Field amplitude over a circular region on the image plane. In one case there is no absorption in the phase plate, and the irradiance would be a small ripple on a great plateau. With the zeroth order attenuated, the contrast increases.

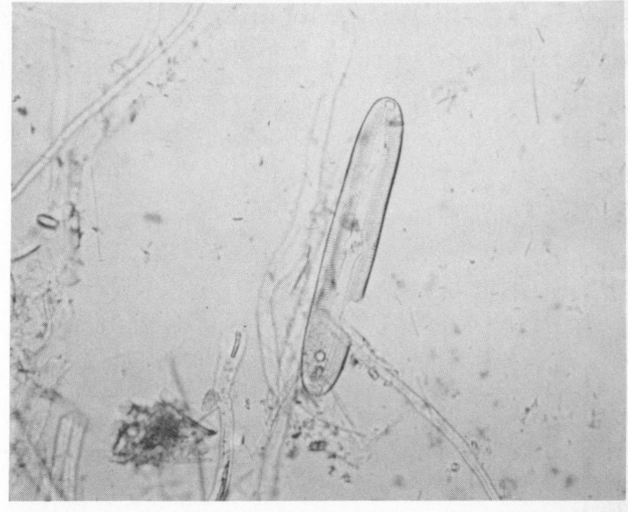

(a)

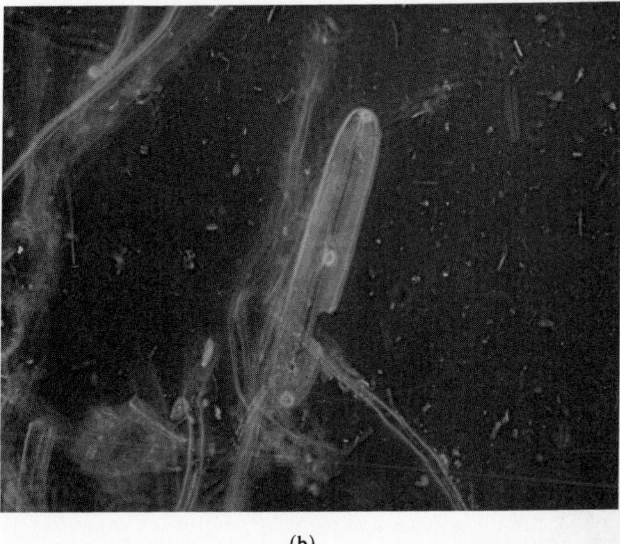

(b)

FIGURE 13.45 (*a*) A conventional photomicrograph of diatoms, fibers, and bacteria. (*b*) A phase photomicrograph of the same scene. (Photos by T. J. Lowery and R. Hawley.)

with an assortment of these phase plates having different absorptions.

In the parlance of modern Optics (the still-blushing bride of communications theory), phase contrast is simply the process whereby we introduce a $\pi/2$ phase shift in the zeroth-order spectrum of the Fourier transform of a phase object (and perhaps attenuate its amplitude as well) through the use of an appropriate spatial filter.

The phase-contrast microscope, which earned Zernike the Nobel Prize in 1953, has found extensive applications (Fig. 13.45), perhaps the most fascinating of which is the study of the life functions of otherwise invisible organisms.

13.2.5 The Dark-Ground and Schlieren Methods

Suppose we go back to Fig. 13.40, where we were examining a phase object, and this time rather than retard and attenuate the central zeroth order, we remove it completely with an opaque disk at S_o. Without the object in place, the image plane will be completely dark—ergo the name ***dark ground***. With the object in position, only the localized diffracted wave will appear at Σ_i to form the image. (This can also be accomplished in microscopy by illuminating the object obliquely so that no direct light enters the objective lens.) Observe that by eliminating the dc contribution, the amplitude distribution (as in Fig. 13.44) will be lowered and portions that were near zero prior to filtering will become negative. Inasmuch as irradiance is proportional to the amplitude squared, this will result in somewhat of a contrast reversal from that which would have been seen in phase contrast (see Section 13.2.3). In general, this technique has not been as satisfactory as the phase-contrast method, which generates a flux-density distribution across the image that is directly proportional to the phase variations induced across the object.

In 1864 A. Toepler introduced a procedure for examining defects in lenses, which has come to be known as the **schlieren** method.* We will discuss it here because of the widespread current usage of the method in a broad range of fluid dynamics studies and furthermore because it is another beautiful example of the application of spatial filtering. Schlieren systems are particularly useful in ballistics, aerodynamics, and ultrasonic wave analysis (Fig. 13.46)—indeed, wherever it is desirable to examine pressure variations as revealed by refractive-index mapping.

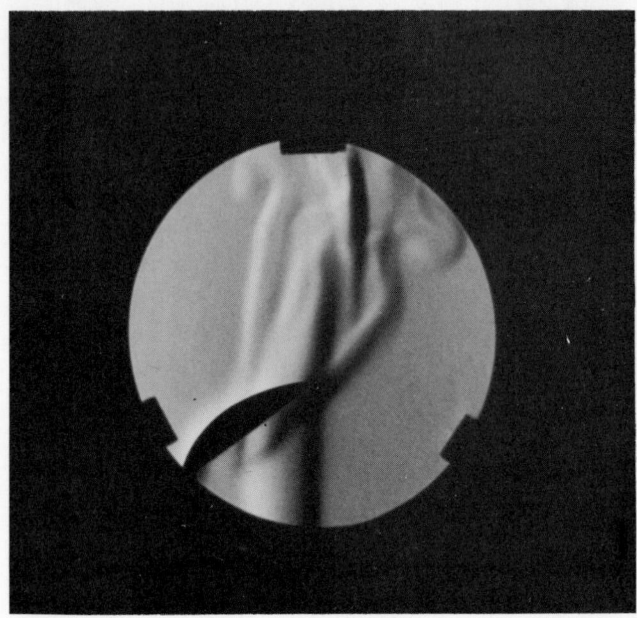

FIGURE 13.46 A schlieren photo of a spoon in a candle flame. (Photo by E. H.)

Suppose that we set up any one of the possible arrangements for viewing Fraunhofer diffraction (e.g., Fig. 10.6 or P.10.5). But now, instead of using an aperture of some sort as the diffracting amplitude object, we insert a phase object, for example, a gas-filled chamber (Fig. 13.47). Again a Fraunhofer pattern is formed in Σ_t, and if that plane is followed by the objective lens of a camera, an image of the chamber is formed on the film plane. We would then photograph any amplitude objects within the test area, but, of course, phase objects would still be invisible. Imagine that we now introduce a knife edge at Σ_t, raising it from below until it obstructs (sometimes only partially) the zeroth-order light and therefore all the higher orders on the bottom side as well. Just as in the dark-ground method, phase objects are then perceptible. Inhomogeneities in the test chamber windows and flaws in the lenses are also noticeable. For this reason and because of the large field of view usually required, mirror systems (Fig. 13.48) have not become commonplace.

Quasimonochromatic illumination is generally made use of when resulting data are to be analyzed electronically, for example, with a photodetector. Sources with a broad spectrum, on the other hand, allow us to exploit the considerable color sensitivity of photographic emulsions, and a number of color schlieren systems have been devised.

*The word *Schlieren* in German means streaks or striae. It's frequently capitalized because all nouns are in German and not because there was a Mr. Schlieren.

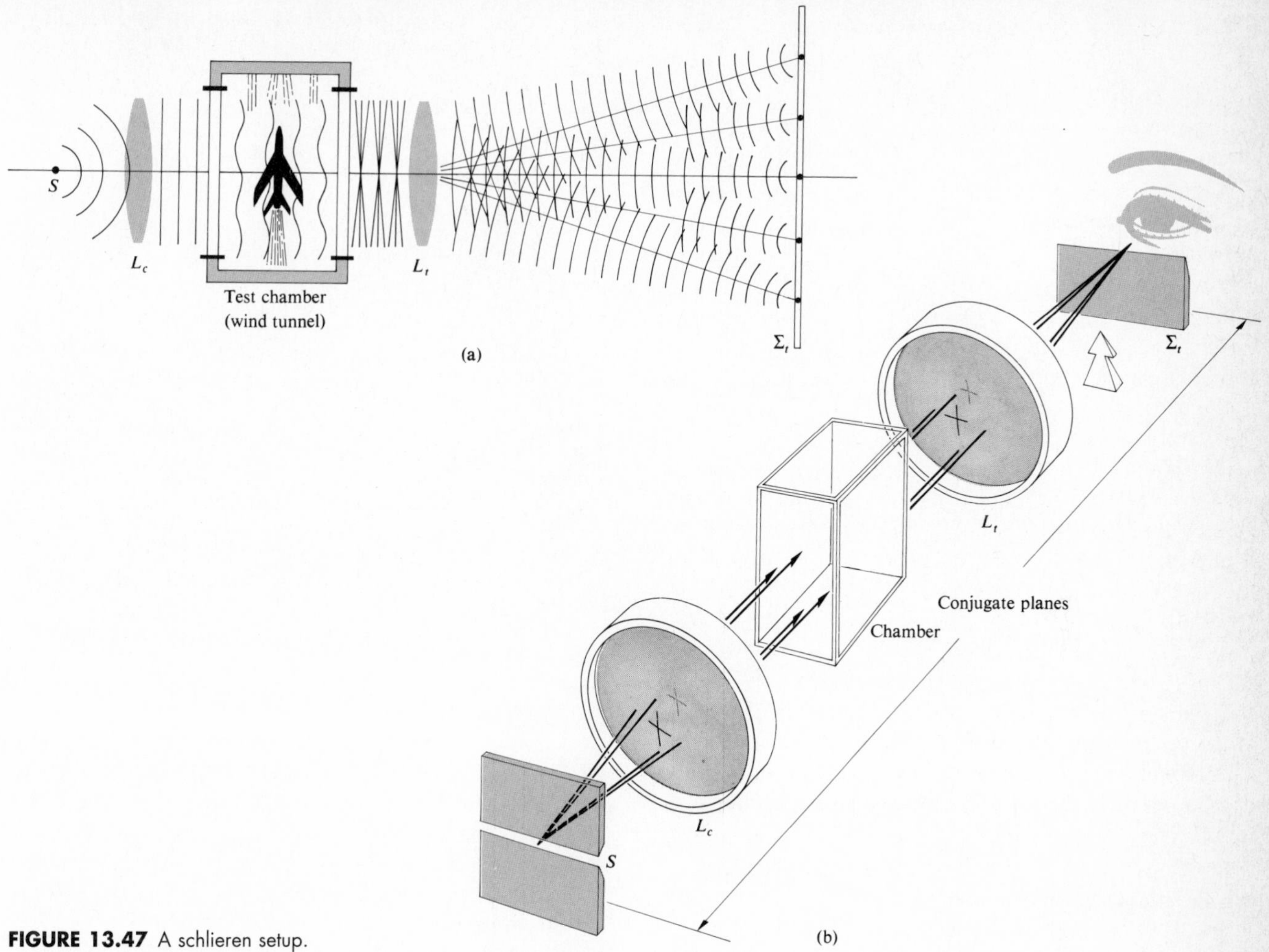

FIGURE 13.47 A schlieren setup.

13.3 Holography

The technology of photography has been with us for a long time, and we've all grown accustomed to seeing the three-dimensional world compressed into the flatness of a scrapbook page. The depthless television pitchman who smiles out of a myriad of phosphorescent flashes, although inescapably there, seems no more palpable than a postcard image of the Eiffel Tower. Both share the severe limitation of being simply irradiance mappings. In other words, when the image of a scene is ordinarily reproduced, by whatever traditional means, what we ultimately see is not an accurate reproduction of the light field that once inundated the object, but rather a point-by-point record of just the square of the field's amplitude. The light reflecting off a photograph carries with it information about the irradiance but nothing about the phase of the wave that once emanated from the object. Indeed, if both the amplitude and phase of the original wave could be reconstructed somehow, the resulting light field (assuming the frequencies are the same) would be indistinguishable from the original. This means that you would then see (and could photograph) the re-

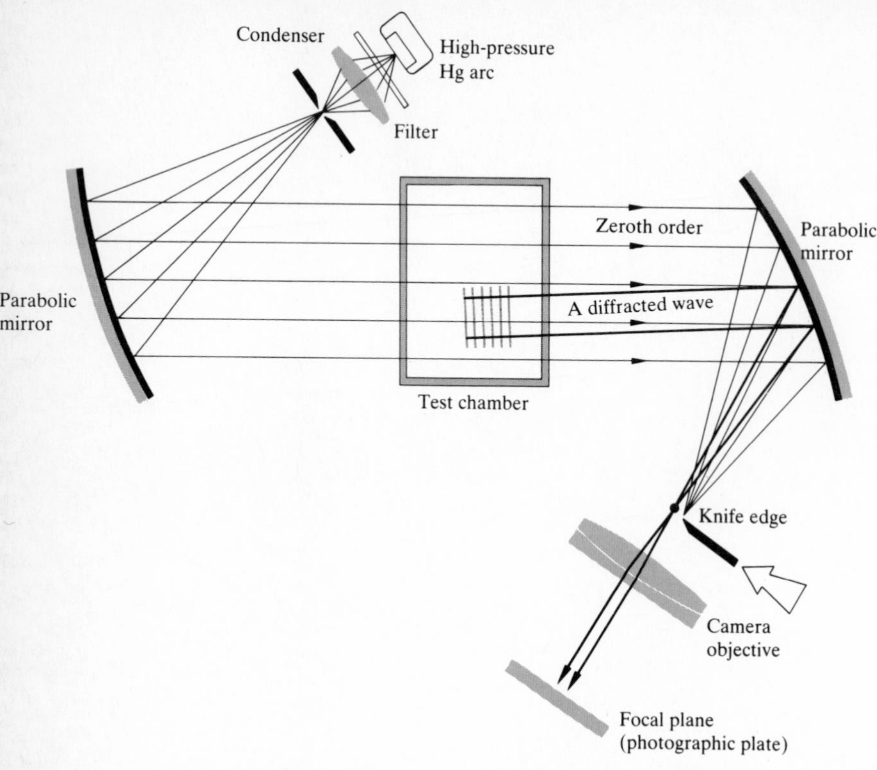

FIGURE 13.48 A schlieren setup using mirrors.

formed image in perfect three-dimensionality, exactly as if the object were there before you, actually generating the wave.

13.3.1 Methods

Dennis Gabor had been thinking along these lines for a number of years prior to 1947, when he began conducting his now famous experiments in holography at the Research Laboratory of the British Thomson–Houston Company. His original setup, depicted in Fig. 13.49, was a two-step lensless imaging process in which he first photographically recorded an interference pattern, generated by the interaction of scattered quasimonochromatic light from an object and a coherent reference wave. The resulting pattern was something he called a **hologram**, after the Greek word *holos*, meaning whole. The second step in the procedure was the *reconstruction* of the optical field or image, and this was done through the diffraction of a coherent beam by a transparency, which was the developed hologram. In a way quite reminiscent of Zernike's phase-contrast technique (Section 13.2.4), the hologram was formed when the unscattered *background* or *reference wave* interfered with the diffracted wave from the small semitransparent object, *S*—which was, in those early days, often a piece of microfilm. The key point is that the interference pattern or hologram contains, by way of the fringe configuration, information corresponding to both the amplitude and phase of the wave scattered by the object.

Admittedly, it's not at all obvious that by now shining a plane wave through the processed hologram one could reconstruct an image of the original object. Suffice it to say for the moment that if the object were very small, the scattered wave would be nearly spherical, and the interference pattern a series of concentric rings (centered about an axis through the object and normal to the plane wave). Except for the fact that the circular fringes would vary gradually in irradiance from one to the next, the resulting flux-density distribution would correspond to a conventional Fresnel zone plate (Section 10.3.5). Recall that a zone plate functions somewhat like a lens in that it diffracts collimated light into a beam converging to a real focal point, P_r. In addition, it produces a diverging wave, which appears to come from the point P_r and constitutes a virtual image. Thus we can imagine, albeit rather simplistically, that each point on an extended object generates its own zone

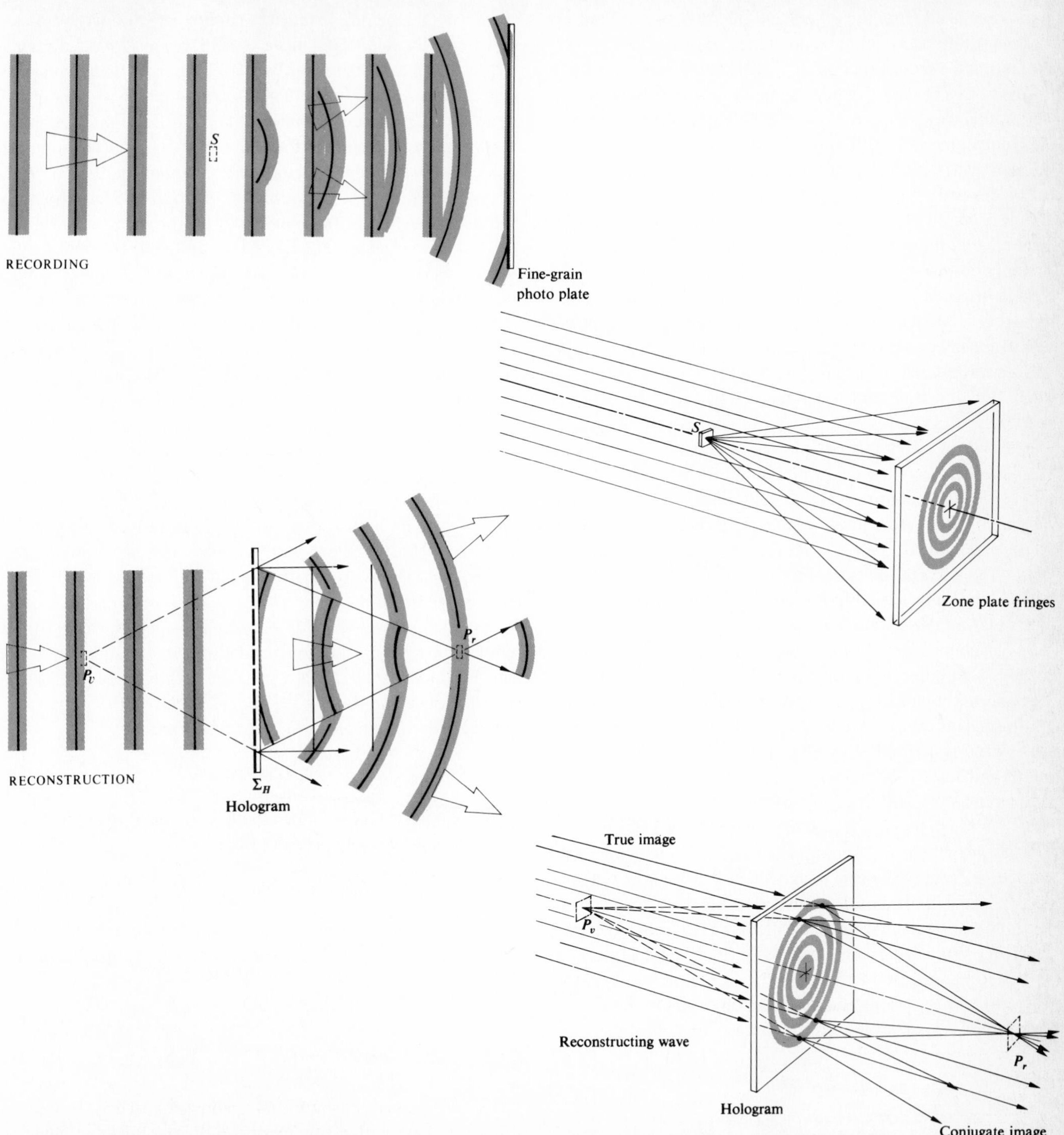

FIGURE 13.49 Holographic (in-line) recording and reconstruction of an image.

plate displaced from the others and that the ensemble of all such partially overlapping zone plates forms the hologram.*

During the reconstruction step, each constituent zone plate forms both a real and virtual image of a single object point, and in this way, point by point, the hologram regenerates the original light field. When the reconstructing beam has the same wavelength as the initial recording beam (which need not necessarily be the case, and quite often isn't), the virtual image is undistorted and appears at the location formerly occupied by the object. Thus it is the virtual image field that actually corresponds to the original object field. As such, the virtual image is sometimes spoken of as the *true image*, while the other is the real or, perhaps more fittingly, the *conjugate image*. In any event, we envision the hologram as a composite of interference patterns, and at least for this very simple configuration, those patterns resemble zone plates. As we will see presently, the sinusoidal grating is an equally fundamental fringe system making up complex holograms.

Gabor's research, which won him the 1971 Nobel Prize in Physics, had as its motivation an improvement in electron microscopy. His work initially generated some interest, but all in all it remained in a state of quasi-unnoticed oblivion for about 15 years. In the early 1960s there was a resurgence of interest in Gabor's **wavefront reconstruction** process and, in particular, in its relation to certain radar problems. Soon, aided by an abundance of the new coherent laserlight and extended by a number of technological advances, holography became a subject of widespread research and tremendous promise. This rebirth had its origin in the Radar Laboratory of the University of Michigan, with the work of Emmett N. Leith and Juris Upatnieks. Among other things, they introduced an improved arrangement for generating holograms, which is illustrated in Fig. 13.50. Unlike Gabor's *in line*-configuration, where the conjugate image was inconveniently located in front of the true image, the two were now satisfactorily separated off-axis, as shown in the diagram. Once again, the hologram is an interference pattern arising from a coherent reference wave and a wave scattered from the object (this type is sometimes referred to as a **side-band Fresnel hologram**). Figure 13.51 shows the equivalent arrangement for producing side-band Fresnel holograms from transparent objects.

What's happening here can be appreciated in two ways—an essentially pictorial, Fourier-optical way and, alternatively, a direct mathematical way. We will look from both perspectives, because they complement each other. First, this is at heart an interference (or, if you like, a diffraction) problem, and we can again return to the notion of the complicated object wavefront being composed of Fourier-component plane waves (Figs. 7.32 and 10.11) traveling in directions associated with the different spatial frequencies of the object's light field, reflected or transmitted. Each one of these Fourier plane waves interferes with the reference wave on the photographic plate and thus preserves the information associated with that particular spatial frequency in the form of a characteristic fringe pattern.

To see how this occurs, examine the simplified two-wave version depicted in Fig. 13.52. At the moment shown, the reference wave happens to have a crest along the face of the film plane, and the scattered object wavelet, coming in at an angle θ, similarly has crests at points A, B, and C. These correspond to points where interference maxima will occur at the moment shown. But as both waves progress to the right, they will remain in-phase at these points, trough will overlap trough, and the maxima will remain fixed at A, B, and C. Similarly, between these points, trough overlaps crest, and minima exist. The relative phase (ϕ) of these two waves, which varies from point to point along the film, can be written as a function of x. Since ϕ changes by 2π as x goes the length of $\overline{AB}$, $\phi/2\pi = x/\overline{AB}$. Notice that $\sin\theta = \lambda/\overline{AB}$, and so getting rid of the specific length $\overline{AB}$, the phase in general becomes

$$\phi(x) = (2\pi x \sin\theta)/\lambda \tag{13.22}$$

If the two waves are assumed to have the same amplitude E_0, the resultant field follows from Eq. (7.17):

$$E = 2E_0 \cos \tfrac{1}{2}\phi \sin(\omega t - kx - \tfrac{1}{2}\phi)$$

and the irradiance distribution, which is proportional to the field amplitude squared, by way of Eq. (3.44), has the form

$$I(x) = \tfrac{1}{2}c\epsilon_0(2E_0 \cos \tfrac{1}{2}\phi)^2 = 2c\epsilon_0 E_0^2 \cos^2 \tfrac{1}{2}\phi$$

or

$$I(x) = 2c\epsilon_0 E_0^2 + 2c\epsilon_0 E_0^2 \cos\phi. \tag{13.23}$$

What we have is a cosinusoidal irradiance distribution across the film plane with a spatial period of $\overline{AB}$ and a spatial frequency ($1/\overline{AB}$) of $\sin\theta/\lambda$.

*See M. P. Givens, "Introduction to Holography," *Am. J. Phys.* **35**, 1056 (1967).

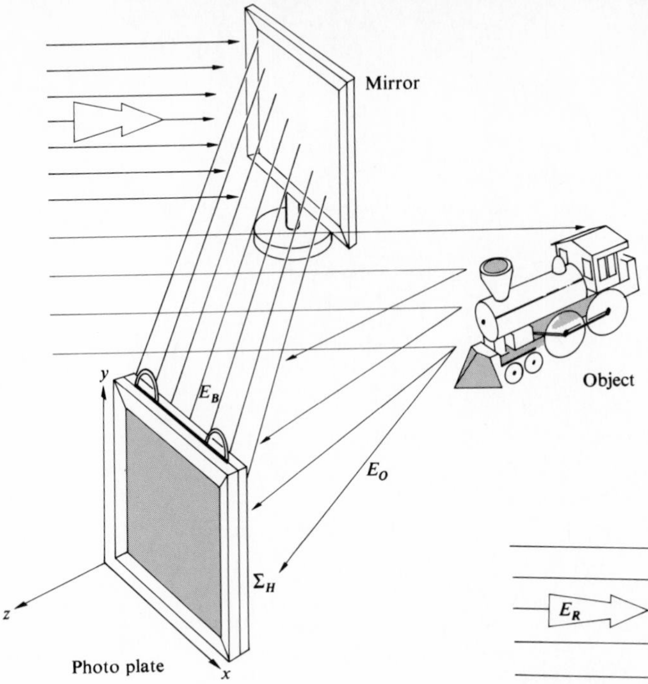

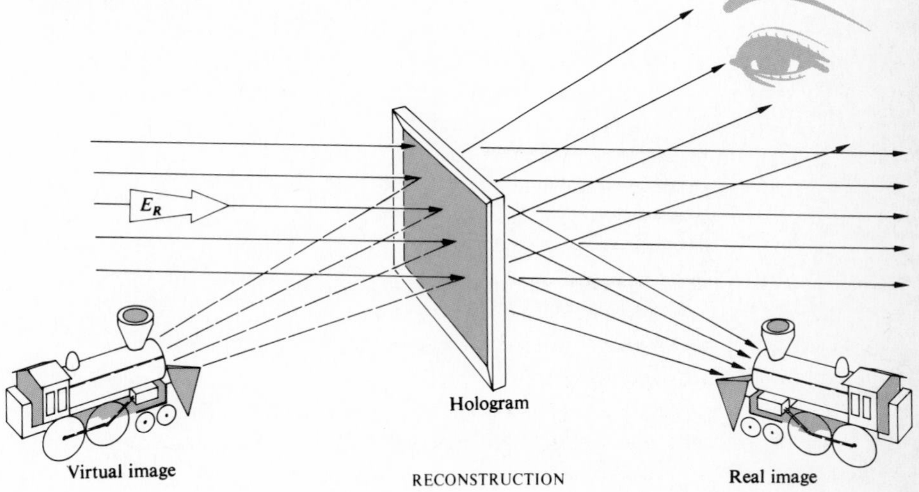

FIGURE 13.50 Holographic (side-band) recording and reconstruction of an image.

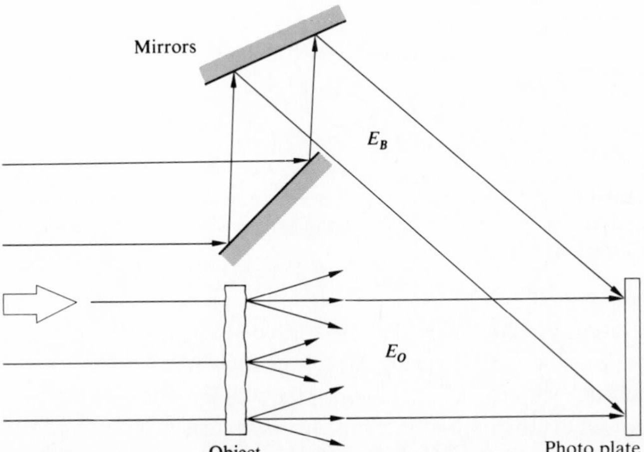

FIGURE 13.51 A side-band Fresnel holographic setup for a transparent object.

Upon processing the film so that the amplitude transmission profile corresponds to $I(x)$, the result is a cosinusoidal grating. When this simple hologram (which essentially corresponds to a structureless object with no information) is illuminated by a plane wave identical to the original reference wave (Fig. 13.52*c*) three beams will emerge: one zeroth and two first order. One of these first-order beams will travel in the direction of the original object beam and corresponds to its reconstructed wavefront.

Now suppose we go one step beyond this most basic hologram and examine an object that has some optical structure. Accordingly, let's use as the object a transparency with a simple periodic structure that has a single spatial frequency—a cosine grating. A slightly idealized representation (which leaves out the weak higher-order terms due to the finite size of the beam and grating) is depicted in Fig. 13.53, which shows the illuminated grating, the three transmitted beams, and the

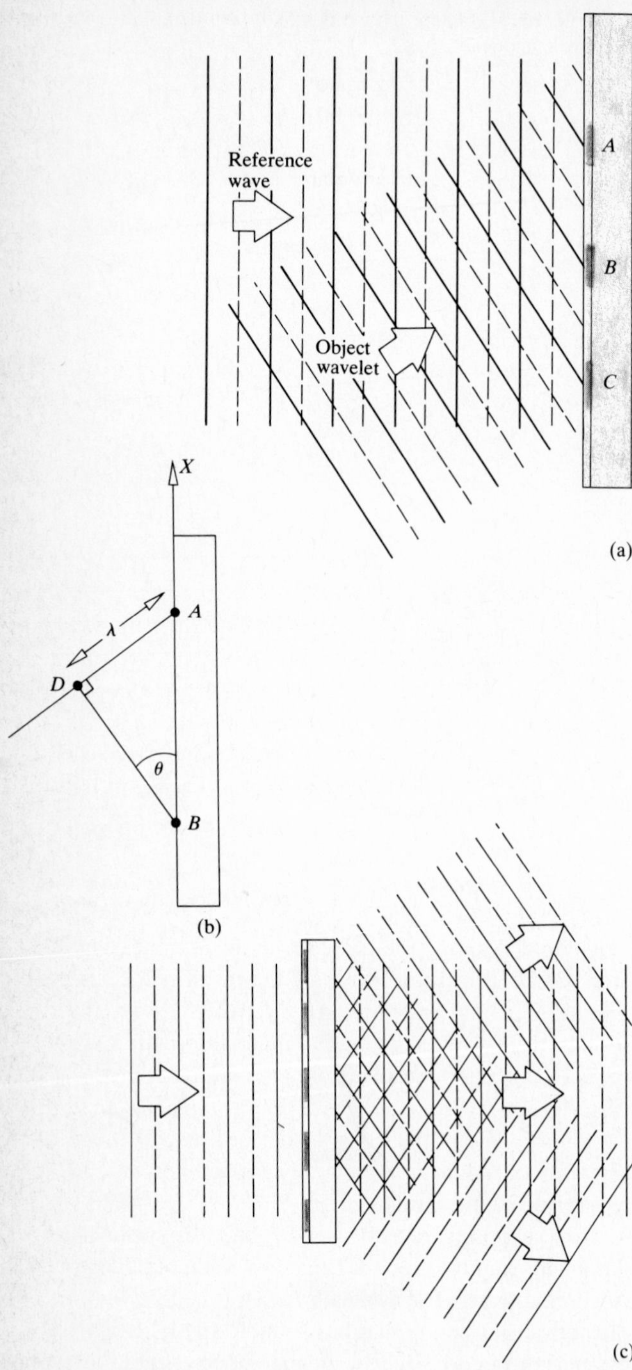

FIGURE 13.52 The interference of two plane waves to create a cosine grating.

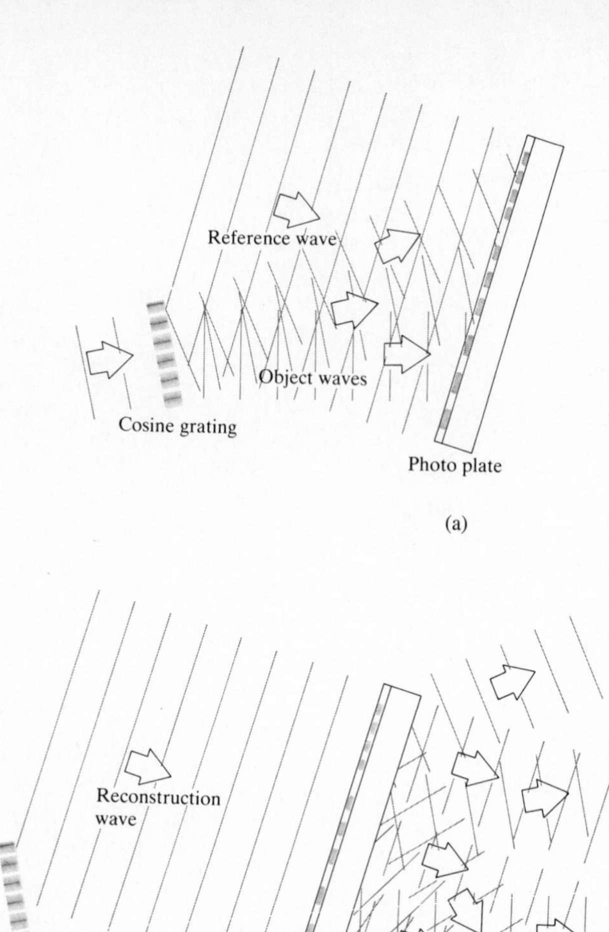

FIGURE 13.53 Notice that there are three regions with different spatial frequencies. Each of these on the reilluminated hologram generates three waves.

reference beam. What results is three slightly different versions of Fig. 13.52, where each of the three transmitted waves makes a slightly different angle (θ) with the reference wave. Consequently, each of the three overlap areas will correspond to a set of cosine fringes of a slightly different spatial frequency, from Eq. (13.22). Again when we play back the resulting hologram, Fig. 13.53*a* and *b*, we have three pieces of business:

the undiffracted wave, the virtual image, and the real image. Observe that it is only where the three beams come together to contribute their spatial frequency content that images of the original grating are formed.

When a still more complex object is used, we can anticipate that the relative phase between the object and reference waves (ϕ) will vary from point to point in a complicated way, thereby modulating the basic carrier signal (Fig. 13.54) produced by two plane waves when no object is present. We can generalize from Fig. 13.53 and conclude that the phase-angle difference ϕ (which varies with θ) is encoded in the configuration of the fringes. Furthermore, had the amplitudes of the reference and object waves been different, the irradiance of those fringes would have been altered accordingly. Thus we can guess that the amplitude of the object wave at every point on the film plane will be encoded in the visibility of the resulting fringes.

The process depicted in Fig. 13.50 can be treated analytically as follows. Suppose that the xy-plane is the plane of the hologram, Σ_H. Then

$$E_B(x, y) = E_{0B} \cos [2\pi\nu t + \phi(x, y)] \qquad (13.24)$$

describes the planar background or reference wave at Σ_H, overlooking considerations of polarization. Its amplitude, E_{0B}, is constant, while the phase is a function of position. This just means that the reference wavefront is tilted in some known manner with respect to Σ_H. For example, if the wave were oriented such that it could be brought into coincidence with Σ_H by a single rotation through an angle of θ about y, the phase at any point on the hologram plane would depend on its value of x. Thus ϕ would again have the form

$$\phi = \frac{2\pi}{\lambda} x \sin \theta = kx \sin \theta$$

being, in that particular case, independent of y and varying linearly with x. For the sake of simplicity, we'll just write it, quite generally, as $\phi(x, y)$ and keep in mind that it's a simple known function. The wave scattered from the object can, in turn, be expressed as

$$E_O(x, y) = E_{0O}(x, y) \cos [2\pi\nu t + \phi_O(x, y)] \qquad (13.25)$$

where both the amplitude and phase are now complicated functions of position corresponding to an irregular wavefront. From the communications-theoretic point of view, this is an amplitude- and phase-modulated carrier wave bearing all of the available information about the object. Note that this information is encoded in spatial rather than temporal variations of the wave. The two disturbances E_B and E_O superimpose and

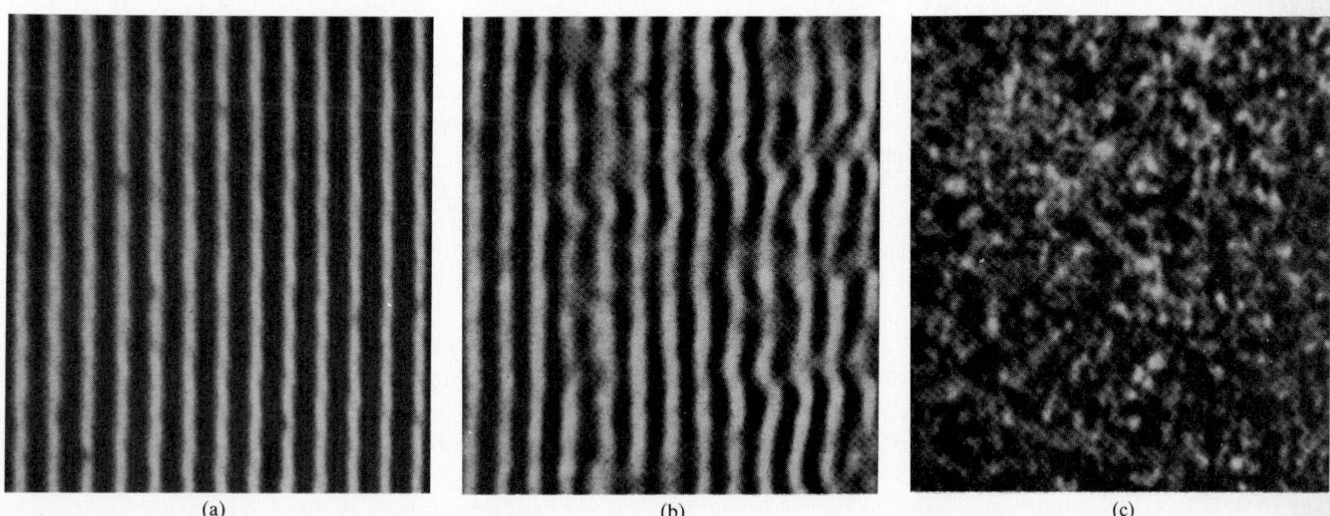

(a) (b) (c)

FIGURE 13.54 Various degrees of modulation of hologram fringes. (Photo courtesy Emmett N. Leith and *Scientific American*.)

interfere to form an irradiance distribution, which is recorded by the photographic emulsion. The resulting irradiance, except for a multiplicative constant, is $I(x, y) = \langle (E_B + E_O)^2 \rangle_T$, which, from Section 9.1, is given by

$$I(x, y) = \frac{E_{0B}^2}{2} + \frac{E_{0O}^2}{2} + E_{0B}E_{0O} \cos (\phi - \phi_O) \qquad (13.26)$$

Observe once again that the phase of the object wave determines the location on Σ_H of the irradiance maxima and minima. Moreover, the contrast or fringe visibility

$$\mathcal{V} \equiv (I_{\max} - I_{\min})/(I_{\max} + I_{\min}) \qquad [12.1]$$

across the hologram plane, which is

$$\mathcal{V} = 2E_{0B}E_{0O}/(E_{0B}^2 + E_{0O}^2) \qquad (13.27)$$

contains the appropriate information about the object wave's amplitude.

Once more, in the parlance of communications theory, we might observe that the film plate serves as both the storage device and detector or mixer. It produces, over its surface, a distribution of opaque regions corresponding to a modulated spatial waveform. Accordingly, the third or difference frequency term in Eq. (13.27) is both amplitude and phase modulated by way of the position dependence of $E_{0O}(x, y)$ and $\phi_O(x, y)$.

Figure 13.54*b* is an enlarged view of a portion of the fringe pattern that constitutes the hologram for a simple, essentially two-dimensional, semitransparent object. Were the two interfering waves perfectly planar (as in Fig. 13.54*a*), the evident variations in fringe position and irradiance, which represent the information, would be absent, yielding the traditional Young's pattern (Section 9.3). The sinusoidal transmission-grating configuration (Fig. 13.54*a*) may be thought of as the carrier waveform, which is then modulated by the signal. Furthermore, we can imagine that the coherent superposition of countless zone-plate patterns, one arising from each point on a large object, have metamorphosed into the modulated fringes of Fig. 13.54*b*. When the amount of modulation is further greatly increased, as it would be for a large, three-dimensional, diffusely reflecting object, the fringes lose the kind of symmetry still discernible in Fig. 13.54*b* and become considerably more complicated. Incidentally, holograms are often covered with extraneous swirls and concentric ring systems that arise from diffraction by dust and the like on the optical elements.

The amplitude transmission profile of the processed hologram can be made proportional to $I(x, y)$. In that case, the *final emerging wave*, $E_F(x, y)$, is proportional to the product $I(x, y)E_R(x, y)$, where $E_R(x, y)$ is the *reconstructing wave* incident on the hologram. Thus if the reconstructing wave, of frequency ν, is incident obliquely on Σ_H, as was the background wave, we can write

$$E_R(x, y) = E_{0R} \cos [2\pi\nu t + \phi(x, y)] \qquad (13.28)$$

The final wave (except for a multiplicative constant) is the product of Eqs. (13.26) and (13.28):

$$E_F(x, y) = \tfrac{1}{2}E_{0R}(E_{0B}^2 + E_{0O}^2) \cos [2\pi\nu t + \phi(x, y)]$$
$$+ \tfrac{1}{2}E_{0R}E_{0B}E_{0O} \cos (2\pi\nu t + 2\phi - \phi_O)$$
$$+ \tfrac{1}{2}E_{0R}E_{0B}E_{0O} \cos (2\pi\nu t + \phi_O) \qquad (13.29)$$

Three terms describe the light issuing from the hologram; the first can be rewritten as

$$\tfrac{1}{2}(E_{0B}^2 + E_{0O}^2)E_R(x, y)$$

and is an amplitude-modulated version of the reconstructing wave. In effect, each portion of the hologram functions as a diffraction grating, and this is again the *zeroth-order*, undeflected, direct beam. Since it contains no information about the phase of the object wave, ϕ_O, it is of little concern here.

The next two or *side-band waves* are the sum and difference terms, respectively. These are the two *first-order waves* diffracted by the grating-like hologram. The first of these (i.e., the sum term) represents a wave that, except for a multiplicative constant, has the same amplitude as the object wave $E_{0O}(x, y)$. Moreover, its phase contains a $2\phi(x, y)$ contribution, which, as you recall, arose from tilting the background and reconstructing wavefronts with respect to Σ_H. It's this phase factor that provides the angular separation between the real and virtual images. Furthermore, rather than containing the phase of the object wave, the sum term contains its negative. Thus it's a wave carrying all of the appropriate information about the object but in a way that is not quite right. Indeed, this is the real image formed in converging light in the space beyond the hologram, that is, between it and the viewer. The negative phase is manifest in an inside-out image something like the pseudoscopic effect occurring when the elements of a photographic stereo pair are interchanged. Bumps appear as indentations, and object points that were in front of and nearer to Σ_H are now imaged nearer to but beyond Σ_H. Thus a point on the original subject closest to the observer appears farthest away in the real image. The scene is turned in on itself along one axis in a way that perhaps must be seen to be appreciated.

For example, imagine you are looking down the holographic conjugate image of a bowling alley. The "back" row of pins, even though partially obscured by the "front" rows, are nonetheless imaged closer to the viewer than is the one-pin. Despite this, bear in mind that it's not as if you were looking at the array from behind. No light from the very backs of the pins was ever recorded—you're seeing an inside-out front view. As a consequence, the conjugate image is usually of limited utility, although it can be made to have a normal configuration by forming a second hologram with the real image as the object.

The difference term in Eq. (13.29), except for a multiplicative constant, has precisely the form of the object wave $E_{0O}(x, y)$. If you were to peer into (not at) the illuminated hologram, as if it were a window looking out onto the scene beyond, you would "see" the object exactly as if it were truly sitting there. You could move your head a bit and look around an item in the foreground in order to see the view it had previously been obstructing. In other words, in addition to complete three-dimensionality, parallax effects are apparent as they are in no other reproducing technique (Fig. 13.55).

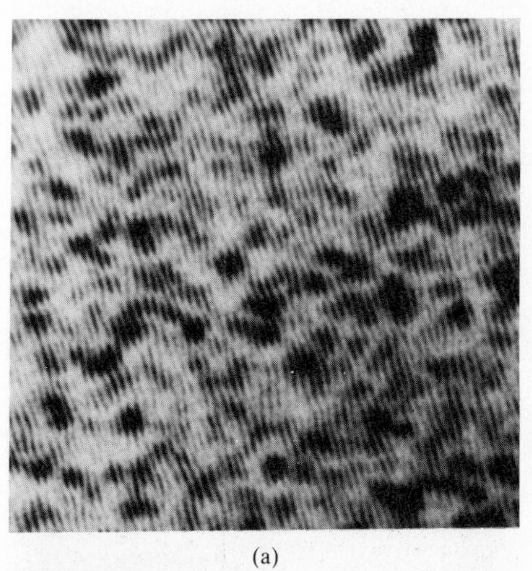
(a)

(b)

(c)

(d)

FIGURE 13.55 Parts (*b*) through (*d*) are three different views photographed from the same holographic image generated by the hologram in (*a*). (Photos from Smith, *Principles of Holography*.)

Imagine that you are viewing the holographic image of a magnifying glass focused on a page of print. As you move your eye with respect to the hologram plane, the words being magnified by the lens (which is itself just an image) actually change, just as they would in "real" life with a "real" lens and "real" print. In the case of an extended scene having considerable depth, your eyes would have to refocus as you viewed different regions of it at various distances. In precisely the same way, a camera lens would have to be readjusted if you were photographing different regions of the virtual image (Fig. 13.56).

Holograms display other extremely important and interesting features. For example, if you were standing close to a window, you could obscure all of it with, say, a piece of cardboard, except for a tiny area through which you could then peer and still see the objects beyond. The same is true of a hologram, since each small fragment of it contains information about the entire object, at least as seen from the same vantage point, and each fragment can reproduce, albeit with diminishing resolution, the entire image.

Figure 13.57 summarizes pictorially much of what's been said so far while also providing a convenient setup for actually making and viewing a hologram. Here the photographic emulsion is shown having some depth, as compared with Fig. 13.52, where it was treated as though it were purely two-dimensional. Of course, any emulsion must certainly have a finite thickness. Typically, it would be about 10 μm thick, as compared with the spatial period of the fringes, which might average around 1 μm or so. Figure 13.58*a* is closer to the point, showing the kind of three-dimensional fringes that actually exist throughout the emulsion. For plane waves, these straight parallel fringe-planes are oriented so as to bisect the angle between the reference and object waves. Realize that all the holograms considered up to now have been viewed by looking through them; they're all **transmission holograms**, and in each case they were made by causing the reference wave and the object wave to traverse the film from the same side.

Something similar happens when the reference and object waves traverse the emulsion from opposite sides, as in Fig.

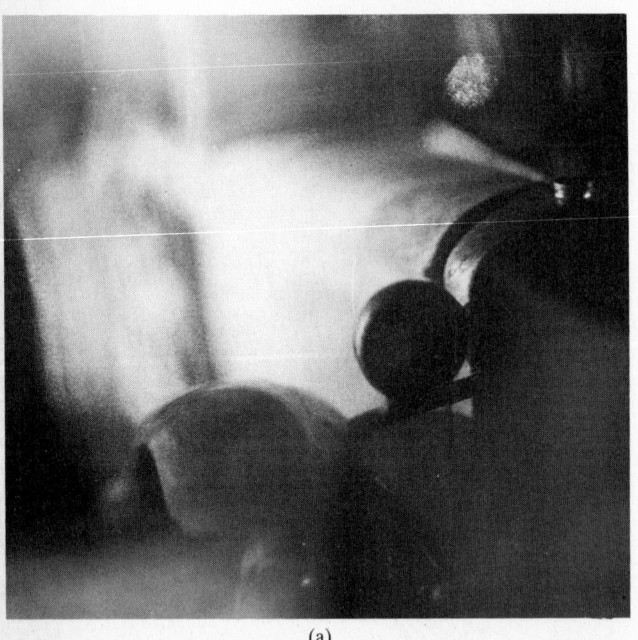

(a)

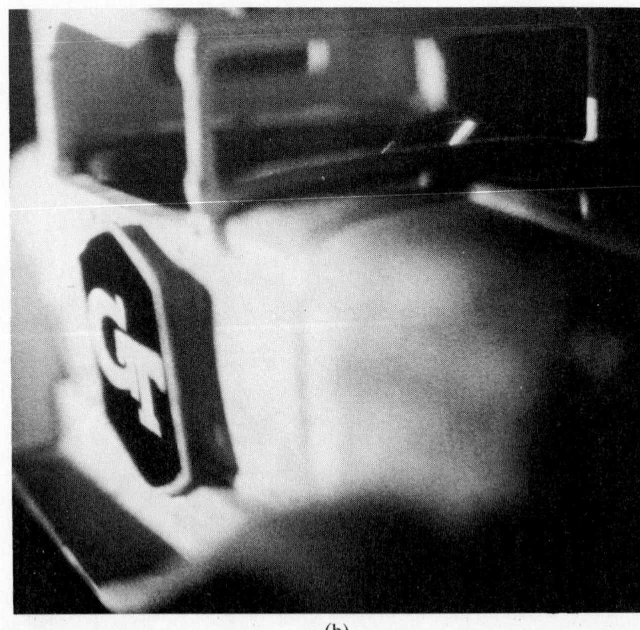

(b)

FIGURE 13.56 A reconstructed holographic image of a model automobile. The camera position and plane of focus were changed between (*a*) and (*b*). (Photos from O'Shea, Callen, and Rhodes, *An Introduction to Lasers and Their Applications*.)

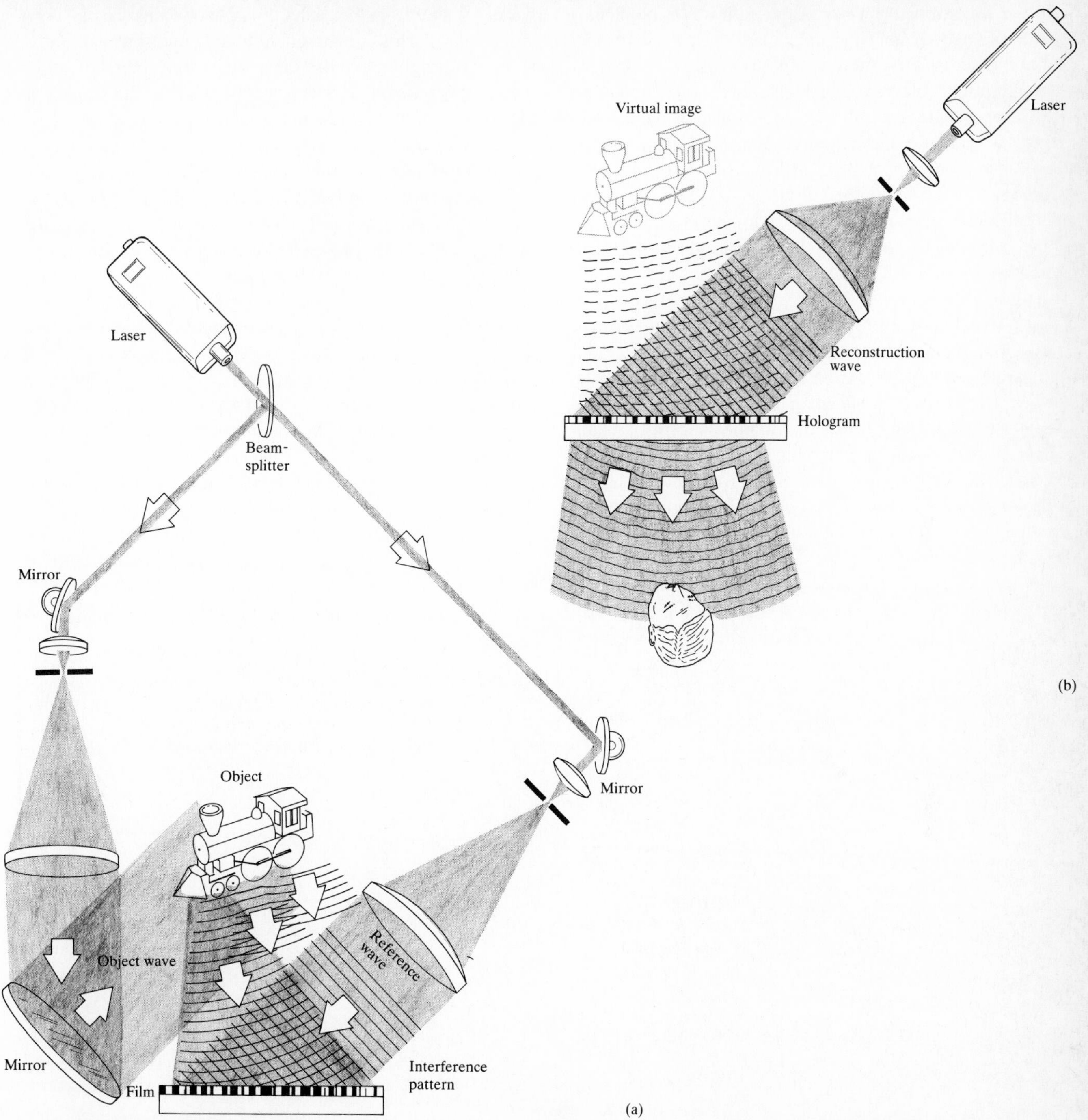

FIGURE 13.57 (*a*) The creation of a transmission hologram of a toy locomotive. (*b*) Replay of a transmission hologram.

13.58*b*. If for simplicity we again let both waves be planar, the resulting pattern can be visualized by sliding two pencils along with the fronts; it should then be clear that the fringes are straight bands (planes) lying parallel to the face of the film plate. When an actual, highly contorted, object wave is made to overlap a planar, coherent, reference wave, these fringes become modulated with the information describing the object. The corresponding three-dimensional diffraction grating is called a **reflection hologram**. During playback it scatters the reilluminating beam back out toward the viewer, and one sees a virtual image behind the hologram (as if looking into a mirror).

The zone-plate interpretation has been applicable to the various holographic schemes we've considered thus far, and this regardless of whether the diffracted wave was of the *near-* or *far*-field variety (i.e., whether we had Fresnel or Fraunhofer holograms, respectively). Indeed, it applies generally where the interferogram results from the superpositioning of the scattered spherical wavelets from each object point and a coherent plane or even spherical reference wave (provided the latter's curvature is different from that of the wavelets). An inherent problem, which these schemes therefore have in common, arises from the fact that the zone-plate radii, R_m, vary as $m^{1/2}$ from Eq. (10.91). Thus the zone fringes are more densely packed farther from the center of each zone lens (i.e., at larger values of m). This is tantamount to an increasing spatial frequency of bright and dark rings, which must be recorded by the photographic plate. The same thing can be appreciated in the cosine-grating representation, where the spatial frequency increases with θ. Since film, no matter how fine-grained, is limited in its spatial frequency response, there will be a cutoff beyond which it cannot record data. All of this represents a built-in limitation on resolution. In contrast, if the mean frequency of the fringes could be made constant, the limitations imposed by the photographic medium would be considerably reduced, and the resolution correspondingly increased. As long as it could record the average spatial fringe frequency, even a coarse emulsion, such as Polaroid P/N, could be used without extensive loss of resolution. Figure 13.59 shows an arrangement that accomplishes just this by having the diffracted object wavelets interfere with a spherical reference wave of about the same curvature. The resulting interferogram is known as a **Fourier-transform** hologram (in this specific instance, it's of the high-resolution *lensless* variety). This scheme is designed to have the reference wave cancel the quadratic (zone-lens type) dependence of the phase with position on Σ_H. But that will occur precisely only for a planar two-dimensional object. In the case of a three-dimensional object

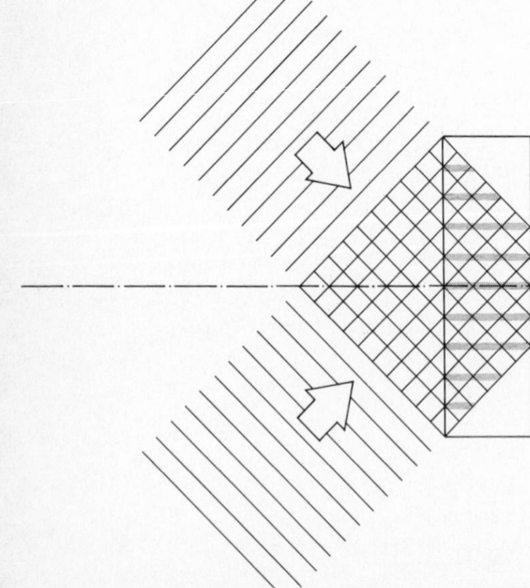

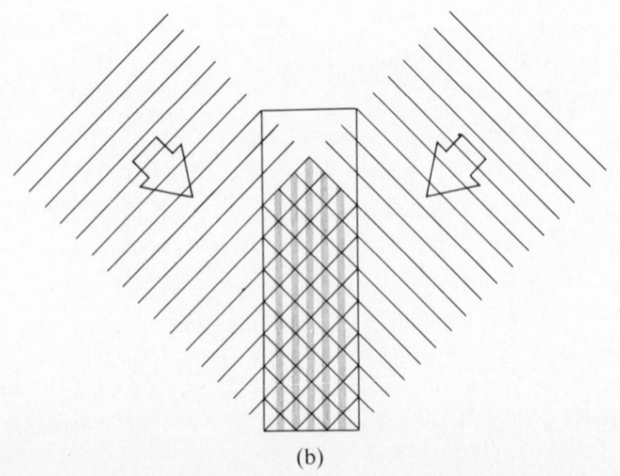

FIGURE 13.58 (*a*) The interference of two plane waves traveling toward the same side to create a transmission hologram. (*b*) The interference of two plane waves traveling toward opposite sides to create a reflection hologram. Refraction has been omitted.

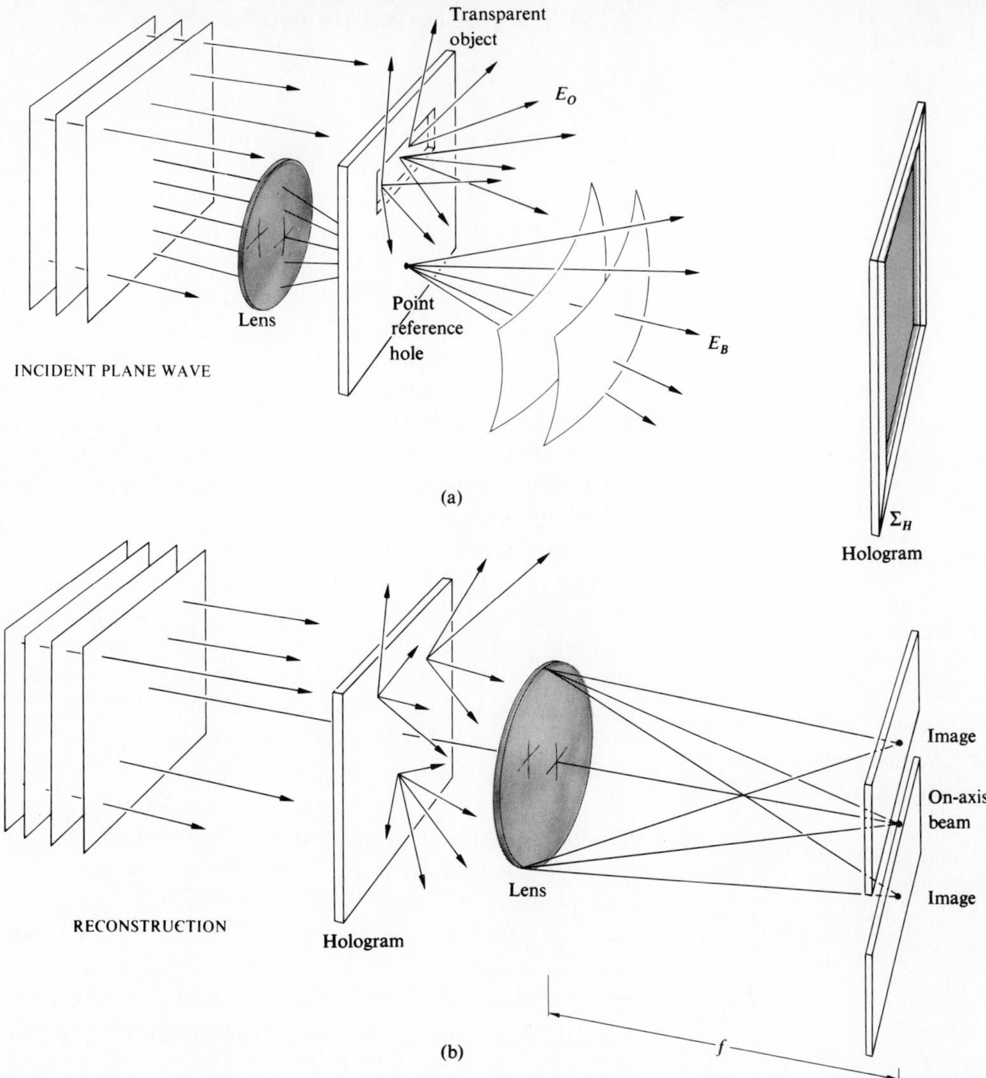

FIGURE 13.59 Lensless Fourier-transform holography (a transparent object).

(Fig. 13.60) this only happens over one plane, and the resulting hologram is therefore a composite of both types, that is, a zone lens and Fourier transform. Unlike the other arrangements, both images generated by a Fourier-transform hologram are virtual, in the same plane, and oriented as if reflected through the origin (Fig. 13.61).

The grating-like nature of all previous holograms is evident here as well. In fact, if you look through a Fourier-transform hologram at a small white-light source (a flashlight in a dark room works beautifully), you see the two mirror images, but they are extremely vague and surrounded by bands of spectral colors. The similarity with white light that has passed through a grating is unmistakable.*

*See DeVelis and Reynolds, *Theory and Applications of Holography*; Stroke, *An Introduction to Coherent Optics and Holography*; Goodman, *Introduction to Fourier Optics*; Smith, *Principles of Holography*; or perhaps *The Engineering Uses of Holography*, edited by E. R. Robertson and J. M. Harvey.

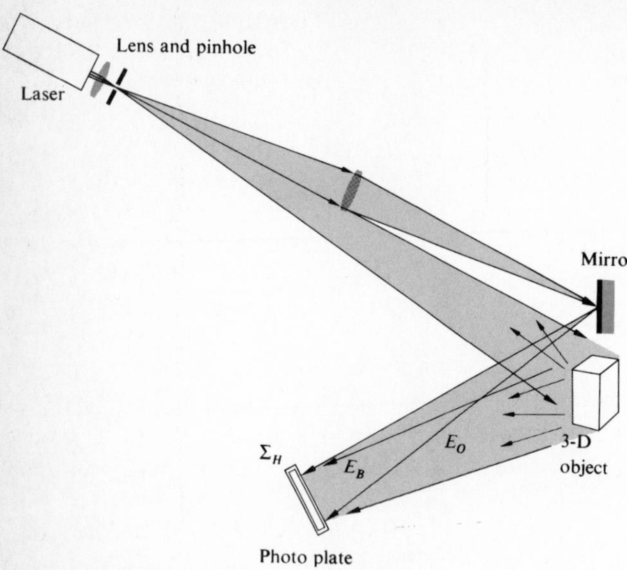

FIGURE 13.60 Lensless Fourier-transform holography (an opaque object).

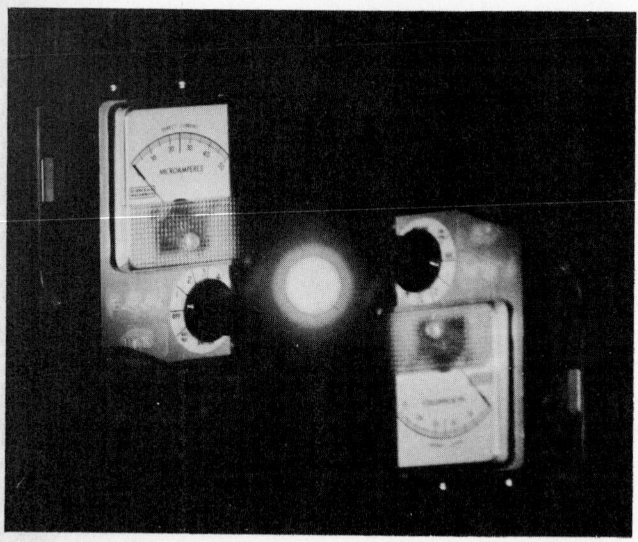

FIGURE 13.61 A reconstruction of a Fourier-transform hologram. [From G. W. Stroke, D. Brumm, and A. Funkhauser, *J. Opt. Soc. Am.* **55**, 1327 (1965).]

13.3.2 Developments and Applications

For years holography was an invention in search of application, that notwithstanding certain obvious possibilities, such as the all too inevitable 3-D billboard. Fortunately, several significant technological developments have in recent times begun what will surely be an ongoing extension of the scope and utility of holography. The early efforts in the field were typified by countless images of toy cars and trains, chess pieces and statuettes—small objects resting on giant blocks of granite. They had to be small because of limited laser power and coherence length, while the ever-present massive granite platform served to isolate the slightest vibrations that might blur the fringes and thereby degrade or obliterate the stored data. A loud sound or gust of air could result in deterioration of the reconstructed image by causing the photographic plate, object, or mirrors to shift several millionths of an inch during the exposure, which itself might last of the order of a minute or so. That was the still-life era of holography. But now, with the use of new, more sensitive films and the short duration ($\approx$40 ns) high-power light flashes from a single-mode pulsed ruby laser, even portraiture and stop-action holography have become a reality* (Fig. 13.62).

Throughout the 1960s and much of the 1970s, the emphasis in the field was on the obvious visual wonders of holography. This continues in the 1980s with the mass production of over a hundred million inexpensive plastic reflection holograms (bonded to credit cards; tucked in candy packages; decorating magazine covers, jewelry, and record albums). The development of a photopolymer that is stable, cheap, and able to produce high-quality images has stimulated the manufacture of even more of these throwaway holograms. Still there is now a widespread recognition of the potential of holography as a nonpictorial instrumentality, and that new direction is finding increasingly important applications.

Volume Holograms

Yuri Nikolayevitch Denisyuk of the Soviet Union, in 1962, introduced a scheme for generating holograms that was conceptually similar to the early (1891) color photographic

*L. D. Siebert, *Appl. Phys. Letters* **11**, 326 (1967), and R. G. Zech and L. D. Siebert, *Appl. Phys. Letters* **13**, 417 (1968).

FIGURE 13.62 A reconstruction of a holographic portrait. (Photo courtesy L. D. Siebert.)

process of Gabriel Lippmann. In brief, the object wave is reflected from the subject and propagates backward, overlapping the incoming coherent background wave. In so doing, the two waves set up a three-dimensional pattern of standing waves, as in Fig. 13.58. The spatial distribution of fringes is recorded by the photoemulsion throughout its entire thickness to form what has become known as a **volume hologram**. Several variations have since been introduced, but the basic ideas are the same; rather than generating a two-dimensional grating-like scattering structure, the volume hologram is a three-dimensional grating. In other words, it's a three-dimensional, modulated, periodic array of phase or amplitude objects, which represent the data. It can be recorded in several media, for example, in thick photoemulsions wherein the amplitude objects are grains of deposited silver; in photochromic glass; with halogen crystals, such as KBr, which respond to irradiation via color-center variations; or with a ferroelectric crystal, such as lithium niobate, which undergoes local alterations in its index of refraction, thus forming what might be called a phase volume hologram. In any event, one is left with a volume array of data, however stored in the medium, which in the reconstruction process behaves very much like a crystal being irradiated by X-rays. It scatters the incident (reconstructing) wave according to Bragg's Law (p. 475). This isn't very surprising, since both the scattering centers and λ have simply been scaled up proportionately.

One important feature of volume holograms is the interdependence [via Bragg's Law, $2d \sin \theta = m\lambda$, Eq. (10.71)] of the wavelength and the scattering angle; that is, only a given color light will be diffracted at a particular angle by the hologram. Another significant property is that by successively altering the incident angle (or the wavelength), a single-volume medium can store a great many coexisting holograms at one time. This latter property makes such systems extremely appealing as densely packed memory devices. For example, an 8-mm-thick hologram has been used to store 550 pages of information, each individually retrievable. In theory a single lithium niobate crystal is capable of easily storing thousands of holograms, and any one of them could be replayed by addressing the crystal with a laserbeam at the appropriate angle. Current research is also focusing on potassium tantalate niobate (KTN) as a potential photorefractive crystal-storage medium. Imagine a 3-D holographic motion picture; a library; or everyone's vital statistics—beauty marks, credit cards, taxes, bad habits, income, life history, and so on, all recorded on a handful of small transparent crystals.

Multicolored reconstructions have been formed using (black and white) volume holographic plates. Two, three, or more different colored and mutually incoherent overlapping laserbeams are used to generate separate, cohabitating, component holograms of the object, and this can be done one at a time or all at once. When these are illuminated simultaneously by the various constituent beams, a multicolored image results.

Another important and highly promising scheme, devised by G. W. Stroke and A. E. Labeyrie, is known as **white-light reflection holography**. Here, the reconstructing wave is an ordinary white-light beam from, say, a flashlight or projector, having a wavefront similar to the original quasimonochromatic background wave. When illuminated on the same side as the viewer, only the specific wavelength that enters the volume hologram at the proper Bragg angle is reflected off to form a reconstructed 3-D virtual image. Thus if the scene were recorded in red laserlight, only red light would presumably be reflected as an image. It is of pedagogical interest to point out, however, that the emulsion may shrink during the fixing process, and if it is not swollen back to its original form chemically (with say triethylnolamine), the spacing of the Bragg planes, d, decreases. That means that at a given angle θ, the reflected wavelength will decrease proportionately. Hence, a scene recorded in He–Ne red might play back in orange or even green when reconstructed by a beam of white light.

If several overlapping holograms corresponding to different wavelengths are stored, a multicolored image will result. The advantages of using an ordinary source of white light to reconstruct full-color 3-D images are obvious and far-reaching.

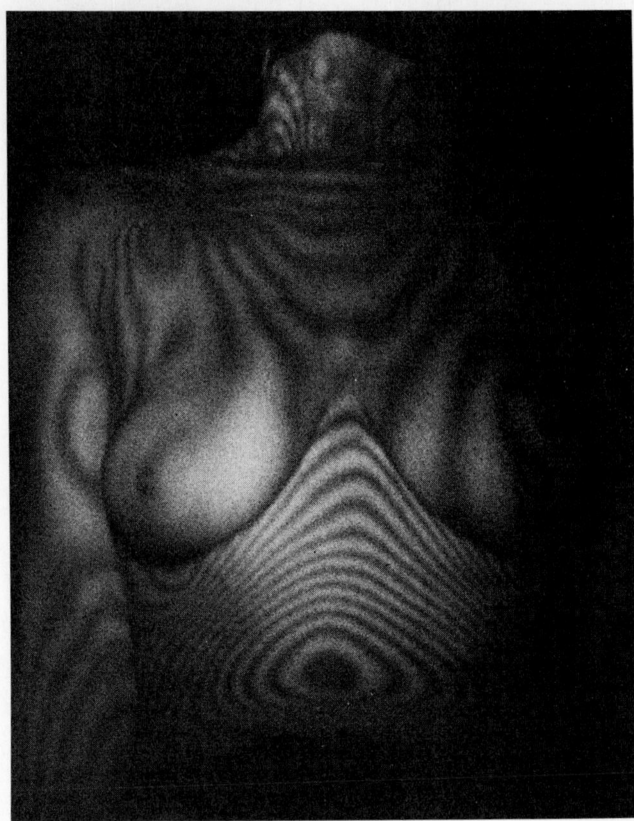

FIGURE 13.63 Double exposure holographic interferogram. [From S. M. Zivi and G. H. Humberstone, "Chest Motion Visualized by Holographic Interferometry," *Medical Research Eng.* p. 5 (June 1970).]

Holographic Interferometry

One of the most innovative and practical of recent holographic advances is in the area of interferometry. Three distinctive approaches have proved to be quite useful in a wealth of nondestructive testing situations where, for example, one might wish to study microinch distortions in an object resulting from strain, vibration, heat, and so on. In the *double exposure* technique, one simply makes a hologram of the undisturbed object and then, before processing, exposes the hologram for a second time to the light coming from the now distorted object. The ultimate result is two overlapping reconstructed waves, which proceed to form a fringe pattern indicative of the displacements suffered by the object, that is, the changes in optical path length (Fig. 13.63). Variations in index such as those arising in wind tunnels and the like will generate the same sort of pattern.

In the *real-time* method, the subject is left in its original position throughout; a processed hologram is formed, and the resulting virtual image is made to overlap the object precisely (Fig. 13.64). Any distortions that arise during subsequent testing show up, on looking through the hologram, as a system of fringes, which can be studied as they evolve in real time. The method applies to both opaque and transparent objects. Motion pictures can be taken to form a continuous record of the response.

The third method is the *time-average* approach and is particularly applicable to rapid, small-amplitude, oscillatory systems. Here the film plate is exposed for a relatively long duration, during which time the vibrating object has executed a number of oscillations. The resulting hologram can be thought of as a superposition of a multiplicity of images, with the effect that a standing-wave pattern emerges. Bright areas reveal undeflected or stationary nodal regions, while contour lines trace out areas of constant vibrational amplitude.

Today holographic testing of mechanical systems is a well-established practice in industry. It continues to serve in a broad range of applications, from noise reduction in automobile transmissions to routine jet engine inspections.

Acoustical Holography

In acoustical holography, an ultra-high-frequency sound wave (ultrasound) is used to create the hologram initially, and a laserbeam then serves to form a recognizable reconstructed image. In one application, the stationary ripple pattern on the surface of a water body produced by submerged coherent transducers corresponds to a hologram of the object beneath (Fig. 13.65). Photographing it creates a hologram that can be illuminated optically to form a visual image. Alternatively, the ripples can be irradiated from above with a laserbeam to produce an instantaneous reconstruction in reflected light.

The advantages of acoustical techniques reside in the fact

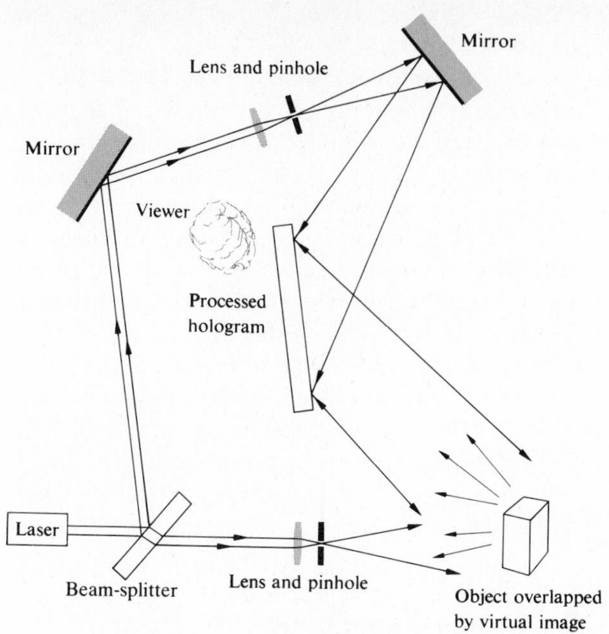

FIGURE 13.64 Real-time holographic interferometry.

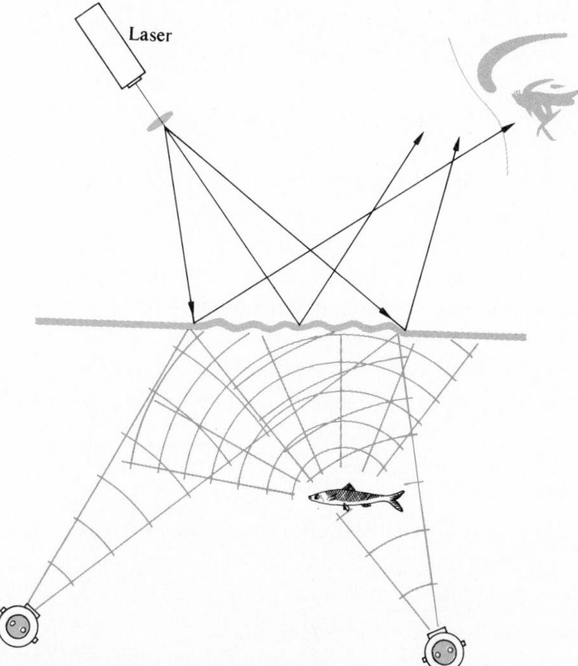

FIGURE 13.65 Acoustical holography.

that sound waves can propagate considerable distances in dense liquids and solids where light cannot. Thus acoustical holograms can record such diverse things as underwater submarines and internal body organs.* In the case of Fig. 13.65, one would see something that resembled an X-ray motion picture of the fish. Figure 13.66 is the image of a penny formed via acoustical holography using ultrasound at a frequency of 48 MHz. In water, that corresponds to a wavelength of roughly 30 μm, and so each fringe contour reveals a change in elevation of $\frac{1}{2}\lambda$ or 15 μm.

Holographic Optical Elements

Evidently, when two plane waves overlap, as in Fig. 13.52, they produce a cosine grating. This suggests the rather obvious notion that holography can be used for nonpictorial purposes, like making diffraction gratings. Indeed, the *holographic optical element* (HOE) is any diffractive device consisting of a "fringe" system (i.e., a distribution of diffracting amplitude or

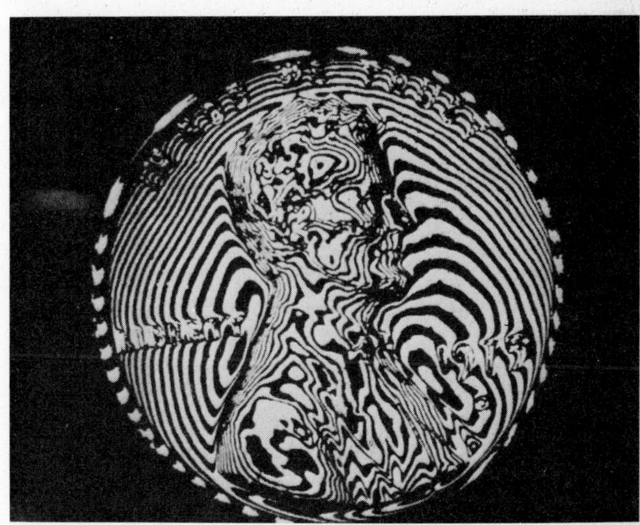

FIGURE 13.66 Interferometric image of a penny via acoustical holography. (Photo courtesy Holosonics, Inc.)

*See A. F. Metherell, "Acoustical Holography," *Sci. Am.* **221**, 36 (October 1969). Refer to A. L. Dalisa et al., "Photoanodic Engraving of Holograms on Silicon," *Appl. Phys. Letters* **17**, 208 (1970), for another interesting use of surface relief patterns.

phase objects) created either directly by interferometry or by computer simulation thereof. Holographic diffraction gratings, both blazed and sinusoidal, are available commercially (with up to around 3600 lines/mm). Although generally less efficient than ruled gratings, they do produce far less stray light, which can be important in many applications.

Suppose we record the interference pattern of a converging beam using a planar reference wave. Upon reilluminating the resulting transmission hologram with a matching plane wave, out will come a recreated converging wave—the hologram will function like a lens (see Fig. 13.49). Similarly, if the reference beam is a diverging wave from a point source and the object is a plane wave, the resulting hologram, reilluminated by the point source, will play back a plane wave. In this way a holographic optical element can perform the tasks of a complex lens with the added benefit of allowing for an inexpensive, lightweight, compact system design.

Holographic optical elements are already in use inside supermarket check-out scanners that automatically read the bar patterns of the Universal Product Code (UPC) on merchandise. A laserbeam passes through a rotating disk composed of a number of holographic lens-prism facets. These rapidly refocus, shift, and scan the beam across a volume of space, ensuring that the code will be read on the first pass across the device. HOEs are used in so-called heads-up displays in airplane cockpits. These allow reflected data to appear on an otherwise transparent screen in front of the pilot's face and yet not obscure the view. They're also in office copy machines and solar concentrators.

As *matched spatial filters*, HOEs are used in optical correlators to spot defects in semiconductors and tanks in reconnaissance pictures. In such cases the HOE is a hologram formed using the Fourier transform of the target (e.g., a picture of a tank or perhaps a printed word) as the object. Suppose the problem is to find a word on a printed page automatically, using an optical computer like that in Fig. 13.32, that is, to cross-correlate the word and the page of words. The target-transform hologram is placed in the transform plane and illuminated with the transform of an entire page of print. The field amplitude emerging from this HOE-filter will then be proportional to the product of the transforms of the page and the word. The transform of this product, generated by the last lens and displayed on the image plane, is the desired cross-correlation (recall the Wiener–Khintchine theorem). If the word is on the page, there will be a high correlation, and a bright spot of light will appear superimposed in the final image everywhere the target word occurs.*

It is possible to synthesize, point by point, a hologram of a fictitious object. In other words, in the most direct approach holograms can be produced by calculating, with a digital computer, the irradiance distribution that would arise were some object appropriately illuminated in a hypothetical recording session. A computer-controlled plotter drawing or cathode ray tube read out of the interferogram is then photographed, thence to serve as the actual hologram. The result upon illumination is a three-dimensional reconstructed image of an object that never had any real existence in the first place. More practically, computer-generated HOEs are now routinely being produced, often to serve as references for optical testing. Since this mating of technologies can in principle generate wavefronts otherwise essentially impossible to produce, the future is very promising.

13.4 Nonlinear Optics

Generally, the domain of *nonlinear optics* is understood to encompass those phenomena for which electric and magnetic field intensities of higher powers than the first play a dominant role. The Kerr Effect (Section 8.11.3), which is a quadratic variation of refractive index with applied voltage, and thereby electric field, is typical of several long-known nonlinear effects.

The usual classical treatment of the propagation of light—superposition, reflection, refraction, and so forth—assumes a linear relationship between the electromagnetic light field and the responding atomic system constituting the medium. But just as an oscillatory mechanical device (e.g., a weighted spring) can be overdriven into nonlinear response through the application of large enough forces, so too we might anticipate that an extremely intense beam of light could generate appreciable nonlinear optical effects.

The electric fields associated with light beams from ordinary or, if you will, traditional sources are far too small for such behavior to be easily observable. It was for this reason, coupled with an initial lack of technical prowess, that the subject had to await the advent of the laser in order that sufficient

*See A. Ghatak and K. Thyagarajan, *Contemporary Optics*, p. 214.

brute force could be brought to bear in the optical region of the spectrum. As an example of the kinds of fields readily obtainable with the current technology, consider that a good lens can focus a laserbeam down to a spot having a diameter of about 10^{-3} inch or so, which corresponds to an area of roughly 10^{-9} m^2. A 200-megawatt pulse from, say, a Q-switched ruby laser would then produce a flux density of 20×10^{16} W/m^2. It follows (Problem 13.27) that the corresponding electric field amplitude is given by

$$E_0 = 27.4\left(\frac{I}{n}\right)^{1/2} \tag{13.30}$$

In this particular case, for $n \approx 1$, the field amplitude is about 1.2×10^8 V/m. This is more than enough to cause the breakdown of air (roughly 3×10^6 V/m) and just several orders of magnitude less than the typical fields holding a crystal together, the latter being roughly about the same as the cohesive field on the electron in a hydrogen atom (5×10^{11} V/m). The availability of these and even greater (10^{12} V/m) fields has made possible a wide range of important new nonlinear phenomena and devices. We shall limit this discussion in the consideration of several nonlinear phenomena associated with passive media (i.e., media that act essentially as catalysts without making their own characteristic frequencies evident). Specifically, we'll consider optical rectification, optical harmonic generation, frequency mixing, and self-focusing of light. In contrast, Stimulated Raman, Rayleigh, and Brillouin Scattering exemplify nonlinear optical phenomena arising in active media that do impose their characteristic frequencies on the lightwave.*

As you may recall, the electromagnetic field of a lightwave propagating through a medium exerts forces on the loosely bound outer or valence electrons. Ordinarily, these forces are quite small, and in a linear isotropic medium the resulting electric polarization is parallel with and directly proportional to the applied field. In effect, the polarization follows the field; if the latter is harmonic, the former will be harmonic as well. Consequently, one can write

$$P = \epsilon_0 \chi E \tag{13.31}$$

where χ is a dimensionless constant known as the electric susceptibility, and a plot of P versus E is a straight line. Quite obviously in the extreme case of very high fields, we can expect that P will become saturated; in other words, it simply cannot increase linearly indefinitely with E (just as in the familiar case of ferromagnetic materials, where the magnetic moment becomes saturated at fairly low values of H). Thus we can anticipate a gradual increase of the ever-present, but usually insignificant, nonlinearity as E increases. Since the directions of $\mathbf{P}$ and $\mathbf{E}$ coincide in the simplest case of an isotropic medium, we can express the polarization more effectively as a series expansion:

$$P = \epsilon_0(\chi E + \chi_2 E^2 + \chi_3 E^3 + \cdots) \tag{13.32}$$

The usual linear susceptibility, χ, is much greater than the coefficients of the nonlinear terms χ_2, χ_3, and so on, and hence the latter contribute noticeably only at high-amplitude fields. Now suppose that a lightwave of the form

$$E = E_0 \sin \omega t$$

is incident on the medium. The resulting electric polarization

$$P = \epsilon_0 \chi E_0 \sin \omega t + \epsilon_0 \chi_2 E_0^2 \sin^2 \omega t + \epsilon_0 \chi_3 E_0^3 \sin^3 \omega t + \cdots \tag{13.33}$$

can be rewritten as

$$P = \epsilon_0 \chi E_0 \sin \omega t + \frac{\epsilon_0 \chi_2}{2} E_0^2(1 - \cos 2\omega t) + \frac{\epsilon_0 \chi_3}{4} E_0^3(3 \sin \omega t - \sin 3\omega t) + \cdots \tag{13.34}$$

As the harmonic lightwave sweeps through the medium, it creates what might be thought of as a polarization wave, that is, an undulating redistribution of charge within the material in response to the field. If only the linear term were effective, the electric polarization wave would correspond to an oscillatory current following along with the incident light. The light thereafter reradiated in such a process would be the usual refracted wave generally propagating with a reduced speed v and having the same frequency as the incident light. In contrast, the presence of higher-order terms in Eq. (13.33) implies that the polarization wave does not have the same harmonic profile as the incident field. In fact, Eq. (13.34) can be likened to a Fourier series representation of the distorted profile of $P(t)$.

*For a more extensive treatment than is possible here, see N. Bloembergen, *Nonlinear Optics*, or G. C. Baldwin, *An Introduction to Nonlinear Optics*.

13.4.1 Optical Rectification

The second term in Eq. (13.34) has two components of great interest. First there is a *dc* or *constant bias polarization* varying as E_0^2. Consequently, if an intense plane-polarized beam traverses an appropriate (piezoelectric) crystal, the presence of the quadratic nonlinearity will, in part, be manifest by a constant electric polarization of the medium. A voltage difference, proportional to the beam's flux density, will accordingly appear across the crystal. This effect, in analogy to its radiofrequency counterpart, is known as **optical rectification**.

13.4.2 Harmonic Generation

The cos $2\omega t$ term [Eq. (13.34)] corresponds to a variation in electric polarization at twice the fundamental frequency (i.e., at twice that of the incident wave). The reradiated light that arises from the driven oscillators also has a component at this same frequency, 2ω, and the process is spoken of as **second-harmonic generation**, or SHG for short. In terms of the photon representation, we can envision two identical photons of energy $h\omega$ coalescing within the medium to form a single photon of energy $h2\omega$. Peter A. Franken and several coworkers at the University of Michigan in 1961 were the first to observe SHG experimentally. They focused a 3-kW pulse of red (694.3 nm) ruby laserlight onto a quartz crystal. Just about one part in 10^8 of this incident wave was converted to the 347.15-nm ultraviolet second harmonic.

Notice that, for a given material, if $P(E)$ is an odd function, that is, if reversing the direction of the **E**-field simply reverses the direction of **P**, the even powers of E in Eq. (13.32) must vanish. But this is just what happens in an isotropic medium, such as glass or water—there are no special directions in a liquid. Moreover, in crystals like calcite, which are so structured as to have what's known as a *center of symmetry* or an *inversion center*, a reversal of all of the coordinate axes must leave the interrelationships between physical quantities unaltered. Thus no even harmonics can be produced by materials of this sort. Third-harmonic generation (THG), however, can exist and has been observed, for example, in calcite. The requirement for SHG that a crystal not have inversion symmetry is also necessary for it to be piezoelectric. Under pressure a piezoelectric crystal [such as quartz, potassium dihydrogen phosphate (KDP), or ammonium dihydrogen phosphate (ADP)] undergoes an asymmetric distortion of its charge distribution, thus producing a voltage. Of the 32 crystal classes, 20 are of this kind and may therefore be useful in SHG. The simple scalar expression [Eq. (13.32)] is actually not an adequate description of a typical dielectric crystal. Things are a good deal more complicated, because the field components in several different directions in a crystal can affect the electric polarization in any one direction. A complete treatment requires that **P** and **E** be related not by a single scalar but by a group of quantities arranged in the particular form of a tensor, namely, the susceptibility tensor.*

A major difficulty in generating copious amounts of second-harmonic light arises from the frequency dependence of the refractive index, that is, dispersion. At some initial point where the incident or ω-wave, generates the second-harmonic or 2ω-wave, the two are coherent. As the ω-wave propagates through the crystal, it continues to generate additional contributions of second-harmonic light, which all combine totally constructively only if they maintain a proper phase relationship. Yet the ω-wave travels at a phase velocity v_ω, which is ordinarily different from the phase velocity, $v_{2\omega}$, of the 2ω-wave. Thus the newly emitted second harmonic periodically falls out-of-phase with some of the previously generated 2ω-waves. When the irradiance of the second harmonic, $I_{2\omega}$, emerging from a plate of thickness ℓ is computed,† it turns out to be

$$I_{2\omega} \propto \frac{\sin^2 [2\pi(n_\omega - n_{2\omega})\ell/\lambda_0]}{(n_\omega - n_{2\omega})^2} \tag{13.35}$$

(see Fig. 13.67). This yields the result that $I_{2\omega}$ has its maximum value when $\ell = \ell_c$, where

$$\ell_c = \frac{1}{4}\frac{\lambda_0}{|n_\omega - n_{2\omega}|} \tag{13.36}$$

This is commonly known as the *coherence length* (although a different name would be better), and it's usually of the order of only about $20\lambda_0$. Despite this, efficient SHG can be accomplished by a procedure known as *index matching*, which negates the undesirable effects of dispersion; in short, one arranges things so that $n_\omega = n_{2\omega}$. A commonly used SHG

*Incidentally, there is nothing extraordinary about this kind of behavior—it comes up all the time. There are inertia tensors, demagnetization coefficient tensors, stress tensors, and so forth.

†See, for example, B. Lengyel, *Introduction to Laser Physics*, Chapter VII. This is a fine elementary treatment.

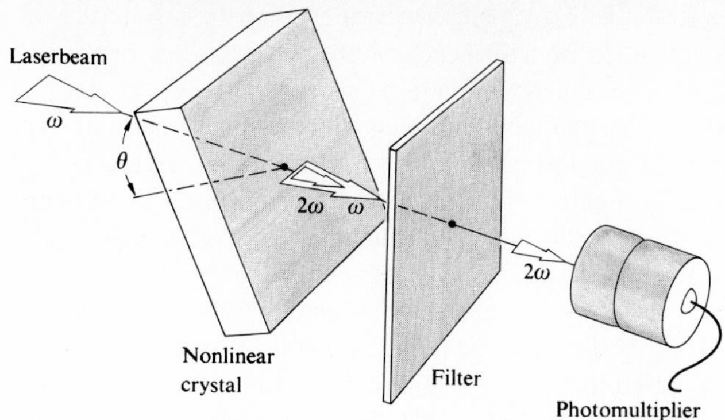

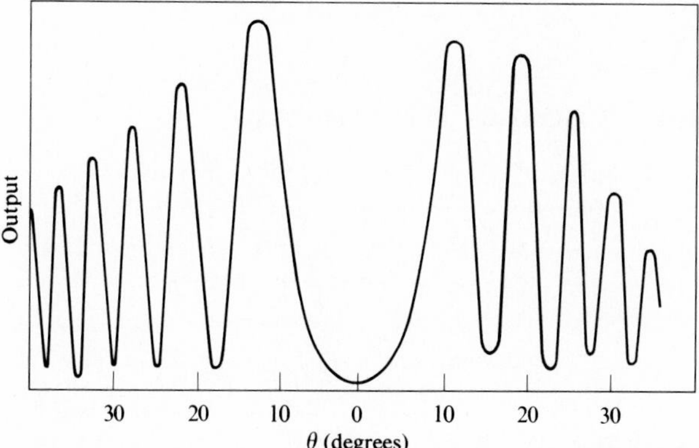

FIGURE 13.67 Second-harmonic generation as a function of θ for a 0.78-mm thick quartz plate. Peaks occur when the effective thickness is an even multiple of ℓ_c. [From P. D. Maker, R. W. Terhune, M. Nisenoff, and C. M. Savage, *Phys. Rev. Letters* **8**, 21 (1962).]

material is KDP. It is piezoelectric, transparent, and also negatively uniaxially birefringent. Furthermore, it has the interesting property that if the fundamental light is a linear polarized *ordinary wave*, the resulting second harmonic will be an *extraordinary wave*. As can be seen from Fig. 13.68, if light propagates within a KDP crystal at the specific angle θ_0 with respect to the optic axis, the index, $n_{0\omega}$, of the ordinary fundamental wave will precisely equal the index of the extraordinary second harmonic $n_{e2\omega}$. The second-harmonic wavelets will then interfere constructively, thereupon increasing the conversion efficiency by several orders of magnitude. Second-harmonic generators, which are simply appropriately cut and oriented crystals, are available commercially, but do keep

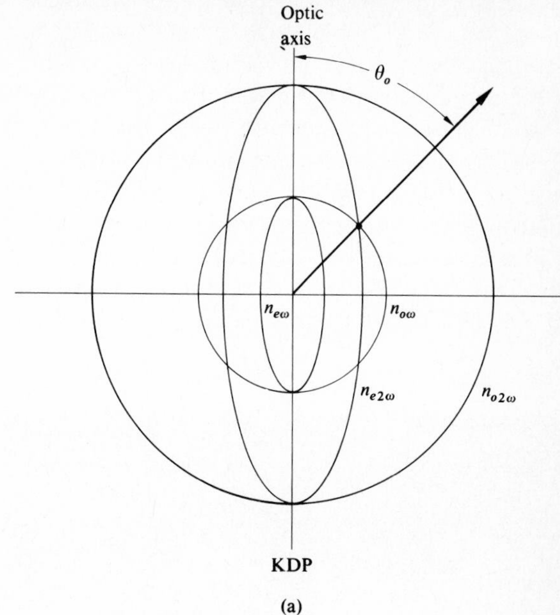

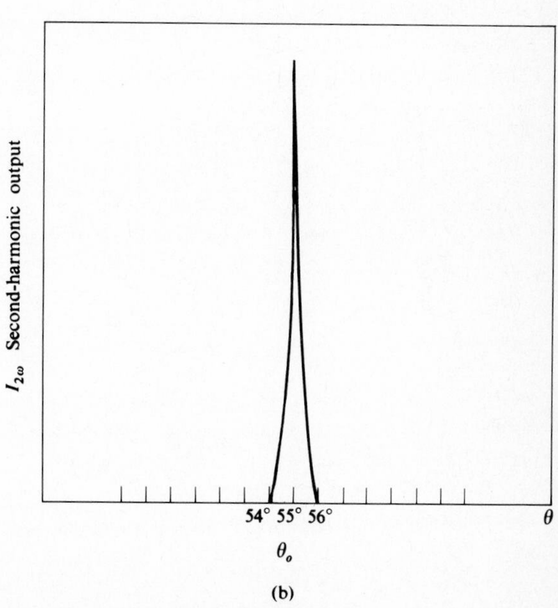

FIGURE 13.68 Refractive index surface for KDP. (*b*) $I_{2\omega}$ versus crystal orientation in KDP. (From Maker et al.)

in mind that θ_0 is a function of λ, and each such device performs at one frequency. Not long ago, a continuous 1-W second-harmonic beam at 532.3 nm was obtained by placing a barium sodium niobate crystal within the cavity of a 1-W 1.06μ laser. The fact that the ω-wave sweeps back and forth through the crystal increases the net conversion efficiency.

Optical harmonic generation soon lost its initial exotic quality and became a routine commercial process by the early 1980s. Still, there continue to be exciting technical accomplishments, such as the 74-cm–diameter harmonic conversion array (Fig. 13.69) built for the Nova laser-fusion program. Its function is to convert upwards of 80% of the infrared (1.05 μm) emission from the neodymium–glass laser into more efficient high-frequency radiation. Because of its great size, the converter is an aligned mosaic of smaller KDP single-crystal panels forming two layers, one behind the other. To generate the second harmonic (green light at 0.53 μm), the array is positioned so that each layer functions independently to produce two overlapping frequency-shifted components. These arise one from each crystal layer and are orthogonally polarized. The third harmonic (blue light at 0.35 μm) is created by reorienting the assembly to the appropriate phase-matching angle so as to shift about two-thirds of the beam energy into the second harmonic as it traverses the first crystal layer. The second layer mixes the remaining IR and the second-harmonic green light to produce third-harmonic blue.

FIGURE 13.69 The KDP frequency converter for the Nova laser. (Photo courtesy Lawrence Livermore National Laboratory.)

13.4.3 Frequency Mixing

Another situation of considerable practical interest involves the *mixing* of two or more primary beams of different frequencies within a nonlinear dielectric. The process can most easily be appreciated by substituting a wave of the form

$$E = E_{01} \sin \omega_1 t + E_{02} \sin \omega_2 t \qquad (13.37)$$

into the simplest expression for P given by Eq. (13.32). The second-order contribution is then

$$\epsilon_0 \chi_2 (E_{01}^2 \sin^2 \omega_1 t + E_{02}^2 \sin^2 \omega_2 t + 2E_{01}E_{02} \sin \omega_1 t \sin \omega_2 t)$$

The first two terms can be expressed as functions of $2\omega_1$ and $2\omega_2$, respectively, while the last quantity gives rise to sum and difference terms, $\omega_1 + \omega_2$ and $\omega_1 - \omega_2$.

As for the quantum picture, the photon of frequency $\omega_1 + \omega_2$ simply corresponds to a coalescing of the two original photons into a new photon, just as it did in the case of SHG, where both quanta had the same frequency. The energy and momentum of the annihilated photons are carried off by the created sum photon. The generation of an $\omega_1 - \omega_2$ difference-photon is a little more involved. Conservation of energy and momentum requires that on interacting with an ω_2-photon, only the higher-frequency ω_1-photon vanishes, thereby creating two new quanta, one an ω_2-photon and the other a difference-photon.

As an application of this phenomenon, suppose we beat,

within a nonlinear crystal a strong wave of frequency ω_p, called the *pump light*, with a weak *signal wave* of lower frequency ω_s, which is to be amplified. Pump light is thereby converted into both signal light and a difference wave, called *idler light*, of frequency $\omega_i = \omega_p - \omega_s$. If the idler light is then made to beat with the pump light, the latter is converted into additional amounts of idler and signal light. In this way both the signal and idler waves are amplified. This is actually an extension into the optical-frequency region of the well-known concept of *parametric amplification*, whose use in the microwave spectrum dates back to the late 1940s. The first *optical-parametric oscillator*, which was operated in 1965, is depicted in Fig. 13.70. The flat parallel end faces of a nonlinear crystal (lithium niobate) were coated to form an optical Fabry–Perot cavity. The signal and idler frequencies (both about 1000 nm) corresponded to two of the resonant frequencies of the cavity. When the flux density of the pumping light was high enough, energy was transferred from it into the signal and idler oscillatory modes, with the consequent buildup of those modes and emission of coherent radiant energy at those frequencies. This transfer of energy from one wave to another within a lossless medium typifies parametric processes. By changing the refractive index of the crystal (via temperature, electric field, etc.), the oscillator becomes tunable. Various oscillator configurations have since evolved, with other nonlinear materials used as well, such as barium sodium niobate. The optical parametric oscillator is a laser-like, broadly tunable source of coherent radiant energy in the IR to the UV.

13.4.4 Self-Focusing of Light

When a dielectric is subjected to an electric field that varies in space, in other words, when there is a gradient of the field parallel to **P**, an internal force will result. This has the effect of altering the density, changing the permittivity, and thereby varying the refractive index, and this in both linear and nonlinear isotropic media. Suppose then that we shine an intense laserbeam with a transverse Gaussian flux-density distribution onto a specimen. The induced refractive-index variations will cause the medium in the region of the beam to function much as if it were a positive lens. Accordingly, the beam contracts, the flux density increases even more, and the contraction continues in a process known as **self-focusing**. The effect can be sustained until the beam reaches a limiting filament diameter (of about 5×10^{-6} m), being totally internally reflected as if it were in a fiberoptic element imbedded within the medium.*

*See J. A. Giordmaine, "Nonlinear Optics." *Phys. Today*, 39 (January 1969).

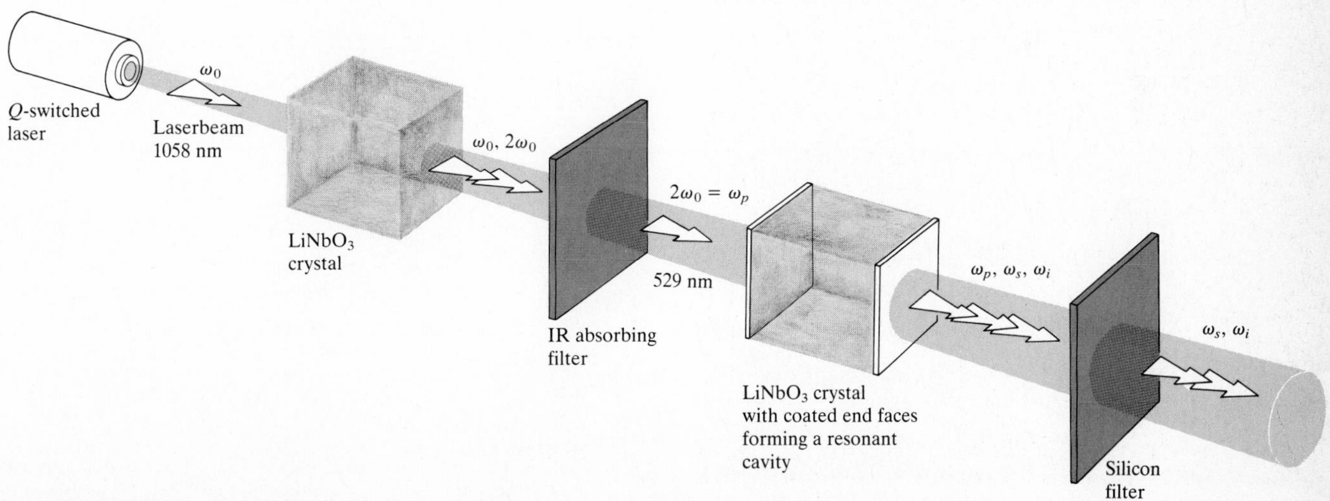

FIGURE 13.70 An optical parametric oscillator. [After J. A. Giordmaine and R. C. Miller, *Phys. Rev. Letters* **4**, 973 (1965).]

PROBLEMS

13.1* After a while, a cube of rough steel (10 cm on a side) reaches equilibrium inside a furnace at a temperature of 400°C. Knowing that its total emissivity is 0.97, determine the rate at which the cube radiates energy from each face.

13.2 A somewhat typical person has a total naked area of about 1.4 m^2 and an average skin temperature of 33°C. Determine the net power radiated per unit area, the irradiance or more precisely the exitance, if the person's total emissivity is 97% and the environment is room temperature (20°C). How much energy does that body radiate per second?

13.3 Suppose that we measure the emitted exitance from a small hole in a furnace to be 22.8 W/cm^2, using an optical pyrometer of some sort. Compute the internal temperature of the furnace.

13.4 The temperature of an object resembling a blackbody is raised from 200 K to 2000 K. By how much does the amount of energy it radiates increase?

13.5* Your average skin temperature is about 33°C. Assuming you radiate as does a blackbody at that temperature, at what wavelength do you emit the most energy?

13.6* What is the wavelength that carries away the most energy when an object resembling a blackbody radiates energy into a room-temperature (20°C) environment?

13.7* The surface temperature of a class O blue-white star is around 40×10^3 K. At what frequency will it radiate most of its energy?

13.8* When the Sun's spectrum is photographed, using rockets to range above the Earth's atmosphere, it is found to have a peak in its spectral exitance at roughly 465 nm. Compute the Sun's surface temperature, assuming it to be a blackbody. This approximation yields a value that is about 400 K too high.

13.9* An object resembling a blackbody emits a maximum amount of energy per unit wavelength in the red end of the visible spectrum (λ = 680 nm). What's its surface temperature?

13.10* The energy per unit area per unit time per wavelength interval emitted by a blackbody at a temperature T is given by

$$I_\lambda = \frac{2\pi hc^2}{\lambda^5}\left[\frac{1}{e^{\frac{hc}{\lambda k_B T}} - 1}\right]$$

At a specific temperature, the total power radiated per unit area of the blackbody is equal to the area under the corresponding I_λ versus λ curve. Use this to derive the Stefan–Boltzmann Law. [*Hint*: To clean up the exponential, change variables in the integral so that

$$x = \frac{hc}{\lambda k_B T}$$

Use the fact that $\int_0^\infty x^n \frac{dx}{e^x - 1} = \Gamma(n+1)\,\zeta(n+1)$ where the gamma function is given by $\Gamma(n+1) = n!$ and the Riemann zeta function for $n = 3$ is $\zeta(4) = \frac{\pi^4}{90}$.]

13.11* In the atomic domain, energy is often measured in electron-volts. Arrive at the following expression for the energy of a light-quantum in eV when the wavelength is in nanometers:

$$\mathscr{E} = \frac{1239.8 \text{ eV}\cdot\text{nm}}{\lambda}$$

What is the energy of a quantum of 600-nm light?

13.12 Figure P.13.12 shows the *spectral irradiance* impinging on a horizontal surface, for a clear day, at sea level, with the Sun at the zenith. What is the most energetic photon we can expect to encounter (in eV and in J)?

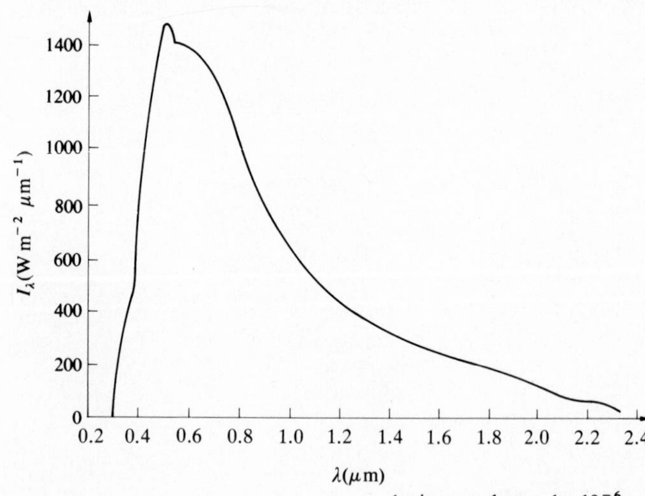

FIGURE P.13.12

13.13* Suppose we have a 100-W yellow lightbulb (550 nm) 100 m away from a 3-cm diameter shuttered aperture. Assuming the bulb to have a 2.5% conversion to radiant power, how many photons will pass through the aperture if the shutter is opened for $\frac{1}{1000}$ s?

13.14 The *solar constant* is the radiant flux density at a spherical surface centered on the Sun having a radius equal to that of the Earth's mean orbital radius; it has a value of 0.133–0.14 W/cm^2. If we assume an average wavelength of about 700 nm, how many photons at most will arrive on each square meter per second of a solar cell panel just above the atmosphere?

13.15 A 50.0-cm^3 chamber is filled with argon gas to a pressure of 20.3 Pa at a temperature of 0°C. All but a negligible number of these atoms are initially in their ground states. A flash tube surrounding the sample energizes 1.0% of the atoms into the same excited state having a mean life of 1.4×10^{-8} s. What is the maximum rate at which photons are subsequently emitted by the gas, of course it falls off with time? Assume both that spontaneous emission is the only mechanism at work and that the medium is an ideal gas.

13.16* Show that for a system of atoms and photons in equilibrium at a temperature T, the ratio of the transition rates of stimulated to spontaneous emission is given by

$$\left[\frac{1}{e^{\frac{h\nu}{k_B T}} - 1}\right]$$

13.17* A system of atoms in thermal equilibrium is emitting and absorbing 2.0-eV light photons. Determine the ratio of the transition rates of stimulated emission to spontaneous emission at a temperature of 300 K. Discuss the implications of your answer. [*Hint*: See the previous problem.]

13.18 Redo the previous problem for a temperature of 30.0×10^3 K and compare the results of both calculations.

13.19* For a system of atoms (in equilibrium) having two energy levels, show that at high temperatures where $k_B T >> \mathscr{E}_j - \mathscr{E}_i$, the number densities of the two states tend to become equal. [*Hint*: Form the ratio of the transition rates for total emission to absorption.]

13.20* Radiation at 21 cm pours down on the Earth from outer space. Its origin is great clouds of hydrogen gas. Taking the background temperature of space to be 3.0 K, determine the ratio of the transition rates of stimulated emission to spontaneous emission and discuss the result.

13.21* The beam ($\lambda = 632.8$ nm) from a He–Ne laser, which is initially 3.0 mm in diameter, shines on a perpendicular wall 100 m away. Given that the system is diffraction limited, how large is the circle of light on the wall?

13.22* Make a rough estimate of the amount of energy that can be delivered by a ruby laser whose crystal is 5.0 mm in diameter and 0.050 m long. Assume the pulse of light lasts 5.0×10^{-6} s. The density of aluminum oxide (Al_2O_3) is 3.7×10^3 kg/m^3. Use the data in the discussion of Fig. 13.6 and the fact that the chromium ions make a 1.79 eV lasing transition. How much power is available per pulse?

13.23 What is the transition rate for the neon atoms in a He–Ne laser if the energy drop for the 632.8 nm emission is 1.96 eV and the power output is 1.0 mW?

13.24* Given that a ruby laser operating at 694.3 nm has a frequency bandwidth of 50 MHz, what is the corresponding linewidth?

13.25* Determine the frequency difference between adjacent axial resonant cavity modes for a typical gas laser 25 cm long ($n \approx 1$).

13.26* A He–Ne c-w laser has a Doppler-broadened transition bandwidth of about 1.4 GHz at 632.8 nm. Assuming $n = 1.0$, determine the maximum cavity length for single-axial mode operation. Make a sketch of the transition linewidth and the corresponding cavity modes.

13.27 Show that the maximum electric-field intensity, E_{max}, that exists for a given irradiance I is

$$E_{max} = 27.4\left(\frac{I}{n}\right)^{1/2} \quad \text{in units of V/m}$$

where n is the refractive index of the medium.

13.28* A He–Ne laser operating at 632.8 nm has an internal beam-waist diameter of 0.60 mm. Calculate the full-angular width, or divergence, of the beam.

13.29 What would the pattern look like for a laserbeam diffracted by the three crossed gratings of Fig. P.13.29?

FIGURE P.13.29

13.30 Make a rough sketch of the Fraunhofer diffraction pattern that would arise if a transparency of Fig. P.13.30*a* served as the object. How would you filter it to get Fig. P.13.30*b*?

(a)

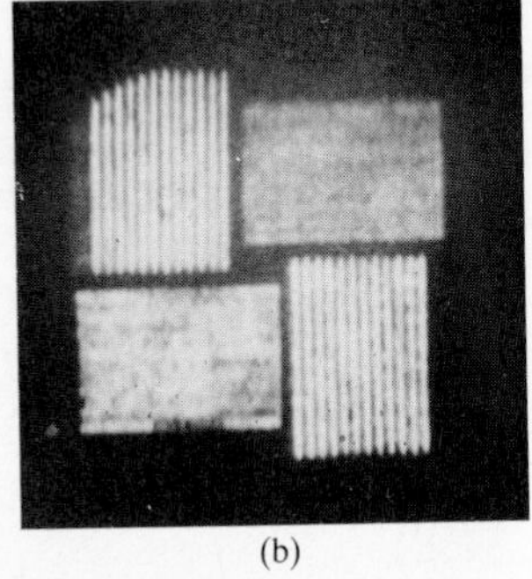
(b)

FIGURE P.13.30 Photos courtesy R. A. Phillips.

13.31 Repeat the previous problem using Fig. P.13.31 instead.

(a)

(b)

FIGURE P.13.31 Photos courtesy R. A. Phillips.

13.32* Repeat the previous problem using Fig. P.13.32 this time.

13.33 Returning to Fig. 13.34, what kind of spatial filter would produce each of the patterns shown in Fig. P.13.33?

13.34 With Fig. 13.33 in mind, show that the transverse magnification of the system is given by $-f_i/f_t$ and draw the appropriate ray diagram. Draw a ray up through the center of the first lens at an angle θ with the axis. From the point where that ray intersects Σ_t, draw a ray downward that passes through the center of the second lens at an angle Φ. Prove that $\Phi/\theta = f_t/f_i$. Using the notion of spatial frequency, from Eq. (11.64), show that k_O at the object plane is related to k_I at the image plane by

$$k_I = k_O(f_t/f_i)$$

What does this mean with respect to the size of the image when $f_i > f_t$? What can then be said about the spatial periods of the input data as compared with the image output?

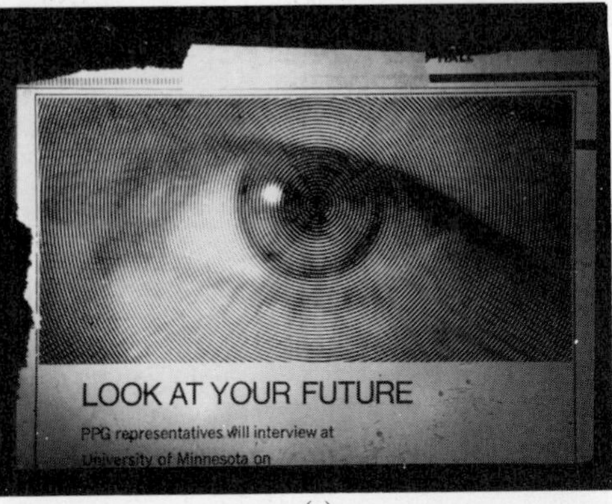

(a)

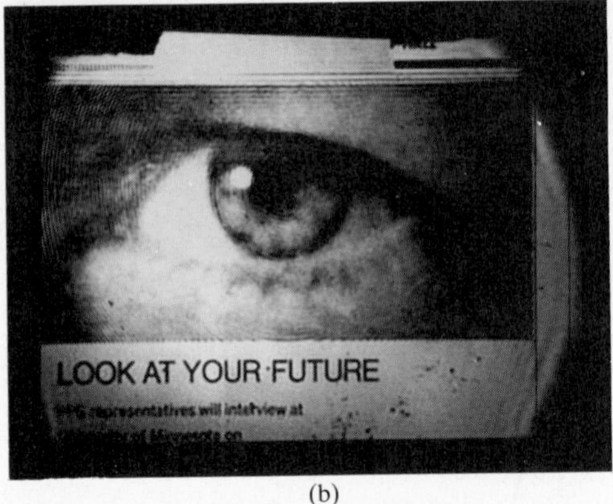

(b)

FIGURE P.13.32 Photos courtesy R. A. Phillips.

13.35 A diffraction grating having a mere 50 grooves per cm is the object in the optical computer shown in Fig. 13.33. If it is coherently illuminated by plane waves of green light (543.5 nm) from a He–Ne laser and each lens has a 100-cm focal length, what will be the spacing of the diffraction spots on the transform plane?

13.36* Imagine that you have a cosine grating (i.e., a transparency whose *amplitude* transmission profile is cosinusoidal) with a spatial period of 0.01 mm. The grating is illuminated by quasimonochro-

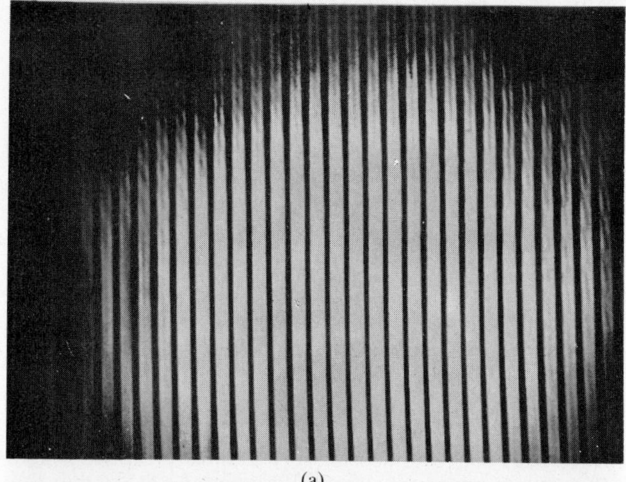

(a)

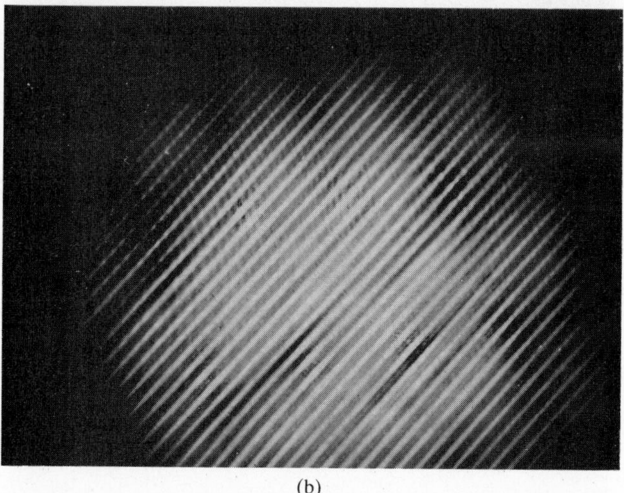

(b)

FIGURE P.13.33 Photos courtesy D. Dutton, M. P. Givens, and R. E. Hopkins.

matic plane waves of $\lambda = 500$ nm, and the setup is the same as that of Fig. 13.33, where the focal lengths of the transform and imaging lenses are 2.0 m and 1.0 m, respectively.

a) Discuss the resulting pattern and design a filter that will pass *only* the first-order terms. Describe it in detail.

b) What will the image look like on Σ_i with that filter in place?

c) How might you pass only the dc term, and what would the image look like then?

13.37 Suppose we insert a mask in the transform plane of the previous problem, which obscures everything but the $m = +1$ diffraction contribution. What will the reformed image look like on Σ_i? Explain your reasoning. Now suppose we remove *only* the $m = +1$ or the $m = -1$ term. What will the re-formed image look like?

13.38* Referring to the previous two problems with the cosine grating oriented horizontally, make a sketch of the electric-field amplitude along y' with no filtering. Plot the corresponding image irradiance distribution. What will the electric field of the image look like if the dc term is filtered out? Plot it. Now plot the new irradiance distribution. What can you say about the spatial frequency of the image with and without the filter in place? Relate your answers to Fig. 11.13.

13.39 Replace the cosine grating in the previous problem with a "square" bar grating, that is, a series of many fine alternating opaque and transparent bands of equal width. We now filter out all terms in the transform plane but the zeroth and the two first-order diffraction spots. These we determine to have relative irradiances of 1.00, 0.36, and 0.36: compare them with Figs. 7.21*a* and 7.22. Derive an expression for the general shape of the irradiance distribution on the image plane—make a sketch of it. What will the resulting fringe system look like?

13.40 A fine square wire mesh with 50 wires per cm is placed vertically in the object plane of the optical computer of Fig. 13.32. If the lenses each have 1.00-m focal lengths, what must be the illuminating wavelength, if the diffraction spots on the transform plane are to have a horizontal and vertical separation of 2.0 mm? What will be the mesh spacing as it appears on the image plane?

13.41* Imagine that we have an opaque mask into which are punched an ordered array of circular holes, all of the same size, located as if at the corners of the boxes of a checkerboard. Now suppose our robot puncher goes mad and makes an additional batch of holes essentially randomly all across the mask. If this screen is now made the object in Problem 13.39, what will the diffraction pattern look like? Given that the ordered holes are separated from their nearest neighbors on the object by 0.1 mm, what will be the spatial frequency of the corresponding dots in the image? Describe a filter that will remove the random holes from the final image.

13.42* Imagine that we have a large photographic transparency on which there is a picture of a student made up of a regular array of small circular dots, all of the same size, but each with its own density, so that it passes a spot of light with a particular field amplitude.

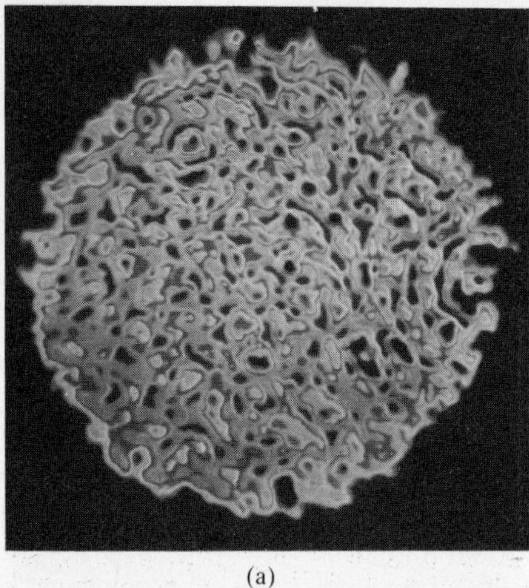

(a)

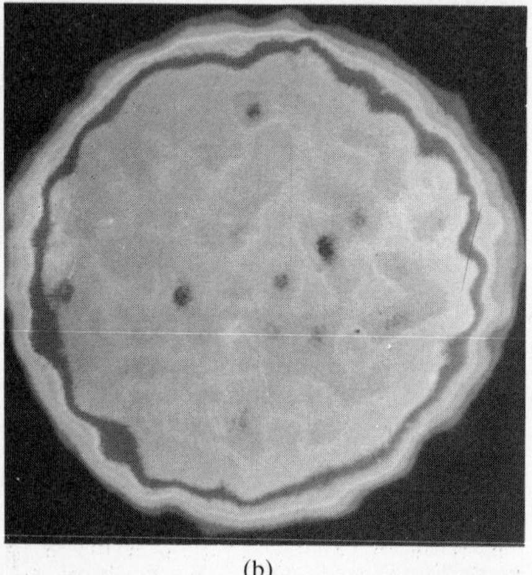

(b)

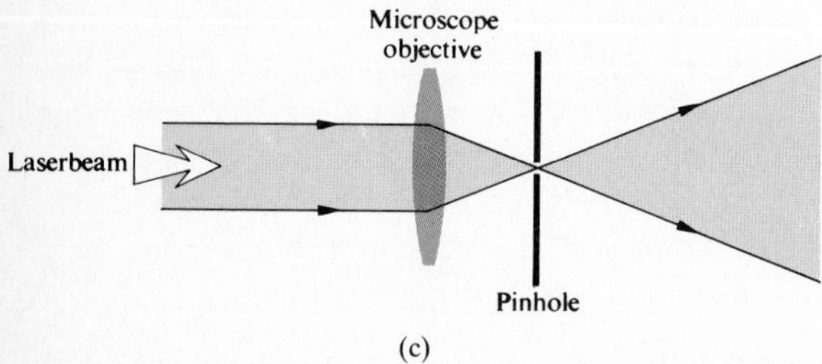

(c)

FIGURE P.13.43 (*a*) and (*b*) A high-power laserbeam before and after spatial filtering. (Photo courtesy Lawrence Livermore National Laboratory.)

Considering the transparency to be illuminated by a plane wave, discuss the idea of representing the electric-field amplitude just beyond it as the product (on average) of a regular two-dimensional array of top-hat functions (Fig. 11.4, p. 516) and the continuous two-dimensional picture function: the former like a dull bed of nails, the latter an ordinary photograph. Applying the frequency convolution theorem, what does the distribution of light look like on the transform plane? How might it be filtered to produce a continuous output image?

13.43* The arrangement shown in Fig. P.13.43 is used to convert a collimated laserbeam into a spherical wave. The pinhole cleans up the beam; that is, it eliminates diffraction effects due to dust and the like on the lens. How does it manage it?

13.44 What would happen to the speckle pattern if a laserbeam were projected onto a suspension such as milk rather than onto a smooth wall?

APPENDIX 1
Electromagnetic Theory

MAXWELL'S EQUATIONS IN DIFFERENTIAL FORM

The set of integral expressions that have come to be known as Maxwell's Equations are

$$\oint_C \mathbf{E}\cdot d\mathbf{l} = -\iint_A \frac{\partial \mathbf{B}}{\partial t}\cdot d\mathbf{S} \qquad [3.5]$$

$$\oint_C \frac{\mathbf{B}}{\mu}\cdot d\mathbf{l} = \iint_A \left(\mathbf{J} + \epsilon\frac{\partial \mathbf{E}}{\partial t}\right)\cdot d\mathbf{S} \qquad [3.13]$$

$$\oiint_A \epsilon\mathbf{E}\cdot d\mathbf{S} = \iiint_V \rho\, dV \qquad [3.7]$$

and

$$\oiint_A \mathbf{B}\cdot d\mathbf{S} = 0 \qquad [3.9]$$

where the units, as usual, are SI.

Maxwell's Equations can be written in a differential form, which is more useful for deriving the wave aspects of the electromagnetic field. This transition can readily be accomplished by making use of two theorems from vector calculus, namely, Gauss's Divergence Theorem,

$$\oiint_A \mathbf{F}\cdot d\mathbf{S} = \iiint_V \nabla\cdot\mathbf{F}\, dV \qquad (A1.1)$$

and Stokes's Theorem

$$\oint_C \mathbf{F}\cdot d\mathbf{l} = \iint_A \nabla\times\mathbf{F}\cdot d\mathbf{S} \qquad (A1.2)$$

Here the quantity **F** is not one fixed vector, but a function that depends on the position variables. It is a rule that associates a single vector, for example, in Cartesian coordinates, $\mathbf{F}(x, y, z)$, with each point (x, y, z) in space. Vector-valued functions of this kind, such as **E** and **B**, are known as vector fields.

Applying Stokes's Theorem to the electric field intensity, we have

$$\oint \mathbf{E}\cdot d\mathbf{l} = \iint \nabla\times\mathbf{E}\cdot d\mathbf{S} \qquad (A1.3)$$

If we compare this with Eq. (3.5), it follows that

$$\iint \nabla\times\mathbf{E}\cdot d\mathbf{S} = -\iint \frac{\partial \mathbf{B}}{\partial t}\cdot d\mathbf{S} \qquad (A1.4)$$

This result must be true for all surfaces bounded by the path C. This can only be the case if the integrands are themselves equal, that is, if

$$\nabla\times\mathbf{E} = -\frac{\partial \mathbf{B}}{\partial t} \qquad (A1.5)$$

A similar application of Stokes's Theorem to **B**, using Eq. (3.13), results in

$$\nabla\times\mathbf{B} = \mu\left(\mathbf{J} + \epsilon\frac{\partial \mathbf{E}}{\partial t}\right) \qquad (A1.6)$$

Gauss's Divergence Theorem applied to the electric field intensity yields

$$\oiint \mathbf{E}\cdot d\mathbf{S} = \iiint \nabla\cdot\mathbf{E}\, dV \qquad (A1.7)$$

If we make use of Eq. (3.7), this becomes

$$\iiint_V \nabla\cdot\mathbf{E}\, dV = \frac{1}{\epsilon}\iiint_V \rho\, dV \qquad (A1.8)$$

and since this is to be true for any volume (i.e., for an arbitrary closed domain), the two integrands must be equal. Consequently, at any point (x, y, z, t) in space–time

$$\nabla\cdot\mathbf{E} = \frac{\rho}{\epsilon} \qquad (A1.9)$$

In the same fashion Gauss's Divergence Theorem applied to the **B**-field and combined with Eq. (3.9) yields

$$\nabla\cdot\mathbf{B} = 0 \qquad (A1.10)$$

Equations (A1.5), (A1.6), (A1.9), and (A1.10) are Maxwell's Equations in differential form. Refer back to Eqs. (3.18) through (3.21) for the simple case of Cartesian coordinates and *free space* ($\rho = J = 0$, $\epsilon = \epsilon_0$, $\mu = \mu_0$).

ELECTROMAGNETIC WAVES

To derive the electromagnetic wave equation in its most general form, we must again consider the presence of some medium. We saw in Section 3.5.1 that there is a need to introduce the *polarization* vector **P**, which is a measure of the overall behavior of the medium, in that it is the resultant electric dipole moment per unit volume. Since the field within the material has been altered, we are led to define a new field quantity, the *displacement* **D**:

$$\mathbf{D} = \epsilon_0 \mathbf{E} + \mathbf{P} \tag{A1.11}$$

Clearly then,

$$\mathbf{E} = \frac{\mathbf{D}}{\epsilon_0} - \frac{\mathbf{P}}{\epsilon_0}$$

The internal electric field **E** is the difference between the field $\mathbf{D}/\epsilon_0$, which would exist in the absence of polarization, and the field $\mathbf{P}/\epsilon_0$ arising from polarization.

For a homogeneous, linear, isotropic dielectric, **P** and **E** are in the same direction and are mutually proportional. It follows that **D** is therefore also proportional to **E**:

$$\mathbf{D} = \epsilon \mathbf{E} \tag{A1.12}$$

Like **E**, **D** extends throughout space and is in no way limited to the region occupied by the dielectric, as is **P**. The lines of **D** begin and end on free, movable charges. Those of **E** begin and end on either *free* charges or bound polarization charges. If no free charge is present, as might be the case in the vicinity of a polarized dielectric or in free space, the lines of **D** close on themselves.

Since in general the response of optical media to **B**-fields is only slightly different from that of a vacuum, we need not describe the process in detail. Suffice it to say that the material will become polarized. We can define a *magnetic polarization* or *magnetization* vector **M** as the magnetic dipole moment per unit volume. In order to deal with the influence of the magnetically polarized medium, we introduce an auxiliary vector **H**, traditionally known as the *magnetic field intensity*

$$\mathbf{H} = \mu_0^{-1}\mathbf{B} - \mathbf{M} \tag{A1.13}$$

For a homogeneous, linear (nonferromagnetic), isotropic medium, **B** and **H** are parallel and proportional:

$$\mathbf{H} = \mu^{-1}\mathbf{B} \tag{A1.14}$$

Along with Eqs. (A1.12) and (A1.14), there is one more *constitutive equation*,

$$\mathbf{J} = \sigma \mathbf{E} \tag{A1.15}$$

Known as *Ohm's Law*, it is a statement of an experimentally determined rule that holds for conductors at constant temperatures. The electric field intensity, and therefore the force acting on each electron in a conductor, determines the flow of charge. The constant of proportionality relating **E** and **J** is the conductivity of the particular medium, σ.

Consider the rather general environment of a linear (nonferroelectric and nonferromagnetic), homogeneous, isotropic medium, which is physically at rest. By making use of the constitutive relations, we can rewrite Maxwell's Equations as

$$\nabla \cdot \mathbf{E} = \rho/\epsilon \qquad [\text{A1.9}]$$

$$\nabla \cdot \mathbf{B} = 0 \qquad [\text{A1.10}]$$

$$\nabla \times \mathbf{E} = -\frac{\partial \mathbf{B}}{\partial t} \qquad [\text{A1.5}]$$

and

$$\nabla \times \mathbf{B} = \mu\sigma\mathbf{E} + \mu\epsilon\frac{\partial \mathbf{E}}{\partial t} \tag{A1.16}$$

If these expressions are somehow to yield a wave equation (2.61), we had best form some second derivatives with respect to the space variables. Taking the curl of Eq. (A1.16), we obtain

$$\nabla \times (\nabla \times \mathbf{B}) = \mu\sigma(\nabla \times \mathbf{E}) + \mu\epsilon\frac{\partial}{\partial t}(\nabla \times \mathbf{E}) \tag{A1.17}$$

where, since **E** is assumed to be a well-behaved function, the space and time derivatives can be interchanged. Equation (A1.5) can be substituted to obtain the needed second derivative with respect to time:

$$\nabla \times (\nabla \times \mathbf{B}) = -\mu\sigma\frac{\partial \mathbf{B}}{\partial t} - \mu\epsilon\frac{\partial^2 \mathbf{B}}{\partial t^2} \tag{A1.18}$$

The vector triple product can be simplified by making use of the operator identity

$$\nabla \times (\nabla \times\ ) = \nabla(\nabla \cdot\ ) - \nabla^2 \tag{A1.19}$$

so that

$$\nabla \times (\nabla \times \mathbf{B}) = \nabla(\nabla \cdot \mathbf{B}) - \nabla^2\mathbf{B}$$

where in Cartesian coordinates

$$(\nabla \cdot \nabla)\mathbf{B} = \nabla^2\mathbf{B} \equiv \frac{\partial^2 \mathbf{B}}{\partial x^2} + \frac{\partial^2 \mathbf{B}}{\partial y^2} + \frac{\partial^2 \mathbf{B}}{\partial z^2}$$

Since the divergence of **B** is zero, Eq. (A1.18) becomes

$$\nabla^2\mathbf{B} = \mu\epsilon\frac{\partial^2 \mathbf{B}}{\partial t^2} - \mu\sigma\frac{\partial \mathbf{B}}{\partial t} = 0 \tag{A1.20}$$

A similar equation is satisfied by the electric field intensity. Following essentially the same procedure as above, take the curl of Eq. (A1.5):

$$\nabla \times (\nabla \times \mathbf{E}) = -\frac{\partial}{\partial t}(\nabla \times \mathbf{B})$$

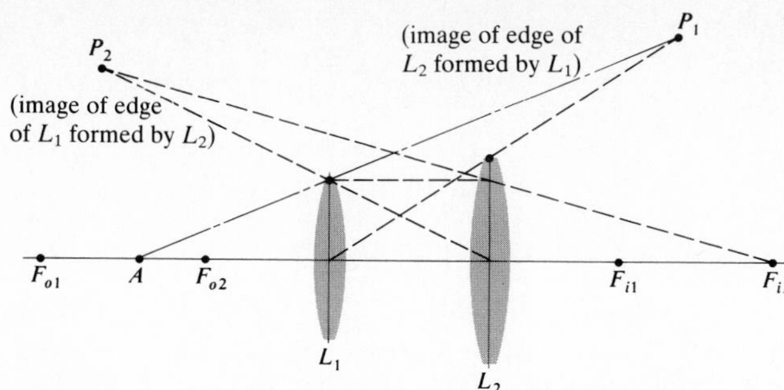

5.40 The A.S. is either the edge of L_1 or L_2. Thus the entrance pupil is either marked by P_1 or P_2. Beyond F_{o1}, P_1 subtends the smaller angle; thus Σ_1 locates the A.S. The image of the A.S. in the lenses to its right, L_2, locates P_3 as the exit pupil.

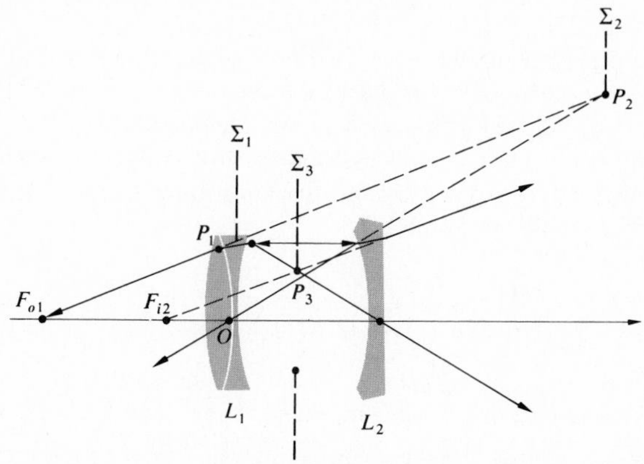

5.41 Draw the chief ray from the tip to L_1 such that when extended it passes through the center of the entrance pupil. From there it goes through the center of the A.S., and then it bends at L_2 so as to extend through the center of the exit pupil. A marginal ray from S extends to the edge of the entrance pupil, bends at L_1 so it just misses the edge of A.S., and then bends at L_2 so as to pass by the edge of the exit pupil.

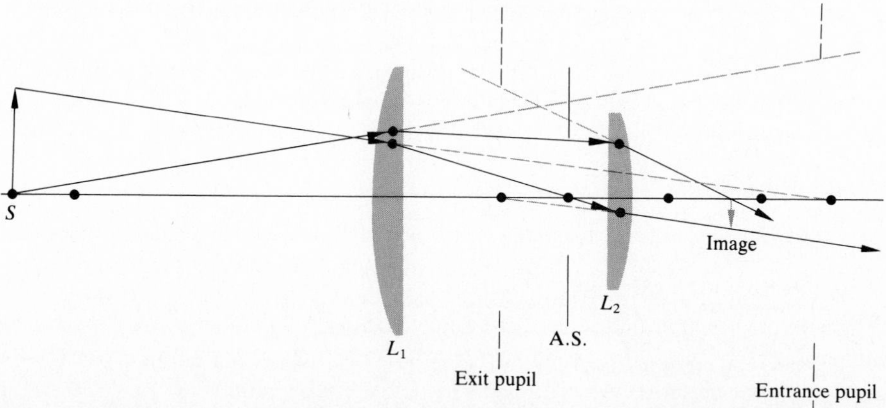

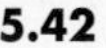

5.42

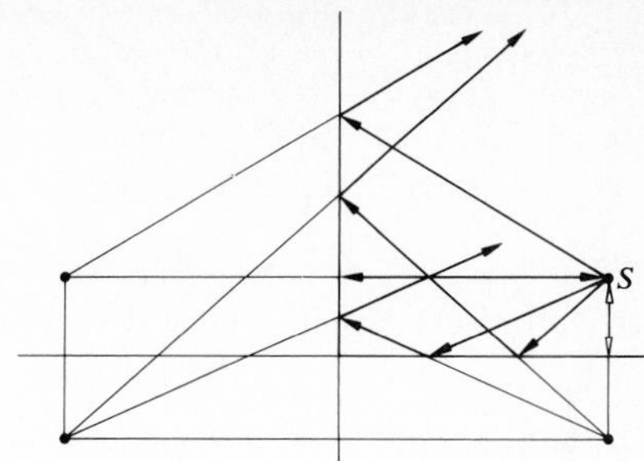

5.43 No—although she might be looking at you.

5.44 The mirror is parallel to the plane of the painting, and so the girl's image should be directly behind her and not off to the right.

5.45 $1/s_o + 1/s_i = -2/R$. Let $R \to \infty$: $1/s_o + 1/s_i = 0$, $s_o = -s_i$, and $M_T = +1$. Image is virtual, same size, and erect.

5.46 From Eq. (5.49), $1/100 + 1/s_i = -2/80$, and so $s_i = -28.5$ cm. Virtual ($s_i < 0$), erect ($M_T > 0$), and minified. (Check with Table 5.5.)

5.48 Image on screen must be real $\therefore$ s_i is +

$$\frac{1}{25} + \frac{1}{100} = -\frac{2}{R}, \qquad \frac{5}{100} = -\frac{2}{R}, \qquad R = -40 \text{ cm}$$

5.49 The image is erect and minified. That implies (Table 5.5) a convex spherical mirror.

5.52 To be magnified and erect the mirror must be concave, and the image virtual; $M_T = 2.0 = s_i/(0.015 \text{ m})$, $s_i = -0.03$ m, and hence $1/f = 1/(0.015 \text{ m}) + 1/(-0.03 \text{ m})$; $f = 0.03$ m and $f = -R/2$; $R = -0.06$ m.

5.53 $M_T = y_i/y_o = -s_i/s_o$, using Eq. (5.50), $s_i = fs_o/(s_o - f)$, and since $f = -R/2$, $M_T = -f/(s_o - f) = -(-R/2)/(s_o + R/2) = R/(2s_o + R)$.

5.56 $M_T = -s_i/25 \text{ cm} = -0.064$; $s_i = 1.6$ cm. $1/25 \text{ cm} + 1/1.6 \text{ cm} = -2/R$, $R = -3.0$ cm.

5.60 $f = -R/2 = 30$ cm, $1/20 + 1/s_i = 1/30$, $1/s_i = 1/30 - 1/20$.

$$s_i = -60 \text{ cm}, \qquad M_T = -s_i/s_o = 60/20 = 3$$

Image is virtual ($s_i < 0$), erect ($M_T > 0$), located 60 cm behind mirror, and 9 inches tall.

5.61 Image rotated through 180°.

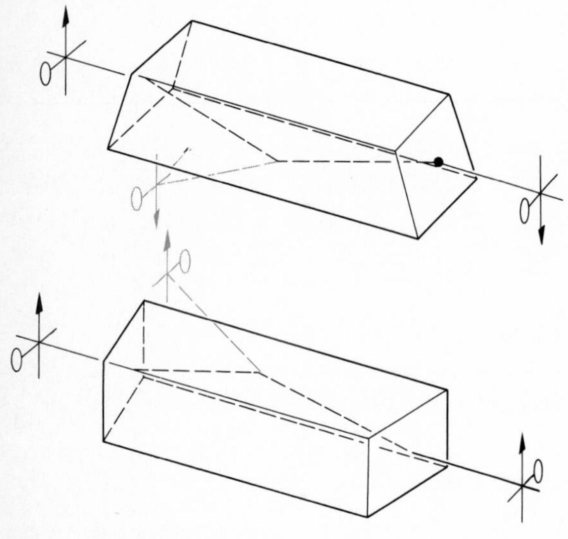

5.62 From Eq. (5.64)

$$\text{NA} = (2.624 - 2.310)^{1/2} = 0.550$$

$$\theta_{\max} = \sin^{-1} 0.550 = 33°22'$$

Maximum acceptance angle is $2\theta_{\max} = 66°44'$. A ray at 45° would quickly leak out of the fiber; in other words, very little energy fails to escape, even at the first reflection.

5.63 Considering Eq. (5.65) (p. 200), $\log 0.5 = -0.30 = -\alpha L/10$, and so $L = 15$ km.

5.64 From Eq. (5.64) (p. 197) NA $= 0.232$ and $N_m = 9.2 \times 10^2$.

5.66 $M_T = -f/x_o = -1/x_o\mathfrak{D}$. for the human eye $\mathfrak{D} \approx 58.6$ diopters.

$$x_o = 230000 \times 1.61 = 371 \times 10^3 \text{ km}$$

$$M_T = -1/3.71 \times 10^6(58.6) = 4.6 \times 10^{-11}$$

$$y_i = 2160 \times 1.61 \times 10^3 \times 4.6 \times 10^{-11} = 0.16 \text{ mm}$$

5.68 $1/20 + 1/s_{io} = 1/4, \qquad s_{io} = 5$ m

$$1/0.3 + 1/s_{ie} = 1/0.6, \qquad s_{ie} = -0.6 \text{ m}$$

$$M_{To} = -5/10 = -0.5$$

$$M_{Te} = -(-0.6)/0.5 = +1.2$$

$$M_{To}M_{Te} = -0.6$$

5.72 Ray 1 in the figure above misses the eye-lens, and there is, therefore, a decrease in the energy arriving at the corresponding image point. This is vignetting.

5.73 Rays that would have missed the eye-lens in the previous problem are made to pass through it by the field-lens. Note how the field-lens bends the chief rays a bit so that they cross the optical axis slightly closer to the eye-lens, thereby moving the exit pupil and shortening the eye relief. (For more on the subject, see *Modern Optical Engineering*, by Smith.)

5.77 $\mathfrak{D}_l - \dfrac{\mathfrak{D}_c}{1 + \mathfrak{D}_c d} = \dfrac{3.2D}{1 + (3.2D)(0.017 \text{ m})} = +3.03D$

or to two figures $+3.0D$. $f_1 = 0.330$ m, and so the far point is 0.330 m − 0.017 m = 0.313 m behind the eye lens. For the contact lens $f_c = 1/3.2 = 0.313$ m. Hence the far point at 0.31 m is the same for both, as it indeed must be.

5.79

(a) The intermediate image-distance is obtained from the lens formula applied to the objective;

$$\frac{1}{27 \text{ mm}} + \frac{1}{s_i} = \frac{1}{25 \text{ mm}}$$

and $s_i = 3.38 \times 10^2$ mm. This is the distance from the objective to the intermediate image, to which must be added the focal length of the eyepiece to get the lens separation; 3.38×10^2 mm + 25 mm = 3.6×10^2 mm.

(b) $M_{To} = -s_i/s_o = -3.38 \times 10^2 \text{ mm}/27 \text{ mm} = -12.5\times$, while the

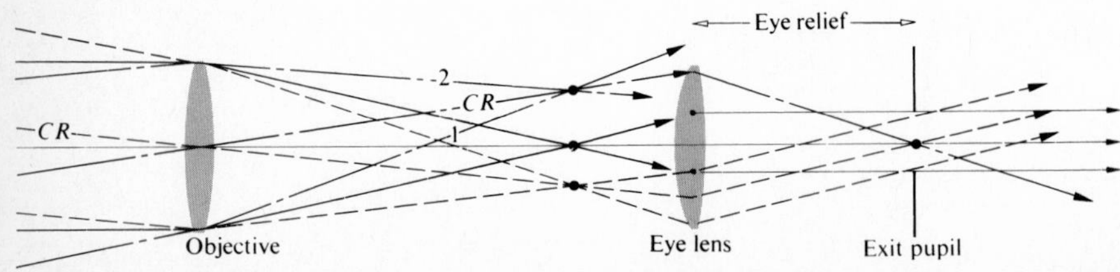

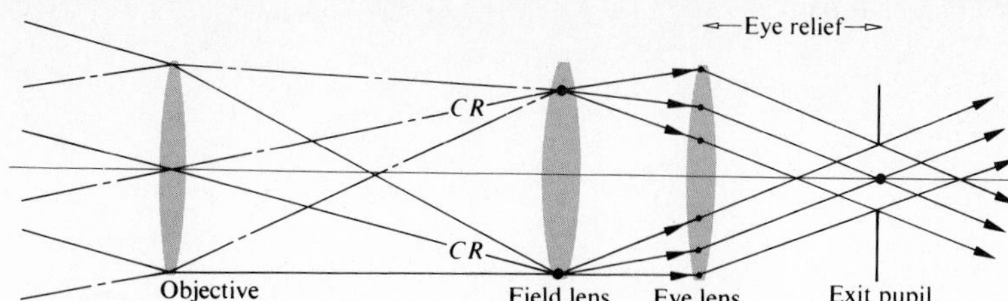

Figure for Solution 5.73

eyepiece has a magnification of $d_o\mathcal{D} = (254 \text{ mm})(1/25 \text{ mm}) = 10.2\times$. Thus the total magnification is MP $= (-12.5)(10.2) = -1.3 \times 10^2$; the minus sign just means the image is inverted.

CHAPTER 6

6.2 From Eq. (6.8),

$$1/f = 1/f' + 1/f' - d/f'f' = 2/f' - 2/3f', \qquad f = 3f'/4$$

From Eq. (6.9), $\overline{H_{11}H_1} = (3f'/4)(2f'/3)/f' = f'/2$

From Eq. (6.10), $\overline{H_{22}H_2} = -(3f'/4)(2f'/3)/f' = -f'/2$

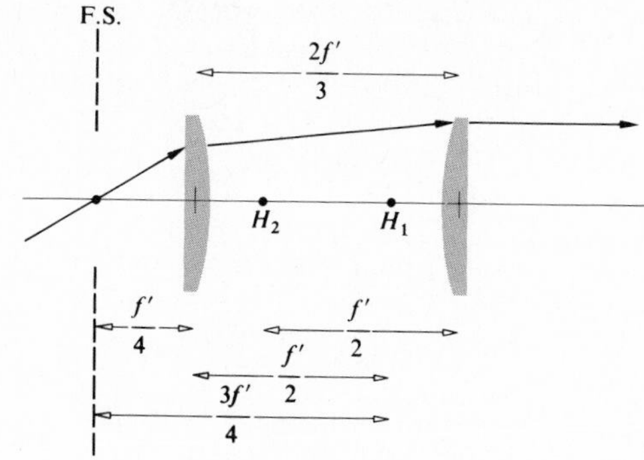

6.3 From Eq. (6.2), $1/f = 0$ when $-(1/R_1 - 1/R_2) = (n_l - 1)d/n_lR_1R_2$. Thus $d = n_l(R_1 - R_2)/(n_l - 1)$.

6.4 $1/f = 0.5[1/6 - 1/10 + 0.5(3)/1.5(6)10]$

$$= 0.5[10/60 - 6/60 + 1/60];\ f = +24$$

$$h_1 = -24(0.5)(3)/10(1.5) = -2.4$$

$$h_2 = -24(0.5)(3)/6(1.5) = -4$$

6.6 $f = \frac{1}{2}nR/(n-1);\ h_1 = +R,\ h_2 = -R$

6.10 $f = 29.6 + 0.4 = 30$ cm; $s_o = 49.8 + 0.2 = 50$ cm; $1/50 + 1/s_i = 1/30$ cm. $s_i = 75$ cm from H_2 and 74.6 cm from the back face.

6.12 From Eq. (6.2),

$$1/f = \tfrac{1}{2}[(1/4.0) - (1/-15) + \tfrac{1}{2}(4.0)/(3/2)(4.0)(-15)]$$

$$= 0.147 \quad \text{and} \quad f = 6.8 \text{ cm}.$$

$h_1 = -(6.8)\frac{1}{2}(4.0)/(-15)(3/2) = +0.60$ cm, while $h_2 = -2.3$. To find the image $1/(100.6) + 1/s_i = 1/(6.8)$; $s_i = 7.3$ cm or 5 cm from the back face of the lens.

6.18 $h_1 = n_{i1}(1 - a_{11})/-a_{12} = (\mathcal{D}_2 d_{21}/n_{t1})f$

$$= -(n_{t1} - 1)d_{21}f/R_2n_{t1}$$

from Eq. (5.64) where $n_{t1} = n_l$;

$$h_2 = n_{t2}(a_{22} - 1)/-a_{12}$$

$$= -(\mathcal{D}_1 d_{21}/n_{t1})f \text{ from Eq. (5.70)}$$

$$= -(n_{i1} - 1)d_{21}f/R_1n_{t1}$$

6.19 $\mathcal{A} = \mathcal{R}_2\mathcal{T}_{21}\mathcal{R}_1$, but for the planar surface

$$\mathcal{R}_2 = \begin{bmatrix} 1 & -\mathcal{D}_2 \\ 0 & 1 \end{bmatrix}$$

and $\mathcal{D}_2 = (n_{t1} - 1)/(-R_2)$ but $R_2 = \infty$

$$\mathcal{R}_2 = \begin{bmatrix} 1 & 0 \\ 0 & 1 \end{bmatrix}$$

which is the unit matrix, hence $\mathcal{A} = \mathcal{T}_{21}\mathcal{R}_1$.

6.20 $\mathcal{D}_1 = (1.5 - 1)/0.5 = 1$

and $\mathcal{D}_2 = (1.5 - 1)/-(-0.25) = 2$

$$\mathcal{A} = \begin{bmatrix} 1 - 2(0.3)/1.5 & -1 + 2(1)(0.3)/(1.5 - 2) \\ 0.3/1.5 & -1(0.3)/1.5 + 1 \end{bmatrix}$$

$$= \begin{bmatrix} 0.6 & -2.6 \\ 0.2 & 0.8 \end{bmatrix}$$

$$|\mathcal{A}| = 0.6(0.8) - (0.2)(-2.6) = 0.48 + 0.52 = 1$$

6.25 See E. Slayter, *Optical Methods in Biology*. $\overline{PC}/\overline{CA} = (n_1/n_2)R/R = n_1/n_2$, while $\overline{CA}/\overline{P'C} = n_1/n_2$. Therefore triangles ACP and ACP' are similar; using the sine law

$$\frac{\sin \measuredangle PAC}{\overline{PC}} = \frac{\sin \measuredangle APC}{\overline{CA}}$$

or

$$n_2 \sin \measuredangle PAC = n_1 \sin \measuredangle APC$$

but $\theta_i = \measuredangle PAC$, thus $\theta_t = \measuredangle APC = \measuredangle P'AC$, and the refracted ray appears to come from P'.

6.26 From Eq. (5.6), let $\cos\varphi = 1 - \varphi^2/2$; then

$$\ell_o = [R^2 + (s_o + R)^2 - 2R(s_o + R) + R(s_o + R)\varphi^2]^{1/2}$$
$$\ell_o^{-1} = [s_o^2 + R(s_o + R)\varphi^2]^{-1/2}$$
$$\ell_i^{-1} = [s_i^2 - R(s_i - R)\varphi^2]^{-1/2}$$

where the first two terms of the binomial series are used,

$$\ell_o^{-1} \approx s_o^{-1} - (s_o + R)h^2/2s_o^3R \quad \text{where } \varphi \approx h/R,$$
$$\ell_i^{-1} \approx s_i^{-1} + (s_i - R)h^2/2s_i^3R$$

Substituting into Eq. (5.5) leads to Eq. (6.40).

6.27

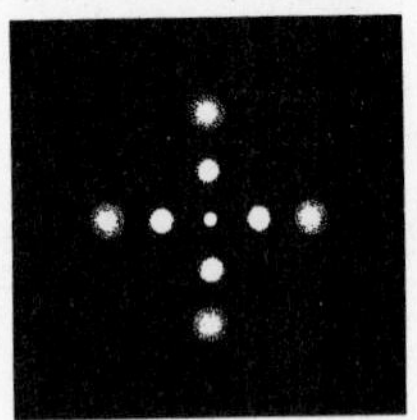

Chapter 7

7.1 $E_0^2 = 36 + 64 + 2\cdot 6\cdot 8\cos\pi/2 = 100$, $E_0 = 10$; $\tan\alpha = \frac{8}{6}$, $\alpha = 53.1° = 0.93$ rad.

$$E = 10\sin(120\pi t + 0.93)$$

7.5 $\dfrac{1\text{ m}}{500\text{ nm}} = 0.2\times 10^7 = 2\,000\,000$ waves

In the glass $\dfrac{0.05}{\lambda_0/n} = \dfrac{0.05(1.5)}{500\text{ nm}} = 1.5\times 10^5$

in air $\dfrac{0.95}{\lambda_0} = 0.19\times 10^7$

total 2 050 000 waves.

$OPD = [(1.5)(0.05) + (1)(0.95)] - (1)(1)$
$OPD = 1.025 - 1.000 = 0.025$ m

$$\frac{\Lambda}{\lambda_0} = \frac{0.025}{500\text{ nm}} = 5\times 10^4 \text{ waves}$$

7.8 $E = E_1 + E_2 = E_{01}\{\sin[\omega t - k(x + \Delta x) + \sin(\omega t - kx)\}$
Since $\sin\beta + \sin\gamma = 2\sin\frac{1}{2}(\beta + \gamma)\cos\frac{1}{2}(\beta - \gamma)$

$$E = 2E_{01}\cos\frac{k\Delta x}{2}\sin\left[\omega t - k\left(x + \frac{\Delta x}{2}\right)\right]$$

7.9 $E = E_0\text{Re}[e^{i(kx+\omega t)} - e^{i(kx-\omega t)}]$
$= E_0\text{Re}[e^{ikx}(e^{i\omega t} - e^{-i\omega t})]$
$= E_0\text{Re}[e^{ikx}2i\sin\omega t]$
$= E_0\text{Re}[2i\cos kx\sin\omega t - 2\sin kx\sin\omega t]$

and $E = -2E_0\sin kx\sin\omega t$. Standing wave with node at $x = 0$.

7.10
$$\frac{\partial E}{\partial x} = -\frac{\partial B}{\partial t}$$

Integrate to get

$$B(x, t) = -\int\frac{\partial E}{\partial x}\,dt = -2E_0k\cos kx\int\cos\omega t\,dt$$
$$= -\frac{2E_0k}{\omega}\cos kx\sin\omega t$$

But $E_0k/\omega = E_0/c = B_0$; thus

$$B(x,t) = -2B_0\cos kx\sin\omega t$$

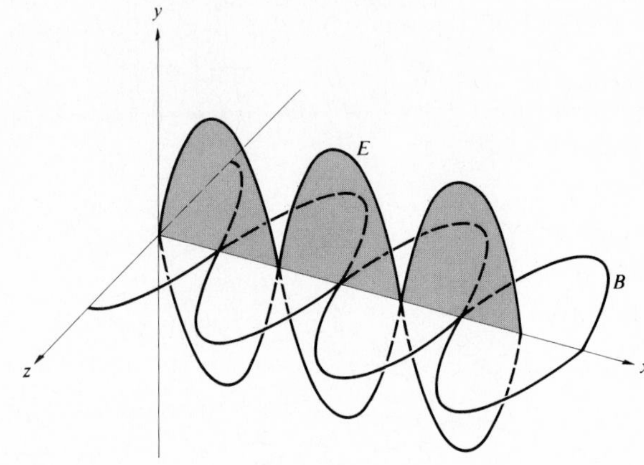

7.15 $E = E_0\cos\omega_c t + E_0\alpha\cos\omega_m t\cos\omega_c t$

$$= E_0\cos\omega_c t + \frac{E_0\alpha}{2}[\cos(\omega_c - \omega_m)t + \cos(\omega_c + \omega_m)t]$$

Audible range $\nu_m = 20$ Hz to 20×10^3 Hz. Maximum modulation frequency $\nu_m(\max) = 20\times 10^3$ Hz.

$$\nu_c - \nu_m(\max) \le \nu \le \nu_c + \nu_m(\max)$$
$$\Delta\nu = 2\nu_m(\max) = 40\times 10^3 \text{ Hz}$$

7.16 $v = \omega/k = ak$, $\quad v_g = d\omega/dk = 2ak = 2v$

7.19
$$v = \sqrt{\frac{g\lambda}{2\pi}} = \sqrt{g/k}$$

$$v_g = v + k\frac{dv}{dk} \qquad [7.38]$$

$$\frac{dv}{dk} = -\frac{1}{2k}\sqrt{\frac{g}{k}} = -\frac{v}{2k}$$

$$v_g = v/2$$

7.21 $v_g = v + k\dfrac{dv}{dk}$ and $\dfrac{dv}{dk} = \dfrac{dv}{d\omega}\dfrac{d\omega}{dk} = v_g\dfrac{dv}{d\omega}$

Since $v = c/n$, $\dfrac{dv}{d\omega} = \dfrac{dv}{dn}\dfrac{dn}{d\omega} = -\dfrac{c}{n^2}\dfrac{dn}{d\omega}$

$$v_g = v - \frac{v_g ck}{n^2}\frac{dn}{d\omega} = \frac{v}{1 + (ck/n^2)(dn/d\omega)} = \frac{c}{n + \omega(dn/d\omega)}$$

7.26 $\omega >> \omega_i$, $\quad n^2 = 1 - \dfrac{Nq_e^2}{\omega^2\epsilon_0 m_e}\sum f_i = 1 - \dfrac{Nq_e^2}{\omega^2\epsilon_0 m_e}$

Using the binomial expansion, we have

$$(1 - x)^{1/2} \approx 1 - \frac{1}{2}x \quad \text{for } x << 1$$

$$n = 1 - Nq_e^2/\omega^2\epsilon_0 m_e 2, \qquad dn/d\omega = Nq_e^2/\epsilon_0 m_e\omega^3$$

$$v_g = \frac{c}{n + \omega(dn/d\omega)}$$

$$= \frac{c}{1 - Nq_e^2/\omega^2\epsilon_0 m_e 2 + Nq_e^2/\epsilon_0 m_e\omega^2}$$

$$= \frac{c}{1 + Nq_e^2/\epsilon_0 m_e\omega^2 2}$$

and $v_g < c$,

$$v = c/n = \frac{c}{1 - Nq_e^2/\epsilon_0 m_e\omega^2 2}$$

Binomial expansion

$$(1 - x)^{-1} \approx 1 + x, \qquad x << 1$$

$$v = c[1 + Nq_e^2/e_0 m_e\omega^2 2]; \qquad vv_g = c^2$$

7.28 $\displaystyle\int_0^\lambda \sin akx \sin bkx\, dx$

$$= \frac{1}{2k}\left[\int_0^\lambda \cos\,[(a-b)kx]k\,dx - \int_0^\lambda \cos\,[(a+b)kx]k\,dx\right]$$

$$= \frac{1}{2k}\frac{\sin(a-b)kx}{a-b}\bigg|_0^\lambda - \frac{1}{2k}\frac{\sin(a+b)kx}{a+b}\bigg|_0^\lambda$$

$= 0 \quad$ if $a \neq b$

Whereas if $a = b$

$$\int_0^\lambda \sin^2 akx\, dx = \frac{1}{2k}\int_0^\lambda (1 + \cos 2akx)k\, dx = \frac{\lambda}{2}$$

The other integrals are similar.

7.29 Even function, therefore $B_m = 0$.

$$A_0 = \frac{2}{\lambda}\int_{-\lambda/a}^{\lambda/a} dx = \frac{2}{\lambda}\left(\frac{\lambda}{a} + \frac{\lambda}{a}\right) = \frac{4}{a}$$

$$A_m = \frac{2}{\lambda}\int_{-\lambda/a}^{\lambda/a} (1)\cos mkx\, dx$$

$$= \frac{2}{mk\lambda}\sin mkx\bigg]_{-\lambda/a}^{\lambda/a}$$

$$A_m = \frac{2}{m\pi}\sin\frac{m2\pi}{a}$$

7.33 $\qquad f'(x) = \dfrac{1}{\pi}\displaystyle\int_0^a E_0 L\frac{\sin kL/2}{kL/2}\cos kx\, dk$

$$= \frac{E_0 L}{\pi 2}\int_0^b \frac{\sin\,(kL/2 + kx)}{kL/2}dk + \frac{E_0 L}{\pi 2}\int_0^b \frac{\sin\,(kL/2 - kx)}{kL/2}dk$$

Let $kL/2 = w$, $(L/2)\,dk = dw$, $kx = wx'$.

$$f'(x) = \frac{E_0}{\pi}\int_0^b \frac{\sin(w + wx')}{w}dw + \frac{E_0}{\pi}\int_0^b \frac{\sin(w - wx')}{w}dw$$

where $b = aL/2$. Let $w + wx' = t$, $dw/w = dt/t$. $0 \le w \le b$ and $0 \le t \le (x' + 1)b$. Let $w - wx' = -t$ in other integral. $0 \le w \le b$ and $0 \le t \le (x' - 1)b$.

$$f'(x) = \frac{E_0}{\pi}\int_0^{(x'+1)b}\frac{\sin t}{t}dt - \frac{E_0}{\pi}\int_0^{(x'-1)b}\frac{\sin t}{t}dt$$

$$f'(x) = \frac{E_0}{\pi}\,\text{Si}\,[b(x' + 1)] - \frac{E_0}{\pi}\,\text{Si}\,[b(x' - 1)], \qquad x' = 2x/L$$

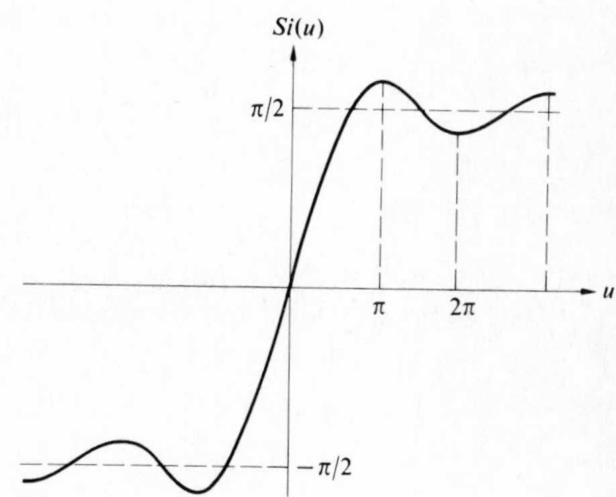

7.34 By analogy with Eq. (7.61),

$$A(\omega) = \frac{\Delta t}{2}E_0 \operatorname{sinc}\,(\omega_p - \omega)\frac{\Delta t}{2}$$

From Table 1 (p. 649) sinc $(\pi/2) = 63.7\%$. Not quite 50% actually,

$$\operatorname{sinc}\left(\frac{\pi}{1.65}\right) = 49.8\%$$

$$\left|(\omega_p - \omega)\frac{\Delta t}{2}\right| < \frac{\pi}{2} \quad \text{or} \quad -\frac{\pi}{\Delta t} < (\omega_p - \omega) < \frac{\pi}{\Delta t}$$

thus appreciable values of $A(\omega)$ lie in a range $\Delta\omega \sim 2\pi/\Delta t$ and $\Delta\nu\,\Delta t \sim 1$. The power spectrum is proportional to $A^2(\omega)$, and $[\text{sinc}\,(\pi/2)]^2 = 40.6\%$.

7.35 $\Delta l_c = c\,\Delta t_c$, $\Delta l_c \approx c/\Delta\nu$. But $\Delta\omega/\Delta k_0 = \bar{\omega}/\bar{k}_0 = c$; thus $|\Delta\nu/\Delta\lambda_0| = \bar{\nu}/\bar{\lambda}_0$,

$$\Delta l_c \approx \frac{c\bar{\lambda}_0}{\Delta\lambda_0\bar{\nu}} \qquad \Delta l_c \approx \bar{\lambda}_0^2/\Delta\lambda_0$$

Or try using the uncertainty principle:

$$\Delta l \approx \frac{h}{\Delta p} \quad \text{where } p = h/\lambda \text{ and } \Delta\lambda_0 << \bar{\lambda}_0$$

7.36

$$\Delta l_c = c\,\Delta t_c = 3 \times 10^8\ \text{m/s}\ 10^{-8}\ \text{s} = 3\ \text{m}$$
$$\Delta\lambda_0 \sim \lambda_0^2/\Delta l_c = (500 \times 10^{-9}\ \text{m})^2/3\ \text{m}$$
$$\Delta\lambda_0 \sim 8.3 \times 10^{-14}\ \text{m} = 8.3 \times 10^{-5}\ \text{nm}$$
$$\Delta\lambda_0/\bar{\lambda}_0 = \Delta\nu/\bar{\nu} = 8.3 \times 10^{-5}/500 = 1.6 \times 10^{-7}$$
$$\approx 1 \text{ part in } 10^7$$

7.37 $\Delta\nu = 54 \times 10^3$ Hz

$$\Delta\nu/\bar{\nu} = \frac{(54 \times 10^3)(10\,600 \times 10^{-9}\ \text{m})}{(3 \times 10^8\ \text{m/s})}$$
$$= 1.91 \times 10^{-9}$$
$$\Delta l_c = c\,\Delta t_c \approx c/\Delta\nu$$
$$\Delta l_c \approx \frac{(3 \times 10^8\ \text{m/s})}{(54 \times 10^3\ \text{Hz})} = 5.55 \times 10^3\ \text{m}$$

7.39 $\Delta l_c = c\,\Delta t_c = 3 \times 10^8 \times 10^{-10} = 3 \times 10^{-2}$ m

$$\Delta\nu \approx 1/\Delta t_c = 10^{10}\ \text{Hz}$$
$$\Delta\lambda_0 \approx \bar{\lambda}_0^2/\Delta l_c \text{ (see Problem 7.35)}$$
$$= (632.8\ \text{nm})^2/3 \times 10^{-2}\ \text{m} = 0.013\ \text{nm}$$
$$\Delta\nu = 10^{15}\ \text{Hz},\ \Delta l_c = c \times 10^{-15} = 300\ \text{nm}$$
$$\Delta\lambda_0 \approx \bar{\lambda}_0^2/\Delta l_c = 1334.78\ \text{nm}$$

Chapter 8

8.1

(a) $\mathbf{E} = \hat{\mathbf{i}}E_0 \cos(kz - \omega t) + \hat{\mathbf{j}}E_0 \cos(kz - \omega t + \pi)$. Equal amplitudes, E_y lags E_x by π. Therefore $\mathcal{P}$-state at 135° or −45°.

(b) $\mathbf{E} = \hat{\mathbf{i}}E_0 \cos(kz - \omega t - \pi/2) + \hat{\mathbf{j}}E_0 \cos(kz - \omega t + \pi/2)$. Equal amplitudes, E_y lags E_x by π. Therefore same as (a).

(c) E_x leads E_y by $\pi/4$. They have equal amplitudes. Therefore it is an ellipse tilted at +45° and is left-handed.

(d) E_y leads E_x by $\pi/2$. They have equal amplitudes. Therefore it is an $\mathcal{R}$-state.

8.2 $\qquad E_x = \hat{\mathbf{i}} \cos \omega t, \qquad E_y = \hat{\mathbf{j}} \sin \omega t$

Left-handed circular standing wave.

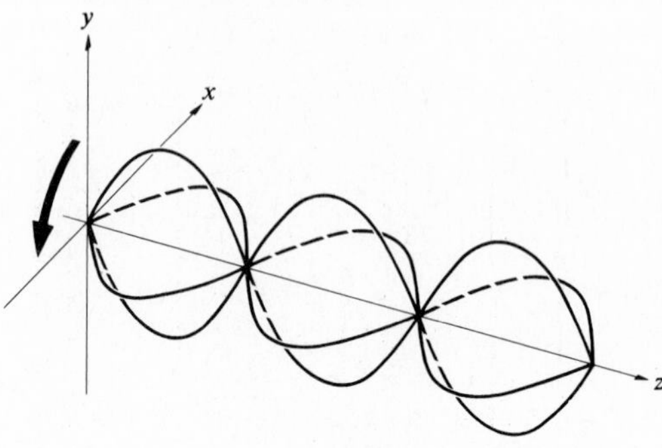

8.3 $\mathbf{E}_{\mathcal{R}} = \hat{\mathbf{i}}E_0 \cos(kz - \omega t) + \hat{\mathbf{j}}E_0 \sin(kz - \omega t)$

$$\mathbf{E}_{\mathcal{L}} = \hat{\mathbf{i}}E_0' \cos(kz - \omega t) - \hat{\mathbf{j}}E_0' \sin(kz - \omega t)$$
$$\mathbf{E} = \mathbf{E}_{\mathcal{R}} + \mathbf{E}_{\mathcal{L}} = \hat{\mathbf{i}}\,(E_0 + E_0') \cos(kz - \omega t)$$
$$+ \hat{\mathbf{j}}(E_0 - E_0') \sin(kz - \omega t).$$

Let $E_0 + E_0' = E_{0x}''$ and $E_0 - E_0' = E_{0y}''$; then $\mathbf{E} = \hat{\mathbf{i}}E_{0x}'' \cos(kz - \omega t) + \hat{\mathbf{j}}E_{0y}'' \sin(kz - \omega t)$. From Eqs. (8.11) and (8.12) it is clear that we have an ellipse where $\varepsilon = -\pi/2$ and $\alpha = 0$.

8.4 $E_{0y} = E_0 \cos 25°$; $E_{0z} = E_0 \sin 25°$;

$$\mathbf{E}(x, t) = (0.91\hat{\mathbf{j}} + 0.42\hat{\mathbf{k}})E_0 \cos(kx - \omega t + \tfrac{1}{2}\pi)$$

8.6 $\mathbf{E} = E_0[\hat{\mathbf{j}} \sin(kx - \omega t) - \hat{\mathbf{k}} \cos(kx - \omega t)]$

8.9 In natural light each filter passes 32% of the incident beam. Half of the incoming flux density is in the form of a $\mathcal{P}$-state parallel to the extinction axis, and effectively none of this emerges. Thus, 64% of the light parallel to the transmission axis is transmitted. In the present problem $32\%I_i$ enters the second filter, and $64\%\,(32\%I_i) = 21\%I_i$ leaves it.

8.20 From the figure (upper right), it follows that

$$I = \tfrac{1}{2}E_{01}^2 \sin^2\theta \cos^2\theta = \frac{E_{01}^2}{8}(1 - \cos 2\theta)(1 + \cos 2\theta)$$
$$= \frac{E_{01}^2}{8}(1 - \cos^2 2\theta) = \frac{E_{01}^2}{8}[1 - (\tfrac{1}{2}\cos 4\theta + \tfrac{1}{2})]$$
$$= \frac{E_{01}^2}{16}(1 - \cos 4\theta) = \frac{I_1}{8}(1 - \cos 4\theta) \qquad \theta = \omega t$$

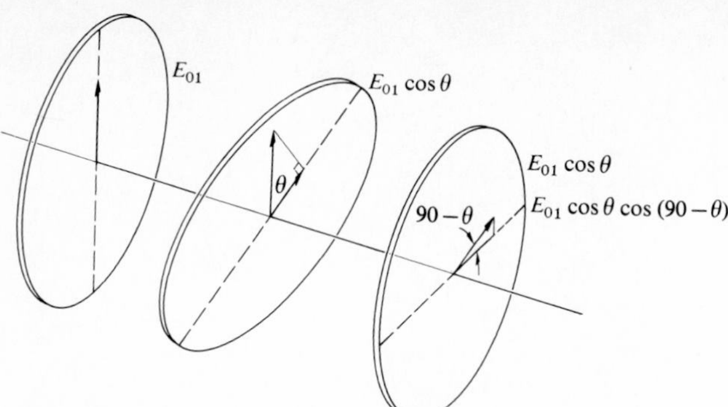

8.21 No. The crystal performs as if it were two oppositely oriented specimens in series. Two similarly oriented crystals in series would behave like one thick specimen and thus separate the *o*- and *e*-rays even more.

8.23 Light scattered from the paper passes through the polaroids and becomes linearly polarized. Light from the upper left filter has its **E**-field parallel to the principal section (which is diagonal across the second and fourth quadrants) and is therfore an *e*-ray. Notice how the letters *P* and *T* are shifted downward in an *extraordinary* fashion. The lower right filter passes an *o*-ray so that the *C* is undeviated. Note that the ordinary image is closer to the blunt corner.

8.24 (a) and (c) are two aspects of the previous problem. (b) shows double refraction because the polaroid's axis is at roughly 45° to the principal section of the crystal. Thus both an *o*- and an *e*-ray will exist.

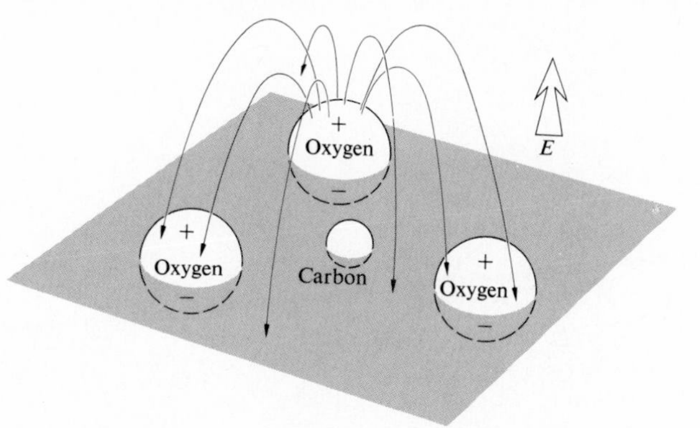

8.25 When **E** is perpendicular to the CO_3 plane the polarization will be less than when it is parallel. In the former case, the field of each polarized oxygen atom tends to reduce the polarization of its neighbors. In other words, the induced field, as shown in the figure, is down while **E** is up. When **E** is in the carbonate plane two dipoles reinforce the third and vice versa. A reduced polarizability leads to a lower dielectric constant, a lower refractive index, and a higher speed. Thus $v_{\parallel} > v_{\perp}$.

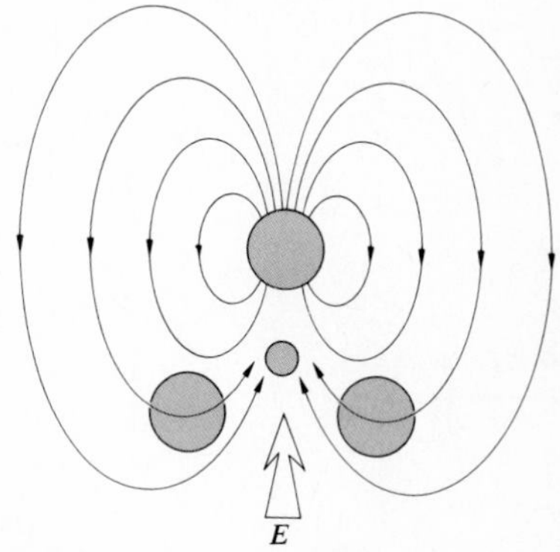

8.33 $n_o = 1.6584$, $n_e = 1.4864$. Snell's Law:

$$\sin\theta_i = n_o \sin\theta_{to} = 0.766$$

$$\sin\theta_i = n_e \sin\theta_{te} = 0.766$$

$$\sin\theta_{to} \approx 0.463, \qquad \theta_{to} \approx 27°35'$$

$$\sin\theta_{te} \approx 0.516, \qquad \theta_{te} \approx 31°4'$$

$$\Delta\theta \approx 3°29'$$

8.35 Calcite $n_o > n_e$. Two spectra will be visible when (b) or (c) is used in a spectrometer. The indices are computed in the usual way, using

$$n = \frac{\sin\frac{1}{2}(\alpha + \delta_m)}{\sin\frac{1}{2}\alpha}$$

where δ_m is the angle of minimum deviation of either beam.

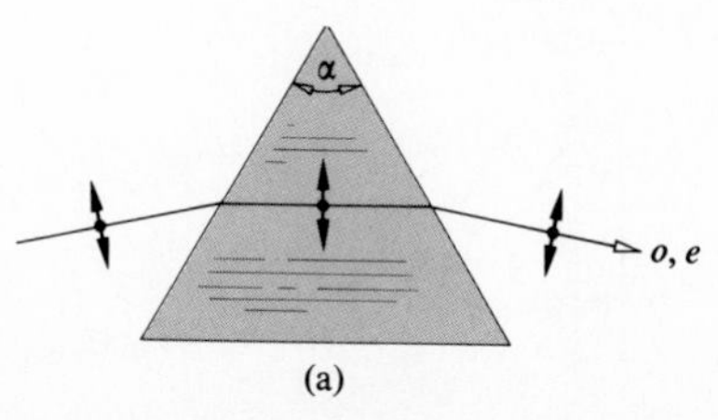

(a)

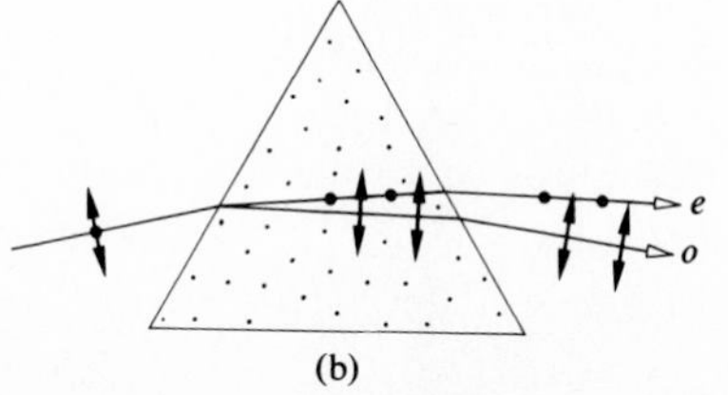

(b)

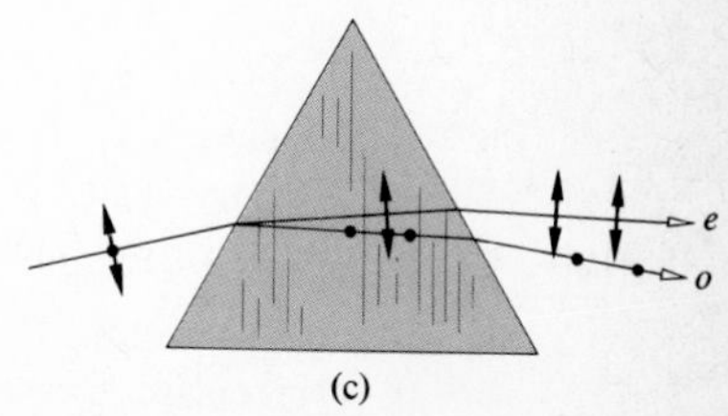

(c)

8.36 E_x leads E_y by $\pi/2$. They were initially in phase and $E_x > E_y$. Therefore the wave is left-handed, elliptical, and horizontal.

8.37 $\sin\theta_c = \dfrac{n_{\text{balsam}}}{n_0} = \dfrac{1.55}{1.658} = 0.935; \; \theta_c \sim 69°$

8.39

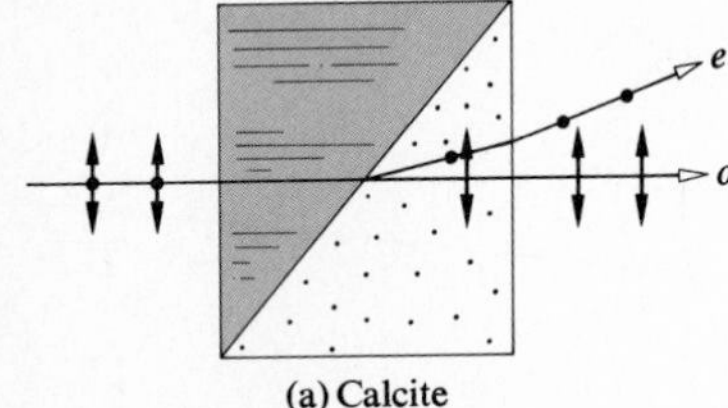

(a) Calcite

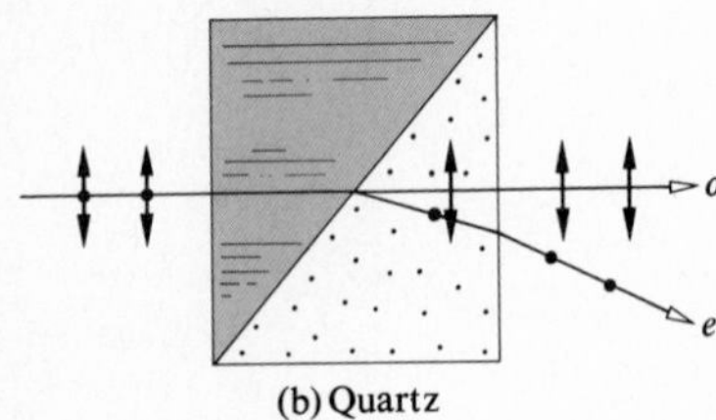

(b) Quartz

(c) Undesired energy in the form of one of the $\mathcal{P}$-states can be disposed of without local heating problems.

(d) The Rochon transmits an undeviated beam (the o-ray), which is therefore achromatic as well.

8.44

$$\Delta\varphi = \frac{2\pi}{\lambda_0} d\,\Delta n$$

but $\Delta\varphi = (1/4)(2\pi)$ because of the fringe shift. Therefore $\Delta\varphi = \pi/2$ and

$$\frac{\pi}{2} = \frac{2\pi\, d\,(0.005)}{589.3 \times 10^{-9}}$$

$$d = \frac{589.3 \times 10^{-9}}{2(10^{-2})} = 2.94 \times 10^{-5}\ \text{m}$$

8.45 The $\mathcal{R}$-state incident on the glass screen drives the electrons in circular orbits, and they reradiate reflected circular light whose **E**-field rotates in the same direction as that of the incoming beam. But the propagation direction has been reversed on reflection, so that although the incident light is in an $\mathcal{R}$-state, the reflected light is left-handed. It will therefore be completely absorbed by the right-circular polarizer. This is illustrated in the figure above.

8.46 Yes. If the amplitudes of the $\mathcal{P}$-states differ. The transmitted beam, in a pile-of-plates polarizer, especially for a small pile.

Figure for Solution 8.45

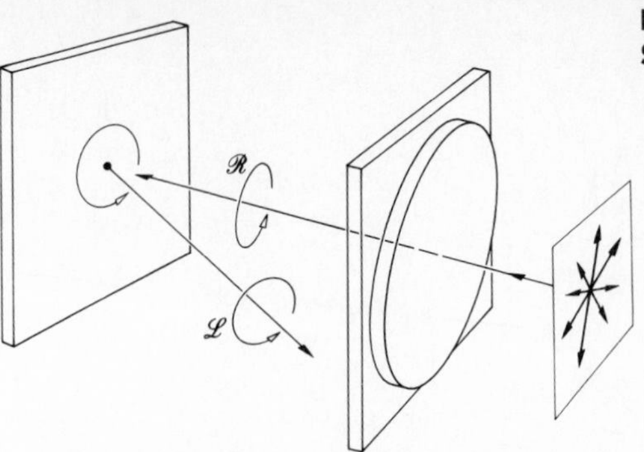

8.48 Place the photoelastic material between circular polarizers with both retarders facing it (as in Fig. 8.51). Under circular illumination no orientation of the stress axes is preferred over any other, and they will thus all be indistinguishable. Only the birefringence will have an effect, and so the isochromatics will be visible. If the two polarizers are different, that is, one an $\mathcal{R}$, the other an $\mathcal{L}$, regions where Δn leads to $\Delta\varphi = \pi$ will appear bright. If they are the same, such regions appear dark.

8.50 $V_{\lambda/2} = \lambda_0/2n_0^3 r_{63}$ [8.44]

$= 550 \times 10^{-9}/2(1.58)^3 5.5 \times 10^{-12}$

$= 10^5/2(3.94) = 12.7$ kV

8.51 $\mathbf{E}_1 \cdot \mathbf{E}_2^* = 0,\; \mathbf{E}_2 = \begin{bmatrix} e_{21} \\ e_{22} \end{bmatrix}$

$$\mathbf{E}_1 \cdot \mathbf{E}_2^* = (1)\,(e_{21})^* + (-2i)(e_{22})^* = 0$$

$$\mathbf{E}_2 = \begin{bmatrix} 2 \\ i \end{bmatrix}$$

$\mathbf{E}_1$ is $\qquad$ $\mathbf{E}_2$ is

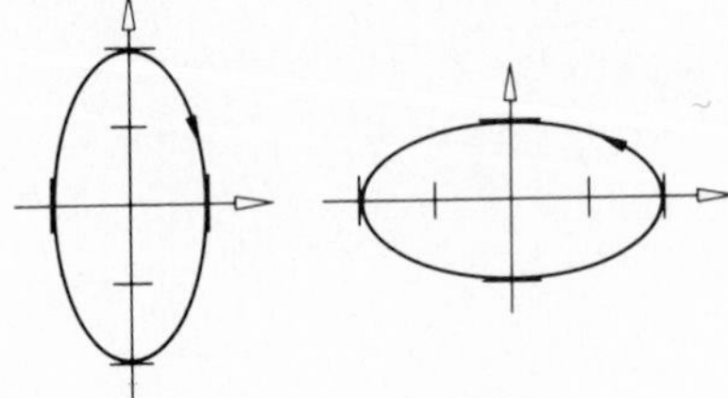

8.57

$$\begin{bmatrix} 1 & 0 & 0 & 0 \\ 0 & 0 & 0 & 1 \\ 0 & 0 & 1 & 0 \\ 0 & -1 & 0 & 0 \end{bmatrix}\begin{bmatrix} 1 & 0 & 0 & 0 \\ 0 & 0 & 0 & -1 \\ 0 & 0 & 1 & 0 \\ 0 & 1 & 0 & 0 \end{bmatrix} = \begin{bmatrix} 1 & 0 & 0 & 0 \\ 0 & 1 & 0 & 0 \\ 0 & 0 & 1 & 0 \\ 0 & 0 & 0 & 1 \end{bmatrix}$$

8.59

$$\begin{bmatrix} 1 & 0 & 0 & 0 \\ 0 & 1 & 0 & 0 \\ 0 & 0 & 0 & -1 \\ 0 & 0 & 1 & 0 \end{bmatrix} \begin{bmatrix} 1 & 0 & 0 & 0 \\ 0 & 1 & 0 & 0 \\ 0 & 0 & 0 & -1 \\ 0 & 0 & 1 & 0 \end{bmatrix} = \begin{bmatrix} 1 & 0 & 0 & 0 \\ 0 & 1 & 0 & 0 \\ 0 & 0 & -1 & 0 \\ 0 & 0 & 0 & -1 \end{bmatrix}$$

$$\begin{bmatrix} 1 & 0 & 0 & 0 \\ 0 & 1 & 0 & 0 \\ 0 & 0 & -1 & 0 \\ 0 & 0 & 0 & -1 \end{bmatrix} \begin{bmatrix} 1 \\ 0 \\ 0 \\ 1 \end{bmatrix} = \begin{bmatrix} 1 \\ 0 \\ 0 \\ -1 \end{bmatrix}$$

$$\begin{bmatrix} 1 & 0 & 0 & 0 \\ 0 & 1 & 0 & 0 \\ 0 & 0 & -1 & 0 \\ 0 & 0 & 0 & -1 \end{bmatrix} \begin{bmatrix} 1 \\ 0 \\ 0 \\ -1 \end{bmatrix} = \begin{bmatrix} 1 \\ 0 \\ 0 \\ 1 \end{bmatrix}$$

$$\begin{bmatrix} 1 & 0 & 0 & 0 \\ 0 & 1 & 0 & 0 \\ 0 & 0 & 0 & 1 \\ 0 & 0 & -1 & 0 \end{bmatrix} \begin{bmatrix} 1 & 0 & 0 & 0 \\ 0 & 1 & 0 & 0 \\ 0 & 0 & 0 & 1 \\ 0 & 0 & -1 & 0 \end{bmatrix} = \begin{bmatrix} 1 & 0 & 0 & 0 \\ 0 & 1 & 0 & 0 \\ 0 & 0 & -1 & 0 \\ 0 & 0 & 0 & -1 \end{bmatrix}$$

8.60

$$\begin{bmatrix} 1 & 0 & 0 & 0 \\ 0 & 1 & 0 & 0 \\ 0 & 0 & 0 & -1 \\ 0 & 0 & 1 & 0 \end{bmatrix} \frac{1}{2}\begin{bmatrix} 1 & 0 & 1 & 0 \\ 0 & 0 & 0 & 0 \\ 1 & 0 & 1 & 0 \\ 0 & 0 & 0 & 0 \end{bmatrix} = \frac{1}{2}\begin{bmatrix} 1 & 0 & 1 & 0 \\ 0 & 0 & 0 & 0 \\ 0 & 0 & 0 & 0 \\ 1 & 0 & 1 & 0 \end{bmatrix}$$

$$\frac{1}{2}\begin{bmatrix} 1 & 0 & 1 & 0 \\ 0 & 0 & 0 & 0 \\ 0 & 0 & 0 & 0 \\ 1 & 0 & 1 & 0 \end{bmatrix} \begin{bmatrix} 1 \\ 0 \\ 0 \\ 0 \end{bmatrix} = \frac{1}{2}\begin{bmatrix} 1 \\ 0 \\ 0 \\ 1 \end{bmatrix}$$

$$\frac{1}{2}\begin{bmatrix} 1 & 0 & 1 & 0 \\ 0 & 0 & 0 & 0 \\ 0 & 0 & 0 & 0 \\ 1 & 0 & 1 & 0 \end{bmatrix} \begin{bmatrix} 1 \\ 0 \\ 0 \\ 1 \end{bmatrix} = \frac{1}{2}\begin{bmatrix} 1 \\ 0 \\ 0 \\ 1 \end{bmatrix}$$

$$\frac{1}{2}\begin{bmatrix} 1 & 0 & 0 & 1 \\ 0 & 0 & 0 & 0 \\ 0 & 0 & 0 & 0 \\ 1 & 0 & 0 & 1 \end{bmatrix} \begin{bmatrix} 1 \\ 0 \\ 0 \\ 1 \end{bmatrix} = \frac{1}{2}\begin{bmatrix} 1 \\ 0 \\ 0 \\ 1 \end{bmatrix}$$

$$\frac{1}{2}\begin{bmatrix} 1 & 0 & 1 & 0 \\ 0 & 0 & 0 & 0 \\ 0 & 0 & 0 & 0 \\ 1 & 0 & 1 & 0 \end{bmatrix} \begin{bmatrix} 1 \\ 0 \\ 0 \\ -1 \end{bmatrix} = \frac{1}{2}\begin{bmatrix} 1 \\ 0 \\ 0 \\ 1 \end{bmatrix}$$

$$\frac{1}{2}\begin{bmatrix} 1 & 0 & 0 & 1 \\ 0 & 0 & 0 & 0 \\ 0 & 0 & 0 & 0 \\ 1 & 0 & 0 & 1 \end{bmatrix} \begin{bmatrix} 1 \\ 0 \\ 0 \\ -1 \end{bmatrix} = \begin{bmatrix} 0 \\ 0 \\ 0 \\ 0 \end{bmatrix}$$

8.62

$$\begin{bmatrix} te^{i\varphi} & 0 \\ 0 & te^{i\varphi} \end{bmatrix}$$

where a phase increment of φ is introduced into both components as a result of traversing the plate.

$$\begin{bmatrix} 1 & 0 \\ 0 & 1 \end{bmatrix} \qquad \begin{bmatrix} 0 & 0 \\ 0 & 0 \end{bmatrix}$$

8.63

$$\begin{bmatrix} t^2 & 0 & 0 & 0 \\ 0 & t^2 & 0 & 0 \\ 0 & 0 & t^2 & 0 \\ 0 & 0 & 0 & t^2 \end{bmatrix} \qquad \begin{bmatrix} 1 & 0 & 0 & 0 \\ 0 & 0 & 0 & 0 \\ 0 & 0 & 0 & 0 \\ 0 & 0 & 0 & 0 \end{bmatrix}$$

8.64

$$V = \frac{I_p}{I_p + I_u} = \frac{(\mathcal{S}_1^2 + \mathcal{S}_2^2 + \mathcal{S}_3^2)^{1/2}}{\mathcal{S}_0}$$

$$I_p = (\mathcal{S}_1^2 + \mathcal{S}_2^2 + \mathcal{S}_3^2)^{1/2}; \qquad I - I_p = I_u$$

$$\mathcal{S}_0 - (\mathcal{S}_1^2 + \mathcal{S}_2^2 + \mathcal{S}_3^2)^{1/2} = I_u$$

$$\begin{bmatrix} 4 \\ 0 \\ 0 \\ 0 \end{bmatrix} + \begin{bmatrix} 1 \\ 0 \\ 0 \\ 1 \end{bmatrix} = \begin{bmatrix} 5 \\ 0 \\ 0 \\ 1 \end{bmatrix}$$

$$5 - (0 + 0 + 1)^{1/2} = I_u$$

CHAPTER 9

9.1 $\mathbf{E}_1 \cdot \mathbf{E}_2 = \frac{1}{2}(\mathbf{E}_1 e^{-i\omega t} + \mathbf{E}_1^* e^{i\omega t}) \cdot \frac{1}{2}(\mathbf{E}_2 e^{-i\omega t} + \mathbf{E}_2^* e^{i\omega t})$,
where Re $(z) = \frac{1}{2}(z + z^*)$.

$$\mathbf{E}_1 \cdot \mathbf{E}_2 = \tfrac{1}{4}[\mathbf{E}_1 \cdot \mathbf{E}_2 e^{-2i\omega t} + \mathbf{E}_1^* \cdot \mathbf{E}_2^* e^{2i\omega t} + \mathbf{E}_1 \cdot \mathbf{E}_2^* + \mathbf{E}_1^* \cdot \mathbf{E}_2]$$

The last two terms are time independent, while

$$\langle \mathbf{E}_1 \cdot \mathbf{E}_2 e^{-2i\omega t} \rangle \to 0 \quad \text{and} \quad \langle \mathbf{E}_1^* \cdot \mathbf{E}_2^* e^{2i\omega t} \rangle \to 0$$

because of the $1/T\omega$ coefficient. Thus

$$I_{12} = 2\langle \mathbf{E}_1 \cdot \mathbf{E}_2 \rangle = \tfrac{1}{2}(\mathbf{E}_1 \cdot \mathbf{E}_2^* + \mathbf{E}_1^* \cdot \mathbf{E}_2)$$

9.2 The largest value of $(r_1 - r_2)$ is equal to a. Thus if $\varepsilon_1 = \varepsilon_2$, $\delta = k(r_1 - r_2)$ varies from 0 to ka. If $a >> \lambda$, cos δ and therefore I_{12} will have a great many maxima and minima and therefore average to zero over a large region of space. In contrast, if $a << \lambda$, δ varies only slightly from 0 to $ka << 2\pi$. Hence I_{12} does not average to zero, and from Eq. (9.17), I deviates little from $4I_0$. The two sources effectively behave as a single source of double the original strength.

9.4 A bulb at S would produce fringes. We can imagine it as made up of a very large number of incoherent point sources. Each of these would generate an independent pattern, all of which would then overlap. Bulbs at S_1 and S_2 would be incoherent and could not generate detectable fringes.

9.7

(a) $(r_1 - r_2) = \pm\frac{1}{2}\lambda$, hence $a \sin\theta_1 = \pm\frac{1}{2}\lambda$ and $\theta_1 \approx \pm\frac{1}{2}\lambda/a = \pm\frac{1}{2}(632.8 \times 10^{-9}\text{ m})/(0.200 \times 10^{-3}\text{ m}) = \pm 1.58 \times 10^{-3}$ rad, or since $y_1 = s\theta_1 = (1.00\text{ m})(\pm 1.58 \times 10^{-3}\text{ rad}) = \pm 1.58$ mm.

(b) $y_5 = s5\lambda/a = (1.00\text{ m})5(632.8 \times 10^{-9})/(0.200 \times 10^{-3}\text{ m}) = 1.582 \times 10^{-2}$ m.

(c) Since the fringes vary as cosine-squared and the answer to (a) is half a fringe width, the answer to (b) is 10 times larger.

9.15 $r_2^2 = a^2 + r_1^2 - 2ar_1 \cos(90 - \theta)$. The contribution to $\cos\delta/2$ from the third term in the Maclaurin expansion will be negligible if

$$\frac{k}{2}\left(\frac{a^2}{2r_1}\cos^2\theta\right) << \pi/2$$

therefore $r_1 >> a^2/\lambda$.

9.16 $E = \frac{1}{2}mv^2; \quad v = 0.42 \times 10^6$ m/s

$\lambda = h/mv = 1.73 \times 10^{-9}; \quad \Delta y = s\lambda/a = 3.46$ mm

9.20 $\Delta y = s\lambda_0/2d\alpha(n - n')$

9.21 $\Delta y = (s/a)\lambda, \quad a = 10^{-2}$ cm, $a/2 = 5 \times 10^{-3}$ cm

9.22 $\delta = k(r_1 - r_2) + \pi$ (Lloyd's mirror)

$\delta = k\{a/2 \sin\alpha - [\sin(90 - 2\alpha)]a/2 \sin\alpha\} + \pi$

$\delta = ka(1 - \cos 2\alpha)/2 \sin\alpha + \pi$

maximum occurs for

$\delta = 2\pi$ when $\sin\alpha\,(\lambda/a) = (1 - \cos 2\alpha) = 2\sin^2\alpha$

First maximum $\alpha = \sin^{-1}(\lambda/2a)$.

9.24 Here $1.00 < 1.34 > 1.00$, hence from Eq. (9.36) with $m = 0$, $d = (0 + \frac{1}{2})(633\text{ nm})/2(1.34) = 118$ nm.

9.27 Eq. (9.37) $m = 2n_f d/\lambda_0 = 10\,000$. A minimum, therefore central dark region.

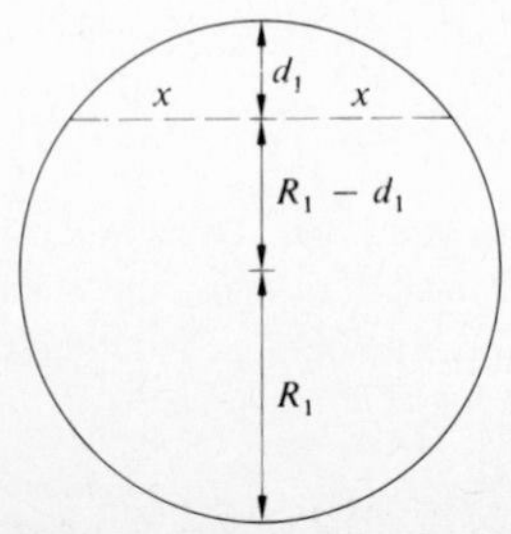

9.28 The fringes are generally a series of fine jagged bands, which are fixed with respect to the glass.

9.29 $x^2 = d_1[(R_1 - d_1) + R_1] = 2R_1d_1 - d_1^2$

Similarly $x^2 = 2R_2d_2 - d_2^2$

$$d = d_1 - d_2 = \frac{x^2}{2}\left[\frac{1}{R_1} - \frac{1}{R_2}\right], \qquad d = m\frac{\lambda_f}{2}$$

As $R_2 \to \infty$, x_m approaches Eq. (9.43).

9.31 $\Delta x = \lambda_f/2\alpha, \qquad \alpha = \lambda_0/2n_f\Delta x$

$\alpha = 5 \times 10^{-5}$ rad $= 10.2$ seconds

9.33 A motion of $\lambda/2$ causes a single fringe pair to shift past, hence $92\,\lambda/2 = 2.53 \times 10^{-5}$ m and $\lambda = 550$ nm.

9.37 $E_t^2 = E_tE_t^* = E_0^2(tt')^2/(1 - r^2e^{-i\delta})(1 - r^2e^{+i\delta})$

$I_t = I_i(tt')^2/(1 - r^2e^{-i\delta} - r^2e^{i\delta} + r^4)$

9.38 (a) $R = 0.80 \therefore F = 4R/(1 - R)^2 = 80$

(b) $\gamma = 4\sin^{-1} 1/\sqrt{F} = 0.448$

(c) $\mathscr{F} = 2\pi/0.448$

(d) $C = 1 + F$

9.39 $$\frac{2}{1 + F(\Delta\delta/4)^2} = 0.81\left[1 + \frac{1}{1 + F(\Delta\delta/2)^2}\right]$$

$F^2(\Delta\delta)^4 - 15.5F(\Delta\delta)^2 - 30 = 0$

9.40 $I = I_{max}\cos^2\delta/2$

$I = I_{max}/2$ when $\delta = \pi/2 \quad \therefore \gamma = \pi$

Separation between maxima is 2π

$$\mathscr{F} = 2\pi/\gamma = 2$$

9.42 At near normal incidence ($\theta_i \approx 0$) Fig. 4.23 *e* indicates that the relative phase shift between an internally and externally reflected beam is π rad. That means a total relative phase difference of

$$\frac{2\pi}{\lambda_f}[2(\lambda_f/4)] + \pi$$

$n_0 < n_1$
$n_1 > n_s$
n_s

or 2π. The waves are in phase and interfere constructively.

9.43 $n_0 = 1 \quad n_s = n_g \quad n_1 = \sqrt{n_g}$

$\sqrt{1.54} = 1.24$

$$d = \frac{1}{4}\lambda_f = \frac{1}{4}\frac{\lambda_0}{n_1} = \frac{1}{4}\frac{540}{1.24}\text{ nm}$$

No relative phase shift between two waves.

9.44 The refracted wave will traverse the film twice, and there will be no relative phase shift on reflection. Hence

$$d = \lambda_0/4n_f = (550\text{ nm})/4(1.38) = 99.6\text{ nm}.$$

CHAPTER 10

10.1 $(R + \ell)^2 = R^2 + a^2$; therefore $R = (a^2 - \ell^2)/2\ell \approx a^2/2\ell$, $\ell R = a^2/2$, so for $\lambda >> \ell$, $\lambda R >> a^2/2 \therefore R = (1 \times 10^{-3})^2 10/2\lambda = 10$ m.

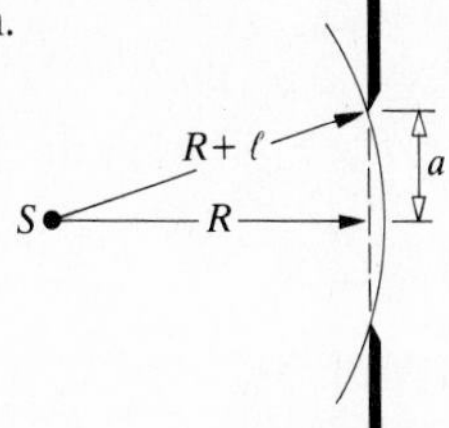

10.2 $E_0/2 = R \sin(\delta/2)$

$E = 2R \sin(N\delta/2)$ chord length

$E = [E_0 \sin(N\delta/2)]/\sin(\delta/2)$

$I = E^2$

10.4 $d \sin\theta_m = m\lambda$, $\quad \theta = N\delta/2 = \pi$

$7 \sin\theta = (1)(0.21) \quad \delta = 2\pi/N = kd\sin\theta$

$\sin\theta = 0.03 \quad \sin\theta = 0.0009$

$\theta = 1.7° \quad \theta = 3$ min.

10.5 Converging spherical wave in image space is diffracted by the exit pupil.

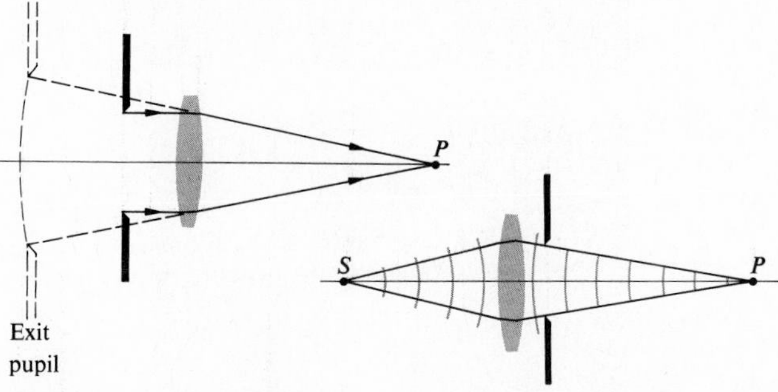

10.6

$\beta = \pm\pi$

$\sin\theta = \pm\lambda/b$

$\theta \approx \pm\lambda/b$

$L\theta \approx \pm L\lambda/b$

$L\theta \approx \pm f_2\lambda/b.$

10.9 $\lambda = (20\text{ cm}) \sin 36.87° = 12$ cm

10.10 $\alpha = \dfrac{ka}{2}\sin\theta, \quad \beta = \dfrac{kb}{2}\sin\theta$

$a = mb,\ \alpha = m\beta,\ \alpha = m2\pi$

$N =$ number of fringes $= a/\pi = m2\pi/\pi = 2m$

10.12 $\alpha = 3\pi/2N = \pi/2$ [10.34]

$$I(\theta) = \frac{I(0)}{N^2}\left(\frac{\sin\beta}{\beta}\right)^2 \quad \text{from Eq. (10.35)}$$

and $I/I(0) \approx \frac{1}{9}$.

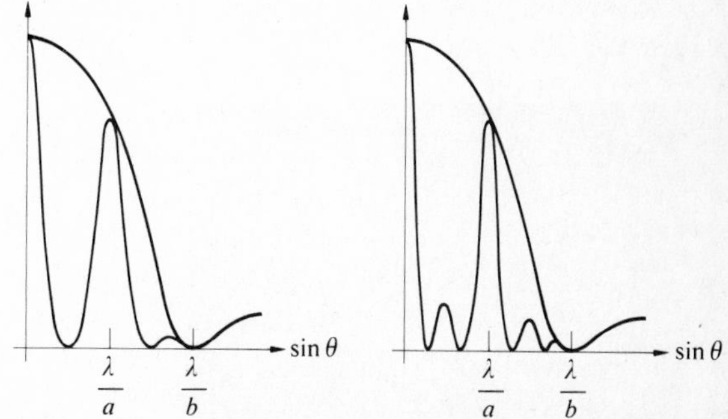

10.15 If the aperture is symmetrical about a line, the pattern will be symmetrical about a line parallel to it. Moreover, the pattern will be symmetrical about yet another line prpendicular to the aperture's symmetry axis. This follows from the fact that Fraunhofer patterns have a center of symmetry.

10.16

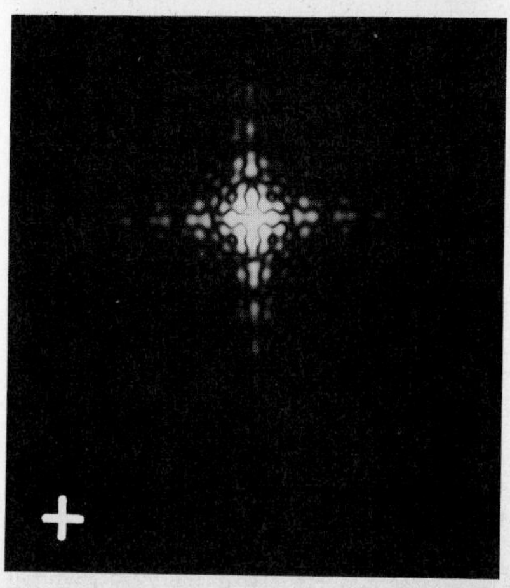

10.17 Three parallel short slits.
10.18 Two parallel short slits.
10.19 An equilateral triangular hole.
10.20 A cross-shaped hole.
10.21 The *E*-field of a rectangular hole.
10.23 From Eq. (10.58), $q_1 \approx 1.22(f/D)\lambda \approx \lambda$.
10.24

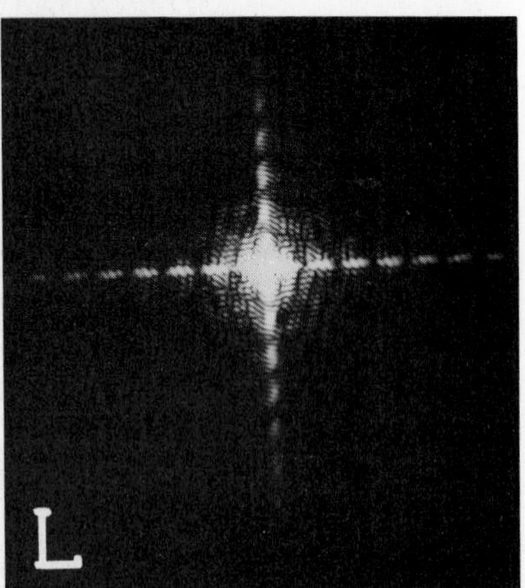

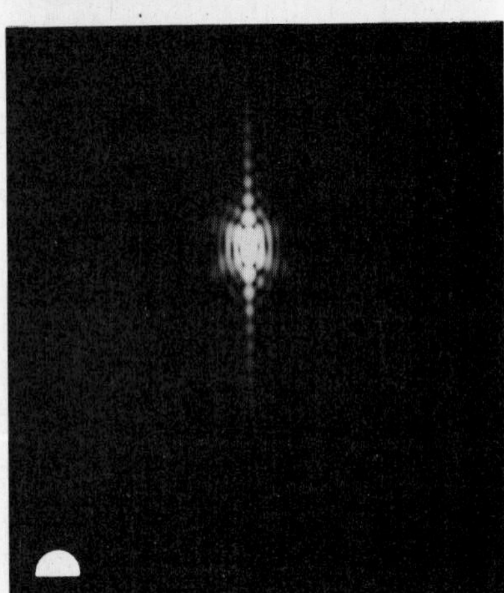

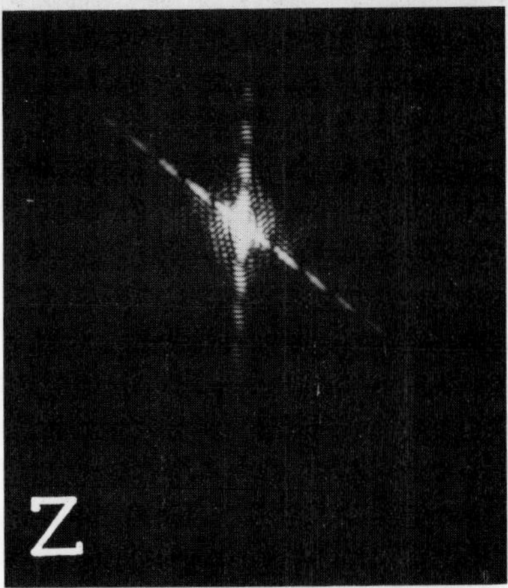

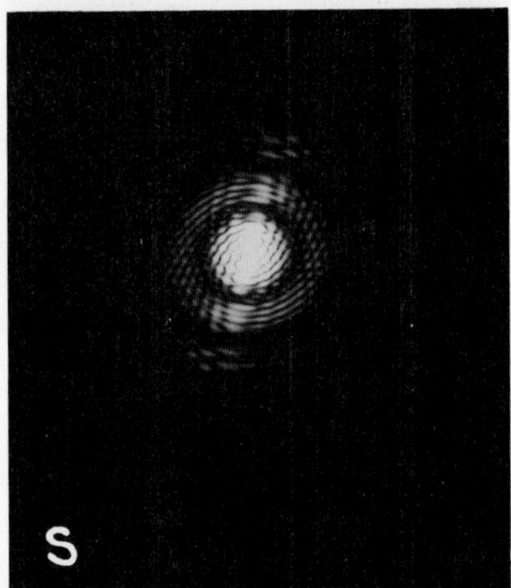

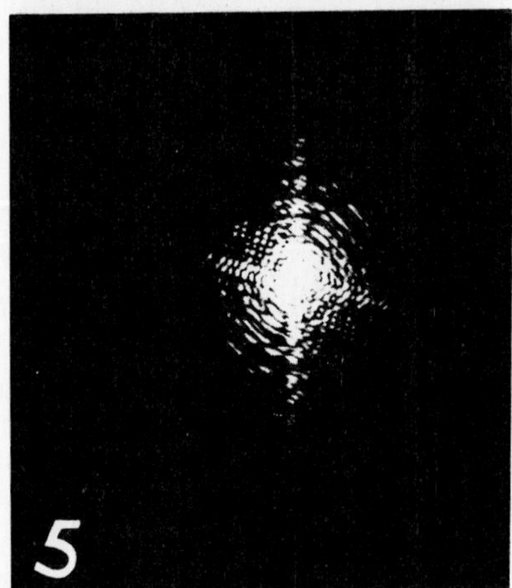

10.27 1 part in 1000. 3 yd ≈ 100 inches. (see Figure on p.675.)

10.32 From Eq. (10.32), where $a = 1/(1000 \text{ lines per cm}) = 0.001$ cm per line (center to center), $\sin\theta_m = 1(650 \times 10^{-9} \text{ m})/(0.001 \times 10^{-2} \text{ m}) = 6.5 \times 10^{-2}$ and $\theta_1 = 3.73°$.

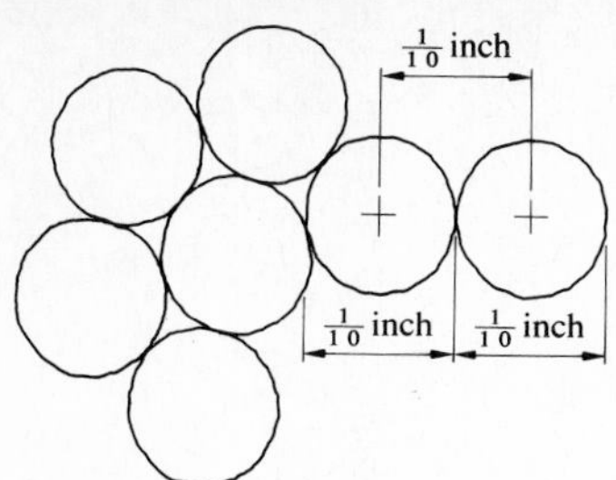

Figure for Solution 10.27

10.35 The largest value of m in Eq. (10.32) occurs when the sine function is equal to one, making the left side of the equation as large as possible, then $m = a/\lambda = (1/10 \times 10^5)/(3.0 \times 10^8 \text{ m/s} \div 4.0 \times 10^{14} \text{ Hz}) = 1.3$, and only the first-order spectrum is visible.

10.37 $\sin\theta_i = n \sin\theta_n$

Optical path length difference $= m\lambda$

$a \sin\theta_m - na \sin\theta_n = m\lambda$

$a(\sin\theta_m - \sin\theta_i) = m\lambda$

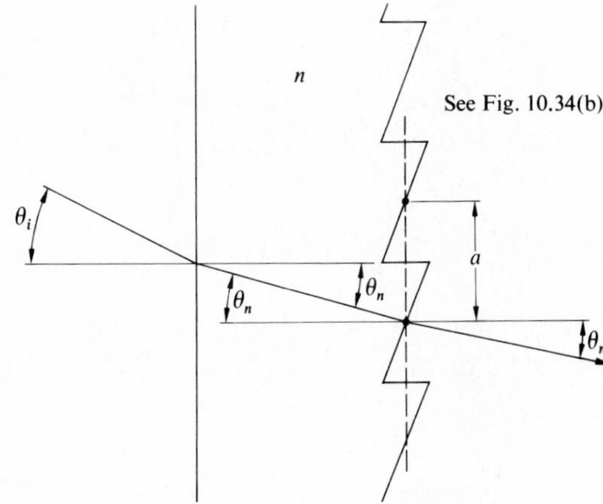

10.38 $\mathcal{R} = mN = 10^6, N = 78 \times 10^3$

$\therefore m = 10^6/78 \times 10^3$

$\Delta\lambda_{\text{fsr}} = \lambda/m = 500 \text{ nm}/(10^6/78 \times 10^3) = 39 \text{ nm}$

$$\mathcal{R} = \mathcal{F}m = \mathcal{F}\frac{2n_f d}{\lambda} = 10^6 \qquad [9.76]$$

$$\Delta\lambda_{\text{fsr}} = \lambda^2/2n_f d = 0.0125 \text{ nm} \qquad [9.78]$$

10.39 $\mathcal{R} = \lambda/\Delta\lambda = 5892.9/5.9 = 999$

$$N = \mathcal{R}/m = 333$$

10.41 $y = L\lambda/d$

$d = 12 \times 10^{-6}/12 \times 10^{-2} = 10^{-4} \text{ m}$

10.43 $A = 2\pi\rho^2 \int_0^{\varphi} \sin\varphi \, d\varphi = 2\pi\rho^2(1 - \cos\varphi)$

$$\cos\varphi = [\rho^2 + (\rho + r_0)^2 - r_l^2]/2\rho(\rho + r_0)$$

$$r_l = r_0 + l\lambda/2$$

Area of first l zones

$$A = 2\pi\rho^2 - \pi\rho(2\rho^2 + 2\rho r_0 - l\lambda r_0 - l^2\lambda^2/4)/(\rho + r_0)$$

$$A_l = A - A_{l-1} = \frac{\lambda\pi\rho}{\rho + r_0}\left[r_0 + \frac{(2l-1)\lambda}{4}\right]$$

10.45

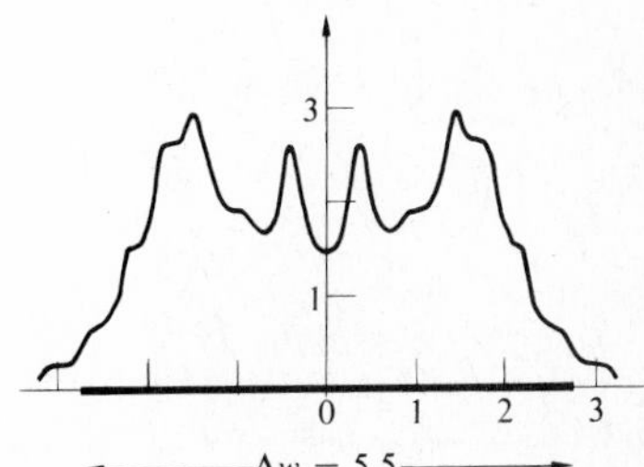

10.46 $I = \frac{I_0}{2}\{[\frac{1}{2} - \mathscr{C}(v_l)]^2 + [\frac{1}{2} - \mathscr{S}(v_l)]^2\}$

$$I = \frac{I_0}{2}\left(\frac{1}{\pi v_1}\right)^2\left[\sin^2\left(\frac{\pi v_1^2}{2}\right) + \cos^2\left(\frac{\pi v_1^2}{2}\right)\right]$$

$$I = \frac{I_0}{2}\left(\frac{1}{\pi v_1}\right)^2$$

10.47 Fringes in both the clear and shadow region [(see M. P. Givens and W. L. Goffe, *Am. J. Phys.* **34**, 248 (1966)].

10.48 $u = y[2/\lambda r_0]^{1/2}$; $\Delta u = \Delta y \times 10^3 = 2.5$

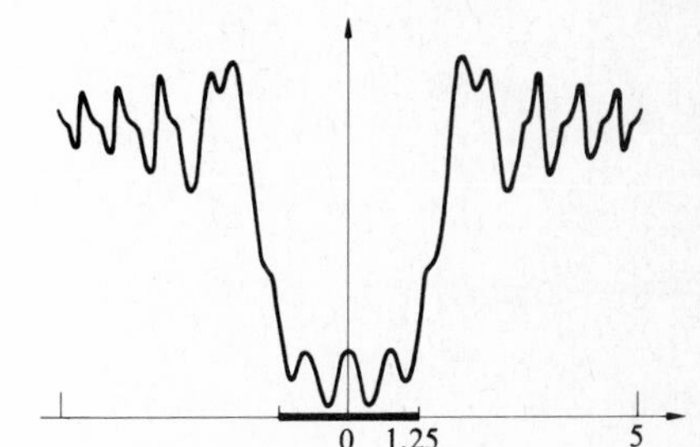

10.49

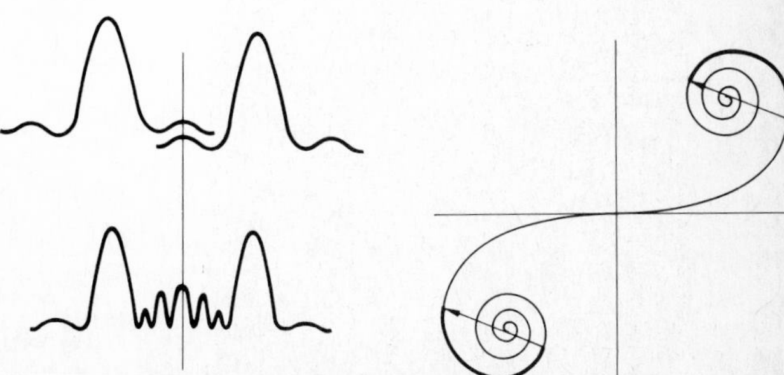

CHAPTER 11

11.1 $E_0 \sin k_p x = E_0(e^{ik_p x} - e^{-ik_p x})/2i$

$$F(k) = \frac{E_0}{2i}\left[\int_{-L}^{+L} e^{i(k+k_p)x}\,dx - \int_{-L}^{+L} e^{i(k-k_p)x}\,dx\right]$$

$$F(k) = -\frac{iE_0 \sin(k+k_p)L}{(k+k_p)} + \frac{iE_0 \sin(k-k_p)L}{(k-k_p)}$$

$$F(k) = iE_0 L[\text{sinc}\,(k-k_p)L - \text{sinc}\,(k+k_p)L]$$

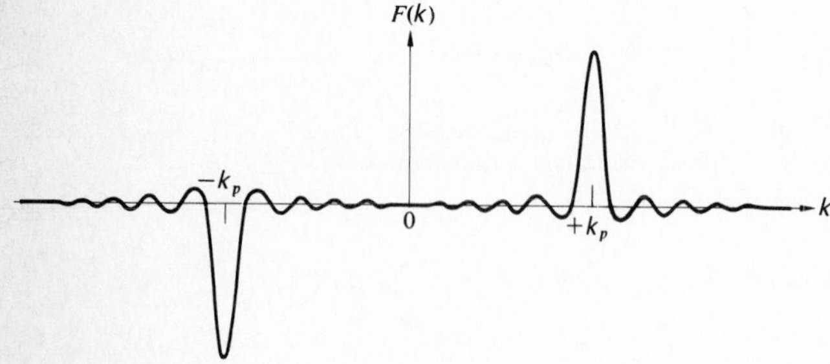

11.3 $\cos^2 \omega_p t = \frac{1}{2} + \frac{1}{2}\cos 2\omega_p t = \frac{1}{2} + \dfrac{e^{2i\omega_p t} + e^{-2i\omega_p t}}{4}$

$$F(\omega) = \tfrac{1}{2}\int_{-T}^{+T} e^{i\omega t}\,dt + \tfrac{1}{4}\int e^{i(\omega+2\omega_p)t}\,dt + \tfrac{1}{4}\int e^{i(\omega-2\omega_p)t}\,dt$$

$$F(\omega) = \frac{1}{\omega}\sin\omega T + \frac{1}{2(\omega+2\omega_p)}\sin(\omega+2\omega_p)T + \frac{1}{2(\omega-2\omega_p)}\sin(\omega-2\omega_p)T$$

$$F(\omega) = T\,\text{sinc}\,\omega T + \frac{T}{2}\,\text{sinc}\,(\omega+2\omega_p)T + \frac{T}{2}\,\text{sinc}\,(\omega-2\omega_p)T$$

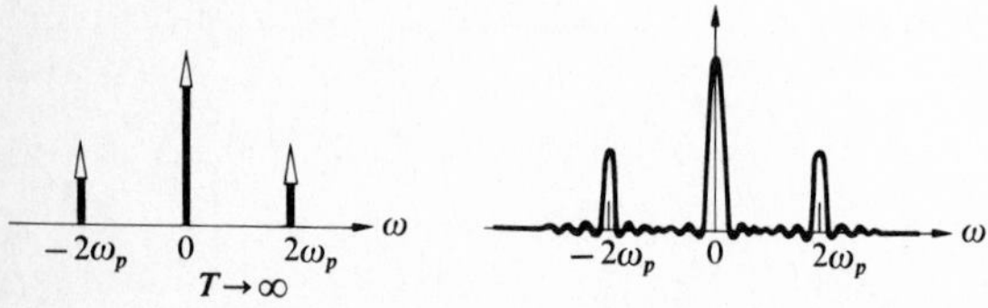

11.6 $\mathscr{F}\{af(x) + bh(x)\} = aF(k) + bH(k)$

11.8 $F(k) = L\,\text{sinc}^2\,kL/2$ at $k = 0$, $F(0) = L$, and $F(\pm 2\pi/L) = 0$.

11.15 $\displaystyle\int_{x=-\infty}^{x=+\infty} f(x)h(X-x)\,dx$

$$= -\int_{x'=+\infty}^{x'=-\infty} f(X-x')h(x')\,dx' = \int_{-\infty}^{+\infty} h(x')f(X-x')\,dx'$$

where $x' = X - x$, $dx = -dx'$.

$$f \circledast h = h \circledast f$$

or

$$\mathscr{F}\{f \circledast h\} = \mathscr{F}\{f\}\cdot\mathscr{F}\{h\} = \mathscr{F}\{h\}\cdot\mathscr{F}\{f\} = \mathscr{F}\{h \circledast f\}.$$

11.17 A point on the edge of $f(x, y)$, for example, at $(x = d, y = 0)$, is spread out into a square 2ℓ on a side centered on $X = d$. Thus it extends no farther than $X = d + \ell$, and so the convolution must be zero at $X = d + \ell$ and beyond.

11.19 $f(x - x_0) \circledast h(x) = \displaystyle\int_{-\infty}^{+\infty} f(x - x_0)h(X - x)\,dx,$

and setting $x - x_0 = \alpha$, this becomes

$$\int_{-\infty}^{+\infty} f(\alpha)h(X - \alpha - x_0)\,d\alpha = g(X - x_0)$$

11.21

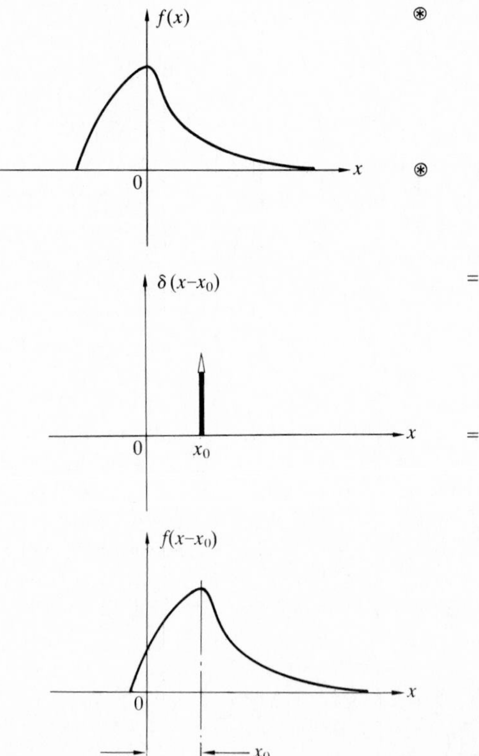

11.24 We see that $f(x)$ is the convolution of a rect-function with two δ-functions, and from the convolution theorem,

$$F(k) = \mathscr{F}\{\text{rect}\,(x) \circledast [\delta(x - a) + \delta(x + a)]\}$$

$$= \mathcal{F}\{\text{rect}\,(x)\} \cdot \mathcal{F}\{[\delta(x-a) + \delta(x+a)]\}$$
$$= a\,\text{sinc}\,\tfrac{1}{2}ka \cdot (e^{ika} + e^{-ika})$$
$$= a\,\text{sinc}\,(\tfrac{1}{2}ka) \cdot 2\cos ka$$

11.25 $f(x) \circledast h(x)$

$$= [\delta(x+3) + \delta(x-2) + \delta(x-5)] \circledast h(x)$$
$$= h(x+3) + h(x-2) + h(x-5)$$

11.28

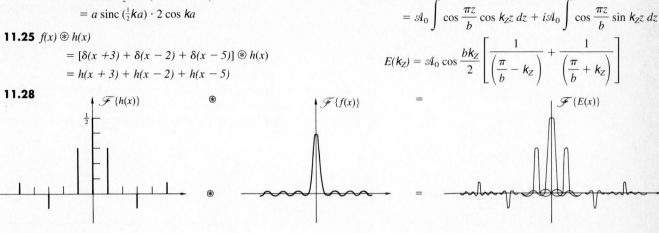

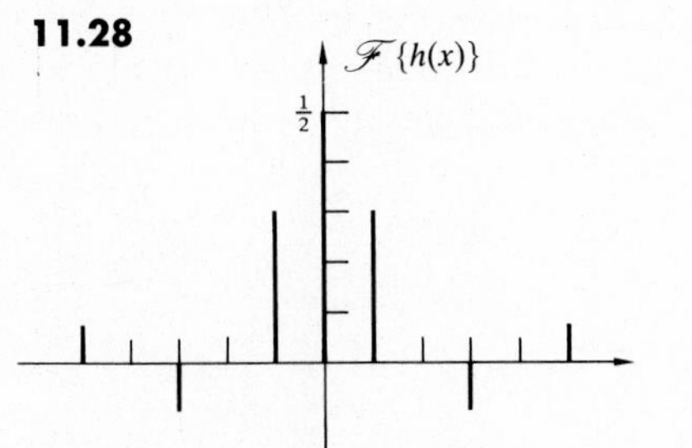

$\circledast$

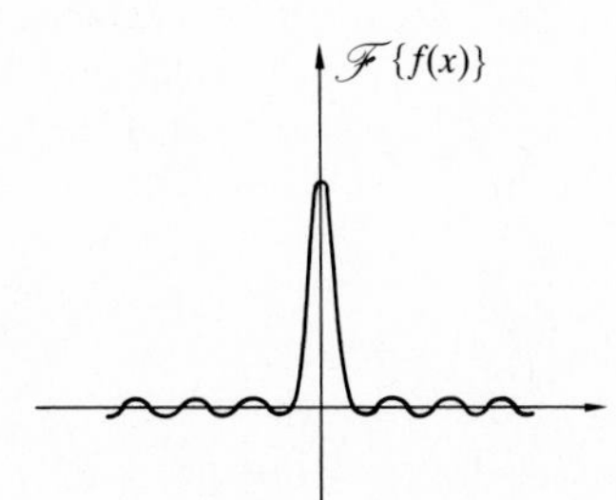

$=$

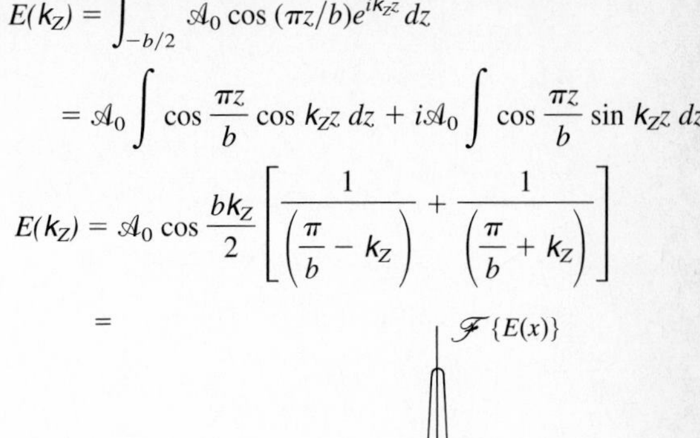

11.29 $\mathcal{A}(y, z) = \mathcal{A}(-y, -z)$

$$E(Y, Z, t) \propto \iint \mathcal{A}(y, z) e^{i(k_Y y + k_Z z)}\,dy\,dz$$

Change Y to $-Y$, Z to $-Z$, y to $-y$, z to $-z$, then k_Y goes to $-k_Y$ and k_Z to $-k_Z$.

$$E(-Y, -Z) \propto \iint \mathcal{A}(-y, -z) e^{i(k_Y y + k_Z z)} dy\,dz$$

$$\therefore E(-Y, -Z) = E(Y, Z)$$

11.30 From Eq. (11.63),

$$E(Y, Z) = \iint \mathcal{A}(y, z) e^{ik(Yy+Zz)/R}\,dy\,dz$$

$$E'(Y, Z) = \iint \mathcal{A}(\alpha y, \beta z) e^{ik(Yy+Zz)/R}\,dy\,dz$$

now let $y' = \alpha y$ and $z' = \beta z$:

$$E'(Y, Z) = \frac{1}{\alpha\beta} \iint \mathcal{A}(y', z') e^{ik[(Y/\alpha)y' + (Z/\beta)z']}\,dy'\,dz'$$

or
$$E'(Y, Z) = \frac{1}{\alpha\beta} E(Y/\alpha, Z/\beta)$$

11.31

$$C_{ff} = \lim_{T\to\infty} \frac{1}{2T} \int_{-T}^{+T} A \sin(\omega t + \varepsilon) A \sin(\omega t - \omega\tau + \varepsilon)\,dt$$

$$= \lim_{T\to\infty} \frac{A^2}{2T} \int [\tfrac{1}{2}\cos(\omega\tau) - \tfrac{1}{2}\cos(2\omega t - \omega\tau + 2\varepsilon)]\,dt$$

since $\cos\alpha - \cos\beta = -2\sin\frac{1}{2}(\alpha+\beta)\sin\frac{1}{2}(\alpha-\beta)$. Thus

$$C_{ff} = \frac{A^2}{2}\cos(\omega\tau)$$

11.32 $E(k_Z) = \displaystyle\int_{-b/2}^{+b/2} \mathcal{A}_0 \cos(\pi z/b) e^{ik_Z z}\,dz$

$$= \mathcal{A}_0 \int \cos\frac{\pi z}{b} \cos k_Z z\,dz + i\mathcal{A}_0 \int \cos\frac{\pi z}{b} \sin k_Z z\,dz$$

$$E(k_Z) = \mathcal{A}_0 \cos\frac{bk_Z}{2}\left[\frac{1}{\left(\dfrac{\pi}{b} - k_Z\right)} + \frac{1}{\left(\dfrac{\pi}{b} + k_Z\right)}\right]$$

Chapter 12

12.1 At low pressures, the intensity emitted from the lamp is low, the bandwidth is narrow, and the coherence length is large. The fringes will initially display a high contrast, although they'll be fairly faint. As the pressure builds, the coherence length will decrease, the contrast will drop off, and the fringes might even vanish entirely.

12.4 Each sine function in the signal produces a cosinusoidal autocorrelation function with its own wavelength and amplitude. All of these are in phase at the zero delay point corresponding to $\tau = 0$. Beyond that origin the cosines soon fall out of phase, producing a jumble where destructive interference is more likely. (The same sort of thing happens when, say, a square pulse is synthesized out of sinusoids—everywhere beyond the pulse all the contributions cancel.) As the number of components increases and the signal becomes more complex—resembling random noise—the autocorrelation narrows, ultimately becoming a δ-spike at $\tau = 0$.

12.6 The irradiance at Σ_0 arising from a point source is

$$4I_0\cos^2(\delta/2) = 2I_0(1 + \cos\delta)$$

For a differential source element of width dy at point S', y from axis, the OPD to P at Y via the two slits is

$$\Lambda = (\overline{S'S_1} + \overline{S_1P}) - (\overline{S'S_2} + \overline{S_2P})$$
$$= (\overline{S'S_1} - \overline{S'S_2}) + (\overline{S_1P} - \overline{S_2P})$$
$$= ay/l + aY/s \text{ from Section 9.3.}$$

The contribution to the irradiance from dy is then

$$dI \propto (1 + \cos k\Lambda)\,dy$$

$$I \propto \int_{-b/2}^{+b/2} (1 + \cos k\Lambda)\, dy$$

$$I \propto b + \frac{d}{ka}\left[\sin\left(\frac{aY}{s} + \frac{ab}{2l}\right) - \sin\left(\frac{aY}{s} - \frac{ab}{2l}\right)\right]$$

$$\begin{aligned} I \propto b + \frac{d}{ka} & [\sin (kaY/s) \cos (kab/2l) \\ & + \cos (kaY/s) \sin (kab/2l) \\ & - \sin (kaY/s) \cos (kab/2l) \\ & + \cos (kaY/s) \sin (kab/2l)] \end{aligned}$$

$$I \propto b + \frac{l2}{ka} \sin (kab/2l) \cos (kaY/s)$$

12.7

$$\mathscr{V} = \frac{I_{\max} - I_{\min}}{I_{\max} + I_{\min}}$$

$$I_{\max} = I_1 + I_2 + 2\sqrt{I_1 I_2}|\tilde{\gamma}_{12}|$$

$$I_{\min} = I_1 + I_2 - 2\sqrt{I_1 I_2}|\tilde{\gamma}_{12}|$$

$$\mathscr{V} = \frac{4\sqrt{I_1 I_2}|\tilde{\gamma}_{12}|}{2(I_1 + I_2)}$$

12.8 When

$$S''S_1O' - S'S_1O' - \lambda/2, 3\lambda/2, 5\lambda/2, \ldots$$

the irradiance due to S' is given by

$$I' = 4I_0 \cos^2 (\delta'/2) = 2I_0(1 + \cos \delta')$$

while the irradiance due to S'' is

$$\begin{aligned} I'' = 4I_0 \cos^2 (\delta''/2) &= 4I_0 \cos^2 (\delta' + \pi)/2 \\ &= 2I_0(1 - \cos \delta') \end{aligned}$$

Hence $I' + I'' = 4I_0$.

12.10 $\theta = \frac{1°}{2} = 0.0087$ rad

$h = 0.32\bar{\lambda}_0/\theta$ using $\bar{\lambda}_0 = 550$ nm

$h = 0.32\,(550 \text{ nm})/0.0087$

$h = 2 \times 10^{-2}$ mm

12.11 $I_1(t) = \Delta I_1(t) + \langle I_1 \rangle$
hence

$$\langle I_1(t + \tau)I_2(t)\rangle = \langle[\langle I_1\rangle + \Delta I_1(t + \tau)][\langle I_2\rangle + \Delta I_2(t)]\rangle$$

since $\langle I_1 \rangle$ is independent of time.

$$\langle I_1(t + \tau)I_2(t)\rangle = \langle I_1\rangle\langle I_2\rangle + \langle \Delta I_1(t + \tau)\, \Delta I_2(t)\rangle$$

if we recall that $\langle \Delta I_1(t)\rangle + 0$. Eq. (12.34) follows by comparison with Eq. (12.32).

12.13 From Eq. (12.22), $\mathscr{V} = 2\sqrt{(10I)I}/(10I + I) = 2\sqrt{10}/11 = 0.57$.

12.15 Using the van Cittert-Zernike theorem, we can find $\tilde{\gamma}_{12}(0)$ from the diffraction pattern over the apertures, and that will yield the visibility on the observation plane: $\mathscr{V} = |\tilde{\gamma}_{12}(0)| = |\text{sinc}\ \beta|$. From Table 1, $\sin u/u = 0.85$ when $u = 0.97$, hence $\pi by/l\lambda = 0.97$, and if $y = \overline{P_1P_2} = 0.50$ mm, then $b = 0.97(l\lambda/\pi y) = 0.97(1.5 \text{ m})(500 \times 10^{-9} \text{ m})/\pi(0.50 \times 10^{-3} \text{ m}) = 0.46$ mm.

12.18 From the van Cittert-Zernike theorem, the degree of coherence can be obtained from the Fourier transform of the source function, which itself is a series of δ-functions corresponding to a diffraction grating with spacing a, where $a \sin \theta_m = m\lambda$. The coherence function is therefore also a series of δ-functions. Hence the $\overline{P_1P_2}$, the slit separation d, must correspond to the location of the first-order diffraction fringe of the source if $\mathscr{V}$ is to be maximum. $a\theta_1 \approx \lambda$, and so $d \approx l\theta_1 \approx \lambda l/a = (500 \times 10^{-9} \text{ m})(2.0 \text{ m})/(500 \times 10^{-6} \text{ m}) = 2.0$ mm.

Chapter 13

13.2 $P/A = \epsilon\sigma(T^4 - T_e^4) = (0.97)(5.6703 \times 10^{-8} \text{ W/m}^2\text{·K}^4) \times (306^4 - 293^4) = I = 76.9 \text{ W/m}^2$. $P = 108$ W.

13.3 $I_e = \sigma T^4$

$$(22.8 \text{ W cm}^2)(10^4 \text{ cm}^2/\text{m}^2) = (5.7 \times 10^{-8} \text{ W m}^{-2}\text{ K}^{-4})T^4$$

$$T = \left[\frac{22.8 \times 10^4}{5.7 \times 10^{-8}}\right]^{1/4} = 1.414 \times 10^3 = 1414 \text{ K}$$

13.4 $T_2^4/T_1^4 = P_2/P_1 = 16 \times 10^{12}/16 \times 10^8 = 1.0 \times 10^4$

13.12 $\lambda(\text{min}) = 300$ nm

$$h\nu = hc/\lambda$$

$$= \frac{(6.63 \times 10^{-34} \text{ J·s})(3 \times 10^8 \text{ m/s})}{300 \times 10^{-9} \text{ m}}$$

$$\mathscr{E} = 6.63 \times 10^{-19} \text{ J} = 4.14 \text{ eV}$$

13.14 $Nh\nu = (1.4 \times 10^3 \text{ W/m}^2)(1 \text{ m}^2)(1 \text{ s})$

$$N = \frac{1.4 \times 10^3(700 \times 10^{-9})}{(6.63 \times 10^{-34})(3 \times 10^8)} = \frac{980 \times 10^{20}}{19.89}$$

$$N = 49.4 \times 10^{20}$$

13.15 Find the number of atoms present. $pV = nRT$; $n = 4.47 \times 10^{-7}$ mol; so there are 2.69×10^{17} atoms and 2.67×10^{15} get excited; the emission rate is $2.67 \times 10^{15}/\tau = 1.92 \times 10^{23}$ photons per second.

13.18 $h\nu/k_BT = 0.774$ and $\dfrac{1}{e^{0.774} - 1} = 0.86$; at the elevated temperature the ratio is substantial and the two modalities are comparable.

13.23 The transition rate must equal $P/h\nu = 3 \times 10^{15}\ \text{s}^{-1}$.

13.27 $I = \frac{1}{2} v \epsilon E_0^2 = \frac{n}{2}\left(\frac{\epsilon_0}{\mu_0}\right)^{1/2} E_0^2$, where $\mu \approx \mu_0$

$$E_0^2 = 2(\mu_0/\epsilon_0)^{1/2}\, I/n \qquad (\mu_0/\epsilon_0)^{1/2} = 376.730\ \Omega$$

$$E_0 = 27.4\,(I/n)^{1/2}$$

13.29

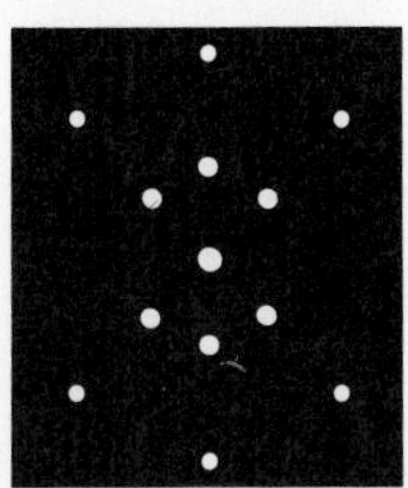

Diffraction pattern

13.30

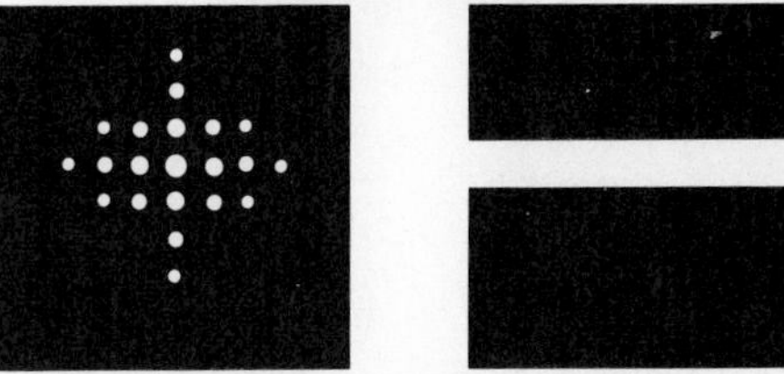

Diffraction pattern Filter

13.31

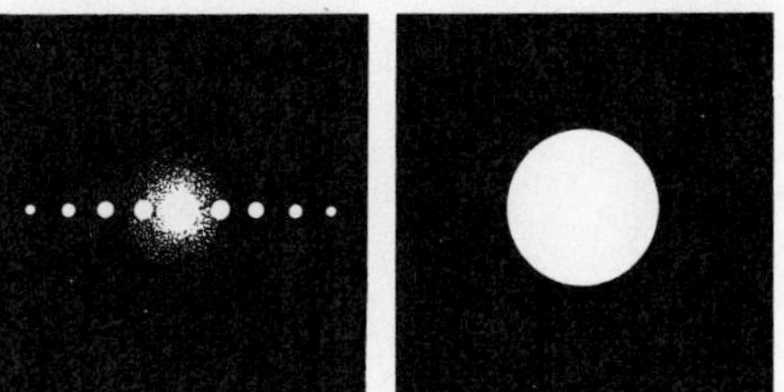

Diffraction pattern Filter

or even better

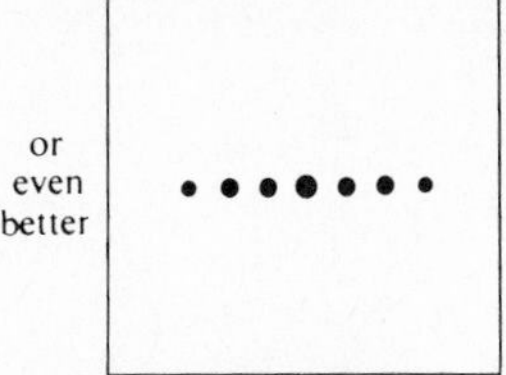

Filter

13.33

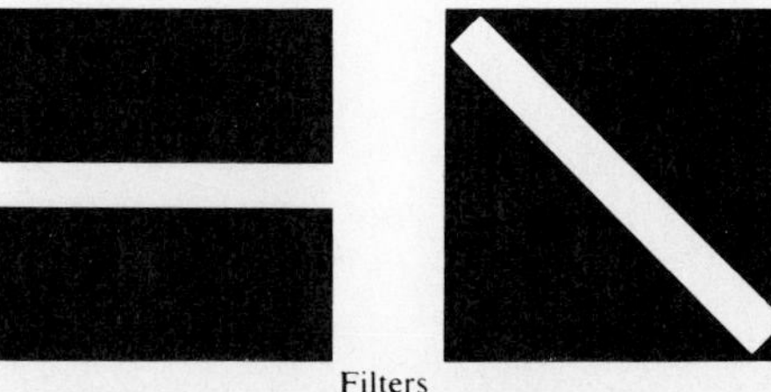

Filters

13.34 From the geometry, $f_t\theta = f_i\,\Phi$: $k_O = k \sin\theta$ and $k_I = k \sin\Phi$, hence $\sin\theta \approx \theta \approx k_O\lambda/2\pi$ and $\sin\Phi \approx \Phi \approx k_I\lambda/2\pi$, therefore $\theta/\Phi = k_O/k_I$ and $k_I = k_O(\Phi/\theta) = k_O(f_t/f_i)$. When $f_i > f_t$ the image will be larger than the object, the spatial periods in the image will also be larger, and the spatial frequencies in the image will be smaller than in the object.

13.35 $a = (1/50)$ cm: $a \sin\theta = m\lambda$, $\sin\theta \approx \theta$, hence $\theta = (5000\text{ m})\,\lambda$, and the distance between orders on the transform plane is $f\theta = 5000\,\lambda f = 2.7$ mm.

13.37 Each point on the diffraction pattern corresponds to a single spatial frequency, and if we consider the diffracted wave to be made up of plane waves, it also corresponds to a single-plane wave direction. Such waves, by themselves, carry no information about the periodicity of the object and produce a more or less uniform image. The periodicity of the source arises in the image when the component plane waves interfere.

13.39 The relative field amplitudes are 1.00, 0.60, and 0.60; hence $E \propto 1 + 0.60\cos(+ky') + 0.60\cos(-ky') = 1 + 1.2\cos ky'$. This is a cosine oscillating about a line equal to 1.0. It varies from +2.2 to −0.2. The square of this will correspond to the irradiance, and it will be a series of tall peaks with a relative height of $(2.2)^2$, between each pair of which there will be a short peak proportional to $(0.2)^2$; notice the similarity with Fig. 11.32.

13.40 $a \sin\theta = \lambda$, here $f\theta = 50\lambda f = 0.20$ cm; hence $\lambda = 0.20/50(100) = 400$ nm. The magnification is 1.0 when the focal lengths are equal, hence the spacing is again 50 wires/cm.

13.44 The inherent motion of the medium would cause the speckle pattern to vanish.

BIBLIOGRAPHY

ANDREWS, C. L., *Optics of the Electromagnetic Spectrum*, Prentice-Hall, Englewood Cliffs, N.J., 1960.

ANTONELLI, CHRISTIAN, FISCHER, GILES, JAMES, and STONER, *Waves and Optics Simulations*, Wiley, New York, 1995.

BAKER, B. B. and E. J. COPSON, *The Mathematical Theory of Huygens' Principle*, Oxford University Press, London, 1969.

BALDWIN, G. C., *An Introduction to Nonlinear Optics*, Plenum Press, New York, 1969.

BARBER, N. F., *Experimental Correlograms and Fourier Transforms*, Pergamon, Oxford, 1961.

BARNOSKI, M., *Fundamentals of Optical Fiber Communications*, Academic Press, New York, 1976.

BARTON, A. W., *A Textbook On Light*, Longmans, Green, London, 1939.

BEARD, D. B. and G. B. BEARD, *Quantum Mechanics With Applications*, Allyn and Bacon, Boston, 1970.

BEESLEY, M., *Lasers and Their Applications*, Taylor and Francis, New York, 1976.

BERAN, M. J. and G. B. PARRENT, JR., *Theory of Partial Coherence*, Prentice-Hall, Englewood Cliffs, N.J., 1964.

BLAKER, J. W. and W. M. ROSENBLUM, *Optics*, Macmillan, New York, 1993.

BLOEMBERGEN, N., *Nonlinear Optics*, Addison-Wesley, Reading, MA, 1991.

BLOOM, A. L., *Gas Lasers*, Wiley, New York, 1968.

BLOSS, D., *An Introduction to the Methods of Optical Crystallography*, Holt, Rinehart and Winston, New York, 1961.

BORN, M. and E. WOLF, *Principles of Optics*, Pergamon, Oxford, 1970.

BOROWITZ, S., *Fundamentals of Quantum Mechanics*, Benjamin, New York, 1967.

BRADDICK, H., *Vibrations, Waves, and Diffraction*, McGraw-Hill, New York, 1965.

BROUWER, W., *Matrix Methods in Optical Instrument Design*, Benjamin, New York, 1964.

BROWN, E. B., *Modern Optics*, Reinhold, New York, 1965.

BUTCHER, P. N. and D. COTTER, *The Elements of Nonlinear Optics*, Cambridge University Press, Cambridge, 1990.

CAJORI, F., *A History of Physics*, Macmillan, New York, 1899.

CATHEY, W., *Optical Information Processing and Holography*, Wiley, New York, 1974.

CHANG, W. S. C., *Principles of Quantum Electronics, Lasers: Theory and Applications*, Addison-Wesley, Reading, Mass., 1969.

COHEN-TANNOUDJI, C., DUPONT-ROC, J. and G. GRYNBERG, *Photons and Atoms*, Wiley, New York, 1989.

COLLIER, R., C. BURCKHARDT, and L. LIN, *Optical Holography*, Academic Press, New York, 1971.

CONRADY, A. E., *Applied Optics and Optical Design*, Dover Publications, New York, 1929.

COULSON, C. A., *Waves*, Oliver and Boyd, Edinburgh, 1949.

CRAWFORD, F. S., JR., *Waves*, McGraw-Hill, New York, 1965.

DAVIS, C. C., *Lasers and Electro-Optics*, Cambridge University Press, Cambridge, 1996.

DAVIS, H. F., *Introduction to Vector Analysis*, Allyn and Bacon, Boston, 1961.

DAVIS, S. P., *Diffraction Grating Spectrographs*, Holt, Rinehart and Winston, New York, 1970.

DENISYUK, Y., *Fundamentals of Holography*, Mir Publishers, Moscow, 1984.

DEVELIS, J. B. and G. O. REYNOLDS, *Theory and Applications of Holography*, Addison-Wesley, Reading, Mass., 1967.

DIRAC, P. A. M., *Quantum Mechanics*, Oxford University Press, London, 1958.

DODD, J. N., *Atoms and Light: Interactions*, Plenum Press, New York, 1991.

DRUDE, P., *The Theory of Optics*, Longmans, Green, London, 1939.

DITCHBURN, R. W., *Light*, Wiley, New York, 1963.

ELMORE, W. and M. HEALD, *The Physics of Waves*, McGraw-Hill, New York, 1969.

FEYNMAN, R. P., *QED*, Princeton University Press, Princeton, N.J., 1985.

FLÜGGE, J., ed., *Die wissenschafliche und angewandte Photographie; Band 1, Das photographische Objektiv*, Springer-Verlag, Wien, 1955.

Fowles, G., *Introduction to Modern Optics*, Holt, Rinehart and Winston, New York, 1968.

FRANÇON, M., *Modern Applications of Physical Optics*, Interscience, New York, 1963.

FRANÇON, M., *Diffraction Coherence in Optics*, Pergamon Press, Oxford, 1966.

FRANÇON, M., *Optical Interferometry*, Academic Press, New York, 1966.

FRANÇON, M., N. KRAUZMAN, J. P. MATHIEU, and M. MAY, *Experiments in Physical Optics*, Gordon and Breach, New York, 1970.

FRANÇON, M., *Optical Image Formation and Processing*, Academic Press, New York, 1979.

FRANK, N. H., *Introduction to Electricity and Optics*, McGraw-Hill, New York, 1950.

FRENCH, A. P., *Special Relativity*, Norton, New York, 1968.

FRENCH, A. P., *Vibrations and Waves*, Norton, New York, 1971.

FROOME, K. D. and L. ESSEN, *The Velocity of Light and Radio Waves*, Academic Press, London, 1969.

FRY, G. A., *Geometrical Optics*, Chilton, Philadelphia, 1969.

GARBUNY, M., *Optical Physics*, Academic Press, New York, 1965.

GASKILL, J., *Linear Systems, Fourier Transforms, and Optics*, Wiley, New York, 1978.

GHATAK, A. K., *An Introduction to Modern Optics*, McGraw-Hill, New York, 1971.

GHATAK, A. and K. THYAGARAJAN, *Contemporary Optics*, Plenum Press, New York, 1978.

GOLDIN, E., *Waves and Photons, An Introduction to Quantum Theory*, Wiley, New York, 1982.

GOLDWASSER, E. L., *Optics, Waves, Atoms, and Nuclei: An Introduction*, Benjamin, New York, 1965.

GOODMAN, J. W., *Introduction to Fourier Optics*, McGraw-Hill, New York, 1968.

GUENTHER, R. D., *Modern Optics*, Wiley, New York, 1990.

HARDY, A. C. and F. H. PERRIN, *The Principles of Optics*, McGraw-Hill, New York, 1932.

HARVEY, A. F., *Coherent Light*, Wiley, London, 1970.

HEAVENS, O. S., *Optical Properties of Thin Solid Films*, Dover Publications, New York, 1955.

HECHT, E., *Optics: Schaum's Outline Series*, McGraw-Hill, New York, 1975.

HERMANN, A., *The Genesis of Quantum Theory (1899–1913)*, MIT Press, Cambridge, Mass., 1971.

HOUSTON, R. A., *A Treatise On Light*, Longmans, Green, London, 1938.

HUNSPERGER, R., *Integrated Optics: Theory and Technology*, Springer-Verlag, Berlin, 1984.

HUYGENS, C., *Treatise on Light*, Dover Publications, New York, 1962 (1690).

IGA, K., *Fundamentals of Laser Optics*, Plenum Press, New York, 1994.

IIZUKA, K., *Engineering Optics*, Springer-Verlag, Berlin, 1987.

INGARD, K. U., *Fundamentals of Waves and Oscillations*, Cambridge University Press, Cambridge, 1988.

JACKSON, J. D., *Classical Electrodynamics*, Wiley, New York, 1962.

JENKINS, F. A. and H. E. WHITE, *Fundamentals of Optics*, McGraw-Hill, New York, 1957.

JENNISON, R. C., *Fourier Transforms and Convolutions for the Experimentalist*, Pergamon, Oxford, 1961.

JOHNSON, B. K., *Optics and Optical Instruments*, Dover Publications, New York, 1947.

JONES, B., et al., *Images and Information*, The Open University Press, Milton Keynes, Great Britain, 1978.

KARIM, M. A., *Electro-Optical Devises and Systems*, PWS-Kent, Boston, 1990.

KEISER, G., *Optical Fiber Communications*, McGraw-Hill, New York, 1991.

KLAUDER, J. and E. SUDARSHAN, *Fundamentals of Quantum Optics*, Benjamin, New York, 1968.

KLEIN, M. V., *Optics*, Wiley, New York, 1970.

KREYSZIG, E., *Advanced Engineering Mathematics*, Wiley, New York, 1967.

LENGYEL, B. A., *Introduction to Laser Physics*, Wiley, New York, 1966.

LENGYEL, B. A., *Lasers, Generation of Light by Stimulated Emission*, Wiley, New York, 1962.

LEVI, L., *Applied Optics*, Wiley, New York, 1968.

LEVINSON, M. D. and S. S. KANO, *Introduction to Nonlinear Laser Spectroscopy*, Academic Press, New York, 1988.

LIPSON, S. G. and LIPSON, H. and D. S. TANNHAUSER, *Optical Physics*, Cambridge University Press, London, (3rd Ed.), 1995.

LONGHURST, R. S., *Geometrical and Physical Optics*, Wiley, New York, 1967.

MACH, E., *The Principles of Physical Optics, An Historical and Philosophical Treatment*, Dover Publications, New York, 1926.

MAGIE, W. F., *A Source Book in Physics*, McGraw-Hill, New York, 1935.

MAIN, I. G., *Vibrations and Waves in Physics*, Cambridge University Press, Cambridge, (3rd Ed.), 1993.

MALACARA, D., *Optical Shoptesting*, Wiley, New York, 1978.

MANDEL, L. and E. WOLF, *Optical Coherence and Quantum Optics*, Cambridge University Press, Cambridge, 1995.

MARION, J. and M. HEALD, *Classical Electromagnetic Radiation*, Academic Press, New York, 1980.

MARTIN, L. C. and W. T. WELFORD, *Technical Optics*, Sir Isaac Pitman & Sons, Ltd., London, 1966.

MATVEEV, A. N., *Optics*, Mir Publishers, Moscow, 1988.

MEYER, C. F., *The Diffraction of Light, X-rays and Material Particles*, University of Chicago Press, Chicago, 1934.

MEYER-ARENDT, J. R., *Introduction to Classical and Modern Optics*, Prentice-Hall, Englewood Cliffs, N.J., 1972.

MEYSTRE, P. and M. SARGENT III, *Elements of Quantum Optics*, Springer-Verlag, Berlin, 1990.

MICKELSON, A. R., *Physical Optics*, Van Nostrand Reinhold, New York, 1992.

MIDWINTER, J., *Optical Fibers for Transmission*, Wiley, New York, 1979.

Military Standardization Handbook—Optical Design, MIL-HDBK-141, 5 October 1962.

MILONNI, P. W. amd J. H. EBERLY, *Lasers*, Wiley, New York, 1988.

MINNAERT, M., *The Nature of Light and Colour in the Open Air*, Dover Publications, New York, 1954.

MÖLLER, K. D., *Optics*, Wiley, New York, 1988.

MORGAN, J., *Introduction to Geometrical and Physical Optics*, McGraw-Hill, New York, 1953.

NEWELL, A. C. and J. V. MOLONEY, *Nonlinear Optics*, Addison-Wesley, Reading, MA, 1992.

NEWTON, I., *Optiks*, Dover Publications, New York, 1952 (1704).

NOAKES, G. R., *A Text-Book of Light*, Macmillan, London, 1944.

NUSSBAUM, A., *Geometric Optics: An Introduction*, Addison-Wesley, Reading, Mass., 1968.

NUSSBAUM, A. and R. PHILLIPS, *Contemporary Optics for Scientists and Engineers*, Prentice-Hall, Englewood Cliffs, N.J., 1976.

OKOSHI, T., *Optical Fibers*, Academic Press, New York, 1982.

O'NEILL, E. L., *Introduction to Statistical Optics*, Addison-Wesley, Reading, Mass., 1963.

O'SHEA, D., W. CALLEN, and W. RHODES, *Introduction to Lasers and Their Applications*, Addison-Wesley, Reading, Mass., 1977.

O'SHEA, D. C., *Elements of Modern Optical Design*, Wiley, New York, 1985.

PALMER, C. H., *Optics, Experiments and Demonstrations*, John Hopkins Press, Baltimore, Md., 1962.

PAPOULIS, A., *The Fourier Integral and Its Applications*, McGraw-Hill, New York, 1962.

PAPOULIS, A., *Systems and Transforms with Applications in Optics*, McGraw-Hill, New York, 1968.

PEARSON, J. M., *A Theory of Waves*, Allyn and Bacon, Boston, 1966.

PERSONICK, S. D., *Optical Fiber Transmission Systems*, Plenum Press, New York, 1981.

PLANCK, M. and M. MASIUS, *The Theory of Heat Radiation*, Blakiston, Philadelphia, 1914.

PRESTON, K., *Coherent Optical Computers*, McGraw-Hill, New York, 1972.

ROBERTSON, E. R. and J. M. HARVEY, eds., *The Engineering Uses of Holography*, Cambridge University Press, London, 1970.

ROBERTSON, J. K., *Introduction to Optics Geometrical and Physical*, Van Nostrand, Princeton, N.J., 1957.

RONCHI, V., *The Nature of Light*, Harvard University Press, Cambridge, Mass., 1971.

ROSSI, B., *Optics*, Addison-Wesley, Reading, Mass., 1957.

RUECHARDT, E., *Light Visible and Invisible*, University of Michigan Press, Ann Arbor, Mich., 1958.

SAFFORD, E. L. JR. and J. A. MCCANN, *Fiberoptics and Lasers*, TAB Books, Blue Ridge Summit, PA, 1988.

SALEH, B. E. A. and M. C. TEICH, *Fundamentals of Photonics*, Wiley, New York, 1991.

SANDBANK, C. P., *Optical Fibre Communication Systems*, Wiley, New York, 1980.

SANDERS, J. H., *The Velocity of Light*, Pergamon, Oxford, 1965.

SARGENT, M., M. SCULLY, and W. LAMB, *Laser Physics*, Addison-Wesley, Reading, Mass., 1974.

SCHAWLOW, A. L., intr., *Lasers and Light; Readings from Scientific American*, Freeman, San Francisco, 1969.

SCHRÖDINGER, E. C., *Science Theory and Man*, Dover Publications, New York, 1957.

SCHROEDER. D. J., *Astronomical Optics*, Academic Press, New York, 1987.

SEARS, F. W., *Optics*, Addison-Wesley, Reading, Mass., 1949.

SHAMOS, M. H., ed., *Great Experiments in Physics*, Holt, New York, 1959.

SHURCLIFF, W. A., *Polarized Light: Production and Use*, Harvard University Press, Cambridge, Mass., 1962.

SHURCLIFF, W. A. and S. S. BALLARD, *Polarized Light*, Van Nostrand, Princeton, N.J., 1964.

SILVAST, W. T., *Laser Fundamentals*, Cambridge University Press, Cambridge, 1996.

SIMMONS, J. and M. GUTTMANN, *States, Waves and Photons: A Modern Introduction to Light*, Addison-Wesley, Reading, Mass., 1970.

SINCLAIR, D. C. and W. E. BELL, *Gas Laser Technology*, Holt, Rinehart and Winston, New York, 1969.

SLAYTER, E. M., *Optical Methods in Biology*, Wiley, New York, 1970.

SMITH, F. and J. THOMSON, *Optics*, Wiley, New York, 1971.

SMITH, H. M., *Principles of Holography*, Wiley, New York, 1969.

SMITH, W. J., *Modern Optical Engineering*, McGraw-Hill, New York, (2nd Ed.), 1990.

Société Française de Physique, ed., *Polarization, Matter and Radiation. Jubilee Volume in Honor of Alfred Kastler*, Presses Universitaires de France, Paris, 1969.

SOMMERFELD, A., *Optics*, Academic Press, New York, 1964.

SOUTHALL, J. P. C., *Introduction to Physiological Optics*, Dover Publications, New York, 1937.

SOUTHALL, J. P. C., *Mirrors, Prisms and Lenses*, Macmillan, New York, 1933.

STARK, H., *Applications of Optical Fourier Transforms*, Academic Press, New York, 1982.

STEWARD, E., *Fourier Optics: An Introduction*, Wiley, New York, (2nd Ed.), 1987.

STONE, J. M., *Radiation and Optics*, McGraw-Hill, New York, 1963.

STROKE, G. W., *An Introduction to Coherent Optics and Holography*, Academic Press, New York, 1969.

STRONG, J., *Concepts of Classical Optics*, Freeman, San Francisco, 1958.

SVELTO, O., *Principles of Lasers*, Plenum Press, New York, 1977.

SYMON, K. R., *Mechanics*, Addison-Wesley, Reading, Mass., 1960.

TATASOV, L., *Laser Age in Optics*, Mir Publishers, Moscow, 1981.

TOLANSKY, S., *An Introduction to Interferometry*, Longmans, Green, London, 1955.

TOLANSKY, S., *Curiosities of Light Rays and Light Waves*, American Elsevier, New York, 1965.

TOLANSKY, S., *Multiple-Beam Interferometry of Surfaces and Films*, Oxford University Press, London, 1948.

TOLANSKY, S., *Revolution in Optics*, Penguin Books, Baltimore, 1968.

TOWNE, D. H., *Wave Phenomena*, Addison-Wesley, Reading, Mass., 1967.

TROUP, G., *Optical Coherence Theory*, Methuen, London, 1967.

VALASEK, J., *Optics, Theoretical and Experimental*, Wiley, New York, 1949.

VAN HEEL, A. C. S., ed., *Advanced Optical Techniques*, American Elsevier, New York, 1967.

VAN HEEL, A. C. S. and C. H. F. VELZEL, *What Is Light?*, McGraw-Hill, New York, 1968.

VASICEK, A., *Optics of Thin Films*, North-Holland, Amsterdam, 1960.

WAGNER, A. F., *Experimental Optics*, Wiley, New York, 1929.

WALDRON, R., *Waves and Oscillations*, Van Nostrand, Princeton, N.J., 1964.

WEBB, R. H., *Elementary Wave Optics*, Academic Press, New York, 1969.

WILLIAMS, W. E., *Applications of Interferometry*, Methuen, London, 1941.

WILLIAMSON, S. and H. CUMMINS, *Light and Color in Nature and Art*, Wiley, New York, 1983.

WILSON, R. G., *Fourier Series and Optical Transform Techniques in Contemporary Optics*, Wiley, New York, 1995.

WOLF, E., ed., *Progress in Optics*, North-Holland, Amsterdam.

WOLF, H. F., ed., *Handbook of Fiber Optics: Theory and Applications*, Garland STPM Press, 1979.

WOOD, R. W., *Physical Optics*, Dover Publications, New York, 1934.

WRIGHT, D., *The Measurement of Color*, Van Nostrand, New York, 1971.

YARIV, A., *Quantum Electronics*, Wiley, New York, 1967.

YOUNG, H. D., *Fundamentals of Optics and Modern Physics*, McGraw-Hill, New York, 1968.

YOUNG, M., *Optics and Lasers*, Springer-Verlag, Berlin, 1986.

ZIMMER, H., *Geometrical Optics*, Springer-Verlag, Berlin, 1970

Index

QMW LIBRARY
(MILE END)